U0166912

自然资源部盐湖资源与环境重点实验室论文集（2018～2020年）

中国盐湖资源与环境研究

主　编　郑绵平
副主编　张永生　孔凡晶

科学出版社
北　京

内 容 简 介

本书收录了近年来自然资源部盐湖资源与环境重点实验室的研究成果，研究领域涉及盐湖地质、矿产、盐化工、环境、生态等学科，研究内容涵盖了盐湖资源与环境面临的热点问题，分为重点实验室及野外台站建设、盐类及其他矿产、盐湖化工、盐湖古气候与生态环境、盐湖资源发展战略五个部分。本书研究成果回应了社会对有关钾、锂等盐湖资源与环境保护等方面的关切。

本书可作为盐湖地质学、盐湖化学、盐湖化学工程与技术、环境科学与工程等专业的研究及产业开发人员、研究生、本科生的参考用书。

审图号：GS 京（2022）1492 号

图书在版编目（CIP）数据

中国盐湖资源与环境研究/郑绵平主编. —北京：科学出版社，2022.11

ISBN 978-7-03-073839-4

Ⅰ. ①中… Ⅱ. ①郑… Ⅲ. ①盐湖－自然资源－生态环境－研究－中国 Ⅳ. ①P942.078

中国版本图书馆 CIP 数据核字（2022）第 220745 号

责任编辑：孟美岑 李亚佩 / 责任校对：何艳萍

责任印制：吴兆东 / 封面设计：北京图阅盛世

科学出版社出版

北京东黄城根北街 16 号

邮政编码：100717

http://www.sciencep.com

北京中科印刷有限公司印刷

科学出版社发行 各地新华书店经销

*

2022 年 11 月第 一 版 开本：880×1230 1/16

2022 年 11 月第一次印刷 印张：35

字数：1 100 000

定价：528.00 元

（如有印装质量问题，我社负责调换）

《中国盐湖资源与环境研究》
编委会名单

主　编：郑绵平

副主编：张永生　孔凡晶

委　员（按姓氏汉语拼音排序）：

卜令忠　陈安东　陈　帅　程贤达　丁　涛
樊　馥　高知睿　巩　鑫　侯献华　黄　琴
李斌凯　李　空　林勇杰　刘春花　刘青枰
雒洋冰　吕苑苑　马黎春　孟凡巍　苗忠英
倪润祥　乜　贞　牛新生　彭　渊　权晓莹
商雯君　水新芳　宋　高　孙明光　王安建
王晨光　王　凯　王利伟　王　林　王琳霖
王　松　王远超　王云生　吴宇靓　邢恩袁
闫丽娟　叶传永　于旭东　余石勇　张洪霞
张会平　张雪飞　赵元艺　郑　洪　仲佳爱

序

我和郑绵平是在 1961 年参加中国科学院西藏科学考察队时相识的，岁月匆匆，想来已 60 余载。数十年来，郑绵平一直从事盐湖研究，他领导的科研队伍已发展壮大成为盐湖资源环境的多学科研究团队。2020 年 12 月，我主持了自然资源部盐湖资源与环境重点实验室学术年会，听取了该团队的研究进展汇报，这本书就是该团队学术成果的总结。

盐湖不但蕴藏着锂、钾、硼、溴、碘等国家急需的战略矿产资源，而且还富含盐藻、卤虫等独特的生物资源。该重点实验室聚焦盐湖钾锂等国家急需的战略矿产资源勘查和综合利用，同时与环境、生物等多学科结合开展综合研究。研究深入实际，重视野外调查，工作区域遍布全国盐类分布区。以西藏扎布耶、当雄错、班戈湖和青海柴达木盆地察尔汗、西台吉乃尔等盐湖野外观测站为依托，取得第一手长期基础观察数据，为盐湖区资源与环境科学研究及开发提供了重要的基础数据。研究成果受到地矿、化工、环境、能源等多部门的关注。

钾盐是粮食的“粮食”，是我国紧缺矿产。郑绵平团队与有关部门密切合作，为我国察尔汗钾盐矿床的发现和陆相成钾理论的创立做出了重要贡献。近十年来，该团队又指导和参与发现柴达木西部早更新世深层砂砾型卤水钾盐，建立承袭成钾新模式，发展了陆相成钾理论。与此同时，他带领团队着力开展我国海相钾、锂调研，在四川海相找钾取得重要的突破。在川东北普光地区发现过去视为呆矿的海相石膏杂卤石钾盐矿床新类型；在云南，对我国古代钾盐矿床成钾时代提出了新的认识，确定该钾盐形成于侏罗纪，提出“两层楼”成矿模式。上述工作对指导我国特提斯海相找钾有重要的意义，为海相找钾开拓了新的思路。我国“十四五”规划提出要大力发展新能源汽车，对盐湖锂资源的需求量将急剧增加。相信《中国盐湖资源与环境研究》一书针对青藏高原锂钾资源和四川盆地锂资源的讨论对扩大锂资源应用前景有重要价值。谨以此为序。

孙鸿烈

2021 年 8 月

前　言

资源与环境是当今世界发展面临的两大主题。现代盐湖是大气圈、岩石圈、水圈、生物圈等的“圈层窗口”，是研究资源与环境国内外面临的重大课题的天然实验室。成盐元素由于其作用的广泛性和重要性，以及近代科学技术的进步，盐类聚集体的研究已突破单学科领域，在宏观上已进入全球乃至行星科学研究领域，在微观上已达到分子和基因层次，盐类科学研究已进入多学科交叉、在广度和深度上大为扩展的崭新时代。综合集成以往单学科成果，揭示其内在规律，形成世界前沿的盐类大科学。

本书收录了自然资源部盐湖资源与环境重点实验室 2018～2020 年的研究成果，以学术性为主，兼顾产业方面的需求，同时涉及矿产资源发展战略及盐湖提锂提钾等方面的内容。重点实验室关注盐湖资源与环境存在的科技问题，聚焦锂钾等战略矿产资源勘查、资源利用等问题，结合环境、化工、生物等多学科交叉，开展盐湖综合研究。重点实验室人员深入实际，重视野外调查，工作区域遍布全国盐类分布区。以西藏扎布耶、当雄错、班戈湖和柴达木察尔汗、西台吉乃尔等盐湖野外观测站为依托，取得第一手长期基础观察数据，为盐湖区资源与环境科学研究及开发提供重要的基础数据支撑。本书以盐类大科学思想为指导，以盐湖及盐类资源与环境问题为主题，涉及盐湖矿产地质、盐湖化工、古气候与环境、资源发展战略等多学科，取得了盐湖科学上的新认识、盐湖资源利用上的新技术及新工艺。

自然资源部科技发展司辛红梅副司长亲临重点实验室指导，中国地质科学院矿产资源研究所作为自然资源部盐湖资源与环境重点实验室的依托单位，陈仁义所长及领导班子在各方面为重点实验室的发展提供了大力支持。以孙鸿烈院士为主任的重点实验室学术委员会同仁为重点实验室建言献策，在重点实验室发展方向、学科建设和人才培养等方面提出了许多宝贵的意见及建议，在此一并表示衷心的感谢。

因所收集的文章源自多种期刊，各源刊的格式标准难免不统一，本着尊重原著的精神，所用物理量单位、符号、图例和参考文献等均尽量保留原文风貌，未做统一标准的处理。

编　者

2021 年 8 月

目　　录

第四部分　盐湖古气候与生态环境

第五部分　盐湖资源发展战略

第一部分

重点实验室及野外台站建设

自然资源部盐湖资源与环境重点实验室 2020 年度运行简况

郑绵平[1,2] 张永生[1,2] 孔凡晶[1,2]

1. 中国地质科学院矿产资源研究所，北京 100037

2. 自然资源部盐湖资源与环境重点实验室，北京 100037

“自然资源部盐湖资源与环境重点实验室”是盐湖与盐类资源环境多学科综合研究的科技平台，主要开展盐湖与盐类资源环境领域基础和应用基础研究，包括盐湖与盐类资源调查评价、综合利用方法技术和工程化、盐类资源环境与盐湖农业、盐湖沉积与环境变化、盐类资源的综合利用、环境保护及优化盐碱地和盐水域等多个交叉学科领域研究，为自然资源保护及开发提供科技支撑。本实验室主要有 4 个研究方向。①盐湖与盐类矿产成矿规律、资源评价理论与方法：对盐湖和盐类矿床钾、硼、锂、铯、铷、溴、碘等国家急需、重点或稀有矿产的成矿规律、找矿方法进行研究；对重点盐类矿产资源综合利用的关键理论与技术方法进行研究，侧重盐类综合利用的物化原理、化工工艺和工程化关键技术，为增加盐类矿产品种和高值化、改变长期以来我国盐类矿产开发中产品单一和低值开发的重大难题提供科技依托。②盐湖环境与全球变化：研究盐湖沉积的气候环境定量化指标，建立青藏高原典型盐湖高精度古气候、古环境演化序列；研究新生代中国盐湖带迁移和时空演化、高原形成及古季风演替的相关性；坚持不同区带盐湖站点的长期科学观测研究，建立我国盐湖区生态环境、盐类和生物资源数据信息系统，引进和完善 GIS 系统，为我国盐湖区环境保护和资源合理开发提供基础资料。③盐湖农业、盐湖与健康：对盐湖及盐碱地分类调查，并相应开展盐生生物种资源调查、筛选、驯化以及新品种的培育研究；开展广义盐湖的生态学研究，探讨气候变化及人类活动对现代盐湖生态系统的演变规律的影响，对生态系统十分脆弱的藏北和柴达木盆地地区的盐湖生态系统容纳能力科学评估，服务于盐湖矿业发展的环境规划，为盐湖生物多样性保护和构建西部盐境生态安全屏障提供依据；开展盐湖水域生态系统的生物群落结构与生物生产力研究，开展盐湖生物产业化发展潜力及其生态风险评估研究，为盐湖农业的健康发展提供理论依据与技术支持。④盐类资源综合利用与产业化：开展盐湖卤水、深部卤水锂钾资源综合利用关键技术创新研究，开发新工艺新技术，为实现盐湖与盐类资源高值化综合利用及产业化提供技术支撑。

“自然资源部盐湖资源与环境重点实验室”更名于 2019 年，其前身相继为“国土资源部盐湖资源与环境重点实验室”（2007 年成立）和“地矿部盐湖资源与环境开放实验室”（1997 年成立），是中国盐湖资源和环境研究的重要科技平台。实验室成员是一支以地质学、化学化工、生物工程学、环境生态学等多学科中青年科技骨干为主的科技团队，重点实验室现有在职研究人员 30 人，其中院士 1 人、研究员/教授级高工 9 人、副研究员（高工）11 人、助理研究员（工程师）9 人；此外还有客座研究员 29 人（院士 2 人）。

2020 年，新冠肺炎疫情在全球爆发，抗疫斗争在全国全面展开。在严峻的疫情形势下，全室科研人员努力拼搏，圆满完成了本年度的各项任务。本年度共承担项目 14 项，其中国家自然科学基金重大研究计划项目 1 项、青年科学基金项目 2 项，国家重点研发计划项目 1 项，国家重点研发计划课题 3 项，中国地质调查项目 2 项，基本科研业务费课题 4 项，研究总经费 11367 万元，年度到款经费 5118.5 万元，工作量饱满。本年度共发表研究论文 19 篇，其中国外 SCI 收录 10 篇，国内 SCI 收录 1 篇，EI 收录 4 篇，中文核心期刊 4 篇。获得发明专利授权 2 项。有 6 人次参加国内学术交流会。在科普方面，2020 年由于受到新冠肺炎疫情的影响，原计划开展的地球日、科技周大型展览等现场活动均没有实施。2020 年科普工作重点放在在线科普活动上，先后开展了线上科普讲座 1 次，制作科普视频 2 部，参与探月中心牵头组织的科普活动 1 次，取得了良好的效果。在日常科普工作中，实验室科普走廊对公众常年开放，普及了盐湖科学知识，倡导科学方法，让公众全面了解重点实验室科研成果及盐湖资源与环境基本知识。

1　年度重要创新进展情况

1.1　四川盆地钾锂找矿新进展

在中国地质调查二级项目“四川盆地东北部锂钾资源综合调查评价”资助下，2020年取得了以下创新进展（项目总指导：郑绵平；项目组成员：张永生、邢恩袁、彭渊、桂宝玲、牛新生、苏奎、左璠璠、林勇杰、商雯君、麻乾坤、崔新宇）。

1.1.1　川宣地1井海相锂钾找矿取得突破

“川宣地1井”位于川东北宣汉县普光镇，是一口“锂钾兼探”综合地质调查井，由郑绵平院士指导的“四川盆地东北部锂钾资源综合调查评价”项目组设计部署，中国地质调查项目与四川巴人新能源有限公司联合资助（总投入3098万元），旨在探测川东北地区深部富锂钾卤水和海相可溶性“新型杂卤石钾盐矿”。该井于2020年8月9日钻至井深3797m完钻，累计取心837.25m，岩心采取率98.09%，取得锂钾找矿双丰收，圆满完成验证井钻探工程地质设计任务。

1.1.1.1　发现嘉陵江组四—五段厚达62.8m的“新型杂卤石钾盐矿”工业矿层

在川宣地1井下三叠统嘉陵江组四—五段井深3000.84～3482.86m，发现海相含钾盐系厚348m（图1），其中钾盐含量（KCl）平均3%、最高达20.5%的“新型杂卤石钾盐矿”累计矿层厚度达62.8m，KCl平均含量为11.97%的高品位工业矿层累计厚度达29.46m，取得四川盆地三叠系海相可溶性固体钾盐找矿的重大突破。结合已有的天然气探井测井解释和全覆盖三维地震资料处理解释结果，川东北宣汉普光地区三叠系海相可溶性“新型杂卤石钾盐矿”潜在资源（KCl）达7亿多吨，有望形成我国首个亿吨级大型海相钾盐资源基地。

(a)

(b)

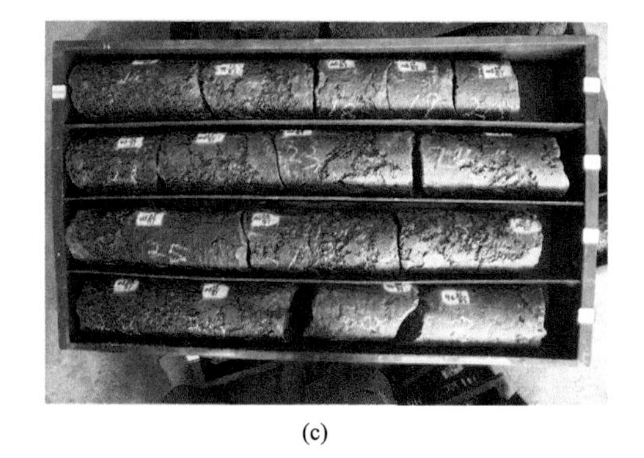

(c)

图1　川宣地1井钻获的海相可溶性“新型杂卤石钾盐矿”矿层岩心

（a）褐红色细-中砾杂卤石碎屑密集分布于石盐基质中；（b）褐红色中-粗砾杂卤石碎屑密集分布于石盐基质中；（c）褐红色粗-巨砾杂卤石碎屑密集分布于石盐基质中

1.1.1.2 发现嘉陵江组四—五段4个含锂钾卤水层（“黑卤”）

川宣地1井测井解释结果显示：在嘉陵江组四—五段井深2958～2960.5m（2.5m）、2963～2965m（2m）、2976.1～2982.8m（6.7m）和3076.85～3098.1m（21.25m）处共发育4个含水层，累计厚度32.45m，测井解释孔隙度2.3%～8.9%，含水饱和度51%～68%，储层主要岩性为裂隙发育的白云岩（图2）。嘉陵江组四—五段井深3008～3090m内泥浆抽样测试结果初步表明，Li、K、B、Br、I均显示明显的高异常值，推断这4个含水层应为富锂钾卤水层。基于四川盆地下三叠统上部嘉陵江组四—五段普遍产出富锂钾的“黑卤”，综合测井解释、岩心编录、泥浆分析结果，基本确定本井嘉陵江组四—五段含水层应为富锂钾的“黑卤”，可作为下一步射孔试水的主力目标层位。

1.1.1.3 发现上三叠统须家河组4个含水层（“黄卤”）

测井解释显示：本井在须家河组陆相砂岩层中，还发育4个含水层，分别位于井深2147.0～2157.7m（10.7m）、2164.0～2172.3m（8.3m）、2425.4～2430.0m（4.6m）、2438.9～2446.6m（7.7m），累计厚度31.3m，测井解释孔隙度8%～20%，含水饱和度72%～96%。基于四川盆地上三叠统须家河组普遍产出富Br、I、含Li、K“黄卤”的区域性分布规律，推断该组4个含水层也应为“黄卤”，可考虑综合利用。

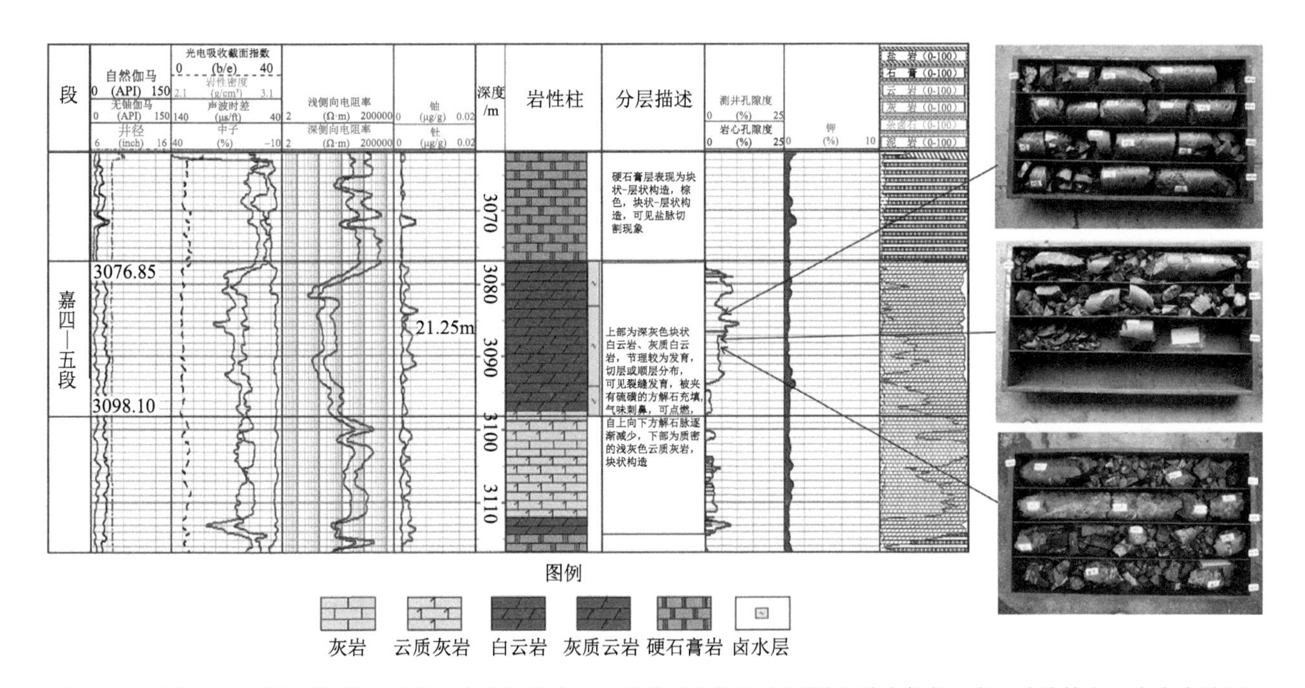

图2 川宣地1井嘉陵江组四—五段主力富锂钾卤水层测井综合解释成果图及裂隙发育、岩心破碎的白云岩卤水储层

1.1.2 宣汉普光核心区富锂钾卤水有利分布区预测

1.1.2.1 建立川宣地1井单井卤水层测井解释模型

利用川宣地1井的岩心编录数据，自然伽马、声波、补偿中子、密度、深侧向测井、浅侧向测井、有效光电吸收截面指数等测井数据，以及泥浆测试Li、K异常数据，采用矿物含量计算、孔隙度计算、含水饱和度计算等方法，综合建立了川宣地1井单井卤水层预测标准技术。识别解释显示：卤水层通常发育在白云岩、灰质白云岩等地层中，测井曲线上表现为低伽马、低电阻、低密度、高声波时差、高中子“三低两高”的基质孔隙发育特征。其中，碳酸盐岩的孔隙型和裂缝—孔隙型储层，双侧向电阻率低于100Ω·m为储卤层。根据建立的测井解释模型和方法，共解释嘉陵江组四—五段和须家河组含水层共8层，累计厚度63.75m。以川宣地1井为标杆，对周围邻井的测井数据进行重新解释标定，得到了更加可靠的储卤层预测结果。

1.1.2.2　井-震标定储卤层横向展布

1）嘉陵江组四—五段测井解释卤水层分布

从普光地区嘉陵江组四—五段综合分析来看，嘉陵江组四—五段上部发育膏盐盆缘滩相和云坪亚相白云岩、石灰岩，是富锂钾卤水储层主要发育段，普光地区卤水储层较发育，除川宣地1井外，普光8井嘉陵江组四—五段测井解释水层31.2m/3层、孔隙度1.7%～3.1%，大湾102井测井解释水层57.7m/7层、孔隙度1.4%～5.7%等，毛坝和分水岭地区基本为石膏和盐岩，卤水储层欠发育。

2）雷口坡组测井解释卤水层分布

结合沉积相展布及测井解释，普光地区雷口坡组一段相比嘉陵江组四—五段，盐岩少，泥质含量高；含水储层致密，为低孔、低渗特征；毛坝地区泥质含量更高，基础孔隙度低，储层不发育，以裂缝型储集空间为主。

3）井-震标定卤水层横向展布

通过对研究区三维地震资料的重新处理，得到了更加清晰的嘉陵江组和雷口坡组的地震剖面，结合前期沉积相展布及测井解释，通过井震标定，利用波阻抗特征参数，对普光地区卤水储层进行反演预测（图3），结果显示在大湾构造的背斜构造核部卤水储层较发育。

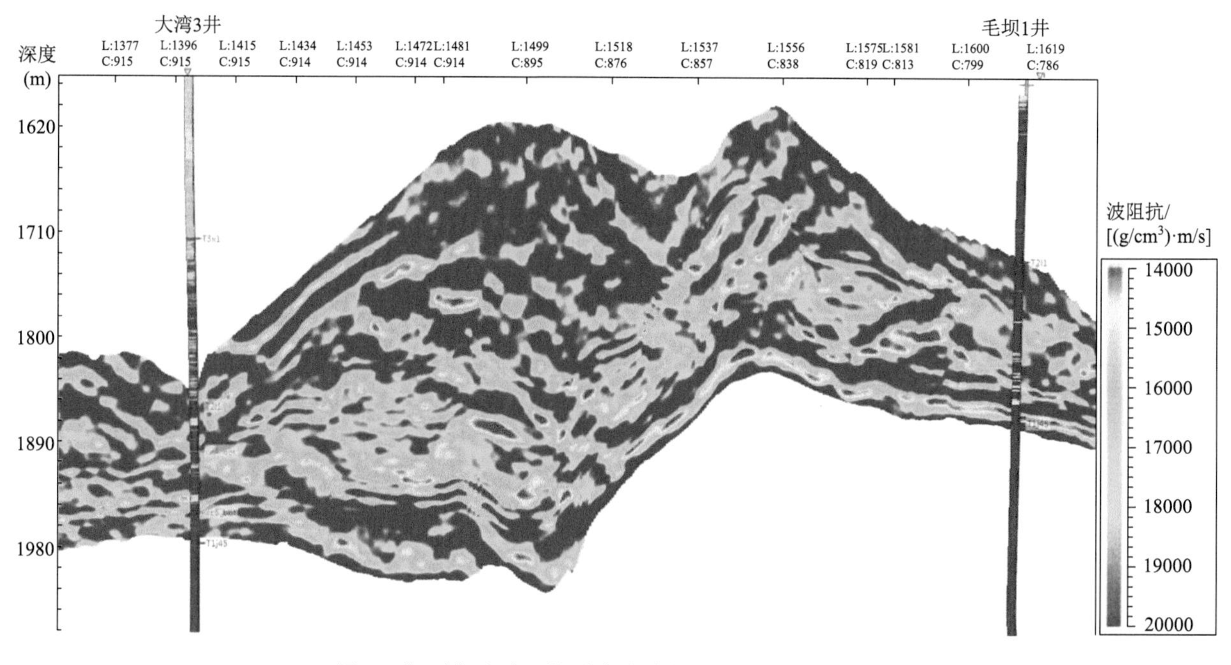

图3　井-震标定波阻抗反演地震剖面卤水储层分布预测

1.1.2.3　富锂钾卤水有利分布区预测

在上述工作基础上，利用单井标定、三维地震反演预测手段，对宣汉普光地区嘉陵江组四—五段富锂钾卤水层进行预测。通过川宣地1井嘉陵江组四—五段卤水层最新测井资料的标定，对已有测井和地震解释成果进行复查，对嘉陵江组四—五段卤水有利区分布有了新的认识，卤水储层主要受控于NE-SW向主干断裂，呈同向展布特征，共划分出2个有利区和1个靶区，卤水有利区面积分别为37km^{2}和100km^2，靶区面积2.4km^2。

1.2　滇西南钾盐找矿新进展

在中国地质调查二级项目“思茅盆地中生代海相钾盐资源调查评价”资助下，2020年取得以下创新成果。（项目总指导：郑绵平。项目组成员：苗忠英、徐建明、张雪飞、宋高、施林峰、吕苑苑、雒洋冰、马妮娜、孔凡晶、马黎春、韩鸿业、杜少荣、娄鹏程、徐其辉。）

1.2.1 构建了镇沅地区侏罗系盐泉的成因模式

镇沅县恩乐镇北部共发现盐泉点10处，其中矿化度大于5g/L的泉点8处。按舒卡列夫分类，其中6处盐泉点属于Cl-Na型水质。6处Cl-Na型盐泉分布在芒磨至考捆之间，5处在长5.5km、宽3.5km、面积约16km^{2}的范围内。其中，5处分布于川河东岸，1处位于川河西岸。调查区盐泉出露位置高程为1100～1250m，在川河两岸，海拔低处出露的盐泉有最高的矿化度，而海拔高处出露的盐泉有最低的矿化度。例如，川河东岸J4号盐泉出露位置的海拔在1100m左右，其矿化度为32.9g/L；J8号盐泉出露位置的海拔接近1250m，其矿化度仅为5.8g/L。地形切面图中盐泉出露位置的海拔与其矿化度之间的关系更加明显。

1.2.2 勐野井组实测剖面对钾盐矿床成因的指示

2020年度对江城县宝藏镇良马河村附近出露的下白垩统勐野井组剖面进行了实际测量。剖面起点坐标为东经101°40′25″，北纬22°39′55″，起点位置岩性为灰白色石英砂岩，层面见流水波痕，具有典型的河流相扒沙河组岩相特征。上覆的勐野井组可划分为三个岩性段，下段主岩性为泥岩、含泥砾粉砂岩、粉砂质泥岩，中段主岩性为粉砂岩、细砂岩、泥质粉砂岩，上段主岩性为泥岩和含泥砾粉砂岩。整段勐野井组产状比较稳定，地层主体倾向为SW向，倾角主体为30°左右。勐野井组上覆第四系河滩。我们所测剖面距勐野井钾盐矿最近距离小于500m，但是从实测剖面构造、岩相、岩性特征等方面根本看不到含盐、含钾的信息，这一客观事实指示：①勐野井钾盐矿床的含钾盐体与勐野井组碎屑岩突变接触，缺少正常的蒸发沉积序列，其成盐、成钾物质应为外源，目前的地质调查成果认为其来自深部中侏罗统源盐层；②通过野外剖面测量，可以获得成盐、成钾沉积和构造背景方面的信息，但是很难直接指示盐体的存在；③目前钾盐勘查地质工作仍要以盐泉、盐溶泥砾岩作为直接找矿标志，以成矿物源、古气候、沉积背景、构造演化研究作为基础理论支撑，采用合理、有效的地球物理手段通过钻探工程实现找矿突破。

1.2.3 MK-3井深层发现了侏罗系轮藻化石

我们在MK-3井2469.9m处的碎屑岩岩心中发现了轮藻化石，共计42枚，其藏卵器已全部石化，但保存完好，经鉴定分别为*Sphaerochara*（球状轮藻）、*Euaclistochara*（真开口轮藻）、*Aclistochara*（开口轮藻）和*Mesochara*（中生轮藻）属。结合前人对此类化石年代学属性的研究成果，可以把MK-3井2469.9m处的地层时代划归中侏罗世晚期—晚侏罗世早期。由于这一层段恰好位于两段盐岩层之间，因此可以推断MK-3井深层含钾盐岩的形成时间可能为中侏罗世晚期—晚侏罗世早期。

1.2.4 景谷拗陷完成二维地震组网测量

景谷拗陷是思茅盆地内可供钾盐资源地质调查开展工作的最大的二级构造单元，其沉积环境相对封闭，沉积地层连续性好，构造条件相对稳定，成钾前景较好。目前盆地内已发现回短、按板井、文卡、凤岗、文晒、香盐、益香共7个岩盐矿床，其中除按板井、益香之外均见钾矿化，但是限于当时钻机的钻探能力，拗陷内更深处的含盐含钾情况未能被揭示，很多探井在岩盐层终孔。为了更加科学、合理、高效地部署钾盐钻探工程，项目组在原有工作基础上追加二维地震探测工作40km，与原有工作共同组建成累计长度120km的二维地震测网。由于积累了原有工作经验，新的地震测量工作野外采集数据的质量有了大幅提升，不仅信噪比有了明显提升，而且现场监控剖面也可以比较好地反映地层结构和构造特征。

1.2.5 盐泉水调查提供新的有利目标区

本年度思茅盆地盐泉水调查结果显示有利的成盐、成钾区有两处，分别为镇沅县恩乐镇复兴村和古城镇黄庄村。目前已在复兴村有利成盐区部署钻探工程 1 处，设计井深 900m。黄庄盐泉的总盐量为 73375mg/L，Br×10^3/Cl 为 0.41，K×10^3/Cl 为 6.35，K×10^3/总盐量为 3.40，钾氯系数/钾盐系数为 1.87。这些参数特征指示盐泉母岩已经达到析钾阶段，因此这一地区是勘查固体钾盐的有利目标区。但是黄庄地区钾盐地质调查工作还有一些基础地质问题需要解决，例如：①前期的地质调查工作认为盐岩赋存于黄庄小背斜轴部灰绿色泥砾岩中，其层位被划分在勐野井组；②盐泉的盐度受大气降水影响较大，与上升泉的特征不相符；③前期调查认为黄庄地区含盐面积仅 1km^2 左右，开发前景较小。因此，后续的地质调查工作还需要加强地层划分对比、含盐地层分布面积、盐泉性质等基础地质问题的研究。

1.2.6 MK-3 井岩心古地磁分析成果

本年度选取了 MK-3 井 200 件岩心样品，进行了古地磁测试分析，测试单位为中国地质科学院地质力学研究所。样品所在地层的倾向数据来自三维地震资料的地质解译，地层倾角由野外采样过程中实际测量，沿与地层倾向正相交的面钻取圆柱体定向样品。测试结果显示：几乎所有样品的剩磁强度在加热到 500℃之后不降反升，同时伴随热退磁曲线在正交矢量投影图上杂乱无章的变化。因此，我们推测这些样品可能发生了轻微的动力变质，形成了新的热不稳定矿物，这些热不稳定矿物在加热到 500℃之后生成新的磁性矿物（如磁铁矿），在冷却过程中重新获得了黏滞剩磁，致使样品的剩磁强度衰减曲线大幅上涨。这与样品所处的地质背景相符，附近发育的 F3 断层是动力变质的一种佐证。样品的磁极性序列揭示 MK-3 井 2447.83～2526.95m 深度段存在 7 个反极性、5 个正极性。正负极性倒转频繁符合侏罗纪和早白垩世的磁性地层特征，但是具体对应哪个时期仍需要对更多岩心样品进行磁性地层测量之后才能确定。

1.3 柴达木深层卤水资源调查新进展

项目组成员为：侯献华、樊馥、苏奎。

（1）在柴达木西部黑北凹地完成黑 ZK10 井选位，该孔钻遇巨厚层大含水量的含钾卤水，求得单孔 KCl 资源量 1300 万 t，柴西深部砂砾孔隙卤水钾盐找矿取得重大进展。

通过收集油田二维地震资料重新处理、解释（图 4、图 5），成功预测了黑 ZK10 孔钻遇目的层深度。①钻孔中揭露含水层累计厚度 681.82m，占孔深的 95%以上，岩性均为含卵石的砾砂、含砾的中粗砂等（图 6、图 7），其含水性、透水性均较好。②经抽水试验，水位 3.23m，大降深单井涌水量 6073m^3/d，最大降深 8.47m，抽卤时间 104h，稳定时间 24h。单位涌水量 750m^3/(d·m)，富水性极强（图 8）。小降深单井涌水量 2989.44m^3/d，降深 5.09m，抽卤时间 32h，稳定时间 8h。经化验分析，KCl 品位 0.68%。③单孔新增 KCl 资源量 1300 万 t。深部砂砾孔隙卤水钾盐找矿取得重大找矿进展。

（2）收集了大浪滩—黑北凹地已有中深钻孔资料，进行了单井相、连井相分析，初步编制了大浪滩—黑北凹地深层砂砾目标储卤层岩相古地理图件。

选择大浪滩—黑北凹地重点地区进行深部砂砾储卤层的岩相古地理编图（1∶10 万），初步识别出扇三角洲相和湖泊相，其中扇三角洲相可划分为扇三角洲平原亚相和扇三角洲前缘亚相，湖泊相可划分为滨湖亚相和浅湖亚相，浅湖亚相又可进一步划分为淡盐湖亚相和盐湖亚相，各亚相特征如下。

扇三角洲平原：扇三角洲平原为扇三角洲的水上部分，大多时间暴露在空气中，故沉积物颜色以红色、褐色等氧化色为主。沉积物主要为红棕色、棕红色、黄褐色含砾粗砂、中砂、细砂，砾石可为粗砾、中砾、细砾不等。自然伽马曲线表现为齿状低值，电测曲线表现为微齿状低值。

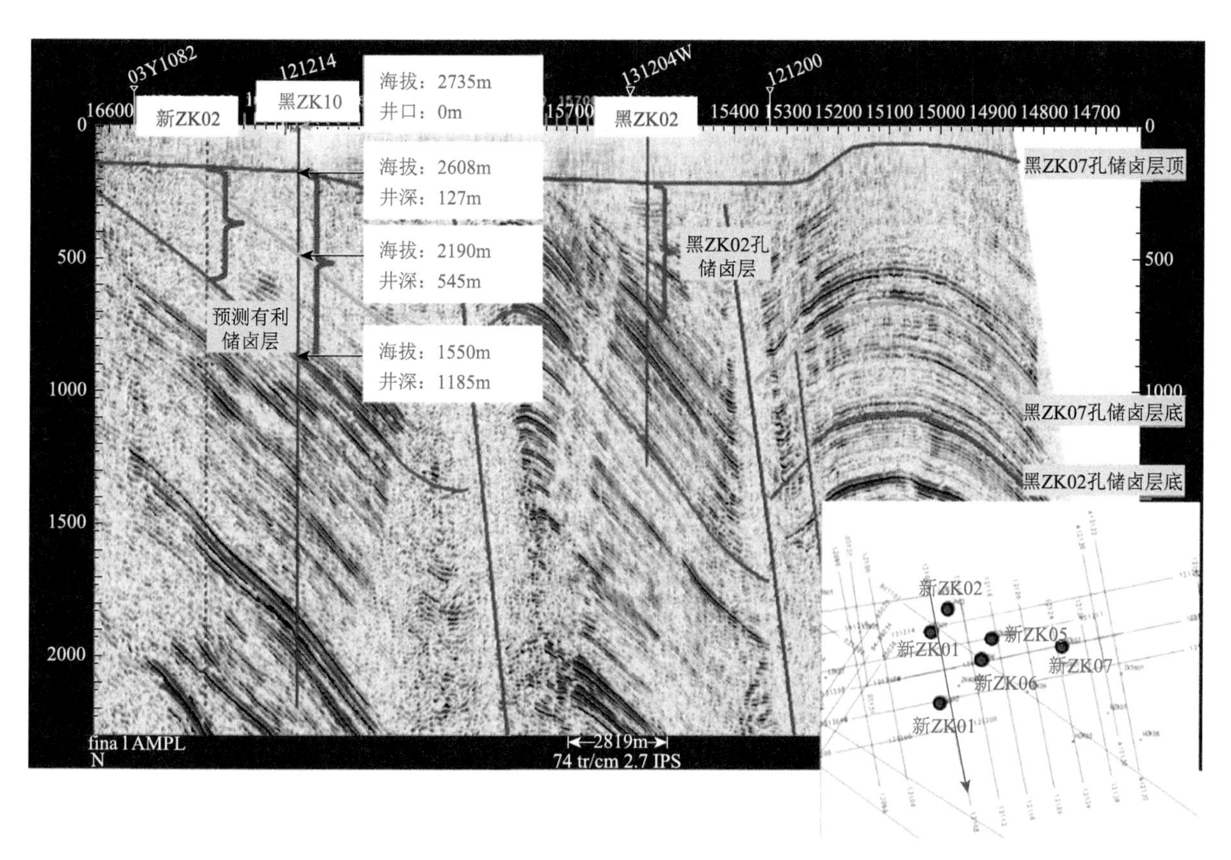

图 4　黑 ZK10 孔砂砾储卤层地震反射特征

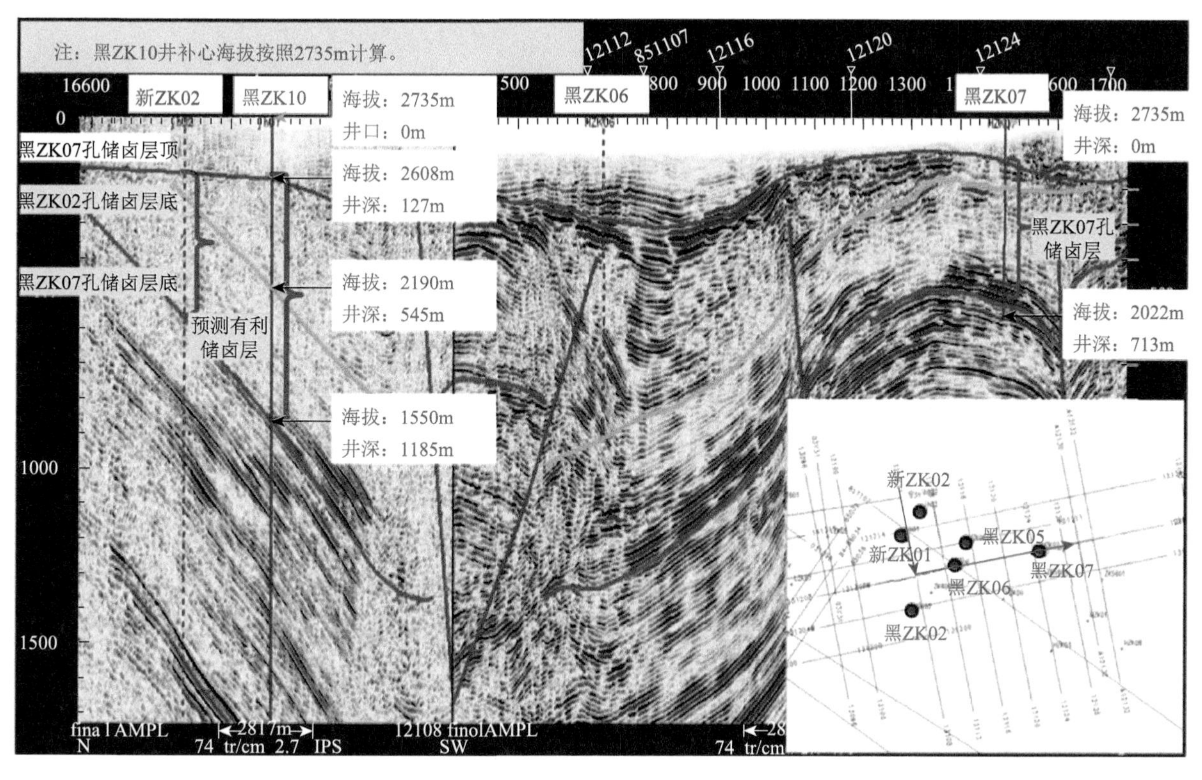

图 5　黑 ZK10 孔砂砾储卤层地震反射特征及深度

图 6　黑 ZK10 孔砂砾储卤层岩性

图 7　黑 ZK10 孔砂砾储卤层岩性特征

图 8　黑 ZK10 孔砂砾孔隙卤水抽水试验

扇三角洲前缘：扇三角洲前缘为扇三角洲的水下部分，多数时间为水下沉积，故沉积物颜色以青灰色、灰白色及深灰色等非氧化色为主。沉积物主要为深灰色、黄色、黄灰色含砾粗、中砂，及灰白色、灰色粗砂、中砂等。自然伽马曲线表现为齿状低值，电测曲线表现为微齿状低值。

滨湖：滨湖水位较浅，沉积物接近水面，时而出露时而淹没，氧化作用强烈，在干旱的气候条件下颜色多呈黄褐色、棕红色等，沉积物主要为黄褐色、棕红色、灰白色黏土粉砂、淤泥粉砂等，或含层状石膏及少量他形石盐颗粒。自然伽马曲线表现为齿状高值，电测曲线表现为微齿状高值。

淡盐湖：淡盐湖始终位于水下，为淡化时期盐湖的沉积。沉积物颜色以灰色、深灰色、灰黑色为主，沉积物主要为灰黑色、灰绿色、深灰色含石膏粉砂质黏土或淤泥，或淤泥粉砂，多含板状石膏层或含芒硝颗粒。自然伽马曲线表现为齿状高值，电测曲线表现为平缓低值。

盐湖：盐湖始终位于水下，为咸化时期的盐湖沉积。沉积物颜色以灰色、深灰色、灰黑色为主，沉积物主要为灰黑色、青灰色含粉砂或淤泥石盐层，石盐多呈他形，半自形-自形均发育，粒径为0.02～0.5cm。自然伽马曲线为平缓低值，电测曲线表现为微齿低值，声波时差曲线为齿状低值。为此，该课题在大浪滩选择具有代表性的钻孔梁ZK01、梁ZK03、梁ZK05、梁ZK10、梁ZK02、梁ZK06，在黑北凹地选择黑ZK01、黑ZK02、黑ZK03、黑ZK04、黑ZK05进行了岩心重新观察描述，并进行了1∶1000单井相划分、连井相研究，编制了该区更新世早期1∶10万岩相古地理编图。

1.4 盐湖卤水及深层卤水提锂工艺取得了一定的进展

项目组成员为：乜贞、卜令忠、王云生、伍倩、余疆江。

1.4.1 加碱法

以西藏扎布耶盐湖卤水为实验材料，开展了一系列加碱实验，进一步优化了提锂工艺。开展冬季成卤加碱提锂实验（扎布耶室内），目的是对比扎布耶冬季成卤在不同加热温度和Na_2CO_3饱和溶液加入量下的析锂效果，确定最优操作条件。实验结论为加入Na_2CO_3对于提高卤水Li^+的析出率有显著效果。原料卤水如不加碱，则无碳酸锂析出；在一定的加热条件下，Na_2CO_3的加入量越大，卤水中Li^+的析出率越高；针对实验用原料卤水，加入Na_2CO_3饱和溶液的比例为4%～8%较合适。

卤水加碱升温析锂实验（北京室内），目的是确定扎布耶冬季成卤在加热温度45℃、不同Na_2CO_3饱和溶液加入量下的析锂效果。实验结论为扎布耶冬季成卤通过加碱升温手段可以得到70%以上的Li_2CO_3混盐。不同成卤加碱40℃升温析锂实验（扎布耶室内），目的是对比考察扎布耶矿区三个车间不同成卤在加热温度40℃、不同Na_2CO_3饱和溶液加入量下的析锂效果。实验结论是加碱法对于提高高Li^+低CO_3^{2-}卤水锂析出率的效果非常理想：采用加碱法，析出的Li_2CO_3品位均接近70%，产量亦有明显提高，但碱液的加入量过大，则会一定程度地降低Li_2CO_3混盐的品位；相同体积的不同原料卤水，按5%的比例加入Na_2CO_3饱和溶液，二车间卤水的析锂效果最好，其次为一车间，三车间最不理想。二车间5000m^3卤水按5%的比例加入碱液，在40℃升温的条件下，可增产30.4t品位达69.9%的Li_2CO_3混盐；Na_2CO_3饱和溶液加入量的较优占比为5%，即一个盛有5000m^3卤水的结晶池需加入Na_2CO_3饱和溶液250m^3，换算成工业纯碱约125t。

实验加碱法对于提高高Li^+低CO_3^{2-}卤水的锂析出率效果显著，单池Li_2CO_3混盐产量和品位均有大幅提高。若要在实际生产中实施加碱法，可在结晶池灌卤时，将Na_2CO_3饱和溶液注入输卤渠，使其与卤水预先混合均匀后再一并灌入结晶池，此加碱方式的提锂效果更好。

1.4.2 扰动法

扰动法原理是阴、阳离子结合需要碰撞，对扎布耶结晶太阳池底部的析锂层进行机械扰动，可大大增加Li^+和CO_3^{2-}的碰撞结合机会，当太阳池内的卤水升至一定温度时，在不破坏过渡层的前提下，对底部析锂层进行适当扰动，可加速Li_2CO_3的结晶沉淀。同时，结晶太阳池底部的下对流层（析锂层）温度、浓度不均一，四周及底部较低，中间较高，机械扰动亦可使析锂层的温度和浓度更加均匀，从而提高锂析出率。

实验结果表明，扰动结晶池底部的析锂层（下对流层），对提高结晶池的单池产量和混盐品位有明显效果。扰动法虽然没有加碱法效果显著，但其适用于所有结晶池，且投资少，可在生产中推广实施。

1.4.3　兑卤法

1）同年度兑卤——同年度冬季尾卤与夏季卤水勾兑

原理是结晶太阳池排出的第一批尾卤是高 Li^+低 CO_3^{2-} 的尾卤，Li^+浓度一般在 1.7～1.8g/L，CO_3^{2-} 浓度在 20g/L 左右，夏季的成卤中 Li^+浓度一般在 1.6g/L 左右，CO_3^{2-} 浓度在 40g/L 以上，二者勾兑可达到提高 CO_3^{2-} 浓度的目的，也利用了尾卤的部分余热，同时减轻了制卤压力。实验结果表明，实验用冬季尾卤（Li^+浓度为 1.93g/L，CO_3^{2-} 浓度为 25.44g/L）和夏季成卤（Li^+浓度为 1.69g/L，CO_3^{2-} 浓度为 53.52g/L）是比较理想的勾兑对象，冬季尾卤和夏季成卤的勾兑比例在 1∶1.25～1∶2 时，得到的 Li_2CO_3 混盐重量和品位均相对较高且基本一致，但其兑卤效果不如夏卤直接加热升温析锂的效果好。

2）跨年度兑卤——上年度夏季卤水与下年度冬季成卤勾兑

原理是将上年度高 CO_3^{2-} 低 Li^+的夏季卤水灌入结晶池，制作太阳池，过冬至下一年度的 2 月份，再与下一年度二三月份的高 Li^+低 CO_3^{2-} 的冬季成卤相兑，取长补短，起到添加 CO_3^{2-}、提高锂析出率的作用，同时还有效地利用了结晶太阳池的部分余热，也缓解了制卤压力。实验结果表明，采用跨年度兑卤法可以显著提高 Li_2CO_3 的析出率；兑卤比例按 1∶0.75 勾兑时，析锂效果最佳，其次为 1∶0.5。与未兑卤的冬季成卤相比，兑卤后 Li_2CO_3 的析出率均提高了 70%以上，品位也都在 70%以上；兑卤法可通过提高冬季成卤中 CO_3^{2-} 的浓度，来增大 Li_2CO_3 的析出率。当向结晶池灌卤时，在关注卤水中 Li^+浓度的同时，也应充分关注 CO_3^{2-} 的浓度；在跨年度兑卤工艺操作中，确保高 CO_3^{2-} 低 Li^+的夏季卤水安全储存过冬，即 CO_3^{2-} 不会因冬季环境气温降低而析出是关键。有效办法是在每年八、九月份生产结束后，向结晶池内灌入夏季卤水，再铺设淡水，构建太阳池，升温过冬。待来年二、三月份结晶池灌卤时，再按一定比例进行冬季成卤和夏季卤水的勾兑。就目前考察的两种提高冬季成卤中 CO_3^{2-} 浓度的方法而言，加碱法因需要外购纯碱，会增加生产成本；而兑卤法则利用冬季成卤的高 CO_3^{2-} 浓度来提高 Li_2CO_3 的析出率，同时兑卤操作还可大大缓解制卤压力，一举多得。

1.5　科普活动在形式上取得新进展

重点实验室科普基地本年度参与开展了 3 次科普活动，包括线上科普讲座、科普视频制作及柴达木盆地火星基地探索。

1.5.1　线上科普讲座

2020 年 9 月 2 日，孔凡晶研究员参加了自然资源部中国地质调查局关于 2020 年全国科技活动周的科普直播讲座。盐是我们生活离不开的日用品，“开门七件事，柴米油盐酱醋茶”，其中盐湖是食盐的主要来源。我国是一个多盐湖国家，盐湖水化学类型多样。盐湖是如何形成的？盐湖有什么资源？我们如何利用和保护好盐湖资源？孔凡晶研究员针对以上问题，以盐湖资源与大盐湖产业为主题，很好地回答了以上问题，并带领公众一起探索丰富多彩的盐湖世界。盐湖的钾盐资源、锂盐资源是我国另一类重要的资源，钾盐是作物钾肥的矿物原料，锂盐是锂电池的矿物原料，孔凡晶研究员呼吁大家利用好保护好盐湖资源。

1.5.2　参加了由探月中心牵头组织的火星基地探索活动

2020 年 8 月 17～24 日，参加了由探月中心组织的火星基地探索活动。柴达木盆地大浪滩干盐湖，是我所联合美国华盛顿大学在青海柴达木大浪滩地区建立的我国第一个具有国际影响力的火星地面模拟试验场，用于火星盐类环境比较行星学研究，取得了丰硕的研究成果，获得了泻盐在自然条件风化过程变化规律的认识，并针对火星盐类的起源及次生过程开展类比研究，在大浪滩干盐滩地区观测到了原生沉积的六水泻盐在暴露于地表风化后形成了硫镁矾，基于水合硫酸镁温湿相图和该区环境条件数据，我们推论高水合硫酸镁在火星条件下也会风化脱水形成硫镁矾，现今火星表面普遍存在的硫镁矾是高水硫酸镁通过次生变化而来。初步建立了盐碱极端环境条件下微生物多样性研究技术方法，获得了类火星盐类环境下微生物多样性的认识，为火星生命探测提供了一种新的思路和途径。

这次活动结合我国“天问”火星探测器的发射，公众对火星兴趣浓厚，孔凡晶研究员作为特邀专家参加了这次活动，对火星模拟试验场盐湖形成、盐湖与生命等相关科学知识进行了讲解，由腾讯公司制作了一部科普视频向公众播放，取得了良好的效果。

1.5.3 2020 年 7～10 月，重点实验室视频制作

为了更好地向公众介绍重点实验室，我们结合自然资源部平台交流活动，制作了介绍重点实验室的视频，为线上进行科普活动提供了重要的作品。

2020 年 11 月，赵元艺、叶传永等编写了画册“给盐湖装上千里眼，把论文写在高原上”，介绍了青藏高原盐湖野外观测站建设情况，工作原理及观测研究成果，观测站未来发展规划等，让读者充分认识了盐湖野外观测站，获得了盐湖相关的科学知识。

2 团队建设与人才培养情况

重点实验室现有在职研究人员 30 人，其中院士 1 人，研究员/教授级高工 9 人，副研究员（高工）11 人，助理研究员（工程师）9 人；此外还有客座研究员 29 人（院士 2 人），协作科研人员 12 人。近两年，科研队伍逐渐壮大，形成了一支以地质学、化学化工、生物工程学、环境生态学等多学科中青年科技骨干为主的科技团队。在读及毕业硕士研究生 30 人，博士研究生 7 人，博士后 4 人。

由郑绵平院士为统领和指导，形成了 3 个主要方向的团队。

2.1 盐湖及盐类矿产资源评价地质矿产团队与项目组

川东北锂钾项目组，成员包括张永生、邢恩袁、牛新生、苏奎、桂宝玲、彭渊、左璠璠等。

滇西南钾盐项目组，成员包括苗忠英、徐建明、齐文、施林峰、张雪飞（兼）、宋高（兼）等。

2.2 盐湖沉积、生态环境与全球变化研究团队与项目组

盐类资源与古气候项目组，成员包括侯献华、陈文西、樊馥、马妮娜、宋高（兼）、陈安东（兼）、张雪飞（兼）、马黎春（兼）、吕苑苑（兼）。

柴达木盐湖农业与生态环境项目组，成员包括孔凡晶、马黎春（兼）、雒洋冰（兼）等。

青藏高原科学观测项目组，成员包括赵元艺、叶传永、刘喜方、卜令忠（兼）、吕苑苑、陈安东、王海雷等。

2.3 盐类先进化工技术团队与项目组

成员包括乜贞、王云生、卜令忠、伍倩、余疆江、雒洋冰等。

3 合作开放与交流情况

由于受到新冠肺炎疫情的影响，拟开展的一系列学术交流活动取消，如拟参加在西班牙召开的第 14 届国际盐湖会议等因疫情取消，推迟到 2021 年 10 月 22 日线上召开。2020 年 11 月在杭州召开了全国矿床会议，本实验室共有 10 余人参加了这次会议，主持了“盐类资源与环境观测、成矿作用与开发利用”专题研讨会，并取得圆满成功。

给青藏高原盐湖装上温湿记录器：野外科学观测研究现状与未来发展

叶传永[1, 2, 3]，郑绵平[1, 2, 3]，赵元艺[1, 2, 3]，卜令忠[1, 2, 3]，余疆江[1, 2, 3]，乜贞[1, 2, 3]，陈文西[1, 2, 3]，侯献华[1, 2, 3]

1. 中国地质科学院矿产资源研究所，北京 100037
2. 自然资源部青藏高原盐湖野外科学观测研究站，北京 100037
3. 自然资源部盐湖资源与环境重点实验室，北京 100037

摘要 野外科学观测研究站是国家科技创新体系的重要组成部分。青藏高原是地球第三极、世界屋脊，也是我国重要的战略资源储备基地。盐湖广泛分布于青藏高原，集大气圈、水圈、生物圈和岩石圈演化的综合信息于一体，备受全世界关注。1990年至今，青藏高原盐湖野外科学观测研究站建成了“一站四点”观测网络，获取重点盐湖区完整、连续、准确、指标齐全的观测数据，积累卤水、盐湖沉积、生物等相关学科野外现象资料，为盐湖学、成矿学、生态学等基础学科的理论创新提供基础资料支撑，为青藏高原隆升机制、极端环境与气候变化等关键科学问题的观测研究提供野外基地，为盐湖资源包括锂铷铯等战略性关键矿产资源、生物资源等安全保护、合理开发、综合利用提供实验场所和平台，培养高原科学新的生长点，保障战略资源和环境的安全可持续发展。为国家培养了一支老中青结合的盐湖野外观测研究队伍，包括国家级人才11人次、高级职称37人次、博士后32名、研究生57名。利用野外观测和现场试验研究数据，形成了多项专利，其中一项被国家知识产权局评为优秀奖。未来，将充分发挥区位和学科优势，稳步提升盐湖野外科学观测研究站条件保障能力、数据汇聚能力、分析挖掘能力以及共享服务能力，努力将盐湖野外科学观测研究站打造成我国推进科技创新和青藏高原保护生态环境、促进生态文明建设的重要科技支撑平台。

关键词 盐湖；野外观测站；青藏高原；生态环境；资源安全

1 引言

野外长期科学观测和定位（定点）试验研究是整个科技工作的重要组成部分，它与野外科学考察和室内实验研究是相辅相成的，其研究成果对编制国家长远规划和区域规划、开发利用自然资源、国土整治和改造、自然灾害防治、生态环境保护、大型工程建设，以及发展具有我国特色的学科等方面具有极其重要的作用[1]。发达国家十分重视生态系统水、土、气、生要素的长期定位观测和联网研究，特别是长期观测和科研样地建设[2-6]。经过几十年的发展，我国基本建成了国家野外科学研究体系，积累了一大批第一手的科学观测研究数据，一批原创性科研成果不断产出[7-12]。

青藏高原盐湖面积大于1km^{2}的有352个，是世界最大的盐湖群，总面积约27136km^2，占全国盐湖总面积的68.35%，成矿最独特，矿种最齐全（含钾、锂、硼、铷、铯等）[13]。青藏高原盐湖分别提供全国85%和80%的钾和锂资源量，是我国钾锂资源的生产基地和后备基地[14]。而且，盐湖及其淡水补给系统，是干旱-半干旱地区人类赖以生存的重要生态地理单元，是地球大气圈、水圈、生物圈和岩石圈演化的窗口（图1），其生态环境脆弱，对气候变化反应尤其敏感。盐湖区气象-水文、生态环境记录是研究区域气候变化与水循环的重要基础数据[15]。盐湖是旱区生态环境变化的灵敏指示器，对气候变化具有敏感性、超前性和预警性。西藏盐湖水位升降变化较复杂，影响高原封闭型湖泊水量均衡的因素有气温、蒸发、降水量、冰川融化等，湖泊水循环模式、能量均衡模式、气温-降水-冰川变化与湖泊水位的关系等很多问题还有待深入研究。通过盐湖科学观测，可以积累长期的气候、水文与水化学动态变化和生态环境基础科学数据，是进行青藏高原盐湖成矿作用、盐湖生态环境、沉积作用等研究的第一手珍贵数据资料。

近 200 年来，国际上发达国家高度重视对重要盐湖（美国大盐湖[16]、以色列的死海[17]等）的野外科学观测[18]。野外观测站从最初的气象、水文等较为单一的观测，发展到地球化学、盐类沉积、水文模型、生态环境等综合观测。所获得的大量基础数据，为盐湖学学科发展和脆弱生态环境预警及宏观决策等提供了不可缺少的支撑。

1990 年，为了开发西藏扎布耶盐湖提锂工艺，中国地质科学院矿产资源研究所科研团队开始在扎布耶盐湖修建野外观测站。到今天，已经初步建成了青藏高原盐湖野外科学观测研究站（以下简称“盐湖野外站”）。盐湖野外站地处高寒、缺氧的青藏高原。2010 年入选国土资源部首批建设野外科学观测基地，成为部属台站，2019 年成为自然资源部首批野外科学观测研究站。目前已经在青藏高原建立了覆盖典型盐湖的“一站四点”观测网络（图 2）。一站：西藏扎布耶中心站；四点：西藏当雄错观测点、西藏班戈湖观测点、青海察尔汗观测点和青海西台吉乃尔观测点。其中，西藏扎布耶盐湖是世界上唯一天然沉积碳酸锂的盐湖；当雄错观测点湖面变化对冰川消退反应灵敏；班戈湖是西藏第一大湖色林错的分蘖湖，值得重点观测研究；青海察尔汗盐湖是我国最大的氯化钾生产基地；西台吉乃尔湖是高镁锂比盐湖，是生产锂盐和硫酸钾的基地。“一站四点”分布在青藏高原典型盐湖成矿带上，在面向国家钾、锂矿产战略需求和盐湖学学科发展长远需要方面，具有明显的领域和区域代表性。

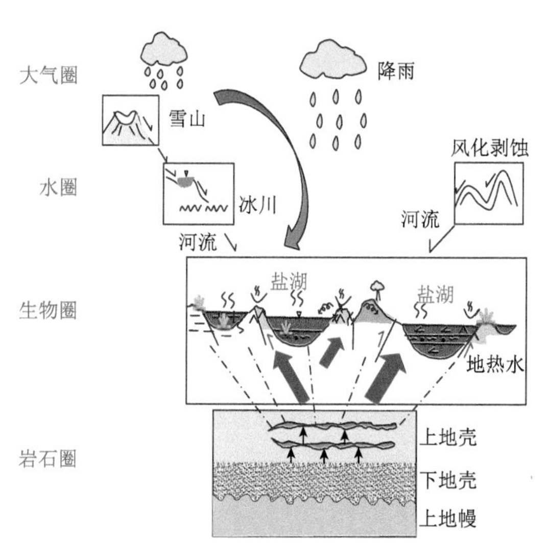

图 1　盐湖圈层演化示意图[4]

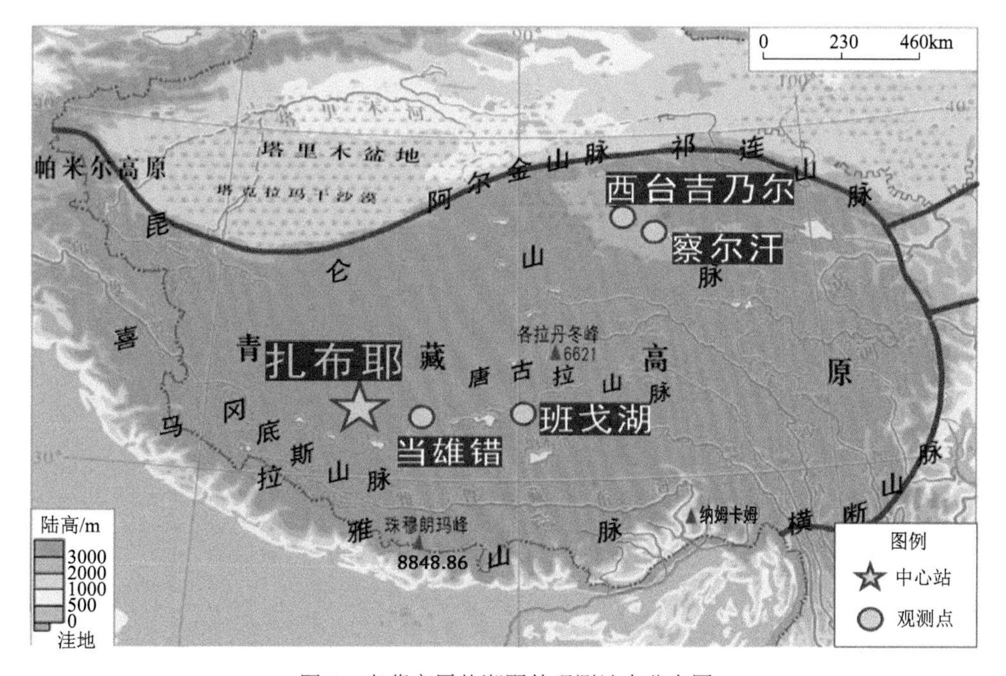

图 2　青藏高原盐湖野外观测站点分布图

2　定位与科学目标

盐湖野外站获取青藏高原盐湖区翔实、连续、系统的原始数据，定位于建设世界一流的盐湖观测数据共享平台，为研究亚洲水塔动态变化提供基础数据，为揭示盐湖关键矿产超常富集、水盐动态平衡、盐湖区气候变化、盐类沉积规律提供观测资料，推动盐湖相关观测技术的创新，引领盐湖学的发展。

科学目标是盐湖形成演化对层圈交互作用的响应。包含以下 4 个方面：恢复高原近代盐湖的盐度、温度、湿度的精细、高分辨率变化过程；重建 13 万年以来多次大湖期与周缘水灾的相关性；预测未来气候、环境变化趋势；揭示盐湖战略性矿产物质来源和富集规律。

3　主要研究内容

围绕上述定位和科学目标，盐湖野外站主要开展两个方面研究。

3.1　水盐动态平衡与成矿规律

制定盐湖水盐动态变化监测数据规范，实时采集气象、水文等数据，传输至数据处理系统，组成盐湖动态变化数据库，建成动态变化数据共享平台。依托盐湖野外站，利用基础观测数据，开展钾、锂、硼、溴、铯等资源的水盐动态平衡规律研究。结合基础地质、水文地质等资料，揭示盐湖中钾、锂、铷、铯、硼等战略关键矿产在岩浆演化-热水运移-地表风化-地表聚集过程中超常富集成矿模式和规律。探讨物源和气候变化对盐湖演化的控制作用。

3.2　湖泊-盐湖沉积与气候条件的耦合关系

综合利用沉积年代学、盐类矿物学、有机地球化学和微体古生物学等观测数据，开展湖泊-盐类层序对气候环境变化的响应研究。通过对第四纪盐湖和现代盐湖沉积的观测与野外试验研究，揭示气候水文变化对湖-河灾害与盐类沉积的控制作用。

4　科研能力与技术平台

盐湖野外站修建了固定建筑物作为工作和生活场所，建筑面积共计 $2200m^2$，配备了相应的盐湖观测仪器设备，现场可完成气象及水文自动观测、化学分析、矿物鉴定等基础工作，具备了在盐湖野外现场开展观测研究工作的基本条件。

4.1　观测内容

对大气圈、水圈、生物圈和岩石圈与盐湖的多圈层交互关系进行全方位观测，具体内容如下。

大气圈：气温、降水量、蒸发量、地温、气压、日照、风向、风速等。

水圈：冰川消退速率、组分含量等；河水与泉水的流量、组分含量、水温、pH 值等；卤水的盐度、组分含量、水位、水温、pH 值等。

生物圈：盐水域（浮游动物/植物/微生物）、盐沼带（植物）的生物种类、形态、数量、生态环境参数等。

岩石圈：盐类沉积物的颜色、粒度、结构、比重、孔隙度、矿物组分、种类等；湖相沉积物的色度、粒度、孔隙度、化学组成、矿物组分、种类等；岩石圈补给的泉水的流速、流量、组分含量、水温、pH 值等。

4.2　观测手段

大气圈：自动气象观测、人工一天三次定时观测、干湿沉降仪等。

冰川：冰缘定点观测、定期采样、遥感解译等。

热水、河水与泉水：定点定时采样观测、流速计等。

卤水：定时定点取样观测等。

生物圈：主要指盐水域、盐沼带，定时定点采样、鉴定。

盐类沉积物：定时定点观测、采样、冷藏、矿物鉴定等。

湖相沉积物：钻探、取心、冷藏、分样、矿物鉴定等。

4.3 观测数据

自 1990 年以来，观测数据涵盖了与盐湖相关的大气圈、水圈、生物圈和岩石圈的各类数据，容量约 200GB，记录本累计约 600 本（册）。

现场利用自动气象观测和人工定时观测采集了大气圈气象参数，包括 8 个参数：气温、降水量、蒸发量、地温、气压、日照、风速、风向。

定点定时采样观测河水、湖水和蒸发实验过程中的卤水，在现场记录河水的参数：流速、流量、水温、pH 值。湖水的参数：盐度、组分、水位、水温、比重、pH 值。蒸发实验卤水的参数：盐度、组分含量、水深、气温、环境湿度、水温、pH 值、比重。在扎布耶中心，对扎布耶盐湖卤水开展天然日晒蒸发实验；在当雄错观测点，对当雄错、拉果错等盐湖湖水开展天然日晒蒸发实验；在班戈湖观测点，对班戈湖、杜佳里湖、朋彦错盐湖卤水蒸发池自然蒸发实验。

以当雄错观测点为基地，研究当雄错、达则错生态系统结构及动态变化规律，布置了 10 个调查点，2011 年 11 月、2012 年 7～8 月对当雄错盐湖水深、水温、气温、酸碱度、透明度及盐度等指标进行了现场观测。

岩石圈的观测数据包括盐类沉积物、湖相沉积物和受岩石圈补给的泉水。根据前面章节设定的观测内容，定点定时采集、观测泉水，在现场记录泉水的参数（流速、流量、水温、组分、pH 值），蒸发实验盐类沉积物的参数（化学组分、矿物结构、矿物种类）。

5 研究成果与科学贡献

自 1990 年以来，中国地质科学院矿产资源研究所盐湖科研团队，通过盐湖野外站这个平台，先后承担了国家重点科技攻关等近 30 项科研项目，合计项目经费 3700 万元，共计发表学术论文 400 余篇，专著 7 部，咨询报告 4 份。以盐湖野外站观测数据为基础，作为第一获奖单位先后获得国家科学技术进步奖二等奖（2 项）（图 3）、部一等奖（3 项）、国家优秀专利奖（1 项）等各级奖项共计 6 项，获发明专利 4 项。

在盐湖野外站定点观测基础上，提出了多级盐湖成矿模式，其中多级中浅盆盐湖成矿模式主要分布于藏北高原，多级深盆盐湖成矿模式主要分布于青海柴达木盆地。

利用野外观测和现场试验研究数据，支撑了盐湖锂、钾资源开发工艺研究，形成了多项专利，为当地资源开发利用提供了科学依据。尤其是基于扎布耶盐湖长期观测和现场试验，研制了仅仅使用当地的太阳能和冷气候资源而开发盐湖锂矿的绿色技术，为世界独有，申报了专利“利用太阳池从碳酸盐型卤水中结晶析出碳酸锂的方法”。该专利 2009 年被国家知识产权局评为优秀奖。该技术已由西藏日喀则扎布耶锂业高科技有限公司实施，开创了我国盐湖卤水提锂之先河，在扎布耶盐湖建设完成了年产 5000t 碳酸锂精矿的生产线，产生了很好的社会经济效益。

6 人才培养与队伍建设

盐湖野外站重视人才梯队建设，形成了一支科研意识超前、技术力量雄厚、观测水平过硬、管理实施高效、老中青结合、专业结构合理的盐湖观测研究队伍，这些人员长期从事盐湖科学观测、野外试验研究等工作。长期在观测站工作的科研人员和技术人员累计超过 100 人。目前，有 40 多人直接参与野外观测和研究工作，其中院士 1 人，研究员和教授级高工 5 人，副研究员和高工 8 人；拥有博士学位者 27 人；另外外聘 12 人，包括气象 4 人、水文 4 人、化验 4 人。

自 1990 年建站以来，培养了中国工程院院士 2 位，国际盐湖协会主席 1 人，国际盐湖学会执委 1 人，中科院“百人计划”2 人，国土资源部“百人计划”2 人，青藏高原青年科技奖 2 人，国土资源部杰出青年科技人才 1 人；研究员/教授 12 人，副研究员/高工 25 人；博士后 32 名、博士 25 名、硕士 32 名。

图 3　青藏高原盐湖科学观测获得的部分奖项

7　开放与交流

盐湖野外站积极开展以我为主的国内外合作与交流。自建站以来，每年都有外单位科研人员到观测站上开展相关科研活动。外单位包括中国科学院系统（青藏高原研究所、青海盐湖研究所、地理科学与资源研究所等）、教育部系统的高校（北京大学、北京师范大学、南京大学等）、自然资源部下属科研院所（中国地质调查局国家地质实验测试中心、中国地质调查局成都地质调查中心等）、其他单位科研机构等。开展的课题有中国地质调查局地质调查项目、国家重点研发计划项目和课题、国家自然科学基金等。

近 5 年来，共接待外单位累计 100 人次，合作开展国家自然科学基金等项目约 20 项，是我国学者在青藏高原盐湖和湖泊研究的主要平台之一。2014 年 7 月 13～18 日，第 12 届国际盐湖会议在河北省廊坊市香河中信国安第一城会议中心成功召开（图 4）。来自中国、俄罗斯、澳大利亚、美国、以色列、巴西、伊朗、克罗地亚、埃及、西班牙、阿根廷及哈萨克斯坦等国家的 300 多名与会专家学者，紧紧围绕“未来盐湖，全球可持续性研究与发展”这一会议主题，对全球变化与盐湖记录、盐湖生态与生物资源、盐类地质学与资源勘查及盐类化工等内容进行了研讨，共同为盐湖未来的科学研究、资源综合利用及保护建言献策。

每个野外站现场都建立了标示说明系统，人们随时可以去参观。多年的地球日，以盐湖观测数据为依托，在观测站现场或者北京开展相关科普活动。

图4 郑绵平院士在2014年第12届国际盐湖会议上做报告

8 发展展望

突出盐湖野外站区域代表性和研究领域代表性的优势，以解决国家重大科学技术问题的能力为目标，特别是青藏高原生态安全屏障建设和我国锂硼钾等战略性矿产资源安全，努力建设自动采集、传输、数据处理的观测系统，最终建成国际一流盐湖野外科学观测研究站。

8.1 总体思路

利用青藏高原区位和盐湖学学科优势，获取长期、连续、高精度的盐湖综合观测数据。阐明青藏高原盐湖水盐动态变化、成矿作用、盐湖极端生态环境以及盐湖沉积对环境的响应。为盐类战略关键矿产资源保障、青藏高原环境保护、重大工程安全提供预警和支撑。

8.2 示范作用

获取和重建高原盐湖气候和水动态的高分辨率数据，为全球变化研究提供重要的依据。观测数据为社会、企业和当地政府决策所用，为地质灾害预警提供数据，为企业生产保驾护航。为我国与周边国家印度、尼泊尔等提供湖泊水文演化资料，为水文地质灾害提供预警。加强与俄罗斯、蒙古国、伊朗、阿根廷、智利、玻利维亚等有盐湖分布的“一带一路”沿线国家的合作交流。

8.3 未来工作重点

持续获取青藏高原重点盐湖高精度、高分辨率的观测数据。打造开放的盐湖科学观测基础设施和数据共享平台。申请极端环境盐湖观测技术专利。取得原创性研究成果，引领世界盐湖科学发展。主办国际盐湖科学观测学术研讨会，开展大型科普活动。

致谢

盐湖野外站自 1990 年成立至今，已有三十余载，从最初的西藏扎布耶观测点到现在基本覆盖青藏高原的“一站四点”观测网络，得到了包括自然资源部（原国土资源部）、中国地质调查局、科技部、国家自然科学基金委员会、西藏自治区自然资源厅、西藏自治区地质矿产勘查开发局等单位的支持，特别是依托单位中国地质科学院矿产资源研究所的大力支持。中国地质科学院矿产资源研究所盐湖研究团队（含聘用人员），不畏高寒缺氧，不分春夏秋冬，一批又一批赶赴盐湖野外站现场，开展仪器维护、读取数据、蒸发实验等工作。在此，对上述有关部门、单位和人员表示衷心的感谢。

参考文献

[1] 孙鸿烈. 发挥优势，提高野外观测试验水平[J]. 中国科学院院刊，1987（1）：5-9.

[2] 杨萍，白永飞，宋长春，等. 野外站科研样地建设的思考、探索与展望[J]. 中国科学院院刊，2020，35（1）：125-135.

[3] Storkey J，Macdonald A J，Poulton P R，et al. Grassland biodiversity bounces back from long-term nitrogen addition[J]. Nature，2015，528（7582）：401-404.

[4] Tilman D，Reich P B，Knops J，et al. Diversity and productivity in a long-term grassland experiment[J]. Science，2001，294（5543）：843-845.

[5] Finlay B J，Esteban G F，Fenchel T. Global diversity and body size[J]. Nature，1996，383（6596）：132-133.

[6] Isbell F，Calcagno V，Hector A，et al. High plant diversity is needed to maintain ecosystem services[J]. Nature，2011，477（7363）：199-202.

[7] 高春东，何洪林. 野外科学观测研究站发展潜力大应予高度重视[J]. 中国科学院院刊，2019，34（3）：344-348.

[8] 杨萍，于秀波，庄绪亮，等. 中国科学院中国生态系统研究网络（CERN）的现状及未来发展思路[J]. 中国科学院院刊，2008（6）：2，555-561.

[9] 杨萍. 需求导向　优势互补　合作共赢　推进野外站联盟建设——野外站联盟建设的进展与展望[J]. 中国科学院院刊，2014，29（5）：636-639，651.

[10] 牛栋，黄铁青，杨萍，等. 中国生态系统研究网络（CERN）的建设与思考[J]. 中国科学院院刊，2006（6）：466-471.

[11] 彭萍，朱立平. 基于野外站网络的青藏高原地表过程观测研究[J]. 科技导报，2017，35（6）：97-102.

[12] 科技部办公厅. 科技部办公厅关于组织填报《国家野外科学观测研究站建设运行实施方案》的通知[EB/OL].（2020-12-29）[2022-06-30]. http://www.most.gov.cn/xxgk/xinxifenlei/fdzdgknr/qtwj/qtwj2020/ 202012/t20201229_160425.html.

[13] 郑绵平，向军，魏新俊，等. 青藏高原盐湖[M]. 北京：北京科学技术出版社，1989.

[14] 郑绵平，张永生，刘喜方，等. 中国盐湖科学技术研究的若干进展与展望[J]. 地质学报，2016，90（9）：2123-2166.

[15] 郑绵平. 论盐湖学[J]. 地球学报，1999（4）：395-401.

[16] Wurtsbaugh W A. The great salt lake ecosystem（Utah，USA）：long term data and a structural equation approach：comment[J]. Ecosphere，2014，5（3）：1-8.

[17] Sirota I，Enzel Y，Mor Z，et al. Sedimentology and stratigraphy of a modern halite sequence formed under Dead Sea level fall[J]. Sedimentology，2021，68（3）：1069-1090.

[18] Wurtsbaugh W A，Miller C，Null S E，et al. Decline of the world's saline lakes[J]. Nature Geoscience，2017，10（11）：816-821.

第二部分

盐类及其他矿产

Progress in the investigation of potash resources in western China*

Mian-ping Zheng[1], Xian-hua Hou[1], Yong-sheng Zhang[1], En-yuan Xing[1], Hong-pu Li[2], Hong-wei Yin[3], Chang-qing Yu[4], Ning-jun Wang[5], Xiao-lin Deng[6], Zhao Wei[6], Zhong-ying Miao[1], Jia-ai Zhong[1], Fan Wang[6], Fu Fan[1], Xue-fei Zhang[1], Xu-ben Wang[7], Tu-qiang Liu[8] and Wei-gang Kong[1]

1. MNR Key Laboratory of Saline Lake Resources and Environments, Institute of Mineral Resources, CAGS, Beijing 100037, China
2. Qaidam Integrated Geological Exploration Institute of Qinghai Province, Golmud 816000, China
3. School of Earth Sciences and Engineering, Nanjing University, Nanjing 210093, China
4. Institute of Geology, Chinese Academy of Geological Sciences, Beijing 100037, China
5. Dazhou Heng Cheng Energy (Group) CO., LTD, Dazhou 635000, China
6. Institute of Geology of China Chemical Geology and Mine Bureau, Zhuozhou 072754, China
7. Chengdu University of Technology, Chengdu 610059, China
8. Sichuan Geological Survey, Chengdu 610081, China

Abstract Through the study of the geological conditions of potash deposits in China from recent years, a new understanding of potash theories has arisen that appropriate Chinese geological features. Important progress and substantial breakthroughs have been gained in the direction and management of potash prospecting. (1) Important breakthroughs in continental potassium prospecting: the "Quaternary gravel type deep potassium rich brine metallogenic model in western Qaidam" ensures Quaternary deep potassium rich brine prospecting will grow new KCl resources by 350 Mt, providing a resource guarantee for meeting the Chinese demand for sylvite. (2) The marine facies potash prospecting shows good prospects: the determination of the new type of Triassic polyhalite potash ore deposits in Sichuan provide an important scientific basis for the establishment of exploration planning and the selection of exploration target areas for polyhalite minerals in the Sichuan Basin; the "two-storey potash deposits model" in southwestern Yunnan has been confirmed, which indicates prospects for the exploration of potash in the deeper Marine facies in southwestern Yunnan are likely to be successful. The discovery of a high concentration of rich bromite salt and potash salt in the Paleogene of the Kuqa depression and the southwestern Tarim region provides strong support for the likelihood large-scale potash deposits exist in these regions.

Keywords Potash deposits theory; Potash basin; Resource quantity; Potassium bearing brine of sandy gravel bed in Qaidam; A new type of polyhalite potash ore in Sichuan; "Two-storey" potash forming model in Yunnan

1 Introduction

Due to the importance of salt resources in ancient and modern times involving human survival and development, before the Common Era, the mining and utilization of halite from salt lakes was recorded in both the Middle East and China; Ancient (Permian) sylvite was first discovered and developed in Germany 200 years ago. Salts science has been evolving for at least 169 years, or ever since Usiglio (1849) first made his complete salt analysis series from Kara-Bugaz's seawater evaporation. Ochsenius was the first to propose the famous "sandbank theory" around marine salt and potash deposits in 1877. He believed that a supply of seawater was necessary for large amounts of evaporate, as halite, which was widely recognized and dominated academia for more than 100

* 本文发表在：China Geology, 2018, 1 (3): 392-401

years. However the "desert theory" (Walther J, 1894) failed to recognize seawater replenishment; during the 1960s–1980s, and foreign scholars have put forward theories based on the background of the giant potash deposits of the "preparation basin-dry salt lake" potash deposits thesis (Ваяшко ММГ, 1965), "salt-forming in the deep water", "bull-eyes type" and "tears type" (Schmalz RF, 1969) and so on into potash theoretical models, and also that in the Mediterranean, for instance, the "dry deep basin" salt-forming thesis (Xu JH, 1980) was also put forward. At present, potash deposits, and Salt Lake brine containing potash, discoveries are distributed on six continents, but this distribution is inconsistent and unevenly distributed (Cocker MD et al., 2016; Prakash S and Verma JP, 2016). World potash resources are primarily produced in evaporative rock deposits in major marine craton basins of the Cambrian through Tertiary salt ages. Salt basins in China are made up of several microcontinents or blocks and orogenic belts, which have particularities but can be compared with other basins. The cratonization era of salt basins in China is also late, and much of this is "paraplatform" and formed by large intermediate blocks among active belts, where the impact of adjacent orogenic belts is powerful, and large continental and marine salt basins are formed in succession by the Cenozoic plateau collision mineralization. The above factors led to particularities in the metallogenic tectonic environment and halite and potash deposits in China's marine salt basins and continental salt basins: salts formed multiple times in different epochs, salts formed by migration and accumulation, the material composition of salts were diverse, salts deformation usually occurred late, salt that is deeply buried, and salt ores that are mainly in a liquid state. Marine salt basins are smaller, and the youngest continental lithium, potash, boron, and other salt resources formed after the late collision of the Tibet Plateau. According to geological characteristics of salt-formations in China, it is necessary to explore prospecting methods for salt resources such as lithium, potash, etc. Based on the development of China's continental potash deposits theory, substantial breakthroughs and some remarkable achievements have been made in guiding the exploration and prospecting work in the western region of the Qaidam Basin in recent years. As the Marine salt basin during the age of potash deposits is so old, and the transformation of the basin so extreme, it is difficult to find potash. However, with the emergence of new knowledge, new technology, and new discoveries, some remarkable progress has been made.

2 The development of the potash forming theory in China

2.1 The progress of the continental potash forming theory in China

In the early days of the founding of the People's Republic of China, under the background dominated by the international marine facies potash deposits theories of 1957 to 1959, Chinese geologists opened up a new way of potash prospecting based on Chinese actual geological conditions, and discovered and evaluated large-scale continental potash deposits such as Qarhan. Yuan JQ et al. (1982) put forward his famous theory involving the "salt-forming environment of deep basins surrounded by high mountains", and pointed out that the main source of potash deposits in the Qaidam Basin in China were medium-acidic magma with large areas of potassium-rich feldspar in the surrounding mountains, and potash bearing deep circulating water around the lake. Zheng MP et al. (1989) proposed the understanding of potash deposits through multistage salt lakes. Zhang PX et al. (1993) further proposed the potash deposits model of "shocking-drying in deep basins surrounded by high mountains". Scholars such as Yang Q et al. (1994) put forward the theory of "deep cycle replenishment" and "adjacent ancient flood lakes replenishment". Based on the recent potash prospecting in the Qaidam Basin, Zheng MP realized that although Qarhan is a mountain with a deep basin geomorphology background, most of its salts are in the multistage salt basin system that migrated from northwest to southeast and gradually enriched potash due to gravity fields and chemical differentiation. The specific process is as follows: the earliest salt-forming deposits of Qaidam were in Shizigou, in the southwest of Qaidam at the

Eocene, which migrated to the East via the North in the Dalangtan at the Oligocene, and moved to Chahansilatu during the middle Neolithic period, and moved to Yiliping at the Pliocene, and moved to Mahai in the early Quaternary; Finally, in the late Pleistocene era, the Qaidam ancient seesaw was inclined to the east due to the pivotal role of the faults in the middle of the Qaidam, which caused potassium-rich brine to accumulate in the 3 lakes area. A large number of drilling sections point to the Qarhan salt lake formation around 47 ka B.P. which eventually formed a large potash deposit (Zheng MP et al., 1989; Huang Q and Han FQ, 2007; Zheng MP et al., 2016). Zheng MP put forward the model of "salt forming in deep basins through multistage lake basins". During this research period, it was also discovered that China's Qaidam Basin underwent 56 Ma of "inherited potash", and it was not until the late Pleistocene period that the large-scale potash rich intercrystalline brine and local potash deposits were formed in the west of Qaidam and the Qarhan playa through reverse S-type migration (Zheng MP et al., 2012). Migration also happened in the lop nur from the northeast to the southwest, and large-scale potassium rich intercrystalline brine appeared in the north depression of lop nur (Wang ML et al., 2001). Migration by the tectonic evolution of salt basins, brought about "inherited potash", and other soluble salt differentiations and accumulations, and are able to provide favorable conditions for "potash deposits forming through inheriting from ancient salt deposits". Theoretical knowledge of "The Continental Potash Deposition" is further developed by the above understanding in our country, which guided an important breakthrough in potash prospecting within potassium rich brines in the gravel layers of western Qaidam (Zheng MP et al., 2013, 2015).

2.2 Progress in the marine potash forming theory in China

Geological structural conditions in China are different from other countries (Canada, Russia, etc.) where giant potash deposits were formed in the background of giant craton. Therefore, some potash theories summarized abroad are difficult to duplicate in China. Important ancient salt basins in China, particularly Marine salt basins, were all developed in the tectonic regions of the above-mentioned Precambrian basements and a protoplatform, especially salt deposits which were mostly developed in a relatively stable continental nucleus. For example, the Yangtze Triassic evaporation basin is located within the structurally stable zone of the Sichuan Basin, which coincides with the basement of the western-middle Sichuan continental nucleus as determined by Wang HZ et al., 2005. In addition, the Late Sinian-Cambrian thick gypsum-salt deposits also developed in this area. The distribution of the Early Ordovician salt basin in northern Shaanxi also overlapped with that of the Ordos continental nucleus. There are vast microcontinents in the salt forming basins of the eastern Tibetan-southwestern Yunnan in China. Compared with large international salt forming basins and potash forming basins, such as the famous Middle Devonian potash basin in Saskatchewan, Canada, which developed in the sub-stable tectonic region on the platform between the Canada Shield and the geosyncline, the south part of the most stable shield (Bear CA, 1973). Also, the Late Devonian potash basin in (Прйпятский) Prypiat of Belarus, which is located in the large graben between the Ukraine Shield and the Belarus Block. They are both in the relatively active sub-stable tectonic region in a stable tectonic zone (Раевсгскийчў BNидр, 1973). The study shows that the geological tectonic setting, which is very favorable for the formation of the potash basin, is the relatively active sub-stable tectonic region in stable tectonic zone and the stable tectonic region in sub-stable tectonic zone (the basement is composed of a continental nucleus or a protoplatform), respectively. Based on this understanding, with regard to a global perspective, in combination with regional geological tectonic settings, from the surface to the spot, deep analysis is used to arrive at the following understanding: (1) The salt forming belt in southwestern Yunnan should be classified as a Tethyan tectonic domain. Especially in the Middle-Late Jurassic period, located north of the equator in the mid-low latitude zone, where a favorable ancient geography and ancient climate environment for salt and potash was formed. The field data of the geological profiles, mine investigations and MES (high-precision method of electromagnetic frequency spectrum detection) indicated that there might be halite layers deep in the Mengye mine (Zheng MP et al., 2010, 2012). The results of Sr isotopic Rb/Sr analysis of halite in the mine indicated that the source of the material was Marine facies, and the salt-forming era might be Jurassic (Zheng MP et al., 2012). Based on the actual regional geological characteristics of marine salt basins in China, core observational, and the interpretative results of a large number of two-dimensional seismic data and drilling verifications, a "two-story potash forming model" was proposed (Zheng MP et al., 2014). (2) The Ordovician salt basin in northern Shaanxi belongs to the late Archean craton, and the salt-bearing layer of the 6th sub-member of the 5th member in Majiagou Formation ($O_2m_5^6$) developed a distribution area of 5.6×10^4 km^2 and an accumulative thickness of 159–200 m of the halite layers. The North China plate within the low latitude area during the Ordovician period, roughly 18° South latitude and 10° West longitude (Wang HH et al., 2016), or near 10°–12° North latitude (Zhao XX et al., 1992; Wu HN et al., 1991), predecessors have been discovered by drilling a thin sylvine layer (Liu Q et al., 1997). Since 2007, a 1 m thickness of the sylvine mineralization segment was found in the Suitan 1 well (KCl content 2%–5.24%) by the China Geological Survey project, located in the local salt forming depression of the most stable ancient continental block in China. Through 16 holes and well-seismic joint calibrations, the local salt forming depression with more than 150 m thick salt layers was found in the northern Shaanxi salt basin, and the northern Shaanxi region where "an uplift is caught in the middle of two depressions, 'W type complex bottom pot', potash forming model" (Zhang YS et al., 2013) was established preliminarily. The potash forming advantageous area was narrowed down to the 2000 km^2 range from 5×10^4 km^2, and the mid-eastern part of the northern Shaanxi as a favorable region for potash prospecting was further focused on (Zhang YS et al., 2014; Gui BL et al., 2017). (3) In Tarim, a typical Paleogene marine salt basin in China, an investigation into Kuqa was performed in the early 1990s. The method of "co-exploration of hydrocarbon and potash" was adopted, based on the detailed and highly accurate review of 70000 m of debris, and 50 m thick sylvine mine layers (in the Paleogene Kumugeliemu Formation, average K content 3.5%) was found in the Yangta 4 Well of the Yangtake structure, which illustrated the high possibility of Paleogene potash forming in the Tarim Basin (Deng XL et al., 2013).

3 Breakthrough of continental potash exploration in China

Qaidam Basin added 2.382×10^7 t of KCl resources in 2017 and the accumulative total of 3.5×10^8 t of KCl resources in 8 years (2010–2017). Under the guidance of the new recognition the "Genetic mode of potassium bearing brine of sandy gravel bed in the western Qaidam Basin", a successive set of potassium bearing brine of sandy gravel bed has been found in Quaternary deep layers of the Dalangtan-Heibei depression and the Kunteyi depression in the western Qaidam Basin (Zheng MP et al., 2015). Liang ZK09, Liang ZK10, Hei ZK01, Hei ZK02, Hei ZK03, Hei ZK04 and Hei ZK05 reveal that the bed is buried approximately 206–900 m deep. The bed has a length greater than 100 km, a width of 8–16 km and an area greater than 1000 km^2 (Fig. 1). The thickness of the bed exposed by the boreholes is between 197.30–692.68 m, with an average thickness of 354.48 m, and the single-well inflow is 338.86–9600 m^3/d. The chemical type of the brine is chloride, and the grade of KCl is between 0.18%–1.56%, with an average grade of 0.52%. KCl resources added in 2017 were 2.382×10^7 t. According to the latest estimation results, KCl resources (level 333–334) in 2010–2017 now total 3.5×10^8 t. The success of this new understanding of the potash forming theory in guiding prospecting illustrates the considerable breakthroughs China has realized in continental potash prospecting.

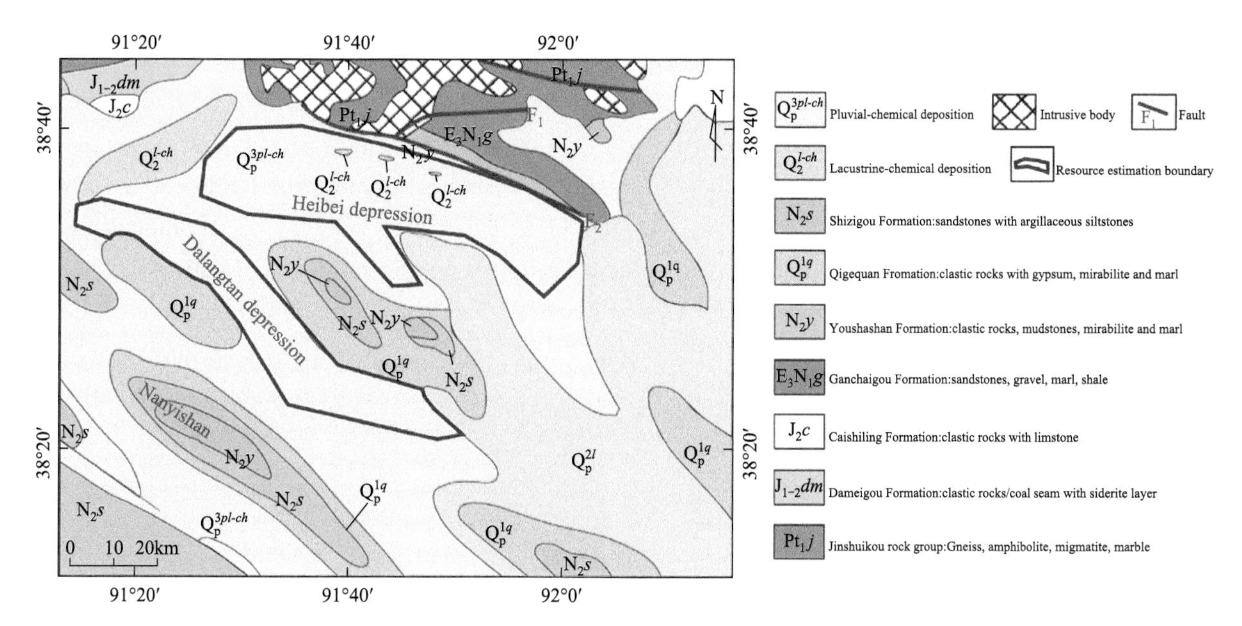

Fig. 1 The Quaternary deep potassium rich brine resource region in the Dalangtan and Heibei depression of the western Qaidam Basin. (modified from Zhang XT, 2007)

4 Recent developments in China in the exploration of marine potash

4.1 The determination of a new type of polyhalite potash ore in the Sichuan Basin

Sichuan Basin is rich in potash resources and although the solid phase and liquid phase of potash deposits are widely distributed, they mainly occur in the Triassic Jialingjiang-Leikoupo Formation. According to the previous survey, the polyhalite displays a prospect that is high in resources. However, most of the Triassic polyhalite in the Sichuan Basin is deeply buried, with the exception of the polyhalite around Huaying mountain fault which is shallow. Difficulties arise with complex structures with regard to exploration and development. Due to the fact that polyhalite with the gypsum, the anhydrite and the dolomite are usually dense massive intergrowths, this kind of polyhalite is not easily dissolved in water. Employing the fracturing method will likely encounter problems such as, hydration and plasticity. Compared with easily soluble potash resources, deep polyhalite mining is rather difficult. The objective conditions, which are difficult to mine at levels far beneath the surface but have scant resources in shallow layers and complex structures, that formed in the past indicate "polyhalite in Sichuan has no industrial value, known as 'dull mine'". However, polyhalite is a kind of high quality slow-release potash fertilizer, which has great strategic significance in terms of a solution to the reserve resources of potash in China. If we provide analysis using these new ideas, and develop scientifically and effectively, an excellent opportunity and a great challenge can be presented to Chinese geosciences and industry (Zheng MP et al., 2010). Under the guidance of Zheng MP, the potash resource surveys project team paid attention to the research on the cause and use of polyhalite 10 years ago, and Professor Anlianying from Chengdu University of Technology arranged to carry out research on the dissolution experiment of polyhalite. At the present time, many aspects and identifications have been obtained (An LY et al., 2008, 2010; Zhao XY et al., 2011). In 2012, the potash research team conducted a survey of polyhalite in the Changshou region of eastern Sichuan. A large number of dispersed grains/gravels of primary polyhalite was found in the salt rocks of the Jialingjiang formation of the Triassic in the Changping 3 well (Wang SL and Zheng MP, 2014), which was the main source of K^+ in the salt. This discovery dictated a new direction for the survey of potash resources in the Sichuan Basin.

In general, potassium sulfate and magnesium sulfate in polyhalite are soluble in water, while gypsum, which is insoluble in water, forms a protective membrane around the periphery of polyhalite particles, making the polyhalite insoluble again in water. As a result, it is difficult for polyhalite to react with water in the interaction with gypsum or dolomite (Fig. 2), however, the dispersed polyhalite in salt rocks is easy to dissolve (Fig. 3). Therefore, it is considered that the latter may be worthy of dissolution in order to exploit it. In 2017, the Hengcheng company successfully achieved the butt well and mining dissolution experiment in dispersed polyhalite bearing salt layers (Fig. 4), indicating that it is anticipated this kind of potash ore layer will be developed and utilized. After the author had been in the field, he further pointed out that the stable distribution and thick layer of polyhalite is a "new type of polyhalite potash ore", which has been unanimously approved by the company and the Sichuan geology and mineral department.

Fig. 2　Petrological and mineralogical characteristics of saline rock with polyhalite.

(a) Water leaching reaction of polyhalite; (b) microscopic features of saline rock with polyhalite (orthogonal light); (c) core photographs of saline rock with polyhalite

Fig. 3　Two types of polyhalite ore.

(a) The sample type of dispersed polyhalite in salt rocks which is easy to dissolve; (b) the sample type of the polyhalite with the anhydrite, being dense massive intergrowths, which is hard to be dissolved in water

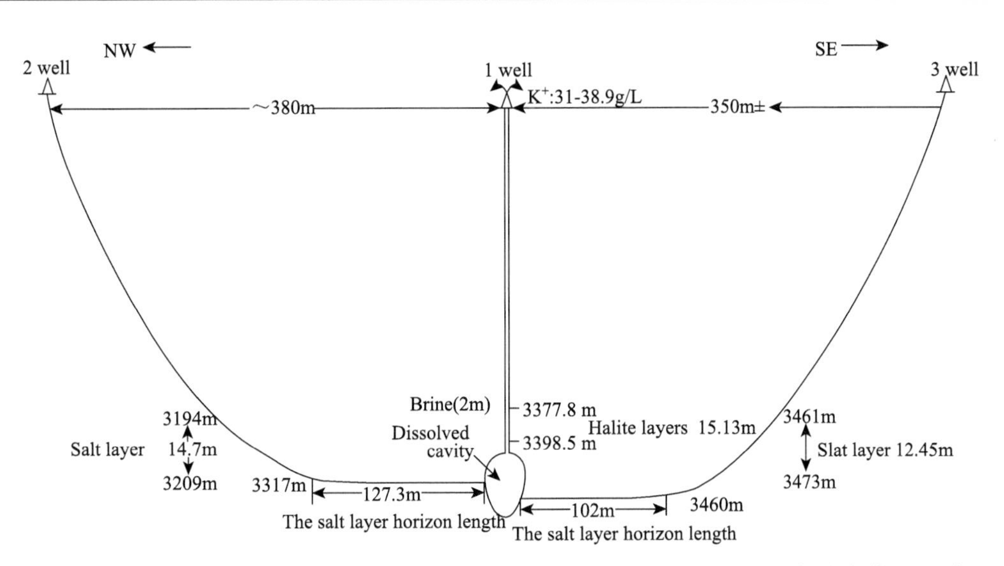

Fig. 4 A schematic diagram of the experiment on the dissolution and mining of the new type of polyhalite potash ore (simplified according to Dazhou Heng Cheng Energy (Group) CO., LTD).

When polyhalite meets water, it reacts as follows:

$$K_2SO_4MgSO_4CaSO_4{\cdot}2H_2O + nH_2O \longrightarrow CaSO_4{\cdot}2H_2O + K_2SO_4 + MgSO_4$$

The distribution of the new type of polyhalite potash ore is stable in the Sichuan Basin. The discovery of a thick and large-scale "new type of polyhalite potash ore" provides an important prospect for slow-release potash fertilizer and a huge amount of polyhalite development in this area. The prospective resources of potash could be in the billions of tons, measured by preliminary data. The determination of a new type of polyhalite potash ore breaks the traditional recognition that polyhalite is a "dull mine", and that the new type of polyhalite potash ore is selected from the traditional, indicating that the new type of polyhalite potash ore can be utilized. This discovery provides a new direction for potash surveys in the Sichuan Basin.

4.2 New evidence of potash forming in Jurassic of Lanping-Simao Basin

Some scholars believe that the potash deposit in the Mengye mine in southwestern Yunnan is mainly continental facies (Li MH et al., 2015), but most scholars believe that the potash deposit in the Mengye mine in southwestern Yunnan is a marine sedimentary environment (Liu Q, 1986; Xu XS and Wu JL, 1983; Shuai KY, 1987; Qu YH, 1997; Qu YH et al., 1998). However, controversy has emerged in the potash age, which directly affects the future direction of potash prospecting within the region. Based on the previous potassium prospecting and research work in southwestern Yunnan (Zheng MP et al., 2014), the strontium isotope analysis of several mine salt samples shows the source of the salt in the Mengye mine of the Simao Basin in Yunnan to be seawater (Miao ZY, 2018), which can be compared with the strontium isotope data of Jurassic seawater. However, the surrounding rocks in contact with the potash ore are typical continental sediments. These provide geochemical and sedimentary constraints for exploring the formation mechanism and distribution law of potash in this region. It is further demonstrated that the "two-storey potash forming model", reveals a good prospect for potash exploration in southwestern Yunnan.

4.3 The prognosis of potash salt resources and the discovery of high-grade bromine in the Kuqa depression of the Tarim Basin

4.3.1 The forecast for potash resources in the Kuqa depression, Tarim Basin

Based on the review and chemical analysis of the anhydrite and salt debris from oil exploration drilling in the Tarim Basin, potassium anomalies and potash mineralization (K = 0.52%, that is, KCl = 1.00%, is anomalous;

K = 1.05%, that is, KCl = 2.00%, is mineralization.) were found in Yangta 4 and Yangta 6 Wells in the south of the Kuqa depression. As shown in Fig. 5, the mineralization of the Yangta 4 Well was the highest and salinity, and the mineralization scale is the largest, with the K content up to 7.83%, while the Cl content generally around 20% to 50%. The lithology is potash bearing salt rocks or argillaceous potash bearing salt rocks (KCl = 14.94%), which can be divided into three mineralization periods: The upper potash mineralization period depth is 5141–5171 m, 30 m thick, and the average content of K is 3.71%. This periods lithology is mainly sylvite; The middle potash mineralization period depth is 5181–5192 m, 11 m thick, and the average content of K is 4.89%. The mid-upper part of this period's lithology is mainly sylvite and potassium bearing halite, and in the mid-lower part of the argillaceous the ingredient is increased. The lower potash mineralization depth is 5196 –5218 m, which is comprised of three layers of interbedded potassium bearing halite and halite layers. The accumulative thickness of the potash bearing salt layers is 9 m, the average content of K is 2.31%, and the cumulative thickness of the mineralization sections is 50 m. The mineralization section of potash bearing salt in Yangta 6 well is one layer, with a K content of 1.07%, a depth of 5078–5079 m and a thickness of 1 m. The main lithology is potassium-bearing halite, and the mineralization is weak.

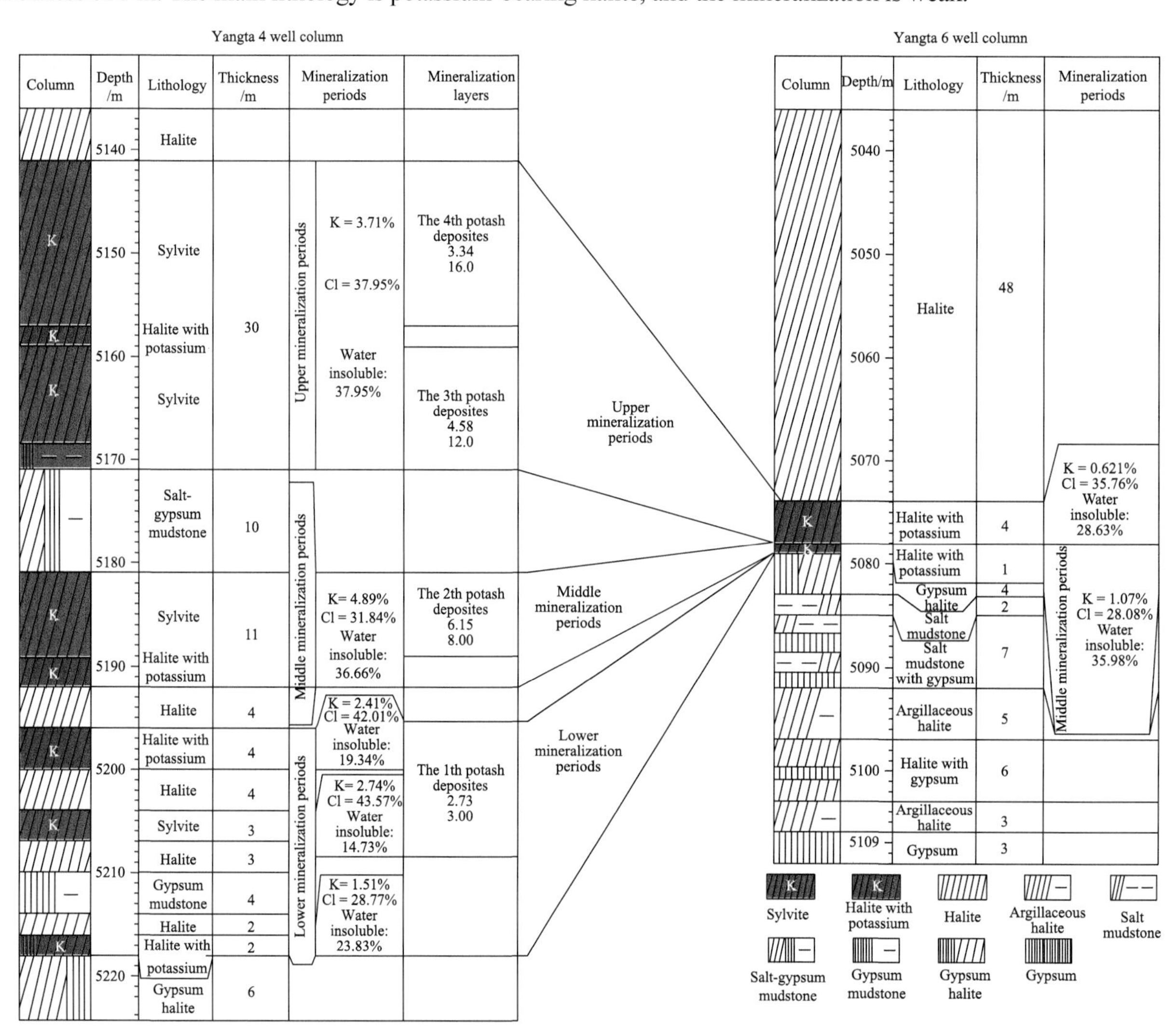

Fig. 5　Comparison of mineralization of potash bearing strata between the Yangta 4 well and the Yangta 6 well in the Kuqa depression, Tarim Basin, China.

According to the well seismic calibration of the Yangta 4 well, wave impedance and gamma joint inversion were carried out for the potash containing target layer of Yangtake area (Fig. 6). The potash bearing strata in Yangtake area was distributed in a North-East South-West direction, and the potash Wells in Yangta 4 and Yangta 6 were used as the boundary in the South, and the potash bearing area was 133 km^2. According to the above parameters of Yangta 4 and Yangta 6 wells, the average proportion of potash ore is 1.993 kg/m^3, and the average thickness of potash ore is 25.5 m in the Yangta 4 and Yangta 6 wells, and the weighted average grade value of ore KCl in Yangta 4 and Yangta 6 wells is 7.11%. The prospective resource of potash was calculated, and the predicted resource quantity of KCl = area × thickness × ore specific gravity × grade = $133\times10^6 m^2\times25.5 m\times 1.993\times10^3 kg/m^3\times7.11\%\times10^{-3} t/kg = 4.8\times10^8 t$.

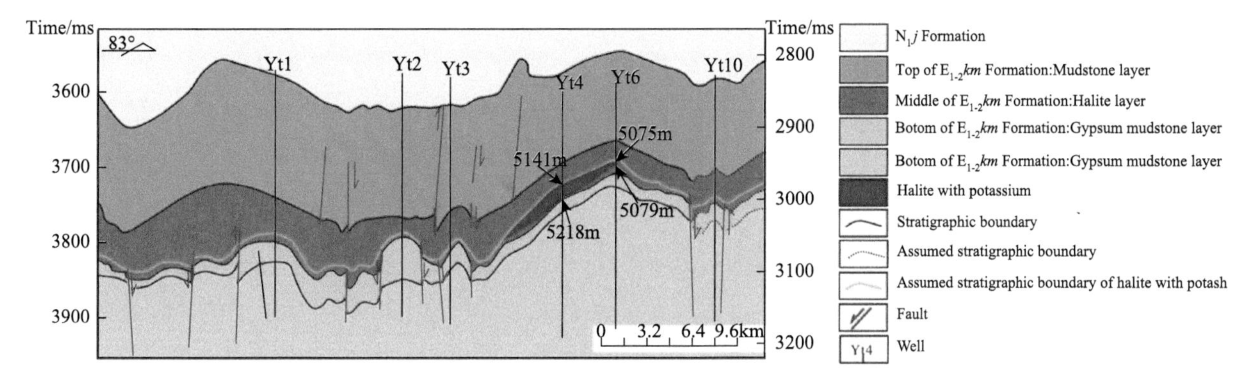

Fig. 6 A section map of the potash-bearing salt layer of the Yangtak tectonic seismic interpretation in the Kuqa depression, Tarim Basin, China.

4.3.2 The recent discovery of high-grade bromine ore in Tarim Basin

In the Kuqa depression and the southwestern Tarim depression of the Tarim Basin (Fig. 7, Fig. 8), the salt debris of oil and gas drilling was reviewed by using the method of "co-exploration of hydrocarbon and potash". After repeated chemical analysis of the debris, high bromine salt debris was found in the deep Paleogene of Keshen 208 well in the Kuqa depression and Wubo 1 well in the southwestern Tarim depression. Within the depth range (4950–5380 m, 6110–6630 m) of Keshen 208 well, the content of bromine in the argillaceous (gypsum)

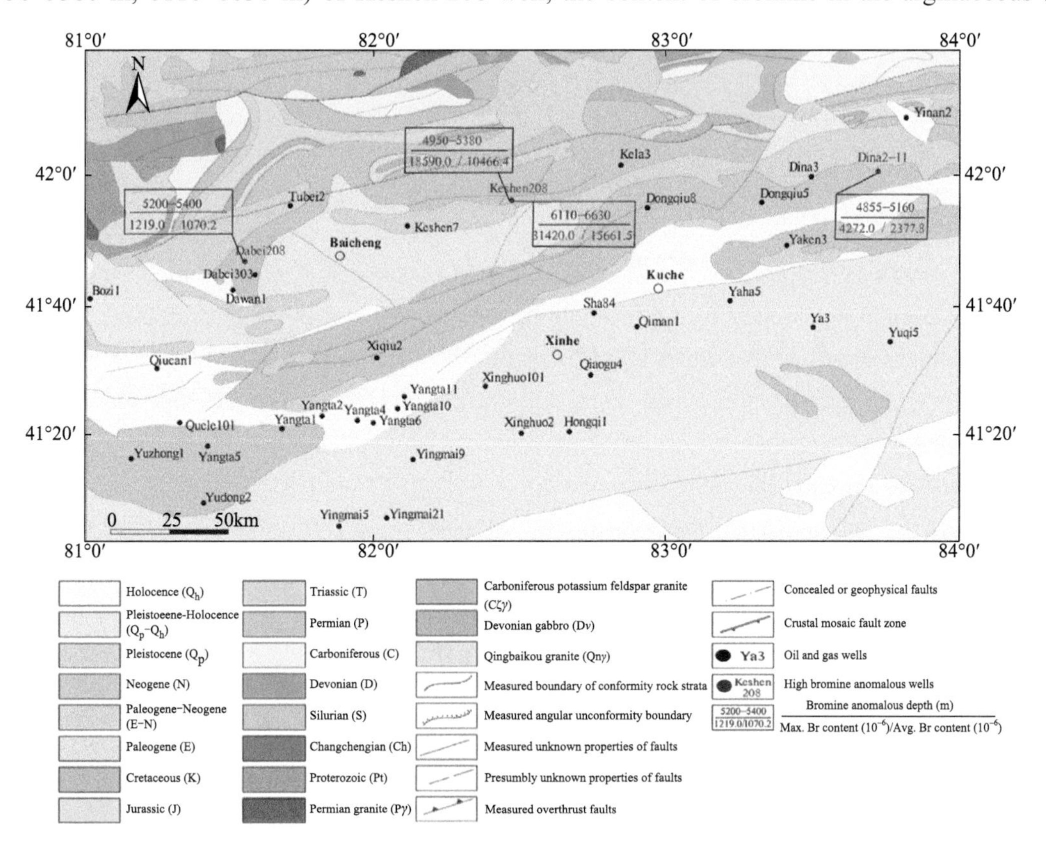

Fig. 7 Well location distribution with abnormal bromine characteristics in the Kuqa depression.

halite and the salt (gypsum) mudstone halite debris is more than 0.60% and 3.00% in some parts. Within the depth range (6430–6676 m) of the Wubo 1 well, the content of bromine in the gypsum (gypsum bearing) halite and the salt anhydrite debris is between 0.68%–0.78% and 1.00% in some areas. This initial discovery of high bromine salt rocks debris reflects the huge potential for bromine exploration in the southwestern Tarim depression and the Kuqa depression, and is of great significance for the prediction of potash resources.

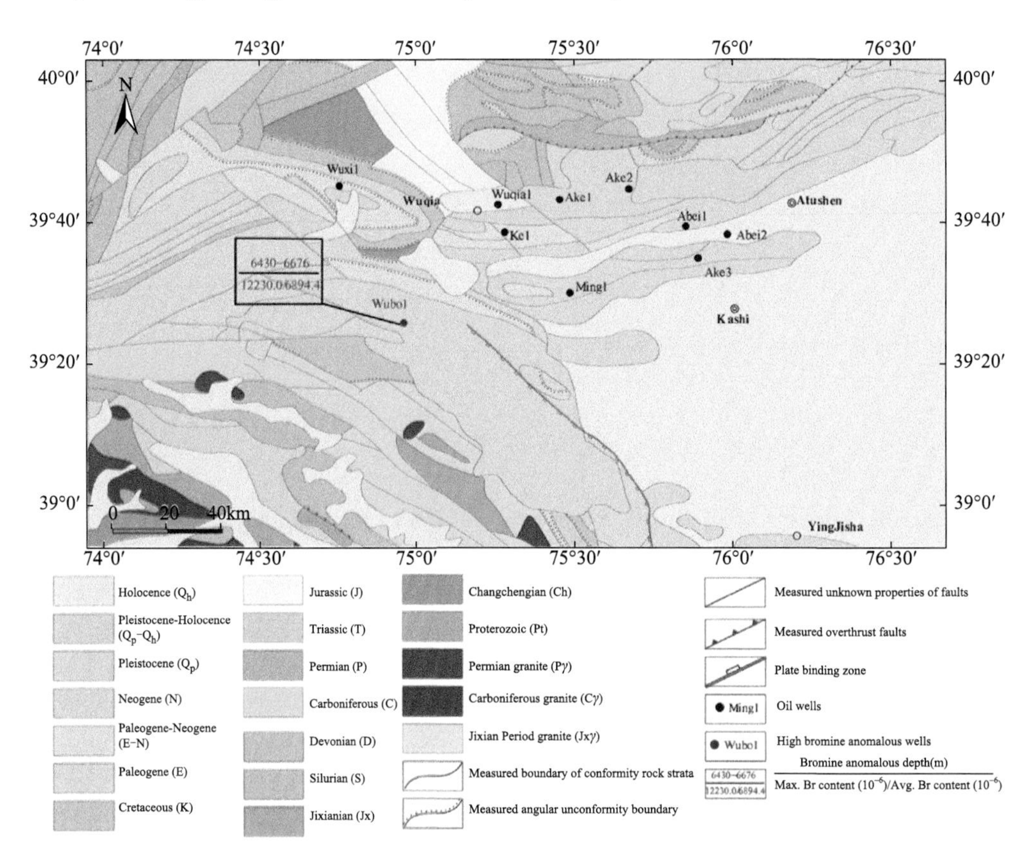

Fig. 8 Well location distribution with abnormal bromine characteristics in the northwest Tarim region.

5 Conclusions

Under the guidance of the new potash formation theory of continental facies in China, the total KCl resource amount in the Quaternary potassium bearing brine in the western Qaidam Basin increased by 350 Mt, providing an important resource guarantee for the new potash salt resource base of Qaidam and future food security. Substantial progress has been made in potash prospecting in marine facies. The determination of a new type of polyhalite potash ore in Sichuan has "activated" a large amount of polyhalite that has long been considered "dull mine", indicates a new direction for potash prospecting in Sichuan Basin. It further proved the era and origin of Mengye potash mine, which indicated the prospect of deep potash explorations in Jurassic. The predicted resource quantity of KCl in the Tarim Kuqa depression is estimated to be 480 Mt, which shows the prospect of potash resources to be high. The discovery of the Paleogene high-bromine salt rocks in the southwestern Tarim depression and Kuqa depression not only increase the available salt minerals in this area, but also anticipates, potash prospecting in the new area of the Tarim Basin.

Acknowledgment

This work was funded by the geological surveys secondary project of CGS of MNR, "Survey and evaluation of potash mineral prospect in western regions of China" (DD20160054). The national Science Foundation of

China and Joint foundation (U1407207), Task of "Potash-rich law, strategic constituency and deep detection technology demonstration in key potash bearing basins" (2017YFC0602806), belonging to the National key research and development plan "Deep Land Resources Exploration and Exploitation". The anonymous reviewers, Dr. Hao Zi-guo, and Dr. Yang Yan revised this manuscript with valuable suggestions and comments, the authors would like to express sincerely thanks to them.

References

An LY, Yin HA, Tang ML. 2010. Feasibility study of leaching mining of deeply buried Polyhalite and Kinetic leaching modeling. Acta Geologica Sinica, 84 (11), 1690-1696.

An LY, Yang ZJ, Liu N, Yin Huian, Tang ML. 2008. Modeling of leaching process of polyhalite in laboratory. Industrial Minerals & Processing, (10), 24-27.

Bear CA. 1973. Geological problems in Saskatchewan potash mining due to peculiar conditions during deposition of potash beds//Coogan A H. Fourth International Symposium on Salt, Vol. 1. Cleveland Ohio: Northern Ohio Geological Society: 101-118.

Cocker MD, Orris GJ, Wynn J. 2016. US Geological Survey assessment of global potash production and resources: a significant advancement for global development and a sustainable future. Geoscience for the public good and global development: toward a sustainable future. Geological Society of America Special Paper, 520, 89-98. doi: 10.1130/9780813725208.

Deng XL, Wei Z, Zhao YH, Wang J. 2013. Significance of the discovery of potash deposits in the Tarim Basin in the Paleozoic. Acta Mineralogica Sinica, 33 (S2), 755-756.

Gui BL, He DF, Zhang YS, Sun YP, Huang JY, Zhang WJ. 2017. Geometry and kinematics of extensional structural wedges. Tectonophysics, 699, 199-212. doi: 10.1016/j.tecto.2017.01.013.

Huang Q and Han FQ. 2007. Salt Lake evolution and paleoclimate fluctuation in Qaidam Basin. Beijing, Science Press, 52-90.

Li MH, Yan MD, Wang ZR, Liu XM, Fang XM, Li J. 2015. The origins of the Mengye potash deposit in the Lanping-Simao Basin, Yunnan Province, Western China. Ore Geology Reviews, 69, 174-186. doi: 10.1016/j.oregeorev.2015.02.003.

Liu Q, Du ZY, Chen YH, Jin RG, Yuan HR, Zhang FG, Zhu YH, Chen YH. 1997. Prospects for potash prospecting in the Ordovician of northern Shaanxi and the Carboniferous of Tarim. Beijing, Atomic Energy Press, 30-37.

Liu Q. 1986. A preliminary study on sedimentary characteristics of salt bearing layer series and potash layers in a certain place of Yunnan province (1968). Collected papers of institute of mineral deposit geology, Chinese Academy of Geological Sciences (18). Geological society of China, 1.

Ваяшко ММГ, 1965. Geochemistry of the formation of potash deposits. China Industry Press: 274-309.

Miao ZY. 2018. New Sr isotope evidence to support the material source of the Mengyejing Potash Deposit in the Simao Basin from Ancient Marine Halite or Residual Sea. Acta Geologica Sinica (English Edition), 92 (2), 866-867. doi: 10.1111/acgs.2018.92.issue-2.

Ochsenius C. 1877. Die Bildung der Steinsalzlager und ihrer Mutterlaugen salze. Pfeffer, Halle, 172.

Раевсгскийчў ВИ ИДР. 1973. МесторожДения калийиыхсолейСССР. изд "Недра".

Prakash S, Verma JP. 2016. Global perspective of potash for fertilizer production. Potassium solubilizing microorganisms for sustainable agriculture. Springer, New Delhi, 327-331.

Qu YH. 1997. Study on the same origin of potassium bearing brine in the Lanping-Simao Basin in China and Khora Basin of Thailand-The favorable layer and area of the potassium deposits in this area. Geology of Chemical Minerals, 19 (2), 81-98.

Qu YH, Yuan PQ, Shuai KY, Zhang Y, Cai KQ, Jia SY, Chen CD. 1998. The regularity and prediction of potash mineralization in the Lanping-Simao Basin. Beijing: Geological Publishing House, 1-114.

Schmalz RF. 1969. Deep-water evaporite deposition: a genetic model. AAPG Bulletin, 53 (4), 798-823.

Shuai KY. 1987. The evolution of Mesozoic and Cenozoic geological tectonic and evaporite formations in Yunnan. Geoscience, 1 (2), 207-229.

Usiglio Par MJ. 1849. Analyse de l'eau de la Mediterannee sur les cotes de Fiance. Etutes sur la composition de l'eau dela Mediterannee et sur I'exploitation des sels quelle contient. Ann. de Chim.et de Phys., 3 (27), 92-107.

Walther J. 1894. Lithogenesis der Genewrt-Beobachtungen uber die Bildung der Gesteine an der heutigen Erdoberflache. Fischer. Jcna: Teil der Einleitung in die Geologie. Fischer. Jcna, III, 535-1055.

Wang HH, Li JH, Zhang HT, Xu L, Li WB. 2016. The absolute paleoposition of the North China Block during the Middle Ordovician. Sci China Earth Sci, 46 (1), 57-66.

Wang HZ, Liu BP, Li ST. 2005. Tectonic division and development stages in China and its adjacent areas//Wang HZ Collected Works. Beijing: Science Press, 359-379.

Wang ML, Liu CL, Jiao PC, Han WT, Song SS, Chen YZ, Yang ZC, Fan WD, Li TQ, Li CH, Feng JX, Chen JZ, Wang XM, Yu ZH, Li YW. 2001. Potash resources of salt lakes in Lop Nur. Geological Publishing House, 29-31.

Wang SL, Zheng MP. 2014. Discovery of Triassic polyhalite in Changshou area of East Sichuan Basin and its genetic study. Mineral Deposits, 33 (5), 1045-1056.

Wu HN, Chang CF, Liu C, Zhong DL. 1991. The Paleozoic to Mesozoic paleomagnetic polar shift curves and the distribution changes of paleolatitude of north and south China block. Journal of Northwest University (Natural Science Edition), 3, 99-107.

Xu JH. 1980. Thin crust plate tectonics and collisional orogeny. Chinese science A, (11), 1081-1089.

Xu XS and Wu JL. 1983. Characteristics of potash deposit in Mengye well, Yunnan, discussion of the genesis of microelement geochemistry and. Journal of the Chinese Academy of Geological Sciences, 5 (1), 17-36.

Yang Q, Wu BH, Wang SZ, Cai KQ, Qian ZH. 1994. Geology of the Qarhan Salt Lake potash deposits. Beijing: Geological Publishing House, 174-191.

Yuan JQ, Huo ZY, Cai KQ. 1982. The advances in the theory of the origin of salt deposits and their influence on the study of mineral deposits. Mineral Deposits, (1), 15-24.

Zhang PX, Zhang BZ, Lowenstein TK, Spencer RJ. 1993. Origin of the ancient anomalous potash evaporites. Beijing: Science China Press, 121-123.

Zhang XT. 2007. Introduction to regional geology of Qinghai Province. Beijing: Geology Press.

Zhang YS, Xing EY, Zheng MP, Su K, Fan F, Gong WQ, Yuan HR, Liu JH. 2014. The discovery of thick-bedded sylvite highly Mineralized Section from the M56 sub-member of ordovician Majiagou formation in Northern Shaanxi Salt Basin and its potash salt-prospecting implications. Acta Geoscientica Sinica, 35 (6), 693-702.

Zhang YS, Zheng MP, Bao HP, Guo Q, Yu CQ, Xing EY, Su K, Fan F, Gong WQ. 2013. Tectonic differentiation of $O_2m_5^6$ deposition stage in salt basin, northern Shaanxi, and its control over the formation of Potassium sags. Acta Geologica Sinica, 87 (1), 101-109.

Zhao XY, An LY, Liu N, Yin HA, Tang ML. 2011. Laboratory simulation of in-situ leaching of Polyhalite. Procedia Earth &Planetary Science, 2 (1), 50-57.

Zhao XX, Coe R S, Liu C, Zhou YX. 1992. New Cambrian and Ordovician paleomagnetic poles for the North China Block and their paleogeographic implications. Journal of Geophysical Research Solid Earth, 97 (B2), 1767-1788. doi: 10.1029/91JB02742.

Zheng MP, Hou XH, Yu CQ, Li HP, Yin HW, Zhang Z, Deng XL, Zhang YS, Guo TF, Wei Z, Wang XB, An LY, Nie Z, Tan XH, Zhang XF, Niu XS. 2015. The leading role of salt formation theory in the breakthrough and important irogress in Potash deposit prospecting. Acta Geoscientica Sinica, 36 (2), 129-139.

Zheng MP, Xiang J, Wei XJ, Zheng Y. 1989. Saline lake on the Qinghai-Xizang (Tibet) Plateau. Science Press, Beijing, 25-353.

Zheng MP, Yuan HR, Zhang YS, Liu XF, Chen WX, Li JS. 2010. Regional distribution and prospects of potash in China. Acta Geologica Sinica, 84 (11), 1523-1553.

Zheng MP, Zhang XF, Hou XH, Wang HL, Li HP, Shi LF. 2013. Geological environments of the late Cenozoic lakes and salt-forming and oil-gas poolforming actions in the Tibetan plateau. Acta Geoscientica Sinica, 34 (2), 129-138.

Zheng MP, Zhang YS, Liu XF, Qi W, Kong FJ, Nie Z, Jia QX, Pu LZ, Hou XH, Wang HL, Zhang Z, Kong WG, Lin YJ. 2016. Progress and prospects of salt lake research in China. Acta Geoscientica Sinica, 90 (9), 2123-2166.

Zheng MP, Zhang Z, Yin HW, Tian XH, Yu CQ, Shi LF, Zhang XF, Yang JX, Jiao J, Wu GP. 2014. A new viewpoint concerning the formation of the Mengyejing potash deposit in Jiangcheng, Yuannan. Acta Geoscientica Sinica, (1), 11-24.

Zheng MP, Zhang Z, Zhang YS, Liu XF, Yin HW. 2012. Potash exploration characteristics in China: understanding and research progress. Acta Geoscientica Sinica, 33 (3), 280-294.

Geothermal-type lithium resources in southern Xizang, China*

Chenguang Wang [1, 2, 3, 4, 5], Mianping Zheng[1, 3, 4], Xuefei Zhang[1, 3, 4], Qian Wu[1, 3, 4], Xifang Liu[1, 3, 4], Jianhong Ren[5, 6] and Shuangshuang Chen[7]

1. MLR Key Laboratory of Saline Lake Resources and Environments, Institute of Mineral Resources, Chinese Academy of Geological Sciences (CAGS), Beijing 100037, China
2. Center for High Pressure Science and Technology Advanced Research, Beijing 100094, China
3. R&D Center for Saline Lakes and Epithermal Deposits, CAGS, Beijing 100037, China
4. Institute of Mineral Resources, CAGS, Beijing 100037, China
5. School of Gemology and Materials Technology, Hebei GEO University, 136 Huaiandong Road, Shijiazhuang 050031, China
6. Gemstone Testing Center of Hebei Geo University, 136 Huaiandong Road, Shijiazhuang 050031, China
7. Guangdong Provincial Key Laboratory of Geodynamics and Geohazards, School of Earth Sciences and Engineering, Sun Yat-Sen University, Guangzhou 510275, China

Abstract High-temperature geothermal water has abundant lithium (Li) resources, and research on the development and utilization of geothermal-type lithium resources around the world are increasing. The Qinghai-Tibetan Plateau contains huge geothermal resources; especially, Li-rich geothermal resources in southern Xizang, southwestern China, are widely developed. The Li-rich geothermal spots in Xizang are mainly distributed on both sides and to the south of the Yarlung Zangbo suture zone. Such resources are often found in the intensely active high-temperature Li-rich geothermal fields and, compared with other Li-rich geothermal fields around the world, the Li-rich geothermal fluid in the Xizang Plateau, southern Xizang is characterized by good quality: the highest reported Li concentration is up to 239 mg/L; the Mg/Li ratio is extremely low and ranges from 0.03 to 1.48 for most of the Li-rich geothermal fluid; the Li/TDS value is relatively high and ranges from 0.25–1.14% compared to Zhabuye Li-rich salt lake (0.19%) and Salar de Uyuni (Bolivia) (0.08–0.31%). Continuous discharge has been stable for at least several decades, and some of them reach industrial grades of salt lake brine (32.74 mg/L). In addition, elements such as boron (B), caesium (Cs), and rubidium (Rb) are rich and can be comprehensively utilized. Based on still-incomplete statistics, there are at least 16 large-scale Li-rich hot springs with lithium concentration of 20 mg/L or more. The total discharge of lithium metal is about 4300 tons per year, equivalent to 25,686 tons of lithium carbonate. Drilling data has shown that the depth is promising and there is a lack of volcanism (nonvolcanic geothermal system). With a background of the partial-melting lower crust caused by the collision of the Indo-Asia continent and based on a comprehensive analysis of the tectonic background of southern Xizang and previous geological, geophysical, and geothermal research, deep molten magma seems to provide a stable heat source for the high-temperature Li-rich geothermal field. The Li-rich parent geothermal fluid rushes to the surface to form hot springs along the extensively developed tectonic fault zones in southern Xizang; some of the Li-rich fluid flows in to form Li-rich salt lakes. However, most of the Li-rich geothermal fluid is remitted to seasonal rivers and has not been effectively exploited, resulting in great waste. With the continuous advance of lithium extraction technologies in Li-rich geothermal fluid, the lithium resource in geothermal water is promising as a new geothermal type of mineral deposit, which can be effectively exploited. This is the first study to undertake a longitudinal analysis on the characteristics, distribution and scale, origin and utilization prospects of Li-rich geothermal resources in southern

* 本文发表在：Acta Geologica Sinica (English Edition), 2021, 95(3): 860-872

Xizang, research that will contribute to a deeper understanding of Li-rich geothermal resources in the area and attract attention to these resources in China.

Keywords Geothermal resources; Lithium; Xizang (Tibet)

1 Introduction

In recent years, with the promotion and popularization of electric vehicles, the demand for lithium resources around the world has been continuously increasing (Park, 2012; Martin et al., 2016). Facing the rising price of the metal, the exploration for lithium and increase in mines has continued to be active worldwide (Liu L J et al., 2019). There are many geothermal fields worldwide, such as the Puga Valley in Kashmir (Chowdhury et al., 1974), the Puna Plateau in Argentina (Kasemann et al., 2004), and the Chilean Lahsen geothermal (Cortecci et al., 2006), revealing the presence of high concentrations of some rare and precious alkali metals such as lithium, rubidium, and caesium (Li, Rb, Cs) and dispersed elements such as arsenic, boron, and bromine (As, B, Br). In particular, the unusual enrichment of lithium in geothermal water is generally accepted as one type of significant and potentially valuable mineral resource for future exploration. Considering that geothermal resources also have huge potential to be green energy, the lithium resources in Li-rich geothermal fluid are receiving increasing attention (Campbell, 2009; Tomaszewska and Szczepański, 2014), and some scholars even point out that lithium resources in geothermal water can reach 2 Mt (Gruber et al., 2011; Kesler et al., 2012).

Lithium resources in salt lakes contribute three-quarters of the world's lithium demand and many scholars consider that geothermal water in Li-rich brine lakes is an important source of lithium (Shcherbakov and Dvorov, 1970; Zheng et al., 1983, 1989, 1990; Campbell, 2009; Munk et al., 2011; Tan et al., 2012; Yu et al., 2013; Ericksen et al., 1978; Luo et al., 2017). Therefore, much research work has been carried out on extracting and utilizing lithium raw materials from Li-rich geothermal water, and several studies have asserted that Li-rich geothermal water will become an effective way to meet the rising demand for the resource (Yanagase et al., 1983; Hano et al., 1992; Jeongeon et al., 2012; Krotscheck and Smith, 2012; Cetiner et al., 2015) because certain high-temperature geothermal waters have unusually high and important economic concentrations of lithium (Campbell, 2009). This is a particularly important lithium resource replenishment considering the current world shortage (Tan et al., 2012).

The Tibetan Plateau, as an eastern extension of the Mediterranean-Himalayan geothermal zone that belongs to one of the world significant representative geothermal zones, has very rich geothermal resources (Fig. 1) and a long history of research process. Most geothermal springs in this region show an unusual enrichment of lithium and other typical resources (e.g. B, Rb, Cs) dissolved in hot water or as geothermal deposits (Tan et al., 2018). Research on Li-rich geothermal resources in China started early when Zheng et al. (1983) discovered a hot spring with high lithium concentration in the southern part of the Bangkog Co in Xizang. They conducted preliminary research on its genesis as early as 1960. In addition, a preliminary estimate of the lithium resources in the Yangbajing geothermal field was also reported, with a resource of 390,000 tons (Zheng and Liu, 1982).

In recent decades, the huge geothermal energy in Xizang's geothermal resources has been effectively exploited (Pang et al., 2013). However, the degree of research on, and exploitation and utilization of the lithium resources are still very low, and they have not attracted their due attention and should be given more attention. With the rapid transformation of China's economy and industrial base, the high-tech industries are developing quickly. China's demand for lithium resources is increasing rapidly year by year, coupled with the continuous expansion of the international lithium demand market and the shortage of Li-rich mineral raw materials, so that any Li-rich water that can be used to extract lithium is an effective solution (Tan et al., 2018). Therefore, in the

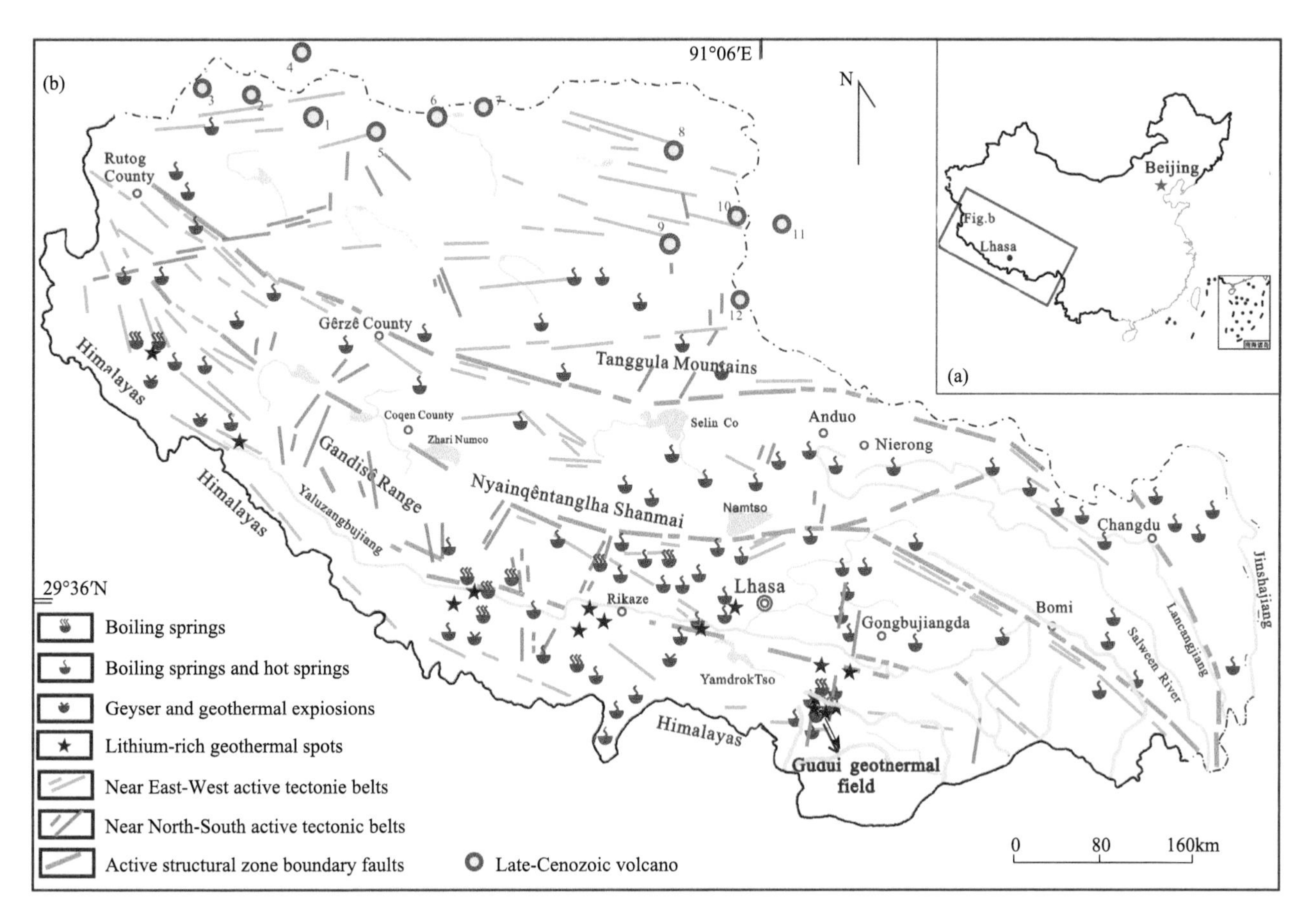

Fig. 1 Distribution of hot spring sites, active tectonics, and Late Cenozoic volcanoes in the Xizangan Plateau (modified from Zheng et al., 1995; Li, 2002; Guo et al., 2014; Luo et al., 2017) and lithium-rich geothermal spots in southern Xizang (after Tong et al., 2000; Wang and Zheng, 2019).

The 12 Late Cenozoic volcanoes are: 1, Shenglidaban; 2, Quanshuigou; 3, Liuhuangdaban; 4, Kaerdaxi; 5, Heishibeihu; 6, Dujianshan; 7, Huangyangling; 8, Yongbohu; 9, Bamaoqiongzong; 10, Kushuihuanbei; 11, Heiguoshan; and 12, Yuyehu (according to Guo et al., 2014). China basemap after China National Bureau of Surveying and Mapping Geographical Information.

context of the exploration, exploitation and market dynamics of lithium resources at home and abroad in recent years, in order to improve the current situation of research and utilization of lithium resources in the high-temperature Li-rich geothermal field in Xizang, based on our team's rich field and other research experiences in the Qinghai-Tibetan Plateau (Zheng and Liu, 1987; Zheng et al., 1983, 1990, 1995) and understanding of the development and utilization status of other high-temperature Li-rich geothermal fields around the world, we propose that Li-rich geothermal water is an important kind of lithium mine. Targeted field and indoor research work have been carried out in response to this concept. Based on this, we consider that with the continuous advancement of extraction technologies, geothermal-type lithium resources have the potential to become a new genetic type of lithium ore deposit.

This article absorbs and summarizes previous relevant research materials (Zheng and Liu, 1982, 1987, 2007; Zheng et al., 1983, 1989, 1990, 1995; Tong et al., 1981, 2000; Zheng et al., 1982; Duo, 2003; Luo et al., 2017; Guo and Wang, 2012; Pang et al., 2013; Tan et al., 2014, 2018; Zheng et al., 2015; Wang R C et al., 2017; Mao, 2018; Guo et al., 2007, 2009, 2010, 2017, 2019a, 2019b; Guo, 2012; Wang and Zheng, 2019; Wang et al., 2019; Xia and Zhang, 2019; Zhang et al., 2019). Together with field investigations and research on the key areas in southern Xizang (e.g., the Gudui geothermal field), we aim to report the characteristics, distribution and scale, origin and utilization prospects of geothermal-type lithium resources in southern Xizang. We also aim to attract attention to Li-rich geothermal systems, strengthen exploration investment and basic research of geothermal-type lithium resources, to find out the accurate lithium resources retained in geothermal water in China, and ultimately, implement rational extraction and utilization of the geothermal-type lithium in the near future.

2 Characteristics of high-temperature lithium-rich hot water in southern Xizang

The collision between India and Eurasia plates that occurred since the Eocene has been resulting in the uplift of the Tibetan Plateau and formation of the Himalayan geothermal belt (Blisniuk et al., 2001; Williams et al., 2001; Duo, 2003; Zhu et al., 2004; Taylor and Yin, 2009; Pan et al., 2012). Many geothermal-active regions occur in the central and southern parts of the Tibetan Plateau (Fig. 1). The high-temperature Li-rich geothermal resources in southern Xizang are mostly distributed on both sides of the Yarlung Zangbo suture zone and associated N-S trending rifts, especially at the intersection of the main and secondary faults (Han, 1990; Wang and Zheng, 2019; Fig. 1). Compared with other high-temperature Li-rich geothermal fields around the world, such as the El Tatio geothermal field in Chile (Cortecci et al., 2006), Clayton Valley (Munk et al., 2011), Rehai geothermal field in SW China (Wang et al., 2016), some active volcanic hydrothermal systems in Yellowstone geothermal field, USA (Guo et al., 2014), and those around Salar de Uyuni, Bolivia (Ericksen et al., 1978), the Li-rich geothermal fields in southern Xizang lack Quaternary volcanic eruptions and even volcanic rocks in the surrounding strata (Fig. 1). In fact, almost all the Late Cenozoic volcanoes in Tibet are located in the north of the country, as shown in Fig. 1. Therefore, the southern Xizang Li-rich geothermal fields all belong to high-temperature non-volcanic geothermal fields. The temperatures of the surface springs are more than 70 ℃, close to the boiling temperature of local water (83 ℃), and the location of the high-temperature hot springs are strongly controlled by small-scale secondary faults, such as Yangbajing geothermal field (Duo, 2003), Yangyi geothermal field (Yuan et al., 2014), Gudui Geothermal field (Fig. 2; Wang and Zheng, 2019) and the Mapamyum geothermal system (Wang et al., 2016). The fluid of the Li-rich hydrothermal systems is almost all pH neutral, and most contain Cl^- as predominant anions and Na^+ as predominant cation, whereas their Ca and Mg concentrations are generally low. The hydrochemical types are mainly Na-Cl type, with surrounding surface water mainly Ca-Mg-SO_4-HCO_3 and Na-Ca-Mg-SO_4-Cl-HCO_3 types (Guo and Wang, 2012; Guo et al., 2014; Wang et al., 2016; Wang and Zheng, 2019; Supplementary data 1).

Lithium concentration in the natural water is extremely low, with less than 0.01 ppm of lithium in the world's major rivers (Morozov, 1969) and most seawater contains no more than 1 ppm lithium (Schwochau, 1984; Kesler et al., 2012). In comparison, the Gudui cold spring water in Xizang only contains 0.017 ppm lithium (Wang, 2017).

The minimum industrial grade of salt-lake brines is LiCl 200 mg/L (equivalent to Li ion 32.74 mg/L), and the cut-off grade is LiCl 150 mg/L (equivalent to Li ion 24.56 mg/L), according to the specifications for salt-lake, salt mineral exploration DZT 0212—2002 issued by the Chinese government. In contrast, the lithium concentration in some geothermal fluid in southern Xizang is extremely high, and some of them reach the available industrial grades. For example, the lithium ion concentration of Jianhaizi hot spring is as high as 239 mg/L (Supplementary data 1). The lithium ion concentration of Zhumosha hot spring is as high as 65.40 mg/L and field investigation found that it develops on the Yarlung Zangbo River. The Moluojiang boiling spring has a lithium ion concentration of 50 mg/L (Supplementary data 1).

In addition, it is reported that more than 60% of the total lithium reserves (about 26.9 Mt) exist in brine, especially in those located in Bolivia, Chile, North America, and China (Cha et al., 2017). However, high concentrations of Mg^{2+} usually coexist with Li^+ in brine, by which the high cost and technical complexity for the recovery of Li^+ from the high Mg/Li ratio brine have led to the situation of lithium resources failing to meet

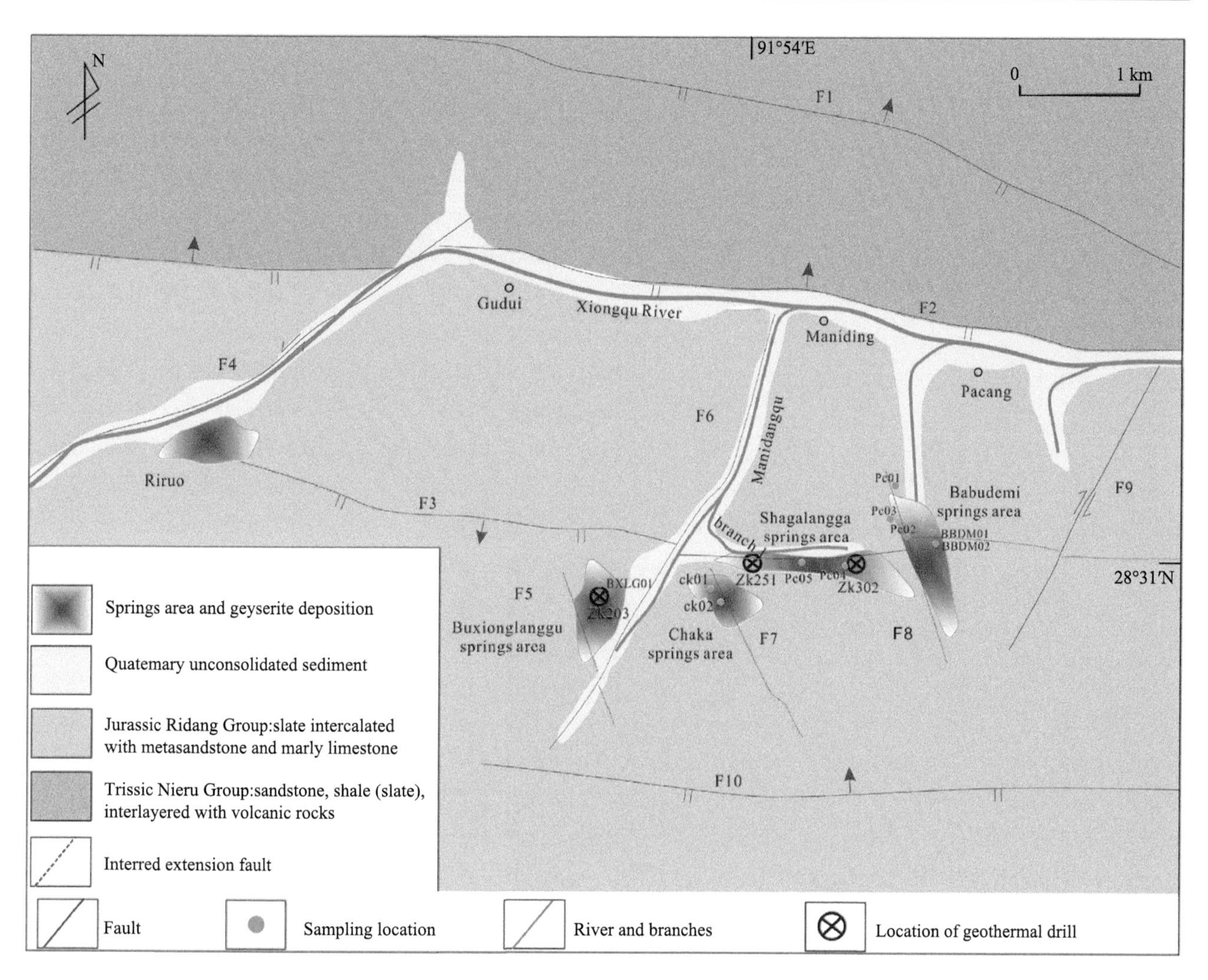

Fig. 2 Simplified geological map and sampling locations of the Gudui Geothermal Field. F5, F6, F7, F8 are extension faults; other properties unknown (modified from Liu et al., 2017; Wang, 2017; Wang and Zheng, 2019).

market demands (Zhao et al., 2017). It is noted that there are plentiful geothermal water resources with high Li/TDS value and low Mg/Li ratio properties in the world (Sun et al., 2019). China's high-temperature Li-rich hot water in southern Xizang has a very low Mg/Li ratio, between 0.03–1.48 (average 0.43; Supplementary data 1); this feature is highly conducive to industrial exploitation. Although the TDS values of most of the Li-rich hot waters are not high, the Li/TDS values are quite high, some as high as 1.14% (Supplementary data 1), which is even higher than Zhabuye (0.19%; communication data from Prof. Liu Xifang), the most Li-rich salt lake in the Qinghai-Tibet Plateau, and at Salar de Uyuni (0.08–0.31%; Schmidt, 2010), the richest Li-bearing salt lake in the world. Moreover, according to Zheng et al. (1983) and Tong et al. (2000), most of the Li-rich geothermal water in southern Xizang has been stably discharged for decades (e.g., Gudui geothermal field for at least 42 years; Fig. 3), and possess stable heat source and lithium source. Nevertheless, so far, most of the Li-rich geothermal water resources have flowed into the surface runoff, causing great waste. In recent years, many boreholes have uncovered the deep geothermal water of different geothermal fields, such as Yangbajing and Gudui, which has even higher lithium concentration than the surface hot springs (personal communication between academicians Zheng Mianping and Duoji), indicating that there is hope for finding more Li-rich geothermal resources in the deep high-temperature Li-rich geothermal field.

Since 1956, extensive geothermal geological surveys have been carried out in Xizang, finding that the hot waters in the Yarlung Zangbo geothermal belt are rich in elements such as B, Li, Cs, and Rb (Zheng et al., 1995). In 1995, during the field investigation in Xizang geothermal fields, Zheng M P discovered the anomalous enrichment

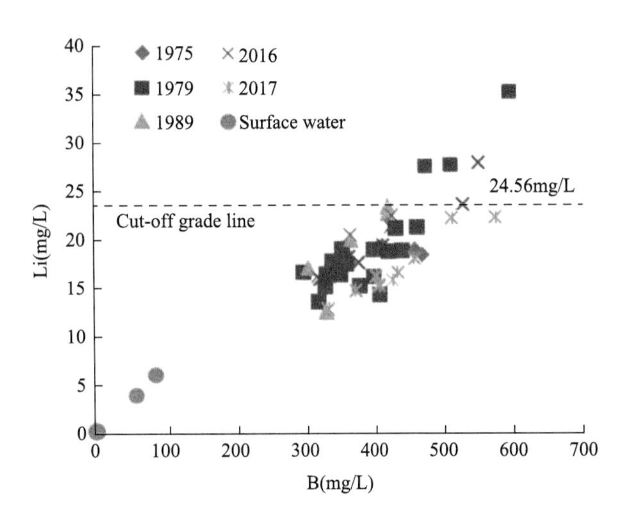

Fig. 3 Trends and relationships of boron and lithium concentration in the Gudui geothermal field from 1975 to 2017 (data from Tong et al., 2000;Wang, 2017; Wang and Zheng, 2019).

of Cs in the sinters, and then proposed a new type of hydrothermal deposit—Cesium-bearing geyserite. The Li-rich hot waters reported in this paper are also developed with high concentrations of B (Fig. 3), Cs, Rb and other resources that can be comprehensively utilized, which can undoubtedly increase the development and utilization value of the high-temperature Li-rich geothermal water in Xizang. In addition, we noticed that some high-temperature Li-rich hot water in Xizang also discharged massive As (approximately 6.6 mg/L in the south bank of the Semi; Zheng et al., 1995); the highest record is 126 mg/L of the Moluojiang hot spring in southern Xizang (Tong et al., 2000), Sb (Semi 1.40 mg/L, Zheng et al., 1995), W (tungsten concentration of Daggyai, Semi and Gudui varies from 289.1 to 1103 μg/L, Guo et al., 2019a), Hg (0.006 mg/L; Zheng et al., 1995) and other harmful elements.

Through the study of the geochemical characteristics and distribution of these harmful elements in high-temperature Li-rich geothermal fields, we can not only effectively control pollution by harmful elements in the Xizang Plateau, but also comprehensively make use of them in the future. In the near future, the value of the exploitation of these metal elements from the high-temperature Li-rich geothermal field will be comparable to that generated by the utilization of geothermal heat.

3 Distribution and scale of high-temperature lithium-rich geothermal resources in southern Xizang

According to genetic type, China's lithium deposits can be divided into endogenous and exogenous. Zheng M P (personal communication) classified Li-containing geothermal fluid into one of the ten kinds of lithium resources according to output background, distribution range, research, typical deposit characteristics and exploited degree. Here we focus on a brief introduction to geothermal-type lithium resources in southern Xizang, China.

China's high-temperature Li-rich geothermal fluid is mainly distributed in Xizang, especially in the south (Fig. 1). Since 1956, after successive large-scale investigations of the geology and geothermal deposits conducted by the Xizang Geology and Mineral Bureau, a Comprehensive Investigation Team of the Chinese Academy of Sciences (CAGS) and the Institute of Mineral Geology of CAGS, the extensive geothermal activities have received increasing attention. The survey found that recent medium- and high-temperature geothermal resources are mainly distributed in the southern part of the Qinghai-Tibetan Plateau, with deposits dominated by siliceous sinter, and the water-chemical type dominated by Na-Cl (Supplementary data 1). Nearly 700 geothermal display areas (points) have been discovered (Zheng et al., 1982; Zhang et al., 1989; Tong et al., 1981, 2000), among which 57 areas are high-temperature geothermal systems (temperature ≥150 ℃). The medium-high temperature geothermal systems generally have a high lithium concentration, especially the hot springs on both sides of the Yarlung Zangbo River, which are rich in B, Cs, Li, Rb and other elements (Zheng et al., 1995). During an in-depth study of the Daggyai geothermal field, Zheng et al. (1990) conducted a preliminary estimate of its resources and pointed out for the first time that the concentration of Cs, Li, Rb and B in many geothermal waters in Xizang reached a single comprehensive utilization index, such as the hot spring in the Semi geothermal field where Li concentration is as high as 35 mg/L. Grimaud et al. (1985) conducted a detailed studied of nearly 300 hot springs in Xizang and found that many geothermal fields produce geothermal water rich in B, Li and Cs. Although many waters in geothermal fields worldwide are enriched with a large amount of Li, many on the Xizangan Plateau show that the anomalous enrichment of elements such as B, Rb, and Cs has reached economic utilization grade (Grimaud et al., 1985). Li Z Q (2002) and Luo Y B et al. (2017) thought that the geothermal fluid activities in Xizang are basically consistent with corresponding activity tectonics. The high-temperature geothermal fluid activity areas in southern Xizang can be divided from west to east into: (1) the Dangreyongco-Dingri; (2) Shenza-Xietongmen; (3) Yadong-Gulu; and (4) Sangri-Cuona geothermal belts (Zheng W et al., 2015; Wang, 2017; Wang and Zheng, 2019), which is basically consistent with Zheng et al. (1983) conclusions about the Xizang Quaternary basin and linear structural sketches. As noted above, when researching the Li-containing hot springs in southern Bangkog Co, Zheng et al. (1983) thought that there was still a large amount of B, Li, potassium (K) and fluorine (F) that were carried out from the ground. On this basis, Zheng et al. (1983) discussed the relationship between Li-rich hot springs and volcanic, epithelial magma.

Based on data from the book *Thermal Springs in Xizang* (Tong et al., 2000), Li-rich geothermal fluid in Xizang was counted, as shown in Supplementary data 1. According to its now-out of date statistics, the total annual Li and B emissions of some high-temperature Li-rich hot springs with flow reports were calculated. The results show that the annual lithium discharge is 4,281 tons, equivalent to 25,686 tons of lithium carbonate, and boron emissions are as high as 91,882 tons (Supplementary data 2), which is enough to illustrate the huge utilization potential. Therefore, the lithium resources carried by geothermal systems can be used to meet the demand in the international market (Tan et al., 2012). Besides, the consensus is that lithium in the Li-rich salt lakes originated from geothermal water (Grimaud et al., 1985; Zheng et al., 1995; Tan et al., 2012), which further illustrates the importance of lithium resources in geothermal water and the urgency of promoting research, extraction and utilization of geothermal lithium resources.

In addition, most of the Xizang Li-rich geothermal water is often accompanied by high-concentration Rb, Cs, and B resources (Supplementary data 1), which further improves the utilization value of the geothermal-type lithium resources. Comprehensive analysis of previous research data, at the boundary condition of about 24.56 mg/L of lithium ion (equivalent to salt lake brine ore), yielded many high-temperature Li-rich hot (boiling) spring areas (screened out from Tong et al., 2000), including Moluojiang, Duoguoqu, Semi, Labulang, the water chemistry characteristics of which are shown in Supplementary data 1 and 2. Through field investigations and research at Buxionglanggu, Shagalangga and Babudemi boiling springs, we found in addition that the Chaka and Riruo hot springs in the Gudui geothermal field are all under the control of tectonic structures being found at the intersection areas of faults (Fig. 2). Plotting the lithium and boron concentration data in geothermal water from the past 42-year history of the Gudui geothermal field (Fig. 3), we found that it has remained basically stable, and that the surface water concentration is significantly lower than that of geothermal water. In addition, we also found that the relationship of boron and lithium concentration between the Gudui geothermal water and surface water is basically linear, indicating the characteristics of binary mixing; this shows that the surface hot spring is the result of mixing between deep Li- and B-rich geothermal fluid and surface water, which is further explained by the existence of a deep parent geothermal fluid (Wang and Zheng, 2019). Generally speaking, the potential of the high-temperature Li-rich geothermal resources further increases its development and utilization value.

4 Origin of high-temperature Li-rich geothermal resources in southern Xizang

Two-thirds of the world's supply of lithium resources relies on extraction from Li-rich salt-lake brines. Most Li-rich hot springs are distributed around the periphery of these salt lake brines, and they play an important role in the formation of continental Li-rich salt lakes, such as the tentatively listed World Heritage Atacama in Chile, Uyuni in Bolivia, Hombre Muerto in Argentina, Silver Peak and Searles in America. Bradley et al. (2013) highlighted the important role of geothermal activity in the basic characteristics of the continental Li-rich brine mineralization model. Munk et al. (2016) further summarized the control effect of geothermal activity on the continental Li-rich salt lake brine as follows: (1) it provides a geothermal fluid source for enhanced leaching of Li from source rocks; (2) it is also likely a direct source of Li from shallow magmatic brines and/or magmatic activity; (3) it may play a role in the concentration of Li through distillation or 'steaming' of thermal waters in the shallow subsurface; (4) thermally driven circulation may be an effective means for advecting Li from source areas. Probably the highly fractionated granites developed in southern Xizang are closely relative to the mineralization of W, Sn, Nb, Ta, Li, Be, Rb, Cs and REEs etc. (Wang R C et al., 2017) to regions of brine accumulation; and (5) it can result in the formation of the Li-rich clay mineral hectorite, which can in turn be a potential source of Li to brines if leaching and transport occur from the clay source. It can be seen that geothermal activity plays an important role in the mineralization of Li-rich salt lake brine. Therefore, in recent years, much research has been conducted on the formation of high-temperature Li-rich geothermal fluids; from the literature, we found that the key factors that control the formation of geothermal fields are the heat source and the passage that allow geothermal water to circulate from depth to the surface.

Geothermal heat has two sources: one is recent active volcanoes. For example, active volcanoes in the Andes of South America have many surrounding geothermal fields; another source is deep molten magma. These geothermal fields are widely developed, such as Salton Sea, California, USA (Brothers et al., 2009; Lachenbruch et al., 1985; Schmitt and Hulen, 2008; Karakas et al., 2017), Mapamyum non-volcanic field (Wang et al., 2016), Yangbajing (Guo et al., 2007), Yangyi (Yuan et al., 2014), and Gudui (Guo et al., 2019a, b). The regional deep faults caused by strong tectonic activities through time constitute mass heat migration channels, which cause the deep heat energy to migrate to shallower strata, forming the geothermal fields and display areas, especially those areas the fractures in different directions cut each other (Wang, 2017). Tectonic and volcanic activities provide channel systems and heat that drive groundwater convective cycles and are necessary for leaching lithium from source rocks (Ericksen et al., 1978). Large geothermal fields are often distributed along large deep fault zones, especially at fault intersections, as noted above, as at Salton Sea, located on the active San Andreas faults (Brothers et al., 2009), superimposed with the subsequent SAF-IF strike-slip fault (Karakas et al., 2017), which is as rich in lithium as Bolivia and Chile's most productive salt lakes (Campbell, 2009); The high-temperature Li-rich geothermal display areas in Xizang are mostly distributed near a deep fault, especially at the intersections of near E-W and N-S deep faults, as shown in Fig. 1. In addition, geothermal fields such as Lake Taupo, New Zealand, Yellowstone, and Geysir in Iceland (Elderfield and Greaves, 1981; Jones et al., 2007; Geilert et al., 2015) also have the above features.

Given suitable heat sources and tectonic conditions, there are two paths for the formation of a high temperature geothermal fluid. One is that atmospheric precipitation or glacial melt water seeps along fault zones. As the depth of infiltration increases, it is closer and closer to the deep molten magma body and then is gradually heated. With buoyancy, the heated water increases and rises constantly along the faults and at last forms hot springs, and finally, densely developed hot springs form geothermal fields; The other way is that atmospheric precipitation or glacial meltwater infiltrates along faults to a certain depth and then mixes with rising hot magmatic fluid to form a deep high-temperature geothermal reservoir. The mixed geothermal reservoir fluid then rises along the fault to the surface to form a hot spring or geothermal field. The geothermal water formed by these two processes has clear differences in water chemistry characteristics.

Chowdhury et al. (1974) pointed out that B, F, As, Li, and Cs are closely related to the final stage of magma crystallization evolution. Based on experimental studies (Fuge, 1977; Webster and Holloway, 1980) and analysis of trace elements in granite (Grimaud et al., 1985), it has been pointed out that magma fluid that is rich in K, Rb, Cs, B, Li, F and Cl can be released in the final stages of granite evolution. From the characteristics of high-temperature Li-rich geothermal water in southern Xizang, most are neutral or alkaline Na-Cl type waters rich in characteristic elements such as B, Li, Rb, Cs and F. What's more, acidic geothermal water is often developed and rich in SO_4^{2-}, which was considered typical for steam-heated acid waters (Guo et al., 2019b). In addition, Cl ion concentration has a good linear relationship with other characteristic ions, as at Gudui (Wang and Zheng, 2019), which indicates that there, deep magmatic fluids are mixed in (Guo et al., 2019a, b). Similarly, other southern Xizang high-temperature Li-rich geothermal fields could also have mixing with deep magma fluids. Many geophysical studies have shown that the existence of a large number of melts in the deep part of southern Xizang (Nelson et al., 1996; Brown et al., 1996; Makovsky et al., 1996; Kind et al., 1996; Chen et al., 1996; Tan et al., 2004; Wei et al., 2010) further supports the above conclusions. All the geothermal fields and hot springs around the world are basically controlled by the above factors, but not all can produce large amounts of Li-rich geothermal water.

Each Li-rich geothermal field has its unique characteristics; the only difference is the source of lithium. Despite in-depth studies, there is still controversy, but basically the consensus supports two sources: (1) water-rock reaction between high temperature geothermal fluid and Li-rich rocks (mostly in volcanic rocks such as tuff, e.g. Vide salt lake, South America, Silver Peak Lake, USA; Bradley et al., 2013); in southern Xizang, highly fractionated granites, such as Ramba leucogranite (Li: 226 ppm; Liu Z C et al., 2014), and Xiaru leucogranite (Li: 2000 ppm; Wang R C et al., 2017) might be examples, which dissolve and bring out lithium from the source rock. Other examples include the Uyuni salt pans in Bolivia, which are surrounded by hot springs containing unusually high lithium concentration associated with Quaternary rhyolite volcanic rocks (Ericksen et al., 1978). Water-rock reaction is considered to be the main source of lithium in brines (Campbell, 2009; Shcherbakov and Dvorov, 1970; Ericksen et al., 1978). Zheng et al. (1983) mentioned the origin of the Li-containing hot springs in Bangkog Co, noticing that water in the volcanic sedimentary beds has a high concentration of lithium and boron and, therefore, considered that the Cenozoic acid volcanic rocks in central and southern Xizang are important sources of boron and lithium in this region (Zheng et al., 1983). Araoka et al. (2014) discovered that the enrichment of lithium in the Nevada Basin was mainly through reactions between high-temperature geothermal fluid associated and Li-rich tuff or volcanic glass using a systematic lithium isotope study of several Li-rich playas (Clayton Valley brine field; Alkali Lake; Columbus Salt Marsh) in Nevada. The lithium concentration of the surrounding geothermal water is as high as 36 ppm (Davis et al., 1986), which indicates the importance of the water-rock reaction (Araoka et al., 2014). (2) High-temperature steam aqueous solution carrying lithium from the late stage of deep molten magma differentiation; because lithium is a moderately incompatible element in the magma system, it usually accumulates in the residual melt during the magma differentiation process (Zhang et al., 1998). Although high-temperature Li-rich geothermal fields are widely developed in southern Xizang, so far there is no related volcanic rock reported, rather they might be related to the intrusive rocks reported by Francheteau et al. (1984) leading to the high heat flow values in southern Xizang (Grimaud et al., 1985). Tan et al. (2014) explained the circulation process of underground geothermal fluid in Xizang's main high-temperature geothermal system through hydrogen and oxygen stable isotope data, the results showing that the formation of the high-temperature geothermal system is mainly due to rapid circulation of groundwater and upwelling of residual magma fluid. And in most cases, the granite outcrops can be observed not far from the hot springs (Grimaud et al., 1985), for example, the Cuonadong tourmaline granite (Liu M L et al., 2019) is developed not far from the Gudui geothermal field to the southeast.

Although many studies have shown the relationship between southern Xizang geothermal fields and deep molten magma, such as Yangbajing, Mapamyum, Daggyai, Semi (Guo et al., 2019a, b), Gudui (Wang and Zheng, 2019) etc., still the accurate source of high lithium concentration is rarely mentioned. Guo et al. (2019a, b) proved that the magma sources contribute to the extraordinary high concentration of As and W in some Li-rich geothermal water. The host rock leaching alone cannot explain the observed lithium anomaly in the Li-rich geothermal systems that have had continuous outflow for decades or longer because a fixed volume of surrounding rock cannot have such a huge amount of metal for hot water leaching. Based on the same considerations, Guo et al. (2019b) estimates of the As discharge flux and the total amount of available As in the Yangbajain system reservoir rocks supporting the existence of a magma chamber have been validated by geophysical and geochemical studies (Brown et al., 1996; Zhao et al., 1998a, b; Zhao et al., 2002). Overall, the results indicate that the estimate of the available As in reservoir rocks is more than two orders of magnitude lower than the estimated total flux, and therefore the As in the Yangbajain geothermal waters cannot be derived exclusively from host rock leaching but should be primarily from the input of magmatic fluid.

Further evidence can be offered by taking the Gudui Li-rich geothermal system as an example, for we consider that the metal lithium in the Gudui geothermal field mainly comes from the deep magmatic hydrothermal system. Therefore, substantial contribution of lithium from underlying magma chambers below southern Xizang is postulated, which are likely attributable to partial melting of the subcontinental lithospheric mantle that has undergone metasomatism by Li-rich fluids derived from subducted oceanic crust and marine sediments (Tian et al., 2018). Therefore, we consider that lithium in geothermal fields is closely related to deep molten magma. It is also easy to capture the direct source with the contribution of magmatic fluid input not being ruled out. Of course, in-depth research on this issue is necessary.

The strongly active high-temperature geothermal fluid in southern Xizang extracts lithium by water-rock reaction when it flows through felsic Li-rich rock rock such as granite or mixes directly with Li-rich magma fluid. Research has shown that high-temperature geothermal fluid dissolves lithium in rocks more than conventional cryogenic fluids (Chagnes and Światowska, 2015). Through the analysis of the trend of lake water composition, Zheng et al. (1990) discovered that the abundances of B, Li, K and Cs in the modern lakes of the Qinghai-Tibetan Plateau are characterized by the outward decline from the Gangdese-Yaluzang as a high value center, which also constitutes the geochemical anomaly areas of B, Li and Cs on the plateau. The origin there is contributed to the diffusion of volcanic geothermal fluid to the surface from deep melts in the western part of the Bangur and Yarlung Zangbo (Zheng et al., 1990). Based on recent researches, we accept that the formation of the Li-rich salt lakes in Xizang has undergone many 'pre-enrichment' processes in the early Li-rich oceanic crust (Zheng M P, unpublished) that partially melted to form Li-rich magma during subduction to complete the initial lithium enrichment. The Li-rich magma gradually cooled and crystallized along the deep and large fractures during the continuous ascending process. In the late stage of crystallization differentiation, lithium was further enriched in the magma hydrothermal fluid. With the further collision of the Indo-Asian continent, under different geological conditions, some of the Li-rich magmatic hydrothermal fluid formed rich tourmaline-spodumene granites or highly fractionated granites, which distribute surrounding the Xizang Semi geothermal field, while part of the Li-rich magmatic hydrothermal fluid can form Li-rich geothermal fluid by mixing with infiltrated surface water and then the mixed fluid rises along the fault zone to reach the surface (Fig. 4, Fig. 5). Some of the geothermal fluid is sent to the salt lakes, and further concentrated by evaporation and enrichment to form the Li-rich salt lakes brines, whereas most flows into surface runoff.

To illustrate the above conclusions, we take Yangbajing and Gudui geothermal fields as examples. Yangbajain is chosen because, among all the hydrothermal areas in Tibet, it has been investigated systematically in the past three decades (Tong et al., 1981; Zhao et al., 1998b; Duo, 2003; Guo et al., 2007, 2009), and the existence of a magmatic heat source and high-temperature reservoir (over 200 ℃) has also been confirmed (Guo et al., 2014). Guo et al. (2010) showed that the deep geothermal fluid is the mixing product of both magmatic and infiltrating snow-melt water, whereas the shallow geothermal fluid is formed by the mixing of deep geothermal fluid with cold groundwater (mixing process and ratios are shown in the Fig. 4). In addition, two groups of high-angle normal NNE-SSW and NE-SW stretching/extensional faults are distributed in front of the Nyenchen Tonglha and Tang mountains near to the geothermal field. These faults were active until the Quaternary and are crucial structures for the occurrence of the field because they provided favorable permeability for the migration and storage of the geothermal fluids at depth. What's more, the deep geothermal fluid is richer in lithium (personal communication with academician Duoji).

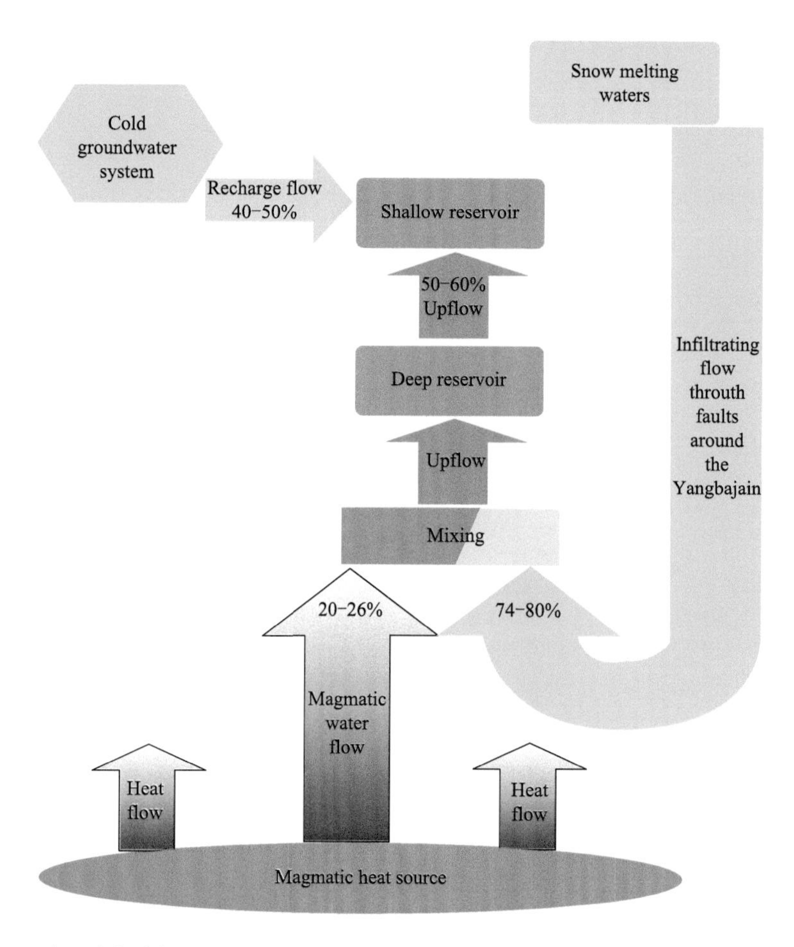

Fig. 4 Conceptual model of the Yangbajing geothermal water system (modified according to Guo et al., 2010).

The reason to select the Gudui geothermal field is that it is characterized by the most intensive hydrothermal activity among all Tibet geothermal areas (Guo et al., 2014), although it is not as well known or systematically researched. Based on its hydrochemistry, Gudui is very likely to be a magmatic hydrothermal system similar to Yangbajain (Guo et al., 2014). In addition, Wang et al. (2019), through hydrogen and oxygen isotope research, suggested that the deep reservoir at Gudui also definitely mixed with magmatic fluid and then, together with strontium isotope data, proposed the 6th-Class Reservoirs Evolution Conceptual Model (6-CRECM) for the geothermal system (Fig. 5). They also found that the lithium concentration is decreasing gradually from deep geothermal fluid, which is mixing more magmatic fluid to shallow fluid, which might provide further clues to prove that lithium comes from deep magmatic fluids (Wang et al., 2019). The geological characteristics of the Gudui area (Fig. 2) and previous studies also show that the strongly developed faults had an important role in the formation of the Gudui Li-rich high-temperature geothermal system (Liu et al., 2017; Wang, 2017; Wang and Zheng, 2019).

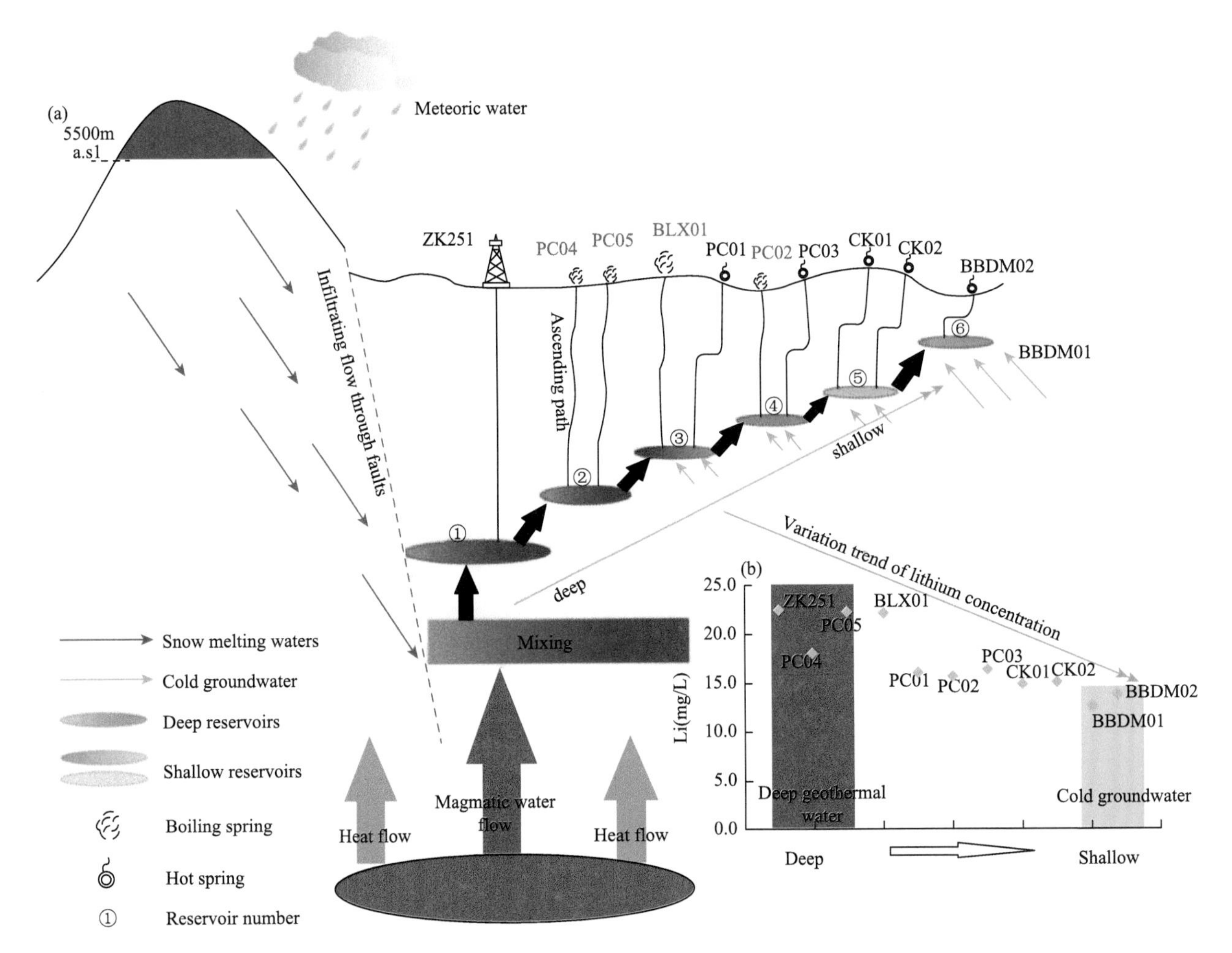

Fig. 5　(a) The 6th-Class Reservoirs Evolution Conceptual Model (6-CRECM) for the Gudui geothermal system; (b) changing trend of lithium concentration from deep geothermal fluid to shallow geothermal fluid of the Gudui geothermal field (modified from Wang et al., 2019).

Taking these two examples into consideration, we think that the deep molten magma of the lower crust provided a stable heat source and lithium source for the high-temperature Li-rich geothermal fields developed in southern Xizang. Li (2002) also thought that the melting layers in the crust in southern Xizang not only provide heat energy for the shallow hydrothermal system, but also supply liquid and metallogenetic elements. The upper fractional melting layer has driven ampliated area's hydrothermal activities in southern Tibet, whereas the near-surface upwelling emplaced melting masses from the lower crust to promote the presenting of some high-temperature geothermal fields (Li, 2002). The deep-produced Li-rich parent geothermal fluid rushes to the surface to form hot springs along the extensively developed tectonic fault zones in southern Xizang, and some of the Li-rich fluid flows into the lakes to form Li-rich salt lakes; however, most of the Li-rich fluid is wasted, flowing into different rivers. More investigations on Li-rich high-temperature geothermal fields are needed to further verify the conclusions reached here.

5　Prospects for the utilization of high temperature Li-rich geothermal resources in southern Xizang

Many worldwide researches have been done on extraction of lithium from liquid minerals, and numerous methods have been proposed, such as solvent extraction, evaporative crystallization, precipitation, ion exchange adsorption, or a combination of methods (Flexer et al., 2018; An et al., 2012; Song et al., 2017), among which the solvent extraction and ion exchange adsorption have received the most attention (Sun et al., 2019), but the recovery of lithium from geothermal fluid resources has rarely been reported to date. Based on the experiences of extracting caesium in the Cs-bearing geyserite deposit in Xizang and the changes in concentration of lithium and other chemical constituents during Li-rich geothermal fluid evaporation and concentration experiments, combined consideration of the topography and geology of the high-temperature Li-rich geothermal area in southern Xizang, we consider the efficiency of simple evaporation concentration is too low to meet current development and utilization needs. Through preliminary exploration and research, we thought that membrane and adsorption methods are still very promising and worth considering, especially with natural non-polluting adsorbents such as aluminum-based or zeolite-based, and even those that combine natural substances and synthetic substances together can be considered. Not only can they be green and pollution-free, but this method can also increase man-made controllability. According to the chemical composition and type of geothermal fluid, appropriate adjustments can be made to improve the efficiency of lithium extraction while reducing energy consumption. So far, our team is still conducting research on the use of different adsorbents to extract lithium resources in Li-rich geothermal water. We are confident that, in the near future, exploitation of high-temperature Li-rich geothermal water in southern Xizang will be enhanced.

Fortunately, Chinese scholars have begun to pay attention to this problem and achieved results. Sun et al. (2019) proposed a green recovery method of lithium from geothermal water based on a novel Li-Fe phosphate electrochemical technique. The recovery rate of Li + is up to 90.65% after eight adsorption-desorption cycles. In addition, only electrical energy is consumed and no organic solvents or other toxic reagents used or produced during the recovery process. All these properties make the above method a green and promising candidate for the recovery of Li^{+} from geothermal water. In addition, Zheng (1999) thought that low-salinity geothermal fluids are also a special ore-forming fluid through a comparative study of the fluids rich in Li, B, Cs and Rb from the Salton Sea, Puga, Taupo and high-temperature geothermal fields developed in Xizang. The study of the low-temperature mineralization that is developing in Xizang will be helpful to improve metallogenic theory and increase ore-forming target areas. Therefore, it is necessary to enhance research on high-temperature Li-rich geothermal resources in southern Xizang.

6 Conclusions

Geothermal-type lithium resources are a new type of lithium resources, which are widely distributed worldwide and have huge potential reserves. Scholars from different countries have carried out research on these resources, which should also be given due attention by Chinese scholars. Our research has discovered geothermal-type lithium reserves in China but further relevant exploration investment and basic research is needed, to realize industrial extraction and utilization of geothermal-type lithium resources. We consider that there will be more efficient and portable methods to extract lithium resources from geothermal water in the near future.

China's geothermal-type lithium resources are mainly distributed in southern Xizang, and strongly controlled by E-W extension structures and the Yarlung Zangbo suture zone. The Cenozoic volcanic activities on both sides of the suture zone and in the southern part of Xizang have not been reported previously, and the magmatism caused by the melting of the upper crust is essential for the development of high-temperature Li-rich geothermal systems. China's high-temperature Li-rich geothermal water is characterised by low Mg/Li ratio, high Li/TDS ratio, and has been continuously and stably discharged for decades; some waters reach industrial grades, and all are associated with high-quality accompanying elements such as B, Cs, Rb that can be comprehensively utilized.

The current high-temperature Li-rich geothermal water in southern Xizang has a large scale of metal lithium emission per year, and the deep potential resources are even more abundant, which has great exploitation and utilization value. With the advancement of technology, methods for extracting lithium from Li-rich geothermal water are gradually diversifying, and will gradually mature. In the near future, utilization of lithium and other resources from the high-temperature Li-rich geothermal water will not only produce its due economic value, but also help to reduce environmental pollution of harmful elements.

In summary, the high-temperature Li-rich geothermal resources widely developed in southern Xizang are a valuable asset worth exploiting, and the near-surface Li-rich geothermal fluid with rare alkali metals is also a modern low-salt hydrothermal ore-forming fluid. The study of its mineralization will help deepen our understanding of low-temperature mineralization and regional prospecting.

Acknowledgements

We are grateful to the earlier generation of geologists Tong Wei, Zhang Zhifei, Zhang Mingtao and Zhu Meixiang for their contributions to the study of geothermal resources in Xizang. They have provided first-hand us with information and laid a good foundation for the development of geothermal science in China. This work was financially supported by the Major Program of the National Natural Science Foundation of China (Project Grant No. 91962219), the National Natural Science Foundation of China (Grant No. U1407207), the Beijing Dizhiguang New Energy Technology Research Institute Co., Ltd, China Geological Survey (Project Grant No. DD20190172) and the National Key Research and Development Program of China (Grant Nos. 2017YFC0602806 and 2017YFC0602802). Susan Turner (Brisbane) assisted with English language.

References

An, J.W., Kang, D.J., Tran, K.T., Kim, M.J., Lim, T., and Tran, T., 2012. Recovery of lithium from Uyuni salar brine. Hydrometallurgy, 117: 64-70.

Araoka, D., Kawahata, H., Takagi, T., Watanabe, Y., Nishimura, K., and Nishio, Y., 2014. Lithium and strontium isotopic systematics in playas in Nevada, USA: constraints on the origin of lithium. Mineralium Deposita, 49 (3): 371-379.

Brown, L.D., Zhao, W., Nelson, K.D., Hauck, M., Alsdorf, D., Ross, A., Cogan, M., Clark, M., Liu, X., and Che, J., 1996. Bright spots, structure, and magmatism in Southern Xizang from indepth seismic reflection profiling. Science, 74: 1688-1690.

Blisniuk, P.M., Hacker, B.R., Glodny, J., Ratschbacher, L., Bi, S., Wu, Z., McWilliams, M.O., and Calvert, A., 2001. Normal faulting in central Tibet since at least 13.5 Myr ago. Nature, 412 (6847): 628-632.

Brothers, D.S., Driscoll, N.W., Kent, G.M., Harding, A.J., Babcock, J.M., and Baskin, R.L., 2009. Tectonic evolution of the Salton Sea inferred from seismic reflection data. Nature Geoscience, 2: 581-584.

Bradley, D., Munk, L., Jochens, H., Hynek, S., and Labay, K., 2013. A preliminary deposit model for lithium brines, U.S. Geological Survey Open-File Report: 2013-1006, 6.

Campbell, M.G., 2009. Battery lithium could come from geothermal waters. New Scientist, 204 (2738): 23.

Cetiner, Z.S., Özgür, D., Göksel, Ö., and Özer, E.P., 2015. Toward utilising geothermal waters for cleaner and sustainable production: potential of li recovery from geothermal brines in turkey. International Journal of Global Warming, 7 (4): 439.

Cha, H., Kim, J., Lee, Y., Cho, J., and Park, M., 2017. Issues and challenges facing flexible lithium-ion batteries for practical application. Small, 1702989.

Chagnes, A., and Światowska, J., 2015. Lithium Process Chemistry. Amsterdam: Elsevier, 1-312.

Chen, L., Booker, J.R., Jones, A.G., Wu, N., Unsworth, M.J., Wei, W., and Tan, H., 1996. Electrically conductive crust in Southern Xizang from INDEPTH magnetotelluric surveying. Science, 274: 1694-1696.

Chowdhury, A.N., Handa, B.K., and Das, A.K., 1974. High lithium, rubidium and cesium contents of thermal spring water, spring sediments and borax deposits in Puga Valley, Kashmir, India. Geochemical Journal, 8: 61-65.

Cortecci, G., Boschetti, T., Mussi, M., Lameli, C.H., Mucchino, C., and Barbieri, M., 2006. New chemical and original isotopic data on waters from El Tatio geothermal field, northern Chile. Geochemical Journal, 39 (6): 547-571.

Davis, J.R., Friedman, I., and Gleason, J.D., 1986. Origin of the lithium-rich brine, Clayton Valley, Nevada. U.S. Geological Survey Bulletin, 1622: 131-138.

Duo, J., 2003. The basic characteristics of the Yangbajing geothermal field-A typical high temperature geothermal system. Engineering Sciences, 5: 42-47 (in Chinese with English abstract).

Ericksen, G.E., Vine, J.D., and Ballón, A.R., 1978. Chemical composition and distribution of lithium-rich brines in salar de Uyuni and nearby salars in southwestern Bolivia. Energy, 3 (3): 355-363.

Elderfield, H., and Greaves, M.J., 1981. Strontium isotope geochemistry of Icelandic geothermal systems and implications for sea water chemistry. Geochimica et Cosmochimica Acta, 45: 2201-2212.

Flexer, V., Baspineiro, C.F., and Galli, C.L., 2018. Lithium recovery from brines: A vital raw material for green energies with a potential environmental impact in its mining and processing. Science Total Environment, 639: 1188-1204.

Francheteau, J., Jaupart, C., Jie, S.X., Kang, W.H., Lee, D.L., Bai, J.C., Wei, H.P., and Deng, H.Y., 1984. High heat flow in Southern Xizang. Nature, 307 (5946): 32-36.

Fuge, R., 1977. On the behaviour of fluorine and chlorine during magmatic differentiation. Contributions to Mineralogy and Petrology, 61 (3): 245-249.

Garrett, D.E., 2004, Handbook of Lithium and Natural Calcium Chloride—Their Deposits, Processing, Uses and Properties (1st ed.): Amsterdam, Boston: Elsevier Academic Press, 476.

Geilert, S., Vroon, P.Z., Keller, N.S., Gudbrandsson, S., Andri, S., and van Manfred, B.J., 2015. Silicon isotope fractionation during silica precipitation from hot-spring waters: Evidence from the Geysir geothermal field, Iceland. Geochimica et Cosmochimica Acta, 164: 403-427.

Grimaud, D., Huang, S., Michard, G., and Zheng, K., 1985. Chemical study of geothermal waters of central Xizang (China). Geothermics, 14 (1): 35-48.

Gruber, P.W., Medina, P.A., Keoleian, G.A., Kesler, S.E., Everson, M.P., and Wallington, T.J., 2011. Global lithium availability—A constraint for electric vehicles? Journal of Industrial Ecology, 15: 760-775.

Guo Q., 2012. Hydrochemistry of high-temperature geothermal system in China: A review[J]. Applied Geochemistry, 27(10): 1887-1898.

Guo Q., and Wang, Y., 2012. Geochemistry of hot springs in the Tengchong hydrothermal areas, Southwestern China. Journal of Volcanology and Geothermal Research, 215-216: 61-73.

Guo, Q., Wang, Y., and Liu, W., 2007. Major hydrogeochemical processes in the two reservoirs of the Yangbajing geothermal field, Xizang, China. Journal of Volcanology and Geothermal Research, 166 (3): 255-268.

Guo, Q., Wang, Y., and Liu, W., 2009. Hydrogeochemistry and environmental impact of geothermal waters from Yangyi of Xizang, China. Journal of Volcanology and Geothermal Research, 180 (1): 9-20.

Guo, Q., Li, Y., and Luo, L., 2019a. Tungsten from typical magmatic hydrothermal systems in China and its environmental transport. Science of the Total Environment, 657: 1523-1534.

Guo, Q., Liu, M., Li, J., Zhang, X., Guo, W., and Wang, Y., 2017. Fluid geochemical constraints on the heat source and reservoir temperature of the Banglazhang hydrothermal system, Yunnan-Xizang Geothermal Province, China. Journal of Geochemical Exploration, 172: 109-119.

Guo, Q., Planer-Friedrich, B., Liu, M., Yan, K., and Wu, G., 2019b. Magmatic fluid input explaining the geochemical anomaly of very high arsenic in some Southern Xizangan geothermal waters. Chemical Geology, 513: 32-43.

Guo, Q.H., Kirk, N.D., and Blaine, M.R., 2014. Towards understanding the puzzling lack of acid geothermal springs in Tibet (China): Insight from a comparison with Yellowstone (USA) and some active volcanic hydrothermal systems. Journal of Volcanology and Geothermal Research, 288: 94-104.

Guo, Q.H., Wang, Y.X., and Liu, W., 2010. O, H, and Sr isotope evidences of mixing processes in two geothermal fluid reservoirs at Yangbajing, Xizang, China. Environmental Earth Sciences, 59: 1589-1597.

Han, T.L., 1990. Relationship of the active structural system to geothermal activity in southern Tibet. Himalayan Geology. Beijing: Geological Publishing House, 45-48 (in Chinese).

Hano, T., Matsumoto, M., and Ohtake, T., 1992. Recovery of lithium from geothermal water by solvent extraction technique. Solvent Extraction and Ion Exchange, 10: 195-206.

Jeongeon, P., Hideki, S., Syouhei, N., and Kazuharu, Y., 2012. Lithium recovery from geothermal water by combined adsorption methods. Solvent Extraction and Ion Exchange, 30: 398-404.

Jones, B., Renaut, R.W, Torfason, H., and Owen, R.B., 2007. The geological history of Geysir, Iceland: A tephrochronological approach to the dating of sinter. Journal of the Geological Society, 164 (6): 1241-1252.

Kasemann, S.A., Meixner, A., Erzinger, J., Viramonte, J.G., Alonso, R.N., and Franz, G., 2004. Boron isotope composition of geothermal fluids and borate minerals from Salar deposits (central Andes/NW Argentina). Journal of South American Earth Sciences, 16: 685-697.

Karakas, O., Dufek, J., Mangan, M.T., Wright, H.M., and Bachmann, O., 2017. Thermal and petrologic constraints on lower crustal melt accumulation under the Salton Sea Geothermal Field. Earth and Planetary Science Letters, 467: 10-17.

Kesler, S.E., Gruber, P.W., Medina, P.A., Keoleian, G.A., and Wallington, T.J., 2012. Global lithium resources: Relative importance of pegmatite, brine and other deposits. Ore Geology Reviews, 48 (5): 55-69.

Kind, R., Ni, J., Zhao, W., Wu, J., Yuan, X., Zhao, L., Sandvol, E., Rees, E.C., Nabelek, J., and Hearn, T., 1996. Evidence from earthquake data for a partially molten crustal layer in southern Xizang. Science, 274: 1692-1694.

Krotscheck, E., and Smith, R.A., 2012. Separation and recovery of lithium from geothermal water by sequential adsorption process with l-MnO_2 and TiO_2. Ion Exchange Letter, 32: 2219-2233.

Lachenbruch, A.H., Sass, J., and Galanis, S., 1985. Heat flow in southernmost California and the origin of the Salton Trough. Journal of Geophysical Research: Solid Earth, 90: 6709-6736.

Li, Z.Q., 2002. Present hydrothermal activities during collisional orogenics of the Xizangan Plateau. Doctor thesis, Beijing: Chinese Academy of Geological Sciences (in Chinese with English abstract).

Liu, L.J., Wang, D.H., Gao, J.Q., Yu, F., and Wang, W., 2019. Breakthroughs of lithium exploration progress (2017-2018) and its significance to China's strategic key mineral exploration. Acta Geologica Sinica, 93 (6): 1479-1488 (in Chinese with English abstract).

Liu, M.L., Guo, Q.H., Wu, G., Guo, W., She, W.Y., and Yan, W.D., 2019. Boron geochemistry of the geothermal waters from two typical hydrothermal systems in Southern Xizang (China): Daggyai and Quzhuomu. Geothermics, 82: 190-202.

Liu, Z.C., Chen, K., and Nan, D., 2017. Hydrochemical characteristics of geothermal water in Gudui, Xizang (Tibet).Geological Review, (s1): 353-354 (in Chinese).

Liu, Z.C., Wu, F.Y., Ji, W.Q., Wang, J.G., and Liu, C.Z., 2014. Petrogenesis of the Ramba leucogranite in the Tethyan Himalaya and constraints on the channel flow model. Lithos, 208-209: 118-136.

Luo, Y.B., Zheng, M. P., and Ren, Y.Q., 2017. Correlation between special salt lakes on the Qinghai-Tibetan Plateau and deep volcanoes-geothermal water. Science and Technology Review, 35 (12): 44-48.

Makovsky, Y., Klemperer, S.L., Ratschbacher, L., Brown, L.D., Li, M., Zhao, W., and Meng, F., 1996. INDEPTH wide-angle reflection observation of P-Wave-to-S-Wave conversion from crustal bright spots in Xizang. Science, 274: 1690-1691.

Martin, G., Rentsch, L., Höck, M., and Bertau, M., 2016. Lithium market research-global supply, future demand and price development. Energy Storage Materials, 6: 171-179.

Morozov, N.P., 1969. Geochemistry of the alkali metals in rivers. Geokhimiya, 6 (3): 729-739.

Munk, L.A., Bradley, D.C., Hynek, S.A., Chamberlain, C.P., 2011. Origin and evolution of Li-rich brines at Clayton Valley, Nevada, USA. Antofagasta: 11th SGA Biennial Meeting, Chile, 217-219.

Munk, L.A., Hynek, S.A., Bradley, D., Boutt, D.F., Labay, K., and Jochens, H., 2016. Lithium brines: a global perspective. Review Economic Geology, 18: 339-365.

Mao, X.P., 2018. Characteristics of temperature distribution and control factors in geothermal field. Acta Geologica Sinica (English Edition), 92 (z2): 96-98.

Nelson, K.D., Zhao, W., Brown, L.D., Kuo, J., Che, J., Liu, X., Klemperer, S.L., Makovsky, Y., Meissner, R., Mechie, J., Kind, R., Wenzel, F., Ni, J., Nabelek, J., Leshou, C., Tan, H., Wei, W., Jones, A.G., Booker, J., Unsworth, M., Kidd, W.S.F., Hauck, M., Alsdorf, D., Ross, A., Cogan, M., Wu, C., Sandvol, E., and Edwards, M., 1996. Partially molten middle crust beneath Southern Xizang: Synthesis of project INDEPTH results. Science, 274: 1684-1688.

Pan, G., Wang, L., Li, R., Yuan, S., Ji, W., Yin, F., Zhang, W., and Wang, B., 2012. Tectonic evolution of the Qinghai-Tibet Plateau. Journal of Asian Earth Sciences, 53: 3-14.

Pang, Z.H., Hu, S.B., and Wang, J.Y., 2013. A roadmap to geothermal energy development in China. Science and Technology Review, (4): 3-10 (in Chinese with English abstract).

Park, J.K., 2012. Principles and applications of lithium secondary batteries. Journal of Solid State Electrochemistry, 17 (8): 2375-2376.

Shcherbakov, A.V., and Dvorov, V.I., 1970. Thermal waters as a source for extraction of chemicals. Geothermics, 2 (2): 1636-1639.

Schwochau, K., 1984. Extraction of metals from sea water. Inorganic Chemistry: Springer, Berlin, Heidelberg, 91-133.

Schmitt, A.K., and Hulen, J.B., 2008. Buried rhyolites within the active, high-temperature Salton Sea geothermal system. Journal of Volcanology and Geothermal Research. 178: 708-718.

Schmidt, N., 2010, Hydrogeological and hydrochemical investigations at the Salar de Uyuni (Bolivia) with regard to the extraction of lithium. FOG-Freiberg Online Geoscience, 26: 1-131.

Song, J.F., Nghiem, L.D., Li, X. M., and He, T., 2017. Lithium extraction from Chinese saltlake brines: opportunities, challenges, and future outlook. Environmental Science-Water Research and Technology, 3: 593-597.

Sun, S., Yu, X.P., Li, M.L., Duoji, Guo, Y.F., and Deng, T.L., 2019. Green recovery of lithium from geothermal water based on a novel lithium iron phosphate electrochemical technique. Journal of Cleaner Production, 247: 119178.

Tan, H., Chen, J., Rao, W., Zhang, W., and Zhou, H., 2012. Geothermal constraints on enrichment of boron and lithium in salt lakes: An example from a river-saltlake system on the northern slope of the eastern Kunlun Mountains, China. Journal of Asian Earth Sciences, 51 (12): 21-29.

Tan, H., Zhang, Y., Zhang, W., Kong, N., Zhang, Q., and Huang, J., 2014. Understanding the circulation of geothermal waters in the Xizangan Plateau using oxygen and hydrogen stable isotopes. Applied Geochemistry, 51: 23-32.

Tan, H., Su, J., Xu, P., Dong, T., and Elenga, H.I., 2018. Enrichment mechanism of Li, B and K in the geothermal water and associated deposits from the Kawu area of the Xizangan plateau: Constraints from geochemical experimental data. Applied Geochemistry, 93: 60-68.

Tan, H.D., Wei, W.B., Unsworth, M., Deng M., Jin S., Booker, J., and Jones, A., 2004. Crustal electrical conductivity structure beneath the Yarlung Zangbo Jiang Suture in the southern Xizang plateau. Chinese Journal of Physics, 47: 780-786 (in Chinese with English abstract).

Taylor, M., and Yin, A., 2009. Active structures of the Himalayan-Tibetan orogen and their relationships to earthquake distribution, contemporary strain field, and Cenozoic volcanism. Geosphere, 5: 199-214.

Tian, S., Hou, Z., Tian, Y., Zhang, Y.J, Hu, W.J., Mo, X.X., Yang, Z.S., Li, Z.Q., and Zhao, M., 2018. Lithium content and isotopic composition of the juvenile lower crust in Southern Xizang. Gondwana Research, 62: 198-211.

Tomaszewska, B., and Szczepański, A., 2014. Possibilities for the efficient utilisation of spent geothermal waters. Environmental Science and Pollution Research, 21 (19): 11409-11417.

Tong, W., Liao, Z.J., Liu, S.B., Zhang, Z.F., You, M.Z., and Zhang, M.T., 2000. Xizangan hot springs. Beijing: Science Press, 1-282 (in Chinese with English abstract).

Tong, W., Zhang, M.T., Zhang, Z.F., Liao, Z.J., You, M.Z., Zhu, M.X., Guo, J.Y., and Liu, S.B., 1981. Geothermy in Xizang, Beijing: Science Press.

Wang, C.G., and Zheng, M.P., 2019. Hydrochemical characteristics and evolution of hot fluids in the Gudui geothermal field in Comei County, Himalayas. Geothermics, 81: 243-258.

Wang, C.G., Zheng, M.P., Zhang, X.F., Xing, E.Y., Zhang, J.Y., Ren, J.H., and Ling, Y., 2019. O, H, and Sr isotope evidence for origin and mixing processes of the Gudui geothermal system, Himalayas, China. Geoscience Frontiers, 11 (4): 1175-1187.

Wang, P., Chen, X., Shen, L., Wu, K.Y., Huang, M, Z., and Xiao, Q., 2016. Geochemical features of the geothermal fluids from the Mapamyum non-volcanic geothermal system (Western Xizang, China). Journal of Volcanology and Geothermal Research, 320: 29-39.

Wang, R.C., Wu, F.Y., Xie, L., Liu, X.C., Wang, J.M., Yang, L., Lai, W., and Liu, C., 2017. A preliminary study of rare-metal mineralization in the Himalayan leucogranite belts, South Xizang. Science China Earth Sciences, 47: 871-880.

Wang, S.Q., 2017. Hydrogeochemical processes and genesis mechanism of high-temperature geothermal system in Gudui, Xizang. Doctor thesis, Beijing: China University of Geosciences (in Chinese with English abstract).

Wang, S.Q., Liu, Z., and Shao, J.L., 2017. Hydrochemistry and H-O-C-S isotopic geochemistry characteristics of geothermal water in Nyemo-Nagqu, Tibet. Acta Geologica Sinica (English Edition), 91 (2): 644-657.

Webster, E.A., and Holloway, J.R., 1980. The partitioning of REE's, Rb and Cp between silicic meh and a Cl fluid. EOS 61, 1152.

Wei, W.B., Jin, S., Ye, G.F., Deng, M., Jing, J.E., Unsworth, M., and Jones, A.G., 2010. Conductivity structure and rheological property of lithosphere in Southern Xizang inferred from super-broadband magmetotulleric sounding. Science in China Series D: Earth Sciences, 53: 189-202.

Williams, H., Turner, S., and Kelley, S., 2001. Age and composition of dikes in Southern Tibet: new constraints on the timing of east-west extension and its relationship to postcollisional volcanism. Geology, 29: 339-342.

Xia, L.Y., and Zhang, Y.B., 2019. An overview of world geothermal power generation and a case study on China—The resource and market perspective. Renewable and Sustainable Energy Reviews, 112: 411-423.

Yanagase, K., Yoshinaga, T., Kawano, K., and Matsuoka, T., 1983. The recovery of lithium from geothermal water in the Hatchobaru area of Kyushu, Japan. Bulletin of Chemistry Society of Japan, 56: 2490-2498.

Yu, J.Q., Gao, C.L., Cheng, A.Y., Liu, Y., Zhang, L., and He, X.H., 2013. Geomorphic, hydroclimatic and hydrothermal controls on the formation of lithium brine deposits in the qaidam basin, northern Xizangan Plateau, China. Ore Geology Reviews, 50 (50): 171-183.

Yuan, J., Guo, Q., and Wang, Y., 2014. Geochemical behaviors of boron and its isotopes in aqueous environment of the Yangbajing and Yangyi geothermal fields, Xizang, China. Journal of Geochemical Exploration, 140: 11-22.

Zhang, L., Chan, L.H., and Gieskes, J.M., 1998. Lithium isotope geochemistry of pore waters from Ocean Drilling Program Sites 918/919, Irminger Basin. Geochimica et Cosmochimica Acta, 62 (14): 2437-2450.

Zhang, L.X., Pang, M.Y., Han, J., Li, Y.Y., and Wang, C.B., 2019. Geothermal power in China: Development and performance evaluation. Renewable and Sustainable Energy Reviews, 116: 109431.

Zhang, Z.F., Shen, M.Z., and Zhao, F.S., 1989. The underground condition of the high temperature hydrothermal system in Xizang. Geothermal Album (Second Series), Beijing: Geological Publishing House, 134-140.

Zhao, M.Y., Ji, Z.Y., Zhang, Y.G., Guo, Z.Y., Zhao, Y.Y., Liu, J., and Yuan, J.S., 2017. Study on lithium extraction from brines based on LiMn2O4/Li1-xMn2O4 by electrochemical method. Electrochimica Acta, 252: 350-361.

Zhao, P., Duo, J., Liang, T.L., Jin, J., and Zhang, H.Z., 1998a. Characteristics of gas geochemistry in Yangbajing geothermal field, Tibet. Chinese Science Bulletin, 43 (21): 1770-1777.

Zhao, P., Jin, J., Zhang, H., Duo, J., and Liang, T., 1998b. Chemical composition of thermal water in the Yangbajing geothermal field, Tibet. Scientia Geologica Sinica, 33 (1): 61-72.

Zhao, W., Zhao, X., Shi, D., Liu, K., Jiang, W., Wu, Z., Xiong, J., and Zhang, Y., 2002. Progress in the study of deep profles (INDEPTH) in the Himalayas and Qinghai-Tibet Plateau. Geological Bulletin of China, 21 (11): 691-700.

Zheng, M.P., 1999. Preliminary discussion of low-salinity hydrothermal fluid mineralization. Chinese Science Bulletin, (S2): 141-143 (in Chinese).

Zheng, M.P., and Liu, W.G., 1982. Lithium-rich magnesium borate deposit discovered in Xizang. Geological Review, 28 (3): 263-266 (in Chinese with English abstract).

Zheng, M.P., and Liu, W.G., 1987. A new lithium mineral—zabuyelite. Acta Mineralogica Sinica, 7 (3): 221-226 (in Chinese with English abstract).

Zheng, M.P., and Liu, X.F., 2007. Lithium Resources in China. Advanced Materials Industry, (8): 13-16 (in Chinese with English abstract).

Zheng, M.P., Liu, W.G., Xiang, J., and Jiang, Z.T., 1983. On saline lakes in Xizang, China. Acta Geologica Sinica, 57 (2): 184-194 (in Chinese with English abstract).

Zheng, M.P., Xiang, J., and Wei, X.J., 1989. Saline Lake on the Qinghai-Xizang (Xizang) Plateau. Beijing: Science Press, 1-431 (in Chinese).

Zheng, M.P., Zheng, Y., and Liu, J., 1990. New discoveries of salt lakes and geothermal deposits in the Qinghai-Tibetan Plateau, Acta Geoscientica Sinica, (1): 151 (in Chinese).

Zheng, M.P., Wang, Q.X., Duoji, Liu, J., Pingcuo, W.J., and Zhang, S.C., 1995. A new type of hydrothermal deposit: Cesium-bearing geyserite in Tibet. Beijing: Geological Publishing House, 1-114.

Zheng, S.H., Zhang, Z.F., Ni, R.L., Hou, F.G., and Shen, M.Z., 1982. Hydrogen-oxygen stable isotope study of geothermal water in Xizang. Acta Scientiarum Naturalium Universitatis Pekinensis, (1): 99-106.

Zheng, W., Tan, H., Zhang, Y., Wei, H.Z., and Dong, T., 2015. Boron geochemistry from some typical Xizangan hydrothermal systems: Origin and isotopic fractionation. Applied Geochemistry, 63: 436-445.

Zhu, D., Pan, G., Mo, X., Duan, L., and Liao, Z., 2004. The age of collision between India and Eurasia. Advances in Earth Science, 19 (4): 564-571 (in Chinese with English abstract).

Hydrochemical characteristics and evolution of hot fluids in the Gudui geothermal field in Comei County, Himalayas*

Chenguang Wang and Mianping Zheng

MLR Key Laboratory of Saline Lake Resources and Environments, Institute of Mineral Resources, Chinese Academy of Geological Sciences, No.26, Baiwanzhuang street, Beijing, China

Abstract The Gudui geothermal field is a typical high-temperature geothermal system characterized by tectonic control and violent hydrothermal explosions in the Himalayas. The geothermal waters are mainly Na-Cl and Na-HCO_3-Cl types. The comprehensive analysis of Na-K, quartz, K-Mg geothermometers and a Na-K-Mg ternary diagram indicate that the reservoir temperature is up to 266.6 ℃. Except for four samples, most geothermal water samples collected from Gudui plot far from the full equilibrium line in Na-K-Mg ternary diagram, suggesting that complete chemical re-equilibrium has not been achieved as these geothermal waters flow upward from reservoirs towards spring vents and possibly mix with cooler waters. The results of geochemical characteristics analysis indicate that Cl, Na, K, SiO_2, B, As, Li, Rb, Cs, and F are the characteristic components of Gudui geothermal waters. The good linear relations between Cl and other characteristic consituents reflect the existence of a parent geothermal liquid (PGL) below Gudui. Comprehensive comparative analysis of the silica-enthalpy diagram and the chloride-enthalpy diagram suggests that the parent geothermal liquid below Gudui has a Cl concentration of 697 mg/L and enthalpy of 1250 J/g. The PGL ascends to the surface through different channels and may cool by conduction of heat to reservoir host rocks, by boiling, or by mixing with cooler groundwater.

Keywords Hydrochemistry; Geothermometry; Characteristic components, Parent geothermal fluid; Gudui geothermal field; Himalayas

1 Introduction

Facing a global energy crisis and environmental problems, people from all over the world pay more and more attention to the exploitation and utilization of clean energy (Capaccioni et al., 2011). As an old energy, geothermal energy is clean, renewable (Smith, 2007; Friedman, 2011; Gluyas et al., 2012; Craig et al., 2013) and the history of geothermal direct-use has been going on for over 2000 years. In recent years, the hydrogeological and geochemical exploration of geothermal fields has become a hot topic all over the world (Delmelle et al., 2000; Dotsika, 2015; Dotsika et al., 2006; Ünsal and Tarcan, 2002; Tassi et al., 2010).

In China, geothermal resources have been found in every province (Liu et al., 2017), however, high-temperature hydrothermal systems are mainly distributed in western Yunnan Province, western Sichuan Province and Southern Tibet in mainland of China, which belong to the Yunnan-Tibet Geothermal belt (YTGB) and are also an important part of the Mediterranean-Himalayas geothermal belt (Guo et al., 2017; Lü et al., 2014). Since the 1970's, the first Tibetan scientific expedition has been carried out by the Chinese Academy of Sciences (Wang et al., 2016). Through the expedition, over 600 geothermal spring sites [Fig. 1(b)] have been found and explored. Most of these geothermal springs are located at the intersections of the east-west structural belts and the north-south structural belts or the boundaries of the active faults [Fig. 1(b)]. The Yadong-Gulu geothermal belt and Cona-Woka geothermal belts are the most representative geothermal

* 本文发表在：Geothermics, 2019, 81 (SEP): 243-258

areas [Fig. 1(b)]. In addition, more than 100 springs in Tibet have reservoir temperatures higher than 150 ℃ and 11 sites had hydrothermal explosions (Liao and Zhao, 1999), among which are the most famous and most successfully utilized high-temperature hydrothermal systems Yangbajing and Yangyi geothermal fields (Duo, 2003; Guo et al., 2007, 2009; Zhao et al., 1998a, b; Du et al., 2005; Guo and Wang, 2012; Shangguan et al., 2005). The potential for power generation from geothermal energy in Tibet is vast but as yet largely untapped.

The Gudui geothermal field is located at the intersection area of approximately North-South Cona-Woka fault belt [Fig. 1(b)] and approximately East-West extending Zhegucuo-Longzi thrust structural zone between the South Tibetan Detachment System (STDS) and the Yarlung Zangbo Suture Zone (YZSZ) [Fig. 1(b)], with an area of 16 km^2 (Wang, 2017; Hou and Cook, 2009). The Gudui geothermal field is characterized by high-temperature geothermal water and the occurrence of intermittent hydrothermal explosions, and is also the largest hydrothermal area in Combai Country, Southeastern Tibet (Tong et al., 2000). According to the investigations by Tong et al. (2000) and Liu et al. (2017), the Gudui geothermal field can be divided into 5 hydrothermal areas, including Buxionglanggu (Fig. 2), Shagalangga (Fig. 2), Bubudemi (Fig. 2), Chaka and Riruo (Fig. 2), in each of which a great number of hot springs discharge to the surface. Of these hydrothermal areas, Buxionglanggu and Shagalangga areas are the most intensely geothermally active with average spring temperatures higher than 50–80℃ (Liu et al., 2017).

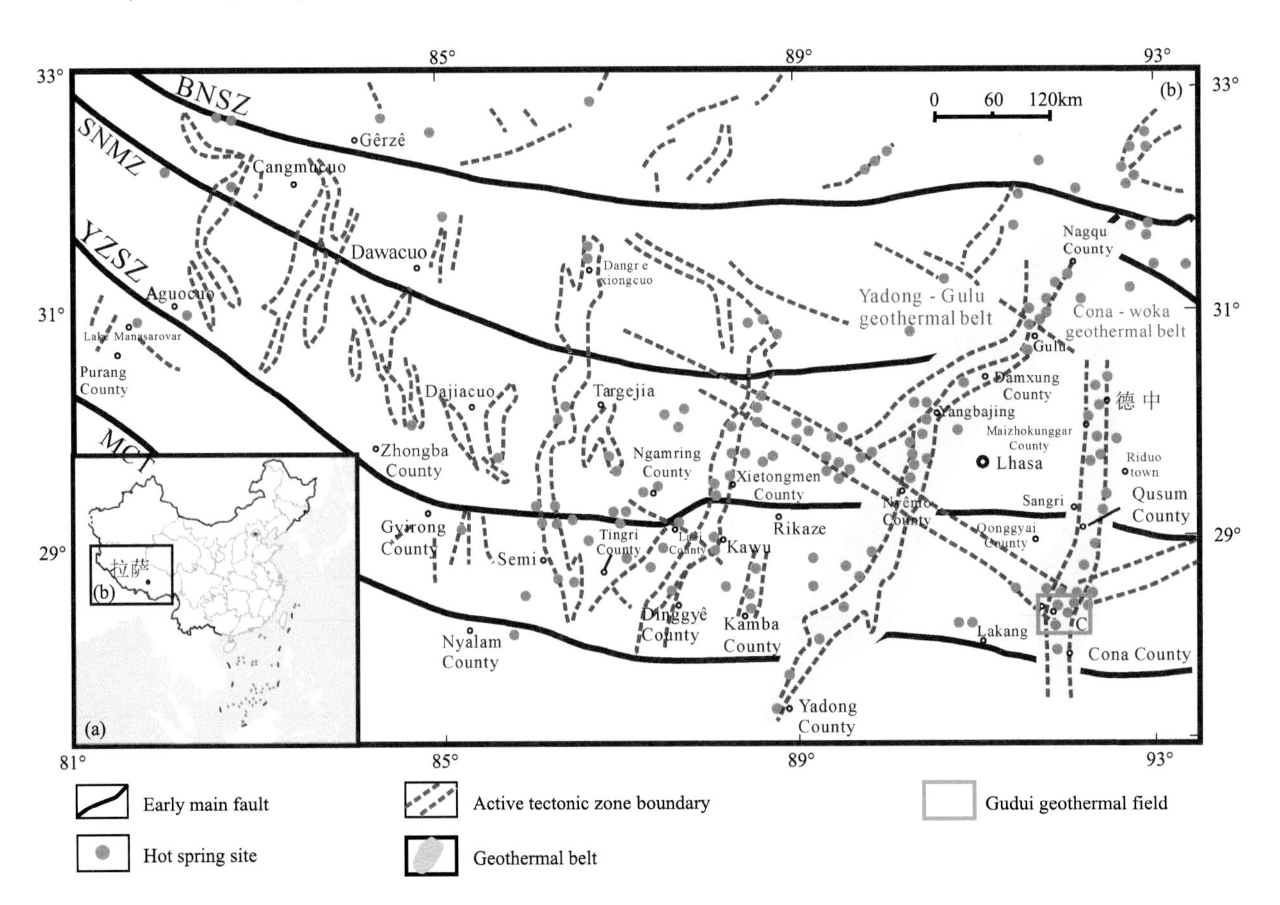

Fig. 1　Distribution of hot spring sites, active tectonics and hydrothermal active belts in the Tibetan Plateau (modified from Li, 2002).

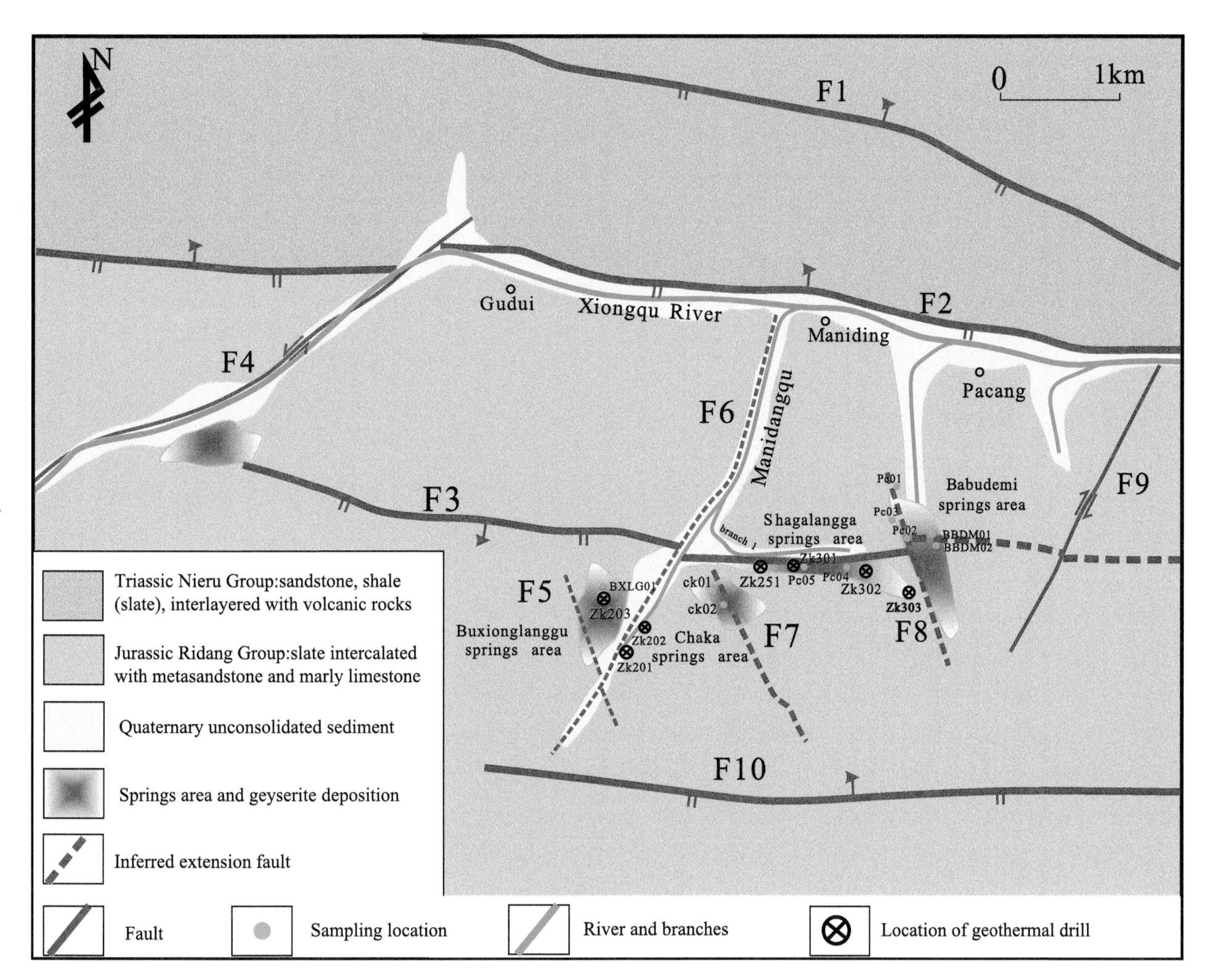

Fig. 2 Simplified geological map and sampling locations of the Gudui geothermal field (F5, F6, F7, F8 are extension faults, other properties are unknown; In addition, although the above figure identifies 7 drill holes, only ZK251, ZK203 and ZK302 are discussed in this paper) (modified from Liu et al., 2017; Wang, 2017; Sun et al., 2018).

Since 2012, under the support of the "High-temperature Geothermal Resources Survey in Key Areas along the Qinghai-Tibet Railway" deployed by the China Geological Survey, the Geothermal Geological Brigade of the Tibet Geological Exploration Bureau has drilled three boreholes, named ZK251, ZK301 and ZK302, in Comei County Gudui geothermal (Fig. 2) and produced geothermal fluids at temperature of 163 ℃, 159 ℃ and 205 ℃ respectively. The geothermal drillhole ZK302 with a temperature of 205 ℃ at 230 m renews the record of the highest temperature at the same depth in a geothermal exploration in China (Hu et al., 2015). Liu et al. (2017) estimated that the total geothermal power generation potential in Gudui is approximately 266 MW and the result of Wang Siqi's (2017) exploration is 115.5 MW, both of which indicate the huge potential for utilizing geothermal resources in the Gudui geothermal field and some researchers even believe that the potential of the Gudui geothermal field may exceed that of the Yangbajing geothermal field, although it is still controversial and requires further in-depth research (Hu et al., 2015). The Gudui area suffers from an acute shortage of power, particularly in the mountainous region of Gudui. The annual average temperature in the study area is about 8.2 ℃ degrees, and it is particularly cold in winter and spring (Wang, 2017). Human and animal heating mostly use fuelwood and cow dung as fuel. Therefore, utilizing geothermal resources cannot only directly improve people's living conditions, but also reduce anthropogenic deforestation and destruction of meadows. However, because of the restrictions on traffic and natural conditions, The Gudui geothermal field is under-researched and undeveloped (Wang, 2017). Hence, in order to develop and utilize geothermal resources in Gudui, there is an urgent need for a detailed investigation of this area and an assessment of its resource potential.

In this study, a detailed geochemical investigation of Gudui geothermal field was conducted on the geothermal waters. This study mainly aims to (i) characterize the geochemical properties and the dissolved constituents of the geothermal waters; (ii) assess the reservoir temperature of hydrothermal waters, and (iii) understand the formation processes and cooling mechanisms of different hot springs in Gudui. The findings will allow us to better understand the geochemical properties of geothermal resources and their evolution in the Himalayan geothermal belt, which is vitally important for policy making to improve hydrothermal resource management and utilization in Tibet.

2 Regional setting and geology

Gudui's hot springs were known as early as the last seventies century and have been investigated by various researchers (Zheng, 1995; Tong et al., 2000; Liu et al., 2017; Wang, 2017). As the largest and most active hydrothermal area in Voca-Cuona geothermal field [Wang, 2017; Fig. 1(b)], the region setting and geology of the Gudui geothermal field are delineated detailedly below:

The Gudui geothermal field is located in Gudui Town, approximately 150 km southeast of Lhasa City, about 90 km west of Cuomei County (28°15′00″–28°45′00″N, 91°30′00″–92°15′00″E, Fig. 1B). The topography of Gudui is high in the south and low in the north, with altitudes varying from 4420 m to 5430 m. It has an average elevation of 4700 m above sea level (asl). Modern climate in Gudui is a typical continental plateau temperate semi-arid monsoon climate with a mean annual temperature of 8.2 ℃ and a mean annual precipitation of 408 mm which is mainly concentrated in June to September (Wang, 2017). The surface waters in the study area are relatively developed, and the main river developed is Xiongqu River, Manidangqu and its tributaries (Fig. 2), which divide the geothermal field into many river valley basins and hot spring areas (Fig. 2). The Xiongqu River is a long-term water current, but the flow changes greatly during the flood season and the dry season, with a range of up to 400 times (Fig. 2). With increasing depth the following sequence of units is found: Quaternary flood alluvium and Quaternary geyserite, Jurassic Ridang Group, Triassic Nieru Group and Cretaceous diorite or diabase (Liao and Zhao, 1999; Du et al., 2005). The tectonic activity in the Gudui region is vigorous and has 14 approximately East-West trending and South-North trending faults (Wang, 2017), among which 10 faults are closely related to geothermal fields, including the early stage East-West trending main faults F1, F2, F3, and F10 (Fig. 2), the late stage North East (F4, F6, F9) (Fig. 2) and North West (F5, F7, F8) faults (Fig. 2). Most of the early stage faults are reverse faults, which have water blocking properties. The late stage faults are the most important structures and serve as the channels for the upward flow of geothermal fluids (Wang, 2017). The location of the hot springs in Gudui tends to be clustered along or near the intersections of faults indicating that the Gudui geothermal field is tectonically controlled. In the study area, magmatic intrusion activity is frequent, but most intrusions are small scale Early Cretaceous diorites and Late Cretaceous diabase (Fig. 2). The strong hydrothermal manifestations in the Gudui geothermal field include hydrothermal explosions [Fig. 3(f)], boiling springs [Fig. 3(k)], boiling mud springs, hot springs [Fig. 3(j)], warm springs [Fig. 3(i)], fumaroles [Fig. 3(c)], steaming ground [Fig. 3(b)], hydrothermal alteration [Fig. 3(l)], and salt sinter [Fig. 3(l)]. The water temperatures of hot springs range from 20 ℃ to more than 83 ℃ (local boiling point corresponding to an average altitude of 4700 m above sea level) (Wang, 2017). Most of the hot springs have a strong sulphurous smell, especially some boiling springs (Tong et al., 2000). During our field work, we found that the main boiling spring outlet at Buxionglanggu exhibits strong intermittent eruptions or hydrothermal explosions [Fig. 3(f)]. The high-temperature water, steam and mud mixture erupts up to tens of meters in the air [Fig. 3(f)], and it showed an eruption at intervals of 10 minutes and each one regularly lasts for 5 minutes.

Shagalangga boiling springs area is characterized by the maximum flow from a single outlet [Fig. 3(e), (g)] which can reach 13.15 L/s and the intense sulfur smell which make it insufferable to approach; In addition, during our fieldwork we discovered the geothermal activity in the Babudemi spring area is significantly weakened, and the host rock has collapsed and covered most of the geothermal area [Fig. 3(j)]. The strongly developed ancient sinter and salt sinter is the most remarkable feature of the Chaka spring area [Fig. 3(i)]. Due to the onset of the rainy season, the Xiongqu River soared and we were unable to reach the Riruo Spring area. The relevant data is cited from previous analysis (Tong et al., 2000; Wang, 2017).

3 Sampling and analysis

3.1 Sample collection

Water samples from 10 sites in Gudui were collected in August 2017. The sampling sites of all water samples are presented in Fig. 2 while other sampling information, in-situ parameters, major characteristics and hydrochemical types are reported in Table 1. The temperature was measured with a hand-held thermometer. The pH of water samples was measured at sampling points using a pH meter that was calibrated in the field prior to sampling. Before determining the pH, the meter was calibrated using pH 7 and pH 4 buffering solutions, respectively. The electrode of the meter was rinsed with distilled water before determining the pH of any subsequent sample to prevent inter-sample contamination. All water samples were filtered through 0.45 μm membranes on site. Samples are stored in 600 mL polyethylene bottles which had been rinsed with deionized water twice before sampling.

3.2 Sample analysis

Standard analytical methods of the Analysis Group, Qinghai Saline Lake Institute and Chinese Academy of Science were employed for the water samples (Lin et al., 2017). For cation analysis, reagent-grade HNO_3 with molar concentration up to 14 M was added to one sample collected at each site to bring the pH below 1. For SiO_2 analysis, the samples were diluted ten-fold using deionized water to prevent SiO_2 in the water from precipitating. The concentrations of Ca^{2+}, Mg^{2+}, Na^+, Li^+ and K^+ were determined by atomic absorption spectrophotometry using a WFX-120 atomic absorption spectrophotometer (Beijing Rayleigh Analytical Instrument Group., Beijing, China) within 2 weeks of sampling, which has a precision of 0.5%. Cl^- was analyzed using the $AgNO_3$ volumetric titration method, with a precision of 0.1%. CO_3^{2-} and HCO_3^- concentrations were determined by titration with 0.1 moL/L HCl, using methyl orange and phenolphthalein as a double indicator, with a precision of 0.3%. SO_4^{2-} was analyzed using the gravimetric method with $BaCl_2$, with a precision of 0.01%. The boron concentration was evaluated by titration with 0.05 mol/L NaOH in the presence of mannitol and using phenolphthalein as the indicator, with a precision of 0.3% (Gao et al., 2012). Total dissolved solids (TDS) were calculated by summing all ions and subtracting half of the HCO_3^-. All analysis was carried out at the Analytical Laboratory of the Research & Development Center of Saline Lake and Epithermal Deposits, Chinese Academy of Geological Sciences (CAGS), Beijing. The hydrochemistry and the charge balances of all water samples were calculated and listed in Table 2. The geothermal water analysis is generally of high analytical accuracy, and the charge balance for most samples is less than or around 10%, especially those samples collected in this study, the charge balance is less than 0.8%.

Fig. 3　Typical hot spring phenomena and surface expressions.

(a) Topography characteristics and the relative location of the Shagalangga and Babudemi spring areas. (b) Hot spring PC01 and the steaming ground. (c) Hot spring PC03 and the use of geothermal water through a water pipe. (d) The characteristics of fault F3 and the location of Buxionglanggu and Chaka spring areas. (e) Intense hydrothermal activity of PC05 and its location relative to F3. (f) The hydrothermal explosion of Buxionglanggu and its strong extrusion of mixed geothermal water, steam and mud. (g) The spring flow of PC05, which is the maximum spring flow of Gudui. (h) The position of spring PC03 and the gully. (i) Chaka spring and the characteristics of salt sinter. (j) Babudimi spring area and the collapsed host rock. (k) Boiling spring. (l) Hydrothermal alteration.

Table 1 Characteristics of water samples from Gudui (temperature in ℃, TDS in mg/L)

NO.	Sample	Name of spring	Spring Type	Sampling location	Hight (m)	T	PH (F)	PH (L)	TDS	Hydrochemical type	Sampling date	References
1	RN1979	Riruo	thermcale	See reference	4440–4540	68.0	na	8.1	3070.0	Na-HCO_3-Cl	July 25, 1979	a
2	RN1989	Riruo	thermcale	See reference	4440–4540	61.0	na	8.3	2460.0	Na-HCO_3-Cl	July 12, 1989	
3	BXLG1975	Buxionglanggu	boiling spring	See reference	4400–4540	86.5	na	8.7	2640.0	Na-Cl	May 31, 1975	
4	BXLG197901	Buxionglanggu	boiling spring	See reference	4400–4540	85.0	na	8.8	3060.0	Na-Cl	July 21, 1979	
5	BXLG197902	Buxionglanggu	boiling spring	See reference	4400–4540	85.0	na	8.7	2960.0	Na-Cl	July 21, 1979	
6	BXLG197903	Buxionglanggu	boiling spring	See reference	4400–4540	45.0	na	8.1	2134.0	Na-HCO_3-Cl	July 21, 1979	
7	CK197901	Chaka	thermcale	91.8924°E, 28.5226°N	4480.0	69.0	na	8.3	2119.0	Na-HCO_3-Cl	July 22, 1979	
8	CK197902	Chaka	thermcale	91.8924°E, 28.5226°N	4480.0	65.0	na	8.3	2196.0	Na-HCO_3-Cl	July 22, 1979	
9	CK197903	Chaka	boiling spring	91.8924°E, 28.5226°N	4480.0	84.0	8.0	8.2	2149.0	Na-HCO_3-Cl	July 11, 1989	
10	CK1989	Chaka	thermcale	91.8924°E, 28.5226°N	4480.0	64.0	na	8.1	2083.0	Na-HCO_3-Cl	July 22, 1979	
11	SGLG197901	Shagalangga	boiling spring	91.8922°E, 28.5217°N	4600.0	85.5	na	8.9	2403.0	Na-Cl	July 24, 1979	
12	SGLG197902	Shagalangga	boiling spring	91.8922°E, 28.5217°N	4600.0	86.0	na	8.9	2330.0	Na-Cl	July 24, 1979	
13	SGLG197903	Shagalangga	boiling spring	91.8922°E, 28.5217°N	4600.0	86.0	na	9.0	2162.0	Na-Cl	July 24, 1979	
14	SGLG197904	Shagalangga	boiling spring	91.8922°E, 28.5217°N	4600.0	84.0	na	9.3	2321.0	Na-Cl	July 24, 1979	
15	SGLG1989	Shagalangga	thermcale	91.8922°E, 28.5217°N	4600.0	70.0	7.5	8.0	1952.0	Na-Cl	July 11, 1989	
16	BBDM1975	Babudemi	boiling spring	91.8923°E, 28.5222°N	4630–4700	86.5	na	8.6	2050.0	Na-Cl	June 1, 1975	
17	BBDM197901	Babudemi	boiling spring	91.8923°E, 28.5222°N	4630–4700	86.0	na	8.8	2091.0	Na-Cl	July 26, 1979	
18	BBDM197902	Babudemi	boiling spring	91.8923°E, 28.5222°N	4630–4700	85.0	na	8.7	2138.0	Na-Cl	July 26, 1979	
19	BBDM197903	Babudemi	boiling spring	91.8923°E, 28.5222°N	4630–4700	86.0	na	9.0	2135.0	Na-Cl	July 26, 1979	
20	BBDM197904	Babudemi	thermcale	91.8923°E, 28.5222°N	4630–4700	47.0	na	8.3	1786.0	Na-HCO_3-Cl	July 26, 1979	
21	BBDM197905	Babudemi	thermcale	91.8923°E, 28.5222°N	4630–4700	76.0	na	8.5	1933.0	Na-Cl	July 26, 1979	
22	BBDM1989	Babudemi	boiling spring	91.8923°E, 28.5222°N	4630–4700	86.5	8.5	8.5	2188.0	Na-Cl	1989	
23	Q001	Gudui	Cold spring	See reference	4630–4700	10.0	6.7	na	826.2	Ca-Mg-SO_4-HCO_3	2017	b
24	Q002	Riruo	thermcale	See reference	4630–4700	64.0	6.5	na	2701.3	Na-Cl-HCO_3	2017	
25	Q004	Babudemi	thermcale	See reference	4630–4700	74.0	7.2	na	2256.7	Na-Cl	2017	
26	Q006	Shagalangga	thermcale	See reference	4630–4700	78.0	8.4	na	2461.9	Na-Cl	2017	
27	H008	Shagalangga	hot spring	See reference	4630–4700	28.0	8.4	na	2356.8	Na-Cl	2017	
28	Q009	Shagalangga	thermcale	See reference	4630–4700	66.0	7.2	na	2346.0	Na-Cl-HCO_3	2017	
29	Q010	Chaka	thermcale	See reference	4630–4700	54.0	7.5	na	2542.7	Na-Cl-HCO_4	2017	
30	Q013	Buxionglanggu	thermcale	See reference	4630–4700	71.0	7.5	na	2998.5	Na-Cl-HCO_6	2017	

Continued

NO.	Sample	Name of spring	Spring Type	Sampling location	Hight (m)	T	PH (F)	PH (L)	TDS	Hydrochemical type	Sampling date	References
31	H011	Surface	surface water	See reference	4630–4700	10.0	8.2	na	835.5	Ca-Mg-SO_4-HCO_3	2017	
32	H014	Surface	surface water	See reference	4630–4700	11.0	8.2	na	1096.2	Na-Ca-Mg-SO_4-Cl-HCO_3	2017	
33	H015	Surface	surface water	See reference	4630–4700	17.0	7.8	na	996.0	Na-Ca-Mg-SO_4-Cl-HCO_4	2017	
34	H016	Surface	surface water	See reference	4630–4700	18.0	7.8	na	545.3	Ca-Mg-SO_4-HCO3	2017	
35	H017	Surface	surface water	See reference	4630–4700	19.0	8.3	na	819.5	Ca-Mg-SO_4-HCO_3	2017	
36	ZK302	Drill water	drill water	See reference	4630–4700	204.0	8.7	na	3324.9	Na-Cl	2017	
37	ZK251	Drill water	drill water	See reference	4630–4700	163.0	8.5	na	2798.3	Na-Cl	2017	
38	ZK203	Drill water	drill water	See reference	4630–4700	175.5	8.7	na	2902.5	Na-Cl-HCO_3	2017	
39	PC01	PC01	thermcale	91.9146°E, 28.5329°N	4449.0	74.0	8.1	na	2182.4	Na-Cl	August 14, 2017	c
40	PC02	PC02	boiling spring	91.9107°E, 28.5311 °N	4605.0	81.0	8.4	na	2206.4	Na-Cl	August 14, 2017	
41	PC03	PC03	thermcale	91.9107°E, 28.5312 °N	4605.0	70.0	nd	na	2255.4	Na-Cl	August 14, 2017	
42	PC04	Shagalangga	boiling spring	91.9108°E, 28.5257 °N	4645.0	85.0	8.2	na	2649.6	Na-Cl	August 14, 2017	
43	PC05	Shagalangga	boiling spring	91.9084°E, 28.5254 °N	4630.0	84.0	8.3	na	2981.1	Na-Cl	August 14, 2017	
44	BLX01	Buxionglanggu	boiling spring	91.8831°E, 28.5197 °N	4488.0	83.0	8.1	na	2996.1	Na-Cl	August 16, 2017	
45	CK01	Chaka	thermcale	91.8925°E, 28.5235 °N	4433.0	70.0	7.4	na	2472.8	Na-HCO_3-Cl	August 16, 2017	
46	CK02	Chaka	thermcale	91.8925°E, 28.5231 °N	4501.0	68.0	7.2	na	2303.3	Na-Cl	August 16, 2017	
47	BBDM01	Babudemi	thermcale	91.9171°E, 28.5333 °N	4520.0	80.0	6.6	na	2316.6	Na-HCO_3-Cl	August 17, 2017	
48	BBDM02	Babudemi	hot spring	91.9171°E, 28.5333 °N	4520.0	20.0	nd	na	2282.6	Na-Cl	August 17, 2017	

Note: PH (F) represents field measured and is for reference only, PH (L) represents indoor measured, and na represents no analysis; nd represents no detected; a = Tong et al. (2000); b = Wang (2017); c = current work in this study; The sampling condition of Wang (2017) may be under the condition of steam loss (maybe not maximum steam loss), which is suggested by an anonymous reviewer. The sampling position of samples 1–6 and 23–38 are only given a coordinate range in the literature. For details, see the corresponding literature

Table 2 Concentrations of major chemical constituents of water samples from Gudui (Note that the unit of all constituents is mg/L.)

No.	Sample no.	HCO_3^-	CO_3^{2-}	SO_4^{2-}	Cl^-	Ca	Mg	Na	K	Sr	SiO_2	HBO_2	As	Li^+	Rb^+	Cs^+	NH_4^+	F	Charge imbalance (%)
1	RN1979	689.0	0.0	125.0	985.0	26.8	11.2	690.0	98.0	2.5	126.0	590.0	0.0	35.0	1.9	9.6	10.8	9.4	−9.8
2	RN1989	609.0	0.0	123.0	718.0	39.7	8.1	600.0	78.0	1.3	125.0	420.0	0.1	22.6	1.4	7.4	8.1	5.5	−5.1
3	BXLG1975	227.0	180.0	182.0	757.0	2.1	2.9	793.8	88.0	0.0	118.0	470.0	0.0	0.0	0.0	0.0	2.7	15.4	−4.7
4	BXLG197901	444.0	157.0	155.0	759.0	1.7	0.1	700.0	85.0	0.5	400.0	510.0	2.4	27.9	1.7	8.6	2.2	26.5	−8.5
5	BXLG197902	459.0	142.0	162.0	746.0	4.6	1.0	700.0	50.0	0.5	362.0	470.0	2.2	27.4	1.6	8.0	3.1	13.0	−8.3
6	BXLG197903	751.0	0.0	121.0	445.0	98.3	17.2	425.0	100.0	1.3	165.0	410.0	0.1	14.2	1.0	4.8	0.0	5.1	−5.2
7	CK197901	534.0	0.0	120.0	556.0	15.4	3.7	500.0	68.0	0.6	198.0	350.0	1.3	16.3	12.5	6.5	6.2	9.5	−7.1
8	CK197902	565.0	0.0	138.0	559.0	20.6	4.0	505.0	68.0	0.6	181.0	400.0	1.3	16.1	1.4	5.6	4.9	7.5	−8.9
9	CK197903	508.0	0.0	128.0	573.0	5.6	3.3	535.0	80.5	0.0	165.0	354.0	1.5	18.9	1.4	5.6	7.1	7.9	−4.4
10	CK1989	435.0	0.0	148.0	549.0	20.6	2.2	510.0	68.0	0.4	198.0	330.0	1.8	12.4	2.4	11.3	5.4	12.4	−4.7
11	SGLG197901	257.0	143.0	165.0	628.0	2.6	0.2	560.0	78.0	0.3	220.0	430.0	2.8	21.1	1.3	5.8	2.1	14.2	−9.2
12	SGLG197902	312.0	84.0	144.0	603.0	2.6	0.2	555.0	78.0	0.3	220.0	440.0	2.4	18.7	1.5	5.6	2.9	15.0	−7.1
13	SGLG197903	260.0	131.0	151.0	591.0	5.9	0.8	535.0	80.0	0.3	215.0	420.0	2.5	18.7	1.4	6.4	2.6	10.5	−8.5
14	SGLG197904	177.0	200.0	115.0	626.0	3.1	0.1	535.0	80.0	0.3	225.0	400.0	2.6	18.9	1.4	6.2	2.1	14.3	−10.2
15	SGLG1989	384.0	0.0	149.0	591.0	14.7	4.3	483.0	78.6	0.0	89.8	304.0	1.0	16.9	1.6	8.0	4.0	9.6	−5.6
16	BBDM1975	134.0	109.0	177.0	576.0	6.9	1.1	450.0	67.0	0.0	151.0	412.0	1.1	19.3	2.2	11.9	3.4	12.5	−11.4
17	BBDM197901	282.0	88.0	144.0	566.0	3.7	0.2	525.0	73.0	0.3	175.0	332.0	2.4	16.3	1.3	6.4	3.1	13.5	−5.8
18	BBDM197902	329.0	54.0	144.0	579.0	4.7	0.2	535.0	73.0	0.3	180.0	360.0	2.4	17.4	1.3	6.4	3.5	12.4	−5.2
19	BBDM197903	260.0	123.0	108.0	594.0	3.8	0.4	535.0	74.0	0.3	182.0	340.0	2.3	17.6	1.4	6.4	2.5	14.3	−6.1
20	BBDM197904	382.0	0.0	124.0	496.0	32.9	1.8	435.0	55.0	0.3	102.0	320.0	1.5	13.4	1.0	5.2	0.0	7.1	−5.7
21	BBDM197905	344.0	0.0	147.0	530.0	6.7	0.3	470.0	63.0	0.5	125.0	380.0	2.1	15.0	1.1	4.8	3.3	12.1	−7.1
22	BBDM1989	360.0	12.4	150.0	608.0	0.2	0.5	558.0	82.4	0.0	180.0	365.0	2.1	19.8	2.4	12.4	3.8	10.2	−2.9
23	Q001	271.0	0.0	346.2	1.7	151.0	46.8	0.9	0.6	1.0	7.3	0.4	0.0	0.0	0.0	0.0	0.0	0.3	−1.1
24	Q002	841.9	0.0	238.5	497.1	131.8	27.3	485.0	60.9	0.7	92.1	321.5	0.0	15.8	0.0	0.0	0.0	5.0	−4.5
25	Q004	392.1	0.0	207.7	538.2	22.5	19.5	450.9	70.5	0.1	151.1	339.9	1.1	16.5	0.0	0.0	0.0	11.0	−6.7
26	Q006	299.8	69.4	180.8	620.5	9.6	17.5	490.3	85.6	0.0	244.4	412.6	2.6	19.3	0.0	0.0	0.0	11.5	−9.0
27	H008	288.3	75.2	214.4	558.8	32.1	17.5	486.2	75.4	0.1	201.1	362.7	2.0	18.1	0.0	0.0	0.0	11.4	−5.6
28	Q009	478.6	0.0	231.6	510.8	38.6	23.4	435.1	74.0	0.1	206.4	318.9	1.6	16.0	0.0	0.0	0.0	10.6	−7.2
29	Q010	593.9	0.0	177.2	565.7	54.6	23.4	456.9	74.0	0.1	207.7	358.3	1.2	18.0	0.0	0.0	0.0	9.5	−7.9

Continued

No.	Sample no.	HCO_3^-	CO_3^{2-}	SO_4^{2-}	Cl^-	Ca	Mg	Na	K	Sr	SiO_2	HBO_2	As	Li^+	Rb^+	Cs^+	NH_4^+	F	Charge imbalance (%)
30	Q013	640.0	0.0	208.0	689.1	9.6	19.5	595.0	93.7	0.1	182.4	525.6	1.7	23.4	0.0	0.0	0.0	11.8	−9.8
31	H011	207.9	6.8	400.6	1.0	109.2	76.0	15.2	2.5	0.8	11.9	3.5	0.0	0.0	0.0	0.0	0.0	0.3	1.6
32	H014	244.5	11.3	292.7	140.6	80.3	52.6	118.0	18.0	0.4	41.1	85.9	0.0	5.9	0.0	0.0	0.0	2.2	−2.5
33	H015	251.4	0.0	277.3	120.0	84.2	51.1	112.2	13.0	0.6	21.3	58.7	0.0	3.8	0.0	0.0	0.0	1.1	0.5
34	H016	196.1	0.0	192.3	5.5	93.2	31.2	11.0	1.4	0.5	11.2	1.8	0.0	0.2	0.0	0.0	0.0	0.2	2.4
35	H017	186.8	0.0	415.4	2.7	118.9	72.1	10.5	1.4	0.4	8.5	2.6	0.0	0.1	0.0	0.0	0.0	0.3	2.3
36	ZK302	344.6	137.9	293.5	866.4	3.3	32.3	773.9	115.9	0.0	159.4	550.2	3.5	27.8	0.0	0.0	0.0	19.0	−4.8
37	ZK251	512.5	108.0	168.2	658.5	17.5	23.4	566.7	86.0	0.1	196.6	425.9	2.5	22.4	0.0	0.0	0.0	10.0	−9.2
38	ZK203	621.6	169.4	233.7	529.6	6.7	30.1	627.5	91.6	0.1	192.7	365.1	2.0	20.4	0.0	0.0	0.0	13.0	−6.5
39	PC01	213.5	90.00	181.1	580.2	23.20	3.80	600.0	62.00	0.0	48.7	403.6	2.6	16.1	0.7	3.4	3.5	5.05	−0.2
40	PC02	73.20	124.8	194.3	641.7	20.40	3.60	630.0	67.00	0.0	72.9	426.9	3.0	15.7	0.8	3.5	3.3	6.04	−0.2
41	PC03	107.4	110.4	199.2	648.5	19.40	3.00	640.0	68.00	0.0	74.3	434.7	3.3	16.5	0.8	3.6	2.8	6.21	−0.2
42	PC04	176.9	132.0	309.5	682.6	23.00	3.40	760.0	77.00	0.0	113	458.0	3.2	18.0	0.9	3.4	4.9	6.57	0.6
43	PC05	244.0	174.0	189.3	808.9	20.40	3.60	840.0	93.00	0.0	218	574.4	4.1	22.2	1.2	5.2	2.9	8.88	−0.4
44	BXLG 01	372.1	138.0	268.4	730.4	19.40	4.00	840.0	78.00	0.0	134	512.3	4.9	22.2	0.9	3.9	5.0	6.53	−0.2
45	CK01	483.1	54.00	151.5	587.1	43.00	6.80	650.0	64.00	0.0	87.7	407.5	2.4	15.0	0.8	3.6	5.1	4.47	0.6
46	CK02	353.8	112.8	107.0	566.6	41.40	5.60	620.0	64.00	0.0	86.1	407.5	2.4	15.2	0.8	3.3	3.8	4.52	0.8
47	BBDM01	536.8	108.0	144.9	460.8	55.40	8.60	590.0	56.00	0.4	26.3	333.8	0.2	12.7	0.8	3.6	3.9	2.73	−0.1
48	BBDM02	317.2	132.0	187.7	529.0	46.60	9.20	610.0	59.00	0.2	33.9	372.6	1.2	14.6	0.7	2.6	0.0	4.51	0.0

Note that for the sake of convenience, B is used to represent HBO_2 in the paper and relevant figures.

4 Results and discussion

4.1 Hydrochemistry

In the Gudui geothermal field, the hydrochemical types of boiling spring waters and thermal samples are mainly Na-Cl, whereas the hot springs and cold springs are mainly Na-HCO_3-Cl, Na-Cl-HCO_3 and Na-HCO_3, the surface waters are Ca-Mg-SO_4-HCO_3 and Na-Ca-Mg-SO_4-Cl-HCO_3 types (Table 1). Most geothermal samples contain Cl^- as the predominant anion (Fig. 4), and Na^+ as the predominant cation (Fig. 4), whereas their Ca and Mg concentrations are generally low (Fig. 4). However, the surface and cold spring waters are just the opposite, the Ca, Mg and SO_4^{2-} concentrations are higher than those of the geothermal samples (Fig. 4). The samples collected from Gudui are also characterized by very high sampling temperature (up to 85 ℃) (Table 1) and SiO_2, B, As, Li, Rb, Cs, F concentrations (Table 2), especially the B and Li content, which is up to 590 mg/L and 35 mg/L, respectively. High concentrations of SiO_2, B, As, Li, Rb, Cs and F in the weak alkaline springs at Gudui were also observed by Zheng (1995), Tong et al. (2000), and Wang (2017). The TDS values for geothermal well, boiling spring, warm spring, hot spring, and surface water are between 2792.26–3324.9 mg/L, 2050.0–3060.0 mg/L, 1786.0–3070.0 mg/L, 2134.0–2356.8 mg/L, and 545.3–1096.2 mg/L (Table 1), respectively. By comparing the TDS values for different natural waters in Gudui, we find that as the temperature decreases, the TDS also gradually decreases. It is interesting that most of the surface waters and cold spring waters have a slightly lower pH and Cl concentration and a higher SO_4^{2-} concentration than the geothermal water samples. The formation of these springs can probably be explained by the theory of Henley and Ellis (1983) that steam or steam and gas escaping from deep hydrothermal systems may be absorbed by superficial groundwater and is perhaps perched above a low permeability cap-rock to form waters enriched in sulfate ion (Fig. 4). The strong sulphurous smell of geothermal water in Gudui also indicates the flashing steam and gas may contain a great deal of H_2S. Since Cl seldom migrates along with flashing steam, the Cl concentration in the surface waters and cold springs is low.

In the shallow subsurface environment, the H_2S dissolved in the groundwater is oxidized to SO_4^{2-}. As a result, the cold springs and surface waters become a little acidic and rich in SO_4^{2-}. However, more work needs to be doned on future samples to confirm the above speculation. Through the above analysis, we can conclude that the Gudui geothermal field is a typical high-temperature geothermal system, which has similar characteristics as the Yangbajing high-temperature field. The change of geothermal water types (from boiling springs to cold springs), hydrochemical types (from Na-Cl type to Na-HCO_3-Cl), and TDS value with decreasing outlet temperature suggests that the geothermal water has mixed with groundwater to different degrees in the process of ascending to the surface in the Gudui geothermal field. In contrast with the hydrochemistry characteristics of geothermal fluids from Shagalangga and Buxionglanggu, we suggest that those below Bubudemi, Chaka, Riruo are likely to have mixed with cold groundwater enriched in Ca and Mg but depleted in Cl, Na, K, SiO_2 and B. From the perspective of the location of the different type springs, we conclude that the Shagalangga and the Buxionglanggu correspond to the central hydrothermal area (high-temperature hydrothermal areas), while the Riruo, Chaka and Bubudemi correspond to the marginal hydrothermal areas (low-temperature hydrothermal areas).

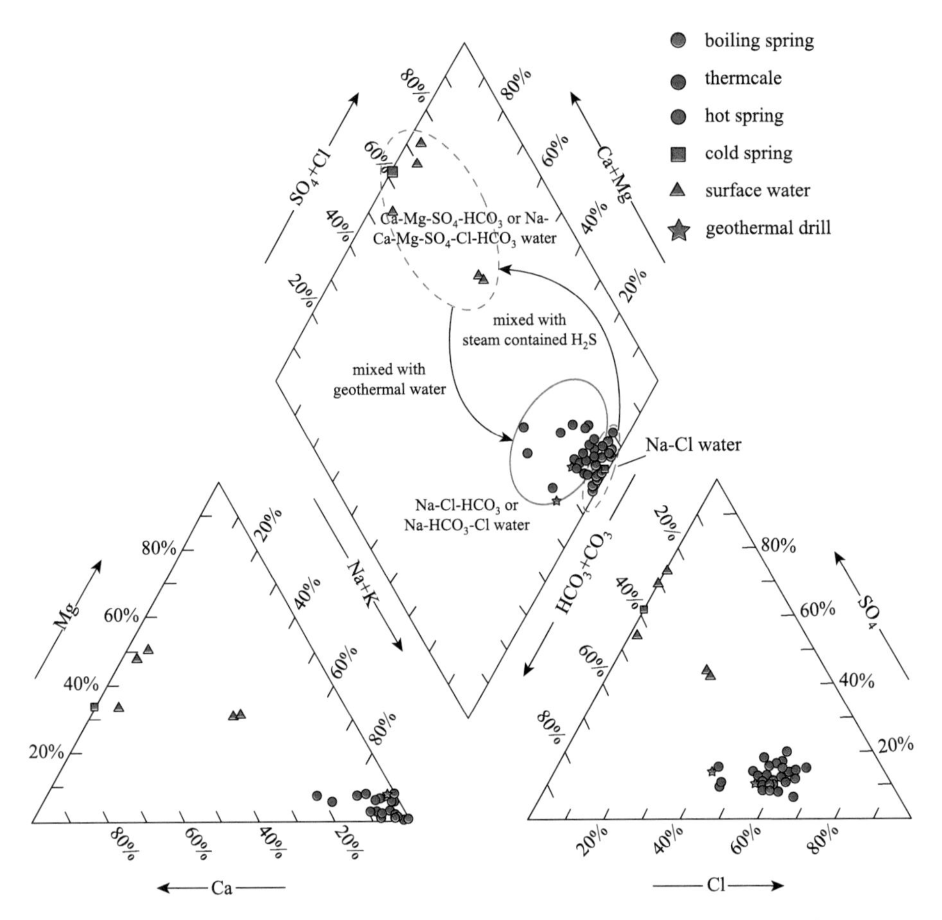

Fig. 4　Piper diagram of cold surface water, thermal water from drills and springs.

4.2　Geothermometry

When using chemical geothermometers (Fournier and Rowe, 1966; Ellis, 1970; Fournier and Truesdell, 1973; Truesdell, 1976; Fournier, 1977) to predict subsurface temperatures in geothermal systems one should keep in mind the various assumptions which are implicit in the methods (Fournier et al., 1974). It is also necessary to be aware of the hydrologic complexities which are commonly present in hot-spring systems (Fournier, 1979). It has been commonly accepted by the international geothermal community and researchers all over the world that the geothermal water ascends to the surface through adiabatic cooling, conductive cooling, most commonly partly adiabatic and partly conductive and mixing with groundwater (Fournier, 1979). However, different cooling processes have a greater influence on the composition variability of geothermal water, which in turn influences the choice of geothermometer for calculation of the geothermal reservoir temperature. As all chemical and isotopic reactions do not proceed at the same rate, different chemical and isotope geothermometers may give different results. If the geothermal water has cooled by conduction, the spring temperature may be much lower than the maximum temperature in the hydrothermal system (just like PC01, PC03, CK01, CK02, BBDM02) and during the relatively slow rate of mass movement to the surface necessary for appreciable conductive cooling, the water composition is likely to change owing to precipitation and water-rock reactions. Waters which start above 100 ℃ and flow at relatively large mass rates directly to the surface will cool adiabatically, and the emerging spring water will be at or slightly above the boiling temperature (just like PC04, PC05, and BLXG01) (Fournier and Rowe, 1966). When initial temperatures are above about 210–230 ℃ (ZK302 reveals that the initial temperature of Gudui geothermal water is at least 204 ℃), silica is likely to precipitate on the way to the surface owing to relatively fast rates of reaction at higher temperatures and the attainment of supersaturation with respect to amorphous silica as the solution cools.

This silica precipitation may coat the channels and prevent other water-rock reactions, particularly those involving Na, K, and Ca. In this situation the Na-K-Ca geothermometer should give higher and more reliable results than the silica geothermometer (Fournier and Rowe, 1966). Provided calcium carbonate precipitates owing to carbon dioxide does not dissipating in time, Na-K geothermometer should give more reliable results in the Gudui geothermal field. Thus, the reservoir temperatures in different hydrothermal areas of Gudui are estimated using the Na-K geothermometer (Giggenbach et al., 1983), the quartz geothermometer without and with maximum steam loss (both Fournier, 1977) and the K-Mg geothermometer (Giggenbach, 1988). Quartz geothermometry was used instead of chalcedony geothermometry, since the solubility of quartz appears to control dissolved silica in a geothermal reservoir at temperatures higher than 120–180 ℃ (Fournier, 1977), which is the case in all the geothermal prospect areas investigated (see Na-K-Mg ternary diagram, Fig. 5). All applied geothermometers assume the attainment of solute/mineral equilibrium in the deep reservoir. Chemical equilibration was tested with Giggenbach (1988) ternary diagram using relative Na/1000, K/100 and $Mg^{1/2}$ concentration (Fig. 5). In this study, we included existing data for the geothermal waters and compared it to the new data from our water sampling campaign in 2017 in order to arrive at a better resolution for the derived geothermal subsurface temperatures estimated by different geothermometers. Furthermore, the reservoir temperatures estimated were compared with measured downhole temperatures from ZK251, ZK203 and ZK302 (data from Wang, 2017). For this, we give a short overview of the wells ZK251, ZK203 and ZK302 in the Gudui geothermal field (after Wang, 2017): ZK251, ZK203 and ZK302 are all non-productive wells drilled in different areas of the Gudui geothermal field. The outflow zone in the Gudui geothermal field lies along a deep, open, young EW trending fault (F3 in Fig. 2) encompassing wells ZK251 and ZK302 with depths of 672 and 400 m and maximum measured temperatures of 163 ℃ and 205 ℃, of which ZK302 has the maximum flow and its total flow of steam and water is up to 93.17 t/h. ZK203 was drilled in the lateral outflow zone with a depth of 705 m and a maximum measured temperature of 173.5 ℃. Temperature and pressure logging curves are shown in Fig. 6. Other relevant detailed information can be found in Wang (2017).

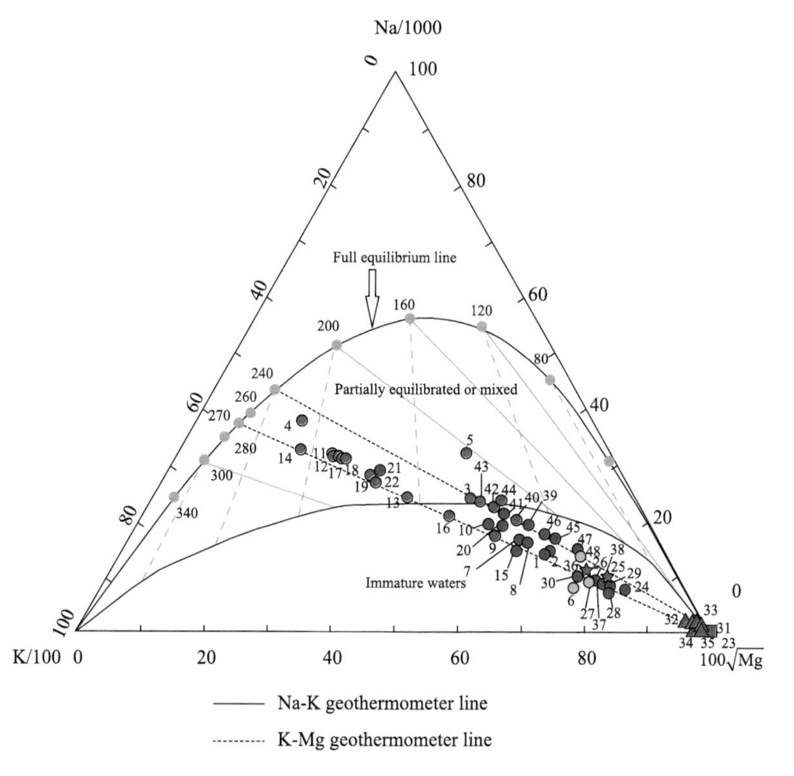

Fig. 5 Triangular diagram Na-K-$Mg_{1/2}$ (from Giggenbach, 1988) for geothermal water samples, geothermal drill water, cold spring water, and surface water from 1975 to 2017. The legend is the same as in Fig. 4; the number label of the symbols is the same as in Table 1, Dashed lines indicate the mixing between geothermal waters and shallow (surface) waters.

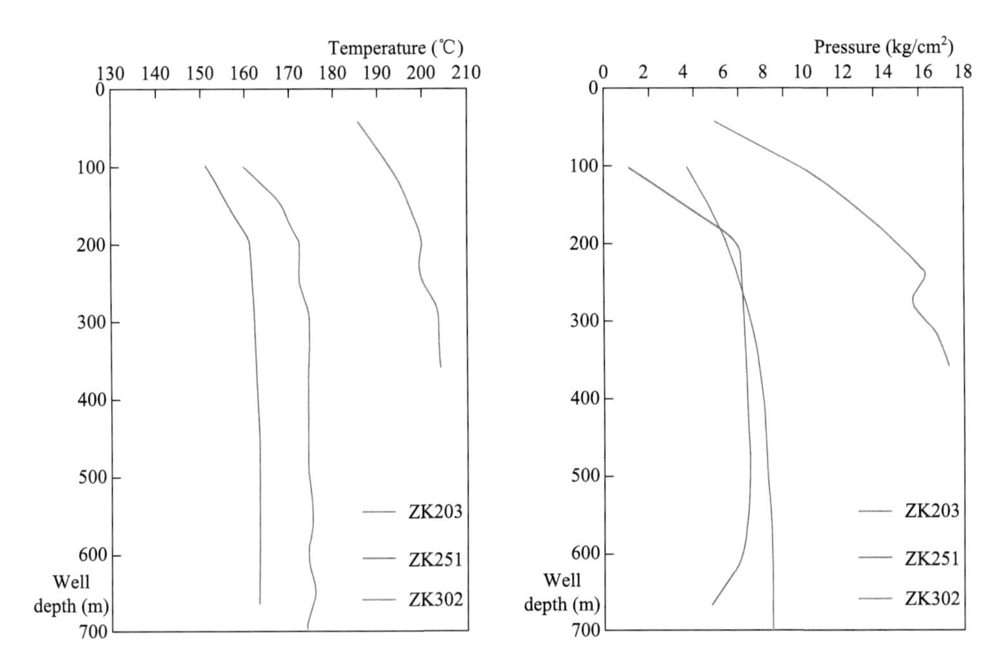

Fig. 6　Temperature and pressure logging curves (from Wang, 2017) for the well ZK203, ZK251 and ZK302.

In our study, we included in detail the chemistry data from Wang (2017) of the wells ZK251, ZK203 and ZK302 (see Tables 1 and 2). Fig. 5 introduces the Na-K-Mg ternary diagram (after Giggenbach, 1988), which can be used to classify hot spring waters as fully equilibrated with rock at given temperatures, partially equilibrated, or immature. This provides an indication of the suitability of the waters for the application of solute geothermometers. The Na-K-Mg ternary diagram (Fig. 5) suggests that hot springs samples labeled 39–48 (new data from our water sampling) plot closely to the boundary between mature waters (partly equilibrated) and immature waters (shallow/mixed waters). For these waters, subsurface temperature estimates can generally be performed with some degree of confidence (Giggenbach, 1988). The diagram also shows that the hot spring labeled 4–5, 11–12, 17–19, and 21–22 are more suitable for geothermometry, because they plot in the partially equilibrated water field. Fig. 5 shows that predecessors' research data and our new data of the Gudui geothermal areas plot on the same dilution tendency and thus are comparable. They refer to the same subsurface temperatures of about 240–270 ℃ for all the samples from the Gudui hot springs. In the following, this result are compared with the subsurface temperatures obtained by silica and cation geothermometry (Table 3). The Na-K, K-Mg and quartz geothermometers yield reservoir temperatures 206.80–266.60 ℃, 142.91–209.38 ℃ and 150.05–232.66 ℃ for the hot springs plotted in the partially equilibrated water field, and 226.46–240.56 ℃, 112.50–142.23 ℃ as well as 74.06–185.90 ℃ for the hot springs plotting closely to the boundary between mature waters and immature waters, respectively. The results for the K-Mg geothermometer are too low to be considered. The reason may be the faster re-equilibration of K-Mg during the upflow of the water, as well as mixing with relatively Mg-rich immature waters. In order to balance all the sampling conditions, the quartz geothermometer was used to calculate the reservoir temperature under the conditions of maximum steam loss and no steam loss. Although the local boiling point is less than 100 ℃, the quartz geothermometer calculation formulas under maximum steam loss conditions may not be applicable. It is better to calibrate the formulas before its application. However, according to the solubility of quartz and temperature curves proposed by Fournier (1977), we found that under the condition of a certain amount of quartz, even if the quartz geothermometer is calibrated, the calculated reservoir temperature with steam loss is still not higher than the calculated result without steam loss. Therefore, this inapplicability does not affect the overall result of the reservoir temperature calculation. Relevant explanations are provided below. The reservoir temperature of Gudui geothermal field estimated by the quartz geothermometer under maximum steam loss is lower than the result estimated by the quartz geothermometer under no steam loss. However, although the reservoir temperature estimated by the quartz geothermometer with no steam loss is consistent with the measured temperatures in ZK251, ZK203 and ZK302, it is too low compared with the Na-K-Mg ternary diagram. It is also known that mixing of geothermal fluids with immature water has a negative effect on the reliability of the silica geothermometers. In contrast, cation geothermometry is less sensitive to mixing and also boiling processes, since it uses ratios rather than absolute abundances of the ions (Gendenjamts, 2003). Besides, the results obtained by the Na-K geothermometer and the Na-K-Mg ternary diagram can confirm each other. Consequently, we can make a conclusion that the Na-K geothermometers provide the most reliable reservoir temperature for the investigated hot spring.

Table 3 Reservoir temperatures (℃) estimated using Na/K, K/Mg and quartz geothermometers

No.	Sample no.	Na/K	SiO_2 (1)	SiO_2 (2)	K/Mg	No.	Sample no.	Na/K	SiO_2 (1)	SiO_2 (2)	K/Mg
1	RN1979	261.95	143.88	150.52	125.44	22	BBDM1989	265.46	162.36	173.14	175.97
2	RN1989	254.15	143.48	150.04	123.38	24	Q002	251.16	128.90	132.74	98.58
3	BXLG1975	240.67	140.65	146.64	143.86	25	Q004	270.73	153.09	161.70	107.35
4	BXLG197901	248.29	210.34	232.25	209.38	26	Q006	281.06	179.56	194.66	114.49
5	BXLG197902	206.80	203.77	224.56	142.91	27	H008	269.89	168.45	180.72	110.77
6	BXLG197903	311.28	157.70	167.37	119.45	28	Q009	278.54	169.91	182.54	106.11
7	CK197901	258.10	167.58	179.64	131.42	29	Q010	273.91	170.26	182.98	106.10
8	CK197902	257.23	162.66	173.51	130.17	30	Q013	271.34	163.08	174.02	115.62
9	CK197903	267.17	157.70	167.37	138.78	36	ZK302	266.75	155.87	165.11	114.42
10	CK1989	256.36	167.58	179.64	139.97	37	ZK251	267.89	167.18	179.14	110.40
11	SGLG197901	260.22	173.50	187.05	190.34	38	ZK203	264.46	166.09	177.78	108.62
12	SGLG197902	261.01	173.50	187.05	190.34	39	PC01	234.94	101.50	101.21	128.04
13	SGLG197903	266.60	172.20	185.41	163.72	40	PC02	237.27	118.38	120.52	131.38
14	SGLG197904	266.60	174.78	188.66	206.62	41	PC03	237.19	119.22	121.48	134.83
15	SGLG1989	274.41	127.72	131.35	133.87	42	PC04	233.35	138.54	144.13	136.86
16	BBDM1975	266.21	153.06	161.66	152.25	43	PC05	240.56	172.98	186.40	142.23
17	BBDM197901	260.06	160.84	171.25	187.55	44	BXLG 01	226.46	146.96	154.23	134.61
18	BBDM197902	258.39	162.36	173.14	187.55	45	CK01	231.07	126.64	130.09	119.99
19	BBDM197903	259.60	162.96	173.88	174.05	46	CK02	234.85	125.80	129.11	122.97
20	BBDM197904	251.75	133.65	138.32	136.25	47	BBDM01	228.17	78.35	74.81	112.50
21	BBDM197905	256.83	143.48	150.04	173.38	48	BBDM02	229.66	87.53	85.33	113.04

SiO_2 (1) represents the temperatures calculated under max. steam loss, SiO_2 (2) represents the temperatures calculated under no steam loss (Fournier, 1977)

However, from the perspective of genesis, the quartz and the K-Mg temperatures may also have a certain geological significance. The comparison of Na-K temperatures, the quartz and the K-Mg temperatures for all geothermal water samples shows that the former give generally higher temperatures than the latter (Table 3; Fig. 7). These differences suggest that the Na-K geothermometer reflects the reservoir temperatures, whereas the quartz or K-Mg geothermometers may record intermediate temperatures (the shallow reservoirs) between the reservoir values and the spring outlet values as the geothermal waters re-equilibrate upon conductive cooling or mixing with cooler Mg-rich waters (Guo and Wang, 2012).

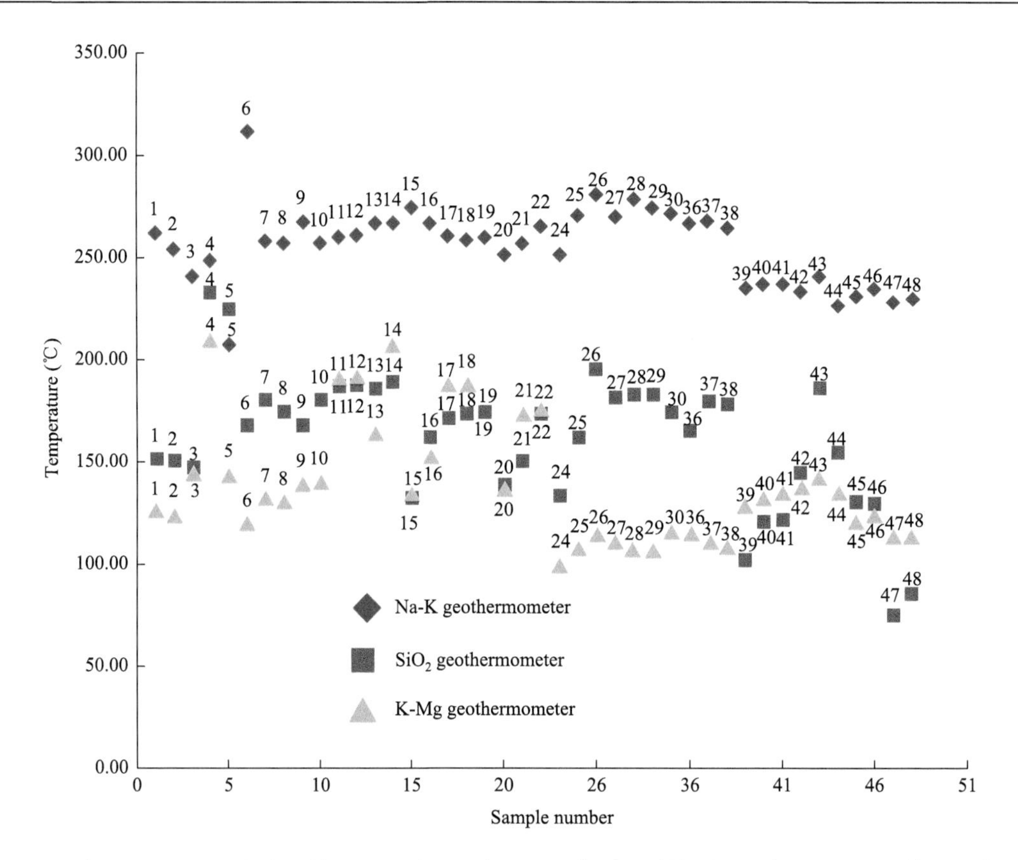

Fig. 7 The reservoir temperature calculated by Na-K, Quartz (no steam loss) and K-Mg geothermometers (the number label of the symbols in this figure is the same as the sample number in Table 1).

Further inspection indicates that the K-Mg temperatures are just slightly lower than the quartz temperatures for most boiling springs and hot springs with water temperatures very close to the boiling point at Gudui, whereas the discrepancies between the temperatures obtained using the K-Mg and quartz geothermometers are much larger for the geothermal springs at low-temperature. For example, there is little difference between the K-Mg temperatures (143.86, 190.34, 190.34, 175.97 and 136.86 ℃, respectively; Table 3) and quartz temperatures (146.64, 187.05, 187.05, 173.14 and 144.13 ℃, respectively; Table 3, Fig. 5) for five boiling springs BXLG1975, SGLG197901, SGLG197902, BBDM1989 and PC04. This means that adiabatic cooling has much more influence on the temperature decrease in geothermal water during its ascent to spring vents than conductive cooling or mixing for those boiling springs or high-temperature hot springs at Gudui. However, for those springs whose sampling temperatures are below the boiling point, conductive cooling or mixing with cooler Mg-rich waters are the major processes responsible for geothermal water temperature decrease when it flows upward to the spring vents, and as a result their K-Mg temperatures and quartz temperatures have significant differences. The triangular diagram Na-K-$Mg^{1/2}$ (Giggenbach, 1988; Fig. 5) documents the disequilibrium situations of most geothermal water samples collected from Gudui, and most of the springs plot far from the full equilibrium line. However, four springs (BXLG197901, SGLG197904, SGLG197901 and SGLG197902) at Gudui, with temperatures of 85 ℃, 84 ℃, 85.5 ℃ and 86 ℃, respectively, plot close to the full equilibrium line, which supports the previous analysis. The result presented in the Na-K-$Mg^{1/2}$ triangular diagram of this study is comparable to those of Wang (2017). Of thirteen water samples collected from the Gudui geothermal field and used by Wang (2017), Sample Q001-Q013, the hot springs and surface water plots in the immature waters area in the triangular diagram, however, warm and the boiling springs plot in the partially equilibrated or mixed area of the triangular diagram, which is consistent with the results of our research.

4.3 Characteristic component analysis

Through the analysis of the line graph for natural water geochemical elements in the Gudui area [Fig. 8(a), (b), (c), (d)], we find that the concentrations of Na, K, Cl, SiO_2, B, As, Li, Rb, Cs and F in the geothermal water are generally high and the change is basically relatively small. However, the concentrations of these constituents have drastically decreased in cold spring Q001 and surface water. On the contrary, cold spring Q001 and surface water contain far more Ca, Mg, and SO_4^{2-} than the geothermal water. This change shows that Na, K, Cl, SiO_2, B, As, Li, Rb, Cs and F may be the characteristic components of geothermal water, which is also consistent with the results of Guo et al. (2012). In addition, the concentrations of these constituents in the geothermal borehole ZK302 water with the highest sampling temperature is the highest of all geothermal water samples which also supports our view. Among these characteristic constituents, Cl is hardly affected by water rock interactions and usually does not enter common rock-forming minerals by adsorption on mineral surfaces (Ellis, 1970; Arnórsson and Andrésdóttir, 1995). It is incompatible in almost all the natural water environments, especially at very high temperatures. Moreover, the Cl concentrations of different waters in Gudui are distinctively different. Therefore, Cl was used as a basic index to compare the hydrochemistry of geothermal waters from different hydrothermal areas of Gudui and to analyze the related hydrogeochemical processes taking place in the reservoirs. Guo et al. (2007) have proved that different sampling time and different laboratories' analysis can make a big difference to a constituent's concentration even for the same sample. Therefore, only the analysis data analysed in recent years is plotted, and is shown in Fig. 9. There are excellent linear relationships for the Gudui samples in the B vs Cl, F vs Cl, K vs Cl, Li vs Cl plots, with squared regression coefficients of 0.9673, 0.9468, 0.957 0.9253, and 0.8788 (Fig. 9), respectively. The surface samples cluster together in these plots around zero without presenting good linear relationship, however, the concentrations of major constituents represented by B, F, K, and Li of the geothermal drill samples are around or higher than the other geothermal samples. It suggests that some geothermal waters from Gudui are formed by mixing between cold groundwater end members and a geothermal fluid end member with B, F, K, and Li as characteristic constituents. In other words, the better linear relations between B, F, K, Li and Cl in the Gudui geothermal waters indicate the existence of a deep parent geothermal liquid (PGL) below Gudui, since these elements are relatively conservative during cooling and mixing. The differences in the ratios of B, F, K, and Li concentrations imply that there are different subsurface temperatures and host rock types in the different hydrothermal areas of Gudui. Affected by the fluid-rock interactions of different types and extent, the B, F, K, and Li concentrations as well as their ratios in the Gudui geothermal waters vary greatly. Although there is a linear relationship between Rb, Cs and Cl, however, what surprises us is that the squared regression coefficients of Rb vs Cl and Cs vs Cl (0.7319 and 0.5386, respectively) are lower than those of B vs Cl, F vs Cl, K vs Cl, Li vs Cl (Fig. 9), which is significantly different from previous results (Guo and Wang, 2012). Systematic research aimed at the mineralization of hot springs silicification was carried out by Zheng (1995), and a new type of hydrothermal deposit—Cesiumbearing Geyserite in Tibet was proposed. Through the research on the Gudui geothermal system, Zheng (1995) discovered the concentration of cesium in some geyserite is as high as 0.1886% (the minimum production-grade of cesium deposit is 0.05–0.06%), which is twice as high as that in Gulu Cs-bearing geyserite (one of the key Cs-bearing geyserite fields), and more than an order of magnitude higher than that in Yangbajing geyserite. Besides, Rb is also enriched to a certain extent but not as greatly as Cs. The relationship between Rb and Cl is a bit better than that of Cs vs Cl (Fig. 9) is also consistent with the above case. Zheng (1995) also suggests that cesium-bearing geyserite is formed in the high-temperature hydrothermal system, which is exactly the case in the Gudui geothermal field.Therefore, the lower squared regression

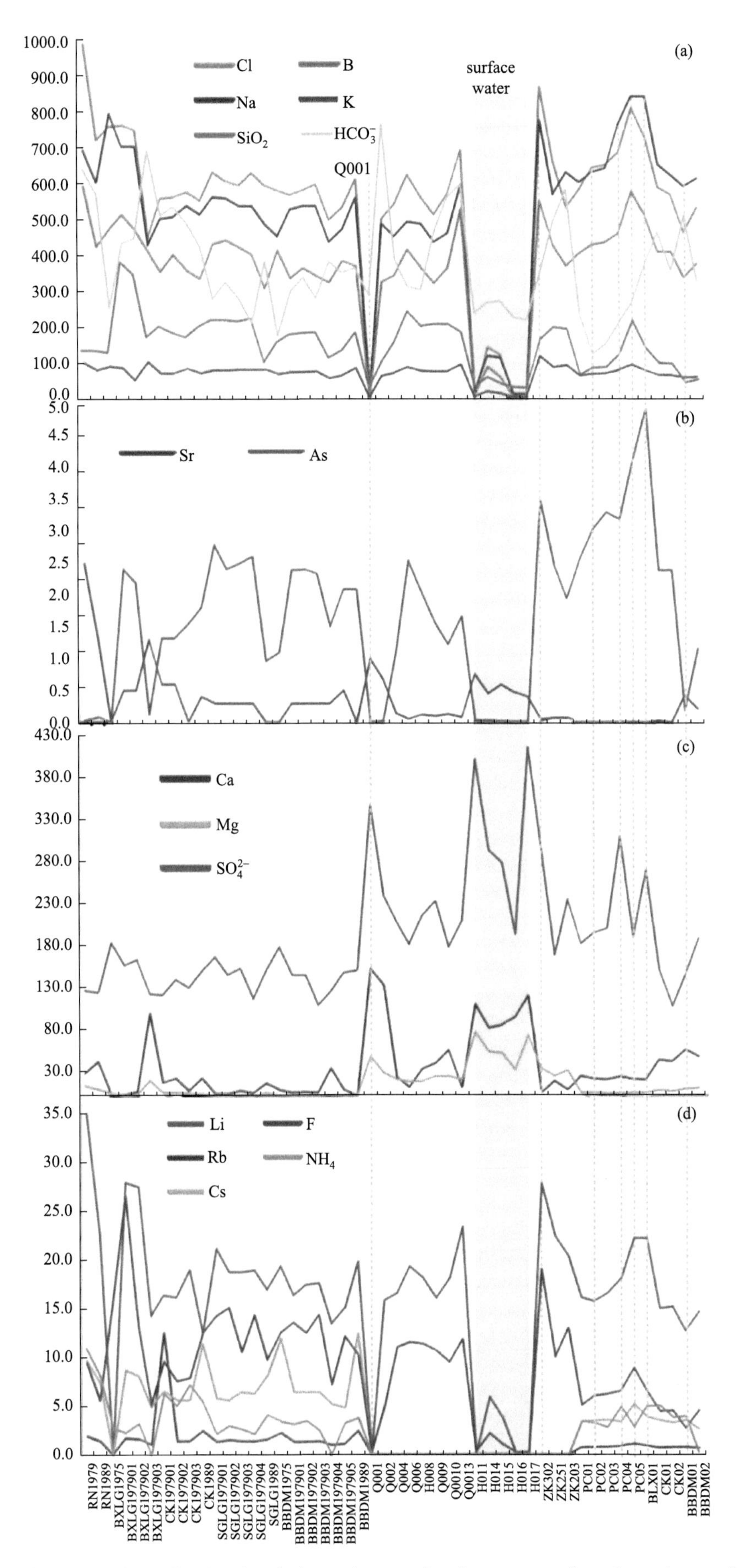

Fig. 8　Line chart of concentrations of major chemical constituents of surface water and geothermal water from 1975 to 2017.

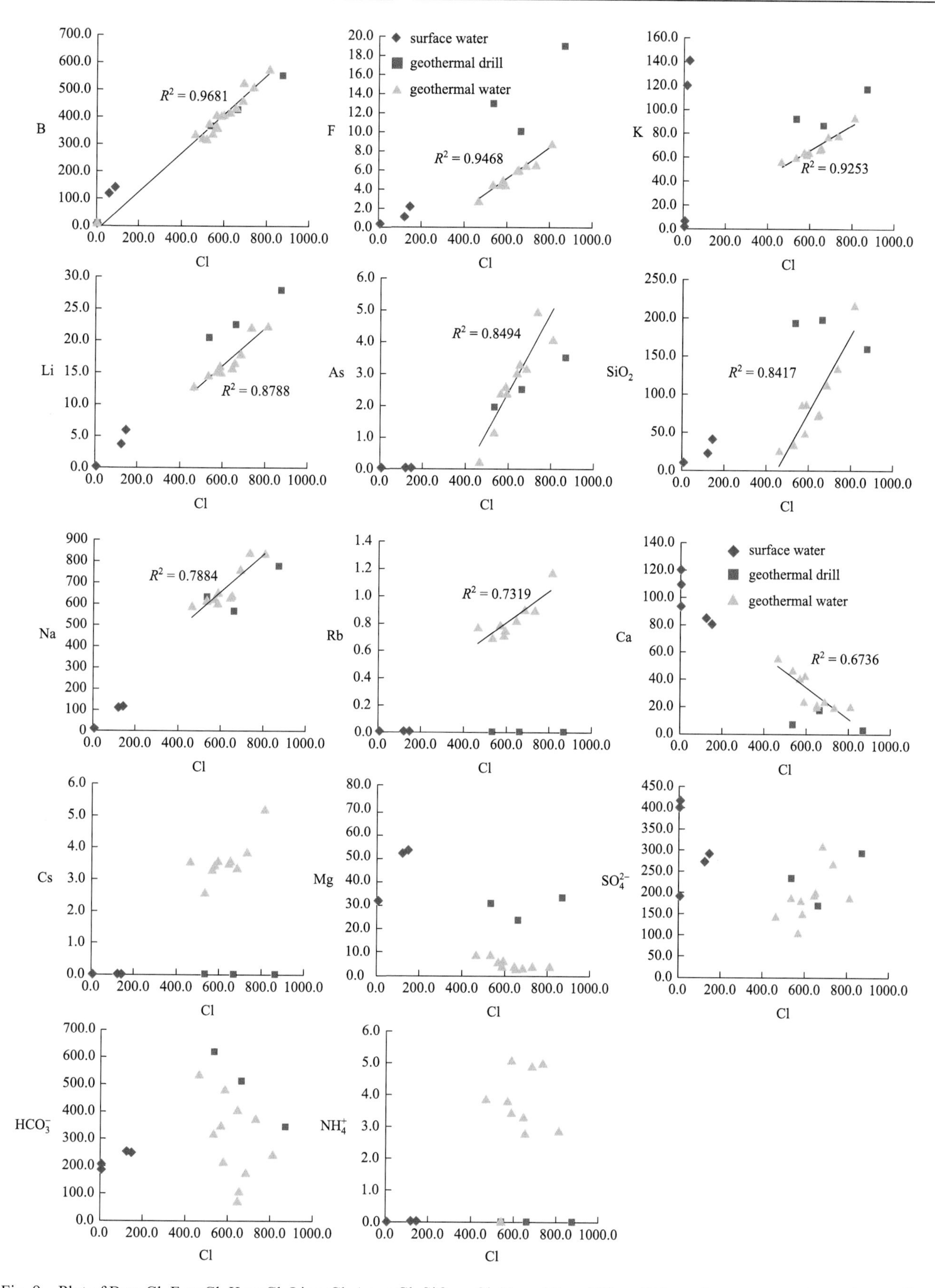

Fig. 9 Plot of B vs Cl, F vs Cl, K vs Cl, Li vs Cl, As vs Cl, SiO_2vs Cl, Na vs Cl, and Rb vs Cl in surface water, geothermal water (in this study) and geothermal drill water samples.

coefficients for Rb vs Cl and Cs vs Cl may be due to their enrichment in geyserite. Na and SiO_2 behave differently from B, F, K, and Li. Although there are linear relationships between Na or SiO_2 and Cl concentrations for the Gudui samples, their squared regression coefficients (0.7884 and 0.8417, respectively) are far lower than those for B vs Cl, F vs Cl, K vs Cl, Li vs Cl (Fig. 9). It reflects that Na and SiO_2 are also characteristic constituents of the Gudui geothermal waters, just like B, F, K, and Li. However, Na and SiO_2 are not conservative constituents and their concentrations in the Gudui geothermal waters may be affected by water-rock reactions (the precipitation of Na and Si as secondary silicate minerals) or decreasing solubilities at lower temperature and pressure (Fournier, 1979; Zheng, 1995; Guo and Wang, 2012), which results in a more scattered distribution of the Gudui data points in the Na vs Cl and SiO_2 vs Cl plots. Not surprisingly, there are no linear relationships in the Ca vs Cl, Mg vs Cl, HCO_3^- vs Cl, and SO_4^{2-} vs Cl plots, since HCO_3^-, SO_4^{2-}, Ca, and Mg are not characteristic constituents of geothermal water. Multiple sources should be responsible for the occurrence of these constituents in the geothermal waters from Gudui, such as dissolution of primary aluminosilicate minerals, mixing with cold groundwater and H_2S oxidation. The lack of linear relationships between Cl and HCO_3^-, SO_4^{2-}, Ca, and Mg may be mainly related to the solubility control of some hydrothermal alteration minerals (Guo and Wang, 2012). Specifically, the concentration of Ca in the geothermal waters can be controlled by the precipitation of calcite (related to loss of CO_2; Fournier, 1979) and some secondary Ca-aluminosilicate minerals (Guo and Wang, 2012).

4.4 Parent geothermal liquid below Gudui

There are hot springs all over the world, on every continent and even under the oceans and seas. It is known that the temperature of rocks within the Earth increases with depth. If water percolates deeply enough into the crust, it will be heated as it comes into contact with hot rocks or mixes with deep thermal fluids. Due to buoyancy, this heated water rises to the surface and forms hot springs. Based on the results of geochemical analysis of Gudui geothermal waters, we believe there should be a Parent Geothermal Liquid (PGL) that is uniform in composition, in chemical equilibrium with the host rocks, and ascending to the surface by different channels to form geothermal springs below the Gudui field.

In order to characterize the PGL of Gudui geothermal field and to delineate its upflow and cooling, it is helpful to understand the hydrochemical characteristics and origin of the Gudui hydrothermal field, which is the most geothermally active and economically valuable system of the Cuona-Woka geothermal belt (Wang, 2017). Considering these aspects, we applied the silica-enthalpy mixing model (Fig. 10) and the chloride-enthalpy mixing model (Fig. 11) to get a deeper insight into the nature of the parent geothermal liquid (PGL) of the Gudui geothermal field. In Fig. 10, we find that the point for ZK302 is below the quartz solubility line. This might probably be due to silica deposition in this high temperature drill water. Fig. 4 and Fig. 5 suggest that almost all the hot spring waters examined are formed by the mixing of hydrothermal water with shallow or surface groundwater. If the water chemistry is a result of simple two component mixing, then the data should plot on a straight line in the silica-enthalpy diagram (Fig. 10).

In the Gudui area, most of the cold and warm spring waters show a relatively linear relation between silica and enthalpy ($R^2 = 0.66$), and can hence be used to estimate the temperature of the PGL in the silica-enthalpy diagram with some degree of reliability. The extrapolation of the line through the data points for the Gudui geothermal field indicate that the temperature of the PGL is about 212 ℃, if no steam separation before mixing is assumed, and 143 ℃, if steam separation takes place before mixing. The PGL temperature estimated by the silica-enthalpy mixing model is too low compared with the results of the Na-K-Mg ternary diagram (240–270 ℃) and Na-K geothermometer (206.80–266.60 ℃), and even lower than the result for the quartz geothermometer (150.05–232.66 ℃). This indicates that the temperature estimated by silica-enthalpy diagram does not represent the true temperature of the PGL.

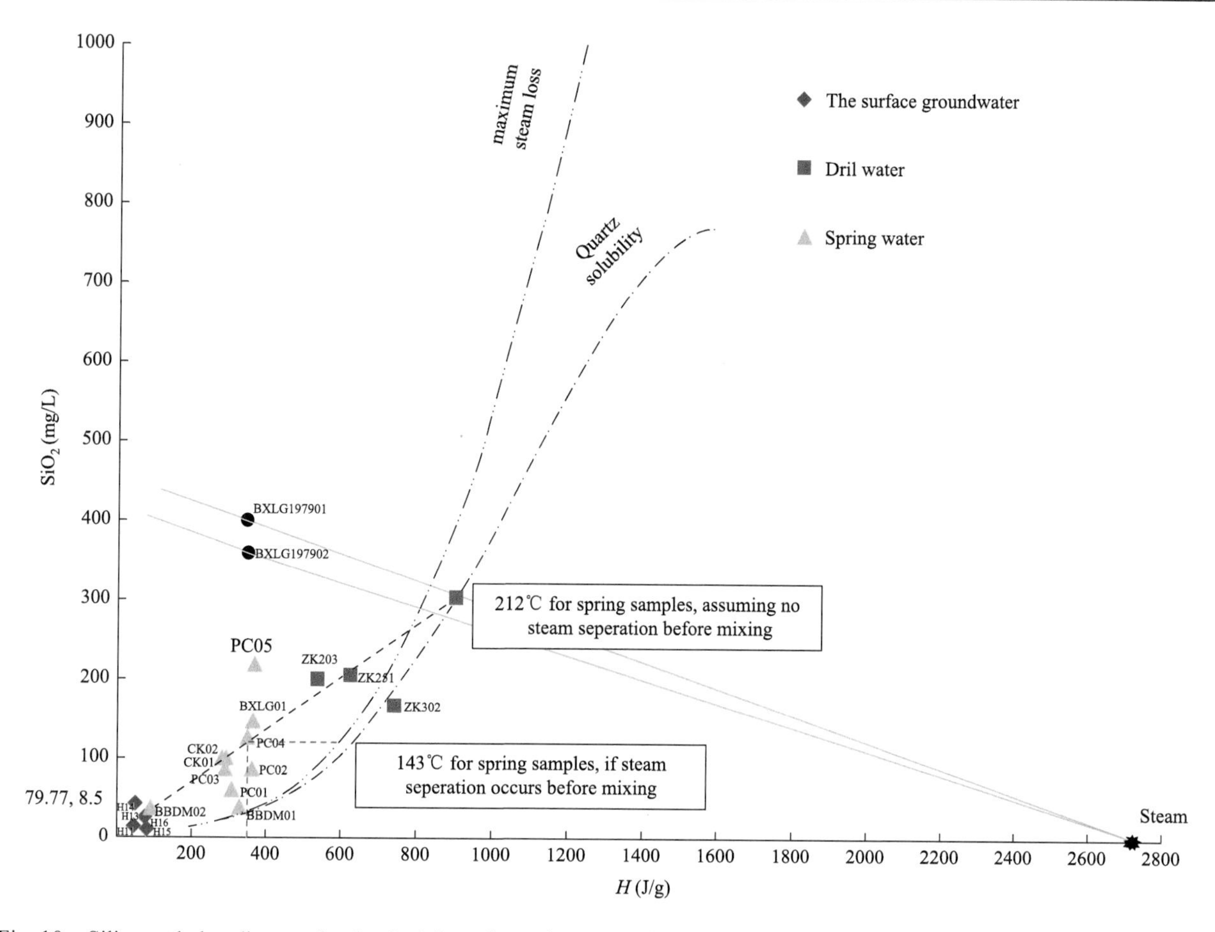

Fig. 10 Silica-enthalpy diagram for the Gudui geothermal waters (after Truesdell and Fournier, 1977). Black dashed line indicate the mixing between geothermal waters and suface waters; Black dots labeled BXLG197901 and BXLG197902 represent typical hot springs with high silica content sampling in 1979.

Therefore, we also applied the chloride-enthalpy mixing model to get the temperature of the PGL (Fig. 11). Enthalpy-chloride diagrams have been widely used to estimate the temperature of parent geothermal liquids and to clarify their cooling mechanisms (Truesdell and Fournier, 1977; Fournier, 1977; Gianelli and Teklemariam, 1993; Sturchio et al., 1996; Guo et al., 2009). In this study, the geothermal water samples collected from Gudui were plotted on an enthalpy chloride diagram (Fig. 11) and divided into two groups: (1) ZK302, ZK251 and ZK203 whose sampling temperatures are 204, 163, 175.5 ℃ respectively, which is significantly higher than the local boiling point (83 ℃). We believe these springs were formed predominantly by adiabatic cooling of geothermal reservoir fluid duo to their high temperature and short distance to PGL. PC04, PC05 and BXLG01 whose sampling temperatures are close to or above the local boiling point (83 ℃). These springs are also formed by adiabatic cooling, since the hot springs with sampling temperatures very close to local boiling point are usually the evidence of geothermal fluid's adiabatic cooling at near-surface (Guo et al., 2017).

Therefore the enthalpy-chloride data points for the geothermal fluid in the reservoirs corresponding to ZK302, ZK251, ZK203, PC04, PC05 and BLX01 should lie along approximately straight lines extending away from the enthalpy of steam at the steam separation temperature. Thus, the straight lines starting from the points of these four samples and radial to a point S with enthalpy value of 2666 J/g (Wang, 2017) and a chloride concentration of 0 mg/L were drawn in Fig. 11. Then the points of the corresponding geothermal reservoir fluids were plotted on these lines based on their Na-K temperatures (Table 3). According to Steam Tables of Pure Waters (Keenan, 1978; Guo et al., 2017), very little error would be introduced using the above method if subsurface boiling took place below 340 ℃ (Fournier, 1979), which is just the case at Gudui. (2) PC01, PC02, PC03, CK01, CK02, BBDM01 and BBDM02 whose sampling temperatures are lower than the local boiling point. These fluids are considered to have been formed mainly by conductive cooling (Fournier, 1979). So the corresponding data points for the geothermal fluids in the reservoirs were plotted in Fig. 11 using their enthalpy values calculated from Na-K temperatures and their chloride concentrations that equal to those of purported conductively cooled spring samples [the chloride content generally will be nearly the same as that of the water in the aquifer feeding the spring (Fournier, 1979)].

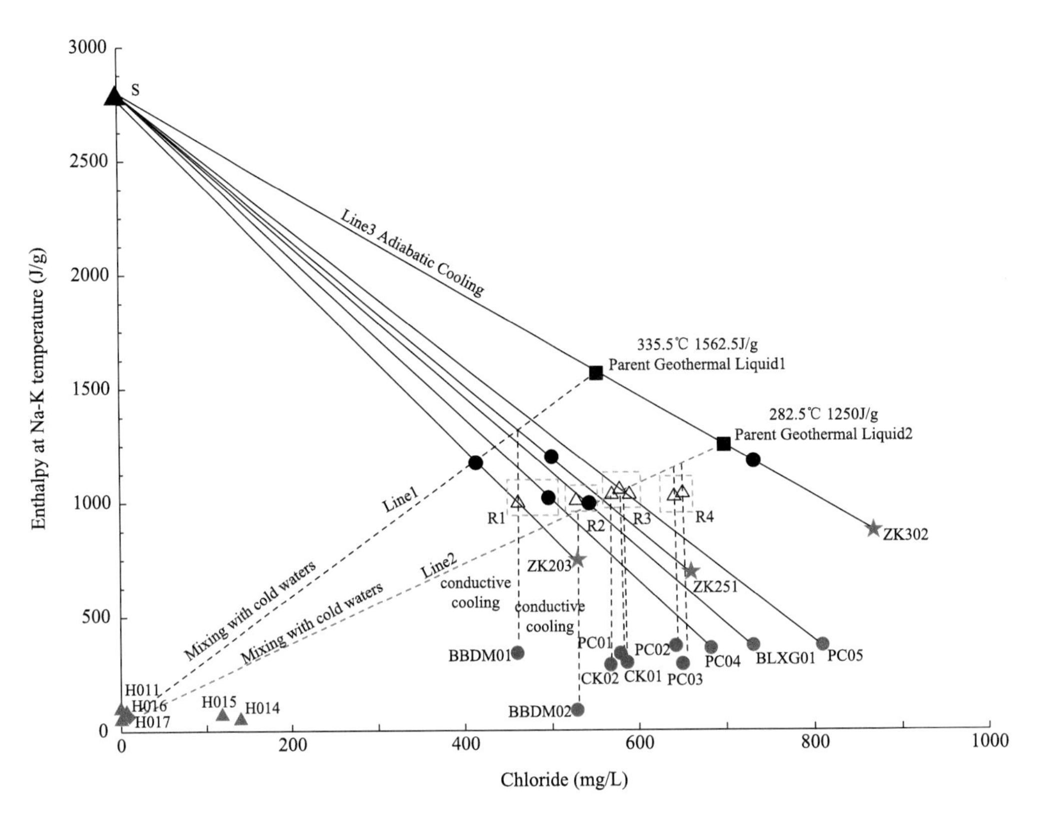

Fig. 11 Plot of enthalpy relative to chloride for geothermal waters from Gudui (Fournier, 1979; Guo and Wang, 2012). Δ: geothermal fluid cooling conductively to form hot springs; ●: geothermal fluid cooling adiabatically to form hot springs; ■: parent geothermal liquid; ▲: steam separated from geothermal fluid. The legend is the same as that in Fig. 4.

In Fig. 11, lines 1 and 3 give a boundary for all the sample points and their corresponding points of geothermal reservoir fluids. The intersection of these two lines represents the Parent Geothermal Liquid 1 below Gudui whose enthalpy value and chloride concentration were estimated to be 1562.5 J/g (corresponding to a temperature of 335.5 ℃) and 567 mg/L, respectively. However, some researchers consider that the results for the Na-K geothermometer are much higher than the temperature measured in reservoir in Tibet (Sun et al., 2015). The main reason for this may be that the partial or complete chemical equilibration has not yet been reached. In order to avoid the influence of this condition on estimating the temperature of the PGL, we select samples BXLG01 and PC05 that are already in partial equilibrium to further estimate the temperature of the PGL. Line 2 passes through the geothermal reservoir corresponding to BXLG01 and PC05 and intersects Line 3. The intersection of these two lines represents the Parent Geothermal Liquid 2 below Gudui whose enthalpy value and chloride concentration were estimated to be 1250 J/g (corresponding to a temperature of 282.5 ℃) and 697 mg/L, respectively. This temperature is slightly higher than the results of the Na-K-Mg ternary diagram (240–270 ℃) and Na-K geothermometer (206.80–266.60 ℃). Therefore, through the above comprehensive comparison, we can conservatively suggest that the Parent Geothermal Liquid 2 (282.5 ℃) may represent the true PGL.

It can also be speculated from Fig. 11 that the PGL below Gudui ascends and emerges at the surface as geothermal springs in different ways: (1) ZK302 located in the centre region of Gudui with the highest temperature and highest chloride concentration, should be formed by the direct upflow of the PGL that undergoes adiabatic cooling, which also defines the right boundary of the springs; (2) The ascending PGL mixes with cold groundwater in an intermediate reservoir (R1, R2, R3, R4, Fig. 11) below Gudui from which the mixed waters keep ascending and cool through two different channels. One channel allows an upflow of geothermal fluid with sufficiently high flow rates to cool adiabatically and to form springs PC04, PC05 and BLXG01, and the other channel is used by a much slower geothermal fluid upflow that cools conductively and forms springs BBDM01, BBDM02, CK01, CK02, PC02, PC03 and PC01. The difference between these reservoirs (R1, R2, R3, R4) is that nearly complete chemical re-equilibration may take place after the PGL arrives at R1, R2 and R3, since the geothermal fluids in R1, R2 and R3 have very similar chloride concentrations and enthalpy values.

It is interesting that the enthalpy vs chloride plot for geothermal waters from Gudui was also reported by Wang (2017). Unlike in this study, seven water samples were used by Wang Siqi, 2017 and only mixing with cooler waters was used to explain the formation of hot springs at Gudui. Moreover, the enthalpy value for the parent geothermal liquid inferred by Wang (2017) are 1051 J/g, a bit lower than that in this study (1250 J/g).

5 Conclusions

The Gudui geothermal field is a representative high-temperature geothermal system in the Himalayas. The high-temperature geothermal waters in the study area are mainly of the Na-Cl type. While the medium-high-temperature geothermal waters are mainly Na-HCO_3-Cl and Na-Cl-HCO_3 types. The reservoir temperature calculation suggests that the Na-K geothermometer reflects the reservoir temperatures (206.80–266.60 ℃), whereas the quartz and K-Mg geothermometers record intermediate temperatures between the reservoir values and the spring outlet values as the geothermal waters ascend. Most geothermal water samples collected from Gudui plot far from the full equilibrium line in Giggenbach's triangular diagram, indicating that conductive cooling or mixing with cooler Mg-rich waters is the major process responsible for the temperature decrease of these geothermal waters when they flow upward to the spring vents. However, some Gudui high temperature geothermal samples plot close to the full equilibrium line, which means that adiabatic cooling also has a significant effect on geothermal water temperature decrease during their ascent to the surface. Geochemical analysis of Gudui geothermal waters shows that Cl Na, K, SiO_2, B, As, Li, Rb, Cs and F can be regarded as the characteristic constituents of the geothermal waters. Further analysis indicates that Cl and other characteristic consituents have good linear relationship, implying the existence of a parent geothermal liquid below Gudui. The plots of enthalpy vs. chloride and silica vs. enthalpy using the Gudui geothermal water samples suggests that the parent geothermal liquid has a Cl^- concentration of 697 mg/L and an enthalpy of 1250 J/g (corresponding to a temperature of 282.5 ℃). Four intermediate reservoirs were additionally identified by the enthalpy-chloride diagram. The PGL below Gudui ascends to the surface to form geothermal springs via different channels. During the ascent, it may cool conductively, adiabatically, or by mixing with cooler shallow groundwaters. The information given by our research shows the hydrochemical characteristics and enhances our understanding of the evolution of hot fluids from the Gudui hydrothermal system. Nevertheless, more investigations on geothermal gases and isotope characteristics in the Gudui geothermal field, are expected to be made for further verifying the conclusions reached in this study.

Acknowledgments

The authors would like to thank Zhijun Xu, Guanglong Zhang for assistance with field investigation, and Dongxin Si for his help in hydrochemical analysis. This research was supported by Projects of China Geological Survey (Grant number: DD20160054). The helpful comments of the editors and the anonymous reviewers are gratefully acknowledged.

References

Arnórsson, S., Andrésdóttir, A., 1995. Processes controlling the distribution of boron and chlorine in natural waters in Iceland. Geochimica Et Cosmochimica Acta 59 (20), 4125-4146.

Capaccioni, B., Vaselli, O., Tassi, F., Santo, A.P., Huertas, A.D., 2011. Hydrogeochemistry of the thermal waters from the Sciacca Geothermal Field (Sicily, southern Italy). Journal of Hydrology 396 (3-4), 292-301.

Craig, J., Absar, A., Bhat, G., Cadel, G., Hafiz, M., Hakhoo, N., Kashkari, R., Moore, J., Ricchiuto, T.E., Thurow, J., Thusu, B., 2013. Hot springs and the geothermal energy potential of Jammu&Kashmir State, N.W. Himalaya, India. Earth-Science Reviews 126, 156-177.

Delmelle, P., Bernard, A., Kusakabe, M., Fischer, T.P., Takano, B., 2000. Geochemistry of the magmatic-hydrothermal system of Kawah Ijen volcano, East Java, Indonesia. Journal of Volcanology and Geothermal Research 97 (1), 31-53.

Dotsika, E., 2015. H-O-C-S isotope and geochemical assessment of the geothermal area of Central Greece. Journal of Geochemical Exploration 150, 1-15.

Dotsika, E., Leontiadis, I., Poutoukis, D., Cioni, R., Raco, B., 2006. Fluid geochemistry of the Chios geothermal area, Chios Island, Greece. Journal of Volcanology and Geothermal Research 154 (3-4), 237-250.

Du, Jianguo, Congqiang, Liu, Bihong, Fu, Ninomiya, Yoshiki, Youlian, Zhang, Chuanyuan, Wang, Hualiu, Wang, Zigang, Sun, 2005. Variations of geothermometry and chemical-isotopic compositions of hot spring fluids in the Rehai geothermal field, southwestern China. Journal of Volcanology and Geothermal Research 142 (3-4), 243-261.

Duo, Ji, 2003. The Basic Characteristics of the Yangbajing Geothermal Field-Typical High Temperature Geothermal System. Engineering Science (in Chinese with English abstract).

Ellis, A.J., 1970. Quantitative interpretation of chemical characteristics of hydrothermal systems. Geothermics 2 (70), 516-528.

Fournier, R.O., 1979. Geochemical and hydrologic considerations and the use of enthalpy-chloride diagrams in the prediction of underground conditions in hot-spring systems. Journal of Volcanology and Geothermal Research 5 (1), 1-16.

Fournier, R.O., 1977. Chemical geothermometers and mixing models for geothermal systems. Geothermics 5 (1), 41-50.

Fournier, R.O., White, D.E., Truesdell, A.H., 1974. Geochemical indicators of subsurface temperature. Part I. Basic assumptions. Journal research of U.S. Geology Survey 2-3.

Friedman, B., 2011. Geothermal Potential Touted. AAPG Explorer, p.14 (16).

Fournier, R.O., Truesdell, A.H., 1973. An empirical Na, K, Ca geothermometer for natural waters. Geochimica Et Cosmochimica Acta 37 (5), 1255-1275.

Fournier, R.O., Rowe, J.J., 1966. Estimation of underground temperatures from the silica content of water from hot springs and wet-steam wells. American Journal of Science 264 (9), 685-697.

Gao, F., Zheng, M., Song, P., Bu, L., Wang, Y., 2012. The 273.15-k-isothermal evaporation experiment of lithium brine from the Zhabei Salt Lake, Tibet, and its Geochemical Significance. Aquatic Geochemistry 18 (4), 343-356.

Gendenjamts, O., 2003. Interpretation of geochemical composition of geothermal fluids from Árskógsströnd, Dalvík, and Hrísey, N-Iceland and in the Khangai area, Mongolia. Report No.10. Geothermal training programme, Reykjavík, Iceland, pp. 219-252.

Gianelli, G., Teklemariam, M., 1993. Water-rock interaction processes in the AlutoLangano geothermal field (Ethiopia). Journal of Volcanology and Geothermal Research 56 (4), 429-445.

Giggenbach, W.F., 1988. Geothermal solute equilibria. Derivation of Na-K-Mg-Ca geoindicators. Geochimica Et Cosmochimica Acta 52 (12), 2749-2765.

Giggenbach, W.F., Gonfiantini, R., Jangi, B.L., Truesdell, A.H., 1983. Isotopic and chemical composition of parbati valley geothermal discharges, North-West Himalaya, India. Geothermics 12 (2-3), 199-222.

Gluyas, J.G., Younger, P.L., Stephens, W.E., 2012. Development of deep geothermal energy resources in the UK. Energy 165 (1), 19-32.

Guo, Q., Liu, M., Li, J., Zhang, X., Guo, W., Wang, Y., 2017. Fluid geochemical constraints on the heat source and reservoir temperature of the Banglazhang hydrothermal system, Yunnan-Tibet Geothermal Province, China. Journal of Geochemical Exploration 172, 109-119.

Guo, Q., Wang, Y., 2012. Geochemistry of hot springs in the Tengchong hydrothermal areas, Southwestern China. Journal of Volcanology and Geothermal Research 215-216, 61-73.

Guo, Q., Wang, Y., Liu, W., 2009. Hydrogeochemistry and environmental impact of geothermal waters from Yangyi of Tibet, China. Journal of Volcanology and Geothermal Research 180 (1), 9-20.

Guo, Q., Wang, Y., Liu, W., 2007. Major hydrogeochemical processes in the two reservoirs of the Yangbajing geothermal field, Tibet, China. Journal of Volcanology and Geothermal Research 166 (3), 255-268.

Henley, R.W., Ellis, A.J., 1983. Geothermal systems ancient and modern: a geochemical review. Earth Science Reviews 19 (1), 1-50.

Hou, Z., Cook, N.J., 2009. Metallogenesis of the Tibetan collisional orogen: A review and introduction to the special issue. Ore Geology Reviews 36 (1), 2-24.

Hu, Xiancai, Guiling, Wang, Xiaoli, Wang, 2015. Geothermal exploration along the Qinghai-Xizang Railway drilled 205℃ geothermal steam. Geothermal Energy 30-30 (in Chinese with English abstract).

Keenan, J.H., 1978. Steam tables: thermodynamic properties of water including vapor, liquid, and solid phases. John Wiley.

Li, Z., 2002. Present hydrothermal activities during collisional orogenics of the Tibetan Plateau. Chinese Academy of Geologecal Sciences (in Chinese with English abstract).

Liao, Z., Zhao, P., 1999. Yunnan-Tibet geothermal belt-Geothermal resources and case histories. Science Press, Beijing (in Chinese with English abstract).

Lin, Y., Zheng, M., Ye, C., 2017. Hydromagnesite precipitation in the Alkaline Lake Dujiali, central Qinghai-Tibetan Plateau: Constraints on hydromagnesite precipitation from hydrochemistry and stable isotopes. Applied Geochemistry 78, 139-148.

Liu, zhao, Chen, kang, Nan, dawa, 2017. Hydrochemical Characteristics of Geothermal Water in Gudui, Xizang (Tibet). Geological Review (s1) 353-354.

Lü, Y.Y., Zheng, M.P., Zhao, P., Xu, R.H., 2014. Geochemical processes and origin of boron isotopes in geothermal water in the Yunnan-Tibet geothermal zone. Science China Earth Sciences 57 (12), 2934-2944.

Shangguan, Z., Zhao, C., Li, H., Gao, Q., Sun, M., 2005. Evolution of hydrothermal explosions at Rehai geothermal field, Tengchong volcanic region, China. Geothermics 34 (4), 518-526.

Smith, T., 2007. A bright future for geothermal energy. GEOExPro 36-46.

Sturchio, N.C., Ohsawa, S., Sano, Y., Arehart, G., Yusa, Y., 1996. Geochemical characteristics of the Yufuin outflow plume, Beppu hydrothermal system, Japan. Geothermics 25 (2), 215-230.

Sun, Hongli, Ma, Feng, Lin, Wenjing, Liu, Zhao, Wang, Guiling, Nan, Dawa, 2015. Geochemical Characteristics and Geothermometer Application in High Temperature Geothermal Field in Tibet. Geological Science and Technology Information 34 (3), 171-177.

Sun, Xiang, Zheng, Youye, Franco Pirajno, T., Campbell, McCuaig, Yu Miao, Xia Shenlan, Qingjie, Song, Huifang, Chang, 2018. Geology, S-Pb isotopes, and $^{40}Ar/^{39}Ar$ geochronology of the Zhaxikang Sb-Pb-Zn-Ag deposit in Southern Tibet: implications for multiple mineralization events at Zhaxikang. Mineralium Deposita 1-24.

Tassi, F., Aguilera, F., Darrah, T., Vaselli, O., Capaccioni, B., Poreda, R.J., Delgado Huertas, A., 2010. Fluid geochemistry of hydrothermal systems in the AricaParinacota, Tarapacá and Antofagasta regions (northern Chile). Journal of Volcanology and Geothermal Research 192, 1-15.

Tong, Wei, Liao, Zhijie, Liu, Shibin, Zhang, Zhifei, You, Maozheng, Zhang, Mingtao, 2000. Tibetan Hot Springs. science press (in Chinese with English contents).

Truesdell, A.H., Fournier, R.O., 1977. Procedure for estimating the temperature of a hotwater component in a mixed water by using a plot of dissolved silica versus enthalpy. Journal Research of U.S. Geology Survey 5, 1.

Truesdell, A.H., 1976. Summary of section III-Geochemical techniques in exploration.Proc.u.n.symp.on the Development&Use of Geothermal Resources San Francisco.

Ünsal, Gemici, Tarcan, G., 2002. Hydrogeochemistry of the Simav geothermal field, western Anatolia, Turkey. Journal of Volcanology and Geothermal Research 116 (3-4), 215-233.

Wang Siqi, 2017. Hydrogeochemical processes and genesis mechanism of high-temperature geothermal system in Gudui, Tibet. China University of Geosciences, Beijing.

Wang, P., Chen, X., Shen, L., Wu, K., Xiao, Q., 2016. Geochemical features of the geothermal fluids from the Mapamyum non-volcanic geothermal system (Western Tibet, China). Journal of Volcanology&Geothermal Research 320, 29-39.

Zhao, P., Dor, J., Liang, T., Jin, J., Zhang, H., 1998a. Characteristics of gas geochemistry in Yangbajing geothermal field, Tibet. Chinese Science Bulletin 43 (21), 1770-1777 (in Chinese with English abstract).

Zhao, P., Jin, J., Zhang, H.Z., 1998b. Chemical composition of thermal water in the Yangbajing geothermal field, Tibet. Scientia Geologica Sinica 33 (1), 61-72 (in Chinese with English abstract).

Zheng, Mianping, 1995. A new type of hydrothermal deposit—Cesium-bearing Geyserite in Tibet. Geological Publishing House (in Chinese with English abstract).

O, H, and Sr isotope evidence for origin and mixing processes of the Gudui geothermal system, Himalayas, China*

Chenguang Wang[1], Mianping Zheng[1], Xuefei Zhang[1], Enyuan Xing[1], Jiangyi Zhang[2], Jianhong Ren[3] and Yuan Ling[1]

1. MLR Key Laboratory of Saline Lake Resources and Environments, Institute of Mineral Resources, CAGS, Beijing 100037, China
2. Hydrochemistry and Environmental Laboratory, Institute of Geology and Geophysics, Chinese Academy of Sciences, Beijing 100029, China
3. State Key Laboratory of Geological Processes and Mineral Resources, China University of Geoscience (Beijing), Beijing 100083, China

Abstract The Gudui geothermal field records the highest temperature at equivalent borehole depths among the mainland hydrothermal systems in mainland of China. Located about 150 km southeast of Lhasa City, the capital of Tibet, the Gudui geothermal field belongs to the Sangri-Cuona rift belt, also known as the Sangri-Cuona geothermal belt, and is representative of the non-volcanic geothermal systems in the Himalayas. In this study, oxygen-18 and deuterium isotope compositions as well as $^{87}Sr/^{86}Sr$ ratios of water samples collected from the Gudui geothermal field were characterized to understand the origin and mixing processes of the geothermal fluids at Gudui. Hydrogen and oxygen isotope plots show both, deep and shallow reservoirs in the Gudui geothermal field. Deep geothermal fluids are the mixing product of magmatic and infiltrating snow-melt water. Calculations show that the magma fluid component of the deep geothermal fluids account for about 21.10–24.04%; magma fluids may also be a contributing source of lithium. The linear relationship of the $^{87}Sr/^{86}Sr$ isotopic ratio versus the 1/Sr plot indicates that shallow geothermal fluids form from the mixing of deep geothermal fluids with cold groundwater. Using a binary mixing model with deep geothermal fluid and cold groundwater as two end-members, the mixing ratios of the latter in most surface hot springs samples were calculated to be between 5% and 10%. Combined with basic geological characteristics, hydrogen and oxygen isotope characteristics, strontium concentration, $^{87}Sr/^{86}Sr$ ratios, and the binary mixing model, we infer the 6th-Class Reservoirs Evolution Conceptual Model (6-CRECM) for the Gudui geothermal system. This model represents an idealized summary of the characteristics of the Gudui geothermal field based on our comprehensive understanding of the origin and mixing processes of the geothermal fluid in Gudui. This study may aid in identifying the geothermal and geochemical origin of the Gudui high-temperature hydrothermal systems in remote Tibet of China, whose potential for geothermal development and utilization is enormous and untapped.

Keywords Geothermal fluid; Oxygen-18 and deuterium isotope $^{87}Sr/^{86}Sr$ ratio; Origin and mixing; Gudui; Himalayas

1 Introduction

Energy resources have, over the course of the 20th century, become increasingly relevant and important to modern society. As a renewable source of energy, geothermal power has attracted increased attention. It is being tapped in over 20 countries around the world, and has become more prevalent in countries like the USA, Iceland, Italy, Germany, Turkey, France, and the Netherlands (Guo and Wang, 2012). Understanding the origin of high-temperature geothermal fluids and its mixing processes are important for geothermal studies (Sheppard and

* 本文发表在：Geoscience Frontiers, 2020, 11 (4): 1175-1187

Lyon, 1984; Marty et al., 1991; Hilton et al., 1993; Clark and Phillips, 2000; Aguilera et al., 2005; Gherardi et al., 2005; Guo et al., 2010). Interactions between high-temperature hydrothermal systems and cold-water systems have been studied globally (Tenu et al., 1981; Delmelle et al., 1998; Brombach et al., 2000; Guo et al., 2010). While China has abundant geothermal resources, high temperature hydrothermal systems are mainly distributed in the western Yunnan Province, western Sichuan Province, and Southern Tibet in mainland of China, which are collectively called the Yunnan-Tibet Geothermal Province and constitute an important part of the Mediterranean-Himalayas geothermal belt (Lü et al., 2014; Guo et al., 2017).

Geothermal exploration suggests that the Gudui geothermal field has the highest recorded temperature at equivalent depths as compared to other hydrothermal systems across the Chinese mainland. Situated in the southeast of Lhasa city, Tibet, western China, the highest down-hole temperature of 205 ℃ was recorded at exploration borehole ZK302 [Figs. 1(a), (b) and 2(b)] at 230 m depth; in comparison, the temperature of the Yangbajing geothermal field is about 130–173 ℃ at the same depth range (Duo, 2003; Hu et al., 2015). The Gudui geothermal field is situated within the Sangri-Cuona rift belt of Tibet [Fig. 1(a)-IV], far from the ocean, on a high plateau surrounded by mountain ranges. A significant number of hydrothermal areas are distributed in the Sangri-Cuona rift belt. Of these, the Gudui geothermal field occupies a total area of around 100 km^2, and is the largest and most active, characterized by copious hydrothermal manifestations, like hot springs (Fig. 2), fumaroles and intensive hydrothermal alteration [Fig. 2(f)] (Wang, 2017). Hydrothermal explosions occur in the southwestern part of the Gudui geothermal field [Fig. 2(e)]. Liu et al. (2017) estimated that its total power generation potential is approximately 266 MW. Wang (2017) predicted an installed capacity of 115.5 MW based on detailed calculations of natural heat release (3.32×10^{14} J) and stored heat (9.83×10^{15} J), with a thermo-electric conversion efficiency of 11% due to the extreme conditions of the Tibetan plateau. Some studies suggest, controversially, that the power generation capacity of the Gudui geothermal field may in the near future greatly exceed that of the Yangbajing geothermal field, whose total installed electricity capacity is only 24.18 MW (Guo et al., 2010; Hu et al., 2015). These estimates for the Gudui geothermal field belie the fact that currently, geothermal utilization is not widely developed in the region. A pilot geothermal power plant is running (Guo et al., 2019a, b), but the hot springs are mainly employed for bathing [Fig. 2(d)], health resorts (balneology), washing wool, a small number of green-houses, and for some spacing heating through water pipelines [Fig. 2(c)]. Currently, Gudui village uses no geothermal power. There are few rivers suitable for hydroelectric power generation. The surface temperature of the hot springs is as high as 86 ℃, making the local hydrothermal resources a promising energy prospect (Wang, 2017). Effective exploitation and utilization of the Gudui geothermal fluids will not only satisfy the local demand for electricity but also promote economic development in the region. However, little is known about the origin and mixing processes in the Gudui geothermal field. Literature in Chinese recording the hydrothermal activities at Gudui were published after an early exploration in 1975 (Tong et al., 2000). Relevant records are not available for the second and third field geologic investigations carried out in July of 1979 and July of 1989 by Zhifei Zhang and Mingtao Zhang, respectively (Tong et al., 2000). Zheng (1995) carried out systematic research aimed at silicification from cesium mineralization at hot springs in Tibet; the concentration of cesium in some Gudui geyserite was as high as 0.12%. The minimum industrial value of cesium deposits is 0.05–0.06%. Liu et al. (2017) presented a brief description of the geological background and hydrochemical characteristics of the Gudui geothermal field. Wang (2017) conducted a relatively detailed study on the geothermal and geological characteristics, hydrochemical types, reservoirs temperatures, etc. of the Gudui geothermal system. No published research focuses on the origin and the specific mixing process of the Gudui geothermal field, which helps discriminate the chemical properties of the geothermal fluid and the source of its recharge. This information has significance for the development of geothermal fields as well as for policy development. Detailed studies of the geothermal water at the Gudui geothermal field are urgently needed for sustainable development. This research paper is a step in that direction.

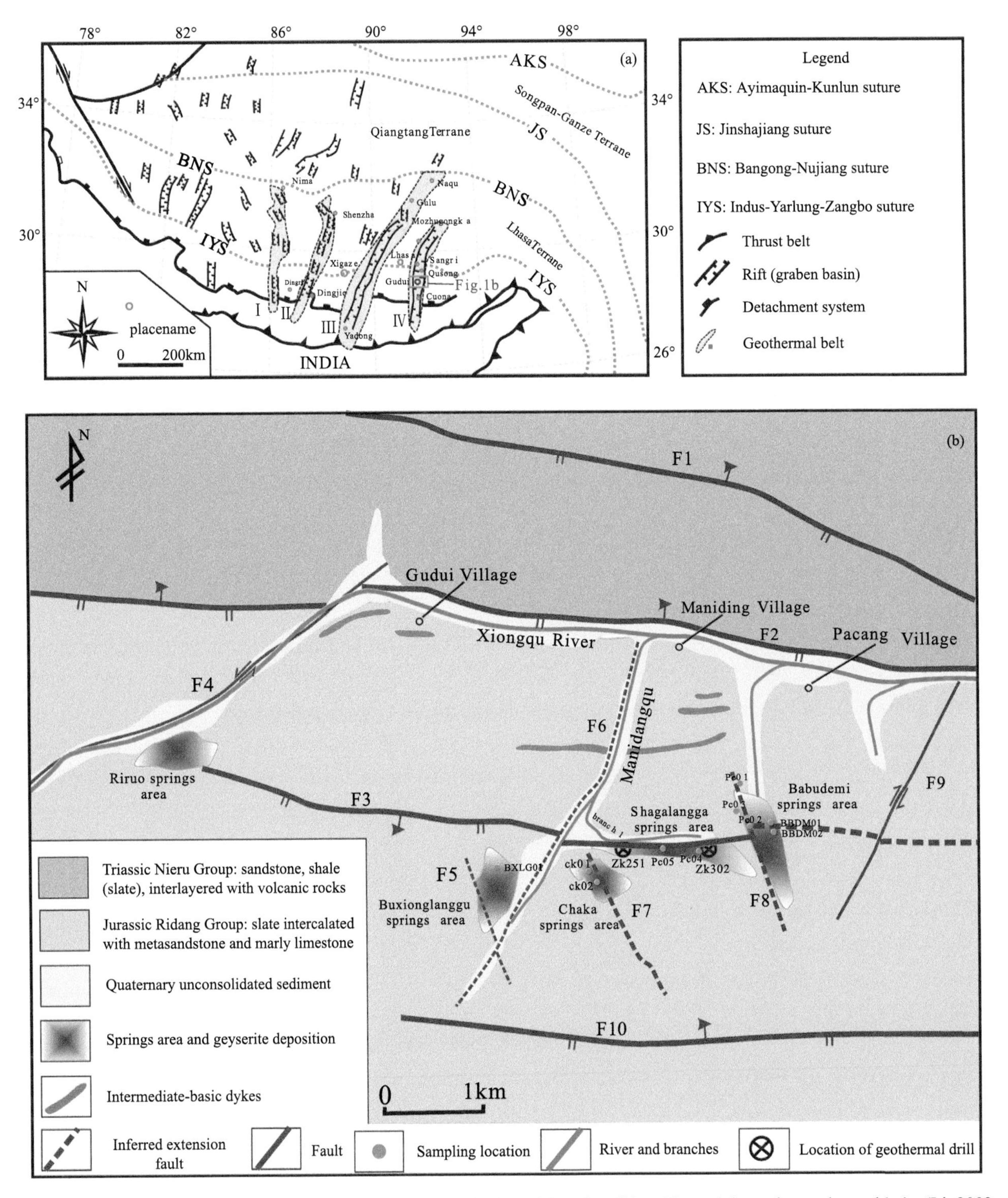

Fig. 1 (a) Simplified map showing the tectonic framework of Tibet with N-S-striking rifts and the main geothermal belts (Li, 2002; Zhang et al., 2015); (b) simplified geological map showing the distribution of water samples from the Gudui geothermal field (Liu et al., 2017; Wang, 2017).

Fig. 2 Field photos of geothermal activity in Gudui

(a, g, h) Fault F3 and the geothermal springs site, taken from different directions and at different scales; (b) the location and surroundings of ZK302; (c) the pipe that transports geothermal water to the rural public bathroom; (d, e) the geothermal eruption or explosion that occurs every 5 min; (f) hydrothermal alteration.

Stable isotope studies play an important role in hydrogeological investigations of both, thermal and non-thermal waters because the isotopes carry a record of fluid origin and processes. Stable O and H isotopes have been widely used in tracing the source of geothermal waters and delineating hydrogeochemical processes occurring in hydrothermal systems (Zheng et al., 1982; Wilkinson et al., 1992; Ghomshei and Clark, 1993; Diamond and Harris, 2001; Majumdar et al., 2005; Guo et al., 2010). In order to identify the source material of soluble components and the mixing process of water from hot springs, information on their isotopic ratios, such as the $^{87}Sr/^{86}Sr$ ratio, is also essential. Strontium consists of four isotopes (84Sr, ^{85}Sr, ^{86}Sr, ^{87}Sr); radiogenic ^{87}Sr is derived from the beta decay of ^{87}Rb, whereas ^{86}Sr is non-radiogenic (Peterman and Wallin, 1999). Pioneering work on $^{87}Sr/^{86}Sr$ research was carried out by Doe et al. (1966), followed by studies on $^{87}Sr/^{86}Sr$ ratios of groundwater (Sunwall and Pushkar, 1979; Nakano et al., 1989; Land et al., 2000; Liotta et al., 2017). In the last two decades, strontium isotope analysis has also been extensively applied in hydrogeochemical studies (Barbieri and Morotti, 2003; Négrel et al., 2007; Nisi et al., 2008). The $^{87}Sr/^{86}Sr$ isotopic ratio frequently acts as a tracer for delineating recharge sources and mixing processes in cold groundwater systems (Lyons et al., 1995; Neumann and Dreiss, 1995; Grobe et al., 2000; Négrel et al., 2001; Barbieri and Morotti, 2003). The $^{87}Sr/^{86}Sr$ isotopic ratio was also applied in hydrogeochemical studies of hydrothermal systems (Elderfield and Greaves, 1981; Graham, 1992; Boschetti et al., 2005; Guo et al., 2010).

In this paper, geochemical investigations using oxygen-18, deuterium and $^{87}Sr/^{86}Sr$ ratios are employed to understand the genesis of the Gudui geothermal fluids and to delineate the mixing processes that occur during their evolution, which is important for management of the geothermal fluid resource.

2 General geology

The Tibetan Plateau experiences intensive hydrothermal activity (Zhang et al., 2015), which is generally regarded as an after product of continental accretion, and as a direct result of the closing of the Tethys Ocean and the India-Asia collision, which also contributes to its uplift (Yin, 2000; Tapponnier et al., 2001). The tectonic units of the Tibetan Plateau tectonic comprise a series of east-west-trending continental blocks: Songpan-Ganzi, Qiangtang, Lhasa, the Tethyan Himalaya, the High Himalaya, and the Lesser Himalaya [Fig. 1(a)] (Tapponnier et al., 2001; Li, 2002; Hou et al., 2004; Zhang et al., 2004, 2015). Hot springs occur locally in these collisional mountain belts (Bucher et al., 2009). Structurally, these sub-belts are separated by the Bangong-Nujiang Suture Zone (BNSZ), the Indus-Yarlung Zangbo Suture Zone (IYZSZ), the South Tibet Detachment System (STDS), the Main Central Thrust Fault (MCT), and the Main Boundary Thrust Fault (MBT) [Fig. 1(a)] (Hou et al., 2004). Many hot springs are located at or close to those faults or suture systems (Kennedy-Bowdoin et al., 2004; Baietto et al., 2005). For example, many boiling geysers or springs develop along the Yaluzangbo River, from the upper reaches to the lower reaches (Zheng et al., 1983). Systems of steeply-dipping faults often border these mountain ranges and occur as belt-internal parallel fault systems, forming fault blocks (Evans et al., 2005). Perpendicular to the above-mentioned sub-belts are nearly north-south trending rifts, attributed to the east-west extension of the Tibetan Plateau (Tapponnier et al., 1981; Zheng et al., 1983; Armijo et al., 1986; Fossen, 2000). From west to east these rifts are the Dangreyongco-Dingri rift belt [Fig. 1(a)-Ⅰ], the Shenza-Xietongmen rift belt [Fig. 1(a)-Ⅱ], the Yadong-Gulu rift belt [Fig. 1(a)-Ⅲ], and the Sangri-Cuona rift belt [Fig. 1(a)-Ⅳ] (Zhang et al., 2015). Geothermal fields, especially for high-temperature systems, are predominantly distributed along these rifts, which may provide channels conducive for the rise of geothermal fluids. The four geothermal belts corresponding to these rifts are the Dangreyongco-Gucuo geothermal belt [Fig. 1(a)-Ⅰ], the Shenza-Dingjie geothermal belt [Fig. 1(a)-Ⅱ], the Naqu-Yadong geothermal belt [Fig. 1(a)-Ⅲ], and the Sangri-Cuona geothermal belt [Fig. 1(a)-Ⅳ; Li, 2002]. This indicates that intensive hydrothermal activities in Tibet are controlled by these rift structures (Zheng et al., 1983 and references therein). The Gudui geothermal field is the most famous of the extended Cuona-Woca geothermal field, a part of the Sangri-Cuona rift belt [Fig. 1(a)-Ⅳ] (Wang, 2017). Bedrock in the Gudui geothermal field includes the Triassic Nieru Group and the Early-Middle Jurassic Ridang Group [Fig. 1(b)]. The Nieru Group consists of quartz sandstone, slate, and a few thin layers of muddy limestone, and is rich in biological fossils (Dong et al., 2017). The lithology of the Ridang Group is coarse-grained metasandstone intercalated with slate, slate intercalated with medium to fine-grained metasandstone, and dark-gray carbonaceous slate intercalated with fine-grained metasandstone and marly limestone (Sun et al., 2016). Intermediate-basic dykes, including gabbro and diorite, are distributed in the southern part of Gudui. According to zircon SHRIMP U-Pb dating results of gabbro and diorite at Gudui by Ren et al. (2014), the age of a surface sample of the gabbro is 133–140.90 Ma, and the age of the diorite is 130±1.70 Ma. They are regarded as the products of the splitting of the Himalayan passive margin which caused the large-scale spreading of the Late Neo-Tethys ocean during the

Late Jurassic-Early Cretaceous. Quaternary sediments with a thickness of several meters are mainly distributed on the banks of the Xiongqu River [Fig. 1(b)] and its tributary terraces, as well as in small basins in the mountains. The sediment types are predominantly Holocene alluvium, proluvium, spring sinter, and glacial deposits (Wang, 2017). Existing drilling data indicates that the Nieru Group, the Ridang Group, and the dykes in these strata constitute the reservoirs, while the Quaternary alluvial deposits are the reservoir caprock. Enhanced hydrothermal activity weakens the caprock. Three groups of high-dip-angle normal or reverse faults intersect with each other in the Gudui geothermal field; these faults are oriented E-W, NE-SW, and NW-SE [Fig. 1(b)]. The Gudui-Longzi fault [F2, Fig. 1(b)] belongs to the Zhegucuo-Longzi thrust tectonic belt (Wang, 2017). The F3 fault [Fig. 1(b)] is mainly responsible for the distribution of geothermal areas in Gudui. Hydrothermal fields with numerous hot springs vents occur where these two active faults converge [Fig. 1(b)]. The June 1, 1806 M_S 7.5 earthquake in Cuona and the December 3,1915 M_S 7.0 earthquake in Sangri indicate the faults in the Sangri-Cuona rift are still active in the Holocene (Wang, 2017). These active faults are crucial structures that enhance local hydrothermal activity by providing favorable permeability pathways for the migration and storage of geothermal fluids at depth.

Intense hydrothermal activities and high subsurface temperatures suggest the presence of a magma chamber below Gudui (Wang, 2017). This inference is based on the existence of a magmatic heat source or water-retaining molten granite below the Yangbajing geothermal field, confirmed by the INDEPTH (International Deep Profiling of Tibet and the Himalayas) project (Zhao et al., 2003) and also reported by a series of articles published in Science (Brown et al., 1996; Chen et al., 1996; Kind et al., 1996; Makovsky et al., 1996; Nelson et al., 1996). At an equivalent depths, the temperature of the Gudui geothermal field is higher than that of the Yangbajing geothermal field. The results of a deep conductivity geophysical exploration confirmed the presence of a low resistivity body beneath the Gudui geothermal field. Tan et al. (2004) and Wei et al. (2010) also reported the existence of a low resistivity body at the southern part of the Cuona-Mozhugongka section (where the study area is located) based on a magnetotelluric survey, and inferred the presence of a local melt or a water-retaining, partially molten rockmass capable of providing a continuous source heat to the Gudui geothermal field. The existence of a magma chamber below Gudui was also reported by Guo et al. (2019a, b). Wang (2017) described, in detail, the geology and hydrogeological characteristics of the Gudui geothermal field.

3 Methods

3.1 Sample collection and basic analysis

Field observations and sampling were conducted in August 2017. During field work, the geothermal areas in Gudui [Fig. 1(b)] were carefully observed for their geological characteristics which included various types of strata, faults, magmatic rocks, and hydrology. Ten water samples including four from boiling springs, five from hot springs, and one cold groundwater sample were collected to obtain their $^{87}Sr/^{86}Sr$ isotopic ratios and $\delta^{18}O$ and δD values. The hot springs sampled were mainly self-flowing springs with slight surface or near-surface evaporation; a heat-resistant glass flask was inserted as deep as possible into the spring head and filled with boiling or hot water, after which the flask was immediately sealed with a plug to prevent vapor-water separation (Tan et al., 2014). When the hot water cooled, it was moved to a small plastic bottle for hydrogen and oxygen isotopic analysis. However, some boiling springs are too violent to allow this kind of operation. In Gudui, a pilot geothermal power plant and excessive geothermal water pumping have resulted in a sharp decrease in the total number of the hot springs; few hot springs are readily available for geochemical sampling.

The geographic distribution of all the sampling sites from this study are shown in Fig. 1(b). All water samples were filtered through 0.45 μm membranes on site. Samples were stored in new 350 mL polyethylene bottles. The bottles were rinsed with deionized water, twice, before sampling. The temperature of the geothermal water was measured in situ using portable thermometers that had been calibrated before use. Total dissolved solids (TDS) were calculated by summing all ions and sub-tracting half of the HCO_3^- at the Analytical Laboratory of the Research & Development Center of Saline Lake and Epithermal Deposits, Chinese Academy of Geological Sciences (CAGS), Beijing. For strontium concentration analysis, reagent-quality $HNO_3$with molar concentration up to 14 M was added to the polyethylene bottles to make the pH of the samples 1 (Guo et al., 2010). The strontium concentration of the samples was measured at the National Research Center of Geoanalysis, at the Chinese Academy of Geological Sciences (CAGS) using a plasma mass spectrometer (PE300Q) within two weeks of sampling. The descriptions of the analytical techniques (including sample preparation) are presented in Chinese standard GB/T 8538–2008.

3.2 Hydrogen and oxygen isotope

δD and δ^{18}O analysis was performed at the Hydrochemistry and Environmental Laboratory, Institute of Geology and Geophysics, Chinese Academy of Sciences. For the measurement of the ^{18}O/^{16}O ratio, the CO_2 equilibration method was employed; for the D/H ratio, H_2 was generated by the Zn-reduction method (Coleman et al., 1982). Isotope ratios of CO_2 and H_2 were measured using a MAT-251 mass spectrometer, and the fractionation factor between CO_2 and water at 25 ℃ was assumed to be 1.0412 (Coplen, 1988). The stable oxygen and hydrogen isotopic values in this study were expressed in usual notation of δ% versus Vienna-Standard Mean Ocean Water (V-SMOW), where $\delta = 1000\ (R_{sample}/R_{SMOW}-1)$, and R = ^{18}O/^{16}O or D/H. The reproducibility is ±0.10% for δ^{18}O and ±0.50% for δD. To determine whether the water from local rivers recharge the Gudui geothermal field, δ^{18}O and δD data from the upper and lower reaches of the small rivers around the geothermal field were collected from previous research (Tan et al., 2014); the local river data is from the Xiongqu upper reaches, Xiongqu lower reaches, Manidang upper reaches, Buxionglanggu, Salanangga, and Babudemi [Fig. 1(b)]. In order to conduct a comprehensive comparison, δ^{18}O and δD data of geothermal water, from the Zangbo River, snow-melt water from Yangbajing, previous study data from Gudui, and the Yaluzangbo River were also collected (Zheng et al., 1982; Guo et al., 2010; Tan et al., 2014; Wang, 2017). Table 1 shows the δ^{18}O and δD compositions of all the water samples used in this paper.

3.3 Strontium isotope analysis

High precision Sr isotopic measurements were carried out at the Nanjing FocuMS Technology Co. Ltd. The strontium samples were purified from a digestion solution using column chemistry where the exchange column combined with BioRad AG50W×8 and Sr Spec resin was used to separate Sr from the sample matrix. The Sr-bearing elution was dried down and re-dissolved in 1.00 mL 2 wt.% HNO_3. Small aliquots were analyzed using Agilent Technologies 7700x quadrupole ICP-MS (Hachioji, Tokyo, Japan) to determine the exact concentration of Sr available. Diluted solution (50 ppb Sr doping with 10 ppb Tl) was introduced into Nu Instruments Nu Plasma II MC-ICP-MS (Wrexham, Wales, UK) by Teledyne Cetac Technologies Aridus II desolvating nebulizer system (Omaha, Nebraska, USA). Raw data of Sr isotopic ratios were corrected for mass fractionation by normalizing to ^{86}Sr/^{88}Sr = 0.1194 with exponential law. International isotopic standards (NIST SRM 987 for Sr) were periodically analyzed to correct instrumental drift. Geochemical reference materials of USGS BCR-2, BHVO-2, AVG-2, STM-2, G-2, RGM-2 were treated as quality control (Weis et al., 2006); the recommended values are 0.705013, 0.703479, 0.703981, 0.703701, 0.709770, and 0.704210, respectively. The strontium isotope results of the above-mentioned USGS Standard Rock Powder (Table 1) is consistent with the USGS recommended value reported by Weis et al. (2006). The 87Sr/$_{86}$Sr ratio measurement errors are under 0.00004, and apply to the last decimal place. Repeated analyses of PC05 sample with the known Sr isotopic value were conducted to ensure instrumental accuracy. All of the above illustrate the accuracy and reliability of the measurement data. Field records, hydrochemical properties, ^{87}Sr/^{86}Sr isotopic ratios, δ^{18}O and δD compositions, reservoir temperatures, and hydrochemical types of all water samples in this paper are listed in Table 1.

Table 1 Hydrochemical properties, $^{87}Sr/^{86}Sr$ isotopic ratios, and $\delta^{18}O$ and δD values of water samples collected in the Gudui geothermal field

No.	Sample type	Sample name	Sample location	Sampling date	Elevation (m)	T	TDS (mg/L)	Sr (μg/L)	Li (mg/L)	$^{87}Sr/^{86}Sr$	Std Error	$\delta^{18}O_{V\text{-}SMOW}$ (‰)	$\delta D_{V\text{-}SMOW}$ (‰)	Reservoir temperatures (℃)	Hydrochemical type
1	USGS Standard Rock Powder	BCR-2	—	—	—	—	—	—	—	0.705045	0.000004	—	—	—	—
2	USGS Standard Rock Powder	BHVO-2	—	—	—	—	—	—	—	0.703525	0.000003	—	—	—	—
3	USGS Standard Rock Powder	AGV-2	—	—	—	—	—	—	—	0.704047	0.000003	—	—	—	—
4	USGS Standard Rock Powder	STM-2	—	—	—	—	—	—	—	0.703706	0.000003	—	—	—	—
5	USGS Standard Rock Powder	G-2	—	—	—	—	—	—	—	0.709771	0.000003	—	—	—	—
6	USGS Standard Rock Powder	RGM-2	—	—	—	—	—	—	—	0.704262	0.000003	—	—	—	—
7	Boiling spring	PC04	Gudui	2017-8-14	4645.00	83.00	2650.00	411.00	18.00	0.709884	0.000004	−16.40	−129.10	233.35	Na-Cl
8	Boiling spring	PC05	Gudui	2017-8-14	4630.00	83.00	2981.00	501.00	22.20	0.710125	0.000003	−16.20	−131.00	240.56	Na-Cl
9	Boiling spring	Parallel samples of PC05	Gudui	—	—	—	—	—	—	0.710092	0.000003	—	—	—	—
10	Boiling spring	BLX01	Gudui	2017-8-16	4488.00	83.00	2996.00	580.00	22.20	0.709591	0.000004	−18.10	−134.40	226.46	Na-Cl
11	Boiling spring	PC02	Gudui	2017-8-14	4605.00	81.00	2206.00	707.00	15.70	0.709646	0.000003	−17.90	−136.50	237.27	Na-Cl
12	Hot spring	PC01	Gudui	2017-8-14	4449.00	74.00	2182.00	566.00	16.10	0.709487	0.000003	−17.60	−139.20	234.94	Na-Cl
13	Hot spring	PC03	Gudui	2017-8-14	4605.00	70.00	2255.00	707.00	16.50	0.709647	0.000002	−17.70	−136.20	237.19	Na-Cl
14	Hot spring	CK01	Gudui	2017-8-16	4433.00	70.00	2473.00	903.00	15.00	0.70955	0.000003	−17.70	−142.00	231.07	Na-HCO_3^-Cl
15	Hot spring	CK02	Gudui	2017-8-16	4501.00	68.00	2303.00	913.00	15.20	0.709546	0.000004	−17.70	−140.40	234.85	Na-Cl
16	Hot spring	BBDM02	Gudui	2017-8-17	4520.00	50.00	2283.00	1198.00	14.60	0.708702	0.000003	−17.20	−138.90	229.66	Na-Cl
17	Cold ground water	BBDM01	Gudui	2017-8-17	4520.00	20.00	2317.00	2559.00	12.70	0.70827	0.000003	−18.70	−144.80	228.17	Na-HCO_3^-Cl
18	Drilling water	ZK302	Gudui	2016	4630-4700	205.00	3325.00	49.00	27.80	—	—	—	—	266.75	Na-Cl
19	Drilling water	ZK251	Gudui	2016	4630-4700	163.00	2798.00	69.00	22.40	—	—	−15.50	−139.00	267.89	Na-Cl

Note that the Std Error represents the error of the $^{87}Sr/^{86}Sr$ ratio. The temperature represents the sampling temperature; the information of ZK302, ZK251 and ZK203 is from Wang (2017); other lithium content data are from this research; the reservoir temperatures are estimated using the Na-K geothermometer (Giggenbach, 1992), and hydrochemical type are cited from Wang and Zheng (2019).

4 Results and discussion

4.1 Hydrochemical properties

Analytical data from Table 1 shows that the water samples from boiling springs have higher TDS values as compared to the hot springs and cold groundwater samples; also, in general, TDS values for drill water are higher than that for the samples from boiling springs. This suggests that dissolved constituents have a positive relationship with temperature, and most of the TDS may have come from the deep (Guo et al., 2010).

The average TDS values from the Gudui geothermal field for drill water, for water samples from boiling springs and hot springs, and cold groundwater are 3008.50, 2708.25, 2306.00, 2283.00 mg/L, respectively. In comparison, the strontium concentration of the high temperature geothermal water in the Gudui geothermal field is much lower than that of the cold groundwater, but much higher than that of the geothermal water in boreholes ZK302 and ZK251 (Table 1). Typically, an increase in the strontium concentration of water is associated with an increase in water temperature, just like in the Yangbajing geothermal field (Zhao et al., 2003; Guo et al., 2010). In fact, Vuataz et al. (1988) had reported this phenomenon for geothermal fluids from Valles Caldera in the Jemez Mountains, New Mexico. They suggest that, in the case of neutral Na-Cl type geothermal waters, the solubility of alkaline-earth cations is partially controlled by the solubility of carbonates, which declines as temperature increases (Vuataz et al., 1988). When hot Na-Cl fluids and relatively cold sedimentary aquifer mix, the strontium concentration of the derived water increases substantially, as does HCO_3^-. The chemistry of the geothermal water from Valles Caldera is consistent with the chemical type of geothermal water in our study area (Table 1). Concentrations of Ca, Mg, and Ba behave in a similar fashion in the Valles Caldera, as well as in the Gudui geothermal field as shown in Fig. 3. Wang (2017) considered that the geochemical composition of geothermal fluid in the study area was controlled by dolomite, through hydrogeochemical simulation of the geothermal water in the Gudui area, indicating that the aforementioned analysis of the relationship between strontium concentration and temperature is applicable to the Gudui geothermal field. The reservoir temperature calculations show that the highest reservoir temperature represented by geothermal drilling borehole ZK251 can reach 267.89 ℃; calculation estimates of the reservoir temperatures for other surface hot springs are also above 200 ℃ (Table 1), which indicate that the Gudui geothermal field is a typical high-temperature geothermal field. Wang and Zheng (2019) presented the general hydrochemical parameters of geothermal waters from Gudui, and proved the existence of a parent geothermal liquid (PGL) below Gudui. High temperature geothermal systems with magmatic heat sources usually host deep parent water that remains in equilibrium with the surrounding rock. The pH of these waters is near neutral, the principal anion is Cl^-, and the principal cation is Na^+ (Li et al., 2015); this is also seen in the case of the Gudui geothermal water (Wang, 2017). Besides, As^{3+} concentrations of the water from the hot springs in Gudui (Tong et al., 2000; Wang, 2017) are unusually high. Many recent reports have claimed that host rock leaching of ubiquitously-occurring arsenic by heated meteoric waters is so efficient that, generally, direct magmatic fluid input is not a prerequisite for high arsenic concentrations in geothermal waters (Stauffer and Thompson, 1984; Webster and Nordstrom, 2003; Birkle et al., 2010; López et al., 2012; Bundschuh and Maity, 2015). However, Guo et al. (2019a, b). conducted a detailed analysis of representative southern Tibetan geothermal water (including samples from the Gudui geothermal field). They suggest a substantial contribution of arsenic from the underlying magma chambers, in addition to host rock leaching, to explain the observed arsenic anomaly. Besides, the distribution of magmatic hydrothermal systems is generally in correspondence to that of high-arsenic geothermal waters in Tibet, implying that magmatic fluid could make a non-negligible contribution to geothermal arsenic. Therefore, partial melts within the crust are detected by low-velocity and/or high-conductivity anomalies in the Gudui area (Tan et al., 2004; Wang, 2017), which serve as the magmatic heat source and partial fluid source of the high-arsenic Gudui geothermal system. We discuss the origin and mixing processes of the geothermal fluid at Gudui using O, H, and Sr isotopes, and further improve our understanding of the process of formation of the hot springs in the Gudui geothermal field.

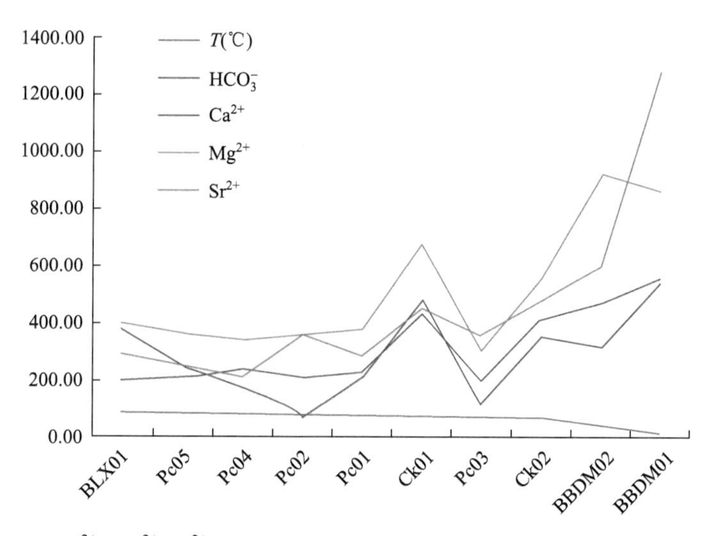

Fig. 3 Variations of HCO_3^-, Ca^{2+}, Mg^{2+}, Sr^{2+} concentrations in geothermal water with temperature decrease as a result of its mixing with cold groundwater.

4.2 Stable oxygen and hydrogen isotopes

The formation of geothermal water was among the earliest topics in hydrology investigated using isotope techniques (Pang, 2006). Craig (1961) showed that geothermal water was mostly meteoric in origin. Truesdell and Hulston (1980) showed that geothermal water can locally be related to groundwater from its isotope composition, with an additional oxygen-18 shift. However, Giggenbach (1992) argued that water from geothermal systems along convergent plate boundaries was a mixture of meteoric and magmatic water. Isotope enrichment of both oxygen-18 and deuterium increases the contribution of magmatic vapours to spring discharges (Pang, 2006). In order to determine the origin of geothermal water at Gudui, the $\delta^{18}O$ and δD values of geothermal water, river water in Gudui and Yangbajing (Zheng et al., 1982; Tan et al., 2014; Wang, 2017), as well as snow melt water from Guo and Wang (2012), and magmatic water from Giggenbach (1992) are plotted in Fig. 4.

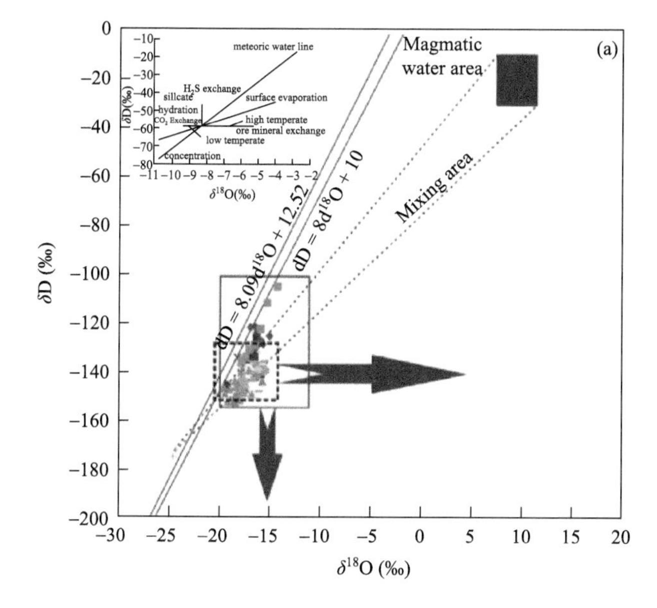

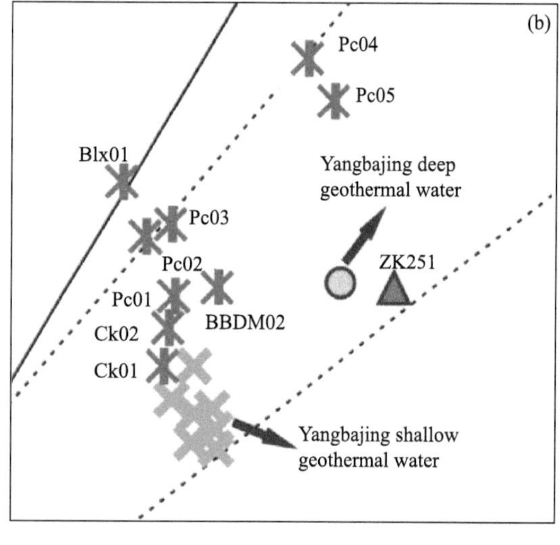

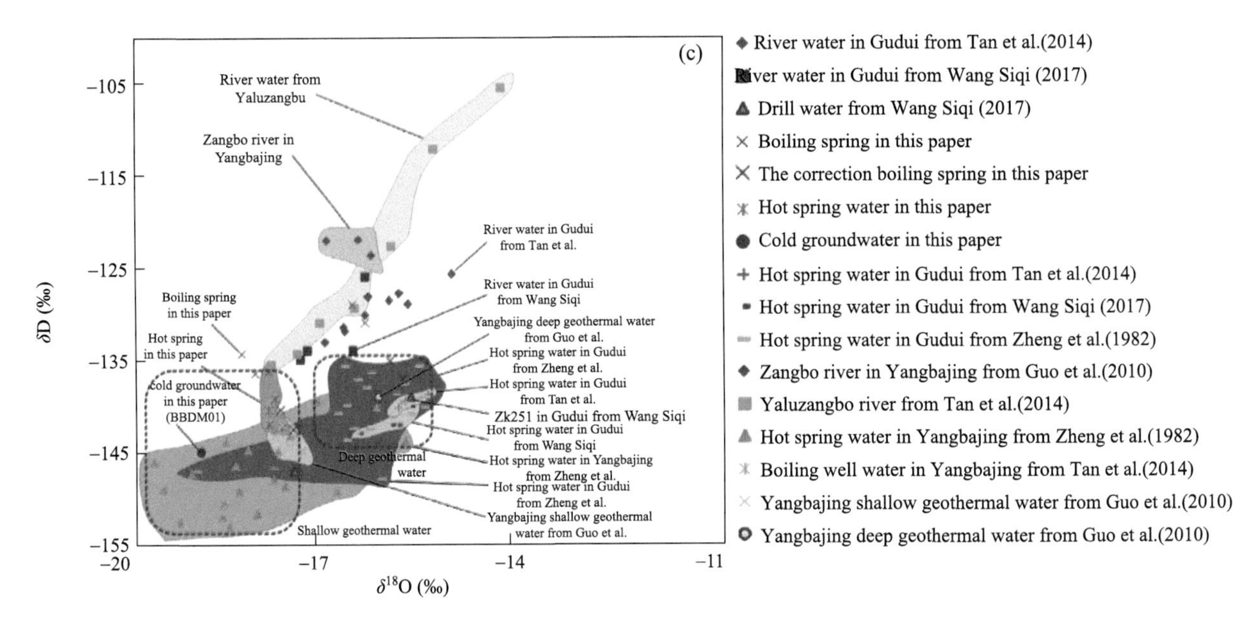

Fig. 4 (a) Plot of δD versus δ^{18}O for all water samples; the magmatic water area was plotted according to Giggenbach (1992); red dotted lines represent different reservoirs; (b) comparison of hydrogen and oxygen isotope plots of geothermal water from Gudui and Yangbajing; (c) an enlarged view of the data in Fig. 4(a).

4.2.1 H and O isotope composition characteristics

The oxygen (δ^{18}O) isotope values of the small rivers in the Gudui geothermal fields are between −17.20‰ and −14.87‰; hydrogen isotope values (δD) are −135‰ to −125.70‰ (Tan et al., 2014; Wang, 2017). This is consistent with water from the Yaluzangbo River that flows across Tibet (δ^{18}O: −17.65‰ to −14.11‰, δD: −135.50‰ to −105.60‰; Tan et al., 2014) as well as the Zangbo River (δ^{18}O: −16.80‰ to −16.10‰, δD: −123.70‰ to −122‰; Guo et al., 2010) which flows through the Yangbajing geothermal field. No distinct variation trend is observed for the δD and δ^{18}O values for rivers in either Gudui or Yangbajing. The Yaluzangbo is the largest river in Tibet, and is mainly affected by different evapo-concentration rates and recharge sources (Tan et al., 2014).

Compared to the water from the Tibetan Plateau rivers, the δD and δ^{18}O values of most of the geothermal water samples in Yangbajing and Gudui were very consistent and notably lower (Zheng et al., 1982; Guo et al., 2010; Tan et al., 2014; Wang, 2017). The trend of the samples from this study, however, approach that of the river water; the four samples from the boiling springs (PC04, PC05, BLX01, PC02) even directly fall in the range of river water. A possible explanation for this could be that the outflow channels from all the springs lead to the local rivers (e.g. the Xiongqu River and the Manidang River in Gudui); thus, the rivers receive almost all the water from the hot springs. The Yaluzangbo River appears to be a discharge tunnel; this is because water from the Gudui geothermal field is much more depleted in D and ^{18}O isotopes than that of the river water. Many boiling geysers or springs develop along the river, from the upper reaches to the lower reaches, indicating negligible recharge of geothermal water by river water. Based on field observations [Fig. 2(b), (e), (g), (h)] and data from previous studies, we speculate that violent water-vapor separation could also cause the deviation in river water data seen at Gudui (Fig. 4) (Wei et al., 1982; Guo et al., 2010; Tan et al., 2014). Fig. 4 suggests that geothermal water at Gudui is slightly heavier than that at Yangbajing, with a variable range of δ^{18}O from −18.78‰ to −15.25‰, δD from −147.90‰ to −127.40‰ for the Gudui geothermal water, and δ^{18}O from −20.60‰ to −15.58‰, δD from −153.00‰ to −139.00‰for the Yangbajing geothermal water. Tan et al. (2014) also found that the oxygen and hydrogen isotopes of geothermal water in the southern Tibet rifts are the heaviest among geothermal water sourced from the Yangbajing-Dangxiong rift basin, Bangong suture and the Yaluzangbo suture [Fig. 1(a)]. It is also noteworthy that the snow-melt water samples lie slightly above the meteoric water line. The deviation of the snow-melt water samples from the meteoric water line may result from isotopic fractionation during the snow melting process. The existence of shallow (at 180–280 m depth) and deep (at a depth of 950–1850 m) reservoirs in the Yangbajing geothermal field have been confirmed by previous studies (Duo, 2003; Guo et al., 2010), and is also expressed in the hydrogen and oxygen isotope compositions [Fig. 4(b)]. The Gudui geothermal water also shows similar characteristics. The plot of the ZK251 data from Wang (2017) is close to that of the deep geothermal water in the Yangbajing geothermal field, which indicates that it can represent the deep reservoir fluid in Gudui. Samples of surface hot springs in this paper are consistent with geothermal water from the shallow reservoirs in Yangbajing, which may indicate that the surface hot springs in Gudui come from shallow reservoirs [Fig. 4(b)]. In addition, the values of hydrogen and oxygen isotopes of the four boiling springs (PC04, PC05, BLX01, PC02) can be corrected as per the calculation of Wei et al. (1982), who reported that the fractionation of isotopes of geothermal water from reservoir to surface indicates a value of 4‰–7‰ for δD, and 0.80‰–1.40‰ for δ^{18}O. The corrected values for PC04 and PC05 plot within the range of the deep reservoir, while values for BLX01 and PC02 fall within the range of the shallow reservoir (Fig. 4). Thus, the hydrogen and oxygen isotope characteristics of the Gudui geothermal water have certain similarities with that of Yangbajing, and also reflect the deep and shallow reservoir characteristics of the Gudui geothermal field.

4.2.2 Recharge source of geothermal water

The average value of the global meteoric precipitation, based on the isotopic relationship between δD and δ^{18}O, plots with a slope of about 8 (Craig, 1961). Fig. 4 shows that the local meteoric water line (LMWL) is very close to the global one, with a slope of 8.09; the equation here is $\delta D = 8.09\ \delta^{18}O + 12.52$ (Wang, 2017). In Fig. 4, all the water samples fall on or close to the LMWL, which indicates that they originate from local precipitation with little or no evapo-concentration; this includes the river water from the Manidang upper reaches, Buxiongnanggu, Salanangga, Babudemi, Xiongqu upper reaches, and the Xiongqu lower reaches in Gudui; the Zangbo River in Yangbajing, and the Yaluzangbo River (Tan et al., 2014); it also includes surface water (Wang, 2017) and a cold groundwater sample. The Gudui river data from Wang (2017) is consistent with that from Tan et al. (2014), and they share considerable overlap with data from the Yaluzangbo River, which indicates that they have the same source of recharge, and may have experienced the same geochemical process, with the exception of a few samples from Tan et al. (2014) which deviated from the LMWL and show a slight evapo-concentration trend.

However, the stable oxygen and hydrogen isotopic compositions of geothermal fluids at Gudui (Table 1) are highly depleted; this might be due to the altitude and distance from the coastline. δ^{18}O and δD values in meteoric water typically decrease from coastal areas to inland areas (Guo et al., 2010), and also decrease with increasing elevation. The Gudui geothermal field is located in an inland area (southern Tibet) with high altitude (altitude varying from 4420 m to 5430 m), and the average elevation is 4700 m above sea level (a.s.l.). The altitudes of the samples used in this study vary between 4433 m and 4645 m, and the average elevation is 4539.60 m. The δ^{18}O and δD values in meteoric waters occurring at Gudui should be very low, even lower than Yangbajing (4300–4500 m; Guo et al., 2010). The very depleted δ^{18}O and δD values of the snowmelt water samples (δ^{18}O: 24.20‰ to −24.50‰, δD: 172‰ to −175‰) in Tibet reported in a previous study (Guo et al., 2010) corroborate this. Thus, meteoric water is an important recharge source for geothermal fluid at Gudui because of its depleted oxygen-18 and deuterium composition, and also because the geothermal water plots very close to the LMWL [Fig. 4(b)] and displays a linear correlation on the δD vs. δ^{18}O chart. Considering the geological conditions at Gudui (Liu et al., 2017; Wang, 2017), it is also likely that the snowmelt water infiltrates the subsurface through the Sangri-Cuona rift belt [Fig. 1(a)], the Zhegucuo-Longzi thrust tectonic belt, and other active faults intersecting each other in different directions in Gudui [Fig. 1(b)]. Wang (2017) calculated that the Gudui geothermal water was mainly recharged by ice and snowmelt water at an altitude of 4800–5200 m. Generally, elevations above 5500 m a.s.l. on the Tibetan Plateau are covered with ice-snow accumulation all year round (Tan et al., 2014). Thus, the recharge elevation of geothermal water is close to the snow line. Besides, field measurements show that the discharge elevation of the geothermal springs at Gudui is no more than 4645 m a.s.l. (Table 1). The differential water head, from the recharge area to the discharge area for the geothermal water circulation, is about 500 m, which may be a major factor driving the rapid circulation of snowmelt water in the Gudui geothermal system. On the contrary, if cold groundwater serves as the main recharge source for geothermal water, no large-scale high-pressure geothermal eruptions with large flux would exist because of the similar elevation from source to discharge. In addition, through communication with the local residents in the Pangcang and Maniding villages [Fig. 1(b)] around geothermal fields, we found that many naturally flowing geothermal geysers used to have low natural flux, but in recent decades, the flux has increased, as in the case of PC05 [Fig. 2(g) and (h)] and BLX01 [Fig. 2(e)]. This phenomenon could be the result of increased melt-water recharge from the global-warming-induced higher rates of melting of ice and snow on the Tibetan Plateau in recent decades (Piao et al., 2010; Wang, 2012). This phenomenon means that the principal recharge source of the Gudui geothermal water should be ice or snowmelt water. However, the geothermal fluids are more enriched in ^{18}O and D than the snowmelt water, which means there must be additional recharge sources (Table 1, Fig. 4); perhaps a water-bearing granitic melt lies at depth supplying heat to the Gudui hydrothermal system. Therefore, magmatic fluids may also contribute to the formation processes of geothermal fluids. All the geothermal water samples fall in the mixing area between snowmelt water and magmatic water [Fig. 4(b)], like the Yangbajing geothermal samples, suggesting that magmatic water contributes to the geothermal fluids that occur at Gudui. Magmatic water contribution, however, is much smaller than that of snow-melt water, as indicated by the distance between the geothermal water samples and the magmatic water area in Fig. 4. The mixing ratio of magmatic water in geothermal fluid can be calculated using a simple model (Pang, 2006):

$$\delta G = \gamma\delta M + (1-\gamma)\delta P$$

where δG, δM, and δP represent the per mil deviations versus SMOW of oxygen-18 or deuterium compositions of geothermal fluid, magmatic water and precipitated meteoric water. γ is the percentage of magmatic water mixed in geothermal fluid. Using the above equation, the mixing ratio of magmatic water in the deep geothermal fluid (well ZK251) was estimated to be between 21.10% and 24.04%. Therefore, snowmelt water and magmatic water mainly charge the deep geothermal fluids in Gudui. As compared to water samples from well ZK251 that represent the deep geothermal fluid, there should be less magmatic water mixed in the water samples from surface hot springs that represent the shallow geothermal fluid due to their longer distance from the magmatic water area, which means that the surface hot springs should get more recharge from non-magmatic water sources. Considering the basic geological characteristics and hydrogeological features of the Gudui geothermal field, there are three possibilities for the recharge of shallow geothermal fluids, (1) by river water; (2) by more snowmelt water as compared to the deep geothermal fluids; and (3) by another source with lower $\delta^{18}O$ and δD values compared to the shallow geothermal water. Fig. 4 shows that the geothermal water has much more depleted $\delta^{18}O$ and δD values than the river water, indicating the recharge source of geothermal water must be located at a higher elevation than the recharge source of the river. Therefore, the river water is not the recharge resource. Considering the high strontium concentrations of surface hot springs, we believe that the snowmelt water with particularly low strontium concentrations should not be the main recharge source of the shallow geothermal fluids. In Fig. 4(c), the cold groundwater sample BBDM01 is located between the snowmelt water samples and the geothermal water in

Gudui. This suggests that the cold groundwater may also be a recharge source for the shallow geothermal fluids. Moreover, the high strontium concentration of cold groundwater also conforms to the characteristics of surface hot springs. Therefore, water from the surface hot springs that represents the shallow reservoir fluid may be a mixture of magmatic water, snow-melt water and cold groundwater.

4.3 Strontium isotopes and mixing ratios

The $^{87}Sr/^{86}Sr$ ratio of the geothermal water at Gudui has a wide spread (Table 1). The concentration of strontium in the deep reservoir (ZK251: 69 μg/L, Table 1) is lower than that of the shallow geothermal water (the surface hot springs; 411.00–1198.00 μg/L), which is notably lower than the cold groundwater BBDM01 (2559.00 μg/L). The mechanism causing this phenomenon has already been mentioned above and will not be repeated here. The cold groundwater sample (BBDM01), and the geothermal water samples from boiling springs and hot springs were all plotted in the $^{87}Sr/^{86}Sr$ isotopic ratio versus 1/Sr plot (Fig. 5). The linear relationship among these samples suggests a binary mixing of component end members (Faure, 1977). Therefore, the strontium isotopic compositions of the water samples indicate that the water from the hot springs PC01, PC02, PC03, BLX01, CK01, CK02, and BBDM02 (the shallow geothermal fluids) were formed as a result of mixing between the boiling springs, PC04 and PC05, and the cold groundwater, BBDM01. In other words, the shallow geothermal water is constituted by the mixing of deep geothermal water and cold groundwater. That is, the deep geothermal water (represented by springs PC04, PC05) was diluted by the cold groundwater during ascent, forming the hot springs at the surface. A binary mixing equation for two endmembers were given by Faure (1986) and Négrel and Roy (1998):

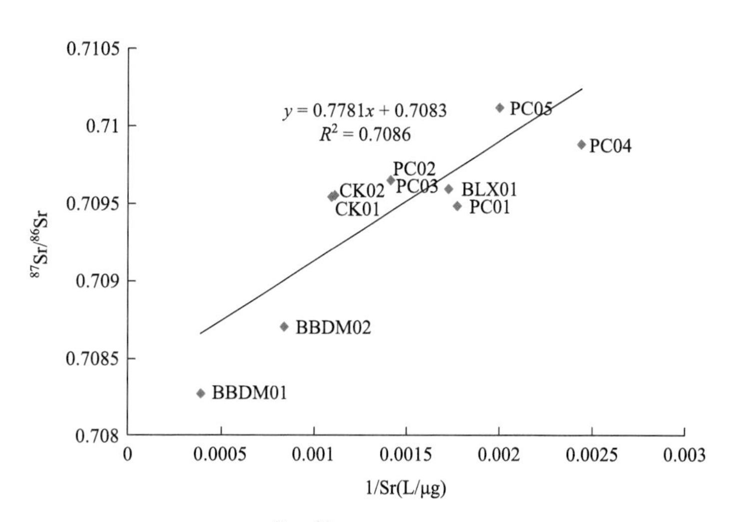

Fig. 5 Plot of the $^{87}Sr/^{86}Sr$ isotopic ratio versus 1/Sr of the geothermal water samples and the cold groundwater sample.

$$R = \frac{C_A R_A (R_B + 1)\alpha + C_B R_B (R_A + 1)(1-\alpha)}{C_A (R_B + 1)\alpha + C_B (R_A + 1)(1-\alpha)}$$

where α is the mixing ratio of endmember A; C_A and C_B are the Sr concentrations of endmembers A and B; R_A and R_B are the $^{87}Sr/^{86}Sr$ isotope ratios of endmembers A and B; and R is the presumed isotope ratio calculated using the predetermined mixing proportions of the two end-members, A and B.

This model can be utilized to calculate the theoretical value of the $^{87}Sr/^{86}Sr$ isotopic ratio of the mixture of the hot springs PC04 and PC05, and the cold groundwater, BBDM01. Thus, the binary mixing model was established. It uses the average value of PC04 and PC05 as endmember A, and the cold groundwater sample BBDM01 as endmember B. The measured $^{87}Sr/^{86}Sr$ isotopic ratios of other hot springs (PC01, PC02, PC03, BLX01, CK01, CK02, and BBDM02) were then compared with those calculated using the mixing model. A theoretical mixing line was plotted according to the calculated $^{87}Sr/^{86}Sr$ isotopic ratios and Sr concentrations of the mixing solutions (Fig. 6). The $^{87}Sr/^{86}Sr$ isotopic ratios versus Sr concentrations of the measured values were also plotted. The water samples from the hot springs cluster closely around the mixing line, supporting the proposed genesis of these geothermal fluids discussed above. This suggests a mixing ratio of 5–10% cold groundwater in most of the geothermal water samples (Fig. 6), except for BBDM02 which is mixed with about 35% cold groundwater. Besides, ZK251 has Sr concentrations of 69 μg/L, which is the lowest among the hot springs samples. Relatively speaking, ZK251 may be more representative of the deep geothermal water, while PC04 and PC05 may represent the deep geothermal fluid mixed with a small amount of cold groundwater with higher Sr concentrations. This shows that the water from the surface hot springs represents the shallow reservoir fluid, which is a mixture of deep geothermal fluid and cold groundwater.

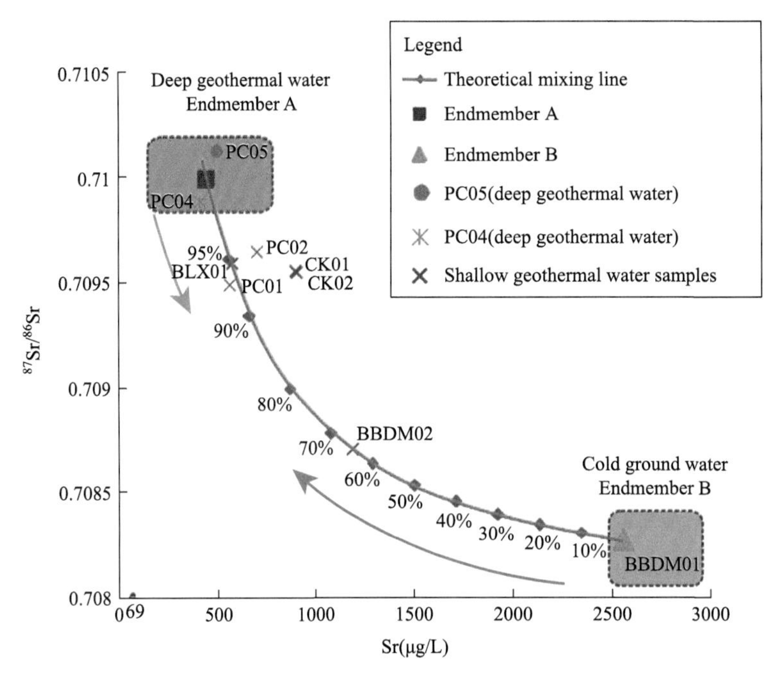

Fig. 6　The $^{87}Sr/^{86}Sr$ isotopic ratio versus Sr concentration for water samples from the hot springs.

The figures on the theoretical mixing line of this binary mixing model represent different values of α. Here, α is the mixing ratio of the geothermal water with average strontium concentrations, and the isotopic ratio of boiling springs PC04 and PC05 (endmember A).

5　Conceptual model of geothermal water circulation

The ongoing collision between the Indian subcontinent and the Asian continent, has in the last 50 Ma, lead to crustal thickening, creating the Tibetan Plateau. The uplift of the Himalayan Mountains and the Tibetan Plateau led to the development of a series of tensile active tectonic belts (Guo and Wang, 2012) such as Sangri-Cuona rift belt [Fig. 1(a)] where the Gudui geothermal field is located. The Gudui geothermal field contains two groups of northeast- and northwest-trending faults [Fig. 1(b)], some among which are tensile active faults that serve as favorable conduits for conducting snowmelt water circulation (Duo, 2003). The average elevation of the central Himalayan Mountains is over 6000 m a.s.l., making them the highest mountains around the geothermal fields in the Tibetan Plateau (Tan et al., 2014). The ice and snowmelt water or precipitation produced at high elevations can easily infiltrate vertically under the influence of gravity–given the significant water-head from the source area to the discharge area–and circulate deeply along the faults developed near Gudui. This was verified by Wang (2017), Guo et al. (2019a, b), and Tan et al. (2014).

The collision of the Indian and Eurasian continental plates has also resulted in the upwelling and heating of fluids from the asthenosphere, which have constantly caused the remelting of the lower crust forming magma, which intrudes upward to the middle or even upper crust along the suture and deep faults (Guillot et al., 1997) [Fig. 1(a)]. Once the deep circulation snowmelt water arrives at the deeper crust and encounters magmatic fluids produced by crustal remelting, considerable magmatic heat energy is absorbed by the cold water and then the heated geothermal water is steadily driven to ascend by convection or buoyancy, forming the deep or shallow reservoirs in Gudui. From the above hydrogen and oxygen isotope characteristics, we know that the Gudui geothermal field comprises of deep and shallow reservoirs. The deep reservoirs are represented by ZK251, PC04, and PC05, and the shallow reservoirs are represented by BXL01, PC01, PC02, PC03, CK01, CK02, and BBDM02. According to the above-mentioned characteristics of the strontium concentration and isotope ratio of the geothermal water in Gudui, we know that the deep geothermal water is characteristic by low strontium concentration and high strontium isotope ratio [Fig. 7(b)], and the cold groundwater is opposite. The binary mixing model calculation of the Gudui geothermal water tells us the hot springs BXL01, PC01, PC02, PC03, CK01, CK02, and BBDM02 are formed by the mixing of the boiling springs PC04 and PC05 with the cold groundwater BBDM01. In addition, the higher strontium concentration and lower isotope ratio represent a higher mixture with cold groundwater [Fig. (6)]. Zhao et al. (2003) conducted systematic studies on strontium isotopes for geothermal systems in Yangbajing, and proved that the water-rock interaction in the reservoir plays a negligible role in producing the $^{87}Sr/^{86}Sr$ ratio of geothermal water at very high reservoir temperatures. Shand et al. (2009) suggested that the difference in the solute $^{87}Sr/^{86}Sr$ ratio is mainly controlled by variations in the initial inputs. Therefore, we assume that the hot springs with similar strontium concentration and isotope ratio are from the same reservoir, and as it mixes with cold groundwater, the strontium concentration increases while the strontium isotope ratio decreases, as does the depth of the reservoirs. Under the conditions of this hypothesis, we obtain the gradual change curve [Fig.7(b)] of strontium concentration and strontium isotope values of the Gudui geothermal samples in this study. Based on this change curve [Fig. 7(b)], we postulate the 6th-Class Reservoirs Evolution Conceptual Model (6-CRECM) for the Gudui geothermal field [Fig. 7(a)]. Fournier (1979) suggested that hot springs emerged at or slightly above their boiling temperature for the prevailing atmospheric pressure conditions, indicating that they flow at relatively fast rates directly to the surface and cool adiabatically. However, the hot springs that emerged below the atmospheric boiling temperature may have cooled adiabatically during vertical movement, and then cooled rapidly by horizontal conduction at shallow horizons. Therefore, based on the surface temperature of the Gudui hot springs in this paper, we can speculate on the evolution of the underground ascending path of different springs [Fig. 7(a)]. In Fig. 7(a), ① and ② represent the reservoirs closest to the deep reservoirs. Due to the hot springs representing the ② reservoir fluids which have higher strontium concentration (Table 1, Fig. 7), the ② reservoir may have mixed with some cold groundwater, and at a shallower depth compared to the ① reservoir, which is represented by ZK251 with the lowest strontium concentration [Fig. 7(a)]. The ③, ④, ⑤, and ⑥ reservoirs represent the shallow reservoirs, and the strontium concentration as well as $^{87}Sr/^{86}Sr$ ratios of the corresponding hot springs are listed in Table 1 and Fig. 7(b). Relatively speaking, the strontium concentration and the isotope ratio of the hot springs representing the ②and ③ reservoirs are slight abnormal. It could be due to the short residence time of the geothermal water in the reservoirs; a shorter residence time suggests disequilibrium; they are marked with gradient color in Fig. 7(a). The equilibrium state in ⑥ reservoir cannot be confirmed with only one sample. However, because it is closest to the surface, it may have a strong convection; therefore, we assume that it has not reached equilibrium. In addition, lithium concentration gradually decreases as the geothermal fluid gradually evolves from deep to shallow (Table 1, Fig. 8), indicating that the source of lithium is the deep geothermal water. It is established that the deep geothermal water is a mixture of magma fluid and snowmelt water. Therefore, lithium is likely to come from the magma component, and the mixing of geothermal fluid and cold groundwater dilutes it.

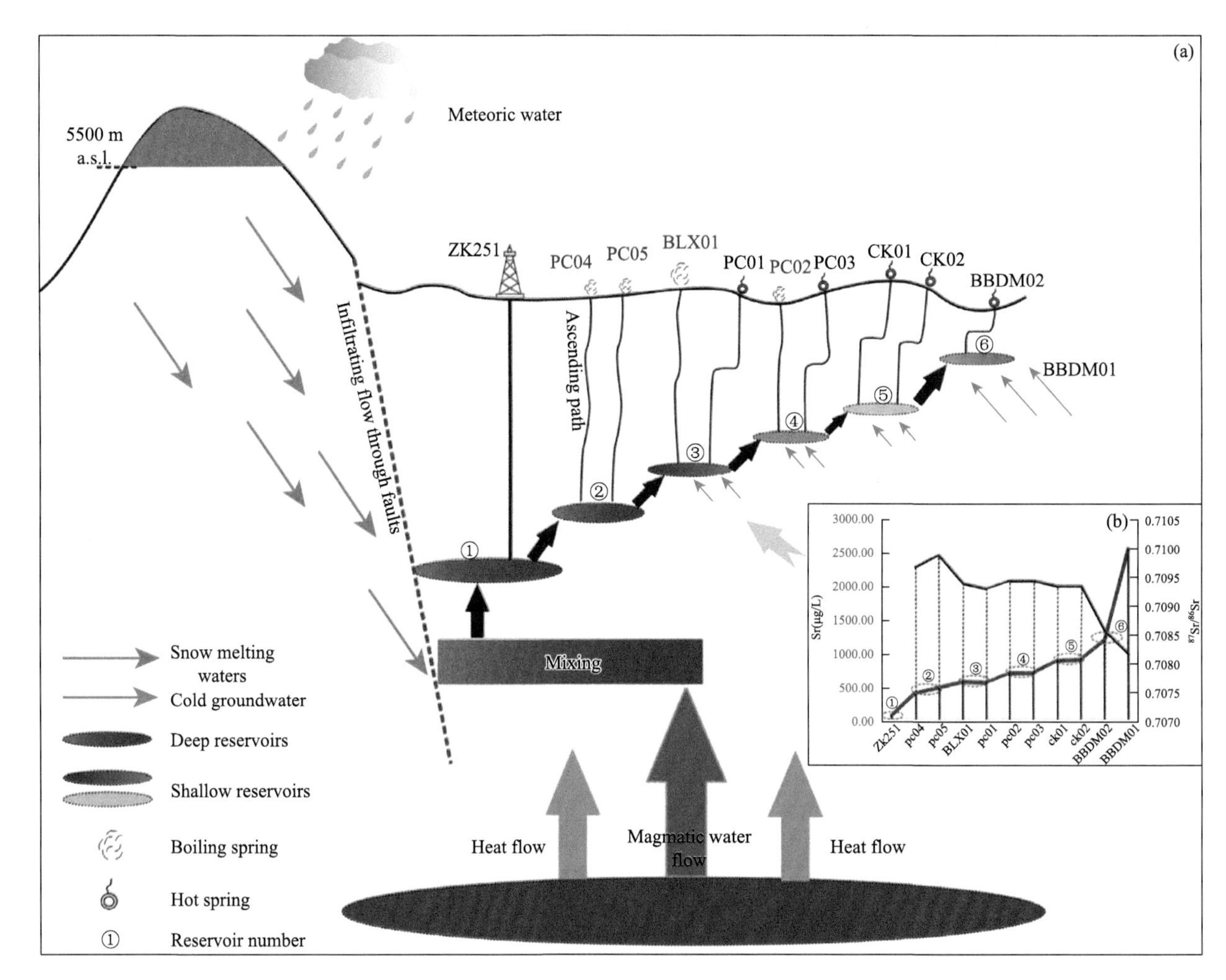

Fig. 7 (a) The 6th-Class Reservoirs Evolution Conceptual Model (6-CRECM) for the Gudui geothermal system; (b) the gradual change curve of strontium concentration and strontium isotope values of the Gudui geothermal samples in this study with the evolution of geothermal water.

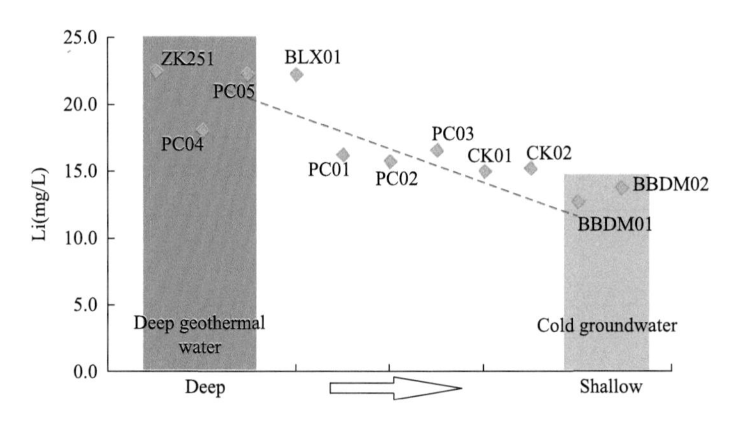

Fig. 8 The change of lithium concentrations in geothermal water with the gradual evolution of geothermal fluid from deep to shallow.

The deep lithium resources in the Gudui geothermal field have great potential; recently, the high-lithium geothermal water discovered by deep drilling in the Yangbajing geothermal field substantiates this viewpoint (personal communication with an academician at the Chinese Academy of Engineering, Duoji). Although we require more in-depth research work to verify these observations, we now have a basic under-standing of the origin, mixing processes, and source of lithium of geothermal water in the Gudui geothermal field.

6 Conclusions

In the Gudui geothermal field, the temperatures, TDS values, and Sr concentrations of cold groundwater and geothermal water are quite different. The strontium concentration in geothermal water is far lower than that in cold groundwater, which is opposite to what has been reported from the Yangbajing geothermal water. Hydrogen and oxygen isotope plotting of the Gudui geothermal water shows the analogous characteristics of deep and shallow reservoirs below the Gudui geothermal field with the Yangbajing geothermal field. The samples PC04, PC05, and ZK251 are from the deep reservoir and the hot springs BXL01, PC01, PC02, PC03, CK01, CK02, and BBDM02 are from shallow reservoir as per the δD versus $\delta^{18}O$ plot. In addition, the hydrogen-oxygen isotope plotting characteristics indicate the contribution of magmatic water and snowmelt water to the deep geothermal fluid. Calculations show the magma fluid component mixed in the deep geothermal fluid accounts for about 21.1–24.04%; magma fluids could be the main source of lithium in geothermal water. In the $^{87}Sr/^{86}Sr$ ratio versus 1/Sr plot (Fig. 5), a linear relationship was detected between the cold groundwater sample (BBDM01) and the hot springs samples from both shallow and deep reservoirs, indicating a binary mixing between the two component endmembers. A mixing model with the deep geothermal water sample (the average of PC04 and PC05) and the cold groundwater sample (BBDM01) as two endmembers was established to calculate the theoretical $^{87}Sr/^{86}Sr$ ratios of the mixing solutions with different end-member mixing ratios. The result indicated that the shallow reservoir is constituted by the mixing of deep geothermal fluid and cold ground-water, and the mixing ratios of cold groundwater in most shallow geothermal water samples range from 5% to 10%. Based on the comprehensive understanding of the basic geological characteristics, hydrogen and oxygen isotope, strontium concentration and strontium isotope characteristics of the Gudui geothermal fluid, we propose a 6th-Class Reservoirs Evolution Conceptual Model (6-CRECM) [Fig. 7(a)] for the Gudui geothermal system, which will help understand the origin and mixing processes of the Gudui geothermal fluid. It may even help local supervisors manage the geothermal water resources at Gudui effectively. Although the conclusions drawn in this paper need to be validated by evidence beyond surface and geochemical investigations, they are suggestive for preliminary identification of heat sources and geological geneses of a hydrothermal system with the highest recorded downhole temperature at an equivalent depth as compared to other geothermal fields in remote Tibet of China, which is vitally important for the efficient management and sustainable exploitation of geothermal resources in Gudui.

Acknowledgements

This work was financially supported by the China Geological Survey (Grant No. DD20160054), the National Natural Science Foundation of China (Grant No. U1407207), and the National Key Research and Development Program of China (Grant No.2017YFC0602802).

References

Armijo, R., Tapponnier, P., Mercier, J.L., Han, T.L., 1986. Quaternary extension in southern Tibet: field observations and tectonic implications. J. Geophys. Res.: Solid Earth 91, 13803-13872.

Aguilera, E., Cioni, R., Gherardi, F., Magro, G., Marini, L., Pang, Z., 2005. Chemical and isotope characteristics of the Chachimbiro geothermal fluids (Ecuador). Geothermics 34, 495-517.

Brown, L.D., Zhao, W., Nelson, K.D., Hauck, M., Alsdorf, D., Ross, A., Cogan, M., Clark, M., Liu, X., Che, J., 1996. Bright spots, structure, and magmatism in southern Tibet from indepth seismic reflection profiling. Science 74, 1688-1690.

Brombach, T., Marini, L., Hunziker, J.C., 2000. Geochemistry of the thermal springs and fumaroles of Basse-Terre Island, Guadeloupe, lesser Antilles. Bull. Volcanol.61, 477-490.

Barbieri, M., Morotti, M., 2003. Hydrogeochemistry and strontium isotopes of spring and mineral waters from Monte Vulture volcano, Italy. Appl. Geochem.18, 117-125.

Boschetti, T., Venturelli, G., Toscani, L., Barbieri, M., Mucchino, C., 2005. The Bagnidi Lucca thermal waters (Tuscany, Italy): an example of Ca-SO_4, waters with high Na/Cl and low Ca/SO_4, ratios. J. Hydrol. (Amst.) 307, 0-293.

Baietto, A., Cadoppi, P., Martinotti, G., Perello, P., Vuataz, F.D., 2005. Mechanisms of up-rise of the hot thermal springs of Valdieri (Argentera Massif, Maritime Alps, northwestern Italy). Geophys. Res. Abstr.7, 03167 https: //doi.org/10.1017/S0009840X00213000.

Bucher, K., Zhang, L., Stober, I., 2009. A hot spring in granite of the Western Tianshan, China. Appl. Geochem.24, 402-410.

Birkle, P., Bundschuh, J., Sracek, O., 2010. Mechanisms of arsenic enrichment in geothermal and petroleum reservoirs fluids in Mexico. Water Res.44, 5605-5617.

Bundschuh, J., Maity, J.P., 2015. Geothermal arsenic: occurrence, mobility and environmental implications. Renew. Sustain. Energy Rev.42, 1214-1222.

Craig, H., 1961. Isotopic variation in meteoric waters. Science 133, 1702-1703.

Coleman, M.L., Shepherd, T.J., Durham, J.J., Rouse, J.E., Moore, G.R., 1982. Reduction of water with zinc for hydrogen isotope analysis. Anal. Chem.54, 993-995.

Coplen, T.B., 1988. Normalization of oxygen and hydrogen isotope data. Chem. Geol. Isot. Geosci.72, 293-297.

Chen, L., Booker, J.R., Jones, A.G., Wu, N., Unsworth, M.J., Wei, W., Tan, H., 1996. Electrically conductive crust in southern Tibet from INDEPTH magnetotelluric surveying. Science 274, 1694-1696.

Clark, I.D., Phillips, R.J., 2000. Geochemical and $_3$He/$_4$He evidence for mantle and crustal contributions to geothermal fluids in the western Canadian continental margin. J. Volcanol. Geotherm. Res.104, 261-276.

Doe, B.R., Hedge, C.E., White, D.E., 1966. Preliminary investigation of the source of lead and strontium in deep geothermal brines underlying the Salton Sea geothermal area. Econ. Geol.61, 462-483.

Delmelle, P., Kusakabe, M., Bernard, A., Fischer, T., Brouwer, S.D., Mundo, E.D., 1998. Geochemical and isotopic evidence for seawater contamination of the hydrothermal system of Taal Volcano, Luzon, the Philippines. Bull. Volcanol.59, 562-576.

Diamond, R.E., Harris, C., 2001. Oxygen and hydrogen isotope geochemistry of thermal springs of the Western Cape, South Africa: recharge at high altitude? J. Afr. Earth Sci. 31, 467-481.

Duo, J., 2003. The basic characteristics of the Yangbajing geothermal field—a typical high temperature geothermal system. Eng. Sci.5, 42-47 (in Chinese with English abstract).

Dong, F.Q., Li, W.Y., Hu, K.W., Huang, R.C., Li, H.E., Dang, H.L., Mao, C.Z., Basangciren, 2017. Study on geochemical characteristics and sedimentary environment of containing radiolarian chert of Triassi Nieru Formation in Gudui area of southern Tibet. Miner. Resour. Geol.31, 166-172 (in Chinese with English abstract).

Elderfield, H., Greaves, M.J., 1981. Strontium isotope geochemistry of Icelandic geothermal systems and implications for sea water chemistry. Geochem. Cosmochim. Acta 45, 2201-2212.

Evans, M.J., Derry, L.A., Francelanord, C., 2005. Hydrothermal flux of metamorphic carbon dioxide from the Central Nepal Himalaya. In: AGU Fall Meeting. AGU Fall Meeting Abstracts.

Faure, G., 1977. Isotope Geology. Wiley, New York.

Fournier, R.O., 1979. Geochemical and hydrologic considerations and the use of enthalpy-chloride diagrams in the prediction of underground conditions in hot-spring systems. J. Volcanol. Geotherm. Res.5, 1-16.

Faure, G., 1986. Principles of Isotope Geology. Wiley, New York.

Fossen, H., 2000. Extensional tectonics in the Caledonides: synorogenic or postorogenic? Tectonics 19, 213-224.

Giggenbach, W.F., 1992. Isotopic shifts in waters from geothermal and volcanic systems along convergent plate boundaries and their origin. Earth Planet. Sci. Lett.113, 495-510.

Graham, I.J., 1992. Strontium isotope composition of Rotorua geothermal waters. Geothermics 21, 165-180.

Ghomshei, M.M., Clark, I.D., 1993. Oxygen and hydrogen isotopes in deep thermal waters from the south meager creek geothermal area, British Columbia, Canada. Geothermics 22, 79-89.

Guillot, S., Sigoyer, J.D., Lardeaux, J.M., Mascle, G., 1997. Eclogitic metasediments from the Tso Morari area (Ladakh, Himalaya): evidence for continental subduction during India-Asia convergence. Contrib. Mineral. Petrol.128, 197-212.

Grobe, M., Machel, H.G., Heuser, H., 2000. Origin and evolution of saline groundwater in the Münsterland Cretaceous Basin, Germany: oxygen, hydrogen, and strontium isotope evidence. J. Geochem. Explor.69-70, 5-9.

Gherardi, F., Panichi, C., Gonfiantini, R., Magro, G., Scandiffio, G., 2005. Isotope systematics of C-bearing gas compounds in the geothermal fluids of Larderello, Italy. Geothermics 34, 442-470.

Guo, Q.H., Wang, Y.X., Liu, W., 2010. O, H, and Sr isotope evidences of mixing processes in two geothermal fluid reservoirs at Yangbajing, Tibet, China. Environ. Earth Sci.59, 1589-1597.

Guo, Q., Planer-Friedrich, B., Liu, M., Yan, K., Wu, G., 2019a. Magmatic fluid input explaining the geochemical anomaly of very high arsenic in some southern Tibetan geothermal waters. Chem. Geol.513, 32-43.

Guo, Q., Wang, Y., 2012. Geochemistry of hot springs in the Tengchong hydrothermal areas, Southwestern China. J. Volcanol. Geotherm. Res.215-216, 61-73.

Guo, Q., Li, Y., Luo, L., 2019b. Tungsten from typical magmatic hydrothermal systems in China and its environmental transport. Sci. Total Environ.657, 1523-1534.

Guo, Q., Liu, M., Li, J., Zhang, X., Guo, W., Wang, Y., 2017. Fluid geochemical constraints on the heat source and reservoir temperature of the Banglazhang hydrothermal system, Yunnan-Tibet Geothermal Province, China. J. Geochem. Explor.172, 109-119.

Hilton, D.R., Hammerschmidt, K., Teufel, S., Friedrichsen, H., 1993. Helium isotope characteristics of Andean geothermal fluids and lavas. Earth Planet. Sci.

Lett.120, 265-282.

Hou, Z.Q., Gao, Y.F., Qu, X.M., Rui, Z.Y., Mo, X.X., 2004. Origin of adakitic intrusives generated during mid-Miocene east-west extension in southern Tibet. Earth Planet. Sci. Lett.220, 139-155.

Hu, X.C., Wang, G.l., Wang, X.l., 2015. Geothermal exploration along the Qinghai-Xizang Railway drilled 205℃ geothermal steam. Geotherm. Energy 3, 30-30 (in Chinese).

Kind, R., Ni, J., Zhao, W., Wu, J., Yuan, X., Zhao, L., Sandvol, E., Rees, E.C., Nabelek, J., Hearn, T., 1996. Evidence from earthquake data for a partially molten crustal layer in southern Tibet. Science 274, 1692-1694.

Kennedy-Bowdoin, T., Silver, E.A., Martini, B.A., Pickles, W.L., 2004. Chemical and structural dynamics of a geothermal system. IEEE Int. Geosci. Remote Sens. Symp. 645-648.

Lyons, W.B., Tyler, S.W., Gaudette, H.E., Long, D.T., 1995. The use of strontium isotopes in determining groundwater mixing and brine fingering in a playa spring zone, Lake Tyrrell, Australia. J. Hydrol.167, 0-239.

Land, M., Ingri, J., Andersson, P.S., Hlander, B., 2000. Ba/Sr, Ca/Sr and $_{87}Sr/_{86}Sr$ ratios in soil water and groundwater: implications for relative contributions to stream water discharge. Appl. Geochem. vol.15, 0-325.

Li, Z.Q., 2002. Present Hydrothermal Activities during Collisional Orogenics of the Tibetan Plateau. D.S. thesis. Chinese Academy of Geologecal Sciences, p.16 (in Chinese with English abstract).

López, D.L., Bundschuh, J., Birkle, P., Armienta, M.A., Cumbal, L., Sracek, O., Cornejo, L., Ormachea, M., 2012. Arsenic in volcanic geothermal fluids of Latin America. Sci. Total Environ.429, 57-75.

Lü, Y.Y., Zheng, M.P., Zhao, P., Xu, R.H., 2014. Geochemical processes and origin of boron isotopes in geothermal water in the Yunnan-Tibet geothermal zone. Sci. China Earth Sci.57, 2934-2944.

Li, H.X., Guo, Q.H., Wang, Y.X., 2015. Evaluation of temperature of parent geothermal fluid and its cooling processes during ascent to surface: a case study in Rehai geothermal field, Tengchong. J. China Univ. Geosci.40, 1576-1584 (in Chinese with English abstract).

Liotta, M., D″Alessandro, W., Arienzo, I., Longo, M., 2017. Tracing the circulation of groundwater in volcanic systems using the $^{87}Sr/^{86}Sr$ ratio: Application to Mt. Etna. J. Volcanol. Geotherm. Res.331, 102-107.

Liu, Z., Chen, K., Nan, dawa, 2017. Hydrochemical characteristics of geothermal water in Gudui, Xizang (Tibet). Geol. Rev. s1, 353-354 (in Chinese).

Marty, B., Gunnlaugsson, E., Jambon, A., Oskarsson, N., Ozima, M., Pineau, F., Torssander, P., 1991. Gas geochemistry of geothermal fluids, the Hengill area, southwest rift zone of Iceland. Chem. Geol.91, 0-225.

Makovsky, Y., Klemperer, S.L., Ratschbacher, L., Brown, L.D., Li, M., Zhao, W., Meng, F., 1996. INDEPTH wide-angle reflection observation of P-wave-to-S-wave conversion from crustal bright spots in Tibet. Science 274, 1690-1691.

Majumdar, N., Majumdar, R.K., Mukherjee, A.L., Bhattacharya, S.K., Jani, R.A., 2005. Seasonal variations in the isotopes of oxygen and hydrogen in geothermal waters from Bakreswar and Tantloi, Eastern India: implications for groundwater characterization. J. Asian Earth Sci.25, 269-278.

Nakano, T., Kajiwara, Y., Farrell, C.W., 1989. Strontium isotope constraint on the genesis of crude oils, oil-field brines and Kuroko ore deposits from the Green Tuff region of northeastern Japan. Geochem. Cosmochim. Acta 53, 2683-2688.

Neumann, K., Dreiss, S., 1995. Strontium 87/strontium 86 ratios as tracers in groundwater and surface waters in Mono Basin, California. Water Resour. Res.31, 3183-3193.

Nelson, K.D., Zhao, W., Brown, L.D., Kuo, J., Che, J., Liu, X., Klemperer, S.L., Makovsky, Y., Meissner, R., Mechie, J., Kind, R., Wenzel, F., Ni, J., Nabelek, J., Leshou, C., Tan, H., Wei, W., Jones, A.G., Booker, J., Unsworth, M., Kidd, W.S.F., Hauck, M., Alsdorf, D., Ross, A., Cogan, M., Wu, C., Sandvol, E., Edwards, M., 1996. Partially molten middle crust beneath southern Tibet: synthesis of project INDEPTH results. Science 274, 1684-1688.

Négrel, P., P., Roy, S., 1998. Chemistry of rainwater in the Massif Central (France): a strontium isotope and major element study. Appl. Geochem.13, 941-952.

Négrel, P., Casanova, J., Aranyossy, J.F., 2001. Strontium isotope systematics used to decipher the origin of groundwaters sampled from granitoids: the Vienne Case (France). Chem. Geol.177, 287-308.

Négrel, P., Guerrot, C., Millot, R., 2007. Chemical and strontium isotope characterization of rainwater in France: influence of sources and hydrogeochemical implications. Isot. Environ. Health Stud.43, 179.

Nisi, B., Buccianti, A., Vaselli, O., Perini, G., Tassi, F., Minissale, A., Montegrossi, G., 2008. Hydrogeochemistry and strontium isotopes in the Arno River Basin (Tuscany, Italy): constraints on natural controls by statistical modeling. J. Hydrol. (Amst.) 360, 166-183.

Peterman, Z.E., Wallin, B., 1999. Synopsis of strontium isotope variations in groundwater at Aspo, southern Sweden. J. Hydrol.14, 939-951.

Pang, Z., 2006. PH dependant isotope variations in arc-type geothermal waters: new insights into their origins. J. Geochem. Explor.89, 306-308.

Piao, S., Ciais, P., Huang, Y., Shen, Z., Peng, S., Li, J., Zhou, L., Liu, H., Ma, Y., Ding, Y., Friedlingstein, P., Liu, C., Tan, K., Yu, Y., Zhang, T., Fang, J., 2010. The impacts of climate change on water resources and agriculture in China. Nature 467, 43-51.

Ren, C., Zhu, l.D., Pan, J.T., Gao, W.X., 2014. Geochemistry and zircon SHRIMP U-Pb dating and their tectonic significance for intermediate-basic dyke in the Gudui region, South Tibet. Acta Geol. Sichuan 4, 496-500 (in Chinese with English abstract).

Sunwall, M.T., Pushkar, P., 1979. The isotopic composition of strontium in brines from petroleum fields of southeastern Ohio. Chem. Geol.24, 189-197.

Sheppard, D.S., Lyon, G.L., 1984. Geothermal fluid chemistry of the Orakeikorako field, New Zealand. J. Volcanol. Geotherm. Res.22, 329-349.

Stauffer, R.E., Thompson, J.M., 1984. Arsenic and antimony in geothermal waters of Yellowstone National Park, Wyoming, USA. Geochem. Cosmochim. Acta 48, 2547-2561.

Shand, P., Darbyshire, D.P.F., Love, A.J., Edmunds, W.M., 2009. Sr isotopes in natural waters: applications to source characterisation and water-rock interaction in contrasting landscapes. Appl. Geochem.24, 0-586.

Sun, X., Zheng, Y., Wang, C., Zhao, Z.Y., Geng, X.B., 2016. Identifying geochemical anomalies associated with Sb-Au-Pb-Zn-Ag mineralization in North Himalaya, southern Tibet. Ore Geol. Rev.73, 1-12.

Truesdell, A.H., Hulston, J.R., 1980. Chapter 5-isotopic evidence on environments of geothermal system. Handb. Environ. Isot. Geochem.179-226.

Tenu, A., Constantinescu, T., Davidescu, F., Nuti, S., Noto, P., Squarci, P., 1981. Research on the thermal waters of the western plain of Romania. Geothermics 10, 1-28.

Tapponnier, P., Mercier, J.L., Proust, F., Andrieux, J., Armijo, R., Bassoullet, J.P., Brunel, M., Burg, J.P., Colchen, M., Dupr′e, B., Girardeau, J., Marcoux, J., Mascle, G., Matte, P., Nicolas, A., Li, T.D., Xiao, X.C., Chang, C.F., Lin, P.Y., Li, G.C., Wang, N.W., Chen, G.M., Han, T.L., Wang, X.B., Den, W.M., Zhen, H.X., Sheng, H.B., Cao, Y.G., Zhou, J., Qiu, H.R., 1981. The Tibetan side of the India-Eurasia collision. Nature 294, 405-410.

Tapponnier, P., Xu, Z.Q., Roger, F., Meyer, B., Arnaud, N., Wittlinger, G., Yang, J.S., 2001. Oblique stepwise rise and growth of the Tibet plateau. Science 294, 1671-1677.

Tong, W., Liao, Z.J., Liu, S.B., Zhang, Z.F., You, M.Z., Zhang, M.T., 2000. Tibetan Hot Springs. Science Press (in Chinese with English abstract).

Tan, H.D., Wei, W.B., Unsworth, Martyn, Deng, M., Jin, S., Booker, John, Jones, Alan, 2004. Crustal electrical conductivity structure beneath the Yarlung Zangbo Jiang suture in the southern Xizang plateau. Chin. J. Phys.47, 685-690 (in Chinese with English abstract).

Tan, H., Zhang, Y., Zhang, W., Kong, N., Zhang, Q., Huang, J., 2014. Understanding the circulation of geothermal waters in the Tibetan Plateau using oxygen and hydrogen stable isotopes. Appl. Geochem.51, 23-32.

Vuataz, F.D., Goff, F., Fouillac, C., Calvez, J.Y., 1988. A strontium isotope study of the VC-1 core hole and associated hydrothermal fluids and rocks from Valles Caldera, Jemez Mountains, New Mexico. J. Geophys. Res.93, 6059.

Wei, K., Lin, R., Wang, Z., 1982. Isotopic composition and tritium content of waters from Yangbajing geothermal area, Xizang (Tibet), China. In: Proc.5th Internat. Conf. Geochronology, Cosmochronology and Isotope Geology. Nikko National Park, Japan.

Wilkinson, M., Crowley, S.F., Marshall, J.D., 1992. Model for the evolution of oxygen isotope ratios in the pore fluids of mudrocks during burial. Mar. Pet. Geol.9, 98-105.

Webster, J.G., Nordstrom, D.K., 2003. Geothermal Arsenic, Arsenic in Ground Water. Springer, pp.101-125.

Weis, D., Kieffer, B., Maerschalk, C., Barling, J., Jeroen, de Jong, Williams, G.A., Hanano, D., Pretorius, W., Mattielli, N., Scoates, J.S., Goolaerts, A., Friedman, R.M., Mahoney, J.B., 2006. High-precision isotopic characterization of USGS reference materials by TIMS and MC-ICP-MS. Geochem. Geophys. Geosyst.7, Q08006. https: //doi.org/10.1029/2006GC001283.

Wei, W.B., Jin, S., Ye, G.F., Deng, M., Jing, J.E., Unsworth, M., Jones, A.G., 2010. Conductivity structure and rheological property of lithosphere in Southern Tibet inferred from super-broadband magnetotelluric sounding. Sci. China Ser. D Earth Sci. 53, 189-202.

Wang, E., 2012. High-altitude salt lake elevation changes and glacial ablation in Central Tibet, 2000-2010. Chin. Sci. Bull.57, 525-534.

Wang, Siqi, 2017. Hydrogeochemical Processes and Genesis Mechanism of High-Temperature Geothermal System in Gudui, Tibet. D.S. Thesis. China University of Geosciences, Beijing (in Chinese with English abstract).

Wang, C.G., Zheng, M.P., 2019. Hydrochemical characteristics and evolution of hot fluids in the Gudui geothermal field in Comei County, Himalayas. Geothermics 81, 243-258. https: //doi.org/10.1016/j.geothermics.2019.05.010.

Yin, A., 2000. Mode of Cenozoic east-west extension in Tibet suggesting a common origin of rifts in Asia during the Indo-Asian collision. J. Geophys. Res. Solid Earth 105, 21745-21759.

Zheng, S.H., Zhang, Z.F., Ni, B.L., Hou, F.G., Shen, M.Z., 1982. Hydrogen and oxygen stable isotope study of geothermal water in Tibet. Acta Sci. Nauralium Univ. Pekin.1, 102-109 (in Chinese with English abstract).

Zheng, M.P., Liu, W.G., Xiang, J., Jiang, X.Z., 1983. On saline lakes in Tibet, China. Acta Geol. Sin.2, 80-90 (in Chinese with English abstract).

Zheng, M.P., 1995. A New Type of Hydrothermal Deposit—Cesium-Bearing Geyserite in Tibet. Geological Publishing House (in Chinese with English abstract).

Zhao, P., Duo, J., Xie, E., Jin, J., 2003. Strontium isotope data for thermal waters in selected high-temperature geothermal fields, China. Acta Petrol. Sin.19, 569-576 (in Chinese with English abstract).

Zhang, P.Z., Shen, Z., Wang, M., Gan, W., Bürgmann, Roland, Molnar, P., Wang, Q., Niu, Z.J., Sun, J.H., Wu, J.C., Sun, H.R., You, X.Z., 2004. Continuous deformation of the Tibetan plateau from global positioning system data. Geology 32, 809.

Zhang, W., Tan, H., Zhang, Y., Wei, H., Dong, T., 2015. Boron geochemistry from some typical Tibetan hydrothermal systems: origin and isotopic fractionation. Appl. Geochem.63, 436-445.

Sedimentary and geochemical characteristics of Triassic new type of polyhalite potassium resources in northeast Sichuan and its genetic study*

Jiaai Zhong[1, 2, 3, 4], Mianping Zheng[1, 2, 3], Yongsheng Zhang[1, 2, 3], Xueyuan Tang[4] and Xuefei Zhang[1, 2, 3]

1. Key Laboratory of Metallogeny and Mineral Assessment, Ministry of Natural Resources, Beijing 100037, China
2. Institute of Mineral Resources Chinese Academy of Geological Sciences, Beijing 100037, China
3. Key Laboratory of Saline Lake Resources and Environment, Ministry of Natural Resources, Beijing 100037, China
4. Sichuan geology and mineral exploration and Development Bureau of four O five geological team, Dujiangyan 611830, Sichuan, China

Abstract Polyhalite has been discovered for years in the Triassic of the Sichuan Basin. However, it is difficult to exploit and utilize such polyhalite because of its deep burial depth and its coexistence with anhydrite or dolomite. Therefore, it has always been regarded as "dead ore". Based on slice identification, X-powder diffraction, chemical analysis, REEs analysis and strontium isotope test on halite samples from the fourth and fifth member of Jialingjiang Formation to Leikoupo Formation of Wells ZK601 and ZK001 in Xuanhan area, Northeast Sichuan Basin, this paper discovers thick layers of granular polyhalite associated with halite and the polyhalite content accounts for 10–30%. These deep polyhalites can be obtained by water-soluble mining and utilized, so they are called "new type polyhalite potash deposits". The deposit is deep buried at 3,000 m underground, and the thickness of a single layer can be more than 30 m. It is stable in regional distribution. The discovery of the "new type polyhalite potash deposits" has "activated" polyhalite, which has been considered as deep "dead ore" and has great significance for potash prospecting in China.

China's marine potassium is mainly developed in the Lower-Middle Triassic Jialingjiang Formation and the Leikoupo Formation of the Sichuan Basin, and abundant potassium salts such as polyhalite and potassium-rich brine have been found[1-7]. In-depth research has been conducted for the Triassic polyhalite in the Sichuan Basin, and the research covers the formation conditions[8, 9], the genesis mechanism[10-12], resource and reserve evaluation, and comprehensive development and utilization. The resource evaluation and exploitation of the shallow-layer polyhalite in the Nongle Village, Qu County, East Sichuan has achieved good results[13-15]. However, the research has been concentrated on the bedded, stratoid, lenticular polyhalite occurring in the anhydrite or interbedding with anhydrite in different thickness. This kind of polyhalite does not coexist with halite, shown in the section, and is mostly deep buried (2,000–5,000 m underground). Since it is difficult to be utilized, it has always been considered as "dead ore". Sichuan BestRed Mining Company has made some progress in in-situ dissolution experiment of deep polyhalite in Guang'an, Sichuan. However, there is still a long way to go to achieve low-cost and efficient industrial mining.

Wang and Zheng[11] found polyhalite in Triassic rock salt in Changshou area, East Sichuan. Afterwards, with the support of the National Geological Survey, the team led by Zheng Mianping conducted in-depth research on the core and well logging data of the potash exploration wells in the Puguang area of Xuanhan County, Sichuan Province and cooperated with Hengcheng company, 405 Geological Brigade, Chengdu University of Technology

* 本文发表在：Scientific Reports, 2020, 10 (1): 13528-13536

and other relevant units to form the synergy effect of innovation. Their findings show that thick layers of granular polyhalite associated with halite are encountered by boreholes such as Hengcheng 2, Hengcheng 3, ZK601, and ZK001, and polyhalite particles of different sizes are distributed in halite matrix deeper than 3,000 m.

Due to the solubility of halite, this kind of deep granular halite can be directly extracted by water-soluble mining and utilized. Academician Zheng Mianping called it "polyhalite grain cemented by halite crystal"—a new type of polyhalite potash deposits (PPD). At present, there are relatively few studies on the characteristics and genesis of the PPD from the fourth and fifth member of Jialingjiang Formation to Leikoupo Formation of the Triassic in the Xuanhan area, and there is a lack of research on its distribution law and prospects for potassium formation. In this paper, the halite samples from the fourth and fifth member of Jialingjiang Formation to the first member of Leikoupo Formation of potash exploration Wells ZK601 and ZK001 in Xuanhan area are systematically collected, and the slice identification, X-powder diffraction, chemical analysis, REEs analysis and strontium isotope test are carried out. The characteristics and genesis of the PPD are studied. At the same time, the well logging data of more than 30 natural gas wells in this area are interpreted, the "new type of polyhalite potash deposits" is identified and its distribution law is studied. Based on this, the prospect of potash in this area is analyzed in order to open up a new direction for the exploration of marine potash in China.

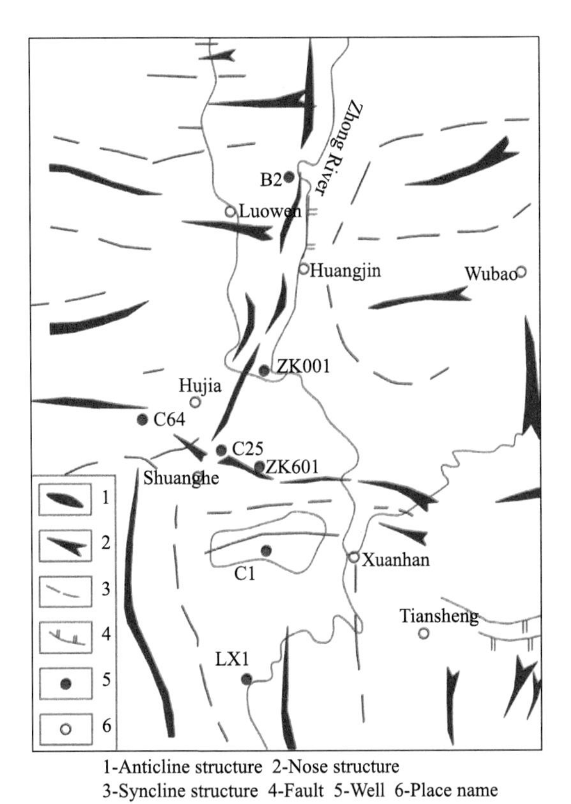

Fig. 1 Structural location map of study area.
Map generated using Geomap4.0
(https: //www.jurassic.com. cn/zh-cn/Products/ Service/145).

Geological background

Puguang gas field is located in the northeast of Sichuan Basin. Tectonically, it belongs to the northwest of the upper Yangtze Block, the hinterland of Sichuan foreland basin, and lies in the transitional zone between the front fault-fold belt of the Dabashan nappe belt and the arcuate fault-fold belt of East Sichuan. The Huangjinkou anticline is narrow and asymmetric and is slightly box shaped, with its axis trending NE 50°–60°. The dip angle of the south east wing is 30°–60°, and that of the north west wing is 20°–35° (Fig. 1).

The strata in the study area are a set of carbonate rock and clastic rock formations of shallow marine and coastal facies, with no magmatism and metamorphism. In this area, the Lower and Middle Jurassic Ziliujing Formation and the strata below are all buried in the hinterland. The salt bearing strata are located in the fourth and fifth members of the Jialingjiang Formation and the first member of Leikoupo Formation of the Triassic in Xuanhan area, with burial depth of about 3,000 m. The salt bearing strata in the study area are a set of evaporites (carbonate-sulfate-chloride combination), which is mainly composed of dolomite, calcareous dolomite, dolomitic limestone, anhydrite and halite, mixed with polyhalite. A small amount of magnesite is seen, and the cumulative thickness is about 300 m.

Results

Profile of salt bearing series. The PPD was developed in the salt forming period of T_1j^5–T_2l^{1-1}. The salt bearing profile consists of carbonate rock, sulfate rock, chloride rock (containing polyhalite), sulfate rock and carbonate rock from the bottom to the top, forming a complete salt bearing profile (Fig. 2). It reflects the deposition process of the salt formation from desalination to salinization, and then gradually desalination. Salt bearing series was developed completely, which is conducive to the preservation of soluble salts.

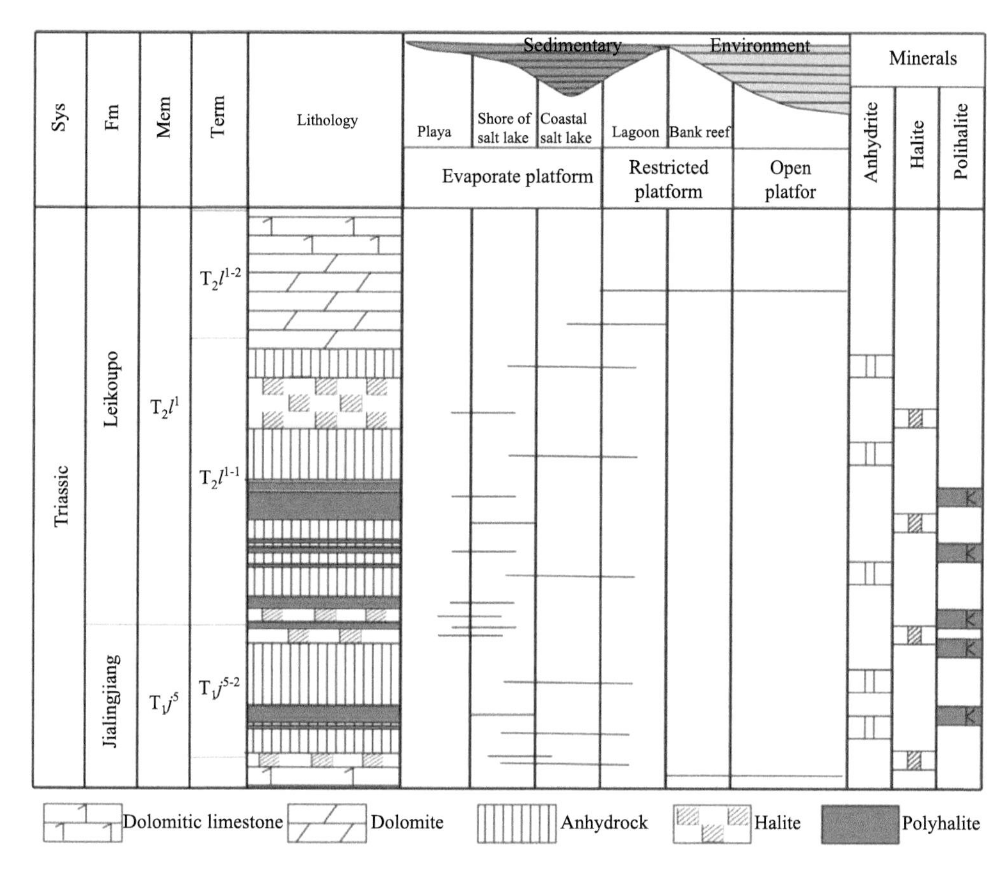

Fig. 2 Profile of polyhalite type potash beds in ZK601.

Cross section characteristics of PPD. According to the results of core drilling and logging, combined with the results from identification of salt bearing series through formation logging, this paper identifies several new type polyhalite potash beds encountered by potash exploration wells and oil and gas drilling wells in the study area and draws three well tie profiles perpendicular to and three ones parallel to the Huangjinkou anticline to depict the distribution of PPD in the study area.

The NW-SE trending section 1 passes through Wells Dawan 102, ZK001, Puguang 6, Puguang 5, Puguang 8 and Laojun 2 (Fig. 3) and is perpendicular to the axis of Huangjinkou anticline. Section 1 is in the middle of Huangjinkou anticline and it passes through the core and southeast wing of the anticline. In this section, the thickness of the first member of Leikoupo formation is basically stable, and the fourth and fifth member of Jialingjiang Formation shows obvious thickening anomaly in the core of anticline, which is caused by the compression and deflection of gypsum salt layer. The thickness of PPD decreases first and then increases from the core to the southeast wing of the anticline. The PPD at the top of T_1j^5 of each well is relatively stable. Multi-layer PPD is encountered by the lower part of Well Dawan-102 in the core of anticline. The PPD in the middle of Well Laojun 2 at the SE end of the section is thicker than that of other wells. The SW-NE trending section 2 passes through Wells ZK601, Puguang 6, Puguang 7 and Puguang 1 (Fig. 3). Section 2 is parallel to the axis of Huangjinkou anticline and is located at its southeast wing. In this section, the thickness of T_2l^1 and T_1j^{4+5} increases first and then decreases from south to north, and it is slightly thick in Well Puguang 7 in the middle and is small in Well ZK601 in the south because T_1j^4 is not drilled through. In this section, the PPD is relatively developed, and its thickness decreases then increases and decreases again from south to north. There are two sedimentary centers in both Well ZK601 and Well Puguang 7.

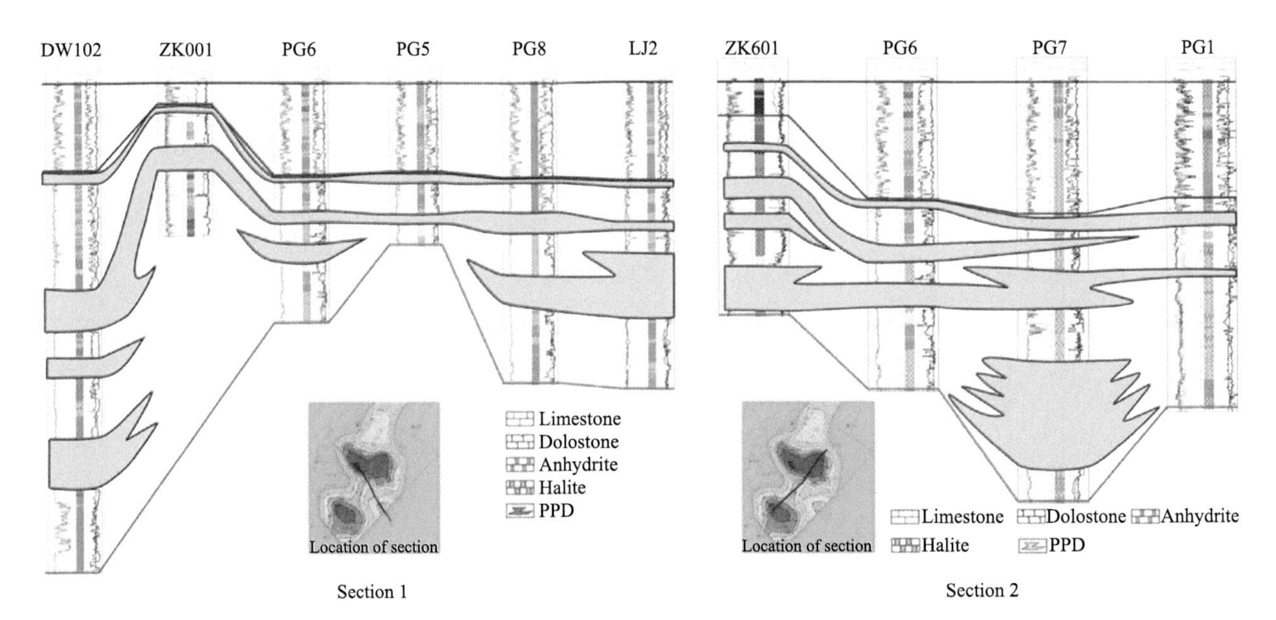

Fig. 3　Well tie profiles of polyhalite bearing halite in the Huangjinkou anticline in Northeast Sichuan Basin.

At the same time, the seismic and geological profile parallel to the core of the anticline is drawn through Wells ZK601, Pg 6, Pg 7 and Pg 1 (Fig. 4). The profile shows that the thickness of T_2l^1–T_1j^{4+5} of Huangjinkou anticline is structurally controlled, and the formation in the core of the anticline is obviously thickened. The thickness change of PPD is consistent with that of T_2l^1–T_1j^{4+5}.

Plane distribution of the PPD. According to the logging results of potash exploration wells and oil and gas wells in Huangjinkou anticline, the thickness data of gypsum salt rock and the thickness of PPD from the fourth member of Jialingjiang Formation to the first member of Leikoupo Formation in more than 30 wells were well collected, and the ratio of gypsum salt rock thickness to formation thickness was calculated. On this basis, the plane distribution showing the ratio of gypsum salt rock thickness to formation thickness of T_2l^1–T_1j^{4+5} of Huangjinkou anticline in Northeast Sichuan Basin [Fig. 5(a)] and plane distribution of PPD thickness [Fig. 5(b)] were plotted, and the research on the distribution law of PDD was conducted.

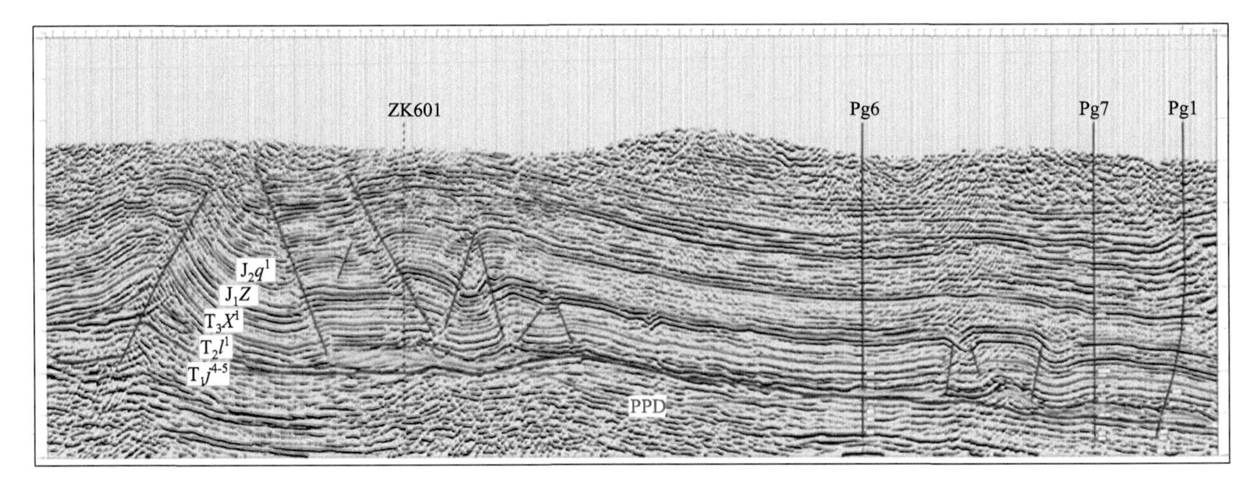

Fig. 4　Seismic and geological profile of Huangjinkou anticline in Northeast Sichuan Basin.

Partly, the ratio of gypsum salt rock thickness to formation thickness reflects the paleo-sedimentary environment, and the plane distribution of gypsum salt rock thickness indicates the paleo-evaporation center. The thickness of gypsum salt rock of T_2l^1–T_1j^{4+5} in the study area is between 31 and 378 m and is mainly in the range of 100–200 m. The ratio of gypsum salt rock thickness to formation thickness is 0.22–0.74, and three high value centers can be seen in the plane distribution, showing local increase and obvious lateral change. There is a certain correlation between Fig. 5(a) and Fig. 5(b). The positions of three high-value centers in Fig. 5(a) are basically consistent with those in Fig. 5(b). The three centers are Well ZK601 of Tuzhu-Liuchi structure in the southwest section of Huangjinkou anticline, Well102 of Maoba-Dawan structure in the northeast section and Well Puguang 8 of Laojunshan structure in the southeast wing of Huangjinkou anticline. The PPD thickness is between 1.3 m (Well Dawan 3) and 84.7 m (Well Dawan 102), and it changes obviously along the anticline dip, showing that it is controlled by the ancient evaporation center and structure.

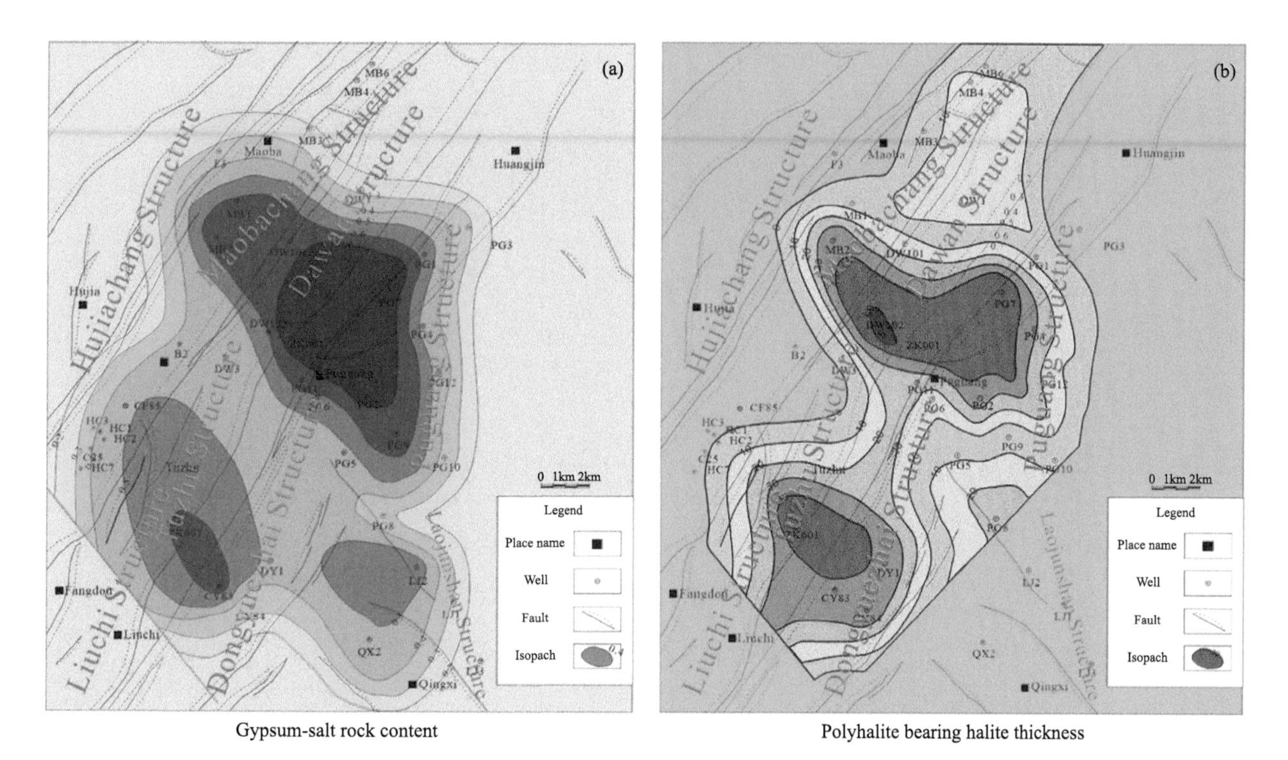

Fig. 5 (a) the ratio of gypsum salt rock thickness to formation thickness of T_2l^1–T_1j^{4+5} of Huangjinkou anticline in Northeast Sichuan Basin; (b) plane distribution of PPD thickness of T_2l^1–T_1j^{4+5} of Huangjinkou anticline in Northeast Sichuan Basin. Map generated using Geomap4.0 (https: //www.jurassic.com.cn/zh-cn/Products/Service/145)

Structure and mineral characteristics of PPD. The PPD is dominated by halite and polyhalite, with a small amount of anhydrite and magnesite. On the core, since more insoluble than halite, the polyhalite occurs as granular particles protruding on the surface of halite. In addition, due to the influence of drilling mud, the polyhalite is yellow-brown and thus easy to be identified. The matrix of PDD consists of light gray transparenttranslucent halite, which dominated by medium-coarse crystal texture. The silt-to boulder-sized polyhalite particles are scattered in the matrix. The polyhalite are mainly granular [Fig. 6(a)], with a few agglomerate and stratoid [Fig. 6(b)]. Polyhalites are mostly micro-fine-grained under the electron microscope. They are colorless, transparent, subhedral-xenomorphic granular, tabular, striped, and bedded (locally assembled ones), and are low relief in the systems of monopolarizer. The interference color order of these crystals are up to second blue, and the polysynthetic twins are common [Fig. 6(c), (d)] under perpendicular polarized light.

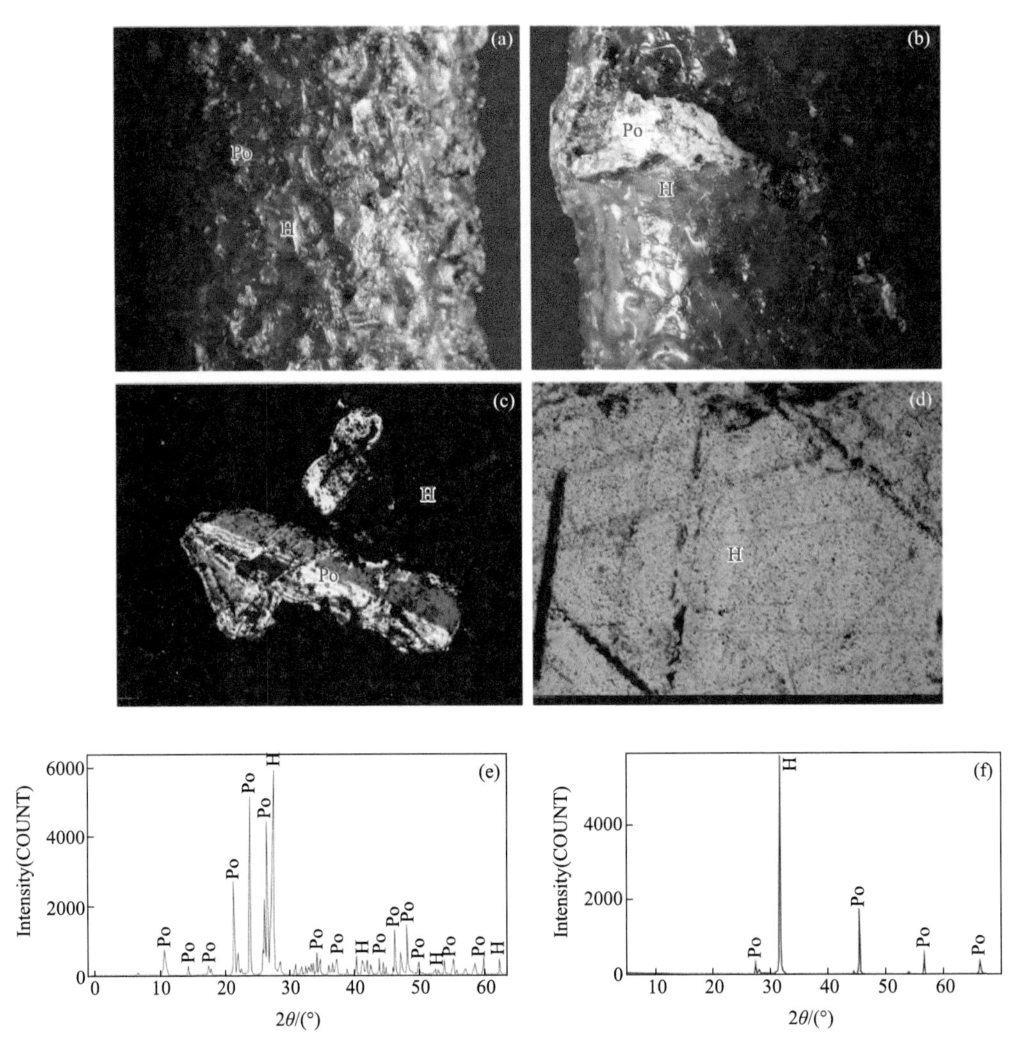

Fig. 6　Core section photograph, microscopic photograph and XRD graph of PPD.

(a, b) Photo of core section of polyhalite bearing halite; (c) microscopic photograph of polyhalite bearing halite, perpendicular polarized light 10×5; (d) microscopic photograph of polyhalite bearing halite, plane polarized light 10×5; (e, f) XRD graph of polyhalite bearing halite

In order to further clarify the types of minerals, X-ray powder diffraction (XRD) was used. The results of XRD analysis show that the sample is mainly composed of halite and polyhalite, and the content of halite and polyhalite in each sample varies, as shown in Fig. 6 (e), (f).

Chemical composition of PPD. A part of PPD samples from Wells ZK601 and ZK001 were analyzed (Table 1). The K^+ content is between 0.79% and 4.37%, with an average value of 1.79%. Some samples have the industrial grade of potassium. The Cl^- content is 34.78–67.32%, with an average value of 53.95%. The Br^- content is 52×10^{-6} to 276×10^{-6}, with an average value of 170×10^{-6}. The ratio of $Br\times10^3/Cl$ is 0.11–0.47, with an average value of 0.32.

Table 1　Chemical composition of polyhalite bearing halite samples

Sample number	Na^+ (%)	K^+ (%)	Ca^{2+} (%)	Mg^{2+} (%)	Cl^- (%)	SO_4^{2-} (%)	Br^- (10^{-6})	$Br\times10^3/Cl$
ZK601-01	33.78	1.14	3.95	0.37	39.16	11.44	121	0.31
ZK601-02	29.52	0.79	9.50	1.60	34.78	25.60	162	0.47
ZK601-07	17.15	1.79	4.39	0.53	58.36	17.77	193	0.33
ZK601-08	17.04	1.87	4.90	0.57	56.19	19.42	163	0.29
ZK601-09	17.78	1.70	4.42	0.53	58.43	17.12	181	0.31
ZK601-10	18.74	1.37	4.11	0.45	59.78	15.52	209	0.35
ZK601-11	18.71	1.96	3.22	0.61	59.98	15.50	192	0.32

Continued

Sample number	Na^+ (%)	K^+ (%)	Ca^{2+} (%)	Mg^{2+} (%)	Cl^- (%)	SO_4^{2-} (%)	Br^- (10^{-6})	Br×10^3/Cl
ZK601-12	21.09	1.25	2.62	0.39	63.65	10.99	255	0.40
ZK601-13	20.53	1.07	1.87	0.32	67.32	8.89	276	0.41
ZK601-23	23.49	4.37	5.97	1.50	39.91	24.76	156	0.39
ZK001-3	37.08	1.14	0.89	0.17	61.02	0.77	192	0.31
ZK001-4	34.31	1.55	1.66	0.54	55.89	6.03	147	0.26
ZK001-5	30.56	2.48	2.79	0.83	51.83	11.45	161	0.31
ZK001-6	33.45	1.47	2.25	0.64	56.92	5.27	132	0.23
ZK001-8	34.41	1.10	1.77	0.43	57.80	4.51	202	0.35
ZK001-9	31.36	2.83	2.68	0.88	49.47	12.82	52	0.11
ZK001-12	28.14	2.55	4.91	1.08	46.63	16.79	96	0.21

The REE characteristics of a PPD. The REE analysis results of some samples from Well ZK601 (Table 2) shows that the ∑REE is low, ranging from 9.24 to 63.54 μg/g, with an average of 34.17 μg/g. The ΣREE of samples is far lower than the average ∑REE of PAAS (184.77 μg/g), indicating that the PPD is rarely affected by terrigenous material in the process of deposition and diagenesis.

Table 2 REE data of the samples in Well ZK601

Sample number	La	Ce	Pr	Nd	Sm	Eu	Gd	Tb	Dy	Ho	Er	Tm	Yb	Lu	ΣREE	LREE/ HREE	$(La/Sm)_N$	$(Gd/Yb)_N$	δEu	δCe
H6	1.61	1.44	1.31	0.83	0.62	0.41	0.46	0.64	0.31	0.42	0.33	0.31	0.24	0.31	9.24	2.06	2.62	1.94	0.76	0.98
H7	3.55	3.02	2.54	2.00	1.23	0.82	0.89	0.85	0.62	0.56	0.62	0.62	0.57	0.62	18.51	2.46	2.88	1.55	0.77	0.99
H8	2.58	2.48	2.30	1.67	1.18	0.68	0.66	0.85	0.47	0.56	0.48	0.62	0.48	0.62	15.60	2.30	2.19	1.37	0.74	1.02
H9	1.61	1.62	1.39	1.00	0.82	0.41	0.42	0.43	0.31	0.42	0.33	0.31	0.33	0.31	9.72	2.39	1.97	1.27	0.66	1.08
H11	2.26	1.73	1.39	1.00	0.56	0.68	0.39	0.43	0.25	0.28	0.29	0.31	0.29	0.31	10.16	3.01	4.00	1.34	1.43	0.95
H12	5.16	3.91	2.87	1.83	0.92	0.54	0.97	1.06	0.84	0.84	0.86	0.93	0.91	0.93	22.57	2.08	5.59	1.06	0.58	0.97
H14	13.23	11.42	8.77	6.83	4.31	1.50	2.74	2.55	2.33	2.23	2.00	1.85	1.91	1.86	63.54	2.63	3.07	1.43	0.42	1.04
H17	9.35	8.03	6.64	5.33	3.18	1.36	1.81	1.49	1.15	1.11	1.14	1.23	1.24	1.24	44.33	3.25	2.94	1.46	0.54	1.00
H21	2.90	2.24	1.72	1.17	0.62	0.41	0.31	0.21	0.22	0.28	0.24	0.31	0.19	0.31	11.12	4.38	4.72	1.61	0.88	0.97
H22	7.10	5.63	5.08	3.33	1.69	0.54	1.00	0.85	0.84	0.70	0.67	0.62	0.62	0.62	29.30	3.95	4.19	1.61	0.40	0.92

The REE distribution model can be used to determine the source of diagenetic fluid[16]. The REE distribution model of sea water is characterized by enrichment of LREE and relative loss of HREE[17]. The values of ΣLREE/ΣHREE of samples range from 2.06 to 4.38, with an average of 3.26, showing the characteristics of distinct REE differentiation, LREE enrichment and HREE depletion, and this is similar to the case in seawater. $(La/Sm)_N$ reflects the fractionation degree of LREE elements, and $(Gd/Yb)_N$ reflects the fractionation degree of HREE. The $(La/Sm)_N$ ranges from 1.97 to 5.59, with an average of 4.10. The $(Gd/Yb)_N$ ranges from 1.06 to 1.94, with an average of 1.43. This shows that the fractionation is moderate for LREE, while not obvious for HREE, reflecting the deep water environment during deposition.

The REE are variable valence elements, especially Ce and Eu, which are prone to change in valence state in different redox environments[18]. In the oxidation environment, Ce^{3+} in water is easily oxidized to Ce^{4+}, resulting in the negative abnormality of Ce in sediment ($\delta Ce<1$). On the contrary, in the reduction environment, Ce^{3+} concentration in water increases, resulting in the positive anomaly of Ce in sediments[19]. In the hydrothermal fluid, the main form of Eu is Eu^{2+} under the condition of high temperature and reduction. When the hydrothermal fluid acts on the sediments, it will lead to the positive anomaly of Eu ($\delta Eu>1$)[20]. Therefore, the Ce and Eu contents in sedimentary rocks can be used to judge the redox condition of water body during sedimentation and whether there was transforming effect by deep fluids in the later stage of diagenesis. The δCe value of the PDD in Well ZK601 is between 0.92 and 1.08, with an average of 0.98, showing weak negative anomaly and reflecting the reduction environment during deposition; the δEu value is between 0.40 and 1.43, with an average of 0.56, showing a medium loss and indicating that there is no influence from hydrothermal fluids during deposition and diagenesis.

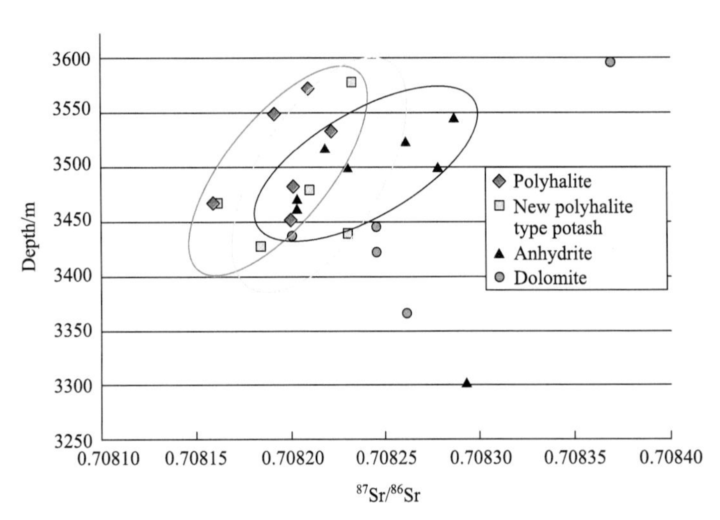

Fig. 7 $^{87}Sr/^{86}Sr$ of different kinds of lithologies in well ZK601.

Strontium isotope characteristics of PPD. The $^{87}Sr/^{86}Sr$ ration of seawater is a function of time in geological history. The longer geological time leads to more ^{87}Sr accumulation and higher $^{87}Sr/^{86}Sr$ ration. The $^{87}Sr/^{86}Sr$ ration is not fractionated by geochemical or evaporation processes and thus retains a fingerprint of its source[21, 22]. Samples of dolomite, anhydrite, polyhalite, and the PPD from salt-bearing strata were collected for strontium isotope test. The $^{87}Sr/^{86}Sr$ rations of samples, ranging from 0.708,16 to 0.708,37, are not significantly different. The $^{87}Sr/^{86}Sr$ rations of polyhalite, the PPD, and anhydrite increase with depth, while this increase trend is not obvious for dolomite (Fig. 7). Another strontium isotopic test was carried out on the manually selected halite and the polyhalite from the PPD at the depth of 3,467.5 m in Well ZK601. The test results are basically the same, with the data 0.708,160 and 0.708,161 respectively, indicating the same source.

The strontium isotopic data of samples were compared with those of Triassic sea water, mantle, crust and other typical domestic rock salt samples (Fig. 8). The strontium isotopic data of the salt bearing strata of Huangjinkou anticline are completely within the range of strontium isotopic data of the Triassic sea water and are far lower than those of the continental rock salt of Qaidam Basin and Dongpu Depression, which shows the marine source of the samples.

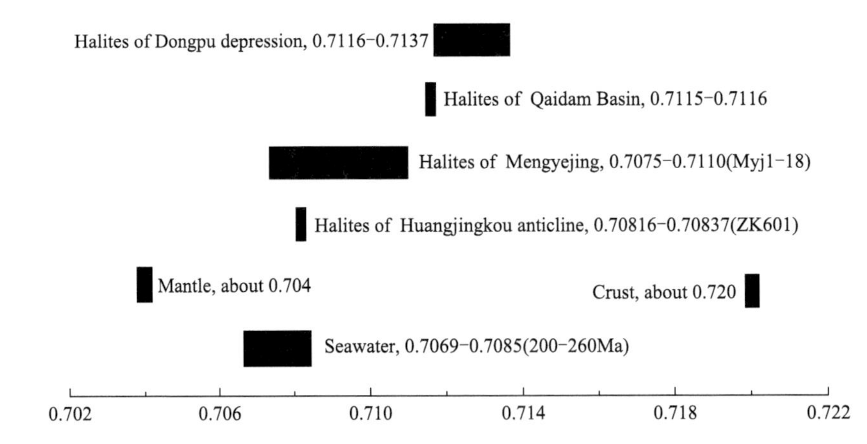

Fig. 8 Range of $^{87}Sr/^{86}Sr$ rations from different strontium sources (mantle and lithosphere data sourced from Kelts[29], seawater data from Burke[30], Hess et al.[31], halite samples in Qaidam Basin and Dongpu Depression are representative of the Land sources, from Tan[32] and Shi[33], Mengyejing data from Zheng et al.[34]).

Discussions

There are three main genesis model of polyhalite in Sichuan Basin: primary sedimentary genesis (the K^+ and

Mg^{2+} rich solution supplied by exotic Ca^{2+} or vice versa), penecontemporaneous metasomatism (the pre deposited gypsum or anhydrite replaced by further enriched brine rich in K^+ and Mg^{2+}), epigenetic metasomatism (gypsum or anhydrite replaced by brine rich in K^+ and Mg^{2+})[23-25].

According to the observation and analysis of core samples, the Triassic evaporite in Northeast Sichuan Basin is simply dominated by anhydrite, halite, polyhalite and magnesite. There are two main forms of polyhalite occurrence: layered, stratoid, lenticular polyhalite coexisting with anhydrite, and agglomerate, granular, dispersed polyhalite coexisting with halite. This paper mainly focuses on the latter—the PPD. Under the microscope, the polyhalites in the PPD are subhedral-xenomorphic granular, tabular, striped and do not coexist with anhydrite, which rules out the third genesis model (anhydrite replaced by brine rich in K^+ and Mg^{2+}).

In the process of evaporation, concentration and crystallization of seawater, bromine is mainly concentrated in the solution without forming a single mineral. However, due to the similar chemical properties of bromine and chloride ions, bromine will enter the chloride minerals in the form of isomorphism. In addition, the concentration of Br^- increases with the increase of the solution concentration, and the concentration of Br^- in the late-crystallized chloride is higher than that in the early-crystallized chloride.

Cheng et al. proposed that the $Br^- \times 10^{-3}/Cl^-$ is 0.11 when salt minerals precipitate after the evaporation and concentration of seawater, about 0.31 when the sylvite precipitates, and 0.45 when the carnallite crystallizes and precipitates[26]. The chemical analysis results of halite samples in the study area shows that the average value of $Br^- \times 10^{-3}/Cl^-$ corresponds to the stage of sylvite precipitation, and the value of $Br^- \times 10^{-3}/Cl^-$ of some samples corresponds to the stage of carnallite precipitation.

The REE content and distribution pattern can reflect the deposition environment of samples. The samples have relatively low total amount of REE and distinct REE differentiation, moderate LREE fractionation, inapparent HREE fractionation, and slight negative anomaly of Ce. This indicates that the Early-Middle Triassic deposition of the Huangjinkou anticline was in deep-water reduction environment and was rarely affected by terrigenous materials. The samples have distinct REE differentiation, enriched LREE, relatively deficient HREE, and the consistence with seawater in terms of REE distribution pattern, which indicates that the sediment source of the salt bearing strata should be marine. The medium loss of δEu indicates that there is no hydrothermal fluid action during deposition and diagenesis.

According to the previous research data, during the salt forming process of the fourth and fifth member of Jialingjiang Formation–the first member of Leikoupo Formation (T_1j^{4+5}–T_2l^1) in the study area, the platform was still connected with the sea water from time to time in the course of regression[27, 28].

According to the $Br^- \times 10^{-3}/Cl^-$ analysis of the PPD in the study area, the paleo-brine could make the deposition of potassium magnesium salt happen. However, in the evaporite minerals, the interbedding of anhydrite, halite, polyhalite and magnesite is common. This indicates that there was dilution by seawater intrusion in the salt forming process. In addition, it is generally believed that the formation of magnesite is related to seawater intrusion into the salt lake, and magnesite is generally contained in the evaporite in the study area, which further indicates that seawater intrusion occurred during the evaporation and concentration of paleo-brine.

Based on the above analysis, it is considered that the formation process of the PPD is as follows: in the stage of salt mineral deposition, the ancient seawater was further evaporated and concentrated. When the stage of sylvite deposition was not completely reached, the seawater rich in Ca^{2+} repeatedly intruded and mixed with seawater rich in K^+ and Mg^{2+}, and the polyhalite precipitated. Therefore, the PPD are primary deposits, which were structurally reformed later.

Conclusions

The PPD in Puguang area of Northeast Sichuan Basin was developed in the salt forming period of the fifth member of Jialingjiang Formation—the first submember of the first member of Leikoupo Formation (T_1j^5–$T_2l^{1\text{-}1}$). The completely salt bearing profile is composed of carbonate rock, sulfate rock, chloride rock, sulfate rock and carbonate rock from the bottom to the top. In PPD, the polyhalite is agglomerate, granular, dispersed and thin-layered, which is beneficial to water-soluble mining.

The PPD was formed before the complete deposition of sylvite. The polyhalite precipitated after the repeatedly intruded Ca^{2+} rich seawater mixed with seawater rich in K^+ and Mg^{2+}. Under the microscope, the polyhalites are subhedral-xenomorphic granular, tabular, and striped. PPD are primary deposits without metasomatism.

In Puguang area, Northeast Sichuan Basin, the PPD is thickly deposited, widely distributed and well preserved. The thickness distribution changes obviously along the anticline trend. The characteristics above show that the PPD is controlled by the ancient evaporation center and structure. The discovery of PPD has activated the polyhalite which has long been considered as a deep useless ore, and PPD has become a large-scale highquality potassium sulfate (K_2SO_4-$MgSO_4$) type potassium ore that can be economically utilized. It has great potential economic value and is expected to become a new strategic base of large-scale marine solid potassium ore in China. The PDD of the Triassic in Sichuan Basin will also become one of the main directions of marine potassium exploration in the future.

Sample collection and test. In this paper, core samples of Triassic PPD were collected from the potash exploration wells in Huangjinkou anticline. The samples were analyzed for petrology and mineralogy, using X-ray powder diffraction, chemical analysis, REEs analysis and strontium isotope test.

The X-ray powder diffraction was completed in the Research Center of Oil and Gas Resources, Northwest Institute of Ecological Environment and Resources, Chinese Academy of Sciences. The REEs test was carried out in Aoshi Analysis and Test (Guangzhou) Co., Ltd., whose laboratory has the test qualification, many years of experimental experience, and the analysis results are reliable.

Method ME-MS61 was used for LA, CE and Y. The results were obtained by four-acid digestion and mass spectrometer based quantitative analysis. Method ME-MS81 was used for other elements. The results were obtained by $LiBO_2$ melting and mass spectrometer based quantitative analysis. MC-ICP-MS (multi-collector inductively coupled plasma mass spectrometer) was used for strontium isotope test. Chemical pretreatment and mass spectrometry was completed in Nanjing FocuMS Technology Co. Ltd. After the rock powder was digested by high-pressure sealed dissolution pellet, Sr was separated from the digestion solution by cation strontium specific resin. In the determination of Sr isotope ratio, $^{86}Sr/^{88}Sr = 0.1194$ was used to calibrate the mass fractionation of the instrument. NIST SRM 987, an international standard for Sr isotopes, was used as an external standard to calibrate the drift of the instrument.

Acknowledgements

This paper was jointly funded by subject of National Key R&D Program of China (subject No. 2017YFC0602806), the Geological Survey Secondary Projects of CGS of MNR (No. DD20190172, Comprehensive Investigation and Evaluation of Lithium and Potash Resources in the Northeastern Sichuan Basin, China), the Sichuan Provincial Geological Prospecting Fund Project (Sichuan Province Xuanhan County Huangjinkou potash general survey) and China Postdoctoral Science Foundation (No. 2017M620885). Thanks to the helpful suggestions from Liu Xifang, Wang Zhengtao from the Institute of Mineral Resources, Chinese Academy of Geological Sciences, and Liu Zhu, Wang Fuming, Pang Bo from Sichuan geology and mineral exploration and Development Bureau of four O five geological team.

References

[1] Wu, Y. L., Zhu, Z. F. & Wang, J. L. Paleogeography of Triassic sedimentary rocks in southwest platform area. *Qinghai Geol.* 3, 126-137 (1983).

[2] Chen, J. Z. Discussion on formation of potassium mineral of Middle and Lower Triassic in the Sichuan Basin. *Geol. Chem. Miner.* 2, 100-103 (1990).

[3] Li, Y. W., Cai, K. Q. & Han, W. T. Origin of potassium-rich brine and the metamorphism of Triassic evaporites in Sichuan Basin. *Geoscience.* 12 (2), 73-79 (1998).

[4] Zheng, M. P. et al. Regional distribution and prospects of potash in China. *Acta Geol. Sin.* 84 (11), 1523-1553 (2010).

[5] Liu, W. Geochemical characteristics of Potassium-rich brine reservoir in the Lower-Middle Trias series of Sichuan Basin. Supervisor: Zhang, C. J., Xu, Z. Q. Chengdu: Chengdu University of Technology. 28-35 (2012) (in Chinese with English abstract).

[6] Li, Z. Geochemical characteristics of potassium-rich brine in Sichuan Basin and its genesis. Supervisor: Zhang, C. J., Xu, Z. Q. Chengdu: Chengdu University of Technology. 7-9 (2014).

[7] Zhen, M. P. *et al.* The leading role of salt formation theory in the breakthroughand important progress in potash deposit prospecting. *Acta Geosci. Sin.* 35 (02), 129-139 (2015).

[8] Han, W. T., Gu, S. Q. & Cai, K. Q. Study on formation conditions of Polyhalite in (K^+, Na^+, Mg^{2+}, Ca^{2+}/Cl^- and SO_4^{2-})-H_2O. *Chin. Sci. Bull.* 1, 362-365 (1982).

[9] Li, Y. W. & Han, W. T. Experimental study on formation conditions of Triassic Parahaline in Sichuan Basin. *Geoscience.* 1 (3-4), 400-411 (1987).

[10] Lin, Y. T. On K_bearing property of the marine Triassic and search for potash salt in Sichuan Basin. *Acta Geol.* 14 (2), 111-121 (1998).

[11] Wang, S. L. & Zheng, M. P. Discovery of Triassic polyhalite in Changshou area of East Sichuan Basin and its genetic study. *Miner. Depos.* 33 (5), 1045-1056 (2014).

[12] Zhong, J. A. et al. Sedimentary characteristics of Deep Polyhalite in HuangJingKou anticline of Northeast Sichuan and its genetic study. *Miner. Depos.* 37 (01), 81-90 (2018).

[13] Pan, Z. H. Genesis of triassic meso-lower mesophane in Nongle, Qu County Sichuan Province. *Build. Mater. Geol.* 1, 6-10 (1988).

[14] Huang, X. Z. & Chen, B. Y. The formation and preservation conditions of the polyhalite deposit in Nongle China. *Nonmetal. Miner. Ind.* 2, 15-19 (1991).

[15] An, L. Y., Yin, H. A. & Tang, M. L. Feasibility study of leaching mining of deeply buried polyhalite and kinetic leaching modeling. *Acta Geol. Sin.* 84 (11), 1690-1696 (2010).

[16] Liang, S. Y., Chen, Y. B., Zhao, G. W., Wang, Y. Q. & Hu, Y. Geochemical characteristics of REEss and their geological significance in the fourth member of the Middle Triassic Leikoupo Formation in Western Sichuan Basin. *Pet. Geol. Exp.* 39 (01), 94-98 (2017).

[17] Bau, M. Rare-earth element mobility during hydrothermal and metamorphic fluid-rock interaction and the significance of the oxidation state of europium. *Chem. Geol.* 93 (3-4), 219-230 (1991).

[18] Murray, R. W., Buchholtz, M. R. & Jones, D. L. REEss as indicators of different marine depositional environment in chert and shale. *Geology* 18 (3), 268-271 (1990).

[19] Hu, Z. et al. Geochemical characteristics of REEss of Huanglong formation dolomites reservoirs in Eastern Sichuan-Northern Chongqing Area. *Acta Geol. Sin.* 83 (6), 782-790 (2009).

[20] Yu, X. Y., Li, P. P., Zou, H. Y., Wang, G. W. & Zhang, Y. REEs geochemistry of dolostones and its indicative significance of the Permian Changxing formation in Yuanba Gasfield Northern Sichuan Basin. *J. Palaeogeogr.* 17 (3), 309-320 (2015).

[21] Miller, E. K., Blum, J. D. & Friedland, A. J. Determiantion of soil exchangeable-cation loss and weathering rates using Sr isotopes. *Nature* 362, 438-441 (1993).

[22] Bailey, S. W., Hornbeck, J. W., Driscoll, C. T. & Gaudette, H. E. Calcium inputs and transport in a base-poor forest ecosystem as interpreted by Sr isotopes. *Water Resour.* Res. 32, 707-719 (1996).

[23] Valyashk, M. G. *Geochemical regularity for the formation of potash deposits* 144 (China Industrial Press, Beijing, 1965).

[24] Peryt, T. M., Pierre, C. & Gryniv, S. P. Origin of polyhalite deposits in the Zechstein (Upper Permian) Zdrada Platform (Northern Poland). *Sedimentology* 45 (3), 565-578 (1998).

[25] Niu, X. Solid Potash mineral characteristics and Genesis Mechanism studies in the Bieletan section of Charhan salt lake. *Chengdu University of Technology* (2014).

[26] Cheng, H. D., Ma, H. Z., Tan, H. B., Xu, J. X. & Zhang, X. Y. Geochemical characteristics of bromide in potassium deposits: review and research perspectives. *Bull. Mineral. Pet. Geochem.* 27 (4), 399-408 (2008).

[27] Zhao, Y. J. et al. The Luzhou-Kaijiang paleouplift control on the formation environments of Triassic salt and potassium of deposits in Eastern Sichuan. *Acta Geol. Sin.* 89 (11), 1983-1989 (2015).

[28] Gong, D. X. *The triassic salt-forming environment, potash-forming conditions and genetic mechanism in Sichuan Basin* (Yi Haisheng. Chengdu University of Technology, Supervisor, 2016).

[29] Kelts, K. Lacaustrine basin analysis and correlation by strontium isotope stratigraphy. in *Abstract of 13th International Sedimentary* (1987).

[30] Burke, W. H. et al. Variation of seawater 87sr/86sr throughout phanerozoic time. *Geology* 10 (10), 516-519 (1982).

[31] Hess, J., Bender, M. L. & Schilling, J. Evolution of ratio of strontium-87 to strontium-86 in seawater from cretaceous to present. *Science* 231 (4741), 979-984 (1986).

[32] Tan, H. B., Ma, H. Z., Li, B. K., Zhang, X. Y. & Xiao, Y. K. Strontium and boron isotopic constraint on the marine origin of the Khammuane Potash deposits in Southeastern Laos. *Chin. Sci. Bull.* 55 (27-28), 3181-3188 (2010).

[33] Shi, Z. S., Chen, K. Y. & He, S. Strotium, sulfur and oxygen isotopic compositions and significance of paleoenvironment of paleogene of Dongpu depression. *Earth Sci. J. China Univ. Geosci.* 30 (04), 430-436 (2005).

[34] Zheng, Z. J., Yin, H. W., Zhang, Z. Z., Zheng, M. P. & Yang, J. X. Strontium isotope characeristics and the origin of salt deposits in Mengyejing, Yunnan Province SW, China. *J. Nanjing Univ. Nat. Sci.* 48 (06), 719-727 (2012).

Distribution of trace elements, Sr-C isotopes, and sedimentary characteristics as paleoenvironmental indicator of the Late Permian Linxi Formation in the Linxi Area, eastern Inner Mongolia*

Linlin Wang[1], Yongsheng Zhang[2], Enyuan Xing[2], Yuan Peng[2] and Dongdong Yu[2]

1. Petroleum Exploration & Production Research Institute, SINOPEC, Beijing 100083, China
2. Institute of Mineral Resources, CAGS, Beijing 100037, China

The Late Permian on the periphery of the Songliao Basin, eastern Inner Mongolia, is an important hydrocarbon source rock system. Its sedimentary environment plays an important role in the evaluation of hydrocarbon prospects in the area. Unfortunately, until now, the interpretation of the sedimentary environment of this area has been controversial. We investigated the Late Permian sedimentary environment by studying the sedimentary characteristics and geochemistry. Based on these investigations, we conclude that the Linxi Formation is mainly composed of clastic sediments, interbedded with limestone lenses, with bioherm limestone at the top of the formation. Inner-layer marine fossils (calcium algae, bryozoans, and sponges) and freshwater and blackish water microfossils (bivalves) are all present, indicative of a typical shallow water sedimentary environment with an open and concussion background. In terms of geochemistry, the formation is relatively light rare Earth enriched, with significant positive Eu anomaly, slight positive La and Y anomaly, weak positive Gd anomaly, and lack of Ce anomaly. The average B/Ga ratio of the mudstone is greater than 3.3, and the average Sr/Ba ratio of the limestone is greater than 1.0. The range of the $^{87}Sr/^{86}Sr$ ratio is from 0.707285 to 0.707953. The range of $\delta^{13}C$ values is from −4.0‰ to 2.4‰. The sediment assemblages, rare Earth elements, trace elements, and $^{87}Sr/^{86}Sr$ and C isotopes of the formation indicate that the Linxi Formation formed in a marine sedimentary environment and occasional marine-terrestrial transitional facies. The formation can be further divided into littoral facies, neritic facies, bathyal facies, and delta front.

1 Introduction

The eastern part of Inner Mongolia in China is located on the junction of the Sino-Korea Plate and the Siberian Plate. During the Permian period, this area experienced an important tectonic transition, and a complex tectonic paleogeographic environment was formed within the sedimentary basin. Since the 20th century, different interpretations have been put forth concerning the paleogeographic environment of the Linxi area of eastern Inner Mongolia during the Late Permian. He et al. interpreted the sedimentary environment of the Linxi Formation, according to the lithology, sedimentary structure, and paleontology of an exposed section in Linxi County. They concluded that the middle-lower parts of the Linxi Formation deposited in the marine environment and the upper part of the Linxi Formation deposited in the terrestrial lake environment[1]. Zhang et al. and Tian et al. concluded that the Linxi Formation formed in a marine sedimentary environment, based on the presence of marine fauna and bioherms in the exposed sections of the Late Permian Linxi Formation in the Linxi area[2, 3]. The other interpretation was that the Late Permian Linxi Formation deposited in terrestrial lake sedimentary environment, according the development of the freshwater bivalve fossils and Angara flora of the Late Permian Linxi Formation in the Linxi area[4-6].

The Late Paleozoic Permian strata in eastern Inner Mongolia are a potential hydrocarbon (oil and gas)

* 本文发表在：Journal of Chemistry, 2020. https: //doi. org/10.1155/2020/7027631

resource on the periphery of Songliao Basin. Clarifying the evolution of the sedimentary environment of the strata is important to locate an Upper Paleozoic hydrocarbon generation horizon and predict hydrocarbon prospects in the area. Meanwhile, the lithofacies of the Linxi Formation record the evolution from filling to closing of the Xing-Meng Trough, which is located on the boundary of the Sino-Korea and Siberian Plates. Although previous studies formed some conclusions concerning the Permian sedimentary environment of eastern Inner Mongolia, especially in the Linxi area, some issues still require clarification. This is mainly due to the fact that the fossil types are not consistent throughout the formation and that determining the sedimentary environment of a complete set of formations based on environmentally meaningful typical fossils, sedimentary structures, or bioconstruction has its limitations. This research study investigates the Late Permian Linxi Formation's sedimentary environment, by analyzing the sedimentary characteristics and the geochemical characteristics of the Linxi Formation in the Linxi area, eastern Inner Mongolia.

2 Regional geological survey

2.1 Structure distribution

The eastern part of Inner Mongolia is located in the Xing-meng orogenic zone. It is bordered in the west and north by the Mongolia-Okhotsk suture zone, which is adjacent to Mongolia and Russia. In the south, it is bordered by the Xar Moron River-Yanji suture zone, which is adjacent to the North China plate. In the east, it is bordered by the Central Sikhote subduction belt near the Pacific Plate. The Linxi area is located east of the Xing-meng orogenic belt and north of the Xar Moron River fault in the area where the Sino-Korea and Siberian Plates meet [Fig. 1 (a)]. The Linxi Formation is mainly distributed in the Keshiketeng Banner and Linxi County. It is an important Permian lithostratigraphic unit in the Upper Permian strata of eastern Xing-meng. The underlying Middle Permian Wujiatun Formation is composed of marine carbonates, volcanic rocks, and clastic sediment, and the overlaying Lower Triassic Xingfulu Formation is composed of terrestrial clastic sediments[7].

2.2 Lithologic characteristics

The Late Permian Linxi Formation can be divided into 5 sections, which we call sections Lin1–5. Section Lin1 is composed of medium-thick gray and gray-green medium-fine-grained sandstone, thin-bedded siltstone, and interbedded siltstone (A1 and A2 in Fig. 2), which is locally interbedded with thin-medium calcarenite or limestone lenses. Section Lin2 is composed of dark gray, grayish-black shale, silty mudstone, and silty slate interbedded with gray-green fine sandstone, and it locally contains visible slump breccia and limestone lenses (A3 and A4 in Fig. 2). Section Lin3 is composed of medium-thick light brown medium-coarse grained sandstone interbedded with thin gray, gray-brown siltstone, argillaceous siltstone, and silty mudstone and local areas of medium-thick gravel coarse grained sandstone and limestone (A5 in Fig. 2). The entire section from bottom to top evolved from sandstone and pebbled sandstone to middle-thick sandstone and upper interbedded sandstone and mudstone. Section Lin4 is composed of dark gray, gray-black shale, silty mudstone, and argillaceous siltstones interbedded with thick gray-green, gray fine grained sandstone, thin-medium limestone, and limestone lenses (A6 and A7 in Fig. 2). The lower part of this section is primarily composed of argillaceous siltstone and silty mudstone. The middle part of the section is dominated by shale, while the upper part of the section is composed of interbedded sandstone and mudstone interbedded with limestone lenses. Section Lin5 is composed of gray argillaceous siltstone, sandy mudstone, light gray medium-fine grained sandstone, tuffaceous sandstone, massive gray bioherm limestone, and framestone (A8 in Fig. 2).

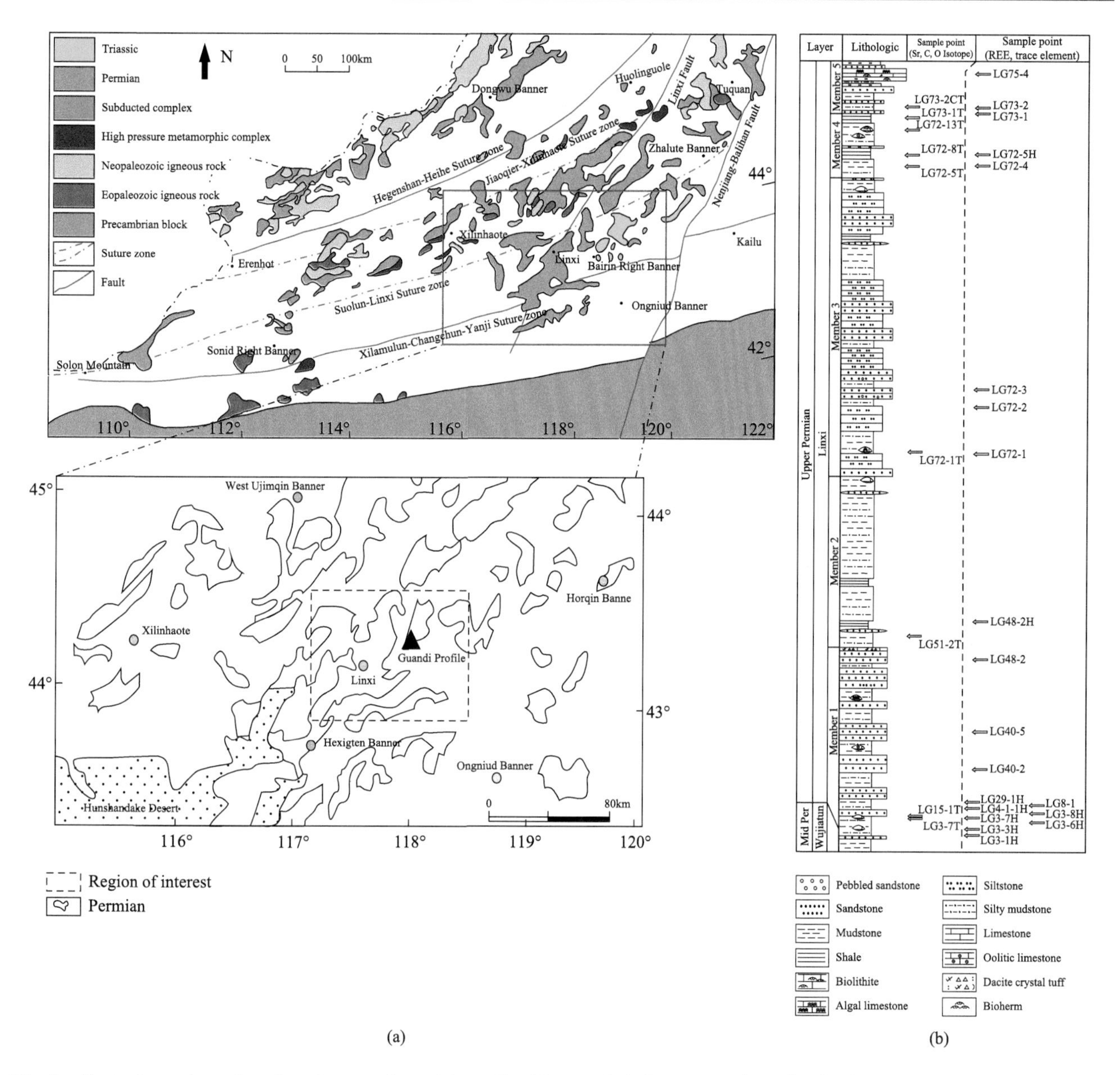

Fig. 1 Tectonics and stratigraphy of eastern Inner Mongolia. (a) tectonic belt, stratum plane distribution, and the location of Guandi profile in the Linxi area; (b) the lithological section and sample distribution of Linxi Formation.

2.3 Sedimentary structures and paleontology fossils

The Linxi Formation contains many types of sedimentary structures and paleontology fossils. In the lower part of section Lin1, wedge-shaped cross bedding, tabular cross bedding, and symmetrical ripple bedding are present in the sandstones, while horizontal bedding (B1 in Fig. 2), current bedding, and small-scale cross bedding are present in the mudstone-siltstones and argillaceous siltstones. The upper part of section Lin1 sandstones contain parallel bedding, oblique bedding, small-scale cross bedding, and current bedding (B2 in Fig. 2), indicating a shallow water concussion environment, containing bivalves *Palaeanodonta* sp., *Palaeomutella* sp., *Anthraconaia* sp., and fragments of plant fossils. The lower part of section Lin2 contains wash-out structures (B3 in Fig. 2), locally developed slump structures, deformation bedding (B4 in Fig. 2), and small current bedding, indicating a relatively deep-water environment. The imbricated structure observed in the lower part of section Lin3 exhibits parallel bedding and tabular cross bedding (B6 in Fig. 2) in the sandstones and current bedding, concussion ripples, bioturbation structures (B5 in Fig. 2), and mud cracks in the mudstones (B7 in Fig. 2), which

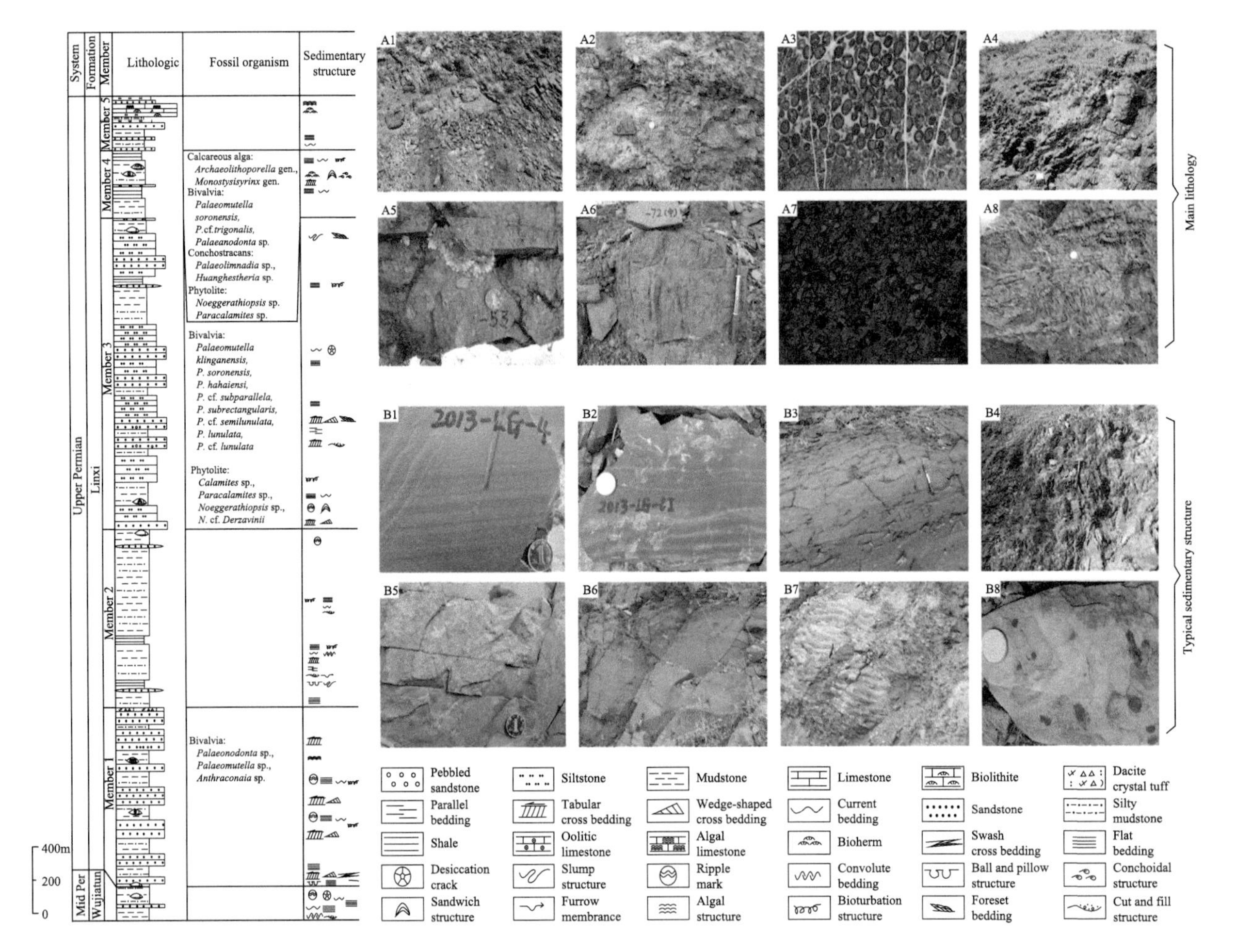

Fig. 2　Sedimentary characteristics of the Late Permian Linxi Formation in the Linxi area of eastern Inner Mongolia.

A1, gray mudstone, section Lin1. A2, gray silty mudstone interbedded with thin yellowish brown silty mudstone, section Lin1. A3, oolitic limestone, light crystal matrix, containing a small amount of oolitic spherulites. A4, slump conglomerate, indicating gravity flow, section Lin2. A5, gray gravel coarse grained sandstone, gravel size 0.2–1.0 cm, subround and subangular gravel that is mainly limestone, section Lin3. A6, laminated limestone with laminated columnar structures, section Lin4. A7, inner clastic granular limestone, particles are mainly inner debris, algae grains, and ooid, section Lin4. A8, framestone with an algae bonding structure, section Lin 5. B1, gray-green silty mudstone with horizontal bedding, section Lin1. B2, gray siltstone with current bedding, section Lin1. B3, lower dark gray mudstone with wash-out structures, unconformably in contact with upper gray sandstone, section Lin 2. B4, gray sandy mudstone with slump and deformation structures, section Lin3. B5, dark gray fine grained sandstone with bioturbation structures, section Lin3. B6, gray fine grained sandstone with tabular cross bedding, section Lin3. B7, gray argillaceous sandstone with ripple bedding and mud cracks, section Lin3. B8, light brown coarse grained gravel sandstone with concretionary structures, section Lin4

is indicative of an intermittently exposed shallow water environment. The middle and upper parts of section Lin3 exhibit lenticular foreset bedding and local contemporaneous slump structures and contain bivalve fossils (*Palaeodoronella soronensis, Palaeodonta* sp., *P.* cf. *trigonalis*, *P.* cf. *semilunulata*, and *P. hahaiensis*) and plant fossil debris, which is indicative of a delta front sedimentary environment. In section Lin4, low angle oblique bedding, graded bedding, concretionary structure (B8 in Fig. 2), and small-scale ripple bedding developed can be seen. The clastic rocks of section Lin4 contain conchostracas (*Huanghestheria* sp., *Palaeolimnadia* sp., and *Pemphicyclus* cf. *trochoides*), bivalve fossils (*Palaeanodonta* sp., *P.* cf. *trigonalis*, and *Palaeomutela soronensis*), plant fragments, and brachiopod fossil fragments. Bryozoans, sponges, and calcareous fossils, including *Monostysisyrinx* gen. and *Archaeolithoporella* gen., can be seen in the limestone, indicating an expansive coastal environment that may have mixed a terrestrial source. The lower part of section Lin5 exhibits wedge-shaped cross bedding and parallel bedding, while the middle and upper parts exhibit visible bioturbation structures, concretionary structures, low angle oblique bedding, and small-scale cross bedding. Bryozoans and sponge bone fossils are present in the limestone[7], indicating a neritic environment.

3 Samples and analytical procedures

The strata discussed in this paper and the samples used for geochemical analysis were primarily collected from the Guandi profile in Linxi County in eastern Inner Mongolia [Fig. 1 (b)]. A total of 33 samples were analyzed. All the samples are fresh and uncontaminated rocks by stripping a weathered layer. Observed under the microscope, the samples were without epigenic dikes or strong recrystallization. Of these samples, 4 of them were limestone of Early Triassic, which were analyzed for Sr-C isotopic ratios.24 of the samples were primarily composed of lenticular limestones from the Linxi Formation. They were analyzed for major, trace, and rare Earth elements and Sr-C isotopic ratios (Tables 1–3). Seven of the samples were mudstone, which were analyzed for major, trace, and rare Earth elements (REEs) (Table 1). The X-ray fluorescence spectroscopy (XRF) was used to determine the oxides of major elements. The LOI (loss on ignition) of the samples was recorded as weight loss after burning at 1000℃. Trace and rare Earth elements were determined by an Element 6000 inductively coupled plasma mass spectrometer (ICP-MS). Both of the tests were performed in the Chinese Academy of Geological Sciences National Geological Experimental Testing Center following the method described by Ryu et al.[15]. The Sr isotopes were analyzed using a British VG354 isotope mass spectrometer following the method described at the Nanjing Institute of Soil Science and Technology Service Center, Chinese Academy of Sciences. The measured value of $^{87}Sr/^{86}Sr$ for American isotope standard NBS987Sr was 0.710236 ± 7 ($n = 10$) with a precision of ±0.005%. The C isotopes were analyzed using MAT-252 isotope mass spectrometer with a precision of ±0.2‰ at Isotope Research Laboratory, Institute of Mineral Resources, Chinese Academy of Geological Sciences.

4 Results

4.1 Rare earth elements and trace elements

Trace element concentrations are reported in Table 1. REE + Y concentrations were normalized to the Post-Archaean Australian Shale (PAAS) composite[16]. The total REE content (ΣREE) of the sediments in the Linxi Formation ranges from 10.07 μg/g to 195.69 μg/g with an average of 76.56 μg/g. These samples of mudstones and lime mudstones with high REE are clearly contaminated by a considerable terrigenous debris component and perhaps masking the anomalies distribution of REE in marine environments. Thus, they are not used any further for palae-oenvironmental reconstructions. We mainly use the trace elements of the carbonates and argillaceous carbonates to discuss the palaeosedimentary environment of the Linxi Formation. Shale-normalised (SN) elemental anomalies were calculated on a linear scale, assuming that differences in concentration between neighbouring pairs are constant, as follows: $La/La^* = La/(3Pr–2Nd)_{PAAS}$, $Ce/Ce^* = Ce/(0.5La + 0.5Pr)_{PAAS}$, $Eu/Eu^* = Eu/ (0.5Sm + 0.5Gd)_{PAAS}$, and $Gd/Gd^* = Gd/(2Tb–Dy)_{PAAS}$.

The ΣREEs of the limestones are between 10.07 μg/g and 91.22 μg/g. Abbreviations are applied to light REE (LREE) and heavy REE (HREE). As can be seen from the PAAS normalized REE distribution pattern [Fig. 3 (a) and Table 1], the limestone samples of Linxi Formation in the Guandi profile of Linxi area display relatively uniform shale-normalised REE + Y patterns [Fig. 3 (a)]. The LREE depleted relative to the HREE with La_{SN}/Yb_{SN} from 0.48 to 0.91, and the limestones display a significant positive Eu anomaly ($(Eu/Eu^*)_{SN} = 1.78$). But significant differences can be noted between the limestones at the bottom of Lin1 and the other limestones of Linxi Formation. The limestones at the bottom of Lin1 show a slight positive La anomaly ($(La/La^*)_{SN} = 1.16$), lack of Ce, Gd, and Y anomaly ($(Ce/Ce^*)_{SN} = 0.99$), ($(Gd/Gd^*)_{SN} = 1.0$), and ($(Y/Ho)_{SN} = 1.07$). The other limestones of Lin1-5 show a distinct positive La anomaly ($(La/La^*)_{SN} = 1.6$), slight positive Y anomaly ($(Y/Ho)_{SN} = 1.25$), weak positive Gd anomaly ($(Gd/Gd^*)_{SN} = 1.08$), weak negative Ce anomaly ($(Ce/Ce^*)_{SN} = 0.93$, except sample LG 40-2), and lack of Ce anomaly ($(Ce/Ce^*)_{SN} = 0.95$) (Table 1).

Table 1 The trace element concentration (in ug/g) of the Linxi Formation in the Guandi profile

Number	LG 3-1H	LG 3-3H	LG 3-6H	LG 3-7H	LG 3-8H	LG 4-1-1H	LG 8-1	LG 29-1H	LG 40-2	LG 40-5	LG 48-2	LG 48-2H	LG 72-1	LG 72-2	LG 72-3	LG 72-4	LG 72-5H	LG 73-1	LG 73-2	LG 75-4
Lithology	Limestone	Limestone	Mudstone	Limestone	Lime mudstone	Mudstone	Bioclastic limestone	Mudstone	Bioclastic limestone	Bioclastic limestone	Lime mudstone	Lime mudstone	Bioclastic limestone	Bioclastic limestone	Bioclastic limestone	Bioclastic limestone	Bioclastic limestone	Bioclastic limestone	Bioclastic limestone	Mudstone
B	18.5	18	94.7	11.5	150	95.7	16.6	57.3	8.34	15.1	33.4	46.7	6.4	14.2	10.8	8.37	13.6	10	3.45	38.1
Ga	4.34	5.28	20.5	3.82	18.9	22.1	6.57	17.1	3.96	0.52	8.1	7.7	1.91	3.15	3.67	3.43	4.35	2.58	0.68	20.1
Rb	27.9	22.4	143	19.7	197	153	39	122	10.7	14.8	41.7	55.6	14.4	24.7	26.9	16.2	15.4	20.8	3.34	116
Sr	1193	950	82.7	1259	351	131	917	58.9	1172	486	692	626	1164	1449	1059	1465	871	1140	3052	250
Ba	166	157	442	70.6	519	461	109	591	49.5	22.2	224	355	86.2	150	204	133	129	95.1	219	581
La	5.39	5.85	33.7	3.38	29.6	37.3	8.14	38.4	7.37	1.95	15.1	14.3	8.96	9.12	7.44	13.4	18.1	7.38	3.74	34.9
Ce	11.8	12.1	70.2	7.3	62.3	77.2	15.9	78.5	15.9	4.06	31.7	25.3	15.4	16	14.3	22.3	35.7	14.1	8.13	60.1
Pr	1.37	1.31	8.15	0.78	7.2	8.98	1.92	9.35	1.96	0.4	3.81	2.99	1.69	1.87	1.51	2.48	4.12	1.52	0.97	7.2
Nd	5.44	5.13	32.2	3.24	26.9	35.7	7.68	37.3	8.42	1.73	15.4	12.6	6.77	8	6.29	9.8	17.1	6.33	4.86	28.1
Sm	1.21	1.16	6.39	0.68	5	7.2	1.55	7.25	2.11	0.37	3.4	2.72	1.54	1.75	1.33	2.18	3.43	1.48	1.41	5.91
Eu	0.43	0.91	1.19	0.31	1.45	1.26	0.53	1.35	0.85	0.13	1.07	0.97	0.54	0.83	0.48	0.74	0.76	0.54	0.73	1.26
Gd	1.29	1.17	6	0.78	5.1	7.17	1.49	6.79	2.41	0.37	3.76	2.83	1.78	1.97	1.39	2.57	3.41	1.68	1.82	5.75
Tb	0.2	0.19	0.97	0.12	0.85	1.09	0.23	1.08	0.38	0.06	0.64	0.44	0.28	0.29	0.21	0.39	0.57	0.26	0.25	0.84
Dy	1.1	0.98	5.61	0.7	4.99	6.1	1.38	5.96	2.22	0.37	3.62	2.75	1.75	1.64	1.33	2.45	3.2	1.6	1.41	5.24
Ho	0.23	0.19	1.15	0.14	1.04	1.24	0.28	1.27	0.45	0.08	0.77	0.58	0.38	0.33	0.27	0.51	0.66	0.35	0.26	1.11
Er	0.65	0.58	3.21	0.41	2.82	3.51	0.81	3.7	1.19	0.24	2.08	1.68	1.09	0.89	0.86	1.57	1.91	1.01	0.67	3.12
Tm	0.09	0.08	0.45	0.05	0.42	0.52	0.13	0.55	0.16	0.03	0.32	0.27	0.16	0.12	0.13	0.23	0.28	0.15	0.09	0.47
Yb	0.6	0.52	3.04	0.36	2.58	3.37	0.8	3.64	1.06	0.26	2.13	1.82	1	0.74	0.83	1.47	1.72	0.92	0.58	3.06
Lu	0.1	0.08	0.49	0.06	0.39	0.5	0.13	0.55	0.15	0.02	0.33	0.26	0.15	0.11	0.12	0.23	0.26	0.13	0.08	0.46
ΣREE	29.90	30.25	172.75	18.3	150.64	191.14	40.97	195.69	44.63	10.07	84.13	69.51	41.49	43.66	36.49	60.32	91.22	37.45	25.00	157.52
LREE	25.64	26.46	151.83	15.69	132.45	167.64	35.72	172.15	36.61	8.64	70.48	58.88	34.90	37.57	31.35	50.90	79.21	31.35	19.84	137.47
HREE	4.26	3.79	20.92	2.62	18.19	23.50	5.25	23.54	8.02	1.43	13.65	10.63	6.59	6.09	5.14	9.42	12.01	6.10	5.16	20.05
La_{SN}/Yb_{SN}	0.66	0.83	0.82	0.69	0.85	0.82	0.75	0.78	0.51	0.55	0.52	0.58	0.66	0.91	0.66	0.67	0.78	0.59	0.48	0.84
(La/La^*)	1.03	1.13	1.07	1.28	0.94	1.09	1.13	1.09	1.23	1.63	1.09	1.47	1.42	1.57	1.47	1.40	1.29	1.44	2.73	1.22
(Ce/Ce^*)	1.00	1.01	0.98	1.04	0.98	0.97	0.93	0.96	0.96	1.06	0.96	0.89	0.91	0.89	0.98	0.89	0.95	0.97	0.98	0.87
(Eu/Eu^*)	1.61	3.66	0.90	1.98	1.35	0.82	1.64	0.90	1.75	1.65	1.40	1.64	1.52	2.08	1.65	1.45	1.04	1.59	2.10	1.02
(Gd/Gd^*)	0.98	0.89	0.98	1.04	0.97	1.02	1.07	0.96	1.02	1.05	0.92	1.11	1.09	1.06	1.15	1.14	0.93	1.09	1.13	1.17
Y/Ho	26.61	31.53	24.87	25.29	24.33	25.00	31.54	24.72	30.89	34.38	25.32	28.97	37.11	36.06	34.15	33.73	27.88	35.43	38.46	27.21

La_{SN}/Yb_{SN} is the ratio of La and Yb by PAAS normalized.

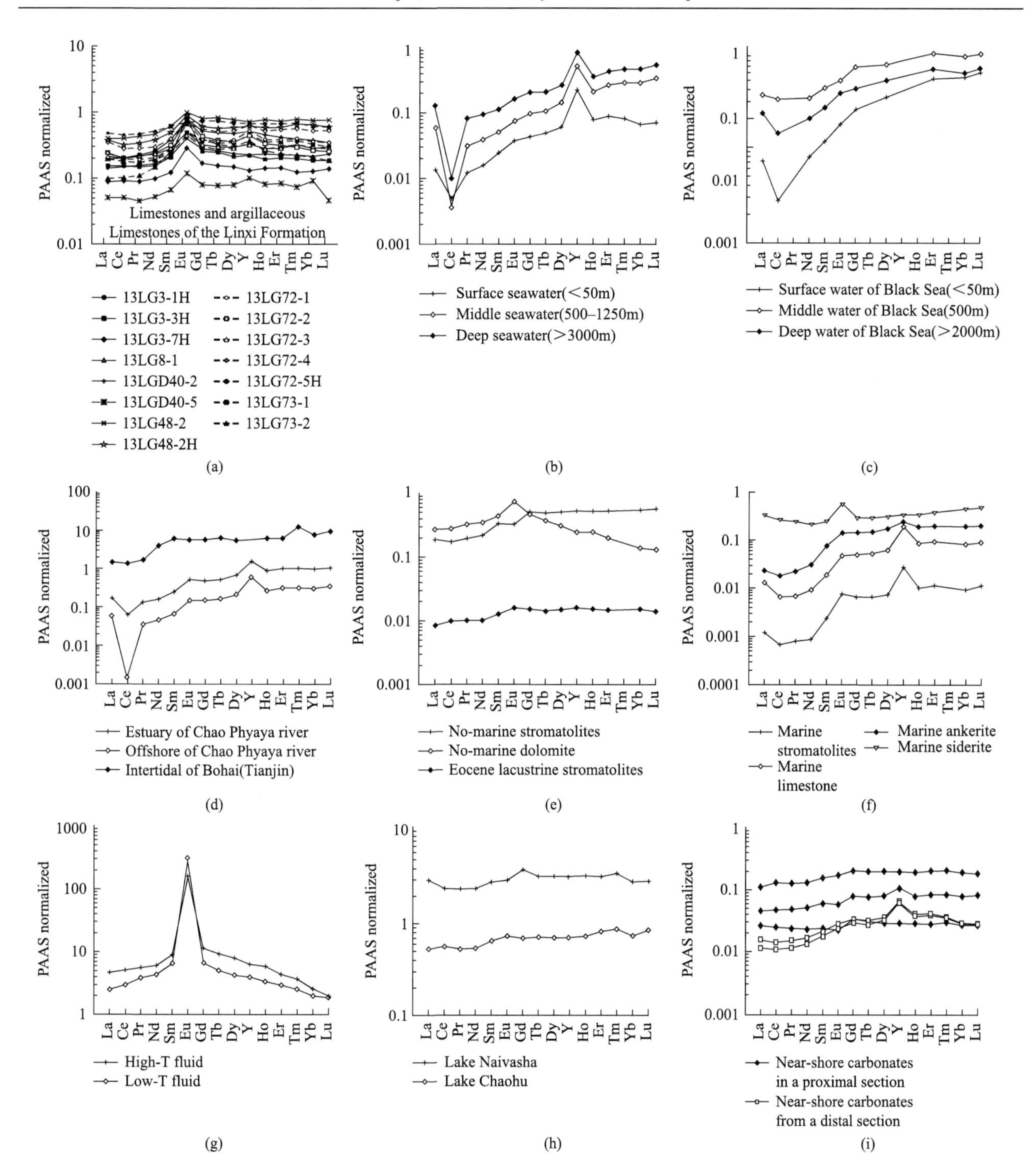

Fig. 3 PAAS standardized comparison chart of the REEs of the Linxi area Guandi profile sediments and the REE of other water bodies, rocks, and minerals.

the REEs of (a) limestone and mudstone in Linxi Formation, (b) seawater from southwestern Pacific[17], (c) seawater in an oxygen-poor environment of the Black Sea[18], (d) the Chao Phraya River estuarine area and distant end[19], seawater form intertidal of Bohai Sea[20], (e) nonmarine carbonate rocks[21], (f) marine carbonate rocks and marine carbonate minerals[22], (g) the high-T and low-T marine hydrothermal fluids form mid-Atlantic ridge[23], (h) water from Lake Naivasha and Chaohu Lake[24, 25], and (i) near-shore carbonates of the Bloeddrif Member, lower Holgat Formation from a proximal section south of the Kuboos Pluton, and a distal section north of the Kuboos Pluton[26]

In addition to REE analysis, other trace element (Sr, Ba, B, and Ga) analyses were conducted on the limestones and mudstones from the Guandi profile in the Linxi area (Table 1). In the Linxi Formation, the Sr content ranges from 58.9 μg/g to 3052 μg/g with an average of 918.43 μg/g. The Ba content ranges from 22.2 μg/g to 591 μg/g with an average of 238.18 μg/g. The Sr/Ba ratios of the limestone range from 1.76 to 23.68 with an average of 8.18. The B content of the mudstones ranges from 18 μg/g to 150 μg/g with an average of 57.21 μg/g, which is enriched compared to the average B content of the crust (7.6 μg/g). The Ga content ranges from 4.34 μg/g to 22.1 μg/g with an average of

12.95 μg/g. The B/Ga ratios of the mudstone range from 1.90 to 7.94 with an average of 4.42.

4.2 Sr and C isotope distribution

The whole rock total Sr content and Sr isotopic data for the Linlin Formation in the Linxi area are presented in Table 2. The C isotopic data for the Linlin Formation are presented in Table 3. The results show that the Sr contents of all nine limestone samples in the Linxi Formation are greater than 850.0 μg/g with a maximum value of 2762.0 μg/g and an average of 1435.1 μg/g. The $^{87}Sr/^{86}Sr$ values of the limestone samples range from 0.707285 to 0.707953 with an average of 0.707624. The $\delta^{13}C$ values of the limestone samples vary from −4.0‰ to 2.4‰, and the average is −1.3‰. As shown in Figure 4, there is no correlation between Sr-C isotope and Mn/Sr, and the Mn/Sr ratio reaches the upper limit of 2–3 for qualified samples recommended by Kaufman et al.[27]. Seven samples have low Mg/Ca ratios (<0.1) and may be presumed to be unaltered. The Sr contents of all of the limestone samples are significantly larger than the minimum Sr content (400 μg/g) suggested by Korte et al. in their Sr isotope deposition studies[28]. In addition, there is no correlation between $^{87}Sr/^{86}Sr$ and Si (representing SiO_2 content) [Fig. 4 (b)], indicating that the Sr isotopic composition is completely derived from the endogenous sedimentary limestone and is not affected by other Si-bearing terrestrial debris and smaller diagenesis events. This indicates that the Sr-C isotopes in the limestone represent the Sr-C isotopes of the sedimentary fluids.

Table 2 Sr isotope values of the limestone from the Linxi area Guandi profile and compared with other parts of the world

$^{87}Sr/^{86}Sr$ (Linxi Formation of Permian in Linxi area, Inner Mongolia)									
Number	LG3-7T	LG15-1T	LG51-2T	LG72-1T	LG72-5T	LG72-8T	LG72-13T	LG73-1T	LG73-2CT
Lithology	Limestone	Oomicrite	Stromatolitic limestone	Limestone	Bioclastic limestone	Bioclastic limestone	Stromatolitic limestone	Limestone	Bioclastic limestone
SiO_2 (%)	13.61	3.11	14.02	18.76	18.48	12.9	11.05	8.93	13.61
MnO (%)	0.30	0.51	0.30	0.23	0.21	0.42	0.51	0.28	0.30
Sr (ug/g)	1275	1639	1604	930.7	859.5	1123	1225	1498	2762
$^{87}Sr/^{86}Sr$	0.707719	0.707593	0.707684	0.707526	0.707758	0.707953	0.707297	0.707802	0.707285
$^{87}Sr/^{86}Sr$ (lacustrine carbonate rocks of Triassic in Linxi area, Inner Mongolia)									
Number	BLS-1	BLS-2	BLS-3	BLS-4					
$^{87}Sr/^{86}Sr$	0.713605	0.714335	0.711949	0.714164					
$^{87}Sr/^{86}Sr$ (marine carbonate and conodonts of Permian/Triassic in the other places of the world)									
Marine limestone of Permian in South China[1]	0.70732	0.70757	0.7076	0.70754	0.70753	0.70757	0.70734	0.70744	0.70744
	0.70741	0.70744	0.70742	0.70745	0.70739	0.70740	0.70749	0.70747	0.70750
	0.70757	0.70758	0.70748	0.70752	0.70755	0.70722	0.70739	0.70767	
Conodonts from Permian/Triassic boundary in Abadeh, Iran[2]	0.707038	0.707079	0.707092	0.707218	0.707225	0.707219	0.707266	0.707392	0.70739
	0.706923	0.706970	0.707010	0.707006	0.70699	0.706945	0.707050	0.707138	0.707046
Marine carbonate of late Permian in upper Yangtze region[3]	0.70743	0.70742	0.70684	0.70697	0.70722	0.70672	0.70755	0.70725	0.70752
Conodonts of Permian/Triassic, United States[4]	0.708111	0.708132	0.708878	0.708841	0.707957	0.708011	0.708093	0.708345	0.707605
	0.707453	0.707629	0.707643	0.707561	0.707506	0.707513	0.707416	0.707461	0.707336
	0.707341	0.707333	0.707388	0.707331	0.707229	0.706914	0.707031	0.706948	0.707247
	0.707406	0.707059	0.707027	0.706913	0.70734	0.707404	0.707373	0.707133	0.707208
	0.707406	0.707059	0.707027	0.706913	0.70734	0.707404	0.707373	0.707133	0.707208
Conodonts of Permian/Triassic, Pakistan[5]	0.707211	0.707388	0.707117	0.70719	0.707168	0.707200	0.707100	0.707102	0.707117
	0.707186	0.707240	0.707130	0.707399	0.707690	0.707322	0.707427	0.707736	0.707499
	0.707568	0.707458	0.707587	0.707469	0.707440	0.707503	0.707308	0.707626	0.707856

1. Data from Tian and Zheng[8]; 2. data come Korte et al. [9, 10]; 3. data come Huang et al. [11]; 4, 5. data from Martin and Macdougall[12].

Table 3 Carbon isotopic data of Permian-Early Triassic carbonate rocks in Linxi area and other regions

C isotope (Linxi Formation of Permian in Linxi area, Inner Mongolia, China)									
Sample number	LG3-7T	LG15-1T	LG51-2T	LG72-1T	LG72-5T	LG72-8T	LG72-13T	LG73-1T	LG73-2CT
$\delta^{13}C$	−4.0	−0.3	−0.4	0.1	2.4	−3.3	−3.2	−0.2	−2.6
C isotope (lacustrine carbonate of Triassic in Linxi area, Inner Mongolia, China)									
Sample number	BLS-2	BLS-3	BLS-4	BLS-5	BLS-6	BLS-7	BLS-8	BLS-9	BLS-11
$\delta^{13}C$	−6.5	−7.4	−9.4	−8.6	−11.0	−9.9	−8.7	−11.4	−8.5
*1C isotope (marine limestone of Permian and lacustrine limestone of Triassic, Inner Mongolia, China)									
$\delta^{13}C$	P_2	−1.15	−0.71	−2.89	T_1	−6.22	−6.22	−12.41	−4.18
*2C isotope (marine carbonate of Permian in Khorat basin, Thailand)									
$\delta^{13}C$	2.06	3.77	4.17	3.01	3.12	2.75	0.88	4.52	−2.78
*3C isotope (marine limestone of the Changhsingian late Permian, Changxing, Zhejiang, China)									
$\delta^{13}C$	−3.0	0.0	−0.4	−0.3	1.0	0.9	3.7	1.9	3.9
	2.3	3.3	3.8	0.5	1.6	2.9	−0.2	2.2	2.8
	−2.5	−1.9	1.7	1.5	1.5	1.9	1.5	2.3	1.9
	2.8	2.7	1.4	2.1	2.0	2.4	2.2	1.6	−1.3
	−3.4	1.8	1.8	2.0	2.5	2.2	1.8	0.5	−3.0

*1. Data from He et al. [1]; *2. data from Du et al. [13]; *3. data from Li[14].

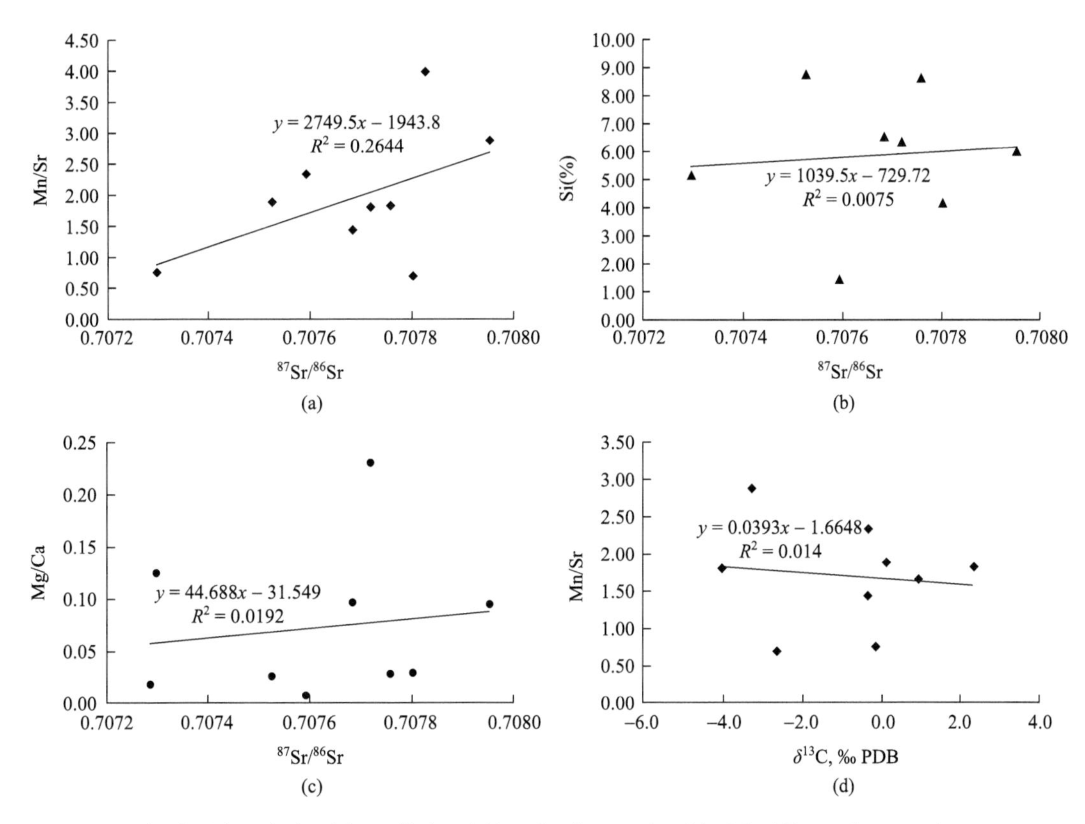

Fig. 4 The relationship to distinguishing the diagenesis of the Linxi Formation samples.
(a) relationship between the $^{87}Sr/^{86}Sr$ values and Mn/Sr values; (b) relationship between the $^{87}Sr/^{86}Sr$ values and Si contents; (c) relationship between the $^{87}Sr/^{86}Sr$ values and Mg/Ca values; (d) relationship between the $\delta^{13}C$ values and Mn/Sr ratios

5 Discussion

5.1 REE distribution and sedimentary environment

The REEs of different water bodies are quite different. The chemical composition of the authigenic minerals and rocks formed by these water bodies can be used to determine the geochemical characteristics of modern and ancient sedimentary media[29-32]. The water involved in the formation of carbonate rocks mainly fall into three categories: seawater, rivers or atmospheric precipitation, and hydrothermal fluid. The behavior of REE in a given type of water body results in distinguishing characteristics. In seawater, the LREE is being depleted relative to the HREE and results in La enrichment [17]. In addition, the REE of seawater in oxygen environment is resulting in Ce depletion[33]. In rivers, lakes, and other freshwater medium, Ce usually exhibits a slight anomaly or no anomaly. The infiltration of weathered leaching water near Eu-rich rocks or Eu bonded to iron hydroxide in water also causes slight positive Eu anomalies[22]. Beyond that, an influx of high-temperature midocean ridge oceanic hydrothermal fluids can also cause positive Eu anomalies.

In order to facilitate a more direct comparison of the REE contents of the different water bodies, PAAS was used to normalize the REE of the ancient marine carbonates and minerals, modern seawater, lakes, lacustrine carbonate rocks, and hydrothermal fluids[17, 18, 20-25] [Fig. 3 (b)–(h)]. Among the rare Earth patterns of the different sediments, water, and minerals, normal seawater has enriched La, negative Ce anomalies, and slightly enriched Gd [Fig. 3 (b) and (c)]. The hydrothermal fluids from the median ridge have obvious positive Eu anomalies [Fig. 3 (g)]. The freshwater and the lacustrine carbonate rocks are characterized by a relatively flat REE patterns without obvious anomaly distribution [Fig. 3 (e) and (h)]. The gulf and coastal waters have positive Ce anomalies and small negative Eu anomalies and show relatively gentle HREE distribution patterns [Fig. 3 (d)]. The marine carbonate rocks and minerals have enriched La, small negative Ce anomalies, and positive Eu anomalies [Fig. 3 (f)]. The samples from a distal section of near-shore marine carbonates display a distinct positive Y anomaly, positive La anomaly, and slight positive Gd anomaly. In contrast, the samples from the proximal section of nearshore marine carbonates display a slight positive Y anomaly, weak positive Gd anomaly, and a lack of La anomaly [Fig. 3 (i)]. It can be noted by the distribution characteristics of REE, except of the Eu, the limestones of Linxi Formation in the Guandi profile of Linxi area display relatively uniform shale-normalised REE + Y patterns [Fig. 3 (a)]. The limestones at the bottom of Lin1 display a slight positive La anomaly and lack of Ce, Gd, and Y anomaly, which are closed to the REE distribution patterns of lake water, no-marine carbonates, and the proximal section of near-shore marine carbonates [Fig. 3 (e), (h), and (i) and Fig. 5]. The middle-upper limestone samples of Lin1 and the limestones of Lin2–5 display a distinct positive La anomaly, slight positive Y anomaly, weak positive Gd anomaly, and lack of Ce anomaly, and its REE patterns resemble those deposits from the distal section of littoral environments, which is dominated by the ocean but with freshwater inputs from continental weathering [Fig. 3 (f) and (i)] or the neritic environments with a terrestrial source mixing. In addition to the above REE of La, Ce, Gd, and Y, almost all the limestones of the Linxi Formation display a distinct positive Eu anomaly. The Eu positive anomalies have no obvious indication of the sedimentary environment. With the lack of Eu positive anomaly in mudstones and no obvious correlation between the concentration of Eu and iron oxide (not shown), we infer that the reason for the positive Eu anomaly is related to the hydrothermal fluid mixing.

5.2 Trace element distribution and sedimentary environment

In addition to REE, other trace element ratios of sedimentary rocks can also be used to determine their sedimentary environments. The Sr in fine-grained marine sediments tends to be enriched relative to Ba, causing the higher Sr/Ba value in the fine-grained sediments. Therefore, the relative contents of Sr and Ba recorded by the fine-grained sediments, especially in mudstones and shales, i.e., Sr/Ba, are positively correlated with the paleosalinity [34-36]. Wei and Algeo conducted a comprehensive study of the Sr/Ba ratios as paleosalinity proxies in ancient shales and mudrocks, and they concluded that the Sr/Ba ratio of freshwater is less than 0.2, the Sr/Ba ratio from 0.2 to 0.5 indicate brackish environments, and ratios greater than 0.5 represent the marine environments. When water depth increases in sedimentary basins, the energy of seawater decreases with the increased clay minerals and fine-grained material. The fine-grained sediment enhances the adsorption of Ba ions and causes a lower Sr/Ba ratio. In general, the Sr/Ba ratios is lower in deep sea or bathyal sediments[36]. In addition, the

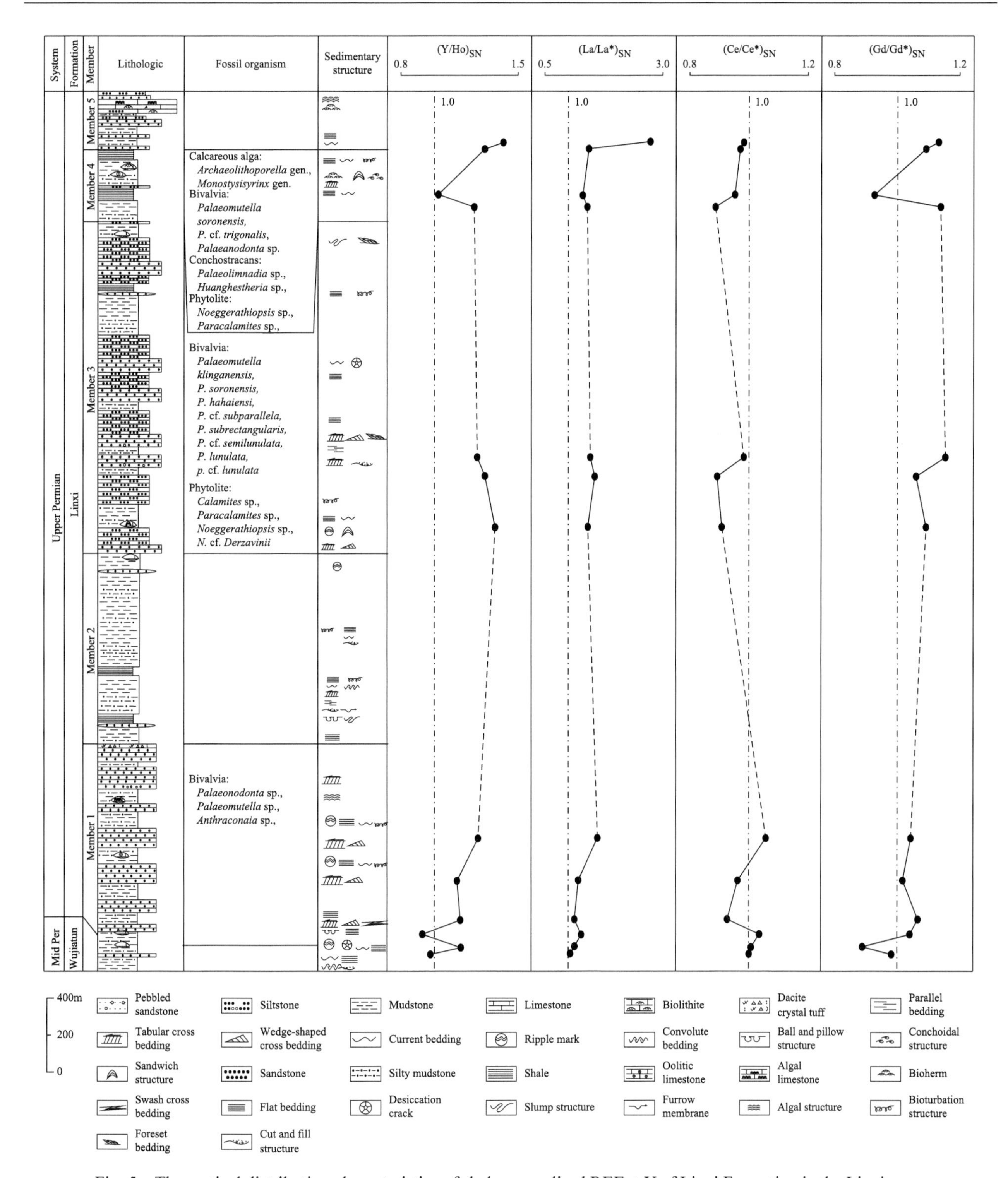

Fig. 5 The vertical distribution characteristics of shale-normalised REE + Y of Linxi Formation in the Linxi area.

incorporation of terrigenous debris in the offshore area can also result in a decrease in Sr/Ba ratios of argillaceous sediments. We compared the Sr and Ba contents of the Permian sedimentary rocks in the other areas of China (Fig. 6 and Table 4) [34, 37, 38]. The Sr/Ba ratios of the marine mudstones of Ordovician Pingliang Formation in southern Ordos Basin range from 0.19 to 2.59 with an average of 0.96. The Sr/Ba ratios of the marine mudstones of the lower Permian Liangshan Formation in Dushan area of Guizhou Province range from 0.14 to 2.79 with an average of 0.91. The Sr/Ba ratios of the mudstones of marine-continental transitional facie of Permian Shanxi Formation in Ordos Basin range from 0.13 to 0.89 with an average of 0.36. The Sr/Ba ratios of the mudstones and lime mudstones in the Linxi Formation of the study area range from 0.1 to 3.09 with an average of 0.93. The Sr/Ba ratios of the mudstones and lime mudstones of Linxi Formation correspond well with the Sr/Ba ratios of the marine mudstones and the mudstones of marine-continental transitional facie in the other areas of China. The Sr/Ba ratios indicate that the Linxi Formation limestone mainly formed in a salt water environment and develop bathyal sea sedimentary environment or have been affected by terrestrial debris (the Sr/Ba ratios of sample LG 29-1H<0.2).

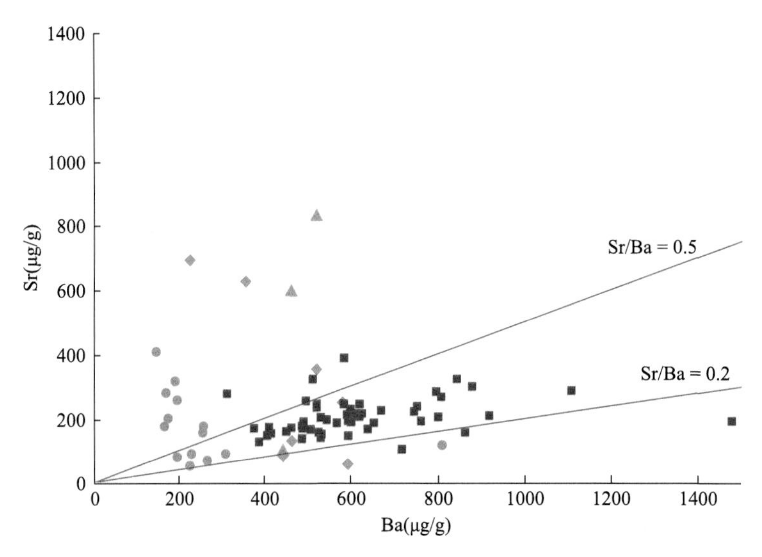

◆ Limestone and mudsone of Linxi Formation in Linxi Area
■ Mudstones of marine-continental transitional facie of Permian in southeastern Ordos Basin
▲ Marine mudstones of Ordovician Pingliang Formation in southern Ordos Basin
● Marine mudstones of the lower Permian Liangshan Formation in Dushan Area, Guizhou

Fig. 6 Relationship between the Sr and Ba contents of the Linxi Formation limestone from the Linxi area Guandi profile in eastern Inner Mongolia and comparison with other regions in China.

Table 4 Sr and Ba data of Permian in other areas of China

Sample	Sr	Ba	Sample	Sr	Ba	Sr	Ba
	124.0	464.5		186.1	652.2	149.8	406.2
	99.7	286.3		285.2	794.3	244.4	519.3
*1Marine mudstones of Ordovician Pingliang Formation in Southern Ordos Basin	102.7	543.9	*3Mudstones of marine-continental transitional facie of Permian Shanxi Formation in Ordos Basin	237.6	521.2	187.9	566.6
	273.4	246.6		171.3	486.5	167.9	637.9
	830.1	319.9		324.0	841.8	170.8	374.1
	598.4	471.1		240.1	749.4	230.6	597.0
	70.9	264.0		269.4	807.4	245.8	583.8
	117.0	808.0		206.3	619.7	323.9	510.8
	53.6	224.0		178.4	486.0	217.1	609.0
*1Marine mudstones of Ordovician Pingliang Formation in Southern Ordos Basin	90.6	307.0		215.2	624.5	162.8	448.9
	90.3	228.0		299.3	877.5	172.8	460.7
	258.0	194.0		225.7	668.1	175.8	409.4
	177.0	164.0		289.0	1108	190.3	600.6
	200.0	173.0		193.2	592.0	192.4	760.8
	157.0	254.0	*3Mudstones of marine-continental transitional facie of Permian Shanxi Formation in Ordos Basin	196.3	544.5	143.2	530.4
	280.0	168.0		202.2	529.4	277.4	310.2
	80.6	195.0		222.3	742.9	150.5	531.4
	316.0	190.0		247.1	619.4	159.3	860.1
*2Marine mudstones of the lower Permian Liangshan Formation in Dushan area of Guizhou Province, China	408.0	146.0		256.3	495.2	194.9	1478
	176.9	255.0		138.9	486.1	130.0	386.7
				105.1	714.9	211.1	917.5
				211.8	589.7	148.1	593.4
				155.1	414.8	158.3	526.0
				190.8	489.2	166.9	505.5
				388.7	583.2	207.3	799.3

*1. Data from Ni et al. [37]; *2. data from Zhang et al. [38]; *3. data from Chen et al. [34]

In addition, B is relatively enriched in marine shales, whereas Ga is relatively enriched in terrestrial shales. Therefore, the B/Ga ratio can be used to distinguish between marine and terrestrial sedimentary environments to a certain degree[36]. Yan et al. conducted a comprehensive study of the Mesozoic biological, mineral, and geochemical characteristics of the Paleocene-Pliocene strata of the Northern Jiangsu basin [39]. Based on a statistical comparison between ancient American marine sediments, modern marine sediments, ancient terrestrial sediments, modern freshwater sediments, modern Chinese marine sediments, and modern lacustrine sediments, they concluded that the B/Ga ratio of terrestrial sediment is less than 3.3, and ratios greater than 3.3 represent sea-land transitional zones and marine sedimentary environments. To more intuitively contrast the content of B and Ga and assess the deposition environment, the data of B and G contents of shale in different regions of the world have been counted and are presented in Table 5 and shown in Fig. 6. The analysis indicates that the B and Ga content of the Changxingjie marine shale in southern China varies from 34.7 μg/g to 141.0 μg/g and from 6.8 μg/g to 25.4 μg/g, respectively, and the average value of B/Ga is 4.79[40]. The B and Ga content of the Permian terrestrial mudstone in Zhungeer Basins of China varies from 14.2 μg/g to 20.0 μg/g and from 4.4 μg/g to 18.5 μg/g, respectively, and with the average values (B/Ga) of 2.27[37]. The B and Ga content of the ancient freshwater argillaceous sediments in the United States varies from 8.0 μg/g to 55.0 μg/g and from 9.0 μg/g to 25.0 μg/g, respectively, and with the average values (B/Ga) of 2.02[41]. The B and Ga content of the ancient marine argillaceous sediment in the United States varies from 65.0 μg/g to 187.0 μg/g and from 10.0 μg/g to 39.0 μg/g, respectively, and with the average values (B/Ga) of 5.77[41]. The B and Ga content of the inner shelf mudstone off the east coast of Inida varies from 80.0 μg/g to 130.0 μg/g and from 11.0 μg/g to 33.0 μg/g, respectively, and with the average values (B/Ga) of 4.84. The B and Ga content of the outer shelf mudstone off the east coast of Inida varies from 90.0 μg/g to 160.0 μg/g and from 16.0 μg/g to 34.0 μg/g, respectively, and with the average values (B/Ga) of 4.81[42]. In the study area, the B and Ga content of Linxi Formation mudstone varies from 33.4 μg/g to 150.0 μg/g and from 7.7 μg/g to 22.1 μg/g, respectively. The B/Ga ratios of Linxi Formation mudstone varies from 1.90 to 7.94 with an average of 4.62 being greater than 3.3, except for the mudstone (Sample LG 75-4, B/Ga, 1.9) at the top of section Lin5, and this corresponds well with the marine mudstone from the other area of the world (Fig. 7). The relative B and Ga contents of the mudstone from the Linxi Formation indicate a predominantly marine sedimentary environment during deposition of the Linxi Formation and that the formation was affected by terrestrial debris at the end of the deposition of section Lin5.

Table 5 B and Ga data of argillaceous sediments in the other places of the world

Sample	B	Ga	Sample	B	Ga	Sample	B	Ga	Sample	B	Ga
Mudstone of Permian in Chao County*1	95.7	22.1	Ancient marine argillaceous sediments, United States*3	95	30	The inner shelf mudstone off the East Coast of Inida*4	120	33	The outer shelf mudstone off the East Coast of Inida*6	160	32
	97.8	20.4		106	29		80	19		160	34
	57.3	20.3		69	18		100	25		110	25
	49.8	20.0		164	25		110	33		110	24
	52.5	22.2		166	18		110	22		100	16
	37	12		178	31		90	11		100	23
	18	12		111	38		120	25		100	21
	8	10		142	39		86	20		130	27
	32	12		87	26		86	20		100	22
	54	17		179	38		110	25		100	20
	34	20		120	28		80	17		120	27
	10	9		187	15		130	28		110	22
Ancient fresh-water argillaceous sediments, United States*2	26	15		127	20		100	20		90	20
	55	25		177	22		130	18		110	27
	26	19		65	10	Terrestrial mudstone of Permian in Junggar Basin*5	14.22	4.41		130	23
	52	18		122	20		29.14	18.53			
				179	20		20.00	9.93			
				74	17						

*1. Data from Cheng[40]; *2, *3. data from Potter et al. [41]; *4, *6. data from Rao and Rao[42]; *5. data from Zhang et al. [37]

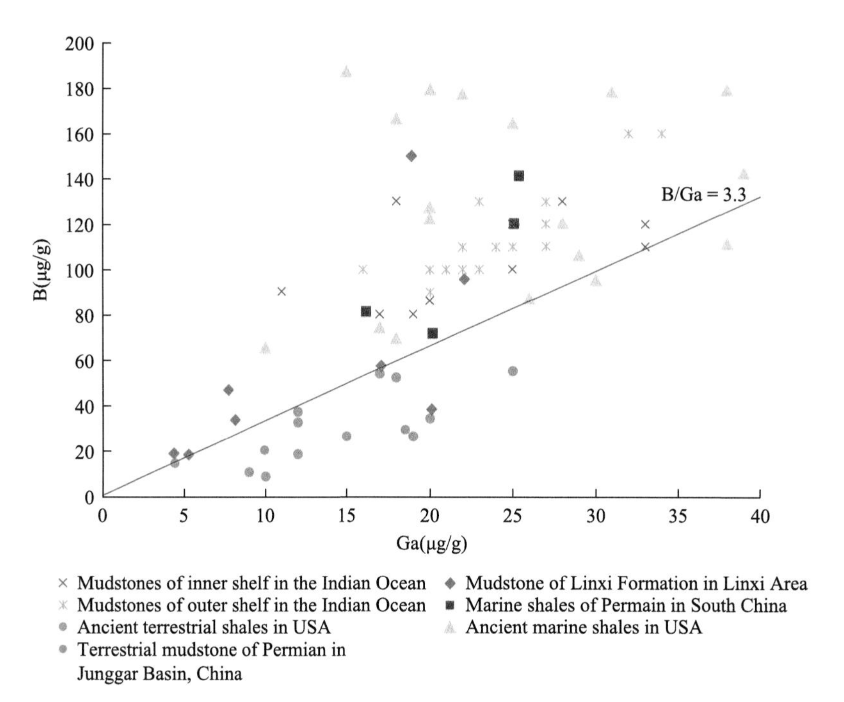

Fig. 7 Relationship between the B and Ga contents of the Linxi Formation mudstone from the Linxi area Guandi profile in eastern Inner Mongolia and comparison with other regions of the world.

5.3 Strontium isotope distribution and sedimentary environment

The stable evolutionary characteristics of Sr isotopes provide a basis for the comparison of Sr isotopes in seawater of the same geological age around the world[43]. In terms of the various sources of Sr, due to the weathering and leaching of crustal Si-Al rocks, lakes, rivers, and groundwater from terrestrial sources transport a large amount of radiogenic ^{87}Sr, which is generated in their catchments (basin), and thus, they have high $^{87}Sr/^{86}Sr$ values (about 0.7115) [44]. However, seawater transports Sr into the mantle at subduction zones, and Sr enters the seawater at midoceanic ridge hydrothermal systems[45]. The seawater also interacts with terrestrial waters that drain into the oceans, resulting in the $^{87}Sr/^{86}Sr$ values of modern seawater being approximately 0.7092[46]. Carbonates and some fossils can record the composition of Sr isotopes in sedimentary media. The composition and evolution of Sr isotopes ($^{87}Sr/^{86}Sr$ values) throughout geologic history have been well documented and reported. Therefore, by comparing the Sr isotopic compositions of endogenous carbonate rocks that have not been altered to the Sr isotopic composition of seawater during that geologic period, we can determine whether the sedimentary strata was deposited in a marine sedimentary environment or a terrestrial sedimentary environment. As shown in Fig. 7, Korte et al. established the background value (0.7073–0.7074) of $^{87}Sr/^{86}Sr$ during the Permian and the Triassic by the analysis of Sr isotopes from brachiopods[9, 10, 28]. Tian and Zheng found that $^{87}Sr/^{86}Sr$ values of Permian marine limestone range from 0.7072 to 0.7077 in South China[8]. Huang et al. found that $^{87}Sr/^{86}Sr$ values of the Late Permian seawater range from 0.7067 to 0.7076 by analyzing the Sr isotopes of marine carbonate rocks from the Upper Permian in the Upper Yangtze region[11]. Martin and Macdougall researched the contents of Sr isotopes in conodonts of Permian/Triassic from the United States and Pakistan, and they found that $^{87}Sr/^{86}Sr$ values of seawater in Permian/Triassic range from 0.7069 to 0.7089 (Table 2)[12]. The results of Sr isotopic analysis show that the $^{87}Sr/^{86}Sr$ values of the Linxi Formation limestone in the Linxi Guandi profile range from 0.70728 to 0.70795 and distribute within the range of Sr isotope composition of Permian seawater (Fig. 8). The $^{87}Sr/^{86}Sr$ values of the four limestone from section Lin1 to the bottom of section Lin3 range from 0.70753 to 0.70776 and fall within the Sr isotope range of the Late Permian Wujiaping and Changxing Period marine carbonates in southern China, indicating a shallow water carbonate sedimentary environment. The $^{87}Sr/^{86}Sr$ values of five limestone samples from sections Lin4 and Lin5 range from 0.70729 to 0.70795. Comparing with the contents of Sr isotope of the Permian marine limestone in other area of China, the $^{87}Sr/^{86}Sr$ values of the lower and upper limestone of section Lin4 have abnormally high values of 0.70795 and 0.70780, respectively, reflecting the fact that terrestrial water or terrestrial debris mixed with the main sediment supply. In addition, the $^{87}Sr/^{86}Sr$ values of one limestone sample each (sample numbers LG72-13T and LG73-2CT) from the top of section Lin4 and the bottom of section Lin5 are consistent with the $^{87}Sr/^{86}Sr$ background values (0.7073–0.7074) of the global Late Permian paleocean and exhibit relatively low $^{87}Sr/^{86}Sr$ values. The interpretation of this phenomenon is that the upper part of the Linxi Formation was deposited during a stage of rift closing and uplift accompanied by intense volcanic eruptions. The eruptions of magmas from a mantle source affected the composition of the Sr isotopes, which resulted in a decrease in the $^{87}Sr/^{86}Sr$ value of the sediments. The tuff interlayer in section Lin5 also supports the occurrence of volcanic activity.

5.4 C isotope distribution and sedimentary environment

The compositions of carbon isotopes in carbonate rocks and their changes can effectively trace sedimentary environments, diagenetic environments, and other geological processes[14, 47-52]. The $\delta^{13}C$ values of all nine limestone samples in the Linxi Formation range from –4.0‰ to 2.4‰ with a variation range of 6.4‰, and 67% of them are in the range of $\delta^{13}C$ values (0 ± 2‰) of normal seawater, indicating a normal marine sedimentary environment in the Linxi area. Compared with the $\delta^{13}C$ of marine Permian carbonate rocks (e.g., eastern Inner Mongolia, Changxing District of Zhejiang and Khorat Basin of Thailand) and Triassic lacustrine limestone in Linxi area (Table 3), the $\delta^{13}C$ values of limestone in the Linxi Formation are in the range of marine limestone but significantly higher than those of the lacustrine limestone (Fig. 9). However, it should be noted that individual sample with a low $\delta^{13}C$ value indicates a further differentiation in the sedimentary environment under normal marine background. For instance, the difference of seawater circulation, atmospheric circulation, and the accordingly production rate of organic carbon may cause changes in carbon isotope composition in limestone. The extensive development of plant debris and bioturbation structures in the Linxi mudstone indicate a shallow water environment with sufficient biomass during the deposition period of the Late Permian Linxi Formation. And so, we infer that the ^{12}C produced by the biological respiration, the oxidation of organic matter, and the freshwater recharge of low-carbon isotopic compositions all contribute to the low carbon isotope value of diagenetic fluids. Moreover, frequent volcanic activity was triggered by strong tectonics at the Late Permian in the study area and the volcanism produced a lot of CO_2. The carbon isotopic compositions of CO_2 (–5‰) may also be one of the reasons for the lower $\delta^{13}C$ value of limestone in the study area.

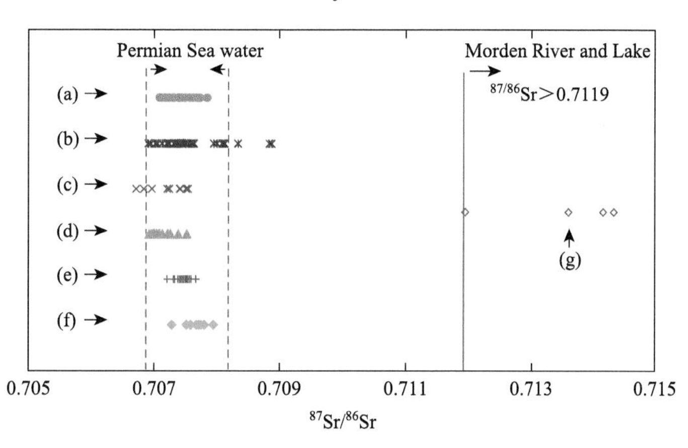

Fig. 8 Sr isotopic compositions of the Linxi Formation limestone from the Linxi area Guandi profile in eastern Inner Mongolia and the comparison with the $^{87}Sr/^{86}Sr$ values of Permian in the other area.

The Sr isotopic compositions of (a) conodonts during the Permian and the Triassic in Pakistan[12], (b) conodonts during the Permian and the Triassic in the United States[12], (c) marine carbonates rocks of Late Permian in Upper Yangtze region[11], (d) conodonts during the Permian and the Triassic in Abadeh, Iran[9, 10], (e) marine limestone of Permian in South China[8], (f) Permian Linxi Formation limestone from the Linxi area in eastern Inner Mongolia, and (g) the lacustrine limestone of Triassic from the Linxi area in eastern Inner Mongolia

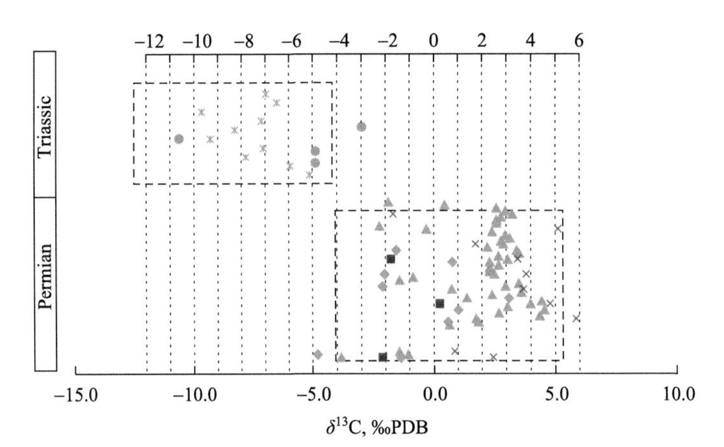

Fig. 9　C isotopic compositions of the Linxi Formation limestone and the lacustrine limestone (Triassic) from the Linxi area in eastern Inner Mongolia and the comparison with the C isotopic of marine limestone (Permian) in the other areas of the world.

5.5　Sedimentary environment and its evolution

These features of trace elements, isotopes, and sedimentary characteristics indicate that the environment of Linxi Formation is closed to the element and isotope compositions characteristic of littoral facies and neritic facies but with freshwater inputs from continental weathering.

Although the characteristics of limestone or shale of Linxi Formation yielded very consistent results, significant differences can be noted between 5 sections of Linxi Formation (Lin1-5) (Fig. 5–Fig. 9). Through the minor differences of trace compositions, lithologic association, sedimentary structures, and paleontology fossils of the Linxi Formation, we made the sedimentary environment dividing in detail. The lithology of section Lin1 of the Linxi Formation is mainly composed of sandy and silty sediments that exhibit symmetrical ripple bedding, horizontal bedding, and other sedimentary structures with directional significance for hydrodynamic conditions. The limestones at the lower section of Lin1 display a slight positive La anomaly, lack of Ce, Gd, and Y anomaly, and the upper section evolves into the characteristic of distinct positive La anomaly and slight positive Y anomaly. These distribution features of REE combining with Sr-C isotope and trace element indicate that the sedimentary environment of section Lin1 evolved from the proximal section of littoral facies with terrigenous clastic mixing to the distal section of littoral facies. Compare with the Lin1 section, the mud content of section Lin2 increased obviously which composed of mudstone, silt mudstone, and shale. The lower part of section Lin2 exhibited slump structures, convolute bedding, and flat bedding; the upper parts developed flat bedding, bioturbation structures, ripple mark structures, and current bedding. These sedimentary structures combining with the distribution of Sr-C isotope and Sr/Ba and B/Ga ratio indicate that the water body deepens during the depositional phase of section Lin2, and the sedimentary environments are interpreted to be bathyal slope to neritic facies. The sandstone content of the section Lin3 increases relatively. The lower parts of section Lin3 exhibited tabular cross bedding, wedge-shaped cross bedding, flat bedding, and ripple mark structures, and these sedimentary structures combining with the distributions of REE, C isotope, and Sr/Ba ratios of the limestones indicated that the sedimentary environment was close to the distal section of littoral facies. The middle and upper parts of section Lin3 developed foreset bedding, desiccation crack, bioturbation structure, slump structure, and current bedding, which indicated the sedimentary environment of delta front and littoral facies. The section Lin4 mainly composed of shale, silty mudstone, argillaceous siltstones, and limestone lenses. The geochemical characteristics of limestones with lower $^{87}Sr/^{86}Sr$ values, slight positive La and Y anomaly, and the sedimentary structures composing of flat bedding, tabular cross bedding, current bedding, bioturbation structure, and the typical marine fossils observed in this section indicate that the sedimentary environment of section Lin4 is mainly neritic facies. Section Lin5 is composed of sandstones and bioherm limestones and develops bryozoans and sponge bone fossils. The geochemical characteristics of the limestones in section Lin5 display a marine distribution pattern, which is displaying as low $^{87}Sr/^{86}Sr$ values and distinct positive La and Y anomaly. Combined with the sedimentary and geochemical characteristics, the sedimentary environment of section Lin5 is interpreted to be neritic facies.

6 Conclusions

The lithologic association of the Linxi Formation is mainly composed of clastic rocks, with interbedded carbonates and lenticular limestone bodies. The upper and lower units of the Linxi Formation limestones consistently show initial $\delta^{13}C$ values, $^{87}Sr/^{86}Sr$ ratios, and the similar REE + Y patterns that are characterized by depletion of LREE relative to HREE, distinct positive La anomaly, minor positive Y anomaly, weak positive Gd anomaly, lack of Ce anomaly, and the same range of Sr/Ba and B/Ga ratio. In general, the sedimentary environment of the Linxi Formation of Late Permian is still dominated by an open marine environment and has freshwater inputs from continental weathering. The marine environments can be divided into 3 major types which are littoral facies, neritic facies, and bathyal facies (Fig. 10). In addition, due to the regional uplift and the accompanying shallow water environment, a delta front environment of sea-land transition zone sediments also developed in the study area in addition to the marine sedimentary environment. The sedimentary environment of the Late Permian Linxi Formation underwent a process from littoral facies to neritic facies (bathyal facies) and then through littoral facies, delta front to neritic facies.

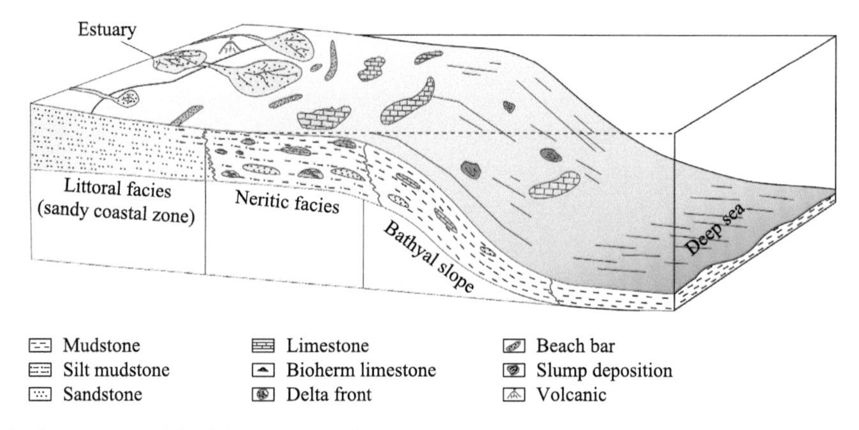

Fig. 10 Sedimentary model of the Late Permian Linxi Formation in the Linxi area of eastern Inner Mongolia.

Acknowledgments

This research was partially supported by the National Key R&D Program of China under grant 2017YFC0603103 and the National Natural Science Foundation of China (no. 41572098).

References

[1] Z. J. He, S. W. Liu, J. S. Ren, and Y. Wang, "Late Permian-early Triassic sedimentary evolution and tectonic setting of the Linxi region, inner Mongolia, "*Regional Geology of China,* vol.16, no.4, pp.403-409, 1997.

[2] Y. Zhang, S. Tian, Z. Li et al., "Discovery of marine fossils in the upper part of the Permian Linxi formation in Lopingian, Xingmeng area, China, "*Chinese Science Bulletin,* vol. 59, no.1, pp.62-74, 2014.

[3] S. Tian, Z. Li, Y. Zhang, Y. Gong, D. Zhai, and M. Wang, "Late carboniferous-Permian tectono-geographical conditions and development in Eastern Inner Mengolia and adjacent areas, "*Acta Geologica Sinica,* vol.90, no.4, pp.688-707, 2016.

[4] H. Yu, "Sedimentary facies and palaeogeography of the Songliao Basin and its peripheral areas during CarboniferousPermian time, "*Sedimentary*

Geology and Tethyan Geology, vol.21, no.4, pp.70-83, 2001.

[5] D. Q. Fang, J. B. Yun, Q. S. Pang, L. H. Zhao, and R. Zhao, "Discussion of carboniferous in the periphery of Songliao Basin, "*Journal of Daqing Petroleum Institute,* vol.28, no.4, pp.93-95, 2004.

[6] R. Zhu, H. Xu, S. Deng, and H. Guo, "Lithofacies palaeogeography of the Permian in northern China, "*Journal of Palaeogeography,* vol.9, no.2, pp.133-142, 2007.

[7] S. G. Tian, Z. S. Li, J. T. Wang, L. P. Zhang, and S. W. Niu, "Carboniferous-Permian tectonic and stratigraphic framework of eastern Inner Mongolia as well as adjacent areas and its formation environment, "*Geological Bulletin of China,* vol.31, no.10, pp.1554-1564, 2012.

[8] J. Tian and Y. Zheng, "The revolution of the isotopic composition of strontium in the Permian Paleo-ocean in South China, "*Acta Sedimentologica Sinica,* vol.13, no.4, pp.125-130, 1995.

[9] C. Korte, H. W. Kozur, M. M. Joachimski, H. Strauss, J. Veizer, and L. Schwark, "Carbon, sulfur, oxygen and strontium isotope records, organic geochemistry and biostratigraphy across the Permian/Triassic boundary in Abadeh, Iran, "*International Journal of Earth Sciences,* vol.93, no.4, pp.565-581, 2004.

[10] C. Korte, T. Jasper, H. W. Kozur, and J. Veizer, "^{87}Sr/^{86}Sr record of Permian seawater, "*Palaeogeography Palaeoclimatology Palaeoecology,* vol.240, no.1-2, pp.89-107, 2006.

[11] S. J. Huang, H. Shi, M. Zhang, L. C. Shen, J. Liu, and W. H. Wu, "Strontium isotope evolution and global sea-level changes of Carboniferous and Permian Marine carbonate, Upper Yangtze Platform, "*Acta Sedimentologica Sinica,* vol.19, no.4, pp.481-487, 2001.

[12] E. E. Martin and J. D. Macdougall, "Sr and Nd isotopes at the Permian/Triassic boundary: a record of climate change, "*Chemical Geology,* vol.125, no.1-2, pp.73-99, 1995.

[13] G. C. Du, H. Cang, S. Q. Hu, Q. R. Cao, and P. P. Gao, "Geochemical characteristics and its paleo-environmental significance of Permian carbonate rocks in Khorat Basin, Thailand, "*Global Geology,* vol.36, no.1, pp.135-143, 2017.

[14] Y. Li, "Carbon and oxygen isotope stratigraphy of the upper Permian changhsingian limestone in meishan section D, changxing, Zhejiang, "*Journal of Stratigraphy,* vol.22, no.1, pp.36-41, 1998.

[15] J. Ryu, A. D. Jacobson, C. Holmden, C. Lundstrom, and Z. Zhang, "The major ion, $\delta^{44/40}$Ca, $\delta^{44/42}$Ca, and $\delta^{26/24}$Mg geochemistry of granite weathering at ph = 1 and t = 25℃: power-law processes and the relative reactivity of minerals, "*Geochimica Et Cosmochimica Acta,* vol. 75, no. 20, pp.6004-6026, 2016.

[16] S. R. Taylor and S. M. Mclennan, "The geochemical evolution of the continental crust, "*Reviews of Geophysics,* vol.33, no.2, pp.241-265, 1995.

[17] J. Zhang and Y. Nozaki, "Rare earth elements and yttrium in seawater: ICP-MS determinations in the east caroline, coral sea, and south Fiji basins of the western south Pacific ocean, "*Geochimica Et Cosmochimica Acta,* vol. 60, no. 23, pp.4631-4644, 1996.

[18] C. R. German, B. P. Holliday, and H. Elderfield, "Redox cycling of rare earth elements in the suboxic zone of the Black Sea, "*Geochimica Et Cosmochimica Acta,* vol. 55, no. 12, pp.3553-3558, 1991.

[19] Y. Nozaki, D. Lerche, D. S. Alibo, and A. Snidvongs, "The estuarine geochemistry of rare earth elements and indium in the Chao Phraya River, Thailand, "*Geochimica Et Cosmochimica Acta,* vol.64, no.23, pp.3983-3994, 2000.

[20] T. Liang, L. Wang, C. Zhang, L. Ding, S. Ding, and X. Yan, "Contents and their distribution pattern of rare earth elements in water and sediment of intertidalite, "*Journal of the Chinese Rare Earth Society,* vol.23, no.1, pp.68-74, 2005.

[21] R. Bolhar and M. Vankranendonk, "A non-marine depositional setting for the northern Fortescue Group, Pilbara Craton, inferred from trace element geochemistry of stromatolitic carbonates, "*Precambrian Research,* vol.155, no.3-4, pp.229-250, 2007.

[22] M. J. V. Kranendonk, G. E. Webb, and B. S. Kamber, "Geological and trace element evidence for a marine sedimentary environment of deposition and biogenicity of 3.45 Ga stromatolitic carbonates in the Pilbara Craton, and support for a reducing Archaean ocean, "*Geobiology,* vol.1, no.2, pp.91-108, 2003.

[23] M. Bau and P. Dulski, "Comparing yttrium and rare earths in hydrothermal fluids from the Mid-Atlantic Ridge: implications for Y and REE behaviour during near-vent mixing and for the Y/Ho ratio of Proterozoic seawater, "*Chemical Geology,* vol.155, no.1-2, pp.77-90, 1999.

[24] S. B. Ojiambo, W. B. Lyons, K. A. Welch, R. J. Poreda, and K. H. Johannesson, "Strontium isotopes and rare earth elements as tracers of groundwater-lake water interactions, Lake Naivasha, Kenya, "*Applied Geochemistry,* vol. 18, no. 11, pp.1789-1805, 2003.

[25] Z. Z. Zhu, Z. L. Wang, B. Gao, and S. L. Wang, "Geochemical characteristics of rare earth elements in Lake Chaohu, East China, "*Geochimica,* vol.35, no.6, pp.639-644, 2006.

[26] H. E. Frimmel, "Trace element distribution in Neoproterozoic carbonates as palaeoenvironmental indicator, "*Chemical Geology,* vol.258, no.3-4, pp.338-353, 2009.

[27] A. J. Kaufman, S. B. Jacobsen, and A. H. Knoll, "The Vendian record of Sr and C isotopic variations in seawater: implications for tectonics and paleoclimate, "*Earth and Planetary Science Letters,* vol.120, no.3-4, pp.409-430, 1993.

[28] C. Korte, H. W. Kozur, P. Bruckschen, and J. Veizer, "Strontium isotope evolution of late Permian and Triassic seawater, "*Geochimica et Cosmochimica Acta,* vol.67, no.1, pp.47-62, 2003.

[29] G. E. Webb and B. S. Kamber, "Rare earth elements in Holocene reefal microbialites: a new shallow seawater proxy, "*Geochimica Et Cosmochimica*

Acta, vol.64, no.9, pp.1557-1565, 2000.

[30] R. Bolhar, B. S. Kamber, S. Moorbath, C. M. Fedo, and M. J. Whitehouse, "Characterisation of early archaean chemical sediments by trace element signatures, "*Earth and Planetary Science Letters,* vol.222, no.1, pp.43-60, 2004.

[31] G. A. Shields and G. E. Webb, "Has the REE composition of seawater changed over geological time? "*Chemical Geology,* vol.204, no.1-2, pp.103-107, 2004.

[32] H. E. Frimmel, "Trace element distribution in Neoproterozoic carbonates as palaeoenvironmental indicator, "*Chemical Geology,* vol.258, no.3-4, pp.338-353, 2009.

[33] Z. X. Ma, B. Li, X. T. Liu, and L. Luo, "Geochemical characteristics and implications for the evolution of sedimentary environments of early cambrian Qingxudong Formation in eastern Guizhou, Southwestern China, "*Geological Science&Technology Information,* vol.34, no.2, pp.71-77, 2015.

[34] H. D. Chen, L. I. Jie, and C. G. Zhang, "Discussion of sedimentary environment and its geological enlightenment of Shanxi Formation in Ordos Basin, "*Acta Petrologica Sinica,* vol.27, no.8, pp.2213-2229, 2011.

[35] N. Adachi, Y. Ezaki, and J. Liu, "The late early cambrian microbial reefs immediately after the demise of archaeocyathan reefs, Hunan Province, South China, "*Palaeogeography, Palaeoclimatology, Palaeoecology,* vol.407, no.4, pp.45-55, 2014.

[36] W. Wei and T. J. Algeo, "Elemental proxies for paleosalinity analysis of ancient shales and mudrocks, "*Geochimica et Cosmochimica Acta,* In press, 2019.

[37] G. W Zhang, S. Tao, D. Z. Tang, Y. B. Xu, Y. Cui, and Q. Wang, "Geochemical characteristics of trace elements and rare earth elements in Permian Lucaogou oil shale, Santanghu Basin, "*Journal of China Coal Society,* vol.42, no.8, pp.2081-2089.

[38] G. T. Zhang, Z. Q. Peng, C. S. Wang, and Z. H. Li, "Geochemical characteristics of the lower Permian Liangshan Formation in Dushan area of Guizhou Province and their implications for the paleoenvironment, "*Geology in China,* vol.43, no.4, pp.1291-1303, 2016.

[39] Q. Yan, G. Zhang, L. Xiang et al., "Marine inundation and related sedimentary environment of Funing group (lower Paleogene), in Jinhu depression, North Jiangsu Plain, "*Acta Geological Sinica,* vol.1, pp.74-84, 1979.

[40] A. Cheng, "Content of B and B/Ga ratio of the Yingying Formation in Chao County, Anhui Province, "*Journal of Stratigraphy,* vol.18, no.4, pp.299-300, 1994.

[41] P. E. Potter, N. F. Shimp, and J. Witters, "Trace elements in marine and fresh-water argillaceous sediments, "*Geochimica Et Cosmochimica Acta,* vol.27, no.6, pp.669-694, 1963.

[42] N. V. N. D. Rao and M. P. Rao, "Trace-element distribution in the continental-shelf sediments off the east coast of India, "*Marine Geology,* vol.15, no.3, pp.43-48, 1973.

[43] C. E. Jones, "Seawater strontium isotopes, oceanic anoxic events, and seafloor hydrothermal activity in the Jurassic and Cretaceous, "*American Journal of Science,* vol. 301, no. 2, pp.112-149, 2001.

[44] B. E. Crowley, J. H. Miller, and C. P. Bataille, "Strontium isotopes ($^{87}Sr/^{86}Sr$) in terrestrial ecological and palaeoecological research: empirical efforts and recent advances in continental-scale models, "*Biological Reviews,* vol. 92, no.1, pp.43-59, 2017.

[45] M. Harris, R. M. Coggon, C. E. Smith-Duque, M. J. Cooper, J. A. Teagle, and D. A. H. Damon, "Channelling of hydrothermal fluids during the accretion and evolution of the upper oceanic crust: Sr isotope evidence from ODP Hole 1256D, "*Earth and Planetary Science Letters,* vol.416, pp.56-66, 2015.

[46] J. M. McArthur, "Recent trends in strontium isotope stratigraphy, "*Terra Nova,* vol.6, no.4, pp.331-358, 2010.

[47] D. Reghellin, H. K. Coxall, G. R. Dickens, and J. Backman, "Carbon and oxygen isotopes of bulk carbonate in sediment deposited beneath the eastern equatorial Pacific over the last 8 million years, "*Paleoceanography,* vol. 30, no.10, pp.1261-1286, 2016.

[48] X. Li, W. Xu, W. Liu et al., "Climatic and environmental indications of carbon and oxygen isotopes from the Lower Cretaceous calcrete and lacustrine carbonates in Southeast and Northwest China, "*Palaeogeography, Palaeoclimatology, Palaeoecology,* vol.385, no.3, pp.171-189, 2013.

[49] M. M. Vieira, A. N. Sial, L. F. De Ros, and S. Morad, "Origin of holocene beachrock cements in Northeastern Brazil: evidence from carbon and oxygen isotopes, "*Journal of South American Earth Sciences,* vol.79, pp.401-408, 2017.

[50] B. Banerjee, S. M. Ahmad, W. Raza, and T. Raza, "Paleoceanographic changes in the Northeast Indian Ocean during middle Miocene inferred from carbon and oxygen isotopes of foraminiferal fossil shells, " *Palaeogeography, Palaeoclimatology, Palaeoecology,* vol.466, pp.166-173, 2017.

[51] Q. Zhang, B. Liang, F. Qin, J. Cao, D. Yong, and J. Li, "Environmental and geochemical significance of carbon and oxygen isotopes of Ordovician carbonate paleokarst in Lunnan, Tarim Basin, "*Environmental Earth Sciences,* vol.75, no.14, pp.1-11, 2016.

[52] D. X. Zhai, Y. S. Zhang, S. G. Tian et al., "The late Permian sedimentary environments of Linxi formation in Xingmeng area: constraints from carbon and oxygen isotopes and trace elements, "*Acta Geoscientica Sinica*, *vol.* 36, no. 3, pp. 333-343, 2015.

Differences in evaporite geochemistry of the greater Ordos Ordovician salt basin, China: evidence from the M_5^6 submember of the Majiagou Formation*

Fu Fan[1], Fan-wei Meng[2], Yong-sheng Zhang[1], Mian-ping Zheng[1], Kui Su[1] and En-yuan Xing[1]

1. Institute of Mineral Resources, Chinese Academy of Geological Sciences, Beijing 100037, China
2. State Key Laboratory of Paleobiology and Stratigraphy, Nanjing Institute of Geology and Paleontology, Chinese Academy of Sciences, Nanjing 210008, China

Abstract Ordovician evaporites are very rare in the global geological record. Thick salt deposits in the Ordovician Majiagou Formation are present in the eastern greater Ordos Basin, China. Potash mineralization layers and a thin industrial grade potash layer occur in the sixth submember of the fifth member of the formation, thus indicating a good potash prospect. The eastern and western salt depressions are separated by an uplift and have been divided into five secondary salt sags (East-1, East-2, West-1, West-2, and West-3). The ratio of the salt layer thickness to that of the whole succession was very low in the West-2 sag and the salt depositional center was the East-2 sag, which implied that seawater entered the basin from the western sags and led to desalination and the formation of deep-water sediments. Analyses indicated that the potash formation conditions were obviously different in the salt sags. Although a thin potash layer was present in the East-1 sag, findings indicated that this area was not favorable for potash accumulation. The geochemical and sedimentary environment of the East-2 sag was similar to that of typical potash basins globally. Rounded sylvine grains occurred between salt crystals. The preservation characteristics indicated that evaporite deposits had reached the KCl bitterns stage. An eastward input of terrestrial materials carried a large amount of clay, and the clay layer protected previously deposited sylvine particles, thereby avoiding further potassium dissolution. Sedimentary, geochemical, and petrological evidences indicate good prospects for potash exploration in the East-2 salt sag.

Keywords Ordos basin; Ordovician; M_5^6submember; Majiagou formation; Sylvine

Introduction

Worldwide, salt deposits are very rare in Ordovician rocks (Ronov et al., 1980; Kovalevych et al., 2006; Meng et al., 2018). Apart from the Ordos basin in central China, massive saline deposits of an Ordovician age have only been reported for the Canadian Arctic basin (Zharkov, 1984). However, the greater Ordos basin contains massive salt deposits (Li et al., 2011). Potassium-magnesium salt deposits are distributed in the eastern part of the northern Shanxi salt basin, which is located in the eastern part of the greater Ordos Basin, thus reflecting the special tectonic and paleogeographic evolution of the area.

The northern Shanxi salt basin is situated to the east of the Yishan slope in the greater Ordos basin (Fig. 1). The Ordovician Majiagou Formation in the greater Ordos basin contains thick salt sediments and can be divided into six members (from lower to upper): Ma1, Ma2, Ma3, Ma4, Ma5, and Ma6 (Meng et al., 2018). Potash mineralization layers and a thin potash layer that reaches an industrial grade occur in the sixth submember of the fifth member of the Majiagou Formation (M_5^6 submember).

* 本文发表在：Carbonates and Evaporites, 2020, 35: 107

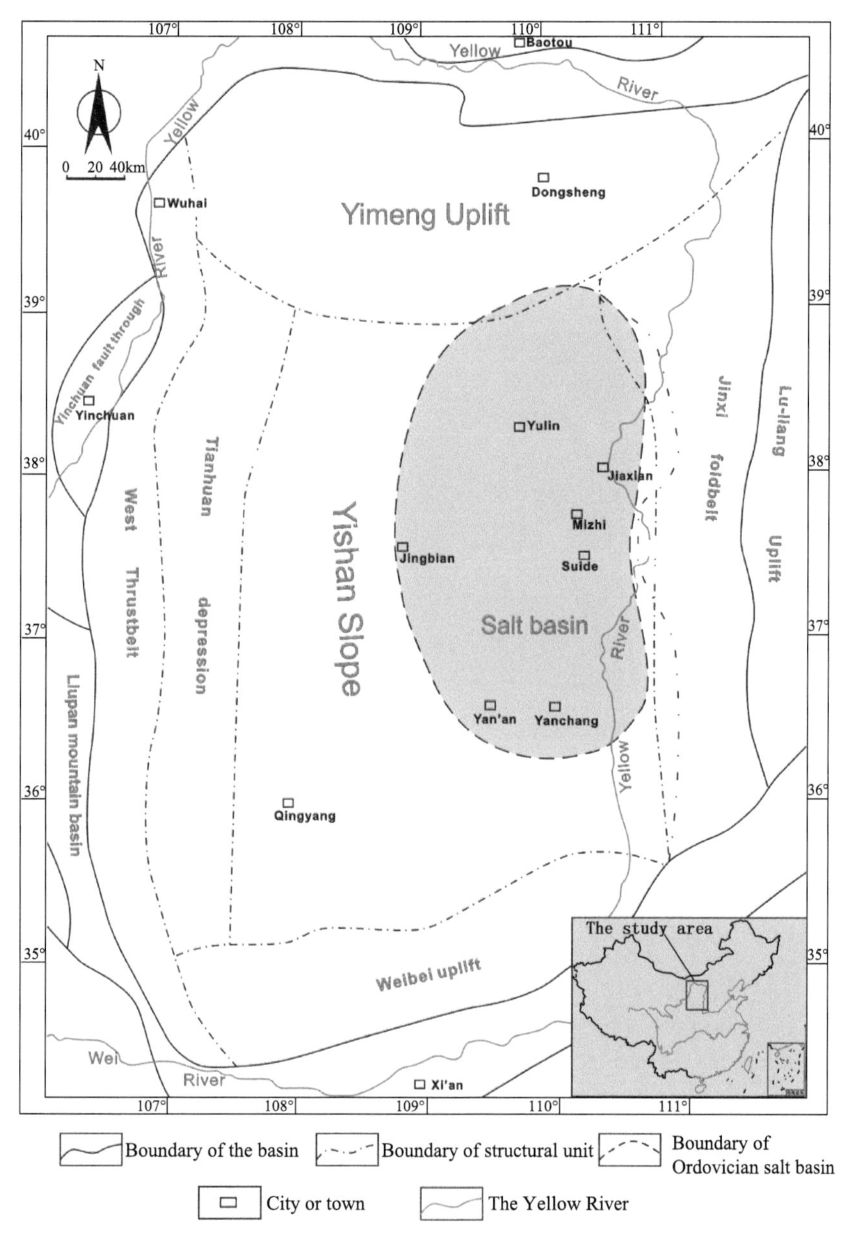

Fig. 1 Location of the study area.

During the Ordovician, the North China block was located at latitude 18° S (Scotese et al., 1979) or 10°–2° N (Zhu et al., 1998; Wu et al., 1990) (Fig. 2). The low-latitude areas at that time may have been favorable for potash deposition. Liu (1997) proposed that the potash minerals in the salt rocks of the Majiagou Formation in the Ordos Basin were derived from Cl-type ancient seawater, which evaporated and concentrated during that period. The lack of a thick layer of K-Mg minerals at the top of the super-thick halite deposits in the Majiagou Formation led Bao (2004) to conclude that the formation of these deposits involved drying out and evaporation followed by recharging and redissolution, whereby recharge by seawater limited the scale of the potash deposits. Chen and Yuan (2010) concluded that the evaporites in the M_5^6 submember were deposited in a deep-water setting in which the previous basin provided a pre-existing brine as the material basis for the potash deposits. However, substantial controversy remains over whether the area had the required conditions for the large-scale accumulation of potash. In recent years, with the growth of oil exploration and salt exploitations, some new wells have been drilled through the evaporites of the Majiagou Formation. These new drillings and core samples have revealed the differentiation of the substructure during the formation of the M_5^6 submember in the northern Shanxi salt basin, which provided the space for the accumulation of potash deposits (Zhang et al., 2013).

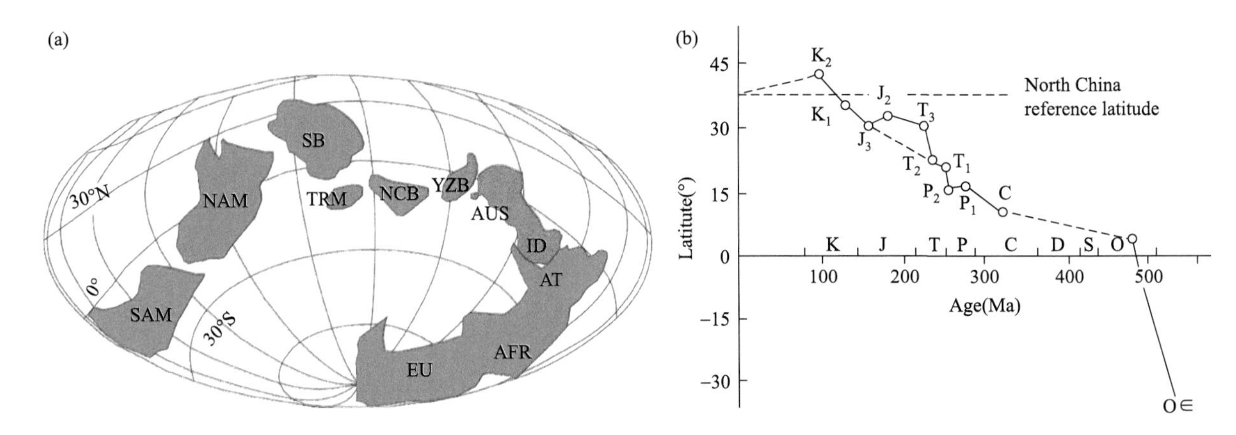

Fig. 2 (a) Reconstruction of Ordovician paleo-continents of the North China, Yangtze, and Tarim blocks (after Zhu et al., 1998). NCB North China block, YZB Yangtze block, TRM Tarim block, SB Siberia, AUS Australia, SAM South America, NAM North America, AFR Africa, ID India, AT Antarctica, EU Paleo-Europe. (b) Paleolatitude distributions of the NCB (after Wu et al., 1990).

In the present study, information from petrological, geochemical, and sedimentological analyses are used to describe the sedimentary characteristics of evaporites and the differences between the various salt sags of the M_5^6submember. We, therefore, aim to provide a basis for studying of the conditions that existed during potash formation in Ordovician age.

Geological background

The basement of the Ordos basin comprises Archean and lower Proterozoic metamorphic rocks. The sedimentary cover contains Proterozoic, Cambrian, Ordovician, Carboniferous, Permian, Triassic, Jurassic, Cretaceous, Tertiary, and Quaternary rocks, but lacks Silurian and Devonian strata. The total thickness of the Proterozoic to Quaternary strata ranges between 5000 and 10,000 m (Yang et al., 2005; Li et al., 2011; Meng et al., 2019).

The Ordovician Majiagou Formation in the northern Shanxi basin contains a great thickness of halite. The lithology includes rhythmic layers with different thicknesses composed of carbonate (limestone and dolomite) and halite, with minor gypsum and dolomitic mudstone interlayers (Yang et al., 2005; Li et al., 2011; Meng et al., 2019).

On the basis of the lithological assemblages and distributions, the Majiagou Formation is divided into the following six lithological members (from lower to upper): Ma1, Ma2, Ma3, Ma4, Ma5, and Ma6. Of these, the Ma1, Ma3, and Ma5 members are evaporite-bearing intervals that contain dolomite and some evaporites, whereas the Ma2, Ma4, and Ma6 members do not contain evaporites (dilution members) and are composed of different types of limestone (Meng et al., 2018). The Ma2 member contains relatively less limestone in some areas. The main evaporite-bearing interval is the Ma5 member, which can be divided into ten submembers on the basis of the lithological assemblages and well log data, whereby salt rocks mainly occur in the Ma_5^4, Ma_5^6, Ma_5^8, and Ma_5^{10} submembers (Fig. 3; Feng et al., 1991, 1998; Guo et al., 2014). The thickness of the Ma5 member varies markedly across the basin; the thickness in the member without evaporites is only 80–133 m, whereas the thickness in the salt depocentre is almost 400 m. The Ma_5^6 submember contains a thin layer that includes potassium minerals. During formation of the Ma_5^6 submember, the salt basin contained two depressions and one uplift. The two depressions contained five secondary salt sags more than 150 m thick (Fig. 4), which are called the West-1, West-2, West-3, East-1, and East-2 salt sags (Zhang et al., 2013). The samples used in this study were obtained from four of these sags (excluding the West-3 sag). Until now, no well drilling data have been available for the West-3 sag.

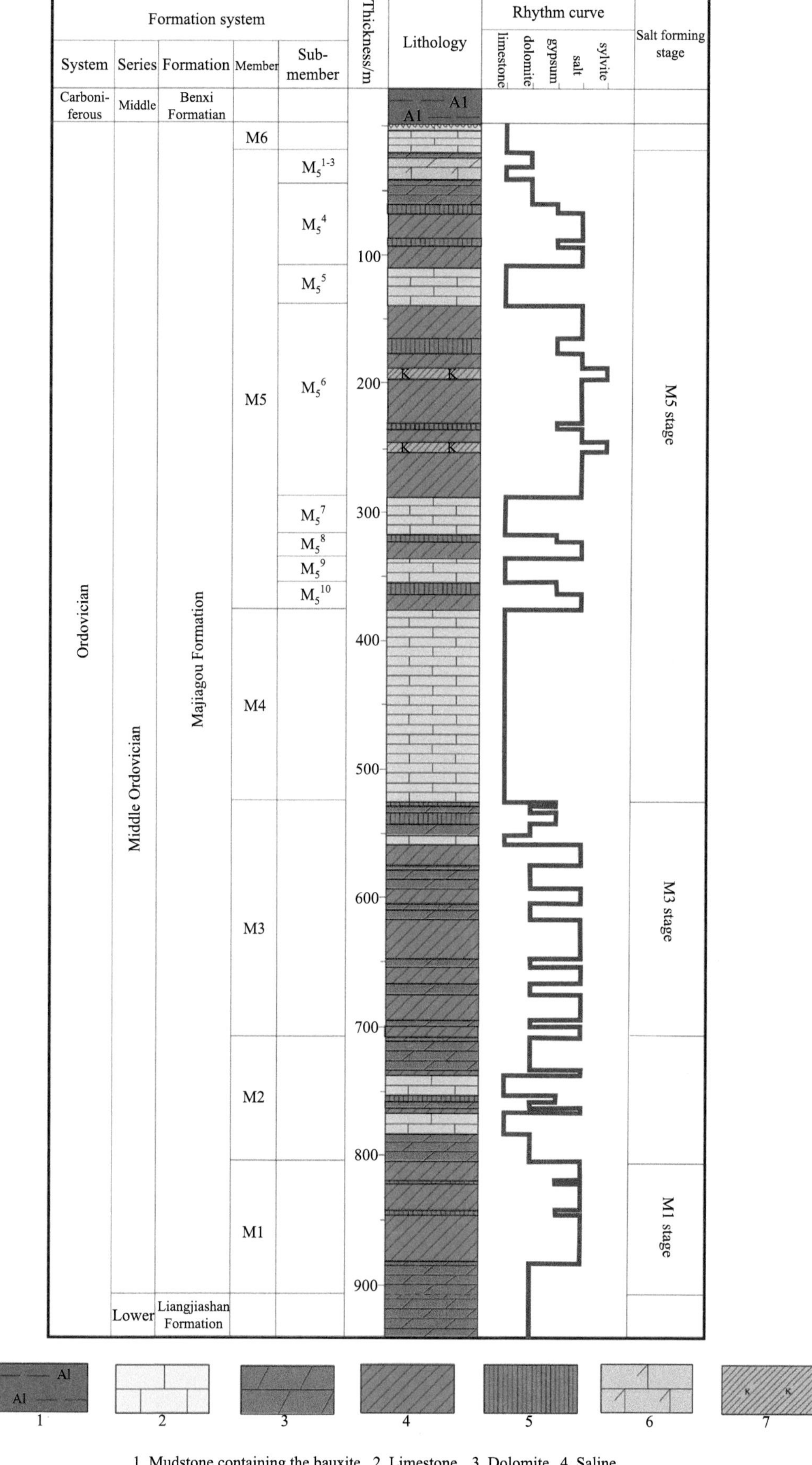

1. Mudstone containing the bauxite 2. Limestone 3. Dolomite 4. Saline
5. Gypsum 6. Dolomitic limstone 7. Saline containing the potassium

Fig. 3 Ordovician strata in the Ordos basin.

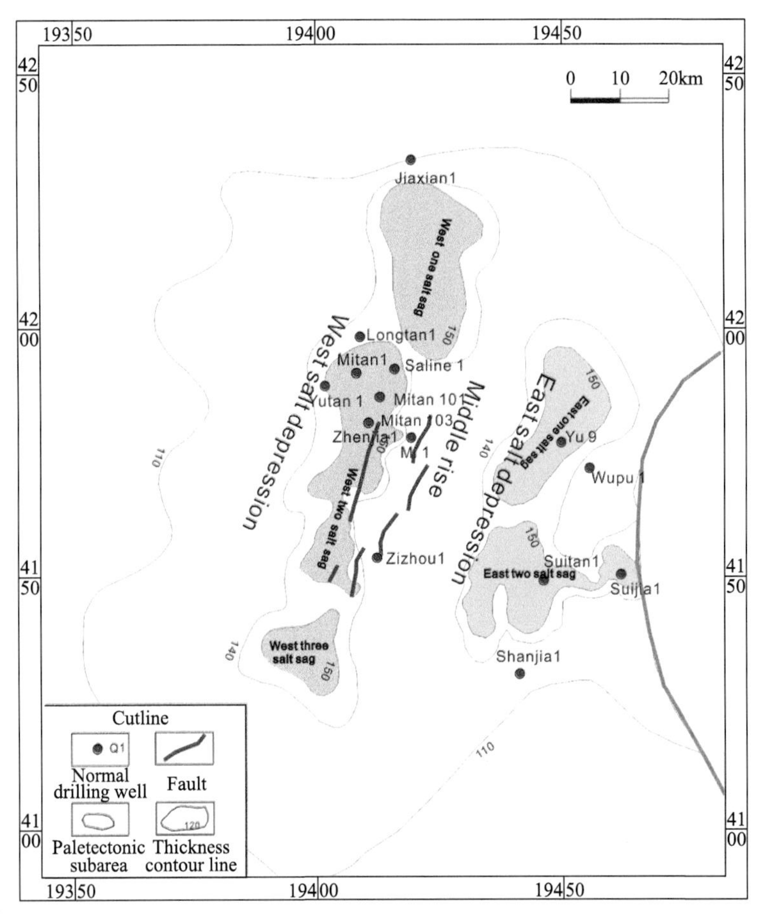

Fig. 4 Paleo-tectonic map of the northern area of the Shanxi salt basin in the M_5^6 stage (after Zhang et al., 2013).

Materials and methods

Sub-samples were obtained from salt rock samples from the Zhenjia-1, Suijia-1, Yutan-1, and Longtan-1 wells. Among them, the stratum with small particle sizes and no obvious recrystallization of rock salt particles with sizes of 0.5 cm were selected for sampling. The Zhenjia-1 well was sampled at depths of 2729.65–2730.66 m, 2738.95–2740.96 m, and 2742.45–2743.62 m with a sampling interval of <1 m. The depth of the salt-bearing member of the Ma_5^6 submember of Suijia-1 well is located at a depth of 2286.11–2404.28 m, and salt samples were taken at an interval of <1 m. The depth of the Ma_5^6 submember in the Yutan-1 well is 2700–2850 m, and the sampling interval was ~1 m. The rest of the drilling sample data was obtained from the drilling data of salt production enterprises and oil fields. The layer interval of the Ma_5^6 submember in the Suitan-1, Yu-9, Mitan-1, Jiaxian-1, Mitan-103, Mitan-101, Wubao-1 wells is 1–2 m. Only a few samples were obtained in Ma_5^6 submember of the Longtan-1 well.

Analyses for K^+, Br^-, and Cl^- contents were performed on the salt rock samples of the Ma_5^6 submember from the Zhenjia-1, Suijia-1, Yutan-1, and Longtan-1 wells. The salt rock samples were crushed and ground to 200 mesh. Pure water was added and fully mixed with the sample at room temperature to partially dissolve the samples, which were then clarified before 1 + 1 hydrochloric acid was added for further dissolution of the samples through boiling. One gram of each sample was weighed and placed into a 100 ml volumetric flask. The K^+ content was determined using an atomic absorption spectrophotometer (AAS, GGX-600) in a 1000 times diluted sample solution. The relative standard deviation was 1.03% and the relative error was 0.82%. The Br^- content was determined by phenol red photo-spectrometry at pH 4.4–5.0 using the chloramines T method as an oxidant to oxidize Br^- to free Br_2, which reacted with phenol red to form tetra-bromophenol red. The relative standard deviation was 15.15%. The Cl^- content was determined by the silver nitrate volumetric method, which is based on the formation of an AgCl precipitate by Ag^+ in Cl^- in a neutral solution. When excessive Ag^+ meets potassium chromate, the solution changes from bright yellow to orange. The relative standard deviation was 4.00% and the relative error was 1.60%. All analyses were conducted at the chemical laboratory of the Salt Lake Centre of the Chinese Academy of Geological Sciences, Beijing, China.

Results and interpretation

Sedimentology and petrology of the salt sags

The samples from the Ma_5^6 submember in the study area mainly contained halite, with lesser quantities of dolomite, anhydrite rock, argillaceous dolomite, argillaceous anhydrite rock, and limestone with gypsiferous mudstone, and a small amount of sylvine (Yang et al., 2005; Li et al., 2011; Meng et al., 2019).

The main color of the halite in the eastern sags was observed to be maroon. Muddy interlayers were common and were mainly greyish-green in color. The cracks in the grey mudstone and dolomite were found to be filled with halite. The salt rocks at different levels were interbedded with muddy deposits and formed rhythmites. The K^+ content in some layers was $>0.09\%$, thus indicating potassium mineralization.

From observations of thin members from the Suijia-1 well in the eastern salt sags, the rock types were identified as halite containing potash and medium- and fine-grained halite, and fine- and medium-grained gypsiferous and argilliferous halite with potash. The main rock types were gypsum, carbonate, mud, halite, and potash. The potash minerals occurred between halite granules [Fig. 5(a)] or inside halite particles [Fig. 5(b)]. The external surfaces of the potash particles shown in Fig. 5 contained large amounts of mud, thus indicating that these particles had experienced dissolution and erosion. The halite particles were rather small in size, and most were xenomorphic, medium-sized and small crystals [Fig. 5(c)], although several showed original chevron crystals [Fig. 5(d)]. These characteristics imply a concentrated brine in shallow environment with some terrigenous inputs during late evaporation stage because the chevron salt is formed in shallow water; however, density-stratified brine in deep water can suppress chevron salt growth at bottom (Hovorka et al., 2007).

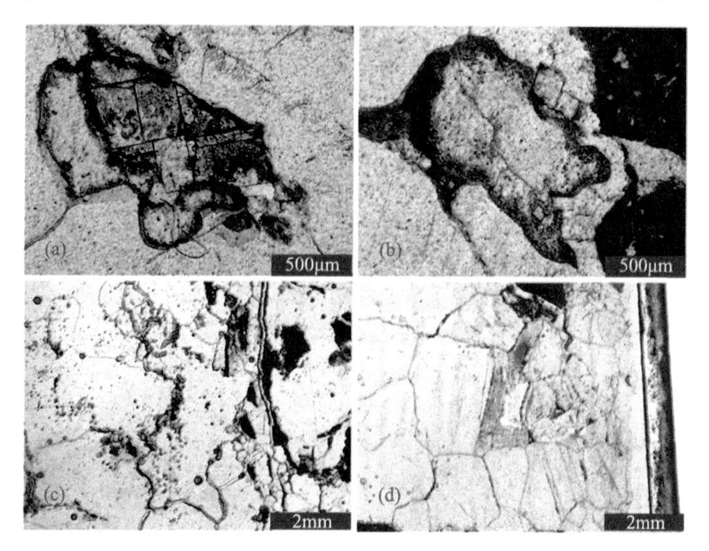

Fig. 5 The petrology characteristics of the Suijia-1 well in the eastern salt depression.

The dominant halites in the western area were observed to be smoky grey in color. These halite crystals were large in size, had a very low potash content, and were rather pure and easily broken. Widespread recrystallization destroyed the original sedimentary structure, which made it difficult to infer the water depth in the western area. However, the smoky grey color of much of the halite indicates reducing conditions under the redox interface. Hovorka et al. (2007) summarized the depth indicators in Permian basin evaporites, and suggested that the characteristics of truncated crystals and original chevron crystals provide a strong indicator of shallow water conditions. On this basis, we infer that the characteristics of truncated sylvine crystals and original chevron crystals in the eastern sags are indicative of a shallow concentrated brine environment. Furthermore, the halite was also commonly of a maroon coloring, which indicates an oxidized sedimentary environment, thus reflecting the shallow water, even for the exposed surface upon the redox interface. Thin-member observations of halite from samples from the Zhenjia-1, Longtan-1, and Yutan-1 wells (west sags) showed that the main rock types were medium-coarse gypsiferous and argilliferous granular halite with medium-coarse halite crystals. Minerals were found to be mostly gypsum, carbonate, minor mud, and halite. The halite particles were large and their surfaces were clean, and displayed a subhedral-xenomorphic texture with mosaic structures between halite particles[Fig. 6(a)]. Contiguous growth was well developed as a result of secondary enlargement and recrystallization [Fig. 6(b)]. The mineral sizes and degree of recrystallization were higher than those of samples from the eastern sags.

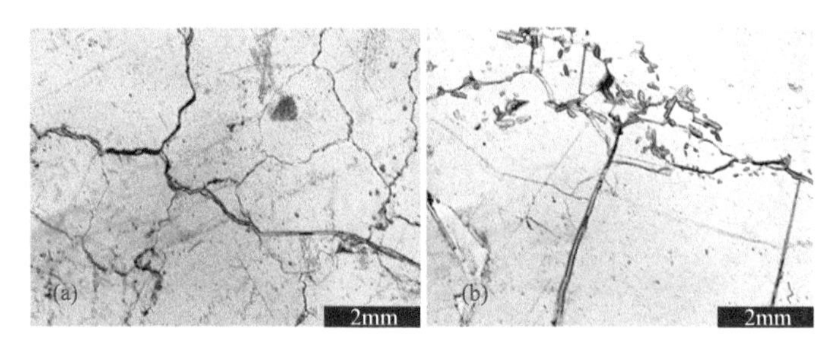

Fig. 6 Petrology characteristics of the western salt depression.

Salt layer thickness in the eastern and western salt sags

Data on the thickness of the salt layers for each of the salt sags are provided in Table 1. The total strata thickness, halite thickness, and ratio of halite thickness to total strata thickness revealed that the strata thickness is high, and the halite layer thickness and the ratio of halite to strata are generally high in the eastern salt sags; however, the halite layer thickness and the ratio of halite to strata differed between sags in the western salt sags. The West-1 salt sag has a small strata and halite thickness, but a high ratio of halite to strata; the West-2 salt sag has a large strata and halite thickness, but a low halite to strata ratio. The differences between the western salt sags are shown in Table 1 and Fig. 7, Fig. 8, and Fig. 9.

Table 1 Thickness of the salt layer in different salt sags

Tectonic Units	Drill well	Thickness of strata (m)	Thickness of the salt layer (m)	Salt/Strata (%)	Single minimum-maximum thickness/m
West-1 salt sag	Jiaxian-1 well	71.1	66.6	93.7	0.62–16.86
	Average	71.1	66.6	93.7	
West-2 salt sag	Mitan-1 well, Mitan-101, Mitan-103, Zhenjia-1 well, Saline-1 well, Longtan-1 well, Yutan-1 well, Zizhou-1 well, Mi-1 well	100.5–167.6	79.9~–140.6	67.1–94.5	0.8–60
	Average	135.8	110.71	81.33	
West-3 salt sag	—	—	—	—	
East-1 salt sag	Wupu-1 well, Yu-9 well	129.3–157.4	127.1–133	98.3–84.5	0.26–10.59
	Average	143.35	130.05	91.4	
East-2 salt sag	Suijia-1 well, Suitan-1 well, Shanjia-1 well	117.6–206.8	100.98–179.2	85.45–92	0.4–14.44
	Average	147.53	129.46	88.05	

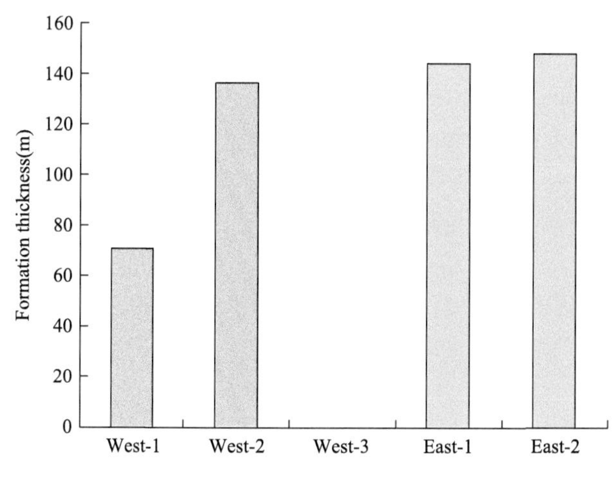

Fig. 7 Thickness ratio of the M_5^6 submember in different salt sags.

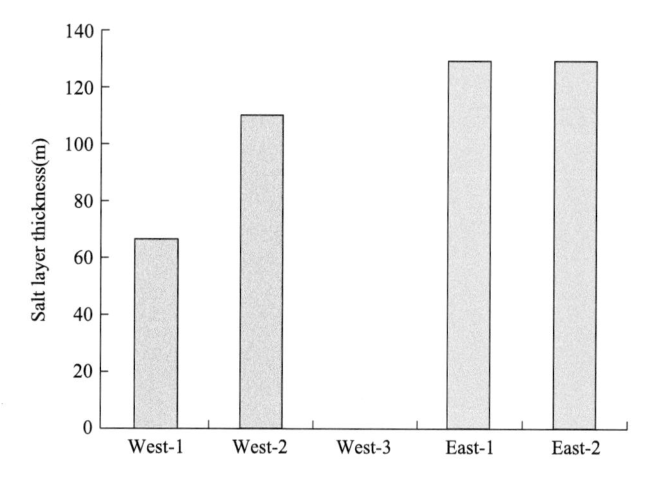

Fig. 8 Thickness ratio of the salt layer of the M_5^6 submember in different salt sags.

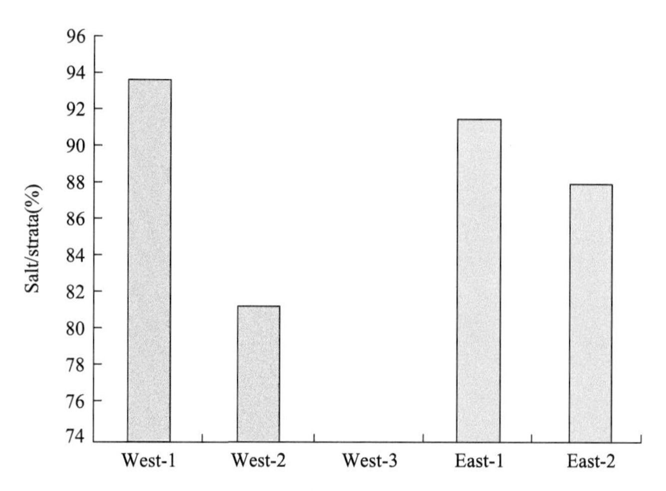

Fig. 9 Salt/strata of the M_5^6 submember in different salt sags.

Geochemical differences between the eastern and western salt sags

Geochemical data for 14 wells in the study area, including the Zhenjia-1, Suijia-1, Longtan-1, and Yutan-1 wells, are summarized in Table 2. The K^+ levels within the Ma_5^6 submember in several wells were between 0.02% and 4.915%. The K^+ contents in the West-1, East-1, and East-2 sags could exceed 0.09% (Table 1), thus indicating that the formation environment was near to the potash deposition stage. Trace element analyses showed that the Br^- contents of the sample from the M_5^6 submember ranged from 100 to 700 ppm. The average Br^- value was between 250 and 300 ppm. The ratio of $Br^-/Cl^- \times 1000$ was between 0.1 and 1.28 (mean 0.46–0.54, Table 2). According to the $Br/Cl \times 10^3$ of the residual brine for a different concentration stage of seawater (Chen, 1972), we infer a marine environment that had reached the stage of potash-carnallite deposition (Table 3).

Table 2 Contrasts among the geochemistry of the salt sags in the northern area of Shanxi

Tectonic Units	Drill well	K^+ (%)	Br^- (%)	$Br^-/Cl^- \times 1000$
West-1 sag	Jiaxian-1 well	0.03–0.57/0.1 (87)	0.03–0.04/0.03 (87)	0.4–0.9/0.54 (87)
	Average	0.1	0.03	0.54
West-2 sag	Mitan-1 well	0.02–0.24/0.08 (91)	0.02–0.07/0.04 (91)	0.3–1.28/0.65 (91)
	Zhenchuan-1 well	—	0.02–0.03\0.02	0.31–0.43/0.368

Continued

Tectonic Units	Drill well	K^+ (%)	Br^- (%)	$Br^-/Cl^- \times 1000$
West-2 sag	Mitan-101 well	0.03–0.12/0.07 (58)	0.03–0.03/0.03 (58)	0.42–0.59/0.48 (58)
	Mitan-103	0.04–0.22/0.07 (56)	0.03–0.03/0.03 (56)	0.45–0.51/0.49 (56)
	Zhenjia-1 well	0.03–0.1/0.05 (88)	0.01–0.04/0.03 (88)	0.17–0.77/0.51 (88)
	Salt 1 well	0.05–0.08/0.07 (8)	0.02–0.23/0.02 (8)	0.30–0.50/0.42 (8)
	Longtan-1 well	0.05–0.13/0.08 (19)	0.02–0.06/0.03 (19)	0.33–1.02/0.48 (19)
	Yutan-1 well	0.04–0.14/0.07 (103)	0.01–0.06/0.03 (103)	0.21–1.06/0.57 (103)
	Average	0.07	0.03	0.51
West-3 sag	—	—	—	—
East-1 sag	Wupu-1 well	0.05–0.34/0.11 (66)	0.01–0.04/0.02 (0.02)	0.1–0.86/0.41 (0.41)
	Yu-9 well	0.02–2.5/0.10 (79)	0.02–0.05/0.03 (79)	0.34–0.86/0.58 (79)
	Average	0.09	0.03	0.50
East-2 sag	Suijia-1 well	0.04–1.24/0.15 (199)	0.01–0.05/0.03 (199)	0.17–0.97/0.50 (199)
	Suide-1 well	0.022–4.915/0.18 (84)	0.01–0.044/0.03 (84)	0.3–0.76/0.51 (84)
	Shanjia-1 well	—	0.01–0.03/0.02	0.03–0.59/0.38
	Average	0.17	0.025	0.46

Notes: The data for the Zhenjia-1, Suijia-1, Longtan-1, and Yutan-1 wells are from the study's test results

Table 3 Characteristics of the Br and $Br^-/Cl^- \times 1000$ at different stages of seawater evolution

Sediment stage	Content of Br^- (10^{-6})		$Br^-/Cl^- \times 1000$	
	Maximum	Minimum	Maximum	Minimum
The saline deposited in the normal seawater	189	68	0.37	0.11
The potash deposited in normal seawater	370	270	0.611	0.445
The carnallite deposited in normal seawater	630	370	10.39	0.611

Geochemical characteristics in the eastern and western salt sags

The geochemical results demonstrated that there are significant differences between the salt-bearing sags in the northern Shanxi basin. The K^+ content in the West-1 and West-2 sags ranged from 0.03% to 0.57% (mean 0.1%) and 0.02% to 0.24% (mean 0.07%), respectively. No data were obtained for the West-3 sag. The K^+ content in the East-1 and East-2 sags ranged from 0.05% to 2.5% (mean 0.09%) and 0.02% to 4.92% (mean 0.17%), respectively. Thus, the mean K^+ contents of the sags were ranked: East-2 > East-1 and West-1 > West-2 (Fig. 10).

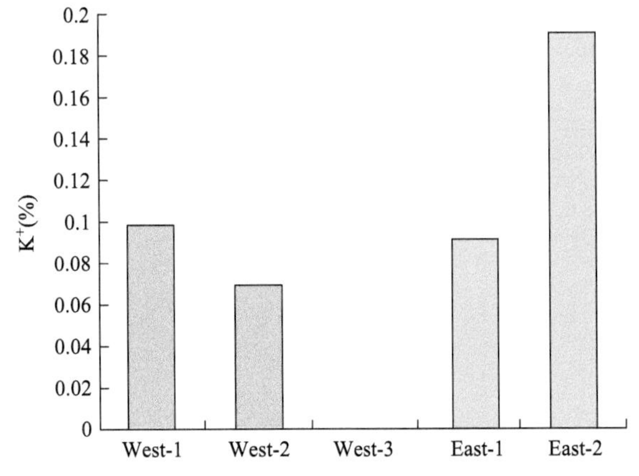

Fig. 10 Contrast of the K^+ content in different salt sags.

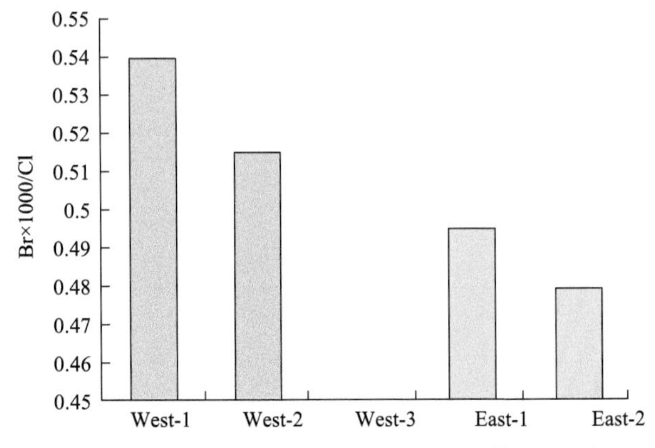

Fig. 11 Contrast of $Br^-/Cl^- \times 1000$ values in different salt sags.

The $Br^-/Cl^- \times 1000$ ratios in the West-1 and West-2 sags ranged from 0.4 to 0.9 (mean 0.54) and 0.17 to 0.57 (mean 0.51), respectively. The ratios in the East-1 and East-2 sags ranged from 0.1 to 0.86 (mean 0.50) and 0.03 to 0.97 (mean 0.46%), respectively. Hence, the $Br^-/Cl^- \times 1000$ ratios in the sags were ranked: West-1>West-2>East-1>East-2 (Fig. 11).

Geochemical profiles in the eastern and western salt sags

The geochemical results from the lower to upper strata in the different salt sags showed that with decreasing depth, the K^+ content initially increased and then decreased. The K^+ levels initially increase, then fall, with the highest value occurring at 2710 m in the West-1 sag (Fig. 12). With decreasing depth, the $Br^-/Cl^- \times 1000$ ratios also decreased, but the changes were less pronounced than those of the K^+ contents in the West-1 sag (Fig. 12).

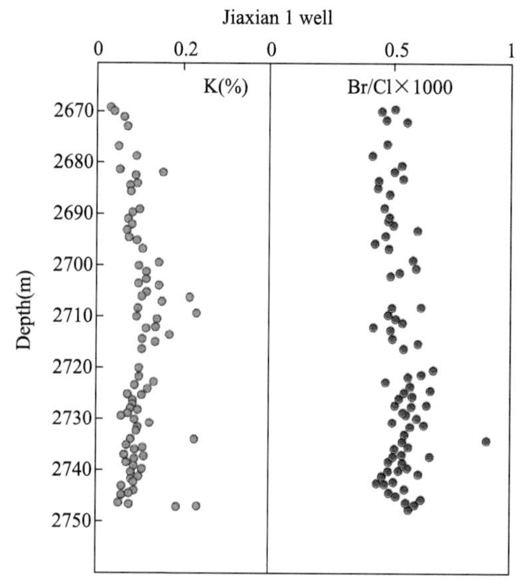

Fig. 12 Change in the trend of K^+ and $Br^-/Cl^- \times 1000$ with depth in the West-1 salt sag.

In the West-2 sag, the curves of K^+ content with depth presented a double peak in several drilling wells. The $Br^-/Cl^- \times 1000$ ratio decreased with decreasing depth and reflected a desalination or a weakened evaporation and concentration trend during brine evolution. There was a weak correlation between the $Br^-/Cl^- \times 1000$ ratio and K^+ content (Fig. 13 and Fig. 14).

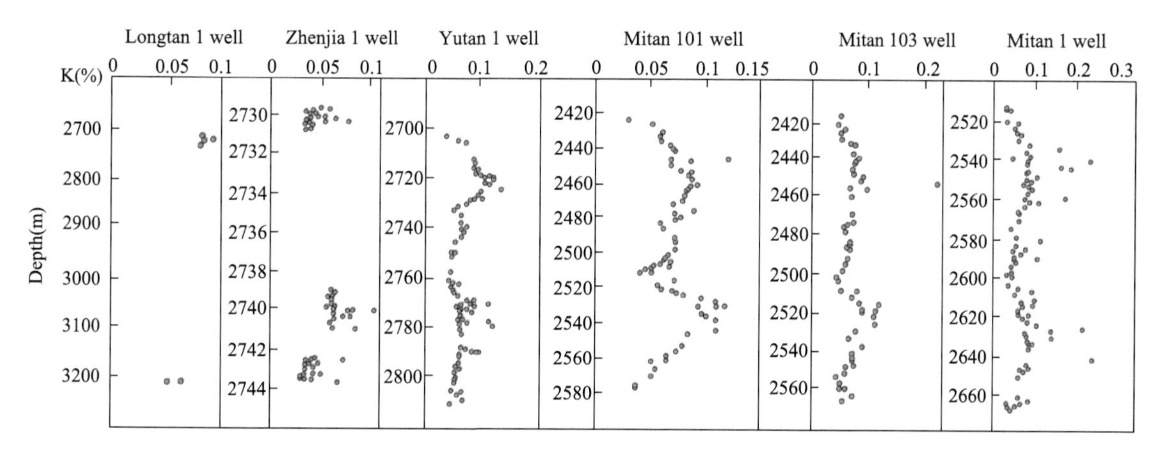

Fig. 13 Change tendency of the K^+ content with depth in the West-2 sag.

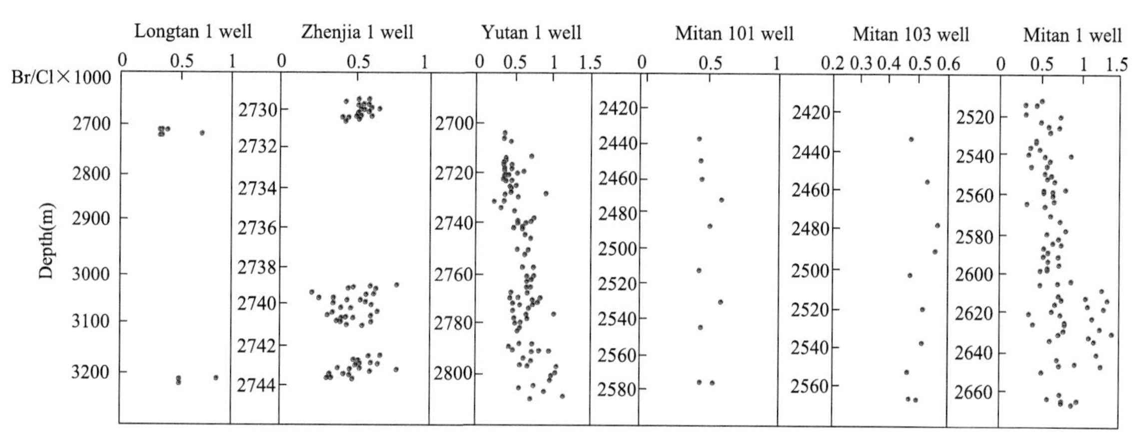

Fig. 14 Change tendency of the $Br^-/Cl^- \times 1000$ content with depth in the West-2 sag.

The curves of K^+ content with depth in the East-1 sag wells also exhibited a double peak. The $Br^-/Cl^- \times 1000$ ratio increased with decreasing depth, but the trend differed to the double peak in K^+ content depth profiles (Fig. 15).

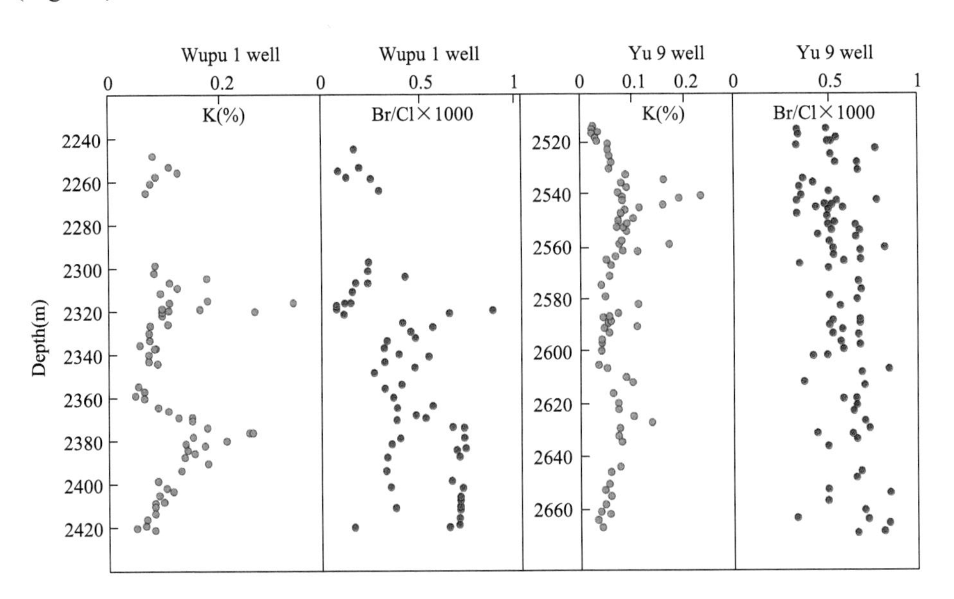

Fig. 15 Change tendency of the K^+ and $Br^-/Cl^- \times 1000$ with depth in the East-1 sag.

The curves of K^+ content with depth in the East-2 sag wells presented either single or double peaks. The single peak of the $Br^-/Cl^- \times 1000$ ratio curve reflected the salinization trend during brine evolution in the East-2 sag (Fig. 16).

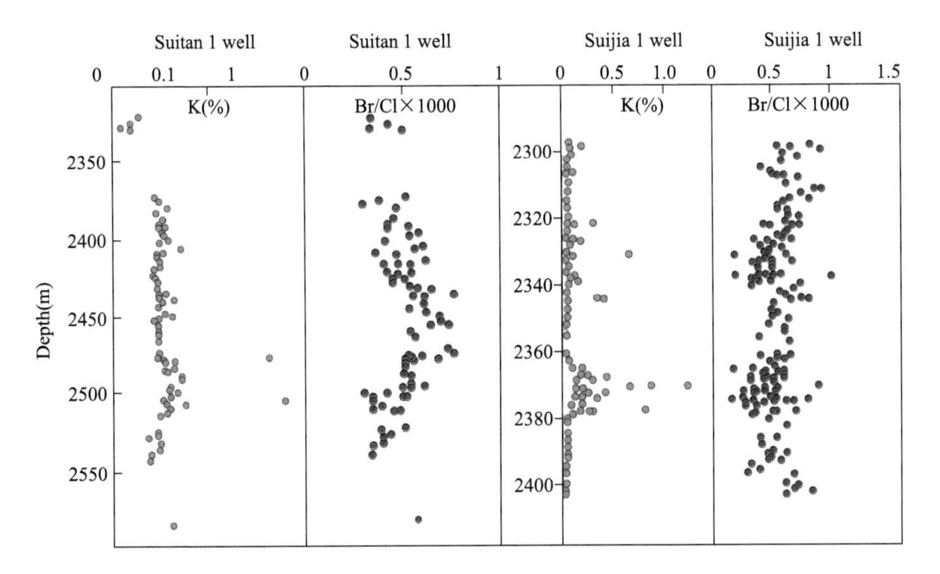

Fig. 16 Change tendency of the K^+ and $Br^-/Cl^- \times 1000$ with depth in the East-2 sag.

Discussion

Differences in potash deposition between salt sags

The results of the geochemical analyses indicate that seawater in the study area had become concentrated to the sylvite and carnallite stage (Meng et al., 2018). The K^+ data from the various secondary sags displayed different patterns of spatial and temporal evolution.

In the West-1 and West-2 sags, the $Br^-/Cl^- \times 1000$ ratios were found to be high, whereas the K^+ contents were generally low. In particular, the potash parameter of the West-2 sag was lowest at the center of the greater Ordos basin. In the depth profiles for the western sags, the curve of the K^+ content exhibited both single and double peaks. The $Br^-/Cl^- \times 1000$ ratio decreased with decreasing depth, thus reflecting desalination or weakened evaporation during brine evolution. Petrological studies of some drilling wells in the West-2 salt sag indicated that the halite consisted of large particles that were dark grey in color. This halite is interpreted as having experienced general recrystallization due to the presence of brine below the redox interface, as well as having experienced desalination in a deep-water environment. The geochemical data also reflected brine desalination that was consistent with the petrological evidence, thereby indicating that conditions were unfavorable for industrial potash deposition.

In the eastern salt sags, thin K^+ ore layers have been reported for the East-1 and East-2 salt sags (Yuan et al., 2010). The K^+ content of samples from the eastern salt sags in the present study were markedly higher than those of sample from the western sags, and the K^+ content of samples from the East-2 sag was slightly higher than those of samples from the East-1 sag.

In depth profiles, the curve of the K^+ content for the East-1 sag presented a double peak, whereas the $Br^-/Cl^- \times 1000$ ratio decreased with decreasing depth, thus indicating brine desalination during brine evolution. The characteristics of brine evolution in the East-1 sag were similar to those in the western sags.

In the East-2 sag, the curve of K^+ content with depth showed a single or double peak, and the $Br^-/Cl^- \times 1000$ ratio initially increased and subsequently decreased with depth, or increased consistently with decreasing depth. The geochemical changes in the East-2 sag were more complex than the other sags. The brine concentration and salinization trends differed between the East-1 sag and the western sags.

Petrological analysis of the East-2 sag demonstrated that halite was mainly maroon in color, and that the muddy interlayers were mostly sage green in color. The residual sylvine particles showed indications of erosion and dissolution. The particle sizes were rather small, and primary cumulate crystals appeared in some samples. The medium and fine sizes of the halite particles indicated enclosed, shallow water conditions above the redox interface during evaporation process.

Classic marine potash deposits worldwide have Br levels that increase with depth in halite sections that contain layers with, or show an initial increase then a subsequent decrease, for example, in the Cassidy Lake Formation of the Windsor Group in the Fundy basin in Canada, in the Michigan basin in the USA, and in the Sakon Nakhon basin in Laos. Such potash layers are mostly pale red in color, which reflects an environment above the redox interface (Qi, 2010; Li et al., 2010; Wittrup and Kyser, 1990; Zhang et al., 2010).

In the East-2 sag, both the geochemical data and sedimentary characteristics were similar to those of classic potassium salt basins; the East-2 salt sag was, therefore, considered to be a favorable setting for the formation of potash deposits.

Mechanism of potash deposition

The Ordos basin, including the northern Shanxi salt basin, was located in an epicontinental sea during the Ordovician period. The area was surrounded by paleo-uplifts that were connected with submarine paleo-uplifts, thereby forming an enclosed and semi-enclosed paleogeographic environment (Meng et al., 2018). With regard to K deposition in epicontinental seas, Usiglio (1849) was the first to propose that halite deposits were formed by evaporation of seawater.

Valyashko (1962) proposed the hypothesis of the prepared basin and dry lake. Chen and Yuan (2010) believed that the North China platform was located in a favorable paleogeographic environment for the formation of thick evaporites. From east to west, in the Jinan-Xuzhou depression, the Shijiazhuang-Lucheng depression, the

Taiyuan-Linfen depression, and the northern area of the Shanxi depression, the thickness of evaporitic rocks increases and salinity also increases, which is similar to the paleogeographic framework proposed for the three-stage deposition and accumulation of potash in the Saskatchewan basin, Canada (Li et al., 2011). The other three adjacent depressions, Jinan-Xuzhou depression, Shijiazhuang-Lucheng, Taiyuan-Linfen depressions possibly provided the pre-existing brine for the Ordovician salt basin as the preparative basins. The distribution was consistent with the K deposition model in a multi-stage basin (Chen and Yuan, 2010).

Feng (1999) considered that seawater entered into the Shanxi basin from the northwest and southeast through the underwater paleo-uplift. In the northern Shanxi basin, Bao et al. (2004) proposed a sedimentary model of drying out and evaporation followed by recharge and redissolution because the basin in which the salt deposits were formed was periodically isolated from the open sea. Low-salinity seawater regularly entered the basin and led to the repeated desalination of the brine in the western sags.

The thickness of the M_5^6 strata, the salt layer, and the salt strata were found to be different in the various salt sags investigated in the present study. The thickest strata and salt layer in the north salt basin were found in the East-2 sag, thus indicating that the eastern sags were the depositional center, whereby the East-1 sag was secondary to the East-2 sag. The western salt sags, on the other hand, exhibited relatively thin strata and salt layer, possibly indicating that the west-2 salt sag was the entry point for seawater from the southwest (Fig. 17).

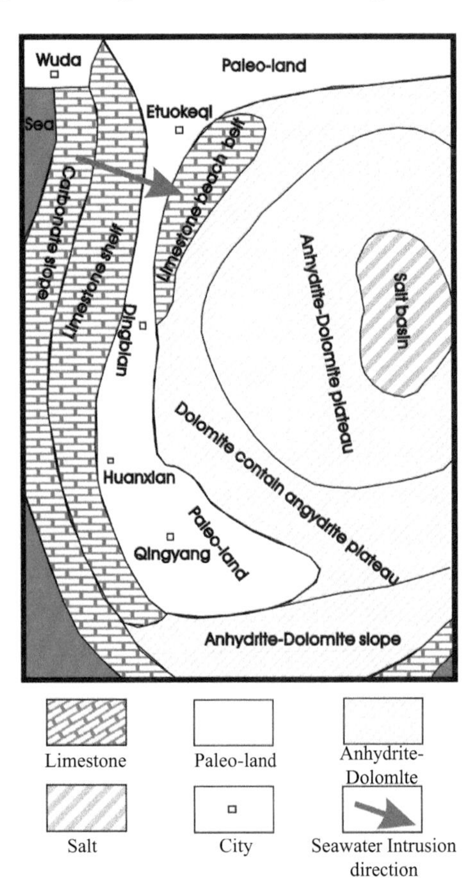

Fig. 17　Middle Ordovician paleogeographic mao of Ordos Basin.

In the western part of the salt basin, the saddle terrain was formed between the Yimeng paleo-land in the north and Qingyang paleo-land in the south. The terrain is relatively low, and it was underwater during the transgression (Liu et al., 1997), which may be an important channel for seawater in the west to invade the salt basin. In addition, from the distribution of lithofacies in the salt basin (Fig. 17), the distribution of lithofacies and sedimentary facies in the evaporation basin is from the outside to the inside, which are successively anhydrite bearing dolomite flat, anhydrite dolomite flat, gypsum salt rock basin, and potassium salt developed near the eastern edge of the salt basin. Only the area near the saddle of Western uplift developed limestone belt (Fig. 17). Research on other potash-bearing salt basins in the world, indicates that potash deposits are mostly distributed far away from the direction of seawater intrusion. The western part of the salt basin near the saddle of the central uplift reflects the development of low-salinity marine sedimentary limestone and the characteristics of potassium deposition in the eastern side of the salt basin, which indicates that the western part is likely to be an important seawater intrusion channel during the evaporation process (Fig. 17).

Moreover, the evolution of the $Br^-/Cl^- \times 1000$ ratios in the West-1 and East-1 sags were consistent with that of the West-2 sag, thereby reflecting the fact that water was well-connected between these sags. The sags were possibly affected by flooding as suggested by the undersaturated seawater under shallow-water conditions. The sedimentology and petrology information clearly revealed the characteristics of the shallow-water conditions in this area. The brine was density-stratified; saturated brine occurred in the deepwater region of the sags and was overlain by unsaturated brine in the upper part of the sags. With evaporation, the upper unsaturated layer would have gradually reached saturation and salts would have deposited together in shallow and deep sags (Galamay et al., 2019).

The geochemistry and sedimentology of the East-2 sag in particular reflected the tendency of the brine to have become concentrated and salinized. This finding indicates an enclosed environment that was only weakly affected by seawater. The thickness of the contour line in Fig. 4 illustrates the middle rise in the north salt basin and the uplift of the terrain between the East-1 and East-2 salt sags. These events possibly led to the enclosed environment of the East-2 salt sag.

The observations from the sampled thin sections demonstrated that sylvite particles grew where salt particles were in contact and formed a disseminated structure. Liu et al. (1997) inferred that the formation of sylvite from an inter-crystal brine indicated a sedimentary environment of a dry saltlake and a residual brine lake. Drill cores containing potash mineralization were observed in the present study to contain many muddy thin interlayers in the eastern salt sags.

Chen and Yuan (2010) conducted chemical analyses of argillaceous sediment in Ordos Ordovician salt basin, believing it originated from freshwater, and characterized the input of roundgenous debris. The authors indicated that periodic movement occurred on the Lishi fault, and that the northern part moved earlier during the final stage of the Late Archean, whereas the southern part moved later. During the Late Triassic, the northern and southern parts connected with each other. Over time, the deposition in the basin changed from being thick in the East and thin in the West from Paleozoic to Middle Triassic to being thin in the East and thick in the west at the end of Triassic (Bai and Sun, 1996). The interlayering of the argillaceous sediment in the East-2 sag was affected by the activation of the north Lishi fracture at an early stage. The thin argillaceous interlayers reflect a stable hydrodynamic environment, an enclosed salt sag, and a weak effect of invading seawater, all of which were favorable circumstances for potash deposition. Microscope observations of thin sections showed that rounded sylvite crystals were present between halite crystals, that sylvite occurred within dissolution embayments of salt particles, and that the mud content was high at the boundaries of the sylvite grains. The erosional characteristics of the sylvite grains indicated that the original potash sediment may have been affected by terrestrial surface water. The increasing mud content at the boundaries of the sylvite particles could have had a protective effect on the sylvine minerals by preventing further dissolution.

The changes in the K^+ contents with depth were markedly different from those of the $Br^-/Cl^- \times 1000$, especially in the West-1 and West-2 sags. The tendency was for an increased $Br^-/Cl^- \times 1000$ ratio with decreasing depth, whereas the K^+ curve shows a double peak. This pattern could be correlated with several sources of material input, including terrigenous materials.

Conclusions

Differences were found in the geochemistry and petrology of the M_5^6 submember between the various salt sags in the study region. We conclude that the West-2 sag was in the direction of seawater invasion to southeast. The western and eastern salt sags exhibit deep-water and desalination characteristics because they were affected by invading seawater, thus indicating unfavorable circumstances for potash preservation. The salinization tendency is apparent in the East-2 sag, and the characteristics of the geochemical and oxidative environment are similar to those of typical potash basins worldwide. The preserved characteristics of the sylvine minerals indicate that the brine evolution reached the dry salt-lake stage in an enclosed environment. The few argillaceous interlayers reflect the input of terrigenous material. The erosion of potash particles and the presence of muds at crystal boundaries reflect fresh-water inputs and the protective effect of argillaceous material for sylvite particles.

The weak correlation between the K^+ content and $Br^-/Cl^- \times 1000$ ratios in the various salt sags reflect that, in addition to seawater, there were many sources of material for the potash deposits.

Acknowledgements

This study was funded by the Ministry of Science and Technology Project (2017YFC0602802), the National Natural Science Foundation of China (Grant Number, 41561144009), Israel Science Foundation (No. 2221/15), and Basic Frontier Scientific Research Program of the Chinese Academy of Sciences (No. ZDBS-LY-DQC021).

References

Bai Y, Sun D (1996) Structural feature and evolution of Lishi fracture zone. Oil Gas Geol 17: 77-80 (**in Chinese with English abstract**)

Bao H, Yang C, Huang J (2004) "Evaporation drying" and "reinfluxing and redissolving": a new hypothesis concerning formation of the Ordovician evaporites in eastern Ordos Basin. J Palaeogeogr 6: 279-288 (**in Chinese with English abstract**)

Chen W, Yuan H (2010) Regional Ore-forming geological conditions of the Ordovician northern Shanxi salt basin. Acta Geol Sin 84: 1565-1575 (**in Chinese with English abstract**)

Feng ZZ, Chen JX, Zhang JS (1991) Lithofacies paleogeography of early Paleozoic in Ordos. Geological Publishing House, Beijing (**in Chinese with English abstract**)

Feng ZZ, Zhang YS, Jin ZK (1998) Type, origin, and reservoir characteristics of dolostones of the Ordovician Majiagou Group, Ordos, North China platform. Sediment Geol 118: 127-140

Galamay AR, Meng FW, Bukowski K, Lyubchak A, Zhang YS, Ni P (2019) Calculation of salt basin depth using fluid inclusions in halite from the Ordovician Ordos Basin in China. Geol Q 63 (3): 619-628

Guo YR, Fu JH, Wei XS, Wei XS, Xu WL, Sun LY, Liu JB, Zhao ZY, Zhang YQ, Gao JR, Zhang YL (2014) Natural gas accumulation and models in Ordovician carbonates, Ordos Basin. NW China Petrol Explor Dev 41 (4): 437-448 (**in Chinese with English abstract**)

Hovorka SD, Holt RM, Powers DW (2007) Depth indicators in Permian Basin evaporites. In: Schreiber BC, Lugli S, Babel M (eds) Evaporites through space and time, vol 85. Geological Society, London, pp 335-364 (special publications 373)

Kovalevych VM, Peryt TM, Zang WL, Vovnyuk SV (2006) Composition of brines in halite-hosted fluid inclusions in the Upper Ordovician, Canning Basin, Western Australia: new data on seawater chemistry. Terra Nova 18: 95-103

Li RX, Guzmics T, Liu XJ, Xie GC (2011) Migration of immiscible hydrocarbons recorded in calcite-hostedfluid inclusions, Ordos Basin: a case study from Northern China. Russ Geol Geophys 52: 1491-1503

Li SP, Ma HZ, Chen YS, Wang SZ, Li WF, Wang QY (2010) Geochemical characteristics of trace elements and ore genesis from potash deposit in Vientiane basin, Laos. Geol Bull China 29: 760-770 (**in Chinese with English abstract**)

Liu Q, Du ZY, Chen YH, Jin RG, Yuan HR, Zhang FG, Zhu YH, Chen YH (1997) Potash prospecting of Ordovician in the north Shanxi and Carboniferous in the Tarim basin. Atomic Energy Press, Beijing, pp 1-237 (**in Chinese with English abstract**)

Meng FW, Zhang YS, Galamy AR, Bukowski K, Ni P, Xing EY, Ji LM (2018) Ordovician seawater composition: evidence from fluid inclusions in halite. Geol Q 62 (2): 344-352

Meng FW, Zhang ZL, Yan XQ, Ni P, Liu WH, Fan F, Xie GW (2019) Stromatolites in Middle Ordovician carbonate-evaporite sequences and their carbon and sulfur isotopes stratigraphy, Ordos Basin, northwestern China. Carbonate Evaporite 34 (1): 11-20

Qi W (2010) The evolution of North America and the formation of potash deposits. Acta Geol Sin 84: 1576-1583 (**in Chinese with English abstract**)

Ronov AB, Khain VE, Balukhovsky AN, Seslavinsky KB (1980) Quantitative analysis of Phanerozoic sedimentation. Sediment Geol 25: 311-325

Scotese CR (1979) Palaeozoic base maps. J Geol 87 (3): 217-277

Usiglio MJ (1849) Étutes sur la composition de l'eau de la Mediterannee et sur l'exploitation des sels qu'elle contient. Ann Chim Phys 27: 172-191

Valyashko MG (1962) The principle of forming of salt deposits (in Russian): Moscow, MGU

Wittrup MB, Kyser TK (1990) The petrogenesis of brine in Devonian potash deposit of western Canada. Chem Geol 82: 103-128

Wu HN, Liu C, Chang CF, Zhong DL (1990) Evolution of the Qinling fold belt and the movement of the north and south China blocks: The evidence of geology and paleomagnetism. Sci Geol Sin 3: 201-214 (**in Chinese with English abstract**)

Yang Y, Li W, Ma L (2005) Tectonic and stratigraphic controls of hydrocarbon systems in the Ordos basin: a multicycle cratonic basin in central China. AAPG Bulletin 89: 255-269

Yuan HR, Zheng MP, Chen WX, Zhang YS, Liu JH (2010) Potash prospects in the Ordovician northern Shanxi Salt Basin. Acta Geol Sin 84: 1554-1563 (**in Chinese with English abstract**)

Zhang X, Ma H, Tan H, Gao D, Li B, Wang M, Tang Q, Yuan X (2010) Preliminary studies of geochemistry and post-depositional change of Dong Tai potash deposit in Laos. Mineral Deposit 29: 713-720 (**in Chinese with English abstract**)

Zhang Y, Zheng MP, Bao HP, Guo Q, Yu CQ, Xing YE, Su K, Fan F, Gong WQ (2013) Tectonic differentiation of O_2 M_5^6 deposition stage in salt basin, northern Shanxi, and its control over the formation of potassium sags. Acta Geol Sin 87 (1): 101-109 (**in Chinese with English abstract**)

Zharkov MA (1984) Paleozoic Salt-bearing Formations of World. Spring-Verlag, Berlin

Zhu RX, Yang ZY, Wu HN, Ma XH, Huang BC, Meng ZF, Fang DJ (1998) Paleomagnetic constraints on the tectonic history of the major blocks of China during Phanerozoic. Sci China (Series D) 41: 1-17

Mineralogical and geochemical characteristics of Triassic Lithium-rich K-bentonite deposits in Xiejiacao section, south China*

Yongjie Lin[1], Mianping Zheng[1], Yongsheng Zhang[1], Enyuan Xing[1], Simon A. T. Redfern [2, 3], Jianming Xu[1], Jiaai Zhong[1] and Xinsheng Niu[1]

1. MNR Key Laboratory of Saline Lake Resources and Environments, Institute of Mineral Resources, Chinese Academy of Geological Sciences, Beijing 100037, China
2. Asian School of the Environment, Nanyang Technological University, Singapore 639798, Singapore
3. Department of Earth Sciences, University of Cambridge, Downing Street, Cambridge CB2 3EQ, UK

Abstract Widespread alteration in the Early-Middle Triassic volcanic ash of the Xiejiacao section, south China, has resulted in significant occurrences of lithium-rich K-bentonite deposits with economic potential. Detailed mineralogical and geochemical investigations of Li-rich K-bentonite deposits from the Xiejiacao section of Guangan city, South China, are presented here. The X-ray diffraction (XRD) data and major element chemistry indicates that the Li-rich K-bentonite deposits contain quartz, clay minerals, feldspar, calcite and dolomite, and the clay minerals are dominated by illite and ordered (R3) illite/smectite (I/S). The concentrations of major and trace elements in Li-rich K-bentonite deposits altered from volcanic ashes are most likely derived from felsic magmas, associated with intense volcanic arc activity. The composition of the clay components suggests that the Li-rich K-bentonite deposits are probably altered from the smectite during diagenesis, whereas smectite is mainly formed by submarine alterations of volcanic materials and subsequently the I/S derived from the volcanogenic smectite illitization. Moreover, accurate determination of the structure in I/S reveals that the temperatures reached by the sedimentary series are around 180 ℃ with a burial depth of ~6000 m. The widely distributed lithium-rich clay deposits strongly indicate widespread eruptions of volcanic ashes in the Early-Middle Triassic, which released huge amounts of volcanic ash. Lithium fixed in the illite and I/S is considered to have leached from the volcanogenic products by a mixed fluid source (i.e., meteoric, porewater and hydrothermal fluids). These Li-rich clay minerals in the marine basin contain economically extractable levels of metal and are a promising new target for lithium exploration.

Keywords Li-rich K-bentonite deposits; Geochemistry; Mineralogy; Early-Middle Triassic; South China

1 Introduction

The increasing demand for lithium in rechargeable batteries, especially for electric vehicles, has attracted a great deal of interest in the search for more potential lithium resources. More recently, Li-rich clays have been now recognized as a significant lithium resource, following the recent assessment of the Li clay deposit in McDermitt/Kings Valley of Nevada, United States and Sonora, Mexico[1]. Although extensive studies have been conducted worldwide on clay deposits of volcanic origin and weathering origin, the global occurrence of lithium-rich clay deposits has rarely been reported, especially in large marine basins [2, 3]. “Mung bean rock” (or “green bean rock”), a type of Li-rich clay deposit, is distributed widely in Early-Middle Triassic strata of south China, over an area of around 700000 km^2, and dozens of centimeters to tens of meters in thickness [4]. It is so called “mung bean rock” because of its green color and often contains siliceous clasts [5, 6]. Several Li-rich clay beds have been found in the outcrop of Early-Middle Triassic strata in Sichuan Basin [7], however, only limited

* 本文发表在：Minerals, 2020, 10 (69)

detailed studies have been published about the mineralogical and geochemical characteristics of mung bean rock. In particular little is known about their formation conditions and the supernormal enrichment mechanism of lithium [2, 6-8].

The Li-rich clay deposits of our study occur and are well-preserved in the Xiejiacao section of Guangan city, South China. Detailed investigations of those minerals are important for obtaining a better understanding of the formation mechanism of Li-rich K-bentonite deposits, particularly of the relationships between the source magmas, altered clay minerals, sedimentary environments, and diagenetic process in these clay deposits. The objectives of this study on Li-rich K-bentonite deposits of the Xiejiacao section are, therefore, to: (1) ascertain if the generic type of Li-rich deposits is volcanic origin or weathering origin; (2) define the clay mineralogy and associated non-clay minerals in detail; (3) characterize the distribution of major and trace elements in the Li-rich clay deposits; and (4) assess the source of the Li-rich K-bentonite deposits and their formation conditions.

2 Geological setting

The Xiejiacao section is situated at Guangan City, Sichuan Province, South China, and tectonically located in the Sichuan Basin, Upper Yangtze Platform (UYP) (Fig. 1). The Yangtze Plate and North China Plate collided diachronically from Late Permian to Triassic, resulting in the closure of the Mianxian-Lueyang paleo-ocean [9, 10]. The palaeogeographic pattern of the Yangtze Plate underwent significant changes in the Triassic. During the Early Triassic to Middle Triassic (Jialingjiang Formation, JF, T_1j), the sedimentary environment of the studied area became restricted due to the westward sea retreat caused by the uplift of the northern margin of the UYP[11]. The whole Sichuan Basin was uplifted due to Episode I of the Indosinian Orogeny by the end of the Middle Triassic, and the sea retreated from the southern part of the basin in this period, thereby exposing the Leikoupo Formation (LF, T_2l) to weathering and erosion for tens of thousands of years. Stratigraphically, the LF conformably overlies the Lower Triassic JF and sits beneath and parallel to the Upper Triassic Xujiahe Formation (XF, T_3xj) [12]. The JF of the UYP, corresponding to the Early Triassic Olenikian stage, is made up of a series of carbonates and evaporites up to 300–1000 m in thickness [13]. The stable UYP was mainly a cratonic carbonate platform during the deposition Middle Triassic LF (corresponding to the Middle Triassic Anisian Stage) and, therefore, limestone, dolomitic limestone, dolomite, gypsiferous dolomite, and gypsum widely formed in this stage [11].

The Li-rich K-bentonite deposits under study, which are composed of a lithology known as mung bean rock, are widely distributed in Early-Middle Triassic strata of the south China, over an area of ~700,000 km^2 and occurring in beds dozens of centimeters to tens of meters in thickness [4]. The Li-rich K-bentonite deposits of the Xiejiacao section is a cemented clay bed ~30 cm thick, exhibiting light green colors and hard characteristics, which has a typical texture of vitric tuff under the microscope (Fig. 2). The Li-rich clay deposits overlies the JF and underlies the LF, and is generally considered to represent the boundary between Early and Middle Triassic. The limestones of the JF, below the bentonitic beds, are rich in foraminifera indicating a late Olenekian stage [16, 17]. However, Zircon U-Pb dating for one sample (XJC-1-R-1) by the laser ablation inductively coupled plasma mass spectrometry (LA-ICP-MS) method yielded one weighted mean $^{206}Pb/^{238}U$ age of 225.9±1.4 Ma (MSWD = 7.1, n = 29) [3], which is approximately Upper Triassic.

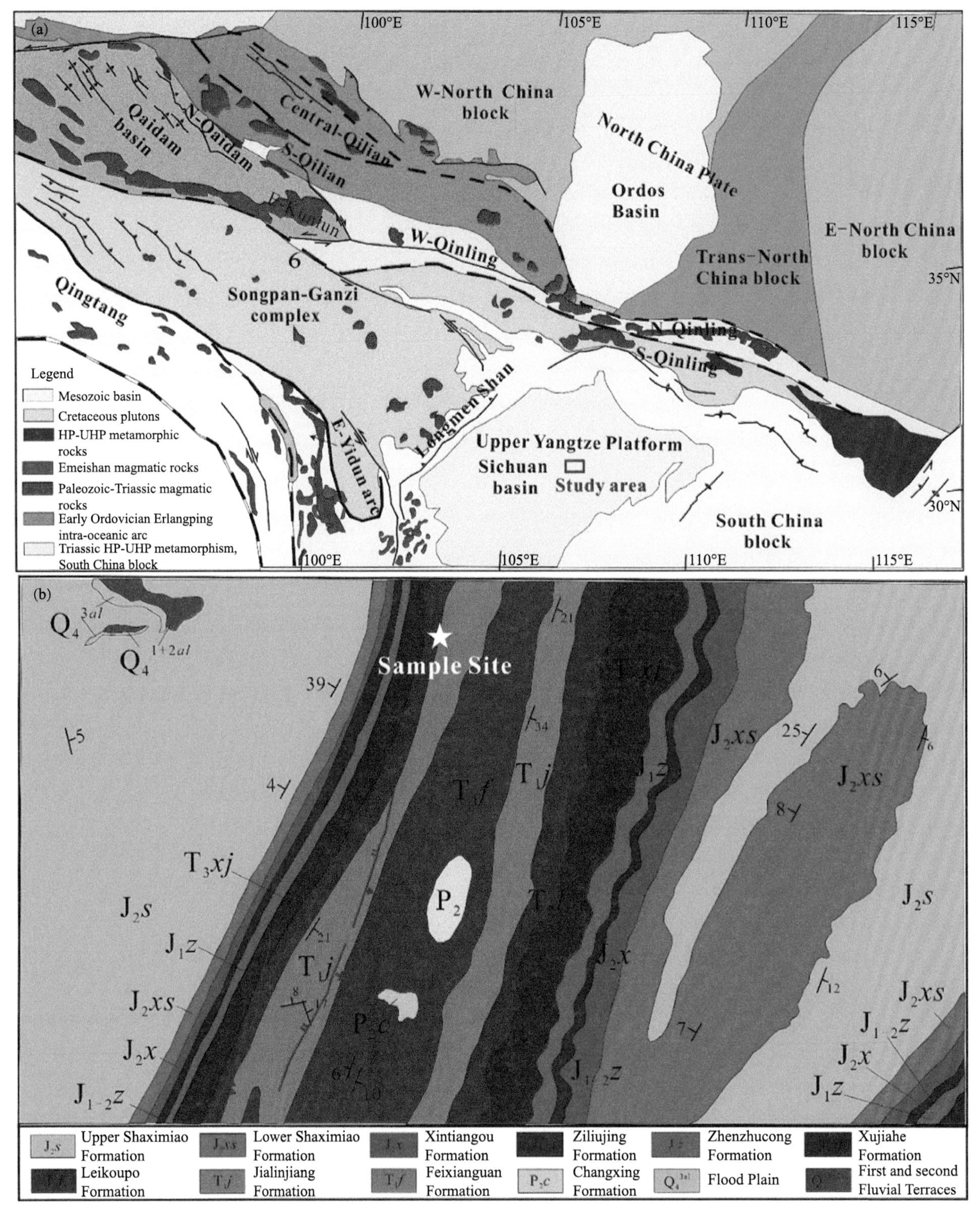

Fig. 1 (a) Simplified regional map showing the locations and tectonics of the Sichuan Basin and its surrounding areas (modified from Enkelmann et al., 2007 [14]). (b) Simplified regional geological map showing the locations of Xiejiacao section in studied area [15].

Fig. 2 (a) Close-up photographs of the Li-rich K-bentonite deposits of Xiejiawan section; (b) a representative hand specimen; (c) microphotograph of Li-rich clays sample S-1 (in plane-polarized light); (d) microphotograph of Li-rich clays sample S-1 (in cross-polarized light).

3 Samples and methods

The Li-rich K-bentonite deposits and JF limestone were collected from the Xiejiacao section, Guangan City, Sichuan Province, in May 2018. One mung bean rock sample (S-1) was collected after removal of the weathered surface crust, which is light green, consolidated and hard, containing elliptical siliceous particles locally (Fig. 2). Three limestone samples were collected from the JF beneath the mung bean rock, and numbered S-2 to S-4. Sun et al. (2018) reported data for three further clay samples (TL-1～3) from the neighboring areas Tongliang section [2], while Ju et al. (2019) published data for one clay sample (XJC-1-R-1) from the same outcrop with this study in the Xiejiacao section [3]. The mineralogy of the whole rock and of the clay minerals on random specimens prepared by the side-loading method was carried out by X-ray diffraction (XRD) a using an X-ray diffractometer (TTR-3, Rigaku Corp, Tokyo, Japan). The XRD was performed at 45 kV and 30 mA with a Cu Kα radiation (λ = 1.54056 Å). The clay mineral fractions (<2 μm) were extracted from the powder samples after centrifugation in distilled water according to the analytical procedure of Jackson et al. (1978) [18]. The XRD analyses were conducted on the air-dried oriented clay sample (AD), ethylene glycol-saturated clay sample (EG), and 550 ℃ heating clay sample, respectively. The mineral compositions and their relative proportions of the bulk rocks and clay minerals in the purified clay samples were obtained using the Clayquan (2016 version) program according to the XRD data of non-oriented powdered samples and oriented clay samples, respectively. The ratio of illite to smectite in the mixed layer illite/smectite (I/S) was calculated by the following formula: $I/(I/S)=\frac{I_{10}(EG)}{I_{10}(550)-I_{10}(EG)}$. Here, I/ (I/S) refers to the mineral contents ratio of illite to mixed layer I/S; I_{10} (EG) indicates the peak area of the (001) peak in the ethylene glycol-saturated clay samples; I_{10} (550) indicates the peak area of (001) for the sample after heating to 550 ℃ [19].

Chemical analyses of all bulk rocks were carried out using X-ray fluorescence spectrometry (XRF) for major elements and inductively coupled plasma mass spectrometry (ICP-MS) for trace elements. Pretreatment procedures were as described by Yang et al. (2007) [20]. Analytical precision was better than 5% for repeated analyses of Chinese national standards GB/T 14506.14–2010 and GB/T 14506.28–2010 [21, 22].

Strontium was separated twice using a cation exchange procedure using 100–200 mesh AG58X8 resin [23] and measured using a Finnigan Triton Thermo ionization mass spectrometer (TIMS) at Nanjing University for isotopic research. The analytical procedure of strontium isotope were described by Hu et al. (2017) [24]. $^{87}Sr/^{86}Sr$ ratios were corrected for mass fractionation by normalizing to $^{86}Sr/^{88}Sr = 0.1194$ with exponential law. Repeated measurements of Sr standard NBS987 yielded $^{87}Sr/^{86}Sr = 0.710259 \pm 0.000015$ (2SD).

4 Results

4.1 Bulk and clay mineralogy

The XRD analyses of the bulk rock, AD-oriented clay sample, EG-saturated clay sample, and 550 ℃ heating clay sample are given in Figure 3. The bulk rock of the Li-rich clay sample (S-1) is composed of clay minerals (37.1%), quartz (33.9%), K-feldspar (19.4%), dolomite (6.6%) and calcite (3.0%). The clay minerals are composed mainly of illite (86%), with minor R3 ordering I/S (13%), and chlorite (1%) (Fig. 3). The air-dried smectite usually show a peak at around 15 Å which changes to around 17 Å upon glycolation. However, the peak at 15–17 Å can only be observed in smectite and partially ordered interstratification I/S, and no peaks at 15–17 Å can be observed in ordered I/S [25]. Therefore, no discrete smectite was found to be present in the samples. The illite is non-expandable and characterized by an intense broad d_{001} peak at around 10 Å, with further peaks at 5 Å and 3.33 Å in the low-angle region, which remain unchanged in all three XRD patterns. In general, the peaks of I/S, overlap with those of illite on air-dried oriented clay samples as well as the 550 ℃ heating clay sample. However, the I/S minerals of S-1 samples are characterized by a strong peak at around 10 Å in the air-dried clay fraction, which collapsed to 9.8966 Å and 10.160 Å after saturation with ethylene glycol and subsequently shrunk to 10 Å after heating at 550 ℃ for 1 h. Due to the presence of only one expandable mineral in the studied sample, the 2θ technique can be reliably used to estimate %S (percent of smectite layers in I/S) [26]. The degree of long-range ordering represented by Reichweite (R) parameters varied consecutively from R0 to R3 via R1 and R2 with increasing illite content [27]. The lowest reliable %S of I/S in Li-rich clay deposits is around 15%, suggesting a high degree of I/S order (R3).

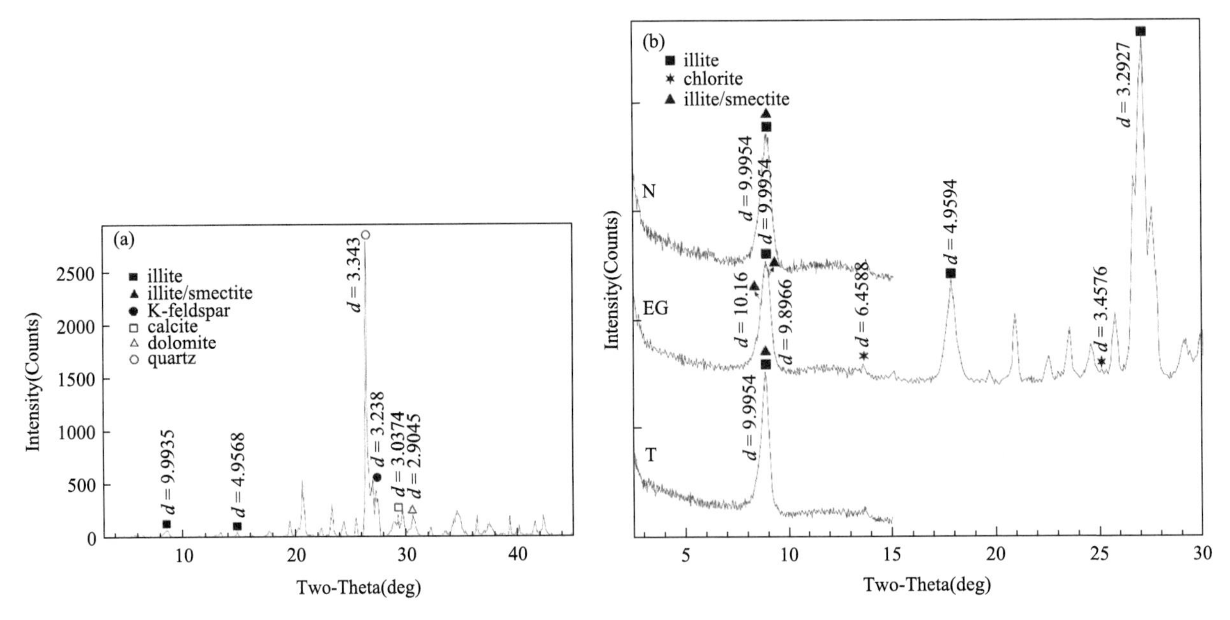

Fig. 3 X-ray diffraction (XRD) patterns of the studied lithium-rich clay sample S-1 from the Xiejiacao section.

(a) Bulk samples; (b) The diffraction pattern of clay minerals. N, Air-dried oriented clay samples; EG, Ethylene glycol-saturated clay samples; T, clay samples post heating to 550 ℃

4.2 Elemental geochemistry

The major and trace element compositions of the Li-rich clay sample (S-1) and limestones samples (S-2, 3, 4) in the studied area are presented in Tables 1 and 2 [3]. The most abundant major component of the limestone sample was CaO (45.68%–54.77%, average = 48.86%), followed by volatiles, measured as loss on ignition (LOI) (38.47–43.92%, average = 41.95%), SiO_2 (0.696%–9.93%, average = 4.075%), and MgO (0.373%–7.20%, average = 3.181%). The results for sample S-1 show that the contents of SiO_2, MgO (4.18%) and K_2O (10.04%) are higher while the Al_2O_3 content (12.41%) is lower than the values reported for the post-Archean Australian shale (PAAS) and north American shale composite (NASC) [28]. The K_2O contents of the ash samples are largely dependent on the presence of I/S and illite, since K is usually fixed in the interlayer sites of illite [29]. The illite typically contains 7% K_2O, and the K_2O content of the ash sample will increase with increased proportion of illite layers in the mixed-layer I/S clays [30-32]. The K_2O contents of the Li-rich clay sample in the Xiejiacao section are obviously higher than those of their limestone host rocks, indicating that K was probably sourced from the porewater and volcanic ashes during early diagenesis. Consequently, a large amount of porewater K is conducive to the formation of mixed-layer R3 ordered I/S clays with smectite layer contents of 15%.

The primitive mantle (PM)-normalized and upper continental crust (UCC)-normalized trace element distributions of our samples are shown in Fig. 4(a), (b)[33]. The most abundant trace element of Li-rich K-bentonite deposits (S-1) is B (867 ppm), followed by Li (321 ppm), Rb (122 ppm), Zr (102 ppm) and Ba (65.4 ppm). The trace elements in Li-rich K-bentonite deposits are characterized as being evidently depleted compared with the PM except for Li, B, Rb and Ba, and depleted compare with the UCC except for Li and B. The clays sample have an overall enrichment of Li and B of 100–1000 times compared to PM, and of 10–100 times compared to UCC. In addition, the clay samples are relatively weakly enriched in high field-strength elements (HFSE), and Zr, Ta, Nb and Hf show a slightly positive anomaly. The trace element compositions of limestone samples (S-3, 4) are characterized as obviously depleted relative to the PM except for Li and B. The limestone sample S-2 has pronounced positive Sr anomalies with a highest Sr value of 1227 ppm.

The rare earth element (REE) data of our sample in the studied area are listed in Table 3 and chondrite-normalized REE distribution patterns are presented in Fig. 4(c). The Eu and Ce anomalies were determined respectively by $\mathrm{Ce/Ce^*} = (\mathrm{Ce_N})/[(\mathrm{La_N}+\mathrm{Pr_N})^{1/2}]$ and $\mathrm{Eu/Eu^*} = (\mathrm{Eu_N})/[(\mathrm{Sm_N}+\mathrm{Tb_N})^{1/2}]$, in which the subscript N denotes normalization of the REE to chondrite [28]. The total REE content of clays sample (S-1) is 73.779 ppm, while limestone samples range from 3.254 ppm to 18.676 ppm with an average of 9.041 ppm. The chondrite-normalize REE distribution patterns of clays and limestone samples show fractionated patterns with a negative Eu anomaly, while the clays sample (S-1) has much larger negative Eu anomalies than the limestone samples. The clay samples have negatively sloping curves with an overall enrichment of light rare earth elements (LREEs) of 10–100 times chondritic, and of heavy rare earth elements (HREEs) a factor of 10. In contrast, the limestone samples have an overall enrichment of LREEs of 1–10 times chondritic, but a deletion of HREEs a factor of 0.01–1. The average (La_N/Yb_N) ratios of limestones and clays samples are 11.1289 and 3.7126 respectively, exhibiting a high degree of fraction between LREEs and HREEs.

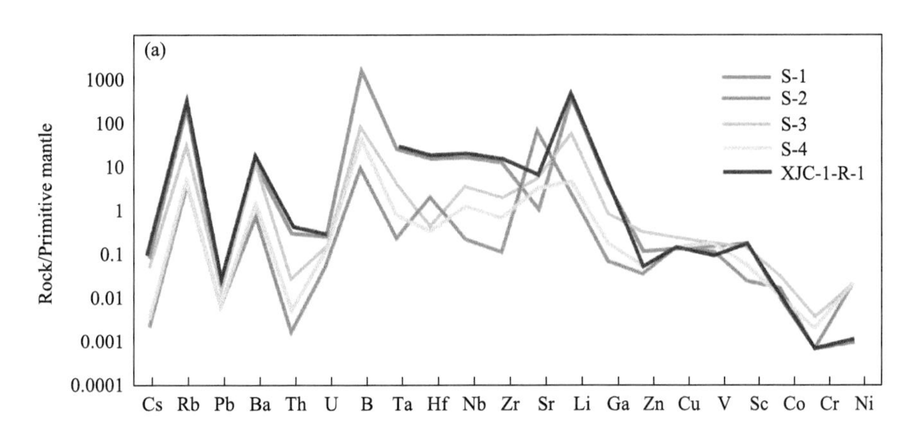

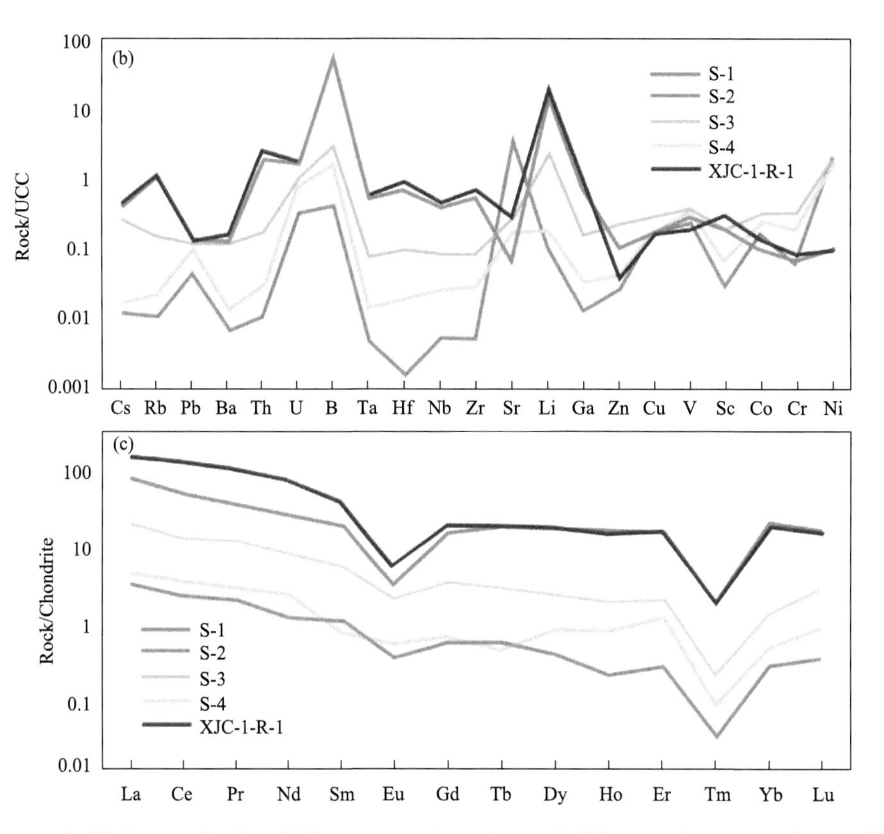

Fig. 4 (a) Primitive-mantle (PM) normalized, and (b) upper continental crust (UCC) normalized trace element distributions, and (c) chondrite-normalized REE distributions of the lithium-rich Kbentonite deposits in Xiejiacao section, South China; Chondrite, PM and UCC normalizing values from Sum and McDonough [33].

Table 1 Major chemical compositions of the collected samples (wt%). Data for sample XJC-1-R-1 are from Ju et al., 2019 [3]; data for TL-1-3 are from Sun et al., 2018 [2]

NO.	SiO_2	Al_2O_3	Fe_2O_3	MgO	CaO	Na_2O	K_2O	MnO	TiO_2	P_2O_5	LOI	FeO	F
S-1	67.73	12.41	0.944	4.18	0.418	0.015	10.04	0.009	0.238	0.029	3.74	0.21	0.261
S-2	0.696	0.154	0.075	0.373	54.77	0.089	0.082	0.005	0.006	0.014	43.46	<0.10	0.044
S-3	9.93	1.82	1.06	1.97	45.68	0.023	0.824	0.041	0.109	0.043	38.47	0.19	0.09
S-4	1.6	0.402	0.357	7.2	46.13	<0.010	0.281	0.015	0.032	0.031	43.92	0.19	0.045
XJC-1-R-1	66.96	14.18	0.60	3.84	0.37	0.06	11.04	0.01	0.23	0.05	3.36	—	—
TL-1	58.52	16.00	0.66	8.11	0.42	<0.01	9.02	<0.01	0.24	0.12	—	—	—
TL-2	62.02	15.00	2.96	3.35	1.36	<0.01	8.44	0.03	0.84	0.38	—	—	—
TL-3	58.24	16.74	0.80	7.29	0.69	<0.01	8.84	<0.01	0.27	0.33	—	—	—

Note: All Fe as Fe_2O_3; LOI is loss on ignition.

Table 2 Concentrations of trace elements and REE of the collected samples (ppm). Data sources are described in the caption for Table 1

NO.	Cs	Rb	Pb	Ba	Th	U	B	Ta	Hf	Nb	Zr	Sr	Li	Ga	Zn	Cu	V	Sc	Co	Cr	Ni
S-1	1.52	122	2.69	65.4	19.8	4.47	867	1.14	4.1	9.63	102	22.4	321	13	7.08	4.1	17.3	2.12	0.97	2.37	2.02
S-2	0.04	1.12	0.89	3.31	0.12	0.9	6	0.01	0.54	0.13	0.9	1227	1.96	0.22	1.75	4.36	15.5	0.31	1.67	2.18	39.7
S-3	0.91	16.3	2.23	62	1.8	2.91	47.2	0.16	0.11	2.09	16.3	92.2	50.1	2.67	17	7.09	23.1	1.99	3.17	11.1	35.9
S-4	0.06	2.34	1.85	6.83	0.33	2.23	25.5	0.03	0.08	0.68	5.2	60.7	3.68	0.53	2.94	4.4	21	0.72	2.35	6.7	35.7
XJC-1-R-1	1.96	140.52	2.86	95.04	31	5.74	—	1.33	5.43	11.94	137.67	101.14	380.02	17.38	2.64	4.09	11.16	3.37	1.38	3.01	1.87
TL-1	11.8	291	5.28	120	29.5	8.02	—	1.55	5.39	12.5	140	26.5	663	22.1	49.1	4.23	4.59	4.03	0.12	—	—
TL-2	12.5	157	34.5	287	21.5	9.81	—	1.5	6.03	17.8	205	50.4	257	0.81	123	35.9	95.3	10.7	8.86	—	—
TL-3	16.8	301	14.2	55.1	16.8	13.3	—	1.59	5.60	13.3	142	21.4	594	0.49	60.7	5.09	4.94	4.4	1.94	—	—

Table 3 Concentrations of rare earth elements of the collected samples (ppm). Data sources are described in the caption for Table 1

NO.	La	Ce	Pr	Nd	Sm	Eu	Gd	Tb	Dy	Y	Ho	Er	Tm	Yb	Lu
S-1	16.3	27.3	2.99	10.3	2.6	0.176	2.75	0.599	3.83	22.3	0.807	2.32	0.461	2.96	0.386
S-2	4.36	6.95	1.01	3.45	0.762	0.114	0.641	0.098	0.553	2.8	0.098	0.309	0.054	0.21	0.067
S-3	0.729	1.3	0.176	0.527	0.155	0.02	0.11	0.02	0.097	0.476	0.012	0.044	0.009	0.046	0.009
S-4	1.02	2.06	0.26	1.01	0.11	0.03	0.131	0.016	0.208	0.998	0.042	0.182	0.023	0.079	0.022
XJC-1-R-1	45.599	99.297	11.286	40.372	6.312	0.28	3.752	0.696	4.506	23.174	0.819	2.555	0.373	2.349	0.348
TL-1	15.9	37.3	4.64	18.5	4.38	0.24	4.61	0.89	6.34	31.2	1.29	3.92	0.62	3.99	0.55
TL-2	79.3	172	20.80	79.3	16.6	1.3	15.4	2.5	15.5	76.8	2.95	8.34	1.23	7.89	1.08
TL-3	26.9	63.6	8.13	33.7	8.98	0.48	10.2	1.86	12.2	61.6	2.39	6.82	1.03	6.38	0.9

4.3 Sr isotopic compositions

The Li-rich K-bentonite sample has a notably higher $^{87}Sr/^{86}Sr$ value of 0.759758 compared to those of 0.708218 to 0.710737 for the limestone samples (Table 4). During the deposition of the JF (Lower Triassic), the $^{87}Sr/^{86}Sr$ ratios of coeval seawater have been previously reported to reach 0.70784 [34], which is approximately consistent with our data. The $^{87}Sr/^{86}Sr$ value of clays sample yield a positive anomaly, which is significantly higher than the amount that could be supplied by river (0.7116), benthic (0.7084) or hydrothermal (0.7037) sources [35]. However, this positive $^{87}Sr/^{86}Sr$ value is close to those of boundary sediments of volcanic origin [36, 37].

Table 4 Sr isotopic compositions of the collected samples

Sample No.	Sample type	$^{87}Sr/^{86}Sr$	Std. error
S-1	Li-rich clays	0.75975786	0.00000577
S-2	Limestones	0.70821791	0.00000562
S-3	limestones	0.71073714	0.00000575
S-4	limestones	0.70876008	0.00000603

5 Discussions

5.1 Source magmas and tectonic settings

The major and trace element compositions of the clay sample (XJC-1-R-1) collected from the same outcrop of the Xiejiacao section and the clay sample (TL-1～3) collected from the Tongliang section in the neighboring area are also presented in Tables 1 and 2 [2, 3]. The REE data of the clay sample (XJC-1-R-1) and (TL-1～3) are also listed in Table 3 and chondrite-normalized REE distribution patterns are presented in Figure 4c. Overall, the major element composition of the bulk rock (S-1) is consistent with the data for the clay sample (XJC-1-R-1) published by Ju et al. (2019) [3], which was collected from the same outcrop in the Xiejiacao section. HFSE, (e.g., Nb, Ti, Zr, Y, Ta) and REEs are widely considered to preserve source characteristics as they usually remain immobile during secondary alteration processes [38, 39]. Therefore, the HFSE contents of bulk rocks are widely used to reveal the parental magmas and tectonic settings of volcanic ashes. The HFSE (Nb, Ta, Hf, Zr), REEs, and TiO_2 are considered important indicators for magmatic origin as they are immobile during diagenesis and weathering [40, 41]. The Al_2O_3/TiO_2 ratio is commonly used as reliable proxy indictor of the provenance, since the contents of Al and Ti remain constant in materials with different degrees of weathering [42, 43]. According to the classification model (Al_2O_3 vs TiO_2), the volcanic ashes corresponding to the Li-rich K-bentonite deposits of the Xiejiacao section (S-1 and XJC-1-R-1) and Tongliang section (TL-1～3) are classified as felsic magmas, with intermediate-acidic to acidic composition [Fig. 5(a)]. The Zr/TiO_2 and Nb/Y ratios are a reliable indicator of alkalinity and differentiation [38, 44]. The Li-rich clay samples of Xiejiacao section and Tongliang section plot in the fields of hyodacitic/dacitic based on the Winchester and Floyd (1977) [44] classification model, suggesting that the Li-rich clays probably originate from rhyodacite/dacite [Fig. 5(b)]. Following the discrimination diagrams after Pearce et al. (1984) [45], the immobile trace element Rb versus Y + Ta show that the Li-rich clays of Xiejiacao section and the Tongliang section lie in the field of within-plate granite, indicating that the parent volcanic eruption were caused by the mid plate volcanoes in the oceanic plate or associated with a mantle plume hotspot [Fig. 5(c)]. Furthermore, magmas that formed by a large fraction of low-temperature melts of felsic continental crust tend to have lower Zr concentration [1, 46]. In addition, magmas with moderate to extreme lithium enrichment are generally considered to have incorporated felsic continental crust [1].

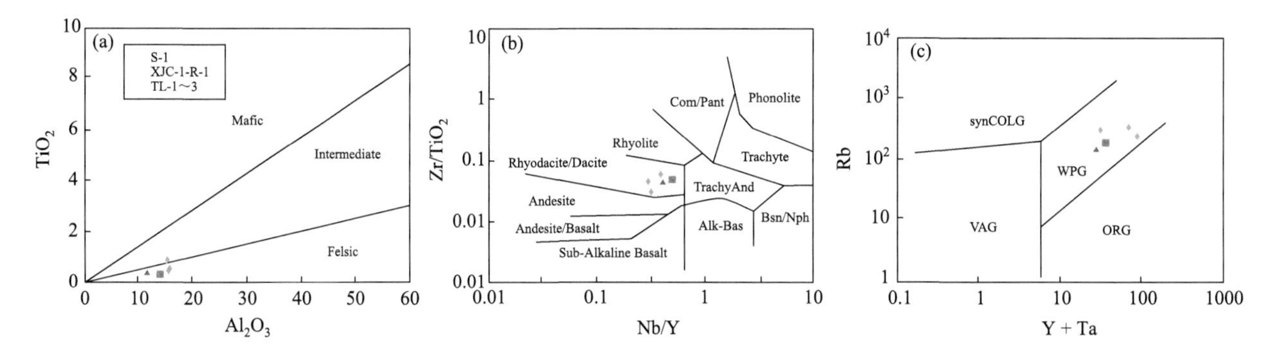

Fig. 5 Li-rich K-bentonite deposits sample plot comparing data from Sun et al. 2018 [2], Ju et al. 2019 [3] and our data.

(a) Plot of TiO_2 and Al_2O_3 (b) Bulk rocks ratios of Nb/Y and Zr/TiO_2 for Li-rich K-bentonite deposits from dashed lines indicate source fields defined by Winchester and Floyd (1977) [44]. (c) Rb versus Y + Ta plots of Li-rich K-bentonite deposits after Pearce et al. (1984) [45]. VAG = Volcanic arc granite; synCOLG = Syn-collision granite; WPG = within-plate granite; ORG = ocean ridge granite

The Sr isotope signatures of samples from marine sections mainly depend on the Sr isotopic composition of seawater and the parent rock of the tuffs [38]. The Sr isotope composition of Li-rich K-bentonite deposits in the Xiejiacao section is 0.759758±0.000012. The Sr isotope composition of underlying limestone samples (S-2~3) of JF range from 0.708218 to 0.710737, which represented the Sr isotopic composition of seawater at that time, far lower than that of the Li-rich clays. Thus, the high $^{87}Sr/^{86}Sr$ values of the Li-rich clays in this study can be attributed to the source parent rock. The increase in the $^{87}Sr/^{86}Sr$ value of clay deposits is mostly related to ignimbrite flare-ups with multiple caldera-forming eruptions, which exhibit a notably high $^{87}Sr/^{86}Sr$ value, corresponding to an increased input from a crustal source and decreased contribution of a mantle source [47]. In general, the rhyolite magmas derived from small proportions (~20%) of felsic continental crust have a high lithium concentration [1]. Therefore, partial melts in the continental crust were initially high in Li, B and other rare metals, which had preconcentrated in the deep stage magma before the volcanic eruption. Overall, the geochemical evidence suggests that the magmatic sources of Li-rich clays are mainly felsic, which most likely derived from a subduction-related volcanism [48, 49].

5.2 Formation mechanism of Li-rich clay

Fresh volcanic ash does not usually contain a significant amount of smectite. In deep-marine environments, volcanic ashes are precursor materials of bentonite, and thus bentonite have a diagenetic origin rather than a sedimentary origin. In this environment, the K, Ca and Mg are readily available and used for the formation of smectite and illite, thus the ash beds are mainly composed of smectite in the early stage and subsequently dominated by illite due to the illitization of smectite [50, 51]. Although the mineralogy and chemistry of the clay minerals that form could be influenced by the variations of burial conditions, the volcanic ash is usually altered into smectite in subaqueous environments [49, 52, 53]. In this study, the clay minerals mainly consist of illite (86%), with minor R3 ordering illite/smectite (13%), and chlorite (1%). The clay minerals of Li-rich clays sample TL-1 and TL-2 from Tongliang section completely comprise I/S, while the sample TL-3 is mainly composed of R3 ordered I/S (68%), illite (24%) and chlorite (5%) [2]. Generally, the I/S in K-bentonite is considered to alter from the smectite during diagenesis, whereas smectite is mainly formed by submarine alteration of volcanic materials [52, 54]. The illitization of smectite begins at about 70–80℃ and leads to a decrease in the percentage of smectite in this process, which is usually characterized by the following reaction sequence: smectite→randomly interstratified illite-smectite mixed layer (I-S)→ordered I-S→illite [55, 56]. Although the composition and permeability of the parent ash seems to be the principal controlling factor in the illitization rates of smectite, it appears that other important factors could have influenced the reaction, such as the burial temperature and time, as well as the chemistry of porewater (especially K^+ content) [56, 57]. The characteristics of mixed layer I/S clay minerals are significantly affected by the deposition facies and porewater K concentrations [58]. The mixed-layer R3 ordered I/S derived from the smectite illitization of a volcanic ash bed observed in this study is probably attributed to deep burial and elevated alteration temperatures [59, 60], while the poorly ordered (R0) mixed-layered I/S is usually attributed to a low burial temperature [61]. The temperatures and depth of post-sedimentary alteration reached by the sediments during the burial diagenesis history could be reliably estimated by the proportion of smectite layers in I/S [58]. The percent of smectite in I/S of the Li-rich K-bentonite is around 15%, suggesting that the temperatures of thermal evolution reached by the sediments are approximately 180 ℃ [55, 62]. Based on the assumption of an average geothermal gradient of 30 ℃/km in the Sichuan basin, the burial depth calculated for the Early-Middle Triassic Li-rich K-bentonites corresponds to around 6 km. The Eu/Eu^* values of eruptive volcanic ashes are initially dependent on the growth of magmatic plagioclase, and the Eu generally tends not to fractionate relative to other REEs during the transport and deposition, except that it can easily be changed in a reducing diagenetic environment. The Eu anomaly, therefore, may be used as a robust proxy to indicate source-magma chemistry [63-65]. Felsic volcanic ashes usually yield a marked significant negative Eu anomaly, and the Li-rich K-bentonite (S-1) shows a negative Eu anomaly ($Eu/Eu^* = 0.47$), lower than that of typical felsic igneous rocks (0.86) [40, 66, 67], which suggests that the Eu is strongly fractionated in a deep burial reducing diagenetic conditions and thus Eu^{3+} becomes more soluble Eu^{2+}. The absence of negative Ce anomalies in the Li-rich clays also indicates that the alteration took place in a suboxic or anoxic environment. The MgO content of Li-rich clays in this study is 4.18%, which is higher than the source magma of felsic volcanic rocks (0.4% to 1.8%) [68]. The Mg fixed in the clay minerals of felsic volcanic ash layers is generally considered to come mainly from the circulation between porewater (e.g., seawater) and the precursor material of Li-rich K-bentonite deposits during the post-sedimentary alteration [69-71]. The ash beds in this study are suggested to have formed in a marine sedimentary setting with a markedly high Mg content. In general, the smectite to chlorite transition is considered to mainly occur in three distinct geological environments: subaerial and submarine hydrothermal systems, sedimentary basins, and regional metamorphic terranes[72]. Smectite and chlorite are ubiquitous products of the diagenesis and low-temperature metamorphism of intermediate to mafic volcanic rocks and volcanogenic sediments[73]. In this study, large amounts of porewater Mg and Fe in alkaline marine environments probably drove the transition of smectite to chlorite [72, 74].

6 Conclusions

The mineralogical and geochemical characteristics of Li-rich clay deposits from the Early-Middle Triassic portion of the Xiejiacao section, south China were investigated in detail in this work. Although the volcanic ashes examined have suffered significant diagenetic changes, geochemical and mineralogical analyses suggest that primary magmatic composition of this bed was mainly rhyodacite/dacitic; trace-element data indicate the parent eruptive centers are characterized by the tectonic settings of a within-plate granite. Moreover, the precise determination of the I/S ratio reveals the thermal evolution of the sedimentary sequences reached approximately

180 ℃ corresponding to a burial depth of around 6 km. Overall, the mineralogical composition of the clay fraction indicates this lithium-rich clay deposit derived from volcanic materials shortly after deposition in deep marine conditions, and the clay fraction mainly contains illite and I/S clays derived from the volcanogenic smectite illitization. The Li in the interlayer space of illite and I/S is probably leached from the eruptive products by a mixed fluid source (i.e. meteoric, porewater and hydrothermal fluids) to become concentrated in the final clay minerals after post-sedimentary alteration.

Acknowledgments

We thank Lisa Stillings (U.S. Geological Survey), Linda Godfrey (Rutgers University) and the two anonymous reviewers for their insightful comments that greatly improved this manuscript.

References

[1] Benson, T.R.; Coble, M.A.; Rytuba, J.J.; Mahood, G. A. Lithium enrichment in intracontinental rhyolite magmas leads to Li deposits in caldera basins. *Nat. Commun.* **2017**, 1, 1-9. [CrossRef] [PubMed]

[2] Sun, Y.; Wang, D.; Gao, Y.; Han, J.; Ma, S.; Fan, X.; Gu, W.; Zhang, L. Geochemical characteristics of lithium-rich mung bean rocks in Tongliang County, Chongqing. *Acta Petrol. Mineral.* **2018**, 37, 395-403. (In Chinese with English Abstract)

[3] Ju, P.; Wang, X.; Wang, Z.; Liu, X.; Zhong, J.; Zhang, Z. Characteristics and geological significance of the Triassic mungbean rocks the Wenquan Town area, northern Chongqing. *Geoscience* **2019**, accepted.

[4] Zhu, Z.; Wang, G. Paleogeography of before and after deposition of green-bean rock (altered tuff) between the early and middle Triassic in the Uper Yangtze platform and its adjacent areas. *Oil Gas Geol.* **1986**, 7, 344-355. (In Chinese with English Abstract)

[5] Chen, Z.; Shen, M.; Zhao, J.; Tang, H. New Analysis of 'Mungbean Rock' Composition. *J. Southwest Pet. Inst.***1999**, 21, 39-42. (In Chinese)

[6] Zhu, L. The Genesis and Characteristics of clay minerals of Green-bean rock in Guizhou Province. *Acta Mineral. Sin.* **1995**, 15, 75-81. (In Chinese with English Abstract)

[7] Ma, S.; Wang, D.; Sun, Y.; Li, C.; Zhong, H. Geochronology and Geochemical Characteristics of Lower-Middle Triassic Clay Rock and Their Significances for Prospecting Clay-Type Lithium Deposit. *Earth Sci.* **2019**, 44, 2-14. (In Chinese with English Abstract)

[8] Sun, Y.; Gao, Y.; Wang, D.; Dai, H.; Gu, W.; Li, J.; Zhang, L. Zircon U-Pb Dating of 'Mung Bean Rock' in the Tongliang Area, Chongqing and Its Geological Significance. *Rock Miner. Anal.* **2017**, 36, 649-658. (In Chinese with English Abstract)

[9] Yin, A.; Nie, S. An indentation model for the North and South China collision and the development of the Tan-Lu and Honam Fault Systems, eastern Asia. *Tectonics* **1993**, 12, 801-813. [CrossRef]

[10] Ratschbacher, L.; Hacker, B.R.; Calvert, A.; Webb, L.E.; Grimmer, J.C.; McWilliams, M.O.; Ireland, T.; Dong, S.; Hu, J. Tectonics of the Qinling (Central China): Tectonostratigraphy, geochronology, and deformation history. *Tectonophysics* **2003**, 366, 1-53. [CrossRef]

[11] Zhong, Y.-J.; Chen, A.-Q.; Huang, K.-K.; Yang, S.; Xu, S.-L. Hydrothermal activity in the fourth member of the Triassic Leikoupo Formation in the Yuanba area, northeast Sichuan, SW China. *Geol.* J. **2019**, 54, 266-277.[CrossRef]

[12] Tan, X.; Li, L.; Liu, H.; Cao, J.; Wu, X.; Zhou, S.; Shi, X. Mega-shoaling in carbonate platform of the Middle Triassic Leikoupo Formation, Sichuan Basin, southwest China. *Sci. China Earth Sci.* **2014**, 57, 465-479.[CrossRef]

[13] Zhu, J. Characteristics and origin of polycrystalline dolomite needles in the Triassic Jialingjiang Formation, Upper Yangtze Platform, southwest China. *Sediment. Geol.* **1998**, 118, 119-126.

[14] Enkelmann, E.; Weislogel, A.; Ratschbacher, L.; Eide, E.; Renno, A.; Wooden, J. How was the Triassic Songpan-Ganzi basin filled? A provenance study. *Tectonics* **2007**, 26, 1-24. [CrossRef]

[15] Wei, Q.; Liang, Z. *Geological Survey Report of the Dianjiang Area, Sichuang Province (Scale 1: 200000 Pieces)*; 107 Geological Brigade of Sichuan Bureau of Geology & Mineral Resources: Chongqing, China, 1980; pp. 1-91. (In Chinese)

[16] Zhou, Z.; Luo, H.; Zhu, Y.; Cai, H.; Xu, B.; Chen, J.; Zhao, Y. Early Trassic Anachronistic Sediments at the Xiejiacao Section of Guangan, Sichuan Province and Their paleoecological Significances: A restudy of a Classical Section. *J. Stratigr.* **2013**, 37, 73-80. (In Chinese with English Abstract)

[17] Yang, H.; Luo, H.; Zhu, Y.; Cai, H.; Xu, B.; Zhou, Z. Early and middle Triassic foraminifera from the Xiejiacao section Guangan, Sichuan Province, China. *Acta Micropalaeontol. Sin.* **2017**, 34, 390-405. (In Chinese with English Abstract)

[18] Jackson, K.; Jonasson, I.; Skippen, G. The nature of metals-sediment-water interactions in freshwater bodies, with emphasis on the role of organic matter. *Earth Sci. Rev.* **1978**, 14, 97-146. [CrossRef]

[19] Wang, C.; Li, M.; Fang, X.; Liu, Y.; Yan, M. Structural characteristic of mixed-layer illite/Smectite clay minerals of the SG-1 core in the western Qaidam basin and its environmental significance. *Quat. Sci.* **2016**, 36, 917-925. (In Chinese with English Abstract)

[20] Yang, X.; Zhu, B.; White, P.D. Provenance of aeolian sediment in the Taklamakan Desert of western China, inferred from REE and major-elemental data. *Quat. Int.* **2007**, 175, 71-85. [CrossRef]

[21] Li, Y.; Xing, Y.; Wang, S. Part 14: Determination of ferrous oxide content. In *Methods for Chemical Analysis of Silicate Rocks*; Standards Press of China: Beijing, China, 2010; pp. 1-3. (In Chinese)

[22] Wang, S.; Yan, M. Part 28: Determination of 16 major and minor elements content. In *Methods for Chemical Analysis of Silicate Rocks*; Standards Press of China: Beijing, China, 2010; pp. 1-6. (In Chinese)

[23] Ling, H.-F.; Feng, H.-Z.; Pan, J.-Y.; Jiang, S.-Y.; Chen, Y.-Q.; Chen, X. Carbon isotope variation through the Neoproterozoic Doushantuo and Dengying Formations, South China: Implications for chemostratigraphy and paleoenvironmental change. *Palaeogeogr. Palaeoclimatol. Palaeoecol.* **2007**, 254, 158-174. [CrossRef]

[24] Hu, Z.; Hu, W.; Wang, X.; Lu, Y.; Wang, L.; Liao, Z.; Li, W. Resetting of Mg isotopes between calcite and dolomite during burial metamorphism: Outlook of Mg isotopes as geothermometer and seawater proxy. *Geochim. Cosmochim. Acta* **2017**, 208, 24-40. [CrossRef]

[25] Lin, X. *X-Ray Diffraction Analysis Technology and Its Geological Application*; Petroleum Industry Press: Beijing, China, 1990.

[26] Środoń, J. X-ray Powder Diffraction Identification of Illitic Materials. *Clays Clay Miner.* **1984**, 32, 337-349. [CrossRef]

[27] Inoue, A.; Lanson, B.; Marques-Fernandes, M.; Sakharov, B.A.; Murakami, T.; Meunier, A.; Beaufort, D. Illite-smectite mixed-layer minerals in the hydrothermal alteration of volcanic rocks: I. One-dimensional XRD structure analysis and characterization of component layers. *Clays Clay Miner.* **2005**, 53, 423-439. [CrossRef]

[28] Taylor, S.R.; McLennan, S.M. *The Continental Crust: Its Composition and Evolution. An Examination of the Geochemical Record Preserved in Sedimentary Rocks*; Blackwell Science: Oxford, UK, 1985.

[29] Hong, H.; Algeo, T.J.; Fang, Q.; Zhao, L.; Ji, K.; Yin, K.; Wang, C.; Cheng, S. Facies dependence of the mineralogy and geochemistry of altered volcanic ash beds: An example from Permian-Triassic transition strata in southwestern China. *Earth Sci. Rev.* **2019**, 190, 58-88. [CrossRef]

[30] Yard, W.N. Mineralogy and petrology of some ordovician K-bentonites and related limestones. *Geol. Soc. Am. Bull.* **1953**, 64, 921-943.

[31] Altaner, S.P.; Hower, J.; Whitney, G.; Aronson, J.L. Model for K-bentonite formation: Evidence from zoned K-bentonites in the disturbed belt, Montana. *Geology* **1984**, 12, 412-415. [CrossRef]

[32] Huff, D.W. K-bentonites: A review. *Am. Mineral.* **2016**, 101, 43-70. [CrossRef]

[33] Sun, S.; McDonough, W.F. Chemical and isotopic systematics of ocean basalts: Implications for mantle composition and processes. In *Magmatism in the Ocean Basins*; Geological Socoety of London Special Publication: London, UK, 1989; pp. 1-345.

[34] Jiang, L.; Cai, C.F.; Worden, R.H.; Li, K.-K.; Xiang, L. Reflux dolomitization of the Upper Permian Changxing Formation and the Lower Triassic Feixianguan Formation, NE Sichuan Basin, China. *Geofluids* **2013**, 13, 232-245. [CrossRef]

[35] Davis, A.C.; Bickle, M.J.; Teagle, D.A.H. Imbalance in the oceanic strontium budget. *Earth Planet. Sci. Lett.* **2003**, 211, 173-187. [CrossRef]

[36] Yang, S.; Hu, W.; Wang, X.; Jiang, B.; Yao, S.; Sun, F.; Huang, Z.; Zhu, F. Duration, evolution, and implications of volcanic activity across the Ordovician-Silurian transition in the Lower Yangtze region, South China. *Earth Planet. Sci. Lett.* **2019**, 518, 13-25. [CrossRef]

[37] Ramkumar, M.; Stuben, D.; Berner, Z.; Schneider, J. $^{87}Sr/^{86}Sr$ anomalies in Late Cretaceous-Early Tertiary strata of the Cauvery basin, south India: Constraints on nature and rate of environmental changes across K-T boundary. *J. Earth Syst. Sci.* **2010**, 119, 1-17. [CrossRef]

[38] Gong, N.; Hong, H.; Huff, W.D.; Fang, Q.; Bae, C.J.; Wang, C.; Yin, K.; Chen, S. Influences of Sedimentary Environments and Volcanic Sources on Diagenetic Alteration of Volcanic Tuffs in South China. *Sci. Rep.* **2018**, 8, 1-12. [CrossRef] [PubMed]

[39] Ge, X.; Mou, C.; Wang, C.; Men, X.; Chen, C.; Hou, Q. Mineralogical and geochemical characteristics of K-bentonites from the Late Ordovician to the Early Silurian in South China and their geological significance. *Geol. J.* **2019**, 54, 514-528. [CrossRef]

[40] He, B.; Zhong, Y.-T.; Xu, Y.-G.; Li, X.-H. Triggers of Permo-Triassic boundary mass extinction in South China: The Siberian Traps or Paleo-Tethys ignimbrite flare-up? *Lithos* **2014**, 204, 258-267. [CrossRef]

[41] Berry, R.W. Eocene and Oligocene Otay-Type Waxy Bentonites of San Diego County and Baja California: Chemistry, Mineralogy, Petrology and Plate Tectonic Implications. *Clays Clay Miner.* **1999**, 47, 70-83. [CrossRef]

[42] Sugitani, K. Anomalously low Al_2O_3/TiO_2 values for Archean cherts from the Pilbara Block, Western Australia—Possible evidence for extensive chemical weathering on the early earth. *Precambrian Res.* **1996**, 80, 49-76. [CrossRef]

[43] Nesbitt, H.W.; Young, G. M. Early Proterozoic climates and plate motions inferred from major element chemistry of lutites. *Nature* **1982**, 299, 715-717. [CrossRef]

[44] Winchester, J.A.; Floyd, P.A. Geochemical discrimination of different magma series and their differentiation products using immobile elements. *Chem. Geol.* **1977**, 20, 325-343. [CrossRef]

[45] Pearce, J.A.; Harris, N.B.W.; Tindle, A.G. Trace Element Discrimination Diagrams for the Tectonic Interpretation of Granitic Rocks. *J. Petrol.* **1984**, 25, 956-983. [CrossRef]

[46] Watson, E.B.; Harrison, T.M. Zircon saturation revisited: Temperature and composition effects in a variety of crustal magma types. *Earth Planet. Sci. Lett.* **1983**, 64, 295-304. [CrossRef]

[47] Ducea, M.N.; Barton, M.D. Igniting flare-up events in Cordilleran arcs. *Geology* **2007**, 35, 1047-1050. [CrossRef]

[48] Su, W.; Huff, W.D.; Ettensohn, F.R.; Liu, X.; Zhang, J.; Li, Z. K-bentonite, black-shale and flysch successions at the Ordovician-Silurian transition, South China: Possible sedimentary responses to the accretion of Cathaysia to the Yangtze Block and its implications for the evolution of Gondwana. *Gondwana Res.* **2009**, 15, 111-130. [CrossRef]

[49] Huff, W.D.; Bermöm, S.M.; Kolata, D.R. Silurian K-bentonites of the Dnestr Basin, Podolia, Ukraine. *J. Geol. Soc.* **2000**, 157, 493-504. [CrossRef]

[50] Mizota, C.; Faure, K. Hydrothermal origin of smectite in volvanic ash. *Clays Clay Miner.* **1998**, 46, 178-182. [CrossRef]

[51] Bohor, T.B.; Triplehorn, M.D. *Tonsteins: Altered Volcanic-Ash Layers in Coal-Bearing Sequences*; Geological Society of America: Boulder, CO, USA, 1993; Volume 285, ISBN 0-8137-2285-3.

[52] Fisher, R.V.; Schmincke, H.-U. *Pyroclastic Rocks*; Springer: Berlin/Heidelberg, Germany, 1984; ISBN 978-3-540-51341-4.

[53] De la Fuente, S.; Cuadros, J.; Fiore, S.; Linares, J. Electron Microscopy Study of Volcanic Tuff Alteration to Illite-Smectite under Hydrothermal Conditions. *Clays Clay Miner.* **2000**, 48, 339-350. [CrossRef]

[54] Lanson, B.; Sakharov, B.A.; Claret, F.; Drits, V.A. Diagenetic smectite-to-illite transition in clay-rich sediments: A reappraisal of X-ray diffraction results using the multi-specimen method. *Am. J. Sci.* **2009**, 309, 476-516. [CrossRef]

[55] Środoń, J.; Clauer, N.; Huff, W.; Dudek, T.; Banaś, M. K-Ar dating of the Lower Palaeozoic K-bentonites from the Baltic Basin and the Baltic Shield: Implications for the role of temperature and time in the illitization of smectite. *Clay Miner.* **2009**, 44, 361-387. [CrossRef]

[56] Lanson, B. Late-Stage Diagenesis of Illitic Clay Minerals as Seen by Decomposition of X-ray Diffraction Patterns: Contrasted Behaviors of Sedimentary Basins with Different Burial Histories. *Clays Clay Miner.* **1998**, 46, 69-78. [CrossRef]

[57] Nadeau, P.H. Burial and Contact Metamorphism in the Mancos Shale. *Clays Clay Miner.* **1981**, 29, 249-259. [CrossRef]

[58] Deconinck, J.F.; Crasquin, S.; Bruneau, L.; Pellenard, P.; Baudin, F.; Feng, Q. Diagenesis of clay minerals and K-bentonites in Late Permian/Early Triassic sediments of the Sichuan Basin (Chaotian section, Central China). *J. Asian Earth Sci.* **2014**, 81, 28-37. [CrossRef]

[59] McCarty, D.K.; Sakharov, B.A.; Drits, V.A. New insights into smectite illitization: A zoned K-bentonite revisited. Am. Mineral. **2009**, 94, 1653-1671. [CrossRef]

[60] Bozkaya, Ö.; Yalçin, H. Geochemistry of mixed-layer illite-smectites from an extensional basin, Antalya unit, southwestern Turkey. *Clays Clay Miner.* **2010**, 58, 644-666. [CrossRef]

[61] Hints, R.; Kirsimäe, K.; Somelar, P.; Kallaste, T.; Kiipli, T. Multiphase Silurian bentonites in the Baltic Palaeobasin. *Sediment. Geol.* **2008**, 209, 69-79. [CrossRef]

[62] Šucha, V.; Kraust, I.; Gerthofferová, H.; Peteš, J.; Sereková, M. Smectite to Illite Conversion in Bentonites and Shales of the East Slovak Basin. *Clay Miner.* **1993**, 28, 243-253. [CrossRef]

[63] Zielinski, R.A. The mobility of uranium and other elements during alteration of rhyolite ash to montmorillonite: A case study in the Troublesome Formation, Colorado, U.S.A. *Chem. Geol.* **1982**, 35, 185-204. [CrossRef]

[64] Sverjensky, D.A. Europium redox equilibria in aqueous solution. *Earth Planet. Sci. Lett.* **1984**, 67, 70-78. [CrossRef]

[65] Püspöki, Z.; Kozák, M.; Kovács-Pálffy, P.; Szepesi, J.; McIntosh, R.; Kónya, P.; Vincze, L.; Gyula, G. Geochemical Records of a Bentonitic Acid-Tuff Succession Related to a Transgressive Systems Tract—Indication of Changes in the Volcanic Sedimentation Rate. *Clays Clay Miner.* **2008**, 56, 23-38. [CrossRef]

[66] Kramer, W.; Weatherall, G.; Offer, R. Origin and correlation of tuffs in the Permian Newcastle and Wollombi Coal Measures, NSW, Australia, using chemical fingerprinting. *Int. J. Coal Geol.* **2001**, 47, 115-135. [CrossRef]

[67] Siir, S.; Kallaste, T.; Kiipli, T.; Hints, R. Internal stratification of two thick Ordovician bentonites of Estonia: Deciphering primary magmatic, sedimentary, environmental and diagenetic signatures. *Est. J. Earth Sci.* **2015**, 64, 140-158. [CrossRef]

[68] Le Maitre, R.W. The Chemical Variability of some Common Igneous Rocks. *J. Petrol.* **1976**, 17, 589-598. [CrossRef]

[69] Huff, W.D.; Bergstrom, S.M.; Kolata, D.R.; Cingolani, C.A.; Astini, R.A. Ordovician K-bentonites in the Argentine Precordillera: Relations to Gondwana margin evolution. *Geol. Soc. Lond. Spec. Publ.* **1998**, 142, 107-126. [CrossRef]

[70] Arslan, M.; Abdioğlu, E.; Kadir, S. Mineralogy, Geochemistry, and Origin of Bentonite in Upper Cretaceous Pyroclastic Units of the Tirebolu Area, Giresun, Northeast Turkey. *Clays Clay Miner.* **2010**, 58, 120-141. [CrossRef]

[71] Zielinski, R.A. Element mobility during alteration of silicic ash to kaolinite-a study of tonstein. *Sedimentology* **1985**, 32, 567-579. [CrossRef]

[72] Schiffman, P.; Staudigel, H. The smectite to chlorite transition in a fossil seamount hydrothermal system: The Basement Complex of La Palma, Canary Islands. *J. Metamorph. Geol.* **1995**, 13, 487-498. [CrossRef]

[73] Bettison-Varga, L. The Role of Randomly Mixed-Layered Chlorite/Smectite in the Transformation of Smectite to Chlorite. *Clays Clay Miner.* **1997**, 45, 506-516. [CrossRef]

[74] Jin, P.; Ou, C.; Ma, Z.; Li, D.; Ren, Y.; Zhao, Y. Evolution of montmorillonite and its related clay minerals and their effects on shale gas development. *Geophys. Prospect. Pet.* **2018**, 57, 344-355. (In Chinese with English Abstract).

Rare earth element and strontium isotope geochemistry in Dujiali Lake, central Qinghai-Tibet Plateau, China: implications for the origin of hydromagnesite deposits*

Yongjie Lin[1], Mianping Zheng[1], Chuanyong Ye[1] and Ian M. Power[2]

1. MNR Key Laboratory of Saline Lake Resources and Environments, Institute of Mineral Resources, Chinese Academy of Geological Sciences, Beijing 100037, China
2. Trent School of the Environment, Trent University, 1600 West Bank Drive Peterborough, Ontario K9L 0G2, Canada

Abstract Rare earth element (REE) and strontium isotope data ($^{87}Sr/^{86}Sr$) are presented for hydromagnesite and surface waters that were collected from Dujiali Lake in central Qinghai-Tibet Plateau (QTP), China. The goal of this study is to constrain the solute sources of hydromagnesite deposits in Dujiali Lake. All lake waters from the area exhibit a slight LREE enrichment (average $[La/Sm]_{PAAS}$ = 1.36), clear Eu anomalies (average $[Eu/Eu^*]_{PAAS}$ = 1.31), and nearly no Ce anomalies. The recharge waters show a flat pattern (average $[La/Sm]_{PAAS}$ = 1.007), clear Eu anomalies (average $[Eu/Eu^*]_{PAAS}$ = 1.83), and nearly no Ce anomalies (average $[Ce/Ce^*]_{PAAS}$ = 1.016). The REE + Y data of the surface waters indicate the dissolution of ultramafic rock at depth and change in the hydrogeochemical characteristics through fluid-rock interaction. These data also indicate a significant contribution of paleo-groundwater to the formation of hydromagnesite, which most likely acquired REE and Sr signatures from the interaction with ultramafic rocks. The $^{87}Sr/^{86}Sr$ data provide additional insight into the geochemical evolution of waters of the Dujiali Lake indicating that the source of Sr in the hydromagnesite does not directly derive from surface water and may have been influenced by both Mg-rich hydrothermal fluids and meteoric water. Additionally, speciation modeling predicts that carbonate complexes are the most abundant dissolved REE species in surface water. This study provides new insights into the origins of hydromagnesite deposits in Dujiali Lake, and contributes to the understanding of hydromagnesite formation in similar modern and ancient environments on Earth.

Keywords Rare earth element; Strontium isotopes; Alkaline lake; Hydromagnesite; Qinghai-Tibet Plateau

1 Introduction

Hydromagnesite $[Mg_5(CO_3)_4(OH)_2 \cdot 4H_2O]$ is a rare magnesium carbonate mineral that has been found globally in numerous localities in a variety of setting including continental lacustrine environments (Braithwaite and Zedef, 1994, 1996; Cangemi et al., 2016; Kazmierczak et al., 2009; Lin et al., 2017a, b; Power et al., 2009, 2014; Renaut and Long, 1989), karst caves (Cañaveras et al., 1999; Fischbeck and Müller, 1971; Hill and Forti, 1986; Northup and Lavoie, 2001), and is commonly associated with ultramafic rocks (Oskierski et al., 2013, 2016; Wilson et al., 2009, 2014; Zedef et al., 2000). The formation mechanism of modern and ancient hydromagnesite in alkaline lake is variable and complex, which more recently has attracted a great deal of interest due to its importance to Mg-carbonate depositional environment and the potential capacity to capture and store CO_2 (Braithwaite and Zedef, 1996; Cangemi et al., 2016; Lin et al., 2017a, b; Power et al., 2009, 2014; Renaut, 1993). The formation mechanisms of hydromagnesite in such deposits have been the subject of continuing debate (Braithwaite and Zedef, 1996; Cangemi et al., 2016; Coshell et al., 1998; Goto et al., 2003; Lin et al., 2017a, b; Power et al., 2009, 2014; Renaut, 1993). Dujiali Lake in the central Qinghai-Tibet Plateau (QTP), China is one of the few modern environments on the Earth's surface that has experienced extensive hydromagnesite precipitation

* 本文发表在：Geochemistry, 2019, 79: 337-346

during the late Quaternary (Lin et al., 2017a, b, 2018). Lin et al. (2017a, b) suggested that supergene formation with authigenic hydromagnesite crystallizing from evaporation of lake water is the dominant precipitation process in Dujiali Lake. In contrast, Yu et al. (2015) carried out evaporation experiments using Dujiali Lake water, which showed that there was no Mg-carbonate precipitation except for northupite [$Na_3Mg(CO_3)_2Cl$]. Consequently, the formation mechanisms of hydromagnesite in Dujiali Lake remain uncertain. In addition, the precipitation of hydromagnesite in alkaline lake requires high Mg/Ca ratios (Chagas et al., 2016; Last and Last, 2012; Power et al., 2014), and the material source and pathway of metallogenic elements in Dujiali Lake is poorly understood.

The REEs constitute a coherent group of elements in geochemical processes due to their equal ionic charges and similar ionic radii (Barrat et al., 2000; Bau, 1991), and exhibit a gradual decrease in their ionic radii with increasing atomic number that leads to systematic changes in chemical properties (Munemoto et al., 2015). Thus, the chemical properties of REEs make them useful as sensitive tracers of geochemical processes in aqueous environments (Lee et al., 2003; Leybourne and Johannesson, 2008; Munemoto et al., 2015; Ojiambo et al., 2003; Willis and Johannesson, 2011). Variation in the strontium isotopic compositions of surface water results from mixing of Sr from various sources with different isotopic ratios (Zieliński et al., 2017). Strontium isotopic compositions ($^{87}Sr/^{86}Sr$) have proved to be a useful tracer for identifying weathering end-members in a watershed due to the fact that radiogenic Sr isotopes ($^{87}Sr/^{86}Sr$) do not fractionate during natural processes (e.g., evaporation, precipitation, biological uptake) in the environment (Blum et al., 2000; Capo et al., 2014; Flockhart et al., 2015; Frost and Toner, 2004; Ojiambo et al., 2003; Peterman et al., 2012; Sahib et al., 2016; Shand et al., 2009). For example, owing to the high Rb and low Sr concentrations in silicates and the conversely low Rb and high Sr concentrations in carbonates, the water draining from a silicate-dominated terrain has higher $^{87}Sr/^{86}Sr$ values than water draining form a carbonate-dominated terrain (Edmond et al., 1996; Elderfield et al., 1990; Huh et al., 1998). Therefore, the $^{87}Sr/^{86}Sr$ values of surface waters can provide information about Sr sources and mixing processes (Semhi et al., 2000; Shand et al., 2009; Tripathy et al., 2010).

Carbonate minerals are found in a wide variety of environments on Earth and are produced by numerous processes over geologic time including precipitation from hydrothermal fluids, microbial metabolism (e.g., stromatolites), or synthesis by organisms like foraminifera, coral, molluscs (Barrat et al., 2000). REE concentrations and strontium isotopic compositions of carbonate minerals are widely used to constrain the origin and formation conditions of carbonate minerals (e.g., Hecht et al., 1999; Himmler et al., 2010; Mantovani et al., 1985; Webb et al., 2009; Zhao et al., 2009).

In this study, we present strontium isotope ($^{87}Sr/^{86}Sr$) and REE data on hydromagnesite and surface waters in the vicinity of Dujiali Lake, central QTP to constrain the nature of the material source of hydromagnesite. To our knowledge, this study is the first comprehensive examination of strontium isotope composition ($^{87}Sr/^{86}Sr$) of hydromagnesite. Analysis of strontium isotopes and REEs in surface waters and hydromagnesite deposits from Dujiali Lake allows direct comparison between the signatures of the waters with those recorded through the precipitation of hydromagnesite.

2 Geological and hydrogeological setting

The Dujiali Lake is located at the center of the Banggong-Nujiang suture zone, central QTP, China (Fig. 1; 32°05′20″ N, 88°42′10″ E, 4524 m above sea level). The Dujiali Lake basin is a downwarped subbasin within a Cenozoic tectonic faulted basin, which derived from the ancient Selin Co basin due to local uplift caused by the tectonic movement. The margins of Dujiali Lake basin are primarily covered by Permian and Cretaceous limestone, Neogene and Paleogene glutenite, Cretaceous andesite, and Jurassic and Cretaceous sandstone, and have multi-step terraces, covered by Quaternary sediments that are a heterogeneous mixture of gravel, sand, silt and clay as well as salt deposits (including hydromagnesite, Zheng et al., 2002; Fig. 2).

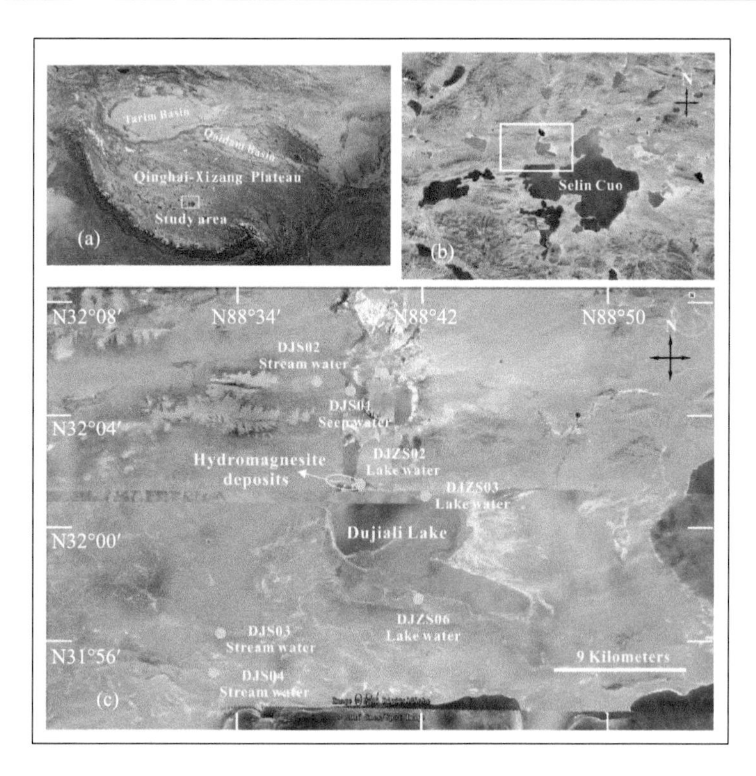

Fig. 1 Satellite images of the study site including Qinghai-Tibetan Plateau (a), Selin Co and Dujiali Lake (b), and Dujiali Lake showing sample locations (c).

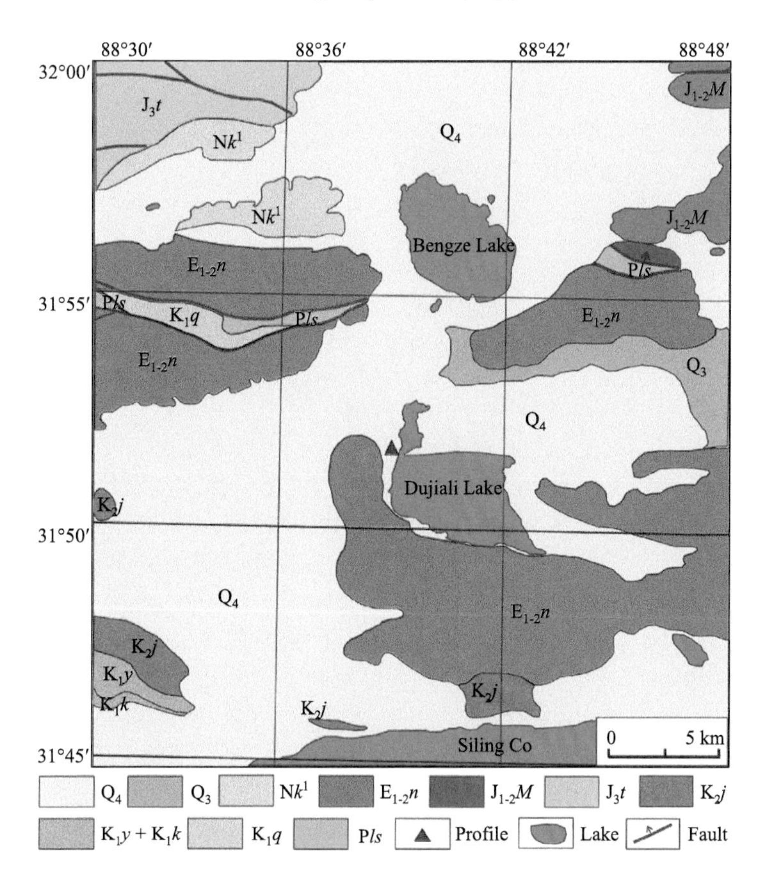

Fig. 2 The surface geology map of the Dujiali Lake area (modified from Qu et al., 2011).

Permian (Pls) and Cretaceous limestone (K_1y and K_1k), Neogene and Paleogene glutenite ($Nk^1$and $E_{1-2}n$), Cretaceous andesite (K_1q), and Jurassic and Cretaceous sandstone ($J_{1-2}M$ and K_2j), and have multi-step terraces, covered by Quaternary sediments (Q_3 and Q_4) that are a heterogeneous mixture of gravel, sand, silt and clay as well as salt deposits (including hydromagnesite, Zheng et al., 2002; Fig. 2)

This region has a cold continental, sub-humid to semi-arid climate with an annual mean temperature of -1 ℃, and annual precipitation and evaporative loss of approximately 200 and 2400 mm, respectively. The Dujiali Lake is a semi-closed endorheic basin and currently a saltwater lake, with an average depth of approximately 3 m and a surface area of ~80 km^2. The lake water has an average pH of 9.37 and total dissolved solids of 1.5–9.3 g/l with Na^+, K^+, Mg^{2+}, SO_4^{2-}, HCO_3^-, CO_3^{2-}, and $B_4O_7^{2-}$ being the dominant ions (Lin et al., 2017a). In addition to precipitation, a seasonal river and groundwater discharge contribute water to the lake.

Fig. 3 (a) Photograph showing the hydromagnesite deposit and profile DJSP01 at Dujiali Lake. (b) Hydromagnesite sediments were bright-white, dry, and clay-like with a weathered surface crust. (c) Photograph showing hydromagnesite directly deposited on Quate.

3 Background and methods

3.1 Field sampling

Hydromagnesite samples were collected from the first lake terrace in the northwest of Dujiali Lake in March 2015 (Lin et al., 2017a, b). The hydromagnesite sediments were distributed in bands following the margins of lake and directly overlay Quaternary sediments. The terrace was subparallel to the land surface and 10 m thick. Seven representative samples were equidistantly taken from the profile (DPSP01) with three samples at different horizons being selected for strontium isotope analysis (Fig. 3). Hydromagnesite sediments were bright-white, dry, and clay-like when sampled (Lin et al., 2017a). In addition, three lake waters were collected from different locations from Dujiali Lake, as well as three stream waters and one seep water were also collected near Dujiali Lake (Fig. 1). The investigated streams are the only rivers contributing to the lake. Water samples were filtered (0.45 μm) in the field into polyethylene bottles that were rinsed with representative water samples. These samples were transferred to the hydrochemistry laboratory at the Institute of Mineral Resources, CAGS within one week and stored in the refrigerator.

3.2 Analytical methods

Strontium was separated by standard cation exchange techniques (Weis et al., 2006) and measured using a Plasma II MC-ICP-MS (Wrexham, UK) at Nanjing FocuMS Technology Co. Ltd for Isotopic and Geochemical Research. Raw data of isotopic ratios were corrected for mass fractionation by normalizing to $^{86}Sr/^{88}Sr = 0.1194$ for Sr with exponential law. The NIST SRM 987 standard gave $^{87}Sr/^{86}Sr$ of 0.710248 (Weis et al., 2006), the maximum difference between duplicates was ± 0.000022 with an average reproducibility ± 0.000012 (± 8, 1 s.d.), the Sr blank measured over the course of the study was 13 pg.

Rare earth element analysis of water samples was performed at Nanjing FocuMS Technology Co. Ltd. by quadrupole ICP-MS (Agilent Technologies 7700x, Hachioji-shi, Japan) following 5× pre-concentration by chelation, using the method of Hall et al. (1995). Briefly, separation and concentration of REEs and Y were achieved using columns containing a CC-1 chelating resin of microporous iminodiacetate. Iminodiacetate resin, Metpac CC-1 (Dionex), was used to concentrate most of the trace metals and to separate them from alkaline and alkaline-earth metals as the resin preferentially retains the trivalent REEs (Hall et al., 1995). The Y and REE isotopes monitored[^{89}Y, ^{139}La, ^{140}Ce, ^{141}Pr, ^{144}Nd, ^{147}Sm, ^{153}Eu, ^{160}Gd, ^{159}Tb, ^{163}Dy, ^{165}Ho, ^{166}Er, ^{169}Tm, ^{174}Yb, and ^{175}Lu] were free of elemental isobaric interferences. The ICP-MS was calibrated and the sample concentrations verified using a series of 5 REE standards of known concentrations (0, 1, 2, 10, 50, 100, 500, and 1000 ng/kg). The precision of the REE measurements was better than ± 10%relative standard deviation for element concentrations exceeding 10 ppm and better than ± 5% for those exceeding 50 ppm. The REE + Y contents of the samples were normalized to post Archean Australian shale (PAAS, Taylor and McLennan, 1985). Y is inserted between Dy and Ho in the REE pattern because of its identical charge and similar ionic radius (Bau and Dulski, 1996). The Eu anomaly was quantified by the term $[Ce/Ce^*]_{PAAS}$, where $Ce^*$is the expected Ce value for a smooth PAAS-normalized REE pattern, such that $Ce^* = 1/2\ (La_{PAAS} + Pr_{PAAS})$ (Bau and Dulski, 1996). The Ce anomaly was determined by the term $[Eu/Eu^*]_{PAAS}$, where $Eu^*$is the expected Eu value for a smooth PAAS-normalized REE pattern, such that $Eu^* = (2/3Sm\ PAAS + 1/3Tb_{PAAS})$ (Bau and Dulski, 1996).

3.3 Speciation modeling

Speciation calculations of REEs in surface waters from Dujiali Lake were performed using the geochemical modeling program Visual MINTEQ (Version 3.1). The Visual MINTEQ database was modified by addition of the 14 naturally occurring REEs and their corresponding inorganic, aqueous complexes with carbonate, sulfate, fluoride, hydroxide, phosphate and chloride ligands. Specifically, infinite dilution stability constants for $LnHCO_3^{2+}$, $LnCO_3^+$, $Ln(CO_3)_2^-$ (Luo and Byrne, 2004), $LnOH^+$ (Klungness and Byrne, 2000), $LnSO_4^+$ (Schijf and Byrne, 2004), and $LnCl_2^+$ (Luo and Byrne, 2001) complexes, where Ln represents any or all of the lanthanides, were added to the database. Although REEs are suggested to closely associate with organic complexes in some natural waters (Ingri et al., 2000; Johannesson et al., 2004), the lack of data limited this study to only inorganic complexes of REEs.

4 Results

4.1 REE + Y concentrations and PAAS-normalized fractionation patterns

The water REE + Y concentrations are presented in Table 1 and Post Archean Australian Shale (PAAS)-normalized REE + Y patterns are shown in Fig. 4. Concentrations of REE + Y in lake waters, stream waters, and hydromagnesite ranged from 1.87–3.43 (avg. = 2.45, SD = 0.7), 2.05–5.08 (avg. = 3.1, 2SD = 1.4), and 0.79–3.64 μg/kg (avg. = 2.11, 2SD = 0.9), respectively. The concentration of REE + Y in seep water was 1.90. Generally, lake waters were HREE-depleted relative to PAAS (i.e., $[La/Yb]_{PAAS}$ ratios of lake water range from

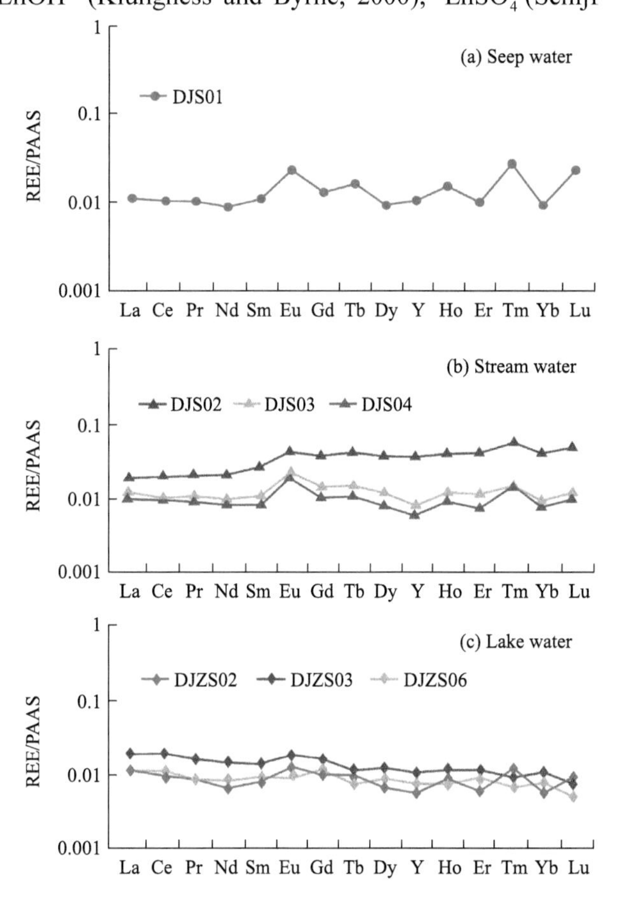

Fig. 4 The PAAS-normalized REE distribution patterns for (a) Seep water; (b) Stream water; (c) Lake water.

Table 1 Rare earth element concentrations in ppb for the surface water samples and ophiolite sample from Dujiali Lake

Sample	Water type	pH	TDS	La	Ce	Pr	Nd	Sm	Eu	Gd	Tb	Dy	Ho	Er	Tm	Yb	Lu	Y	ΣREE	$(La/Yb)_{PAAS}$	$(Eu/Eu^*)_{PAAS}$	$(Ce/Ce^*)_{PAAS}$
DJS01	Seep water	7.50	566	0.43	0.78	0.09	0.31	0.06	0.03	0.06	0.01	0.04	0.02	0.03	0.01	0.03	0.01	0.28	1.90	1.20	1.83	0.90
DJS02	Stream water	7.40	424	0.75	1.66	0.18	0.69	0.15	0.05	0.16	0.03	0.17	0.04	0.12	0.02	0.12	0.02	0.92	4.18	0.48	1.40	1.04
DJS03	Stream water	8.20	699	0.40	0.81	0.10	0.32	0.07	0.03	0.06	0.01	0.06	0.01	0.03	0.01	0.03	0.01	0.23	1.93	1.16	1.83	0.95
DJS04	Stream water	7.80	319	0.47	0.79	0.09	0.29	0.05	0.02	0.05	0.01	0.04	0.01	0.02	0.01	0.02	0.01	0.17	1.88	1.58	2.27	0.90
DJZS02	Lake water	9.00	1478	0.42	0.77	0.08	0.26	0.04	0.01	0.04	0.01	0.03	0.01	0.02	0.01	0.02	0.00	0.15	1.71	2.16	1.50	0.98
DJZS03	Lake water	9.50	6176	0.74	1.51	0.13	0.47	0.08	0.02	0.08	0.01	0.06	0.01	0.03	0.00	0.03	0.00	0.26	3.16	1.96	1.35	1.10
DJZS06	Lake water	9.70	4882	0.41	0.87	0.08	0.27	0.05	0.01	0.05	0.01	0.04	0.01	0.03	0.00	0.02	0.00	0.19	1.82	1.37	1.08	1.13
O-1	Ophiolite	=	=	53, 640	104, 960	10, 990	49, 600	11, 060	2850	13, 000	1870	10, 330	2160	6100	830	4930	730	64, 750	337, 800		1.447	0.461
DJSP0102	Hydromagnesite	=	=	-	0.61	1.17	0.14	0.66	0.13	0.03	0.08	0.02	0.09	0.03	0.05	0.01	0.09	0.01	0.52	3.63	0.51	2.35
DJSP0106	Hydromagnesite	=	=	0.33	0.58	0.07	0.26	0.06	0.01	0.05	0.01	0.03	0.01	0.03	0.01	0.04	0.01	0.30	1.76	0.63	1.41	0.90
DJSP0110	Hydromagnesite	=	=	0.32	0.57	0.07	0.29	0.05	0.01	0.04	0.01	0.06	0.01	0.03	0.01	0.04	0.00	0.36	1.85	0.58	2.04	0.87
DJSP0114	Hydromagnesite	=	=	0.24	0.45	0.05	0.20	0.05	0.01	0.03	0.01	0.05	0.01	0.03	0.00	0.02	0.01	0.31	1.47	0.90	3.13	0.92
DJSP0118	Hydromagnesite	=	=	0.37	0.59	0.08	0.28	0.05	0.01	0.05	0.01	0.07	0.01	0.04	0.01	0.05	0.01	0.43	2.05	0.60	1.83	0.78
DJSP0122	Hydromagnesite	=	=	0.60	1.15	0.13	0.55	0.09	0.02	0.06	0.01	0.07	0.02	0.03	0.01	0.03	0.01	0.37	3.13	1.54	2.39	0.96
DJSP0126	Hydromagnesite	=	=	0.13	0.24	0.03	0.12	0.02	0.01	0.02	0.00	0.02	0.00	0.02	0.00	0.02	0.00	0.16	0.79	0.50	3.56	0.92

Note: The data for the ophiolite sample O-1 were cited from Qu et al. (2011), and the data for hydromagnesite sample were cited from Lin et al. (2017b).

The rare earth element concentrations in μg/l for the surface water samples and in μg/kg for the solid samples, respectively.

The La/Yb value, Eu anomaly and negative Ce anomaly of Ophiolite sample were normalized to Upper crust composition.

Dashes (-) denote no data.

1.371 to 2.156). Seep water was slightly HREE-depleted relative to PAAS with a $[La/Yb]_{PAAS}$ ratios of 1.201 and two stream waters (DJS03 and DJS04) were also HREE-depleted relative to PAAS with $[La/Yb]_{PAAS}$ ratios of 1.156 and 1.582. In contrast with those two stream waters (DJS03 and DJS04), the stream (DJS02) was HREE-enriched relative to PAAS with a $[La/Yb]_{PAAS}$ ratio of 0.477. Lake waters had slightly positive Ce anomalies with $[Ce/Ce^*]_{PAAS}$ ranging from 0.98 to 1.13 (average = 1.07). Two stream water samples showed slightly negative Ce anomalies with $[Ce/Ce^*]_{PAAS}$ ratios of 0.71 and 0.89 (DJS03 and DJS04), while the stream water collected from the north part of the lake had a slightly positive Ce anomaly ($[Ce/Ce^*]_{PAAS}$ = 1.036). Seep water collected from the north area of the lake had a slightly Ce negative anomaly ($[Ce/Ce^*]_{PAAS}$ = 0.90). All waters collected from Dujiali Lake area exhibited Eu anomalies, as the average $[Eu/Eu^*]_{PAAS}$ value 1.31 for lake water, 1.83 for stream water and 1.83 for seep water. In summary, all sampled lake waters in the Dujiali Lake area exhibited a slight LREE enrichment, and variable Eu anomalies and nearly no Ce anomalies, while the surface recharge waters showed flat (or slightly LREE enriched, as in the case of DJS02) patterns, with obvious Eu anomalies and nearly no Ce anomalies.

4.2 REE inorganic species

Although REEs may associate with organic complexes in some natural waters (Ingri et al., 2000; Johannesson et al., 2004), the solution complexation is still poorly understood (Donat et al., 1994; Johannesson et al., 2004). Our focus was the influence of dissolved inorganic species on REEs given the saline water chemistry of Dujiali Lake. Average speciation modeling results for the surface waters from Dujiali Lake are plotted in Fig. 5 with the percentage of each aqueous REE species plotted against the REE atomic number. Speciation modeling predicts that the dominant dissolved REE inorganic species are carbonate complexes [$Ln(CO_3)_2^-$ + $LnCO_3^+$] in surface waters. The proportion of dissolved REEs present in other inorganic ions such as Ln^{3+}, $LnSO_4^+$ and $LnHCO_3^{2+}$ are less than 3% in seep water and stream water. The carbonate complexes such as $Ln(CO_3)_2^-$ are predicted to dominate all of the REEs in the surface waters. The only exception is La, Ce, Pr and Nd in seep water, La, Ce and Pr in stream water where the carbonate complex $LnCO_3^+$ is predicted to account for an equal or greater concentration than the corresponding carbonate complex $Ln(CO_3)_2^-$.

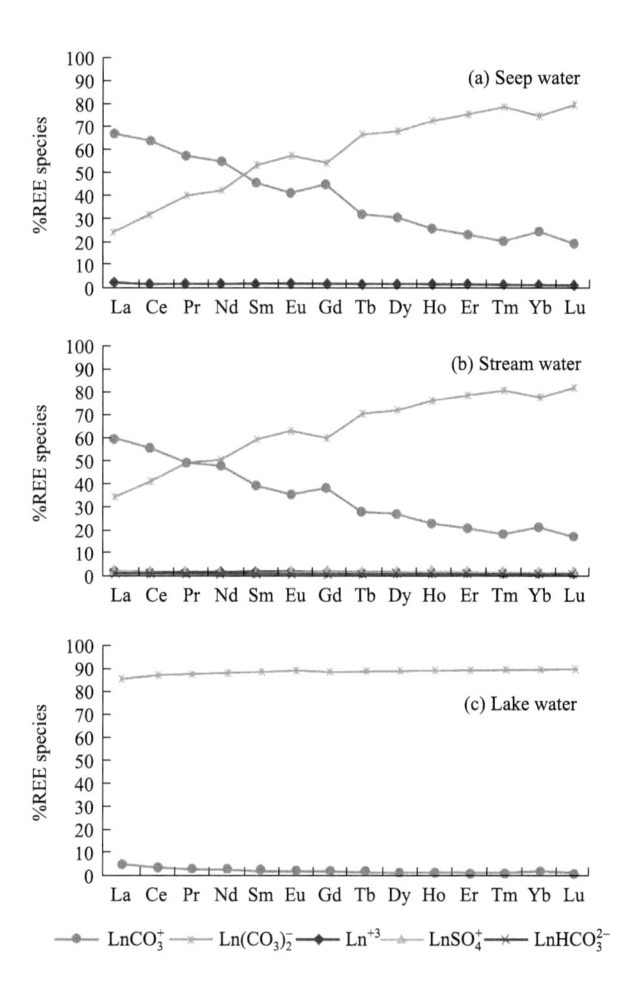

Fig. 5 Results from aqueous speciation calculations for average compositions of surface water samples: (a) Seep water; (b) Stream water; (c) Lake water. Ln refers to any lanthanide element.

4.3 Strontium isotopic compositions

The $^{87}Sr/^{86}Sr$ ratios of hydromagnesite, Dujiali Lake water, and recharge waters are presented in Table 2 and Fig. 6. The $^{87}Sr/^{86}Sr$ values of the hydromagnesite samples range from 0.710045 to 0.710060. The seep water has $^{87}Sr/^{86}Sr$ values of 0.709325, while the $^{87}Sr/^{86}Sr$ values of stream water ranges from 0.708964 to

0.710814. The $^{87}Sr/^{86}Sr$ values of lake waters vary from 0.708873 to 0.709353. In general, seep water from the north of the lake, and stream water around the lake have slightly more radiogenic $^{87}Sr/^{86}Sr$ ratios than the Dujiali Lake water, while the $^{87}Sr/^{86}Sr$ values of hydromagnesite are slightly more radiogenic than those of the water samples.

Table 2 Sr isotopic compositions of surface waters and hydromagnesite from Dujiali Lake

Sample No.	Sample type	$^{87}Sr/^{86}Sr$	Std. error
DJS01	Seep water	0.709325	0.000005
DJS02	Stream water	0.710814	0.000007
DJS03	Stream water	0.710363	0.000005
DJS04	Stream water	0.708964	0.000005
DJZS02	Lake water	0.709353	0.000005
DJZS03	Lake water	0.708873	0.000008
DJZS06	Lake water	0.709004	0.000005
DJSP0101	Hydromagnesite	0.710060	0.000003
DJSP0104	Hydromagnesite	0.710045	0.000004
DJSP0107	Hydromagnesite	0.710057	0.000004

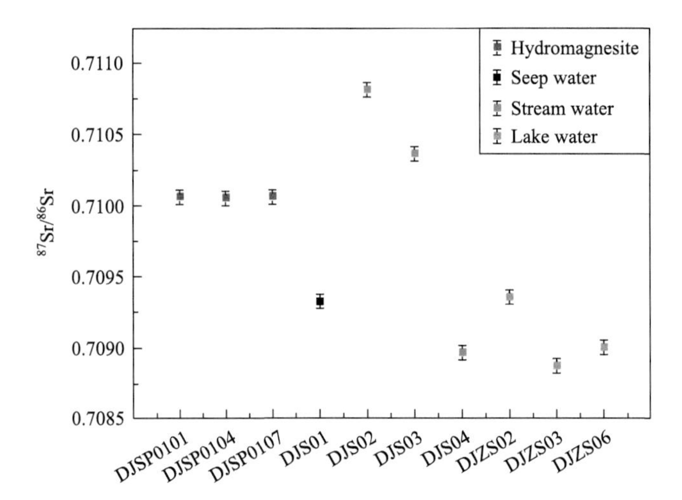

Fig. 6　The $^{87}Sr/^{86}Sr$ ratios of hydromagnesite and surface waters from the Dujiali Lake area.

5　Discussion

5.1　Enrichment mechanism of REEs in Dujiali Lake water

Chemistry of lake and surface water inputs including major solute compositions, total dissolved solids (TDS) concentrations, and pH, was previously reported and interpreted by Lin et al. (2017a). Briefly, the chemical composition of surface water around Dujiali Lake evolved from the rock-weathering-type waters (seep water and stream water; Ca-Mg-HCO_3water type) to more concentrated sodic waters (lake water; Na-SO_4-Cl water type) due to evaporation. The pH and presence of colloids and complexes are considered to play major roles in controlling REE concentrations in natural waters (Åström and Corin, 2003; Protano and Riccobono, 2002; Sholkovitz, 1995; Tabaksblat, 2002). In this study, the carbonate complexes [$Ln(CO_3)_2^-$] are predicted to dominate all of the REEs in the surface waters. The only exceptions are La, Ce, Pr and Nd in seep water, and La, Ce and Pr in stream water where the carbonate complexes $LnCO_3^+$ are predicted to account for equal or higher concentrations than the corresponding carbonate complex, $Ln(CO_3)_2^-$. These results suggest that the $LnCO_3^+$ complexes predominate in LREEs and become less important through the lanthanide series. This trend is balanced by increases in $Ln(CO_3)_2^-$ concentrations.

Dujiali Lake water with high pH values has carbonate complexes [$Ln(CO_3)_2^-$] as the dominant aqueous species comprising more than 95% of each dissolved REE, and exhibits a slightly LREE enrichment. Although the chloride concentration is high in surface waters, the modeling results show that REE-chloride complexes are very rare. The free ion (Ln^{3+}) in solution constitutes approximately 1% in seep water and stream water, and is even negligible in lake water as the importance of the carbonate complexes increases, which supports previous works that suggested that the free ion (Ln^{3+}) should be present predominantly in low pH-values water, such as acidic Lake Colour in Canada (Johannesson and Lyons, 1995; Lee and Byrne, 1992; Wood, 1990).

The stability constants for the $Ln(CO_3)_2^-$ species are much greater than those for $LnCO_3^+$ species. Furthermore, $Ln(CO_3)_2^-$ complexes with HREEs are approximately two orders of magnitude more stable than with the LREEs (Johannesson et al., 1994). In this study, all lake water samples exhibited a slight LREE enrichment (average $[La/Sm]_{PAAS} = 1.36$), while the surface recharge water sample showed a flat pattern (average $[La/Sm]_{PAAS} = 1.007$). We suggest that the LREE enrichment of lake water is the result of complexation of the REEs with carbonate ions (e.g., Johannesson and Lyons, 1994). The larger stability constants of the LREE-carbonate species as compared to similar HREE species result in increasing stability of the LREEs in lake waters. Sulfate was not found to be an important complexes of REEs in the lake waters due to the overwhelming effect of carbonate complexation given the high alkalinity of the lake water (Johannesson et al., 1994; Johannesson and Lyons, 1994). As such, the percentage of REEs complexed with carbonate ions, especially as $Ln(CO_3)_2^-$, in alkaline lake water is very high (greater than 90% in this study); a much greater proportion in comparison to REEs in seawater (Johannesson et al., 1994; Millero, 1992; Och et al., 2014). Other REE species including sulfate and free metal species play a significant role in seawater (Millero, 1992); however, the shale-normalized values of the REEs in alkaline lake waters are greater than seawater values where the alkalinity is lower than in lake waters (average pH value Dujiali Lake was 9.37) (Broecker and Peng, 1982).

5.2 Material sources to hydromagnesite deposits

The mineralogy of lake sediments provides essential information of paleoenvironmental changes given that the composition of detritus is mainly influenced by the sediment provenance, the tectonic setting, and the weathering and transportation processes within the watershed (Last and Smol, 2001; Och et al., 2014). The main pathway for REEs entering into a lake is as detritus because they are relatively immobile during dissolution of source rocks (Och et al., 2014). Consequently, REEs in lake sediments are expected to exhibit normalized patterns similar to the REE signature of the source rocks (Jin et al., 2006; Piper and Bau, 2013). The REE + Y patterns of the Dujiali Lake hydromagnesite deposits exhibit slight HREEs enrichments, slight Ce negative anomaly, and a consistently positive Eu anomaly (Lin et al., 2017b), which is significantly different from the REE + Y patterns of surface water. In addition, the REE + Y patterns of ultramafic rocks in the surrounding area exhibit a pronounced positive Eu anomaly and negative Ce anomaly normalized to Upper crust composition (Qu et al., 2011). It is generally understood that the Eu anomalies in waters are inherited from host rocks, which can be used as a tracer for material source (Chen et al., 2017). The positive Eu anomalies of water samples in this study are likely attributed to preferential dissolution of ultramafic rocks, which reflects the variable influence of fluid-rock interaction. Other examples of hydromagnesite deposits that have formed through ophiolite weathering include the hydromagnesite-magnesite playas in northern British Columbia, Canada (Power et al., 2014) and the hydromagnesite microbialites in Turkey (Braithwaite and Zedef, 1996). The ΣREE data of different surface water samples in this study are relatively homogeneous but not consistent with TDS, which suggest that REEs likely have different sources than that of the major ions in these waters (Chen et al., 2017).

Strontium isotopes ($^{87}Sr/^{86}Sr$) are a valuable tool for determining sources and elucidating mixing relationships in weathering regimes, and quantifying end-member mixing processes (Bakari et al., 2013; Chamberlain et al., 2005; Dowling et al., 2013; Lyons et al., 2002). The $^{87}Sr/^{86}Sr$ values of the hydromagnesite samples were slightly more radiogenic than the surface water samples, except for two of the three stream water samples (DJS02 and DJS03), while the surface waters, with exception of stream water DJS04, have slightly more radiogenic average $^{87}Sr/^{86}Sr$ value than lake water. In general, the strontium isotope systematics of the major rivers in the world can largely be described by mixing between strontium derived from limestone with low $^{87}Sr/^{86}Sr$ ratios and strontium derived from silicate with radiogenic $^{87}Sr/^{86}Sr$ ratios, which leads to a global average $^{87}Sr/^{86}Sr$ ratio of 0.7119 (Palmer and Edmond, 1992). Sr originating from silicate rock has higher radiogenic $^{87}Sr/^{86}Sr$ ratios (>0.710), whereas Sr sourced from carbonate rock has lower radiogenic $^{87}Sr/^{86}Sr$ ratios (Christian et al., 2011; Clow et al., 1997; France-Lanord and Derry, 1999; Graustein and Armstrong, 1983; Katz and Bullen, 1996; Miller et al., 1993). In addition, Mg^{2+}/Ca^{2+} ratios are highest in dolostone aquifers (0.085–1.07), the second in silicate rocks, but lowest in limestone aquifer with the value from 0.01 to 0.26 (Karst Research Group, 1978). As shown in Fig. 7, the $^{87}Sr/^{86}Sr$ ratios of weathering end-members of silicate rock, dolostone, and limestone, the surface recharge water located in the vicinity of the dissolution line of carbonate rock, suggesting Sr in the surface recharge water mainly originated from dissolution of carbonate rock. It can be concluded that the surface recharge water chemistry is mainly the results of interaction with carbonate rock (e.g., limestone). However, the hydromagnesite has higher radiogenic $^{87}Sr/^{86}Sr$ ratios than the surface recharge water. This finding implies that the present inflow of Sr into Dujiali Lake may be a mixture of modern surface waters and that the Sr in the hydromagnesite samples is not simply a mixture of these waters, but that paleo- groundwater was likely a contributor as well. Our previous investigation suggested that hydromagnesite may have precipitated from waters influenced by both Mg-rich hydrothermal fluids and meteoric water (Lin et al., 2017b).

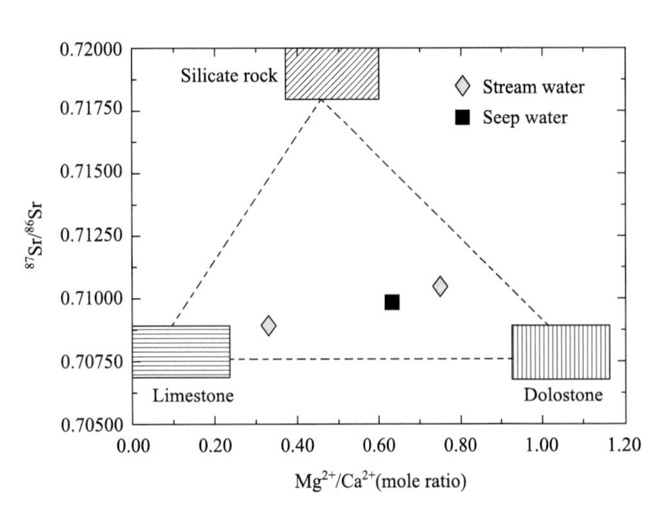

Fig. 7 Plot of $^{87}Sr/^{86}Sr$ vs. Mg/Ca in Lake Dujiali area from different aquifers (Pu et al., 2012).

It is widely understood that the deposition of hydromagnesite in alkaline lakes requires high Mg/Ca ratios (Chagas et al., 2016; Last and Last, 2012; Lin et al., 2017b; Müller et al., 1972; Power et al., 2014). The Mg/Ca ratios of surface waters in this study are relatively homogeneous but not consistent with TDS, which suggest that REEs most likely have different sources with Mg^{2+}. The saturation indices of various carbonate minerals show that Dujiali Lake water is undersaturated with respect to hydromagnesite with an average saturation index of −1.51 (Lin et al., 2017a). The only Mg-carbonate mineral that precipitated from Dujiali Lake water in a field evaporation experiment was northupite [$Na_3Mg(CO_3)_2Cl$] (Yu et al., 2015). Thus, the presentday lake chemistry does not readily lead to precipitation of hydromagnesite even when evaporated. It is reasonable to consider that part of the material source of hydromagnesite may be paleo-groundwater, which can obtain REE and Sr signatures through the interaction with ultramafic rocks.

6 Conclusion

Strontium isotopic compositions and rare earth element concentrations of carbonate sediments and surface waters from Dujiali Lake in QTP, China provide new insights into the origins of hydromagnesite. Speciation modeling predicts that the vast majority of REEs are complexed with carbonate ions [$Ln(CO_3)_2^-$], typically in amounts greater than 99% for each of the REEs. The enrichment of LREEs in lake waters is primarily due to the preferential formation of stronger carbonate complexes with the LREEs. Overall, the REE results and $^{87}Sr/^{86}Sr$ data support our previous investigations where the hydrogeochemistry of surface waters, trace elements content of hydromagnesite deposits were employed to constrain hydromagnesite precipitation, suggesting that paleo-groundwater is also a minor contributor to the hydromagnesite composition and partial REE and Sr signatures from the ultramafic rocks. Combining the results of trace elements including REEs and strontium isotopes leads to a more detailed understanding of the solute sources of hydromagnesite in Dujiali Lake and provides insights into the geochemical processes that drive Mg-carbonate deposition near the Earth's surface.

Acknowledgments

We thank Professor Martin Dietzel and the three anonymous reviewers for their insightful comments that greatly improved this manuscript. This research was supported by subject of The National Key Research and Development Program of China (Grant Number: 2017YFC0602806), National Natural Science Foundation of China (Grant Number: 41603048, U1407207, and 41473061), and Projects of China Geological Survey (Grant Number: DD20160025 and DD20160054).

References

Åström, M., Corin, N., 2003. Distribution of rare earth elements in anionic, cationic and particulate fractions in boreal humus-rich streams affected by acid sulphate soils. Water Res. 37, 273-280 doi: 10.10.1016/S0043-1354 (02) 00274-9.

Bakari, S. S., Aagaard, P., Vogt, R. D., Ruden, F., Johansen, I., Vuai, S. A., 2013. Strontium isotopes as tracers for quantifying mixing of groundwater in the alluvial plain of a coastal watershed, south-eastern Tanzania. J. Geochem. Explor. 130, 1-14. https: //doi.org/10.1016/j.gexplo.2013.02.008.

Barrat, J. A., Boulègue, J., Tiercelin, J. J., Lesourd, M., 2000. Strontium isotopes and rareearth element geochemistry of hydrothermal carbonate deposits from Lake Tanganyika, east Africa. Geochim. Cosmochim. Acta 64, 287-298. https: //doi.org/10.1016/S0016-7037 (99) 00294-X.

Bau, M., 1991. Rare-earth element mobility during hydrothermal and metamorphic fluidrock interaction and the significance of the oxidation state of europium. Chem. Geol. 93, 219-230.

Bau, M., Dulski, P., 1996. Distribution of yttrium and rare-earth elements in the Penge and Kuruman iron-formations, Transvaal Supergroup South Africa. Precambrian Res. 79 (1-2), 37-55. https: //doi.org/10.1016/0301-9268 (95) 00087-9.

Blum, J. D., Taliaferro, E. H., Weisse, M. T., Holmes, R. T., 2000. Changes in Sr/Ca Ba/Ca and $^{87}Sr/^{86}Sr$ ratios between trophic levels in two forest ecosystems in the northeastern U.S.A. Biogeochemistry 49 (1), 87-101. https: //doi.org/10.1023/A: 1006390707989.

Braithwaite, C. J. R., Zedef, V., 1994. Living hydromagnesite stromatolites from Turkey. Sediment. Geol.92, 1-5. https: //doi.org/10.1016/0037-0738 (94) 90051-5.

Braithwaite, C. J. R., Zedef, V., 1996. Hydromagnesite stromatolites and sediments in an alkaline lake Salda Golu, Turkey. J. Sediment. Petrol. 66 (5), 991-1002. https: //doi. org/10.1306/D426845F-2B26-11D7-8648000102C1865D.

Broecker, W. S., Peng, T. H., 1982. Tracers in the Sea. Lamont-Doherty Geological Observatoryhttps: //doi.org/10.1016/0016-7037 (83) 90075-3.

Cañaveras, J. C., Hoyos, M., Sanchez-Moral, S., Sanz-Rubio, E., Bedoya, J., Soler, V., Groth, I., Schumann, P., Laiz, L., Gonzalez, I., Saiz-Jimenez, C., 1999. Microbial communities associated with hydromagnesite and needle-fiber aragonite deposits in a karstic cave (Altamira, Northern Spain). Geomicrobiol. J.16, 9-25. https: //doi.org/10.1080/014904599270712.

Cangemi, M., Censi, P., Reimer, A., D'Alessandro, W., Hause-Reitner, D., Madonia, P., Oliveri, Y., Pecoraino, G., Reitner, J., 2016. Carbonate precipitation in the alkaline lake Specchio di Venere (Pantelleria Island Italy) and the possible role of microbial mats. Appl. Geochem. 67, 168-176. https: // doi.org/10.1016/j.apgeochem.2016.02. 012.

Capo, R. C., Stewart, B. W., Rowan, E. L., Kolesar Kohl, C. A., Wall, A. J., Chapman, E. C., Hammack, R. W., Schroeder, K. T., 2014. The strontium isotopic evolution of Marcellus Formation produced waters, southwestern Pennsylvania. Int. J. Coal Geol. 126, 57-63. https: //doi.org/10.1016/ j.coal.2013.12.010.

Chagas, A. A. P., Webb, G. E., Burne, R. V., Southam, G., 2016. Modern lacustrine microbialites: towards a synthesis of aqueous and carbonate geochemistry and mineralogy. Earth-Sci. Rev.162, 338-363. https: //doi.org/10.1016/j.earscirev.2016.09.012.

Chamberlain, C. P., Waldbauer, J. R., Jacobson, A. D., 2005. Strontium, hydrothermal systems and steady-state chemical weathering in active mountain belts.

Earth Planet. Sci. Lett. 238, 351-366. https: //doi.org/10.1016/j.epsl.2005.08.005.

Chen, L., Ma, T., Du, Y., Xiao, C., 2017. Dissolved rare earth elements of different waters in Qaidam Basin, Northwestern China. Procedia Earth Planet. Sci.17, 61-64. https: //doi.org/10.1016/j.proeps.2016.12.031.

Christian, L. N., Banner, J. L., Mack, L. E., 2011. Sr isotopes as tracers of anthropogenic influences on stream water in the Austin Texas, area. Chem. Geol. 282 (3-4), 84-97. https: //doi.org/10.1016/j.chemgeo.2011.01.011.

Clow, D. W., Mast, M. A., Bullen, T. D., Turk, J. T., 1997. Strontium 87/strontium 86 as a tracer of mineral weathering reactions and calcium sources in an alpine/subalpine watershed, Loch Vale Colorado. Water Resour. Res.33 (6), 1335-1351. https: //doi. org/10.1029/97WR00856.

Coshell, L., Rosen, M. R., Mcnamara, K. J., 1998. Hydromagnesite replacement of biomineralized aragonite in a new location of Holocene stromatolites Lake Walyungup, western Australia. Sedimentology 45 (6), 1005-1018. https: //doi.org/10.1046/j. 1365-3091.1998.00187.x.

Donat, J. R., Lao, K. A., Bruland, K. W., 1994. Speciation of dissolved copper and nickel in South San Francisco Bay: a multi-method approach. Anal. Chim. Acta 284, 547-571. https: //doi.org/10.1016/0003-2670 (94) 85061-5.

Dowling, C. B., Lyons, W. B., Welch, K. A., 2013. Strontium isotopic signatures of streams from Taylor Valley, Antarctica, revisited: the role of carbonate mineral dissolution. Aquat. Geochem. 19, 231-240. https: //doi.org/10.1007/s10498-013-9189-4.

Edmond, J. M., Palmer, M. R., Measures, C. I., Brown, E. T., Huh, Y., 1996. Fluvial geochemistry of the eastern slope of the northeastern Andes and its foredeep in the drainage of the Orinoco in Colombia and Venezuela. Geochim. Cosmochim. Acta 60 (16), 2949-2974. https: //doi.org/ 10.1016/0016-7037 (96) 00142-1.

Elderfield, H., Upstill-Goddard, R., Sholkovitz, E. R., 1990. The rare earth elements in rivers, estuaries, and coastal seas and their significance to the composition of ocean waters. Geochim. Cosmochim. Acta 54 (4), 971-991. https: //doi.org/10.1016/0016-7037 (90) 90432-K.

Fischbeck, R., Müller, G., 1971. Monohydrocalcite, hydromagnesite, nesquehonite, dolomite, aragonite, and calcite in speleothems of the Fränkische Schweiz, Western Germany. Contrib. Mineral. Petrol.33, 87-92. https: //doi.org/10.1007/BF00386107.

Flockhart, D. T. T., Kyser, T. K., Chipley, D., Miller, N. G., Norris, D. R., 2015. Experimental evidence shows no fractionation of strontium isotopes ($^{87}Sr/^{86}Sr$) among soil, plants, and herbivores: implications for tracking wildlife and forensic science. Isotopes Environ. Health Stud. 51 (3), 372-381. https:// doi.org/10.1080/10256016.2015. 1021345.

France-Lanord, C., Derry, L. A., 1999. The strontium isotopic budget of Himalayan Rivers in Nepal and Bangladesh. Geochim. Cosmochim. Acta 63 (13-14), 1905-1925. https: //doi.org/10.1016/S0016-7037 (99) 00081-2.

Frost, C. D., Toner, R. N., 2004. Strontium isotopic identification of water-rock interaction and ground water mixing. Ground Water 42 (3), 418-432. https: // doi.org/10.1111/j. 1745-6584.2004.tb02689.x.

Goto, A., Arakawa, H., Morinaga, H., Sakiyama, T., 2003. The occurrence of hydromagnesite in bottom sediments from Lake Siling, central Tibet: implications for the correlation among $\delta^{18}O$, $\delta^{13}C$ and particle density. J. Asian Earth Sci.21, 979-988.

Graustein, W. C., Armstrong, R. L., 1983. The use of strontium-87/strontium-86 ratios to measure atmospheric transport into forested watersheds. Science 219 (4582), 289-292. https: //doi.org/10.1126/science.219.4582.289.

Hall, G. E. M., Vaive, J. E., Mcconnell, J. W., 1995. Development and application of a sensitive and rapid analytical method to determine the rare-earth elements in surface waters. Chem. Geol. 120, 91-109.

Hecht, L., Freiberger, R., Gilg, H. A., Grundmann, G., Kostitsyn, Y. A., 1999. Rare earth element and isotope (C, O, Sr) characteristics of hydrothermal carbonates: genetic implications for dolomite-hosted talc mineralization at Gopfersgrun (Fichtelgebirge Germany). Chem. Geol. 155, 115-130. https: //doi.org/10.1016/S0009-2541 (98) 00144-2.

Hill, C., Forti, P., 1986. Cave Minerals of the World. National Speleological Society.

Himmler, T., Bach, W., Bohrmann, G., Peckmann, J., 2010. Rare earth elements in authigenic methane-seep carbonates as tracers for fluid composition during early diagenesis. Chem. Geol. 277, 126-136. https: //doi.org/10.1016/j.chemgeo.2010.07. 015.

Huh, Y., Tsoi, M. Y., Zaitsev, A., Edmond, J. M., 1998. The fluvial geochemistry of the rivers of Eastern Siberia: I. Tributaries of the Lena River draining the sedimentary platform of the Siberian Craton. Geochim. Cosmochim. Acta 62 (10), 1657-1676. https: //doi.org/10.1016/S0016-7037 (98) 00107-0.

Ingri, J., Widerlund, A., Land, M., Gustafsson, Ö., Andersson, P., Öhlander, B., 2000. Temporal variations in the fractionation of the rare earth elements in a Boreal river; the role of colloidal particles. Chem. Geol. 166, 23-45. https: //doi.org/10.1016/S0009-2541 (99) 00178-3.

Jin, Z., Li, F., Cao, J., Wang, S., Yu, J., 2006. Geochemistry of Daihai Lake sediments Inner Mongolia, north China: implications for provenance, sedimentary sorting, and catchment weathering. Geomorphology 80, 147-163 doi: 10.016/j.geomorph.2006.02.006.

Johannesson, K. H., Lyons, W. B., 1995. Rare-earth element geochemistry of Colour Lake, an acidic freshwater lake on Axel Heiberg Island Northwest Territories, Canada. Chem. Geol. 119, 209-223. https: //doi.org/10.1016/0009-2541 (94) 00099-T.

Johannesson, K. H., Lyons, W. B., 1994. The rare earth element geochemistry of Mono Lake water and the importance of carbonate complexing. Limnol. Oceanogr. 39, 1141-1154. https: //doi.org/10.4319/lo.1994.39.5.1141.

Johannesson, K. H., Lyons, W. B., Bird, D. A., 1994. Rare earth element concentrations and speciation in alkaline lakes from the western U.S.A. Geophys. Res. Lett. 21, 773-776. https: //doi.org/10.1029/94GL00005.

Johannesson, K. H., Tang, J., Daniels, J. M., Bounds, W. J., Burdige, D. J., 2004. Rare earth element concentrations and speciation in organic-rich blackwaters of the Great Dismal Swamp, Virginia, USA. Chem. Geol. 209 (3-4), 271-294. https: //doi.org/10. 1016/j.chemgeo.2004.06.012.

Karst Research Group, 1978. Karst Research of China. Science Press, Beijing.

Katz, B. G., Bullen, T. D., 1996. The combined use of $^{87}Sr/^{86}Sr$ and carbon and water isotopes to study the hydrochemical interaction between groundwater and lakewater in mantled karst. Geochim. Cosmochim. Acta 60 (24), 5075-5087. https: //doi.org/10.1016/S0016-7037 (96) 00296-7.

Kazmierczak, J., Altermann, W., Kremer, B., Kempe, S., Eriksson, P. G., 2009. Mass occurrence of benthic coccoid cyanobacteria and their role in the production of Neoarchean carbonates of South Africa. Precambrian Res. 173, 79-92. https: //doi. org/10.1016/j.precamres.2009.02.002.

Klungness, G. D., Byrne, R. H., 2000. Comparative hydrolysis behavior of the rare earths and yttrium: The influence of temperature and ionic strength. Polyhedron 19, 99-107. https: //doi.org/10.1016/S0277-5387 (99) 00332.

Last, F. M., Last, W. M., 2012. Lacustrine carbonates of the northern Great Plains of Canada. Sediment. Geol. 277-278, 1-31. https: //doi.org/ 10.1016/j.sedgeo.2012.07. 011.

Last, W. M., Smol, J. P., 2001. Tracking Environmental Change Using Lake Sediments. Volume 2: Physical and Geochemical Methods. https: //doi.org/ 10.1007/0-306-47670-3.

Lee, J. H., Byrne, R. H., 1992. Examination of comparative rare earth element complexation behavior using linear free-energy relationships. Geochim. Cosmochim. Acta 56, 1127-1137. https: //doi.org/10.1016/0016-7037 (92) 90050-S.

Lee, S. G., Lee, D. H., Kim, Y., Chae, B. G., Kim, W. Y., Woo, N. C., 2003. Rare earth elements as indicators of groundwater environment changes in a fractured rock system: evidence from fracture-filling calcite. Appl. Geochem.18, 135-143. https: //doi.org/10. 1016/S0883-2927 (02) 00071-9.

Leybourne, M. I., Johannesson, K. H., 2008. Rare earth elements (REE) and yttrium in stream waters, stream sediments, and Fe-Mn oxyhydroxides: fractionation, speciation, and controls over REE + Y patterns in the surface environment. Geochim. Cosmochim. Acta 72, 5962-5983. https: //doi.org/ 10.1016/j.gca.2008.09.022.

Lin, Y., Zheng, M., Ye, C., 2017a. Hydromagnesite precipitation in the Alkaline Lake Dujiali, central Qinghai-Tibetan Plateau: constraints on hydromagnesite precipitation from hydrochemistry and stable isotopes. Appl. Geochem. 78, 139-148. https: //doi. org/10.1016/j.apgeochem.2016.12.020.

Lin, Y., Zheng, M., Ye, C., Power, I. M., 2018. Thermogravimetric analysis-mass spectrometry (TGA-MS) of hydromagnesite from Dujiali Lake in Tibet China. J. Therm. Anal. Calorim.133 (3), 1429-1437. https: //doi.org/10.1007/s10973-018-7197-8.

Lin, Y., Zheng, M., Ye, C., Power, I. M., 2017b. Trace and rare earth element geochemistry of Holocene hydromagnesite from Dujiali Lake, central Qinghai-Tibetan Plateau, China. Carbonates Evaporites 1-15. https: //doi.org/10.1007/s13146-017-0395-9.

Luo, Y. R., Byrne, R. H., 2001. Yttrium and Rare Earth Element Complexation by Chloride Ions at 25°C. Journal of Solution Chemistry 30, 837-845.

Luo, Y. R., Byrne, R. H., 2004. Carbonate complexation of yttrium and the rare earth elements in natural waters. Geochimica et Cosmochimica Acta 68, 691-699. https: //doi.org/10.1016/S0016-7037 (03) 00495.

Lyons, W. B., Nezat, C. A., Benson, L. V., Bullen, T. D., Graham, E. Y., Kidd, J., Welch, K. A., 2002. Strontium isotopic signatures of the streams and lakes of Taylor Valley, Southern Victoria Land Antarctica: chemical weathering in a polar climate. Aquat. Geochem.8, 75-95. https: //doi.org/10.1023/A: 1021339622515.

Mantovani, M. S. M., Marques, L. S., De Sousa, M. A., Civetta, L., Atalla, L., Innocenti, F., 1985. Trace element and strontium isotope constraints on the origin and evolution of Paraná Continental Flood Basalts of Santa Catarina State (Southern Brazil). J. Petrol. 26, 187-209. https: //doi.org/10.1093/ petrology/26.1.187.

Miller, E. K., Blum, J. D., Friedland, A. J., 1993. Determination of soil exchangeable-cation loss and weathering rates using Sr isotopes. Nature 362, 438-441. https: //doi.org/10.1038/362438a0.

Millero, F. J., 1992. Stability constants for the formation of rare earth-inorganic complexes as a function of ionic strength. Geochim. Cosmochim. Acta 56, 3123-3132. https: //doi.org/10.1016/0016-7037 (92) 90293-R.

Müller, G., Irion, G., Förstner, U., 1972. Formation and diagenesis of inorganic Ca-Mg carbonates in the lacustrine environment. Naturwissenschaften 59, 158-164. https: //doi.org/10.1007/BF00637354.

Munemoto, T., Ohmori, K., Iwatsuki, T., 2015. Rare earth elements (REE) in deep groundwater from granite and fracture-filling calcite in the Tono area, central Japan: prediction of REE fractionation in paleo-to present-day groundwater. Chem. Geol. 417, 58-67. https: //doi.org/10.1016/j.chemgeo. 2015.09.024.

Northup, D. E., Lavoie, K. H., 2001. Geomicrobiology of caves: a review. Geomicrobiol. J. 18, 199-222. https: //doi.org/10.1080/01490450152467750.

Och, L. M., Müller, B., Wichser, A., Ulrich, A., Vologina, E. G., Sturm, M., 2014. Rare earth elements in the sediments of Lake Baikal. Chem. Geol. 376, 61-75. https: //doi.org/10.1016/j.chemgeo.2014.03.018.

Ojiambo, S. B., Lyons, W. B., Welch, K. A., Poreda, R. J., Johannesson, K. H., 2003. Strontium isotopes and rare earth elements as tracers of groundwater-lake water interactions, Lake Naivasha Kenya. Appl. Geochem. 18, 1789-1805. https: //doi.org/10.1016/S0883-2927 (03) 00104-5.

Oskierski, H. C., Dlugogorski, B. Z., Jacobsen, G., 2013. Sequestration of atmospheric CO_2 in chrysotile mine tailings of the Woodsreef Asbestos Mine Australia: quantitative mineralogy, isotopic fingerprinting and carbonation rates. Chem. Geol. 358, 156-169. https: //doi.org/10.1016/j.chemgeo. 2013.09.001.

Oskierski, H. C., Dlugogorski, B. Z., Oliver, T. K., Jacobsen, G., 2016. Chemical and isotopic signatures of waters associated with the carbonation of ultramafic mine tailings, Woodsreef Asbestos Mine Australia. Chem. Geol. 436, 11-23. https: //doi.org/10. 1016/j.chemgeo.2016.04.014.

Palmer, M. R., Edmond, J. M., 1992. Controls over the strontium isotope composition of river water. Geochim. Cosmochim. Acta 56 (5), 2099-2111. https: //doi.org/10. 1016/0016-7037 (92) 90332-D.

Peterman, Z. E., Thamke, J., Futa, K., Preston, T., 2012. Strontium isotope systematics of mixing groundwater and oil-field brine at Goose Lake in northeastern Montana USA. Appl. Geochem. 27, 2403-2408. https: //doi.org/10.1016/j.apgeochem.2012.08.004.

Piper, D. Z., Bau, M., 2013. Normalized rare earth elements in water sediments, and wine: identifying sources and environmental redox conditions. Am. J. Anal. Chem. 04, 69-83. https: //doi.org/10.4236/ajac.2013.410A1009.

Power, I. M., Wilson, S. A., Harrison, A. L., Dipple, G. M., McCutcheon, J., Southam, G., Kenward, P. A., 2014. A depositional model for hydromagnesite-magnesite playas near Atlin, British Columbia Canada. Sedimentology 61, 1701-1733. https: //doi.org/10.1111/sed.12124.

Power, I. M., Wilson, S. A., Thom, J.M., Dipple, G. M., Gabites, J. E., Southam, G., 2009. The hydromagnesite playas of Atlin, British Columbia Canada: a biogeochemical model for CO_2sequestration. Chem. Geol. 260, 302-316. https: //doi.org/10.1016/j. chemgeo.2009.01.012.

Protano, G., Riccobono, F., 2002. High contents of rare earth elements (REEs) in stream waters of a Cu-Pb-Zn mining area. Environ. Pollut. 117, 499-514. https: // doi.org/10. 1016/S0269-7491 (01) 00173-7.

Pu, J., Yuan, D., Zhang, C., Zhao, H., 2012. Tracing the sources of strontium in karst groundwater in Chongqing China: a combined hydrogeochemical approach and strontium isotope. Environ. Earth Sci. 67 (8), 2371-2381. https: //doi.org/10.1007/s12665-012-1683-2.

Qu, Y., Wang, Y., Duan, J., 2011. Geological Survey Reports of the Regional People's Republic of China (Scale 1: 250000. Duoba Pieces). Chinese Ed. China University of Geosciences Press, Wuhan.

Renaut, R. W., 1993. Morphology, distribution, and preservation potential of microbial mats in the hydromagnesite-magnesite playas of the Cariboo Plateau, British Columbia, Canada. Hydrobiologia 267, 75-98. https: //doi.org/10.1007/BF00018792.

Renaut, R. W., Long, P. R., 1989. Sedimentology of the saline lakes of the Cariboo Plateau, Interior British Columbia Canada. Sediment. Geol. 64, 239-264. https: //doi.org/10. 1016/0037-0738 (89) 90051-1.

Sahib, L. Y., Marandi, A., Schüth, C., 2016. Strontium isotopes as an indicator for groundwater salinity sources in the Kirkuk region, Iraq. Sci. Total Environ. 562, 935-945. https: //doi.org/10.1016/j.scitotenv.2016.03.185.

Schijf, J., Byrne, R. H., 2004. Determination of SO4β1 for yttrium and the rare earth elements at I = 0.66 m and t = 25°C-implications for YREE solution speciation in sulfate-rich waters. Geochimica et Cosmochimica Acta 68, 2825-2837. https: //doi. org/10.1016/j.gca.2003.12.003.

Semhi, K., Clauer, N., Probst, J. L., 2000. Strontium isotope compositions of river waters as records of lithology-dependent mass transfers: the Garonne river and its tributaries (SW France). Chem. Geol. 168 (3-4), 173-193. https: //doi.org/10.1016/S0009-2541 (00) 00226-6.

Shand, P., Darbyshire, D. P. F., Love, A. J., Edmunds, W. M., 2009. Sr isotopes in natural waters: applications to source characterisation and water-rock interaction in contrasting landscapes. Appl. Geochem. 24 (4), 574-586. https: //doi.org/10.1016/j. apgeochem.2008.12.011.

Sholkovitz, E. R., 1995. The aquatic chemistry of rare earth elements in rivers and estuaries. Aquat. Geochem. 1, 1-34. https: //doi.org/ 10.1007/ BF01025229.

Tabaksblat, L. S., 2002. Specific features in the formation of the mine water microelement composition during ore mining. Water Resour. 29, 333-345. https: // doi.org/10. 1023/A: 1015640615824.

Taylor, S. R., McLennan, S. M., 1985. The Continental Crust: Its Composition and Evolution. An Examination of the Geochemical Record Preserved in Sedimentary Rocks. https: //doi.org/10.1017/S0016756800032167.

Tripathy, G. R., Goswami, V., Singh, S. K., Chakrapani, G. J., 2010. Temporal variations in Sr and $^{87}Sr/^{86}Sr$ of the Ganga headwaters: estimates of dissolved Sr flux to the mainstream. Hydrol. Process. 24 (9), 1159-1171. https: //doi.org/10.1002/hyp. 7572.

Webb, G. E., Nothdurft, L. D., Kamber, B. S., Kloprogge, J. T., Zhao, J. X., 2009. Rare earth element geochemistry of scleractinian coral skeleton during meteoric diagenesis: a sequence through neomorphism of aragonite to calcite. Sedimentology 56, 1433-1463. https: //doi.org/10.1111/j.1365-3091.2008.01041.x.

Weis, D., Kieffer, B., Maerschalk, C., Barling, J., De Jong, J., Williams, G. A., Hanano, D., Pretorius, W., Mattielli, N., Scoates, J.S., Goolaerts, A., Friedman, R.M., Mahoney, J.B., 2006. High-precision isotopic characterization of USGS reference materials by TIMS and MC-ICP-MS. Geochemistry. Geophys. Geosyst.7. https: //doi.org/10.1029/2006GC001283.

Willis, S. S., Johannesson, K. H., 2011. Controls on the geochemistry of rare earth elements in sediments and groundwaters of the Aquia aquifer, Maryland USA. Chem. Geol. 285, 32-49. https: //doi.org/10.1016/j.chemgeo.2011.02.020.

Wilson, S. A., Dipple, G. M., Power, I. M., Thom, J. M., Anderson, R. G., Raudsepp, M., Gabites, J. E., Southam, G., 2009. Carbon dioxide fixation within mine wastes of ultramafic-hosted ore deposits: examples from the Clinton Creek and Cassiar Chrysotile deposits Canada. Econ. Geol. 104, 95-112. https: // doi.org/10.2113/gsecongeo.104. 1.95.

Wilson, S. A., Harrison, A. L., Dipple, G. M., Power, I. M., Barker, S. L. L., Ulrich Mayer, K., Fallon, S. J., Raudsepp, M., Southam, G., 2014. Offsetting of CO_2emissions by air capture in mine tailings at the Mount Keith Nickel Mine Western Australia: rates, controls and prospects for carbon neutral mining. Int. J. Greenh. Gas Control 25, 121-140. https: //doi.org/10.1016/j.ijggc.2014.04.002.

Wood, S. A., 1990. The aqueous geochemistry of the rare-earth elements and yttrium. Chem. Geol. 82, 159-186. https: //doi.org/10.1016/0009-2541 (90) 90080-Q.

Yu, J., Zheng, M., Wu, Q., Wang, Y., Nie, Z., Bu, L., 2015. Natural evaporation and crystallization of Dujiali salt lake water in Tibet. Chem. Ind. Eng. Prog. 34,

4172-4178.

Zedef, V., Russell, M. J., Fallick, A. E., Hall, A. J., 2000. Genesis of vein stockwork and sedimentary magnesite and hydromagnesite deposits in the ultramafic terranes of southwestern Turkey: a stable isotope study. Econ. Geol. 95, 429-445. https: //doi. org/10.2113/gsecongeo.95.2.429.

Zhao, Y. Y., Zheng, Y. F., Chen, F., 2009. Trace element and strontium isotope constraints on sedimentary environment of Ediacaran carbonates in southern Anhui South China. Chem. Geol. 265, 345-362. https: //doi.org/10.1016/j.chemgeo.2009.04.015.

Zieliński, M., Dopieralska, J., Belka, Z., Walczak, A., Siepak, M., Jakubowicz, M., 2017. The strontium isotope budget of the Warta River (Poland): between silicate and carbonate weathering, and anthropogenic pressure. Appl. Geochem. 81, 1-11. https: //doi.org/10.1016/j.apgeochem.2017.03.014.

Zheng, X. Y., Zhang, M. G., Xu, C., Li, B. X., 2002. The Saline lakes of China. Science Press, Beijing, pp. 49-51 (in Chinese).

蒸发岩中硫同位素的地球化学特征及其沉积学意义

——以思茅盆地 MZK-3 井为例*

苗忠英[1]，郑绵平[1]，张雪飞[1]，张震[1]，高运志[2]

1. 中国地质科学院矿产资源研究所，自然资源部盐湖资源与环境重点实验室，北京 100037
2. 中国地质大学（北京），北京 100083

内容提要 思茅盆地目前是中国唯一的含古代固体钾盐矿床的沉积盆地，其钾盐形成时代、物源特征、海侵方向等多年来一直存在争议。本文依据海相硬石膏的形成条件、存在形式、同位素分馏机理，重点分析了盆地内 MZK-3 井蒸发岩硫同位素的地球化学特征。结果表明：①岩盐中的硬石膏在蒸发盆地析岩盐阶段即可形成，单独成层的硬石膏是由原始沉积的石膏经历了沉积埋藏升温进而脱水而成；②岩盐中硬石膏的硫同位素值具有“双峰”特征，分别为 14‰～16‰和 8‰～10‰或 6‰～8‰，这体现了硫酸盐的双重来源——原始海水中的硫酸盐和陆源淡水输入的硫酸盐或火山活动提供的硫源；③硬石膏层的硫同位素在区域上具有对比性，结合 $^{87}Sr/^{86}Sr$ 值的特征，认为其代表了海相的沉积环境；④硬石膏层的硫同位素值平面上由南向北降低，可能反映了在此方向陆源淡水或火山活动对蒸发岩盆地的影响逐渐增强，进而说明这可能也是海侵的方向。可见对硬石膏硫同位素的研究，不仅在沉积学上能揭示物源、沉积环境、海侵方向等信息，更能对研究区钾盐矿床勘查提供有益的参考。

关键词 硫同位素；蒸发岩；地球化学；钾盐；思茅盆地

硫元素在自然界中分布较广，在地壳中的相对含量为 0.048%，在所有元素中排在第 14 位（Seal，2006）。其存在形式有游离态和化合态两种：游离态主要以单质硫的形式存在于火山口附近或地壳岩层里；化合态主要以硫化物、二氧化硫、亚硫酸盐和硫酸盐的形式出现；煤、石油和蛋白质里也可有少量的硫元素以有机化合物的形式出现（Jiang Jigang et al.，1988；Song Jinming，1990；Hu Wenxuan et al.，1991；Zhu Guangyou et al.，2006；Hu Guyue et al.，2013；Nie Fei et al.，2015）。

蒸发岩中的硫元素主要以硫酸盐的形式出现，相应的矿物主要为石膏和硬石膏。通过对石膏或硬石膏硫同位素值的分析，有学者指出：异常高的硫同位素值指示沉积环境水体分层，而且硫酸盐还原菌活性强（Li Renwei et al.，1989；Wang Chunlian et al.，2013）；蒸发岩硫同位素在同一时代地层内变化较小，可用于地层对比（Huang Jianguo et al.，1989；Tabakh et al.，1998）；硫同位素垂直于地层方向的韵律性变化指示蒸发盆地的封闭性较差，经常有海水或淡水补给（Huang Jianguo et al.，1989；Zhao Haitong et al.，2018），沿此方向的突然变化指示了突发性古气候事件及由此产生的沉积环境氧化还原条件突变（Zhang Tonggang et al.，2003）；通过全球海相蒸发岩硫同位素值对比及其同位素分馏机理的分析，可构建硫同位素年代地层曲线（Holser et al.，1966；Cortecci et al.，1981）。此外，有些学者通过对石膏硫同位素值的分析，探讨了金属矿床的类型和成因机理（Yan Yuehua et al.，2002；Li Yanhe et al.，2013；Hu Guyue et al.，2013；Ren Shunli et al.，2018）。

兰坪-思茅盆地内海相蒸发岩发育较广，上三叠统、下—中侏罗统以及下白垩统内都存在，但是公开发表的硫同位素数据相对较少。利用蒸发岩硫同位素特征探讨物源、沉积环境和古气候更是鲜有报道。本文拟通过对 MZK-3 井钻获岩心中硬石膏夹层、岩盐和钾石盐中硬石膏晶体硫同位素地球化学特征的分析，丰富研究区的硫同位素数据、解析成盐物质的来源和沉积环境的物理化学条件。

* 本文发表在：地质学报，2019，93（5）：1166-1179

1 地质背景

兰坪-思茅盆地处于印支地块北部（曲一华等，1998），盆地边界由西部的澜沧江大断裂和东部的金沙江-哀牢山大断裂控制。在盆地内的次级断裂控制下，可进一步划分为 5 个隆起和 5 个拗陷共计 10 个二级构造单元。盆地内发育的含盐带（Yang Jianxu et al.，2013）主体分布在除大渡岗拗陷以外的其余 4 个拗陷中（图 1）。

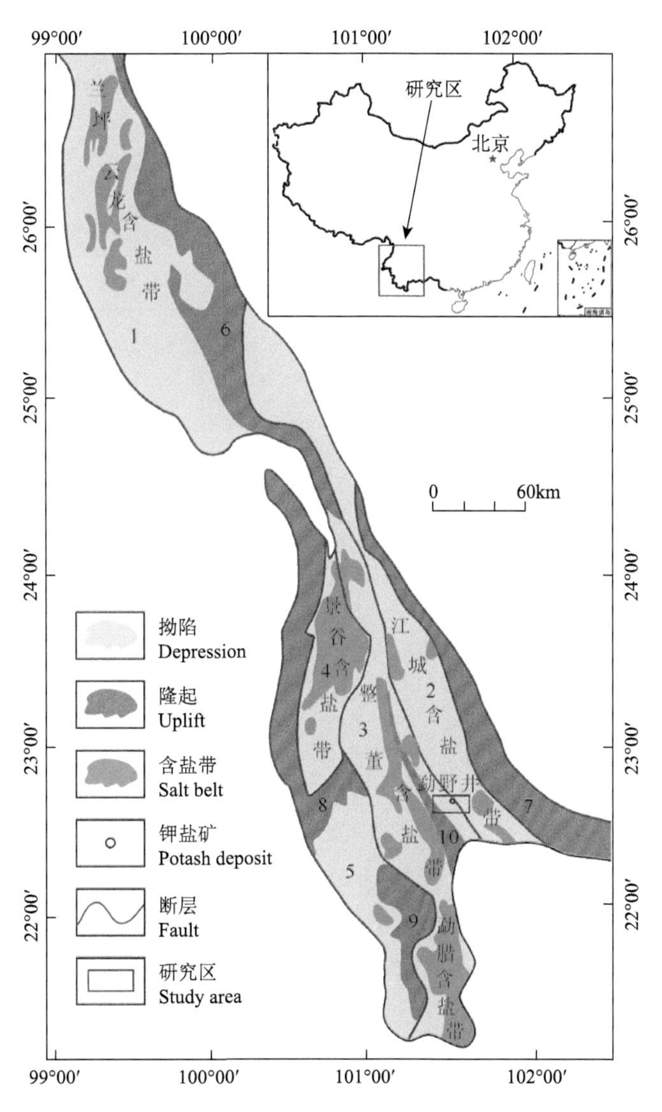

图 1 兰坪-思茅盆地构造单元划分及主要盐产地（据 Yang Jianxu et al.，2013 修改）

Fig. 1 Tectonic units and main production area of salt in the Lanping-Simao basin (after Yang Jianxu et al., 2013).

1-兰坪-云龙拗陷；2-江城拗陷；3-磨黑-勐腊拗陷；4-景谷拗陷；5-大渡岗拗陷；6-乔后隆起；7-墨江隆起；8-景洪隆起；9-勐远隆起；10-无量山隆起

1-Lanping-Yunlong depression；2-Jiangcheng depression；3-Mohe-Mengla depression；4-Jinggu depression；5-Dadugang depression；6-Qiaohou uplift；7-Mojiang uplift；8-Jinghong uplift；9-Mengyuan uplift；10-Wuliangshan uplift

盆地内的主要沉积盖层为中—新生界（图 2）。三叠系缺失下统，中—上统以发育灰色、灰黑色泥岩、页岩、泥灰岩、灰岩为特征。侏罗系呈“红-杂-红”三分特征。其中，下侏罗统主要为紫红色泥岩和粉砂岩，夹砂岩和膏岩；中侏罗统除了紫红色泥岩—粉砂岩等细碎屑岩外，还有灰绿色膏岩、灰岩、泥灰岩与之互层；上侏罗统为紫红色泥岩—粉砂岩夹细砂岩。下白垩统以发育河、湖相砂岩为特征；下白垩统勐野井组是本区的产盐、产钾地层，以发育泥砾岩为特征。古近系等黑组和勐腊组发育的规模较小，以紫红色泥岩、砂岩、砾岩为特征。

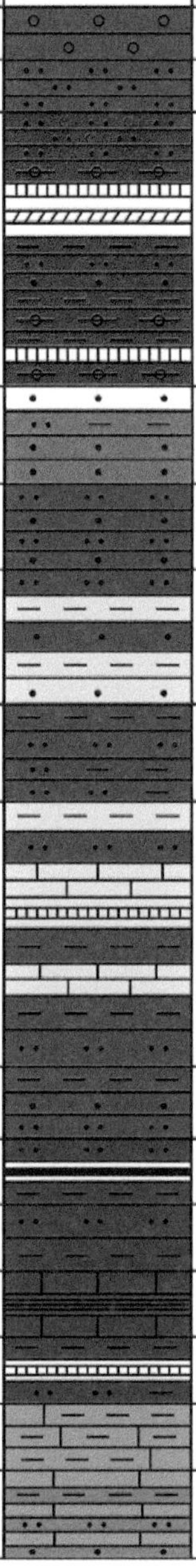

地层				岩性	描述
界	系	统	组		
新生界	古近系	渐新统	勐腊组		紫红色砾岩、砂砾岩、中-细粒砂岩夹粉砂岩
		始新统	等黑组		紫红色粉砂岩、泥质岩夹细砂岩，局部夹灰绿色泥岩
中生界	白垩系	下白垩统	勐野井组		上段：上亚段为棕红色钙泥质粉砂岩、粉砂质泥岩；下亚段为棕红色、杂色泥砾岩夹石膏、泥岩、泥灰岩等，深部为各类蒸发岩 中段：棕红色泥质岩、粉砂岩夹少量细砂岩 下段：棕红色、杂色泥砾岩夹泥质岩，泥灰岩、石膏
			虎头寺组		灰紫色、浅紫红色，顶部灰白色细粒石英砂岩，局部夹粉砂质泥岩，顶部白色砂岩中含铜
			南新组		上部以紫红色粉砂岩、泥岩为主，夹砂岩；下部紫红色细砂岩夹粉砂岩及少量砂砾岩，有时底部为砾岩
			景星组		上段暗紫红色粉砂岩、泥岩，夹少量灰绿色泥质岩及灰白色、紫红色细砂岩；下段灰白色、灰绿色石英砂岩与杂色泥质岩、粉砂岩的韵律性互层，以砂岩为主
	侏罗系	上侏罗统	坝注路组		紫红色泥岩、粉砂质泥岩、泥质粉砂岩、粉砂岩夹细砂岩
		中侏罗统	和平乡组		上段杂色泥质岩、粉砂岩夹砂岩、泥灰岩，上部介壳泥灰岩、燧石条带灰岩及泥砾岩、石膏 下段紫红色细碎屑岩，底部砂砾岩、砾岩，顶部介壳泥质灰岩及灰绿色泥质岩
		下侏罗统	张科寨组		紫红色泥质岩、粉砂岩夹砂岩及膏岩、泥砾岩
	三叠系	上三叠统	麦初箐组		灰黑色粉砂岩、泥岩，中上部夹薄煤层
			挖鲁八组		灰黑色粉砂岩、泥岩
			三合洞组		灰岩及灰黑色页岩
			歪古村组		杂色砂泥岩夹蒸发岩，底部含砾粗砂岩、砾岩
		中三叠统	臭水组		灰色钙质泥岩夹泥质灰岩、泥灰岩
			黄竹林组		下部为灰紫色砂岩、粉砂岩夹灰岩，底部为砾岩，上部为灰岩夹泥岩
下伏地层			二叠系或石炭系		

图 2　兰坪-思茅盆地岩性柱状图（据曲一华等，1998 修改）

Fig. 2　Generalized stratigraphic column of the Mesozoic in the Lanping-Simao basin (after Qu Yihua et al., 1998).

特提斯的构造演化对盆地沉积盖层的岩性、岩相特征影响强烈。泥盆纪—晚三叠世，古特提斯从形成、发展到萎缩、消亡，盆地内的沉积盖层主要为海相碎屑岩和灰岩（Tan Fuwen et al.，1999；Mou Chuanlong et al.，2005；Yin Fuguang et al.，2006）。侏罗纪至今，盆地内以发育陆相红层为特征，主要岩性为河湖相砂岩、页岩、砾岩和煤层（曲一华等，1998；Yin Fuguang et al.，2006）。中侏罗统内出现海相灰岩和双壳类化石，反映了中特提斯的一次海侵（曲一华等，1998）。陆相下白垩统勐野井组内见海相石盐和钾石盐且形成了矿床，反映了中特提斯闭合后的残留海在极端干旱气候条件下侧向迁移析盐（Miao Zhongying et al.，2017）或中特提斯海侵成盐（Wang Licheng，2014）。

2 样品来源和分析方法

2.1 样品来源

样品来源于 MZK-3 井，其地理位置为 22°40′12.00″N，101°37′23.80″E，处于云南省普洱市江城县宝藏乡境内（图 3）。井口周围出露地层主要为南新组和勐野井组，也有少量景星组、虎头寺组和新近系。开孔层位为勐野井组，井深 410.4m 钻遇虎头寺组，井深 470.1m 钻遇南新组，1000.1m 终孔，景星组未钻穿。蒸发岩出现在 179.3～249.3m 深度段，主要岩性有硬石膏、石盐和钾石盐。针对硫同位素的分析，我们自上而下选取了 7 个不同岩相特征的样品（表 1）。

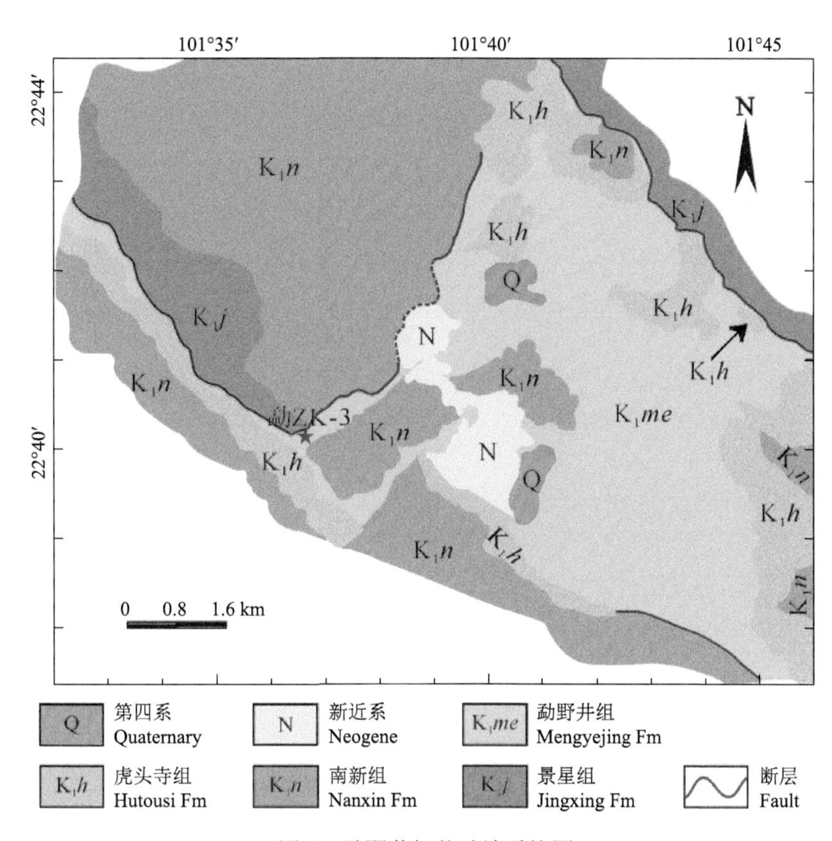

图 3 勐野井钾盐矿地质简图

（据 Zheng Zhijie et al.，2012 修改）

Fig. 3 Geologic map of the Mengyejing potash deposit.

(after Zheng Zhijie et al., 2012)

表 1 思茅盆地 MZK-3 井蒸发岩的取样位置及岩性特征

Table 1 Sampling location and lithology characteristics of evaporites from well MZK-3in the Simao basin

序号	样品编号	深度（m）	岩性	岩性描述
1	MZK-3-6-1	178.5	红褐色硬石膏	红褐色细晶结构，晶体粒径为 0.1～0.25mm，厚层状构造，渗透性好，表面见重结晶针状石膏
2	MZK-3-6-2	179.2	红褐色硬石膏	同上
3	MZK-3-7-1	181.7	黑色岩盐	黑色中-粗晶结构，晶体粒径为 0.25～1.0mm，块状构造
4	MZK-3-8-1	206.2	浅灰色石盐	浅灰色中-粗晶结构，晶体粒径为 0.25～1.0mm，块状构造，较纯净，含有钾石盐
5	MZK-3-8-5	213.0	浅灰色钾石盐	同上
6	MZK-3-10-2	232.5	黑色岩盐	黑色中-粗晶结构，晶体粒径为 0.25～1.0mm，块状构造
7	MZK-3-12-1	248.2	黑色岩盐	同上

2.2 分析方法

2.2.1 薄片鉴定

使用加拿大树胶作为黏结剂将蒸发岩样品固定在载玻片上，使用饱和盐水将样品磨制成 0.03mm 厚，黏结盖玻片后备偏光显微镜下观察。偏光显微镜型号为 Leica 2700P，LED 光源，图像采集系统为 LAS 成像模块。

2.2.2 扫描电镜分析

本次研究使用的扫描电镜型号为 TM3030，能谱仪型号为 XFlash MIN SVE。具体方法为：先将导电胶带黏结到样品柱上，再将粒径为 3～5mm 的全岩样品黏结在胶带上，然后将样品柱放到样品支架上调节高度，使样品表面的高度与标尺之间的距离约为 1mm，最后将样品支架放入样品室开始检测。

2.2.3 X 射线衍射分析

将粉碎至 200 目的样品放入带凹槽的载玻片，用盖玻片将样品压实备分析。XRD 仪器型号为 Rigaku MiniFlex600，辐射为 Cu 靶，管电压/电流为 40kV/15mA，步长 0.02°，扫描角度为 3°～70°。借助 MDI Jade 软件完成矿物鉴定。

2.2.4 化学分析

岩盐样品化学分析在自然资源部盐湖资源与环境重点实验室完成。使用原子吸收分光光度法测定 Na^+、K^+、Mg^{2+}、Ca^{2+}的含量，仪器型号为 GGX-800；使用硫酸钡重量法测定硫酸根的含量；使用硝酸银容量法测定氯根的含量；使用酚红光度法测定溴的含量；使用酸碱滴定容量法测定碳酸根、重碳酸根的含量。更具体的实验步骤可参考文献（Lin et al.，2017）。

2.2.5 稳定硫同位素

首先将全岩样品直接粉碎至 200 目，然后称取约 15mg 样品，利用碳酸钠和氧化锌半熔法将样品中的硫元素转化提取为纯净的硫酸钡。将硫酸钡、五氧化二钒和石英砂按 2∶7∶7 的质量比混合均匀并研磨至 180 目。当预处理制样装置真空度达 2.0×10^{-2}Pa 时，加热样品至 980℃进行氧化反应生成二氧化硫。用液氮冷冻法收集、纯化二氧化硫，用 Delta V Plus 气体同位素质谱仪分析硫同位素的组成。测量结果以 CDT 为标准，记为 $\delta^{34}S_{V\text{-}CDT}$。采用的硫同位素标准参考物质为硫化银 GBW-04414 和 GBW-04415，分析精度优于±0.2‰（Liu Hanbin et al.，2013）。

3 分析结果

3.1 化学成分

样品中主要的阳离子为 Na^+、K^+、Mg^{2+}和 Ca^{2+}，主要阴离子为 Cl^-、Br^-、SO_4^{2-} 和 HCO_3^-（表 2）。MZK-3-6-1 号样品中未检测出 Br^-，同样岩性的样品 MZK-3-6-2 中 Br^-含量达到 300×10^{-6}。岩盐中 Br^-相对含量的变化范围为 10×10^{-6}～400×10^{-6}，含钾石盐的样品（MZK-3-8-1 和 MZK-3-8-5）Br^-含量明显高于纯石盐样品。

岩盐样品的 $Br^-\times10^3/Cl^-$系数为 0.02～0.71，平均值为 0.22。与相邻的呵叻盆地岩盐样品相比（Gao Xiang et al.，2012），这个参数值明显偏低。但是 $Br^-\times10^3/Cl^-$系数与 K^+的含量呈明显的指数正相关，其相关系数 r^2 可达 0.9467（图 4）。这反映 Br^-和 K^+在蒸发岩形成晚期同步相对富集的特征。

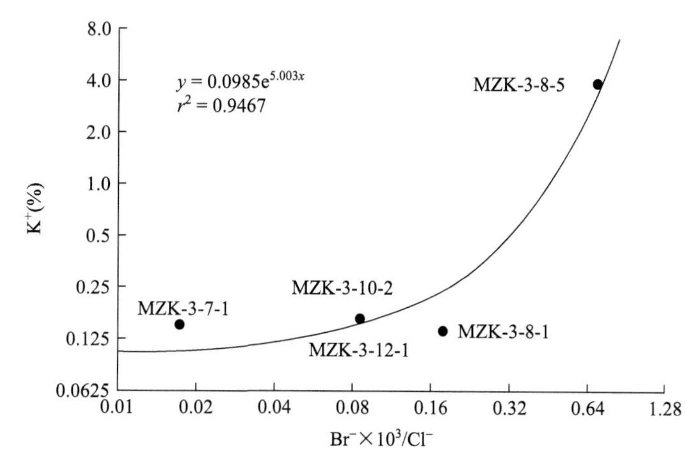

图 4 思茅盆地 MZK-3 井蒸发岩 $Br^-\times10^3/Cl^-$系数与 K^+的相对含量之间呈正相关

Fig. 4 Positive correlation between the $Br^-\times10^3/Cl^-$ coefficient and the relative content of K^+ of evaporate from well MZK-3 in the Simao basin.

表 2 思茅盆地 MZK-3 井蒸发岩化学分析结果

Table 2 Chemical analysis results of evaporite from well MZK-3 in the Simao basin

样品编号	不同离子的相对含量（%）							水不溶物（%）	KCl（%）	Br^-（$\times10^{-6}$）	$\delta^{34}S_{V-CDT}$（‰）
	Na^+	K^+	Mg^{2+}	Ca^{2+}	Cl^-	HCO_3^-	SO_4^{2-}				
MZK-3-6-1	1.03	0.08	0.03	3.46	2.08	0.28	7.21	85.37	0.15	0	13.7
MZK-3-6-2	0.48	0.04	0.11	1.67	1.2	0.24	3.58	91.52	0.08	300	14.2
MZK-3-7-1	35.96	0.15	0.01	0.6	56.82	0.06	0.53	0.12	0.29	10	16.0
MZK-3-8-1	33.44	0.14	0.01	0.54	54.98	0.12	0.82	0.04	0.27	100	15.4
MZK-3-8-5	35.09	3.8	0.01	0.24	56.66	0.34	1.88	1.16	8.69	400	8.9
MZK-3-10-2	36.43	0.16	0.01	0.42	57.5	0.15	1.28	0.11	0.12	50	7.0
MZK-3-12-1	39.29	0.16	0	0.38	57.66	0.12	0.53	0.23	0.31	50	6.7

所有样品中SO_4^{2-}的相对含量为 0.53%～7.21%，平均值为 2.26%。虽然硬石膏中SO_4^{2-}的相对含量较高，但是水不溶物的相对含量均大于 85%。这可能与采用的实验方法有关。使用水溶法测定硬石膏样品中SO_4^{2-}的相对含量时，由于其较低的溶解度可导致上述现象发生。岩盐样品中SO_4^{2-}的相对含量为 0.53%～1.88%，平均值为 1.01%。有趣的是，岩盐样品中SO_4^{2-}的相对含量与 Ca^{2+}的相对含量负相关，而与 K^+的相对含量正相关（图 5）。这种明显与蒸发岩沉积序列违背的现象，可能说明原始卤水中有相对富集SO_4^{2-}的水源混入或者是火山硫源混入后氧化为SO_4^{2-}。

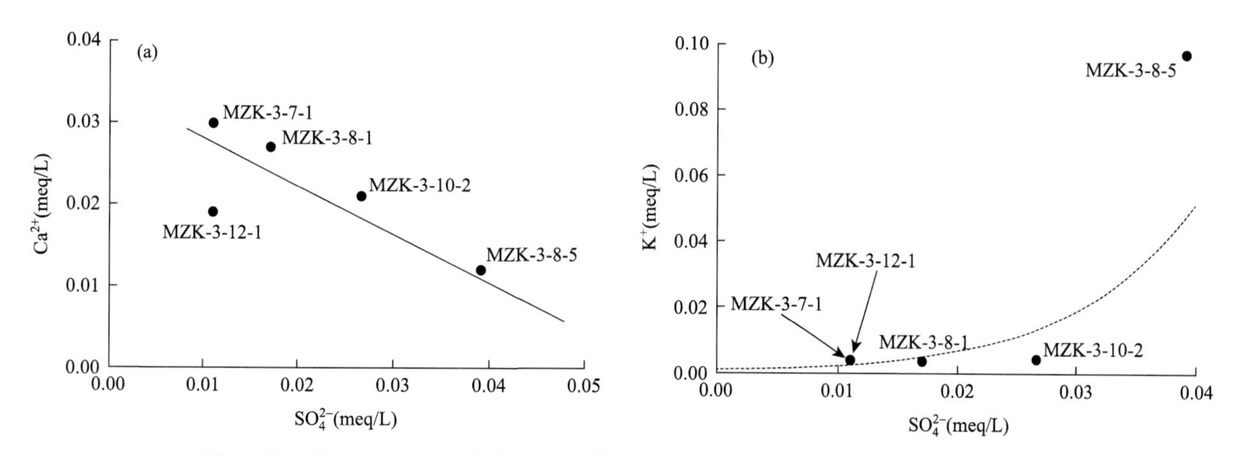

图 5 思茅盆地 MZK-3 井蒸发岩SO_4^{2-}的相对含量与 Ca^{2+}和 K^+相对含量之间的关系

Fig. 5 Relationship of relative content between SO_4^{2-} and Ca^{2+} and K^+ of evaporate from well MZK-3in the Simao basin.

3.2 矿物成分

单偏光下硬石膏晶体呈粒状-短柱状，显淡蓝色、紫色、浅红色，两组解理发育，解理角接近 90°，正低-中凸起；正交偏光下，硬石膏晶体干涉级序较高，显紫色、蓝色和绿色，平行消光[图 6（a）、（b）]。有少量他形方解石晶体充填在硬石膏晶体之间，在正交光下呈高级白干涉色，两组解理夹角近 60°。

图 6　思茅盆地 MZK-3 井蒸发岩偏光显微照片

Fig. 6　Polarization photomicrograph of evaporate from well MZK-3in the Simao basin.

（a）MZK-3-6-1 硬石膏样品单偏光镜下照片；（b）为与（a）相同视域的正交偏光镜下照片；（c）MZK-3-8-5 岩盐单偏光镜下照片；（d）MZK-3-8-1 岩盐单偏光镜下照片

(a) Polarization photomicrograph of the anhydrite with number MZK-3-6-1; (b) cross polarization micrograph is same field of vision with (a); (c) polarization photomicrograph of the halite with number MZK-3-8-5; (d) polarization photomicrograph of the halite with number MZK-3-8-1

单偏光下石盐晶体无色透明，而钾石盐晶体因含有 Fe^{3+}而略带红色；此外，由于二者折射率的差异（$N_{石盐}=1.5443$；$N_{钾石盐}=1.4902$），提升镜筒时贝克线向石盐晶体移动。石盐晶体多为他形；晶间钾石盐晶体呈他形，而晶内钾石盐晶体的自形程度较高[图 6（c）、（d）]。

我们通过泥粉晶 X 衍射定性分析样品的矿物成分结果显示（图 7），MZK-3-6-1 和 MZK-3-6-2 号样品的主要矿物成分是硬石膏，与野外和室内薄片鉴定以及化学分析的结果一致，但是样品中少量的方解石矿物在谱图中没有得到显示，可能与其相对含量低有关；岩盐样品的主要矿物成分是石盐（MZK-3-7-1），部分样品含钾石盐（MZK-3-8-5）；个别样品在野外通过钾试剂（亚硝酸铜铅钠）鉴定出含钾离子，但是谱图中并未体现出含钾矿物（MZK-3-8-1）。岩盐中含硫矿物在谱图中没有体现。

我们进一步分析了样品的扫描电镜图像和能谱特征（图 8）。结果显示硬石膏电镜图像呈粒状或短柱状，能谱的出峰位置主要在 O、S、Ca 元素处，其相对含量符合硬石膏的特征。

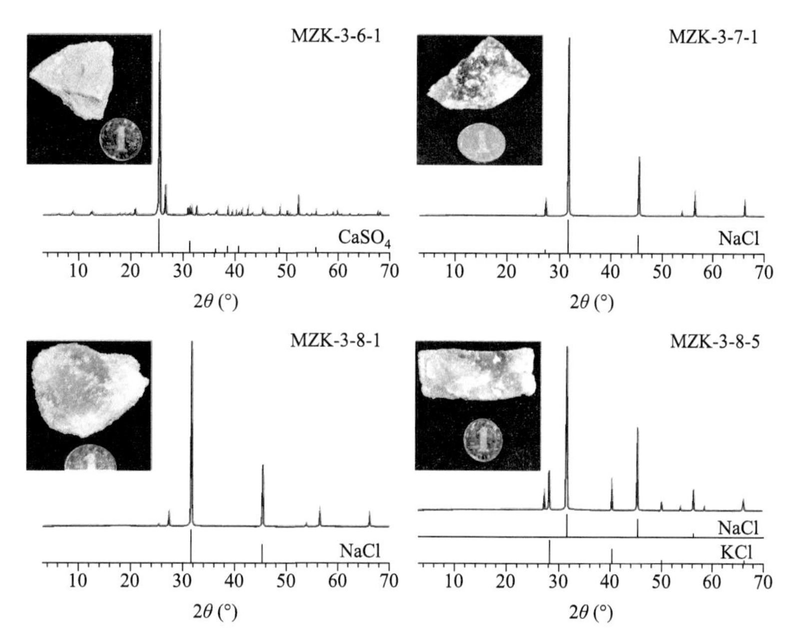

图 7　思茅盆地 MZK-3 井蒸发岩样品泥粉晶 X 衍射谱图

Fig. 7　X-ray powder diffraction curve of the evaporite from well MZK-3in the Simao basin.

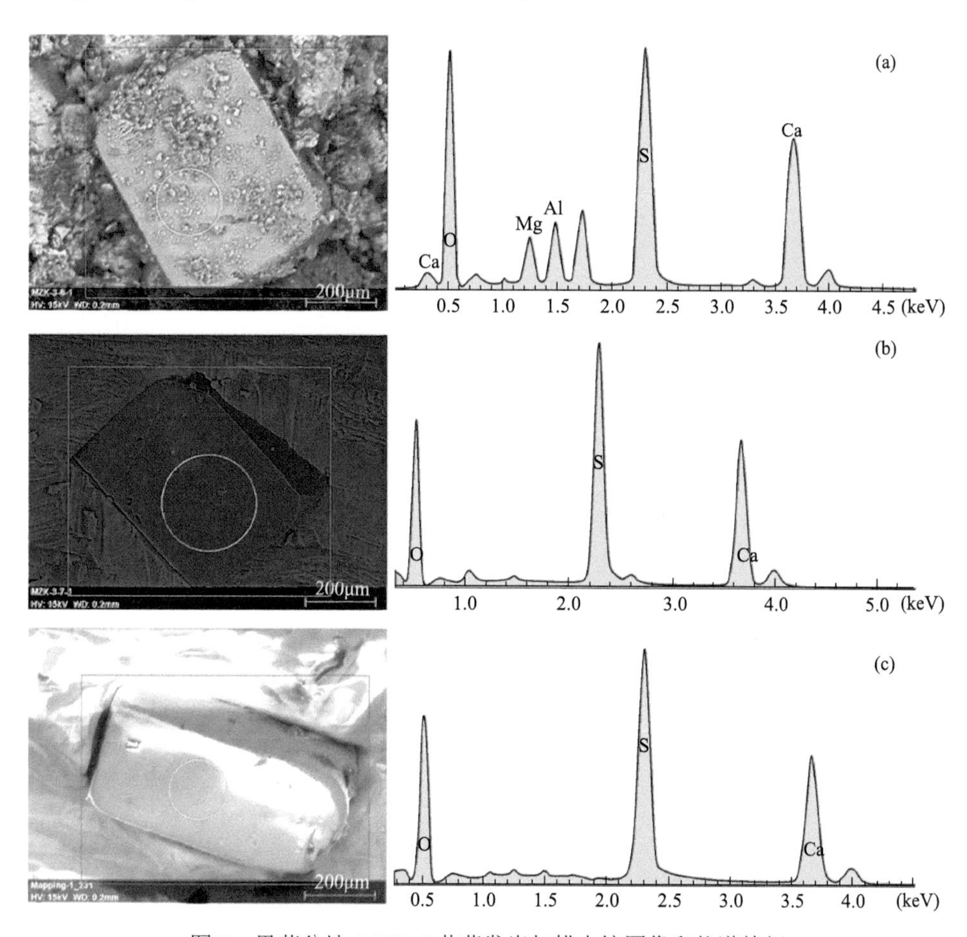

图 8　思茅盆地 MZK-3 井蒸发岩扫描电镜图像和能谱特征

Fig. 8　The SEM images and energy spectrum of evaporate from well MZK-3 in the Simao basin.

（a）MZK-3-6-1 硬石膏样品中的晶体；（b）MZK-3-7-1 岩盐中的硬石膏晶体；（c）MZK-3-12-1 岩盐中的硬石膏晶体

(a) Crystal in the anhydrite with sample number MZK-3-6-1; (b) and (c) anhydrite crystal in the halite with sample number MZK-3-7-1and MZK-3-12-1, respectively

通过上述分析，我们确定了样品的基本化学和矿物组成以及硬石膏在岩盐中客观存在的事实，为进一步分析硫同位素地球化学特征及其地质意义奠定了物质基础。

3.3 稳定硫同位素

前文分析表明样品中的硫元素以硬石膏的形式存在，其同位素值（$\delta^{34}S_{V\text{-}CDT}$）为6.7‰～16.0‰，平均值为11.7‰（表2）。最大值和最小值均出现在岩盐样品中（图9），两者相差9.3‰，说明其物质来源或沉积环境波动较大；硬石膏样品中的硫同位素比较稳定，在14.0‰±0.3‰范围内波动。

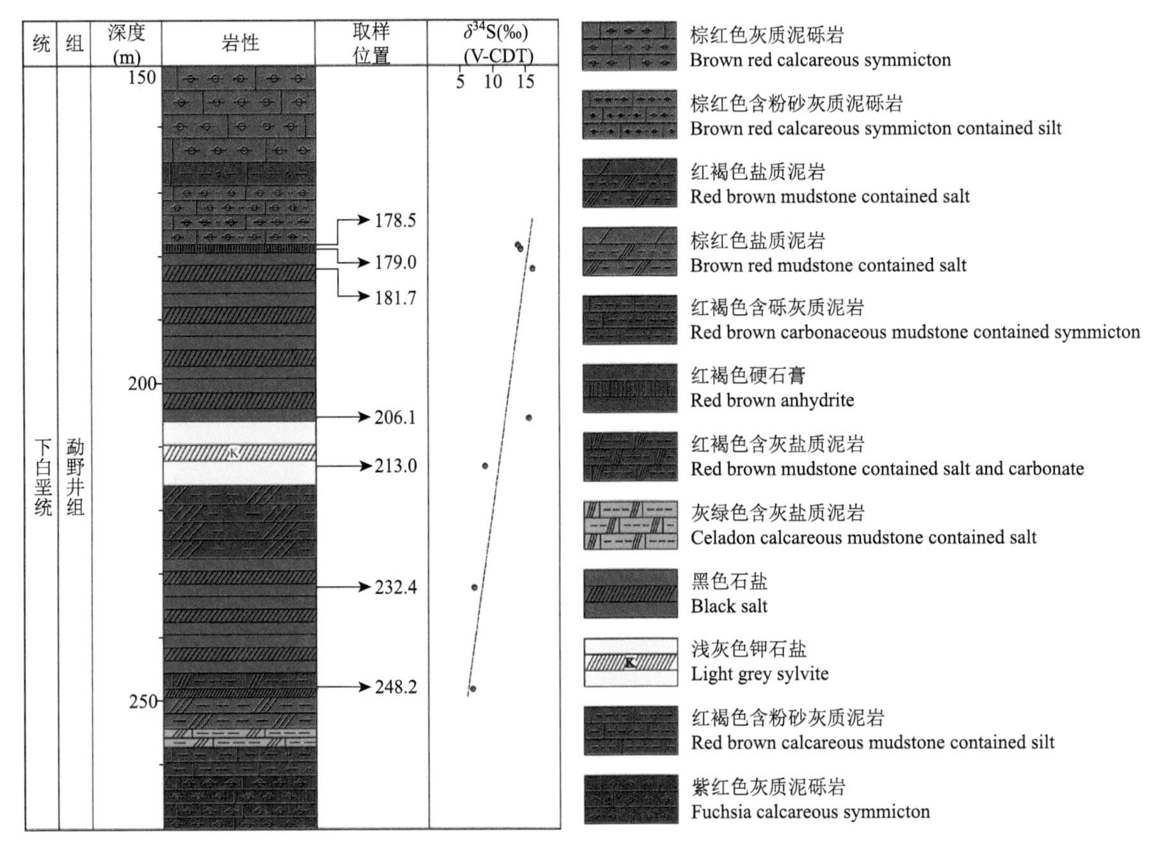

图9　思茅盆地MZK-3井蒸发岩硫同位素在地层中的分布特征

Fig. 9　Distribution of the $\delta^{34}S_{V\text{-}CDT}$ values of evaporate from well MZK-3 in the Simao basin.

岩盐样品$\delta^{34}S_{V\text{-}CDT}$值与$Na^+$浓度负相关，说明原始卤水相对富集重硫同位素，轻硫同位素主要源于后期外源补给，与Ca^{2+}浓度正相关也是这种认识的佐证[图10（a）、（b）]。K^+的浓度和溴氯系数间接反映了卤水的浓缩阶段，而$\delta^{34}S_{V\text{-}CDT}$值与它们的相关性不明显，推测也是外源硫干扰的结果。$\delta^{34}S_{V\text{-}CDT}$值与$SO_4^{2-}$浓度负相关[图10（c）]，间接反映了补给硫具有轻硫同位素的特征。HCO_3^-可能源于样品中微量碳酸盐的溶解，由于其相对含量较低，故与$\delta^{34}S_{V\text{-}CDT}$值的关系[图10（d）]较难反映卤水浓缩阶段和物源特征。

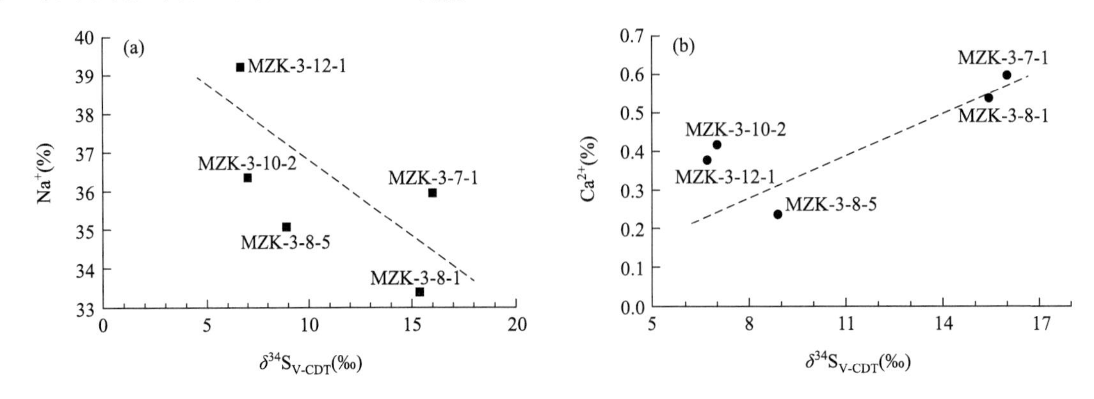

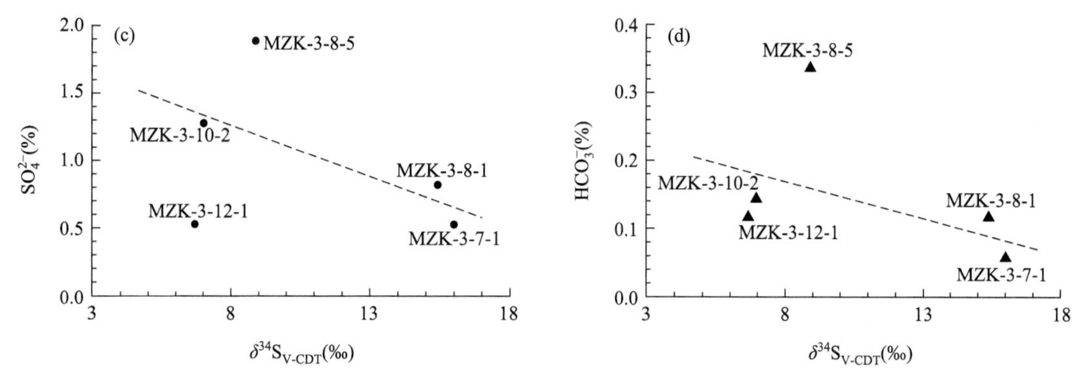

图10 思茅盆地MZK-3井岩盐硫同位素值与化学成分之间的关系

Fig. 10 Relationship between the δ^{34} S_{V-CDT} values and chemical composition of halite from well MZK-3in the Simao basin.

4 讨论

4.1 蒸发岩的埋藏史

本次研究的样品 $^{87}Sr/^{86}Sr$ 平均值为0.7081（Miao，2018），明显低于现代海水，结合伴生矿物和围岩特征，认为它们是物源为海水的蒸发岩。作为一类特殊的沉积岩，其中的硬石膏多以石膏的形式从原始饱和卤水中析出，而石膏向硬石膏转化是一个吸热脱水的过程，其能量来源主要为太阳能和地热能。

理论上，石膏转化为硬石膏的温度下限为42～58℃（Ossorio et al.，2014），但是溶液中实现这一转化的温度往往在90℃以上。例如，在盐溶液中，石膏向硬石膏转化的最小温度为90.5℃；在纯硫酸盐溶液中，这一转换温度至少为97℃（Ostroff，1964）。根据卫星遥感资料获得的现代盐湖表面水体最高温度通常为22.5～28℃。例如，美国犹他州大盐湖最高水面温度为26℃（Crosman et al.，2009）；非洲中南部Malawi湖最高水面温度为28℃（Chavula et al.，2009）；瑞典的Vänern和Vättern湖最高水面温度为22.5℃（Reinart et al.，2008）；伊朗的Urmia湖最高水面温度为26℃（Sima et al.，2013）。湖水若产生盐梯度，进而引发太阳池效应，理论上湖底保温层的温度可达97℃，人工模拟这一效应的温度通常为50～75℃（Srinivasan，1990）。可见自然条件下，盐度未饱和的湖水很难达到直接析出硬石膏的温度条件。

如果溶液中饱含NaCl和$CaSO_4$，石膏向硬石膏转化的温度可低至18℃（Ossorio et al.，2014）。本文所分析的岩盐样品中，硬石膏呈完整单晶分散在石盐晶体之间，说明它们在析出时，溶液是饱含NaCl和$CaSO_4$的。在这种条件下，温度低至18℃即可实现石膏向硬石膏的转化。通过现代盐湖表面水体温度的分析可知，这一温度条件较易达到，所以岩盐中的硬石膏应该是原始沉积成因，不能反映蒸发岩的沉积埋藏史。

单独成层的石膏向硬石膏转化满足温度下限条件即可。当地表温度条件不能满足其转化要求时，深埋过程中地热能的作用将促使这种转化的发生，而地层温度又是深度的函数，所以我们可以根据石膏脱水的温度条件和古地温梯度估算蒸发岩在地质历史中最小的埋藏深度。

思茅盆地现今的地温梯度为18.5～28.4℃/km（Wang Jiyang et al.，1990），古地温资料还未见报道。由于晚三叠世以来，思茅地块拼贴到了华南板块之上，与楚雄盆地不仅位置相邻而且构造背景相近，故可参考其古地温资料。研究表明（Wang Guoli et al.，2005）：楚雄盆地晚三叠世—早白垩世的古地温梯度<25℃/km；古近纪—新近纪的古地温梯度不低于30℃/km；现今的地温梯度为25℃/km；晚三叠世以来经历的最高古地温梯度为40℃/km。所以，研究区石膏—硬石膏转化的深度至少为1050～1450m。硬石膏样品现今的埋藏深度仅为178.5～179.2m，所以蒸发岩至少抬升了871.5～1271.5m。抬升的时期很可能为始新世，此时新特提斯洋闭合、印度板块与欧亚大陆碰撞导致地壳隆升，先期的沉积岩逐渐抬升遭受剥蚀。晚始新世—渐新世的勐腊组不整合于下伏的下白垩统勐野井组之上也是地层抬升剥蚀的有力证据。

4.2　硫同位素的示源和古环境意义

沉积盆地汇集了不同来源的硫元素，由于重硫同位素 ^{34}S 具有在高价态化合物中富集的趋向，即 $S^{2-}\rightarrow S^{0}\rightarrow SO_2\rightarrow SO_3^{2-}\rightarrow SO_4^{2-}$ 依次富集重硫同位素，故可以根据沉积物中 $\delta^{34}S_{V\text{-}CDT}$ 值的特征反推硫元素的来源。蒸发盆地内硫酸盐具有较低的 $\delta^{34}S_{V\text{-}CDT}$ 值，可能反映母源主要为黑色页岩，母源中硫元素以 S^{2-}的形式存在；较高的 $\delta^{34}S_{V\text{-}CDT}$ 值可能反映了母源主要为先期沉积的蒸发岩，母源中硫元素以 SO_4^{2-} 的形式存在；处于这两者之间的 $\delta^{34}S_{V\text{-}CDT}$ 值可能反映了母源主要为大气降水，母源中的硫元素以 SO_3^{2-} 的形式存在。蒸发盆地内硫酸盐异常高的 $\delta^{34}S_{V\text{-}CDT}$ 值反映沉积环境水体具有理化性质分层、底部为还原环境、微生物硫酸盐还原反应强烈等特征。

陆相蒸发盆地由于规模较小，所以硫酸盐的封闭性相对较弱，$\delta^{34}S_{V\text{-}CDT}$ 值的变化特征受控因素较多，既有物源的影响，又有古气候的影响，还有微生物还原的影响，所以根据其中硫酸盐 $\delta^{34}S_{V\text{-}CDT}$ 值的特征判识物源难度较大，但是可以识别出是否遭受硫酸盐还原菌的影响。例如，东濮凹陷和江陵凹陷古近系硬石膏 $\delta^{34}S_{V\text{-}CDT}$ 平均值分别为 34.0‰（Li Renwei et al.，1989）和 29.4‰（Wang Chunlian et al.，2013），据此不能区分硫酸盐母源是由陆源淡水携带进入沉积盆地还是源自海侵水体，但是由于这些数据明显高于同一地质时期海水的 $\delta^{34}S_{V\text{-}CDT}$ 平均值，所以可以推断硬石膏遭受了硫酸盐还原菌的强烈作用；伴生的黑色泥岩、页岩说明沉积环境氧化还原电位较低，利于硫酸盐还原菌的生存，是上述推断的有利证据。氧化还原电位较高的第四系盐湖或干盐湖中不具备硫酸盐还原菌的生存条件，其中未遭受过微生物还原影响的硫酸盐 $\delta^{34}S_{V\text{-}CDT}$ 值最小者为 3.9‰左右（Peng Licai et al.，1999），最大者可达 19.9‰（Fan Qishun et al.，2009），可能反映了母源区的 $\delta^{34}S_{V\text{-}CDT}$ 值特征。

海相沉积盆地不仅规模大，而且由于硫元素在海洋中混合均匀的时间远小于其存留时间，所以同一地质时期海相蒸发岩 $\delta^{34}S_{V\text{-}CDT}$ 值相差不大，也是同期海水 $\delta^{34}S_{V\text{-}CDT}$ 值的反映（Holser et al.，1966）。

思茅盆地的蒸发岩主体发育于“红层”中，反映沉积环境的氧化还原电位较高，不利于硫酸盐还原菌的生存，所以决定蒸发岩中硬石膏硫同位素特征的主控因素是物源输入。区域上硬石膏样品 $\delta^{34}S_{V\text{-}CDT}$ 的平均值可对比：研究区的平均值为 14.0‰；呵叻盆地马哈萨拉堪组的平均值为 16.2‰（Li et al.，2018），沙空那空盆地贡塔组的平均值为 15.0‰（Zhang Hua et al.，2014）；兰坪盆地的平均值为 14.8‰（Wang Licheng et al.，2014）。这种区域内稳定的硫同位素值体现了蒸发岩沉积期盆地规模较大、物源为海水的特征。此外，岩心样品硬石膏的硫同位素还呈现出由南向北逐渐变轻的趋势，即呵叻→沙空那空→思茅盆地硬石膏 $\delta^{34}S_{V\text{-}CDT}$ 值逐渐减小（图 11）。这种变化特征可能反映了陆源淡水或火山活动对硫酸盐沉积的影响程度在逐渐加强。河流携带同位素较轻的硫酸盐或者火山活动提供同位素较轻的硫元素汇入蒸发岩盆地，对呵叻→沙空那空→思茅盆地的影响由弱到强，从而导致硬石膏的 $\delta^{34}S_{V\text{-}CDT}$ 值逐渐减小。关于火山活动对蒸发岩影响的可能性，可以通过硼酸盐矿物的出现、异常低的 $\delta^{11}B$ 值和钾石盐中异常高的 Br 含量来证实（曲一华等，1998；Cheng Huaide et al.，2008；Zhang Congwei et al.，2011）。

本次分析的岩盐样品中硬石膏 $\delta^{34}S_{V\text{-}CDT}$ 的平均值为 10.8‰。勐野井钾盐矿岩盐样品中硬石膏 $\delta^{34}S_{V\text{-}CDT}$ 的平均值为 10.6‰（Xu Jianxin，2008）。这说明研究区内，岩盐样品中硬石膏 $\delta^{34}S_{V\text{-}CDT}$ 的值普遍低于硬石膏层的 $\delta^{34}S_{V\text{-}CDT}$ 值是客观事实。我们进一步分析这些样品的硫同位素分布特征，结果显示岩盐中硬石膏的 $\delta^{34}S_{V\text{-}CDT}$ 值具有双峰特征（图 12）。其中的一个峰 $\delta^{34}S_{V\text{-}CDT}$ 值为 14‰～16‰，与硬石膏层的 $\delta^{34}S_{V\text{-}CDT}$ 值相近，体现了原始海水的硫同位素特征；另一个峰 $\delta^{34}S_{V\text{-}CDT}$ 值为 8‰～10‰或 6‰～8‰，明显低于硬石膏层的 $\delta^{34}S_{V\text{-}CDT}$ 值。蒸发岩结晶析盐实验显示，不同析盐阶段石膏的 $\delta^{34}S_{V\text{-}CDT}$ 值最多相差约 2‰，且析出之后若无硫酸盐还原反应的发生将不再发生同位素的分馏。所以岩盐中硬石膏较低的 $\delta^{34}S_{V\text{-}CDT}$ 值应是原始物源特征的反映。它说明有陆源淡水携带同位素较轻的硫酸盐汇入盐盆地，或火山活动产生同位素较轻的硫元素（或硫化物）汇入沉积盆地后再氧化成硫酸盐，进而影响了硬石膏的 $\delta^{34}S_{V\text{-}CDT}$ 特征。

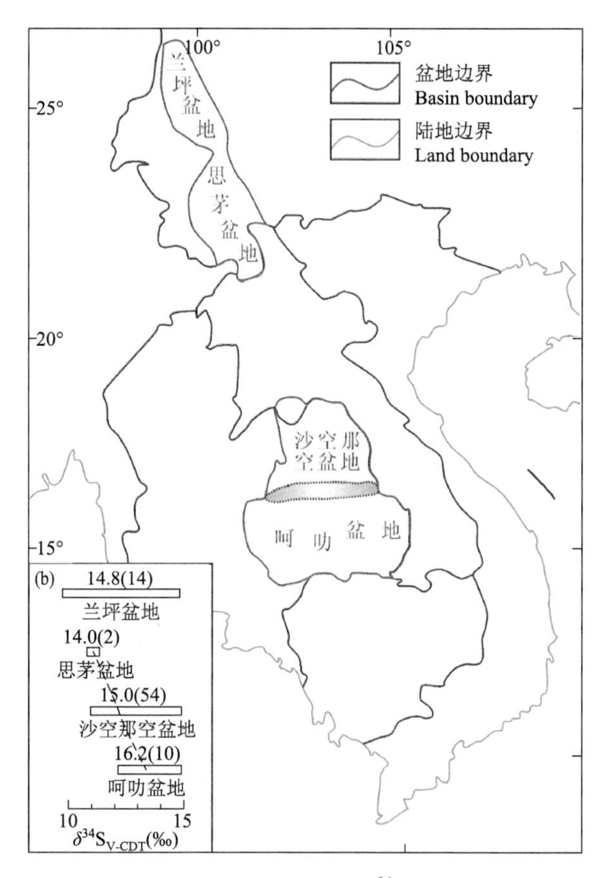

图11　思茅盆地MZK-3井蒸发岩硬石膏$\delta^{34}S_{V\text{-}CDT}$值区域上的变化特征

Fig. 11　Characteristics of the $\delta^{34}S_{V\text{-}CDT}$values of the anhydrite of evaporate from well MZK-3in the Simao basin.

呵叻盆地数据源于Li et al.，2018；沙空那空盆地数据源于Zhang Hua et al.，2014；兰坪盆地数据源于Wang Licheng et al.，2014

$\delta^{34}S_{V\text{-}CDT}$ values of the K horat basin after Li et al.，2018；$\delta^{34}S_{V\text{-}CDT}$ values of the Sakon Nakon basin after Zhang Hua et al.，2014；$\delta^{34}S_{V\text{-}CDT}$ values of the Lanping basin after Wang Licheng et al.，2014

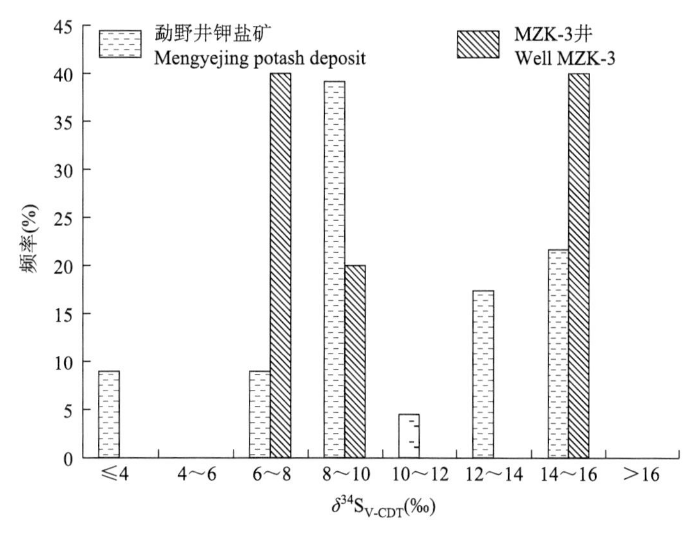

图12　思茅盆地MZK-3井岩盐中硬石膏硫同位素值的分布特征

Fig. 12　Distribution characteristics of the $\delta^{34}S_{V\text{-}CDT}$ values of the anhydrite in the halite from well MZK-3 in the Simao basin.

勐野井钾盐矿硬石膏硫同位素数据来自Xu Jianxin，2008

$\delta^{34}S_{V\text{-}CDT}$ values of the Mengyejing potash deposit after Xu Jianxin，2008

MZK-3井蒸发岩段硬石膏$\delta^{34}S_{V\text{-}CDT}$值由深向浅逐渐增大（图9），反映了外源硫的影响逐渐减小。卤水蒸发浓缩至析盐（NaCl）阶段，相应的古气候条件一定是异常干旱，从而导致经水体携带进入蒸发盆地的外源硫的输入量较低。但是，由于这一阶段卤水中$CaSO_4$含量也较低，所以很少量的外源硫输入就可以改变原始硫酸根的相对含量，进而对$\delta^{34}S_{V\text{-}CDT}$值产生较大影响——表现为有外源输入时，$\delta^{34}S_{V\text{-}CDT}$值较低；

无外源输入时，$\delta^{34}S_{V\text{-}CDT}$值保持原始物源特征。卤水蒸发浓缩至析膏盐阶段，尽管可能相对于析盐阶段的古气候条件略微湿润，有相对较多的外源硫补给蒸发盆地，但是相对于盆地内原始硫酸盐，这些补给的含量较低，所以对$\delta^{34}S_{V\text{-}CDT}$值的影响有限。这可能就是我们看到的硬石膏层$\delta^{34}S_{V\text{-}CDT}$值相对较高且稳定、石盐中硬石膏$\delta^{34}S_{V\text{-}CDT}$值出现双峰的原因。

4.3 硫同位素在钾盐成矿研究中的意义

思茅、沙空那空、呵叻盆地岩心样品硬石膏硫同位素特征可对比，说明蒸发岩沉积期，它们是相关联的蒸发岩盆地系，而且硫同位素如此大范围的可比性反映了一种海相的沉积环境。

硫同位素平面上的变化特征反映了陆源淡水或火山活动对蒸发岩盆地影响的相对强弱。同位素较轻者受陆源淡水或火山活动的影响相对较强，反之则弱。这可能也间接反映了海水补给蒸发岩盆地的方向——沿着海水补给方向，陆源淡水或火山活动对蒸发岩盆地的影响逐渐增强，硬石膏硫同位素逐渐变轻。呵叻→沙空那空→思茅盆地硬石膏硫同位素逐渐变轻，所以海侵的方向可能是由呵叻经沙空那空向思茅盆地。这还可能间接反映了蒸发盆地系中卤水的迁移方向——由预备盆地向终极析钾盆地方向外源影响逐渐减弱，硫同位素值逐渐升高，即浓缩卤水沿思茅→沙空那空→呵叻盆地方向迁移。

对蒸发岩埋藏史的研究揭示了地层普遍遭受抬升剥蚀的客观事实。这一事实符合研究区现今地层逆冲推覆、飞来峰和断层发育等构造特征，也符合印度板块与欧亚大陆碰撞、青藏高原隆升、印支地块走滑拉分这一区域地质特征。地层的抬升，一方面使先期沉积埋深的钾盐矿床抬升接近地表，利于现今开采利用，另一方面未被抬升的深部依然是钾盐勘查的有利目标。我们在平面选点勘查钾盐的同时，也不应轻易放弃对深部资源的勘查。

5 结论

（1）MZK-3 井蒸发岩地质历史中的埋藏深度至少为 1050～1450m，约从始新世开始受印度板块与欧亚大陆碰撞的影响开始抬升，至今抬升了 871.5～1271.5m。

（2）硬石膏的硫同位素值可在区域上对比，反映了一种海相的沉积环境。岩盐中硬石膏的硫同位素值具有“双峰”特征，分别为 14‰～16‰和 6‰～8‰或 8‰～10‰。这种同位素的差异超出了蒸发岩结晶析出产生同位素分馏的限度，必然是外源输入影响的结果。

（3）自南向北硬石膏的硫同位素值逐渐降低，反映了陆源淡水或火山活动对其影响逐渐增强，即海水的影响逐渐减弱，进而说明海侵的方向可能与之相同。当然，若蒸发岩是由浓缩卤水侧向迁移蒸发析出，则硫同位素值由小向大的变化可能反映了卤水的迁移方向。

References

Chavula G, Brezonik P, Thenkabail P, Johnson T, Bauer M. 2009. Estimating the surface temperature of Lake Malawi using AVHRR and MODIS satellite imagery. Physics and Chemistry of the Earth, 34: 749-754.

Cheng Huaide, Ma Haizhou, Tan Hongbing, Xu Jianxin, Zhang Xiying. 2008. Geochemical characteristics of bromide in potassium deposits: Review and research perspectives. Bulletin of Mineralogy, Petrology and Geochemistry, 27 (4): 399-408 (in Chinese with English abstract).

Cortecci G, Reyes E, Berti G, Casati P. 1981. Sulfur and oxygen isotopes in Italian marine sulfates of Permian and Triassic ages. Chemical Geology, 34: 65-79.

Crosman E T, Horel J D. 2009. MODIS-derived surface temperature of the Great Salt Lake. Remote Sensing of Environment, 113: 73-81.

Fan Qishun, Ma Haizhou, Tan Hongbing, Li Tingwei. 2009. Geochemistry characteristics of sulfur isotope in oilfield brine of the western Qaidam basin. Bulletin of Mineralogy, Petrology and Geochemistry, 28 (2): 137-142 (in Chinese with English abstract).

Gao Xiang, Cai Keqin, Li Dairong, Peng Qiang, Fang Qinfang, Qin Hong. 2012. Mineralogical and geochemical characteristics and genesis of the potassiummagnesium salt deposit in Khammouan Province, Laos. Acta Petrologica et Mineralogica, 31 (4): 578-588 (in Chinese with English abstract).

Holser W T, Kaplan I R. 1966. Isotope geochemistry of sedimentary sulfates. Chemical Geology, 1: 93-135.

Hu Guyue, Li Yanhe, Zeng Pusheng. 2013. The role of halosalt in mineralization of the Jinding Pb-Zn deposit: Evidence from sulfur and strontium isotopic compositions. Acta Geologica Sinica, 87 (11): 1694-1702 (in Chinese with English abstract).

Hu Wenxuan, Hu Shouxi, Zhao Yuchen. 1991. Sedimentary genesis of anhydrite deposits in the bolcanicseries and their relation to the pyrite depositsin Xiangshan district Anhui Province. Geoscience, 5 (2): 164-173 (in Chinese with English abstract).

Huang Jianguo, Liu Shiwan. 1989. Sulfur isotope distribution of Triassic evaporite and its geological significance in Sichuan Basin. Acta Sedimentologica Sinica, 7 (2): 105-110 (in Chinese with English abstract).

Jiang Jigang, Sheng Guoying, Fu Jiamo. 1988. Discovery of immature high sulfur crude oil and its significance in gypsum-salt bearing sedimentary basin. Petroleum Geology &Experiment, 10 (4): 337-343 (in Chinese with English abstract).

Li M H, Yan M D, Fang X M, Zhang Z J, Wang Z R, Sun S R, Li J, Liu X M. 2018. Origins of the Mid-Cretaceousevaporite deposits of the Sakhon Nakhon basin in Laos: Evidence from the stable isotopes of halite. Journal of Geochemical Exploration, 184: 209-222.

Li Renwei, Xin Maoan. 1989. Origin of evaporites of Dongpu basin. Acta Sedimentologica Sinica, 7 (4): 141-148 (in Chinese with English abstract).

Li Yanhe, Xie Guiqing, Duan Chao, Han Dan, Wang Chengyu. 2013. Effect of sulfate evaporate salt layer over the formation of skarn-type iron ore. Acta Geologica Sinica, 87 (9): 1324-1334 (in Chinese with English abstract).

Lin Y J, Zheng M P, Ye C Y. 2017. Hydromagnesite precipitation in the Alkaline Lake Dujiali, central Qinghai-Tibetan plateau: Constraints on hydromagnesite precipitation from hydrochemistry and stable isotopes. Applied Geochemistry, 78: 139-148.

Liu Hanbin, Jin Guishan, Li Junjie, Han Juan, Zhang Jianfeng, Zhang Jia, Zhong Fangwen, Guo Dongqiao. 2013. Determination of stable isotope composition in uranium geological samples. World Nuclear Geoscience, 30 (3): 174-179 (in Chinese with English abstract).

Miao Z Y. 2018. New Sr isotope evidence to support the material source of the Mengyejing potash deposit in the Simao basin from ancient marine halite or residual sea. Acta Geologica Sinica (English Edition), 92 (2): 866-867.

Miao Zhongying, Zhang Zhen, Zheng Mianping, Niu Xinsheng, Zhang Xuefei. 2017. Tectonic evolution of eastern Tethys and formation of evaporite in Lanping-Simao basin, southwest China. Acta Geoscientia Sinica, 38 (6): 883-896 (in Chinese with English abstract).

Mou Chuanlong, Tan Qinyin, Yu Qian. 2005. The permian sequence stratigraphic framework and source-reservoir caprock associations in the Simao basin, Yunnan. Sedimentary Geology and Tethyan Geology, 25 (3): 33-37 (in Chinese with English abstract).

Nie Fei, Dong Guochen, Mo Xuanxue, Zhao Zhidan, Wang Peng, Cui Ziliang, Fan Wenyu, Liu Shusheng. 2015. The characteristics of sulfur and lead isotopic compositions of the Xiyi Pb-Zn deposit in Baoshan block, western Yunnan. Acta Petrologica Sinica, 31 (5): 1327-1334 (in Chinese with English abstract).

Ossorio M, Van Driessche A E S, Pérez P, García-Ruiz J M, Du Yuansheng. 2014. The gypsum-anhydrite paradox revisited. Chemical Geology, 386: 16-21.

Ostroff A G. 1964. Conversion of gypsum to anhydrite in aqueous salt solutions. Geochimica et Cosmochimica Acta, 28: 1363-1372.

Peng Licai, Yang Ping, Pu Renlong. 1999. The sulfur isotope composition of sulfate rock deposited in continental brackish lakes and its geological significance. Bulletin of Mineralogy, Petrology and Geochemistry, 18 (2): 31-34 (in Chinese with English abstract).

Qu Yihua, Yuan Pinquan，Shuai Kaiye, et al. 1988. Metallogenic regularity and prediction of potash in Lanping-Simao Basin. Beijing: Geology Press.

Reinart A, Reinhold M. 2008. Mapping surface temperature in large lakes with MODIS data. Remote Sensing of Environment, 112: 603-611.

Ren Shunli, Li Yanhe, Zeng Pusheng, Qiu Wenlong, Fan Changfu, Hu Guyue. 2018. Effect of sulfate evaporate salt layer in mineralization of the Huize and Maoping lead-zinc deposits in Yunnan: evidence from sulfur isotope. Acta Geologica Sinica, 92 (5): 1041-1055 (in Chinese with English abstract).

Seal R R. 2006. Sulfur isotope geochemistry of sulfide minerals. Reviews in Mineralogy &Geochemistry, 61 (1): 633-677.

Sima S, Ahmadalipour A, Tajrishy M, Du Yuansheng. 2013. Mapping surface temperature in a hyper-saline lake and investigating the effect of temperature distribution on the lake evaporation. Remote Sensing of Environment, 136: 374-385.

Song Jinming. 1990. Species of sulfur in marine environments. Environmental Chemistry, 9 (6): 59-64 (in Chinese with English abstract).

Srinivasan J. 1990. Performance of a small solar pond in the tropics. Solar Energy, 45: 221-230.

Tabakh M E, Schreiber B C, Utha-Aroon C, Coshell L, Warren J K. 1998. Diagenetic origin of Basal anhydrite in the Cretaceous Maha Sarakham salt: Khorat plateau, NE Thailand. Sedimentology, 45: 579-594.

Tan Fuwen, Xu Xiaosong, Yin Fuguang, Li Xingzhen. 1999. Upper carboniferous sediments in the Simao region, Yunnan and their tectonic settings. Sedimentary Facies and Palaeogeography, 19 (4): 26-34 (in Chinese with English abstract).

Wang Chunlian, Liu Chenglin, Xu Haiming, Wang Licheng, Shen Lijian. 2013. Sulfur isotopic composition of sulfate and its geological significance of Member 4 of Palaeocene Shashi Formation in Jiangling depression of Hubei Province. Journal of Jilin University (Earth Science Edition), 43 (3): 691-703 (in Chinese with English abstract).

Wang Guoli, Cai Liguo, Wang Jiyang, Shi Xiaobin. 2005. Paleo-geothermal field and tectonic-thermal evolution in the Chuxiong basin of China. Petroleum Geology &Experiment, 27 (1): 28-31, 38 (in Chinese with English abstract).

Wang Jiyang, Huang Shaopeng. 1990. Compilation of heat flow data in the China continental area (2rd edition). Seismology and Geology, 12 (4): 351-366 (in Chinese with English abstract).

Wang Licheng, Liu Chenglin, Fei Mingming, Shen Lijian, Zhang Hua. 2014. Sulfur isotopic composition of sulfate and its geological significance of the Yunlong Formation in the Lanping basin, Yunnan Province. China Mining Magazine, 23 (12): 57-65 (in Chinese with English abstract).

Wang Licheng, Liu Chenglin, Shen Lijian, Bo Ying. 2018. Research advances in potash forming of the Simao basin, eastern Tethyan Realm. Acta Geologica Sinica, 92 (8): 1707-1723 (in Chinese with English abstract).

Xu Jianxin. 2008. Geochemistry and genesis of Mengyejing potash deposits, Yunnan. Qinghai Institute of Salt Lakes, Chinese Academy of Sciences, Xining (in Chinese with English abstract).

Yan Yuehua, Wu Yi. 2002. Archean evaporites of the North China Craton: Evidence on major and trace elements and S-isotope. Acta Petrologica Sinica, 18 (4): 531-538 (in Chinese with English abstract).

Yang Jianxu, Yin Hongwei, Zhang Zhen, Zheng Mianping. 2013. Geologic settings of the potassium formations in the Lanping-Simao basin, Yunnan Province. Geotectonica et Metallogenia, 37 (4): 633-640 (in Chinese with English abstract).

Yin Fuguang, Pan Guitang, Wan Fang, Li Xingzhen, Wang Fangguo. 2006. Tectonic facies along the Nujiang-Lancangjiang-Jinshajiang orogenic belt in southwestern China. Sedimentary Geology and Tethyan Geology, 26 (4): 33-39 (in Chinese with English abstract).

Zhang Congwei, Gao Donglin, Zhang Xiying, Tang Qiliang, Shi Lin. 2011. Comparasion of geochemistry characteristics in palaeocene salt-bearing strata of Lanping-Simao basin and Chuxiong basin. Journal of Salt Lake Research, 19 (3): 8-14 (in Chinese with English abstract).

Zhang Hua, LiuChenglin, Wang Licheng, Fang Xiaomin. 2014. Characteristics of evaporites sulfur isotope from potash deposit in Thakhek basin, Laos, and its implication for potash formation. Geological Review, 60 (4): 851-857 (in Chinese with English abstract).

Zhang Tonggang, Chu Xuelei, Feng Lianjun, Zhang Qirui, Guo Jianping. 2003. The effects of the Neoproterozoic Snowball Earth on carbon and sulfur isotopic compositions in seawater. Acta Geoscientia Sinica, 24 (6): 487-493 (in Chinese with English abstract).

Zhao Haitong, Zhang Yongsheng, Xing Enyuan, Wang Linlin, Yu Dongdong, Shang Wenjun, Gui Baoling, Li Kai. 2018. Sulfur isotopic characteristics of evaporite in the Middle Ordovician Mawu Member in the salt basin of northern Shaanxi and its paleoenvironment significance. Acta Geologica Sinica, 92 (8): 1680-1692 (in Chinese with English abstract).

Zheng Zhijie, Yin Hongwei, Zhang Zhen, Zheng Mianping, Yang Jianxu. 2012. Strontium isotope characteristics and the origin of salt deposits in Mengyejing, Yunnan Province, SW China. Journal of Nanjing University: Nat Sci Ed, 48 (6): 719-727 (in Chinese with English abstract).

Zhu Guangyou, Zhang Shuichang, Liang Yingbo, Dai Jinxing. 2006. Stable sulfur isotopic composition of hydrogen sulfide and its genesis in Sichuan basin. Geochimica, 35 (4): 432-442 (in Chinese with English abstract).

参考文献

程怀德, 马海州, 谭红兵, 许建新, 张西营. 2008. 钾盐矿床中 Br 的地球化学特征及研究进展. 矿物岩石地球化学通报, 27 (4): 399-408.

樊启顺, 马海州, 谭红兵, 李廷伟. 2009. 柴达木盆地西部油田卤水的硫同位素地球化学特征. 矿物岩石地球化学通报, 28 (2): 137-142.

高翔, 蔡克勤, 李代荣, 彭强, 方勤方, 秦红. 2012. 老挝甘蒙省钾镁盐矿床含矿段的矿物学和地球化学特征及成因. 岩石矿物学杂志, 31 (4): 578-588.

胡古月, 李延河, 曾普胜. 2013. 膏盐在金顶铅锌矿成矿中的作用: 硫和锶同位素证据. 地质学报, 87 (11): 1694-1702.

胡文瑄, 胡受奚, 赵玉琛. 1991. 安徽向山地区火山岩层中硬石膏的沉积成因特征及其与硫铁矿的关系. 现代地质, 5 (2): 164-173.

黄建国, 刘世万. 1989. 四川盆地三叠纪蒸发岩地层硫同位素的分布. 沉积学报, 7 (2): 105-110.

江继纲, 盛国英, 傅家谟. 1988. 膏盐沉积盆地中未成熟高硫原油的发现及意义. 石油实验地质, 10 (4): 337-343.

李任伟, 辛茂安. 1989. 东濮盆地蒸发岩的成因. 沉积学报, 7 (4): 141-148.

李延河, 谢桂青, 段超, 韩丹, 王成玉. 2013. 膏盐层在矽卡岩型铁矿成矿中的作用. 地质学报, 87 (9): 1324-1334.

刘汉彬, 金贵善, 李军杰, 韩娟, 张建锋, 张佳, 钟芳文, 郭东侨. 2013. 铀矿地质样品的稳定同位素组成测试方法. 世界核地质科学, 30 (3): 174-179.

苗忠英, 张震, 郑绵平, 牛新生, 张雪飞. 2017. 东特提斯构造演化与兰坪-思茅盆地蒸发岩的形成. 地球学报, 38 (6): 883-896.

牟传龙, 谭钦银, 余谦. 2005. 云南思茅盆地二叠纪层序格架与生储盖研究. 沉积与特提斯地质, 25 (3): 33-37.

聂飞, 董国臣, 莫宣学, 赵志丹, 王鹏, 崔子良, 范文玉, 刘书生. 2015. 云南保山西邑铅锌矿床硫铅同位素地球化学特征研究. 岩石学报, 31 (5): 1327-1334.

彭立才, 杨平, 濮人龙. 1999. 陆相咸化湖泊沉积硫酸盐岩硫同位素组成及其地质意义. 矿物岩石地球化学通报, 18 (2): 31-34.

曲一华, 袁品泉, 帅开业, 张瑛, 蔡克勤, 贾疏源, 陈朝德. 1998. 兰坪-思茅盆地钾盐成矿规律及预测. 北京: 地质出版社.

任顺利, 李延河, 曾普胜, 邱文龙, 范昌福, 胡古月. 2018. 膏盐层在云南会泽和毛坪铅锌矿成矿中的作用: 硫同位素证据. 地质学报, 92 (5): 157-171.

宋金明. 1990. 海洋环境中硫的存在形式. 环境化学, 9 (6): 59-64.

谭富文, 许效松, 尹福光, 李兴振. 1999. 云南思茅地区上石炭统沉积特征及其构造背景. 岩相古地理, 19 (4): 26-34.

汪集旸, 黄少鹏. 1990. 中国大陆地区大地热流数据汇编 (第二版) . 地震地质, 12 (4): 351-366.

王春连, 刘成林, 徐海明, 王立成, 沈立建. 2013. 湖北江陵凹陷古新统沙市组四段硫酸盐硫同位素组成及其地质意义. 吉林大学学报 (地球科学版), 43 (3): 691-703.

王国力, 蔡立国, 汪集旸, 施小斌. 2005. 楚雄盆地构造-热演化与古地温场研究. 石油实验地质, 27 (1): 28-31, 38.

王立成, 刘成林, 费明明, 沈立建, 张华. 2014. 云南兰坪盆地云龙组硫酸盐硫同位素特征及其地质意义. 中国矿业, 23 (12): 57-65.

王立成, 刘成林, 沈立建, 伯英. 2018. 东特提斯域思茅盆地钾盐成矿研究进展. 地质学报, 92 (8): 161-177.

许建新. 2008. 云南勐野井钾盐矿床地球化学与成因研究. 西宁: 中国科学院盐湖研究所.

阎月华, 吴毅. 2002. 内蒙千里山太古代蒸发岩: 常量元素、微量元素及硫同位素证据. 岩石学报, 18 (4): 531-538.
杨尖絮, 尹宏伟, 张震, 郑绵平. 2013. 滇西兰坪-思茅盆地成钾地质条件分析. 大地构造与成矿学, 37 (4): 633-640.
尹福光, 潘桂棠, 万方, 李兴振, 王方国. 2006. 西南 "三江" 造山带大地构造相. 沉积与特提斯地质, 26 (4): 33-39.
张从伟, 高东林, 张西营, 唐启亮, 时林. 2011. 兰坪-思茅盆地与楚雄盆地古新统含盐系地球化学特征对比. 盐湖研究, 19 (3): 8-14.
张华, 刘成林, 王立成, 方小敏. 2014. 老挝他曲盆地钾盐矿床蒸发岩硫同位素特征及成钾指示意义. 地质论评, 60 (4): 851-857.
赵海彤, 张永生, 邢恩袁, 王琳霖, 于冬冬, 商雯君, 桂宝玲, 李凯. 2018. 陕北盐盆中奥陶统马五段蒸发岩硫同位素特征及其古环境意义. 地质学报, 92 (8): 134-146.
郑智杰, 尹宏伟, 张震, 郑绵平, 杨尖絮. 2012. 云南江城勐野井盐类矿床 Sr 同位素特征及成盐物质来源分析. 南京大学学报 (自然科学版), 48 (6): 719-727.
张同钢, 储雪蕾, 冯连君, 张启锐, 郭建平. 2003. 新元古代 "雪球" 事件对海水碳、硫同位素组成的影响. 地球学报, 24 (6): 487-493.
朱光有, 张水昌, 梁英波, 戴金星. 2006. 四川盆地 H2S 的硫同位素组成及其成因探讨. 地球化学, 35 (4): 432-442.

Geochemistry of sulfur isotope in evaporite and its sedimentology significance: a case study from the well MZK-3in the Simao basin, southwestern China

Zhongying Miao[1], Mianping Zheng[1], Xuefei Zhang[1], Zhen Zhang[1] and Yunzhi Gao[2]

1. M LR Key Laboratory of Saline Lake Resources and Environments, Instituteof Mineral Resources, CAGS, Beijing 100037
2. China University of Geosciences, Beijing 100083

Abstract The Simao basin is currently the sole sedimentary basin which hosts ancient solid potash deposit in China. But it forming time, provenance features and transgressive direction have long been controversial. Based on formation conditions, occurrence forms and isotope fractionation mechanism of marine anhydrite, this paper analyzed the sulfur isotope geochemical characteristics of evaporite in the well MZK-3 in detail. The results are as follows: (1) anhydrite in halite can be formed from the precipitation of salt in evaporite basins. Independent stratiform anhydrite can be formed from the dehydration of original deposited gypsum due to high temperature from deep burial. (2) The sulfur isotope of anhydrite in halite is characterized by "double peaks", with $\delta^{34}S$ values of 14‰～16‰ and 8‰～10‰ or 6‰～8‰, respectively. This may reflect the dual sources of sulphate: one from seawater and the other from fresh water or volcanic hydrothermal fluid. (3) $\delta^{34}S$ values of the stratiform anhydrite can be contrasted regionally. $^{87}Sr/^{86}Sr$ values of the samples may represent a marine environment. (4) $\delta^{34}S$ values of the stratiform anhydrite decease from south to north, reflecting that relative influence of fresh water or hydrothermal fluid on evaporite basin correspondingly increases from south to north, and thus imply the transgression direction. Therefore, the study of anhydrite not only reveals source, sedimentary environment and trangressive direction, but also provides a useful reference for investigation of potash deposits in the study area.

Keywords Sulfur isotope; Evaporite; Geochemistry; Potash deposit; Simao basin

罗布泊富钾卤水矿床地球化学空间分布特征*

王凯[1,2]，孙明光[1]，马黎春[1]，汤庆峰[3]，颜辉[4]，张瑜[1]

1. 中国地质科学院矿产资源研究所，自然资源部成矿作用与资源评价重点实验室，北京 100037
2. 中国地质大学（北京），北京 100083
3. 北京市理化分析测试中心，北京 100089
4. 国投新疆罗布泊钾盐有限责任公司，新疆哈密 839000

内容提要 罗布泊是我国最重要的成盐成钾盆地之一，富钾卤水资源丰富。在 14 年的工业开采过程中，不同矿层卤水的地球化学特征持续处于变化中。为了加强对不同矿区富钾卤水的地球化学空间分布特征的认识，合理开发利用卤水资源，本文应用克里金插值法，对 2017 年采集的 89 个常观孔卤水样品进行了空间分析，绘制了 K^+、Cl^-、Na^+、Mg^{2+}、Li^+、B^{3+}、SO_4^{2-} 等值线图，并对其空间分布特征进行了综合分析，结果显示：K^+、Cl^-、Mg^{2+}、Li^+、B^{3+}等离子具有较为一致的分布趋势，其高值区主要分布在罗北凹地及腾龙矿区北部，低值区主要分布于罗北凹地北部及矿区西南部，而 SO_4^{2-} 分布趋势则相反；由于受周缘地区淡水补给的影响，矿区西南部、北部及东南部出现淡化现象，但是淡水只影响到 W_1、W_2 层，W_3 层未见明显淡化现象；2017 年罗布泊矿区 K^+平均含量为 8.83g/L，除受淡水影响区域外，矿区内大部分地区样品 KCl 品位均高于 1%的工业单独开采品位，与开采前 KCl 品位相比，目前罗北凹地 W_1 储层 KCl 平均品位约下降了 9.7%。

关键词 罗布泊；富钾卤水；空间分布；地球化学；卤水矿床

罗布泊位于新疆塔里木盆地最东端，是我国除柴达木盆地以外最重要的成盐成钾盆地，已探明的钾盐资源量（折合成氯化钾储量）达 1.45 亿吨（Wang Mili，2001），是我国最大的陆相卤水钾盐矿床之一。罗布泊富钾卤水主要为硫酸镁亚型卤水，卤水矿化度范围为 247～385g/L，其中罗北凹地矿化度最高，平均为 367g/L，卤水中钾离子（折合成氯化钾）的品位变化于 1.0%～1.45%（Sun Mingguang et al.，2018），高于 1%的工业品位，并且伴生石盐、钙芒硝等固体矿床。从 21 世纪初，国投罗布泊钾盐公司在罗布泊地区设厂开采富钾卤水至今，已达到年产能 130 万吨，是世界上最大的硫酸钾生产基地。目前我国在罗布泊地区有规模的开采富钾卤水已有 14 年，在富钾卤水的成矿条件、储层特征、地球化学特征及开采工艺等方面取得了一定的进展（Wang Mili.，2001，2006；Chen Yongzhi et al.，2001；Liu Chenglin et al.，2002；Gu Xinlu et al.，2003，2009；Lin Jingxing et al.，2005；Ma Lichun et al.，2010a，2010b），但在宏观上对于不同储层卤水的地球化学空间分布研究较少（Sun Mingguang et al.，2018）。本文将根据 2017 年采集的矿区 89 个常观孔卤水地球化学数据，对不同矿层的地球化学空间分布特征进行系统研究，揭示罗布泊钾盐矿床连续开采 14 年后富钾卤水地球化学的空间分布特征，为后期合理开发罗布泊卤水资源及可持续利用提供理论基础。

1 研究区背景

由于受构造运动影响，罗布泊盐湖成为塔里木盆地的汇水中心（Lin Changsong et al.，2011），目前已无地表水系补给，主要接受来自库鲁克塔格山、北山的基岩裂隙水及阿尔金山等的地下水补给，同时接受阿奇克谷地、孔雀河三角洲和塔里木河冲积平原的侧向补给，通过流水的不断溶滤、蒸发，最终到达罗布泊地区，水体矿化度逐渐升高，区域地质简图见图 1。

* 本文发表在：地质学报，2020，94（4）：1183-1191

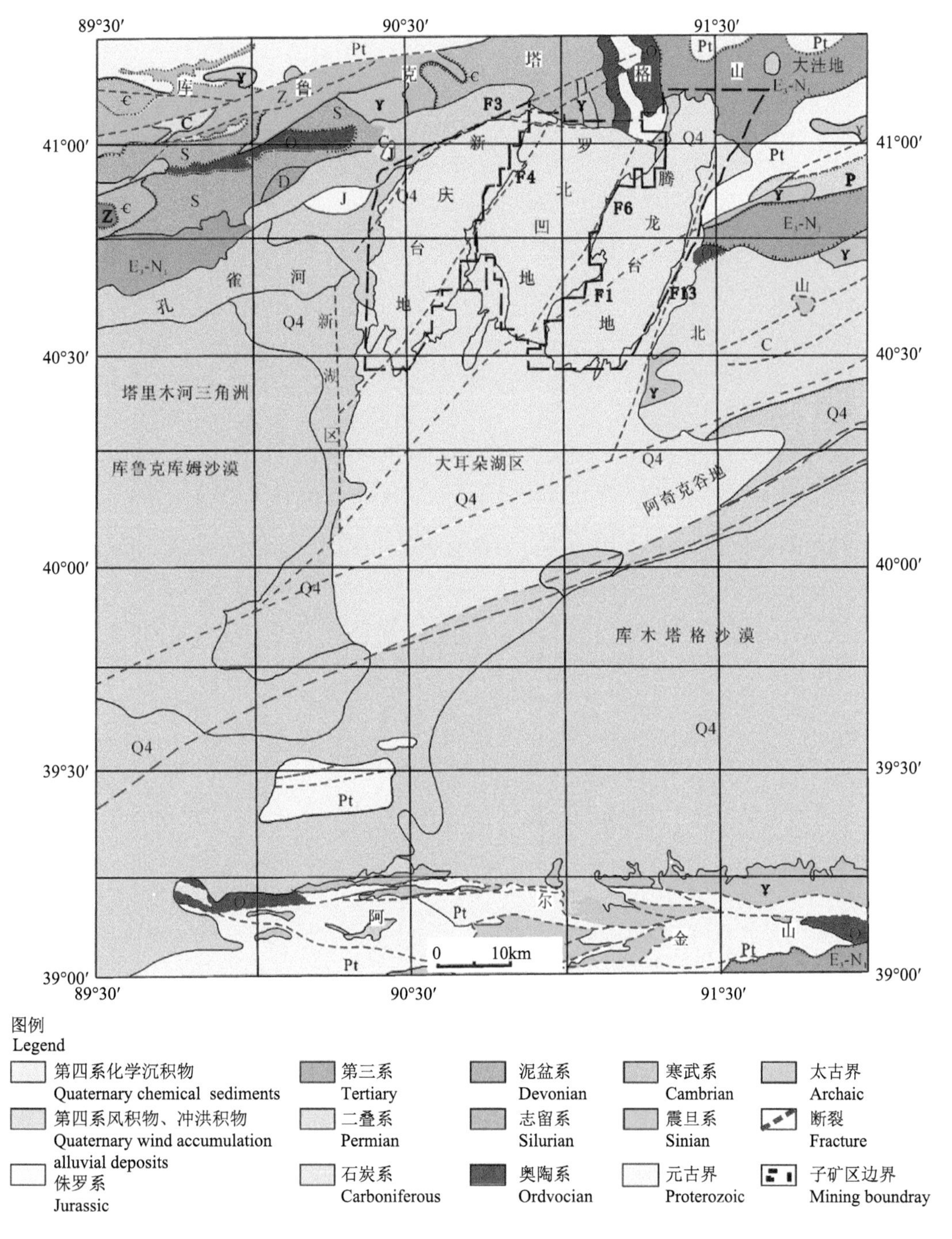

图 1 新疆罗布泊地区区域地质简图

Fig. 1 Simplified geological map in Lop Nor region, XinJiang Province.

本研究区位于罗布泊盆地北部，为罗布泊盆地的主要成盐聚钾区。研究区包含三个子矿区，分别为新庆矿区、罗北凹地矿区和腾龙矿区，见图 1。由于受新构造运动影响，罗布泊盆地内发育有多条断裂及隐伏断裂。其中新庆矿区和罗北凹地矿区被 F_4 断裂所分隔，罗北凹地矿区与腾龙矿区被 F_6 断裂所分隔。F_1 断裂为区域压扭性断裂，从腾龙矿区穿过并将腾龙矿区分割为南北两部分，见图 1。通过罗布泊地区已实施的多孔钻孔资料可知，罗布泊地区富钾卤水的储层主要为钙芒硝层，少量为石盐、石膏和砂砾岩层。由于受断裂的影响，目前所揭露的各个矿区富钾卤水层数量略有不同。罗北凹地共有七个卤水层，包括一层潜水层以及六层承压水层（Tang Pinghui et al.，2008），受技术条件及经济条件限制，目前只开采潜水层及前三层承压水层；新庆矿区揭露有两层，均为承压水层（Fan Yao，2011）；腾龙矿区揭露有三层富钾卤水层，包括一个潜水层和两个承压水层（Yang Li，2009），其中潜卤水层仅分布在 F_1 断裂北部，南部只含有承压水层。由于目前矿区布设的 89 个常观孔只用于监测主开采层，即最浅的三个卤水矿层（W_1、W_2、

W_3），因此本文将根据实际采集的卤水样品及分析数据，讨论这三个矿层的地球化学空间分布特征并进行综合分析。

2 研究方法

2.1 采样方法

2017 年 7～9 月，采集了矿区内 89 个观测孔的卤水样品，并利用测绳、密度计及 pH 计等现场测定卤水埋深、卤水密度以及 pH，同时利用 GPS 进行地理定位和高程测量。利用 500mL 取样瓶进行取样保存，每个观测孔取样两瓶，现场完成密度及 pH 测量后迅速用胶带密封，以防卤水在极端干旱气候条件下蒸发或在运输过程中泄露。

2.2 分析方法

所取卤水样品均送往中国地质科学院国家地质实验测试中心进行主量元素 Cl^-、Na^+、K^+、SO_4^{2-}、Mg^{2+}、Ca^{2+}、CO_3^{2-}、HCO_3^- 及微量元素 Li^+、B^{3+}、Br^-、I^-、Rb^+、Cs^+、Sr^{2+}的测试分析。其中 Na^+、K^+、Mg^{2+}、Ca^{2+}、Sr^{2+}、Rb^+、Li^+、Cs^+采用原子吸收分光光度法测试（误差小于 2%）；Br^-、I^-、CO_3^{2-}、HCO_3^-、Cl^-采用常规滴定法（误差小于 0.2%～0.3%）测试；SO_4^{2-}、B^{3+}采用可见光光度仪测试（误差小于 1%）。

2.3 制图方法

针对主微量元素的测试结果，利用 Surfer 软件克里金插值法分别绘制出不同矿区不同储层的卤水的地球化学空间分布图。

3 结果

根据 2017 年采集的矿区卤水样品及主微量元素化学分析结果，矿区内卤水矿化度较高，变化于 226.09～393.98g/L 之间，其中 Cl^-含量变化于 113～190g/L，平均含量为 171.15g/L；K^+含量为 4.40～10.9g/L，平均含量为 8.84g/L；Na^+含量的最大值为 119g/L，最小值为 60.7g/L；SO_4^{2-} 含量变化于 15.7～100g/L 之间，平均含量为 58.23g/L；Mg^{2+}含量的最小值为 8.68g/L，最大值为 33g/L，平均含量达到 21.83g/L。由于高浓度卤水中 Ca^{2+}含量较低，平均值仅为 0.16mg/L，HCO_3^- 平均含量仅为 110mg/L，而几乎不含 CO_3^{2-}，微量元素中 Br^-、I^-、Rb^+、Cs^+、Sr^{2+}平均含量分别为 9.94mg/L、0.19mg/L、1.39mg/L、0mg/L、3.51mg/L，均低于其工业品位，因此本文不予讨论；硼（折算成 B_2O_3）含量范围为 277.3～755.6mg/L，均大于综合利用品位 150mg/L（Sun Mingguang et al.，2018），Li^+平均含量为 18.57mg/L，虽低于 200mg/L 的工业品位，但在分离掉其他盐类后的饱和氯化镁母液卤水中，锂浓度可以富集十倍以上（Zhi Hongjun，2012），可进行综合利用，因此本文对于硼、锂的分布特征也做一定讨论分析。

4 讨论

依据瓦里亚什科水化学分类方案（1965），目前采集的三个子矿区的卤水样品，水化学类型均为硫酸镁亚型。为了更形象地描述富钾卤水矿床各离子的空间分布特征，本文使用 Surfer 软件克里金插值法对 Cl^-、Na^+、K^+、SO_4^{2-}、Mg^{2+}、Li^+、B^{3+}空间分布特征进行了分析，绘制了空间分布图。

4.1 氯离子空间分布特征

图 2 为罗布泊富钾卤水中氯离子含量分布图。W_1 储层中氯离子含量最高值为 187g/L，最低值为 113g/L，

平均含量为 172g/L。由图 2（a）可知，W_1 储层整体由东南向中部的罗北凹地呈梯度增高趋势，高值区主要位于罗北凹地矿区，并向腾龙台地北部延伸。腾龙矿区以 F_1 断裂为界，氯离子含量呈明显的北高南低态势，南部可能受北山基岩裂隙水补给影响。在罗北凹地北部有一明显低值区，可能是受库鲁克塔格山前裂隙水补给所致。W_2 储层氯离子含量最高值为 190g/L，最小值为 147g/L，平均含量为 179.5g/L，由图 2（b）可以明显的看出，W_2 储层氯离子含量从东西两侧台地向中部的罗北凹地逐渐升高，同时在腾龙矿区北部形成一高值区，与 W_1 储层类似，且高值区分布处于 TDS 340～360g/L 之间，符合卤水浓缩的一般规律。在罗北凹地北部有一片近南北向分布的低值区，显示有来自北部山区淡水补给。新庆矿区氯离子呈阶梯状自东南向西北升高，这可能与西部塔里木河冲积平原及孔雀河三角洲的侧向补给有关；W_3 储层氯离子含量最高值为 185g/L，最低值为 119g/L，平均含量为 164g/L，其分布趋势与 W_1、W_2 类似，见图 2（c），高浓度区主要分布于罗北凹地矿区及腾龙矿区北部，低值区主要分布于东西两翼南部，但由于矿层深度增加，W_3 储层几乎不受北部山前淡水补给的影响。

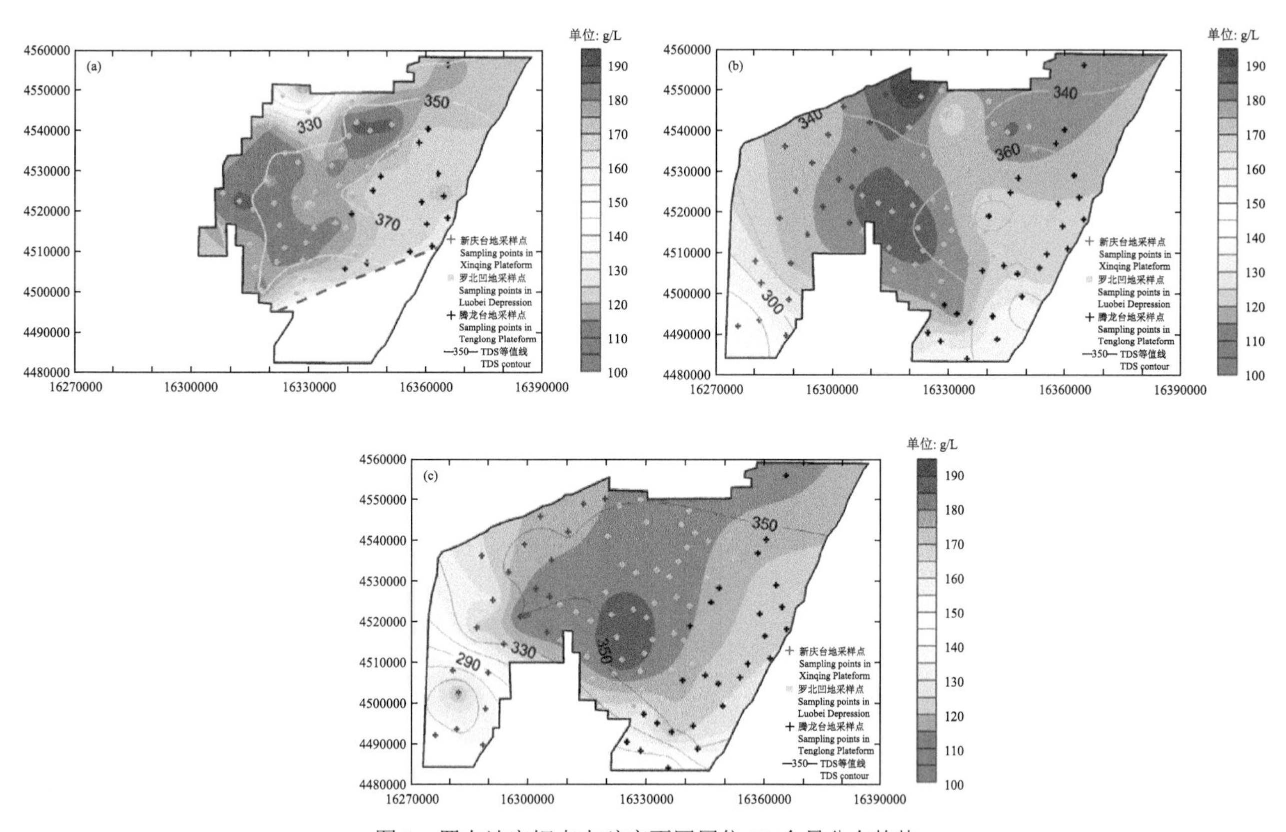

图 2　罗布泊富钾卤水矿床不同层位 Cl^- 含量分布趋势

Fig. 2　The distribution trend of Cl^- content for different layers in Lop Nor potash-rich brine deposit.

（a）W_1 储层；（b）W_2 储层；（c）W_3 储层
(a) layer W_1; (b) layer W_2; (c) layer W_3

4.2　钾离子空间分布特征

图 3 为罗布泊富钾卤水中 K^+ 分布情况。W_1 储层中钾离子最高含量达到 10.9g/L，最低值为 5.8g/L，平均含量为 8.9g/L。由图 3（a）可知，W_1 层位与 Cl^- 分布趋势十分一致，除矿区北部、西南及东南部受淡水补给影响外，整体自西南向东北呈条带状分布，最高值位于罗北凹地中部地区，与 Cl^- 高值区的中心区域吻合度很好；W_2 储层钾离子含量的最高值为 10.3g/L，最低值为 6.27g/L，平均含量为 8.88g/L。钾离子含量的高值区主要分布于罗北凹地中南部及腾龙矿区中东部，和 TDS 分布趋势较为一致，均位于矿化度的高值区（TDS＞360g/L），见图 3（b）。W_3 储层钾离子含量最高值为 10.3g/L，最低值为 4.4g/L，平均含量为 8.3g/L，其分布趋势和 W_1 类似，见图 3（c），钾离子主要富集在罗北凹地矿区，主要的低值区位于新庆矿区西南部和腾龙矿区东南部。

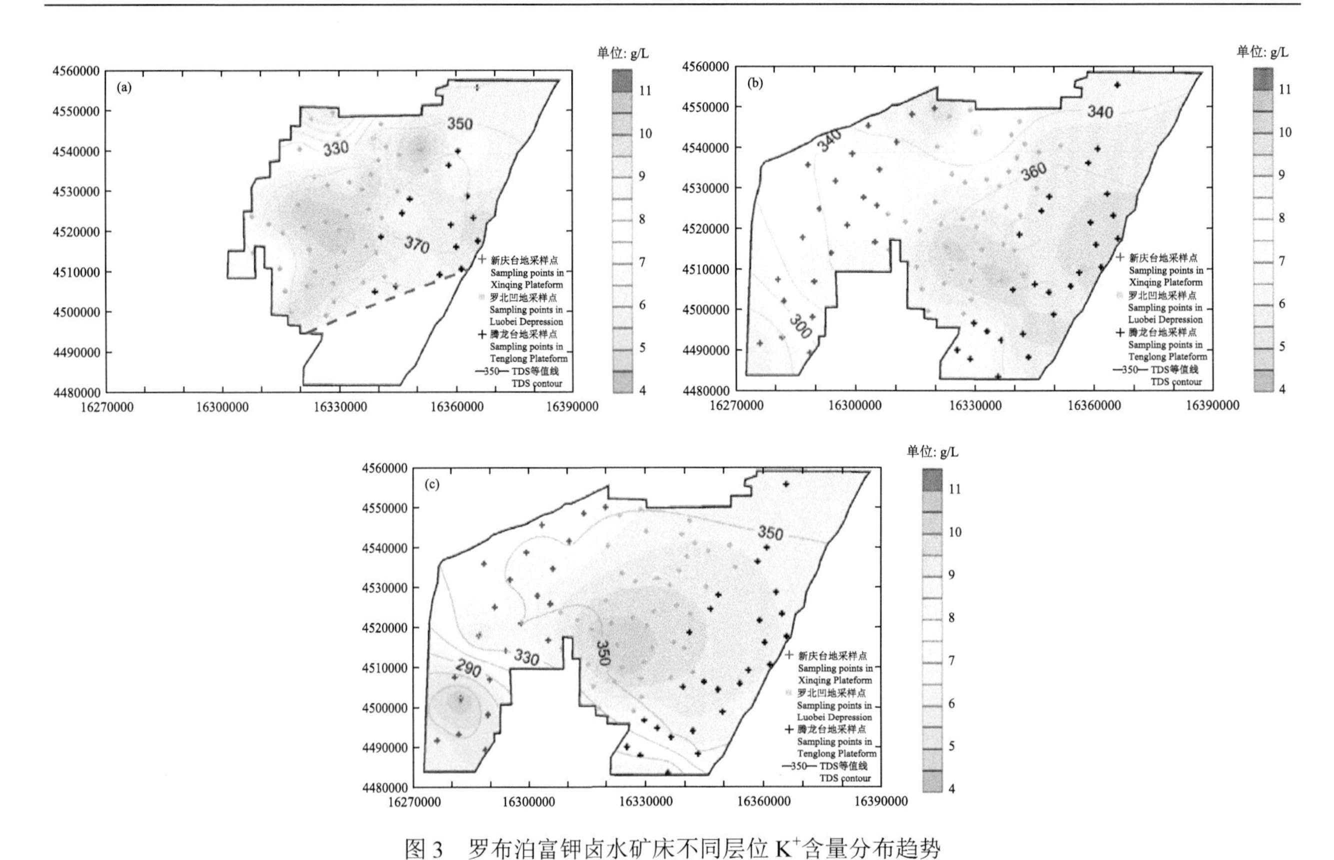

图 3　罗布泊富钾卤水矿床不同层位 K^+含量分布趋势

Fig. 3　The distribution trend of K^+ content for different layers in Lop Nor potash-rich brine deposit.

（a）W_1 储层；（b）W_2 储层；（c）W_3 储层

(a) layer W_1; (b) layer W_2; (c) layer W_3

4.3　镁离子空间分布特征

图 4 为罗布泊矿区各个层位的镁离子分布图。W_1 层位镁离子含量的最大值为 31g/L，最小值为 10g/L，平均含量为 21.4g/L。由图 4（a）可知，W_1 层位镁离子含量从盆地外围向中心呈浓缩趋势，至罗北凹地矿区中部达到最大值，和罗北凹地 K^+高值区域较为吻合。而矿区北部、西部和东南部的低值区则是由于淡水补给所致；W_2 储层中镁离子含量最大值为 27g/L，最小值为 10g/L，平均含量为 20.7g/L。镁离子含量高值区域主要位于腾龙矿区东南部，和 TDS 分布趋势较为一致，均位于高矿化度区域（TDS＞360g/L），见图 4（b）。W_3 储层中镁离子含量最高值为 28g/L，最小值为 11g/L，平均含量为 21.4g/L。与 W_2 储层分布趋势类似，见图 4（c），从西往东镁离子含量呈递增趋势，至腾龙矿区中南部达到最大值。

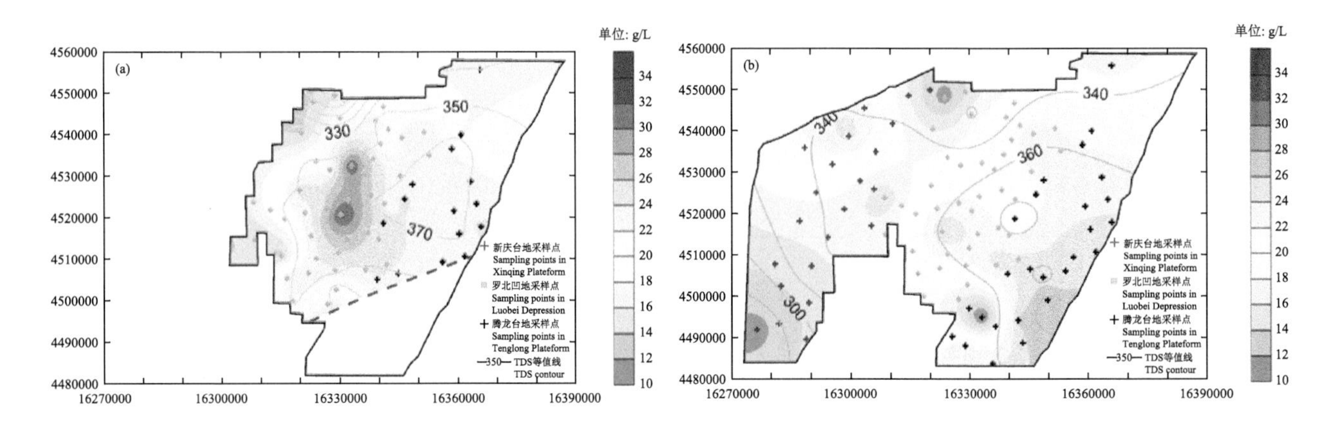

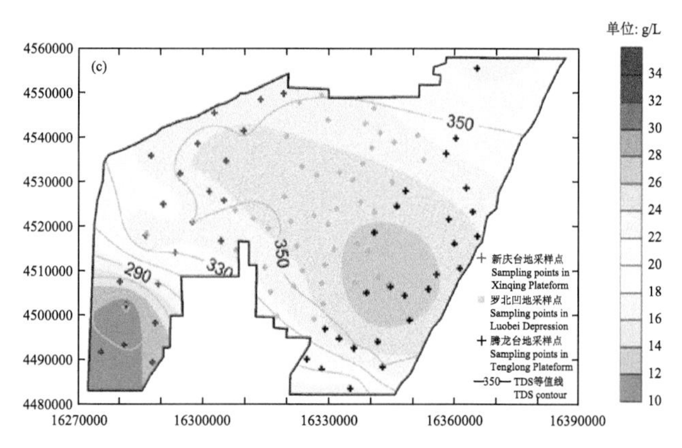

图 4 罗布泊富钾卤水矿床不同层位 Mg^{2+}含量分布趋势

Fig. 4 The distribution trend of Mg^{2+} content for different layers in Lop Nor potash-rich brine deposit.

（a）W_1 储层；（b）W_2 储层；（c）W_3 储层

(a) layer W_1; (b) layer W_2; (c) layer W_3

4.4 锂离子空间分布特征

图 5 为罗布泊矿区各个层位 Li^+分布图。W_1 层位 Li^+含量最大值为 28.4mg/L，最小值为 7.44mg/L，平均含量为 19.70mg/L。由图 5（a）可知，W_1 层位 Li^+主要富集在罗北凹地矿区中部地区，和 Mg^{2+}分布趋势较为一致，从盆地外围向中心呈浓缩趋势。W_2 层位 Li^+含量最大值为 25.3mg/L，最小值为 7.44mg/L，平均含量为 17.15mg/L。其高值区主要位于矿区的中东部地区，而西部的新庆矿区 Li^+含量普遍较低，见图 5（b）；W_3 层位 Li^+含量最大值为 25.2mg/L，最小值为 4.71mg/L，平均含量为 5.13mg/L。且 Li^+高值区主要位于罗北凹地矿区，并呈近同心圆状向外围递减，见图 5（c）。三个矿层 Li^+含量和 Mg^{2+}分布趋势较为一致，见图 4、图 5。

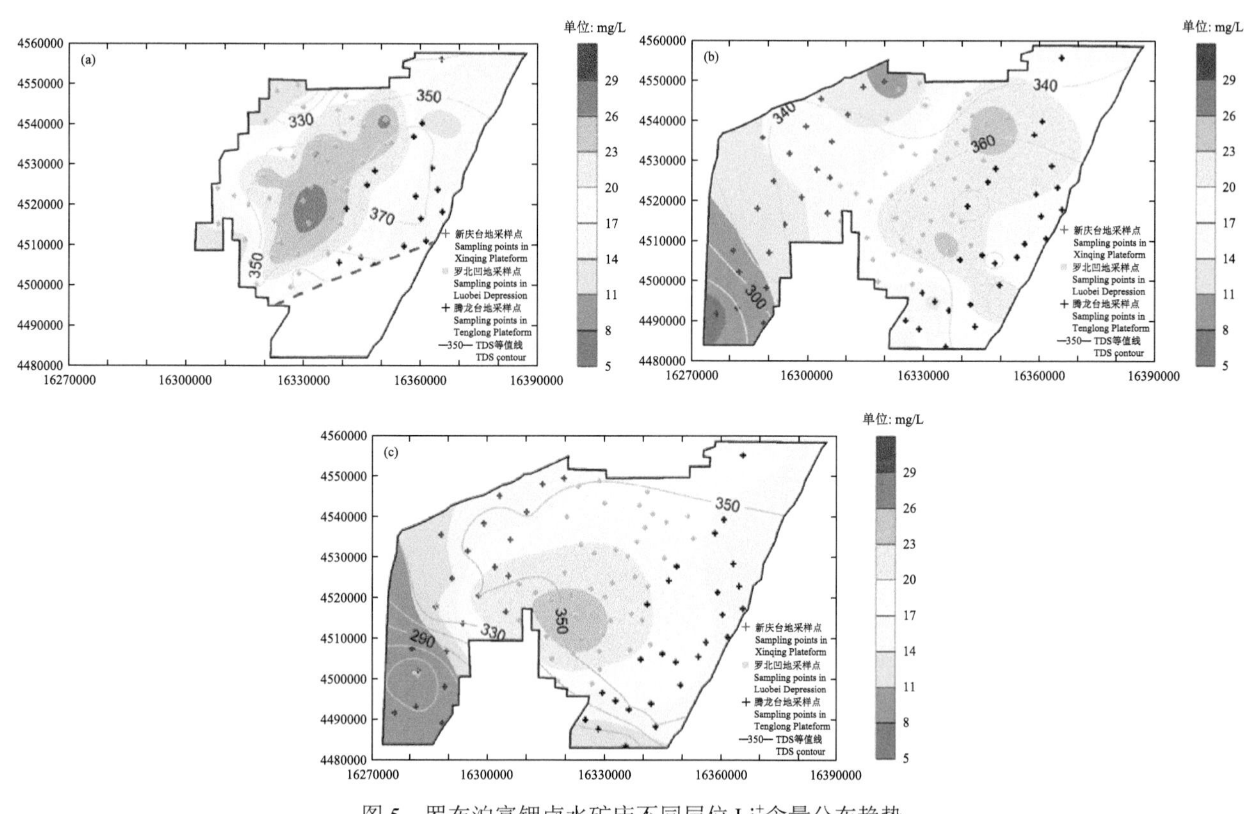

图 5 罗布泊富钾卤水矿床不同层位 Li^+含量分布趋势

Fig. 5 The distribution trend of Li^+ content for different layers in Lop Nor potash-rich brine deposit.

（a）W_1 储层；（b）W_2 储层；（c）W_3 储层

(a) layer W_1; (b) layer W_2; (c) layer W_3

4.5　硼离子空间分布特征

图 6 为罗布泊矿区各个层位 B^{3+}分布图。W_1 层位 B^{3+}含量最大值为 122mg/L，最小值为 39.8mg/L，平均含量为 78.16mg/L。由图 6（a）可知，B^{3+}主要位于罗北凹地矿区中部，并向周围呈梯度递减。W_2 层位 B^{3+}含量最大值为 100mg/L，最小值为 39.8mg/L，平均含量为 71.1mg/L。由图 6（b）可知，B^{3+}主要富集在罗北凹地及腾龙矿区，新庆矿区 B^{3+}含量较低；W_3 层位 B^{3+}含量最大值为 93.3mg/L，最小值为 39.8mg/L，平均含量为 62.9mg/L。与 W_1 储层类似，B^{3+}含量主要富集在罗北凹地矿区，并向周围呈梯度递减，见图 6（c）。总体来说 B^{3+}表现出了与 Li^+和 Mg^{2+}相同的分布趋势，见图 4、图 5、图 6，这也体现了盐湖卤水终端成矿元素的富集规律。

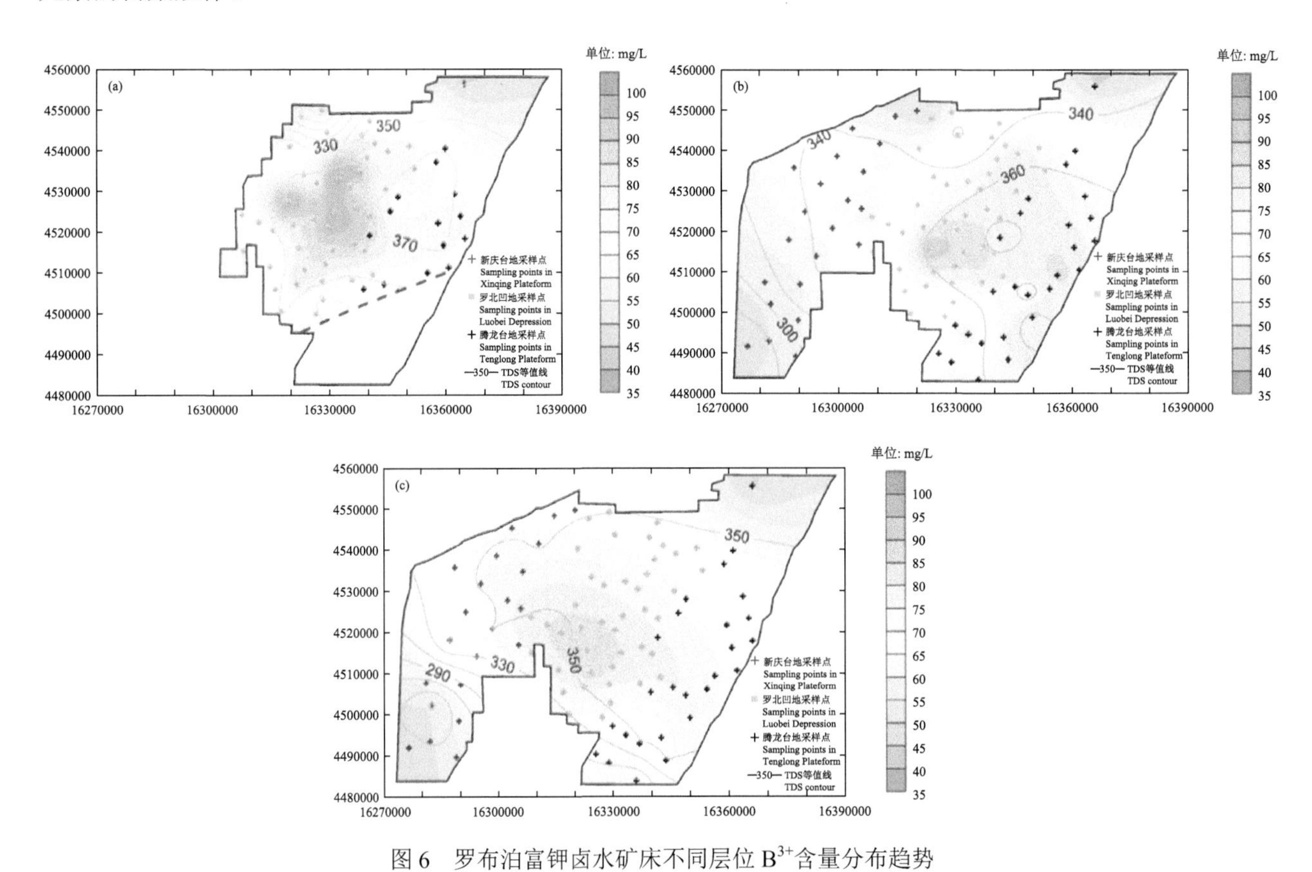

图 6　罗布泊富钾卤水矿床不同层位 B^{3+}含量分布趋势

Fig. 6　The distribution trend of B^{3+} content for different layers in Lop Nor potash-rich brine deposit.

（a）W_1 储层；（b）W_2 储层；（c）W_3 储层

(a) layer W_1; (b) layer W_2; (c) layer W_3

4.6　硫酸根离子空间分布特征

图 7 为罗布泊富钾卤水矿床中 SO_4^{2-} 的分布情况。W_1 储层 SO_4^{2-} 含量的最大值为 94g/L，最小值为 15.7g/L，平均含量为 56g/L。由图 7（a）可知，W_1 储层从西北往东南呈递增趋势，至腾龙矿区达到最大值，而低值区主要位于罗北凹地矿区大部及腾龙矿区北部地区，其总体趋势与 Cl^-相反，但与 TDS 分布趋势相一致。W_2 储层中 SO_4^{2-} 含量最大值为 100g/L，最低值为 15.7g/L，平均离子含量为 57.5g/L，且 SO_4^{2-} 含量的最大值位于腾龙台地东南部，这与 W_1 储层分布特征类似，见图 7（a）、图 7（b），高值区主要分布于 TDS＞360g/L 区域，说明硫酸盐型卤水在高矿化阶段，对于矿化度依然有重要贡献。W_3 储层中 SO_4^{2-} 含量最大值为 75.5g/L，最小值为 15.7g/L，平均含量为 52.8g/L。其分布趋势与 W_3 储层 K^+分布趋势相反，SO_4^{2-} 含量最大值位于腾龙矿区中南部，而罗北凹地矿区 SO_4^{2-} 含量较低，见图 7（c），但新庆矿区南部的低值区域则是由于西部干三角洲的侧向补给所致。

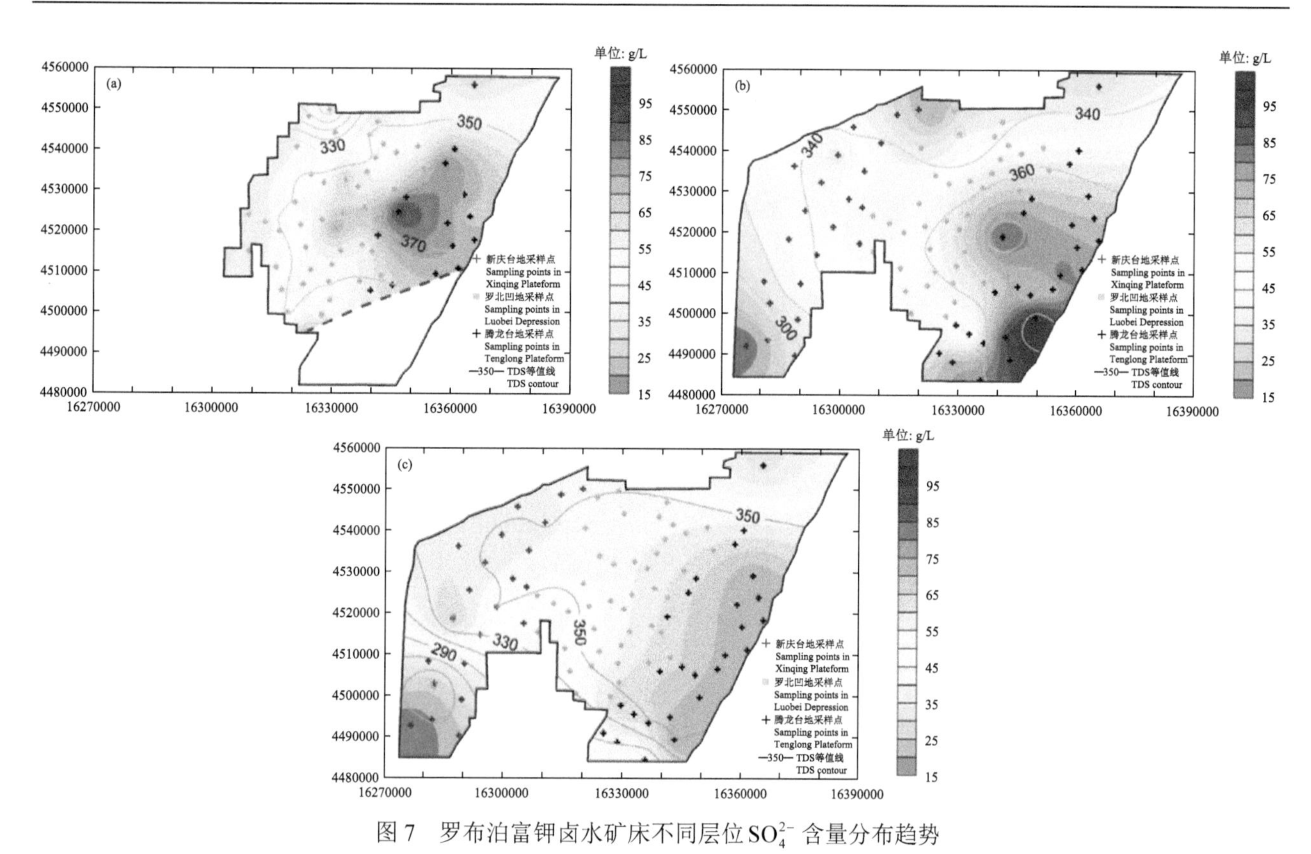

图 7 罗布泊富钾卤水矿床不同层位 SO_4^{2-} 含量分布趋势

Fig. 7 The distribution trend of SO_4^{2-} content for different layers in Lop Nor potash-rich brine deposit.

(a) W_1 储层；(b) W_2 储层；(c) W_3 储层

(a) layer W_1; (b) layer W_2; (c) layer W_3

4.7 钠离子空间分布特征

图 8 揭示了钠离子含量在罗布泊各个矿区不同层位的分布情况。W_1 储层的钠离子含量最高值为112g/L，最小值为60.7g/L，平均含量为98g/L。W_1 储层钠离子含量高值区位于罗北凹地的西北部及腾龙台地的中北部，与 TDS 分布趋势类似，矿区北部山前及腾龙台地南部可能有淡水补给，形成一南北向分布的低值区，见图 8（a）；W_2 储层中钠离子含量最高值为119g/L，最低值为83.5g/L，平均含量为99.7g/L。由图 8（b）可知，W_2 储层的钠离子从西南往东北呈递增趋势，仅在罗北凹地两侧的断裂带附近形成局部高值区。罗北凹地矿区北部、新庆矿区西南部及腾龙矿区南部同样由于淡水补给出现了几个低值区域；W_3 储层中钠离子含量最高值为103g/L，最低值为68.7g/L，平均值为92.2g/L。与 W_1、W_2 储层分布趋势类似，W_3 储层从西南往东北呈明显的递增趋势，且钠离子含量高值区主要分布在腾龙矿区北部，见图 8（c），新庆矿区南部 Na^+ 含量明显偏低，这和矿区接受西部塔里木河及孔雀河三角洲的侧向补给相关，W_3 储层没有受到北部淡水补给的影响。三个矿层钠离子的分布趋势与 TDS 分布趋势较为一致，反映了卤水演化的一般规律。

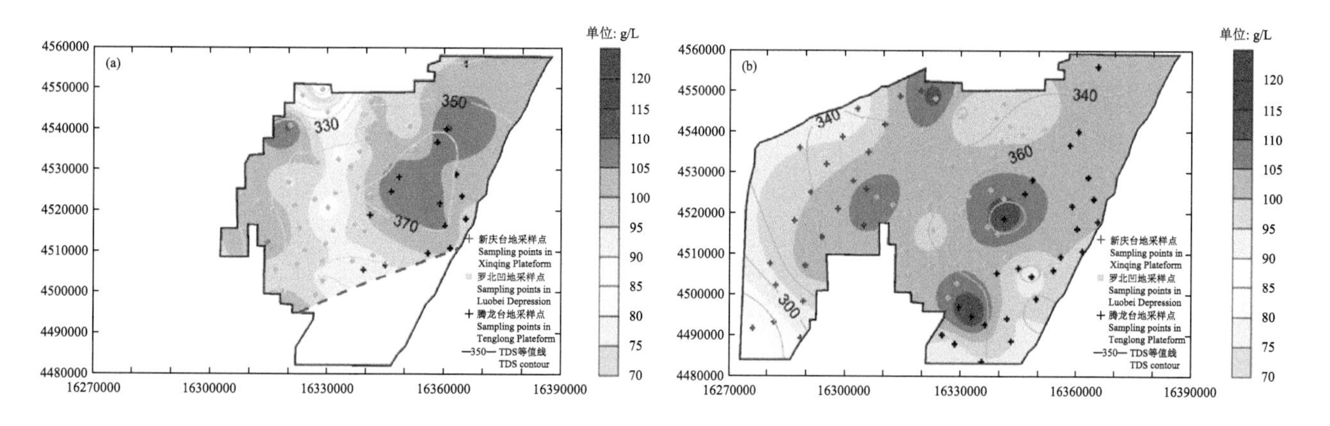

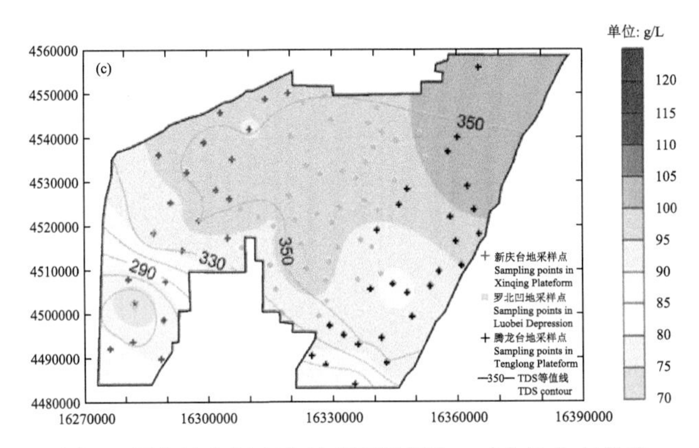

图 8　罗布泊富钾卤水矿床不同层位 Na^+ 含量分布趋势

Fig. 8　The distribution trend of Na^+ content for different layers in Lop Nor potash-rich brine deposit.

（a）W_1 储层；（b）W_2 储层；（c）W_3 储层

(a) layer W_1; (b) layer W_2; (c) layer W_3

4.8　开采前后化学特征对比

1996～1997 年，中国地质科学院矿产资源研究所及新疆地矿局第三地质大队对罗布泊富钾卤水矿床实施了系统的勘探工作，获取了大量的钻孔和卤水样品，为研究该矿床开采前的卤水地球化学特征提供了一定基础（Wang Mili，2001），但当时的工作区域仅限于罗北凹地，未对新庆及腾龙台地开展任何工作，因此，开采前后的对比仅限于罗北凹地子矿区。且早期的采样均是通过浅坑或浅钻获取的 2m 以浅的卤水样品，只能反映罗北凹地 W_1 层顶部的卤水化学特征，其具体化学统计特征见表 1。

表 1　罗北凹地开采前卤水化学统计特征（王弭力，2001）

Table 1　Chemical statistic characteristics of brine before mining inLuobei depression（Wang Mili，2001）

分析项目	盐度（$g·L^{-1}$）	KCl（%）	Mg^{2+}（$g·L^{-1}$）	Cl^-（$g·L^{-1}$）	SO_4^{2-}（$g·L^{-1}$）	B^{3+}（$mg·L^{-1}$）	Li^+（$mg·L^{-1}$）
最大值	410.10	1.82	26.95	189.32	88.95	194.00	44.58
最小值	298.50	0.88	8.45	159.55	13.39	25.60	4.17
平均值	353.46	1.4	17.42	179.95	44.03	67.22	18.17

由表 1 可知，开采前罗北凹地 W_1 层顶部卤水矿化度范围为 298.50～410.10g/L，而 2017 年罗北凹地 W_1 层矿化度范围为 239.57～388.20g/L，指示矿区发生了不同程度的淡化现象；KCl 品位开采前为 0.88%～1.82%，平均值为 1.4%，开采后 KCl 品位为 1.07%～1.56%，平均值为 1.26%，KCl 平均值下降了 9.71%，而 SO_4^{2-}、Mg^{2+}、B^{3+}、Li^+等离子参考意义不大，一是两期的数据所采用的测试方法不同，导致测试结果具有明显差异性，二是早期采集的数据均来源于近地表层，容易受大气降水及周缘季节性洪流的影响而不十分稳定，不能完全代表 W_1 层，因此只能作为一般性参考。但从长期的监测数据来看，罗布泊富钾卤水矿床无论是矿化度还是含钾品位经过十几年的开采均呈下降趋势。

5　结论

本文应用 Surfer 软件克里金插值法，对 2017 年采集的 89 个常观孔的卤水样品进行了空间分析，绘制了不同矿层主量元素 K^+、Na^+、SO_4^{2-}、Cl^-、Mg^{2+}以及微量元素 Li^+、B^{3+}的等值线分布图，揭示了罗布泊钾盐矿床开采 14 年后地球化学的空间分布趋势，通过综合对比分析，主要得出以下结论：

（1）W_1、W_2、W_3 三个层位的 Cl^-、K^+、Mg^{2+}、Li^+、B^{3+}总体具有较为一致的分布特征：离子含量高值区域主要富集在罗北凹地矿区及腾龙矿区中北部地区，离子含量低值区主要分布于矿区西南部和东南部。而 SO_4^{2-} 分布趋势整体与 Cl^-等离子相反，其高值区主要位于腾龙矿区东南部，这符合卤水演化的一般规律。

（2）罗布泊钾盐矿区由于受北山裂隙水、孔雀河、塔里木河冲积平原以及库鲁克塔格山区基岩裂隙水补给导致矿区东南部、西南部以及矿区北部出现有淡化现象，其淡化范围和趋势还需开展长期的观测和模拟实验，以评估对钾盐矿床开发的影响。其中由于受深度的控制，北部山前裂隙水只影响了 W_1、W_2 储层，W_3 储层未见明显变化。

（3）罗布泊钾盐矿床经历了 14 年的连续开采，2017 年其 K^+含量变化于 4.4～10.9g/L，平均含量为 8.83g/L，除少部分受淡水影响的区域外，矿区内大部分区域卤水的 K^+含量（折算成 KCl）均高于 1%的工业单独开采品位；与开采之前相比，罗北凹地 W_1 储层 KCl 平均品位约下降了约 9.7%。

References

Cheng Yongzhi, Wang Mili, Yang Zhichen, Liu Chenglin, Jiao Pengcheng. 2001. The making of potash-bearing salts mixtures through the processing of magnesium sulfate sub-type brine in Lop Nur saline lake, Xinjiang. Acta Geoscientical Sinica, 22 (5): 465-470 (in Chinese with English abstract).

Fan Yao. 2011. Study on chemical characteristics and evolutionary mechanism of Xinqing underground brine in ruoqiang County, Xinjiang (in Chinese with English abstract).

Gu Xinlu, Zhao Zhenhong, Li Qinghai, Chang Zhiyong, Chen Debin, Liu Yuxin. 2003. Analysis of the developing prospect of the unconfined brine kalium mine in the north hollow of the Lop Nur region. HYdrogeology &Engineering Geology, 30 (02): 32-36 (in Chinese with English abstract).

Gu Xinlu, Lu Chengxin, Zeng Yonggang, Song Wenjie. 2009. Quaternary salt mineral of containing potassium occurrence character analyzing in Luo Nor Tonlon Terrace. Xinjiang Geology, 27 (04): 346-349 (in Chinese with English abstract).

Han Fangfang, Shi Deyang, Sun Zhongwei. 2013. A brief talk on the structure and water control in the northern area of Lop Nor, Xinjiang. West-China Exploration Engineering, 25 (8): 174-176 (in Chinese with English abstract).

Hu Dongsheng, Zhang Huajing. 2004. Lake-evaporated salt resources and the environmental evolution in the Lop Nur region. Journal of Glaciology and Geocryology, 26 (2): 212-218 (in Chinese with English abstract).

Lin Jingxing, Zhang Jing, Ju Yuanjing, Wang Yong, Lin Fang, Zhang Junpai, Wang Shaofang, Wei Mingrui. 2005. The lithostratigraphy, magnetostratigraphy, and climatostratigraphy in the Lop Nur region, Xinjiang. Journal of Stratigraphy, 29 (4): 317-322 (in Chinese with English abstract).

Lin Changsong, Li Sitian, Liu Jingyan, Qian Yixiong, Luo Hong, Chen Jianqiang, Peng Li, Riu Zhifeng. 2011. Tectonic framework and paleogeographic evolution of the tarim basin during the Paleozoic major evolutionary stages. Acta Petrologica Sinica, 27 (1): 210-218 (in Chinese with English abstract).

Liu Chenglin, Wang Mili, Jiao Pengcheng, Li Yongzhi, Li Shude. 2002. Formation of pores and brine reserving mechanism of the aquifers in Quaternary potash deposits in Lop Nur Lake, Xinjiang. Geological Review, 48 (4): 437-444 (in Chinese with English abstract).

Ma Lichun, Lowenstein T K, Li Baoguo, Jiang Pingan, Liu Chenglin, Zhong Junpin, Sheng Jiandong, Qiu Honglie, Wu Hongqi. 2010a. Hydrochemical characteristics and brine evolution paths of Lop Nor Basin, Xinjiang Province, western China. Applied Geochemistry, 25 (11): 1770-1782.

Ma Lichun, Liu Chenglin, Jiao Pengcheng, Chen Yongzhi. 2010b. A preliminary discussion on geological conditions and indicator pattern of potash deposits in typical playas of Xinjiang. Mineral Deposits, 29 (04): 593-601 (in Chinese with English abstract).

Sun Mingguang, Ma Lichun. 2018. Potassium-rich brine deposit in Lop Nor basin, XinJiang, China. Scientific Reports.

Tang Pinghui, Shen Tiande, Gu Xinlu. 2008. Study on the occurrence of liquid potash deposits in Luobei depression, Lop Nor, Xinjiang. Tianshan Colloquium on Geological and Mineral Resources (in Chinese with English abstract).

Wang Mili. 2001. Saline lake potash resources in the Lop Nur, Xinjiang (in Chinese with English abstract).

Wang Mili, Liu Chenglin, Jiao Pengcheng. 2006. Investigation and scientific research progress and exploitation present situation of Lop Nur salty lake potash deposits, Xinjiang, China. Geological Review, 52 (6): 757-764 (in Chinese with English abstract).

Yang Li. 2009. Study on the distribution and evolution of underground brine chemical field in Teng long platform, Xinjiang (in Chinese with English abstract).

Zhi Hongjun. 2012. Development and utilization of lithium, boron and other resources in Lop Nor, Xinjiang. Xinjiang Nonferrous Metals, 35 (6): 49-51 (in Chinese with English abstract).

参考文献

陈永志，王弭力，杨智琛，刘成林，焦鹏程. 2001. 罗布泊硫酸镁亚型卤水制取钾混盐工艺试验研究. 地球学报, 22 (5): 465-470.

范尧.2011. 新疆若羌县新庆地下卤水水化学特征及演化机制研究. 中国地质大学（武汉）.

顾新鲁，赵振宏，李清海，常志勇，陈德斌，刘豫新. 2003. 罗布泊地区罗北凹地潜卤水钾矿床成因与开发前景. 水文地质工程地质, 30 (02): 32-36.

顾新鲁，陆成新，曾永刚，宋文杰. 2009. 罗布泊腾龙台地第四系含钾盐类矿物赋存特征分析. 新疆地质, 27 (04): 346-349.

韩芳芳，师德扬，孙忠伟. 2013. 浅谈新疆罗布泊北部地区构造及其控水作用. 西部探矿工程, 25 (8): 174-176.

胡东生，张华京. 2004. 罗布泊荒漠地区湖泊蒸发盐资源的形成及环境演化. 冰川冻土, 26 (2): 212-218.

林景星，张静，剧远景，王永，林防，张俊牌，王绍芳，魏明瑞. 2005. 罗布泊地区第四纪岩石地层、磁性地层和气候地层. 地层学杂志, 29 (4): 317-322.

林畅松，李思田，刘景彦，钱一雄，罗宏，陈建强，彭莉，芮志峰. 2011. 塔里木盆地古生代重要演化阶段的古构造格局与古地理演化. 岩石学报, 27 (1): 210-218.
刘成林，王弭力，焦鹏程，陈永志，李树德. 2002. 罗布泊第四纪卤水钾矿储层孔隙成因与储集机制研究. 地质论评, 48 (4): 437-444.
马黎春，刘成林，焦鹏程，陈永志. 2010. 新疆典型干盐湖成钾条件对比与指标模型初探. 矿床地质, 29 (04): 593-601.
M. г. 瓦里亚什科，范立. 1965. 钾盐矿床形成的地球化学规律. 中国工业出版社.
唐平辉，谌天德，顾新鲁. 2008. 新疆罗布泊罗北凹地液体钾盐矿床赋存规律的研究. 天山地质矿产资源学术讨论会.
王弭力. 2001. 罗布泊盐湖钾盐资源. 地质出版社.
王弭力，刘成林，焦鹏程. 2006. 罗布泊盐湖钾盐矿床调查科研进展与开发现状. 地质论评, 52 (6): 757-764.
杨丽. 2009. 新疆腾龙台地地下卤水化学场的分布特征与演化研究. 中国地质大学（武汉）.
支红军. 2012. 新疆罗布泊卤水锂铯硼等资源开发利用. 新疆有色金属, 35 (6): 49-51.

Spatial variability in the geochemical characteristics of the K-rich brines in the Lop Nor

Kai Wang[1, 2], Mingguang Sun[1], Lichun Ma[1], Qingfeng Tang[3], Hui Yan[4] and Yu Zhang[1]

1. M NR Key Laboratory of Metallogeny and Mineral Assessment, Institute of Mineral Resources, CAGS, Beijing 100037, China
2. China University of Geosciences (Beijing), Beijing 100083, China
3. Beijing Center for Physical & Chemical Analysis, Beijing 100089, China
4. SDIC Xinjiang Lop Nor Potash Co., Ltd, Hami, Xinjiang 839000, China

Abstract Lop Nor, containing abundant potassium-rich brine, is one of the most important salt-forming basin in China. Past 14 years of industrial mining show changing geochemical characteristics of brines in different ore layers. In order to estimate the spatial geochemical distribution of potassium-rich brines in different mining areas and utilize brine water resources, we have used the Kriging interpolation method to analyse the spatial distribution of 89 brine samples collected in 2017, and drawn the contour maps of K^+, Cl^-, Na^+, Mg^{2+}, Li^+, B^{3+}, SO_4^{2-} . The results show that K^+, Cl^-, Mg^{2+}, Li^+, B^{3+} have a relatively consistent distribution trend. The high value area is mainly located in the Luobei depression and the northern part of the Tenglong platform, and the low value areas are located in the north of the Luobei depression and southwest of the mining area. However, the SO_4^{2-} distribution trend is reversed. Due to the influence of freshwater supply from the surrounding areas, desalination occurs in the southwest, north and southeast of the mining area; however, only W_1 and W_2 layers were affected, and the W_3 layer shows no obvious desalination. The average K^+ content of Lop Nor is 8.83g/L. Except for the area affected by fresh water, the KCl grade in most areas was higher than the industrial separate mining grade of 1%. Compared with KCl grade before mining, the average grade of KCl in W_1 layer of Luobei depression is about 9.7%.

Keywords Lop Nor; Potassium-rich brine; Spatial distribution; Geochemistry; Brine deposits

思茅盆地下白垩统蒸发岩硫同位素地球化学特征及其钾盐成矿意义*

苗忠英[1]，郑绵平[1]，张雪飞[1]，张震[1]，刘建华[1]，高运志[2]，翟雪峰[3]

1. 中国地质科学院矿产资源研究所，自然资源部盐湖资源与环境重点实验室，北京 100037
2. 中国地质大学（北京），北京 100083
3. 中国煤炭地质总局第四水文地质队，河北邯郸 056006

摘要 关于思茅盆地下白垩统勐野井组蒸发岩主要物源为海水的认识争议很少，但是关于其成矿时代和成矿模式的认识还有争议，关于陆源淡水对蒸发岩物质成分的影响还缺乏探讨。本文主要通过分析盆地内L2井27件蒸发岩样品的化学成分和硫同位素地球化学特征，结合邻区已发表的硫同位素数据，探讨了蒸发岩的物质来源、陆源淡水对蒸发岩物质成分的影响、成盐时代以及可能的钾盐成矿模式。结果表明：①思茅盆地蒸发岩受陆源淡水和火山热液补给，其中陆源淡水补给使蒸发岩硫同位素明显低于同一地质时期的其他海相样品；②海水可能自现今盆地北西方向补给，一级周期上海水补给存在两次，二级周期上海水补给至少存在七次；③物源海水的时代为中侏罗世，沉积析盐的时代为早白垩世晚期，可能的钾盐成矿模式为中侏罗世海水侧向迁移成矿。这些结果对解释思茅盆地及邻区海相蒸发岩异常低的硫同位素值、高硫同位素值与中侏罗世海水相当以及钾盐成矿缺失“碳酸盐岩相和硫酸盐岩相”有重要的意义。

关键词 思茅盆地；蒸发岩；钾盐；硫同位素；成矿作用；地球化学

思茅盆地下白垩统蒸发岩主要发育在勐野井组红褐色泥砾岩层内，含硫矿物主要以层状、透镜状、结核状、分散单晶状硬石膏或是脉状、纹层状、块状石膏的形式出现，另有少量以黄铁矿的形式出现在与盐岩伴生的灰黑色泥砾岩中（曲一华等，1998；许建新，2008）。

沉积岩中硬石膏硫同位素在研究矿床成因（黄作良等，1996；Tabakh et al.，1998；李延河等，2013；Li et al.，2018）、地层划分和对比（黄建国和刘世万，1990；梁汉东和丁悌平，2004）、沉积环境（史忠生等，2004；王春连等，2013）、成矿物质来源（牛新生等，2014；张华等，2014）、古气候条件和古海洋环境（张同钢等，2004）、地质历史时期全球环境变化（张同钢等，2003）、沉积岩层定年（Holser and Kaplan，1966；Claypool et al.，1980；Cortecci et al.，1981；Paytan et al.，2004）、沉积物生物化学反应（Canfield and Thamdrup，1994）等方面有较好的应用效果。尤其是海相蒸发岩的硫同位素，在判识物源、生物化学反应、地层对比和定年等方面的应用效果更好。这主要归因于“硫酸根在海水中混合均匀的时间远小于海洋与大陆之间硫元素循环的时间；若未发生生物化学反应，硫酸盐蒸发析出之后硫同位素值较稳定（Holser and Kaplan，1966）”。但是，作为目前为止中国唯一海相固体钾盐产地的思茅盆地，类似的研究和应用还非常少。

本文拟通过对思茅盆地磨黑地区L2井蒸发岩样品的化学成分和硬石膏硫同位素值的分析，结合勐野井钾盐矿和沙空那空、呵叻盆地已发表的硬石膏硫同位素数据，探讨研究区钾盐成矿的物质来源、海水补给的方向和次数以及钾盐成矿可能的模式，以期为科学合理地解释矿床成因、简洁高效地部署地质调查工作提供依据。

1 地质背景

思茅盆地位于印支地块北部（Metcalfe，2013），大地构造上属于三江造山带（潘桂棠等，2009），盆

* 本文发表在：地球学报，2019，40（2）：279-290

地东、西边界分别为金沙江-哀牢山断裂带和澜沧江断裂带（曲一华等，1998），其沉积基底为前寒武系—下古生界绿片岩相变质杂岩（殷鸿福等，1999），沉积盖层主要为中—新生界（谭富文，2002；尹福光等，2006）。其中，三叠系缺失下统，中—上三叠统主要发育海相碎屑岩和碳酸盐岩（曲一华等，1998；尹福光等，2006）；下侏罗统主要发育海湾潮坪—潟湖相细碎屑岩（曲一华等，1998；郑绵平等，2014），中侏罗统有海侵的特征（廖宗廷和陈跃昆，2005），上侏罗统主岩性为陆相红色细碎屑岩（曲一华等，1998）；白垩系为一套典型河湖相砂岩、页岩、砾岩（陈跃昆等，2004；尹福光等，2006）。

思茅盆地可划分出四个隆起和四个拗陷共计八个二级构造单元，包含江城、景谷、整董、勐腊四个含盐带（图 1）。其中，江城含盐带勐野井钾盐矿是我国目前发现的唯一的海相固体钾盐矿床，整董含盐带磨黑盐矿在产食盐。这两处产盐地的产层均为下白垩统顶部的勐野井组。根据 1∶20 万区域地质调查资料，思茅盆地内勐野井组整体岩性特征表现为：上段为紫红色、棕红色及杂色泥砾岩、泥岩、粉砂质泥岩、泥质粉砂岩、粉砂岩夹细砂岩；中段为紫红色、棕红色钙质泥岩、粉砂岩、粉砂质泥岩、泥质粉砂岩夹细砂岩；下段为紫红色粉砂岩、钙质泥岩、杂色泥砾岩。勐野井地区钾石盐矿床产自勐野井组上段，中段无蒸发岩，下段含硬石膏和石盐；磨黑地区岩盐仅产自勐野井组上段，厚度可达 126.2m。

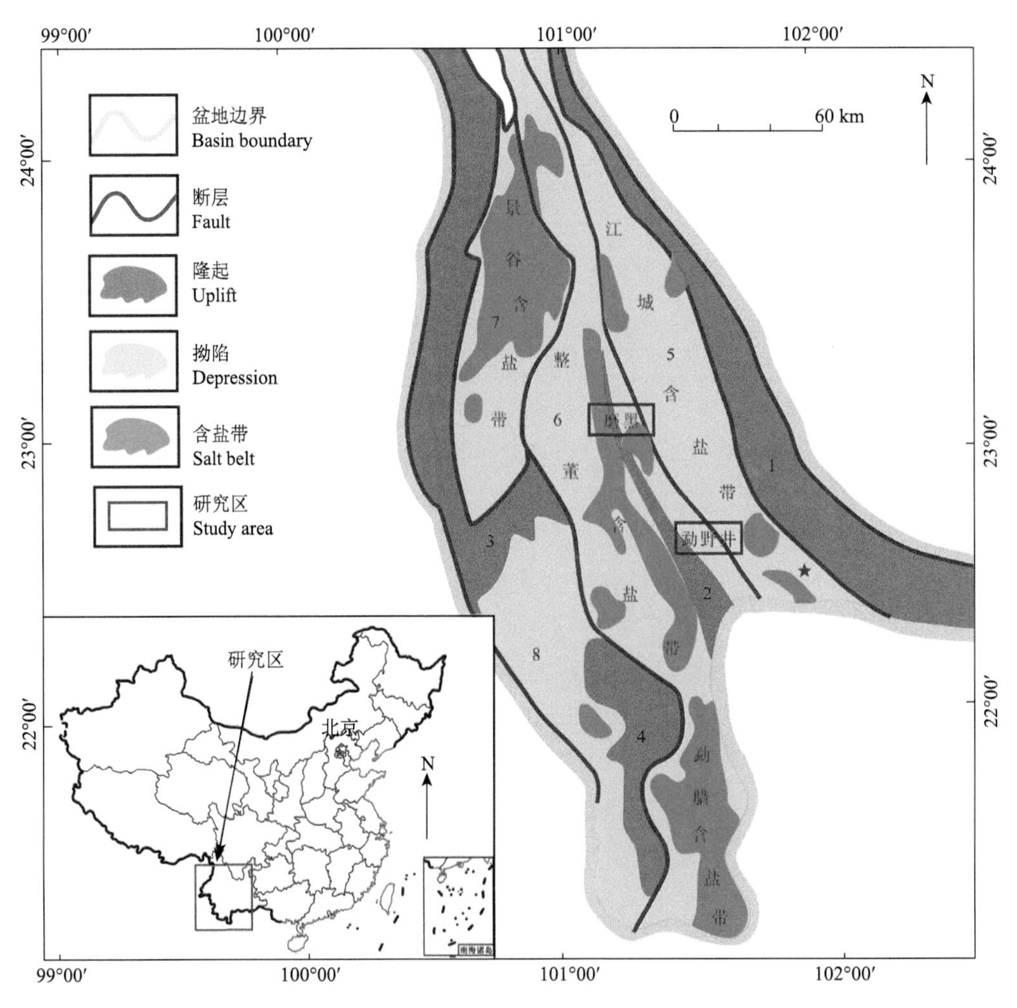

图 1　思茅盆地构造单元划分及含盐带分布特征

Fig. 1　Tectonic units and salt-bearing belt in the Simao basin.

1-墨江隆起；2-无量山隆起；3-景洪隆起；4-勐远隆起；5-江城拗陷；6-磨黑—勐腊拗陷；7-景谷拗陷；8-大渡岗拗陷

1-Mojiang uplift; 2-Wuliangshan uplift; 3-Jinghong uplift; 4-Mengyuan uplift; 5-Jiangcheng depression; 6-Mohe–Mengla depression; 7-Jinggu depression; 8-Dadugang depression

2 样品来源和分析方法

2.1 样品来源

本文分析的样品来自磨黑盐矿外围的 L2 井，其地理坐标为 23°9′1.5″N，101°9′28.7″E（图 2）。井口周围出露的地层为勐野井组，周边还有下白垩统南新组、景星组和侏罗系坝注路组、和平乡组以及上三叠统路马组出露。主体发育北西-南东向逆断层，倾角 40°～60°。

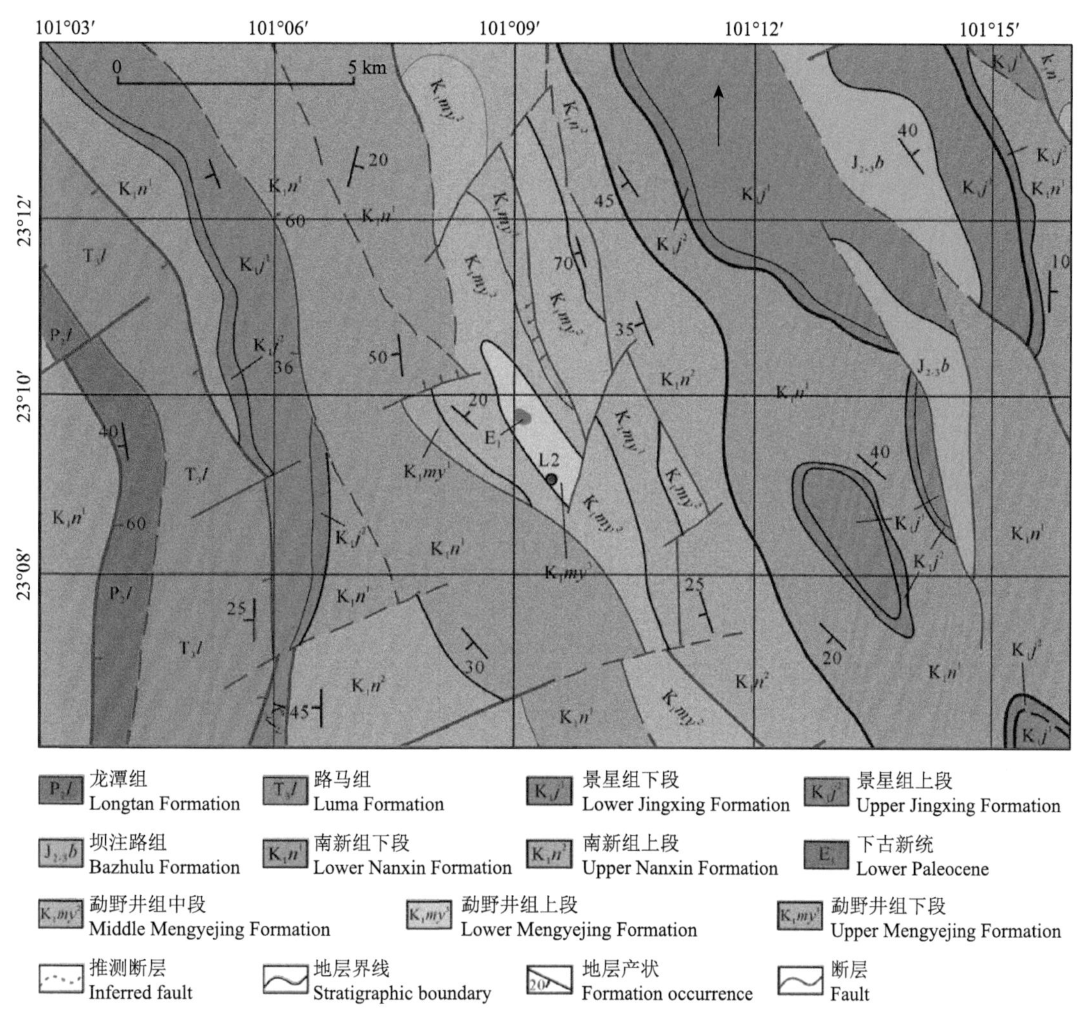

图 2 L2 井位置及磨黑地区地质图

Fig. 2 Location of well L2 and geological map around the Mohei area.

L2 井揭露的地层主要为勐野井组，南新组未钻穿（图 3）。勐野井组岩性具三分特征：上段主要为紫红色泥砾岩夹盐岩；中段主要为紫红色含泥粉砂岩夹盐岩；下段主要为浅紫红色粉砂岩和含泥粉砂岩。钻遇的南新组主要为灰色、灰白色细-中砂岩。

本次研究共取蒸发岩样品 27 件，分布在井深 31.4～631.8m 范围内，主要岩性有硬石膏岩、泥砾质石盐岩、含泥砾石盐岩和石盐岩。其中硬石膏的存在形式主要有薄层状、透镜体、结核状和分散单晶状（表 1）。结核状硬石膏存在于石盐岩中，为了排除石盐岩中分散单晶状硬石膏对其同位素分析结果的影响，先用清水将结核表面的岩盐溶掉，然后在 60℃恒温条件下烘干 48h 备分析；薄层状硬石膏分布于石盐岩层顶部；透镜状硬石膏主要存在于泥砾岩层中。

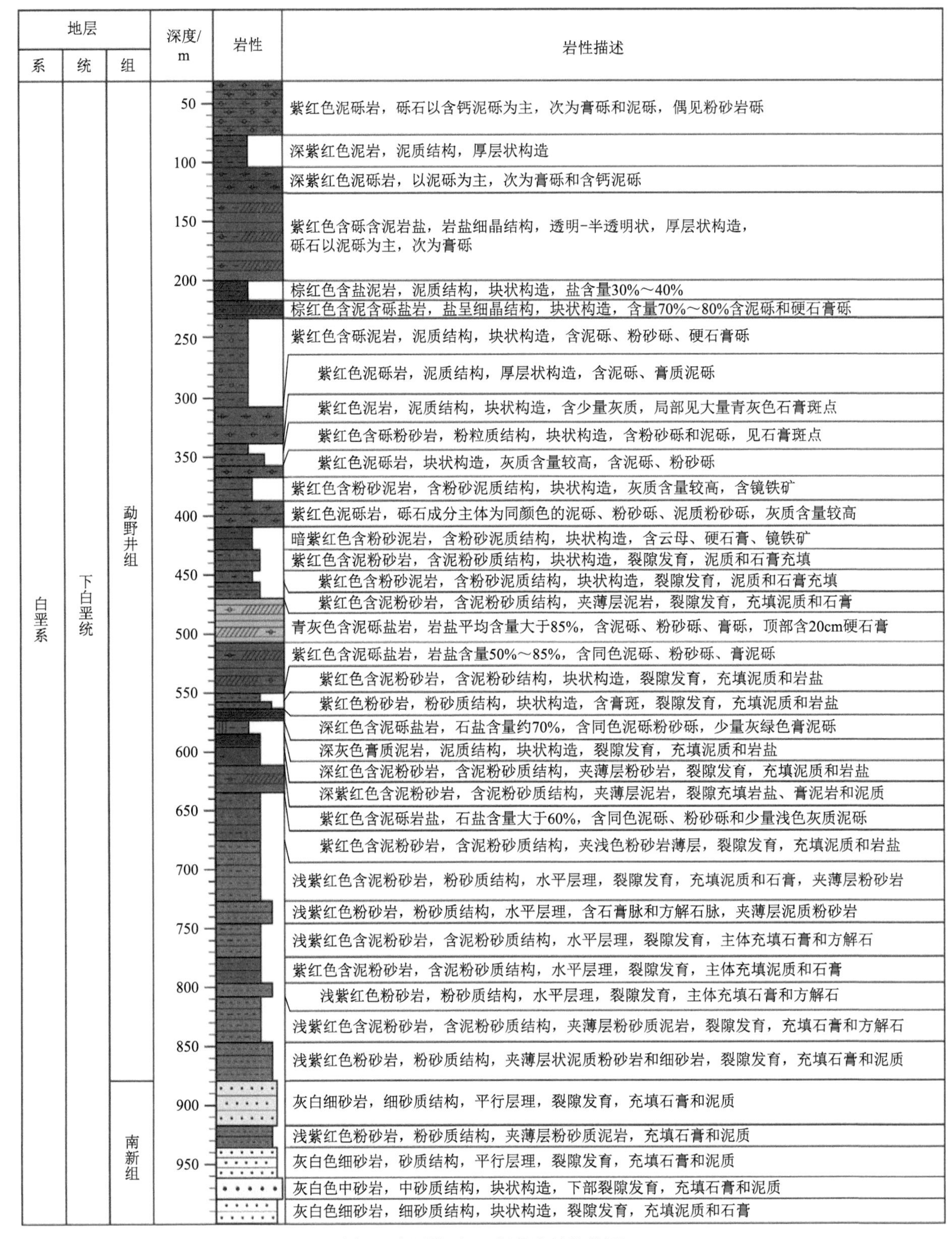

图 3　磨黑地区 L2 探井岩性柱状图

Fig. 3　Stratigraphic column of well L2 in the Mohei area.

表 1　L2 井蒸发岩主要化学成分及硬石膏硫同位素特征

Table 1　Main chemical constituents and sulfur isotopes of halite in well L2

序号	样品编号	岩性	深度/m	$\delta^{34}S_{V\text{-}CDT}$/‰	主离子的相对含量/%							$Br^-/10^{-6}$	$CaCl_2$/%	$CaSO_4$/%
					Na^+	K^+	Mg^{2+}	Ca^{2+}	Cl^-	HCO_3^-	SO_4^{2-}			
1	L2-2-1	硬石膏透镜体	31.5	10.6	0.02	0.02	0.02	6.54	0.36	0.27	14.81	0.00	0.48	20.99
2	L2-6-2	硬石膏透镜体	64.6	12.7	0.04	0.07	0.07	0.66	0.67	0.32	0.84	0.00	0.57	1.19

续表

序号	样品编号	岩性	深度/m	$\delta^{34}S_{V\text{-}CDT}$/‰	主离子的相对含量/%							$Br^-/10^{-6}$	$CaCl_2$/%	$CaSO_4$/%
					Na^+	K^+	Mg^{2+}	Ca^{2+}	Cl^-	HCO_3^-	SO_4^{2-}			
3	L2-9-2	硬石膏透镜体	103.8	11.2	0.11	0.06	0.03	6.60	0.48	0.24	15.07	0.00	0.40	21.36
4	L2-12-1	含砾含泥盐岩	145.1	13.1	32.00	0.04	0.04	1.72	50.34	0.24	2.06	45.00	1.51	2.92
5	L2-14-1	含泥含砾盐岩	155.2	10.8	20.49	0.09	0.11	1.92	33.81	0.37	1.81	34.99	2.89	2.57
6	L2-19-1	含砾含泥盐岩	199.0	13.0	24.79	0.06	0.04	2.08	40.04	0.37	2.50	39.99	2.53	3.55
7	L2-28-1	膏质盐岩	231.8	13.0	19.20	0.06	0.03	1.06	30.97	0.31	0.21	30.00	2.04	0.30
8	L2-28-1-R	岩盐中的硬石膏砾	231.8	18.1	0.09	0.06	0.03	0.40	0.54	0.44	0.28	0.00	0.38	0.40
9	L2-38-2	硬石膏透镜体	296.5	10.4	0.11	0.01	0.02	6.06	0.33	0.21	13.86	0.00	0.24	19.64
10	L2-X-2	薄层状硬石膏	470.2	17.1	7.10	0.01	0.04	5.68	10.47	0.23	13.79	10.00	3.46	19.03
11	L2-62-1	含砾盐岩	471.1	13.4	33.09	0.08	0.04	2.08	53.45	0.18	1.84	79.98	0.00	2.61
12	L2-64-2-2	青灰色盐岩	495.8	12.0	27.29	0.14	0.04	2.08	44.45	0.23	1.89	89.97	3.36	2.68
13	L2-64-3	青灰色盐岩	499.2	12.8	32.18	0.10	0.04	2.32	52.74	0.28	1.17	89.96	4.73	1.66
14	L2-66-1-1	青灰色盐岩	503.3	15.9	35.28	0.06	0.03	2.22	56.72	0.20	2.01	89.96	3.54	2.85
15	L2-66-1-2	青灰色盐岩	503.5	18.0	30.18	0.07	0.02	1.94	48.33	0.16	1.76	89.96	2.70	2.50
16	L2-66-2-2	青灰色盐岩	505.7	12.5	30.59	0.11	0.04	1.70	48.00	0.29	2.47	89.97	1.13	3.50
17	L2-68-2	泥砾质盐岩	514.2	12.8	28.10	0.06	0.04	1.68	44.29	0.21	2.42	79.99	1.42	3.43
18	L2-68-3	泥砾质盐岩	517.5	12.1	27.99	0.05	0.05	1.92	44.37	0.24	2.30	49.99	1.81	3.27
19	L2-70-1	含泥砾盐岩	523.2	12.1	21.90	0.07	0.06	1.60	35.47	0.28	1.60	45.00	2.33	2.26
20	L2-70-3	含泥砾盐岩	532.3	13.0	27.89	0.06	0.06	2.42	45.49	0.32	1.89	49.99	3.80	2.68
21	L2-72-2	泥砾质盐岩	544.0	11.5	27.10	0.06	0.13	0.68	41.44	0.34	2.17	45.00	0.00	1.93
22	L2-72-3	泥砾质盐岩	545.4	12.2	20.39	0.08	0.16	1.54	33.56	0.29	1.35	44.99	2.44	1.91
23	L2-79-1	泥砾质盐岩	571.5	10.5	21.40	0.06	0.08	1.50	34.00	0.26	2.11	50.00	1.48	2.99
24	L2-89-1	含泥砾盐岩	613.4	10.6	23.19	0.05	0.29	3.30	40.64	0.22	2.27	49.98	6.31	3.22
25	L2-89-2	含泥砾盐岩	615.8	11.0	24.39	0.05	0.10	1.14	38.57	0.31	1.68	44.99	0.94	2.38
26	L2-91-1	含泥砾盐岩	628.6	11.6	24.70	0.06	0.23	1.76	39.88	0.29	2.80	44.99	1.37	3.97
27	L2-91-2	含泥砾盐岩	631.8	11.5	21.69	0.06	0.09	1.88	36.07	0.38	1.70	39.99	2.90	2.40

2.2 分析方法

2.2.1 化学分析

盐岩样品化学分析在自然资源部盐湖资源与环境重点实验室完成。使用原子吸收分光光度法测定Na^+、K^+、Mg^{2+}、Ca^{2+}的含量，仪器型号为GGX-800，硫酸钡重量法测定硫酸根的含量，硝酸银容量法测定氯根的含量，酚红光度法测定溴的含量，酸碱滴定容量法测定碳酸根、重碳酸根的含量。更具体的实验步骤可参考文献Lin等（2017）。

2.2.2 硫同位素分析

首先将待测样品经碳酸钠-氧化锌半熔法转化为硫酸钡，再用五氧化二矾法将硫酸钡转化为二氧化硫，纯化并收集二氧化硫，使用Delta V Plus气体同位素质谱仪分析硫同位素的组成，测量结果以CDT为标准，记为$\delta^{34}S_{V\text{-}CDT}$，采用的硫同位素标准参考物质为硫化银GBW-04414和GBW-04415，分析精度优于±0.2‰。更具体的实验步骤可参考文献刘汉彬等（2013）。

3 结果

3.1 化学成分

样品中主要的阳离子为Na^+、K^+、Mg^{2+}和Ca^{2+}，主要阴离子为Cl^-、Br^-、SO_4^{2-}和HCO_3^-（表1）。除了

薄层状硬石膏 L2-X-2 中检测出 $10\times10^{-6}$$Br^-$外，其他硬石膏样品中均未检测出 Br^-。盐岩中 Br^-相对含量的变化范围为 30×10^{-6}～90×10^{-6}，平均含量为 58×10^{-6}，明显低于勐野井钾盐矿的样品。阳离子主要为 Na^+ 和 Ca^{2+}，表现在岩性上为石盐岩和硬石膏岩。K^+含量很低，通过假定盐计算得到 KCl 的相对含量不超过 0.3%。此外，通过假定盐计算还得到 $CaCl_2$ 的相对含量为 0.24%～6.31%，平均值为 2.19%。

3.2 硫同位素

从表 1 中可以看出 L2 井硬石膏的硫同位素值比较离散，最小值为 10.4‰，最大值为 18.1‰，平均值为 12.7‰。全部 27 件样品的硫同位素值以 12‰～14‰为主峰，10‰～12‰为次峰（图 4）。

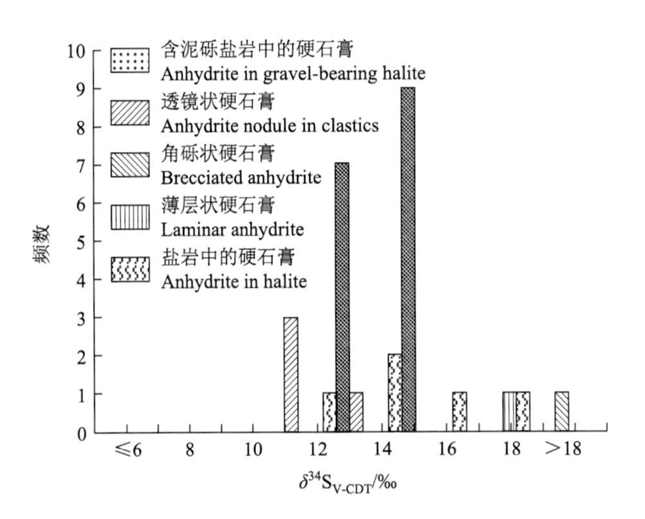

图 4　L2 井蒸发岩硫同位素值整体分布特征

Fig. 4　The overall distribution characteristics of sulfur isotope values of evaporites in well L2.

硬石膏硫同位素值随其赋存形式改变而发生变化。以泥砾岩为围岩的透镜体状硬石膏，其硫同位素值主体分布在 10‰～12‰区间段内；含泥砾石盐岩和石盐岩中分散单晶状的硬石膏，其硫同位素值主体分布在 12‰～14‰区间段内；存在于石盐岩中的角砾状硬石膏，其硫同位素值分布在 18‰～20‰区间段内；石盐岩层顶部直接与之接触的薄层状硬石膏，其硫同位素值分布在 16‰～18‰区间段内（图 5）。

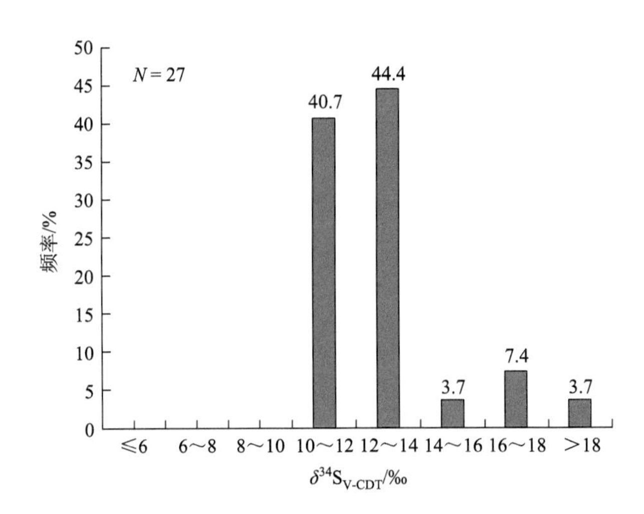

图 5　L2 井不同类型硬石膏的硫同位素值分布特征

Fig. 5　The distribution characteristics of sulfur isotope values of different types anhydrite in well L2.

4 讨论

4.1 思茅盆地蒸发岩硫同位素的示源意义

由于硫酸根在海水中混合均匀的时间远小于其存留时间，所以同一地质时期海水中硫酸根的硫同位素值比较稳定，若沉积析出之后未遭受生物化学作用的改造，则这种稳定的同位素特征会在沉积析出的硬石膏中得到继承，并可在区域上进行对比（Holser and Kaplan，1966；Canfield and Thamdrup，1994）。而陆相湖盆硫酸盐的封闭性相对较差，所以硫同位素值的变化幅度也较大。例如，第四纪盐湖或干盐湖中硫酸盐的 $\delta^{34}S_{V\text{-}CDT}$ 值最小者为 3.9‰左右（彭立才等，1999），最大者可达 19.9‰（樊启顺等，2009）。

思茅盆地蒸发岩溴、铷、锶、硼含量及锶同位素地球化学特征显示其物源为海水（郑智杰等，2012；杨尖絮等，2013；Miao，2018），且围岩全部为“红层”，即沉积环境氧化还原电位较高，很难发生硫酸盐微生物还原反应，所以应用硫同位素特征探讨其物源信息具备了充分的条件。

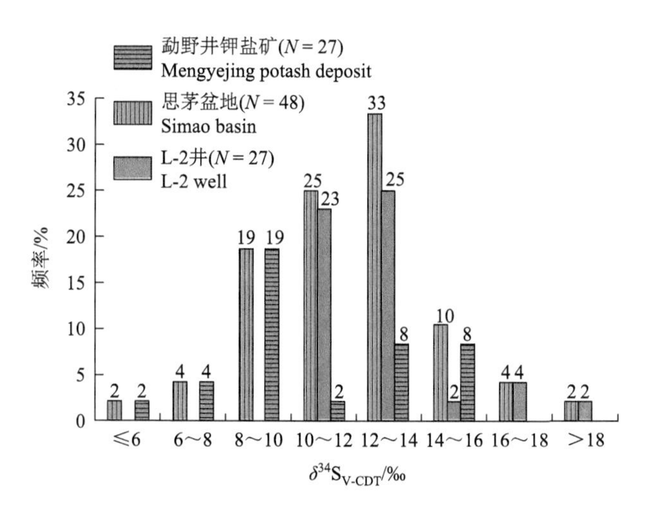

图 6 思茅盆地硬石膏硫同位素值的分布特征

Fig. 6 The distribution characteristics of sulfur isotope values of anhydrite in the Simao basin.

除了本文报道的 L2 井蒸发岩硫同位素数据外，许建新（2008）在研究中也分析了勐野井钾盐矿盐岩样品的硫同位素特征，其最小值为 3.9‰，最大值为 15.5‰，平均值为 10.6‰，主峰值区间呈“双峰”特征，低值区主峰为 8‰～10‰，高值区主峰分别为 12‰～14‰和 14‰～16‰（图 6）。

统计 L2 井和勐野井钾盐矿的硫同位素数据可以发现，逾 83%的蒸发岩样品硫同位素值小于 14‰，逾 50%的蒸发岩样品硫同位素值小于 12‰，明显低于晚三叠世以来海相硬石膏硫同位素的最低值。这说明盆地内以海水为母源的蒸发岩在沉积析盐期，很可能受到具有轻硫同位素特征的外源硫补给的影响，L2 井岩盐中硫同位素值与硫酸根含量负相关是这种认识的有利证据（图 7）。

这种外源补给可能来自陆源水体，也可能来自深部热液。首先，勐野井钾盐矿中出现的硼酸盐矿物及其硼同位素特征反映了热液是蒸发岩的物源之一；利用化学分析的结果假定盐计算能够得出 L2 井蒸发岩中含有 $CaCl_2$。这说明除了海水外，热液确实对蒸发岩的物源补给有贡献。其次，可以从区域上热液补给的强度和硫同位素的变化特征来推断热液补给是否影响蒸发岩的硫同位素特征。沙空那空和呵叻盆地位于思茅盆地南部，同一时代的蒸发岩内发育溢晶石矿物（Li et al.，2018），而思茅盆地仅能通过假定盐计算发现含有 $CaCl_2$，这说明热液对沙空那空和呵叻盆地的影响强于思茅盆地。可是，受热液影响相对较强的沙空那空和呵叻盆地中硬石膏的硫同位素值明显大于思茅盆地。例如，单独成层的硬石膏在呵叻盆地内硫同位素值的主峰为 14.0‰～16.0‰，在沙空那空盆地内的主峰为 14.0‰～15.0‰，在思茅盆地内的主峰为 10.0‰～12.0‰（图 8）；在呵叻盆地盐岩中硬石膏的硫同位素值主峰为 16.0‰～18.0‰，在沙空那空盆地内硫同位素值的主峰为 14.0‰～16.0‰，在思茅盆地内

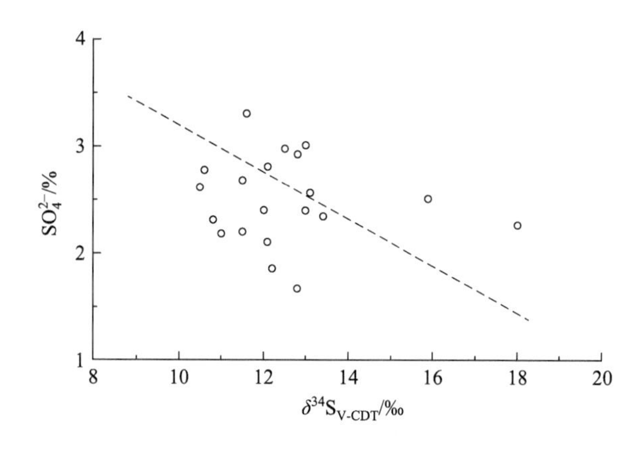

图 7 L2 井岩盐中硫同位素值与硫酸根含量负相关

Fig. 7 Sulfur isotope values in halite of L2 well exhibiting negative correlation with sulfate content.

硫同位素值的主峰为 8.0‰～10.0‰和 12.0‰～14.0‰（图 9），因此可以推断热液补给对研究区蒸发岩硫同位素特征的影响较弱。

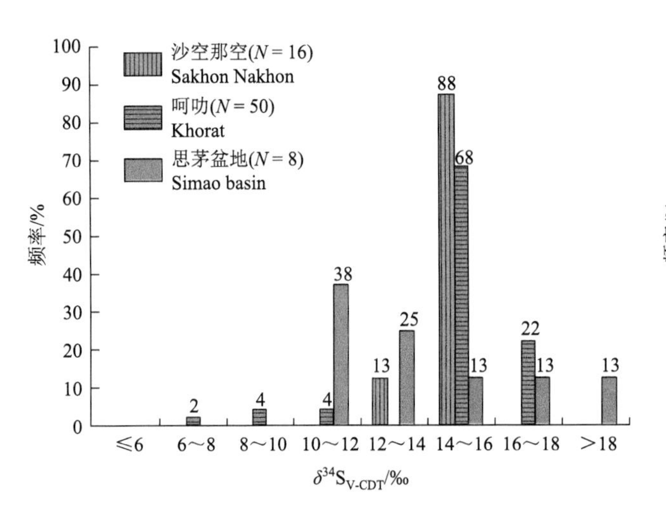

图 8　思茅—沙空那空—呵叻盆地硬石膏层硫同位素值分布特征

Fig. 8　The distribution characteristics of sulfur isotope values of stratiform anhydrite in the Simao–Sakhon Nakhon–Khorat basins.

呵叻盆地数据来自 Tabakh et al.，1999；沙空那空盆地数据来自张华等，2014 和 Li et al.，2018

Khorat basin data after Tabakh et al., 1999; Sakhon Nakhon basin data after ZHANG et al., 2014; Li et al., 2018

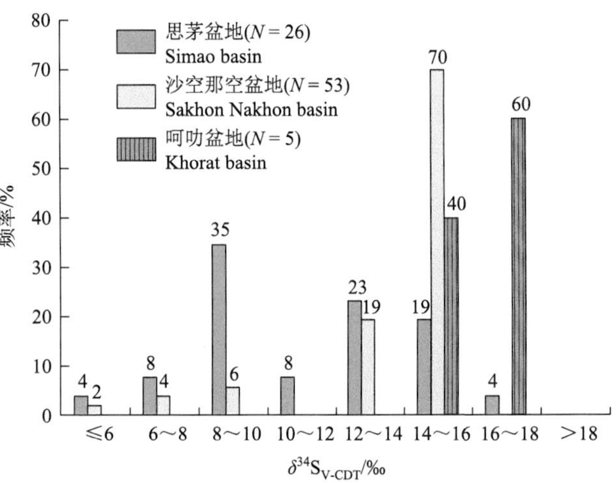

图 9　思茅—沙空那空—呵叻盆地盐岩中硬石膏硫同位素值的分布特征

Fig. 9　The distribution characteristics of sulfur isotope values of anhydrite from the halite in the Simao–Sakhon Nakhon–Khorat basins.

呵叻盆地数据来自 Tabakh et al.，1999；沙空那空盆地数据来自 Li et al.，2018；思茅盆地部分数据来自许建新，2008

Khorat basin data after Tabakh et al., 1999; Sakhon Nakhon basin data after Li et al., 2018; part data of Simao basin after XU, 2008

与蒸发岩相伴生的泥砾是陆源淡水补给的有力证据，特别是勐野井组盐岩中泥砾的稀土元素和微量元素地球化学特征也显示其母质来自陆地（石海岩等，2014），其汇入蒸发盆地应主要以陆源淡水（河流或季节性洪水）作为搬运载体。根据邻区呵叻盆地内同一时代的陆相碎屑岩中硬石膏结核的硫同位素值为 6.4‰～10.9‰（Tabakh et al.，1999），以及 L2 井泥砾岩层中硬石膏透镜体较低的硫同位素值可以推断：补给蒸发盆地的陆源淡水中硫酸根应该具有较轻的硫同位素。而思茅盆地内硫同位素值较低的蒸发岩样品多与泥砾相关。例如，L2 井含泥砾（泥砾质）盐岩中硬石膏硫同位素值全部低于 14‰，L2 井以泥砾岩为围岩的硬石膏样品其硫同位素值主体分布在 10‰～12‰，勐野井钾盐矿含泥砾盐岩中硬石膏的硫同位素值全部低于 10‰，可见思茅盆地蒸发岩较轻的硫同位素特征主要受陆源淡水补给的影响。

既然较低的硫同位素值是陆源淡水补给所致，较高的硫同位素值是母源海水的特征，那么可以通过硫同位素值的高低变化来反演思茅盆地内海水补给的次数。以 L2 井为例，海水补给在一级周期上存在两次，对应着盆地内发育的两套盐岩；在二级周期上至少存在七次，每一次补给都会使硬石膏硫同位素值产生由小向大的变化；勐野井组沉积期间，海水对沉积盆地的影响逐渐减弱，表现为硬石膏硫同位素值整体呈现出由大向小变化的趋势（图 10）。

4.2　思茅盆地蒸发岩硫同位素的地质年代意义

全球晚三叠世以来海相硬石膏硫同位素值变化曲线特征表现为（图 11）：晚三叠世至早白垩世早期（205～125Ma），海相硬石膏的硫同位素值由 16‰增大至 18‰；随后，至早白垩世阿普特期与阿尔必期之交（约 113Ma），海相硬石膏硫同位素值由 18‰快速降低至 14‰，是晚三叠世以来的最低值；随后，至古近纪海相硬石膏硫同位素值逐渐升高至 21‰左右；古近纪至今，海相硬石膏的硫同位素值稳定在 21‰左右（Holser and Kaplan，1966；Claypool et al.，1980；Paytan et al.，2004）。

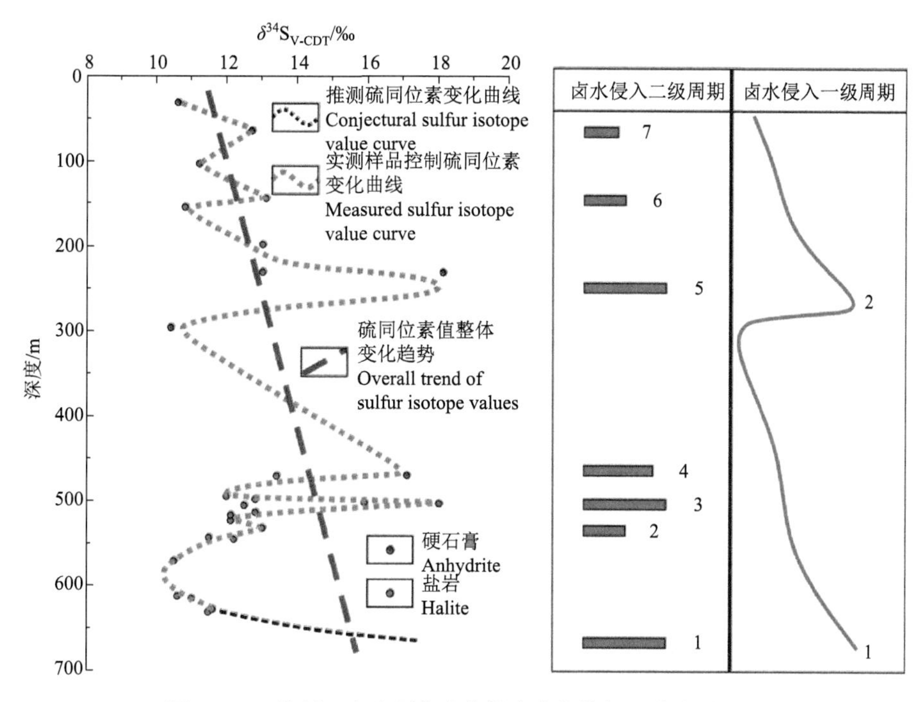

图10　L2井硬石膏硫同位素值纵向变化特征及其地质意义

Fig. 10　Variation characteristics of sulfur isotope values with depth and its geological significance in well L2.

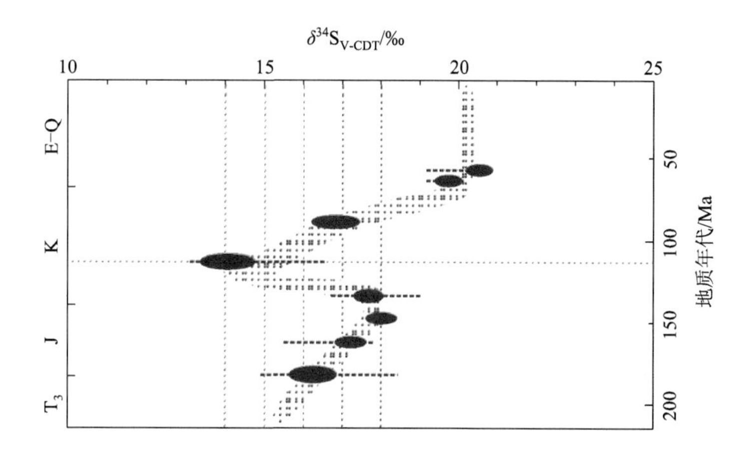

图11　晚三叠世以来海相硬石膏硫同位素值变化特征示意图（据Claypool et al.，1980修改）

Fig. 11　Schematic diagram of the variation of sulfur isotope values of marine anhydrite since the Late Triassic (after Claypool et al., 1980).

思茅盆地内蒸发岩硫同位素值较小的样品受陆源淡水补给的影响相对较大，进而削弱了其年代学意义。硫同位素值较大的样品可近似代表地质时期原始海水的硫同位素特征。其中，编号为L2-28-1-R（$\delta^{34}S_{V-CDT}$值为18.1‰）的样品为盐岩中的硬石膏砾，它很可能是析盐阶段外源输入或者是成岩阶段后构造活动将先期形成的硬石膏裹挟进盐岩，因此其作为原始卤水硫同位素特征的代表性不强；编号为L2-66-1-2的样品为青灰色盐岩，硬石膏呈分散单晶状存在于盐岩中，可代表原始卤水的硫同位素特征，其值为18.0‰（$^{87}Sr/^{86}Sr$为0.7082）；编号为L2-X-2的样品为薄层状硬石膏，尽管其析出时的卤水盐度较析盐期低，但是由于此时$CaSO_4$在卤水中的相对含量较高，导致陆源淡水对其硫同位素特征的影响相对较弱，故其现今的硫同位素值可代表原始卤水的硫同位素特征，其值为17.1‰（$^{87}Sr/^{86}Sr$为0.7084）。所以思茅盆地蒸发岩原始物源硫同位素值主体很可能为17.0‰～18.0‰，与中—晚侏罗世或晚白垩世海水的硫同位素特征相近。思茅盆地勐野井组凝灰岩夹层的SHRIMP锆石U-Pb年龄为110～100Ma（Wang et al.，2015），根据2015年的国际年代地层表，这套地层属于下白垩统顶部（Cohen et al.，2013），即盆地内蒸发岩的原始物源应为中—晚侏罗世的海水。而盆地内上侏罗统为典型的陆相红层，所以蒸发岩的原始物源很可能为中侏罗世的海水。

中侏罗世海水侧向迁移成盐模式可能发生在由羌塘、昌都、兰坪—思茅、呵叻盆地构成的蒸发盆地系。中侏罗世，由羌塘盆地向呵叻盆地方向相对海拔逐渐升高，呵叻盆地高出海平面，地层缺失或发育陆相沉积，其余盆地均发育海相沉积，构成了蒸发盆地系的预备盆地。这种认识主要源自对各盆地相同时代沉积岩主岩性的对比分析。羌塘盆地中侏罗统主要由色哇组、布曲组和夏里组构成。其中，色哇组上部为泥晶灰岩、粉砂质泥岩夹细砂岩条带，下部为钙质页岩夹粉砂岩条带；布曲组主要为灰黑色巨厚生物碎屑灰岩、灰色厚层结晶灰岩夹生物碎屑灰岩及深灰色厚层白云质泥晶灰岩、角砾状灰岩；夏里组主要为灰绿色和紫色砂岩、粉砂岩、钙质或泥质粉砂岩，夹数层鲕状灰岩和泥岩，局部夹细砾岩及石膏层，它们体现了典型海相地层的特征（白生海，1989；吴珍汉等，2014）。昌都盆地中侏罗统有广泛的海相沉积，可与羌塘盆地、兰坪—思茅盆地相连接，其中的东大桥组含海相介壳类化石（杜德勋等，1997）。兰坪—思茅盆地中侏罗统在盆地西部为和平乡组、在盆地东部为花开佐组，含海相双壳类化石，主要岩性呈红色、灰色相间的“杂色岩组”，下部以红色碎屑岩为主，向上灰绿色层增加，顶部为盐溶泥砾岩、膏盐岩（曲一华等，1998）。呵叻盆地侏罗系的划分还存在争议。有学者认为呵叻盆地侏罗系缺失中—下侏罗统（王俊等，2017）；有学者认为呵叻盆地侏罗系缺失下侏罗统，中—上侏罗统由 Phu Kradung 组构成（沙金庚等，2012）；还有学者认为呵叻盆地侏罗系上、中、下统分别由 Sao Khua 组、Phra Wihan 组、Phu Kradung 组构成（吴根耀，1991；Tabakh et al.，1998；胡双全等，2017）。尽管这些划分方案存在差异，但是它们有一个共同的特征：中侏罗世的呵叻盆地为陆相环境——因为除了地层缺失这一特征外，上述各岩性组均包含恐龙化石或恐龙足迹化石（沙金庚等，2012）。

晚侏罗世—早白垩世早期，海水逐渐向羌塘盆地方向退缩，其他盆地均为陆相沉积，而羌塘盆地直至早白垩世早期还发育海相地层（Tabakh et al.，1998；朱丽霞等，2012）。随着中特提斯的闭合，羌塘盆地逐渐隆升，当其相对高度大于同一蒸发盆地系的其他盆地时，残留的中侏罗世古海水发生侧向迁移。早白垩世晚期约 110Ma，首先在思茅盆地蒸发浓缩析出钾石盐；随着中特提斯闭合影响的增强，蒸发盆地系之间的相对高差进一步增大，残留古海水继续侧向迁移，晚白垩世早期 92～65.5Ma（呵叻盆地盐岩夹层的碎屑岩地层古地磁年龄据 Zhang et al.，2018）在呵叻盆地蒸发浓缩析出钾石盐、钾镁盐（图 12）。

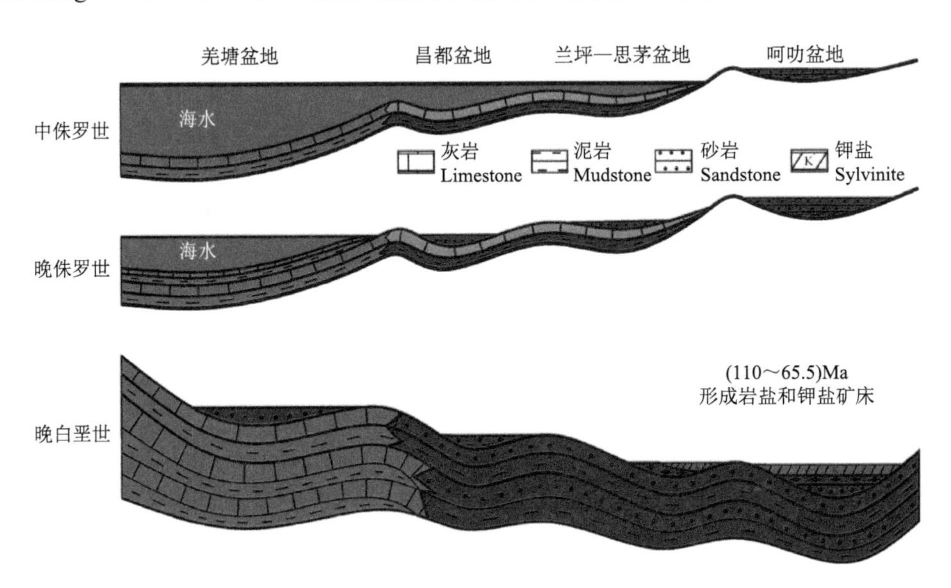

图 12 中侏罗世海水侧向迁移成钾示意图

Fig. 12 Schematic diagram of deposition of potassium showing that material source originated from Middle Jurassic water.

中侏罗世，思茅盆地处于上述蒸发盆地系边缘，利于发育潟湖环境，海水也可能直接在近岸封闭环境蒸发浓缩形成盐岩。例如，勐野井钾盐矿附近的 MK-1 井在 2253～2323m 深度段的中侏罗统花开佐组发现有石盐沉积，并见部分钾石盐充填在石盐裂隙中（郑绵平等，2014）。

5 结论

通过对思茅盆地已有的蒸发岩样品化学成分和硫同位素地球化学特征分析，本项研究对盆地内蒸发岩的成盐物质来源、成盐时代和成盐模式取得了以下几点初步认识。

（1）思茅盆地蒸发岩的物质来源主要为海水。火山热液和陆源淡水对蒸发岩的物质组成也有影响，但是对其中硫同位素影响较大的是陆源淡水。

（2）海水和陆源淡水对蒸发盆地的补给交替进行。在一级周期上，海水补给的次数有两次；在二级周期上，海水补给盆地的次数至少为七次。

（3）海相硬石膏硫同位素年代学意义揭示思茅盆地蒸发岩的物源可能为中侏罗世或晚白垩世的海水。根据相邻盆地的岩性特征、构造演化特征和钾盐成矿时代差异，提出了中侏罗世残留海水侧向迁移的成盐模式。

参考文献

白生海. 1989. 青海西南部海相侏罗纪地层新认识[J]. 地质论评, 35 (6): 529-536.

陈跃昆, 廖宗廷, 魏志红, 李明辉. 2004. 兰坪-思茅中生代盆地的特征及构造演化[J]. 石油实验地质, 26 (3): 219-222, 228.

杜德勋, 罗建宁, 李兴振. 1997. 昌都地块沉积演化与古地理[J]. 岩相古地理, 17 (4): 1-17.

樊启顺, 马海州, 谭红兵, 李廷伟. 2009. 柴达木盆地西部油田卤水的硫同位素地球化学特征[J]. 矿物岩石地球化学通报, 28 (2): 137-142.

胡双全, 杜贯超, 甄胜利. 2017. 泰国呵叻盆地二叠系 Pha Nok Khao 组碳酸盐岩沉积相特征[J]. 沉积学报, 35 (4): 789-796.

黄建国, 刘世万. 1990. 四川盆地东部海相三叠系稳定同位素地球化学及地质解释[J]. 岩相古地理, 4 (4): 16-25.

黄作良, 莫珉, 祖恩东. 1996. 辽宁砖庙矿区硼矿床硬石膏的发现及成因讨论[J]. 现代地质, 10 (3): 350-355.

李延河, 谢桂青, 段超, 韩丹, 王成玉. 2013. 膏盐层在矽卡岩型铁矿成矿中的作用[J]. 地质学报, 87 (9): 1324-1334.

梁汉东, 丁悌平. 2004. 中国煤山剖面二叠/三叠系事件界线地层中石膏的负硫同位素异常[J]. 地球学报, 25 (1): 33-37.

廖宗廷, 陈跃昆. 2005. 兰坪-思茅盆地原型的性质及演化[J]. 同济大学学报 (自然科学版), 33 (11): 1527-1531.

刘汉彬, 金贵善, 李军杰, 韩娟, 张建锋, 张佳, 钟芳文, 郭东侨. 2013. 铀矿地质样品的稳定同位素组成测试方法[J]. 世界核地质科学, 30 (3): 174-179.

牛新生, 刘喜方, 陈文西. 2014. 西藏北羌塘盆地多格错仁地区盐泉水化学特征及其物质来源[J]. 地质学报, 88 (6): 1003-1010.

潘桂棠, 肖庆辉, 陆松年, 邓晋福, 冯益民, 张克信, 张智勇, 王方国, 邢光福, 郝国杰, 冯艳芳. 2009. 中国大地构造单元划分[J]. 中国地质, 36 (1): 1-28.

彭立才, 杨平, 濮人龙. 1999. 陆相咸化湖泊沉积硫酸盐岩硫同位素组成及其地质意义[J]. 矿物岩石地球化学通报, 18 (2): 31-34.

曲一华, 袁品泉, 帅开业, 张瑛, 蔡克勤, 贾疏源, 陈朝德. 1998. 兰坪-思茅盆地钾盐成矿规律及预测[M]. 北京: 地质出版社.

沙金庚, MEESOOK A, NGUYEN X K. 2012. 泰国、老挝南部和越南中部的非海相白垩纪双壳类生物地层学[J]. 地层学杂志, 36 (2): 382-399.

石海岩, 马海州, 苗卫良, 李永寿, 张西营. 2014. 云南思茅盆地江城上白垩统勐野井组稀土微量元素特征及地质意义[J]. 地球化学, 43 (4): 415-427.

史忠生, 陈开远, 史军, 何胡军, 柳保军. 2004. 东濮盐湖古近系硫酸盐硫同位素组成及地质意义[J]. 石油勘探与开发, 31 (6): 44-46.

谭富文. 2002. 云南思茅三叠纪弧后前陆盆地的沉积特征[J]. 沉积学报, 20 (4): 560-567.

王春连, 刘成林, 徐海明, 王立成, 沈立建. 2013. 湖北江陵凹陷古新统沙市组四段硫酸盐硫同位素组成及其地质意义[J]. 吉林大学学报 (地球科学版), 43 (3): 691-703.

王俊, 鲍志东, 吴义平, 杨益春, 何陵沅. 2017. 泰国呵叻高原盆地油气成藏组合与资源潜力[J]. 西南石油大学学报 (自然科学版), 39 (4): 57-70.

吴根耀. 1991. 中国滇西-泰国地区侏罗纪—第四纪盆地发育及其对比研究[J]. 地质科学, (4): 359-368.

吴珍汉, 高锐, 卢占武, 叶培盛, 陆露, 殷裁云. 2014. 羌塘盆地结构构造与油气勘探方向[J]. 地质学报, 88 (6): 1130-1144.

许建新. 2008. 云南勐野井钾盐矿床地球化学与成因研究[D]. 西宁: 中国科学院盐湖研究所.

杨尖絮, 尹宏伟, 张震, 郑绵平. 2013. 滇西兰坪-思茅盆地成钾地质条件分析[J]. 大地构造与成矿学, 37 (4): 633-640.

殷鸿福, 吴顺宝, 杜远生, 彭元桥. 1999. 华南是特提斯多岛洋体系的一部分[J]. 地球科学——中国地质大学学报, 24 (1): 1-12.

尹福光, 潘桂棠, 万方, 李兴振, 王方国. 2006. 西南 "三江" 造山带大地构造相[J]. 沉积与特提斯地质, 26 (4): 33-39.

张华, 刘成林, 王立成, 方小敏. 2014. 老挝他曲盆地钾盐矿床蒸发岩硫同位素特征及成钾指示意义[J]. 地质论评, 60 (4): 851-857.

张同钢, 储雪蕾, 冯连君, 张启锐, 郭建平. 2003. 新元古代 "雪球" 事件对海水碳、硫同位素组成的影响[J]. 地球学报, 24 (6): 487-493.

张同钢, 储雪蕾, 张启锐, 冯连君, 霍卫国. 2004. 扬子地台灯影组碳酸盐岩中的硫和碳同位素记录[J]. 岩石学报, 20 (3): 717-724.

郑绵平, 张震, 尹宏伟, 谭筱虹, 于常青, 施林峰, 张雪飞, 杨尖絮, 焦建, 武国朋. 2014. 云南江城勐野井钾盐成矿新认识[J]. 地球学报, 35 (1): 11-24.

郑智杰, 尹宏伟, 张震, 郑绵平, 杨尖絮. 2012. 云南江城勐野井盐类矿床 Sr 同位素特征及成盐物质来源分析[J]. 南京大学学报 (自然科学版), 48 (6): 719-727.

朱丽霞, 谭富文, 付修根, 陈明, 冯兴雷, 曾胜强. 2012. 北羌塘盆地晚中生代地层: 早白垩世海相地层的发现[J]. 沉积学报, 30 (5): 825-833.

References

BAI Hai-sheng. 1989. New recognition of the marine Jurassic strata in southwestern Qinghai[J]. Geological Review, 35 (6): 529-536 (in Chinese with English abstract).

CANFIELD D E, THAMDRUP B. 1994. The Production of 34S-Depleted Sulfide during bacterial disproportionation of elemental sulfur[J]. Science, 266: 1973-1975.

CHEN Yao-kun, LIAO Zong-ting, WEI Zhi-hong, LI Ming-hui. 2004. Characteristics and tectonic evolution of the Lanping-Simao mesozoic basin[J]. Petroleum Geology & Experiment, 26 (3): 219-222, 228 (in Chinese with English abstract).

CLAYPOOL G E, HOLSER W T, KAPLAN I R, SAKAI H, ZAK I. 1980. The age curves of sulfur and oxygen isotopes in marine sulfate and their mutual interpretation[J]. Chemical Geology, 28: 199-260.

COHEN K M, FINNEY S C, GIBBARD P L, FAN J X. 2013. The ICS International Chronostratigraphic Chart[J]. Episodes, 36: 199-204.

CORTECCI G, REYES E, BERTI G, CASATI P. 1981. Sulfur and oxygen isotopes in Italian marine sulfates of Permian and Triassic ages[J]. Chemical Geology, 34: 65-79.

DU De-xun, LUO Jian-ning, LI Xing-zhen. 1997. Sedimentary evolution and palaeogeography of the Qamdo Block in Xizang[J]. Journal of Palaeogeography, 17 (4): 1-17 (in Chinese with English abstract).

FAN Qi-shun, MA Hai-zhou, TAN Hong-bing, LI Ting-wei. 2009. Geochemistry characteristics of sulfur isotope in oilfield brine of the Western Qaidam Basin[J]. Bulletin of Mineralogy, Petrology and Geochemistry, 28 (2): 137-142 (in Chinese with English abstract).

HOLSER W T, KAPLAN I R. 1966. Isotope geochemistry of sedimentary sulfates[J]. Chemical Geology, 1: 93-135.

HU Shuang-quan, DU Gui-chao, ZHEN Sheng-li. 2017. Facies characteristics of a permian carbonate platform of Pha Nok Khao Formation from Khroat Basin, Thailand[J]. Acta Sedimentologica Sinica, 35 (4): 789-796 (in Chinese with English abstract).

HUANG Jian-guo, LIU Shi-wan. 1990. Stable isotope geochemistry and geological interpretation of Triassic marine carbonate rocks in eastern Sichuan Basin[J]. Journal of Palaeogeography, 4 (4): 16-25 (in Chinese with English abstract).

HUANG Zuo-liang, MO Min, ZU En-dong. 1996. Discovery and geological significances of anhydrite on boron deposits in Zhuan Miao Area, Liaoning[J]. Geoscience, 10 (3): 350-355 (in Chinese with English abstract).

LI Ming-hui, YAN Mao-du, FANG Xiao-min, ZHANG Zeng-jie, WANG Zheng-rong, SUN Shu-rui, LI Jiao, LIU Xiao-ming. 2018. Origins of the Mid-Cretaceous evaporite deposits of the Sakhon Nakhon Basin in Laos: Evidence from the stable isotopes of halite[J]. Journal of Geochemical Exploration, 184: 209-222.

LI Yan-he, XIE Gui-qing, DUAN Chao, HAN Dan, WANG Cheng-yu. 2013. Effect of sulfate evaporate salt layer over the formation of Skarn-Type iron ores[J]. Acta Geologica Sinica, 87 (9): 1324-1334 (in Chinese with English abstract).

LIANG Han-dong, DING Ti-ping. 2004. Evidence of extremely light gypsum from the Permian-Triassic (P/T) event boundary at the Meishan Section of south China[J]. Acta Geoscientica Sinica, 25 (1): 33-37 (in Chinese with English abstract).

LIAO Zong-ting, CHEN Yao-kun. 2005. Nature and evolution of Lanping-Simao Basin prototype[J]. Journal of Tongji University (Natural Science), 33 (11): 1527-1531 (in Chinese with English abstract).

LIN Yong-jie, ZHENG Mian-ping, YE Chuan-yong. 2017. Hydromagnesite precipitation in the Alkaline Lake Dujiali, central Qinghai-Tibetan Plateau: Constraints on hydromagnesite precipitation from hydrochemistry and stable isotopes[J]. Ap plied Geochemistry, 78: 139-148.

LIU Han-bin, JIN Gui-shan, LI Jun-jie, HAN Juan, ZHANG Jian-feng, ZHANG Jia, ZHONG Fang-wen, GUO Dong-qiao. 2013. Determination of stable isotope composition in uranium geological samples[J]. World Nuclear Geoscience, 30 (3): 174-179 (in Chinese with English abstract).

METCALFE I. 2013. Gondwana dispersion and asian accretion: Tectonic and palaeogeographic evolution of eastern Tethys[J]. Journal of Asian Earth Sciences, 66: 1-33.

MIAO Zhong-ying. 2018. New Sr isotope evidence to support the material source of the Mengyejing Potash Deposit in the Simao Basin from ancient marine halite or residual sea[J]. Acta Geologica Sinica (English Edition), 92 (2): 866-867.

NIU Xin-sheng, LIU Xi-fang, CHEN Wen-xi. 2014. Hydrochemical characteristic and origin for salt springs water in Dogai Coring Area of North Qiangtang Basin, Tibet[J]. Acta Geologica Sinica, 88 (6): 1003-1010 (in Chinese with English abstract).

PAN Gui-tang, XIAO Qing-hui, LU Song-nian, DENG Jin-fu, FENG Yi-min, ZHANG Ke-xin, ZHANG Zhi-yong, WANG Fang-guo, XING Guang-fu, HAO Guo-jie, FENG Yan-fang. 2009. Subdivision of tectonic units in China[J]. Chinese Geology, 36 (1): 1-28 (in Chinese with English abstract).

PAYTAN A, KASTNER M, CAMPBELL D, THIEMENS M H. 2004. Seawater sulfur isotope fluctuations in the Cretaceous[J]. Science, 304: 1663-1665.

PENG Li-cai, YANG Ping, PU Ren-long. 1999. The sulfur isotope composition of sulfate rock deposited in continental brackish lakes and its geological significance[J]. Bulletin of Mineralogy Petrology and Geochemistry, 18 (2): 31-34 (in Chinese with English abstract).

QU Yi-hua, YUAN Pin-quan, SHUAI Kai-ye, ZHANG Ying, CAI Ke-qin, JIA Shu-yuan, CHEN Chao-de. 1998. Potash-forming Rules and Prospect of Lower Tertiary in Lanping-Simao Basin, Yunnan[M]. Beijing: Geology Publishing House (in Chinese).

SHA Jin-geng, MEESOOK A, NGUYEN X K. 2012. Non-marine Cretaceous bivalve biostratigraphy of Thailand, southern Lao PDR and central Vietnam[J]. Journal of Stratigraphy, 36 (2): 382-399 (in Chinese with English abstract).

SHI Hai-yan, MA Hai-zhou, MIAO Wei-liang, LI Yong-shou, ZHANG Xi-ying. 2014. Characteristics and geological significances of rare earth and trace elements from Upper Cretaceous Mengyejing Formation of Simao Basin in Jiangcheng County, Yunnan Province[J]. Geochimica, 43 (4): 415-427 (in Chinese with English abstract).

SHI Zhong-sheng, CHEN Kai-yuan, SHI Jun, HE Hu-jun, LIU Bao-jun. 2004. Sulfur isotopic composition and its geological significance of the Paleogene sulfate rock deposited in Dongpu Depression[J]. Petroleum Exploration and Development, 31 (6): 44-46 (in Chinese with English abstract).

TABAKH M E, SCHREIBER B C, UTHA-AROON C, COSHELL L, WARREN J K. 1998. Diagenetic origin of basal anhydrite in the Cretaceous Maha

Sarakham salt: Khorat Plateau, NE Thailand[J]. Sedimentology, 45: 579-594.

TABAKH M E, UTHA-AROON C, SCHREIBER B C. 1999. Sedimentology of the Cretaceous Maha Sarakham evaporites in the Khorat Plateau of northeastern Thailand[J]. Sedimentary Geology, 123: 31-62.

TAN Fu-wen. 2002. The sedimentary characteristics of Simao Triassic Rear Arc Foreland Basin, Yunnan Province[J]. Acta Sedimentologica Sinica, 20 (4): 560-567 (in Chinese with English abstract).

WANG Chun-lian, LIU Cheng-lin, XU Hai-ming, WANG Li-cheng, SHEN Li-jian. 2013. Sulfur isotopic composition of sulfate and its geological significance of member 4 of palaeocene Shashi formation in Jiangling Depression of Hubei Province[J]. Journal of Jilin University (Earth Science Edition), 43 (3): 691-703 (in Chinese with English abstract).

WANG Jun, BAO Zhi-dong, WU Yi-ping, YANG Yi-chun, HE Ling-yuan. 2017. Play and hydrocarbon potential of the Khorat Plateau Basin in Thailand[J]. Journal of Southwest Petroleum University (Science & Technology Edition), 39 (4): 57-70 (in Chinese with English abstract).

WANG Li-cheng, LIU Cheng-lin, FEI Ming-ming, SHEN Li-jian, ZHANG Hua, ZHAO Yan-jun. 2015. First SHRIMP U-Pb zircon ages of the potash-bearing Mengyejing Formation, Simao Basin, southwestern Yunnan, China[J]. Cretaceous Research, 52: 238-250.

WU Gen-yao. 1991. Development of Jurassic-Quaternary basins in western Yunnan, China, and Thailand: A comparative study[J]. Chinese Journal of Geology, (4): 359-368 (in Chinese with English abstract).

WU Zhen-han, GAO Rui, LU Zhan-wu, YE Pei-sheng, LU Lu, YIN Cai-yun. 2014. Structures of the Qiangtang Basin and its significance to oil-gas exploration[J]. Acta Geologica Sinica, 88 (6): 1130-1144 (in Chinese with English abstract).

XU Jian-xin. 2008. Geochemistry and genesis of Mengyejing Potash Deposits, Yunnan[D]. Xining: Qinghai Institute of Salt Lakes, Chinese Academy of Sciences (in Chinese with English abstract).

YANG Jian-xu, YIN Hong-wei, ZHANG Zhen, ZHENG Mian-ping. 2013. Geologic settings of the potassium formations in the Lanping-Simao Basin, Yunnan Province[J]. Geotectonica et Metallogenia, 37 (4): 633-640 (in Chinese with English abstract).

YIN Fu-guang, PAN Gui-tang, WAN Fang, LI Xing-zhen, WANG Fang-guo. 2006. Tectonic facies along the Nujiang-Lancangjiang-Jinshajiang orogenic belt in southwestern China[J]. Sedimentary Geology and Tethyan Geology, 26 (4): 33-39 (in Chinese with English abstract).

YIN Hong-fu, WU Shun-bao, DU Yuan-sheng, PENG Yuan-qiao. 1999. South China defined as part of tethyan archipelagic ocean system[J]. Earth Science-Journal of China University of Geosciences, 24 (1): 1-12 (in Chinese with English abstract).

ZHANG Da-wen, YAN Mao-du, FANG Xiao-min, YANG Yi-bo, ZHANG Tao, ZAN Jin-bo, ZHANG Wei-lin, LIU Cheng-lin, YANG Qian. 2018. Magnetostratigraphic study of the potash-bearing strata from drilling core ZK2893 in the Sakhon Nakhon Basin, eastern Khorat Plateau[J]. Palaeogeography, Palaeoclimatology, Palaeoecology, 489: 40-51.

ZHANG Hua, LIU Cheng-lin, WANG Li-cheng, FANG Xiao-min. 2014. Characteristics of evaporites sulfur isotope from potash deposit in Thakhek Basin, Laos, and its implication for potash formation[J]. Geological Review, 60 (4): 851-857 (in Chinese with English abstract).

ZHANG Tong-gang, CHU Xue-lei, FENG Lian-jun, ZHANG Qi-rui, GUO Jian-ping. 2003. The effects of the Neoproterozoic Snowball Earth on carbon and sulfur isotopic compositions in seawater[J]. Acta Geoscientia Sinica, 24 (6): 487-493 (in Chinese with English abstract).

ZHANG Tong-gang, CHU Xue-lei, ZHANG Qi-rui, FENG Lian-jun, HUO Wei-guo. 2004. The sulfur and carbon isotopic records in carbonates of the Dengying Formation in the Yangtze Platform, China[J]. Acta Petrologica Sinica, 20 (3): 717-724 (in Chinese with English abstract).

ZHENG Mian-ping, ZHANG Zhen, YIN Hong-wei, TAN Xiao-hong, YU Chang-qing, SHI Lin-feng, ZHANG Xue-fei, YANG Jian-xu, JIAO Jian, WU Guo-peng. 2014. A New Viewpoint concerning the Formation of the Mengyejing Potash Deposit in Jiangcheng, Yunnan[J]. Acta Geoscientica Sinica, 35 (1): 11-24 (in Chinese with English abstract).

ZHENG Zhi-jie, YIN Hong-wei, ZHANG Zhen, ZHENG Mian-ping, YANG Jian-xu. 2012. Strontium isotope characteristics and the origin of salt deposits in Mengyejing, Yunnan province, SW China[J]. Journal of Nanjing University: Nat Sci Ed, 48 (6): 719-727 (in Chinese with English abstract).

ZHU Li-xia, TAN Fu-wen, FU Xiu-gen, CHEN Ming, FENG Xing-lei, ZENG Sheng-qiang. 2012. Strata of the Late Mesozoic in the north of Qiangtang Basin: A discovery of the Early Cretaceous marine strata[J]. Acta Sedimentologica Sinica, 30 (5): 825-833 (in Chinese with English abstract).

Sulfur isotope geochemistry of the lower cretaceous evaporite and its significance for potash mineralization in the Simao Basin, southwest China

Zhong-ying Miao[1], Mian-ping Zheng[1], Xue-fei Zhang[1], Zhen Zhang[1], Jian-hua Liu[1], Yun-zhi Gao[2] and Xue-feng Zhai[3]

1. MNR Key Laboratory of Saline Lake Resources and Environments, Institute of Mineral Resources, Chinese Academy

of Geological Sciences, Beijing 100037, China
2. China University of Geosciences (Beijing), Beijing 100083, China
3. No.4 Hydrogeological Party of China National Administration of Coal Geology, Handan, Hebei 056006, China

Abstract Scientists believe that primary material sources of the Mengyejing potash deposit are seawater but there are still disputes about the metallogenic epoch and metallogenic model. In addition, the influence of terrestrial freshwater on the composition of evaporite is still lacking of discussion. In order to explore the material source of evaporite, the effect of freshwater on the evaporite composition, the age of material source, the metallogenic epoch, and the probable potassium metallogenic model, the authors analyzed composition and sulfur isotope of 27 evaporite samples from well L2 in the Simao basin, and also considered the published sulfur isotope data of evaporite from adjacent basins. Some conclusions have been reached: (1) Material sources of the evaporite in the Simao basin were recharged by freshwater and volcanic hydrothermal fluids except seawater. The data of sulfur isotope of the evaporite were significantly lower than that of other marine samples in the same geological epoch, it was caused by freshwater flowing into basin. (2) Seawater flowed northwestward into the Simao basin. There existed two times of seawater recharge in the first-order period and at least seven times in the second-order period. (3) The epoch of material source seawater is the Middle Jurassic, and the salting out age is the late Early Cretaceous. The probable potassium mineralization model was as follows: "Middle Jurassic seawater of lateral migration evaporated and crystallized in the late Early Cretaceous". These results are of great significance in explaining the abnormally low sulfur isotope values and high sulfur isotope values of marine evaporites in the Simao basin and its adjacent areas, which are comparable to those of Middle Jurassic seawater, and the absence of "carbonate facies and sulphate facies" in potassium mineralization.

Keywords Simao basin; Evaporite; Potash; Sulfur isotope; Mineralization; Geochemistry

川东北宣汉地区新型杂卤石钾盐矿的地球化学特征及其意义*

商雯君[1,2]，张永生[1,2]，李空[3]，邢恩袁[1,2]，桂宝玲[1,2]，彭渊[1,2]，赵海彤[1,2]

1. 中国地质科学院矿产资源研究所，北京 100037
2. 国土资源部盐湖资源与环境重点实验室，北京 100037
3. 中国人民武装警察部队黄金部队第五支队，陕西西安 710199

摘要 新型杂卤石钾盐矿分布于川东北宣汉地区早—中三叠世蒸发岩层中，以杂卤石碎屑不均匀地分布于石盐层为特征。笔者对 HC3 井新型杂卤石钾盐矿段进行连续取样，并进行水溶化学实验，根据其可溶部分的主量、微量元素特征发现：①溶液中的 (K + Mg)/Ca 值平均为 1.24，略小于杂卤石中的 (K + Mg)/Ca 值（1.27），整体损失率为 2.36%，远小于 $CaSO_4$ 组分在杂卤石中的含量（48%），表明杂卤石在石盐基质溶解形成的 NaCl 溶液中有较好的溶解度，有利于溶采；②$Br\times10^3/Cl$ 值分布在 0.16～0.44，显示新型杂卤石钾盐矿的蒸发阶段位于正常石盐阶段内，整体较高，蒸发阶段稳定无剧烈波动，仅顶部出现快速淡化，结合前人对杂卤石的成因研究，笔者认为石盐层不具备形成杂卤石的条件，杂卤石碎屑是一种“外来物”；③含硬石膏碎屑中，(K + Mg)/Ca 值随着 $Br\times10^3/Cl$ 值的增加而增高，表明碎屑中杂卤石的含量随成盐卤水浓缩程度的升高而增加。

关键词 地球化学；杂卤石；新型杂卤石钾盐矿；矿床成因；宣汉地区

杂卤石是一种常见的含钾、镁、钙的硫酸盐矿物（$K_2SO_4\cdot MgSO_4\cdot 2CaSO_4\cdot 2H_2O$），由德国化学家 Stromeyer 于 1818 年首次发现，通常出现在蒸发沉积岩序列（Stromeyer，1818），是可溶性钾盐的重要替代资源。目前，在中国四川盆地发现大量杂卤石岩，主要呈透镜状、似层状或层状赋存于膏岩层中（黄宣镇，1996），仅川东地区的杂卤石（折合 K_2O）远景资源量就超过百亿吨（金锋，1989）。但由于杂卤石和硬石膏较难溶于水，加之四川盆地杂卤石岩大多埋深较深（目前仅农乐地区发现浅层杂卤石）、盆地构造复杂，使其难以开发，因此杂卤石长期仅作为寻找钾盐矿床的重要标志和线索，并未得到很好的利用。近期，在川东北宣汉地区的钻井中发现大量赋存于盐层中的碎屑颗粒状杂卤石，基于石盐易溶的特性，这类杂卤石可利用溶采法开采，具有较大的经济潜力。盐层中的杂卤石碎屑含量约 20%，整体 $w(K)$高达 10%，超过钾的工业品位，被定名为新型杂卤石钾盐矿（郑绵平等，2018）。本文利用 HC3 井中新型杂卤石钾盐矿段的水溶化学主量、微量元素实测分析数据，探讨了杂卤石的可利用性和沉积阶段，并浅析其成因。

1 区域地质

1.1 四川盆地早—中三叠世成盐背景

三叠纪早期，随着扬子地块向北漂移与欧亚大陆聚合，周边山系和海隆陆续抬升，中部基地相对下降，四川盆地的雏形就此形成，在此过程中四川盆地总体上处于浅水海退环境，其间发生多次海退海进事件，形成了大量蒸发岩与碳酸岩的交互沉积层，分布面积约达 $1.8\times10^5km^2$（郑绵平等，2010），至中三叠世末，扬子地台与华北大陆拼接，四川盆地结束了海相沉积的历史（蔡克勤等，1986），整个过程中，四川盆地经历了从广海盆地相—台棚浅海相—局限海相—潟湖相—盐湖相—湖泊—三角洲—河流沉积的演化过程。早—中三叠世，局限盆地环境和干旱的气候使四川盆地沉积了厚达 500～1700m 的膏盐层，杂卤石广泛分布其中（吴应林等，1983；李亚文等，1998；林耀庭等，1998，2004），共形成 6 个成盐期，分别是早三叠世嘉陵江组二段（T_1j^2）、四段（T_1j^4）、五段至中三叠世雷口坡组一段一亚段（T_1j^5-T_2l^{1-1}），雷口坡组一段三亚段（T_2l^{1-3}）

* 本文发表在：矿床地质，2020，39（2）：369-380

以及雷口坡组三段（T_2l^3）、四段（T_2l^4）等。主要含盐岩系地层为下三叠统嘉陵江组五段至中三叠统雷口坡组一段的底部（仲佳爱等，2018），岩性以白云岩、膏岩、盐岩为主，夹石灰岩、杂卤石。

研究区处于四川盆地的石膏-杂卤石沉积区[图 1（a）]。

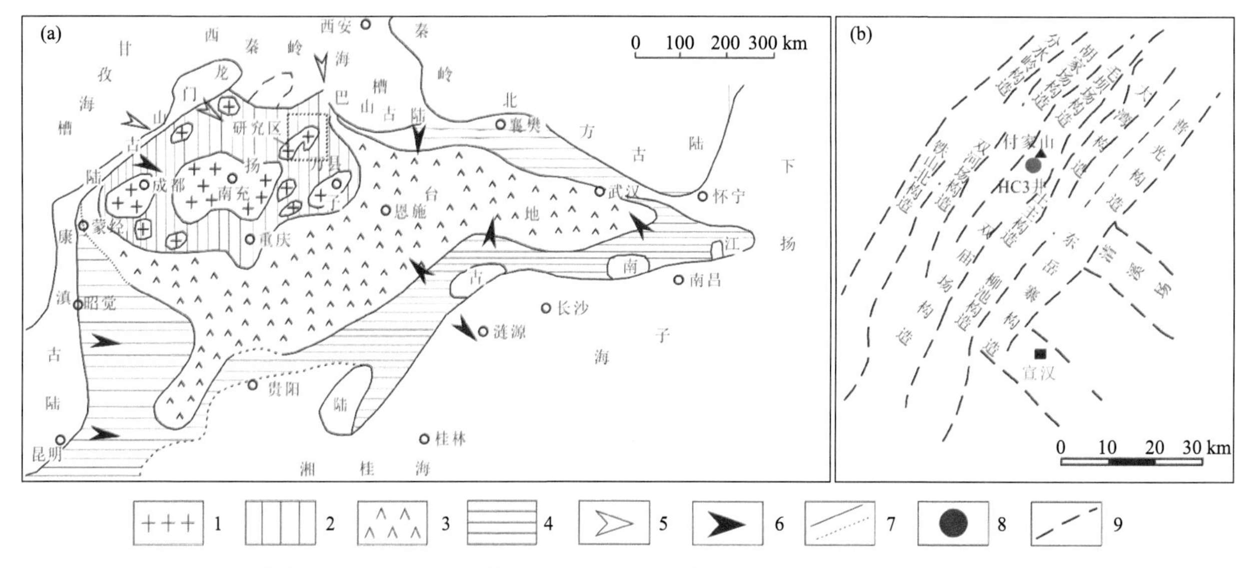

图 1　四川盆地早—中三叠世沉积特征（a，根据林耀庭，1995）和宣汉地区大地构造简图（b）

Fig. 1　Sedimentary characteristics of Sichuan basin (a, modified after Lin, 1995) and geotectonic schematic diagram of the Hanxuan area (b).

1-盐盆（体）；2-石膏-杂卤石沉积区；3-石膏沉积区；4-碳酸盐岩、碎屑岩沉积区；5-海水补给方向；6-陆源补给方向；7-实测或推测的岩相界线；8-HC3 井；9-断层

1-Salt basin (body); 2-Gypsum-polyhalite deposit area; 3-Gypsum deposit area; 4-Carbonate and clastic sedimentary areas; 5-The direction of seawater; 6-The direction of terrestrial water; 7-Measured or inferred lithofacies boundaries; 8-HC3 well; 9-Fault

1.2　HC3 井盐层特征

HC3 井位于黄金口构造带付家山背斜[图 1（b）]，钻遇地层层序正常，开孔地层为侏罗系中统上沙溪庙组，完钻地层为下三叠统嘉陵江组嘉五段二亚段（未钻穿）。

该井钻遇盐岩主要层位为嘉陵江组嘉五段二亚段（T_1j^{5-2}），雷口坡组未见石盐，盐岩有 3 层以上，厚度变化较大，由无色、暗红色、黑色石盐组成，晶体颗粒结晶较好。T_1j^{5-2} 埋深 3430～3476.3m，厚 46.3m，顶部为灰白色膏质盐岩与无色岩盐互层，与上覆地层雷一段底浅灰色含膏灰岩分层。中—下部分为新型杂卤石钾盐矿（石盐段岩心外侧可见覆盖有黄色泥壳的杂卤石，可与硬石膏明显区分），累计厚度 12m，主要由石盐岩及杂卤石碎屑组成（图 2）。

2　样品及测试方法

2.1　样品特征

样品为 HC3 井新型杂卤石钾盐矿，由四川达州市恒成能源（集团）有限责任公司提供。HC3 井新型杂卤石钾盐矿段主要由无色、暗红色、黑色石盐、杂卤石组成，含少量硬石膏，微量菱镁矿、黏土及石英。石盐基质主体部分为红色[图 3（a）]，顶底板逐渐过渡为黑色[图 3（b）]，厚分别为 1.0m 和 1.15m，向上（下）过渡为硬石膏岩，菱镁矿含量逐渐增加，有明显的变形揉皱现象。颜色较深的暗红色石盐基质含杂卤石碎屑成分较高，与红色较纯的石盐互层，部分杂卤石碎屑有定向排列的趋势，与地层倾向一致，表明在地层发生高角度倾斜的过程中，杂卤石碎屑颗粒已经存在于石盐层中。杂卤石碎屑呈粉屑状、蠕虫状、不规则团块状或条带状不均匀地分布于石盐基质中，碎屑含量超过 20%。杂卤石晶体直径变化较大[图 4（a）、（b）]，未见与石盐交代，常见杂卤石交代硬石膏[图 4（e）、（f）]，值得注意的是发现硬石膏交代并包裹杂卤石的现象[图 4（c）、（d）]。扫描电镜下，杂卤石与硬石膏不易区分，主要依靠能谱划分[图 4（g）、（h）]。

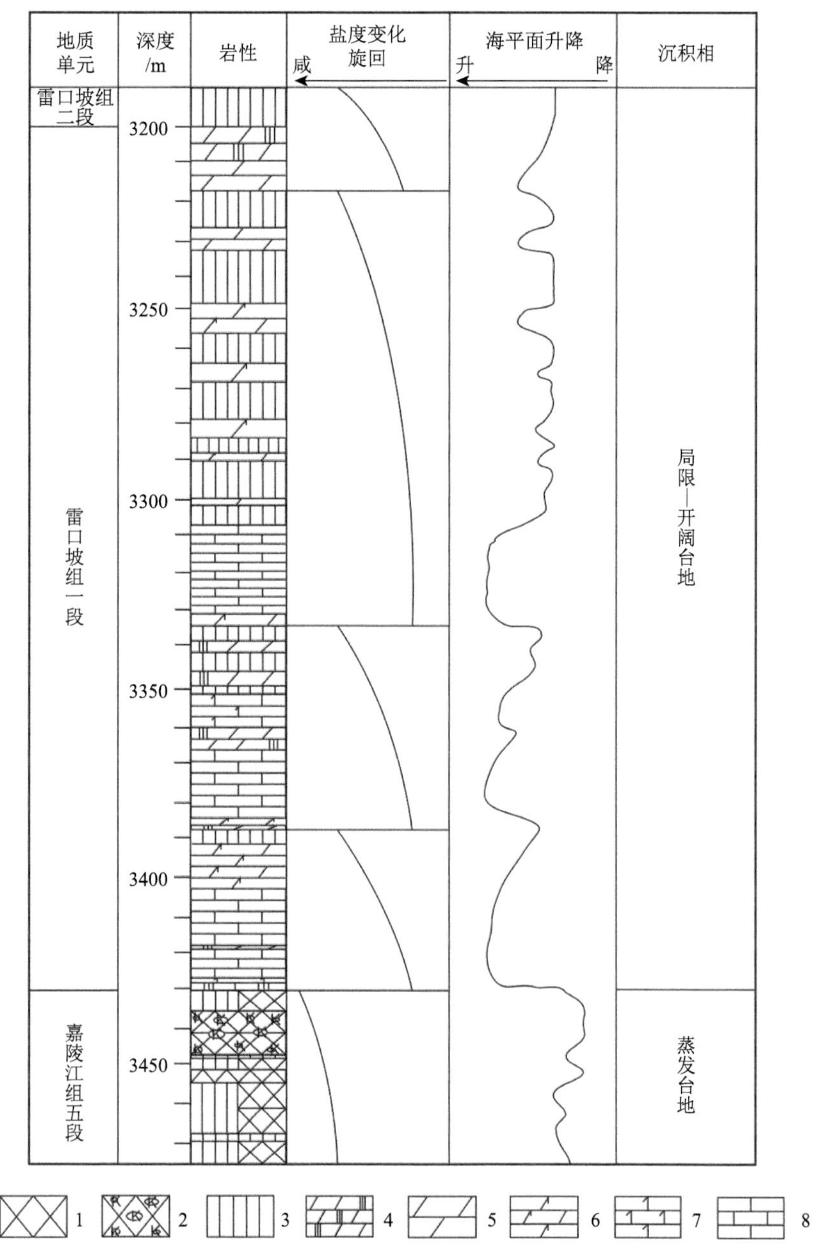

图 2　HC3 井嘉陵江组五段—雷口坡组一段岩性特征及新型杂卤石钾盐矿矿层特征

Fig. 2　Lithological characteristics of the 5th member of Jialingjiang Formation and the 1st member of Leikoupo Formation of HC3 well and characteristics of the new type of polyhalite potassium ore deposit.

1-石盐岩；2-新型杂卤石钾盐矿；3-硬石膏岩；4-（含膏）膏质云岩；5-白云岩；6-含灰云岩；7-云质灰岩；8-灰岩

1-Halite; 2-The new type of polyhalite potassium ore deposit; 3-Anhydrite; 4-Gypsiferous dolomite; 5-Dolomite; 6-Calcitic dolomites; 7-Dolomitic limestone; 8-Limestone

图 3　新型杂卤石钾盐矿岩心照片

Fig. 3　Photos of the new type of polyhalite potassium ore deposit.

（a）杂卤石赋存基质为“红盐”，样品埋深 3441m；（b）顶板黑色石盐，样品埋深 3433m；H-石盐；Pol-杂卤石

(a) The matrix in which polyhalite occurs as “red salt”, at 3441 m; (b) Roof black halite，at 3433; m H-Hailte; Pol-Polyhalite

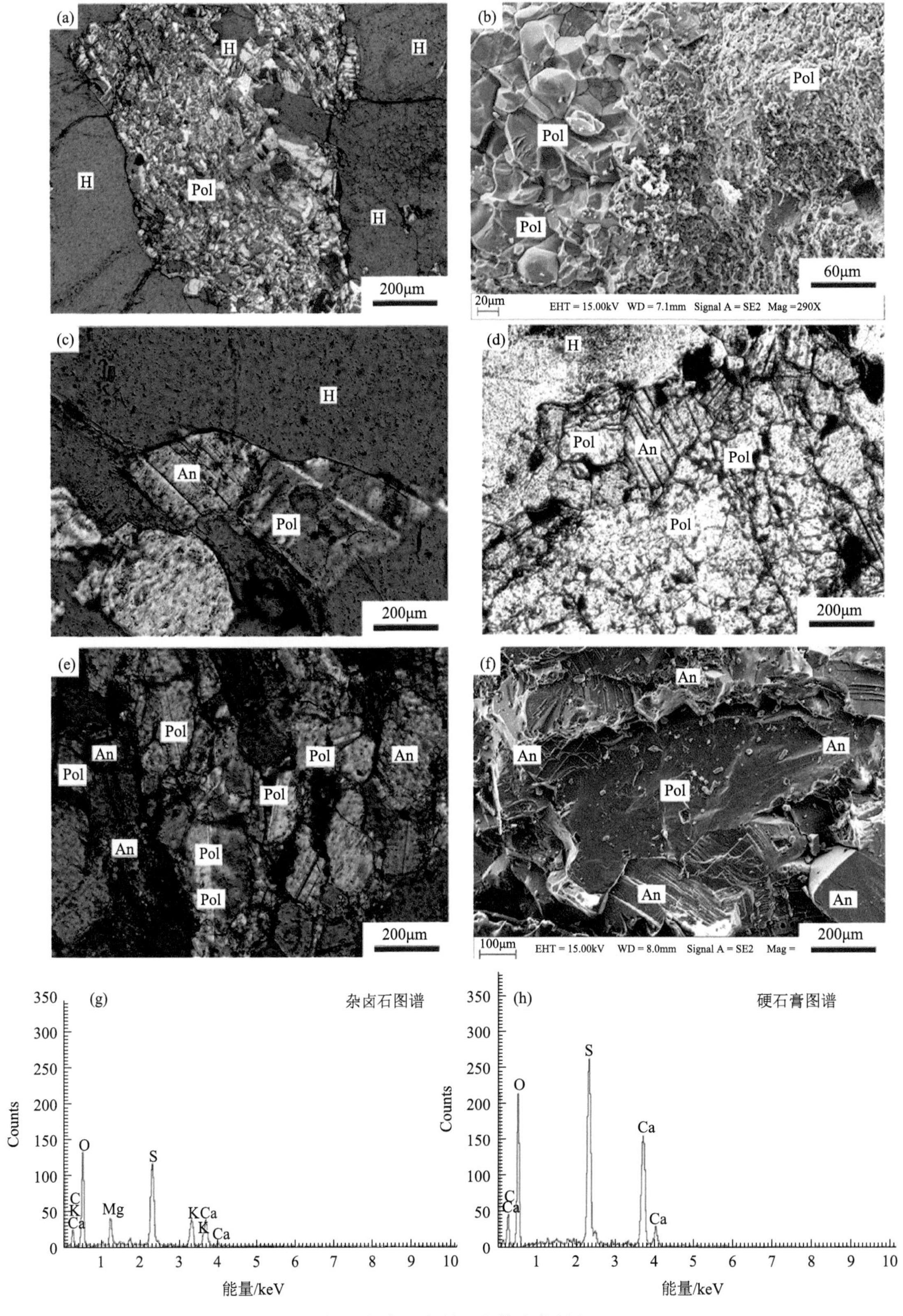

图 4　杂卤石和硬石膏的交代特征

Fig. 4　Metasomatic features of polyhalite and anhydrite.

（a）石盐基质中的杂卤石团块（正交偏光下）；（b）杂卤石颗粒大小不一致（扫描电镜下）；（c）杂卤石团块边缘硬石膏和杂卤石互相交代（正交偏光下）；（d）杂卤石团块边缘杂卤石交代硬石膏（偏光下）；（e）杂卤石被硬石膏交代并被硬石膏包围（正交偏光下）；（f）杂卤石被硬石膏交代并被硬石膏包围（扫描电镜下）；（g）杂卤石图谱（扫描电镜中）；（h）硬石膏图谱（扫描电镜中）；H-石盐；An-硬石膏；Pol-杂卤石

(a) Polyhalite mass in halite matrix (PLM); (b) The inconsistent particle sizes of polyhalite (SEM); (c) Metasomatism between anhydrite and polyhalite at the edges of polyhalite clumps (PLM); (d) The polyhalide of aggregation edge metasomatism anhydrite (crossed nicols); (e) Polyhalite replaced and surrounded by anhydrite (SEM); (f) Polyhalite replaced and surrounded by anhydrite (SEM); (g) Polyhalite spectrum; (h) Anhydrite spectrum; H-Halite; An-Anhydrite; Pol-Polyhalite

笔者对新型杂卤石钾盐矿每隔 20cm 进行连续取样，并在杂卤石碎屑的含量或石盐特征发生明显变化部位加密取样，在新鲜面敲下大约 1cm^3 大小的含杂卤石碎屑的石盐样品，共有 57 件样品进行水溶化学实验。由于杂卤石碎屑的含量在石盐基质中的含量并不均匀，因此选择敲样的部位，其杂卤石碎屑的含量应大致等于这 20cm 岩心中（或杂卤石碎屑的含量和石盐特征发生明显变化的岩心段）杂卤石碎屑在石盐基质中的平均含量。

2.2 测试方法

水溶化学实验在中国地质科学院盐湖资源与环境重点实验室完成，实验前将挑选样品磨成 200 目粉末，称重 1g 加 50ml 蒸馏水，充分搅拌，放置 12 个小时，定容 100ml。

测试方法：溶液中的 Li^+、Na^+、K^+、Rb^+、Cs^+、Mg^{2+}、Ca^{2+}、Sr^{2+}采用原子吸收分光光度法测量，Cl^-采用硝酸银容量法滴定；Br^-采用酚红比色法；碱度（HCO_3^-、OH^-、CO_3^{2-}）使用容量法分析；SO_4^{2-} 利用氯化钡重量法。

3 实验结果

3.1 主量元素特征

溶液主要由 Na^+和 Cl^-组成（除 H^+、OH^-外），除此之外，还含有较高的 K^+、Ca^{2+}、Mg^{2+}、SO_4^{2-} 等离子，其中，$w(K^+)$ 最小值为 0.09%，最大值为 7.9%，平均为 3.31%；$w(Mg^{2+})$ 最小值为 0.07%，最大值为 2.76%，平均为 0.9%；$w(Ca^{2+})$ 最小值为 0.45%，最大值为 5.80%，平均为 3.66%；$w(SO_4^{2-})$ 最小值为 1.63%，最大值为 32.64%，平均为 14.67%；$w(HCO_3^-)$ 最小值为 0.13%，最大值为 0.27%，平均为 0.23%（表 1）。

表 1 HC3 井岩盐水溶化学分析结果

Table 1 Rock salt water soluble chemical composition of HC3 well

序号	样品编号	w(B)/%							w(B)/10^{-6}		Br×10^3/Cl	水不溶物	(K + Mg)/Ca
		Na^+	K^+	Mg^{2+}	Ca^{2+}	Cl^-	SO_4^{2-}	HCO_3^-	$Br^-_{原始}$	$Br^-_{校正}$			
1	HC3-24-5	31.98	0.13	0.09	3.02	48.81	9.31	0.22	80.00	91.71	0.16	7.50	0.08
2	HC3-24-6	38.48	0.14	0.07	0.76	58.20	2.16	0.22	100.00	103.46	0.17	2.02	0.39
3	HC3-24-7-1	29.50	0.09	0.08	3.20	45.75	7.85	0.15	80.00	90.26	0.17	14.39	0.06
4	HC3-24-7-2	37.90	0.25	0.09	0.45	58.23	1.63	0.23	120.00	123.28	0.21	3.00	1.55
5	HC3-24-8-1	34.99	1.09	0.34	1.26	52.98	6.14	0.15	120.00	131.83	0.23	3.28	1.29
6	HC3-24-8-2	31.48	1.85	0.69	3.14	49.71	10.60	0.23	140.00	167.69	0.28	5.82	0.87
7	HC3-24-9-1	36.19	1.00	0.29	1.03	55.79	4.33	0.23	160.00	171.82	0.29	1.71	1.61
8	HC3-24-9-2	28.49	3.00	0.93	4.24	46.31	15.17	0.34	160.00	209.64	0.35	5.11	1.01
9	HC3-24-10-2	31.39	1.76	0.45	3.84	50.96	10.14	0.18	160.00	191.32	0.31	3.37	0.60
10	HC3-24-11-1	28.19	2.24	0.70	3.98	44.88	12.89	0.21	160.00	200.05	0.36	8.74	0.78
11	HC3-24-11-2	32.29	1.74	0.46	3.18	52.21	8.08	0.21	180.00	208.50	0.34	3.28	0.74
12	HC3-24-12	31.09	2.05	0.59	3.86	49	11.87	0.17	180.00	220.97	0.37	6.69	0.72
13	HC3-24-13	33.00	1.82	0.44	2.96	53.12	7.92	0.13	180.00	207.54	0.34	2.19	0.80
14	HC3-25-1-1	26.19	4.20	1.60	4.54	40.41	21.40	0.29	140.00	205.97	0.35	4.56	1.36
15	HC3-25-1-2	27.40	2.76	0.87	5.06	43.64	16.03	0.20	140.00	186.47	0.32	8.70	0.75
16	HC3-25-2	28.39	2.09	0.61	4.88	44.89	13.66	0.26	140.00	178.34	0.31	20.34	0.58
17	HC3-25-3-1	31.78	2.29	0.68	3.44	49.53	11.34	0.16	160.00	194.91	0.32	3.74	0.91
18	HC3-25-3-2	31.79	2.30	0.61	3.76	50.43	10.75	0.20	160.00	194.22	0.32	2.36	0.82
19	HC3-25-4	29.49	2.80	0.86	4.16	45.23	15.12	0.26	160.00	208.33	0.35	6.67	0.94

续表

序号	样品编号	w(B)/%							w(B)/10^{-6}		Br×10^3/Cl	水不溶物	(K + Mg)/Ca
		Na^+	K^+	Mg^{2+}	Ca^{2+}	Cl^-	SO_4^{2-}	HCO_3^-	$Br^-_{原始}$	$Br^-_{校正}$			
20	HC3-25-5	29.90	2.69	0.82	3.84	48.65	12.69	0.23	160.00	200.68	0.33	1.93	0.97
21	HC3-25-6-1	31.50	2.28	0.64	2.86	51.15	9.61	0.26	180.00	213.40	0.35	1.39	1.12
22	HC3-25-6-2	29.90	2.57	0.79	3.46	47.76	12.28	0.22	160.00	198.31	0.34	2.34	1.04
23	HC3-25-6-3	29.79	2.85	0.89	3.90	46.66	14.14	0.23	160.00	205.15	0.34	3.99	1.02
24	HC3-25-7	28.19	2.62	0.85	4.86	45.78	14.06	0.24	160.00	206.80	0.35	6.80	0.75
25	HC3-25-8	29.29	2.53	0.84	4.54	48.64	13.07	0.26	160.00	203.15	0.33	4.47	0.79
26	HC3-25-9-1	31.84	2.26	0.62	3.38	50.97	9.99	0.24	180.00	215.54	0.35	2.98	0.92
27	HC3-25-9-2	29.99	2.65	0.81	4.14	48.28	13.53	0.24	160.00	203.48	0.33	3.24	0.89
28	HC3-25-10-1	31.69	2.28	0.66	3.06	51.32	9.91	0.23	160.00	190.79	0.31	1.58	1.04
29	HC3-25-10-2	31.00	2.70	0.80	3.10	50.08	11.11	0.24	160.00	195.00	0.32	1.34	1.22
30	HC3-25-11-1	27.59	3.60	1.09	4.24	43.27	16.90	0.26	160.00	216.48	0.37	1.96	1.18
31	HC3-25-11-2	29.30	3.40	1.05	4.00	45.79	15.21	0.13	140.00	183.70	0.31	2.45	1.15
32	HC3-25-12-1	29.89	2.8	0.92	4.04	46.26	15.26	0.28	140.00	182.53	0.30	4.5	0.99
33	HC3-25-12-2	25.49	5.4	1.6	4.2	41.1	20.59	0.23	140.00	205.94	0.34	3.07	1.76
34	HC3-25-13	24.79	5.6	1.62	4.4	40.21	21.91	0.22	140.00	211.32	0.35	2.21	1.73
35	HC3-25-14	30.99	2.6	0.84	3	48.21	12.34	0.22	160.00	197.53	0.33	2.82	1.24
36	HC3-25-15	20.39	6.7	2.14	4.78	32.37	26.82	0.24	140.00	236.01	0.43	6.94	1.95
37	HC3-26-2-1	24.49	5.2	1.42	4.68	39.67	20.99	0.24	140.00	207.50	0.35	4.76	1.49
38	HC3-26-2-2	14.7	7.7	2.76	5.58	22.77	32.64	0.22	100.00	195.69	0.44	14.02	1.95
39	HC3-26-4-1	25.59	4.2	1.18	4.6	40.38	18.5	0.26	120.00	168.40	0.30	6.67	1.24
40	HC3-26-4-2	26.19	3.5	0.88	3.92	42.51	14.88	0.26	140.00	182.86	0.33	7.08	1.20
41	HC3-26-5-2	25.49	4.2	1.14	3.92	42.34	15.03	0.26	140.00	185.55	0.33	6.86	1.46
42	HC3-26-6-1	33.49	2.8	0.42	1.62	54.26	5.43	0.26	180.00	201.18	0.33	0.07	2.37
43	HC3-26-6-2	28.59	4.1	1.12	3.52	44.83	15.55	0.24	140.00	185.50	0.31	0.15	1.59
44	HC3-26-7	18	6.8	2.26	5.06	28.29	28.21	0.26	100.00	174.19	0.35	11.3	1.89
45	HC3-26-8-1	29.79	4.4	0.88	3.18	47.86	13	0.21	160.00	204.26	0.33	0.34	1.78
46	HC3-26-8-2	30.6	3.9	0.82	2.62	49.47	11.36	0.26	160.00	197.43	0.32	2.44	2.00
47	HC3-26-10	32.98	3.7	0.48	1.7	53.72	6.24	0.26	180.00	205.43	0.34	0.43	2.90
48	HC3-26-11-1	29.19	4.2	1.04	3.22	46.08	14.7	0.22	140.00	182.72	0.30	0.57	1.75
49	HC3-26-11-2	17	7.9	2.48	5.8	28.11	31.28	0.24	100.00	191.20	0.36	9.48	1.87
50	HC3-26-12	23.99	6	1.7	4.9	38.07	23.68	0.24	120.00	189.04	0.32	1.56	1.65
51	HC3-26-13	26.99	4.3	0.94	3.24	44.12	13.3	0.27	140.00	179.60	0.32	8.85	1.76
52	HC3-26-14	29.69	4.1	0.84	2.9	48.21	12.38	0.22	160.00	201.11	0.33	0.3	1.84
53	HC3-26-15-2	27.39	5.1	1.16	3.98	43.58	17.61	0.24	160.00	222.50	0.37	4.24	1.67
54	HC3-26-16	24.29	5.2	1.36	4.8	38.25	21.41	0.13	120.00	178.84	0.31	10.61	1.40
55	HC3-26-17-1	26.59	4	0.3	3.74	44.12	10.5	0.24	160.00	197.00	0.36	14.44	1.23
56	HC3-26-19	25.99	3.9	0.24	4.14	42.7	11.03	0.26	140.00	174.06	0.33	18.25	1.07
57	HC3-26-20	24.8	4.5	0.6	4.7	37.9	18.27	0.27	140.00	195.37	0.37	17.43	1.15
	平均	28.73	3.3	0.90	3.66	45.68	14.07	0.23	145.96	188.97	0.32	7.50	1.24

K^+、Mg^{2+}、Ca^{2+}及SO_4^{2-}等离子的平均含量较高，而不溶物含量较低，平均仅5.48%，表明溶液中溶解的硫酸盐较多，平均可达21.94%（忽略石盐晶格包裹体中的K^+、Mg^{2+}、Ca^{2+}、SO_4^{2-}等离子），杂卤石或硬石膏等难溶硫酸盐在水溶化学实验过程中发生溶解。

3.2 溴氯特征

溴氯系数是判断卤水蒸发沉积阶段的重要地球化学指标，石盐矿物中的溴氯系数（$Br\times10^3/Cl$）常被用来反映盐湖卤水的浓缩程度和沉积阶段（刘群等，1987；Walter et al.，1990；林耀庭，1995；Warren，2006；Gupta et al.，2012；王淑丽等，2014），也常作为钾盐找矿的地化指标。溴几乎仅以氯的类质同象存在于氯化型矿物中，不含氯的碳酸盐、硫酸盐、硼酸盐等矿物中几乎不含溴，忽略盐类矿物包裹体中含有的痕量溴以及盐类沉积物中的黏土可吸附的痕量溴（程怀德等，2008），本文假设在新型杂卤石钾盐矿中，Br 离子几乎全部存在于石盐晶格或包裹体中，硫酸盐的溶解不会增加溶液中的 Br 离子，但会影响溶液中离子的总量，导致 Br 偏小。

通过除去溶液中的 Ca^{2+}、Mg^{2+}、K^{+}、SO_4^{2-} 校正石盐中 Br 的真实浓度，得到 $w(Br^-_{校正})$为 90.26×10^{-6}～236.01×10^{-6}，平均 188.98×10^{-6}；$Br\times10^3/Cl$ 分布在 0.16～0.44 之间，平均 0.32，HC3 井新型杂卤石钾盐矿段中的 $w(Br^-_{校正})$曲线与 $Br\times10^3/Cl$ 曲线相符（图 5），说明实验和校正结果可靠，可以反映杂卤石位于正常石盐沉积阶段，尚未达到钾镁盐析出阶段，但整体较高，且个别样品已达钾盐析出阶段的底部（图 5）。

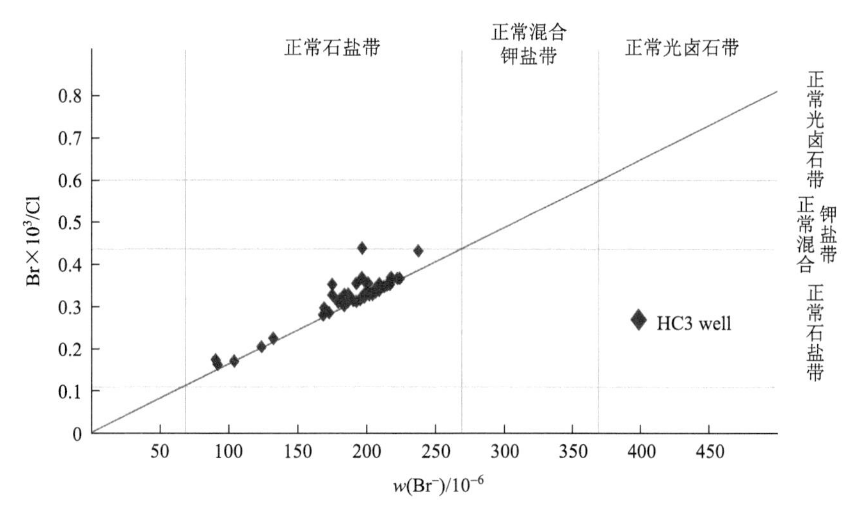

图 5　新型杂卤石钾盐矿中 $w(Br^-)$与 $Br\times10^3/Cl$ 的关系图（底图数据来源于 Valyashko，1956；Lin，1995）

Fig. 5　The characteristics of $w(Br^-)$ and $Br\times10^3/Cl$ of the new type of polyhalite potassium ore deposit (base map data after Valyashko, 1956; Lin, 1995).

4　讨论

4.1　新型杂卤石钾盐矿的可利用性

新型杂卤石钾盐矿的可溶成分为石盐、杂卤石及少量硬石膏，其中石盐为易溶盐，杂卤石和硬石膏等硫酸盐为难溶盐。常温下，随着时间的增加，200 目的杂卤石粉末在纯水中逐渐溶解，这个时间长达数天，其中 Ca 浓度随时间增加未有改变，基本维持在比较低的浓度水平上（安莲英等，2004）。在纯水中，杂卤石溶解形成了难溶的石膏并包裹杂卤石团块，导致其不能继续溶解，这种现象称为“枸溶性”，这也是造成产于膏岩层中杂卤石难以溶采的原因。而在新型杂卤石钾盐矿中，溶解的硫酸盐总量平均达 21.94%，不溶物仅占 5.48%，且溶液 Ca 浓度较高，表明杂卤石或硬石膏等难溶硫酸盐在石盐溶解形成的 NaCl 溶液中具有较高的溶解度。基于杂卤石（$K_2SO_4\cdot MgSO_4\cdot 2CaSO_4\cdot 2H_2O$）和硬石膏（$CaSO_4$）的组成元素，本文利用$(K+Mg)/Ca$ 值指示样品中溶解杂卤石和硬石膏的情况，杂卤石中$(K+Mg)/Ca$ 值为 1.27，硬石膏溶解会降低溶液中的$(K+Mg)/Ca$值，反之，杂卤石溶解形成的 Ca 和SO_4^{2-}沉淀会增加溶液中的$(K+Mg)/Ca$值。实验结果表明溶液中的$(K+Mg)/Ca$值平均为 1.24，略小于杂卤石中的$(K+Mg)/Ca$值，可能由于在本次水溶化学实验流程所制约的条件下，杂卤石溶解形成的 Ca 和SO_4^{2-}离子发生少量沉淀导致，整体损失率为 2.36%，远小于 $CaSO_4$

组分在杂卤石中的含量（48%），新型杂卤石钾盐矿中杂卤石的整体溶解性较好，石膏沉淀较少。但杂卤石碎屑中杂卤石与硬石膏的占比并不同，尤其是HC3-24-3、HC3-24-6、HC3-24-7-1这几个样品，(K + Mg)/Ca值远小于1.27，溶解的难溶组分主要为硬石膏。

基于杂卤石的溶解度大于硬石膏的原理，笔者认为杂卤石-硬石膏（石膏）在NaCl溶液的溶解过程中，溶液的K^+、Na^+、Mg^{2+}、Ca^{2+}/Cl^-、SO_4^{2-}、H_2O体系如未达到Ca（或SO_4^{2-}）饱和，石膏可继续导致溶液中Ca远大于K、Mg含量；反之，当溶液的K^+、Na^+、Mg^{2+}、Ca^{2+}/Cl^-、SO_4^{2-}、H_2O体系超过Ca（或SO_4^{2-}）饱时，则导致石膏沉淀，杂卤石、硬石膏在NaCl溶液中的溶解情况还需要更进一步的实验研究。总之，在新型杂卤石钾盐矿中，其石盐基质溶解的产物本身就是杂卤石的良好溶剂，可降低杂卤石碎屑在溶矿过程中因难溶发生沉淀而导致的损失率，对溶采技术具有一定的科学指导意义，但新型杂卤石钾盐矿是否具有较高的溶采回收率还需进一步的研究和实践证明。

4.2　新型杂卤石钾盐矿的沉积阶段及对杂卤石含量的影响

$w(Br^-)$曲线及$Br\times10^3/Cl$值曲线稳定平缓，卤水无明显淡化或浓缩的迹象，仅沉积后期$w(Br^-)$及$Br\times10^3/Cl$值快速降低，表明此时盆地逐渐淡化直至顶板硬石膏层出现，石盐层在后期地质构造中没有发生明显层位上的变化。此外，通过$w(Br^-)$、$Br\times10^3/Cl$值和岩心特征的关系可以发现，新型杂卤石钾盐矿层底部“黑盐”中的$Br\times10^3/Cl$值特征稳定与“红盐”相似，但顶部“黑盐”的Br特征反映盆地逐渐淡化，表明石盐颜色与蒸发阶段无关（图6），可能是由蒸发环境改变造成。除此之外，底部“黑盐”与“红盐”的杂卤石含量较高，顶部“黑盐”中的硬石膏占比较高，表明石盐颜色并不影响杂卤石碎屑中的成分，而卤水浓缩阶段对杂卤石碎屑的成分可能造成一定影响。

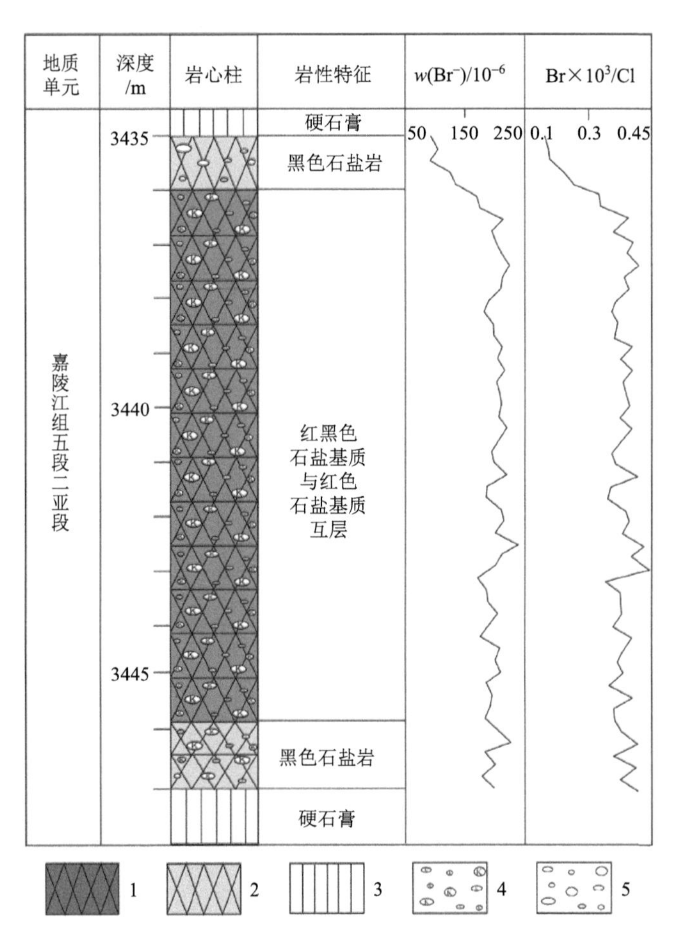

图6　HC3井新型杂卤石钾盐矿段岩心柱状简图及Br特征

Fig. 6　Core histogram and characteristics of Br from the new type of polyhalite potassium ore deposit in HC3 well.

1-红色石盐岩；2-黑色石盐岩；3-硬石膏岩；4-杂卤石碎屑；5-硬石膏碎屑

1-Red halite; 2-Black halite; 3-Anhydrite; 4-Clastic polyhalite; 5-Clastic anhydrite

水溶化学结果显示，蒸发阶段较低时（$w(Br^-)<100\times10^{-6}$；$Br\times10^3/Cl<0.20$），碎屑中的硬石膏含量较多，随着蒸发阶段的升高，碎屑中杂卤石所占比例逐渐升高（图 7），可能由于盆地卤水对杂卤石-石膏碎屑的交代，即随着卤水中 K、Mg 的不断富集，造成碎屑中杂卤石含量的升高，或在后期成岩过程中，石盐释放的富 K、Mg 晶间卤水，甚至钾镁盐在高温高压条件下交代碎屑中的硬石膏，形成杂卤石。

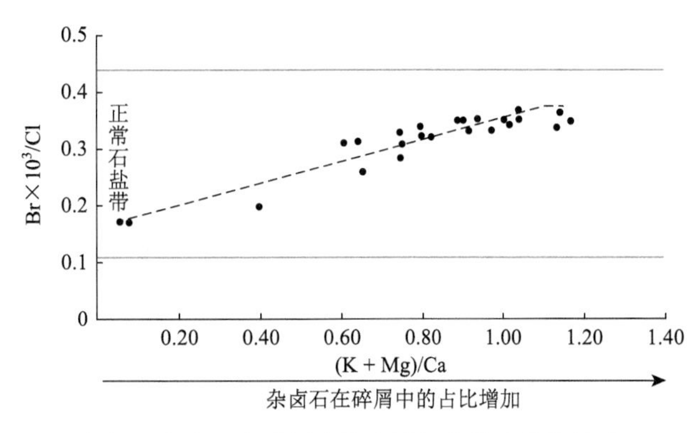

图 7　杂卤石在碎屑中的含量与蒸发阶段的关系

Fig. 7　The relationship between the proportion of polyhalite in detritus and the evaporation stage.

4.3　新型杂卤石钾盐矿成因浅析

蒸发实验结果表明，杂卤石可以在 K^+、Na^+、Mg^{2+}、Ca^{2+}/Cl^-、H_2O 六元体系达到一定平衡时直接形成（韩蔚田等，1982），也可以由含 K、Mg 组分较高的浓卤水与石膏（硬石膏）等硫酸钙盐接触交代形成（Freyer et al.，2003；Wollmann et al.，2008，2009；Wollmann，2010）。前者为原生沉积型杂卤石，后者为交代型杂卤石，其中交代型杂卤石又可分为以下 2 种：①准同生交代型杂卤石，即在石盐蒸发阶段后期，继续浓缩而含 K、Mg 较高的浓卤水与较早沉淀的石膏（硬石膏）等硫酸钙盐发生反应形成的杂卤石（袁见齐，1974）；②后生交代型杂卤石，富钾镁的卤水通过构造裂隙运移并与石膏（硬石膏）等硫酸钙盐接触发生交代作用而形成的杂卤石（袁见齐，1974；廖林志等，1984）。

根据杂卤石的成因，交代型杂卤石主要赋存于膏盐等硫酸钙盐中，而石盐中形成杂卤石则需满足以下条件：在石盐沉积过程中，强烈的蒸发浓缩作用使卤水中的 K、Mg 离子逐渐富集，此时富 Ca 流体的少量补给（达到杂卤石析出的范围）可以形成原生杂卤石，过量补给则导致石膏的出现（赵德钧等，1987）。这种类型的杂卤石往往出现在石盐层的中上部，且厚度较薄、稳定、延伸远，杂卤石质地较纯无交代现象，具有微层理或韵律结构，是卤水持续蒸发浓缩的产物（袁见齐，1974；赵德钧等，1987；王弭力，1982），具有 $w(Br^-)$及 $Br\times10^3/Cl$ 持续增加的特征，中国古近纪现代盐湖中可见此类型杂卤石，如大汶口及潜江等地的膏盐盆中。在新型杂卤石钾盐矿中，杂卤石呈碎屑颗粒状分布于整个石盐层中，可见厚度较大的条带状杂卤石，出现杂卤石交代硬石膏以及杂卤石被硬石膏交代并包裹的复杂交代关系，与原生沉积型杂卤石的特征相差甚远，且 $w(Br^-)$及 $Br\times10^3/Cl$ 比值分布稳定除顶部外没有明显变化，不具备原生杂卤石的形成条件，因此，笔者认为新型杂卤石钾盐矿中的杂卤石并非在石盐沉积过程中形成。杂卤石常见被硬石膏交代并包裹的现象，表明杂卤石曾被较淡流体甚至淡水溶解，在成岩作用后显然不可能发生这种情况，这点进一步佐证了新型杂卤石钾盐矿地球化学特征所得到的结论。此外，这种杂卤石分布于石盐层中而非膏岩等硫酸盐矿物中，也并非一般的交代作用形成，是一种新类型成因的杂卤石矿。四川盆地尤其是盆地中部和东部地区存在大量与硬石膏互层的杂卤石，笔者倾向于认为新型杂卤石钾盐矿中的杂卤石碎屑与该类型杂卤石同源。

综上，笔者认为新型杂卤石钾盐矿的形成过程包括：已经形成的未固结-半固结杂卤石-石膏层经历某种作用发生破碎，并被较淡流体搬运（或原地堆积）进入成盐盆地，最后被石盐胶结，在此过程中，碎屑中的石膏被盆地内的富 K、Mg 卤水或后期晶间富 K、Mg 卤水交代，造成杂卤石的富集。

5　结论

（1）在新型杂卤石钾盐矿中，其石盐基质的溶解产物本身就是杂卤石良好的溶剂，对新型杂卤石钾盐矿的溶采回收率及成本有正面影响。

（2）新型杂卤石钾盐矿的石盐蒸发阶段不具备杂卤石形成的条件，杂卤石碎屑不是在石盐沉积阶段形成的，而是“外来物”。

（3）杂卤石与硬石膏在碎屑中的含量受卤水蒸发阶段的影响，富 K、Mg 流体交代碎屑中的石膏（硬石膏）形成杂卤石，且蒸发阶段较低时（$w(Br^-)<100\times10^{-6}$；$Br\times10^3/Cl<0.20$）不利于硬石膏向杂卤石转化，当卤水浓缩达到石盐沉积的上部，有利于杂卤石的优化。

References

An L Y, Yin H, Tang M L, Gu J and Yao L P. 2004. Study on the dissolving behavior of polyhalite in water[J]. Journal of Mineralogy and Petrology, 24 (4): 108-110 (in Chinese with English abstract).

Cai K Q and Yuan J Q. 1986. Mineralization conditions and prospecting direction of Triassic sylvite in Sichuan Province[J]. Chemical Geology, (2): 1-9 (in Chinese with English abstract).

Cheng H D, Ma H Z, Tan H B, Xu J X and Zhang X Y. 2008. Geochemical characteristics of bromide in potassium deposits: Review and research perspectives[J]. Bulletin of Mineralogy, Petrology and Geochemistry, 27 (4): 399-408 (in Chinese with English abstract).

Freyer D and Voigt W. 2003. Crystallization and phase stability of $CaSO_4$ and $CaSO_4$-based salts[J]. Monatshefte für Chemie, 134: 693-719.

Gupta I, Wilson A M and Rostron B J. 2012. Cl/Br compositions as indicators of the origin of brines: Hydrogeologic simulations of the Alberta Basin, Canada[J]. Geological Society of America Bulletin, 124 (1-2): 200-212.

Han W T, Gu S Q and Cai K Q. 1982. On the formative conditions of polyhalite in the six-component system K^+, Na^+, Mg^{2+}, Ca^{2+}/Cl^-, SO_4^{2-}-H_2O[J]. Chinese Science Bulletin, 27: 362-365 (in Chinese with English abstract).

Huang X Z. 1996. The first polyhalite deposit of China[J]. Yunnan Geology, 15 (1): 52-61 (in Chinese with English abstract).

Jin F. 1989. Enlightenment to China from the exploitation and utilization of the resources of halite abroad[J]. Chemical Mining Technology, (3): 31-34 (in Chinese).

Li Y W, Cai K Q and Hai T W. 1998. Origin of potassium-rich brine and the metamorphism of Triassic evaporites in Sichuan Basin[J]. Geoscience, 12 (2): 73-79 (in Chinese with English abstract).

Liao Z L Huang F H, Liu Q C and Hu C S. 1984. The characteristics and origin of polyhalite in gypsum-anhydrite deposits in Longuio area, eastern Sichuan[J]. Journal of Mineralogy and Petrology, (1): 94-101 (in Chinese with English abstract).

Lin Y T. 1995. Geochemical behaviour of bromine and its application to prospection for potash resource in Sichuan[J]. Geology of Chemical Minerals, 17 (3): 175, 181 (in Chinese with English abstract).

Lin Y T and Yin S M. 1998. Distribution, genesis and significance of shallow-seated polyhalite ore in Quxian, Sichuan[J]. Acta Geologica Sichuan, 18 (2): 121-125 (in Chinese with English abstract).

Lin Y T and He J Q. 2004. The Characters and genies meanings of the shallow polyhalide potash deposit in Huaying mountain, Sichuan[J]. Chemical Mineral Geology, 26 (3): 145-149 (in Chinese with English abstract).

Liu Q, Chen Y H, Li Y C, Lan Q C, Yuan H R and Yan D L. 1987. China Mesozoic and Cenozoic terrigenous clastic-chemical rock type salt deposits[M]. Beijing: Science and Technology Press. 14-15, 52 (in Chinese).

Peng D H, Ma H Z, Tang H B, Yu J X and Zhang X Y. 2008. Geochemical characteristics of Bromide in potassium deposits: Review and research perspectives[J]. Bulletin of Mineralogy, Petrology andGeochemistry, 27 (4): 399-408 (in Chinese with Englishab stract).

Stromeyer F. 1818. De Polyhalite, nova e salium classe fossilium specie[A]. In: Göttingische gelehrte Anzeigen: unter der Aufsicht der Königlichen Gesellschaft der Wissenschaften, Stück[C]. 209 (31): 2081-2084.

Valyashko M G. 1956. Geochemistry of bromide in the processes of salt deposition and the use of the bromide content as a genetic and prospecting tool[J]. Geochemistry USSR, (6): 570-587.

Walter L M, Stueber A M and Huston T J. 1990. Br-Cl-Na systematics in Illinois basin fluids: Constraints on fluid origin and evolution[J]. Geology, 18 (4): 315-318.

Wang M L. 1982. The geological significance of polyhalite in depression Q[J]. Geological Review, 28 (1): 28-37 (in Chinese with English abstract).

Warren J K. 2006. Evaporites: Sediments, resources and hydrocarbons[M]. Berlin-Heidelberg: Springer Verlag.

Wang S L and Zheng M P. 2014. Discovery of Triassic polyhalite in Changshou area of East Sichuan Basin and its genetic study[J]. Mineral Deposits, 33 (5): 1045-1056 (in Chinese with English abstract).

Wollmann G, Freyer D and Voigt W. 2008. Polyhalite and its analogous triple salts[J]. Monatshefte für Chemie, 139: 739-745.

Wollmann G, Seidel J and Voigt W. 2009. Heat of solution of polyhalite and its analogues at T = 298.15 K[J]. Journal of Chemical Thermodynamics, 41: 484-488.

Wollmann G. 2010. Crystallization fields of polyhalite and its heavy metal analogues[D]. Ph. D. thesis. University Bergakademie Freiberg.

Wu Y L, Zhu Z F and Wang J L. 1983. Study on petrographic paleogeography of the Triassic salt-forming period in the southwestern platform area[J]. Qinghai Geology, (3): 126-137 (in Chinese).

Yuan J Q. 1974. A brief introduction to the information of foreign heterohalite[J]. Chemical Mining Technology, (6): 47-58 (in Chinese).

Zhao D J, Han W T, Cai K Q and Gao J H. 1987. The study of polyhalite genesis and its significance of potash-finding in Dawenkou depression, Shandong Province[J]. Earth Science—Journal of Wuhan College of Geology, 12 (4): 349-356 (in Chinese with English abstract).

Zheng M P, Yuan K R, Zhang Y S, Liu X F, Chen W X and Li J S. 2010. Regional distribution and prospects of potash in China[J]. Acta Geologica Sinica, 84 (11): 1523-1553 (in Chinese with English abstract).
Zheng M P, Zhang Y S, Shang W J, Xing E Y, Zhong J A, Gui B L and Peng Y. 2018. Discovery of a new type of polyhalite potassium ore in Puguang region, northeastern Sichuan[J]. Geology in China, 45 (5): 1074-1075 (in Chinese with English abstract).
Zhong J A, Zheng M P, Tang X Y, Liu Z, Wang F M, Cai C and Pang B. 2018. Sedimentary characteristics and genetic study of deep polyhalite in Huangjingkou anticline of northeast Sichuan[J]. Mineral Deposits, 37 (1): 81-90 (in Chinese with English abstract).

参考文献

安莲英, 殷辉安, 唐明林, 顾娟, 尧丽萍. 2004. 杂卤石溶解性能的研究[J]. 矿物岩石, 24 (4): 108-110.
蔡克勤, 袁见齐. 1986. 四川三叠系钾盐成矿条件和找矿方向[J]. 化工地质, (2): 1-9.
程怀德, 马海州, 谭红兵, 许建新, 张西营. 2008. 钾盐矿床中 Br 的地球化学特征及研究进展[J]. 矿物岩石地球化学通报, 27 (4): 399-408.
韩蔚田, 谷树起, 蔡克勤. 1982. K^{+}、Na^{+}、Mg^{2+}、Ca^{2+}/Cl^{-}、SO_4^{2-}-H_2O 六元体系中杂卤石形成条件的研究[J]. 科学通报, (6): 362-365.
黄宣镇. 1996. 中国首例杂卤石矿床[J]. 云南地质, 15 (1): 52-61.
金锋. 1989. 国外杂卤石资源开发利用对我国的启示[J]. 化工矿山技术, (3): 31-34.
李亚文, 蔡克勤, 韩蔚田. 1998. 四川盆地三叠系蒸发岩的变质作用与富钾卤水的成因[J]. 现代地质, 12 (2): 73-79.
廖林志, 黄馥华, 刘铅超, 胡朝书. 1984. 川东农乐石膏-硬石膏矿床中的杂卤石[J]. 矿物岩石, (1): 94-101.
林耀庭. 1995. 溴的地球化学习性及其在四川找钾工作中的应用[J]. 化工矿产地质, 17 (3): 175-181.
林耀庭, 尹世明. 1998. 四川渠县浅层杂卤石矿分布特征及其成因和意义[J]. 四川地质学报, 18 (2): 121-125.
林耀庭, 何金权. 2004. 四川华蓥山浅层杂卤石钾矿地质特征及其成因意义[J]. 化工矿产地质, 26 (3): 145-149.
刘群, 陈郁华, 李银彩, 蓝庆春, 袁鹤然, 阎东兰. 1987. 中国中、新生代陆源碎屑-化学岩型盐类沉积[M]. 北京: 北京科学技术出版社. 14-15, 52.
王弭力. 1982. Q 凹陷杂卤石的地质意义[J]. 地质论评, 28 (1): 28-37.
王淑丽, 郑绵平. 2014. 川东盆地长寿地区三叠系杂卤石的发现及其成因研究[J]. 矿床地质, 33 (5): 1045-1056.
吴应林, 朱忠发, 王吉礼. 1983. 西南地台区三叠纪成盐期岩相古地理研究[J]. 青海地质, (3): 126-137.
袁见齐. 1974. 国外杂卤石资料简介[J]. 化工矿山技术, (6): 47-58.
赵德钧, 韩蔚田, 蔡克勤, 高建华. 1987. 大汶口凹陷下第三系含盐段杂卤石的成因及其找钾意义[J]. 地球科学, 12 (4): 349-356.
郑绵平, 袁鹤然, 张永生, 刘喜方, 陈文西, 李金锁. 2010. 中国钾盐区域分布与找钾远景[J]. 地质学报, 84 (11): 1523-1553.
郑绵平, 张永生, 商雯君, 邢恩袁, 仲佳爱, 桂宝玲, 彭渊. 2018. 川东北普光地区发现新型杂卤石钾盐矿[J]. 中国地质, 45 (5): 1074-1075.
仲佳爱, 郑绵平, 唐学渊, 刘铸, 王富明, 蔡策, 庞博. 2018. 川东北黄金口背斜三叠系深部杂卤石特征及成因探讨[J]. 矿床地质, 37 (1): 81-90.

Geochemical characteristics of a new type of polyhalite potassium ore deposits in Xuanhan area, northeast Sichuan, and their significance

Wenjun Shang[1,2], Yongsheng Zhang[1,2], Kong Li[3], Enyuan Xing[1,2], Baoling Gui[1,2], Yuan Peng[1,2] and Haitong Zhao[1,2]

1. Institute of Mineral Resources, CAGS, Beijing 100037, China
2. MNR Key Laboratory of Saline Lake Resources and Environments, Beijing 100037, China
3. No.5 Gold Geological Party of CAPF, Xi'an 710199, Shaanxi, China

Abstract The new type of polyhalite potassium ore deposits are distributed in the Early-Middle Triassic evaporative strata in Xuanhan area, northeast Sichuan. They are characterized by the uneven distribution of the polyhalite debris in the rock salt layer. In this paper, the main target strata of HC3 well was sampled continuously and tested by water. The major and trace elements of the soluble part of the new type of polyhalite potassium ore deposit have the following features: ①The average ratio of (K + Mg)/Ca in the solution is 1.24, slightly less than the ratio of (K + Mg)/Ca in the polyhalide (1.27), and the overall loss rate is 2.36%, far less than the content of $CaSO_4$ in the polyhalide (48%), indicating that the solubility of the polyhalites in the solution of NaCl dissolved by the halite matrix was good, which had a positive significance for solution mining. ②$Br \times 10^3/Cl$ ratios are in the range of 0.16～0.44, imply that the new type of polyhalite potassium ore deposit was at the normal rock salt phase evaporation stage. The evaporation stage was stable without drastic fluctuation，with only quick desalination at the top. In combination with previous research on the origin of polyhalite， the authors hold that it was a kind of "foreign material". ③The (K + Mg)/Ca of the new type of polyhalite potassium ore deposit can indicate the proportion of polyhalite and anhydrite in the detritus, and its value tends to increase with the increase of the value of $Br \times 10^3/Cl$, indicating that the content of the polyhalite in the detritus increased with the increase of the degree of concentration of the brine.

Keywords Geochemistry; Polyhalite; New type of polyhalite potassium ore deposit; Ore genesis; Xuanhan area

上黑龙江盆地虎拉林早白垩世岩体锆石 U-Pb 年代学、Hf 同位素及地球化学特征研究*

巩鑫[1,2,3]，赵元艺[1]，水新芳[1]，程贤达[1,2]，王远超[1,2,4]，刘璇[1,2]，谭伟[1,2]

1. 中国地质科学院矿产资源研究所，北京，100037
2. 中国地质大学地球科学与资源学院，北京，100083
3. 贵州省有色金属和核工业地质勘查局地质矿产勘查院，贵阳，550005
4. 中国人民武装警察部队黄金第十二支队，成都，610000

内容提要 上黑龙江盆地虎拉林金矿床位于兴蒙造山带东段大兴安岭北部额尔古纳微地块北缘，夹持于蒙古鄂霍茨克缝合带与得尔布干断裂之间，其矿体主要赋存于隐爆角砾岩中，与早白垩世岩浆活动形成的花岗斑岩、石英斑岩关系密切。本文采用 LA-ICP-MS 锆石 U-Pb 定年方法，获得花岗斑岩岩体结晶年龄为 141.7±1.1Ma（MSWD = 0.086）、石英斑岩（2 件）岩体结晶年龄为 144.9±0.56Ma（MSWD = 0.580）和 142.6±0.74Ma（MSWD = 0.077），两者均为早白垩世岩浆活动产物。两岩体锆石 $\varepsilon_{Hf}(t)$ 分别介于−4.14～0.16、−2.25～1.37，且大部分数据位于球粒陨石演化线之下，两阶段模式年龄（t_{DM2}）分别为 987.08～1259.04Ma、908.38～1138.56Ma。岩石地球化学特征显示，花岗斑岩 $w(SiO_2)$ 为 66.21%～66.70%，$w(Al_2O_3)$ 为 15.18%～15.45%，$w(Na_2O + K_2O)$ 为 7.56%～8.17%，铝饱和指数 A/CNK 为 1.03～1.06，显示弱过铝质高钾钙碱性-钾玄岩序列；石英斑岩 $w(SiO_2)$ 为 66.34%～68.25%，$w(Al_2O_3)$ 为 13.77%～15.05%，$w(Na_2O + K_2O)$ 为 6.76%～9.06%，在 SiO_2-K_2O 图解中，落入钾玄岩序列范围内。花岗斑岩、石英斑岩 ΣREE 分别为 71.07×10^{-6}～97.92×10^{-6}、135.53×10^{-6}～156.15×10^{-6}，$(La/Yb)_N$ 分别为 24.87～26.81、31.33～40.62，均表现轻稀土（LREE）相对富集、重稀土（HREE）相对亏损的右倾曲线特征，且轻、重稀土分馏明显；δEu 分别为 0.65～0.88、0.41～0.50，后者较前者铕负异常明显；两岩体均相对富集 Rb、K 等大离子亲石元素（LILEs）、LREE 及 U 等不相容元素，强烈亏损 Nb、Ta、Ti、P 等高场强元素（HFSEs）。岩石成因类型判别图显示上黑龙江盆地虎拉林早白垩世侵入岩为 I 型花岗岩，形成于蒙古-中朝大陆与西伯利亚大陆后碰撞环境，由挤压转为伸展的构造背景下，可能是蒙古-鄂霍茨克造山带陆陆碰撞期间加厚古老地壳拆沉、部分熔融，并受到幔源物质或新生地壳熔融混染的产物，进而限制了蒙古-鄂霍茨克洋至少在 144.9Ma 以前已经完全闭合。

关键词 锆石 U-Pb 年代学；Hf 同位素；地球化学；早白垩世；虎拉林；上黑龙江盆地

大兴安岭地区是我国东北部重要的银、铅、锌、铜及钼等多金属成矿带，带内多数矿床与中生代岩浆活动关系密切，成矿类型以斑岩型、夕卡岩型及浅成低温热液矿床为主（Zhao Yiming et al.，1997；Chen Yanjing et al.，2007；Wu Guang et al.，2007；Zhou Zhenhua et al.，2012；Ouyang Hegen et al.，2013；Li Yun et al.，2017；Li Xiangwen et al.，2018）。上黑龙江盆地位于大兴安岭最北侧，区域内断裂构造发育、岩浆活动频繁，具有加里东期、海西期及燕山期等多期次侵入岩的存在（Wu Guang，2005；Sui Zhenmin et al.，2006；Wu Guang et al.，2010）。燕山期岩浆活动可分为中—晚侏罗世及早白垩世两期，其中早白垩世岩浆活动与区域成矿关系密切，主要岩石类型有花岗闪长岩、闪长岩、花岗斑岩、石英斑岩等，现已发现区域内砂宝斯、二十一站、龙沟河、洛古河、十五里桥及宝兴沟矿床（点）均形成于晚侏罗世—早白垩世（Wu Guang et al.，2008，2014；Liu Jun et al.，2013a；Sun Qi et al.，2016；Sun Yanfeng，2015；Li Xiangwen，2015；Li Ruihua et al.，2018）。

虎拉林金矿床位于上黑龙江盆地西侧，隶属于内蒙古额尔古纳市恩和哈达镇管辖，自 1998 年发现 Au

---

* 本文发表在：地质学报，2020，94（2）：553-572

异常以来，前人在该地区开展过大量的基础地质研究工作，通过对矿石中钾长石进行 ^{40}Ar-^{39}Ar 阶段升温法测定了矿床成矿年龄为 136.34±0.36Ma（Zhang Yong et al.，2003），查明了矿床地质、矿石、围岩蚀变及金矿赋存等基本特征，矿床资源量计算结果显示已达小型矿山规模，同时具有中型矿山潜力（Ding Qingfeng et al.，2006；Song Guibin et al.，2007；Liu Guige et al.，2009；Liu Jimin et al.，2012）。由于矿床特殊的地理位置及地表覆盖严重等因素，使前人均未对矿床内相关岩体进行相应的研究工作，这不仅制约了成岩作用与成矿作用关系的探讨，而且严重制约着对研究区岩浆活动与区域构造演化、深部演化过程及岩浆源区的认识。

为厘清虎拉林金矿床岩体形成时代、来源、类型及构造背景、形成机制等问题，本文在前人研究基础之上，对矿床进行详尽的野外地质研究，选取代表性岩体样品进行详细的定年及地球化学研究，在此基础上确定其成岩时代，探讨岩石成因类型及形成构造背景，为分析成岩与成矿的关系及剖析成矿动力学机制提供理论基础。

1　地质背景

1.1　区域地质

兴蒙造山带东段北与西伯利亚克拉通以蒙古-鄂霍茨克缝合带相隔，南以赤峰-开源断裂与华北克拉通相毗邻。造山带先后受到古亚洲洋、蒙古-鄂霍茨克洋、古太平洋等板块构造运动影响，使其内部结构复杂；造山带自西北至南东以塔源-喜桂图断裂、贺根山-黑河断裂、佳木斯-牡丹江断裂为界依次划分为额尔古纳地块、兴安地块、松辽地块及佳木斯地块（Wu Fuyuan et al.，2002，2011；Wu Guang et al.，2012；Zhao Shuo et al.，2014；Ji Zheng et al.，2018）［图 1（a）］。上黑龙盆地即位于兴蒙造山带额尔古纳地块最北部，呈东西向展布，延伸约 250km，南北宽约 60km，南北分别紧邻得尔布干断裂带及蒙古-鄂霍茨克缝合带，被前人解释为蒙古-鄂霍茨克造山带东南缘的前陆盆地（Li Jinyi et al.，2004a；Wu Guang et al.，2008），该处是古亚洲洋构造域、环太平洋构造域与蒙古-鄂霍茨克构造域强烈叠加、复合、转换的关键部位，特殊的地理位使其区内构造演化复杂、断裂构造发育，从而使得该地区成为重要的贵金属、有色金属及黑色金属矿集区（Wu Guang et al.，2007；Zhou Zhenhua et al.，2012；Wang Tianhao et al.，2014；Che Hewei et al.，2015）。

上黑龙江盆地内地层较发育，主要由中元古兴华渡口群（Pt_1xh）片岩、片麻岩及大理岩、下寒武统额尔古纳河组（Є*e*）大理岩、板岩及早古生代花岗岩类等组成基底（Wu Guang et al.，2002），上层零星发育下泥盆统泥鳅河组（D_1n）灰岩、下白垩统九峰山组（K_1j）砂岩及砾岩，盆地盖层为早—中侏罗世沉积建造及晚侏罗世—早白垩世火山岩建造，自下而上依次为绣峰组（J_2x）、二十二站组（J_2er）及漠河组（J_2m），主要由砾岩、砂岩、粉砂岩及泥岩等构成；火山岩主要由凝灰岩、中酸性火山岩及喷发相火山碎屑岩等构成，自下而上主要包括塔木兰沟组（J_3t）、上库力组（K_1s）及伊列克得组（K_1y）（Regional Geology of Heilongjiang Province，1993；Geological Survey Institute of Inner Mongolia，2003；Wu Guang et al.，2008；Li Ruihua et al.，2018）。

随着早古生代期间额尔古纳地块与兴安地块碰撞拼合完成，后持续受到古亚洲洋洋壳俯冲、微地块碰撞拼合的影响，直至中生代蒙古-鄂霍茨克洋盆闭合，蒙古-华北大陆与西伯利亚大陆发生碰撞拼接，额尔古纳地块长时间处于挤压-伸展等构造活动强烈环境（Sui Zhenmin，2007；She Hongquan et al.，2012；Tang Jie，2016）。在强烈的构造运动作用下，形成了以北东向、北北东向韧性剪切变形带和断裂为主体，北西向韧性和脆性断裂次之的断裂体系，构成了本区网状断裂构造格局［图 1（b）］。区域褶皱构造发育简单。

区域受古生代古亚洲洋构造域与中—新生代蒙古-鄂霍茨克构造域及太平洋构造域叠加影响（Wu Fuyuan et al.，2011；Chen Yanjing et al.，2012；She Hongquan et al.，2012；Chai Mingchun et al.，2018），使得区域内岩浆活动频繁，侵入岩发育。

其中燕山期岩浆活动与区域成矿关系密切，以中酸性侵入岩为主，岩性以花岗岩、花岗斑岩、二长花岗岩、花岗闪长岩、石英闪长岩及石英闪长斑岩为主，均以岩株、岩枝及岩基产出。

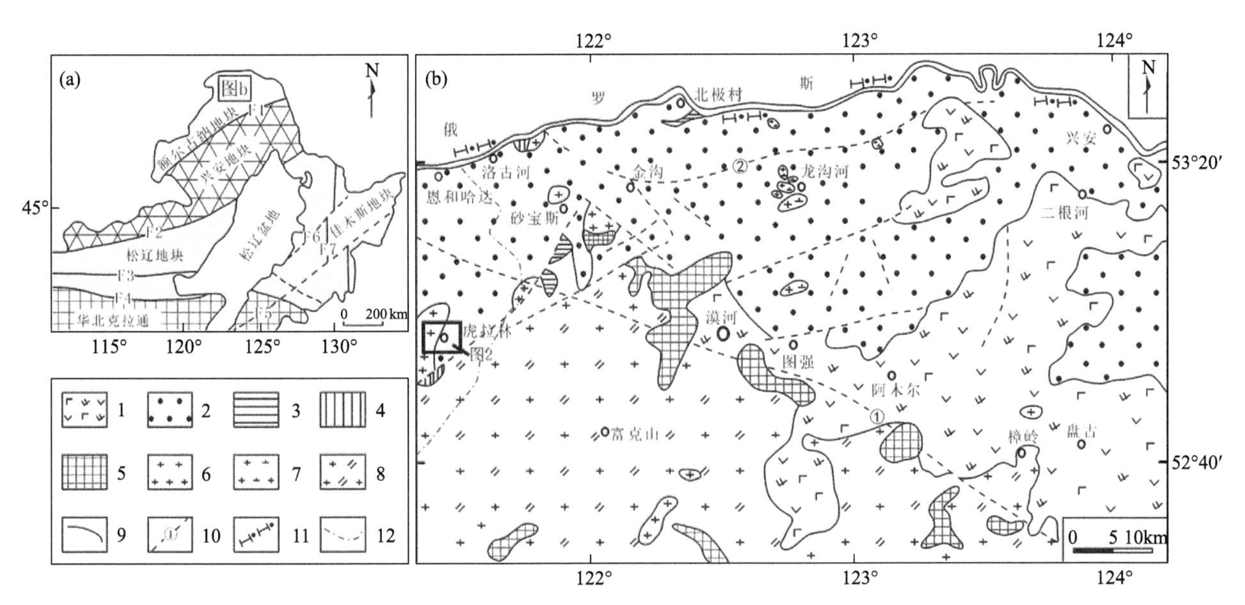

图 1 上黑龙江盆地西段大地构造位置（a，据 Liu Jun et al.，2013a）及区域地质简图（b，据 Wu Guang et al.，2007 修改）

Fig. 1 Geotectonic units (a, after Liu Jun et al., 2013a) and sketch regional geological map (b, modified after Wu Guang et al, 2007) of western part of Upper Heilongjiang basin.

1-中生代火山岩；2-中生代沉积岩；3-古生代沉积岩；4-古生代变质岩；5-元古代结晶基底；6-中生代花岗岩；7-中生代花岗闪长岩；8-古生代花岗岩；9-地质界线；10-断层及编号；11-国界线；12-省界线；F1-塔源-喜桂图断裂；F2-贺根山-黑河断裂；F3-西拉木伦-长春断裂；F4-赤峰-开源断裂；F5-敦化-密山断裂；F6-依兰-伊通断裂；F7-佳木斯-牡丹江断裂；①-西林吉-塔河断裂；②-漠河韧性剪切带

1-Mesozoic volcanic rocks; 2-Mesozoic sedimentary rocks; 3-Paleozoic sedimentary rocks; 4-Paleozoic met-amorphic rocks; 5-Granulitite basement; 6-Mesozoic granite; 7-Mesozoic granodiorite; 8-Paleozoic granite; 9-geological boundary; 10-fault and its serial number; 11-national boundary; 12-provincial boundary; F1-Tayuan-Xiguitu fault; F2-Hegenshan-Heihe fault; F3-Xilamulun-Changchun fault; F4-Chifeng-Kaiyuan fault; F5-Dunhua-Mishan fault; F6-Yilan-Yitong fault; F7-Jiamusi-Mudanjiang fault; ①-Xilinji-Tahe faults; ②-Mohe ductile shear zone

1.2 矿床地质

虎拉林矿床位于额尔古纳微地块北部，区内地层出露简单，仅存在中侏罗统绣峰组（J_2x）及第四系盖层（Q^{al}）。中侏罗统绣峰组为一套陆源碎屑砂岩，岩性主要为含砾砂岩、中粒砂岩及含岩屑长石石英砂岩，磨圆度整体较差，成分主要由石英、长石及碎屑构成，岩石蚀变强烈，主要分布在矿床东南部（Ding Qingfeng et al.，2006；Liu Jimin et al.，2012）。第四系盖层以冲积物为主，主要由浮土、砂岩、角砾岩、花岗斑岩、石英斑岩及花岗岩转块构成，多分布在矿床沟谷及坡度较缓地带。

矿床内构造以角砾岩筒、断裂构造及节理为主，均为重要的控岩、控矿构造。断裂构造发育，规模较小，均以 SN 向发育，控制着矿体规模及走向；角砾岩筒构造呈环状，直径 250～400m，是次火山岩活动中心，与成矿关系密切；节理发育，主要呈 NW、NW 向，规模小，其内黄铁矿脉发育，是 Au 主要赋存载体；褶皱构造不发育（Zhang Yong et al.，2003；Song Guibin et al.，2007）。

矿床内岩浆活动强烈，本次研究发现至少存在两期次岩浆活动，即晚二叠世（另文发表）及早白垩世岩浆侵入活动，后者与成矿关系密切。侵入岩体主要由晚二叠世似斑状花岗岩、早白垩世花岗斑岩及石英斑岩构成，其中早白垩世花岗斑岩多呈岩基产出，少量呈岩株产出穿插于侏罗系砂岩中；早白垩世石英斑岩多呈岩株被包裹于早白垩世花岗斑岩中，或穿插于晚二叠世花岗岩中；晚二叠世花岗岩呈岩株被包裹于早白垩世花岗斑岩或侏罗系砂岩中。

2 样品特征及分析方法

2.1 样品特征及岩相学

本次共在上黑龙江盆地虎拉林金矿床虎拉沟采集 9 件样品，岩性分别为花岗斑岩（3 件）、石英斑岩（6 件），具体描述如下。

灰黑色花岗斑岩（HLL1-4），取样点位 52°55′27.53″N，121°22′41.65″E[图 2、图 3（a）]。灰黑色，斑

状结构，块状构造，可见有肉红色巨斑晶钾长石。斑晶约占20%，主要为钾长石、石英、斜长石及黑云母等，基质多为隐晶质。其中钾长石斑晶约为40%，呈板状、长柱状，粒径0.5～11.4mm，具卡式双晶结构；石英斑晶约为35%，颗粒大，形状不规则，粒径0.40～4.2mm，表面见有裂纹；斜长石斑晶约为15%，呈长条状、板状，粒径0.2～2.6mm，见有聚片双晶、卡纳复合双晶结构；黑云母含量约为6%，为片状，粒径0.5～2.3mm，蚀变强烈；黄铁矿呈粒状分布于岩石中[图3（d）、（e）]。岩石蚀变主要为高岭土化、绢英岩化等[图3（c）]。

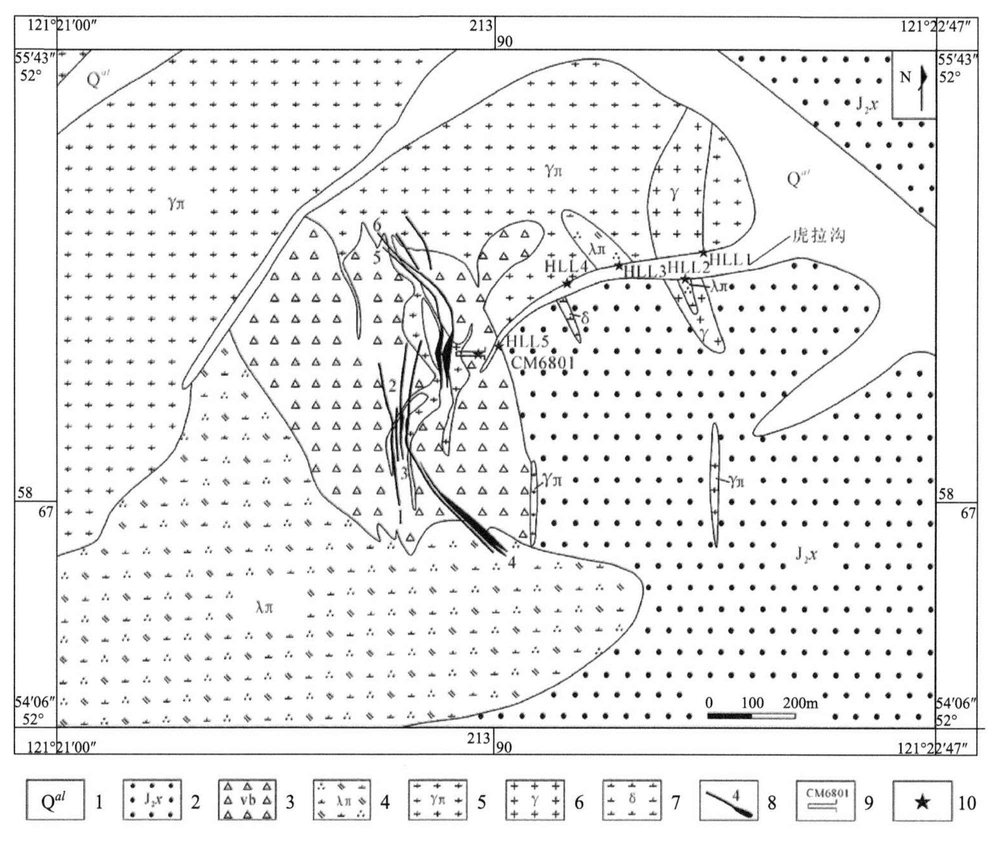

图2　上黑龙江盆地虎拉林金矿床地质简图（据 Ding Qingfeng et al.，2006 修改）

Fig. 2　The sketch regional geological map of Hulalin gold deposits in Upper Heilongjiang basin (modified after Ding Qingfeng et al., 2006).

1-第四系冲积物；2-侏罗系中统绣峰组砂岩；3-隐爆角砾岩；4-早白垩世石英斑岩；5-早白垩世花岗斑岩；6-晚二叠世花岗岩；7-闪长岩；8-矿体及编号；9-坑道及编号；10-取样点位

1-Quaternary flood accumulation; 2-Middle Jurassic Xiufeng formation; 3-Crytoexplosive breccia; 4-Early Cretaceous Quartzite porphyry; 5-Early Cretaceous Granite porphyry; 6-LatePermian Granite; 7-Diorite; 8-Ore body and its number; 9-tunnel and its number; 10-sampling point

图 3 上黑龙江盆地虎拉林早白垩世侵入岩岩体及显微照片

Fig. 3 Early Cretaceous intrusion rock pictures and photomicographs of Huilalin in Upper Heilongjiang Basin.

（a）-花岗斑岩野外露头（HLL1 点）；（b）-石英斑岩野外露头（HLL2 点）；（c）-花岗斑岩（HLL1-4）；（d）-斑状结构（HLL1-4），正交偏光；（e）-斑状结构（HLL1-5），正交偏光；（f）-石英斑岩（HLL2-1）；（g）-斑状结构（HLL2-1），正交偏光；（h）-斑状结构（HLL3-1），正交偏光；Bt-黑云母；Kfs-钾长石；Pl-斜长石；Py-黄铁矿；Qtz-石英

(a)-field photos of granite porphyry (HLL1position); (b)-field photos of quartz porphyry (HLL2position); (c)-granite porphyry (HLL1-4); (d)-porphyritic structure (HLL1-4), cross-polarized light; (e)-porphyritic structure (HLL1-5), cross-polarized light; (f)-quartz porphyry (HLL2-1); (g)-porphyritic structure (HLL2-1), cross-polarized light; (h)-porphyritic structure (HLL3-1), crosspolarized light; Bt-biotite; Kfs-K-feldspar; Pl-plagioclase; Py-pyrite; Qtz-quartz

褐黄色石英斑岩（HLL2-1、HLL3-1）：取样点位分别为 52°55′29.76″N，121°22′29.99″E、52°55′28.26″N，121°22′14.35″E[图 2、图 3（b）]。褐黄色，斑状结构，块状构造。斑晶约占 25%，以石英，少量钾长石等为主，基质多为隐晶质。石英斑晶约为 85%，多呈自形粒状，颗粒较大，粒径 0.1～2.4mm，表面具碎裂结构；钾长石斑晶约为 10%，呈板状、长柱状，粒径 0.1～0.9mm，偶可见有卡式双晶结构[图 3（g）、（h）]。岩石蚀变以钾化、绢英岩化为主[图 3（f）]。

2.2 LA-ICP-MS 锆石 U-Pb 和 Hf 同位素测试

将野外采集样品破碎，按重力和磁选方法进行分选，最后在双目镜下根据锆石颜色、自形程度、形态特征初步分选，挑选具有代表性的锆石；将挑选出的锆石用环氧树脂制靶、打磨及抛光。锆石的分选工作、阴极发光（CL）图像及透、反射电子像（BSE）均在北京锆年领航科技有限公司完成，后两步工作选用仪器为日本 JEOL 公司生产的 JSM6510 型扫描电子显微镜。综合分析锆石反射电子像及阴极发光图像，识别锆石内部结构及晶体形貌，以选取测试最佳点。LA-ICP-MS 锆石 U-Pb 定年测试分析在中国地质科学院矿产资源研究所自然资源部成矿作用与资源评价重点实验室完成，所用仪器为德国 Finnigan Neptune 型 LA-ICP-MS 及与之相配套的 New Wave UP213nm 激光剥蚀系统。本次测试采用激光束斑直径为 30μm，采用 He 作为剥蚀气体，采用标准锆石 GJ-1 作为外部年龄标准进行 U、Pb 同位素分馏矫正，详细的分析流程及原理见文（Hou Kejun et al.，2009）。测试数据采用 ICPMSDateCal8.0 程序处理，操作流程及原理详见文（Liu Yongsheng et al.，2008）；年龄计算及图件生成均采用国际标准程序 Isoplot（ver3.0）（Ludwig，2003）进行处理。

在锆石 LA-ICP-MS U-Pb 同位素分析的基础上，对虎拉林花岗斑岩及石英斑岩 3 件样品进行锆石原位 Hf 同位素分析，该测试在中国地质科学院矿产资源研究所自然资源部成矿作用与资源评价重点实验室进行。测试所用仪器为配有 193nm 激光取样系统的 Finnigan Neptune 型多接收电感耦合等离子质谱仪（LA-MC-ICPMS）及 Newwave UP213 激光剥蚀系统，试验过程采用 He 作为剥蚀物质载气，激光束斑直径为 55μm，剥蚀时间为 30s，样品测定以国际标准锆石 GJ-1 作为外标。详细测试流程及仪器运行过程见文（Hou Kejun et al.，2007）。

2.3 主微量元素分析方法

野外采集样品清洗后，由北京锆年领航科技有限公司粉碎至 200 目后，送至核工业北京地质研究院分析测试研究中心进行主微量分析测试，主量元素采用 X 射线荧光光谱法（XRF），测试仪器为 PW2404 顺序扫描 X 射线荧光光谱仪，SiO_2、Al_2O_3、MgO、Na_2O 检出限为 0.015%，CaO、K_2O、TiO_2 检出限为 0.01%，TFe_2O_3、MnO、P_2O_5 检出限为 0.005%，测试流程参考 GB/T 14506.14—2010；微量和稀土元素测定采用电感耦合等离子体质谱仪，对于含量大于 20×10^{-6} 的元素，误差为±5%；含量小于 20×10^{-6} 的元素，误差为±10%，测试流程参考 GB/T 14506.30—2010。

3 测试结果

3.1 锆石 LA-ICP-MS 年代学

3.1.1 灰黑色花岗斑岩（样品编号 HLL1-4）

锆石形态均呈自形晶—半自形晶，少量锆石边部缺失，粒径 50～210μm，多呈长柱状、短柱状，少量呈锥状及浑圆状，长宽比介于 1∶1～3.5∶1，大部分呈无色透明，部分呈褐黑色，少量锆石发育有增生边结构。阴极发光（CL）图像显示，锆石具有清晰的生长韵律环带[图 4（a）]；挑选 17 颗结晶较好的锆石对其进行测试，共 17 个测试点，锆石 Th 含量介于 42.09×10^{-6}～403.80×10^{-6}，平均为 131.76×10^{-6}；U 含量介于 173.84×10^{-6}～1774.45×10^{-6}，平均为 409.83×10^{-6}，Th/U 值介于 0.23～0.50，均大于 0.1（表 1），且锆石 Ce 显示强烈正异常[图 5（a）]，据以上特征表明锆石为岩浆成因（Qu Xiaoming et al.，2006；Zhao Zhidan et al.，2018）。17 个测点中包括 2 个早期捕获锆石（HLL1-4-4、HLL1-4-10），$^{206}Pb/^{238}U$ 年龄分别为 251Ma±7Ma、252Ma±4Ma。其余 15 个测点均落在谐和线及其附近[图 6（a）]，其 $^{206}Pb/^{238}U$ 加权平均值为 141.7Ma±1.1Ma（MSWD＝0.086），这一年龄代表该岩体的岩浆结晶年龄。

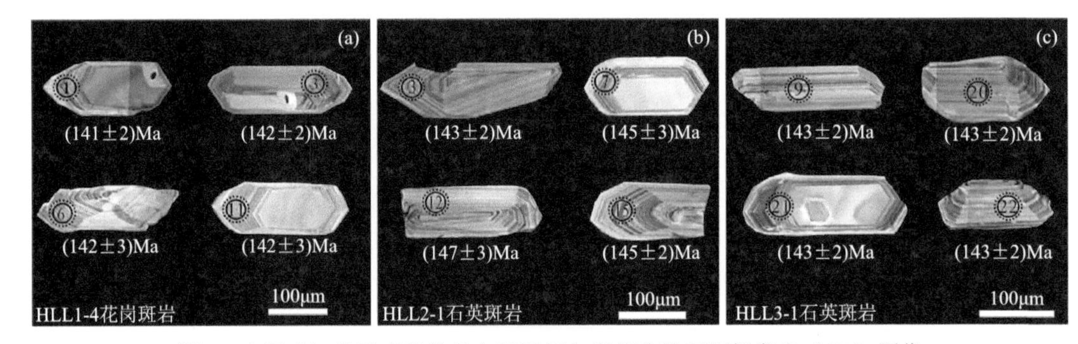

图 4 上黑龙江盆地虎拉林早白垩世侵入岩部分锆石阴极发光（CL）图像

Fig. 4 Cathodoluminescenc (CL) images of zircons selected for analysis from the Early Cretaceous intrusion rock in Hulalin of Upper Heilongjiang basin.

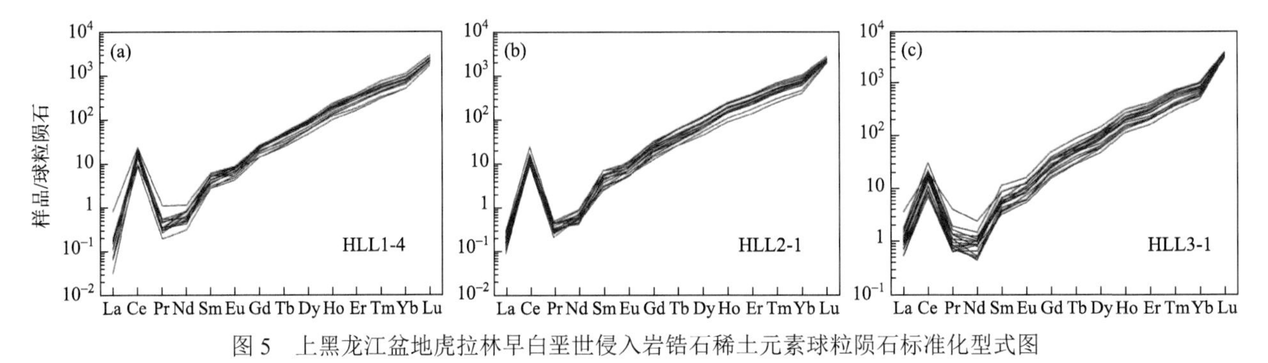

图 5 上黑龙江盆地虎拉林早白垩世侵入岩锆石稀土元素球粒陨石标准化型式图

Fig. 5 Chondrite-normalized REE patterns for zircons of the Early Cretaceous intrusion rock in Hulalin of Upper Heilongjiang basin.

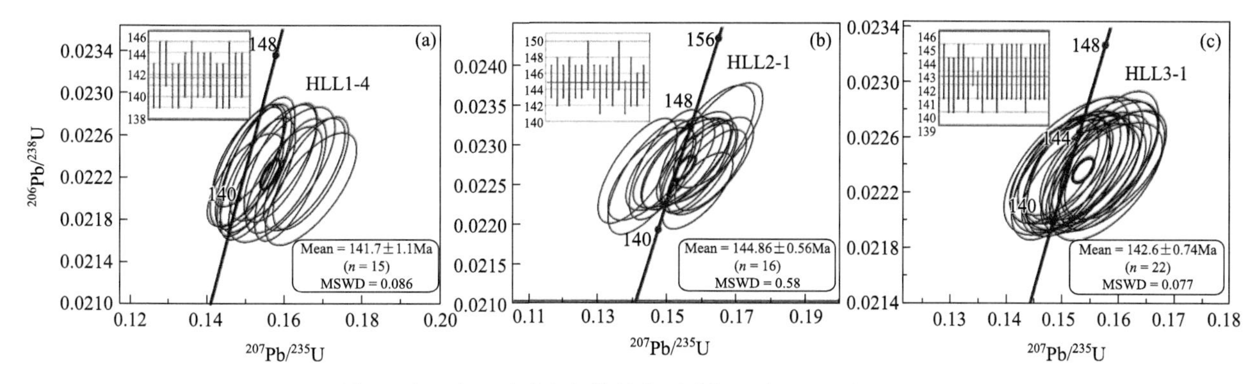

图 6 上黑龙江盆地虎拉林早白垩世侵入岩 U-Pb 谐和图解

Fig. 6 Zircon U-Pb Concordia diagrams the Early Cretaceous intrusion rock in Hulalin of Upper Heilongjiang basin.

表 1 上黑龙江盆地虎拉林矿床早白垩世侵入岩锆石 LA-ICP-MS U-Pb 分析结果表

Table 1 LA-ICP-MS zircon U-Pb analyses ofthe Early Cretaceous intrusion rock in Hulalin of Upper Heilongjiang basin

样品号及分析点号	含量（$\times10^{-6}$）			Th/U	同位素比值±1σ			年龄（Ma）±1σ			谐和度(%)
	Pb	Th	U		$^{207}Pb/^{206}Pb$	$^{207}Pb/^{235}U$	$^{206}Pb/^{238}U$	$^{207}Pb/^{206}Pb$	$^{207}Pb/^{235}U$	$^{206}Pb/^{238}U$	
HLL1-4 花岗斑岩											
HLL1-4-1	27.55	149.32	435.52	0.34	0.0501±0.0020	0.1530±0.0064	0.0222±0.0004	201±66	145±6	141±2	97
HLL1-4-2	20.73	109.56	314.05	0.35	0.0547±0.0021	0.1666±0.0064	0.0222±0.0004	400±57	156±6	142±2	90
HLL1-4-3	22.24	117.10	348.07	0.34	0.0516±0.0020	0.1588±0.0063	0.0223±0.0004	268±60	150±5	142±2	95
HLL1-4-4	26.61	92.08	228.93	0.40	0.0548±0.0023	0.2983±0.0063	0.0396±0.0004	403±56	265±11	251±7	94
HLL1-4-5	22.30	117.77	363.91	0.32	0.0489±0.0021	0.1497±0.0065	0.0222±0.0003	147±77	142±6	141±2	99
HLL1-4-6	22.90	120.28	361.17	0.33	0.0522±0.0021	0.1596±0.0064	0.0222±0.0003	293±65	150±6	141±2	93
HLL1-4-7	18.91	97.03	314.72	0.31	0.0499±0.0021	0.1521±0.0063	0.0223±0.0004	188±63	144±6	142±3	98
HLL1-4-8	22.88	115.86	366.54	0.32	0.0489±0.0020	0.1501±0.0062	0.0223±0.0004	141±66	142±6	142±2	99
HLL1-4-9	26.52	137.41	370.69	0.37	0.0538±0.0022	0.1649±0.0069	0.0222±0.0004	361±66	155±6	142±2	91
HLL1-4-10	159.89	403.80	1774.45	0.23	0.0534±0.0009	0.2944±0.0064	0.0398±0.0007	347±23	262±5	252±4	95
HLL1-4-11	9.27	42.09	173.84	0.24	0.0538±0.0032	0.1592±0.0090	0.0222±0.0004	364±93	150±8	142±3	94
HLL1-4-12	37.32	225.43	454.20	0.50	0.0470±0.0020	0.1508±0.0063	0.0224±0.0003	133±71	143±6	143±2	99
HLL1-4-13	17.72	85.10	283.09	0.30	0.0547±0.0026	0.1666±0.0077	0.0221±0.0004	399±75	156±7	141±2	89
HLL1-4-14	15.06	72.88	240.76	0.30	0.0535±0.0022	0.1613±0.0061	0.0222±0.0004	348±54	152±5	141±2	92
HLL1-4-15	15.13	74.62	248.29	0.30	0.0512±0.0024	0.1569±0.0074	0.0223±0.0004	251±78	148±6	142±2	96
HLL1-4-16	17.36	90.26	286.07	0.32	0.0500±0.0020	0.1529±0.0064	0.0223±0.0004	195±62	144±6	142±3	98
HLL1-4-17	31.99	189.27	402.80	0.47	0.0489±0.0018	0.1510±0.0059	0.0223±0.0004	141±60	143±5	142±2	99
HLL2-1 石英斑岩											
HLL2-1-1	23.90	123.70	309.74	0.40	0.0513±0.0023	0.1595±0.0072	0.0227±0.0004	255±72	150±6	145±2	96
HLL2-1-2	11.72	51.22	194.64	0.26	0.0515±0.0027	0.1616±0.0091	0.0230±0.0005	262±89	152±8	147±3	96
HLL2-1-3	16.60	75.60	283.35	0.27	0.0522±0.0024	0.1590±0.0067	0.0224±0.0003	292±70	150±6	143±2	95
HLL2-1-4	18.37	79.32	284.05	0.28	0.0500±0.0027	0.1558±0.0088	0.0228±0.0004	197±97	147±8	145±3	98
HLL2-1-5	13.49	62.20	207.70	0.30	0.0470±0.0027	0.1456±0.0083	0.0226±0.0004	50±92	138±7	144±2	95
HLL2-1-6	14.33	64.99	239.51	0.27	0.0493±0.0029	0.1544±0.0093	0.0228±0.0004	160±106	146±8	145±2	99
HLL2-1-7	11.66	47.97	202.28	0.24	0.0488±0.0025	0.1497±0.0071	0.0228±0.0005	138±72	142±6	145±3	97
HLL2-1-8	15.93	69.73	273.33	0.26	0.0495±0.0025	0.1545±0.0078	0.0228±0.0004	170±85	146±7	145±2	99
HLL2-1-9	14.46	58.82	261.01	0.23	0.0513±0.0024	0.1625±0.0083	0.0227±0.0004	256±83	153±7	145±3	94
HLL2-1-10	11.19	46.03	217.80	0.21	0.0493±0.0031	0.1509±0.0085	0.0228±0.0004	163±97	143±7	145±2	98
HLL2-1-11	13.08	54.19	234.03	0.23	0.0505±0.0024	0.1584±0.0076	0.0228±0.0003	219±83	149±7	145±2	97

续表

样品号及分析点号	含量（$\times10^{-6}$）			Th/U	同位素比值$\pm1\sigma$			年龄（Ma）$\pm1\sigma$			谐和度（%）
	Pb	Th	U		^{207}Pb/^{206}Pb	^{207}Pb/^{235}U	^{206}Pb/^{238}U	^{207}Pb/^{206}Pb	^{207}Pb/^{235}U	^{206}Pb/^{238}U	
HLL2-1-12	19.06	81.45	265.99	0.31	0.0515±0.0026	0.1648±0.0085	0.0231±0.0004	263±83	155±7	147±3	95
HLL2-1-13	16.14	70.82	282.50	0.25	0.0507±0.0021	0.1584±0.0068	0.0228±0.0004	227±67	149±6	145±2	97
HLL2-1-14	12.24	53.07	214.73	0.25	0.0472±0.0026	0.1452±0.0081	0.0226±0.0005	62±84	138±7	144±3	95
HLL2-1-15	25.26	119.45	328.14	0.36	0.0522±0.0023	0.1626±0.0069	0.0227±0.0003	296±69	153±6	145±2	94
HLL2-1-16	11.96	53.64	211.78	0.25	0.0460±0.0025	0.1415±0.0075	0.0227±0.0004	3±77	134±7	145±3	92
HLL3-1 石英斑岩											
HLL3-1-1	60.83	324.01	965.91	0.34	0.0486±0.0017	0.1504±0.0056	0.0224±0.0002	127±68	142±5	143±1	99
HLL3-1-2	51.45	267.45	831.50	0.32	0.0493±0.0017	0.1527±0.0061	0.0224±0.0003	163±65	144±5	143±2	98
HLL3-1-3	35.94	167.74	695.20	0.24	0.0512±0.0022	0.1574±0.0068	0.0223±0.0003	249±78	148±6	142±2	95
HLL3-1-4	29.57	166.06	452.43	0.37	0.0494±0.0026	0.1519±0.0082	0.0224±0.0003	169±97	144±7	143±2	99
HLL3-1-5	53.05	273.16	791.90	0.34	0.0493±0.0018	0.1516±0.0057	0.0224±0.0003	162±65	143±5	143±2	99
HLL3-1-6	68.09	348.88	1097.98	0.32	0.0484±0.0017	0.1485±0.0052	0.0224±0.0003	118±61	141±5	143±2	98
HLL3-1-7	42.79	234.56	689.11	0.34	0.0508±0.0018	0.1558±0.0054	0.0224±0.0003	233±58	147±5	143±2	97
HLL3-1-8	84.60	499.64	923.46	0.54	0.0489±0.0015	0.1508±0.0047	0.0224±0.0003	144±52	143±4	143±2	99
HLL3-1-9	34.27	157.19	671.50	0.23	0.0506±0.0024	0.1531±0.0069	0.0223±0.0003	222±82	145±6	142±2	98
HLL3-1-10	35.92	182.17	643.99	0.28	0.0505±0.0019	0.1555±0.0062	0.0223±0.0003	220±70	147±5	142±2	96
HLL3-1-11	30.69	134.93	592.26	0.23	0.0493±0.0024	0.1511±0.0070	0.0224±0.0003	160±85	143±6	143±2	99
HLL3-1-12	41.61	200.63	706.17	0.28	0.0492±0.0019	0.1521±0.0062	0.0224±0.0003	157±72	144±5	143±2	99
HLL3-1-13	33.57	155.27	550.39	0.28	0.0518±0.0023	0.1580±0.0067	0.0223±0.0003	277±72	149±6	142±2	95
HLL3-1-14	156.66	979.15	1874.04	0.52	0.0494±0.0011	0.1525±0.0040	0.0223±0.0002	167±41	144±4	142±1	98
HLL3-1-15	23.01	91.13	508.24	0.18	0.0498±0.0024	0.1518±0.0071	0.0223±0.0003	183±85	143±6	142±2	99
HLL3-1-16	42.94	208.55	780.17	0.27	0.0504±0.0019	0.1543±0.0055	0.0224±0.0003	213±62	146±5	143±2	97
HLL3-1-17	58.47	297.95	711.80	0.42	0.0516±0.0020	0.1582±0.0059	0.0224±0.0003	265±59	149±5	143±2	95
HLL3-1-18	55.73	294.94	878.79	0.34	0.0489±0.0020	0.1496±0.0058	0.0223±0.0003	142±7	142±5	142±2	99
HLL3-1-19	31.39	157.54	590.35	0.27	0.0508±0.0018	0.1570±0.0058	0.0224±0.0003	232±64	148±5	143±2	96
HLL3-1-20	39.83	207.59	625.01	0.33	0.0523±0.0021	0.1594±0.0060	0.0224±0.0003	297±64	150±5	143±2	95
HLL3-1-21	43.40	204.53	821.96	0.25	0.0512±0.0018	0.1583±0.0058	0.0224±0.0003	251±61	149±5	143±2	95
HLL3-1-22	26.72	120.54	502.33	0.24	0.0502±0.0025	0.1550±0.0079	0.0224±0.0003	205±92	146±7	143±2	97

3.1.2　石英斑岩（样品编号 HLL2-1、HLL3-1）

锆石均呈自形晶—半自形晶，少量锆石边部缺失，粒径 40～280μm，多呈长柱状及短柱状，少量呈浑圆状，长宽比介于 1.2∶1～4∶1，大部分呈无色透明。阴极发光（CL）图像显示，锆石均具有清晰的生长韵律环带[图 4（b）、（c）]；样品 HLL2-1、HLL3-1 分别挑选 16 颗、22 颗结晶较好的锆石进行测试，分别共 16 个、22 个测试点，前者锆石 Th 含量介于 46.03×10^{-6}～123.70×10^{-6}，平均为 69.51×10^{-6}；U 含量介于 194.64×10^{-6}～328.14×10^{-6}，平均为 250.66×10^{-6}，Th/U 比值介于 0.21～0.40，均大于 0.1（表 1）；后者锆石 Th 含量介于 91.13×10^{-6}～979.15×10^{-6}，平均为 257.89×10^{-6}；U 含量介于 452.43×10^{-6}～1874.04×10^{-6}，平均为 768.39×10^{-6}，Th/U 比值介于 0.18～0.54，均大于 0.1（表 1），锆石 Ce 均显示强烈正异常[图 5（b）、（c）]，据以上特征表明两样品锆石均为岩浆成因（Qu Xiaoming et al.，2006；Zhao Zhidan et al.，2018）。两样品锆石测点均落在谐和线及其附近[图 6（b）、（c）]，显示出很好的谐和性，其 ^{206}Pb/^{238}U 加权平均值分别为 144.9±0.56Ma（MSWD＝0.580）、142.6±0.74Ma（MSWD＝0.077），该年龄代表该岩体的岩浆结晶年龄。

3.2 锆石 Hf 同位素特征

虎拉林花岗斑岩及石英斑岩 3 件样品锆石原位 Hf 同位素分析结果见表 2。花岗斑岩（HLL1-4）所测 10 颗锆石的 $^{176}Yb/^{177}Hf$、$^{176}Lu/^{177}Hf$ 值分别介于 0.025279～0.044370、0.000478～0.000894，$^{176}Hf/^{177}Hf$ 值为 0.282655～0.282776，加权平均值为 0.282727±0.000025（2σ，$n = 10$），$\varepsilon_{Hf}(t)$ 值大部分为负值，介于−4.14～0.16，均值为−1.59，Hf 同位素单阶段模式年龄（t_{DM1}）和两阶段模式年龄（t_{DM2}）分别变化于 667.66～836.42Ma、987.08～1259.04Ma。石英斑岩（HLL2-1、HLL3-1）所测 20 颗锆石的 $^{176}Yb/^{177}Hf$、$^{176}Lu/^{177}Hf$ 值分别介于 0.011255～0.024552、0.000417～0.000857，$^{176}Hf/^{177}Hf$ 值为 0.282708～0.282811，加权平均值为 0.282747±0.000023（2σ，$n = 20$），$\varepsilon_{Hf}(t)$ 值大部分为负值，介于−2.25～1.37，均值为−0.89，Hf 同位素单阶段模式年龄（t_{DM1}）和两阶段模式年龄（t_{DM2}）分别变化于 618.59～762.04Ma、908.38～1138.56Ma。

表 2 上黑龙江盆地虎拉林矿床早白垩世侵入岩锆石 Hf 同位素分析结果表

Table 2 LA-ICP-MS zircon Hf analyses of the Early Cretaceous intrusion rock in Hulalin of Upper Heilongjiang basin

测点号	年龄（Ma）	$^{176}Yb/^{177}Hf$	2σ	$^{176}Lu/^{177}Hf$	2σ	$^{176}Hf/^{177}Hf$	2σ	f_s	$\varepsilon_{Hf}(t)$	t_{DM1}（Ma）	t_{DM2}（Ma）
HLL1-4 花岗斑岩											
HLL1-4-01	141	0.029025	0.000207	0.000894	0.000004	0.282702	0.000021	−0.97	−2.48	777.26	1156.39
HLL1-4-02	142	0.025279	0.000162	0.000719	0.000003	0.282763	0.000023	−0.98	−0.31	687.50	1016.60
HLL1-4-03	141	0.033824	0.000182	0.000644	0.000005	0.282776	0.000021	−0.98	0.16	667.66	987.08
HLL1-4-04	142	0.036665	0.000186	0.000772	0.000002	0.282715	0.000022	−0.98	−2.01	755.89	1124.88
HLL1-4-05	142	0.030913	0.000325	0.000833	0.000004	0.282773	0.000027	−0.97	0.03	676.07	995.69
HLL1-4-06	142	0.029343	0.000413	0.000614	0.000007	0.282694	0.000027	−0.98	−2.77	782.71	1172.11
HLL1-4-07	141	0.028775	0.000166	0.000493	0.000004	0.282740	0.000025	−0.99	−1.13	715.62	1067.79
HLL1-4-08	141	0.033875	0.000064	0.000635	0.000011	0.282713	0.000029	−0.98	−2.08	756.04	1129.28
HLL1-4-09	142	0.039722	0.000114	0.000594	0.000004	0.282655	0.000028	−0.98	−4.14	836.42	1259.04
HLL1-4-10	142	0.044370	0.000487	0.000478	0.000001	0.282738	0.000028	−0.99	−1.19	717.91	1071.24
HLL2-1、HLL3-1 石英斑岩											
HLL2-1-01	143	0.023377	0.000103	0.000520	0.000013	0.282712	0.000021	−0.98	−2.13	755.79	1130.65
HLL2-1-02	145	0.019134	0.000165	0.000417	0.000004	0.282795	0.000021	−0.99	0.80	638.37	942.35
HLL2-1-03	145	0.017311	0.000053	0.000511	0.000010	0.282730	0.000025	−0.98	−1.48	730.08	1088.22
HLL2-1-04	145	0.021597	0.000070	0.000586	0.000005	0.282750	0.000023	−0.98	−0.79	704.01	1044.43
HLL2-1-05	145	0.023907	0.000244	0.000545	0.000011	0.282754	0.000024	−0.98	−0.65	697.82	1035.42
HLL2-1-06	145	0.018528	0.000244	0.000461	0.000010	0.282750	0.000022	−0.99	−0.78	701.54	1043.39
HLL2-1-07	145	0.014319	0.000091	0.000576	0.000004	0.282769	0.000021	−0.98	−0.12	677.22	1001.53
HLL2-1-08	147	0.020578	0.000464	0.000602	0.000002	0.282741	0.000024	−0.98	−1.10	716.73	1063.27
HLL2-1-09	145	0.018155	0.000052	0.000795	0.000003	0.282735	0.000027	−0.98	−1.32	729.19	1079.76
HLL2-1-10	145	0.015342	0.000068	0.000788	0.000004	0.282768	0.000028	−0.98	−0.13	681.60	1003.76
HLL3-1-01	142	0.015186	0.000486	0.000580	0.000004	0.282709	0.000021	−0.98	−2.24	761.29	1138.56
HLL3-1-02	143	0.011255	0.000101	0.000509	0.000002	0.282723	0.000022	−0.98	−1.74	740.24	1105.88
HLL3-1-03	143	0.014563	0.000395	0.000604	0.000004	0.282708	0.000025	−0.98	−2.25	762.04	1138.53
HLL3-1-04	143	0.017548	0.000247	0.000623	0.000012	0.282749	0.000026	−0.98	−0.82	706.09	1048.09
HLL3-1-05	142	0.014885	0.000350	0.000585	0.000005	0.282746	0.000021	−0.98	−0.91	708.72	1053.84
HLL3-1-06	142	0.012494	0.000324	0.000423	0.000002	0.282784	0.000021	−0.99	0.42	653.28	968.15
HLL3-1-07	142	0.014877	0.000224	0.000468	0.000004	0.282711	0.000025	−0.99	−2.14	755.19	1131.64
HLL3-1-08	143	0.016723	0.000121	0.000577	0.000004	0.282811	0.000027	−0.98	1.37	618.59	908.38
HLL3-1-09	143	0.024285	0.000138	0.000578	0.000006	0.282751	0.000021	−0.98	−0.74	701.74	1042.17
HLL3-1-10	143	0.024552	0.000144	0.000857	0.000017	0.282746	0.000025	−0.97	−0.93	714.61	1056.12

3.3 地球化学特征

3.3.1 主量元素

虎拉林花岗斑岩、石英斑岩主量元素分析结果见表 3。花岗斑岩 SiO_2、Al_2O_3、MgO、CaO、全碱（$Na_2O + K_2O$）的含量分别为 66.21%～66.70%、15.18%～15.45%、1.36%～1.49%、2.27%～2.41%、7.56%～8.17%，LOI 均值为 2.80%，K_2O/Na_2O 值为 1.20～1.48，$Mg^{\#}$值为 51～52。样品碱度率（AR）为 2.50～2.69，里特曼指数（σ）为 2.35～2.80（均小于 3.3），在 TAS 图解上[图 7（a）]，样品落在石英二长岩区域，在 SiO_2-K_2O 图解[图 7（b）]中，样品主要落入高钾钙碱性-钾玄岩序列范围内，显示由高钾钙碱性序列向钾玄岩序列过渡。样品铝饱和指数 A/CNK 均小于 1.05，显示样品属弱过铝质系列。花岗斑岩 LOI 均值相对偏高，表明岩石经历了弱蚀变，与手标本现象一致。石英斑岩 SiO_2、Al_2O_3、MgO、CaO、全碱（$Na_2O + K_2O$）的含量分别为 66.34%～68.25%、13.77%～15.05%，0.31%～0.36%、0.07%～0.08%、6.76%～9.06%，LOI 均值 4.86%，K_2O/Na_2O 值为 37.03～45.59，$Mg^{\#}$值为 13～17。样品碱度率（AR）为 2.62～4.15，里特曼指数（σ）为 1.79～3.17（均>3.3），在 TAS 图解上[图 7（a）]，大部分样品落在石英二长岩区域，在 SiO_2-K_2O 图解[图 7（b）]中，样品落入钾玄岩序列范围内。石英斑岩中 Na_2O 整体较低，K_2O 相对较高，以及 LOI 偏高均表明石英斑岩经历了弱钾化、绢英岩化蚀变，与手标本现象一致。

表 3　上黑龙江盆地虎拉林早白垩世侵入岩体主量（%）、稀土及微量元素（10^{-6}）分析结果

Table 3　The data of major (%), rare earth and trace element (10^{-6}) of the Early Cretaceous intrusion rock in Hulalin of Upper Heilongjiang basin

样品号	HLL1-4	HLL1-5	HLL1-6	HLL2-1	HLL2-2	HLL2-3	HLL3-1-1	HLL3-1-2
岩性		花岗斑岩				石英斑岩		
SiO_2	66.70	66.58	66.21	68.25	66.81	66.34	67.56	67.43
TiO_2	0.39	0.39	0.38	0.43	0.44	0.42	0.40	0.41
Al_2O_3	15.24	15.18	15.45	14.74	13.83	13.77	15.02	15.05
Fe_2O_3	1.53	1.61	1.38	2.64	3.90	4.22	3.95	3.94
FeO	1.07	1.07	1.01	0.24	0.26	0.26	0.20	0.24
MnO	0.041	0.039	0.036	0.004	0.004	0.004	0.004	0.004
MgO	1.43	1.49	1.36	0.31	0.34	0.34	0.36	0.36
CaO	2.41	2.27	2.37	0.07	0.07	0.07	0.07	0.08
Na_2O	3.44	3.22	3.30	0.22	0.20	0.22	0.15	0.16
K_2O	4.12	4.58	4.87	8.84	8.25	8.22	6.61	6.62
P_2O_5	0.14	0.12	0.13	0.05	0.10	0.13	0.08	0.08
LOI	2.80	2.77	2.82	3.60	5.25	5.42	5.02	5.02
Total	99.30	99.32	99.32	99.40	99.44	99.42	99.42	99.40
TFe_2O_3	2.72	2.80	2.50	2.91	4.19	4.51	4.17	4.21
$Mg^{\#}$	51.02	51.32	51.87	17.47	13.78	13.03	14.64	14.55
K_2O/Na_2O	1.20	1.42	1.48	40.18	41.25	37.03	45.59	40.86
σ	2.35	2.51	2.80	3.17	2.88	2.93	1.79	1.82
DI	80.20	80.65	81.11	90.55	89.20	88.87	85.99	85.93
La	19.40	19.80	14.80	37.00	36.40	32.40	32.10	32.90
Ce	38.40	41.40	28.30	69.00	66.50	59.70	60.90	61.00

续表

样品号	HLL1-4	HLL1-5	HLL1-6	HLL2-1	HLL2-2	HLL2-3	HLL3-1-1	HLL3-1-2
Pr	4.89	5.36	3.35	8.12	7.57	7.15	6.85	7.11
Nd	19.10	20.20	18.20	28.50	28.00	25.10	24.80	26.10
Sm	3.27	3.70	2.31	4.83	4.64	4.19	3.91	4.30
Eu	0.85	0.84	0.43	0.63	0.67	0.58	0.49	0.54
Gd	2.42	2.86	1.56	3.55	3.15	2.82	3.12	3.21
Tb	0.29	0.37	0.21	0.44	0.40	0.32	0.39	0.38
Dy	1.41	1.69	0.93	1.88	1.56	1.60	1.69	1.70
Ho	0.26	0.24	0.12	0.22	0.20	0.17	0.25	0.24
Er	0.67	0.75	0.40	0.98	0.75	0.76	0.79	0.77
Tm	0.07	0.09	0.05	0.08	0.09	0.09	0.13	0.11
Yb	0.51	0.54	0.37	0.80	0.61	0.54	0.65	0.61
Lu	0.07	0.09	0.05	0.13	0.13	0.10	0.11	0.08
ΣREE	91.59	97.92	71.07	156.15	150.66	135.53	136.18	139.05
ΣLREE	85.91	91.30	67.39	148.08	143.78	129.12	129.05	131.95
ΣHREE	5.69	6.62	3.68	8.07	6.88	6.41	7.13	7.10
LREE/HREE	15.11	13.80	18.30	18.34	20.90	20.16	18.11	18.58
$(La/Yb)_N$	25.86	24.87	26.81	31.33	40.52	40.62	33.47	36.45
δCe	0.91	0.93	0.91	0.90	0.90	0.89	0.93	0.90
δEu	0.88	0.76	0.65	0.44	0.50	0.49	0.41	0.42
V	39.00	50.60	28.10	30.70	32.70	34.60	34.10	30.20
Cr	28.90	31.30	16.20	22.80	30.20	26.80	24.20	22.60
Co	6.11	6.54	3.32	2.18	0.61	1.02	4.65	4.26
Ni	9.52	10.70	5.74	3.47	2.08	2.32	7.74	7.20
Ga	21.20	23.60	13.50	17.10	18.40	18.20	19.60	19.30
Rb	123.0	149.0	84.90	248.0	237.0	240.0	199.0	187.0
Ba	1061	1106	857	1219	1157	1171	1010	967
Th	4.48	4.96	3.30	6.62	6.96	10.80	7.26	6.99
U	1.88	1.98	1.26	1.62	1.47	1.55	1.50	1.47
Ta	0.30	0.32	0.16	0.46	0.39	0.46	0.43	0.39
Nb	5.11	5.28	2.94	6.62	6.26	6.56	5.88	5.90
Sr	754.0	738.0	482.0	146.0	159.0	176.0	228.0	216.0
Zr	156.0	132.0	151.0	133.0	115.0	126.0	149.0	185.0
Hf	3.76	3.40	3.86	3.24	2.79	3.05	3.82	4.50
Y	6.27	6.62	4.00	6.40	5.55	4.91	6.69	6.31
Zr/Hf	41.49	38.82	39.12	41.05	41.22	41.31	39.01	41.11
Nb/Ta	17.03	16.50	18.38	14.42	15.97	14.26	13.71	15.21
Rb/Sr	0.16	0.20	0.18	1.70	1.49	1.36	0.87	0.87

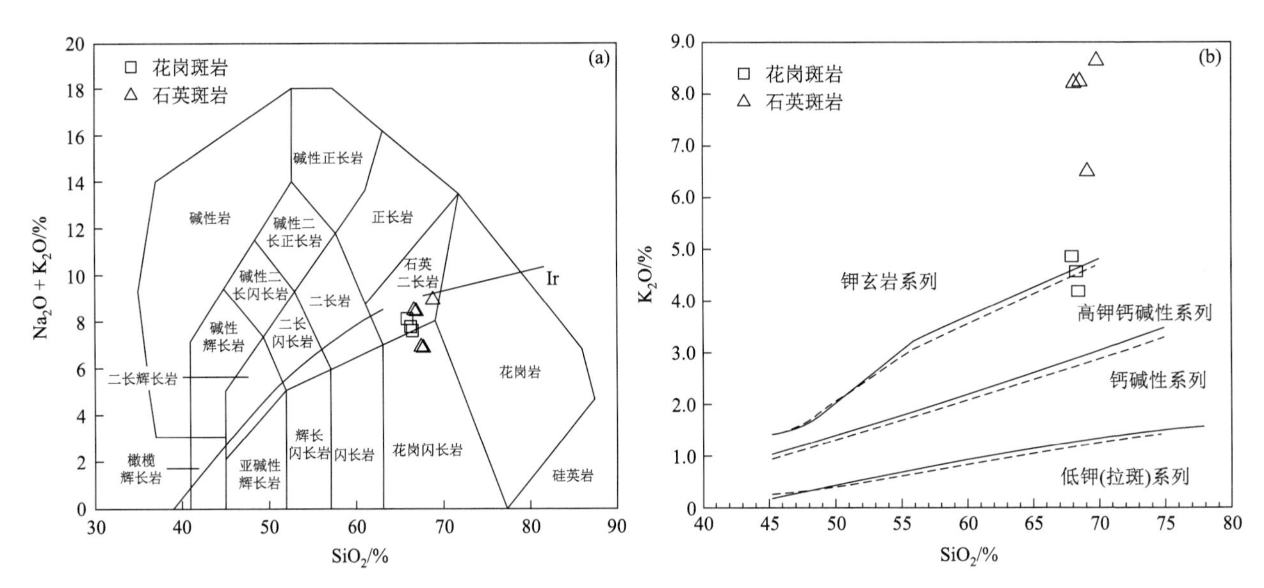

图 7 上黑龙江盆地虎拉林早白垩世侵入岩体 TAS 图解（a）（据 Middlemost，1994）和 SiO_2-K_2O 图解（b）（实线据 Peccerillo and Taylor，1976；虚线据 Middlemost，1985）

Fig. 7 TAS diagrams (a) (after Middlemost, 1994) and SiO_2-K_2O diagram (b) (solid line after Peccerillo and Taylor, 1976; dash line after Middlemost, 1985) of the Early Cretaceous intrusion rock in Hulalin of Upper Heilongjiang basin.

3.3.2 微量元素

虎拉林花岗斑岩及石英斑岩球粒陨石标准化稀土元素配分曲线[图 8（a）]显示相类似的配分型式，均表现为轻稀土（LREE）相对富集，重稀土（HREE）相对亏损的右倾曲线。花岗斑岩 ΣREE、ΣLREE、ΣHREE、LREE/HREE、$(La/Yb)_N$ 分别为 $71.07\times10^{-6}\sim97.92\times10^{-6}$、$67.39\times10^{-6}\sim91.30\times10^{-6}$、$3.68\times10^{-6}\sim6.62\times10^{-6}$、13.80～18.30、24.87～26.81；石英斑岩 ΣREE、ΣLREE、ΣHREE、LREE/HREE、$(La/Yb)_N$ 分别为 $135.53\times10^{-6}\sim156.15\times10^{-6}$、$129.05\times10^{-6}\sim148.08\times10^{-6}$、$6.41\times10^{-6}\sim8.07\times10^{-6}$、18.11～20.90、31.33～40.62；花岗斑岩及石英斑岩 δEu 分别为 0.65～0.88、0.41～0.50。整体上花岗斑岩稀土元素含量相对低于石英斑岩含量，且后者 Eu 负异常较前者明显。

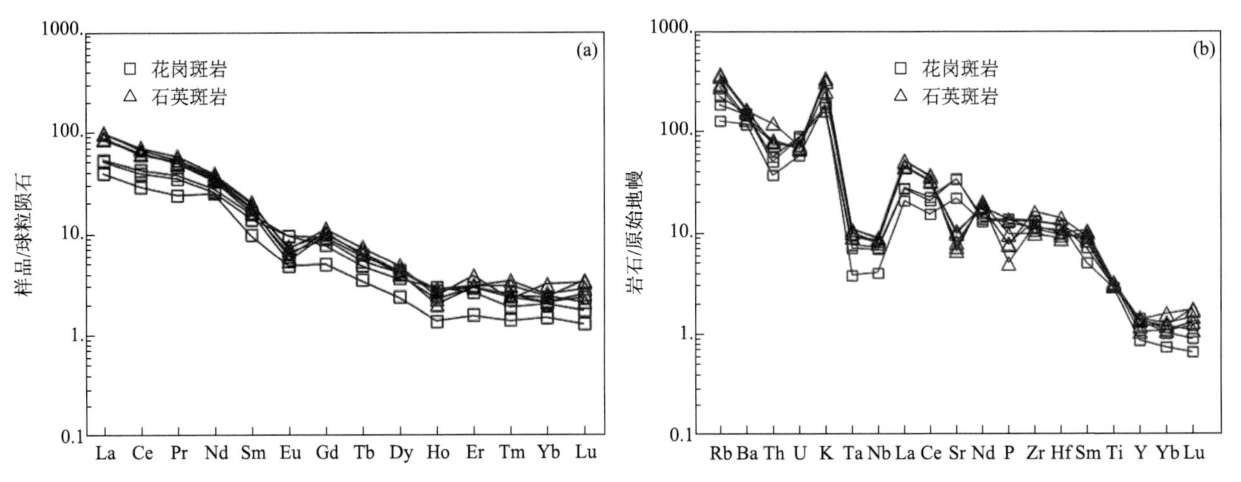

图 8 上黑龙江盆地虎拉林早白垩世侵入岩体球粒陨石标准化稀土元素配分图（a）（标准化值据 Boynton，1984）和原始地幔标准化微量元素蛛网图（b）（标准化值据 Sun and McDonough，1989）

Fig. 8 Chondrite normalized REE patterns. (a) (normalized after Boynton, 1984) and primitive mantle-normalized trace element spider diagrams (b) (normalized after Sun and McDonough，1989) of the Early Cretaceous intrusion rock in Hulalin of Upper Heilongjiang basin.

在原始地幔标准化微量元素蛛网图[图 8（b）]中，两种岩性仍显示出相似的特征，均相对富集大离子亲石元素（LILE）（如Rb、K等）、LREE及不相容元素（如U），强烈亏损高场强元素（HFSE）（如Nb、Ta、Ti、P等），指示母岩浆在上升演化过程中发生了显著的分离结晶作用。如Nb、Ta、Ti的亏损指示富钛矿物的分离，而Sr、P的亏损指示斜长石、磷灰石的分离结晶作用（Zhang Zhaochong et al.，2009；Liu Jun et al.，2013b）。

4 讨论

4.1 形成时代

虎拉林岩体主要为花岗斑岩及石英斑岩等，其锆石CL图像均显示具有清晰的生长韵律环带，较高的Th/U值，锆石Ce显示强烈正异常，均表现出岩浆锆石特征，其测试年龄可代表岩体形成年龄（Gao Yuan et al.，2013）。花岗斑岩（1件）及石英斑岩（2件）$^{206}Pb/^{238}U$加权平均值分别为141.7±1.1Ma（MSWD＝0.086）、144.9±0.56Ma（MSWD＝0.580）、142.6±0.74Ma（MSWD＝0.077），认为在误差范围内岩体形成年龄一致，均为早白垩世。因此确定虎拉林岩体形成年龄为141.7～144.86Ma，是燕山晚期早白垩世岩浆活动的产物。

4.2 物质来源

一般认为，由俯冲板片部分熔融的岩浆具有较高的$Mg^{\#}$（＞50）（Xu Jifeng et al.，2002），而由古老地壳部分熔融的岩浆具有较低的$Mg^{\#}$（＜50）（Xiong Xiaolin et al.，2003；Zhao Zhenhua et al.，2009），因此$Mg^{\#}$可以作为判断岩浆来源的指示剂。虎拉林花岗斑岩、石英斑岩岩体$Mg^{\#}$分别为51～52、13～17，暗示岩体具有古老地壳部分熔融特征；微量元素因稳定的地球化学性质而成为判断岩浆物质来源的重要有效指示剂（Zhang Jing et al.，2010），如La-La/Yb与Yb-La/Y图解、Nb/Ta、Zr/Hf及Rb/Sr等比值（Taylor and McLennan，1985；Cai Hongming et al.，2010；Zhao Di et al.，2018）。La-La/Yb与Yb-La/Y图解显示早白垩世侵入岩体源区具有壳幔混源的性质[图9（a）、（b）]；花岗斑岩、石英斑岩Nb/Ta分别为16.50～18.38、13.71～15.97，花岗斑岩、石英斑岩Zr/Hf分别为38.82～41.49、39.01～41.31，与原始地幔Nb/Ta、Zr/Hf标准值（～18、～37）、原始地壳Nb/Ta、Zr/Hf标准值（～11、～33）相比，显示岩浆均来自壳幔混合源区；花岗斑岩、石英斑岩Rb/Sr分别为0.16～0.20、0.87～1.70，与原始地幔Rb/Sr（0.03～0.047）、原始地壳Rb/Sr（＞0.5）（Sun and McDonough，1989）相比，显示岩浆来源于壳幔混合源区。

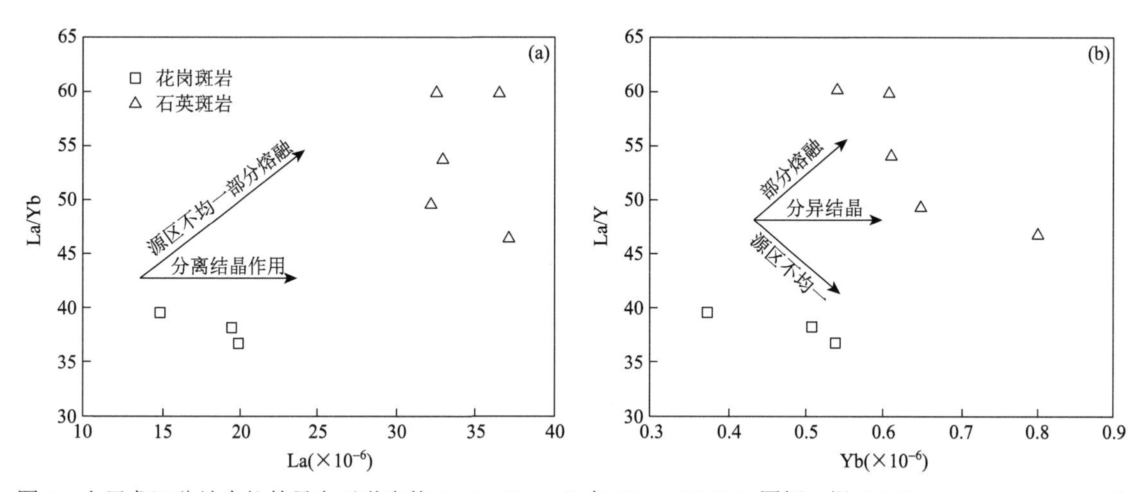

图9 上黑龙江盆地虎拉林早白垩世岩体La-La/Yb（a）与Yb-La/Y（b）图解（据Cai Hongming et al.，2010）

Fig. 9 La-La/Yb (a) and Yb-La/Y (b) diagrams for the Early Cretaceous intrusion rock in Hulalin of Upper Heilongjiang basin (after Cai Hongming et al., 2010).

锆石相对其他矿物具有较好的稳定性，锆石Lu-Hf同位素具有较高的封闭体系温度，从而能为分析岩浆源区及性质提供重要的信息（Wang Tianhao et al.，2014；Che Hewei et al.，2015；Tang Jie，2016）。虎拉林花岗斑岩、石英斑岩锆石Hf同位素测试结果中$^{176}Lu/^{177}Hf$均小于0.0002（表2），表明锆石自结晶以

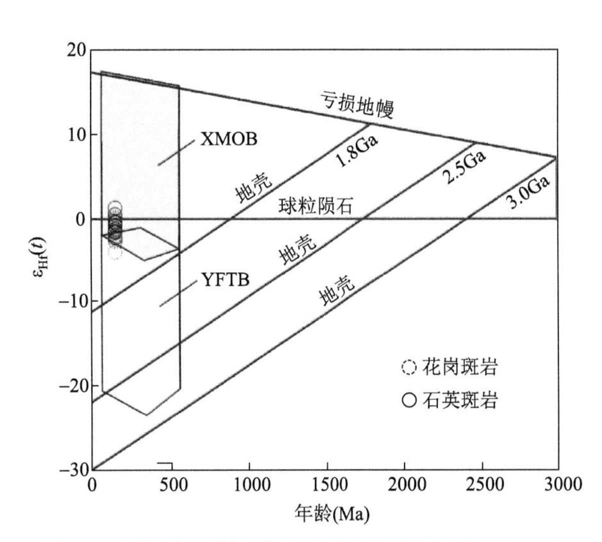

图 10　上黑龙江盆地虎拉林早白垩世岩体锆石 $\varepsilon_{Hf}(t)$-t 图解（据 Wang Tianhao et al.，2014）

Fig. 10　$\varepsilon_{Hf}(t)$-t diagram of zircons of the Early Cretaceous intrusion rock in Hulalin of Upper Heilongjiang Basin (after Wang Tianhao et al., 2014).

XMOB-兴蒙造山带；YFTB-燕山褶皱带

XMOB-XingMeng orogenic Belt；YFTB-YanShan fold belt

来放射成因的 Hf 积累很少，其 $^{176}Hf/^{177}Hf$ 可以代表岩体结晶时的初始 $^{176}Hf/^{177}Hf$（Sun Lixin et al.，2012）。一般而言，岩石中正 $\varepsilon_{Hf}(t)$值表明岩浆来源于亏损地幔或从亏损地幔中新增生的年轻地壳物质部分熔融，负 $\varepsilon_{Hf}(t)$值则代表岩浆来源古老地壳重熔（Wu Fuyuan et al.，2007）。

虎拉林花岗斑岩所测 10 颗锆石 Hf 同位素组成相对比较均一，$^{176}Hf/^{177}Hf$ 值为 0.282655～0.282776，$\varepsilon_{Hf}(t)$介于−4.14～0.16，均值为−1.59。在 $\varepsilon_{Hf}(t)$-t 图解中，除两个数据点分布在球粒陨石演化线之上其余均分布在演化线之下（图 10），样品 Hf 同位素两阶段模式年龄（t_{DM2}）为 987.08～1259.04Ma。石英斑岩所测 20 颗锆石 Hf 同位素组成相对比较均一，$^{176}Hf/^{177}Hf$ 值为 0.282708～0.282811，$\varepsilon_{Hf}(t)$介于−2.25～1.37，均值为−0.89。在 $\varepsilon_{Hf}(t)$-t 图解中，大部分数据点分布在球粒陨石演化线之下（图 10），样品 Hf 同位素两阶段模式年龄（t_{DM2}）为 908.38～1138.56Ma。

综合上述特征表明，虎拉林花岗斑岩、石英斑岩的岩浆均来自中—新元古代古老地壳的熔融，且存在幔源物质或从亏损地幔中新增生地壳物质部分熔融的参与，两种岩体为同源岩浆结晶分异作用的结果。

4.3　成因类型

虎拉林花岗斑岩、石英斑岩形成年龄在误差范围内一致，为早白垩世同期岩浆活动产物。花岗斑岩具有富铝（15.18%～15.45%）、富碱（7.56%～8.17%）特征，并且具有较高的分异系数（DI = 80.20～81.11），为岩浆高程度结晶分异作用产物。根据花岗斑岩样品碱度率 AR（2.50～2.69）、里特曼指数 σ（2.35～2.80）、铝饱和指数 A/CNK 均小于 1.05 等特征，推断该花岗斑岩属于弱铝、高钾钙碱性-钾玄岩系列。石英斑岩同样具有富铝（13.77%～15.05%）、富碱（6.76%～9.06%）特征，且具有高分异系数（DI = 85.93～90.55），为岩浆高度结晶分异作用产物。根据石英斑岩样品碱度率 AR（2.62～4.15）、里特曼指数 σ（1.79～3.17），推断该石英斑岩属于钾玄岩系列。两种斑岩稀土元素、微量元素均具有相似的特征，均相对富集大离子亲石元素（LILEs）、LREE 及不相容元素，强烈亏损高场强元素（HFSEs）。推断虎拉林花岗斑岩与石英斑岩早白垩世侵入岩体可能为同源同期岩浆不同演化阶段高程度分离结晶作用的结果。

在 10000Ga/Al-Y 图解中，两种侵入岩样品点均落在 I 型和 S 型花岗岩区[图 11(a)，Chappell and White，1992]；花岗岩中 P_2O_5 含量是判别岩石类型的一个有利指标（Wang Yinhong et al.，2015），两种侵入岩样品中 P_2O_5 为 0.08～0.14，均明显低于典型的 S 型花岗岩，且 $\omega(P_2O_5)$随着分异作用的进行具有递减趋势[图 11（b）]，而在 S 型花岗岩中，P_2O_5 随着 SiO_2 含量的递增而呈现递增或基本不变的趋势（Wu Fuyuan et al.，2003；Li Xianhua et al.，2007），暗示虎拉林早白垩世花岗岩具有 I 型花岗岩特征；I 型岩浆结晶早期可形成富含 Th 和 Y 的矿物，且 Th、Y 与 Rb 具有正相关关系（Wu Fuyuan et al.，2003；Wang Yinhong et al.，2015；Wang Qisong et al.，2019）。虎拉林早白垩世侵入岩 Th、Y 含量分别为 3.30×10^{-6}～10.80×10^{-6}、4.00×10^{-6}～6.69×10^{-6}，且与 Rb 均呈现正相关趋势[图 11（c）、（d）]。综述上述特征，推断虎拉林早白垩世侵入岩为 I 型花岗岩。

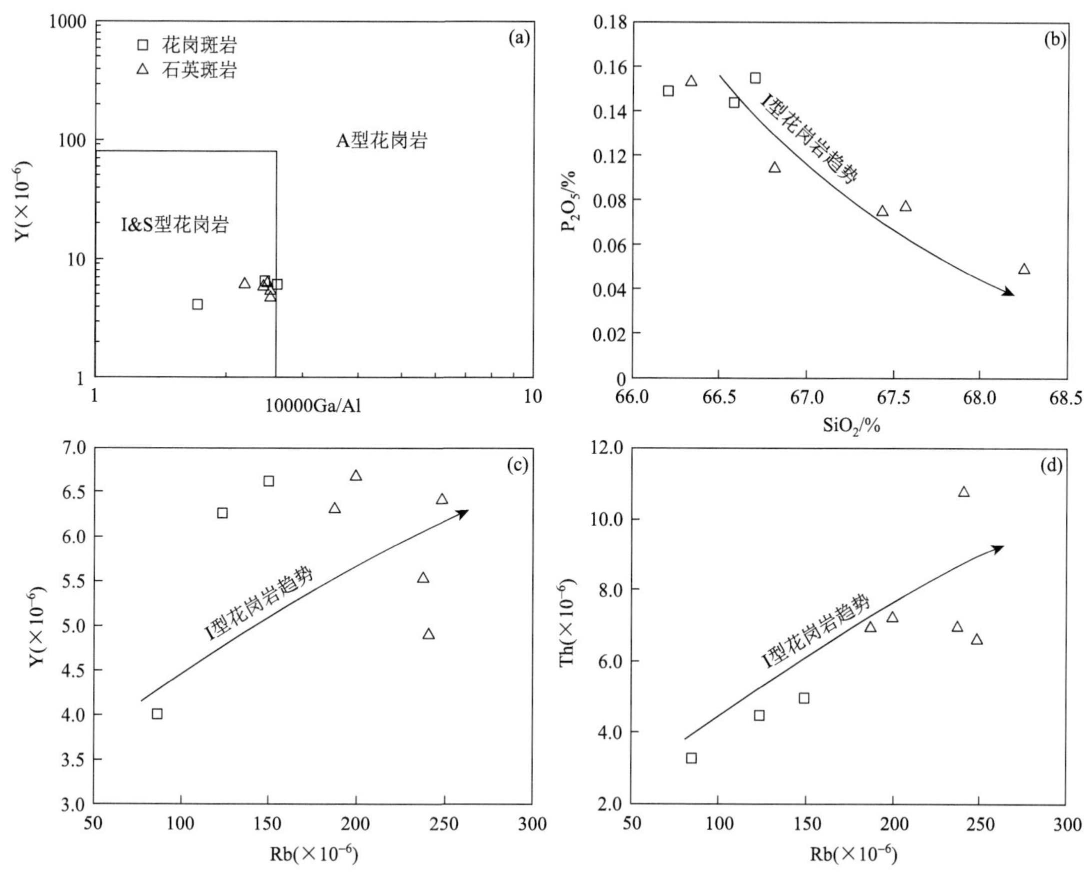

图 11 上黑龙江盆地虎拉林早白垩世侵入岩岩石类型判别图[（a）据 Whalen et al.，1987；（b），（c），（d）据 Li Xianhua et al.，2007]

Fig. 11 Granites discrimination diagram of the Early Cretaceous intrusion rock in Hulalin of Upper Heilongjiang Basin [(a) after Whalen et al., 1987; (b), (c), (d) after Li Xianhua et al., 2007].

4.4 构造背景及形成机制

在花岗岩类(Yb + Nb)-Rb、Yb-Nb 构造环境判别图中，虎拉林早白垩世侵入岩体样品点全部落在火山弧型与同碰撞型区域及两者交界区[图 12（a）、（b）]。经前人证明，落在火山弧型、同碰撞型花岗岩范围内并不一定代表该期岩浆形成于火山弧及同碰撞环境，而仅暗示其源区具有类似火山弧、同碰撞构造环境，对岩体形成确切构造环境需要做出进一步分析（Zhang Zhaochong et al.，2009）。

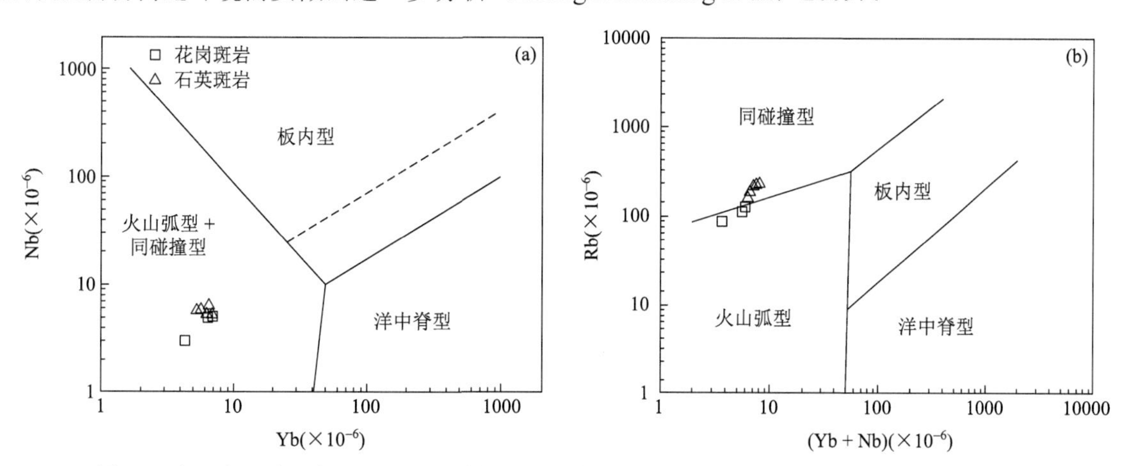

图 12 上黑龙江盆地虎拉林早白垩世侵入岩体构造环境判别图（底图据 Pearce et al.，1984）

Fig. 12 Diagram of tectonic setting of the Early Cretaceous intrusion rock in Hulalin of Upper Heilongjiang basin (after Pearce et al., 1984).

（a）Ta-Yb 判别图；（b）Nb-Y 判别图

(a) diagram of Ta-Yb; (b) diagram of Nb-Y

岩石类型及地球化学特征决定其源区的形成过程，而源区及成岩过程与其形成的构造环境有紧密的联系（Zhang Zhaochong et al.，2007，2009）。虎拉林岩体中花岗斑岩、石英斑岩，均属于高钾钙碱性-钾玄岩 I 系列（图 11），这与前人认为形成于后碰撞阶段岩石组合类型相符（Wu Guang et al.，2005；Zhang Yanlong et al.，2008；Gao Yang et al.，2009），微量元素蛛网图[图 8（b）]中均相对富集大离子亲石元素（LILEs），亏损高场强元素（HFSEs），强烈亏损 Nb、Ta、Ti、P，其特征也与后碰撞侵入岩特征一致（Wu Guang et al.，2005）。岩石 La-La/Yb 与 Yb-La/Y、Nb/Ta（图 9）、Zr/Hf 及 Rb/Sr 等图解及比值与锆石 Hf 同位素（图 10）均表明虎拉林早白垩侵入岩体来自于中—新元古代古老地壳的熔融，且存在幔源物质或从亏损地幔中新增生地壳物质部分熔融的参与，暗示存在幔源岩浆提供热量及参与熔融下地壳物质作用，表明该时期地壳减薄、地幔软流圈物质上涌，同样具有后碰撞环境特征。上黑龙江盆地区域内下白垩统九峰山组下部以砂岩、粉砂岩及砾岩为主，上部以块状粗砾岩为主，沉积环境由湖泊相演变为河流相，整体向陆相沉积环境转变，暗示了在早白垩世，上黑龙江盆地整体处于盆地衰亡期，从沉积地层方面也印证了在早白垩世区域内处于后碰撞构造环境（Geological Survey Institute of Inner Mongolia，2003；Zhao Huanli et al.，2011）。

上黑龙江盆地位于兴蒙造山带东段的大兴安岭地区，自古生代至中生代先后遭受古亚洲洋构造域、古太平洋构造及蒙古-鄂霍茨克构造域的叠加影响，使得该地区具有复杂的地质构造背景（Wu Fuyuan et al.，2011；Sun Deyou et al.，2013；Xu Wenliang et al.，2013a）。鉴于古亚洲洋在二叠纪已完全闭合（Shao Jian et al.，1999；Wu Guang et al.，2008，2010；Li Ruihua et al.，2018），研究区早白垩世侵入岩无法与古亚洲洋的俯冲、碰撞作用相联系。研究区北与蒙古-鄂霍茨克造山带相毗邻，已有区域构造资料表明，该造山带是由晚古生代—中生代期间蒙古-鄂霍茨克洋自西向东呈剪刀式闭合而形成的（Zorin，1999；Parfenov et al.，2001；Li Jinyi et al.，2004a；Xu Wenliang et al.，2013b；Li Yu et al.，2015）。受蒙古-鄂霍茨克洋闭合影响，蒙古-中朝大陆和西伯利亚大陆相碰撞，使得上黑龙江盆地在中侏罗世末期—晚侏罗世早期形成漠河推覆构造和逆冲断层，致使该处地壳缩短、增厚；早白垩世研究区发生左行韧性走滑剪切作用，致使额尔古纳地块向东逃逸，构造背景由挤压转变为伸展环境（Li Jinyi et al.，1999，2004b；Wu Fuyuan et al.，2002；Wu Guang et al.，2008；Tang Jie，2016）。随后岩石圈由先前的增厚、拆沉转为伸展减薄，导致软流圈地幔上隆，幔源岩浆上涌导致新生地壳及原有古老地壳部分熔融，形成早白垩世侵入岩的岩浆源区，随着岩浆上涌遇环境变化形成侵入岩体。虎拉林早白垩世侵入岩形成构造环境及形成机制与前人已发表有关额尔古纳地块早白垩世 A、I 型花岗岩具有相同的构造背景及形成机制（Wu Fuyuan et al.，2002；Tang Jie，2016）。

因此，虎拉林早白垩世侵入岩形成于蒙古-中朝大陆与西伯利亚大陆后碰撞环境，由挤压转为伸展的构造背景下，可能是蒙古-鄂霍茨克造山带陆陆碰撞期间加厚地壳拆沉、部分熔融，并受到幔源物质或从亏损地幔中新增生地壳部分熔融物质混染的产物。

5 结论

（1）虎拉林花岗斑岩岩体结晶年龄为 141.7±1.1Ma（MSWD = 0.086）；石英斑岩岩体结晶年龄为 144.9±0.56Ma（MSWD = 0.580）、142.6±0.74Ma（MSWD = 0.077），两岩体均为早白垩世岩浆活动产物。

（2）虎拉林花岗斑岩、石英斑岩 $\varepsilon_{Hf}(t)$均值分别为−1.59、−0.89，且大部分数据点均分布在球粒陨石演化线之下，两阶段模式年龄（t_{DM2}）分别为 987.08～1259.04Ma、908.38～1138.56Ma，显示两种岩体均自于中—新元古代古老地壳的熔融，且存在幔源物质或从亏损地幔中新增生地壳部分熔融物质的参与。

（3）虎拉林花岗斑岩、石英斑岩均为富铝、富碱性高分异 I 型花岗岩，属于高钾钙碱性向钾玄岩过渡序列；均具富集轻稀土、相对亏损重稀土右倾模式，及富集 Rb、K 等大离子亲石元素（LILEs）、LREE 及 U 等不相容元素，强烈亏损 Nb、Ta、Ti、P 等高场强元素（HFSEs）等特征，石英斑岩负铕异常较花岗斑岩显著。

（4）根据虎拉林花岗斑岩、石英斑岩岩石类型及地球化学特征，推测岩体形成蒙古-中朝大陆与西伯利亚大陆后碰撞环境，由挤压转为伸展的构造背景下，可能是蒙古-鄂霍茨克造山带陆陆碰撞期间加厚古老地壳拆沉、部分熔融，并受到幔源物质或新生地壳熔融混染的产物。进而限制了蒙古-鄂霍茨克洋至少在 144.9Ma 以前已经完全闭合。

致谢

野外工作得到内蒙古牙克石（满归）北部原始森林管护局及武警黄金第三支队大力支持及帮助；样品锆石 LA-ICP-MS 测试、Hf 同位素测试得到了中国地质科学院矿产资源研究所侯可军副研究员热情帮助及指导；岩石主、微量测试得到核工业北京地质研究院分析测试研究中心大力支持。地质学报编辑部及匿名审稿专家对本文提出大量宝贵意见。在此一并致以诚挚的感谢！

References

Bureau of Geology and Mineral Resoures of HeiLongijiang Province. 1993. Regional Geology of Heilongjiang Province. Beijing: Geological Publishing House: 5-438.

Boynton W V. 1984. Geochemistry of the rare earth elements: Meteorite studies. Developments in Geochemistry, 2: 63-114.

Cai Hongming, Zhang Hongfei, Xu Wangchun, Shi Zhangliang, Yuan Honglin. 2010. Petrogenesis of Indosinian volcanic rocks in Songpan-Garze fold belt of the northeastern Tibetan Plateau: New evidence for lithospheric delamination. Science China (Earth Sciences), 40 (11): 1518-1532. (in Chinese with English abstrcat).

Chai Mingchun, Zhao Guoying, Qin Xiaofeng, Wang Quan, Gao Su, Cao Kun. 2018. LA-ICP-MS U-Pb Ages and Geochemical Characteristics of the Intermediate-Acidic Intrusions in Shibazhan-Hanjiayuan area, Great Hinggan Mountains, and Its Geological Significance. Geological Review, 64 (3): 569-583. (in Chinese with English abstrcat).

Chappell B W, White A J R. 1992. I-and S-type granites in the lachlan fold belt. Transactions of the Royal Society of Edinburgh: Earth Sciences, 83 (1-2): 1-26.

Che Hewei, Zhou Zhenhua, Ma Xinghua, Ouyang Hegen, Liu Jun. 2015. Geochemical Characteristics, Zircons U-Pb Ages and Hf Isotopic Composition of the Dacite Porphyry from Zhengguang Au Deposit in Northern Great Xing'an Rang. Acta Geologica Sinica, 89 (8): 1417-1436. (in Chinese with English abstrcat).

Chen Yanjing, Chen Huayong, Zaw Khin, Pirajno Franco and Zhang Zengjie. 2007. Geodynamic settings and tectonic model of skarn gold deposits in China: An overview. Ore Geology Reviews, 31 (1): 139-16.

Chen Yanjing, Zhang Cheng, Li Nuo, Yang Yongfei, Deng Ke. 2012. Geology of the Mo Deposits in Northeast Chins. Journal of Jilin University (Earth Science Edition), 42 (5): 1223-1268. (in Chinese with English abstrcat).

Ding Qingfeng, Sun Fengyue, Liang Haijun. 2006. Discussion on geological character istics and metallogeny of Hulalin Gold Deposit in Inner Mongolia. Gold, 27 (10): 6-12. (in Chinese with English abstrcat).

Gao Yang, Zhang Zhaochong, Yang Tiezheng. 2009. Geologygeochemistry and petrogenesis of Late Hercynian granites in Baoshan area, Heilongjiang Province. Acta Petrologica et Mineralogica, 28 (5): 433-449. (in Chinese with English abstrcat).

Gao Yuan, Zheng Changqing, Yao Wenhua, Wang Hao, Li Juan, Shi Lu, Cui Fanghua, Gao Feng, Zhang Xingxing. 2013. Geochemistry and ziron U-Pb geochronology of the Luotuobozi Pluton in the Haduohe Area in the Northern Daxing'anling. Acta Geologica Sinica, 87 (9): 1293-1310. (in Chinese with English abstrcat).

Hou Kejun, Li Yanhe, Zou Tianren, Qu Xiaoming, Shi Yuruo, Xie Guiqing. 2007. Laser ablation-MC-ICP-MS technique for Hf isotope microanalysis of zircon and its geological applications. Acta Petrologica Sinica, 23 (10): 2595-2604. (in Chinese with English abstract).

Hou Kejun, Li Yanhe, Tian Yourong. 2009. In situ U-Pb zircon dating using laser ablation-multi ion counting-ICP-MS. Mineral Deposits, 28 (4): 481-492. (in Chinese with English abstrcat).

Ji Zheng, Ge Wenchun, Yang Hao, Bi Junhui, Yu Qian, Dong Yu. 2018. The Late Triassic Andean-type andesite from the central Great Xing'an Range: Products of the southward subduction of the Mongol-Okhotsk oceanic plate. Acta Petrologica Sinica, 34 (10): 2917-2930. (in Chinese with English abstrcat).

Li Jinyi, He Zhengjun, Mo Shenguo, Zheng Qingdao. 1999. The Late Mesozoic orogenic processes of Mongolia-Okhotsk orogen: evidence from field investigations into deformation of the Mohe area, NE China. Global Geology, 2 (2): 172-178.

Li Jinyi, He Zhengjun, Mo Shenguo, Sorokin A A. 2004a. The age of conglomerates in the lower part of the Xiufeng Formation in the northern DaHinggan Mountains, NE China, and their tectonic implications. Geological Bulletin of China, 23 (2): 120-129. (in Chinese with English abstract).

Li Jinyi, He Zhengjun, Mo Shenguo, Sun Guihua, Chen Wen. 2004b. The timing of crustal sinistral strike-slip movement in the northern Great Khing'an ranges and its constraint on reconstruction of the crustal tectonic evolution of NE China and adjacent areas since the Mesozoic. Earth Science Frontiers, 11 (3): 157-167. (in Chinese with English abstract).

LI Xiangjun. 2015. Metallogenic regularities of gold deposits in upperHeilongjiang metallogenic belt and its prospecting. Changchun: Jilin University, 1-78.

Li Ruihua, Zhang Han, Sun Fengyue, Wu Guang, Zhang Yuting, Wang Shuo. 2018. Geochronology and geochemistry of the Ershiyizhan intrusive rocks in the northern Great Hinggan Range, and its prospecting implications. Acta Petrologica Sinica, 34 (6): 1725-1740. (in Chinese with English abstrcat).

Li Xianhua, Li Wuxian, Li Zhengxiang. 2007. On the genetic classification and tectonic implications of the Early Yanshanian granitoids in the Nanling Range, South China. Chinese Science Bulletin, 52 (14): 1873-1885.

Li Xiangwen, Yang Yanchen, Wang Xianzhong, Liu Zhijie, Gong Weiguo. 2012. Geologic characteristics and metallogenic tectonic environment of the Baoxinggou gold deposit in Tahe, Heilongjiang Province. Journal of Jilin University (Earth Science Edition), 42 (6): 1700-1710. (in Chinese with English abstract).

Li Xiangwen, Zhang Zhiguo, Wa Keyong, Sun Jiapeng, Yang Jibo, Yang He. 2018. Characteristics of ore-forming fluid and genesis of Baoxinggou gold desposit in North of Great Xing'an Range. Journal of Jilin University (Earth Science Edition), 48 (4): 1071-1084. (in Chinese with English abstrcat).

Li Yu, Ding Leilei, Xu Wenliang, Wang Feng, Tang Jie, Zhao Shuo, Wang Zijin. 2015. Geochronology and geochemistry of muscovite granite in Sunwu area, NE China: Implications for the timing of closure of the Mongol-Okhotsh Ocean. Acta Petrologica Sinica, 31 (1): 56-66. (in Chinese with English abstract).

Li Yun, Hou Xiaoyu, Yang Bo, Zhao Yuanyi, Chen Long. 2017. Indo-Chinese Magmatic Activity in the Duobaoshan District of Heilongjiang Province and Its Geological Significance. Acta Geologica Sinica (English Edition), 91 (6): 2058-2077.

Liu Guige, Wang Ende, Li Jiemei, Li Yanqiu, Chen Yonggan, Chang Chunjiao, Wang Zhihua, Chao Yinyin. 2009. Petrochemical characteristics and genesis of Hulalin gold deposit. Gold Science and Technology, 17 (5): 1-5. (in Chinese with English abstrcat).

Liu Jimin, Zhang Chunge, Wang Xiaoyong, Chang Lihai. 2012. The ore and gold minerals characteristics of Hulalin gold deposit in North of Daxing'anling. Gold Science and Technology, 20 (1): 32-36. (in Chinese with English abstrcat).

Liu Yongsheng, Hu Zhaochu, Gao Shan, Gunther Detlef, Xu Jaun, Gao Changgui, Chen Haihong. 2008. In situ analysis of major and trace elements of anhydrousminerals by LA-ICP-MS without applying an internal standard. Chemical Geology, 257 (1-2): 34-43.

Liu Jun, Wu Guang, Qiu Huaning, Gao Dezhu, Yang Xinsheng. 2013a. $^{40}Ar-^{39}Ar$ Dating of Gold-Bearing Quartz Vein from the Shabaosi Gold Deposit at the Northern End of the Great Xing'an Range and Its Tectonic Signficance. Acta Geologica Sinica, 87 (10): 1570-1579. (in Chinese with English abstrcat).

Liu Jun, Mao Jingwen, Wu Guang, Luo Dafeng, Wang Feng, Zhou Zhenhua, Hu Yanqing. 2013b. Zircon U-Pb dating for the mangmatic rocks in the Chalukou Porphyry Mo deposit in the Northern Great Xing'an Range, China, and Its Geological Significance. Acta Geologica Sinica, 87 (2): 208-226. (in Chinese with English abstrcat).

Ludwig K R. 2003. User's manual for Isoplot 3.00: A geochronological tookit for Microsoft Excel. Berkeley Geochronology Center Special Pubication, 4: 1-70.

Middlemost E A K. 1985. Magmas and Magmatic Rocks: An Introduction to Igneous Petrology. London: Longman, 1-266.

Middlemost E A K. 1994. Naming materials in the magma/igneous rock system. Earth Science Reviews, 37 (3-4): 215-224.

Ouyang Hegen, Mao Jingwen, Santosh M, Zhou Jie, Zhou Zhenhua, Wu Yue, Hou Lin. 2013. Geodynamic setting of Mesozoic magmatism in ne china and surrounding regions: Perspectives from spatio-temporal distribution patterns of ore deposits. Journal of Asian Earth Sciences, 78: 222-236.

Parfenov L M, Popeko L I, Tomurtogoo O. 2001. Problems of tectonics of the Mongol-Okhotsk orogenic belt. Geology of the Pacific Ocean, 16 (5): 797-830.

Pearce J A, Harris N B W, Tindle A G. 1984. Trace element discrimination diagrams for the tectonic interpretation of granitic rocks. Journal of Petrology, 25 (4): 956-983.

Peccerillo R, Taylor S R. 1976. Geochemistry of Eocene calcalkaline volcanic rocks from the Kastamonu area, northem Turkey. Contributions to Mineralogy and Petrology, 58 (1): 63-81.

Shao Jian, Zhang Luqiao, Mu Baolei. 1999. Magmatism in the Mesozoic extending orogenic process of Da Hinggan Mountains. Earth Science Frontiers, 6 (4): 339-346. (in Chinese with English abstrcat).

Qu Xiaoming, Hou Zengqian, Mo Xuanxue, Dong Guochen, Xu Wenyi, Xin Hongbo. 2006. Relationship between Gangdisi porphyry copper deposits and uplifting of southern Tibet plateau: Evidence from multistage zircon of ore-bearing porphyries. Mineral Deposits, 25 (4): 388-400. (in Chinese with English abstrcat).

She Hongquan, Li Jinwen, Xiang Anping, Guan Jidong, Yang Yuncheng, Zhang Dehui, Tan Gang, Zhang Bin. 2012. U-Pb ages of the zircons from primary rocks in middle-northern Daxinganling and its implications to geotectonic evolution. Acta Petrologica Sinica, 28 (2): 571-594. (in Chinese with English abstrcat).

Song Guibin, Song Bingjian, Yue Jihang. 2007. Discussion on geological features and ore-forming rule of Hulalin Gold Deposit in Eerguna Neimenggu. Gold Science and Technology, 15 (5): 30-35. (in Chinese with English abstrcat).

Sui Zhenmin, Ge Wenchun, Wu Fuyuan, Xu Xuechun, Wang Qinghai. 2006. U-Pb chronology in zircon from Harabaqi granitic pluton in northeastern Daxing'anling area and its origin. Global Geology, 25 (3): 229-236. (in Chinese with English abstrcat).

Sui Zhenmin, Ge Wenchun, Wu Fuyuan, Zhang Jiheng, Xu Xuechun, Cheng Ruiyu. 2007. Zircon U-Pb ages, geochemistry and its petrogenesis of Jurassic granites in northeastern part of the Da Hinggan Mts. Acta Petrologica Sinica, 23 (2): 461-480. (in Chinese with English abstrcat).

Sun Deyou, Gou Jun, Wang Tianhao, Ren Yunsheng, Liu Yongjiang, Guo Hongyu, Liu Xiaoming, Hu Zhaochu. 2013. Geochronological and geochemical constraints on the Erguna massif basement, NE China-subduction history of the Mongol-Okhotsk oceanic crust, International Geology Review, 55 (14): 1801-1816.

Sun Lixin, Ren Bangfang, Zhao Fengqing, Peng Lina. 2012. Zircon U-Pb ages and Hf isotope characteristics of Tiapingchuan large porphyritic granite pluton of

Erguna Massifin the Great Xing'an Range. Earth Science Frontiers, 19 (5): 114-122. (in Chinese with English abstract).

Sun Qi, Ren Yunsheng, Li Dexin, Yang Qun. 2016. Isotopic geochemistry and metallogenic material source of Shabaosi Gold Deposit in Heilongjiang Province. Gold, 37 (5): 10-15. (in Chinese with English abstrcat).

Sun S S, McDonough W F. 1989. Chemical and isotopic systematics of oceanic basalts: Implications for mantle composition and processes. In: Saunders AD and Norry MJ (eds.). Magmatism in the Ocean Basins. Geological Society, London, Special Publication, 42 (1): 313-345.

Sun Yanfeng. 2015. Geological Characteristics and Metallogenesis of the Shiwuliqiao Gold Deposit in Tahe, Heilongjiang province. Beijing: China University of Geosciences, 1-68. (in Chinese with English abstrcat).

Taylor S R, McLennan S M. 1985. The Continental Crust: Its Composition and Evolution: An Examination of the Geochemical Record Preserved in Sedimentary Rocks. Oxford, London: Blackwell Scientific Publication, 1-301.

Tang Jie. 2016. Geochronology and geochemistry of the Mesozoic igneous rocks in the Erguna Massif, NE China: Constraints on the tectonic evolution of the Mongol-Okhotsk suture zone. Changchun: Jilin University, 1-205. (in Chinese with English abstrcat).

Wang Tianhao, Zhang Shuiyi, Sun Deyou, Gou Jun, Ren Yunsheng, Wu Pengfei, Liu Xiaoming. 2014. Zircon U-Pb ages and Hf isotopic characteristics of Mesozoic granitoids from southern Manzhouli, Inner Mongolia. Global Geology, 33 (1): 26-38. (in Chinese with English abstrcat).

Wang Yinhong, Zhang Fangfang, Liu Jiajun, Xue Chunji, Wang Jianping, Liu Bin, Lu Weiwei. 2015. Petrogenesis of granites in Baishan molybdenum deposit, eastern Tianshan, Xinjiang: Zircon U-Pb geochronology, geochemistry, and Hf isotope constraints. Acta Petrologica Sinica, 31 (7): 1962-1976. (in Chinese with English abstract).

Wang Qisong, Zhang Jing, Wang Su, Yu Lidong, Xiao Bing. 2019. Petrogenesis and tectonic setting of quartz porphyry in Mazhuangshan gold deposit, Eastern Tianshan Orogen: Evidence from geochemistry, zircon U-Pb geochronology and Sr-Nd-Hf isotopes. Acta Petrologica Sinica, 35 (5): 1503-1518. (in Chinese with English abstract).

Whalen J B, Currie K L, Chappell B W, 1987. A-type granites: geochemical characteristics, discrimination and petrogenesis. Contributions to Mineralogy and petrology, 95 (4): 407-419.

Wu Fuyuan, Sun Deyou, Li Huimin, Jahn Bor-ming, Wilde S. 2002. A-type granites in northeastern China: Age and geochemical constraints on their petrogenesis. Chemical Geology, 187 (1-2): 143-173.

Wu Fuyuan, Jahn B M, Wilde S A, Lo Chinghua, Yui T F, Lin Qiang, Ge Wenchun, Sun Deyou. 2003. Highly fractionated Itype granites in NE China (I): Geochronology and petrogenesis. Lithos, 66 (3-4): 241-273.

Wu Fuyuan, Li Xianhua, Zheng Yongfei, Gao Shan. 2007. Lu-Hf isotopicsystematic and their applications in petrology. Acta Petrologica Sinica, 23 (2): 185-220. (in Chinese with English abstract).

Wu Fuyuan, Sun Deyou, Ge Wenchun, Zhang Yanbin, Grant Matthew L, Wilde Simon A, Jahn Bor-Ming. 2011. Geochronology of the Phanerozoic granitoids in northeastern China. Journal of Asian Earth Sciences, 41 (1): 1-30.

Wu Guang, Zhu Qun, Zhao Caisheng. 2002. Types and geological characteristics of gold-copper deposits in upper Heilongjiang depression, North Daxinganling. Mineral Deposits, 21 (S1): 261-264. (in Chinese with English abstract).

Wu Guang. 2005. Metallogenic Setting and Metallogenesis of Nonferrous-precious Metals in Northern Da Hinggan Moutain. Changchun: Jilin University, 1-221. (in Chinese with English abstrcat).

Wu Guang, Sun Fengyue, Zhao Caisheng, Li Zhitong, Zhao Ailin, Pang Qingbang, Li Guangyuan. 2005. The discovery of Post-Paleozoic collision granites in the northern margin of Erguna block and its geological significance. Chinese Science Bulletin, 50 (20): 2278-2288. (in Chinese with English abstract).

Wu Guang, Sun Fengyue, Zhao Caisheng, Ding Qingfeng, Wang Li. 2007. Fluid inclusion study on gold deposits in northwestern Erguna metallogenic belt China. Acta Petrologica Sinica, 23 (9): 2227-2240. (in Chinese with English abstrcat).

Wu Guang, Chen Yanjing, Sun Fengyue, Li Jingchun, Li Zhitong, Wang Xijin. 2008. Geochemistry of the Late Jurassic granitolds in the northern end area of Da Hinggan Mountains and their geological and prospecting implications. Acta Petrological Sinica, 24 (4): 899-910. (in Chinese with English abstrcat).

Wu Guang, Chen Yanjing, Sun Fengyue, Zhang Zhe, Liu Ankun, Li Zhitong. 2010. Geochemistry and genesis of the Late Jurassic granitoids at northern Great Hinggan Range: Implications for exploration. Acta Geologica Sinica, 84 (2): 321-332.

Wu Guang, Chen Yuchuang, Chen Yanjing, Zeng Qingtao. 2012. Zircon U-Pb ages of the metamorphic supracrustal rocks of the Xinghuadukou Group and granitic complexes in the Argun massif of the northern Great Hinggan Range, NE China, and their tectonic implications. Journal of Asian Earth Sciences, 49 (4): 214-233.

Wu Guang, Wang Guorui, Liu Jun, Zhou Zhenhua, Li Tiegang, Wu Hao. 2014. Metallogenic series and ore-forming pedigree of main ore deposits in northern Great Xing'an Range[J]. Mineral Deposits，33(6): 1127-1150.

Xiong Xiaolin, Li Xianhua, Xu Jifeng, Li Wuxian, Zhao Zhenhua, Wang Qiang, Chen Xiaoming. 2003. Extremely high-Na adakite-like magmas derived from alkali-rich basaltic underplate: The late cretaceous Zhantang andesites in the Huichang Basin, SE China. Geochemical Journal, 37 (2): 233-252.

Xu Jifeng, Shinjo R, Defant M J, Wang Qiang, Rapp R P. 2002. Origin of Mesozoic adakitic intrusive rocks in the Ningzhen area of East China: Partial melting of delaminated lower continental crust? Geology, 30 (12): 1111-1114.

Xu Wenliang, Pei Fuping, Wang Feng, Meng En, Ji Weiqiang, Yang Debin, Wang Wei. 2013a. Spatial-temporal relationships of Mesozoic volcanic rocks in NE China: Constraints on tectonic overprinting and transformations between multiple tectonic regimes. Journal of Asian Earth Sciences, 74 (18): 167-193.

Xu Wenliang, Wang Feng, Pei Fuping, Tang Jie, Xu Meijun, Wang Wei. 2013. Mesozoic tectonic regimes and regional ore-forming background in NE China: Constraints from spatial and temporal variations of Mesozoic volcanic rock associations. Acta Petrologica Sinica, 29 (2): 339-353. (in Chinese with English abstract).

Zhang Zhaochong, Zhou Gang, Yan Shenghao, Chen Bolin, He Yongkang, Chai Fnegmei, He Lixin. 2007. Geology and Geochemistry of the Late Paleozoic Volcanic Rock of the South Margin of the Altai Mountains and Implications for Tectonic Evolution. Acta Geologica Sinica, 81 (3): 344-358. (in Chinese with English abstrcat).

Zhao Zhaochong, Dong Shuyun, Huang He, Ma Letian, Zhang Dongyang, Zhang Shu, Xue Chunji. 2009. Geology and geochenmistry of the Permian intermediate-acid intrusions in the southwestern Tianshan, Xinjiang, China: implications for petrogenesis and tectonics. Geological Bulletin of China, 28 (12): 1827-1839. (in Chinese with English abstrcat).

Zhang Jing, Deng Jun, Li Shihui, Yan Ni, Yang Liqiang, Ma Nan, Wang Qingfei, Gong Qingjie. 2010. Petrological characteristics of magmatites and their relationship with gold mineralization in the Chang'an gold deposit in southern Ailaoshan metallogenic belt. 26 (6): 1740-1750. (in Chinese with English abstrcat).

Zhang Yong, Liu Yinchun, Wang Keqiang, Wang Zhihua, Liu Guige. 2003. Characteristics of explosionbreccias type gold deposit in Hulalin, Inner Mongolia. Gold Geology, 9 (4): 13-18. (in Chinese with English abstrcat).

Zhang Yanlong, Ge Wenchun, Liu Xiaoming, Zhang Jiheng. 2008. Isotopic Characteristics and Its Significance of the Xinlin Town Pluton, Great Hinggan Mountains. Journal of Jilin University (Earth Science Edition), 38 (2): 177-185. (in Chinese with English abstrcat).

Zhao Huanli, Liu Xuguang, Liu Haiyang, Zhu Chunyan. 2011. Petrological evidence of Paleozoic marine basin closure in Duobaoshan of Heilongjiang. Global Geology, 30 (1): 18-27. (in Chinese with English abstrcat).

Zhao Di, Ji Zheng, Yang Hao, He Yue, Jing Yan, Wang Qinghai, Chen Huijun. 2018. Petrogenesis of Late Carboniferous granites in Boketu region of northern Great Xing'an Range: evidence from zircon U-Pb age, geochemistry and Lu-Hf isotopes. Global Geology, 37 (3): 712-723.

Zhao Shuo, Xu Wenliang, Wang Wei, Tang Jie, Zhang Yihan. 2014. Geochronology and Geochemistry of Middle-Late Ordovician Granites and Gabbros in the Erguna Region, NE China: Implications for the Tectonic Evolution of the Erguna Massif. Journal of Earth Science, 25 (5): 841-853.

Zhao Yiming, Zhang Dequan. Metallogeny and Prospective evaluation of copper-polymetallic deposits in the Da Hinggan mountain and its adjacent regions. Beijing: Seismological Press, 1-318. (in Chinese with English abstrcat).

Zhao Zhenhua, Xiong Xiaolin, Wang Qiang, Bai Zhenghua, Qiao Yulou. 2009. Late Paleozoic underplating in North Xinjiang: Evidence from shoshonites and adakites. Gondwana Research, 16 (2): 216-226.

Zhao Zhidan, Liu Dong, Wang Qing, Zhu Dicheng, Dong Guochen, Zhou Su, Mo Xuanxue. 2018. Zircon trace elements and their use in probing deep processes. Earth Science Frontiers, 25 (6): 124-135. (in Chinese with English abstrcat).

Zhou Zhenhua, Mao Jingwen, Lyckberg P. 2012. Geochronology and isotopicgeochemistry of the A-type granites from the Huanggang Sn-Fe deposit, southern Great Hinggan Range, NE China: Implication for their origin and tectonic setting. Journal of Asian Earth Science, 49 (1): 272-286.

Zorin Y A. 1999. Geodynamics of the western part of the Mongolia-Okhotsk collisional belt, Trans-Baikal region (Russia) and Mongolia. Tectonophysics, 306 (1): 33-56.

参考文献

蔡宏明, 张宏飞, 徐旺春, 时章亮, 袁洪林. 2010. 松潘带印支期岩石圈拆沉作用新证据: 来自火山岩岩石成因的研究. 中国科学 (地球科学), 40 (11): 1518-1532.

柴明春, 赵国英, 覃小锋, 王泉, 高溯, 曹昆. 2018. 大兴安岭十八站-韩家园地区中酸性侵入岩 LA-ICP-MS 锆石 U-Pb 年龄、地球化学特征及其地质意义. 地质论评, 64 (3): 569-583.

车合伟, 周振华, 马星华, 欧阳荷根, 刘军. 2015. 大兴安岭北段争光金矿应安斑岩地球化学特征、锆石 U-Pb 年龄及 Hf 同位素组成. 地质学报, 89 (8): 1417-1436.

陈衍景, 张成, 李诺, 杨永飞, 邓轲. 2012. 中国东北钼矿床地质. 吉林大学学报 (地球科学版), 42 (5): 1223-1268.

丁清峰, 孙丰月, 梁海军. 2006. 内蒙古虎拉林金矿矿床地质特征及成因探讨. 黄金, 27 (10): 6-12.

高阳, 张招崇, 杨铁铮. 2009. 黑龙江宝山一带海西晚期强过铝花岗岩地质地球化学及岩石成因. 岩石矿物学杂志, 28 (5): 433-449.

高源, 郑常青, 姚文贵, 王浩, 李娟, 施璐, 崔芳华, 高峰, 张行行. 2013. 大兴安岭北段哈多河地区骆驼脖子岩体地球化学和锆石 U-Pb 年代学. 地质学报, 87 (9): 1293-1310.

黑龙江省地质矿产局. 1993. 黑龙江省区域地质志. 北京: 地质出版社, 5-438.

侯可军, 李延河, 邹天人, 曲晓明, 石玉若, 谢桂青. 2007. LA-MCICP-MS 锆石 Hf 同位素的分析方法及地质应用. 岩石学报, 23 (10): 2595-2604.

侯可军, 李延河, 田有荣. 2009. LA-MC-ICP-MS 锆石微区原位 UPb 定年技术. 矿床地质, 28 (4): 481-492.

纪政, 葛文春, 杨浩, 毕君辉, 于倩, 董玉. 2018. 大兴安岭中段晚三叠世安第斯型安山岩: 蒙古-鄂霍茨克大洋板片南向俯冲作用的产物. 岩石学报, 34 (10): 2917-2930.

李锦轶, 和政军, 莫申国, Sorokin A A. 2004a. 大兴安岭北部绣峰组下部砾岩的形成时代及其大地构造意义. 地质通报, 23 (2): 120-129.

李锦轶, 莫申国, 和政军, 孙桂华, 陈文. 2004b. 大兴安岭北段地壳左行走滑运动时代及其对中国东北及邻区中生代以来地壳构造演化重建的制约. 地学前缘, 11 (3): 157-167.

李睿华, 张晗, 孙丰月, 武广, 张宇婷, 王硕. 2018. 大兴安岭北段二十一站岩体年代学、地球化学及其找矿意义. 岩石学报, 34 (6): 1725-1740.

李向文, 杨言辰, 王献忠, 刘智杰, 公维国. 2012. 黑龙江省塔河县宝兴沟金矿床地质特征及成矿构造环境. 吉林大学学报 (地球科学版), 42 (6): 1700-1710.

李向文, 张志国, 王可勇, 孙加鹏, 杨吉波, 杨贺. 2018. 大兴安岭北段宝兴沟金矿床成矿流体特征及矿床成因. 吉林大学学报 (地球科学版), 48 (4): 1071-1084.

李向文. 2015. 上黑龙江成矿带金矿床成矿规律与找矿预测研究. 长春: 吉林大学, 1-78.

李宇, 丁磊磊, 徐文良, 王枫, 唐杰, 赵硕, 王子进. 2015. 孙吴地区中侏罗世白云母花岗岩的年代学与地球化学: 对蒙古-鄂霍茨克洋闭合时间的限定. 岩石学报, 31 (1): 56-66.

刘桂阁, 王恩德, 李杰美, 李艳秋, 陈勇敢, 常春郊, 王治华, 朝银银. 2009. 虎拉林金矿岩石地球化学特征及成因. 黄金科学技术, 17 (5): 1-5.

刘吉民, 张纯歌, 王晓勇, 常立海. 2012. 大兴安岭北部虎拉林金矿床矿石及金矿物特征. 黄金科学技术, 20 (1): 32-36.

刘军, 武广, 邱华宁, 高德柱, 杨鑫生. 2013a. 大兴安岭北部砂宝斯金矿床含金石英脉 ^{40}Ar-^{39}Ar 年龄及其构造意义. 地质学报, 27 (10): 1570-1579.

刘军, 毛景文, 武广, 罗大锋, 王峰, 周振华, 胡妍青. 2013b. 大兴安岭北部斑岩钼矿床岩浆岩锆石 U-Pb 年龄及其地质意义. 地质学报, 87 (2): 208-226.

内蒙古自治区地质调查院. 2003. 区域地质调查报告 (莫尔道嘎镇幅)

曲晓明, 侯增谦, 莫宣学, 董国臣, 徐文艺, 辛洪波. 2006. 冈底斯斑岩铜矿与南部青藏高原隆升之间关系: 来自含矿斑岩中多阶段锆石的证据. 矿床地质, 25 (4): 388-400.

邵济安, 张履桥, 牟保磊. 1999. 大兴安岭中生代伸展造山过程中的岩浆作用. 地学前缘, 6 (4): 339-346.

佘宏全, 李进文, 向安平, 关继东, 杨郧城, 张德会, 谭刚, 张斌. 2012. 大兴安岭中北段原岩锆石 U-Pb 测年及其与区域构造演化关系. 岩石学报, 28 (2): 571-594.

宋贵斌, 宋丙剑, 岳继航. 2007. 内蒙古额尔古纳市虎拉林金矿床地质特征及成矿规律探讨. 黄金科学技术, 15 (5): 30-35.

隋振民, 葛文春, 吴福元, 徐学纯, 王清海. 2006. 大兴安岭东北部哈拉巴奇花岗岩体锆石 U-Pb 年龄及其成因. 世界地质, 25 (3): 229-236.

隋振民, 葛文春, 吴福元, 张吉衡, 徐学纯, 程瑞玉. 2007. 大兴安岭东北部侏罗纪花岗质岩石的锆石 U-Pb 年龄、地球化学特征及成因. 岩石学报, 23 (2): 461-480.

孙立新, 任邦方, 赵凤清, 彭丽娜. 2012. 额尔古纳地块太平川巨斑状花岗岩的锆石 U-Pb 年龄和 Hf 同位素特征. 地学前缘, 19 (5): 114-122.

孙琦, 任云生, 李德新, 李德新, 杨群. 2016. 黑龙江省砂宝斯金矿床同位素地球化学特征与成矿物质来源. 黄金, 37 (5): 10-15.

孙彦峰. 2015. 黑龙江省塔河县十五里桥金矿床地质特征及成因探讨. 北京: 中国地质大学硕士论文, 1-68.

唐杰. 2016. 额尔古纳地块中生代火成岩的年代学与地球化学: 对蒙古-鄂霍茨克缝合带构造演化的制约. 长春: 吉林大学博士论文, 1-205.

王天豪, 张书义, 孙德有, 苟军, 任云生, 武鹏飞, 柳小明. 2014. 满洲里南部中生代花岗岩的锆石 U-Pb 年龄及 Hf 同位素特征. 世界地质, 33 (1): 26-38.

王银宏, 张方方, 刘家军, 薛春纪, 王建平, 刘彬, 路魏魏. 2015. 东天山白山钼矿区花岗岩的岩石成因: 锆石 U-Pb 年代学、地球化学及 Hf 同位素约束. 岩石学报, 31 (7): 1962-1976.

王琦崧, 张静, 王肃, 于立栋, 肖兵. 2019. 东天山马庄山金矿区赋矿石英斑岩的岩石成因和构造背景; 元素地球化学、U-Pb 年代学和 Sr-Nd-Hf 同位素约束. 岩石学报, 35 (5): 1503-1518.

吴福元, 李献华, 郑永飞, 高山. 2007. Lu-Hf 同位素体系及其岩石学应用. 岩石学报, 23 (2): 185-220.

武广. 2005. 大兴安岭北部区域成矿背景与有色、贵金属矿床成矿作用. 长春: 吉林大学, 1-221.

武广, 朱群, 赵财胜. 2002. 大兴安岭北部上黑龙江拗陷区金铜矿床类型及地质特征. 矿床地质, 21 (S1): 261-264.

武广, 孙丰月, 赵财胜, 李之彤, 赵爱琳, 庞庆帮, 李广远. 2005. 额尔古纳地块北缘早古生代后碰撞花岗岩的发现及其地质意义. 科学通报, 50 (20): 2278-2288.

武广, 孙丰月, 赵财胜, 丁清峰, 王力. 2007. 额尔古纳成矿带西北部金矿床流体包裹体研究. 岩石学报, 23 (9): 2227-2240.

武广, 陈衍景, 孙丰月, 李景春, 李之彤, 王希今. 2008. 大兴安岭北段晚侏罗世花岗岩类地球化学及其地质找矿意义. 岩石学报, 24 (4): 899-910.

武广, 王国瑞, 刘军, 周振华, 李铁刚, 吴昊. 2014. 大兴安岭北部主要金属矿床成矿系列和区域矿床成矿谱系. 矿床地质, 33(6): 1127-1150.

许文良, 王枫, 裴福萍, 孟恩, 唐杰, 徐美君, 王伟. 2013. 中国东北中生代构造体制与区域成矿背景: 来自中生代火山岩组合时空变化的制约. 岩石学报, 29 (2): 339-353.

张招崇, 周刚, 闫升好, 陈柏林, 贺永康, 柴凤梅, 何立新. 2007. 阿尔泰山南缘晚古生代火山岩的地质地球化学特征及其对构造演化的启示. 地质学报, 81 (3): 344-358.

张招崇, 董书云, 黄河, 马乐天, 张东阳, 张舒, 薛春纪. 2009. 西南天山二叠纪中酸性侵入岩的地质学和地球化学: 岩石成因和构造背景. 地质通报, 28 (12): 1827-1839.

张静, 邓军, 李士辉, 燕旎, 杨立强, 马楠, 王庆飞, 龚庆杰. 2010. 哀牢山南段长安金矿床岩浆岩的岩石学特征及其与成矿关系探讨. 岩石学报, 26 (6): 1740-1750.

张勇, 刘荫椿, 王科强, 王治华, 刘桂阁. 2003. 内蒙古虎拉林爆破角砾岩型金矿床特征. 黄金地质, 9 (4): 13-18.

张彦龙, 葛文春, 柳小明, 张吉衡. 2008. 大兴安岭新林镇岩体的同位素特征及其地质意义. 吉林大学学报 (地球科学版), 38 (2): 177-185.

赵焕利, 刘旭光, 刘海洋, 朱春艳. 2011. 黑龙江多宝山古生代海盆闭合的岩石学证据. 世界地质, 30 (1): 18-27.

赵迪, 纪政, 杨浩, 和越, 景妍, 王清海, 陈会军. 2018. 大兴安岭北段博克图地区晚石炭世花岗岩的成因: 来自锆石 U--Pb 年龄、地球化学及 Lu--Hf 同位素证据. 世界地质, 37 (3): 712-723.

赵一鸣, 张德全. 1997. 大兴安岭及其邻区铜多金属矿床成矿规律与远景评价. 北京: 地震出版社, 1-318.

赵志丹, 刘栋, 王青, 朱弟成, 董国臣, 周肃, 莫宣学. 2018. 锆石微量元素及其揭示的深部过程. 地学前缘, 25 (6): 124-135.

Zircon U-Pb chronology, Hf isotope and geochemistry studies of the Early Cretaceous rock mass in the Hulalin of Upper Heilongjiang basin

Xin Gong[1, 2, 3], Yuanyi Zhao[1], Xinfang Shui[1], Xianda Cheng[1, 2], Yuanchao Wang[1, 2, 4], Xuan Liu[1, 2] and Wei Tan[1, 2]

1. Institute of Mineral Resources, Chinese Academy of Geological Sciences, Beijing 100037
2. School of Earth Sciences and Resources, China University of Geosciences (Beijing), Beijing 100083
3. Geological and Mineral Exploration Institute, Guizhou Bureau of Geological exporation for Non-ferrous Metals and Nuclear Industry, Guiyang 550005
4. 12th Gold Team of Chinese People's Armed Police, Chengdu 610000

Abstract The Hulalin gold deposit in the Upper Heilongjiang Basin is located in the northern margin of the Erguna micro-block in the northern Daxinganling, east of the Xingmeng orogenic belt, and is clamped between the Mongolian Okhotsk suture zone and the Deerbugan fault. The body mainly exists in the cryptoexplosive breccia, which is closely related to the granite porphyry and quartz porphyry formed by the Early Cretaceous magmatism. In this paper, the LA-ICP-MS zircon U-Pb dating method was used to obtain the crystallization age of granite porphyry rock mass were 141.7±1.1 Ma (MSWD = 0.086), quartz porphyry rock mass (2 pieces) were 144.86±0.56Ma (MSWD = 0.580) and 142.6±0.74Ma (MSWD = 0.077), both of which are products of early Cretaceous magma activity. The zircon $\varepsilon_{Hf}(t)$ of granite porphyry and quartz porphyry are −4.14～0.16, −2.25～1.37, and most of the data is below the chondrite line. The Two-stage model ages (t_{DM2}) of this two kinds of rock masss are 987.08～1259.04Ma, 908.38～1138.56Ma. The geochemical characteristics of the rock show that the granite porphyry $w(SiO_2)$ is 66.21%～66.70%, $w(Al_2O_3)$ is 15.18%～15.45%, $w(Na_2O + K_2O)$ is 7.56%～8.17%, aluminum saturation index A/CNK is 1.03～1.06, showing weak aluminum high potassium calc-alkaline-potassium rock sequence. Quartz porphyry $w(SiO_2)$ is 66.34%～68.25%, $w(Al_2O_3)$ is 13.77%～15.05%, $w(Na_2O + K_2O)$ is 6.76%～9.06%. In the SiO_2-K_2O diagram, it falls into the sequence of potassium basalt. The REE of granite porphyry and quartz porphyry are 71.07×10^{-6}～97.92×10^{-6}, 135.53×10^{-6}～156.15×10^{-6} respectively, and $(La/Yb)_N$ is 24.87～26.81, 31.33～40.62 respectively, all of which show the relative enrichment of light rare earth (LREE) and the right declination of heavy rare earth (HREE), and the light and heavy rare earth fractionation is obvious. The δEu is 0.65～0.88 and 0.41～0.50 respectively. The latter is more abnormal than the former. Both rock masses are relatively rich in incompatible elements such as Rb, K and other large ion lithophile elements (LILEs), LREE and U, and strongly deplete Nb, Ta, Ti, P and other high field strength elements (HFSEs). The rock genetic type discriminant map shows that the Early Cretaceous intrusive rocks of the Hulalin in the Upper Heilongjiang Basin are I-type granites, which are formed in the Mongolia-China and North Korea continent and the Siberian continent, colliding with the Late collision environment, from the extrusion to the extension of the tectonic setting. It may be a product of thickening of the crust during the ancient land-continent collision of the Mongolia-Okhotsk orogenic belt, partial melting, and by mantle source material or new crust melting mixed contamination products. This further restricted the Mongolian-Okhotsk ocean from being completely closed until at least 144.9Ma.

Keywords Zircon U-Pb chronology; Hf isotope; Geochemistry; Early Cretaceous; Hulalin; Upper Heilongjiang basin area

漠河盆地二十二站组砂岩年代学、地球化学及其地质意义*

王远超[1,2,3]，赵元艺[1]，刘春花[1]，水新芳[1]，程贤达[2,4]，巩鑫[2,5]，刘璇[2]，谭伟[2]，洪骏男[6]

1. 中国地质科学院矿产资源研究所，北京 100037
2. 中国地质大学（北京）地球科学与资源学院，北京 100083
3. 中国人民武装警察部队黄金第十二支队，成都 610000
4. 中国地质调查局西安矿产资源调查中心，西安 710100
5. 贵州省有色金属和核工业地质勘查局地质矿产勘查院，贵阳 550005
6. 黑龙江省齐齐哈尔地质勘查总院，齐齐哈尔 161000

内容提要 漠河盆地位于蒙古-鄂霍茨克褶皱带中的额尔古纳微板块的北缘，地处西伯利亚板块与华北板块碰撞拼合部位。二十二站组是漠河盆地中生代沉积地层之一，前人对其形成时代和物源进行了探讨，但仍存在很大争议，本文在前人研究的基础上，通过碎屑锆石年代学和岩石地球化学再次厘定其形成时代，并对物源及源区大地构造背景进行探讨。碎屑锆石年代学研究表明，二十二站组碎屑锆石大部分为岩浆结晶锆石，少部分锆石颗粒为增生-混合型锆石，显示出经历了后期构造-热事件改造。此外，少部分锆石颗粒磨圆好，显示出其经历了多次搬运、沉积过程的特征，从而指示早先形成的古老沉积岩为二十二站组提供了物源。获得最年轻的锆石年龄为 134±1Ma，结合前人区域地质调查报告中发现了 J_3-K_1 时期的古生物化石，将二十二站组的沉积下限限定为早白垩世早期，同时也说明了研究区存在早白垩世早期火成岩物源。主、微量元素构造环境判别及物源分析揭示二十二站组物源主要为活动大陆边缘及大陆岛弧环境的上地壳长英质、安山质源区，并混有下地壳深部物质（基性岩）。锆石 LA-ICP-MS U-Pb 定年结果表明，二十二站组碎屑锆石有随着时代变新锆石保存数量增多的趋势，其年龄分布整体上可被划分为四个时期：新太古代（2711±10Ma，$N=1$），说明额尔古纳地块存在新太古代的基底信息；中元古代—古元古代（2428～1238Ma，$N=11$），指示兴华渡口岩群为二十二站沉积物提供了部分物源；新元古代（921～561Ma，$N=7$），是晋宁期古亚洲洋向额尔古纳-兴安地块俯冲形成大陆岩浆弧（活动大陆边缘）构造事件在研究区的记录；中生代—晚古生代（540～134Ma，$N=280$），是蒙古-鄂霍茨克洋俯冲、闭合过程中形成的花岗质岩浆在研究区的物质记录，且显生宙花岗岩质岩浆为二十二站组提供了最为丰富的物源。

关键词 碎屑锆石 U-Pb 年代学；地球化学；砂岩；物源分析；二十二站组；漠河盆地

漠河盆地位于中国东北边陲黑龙江省最北端，额尔古纳地块北缘[图 1（a）]，坐标为东经 121°07′～125°45′，北纬 52°00′～54°00′，呈东西展布，长约 300km，宽约 80km，国内面积约 24000km^2。盆地北部为蒙古-鄂霍茨克造山带，地层侵入中生代火成岩，是中国学者直接了解蒙古-鄂霍茨克造山带演化的重要窗口（He Zhengjun et al.，2003）。

漠河盆地整体的地质工作程度相对较低，植被覆盖严重，地层出露较差，1993 年《黑龙江区域地质志》和 2014 年《开库康幅、塔河县幅、新街基幅 1∶25 万区域地质调查报告》中对其进行过地质调查工作，近些年来，其构造属性、地层年代学和物源是研究的重点。前人对漠河盆地的构造属性问题主要有 3 种观点：前陆盆地（He Zhengjun et al.，2003；Zhang Shun et al.，2003），前陆盆地的陆相磨拉石部分或磨拉石盆地（Hou Wei et al.，2010b；Wu Genyao et al.，2006），挤压背景下形成的挤压挠曲盆地或山间盆地（He Zhonghua et al.，2008a）。漠河盆地二十二站组主要出露于漠河前陆盆地东段，呈东西向带状分布①，是漠河盆地最重要的地层之一。关于二十二站地层时代归属和物源组成存在诸多争议。尤其是关于沉积时代争

* 本文发表在：地质学报，2020（3）：869-887

议最大，有些学者认为其沉积时代为中侏罗世（Qiherige，1995；He Zhengjun et al.，2003；Hou Wei，2006；Wang Jian，2007；He Zhonghua et al.，2008a，2008b），有些学者认为晚侏罗世（Sun Guangrui et al.，2002；Wu Heyong et al.，2003b；Xin Renchen et al.，2003；Zhang Shun et al.，2003；Yang Jianguo et al.，2006；Li Chunlei，2007；Hou Wei et al.，2010a，2010b，2010c；Sun Qiushi，2013；Li Liang et al.，2017），有些学者认为中-晚侏罗世（Sun Guangrui et al.，2002）。最新的锆石 LA-ICP-MS U-Pb 同位素测年（Li Liang et al.，2017；Zhen Miao，2017）所得的结果也并不一致。Zhen Miao（2017）测得最老的锆石年龄为 348±3.2Ma，并说明了母源区存在古生代地质体，而 Li Liang 等（2017）仅仅获得中生代碎屑锆石三个年轻的年龄峰值（158Ma，190Ma，210Ma），而并未获得更老的锆石年龄。关于二十二站组的物源也存在较大争议，许多学者认为二十二站组的物源来自北部的蒙古-鄂霍茨克造山带与南部的额尔古纳地块（He Zhengjun et al.，2003；Hou Wei，2006；Li Chunlei，2007；He Zhonghua et al.，2008a，2008b；Hou Wei et al.，2010a，2010b，2010c；Li Liang et al.，2017）；也有学者认为盆地沉积物主要来自南侧的陆块抬升基底，即古元古界兴华渡口群与寒武系兴隆群、古生代同碰撞和后碰撞花岗质岩石及早中生代中酸性火成岩（He Zhonghua et al.，2008a）；也有学者认为盆地的物源除来自南北两侧外，盆地内部也提供了部分物源（Sun Qiushi，2013）。

对碎屑锆石的研究可以确定沉积物形成下限并分析其物源组成，但是运用碎屑锆石研究首先需要关注的基本问题是分析的数据量和锆石颗粒的选择（Guo Pei et al.，2017）。Anderson 使用标准二项概率公式，认为若要使占颗粒数 5%的年龄成分得到识别的概率达到 95%，则至少需要 60 个分析数据（Anderson，2002）。因此，为避免前人因碎屑锆石测试数据量不足而造成结论不一致的情况，本文以二十二站组中分布最广的砂岩为研究对象，重新厘定其沉积时代，并进一步揭示其沉积物来源及大地构造环境，为深入认识盆地构造属性和地球动力学演化提供证据。同时，黑龙江省齐齐哈尔地质勘查总院正在二十一站矿区进行找矿工作，二十二站组是赋矿岩体的围岩，矿区二十二站组一段、二段、三段均有出露，对其进行研究将对下一步的找矿工作具有现实的指导意义。

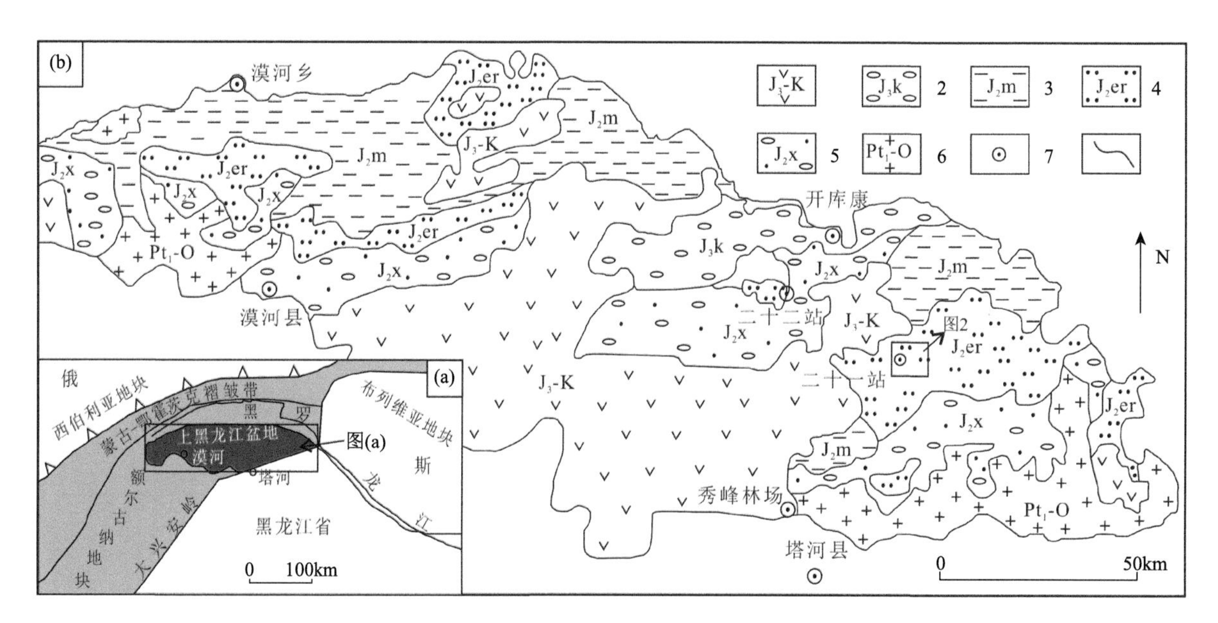

图 1　漠河盆地大地构造位置图（a，据 Li Xiangwen et al.，2017 修改）、漠河盆地地质简图（b，据 Zhao Liguo et al.，2014 修改）

Fig. 1　Geotectonic location map of Mohe Basin (a，modified after Li Xiangwen et al.，2017)，Generalized geological map of Mohe Basin (after Zhao Liguo et al.，2014).

1-晚侏罗世—早白垩世火山地层；2-上侏罗统开库康组；3-中侏罗统漠河组；4-中侏罗统二十二站组；5-中侏罗统绣峰组；6-古元古代—奥陶纪盆地基底；7-主要城镇；8-地质界线

1-Late Jurassic–Early Cretaceous volcanic rock reservoir; 2-Upper Jurassic Kaikukang Formation;3-Middle Jurassic Mohe Formation;4-Middle Jurassic Ershierzhan Formation; 5-Middle Jurassic Xiufeng Formation; 6-Palaeoproterozoic–Ordovician basin basement;7-Main towns;8-geological boundary

1 地质背景

1.1 构造

漠河盆地大地构造上位于蒙古-鄂霍茨克褶皱带中的额尔古纳微板块的北缘[图 1（a）]，地处西伯利亚板块与华北板块碰撞拼合部位，盆地西部延入蒙古国境内，北面及东面过黑龙江延入俄罗斯境内（Zhang Shun et al.，2003；Wu Genyao et al.，2006），在现今的大地构造格局上占有重要位置。漠河盆地被划分为洛古河拗陷、阿木尔拗陷、腰站拗陷、二十二站隆起和额木尔河推覆带五个二级构造单元（Miao Zhongying et al.，2014）。漠河盆地在蒙古-鄂霍茨克洋侏罗纪由西向东逐渐俯冲的大地构造背景下，经历了原型盆地形成，挤压回返，火山断陷盆地形成和抬升萎缩四个阶段，盆地西部具有挤压推覆，中部和东部具有拉张断陷的构造特点（Zhang Shun et al.，2003）。漠河盆地具有双层结构特征：下部的盆地受南北向发育的大型逆冲推覆构造控制，呈东西向展布，主要充填物为晚侏罗世碎屑岩，上部的盆地是早白垩世形成的受北东向伸展断层控制的小型断陷盆地，叠加在早期近东西向分布的构造带之上，主要充填物为火山碎屑岩和火山岩（He Zhonghua et al.，2008a）。

1.2 地层

漠河盆地基底主要为结晶变质岩和花岗岩，由古元古界兴华渡口岩群和寒武系兴隆群，泥盆系泥鳅河组、霍龙门组以及古生代花岗岩组成。盖层主要为上侏罗陆相碎屑岩和白垩纪火山岩，为晚侏罗世南北挤压的前陆盆地叠加早白垩世伸展的火山断陷盆地（He Zhonghua et al.，2008b；Li Chunlei，2007；Sun Qiushi，2013；Li Liang et al.，2017），盖层地层由下至上地层层序为上侏罗统绣峰组、二十二站组、额木尔河组和开库康组，下白垩统塔木兰沟组、上库力组和伊力克得组，新近系中新统—上新统金山组和第四系（Wu Heyong et al.，2004；Hou Wei et al.，2010b）。上侏罗统绣峰组、二十二站组、额木尔河组和开库康组沉积厚度大，分布范围广，是漠河盆地油气勘探主要目的层段（Wu Heyong et al.，2003a）。二十二站组为陆源细碎屑沉积，以灰黑色、灰绿色粉、细、中、粗粒长石岩屑砂岩、泥质粉砂岩互层为主，局部夹含砾砂岩、泥岩及煤线，属于河流冲积相、河漫滩相、沼泽相沉积建造，产淡水动物化石及植物化石（Du Tiantian et al.，2019；Du Bingying et al.，2019）。该组地层为近源快速沉积成岩，与下伏绣峰组和上覆额尔木河组均为整合接触（Du Tiantian et al.，2019）。二十二站组省内典型剖面为塔河县开库康乡嫩漠公路二十二站后山剖面（Du Bingying et al.，2019）。二十二站组主要出露于漠河盆地东段（腰站盆地）的中间部位，呈东西向带状分布，出露面积 1379.43km^2，占地层总面积的 14.71%①。

2 样品特征及测试方法

2.1 样品特征

本文样品采自于塔河县二十一站矿区探槽 D50TC1[图 2（a）]和钻孔 ZK30-1[图 2（b）]，GPS 坐标分别为 N52°50′59″ E124°53′40″和 N52°50′59″ E124°53′28″。所采集样品均隶属二十二站组地层，探槽中砂岩手标本及镜下特征[图 3（c）、（d）]与钻孔中砂岩手标本及镜下特征[图 3（e）、（f）]一致，岩性为中粗粒长石砂岩。岩石主要由碎屑物质和填隙物组成。碎屑物质主要为石英、长石、岩屑。其中石英以单晶石英为主，少部分为结核状石英颗粒，石英粒径 0.25～1mm 不等，含量 40%；长石基本上发生了黏土化和绢云母化，原始晶型需仔细辨认，长石粒径 0.25～1mm 不等，含量约 25%；岩屑大部分也发生了黏土化和绢云母化，岩屑母岩类型很难辨认，疑似火成岩岩屑[图 2（d）]，岩屑粒径 0.25～1mm 不等，含量约 10%。填隙物主要由黏土杂基、胶结物和自生矿物组成。黏土杂基，泥状结构，含量约为 5%；胶结物含量约 5%，自生矿物主要表现为部分石英具自生加大边结构，分布在石英碎屑较富集处，含量为 15%。

2.2 锆石 LA-ICP-MS U-Pb 测年

锆石样品的挑选、制靶和阴极发光（CL）照像在北京锆年领航科技有限公司实验室完成。在双目镜（OLYMPUS）下挑选一定数量的锆石颗粒粘在双面胶上，然后以低温环氧树脂浇铸，过夜烘干，待环氧树脂充分固化后用砂纸打磨，在偏光显微镜下查看确定磨到锆石最大截面为止，然后在自动抛光机上对锆石表面进行抛光，直至锆石表面达到镜面效果。依次用酒精、洗涤剂和水在超声波中对靶表面进行清洗。制靶完成后先在偏光显微镜下按最宽视域依次进行透射光和反射光照相，然后对锆石靶进行镀金处理，在配备了英国 Gatan 阴极荧光探头的日本电子 JSM6510 型电子显微镜进行阴极发光照相。在进行锆石年龄测试之前，通过透射光、反射光和 CL 图像仔细观察研究锆石晶体形态与内部生长层的分布和结构，选择锆石最佳测试点位。锆石 U-Pb 年龄测定在自然资源部成矿作用与资源评价重点实验室完成，所用仪器 Finnigan Neptune 型 LA-ICP-MS 及与之配套的 Newwave UP213 激光剥蚀系统。激光剥蚀束斑直径为 30μm，激光脉冲为 10Hz，能量为 32～36mJ，激光剥蚀样品的深度为 20～40μm，采样方式为单点剥蚀，测试时锆石年龄采用国际标准锆石 GJ-1 作为外标，元素含量采用 NIST SM610 作为外标，^{29}Si 作为内标元素（锆石中 SiO_2 的质量分数为 32.8%），进行锆石 U-Pb 定年和锆石成分测试；测试中采用氦气作为剥蚀物质的载气。每测五个点之后插入一次标样测点，以便及时校正。普通铅校正采用 Anderson（2002）推荐的方法；采用 ICPMSDataCal（V3.7）程序对同位素比值数据进行处理，详细的仪器操作条件和数据处理方法见 Liu Yongsheng et al.（2010）。年龄计算、协和图及概率密度图的绘制采用 ISOPLOT 程序（Ludwig，2003）进行绘制。

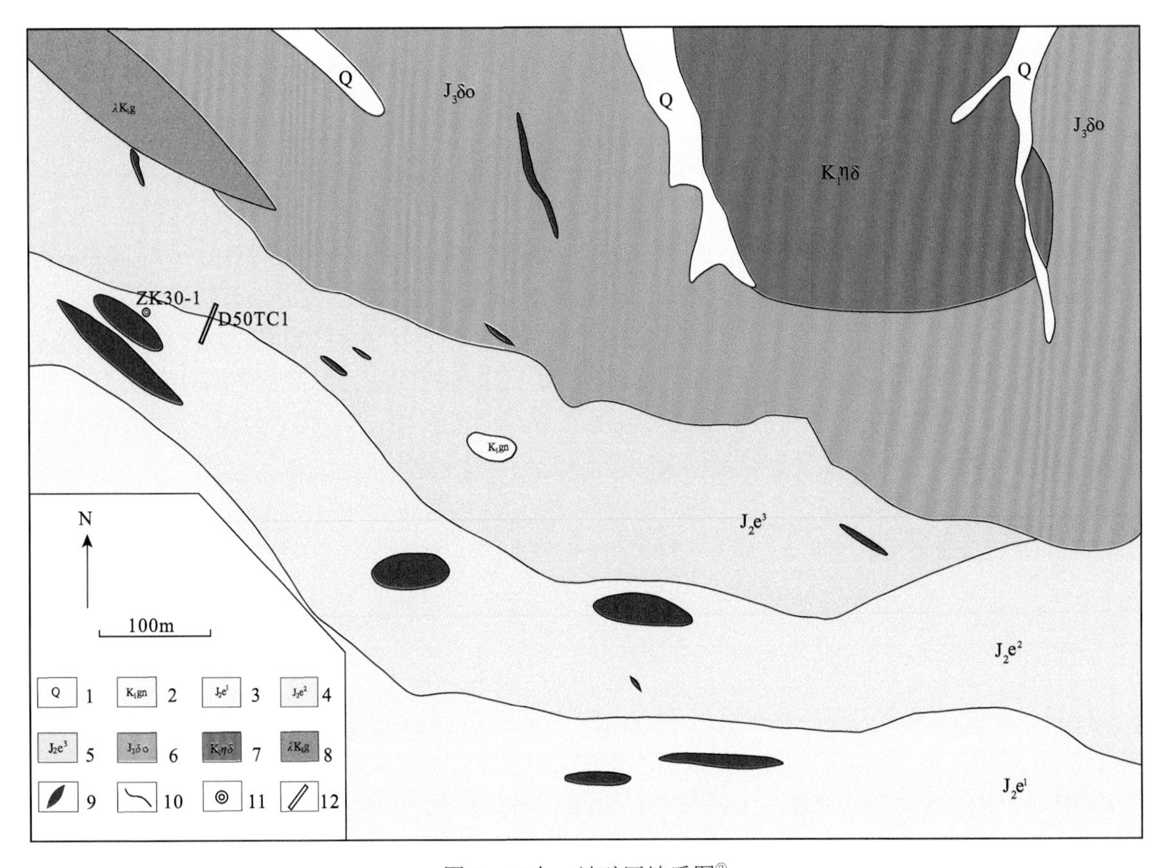

图 2 二十一站矿区地质图②

Fig. 2 Ershiyizhan mining area geological map showing sample locations.

1-第四系；2-光华组；3-二十二站组一段；4-二十二站组二；5-二十二站组三段；6-石英闪长岩；7-二长岩；8-潜火山岩；9-岩脉；10-地质界线；11-钻孔；12-探槽

1-Quaternary system; 2-Guanghua Formation; 3-1th Member of Ershierzhan Formation; 4-2th Member of Ershierzhan Formation; 5-3th Member of Ershierzhan Formation; 6-quartz diorite; 7-monzonite; 8-subvolcanic rock; 9-rock vein; 10-geological boundary;11-drilling hole; 12-exploratory trench

2.3 岩石地球化学分析

岩石地球化学分析样品在北京锆年领航科技有限公司实验室制备。样品经仔细挑选，洗净、晒干、磨成200目粉末。岩石的主量元素、微量元素和稀土元素分析在核工业北京地质研究院分析测试研究中心完成。主量元素使用X-射线荧光光谱仪（飞利浦PW2404）完成，其中Al_2O_3、SiO_2、MgO、Na_2O检测限为0.015%，CaO、K_2O、TiO_2检测限为0.01%，$Fe_2O_3^T$、MnO、P_2O_5检测限为0.005%；FeO用容量法完成（检测限为0.1%）。微量元素及稀土元素使用电感耦合等离子体质谱FininganMAT HR-ICP-MS（Element I型）完成，分析精度优于5%。

3 LA-ICP-MS U-Pb锆石定年

3.1 锆石特征

整体来看，4个样品中碎屑锆石形态特征非常相似，大部分锆石晶型完好，呈棱角状、次棱角柱状，少部分锆石呈次圆状，个别锆石呈浑圆状，粒度为70×100μm～100×180μm，长宽比一般为1.2∶1，少量为1.5∶1。其中破碎的不完整锆石仍能保持棱角状，反映锆石的近距离搬运。大部分锆石颗粒CL图像均显示比较清晰的韵律震荡环带特征（图3），说明所研究样品中大部分被测锆石为岩浆结晶锆石（Connelly，2000；Li Yun et al.，2017）。极少部分锆石由暗色核部和亮色边构成（图3），这是由原生岩浆结晶锆石增生或重结晶所致，因此构成了一个锆石内部以岩浆成因为主，晶体边部以变质成因为主的复合成因锆石。这类锆石属于增生-混合型锆石，指示后期构造岩浆作用对早先形成锆石的改造，从而反映了二十二站组沉积物经历了多期构造热事件。不同结构特征的锆石反映了锆石来源的复杂，具碎屑锆石特点。且绝大部分锆石（98.7%）的Th/U比值都大于0.4。只有39颗锆石的Th/U比值小于0.4，其中8颗锆石Th/U≤0.1。一般认为Th/U比值大于0.4的锆石为岩浆成因，小于0.1的为变质成因（Corfu et al.，2003；Wu Yuanbao et al.，2004；Hermann et al.，2001）。因此可以认为本次测试的锆石绝大部分为岩浆成因。

3.2 锆石U-Pb年龄结果

由于$^{207}Pb/^{206}Pb$年龄误差较大，对于年轻的锆石使用$^{206}Pb/^{238}U$年龄，而对于古老的锆石使用$^{207}Pb/^{206}Pb$年龄，这样获得的结果更可靠（Compston et al.，1992）。对于锆石年龄大于1000Ma的数据，本文采用其$^{207}Pb/^{206}Pb$年龄值，小于1000Ma的数据则采用$^{206}Pb/^{238}U$年龄，大部分以满足谐和度大于90%且小于110%为标准遴选出有效年龄数据，但是由于个别锆石偏小，测试过程中信号较弱，从而导致其协和度低于90%，这样的锆石年龄本文仍然作为有效年龄使用。样品D50TC1-02、D50TC1-09、D50TC1-11、ZK30-1-28分别获得46个、68个、99个和86个有效数据，总体数目符合Anderson（2002）提出的使占颗粒数5%的年龄成分得到识别的概率达到95%所需的锆石数目。所有锆石LA-ICP-MS U-Pb同位素分析数据及年龄结果见附件1。图4为样品中代表性碎屑锆石CL图像及对应年龄值。图5展示了四个样品锆石分析的U-Pb年龄谐和曲线图和频谱图。另外，本次测试在选择具亮边锆石颗粒测点时，尽量选其靠中心部位，以避免后期热液活动对锆石的影响。

本次研究4件样品的锆石CL图像（图4）显示，测年值＞543Ma的锆石颗粒的磨圆度较高，说明沉积物经历了长距离或多次搬运，是漠河盆地岩石中记录到的前寒武系剥蚀区的物质信息。本次测试共获得299个锆石年龄数据，由图5可知，各个样品碎屑锆石年龄分布特征基本一致。综合4件样品锆石年龄分布特征，锆石年龄范围大，在138±2Ma～2711±10Ma之间(附件1)。锆石年龄主要分布于侏罗纪($N=126$)，其次是三叠纪（$N=28$）、二叠纪（$N=21$）、石炭纪（$N=41$）、泥盆纪（$N=22$）、志留纪（$N=8$）、奥陶纪（$N=22$）、寒武纪（$N=9$），震旦纪（$N=1$）、南华纪（$N=3$）、青白口纪（$N=3$）、蓟县纪（$N=3$）和古元古代（$N=8$）仅少量分布，主峰期为140～250Ma，其次为210～370Ma，420～500Ma，最显著的年龄峰值为190Ma左右。这些锆石年龄数据特征说明碎屑锆石的多来源性，且这些锆石大多具有较好的谐和

图 3　二十一站矿区野外照片、手标本及显微镜照片.

Fig. 3　Photos of outcrops，hand specimens and photomicrographs in the Ershiyizhan deposit.

（a）野外探槽照片；（b）野外钻孔岩心照片；（c）探槽中二十二站组砂岩手标本；（d）探槽中二十二站组砂岩手标本显微照片（正交偏光）（e）钻孔中二十二站组砂岩岩心；（f）钻孔中二十二站组砂岩岩心显微照片（正交偏光）；Q-石英；Pl-斜长石；Lv-岩屑

(a) Photographs of field trenches; (b) Photographs of drill hole cores in the field; (c) the Ershierzhan Formation sandstone Hand specimen in the trench; (d) Micrographs of the Ershierzhan Formation sandstone in the trenches (perpendicular polarized) (e) the Ershierzhan Formation sandstone in the drill hole; (f) the Micrograph of the Ershierzhan Formation sandstone in the drill hole (perpendicular polarized); Q-quartz; Pl-plagioclase; Lv-igneous rock cuttings

性[图 5（a）、（c）、（e）、（g）]，谐和线上的锆石年龄即可代表其真实的形成年龄。值得关注的是，具有年龄古老的锆石，比如 1999±10Ma，684±7Ma，746±8Ma，2301±29Ma，2210±15Ma，2003±9Ma，2002±11Ma，这些锆石颗粒磨圆度较高，表明其可能经历了一次或多次的搬运和再循环过程，为古老沉积岩系的再循环锆石，说明了漠河盆地二十二站组沉积物中记录了源自元古代剥蚀区的物质信息。所有测试数据中，有一颗锆石年龄为 2711Ma±10Ma，谐和度 97%，是本次测试捕获到的最老年龄信息，说明二十二站组沉积物中存在新太古界的地壳物质。另外，所有样品中均未发现来自新生代（Cz）的锆石，说明二十二站组沉积物中没有新生代的物质来源。

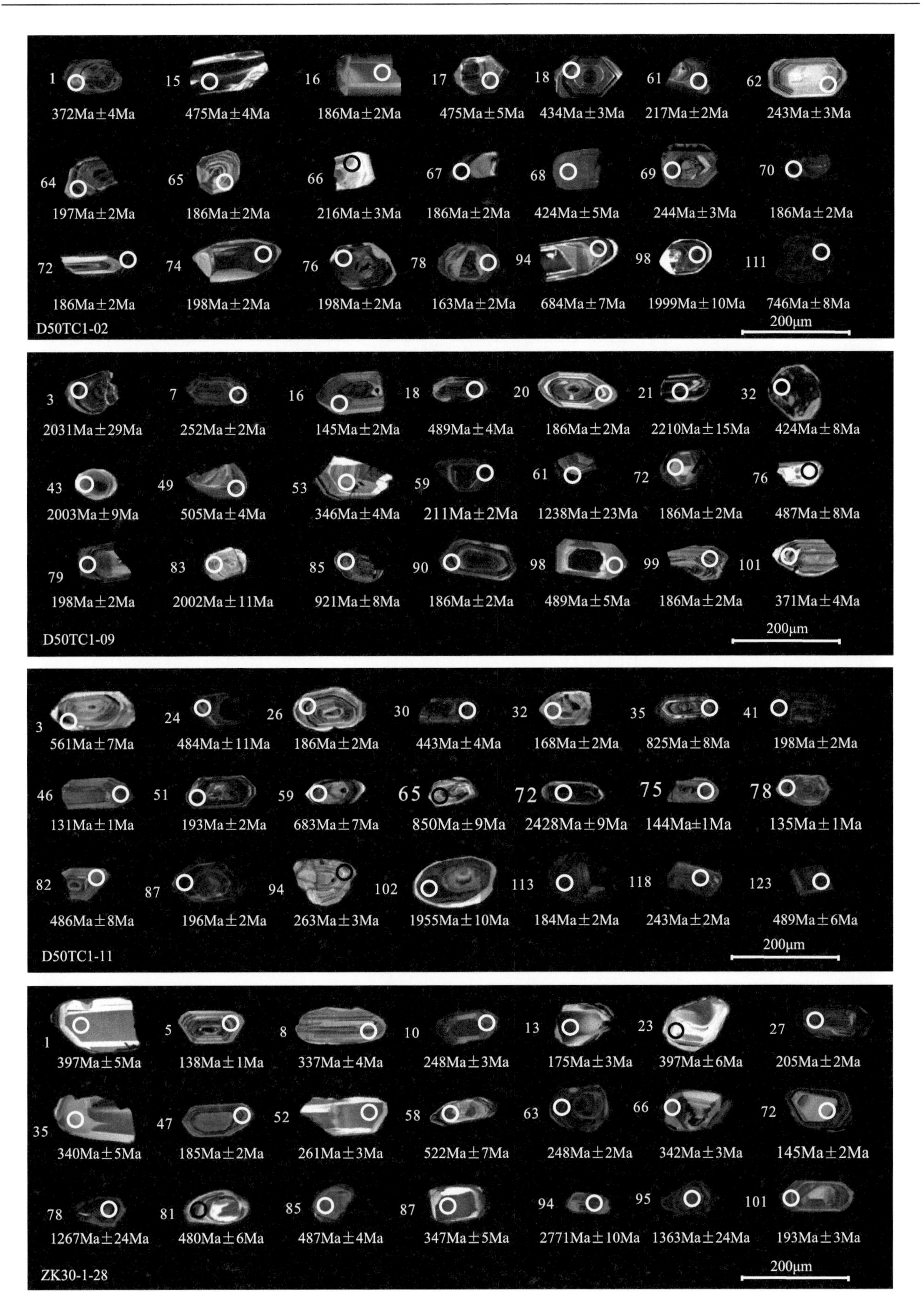

图 4 二十二站组砂岩代表性碎屑锆石阴极发光图像及锆石年龄

Fig. 4 CL images and zircon U-Pb ages of the representative detrital zircons from Ershierzhan Formation sandstone (the circles show the number of spot analysis and the age result is indicated below).

4　岩石地球化学特征

4.1　主量元素

二十二站组长石砂岩主量元素分析见表 1。长石砂岩测得 SiO_2 含量为 68.73%～73.94%，平均 70.14%；Al_2O_3 含量为 12.73%～15.11%。岩石的 SiO_2/Al_2O_3 比值能够反映沉积岩的成熟度（Taylor and McLennan，1985），二十二站组长石砂岩 SiO_2/Al_2O_3 比值范围为 4.45～5.61，平均值为 4.99，显示出成熟度中等的特征。Fe_2O_3/K_2O 比值反映了风化过程中砂岩中铁、镁矿物的稳定程度。本次测试样品的 Fe_2O_3/K_2O 比值范围为 0.12～0.25，平均值为 0.20，Liliang et al.（2017）测得样品的 Fe_2O_3/K_2O 比值范围为 0.41～1.03，平均值为 0.78，表明本次采集样品经历了一定的风化淋滤作用，造成砂岩中铁镁矿物的分解、流失，同时也表明砂岩中铁镁矿物的稳定程度较低。在砂岩类别判别图解中[图 6（a）]，样品点大部分均落入长石砂岩或杂砂岩与长石砂岩接触区域附近，与野外观察结果及镜下特征一致，本次研究将其定名为长石砂岩。

4.2　稀土元素和微量元素

所采样品稀土和微量元素数据见表 1。稀土总量变化范围 122.27×10^{-6}～135.8×10^{-6}，平均值为 142.6×10^{-6}，整体上看，稀土元素含量变化范围较小。在稀土元素球粒陨石标准化图解上[图 6（b）]，本次所测岩石样品与前人测试表现出一致的趋势，表明它们基本上来自相同的物源，均呈现一定程度的轻、重稀土元素分馏，总体表现为轻稀土元素富集，重稀土元素相对亏损的特征，LREE/HREE 平均比值为 10.19，Eu/Eu^* 比值范围为 0.54～0.66，平均值为 0.61，具明显的负 Eu 异常，$(La/Yb)_N$ = 12.62～13.43，无 Ce 异常。

样品全球平均大陆上地壳（UCC）标准化稀土元素配分曲线较为平缓，本次测试样品的 $(Eu/Eu^*)_{UCC}$ 和 $(La/Yb)_{UCC}$ 值与前人样品数据（Liliang et al.，2017）相近（表 1），总体上表现出轻微的轻稀土富集，大部分呈现出弱的 Eu 负异常，但个别样品呈现出 Eu 正异常的特征。这些特征表明，测试砂岩样品与全球平均大陆上地壳的稀土元素含量非常接近[图 6（c）]（Bhatia，1985）。

特征微量元素原始地幔标准化微量元素蛛蛛图中规律明显。整体而言，亏损高场强元素 Nb、Ta，以及大离子亲石元素 Ba 和 Sc，但是本次样品呈现出轻微的 Ba 富集，富集高场强元素 Th、La、Ce、Nd、Sm 和大离子亲石元素 Rb、K，但 U、Th、La、Ce 等轻稀土元素富集不明显。

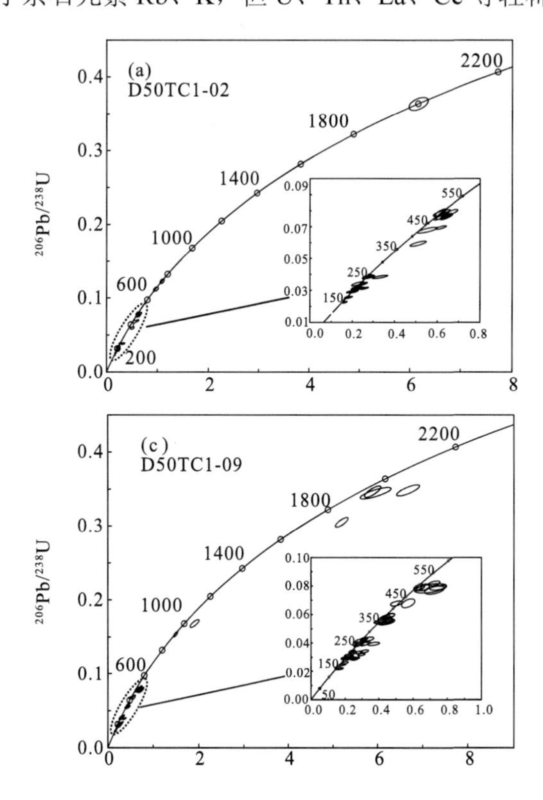

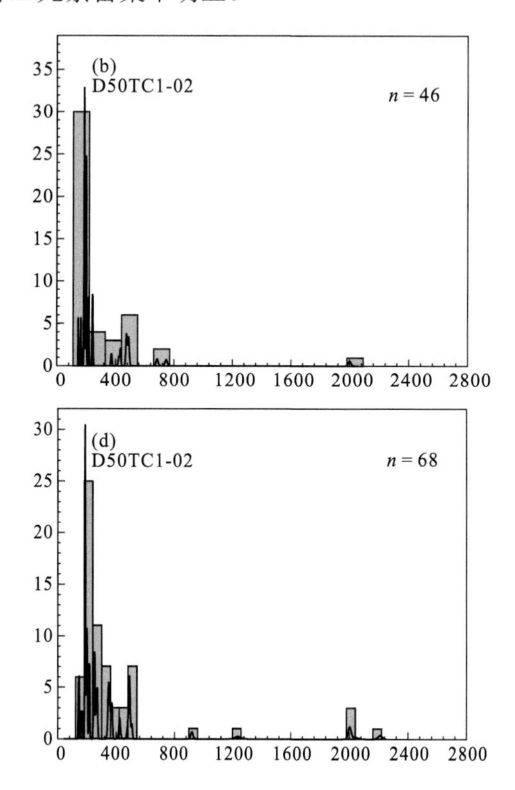

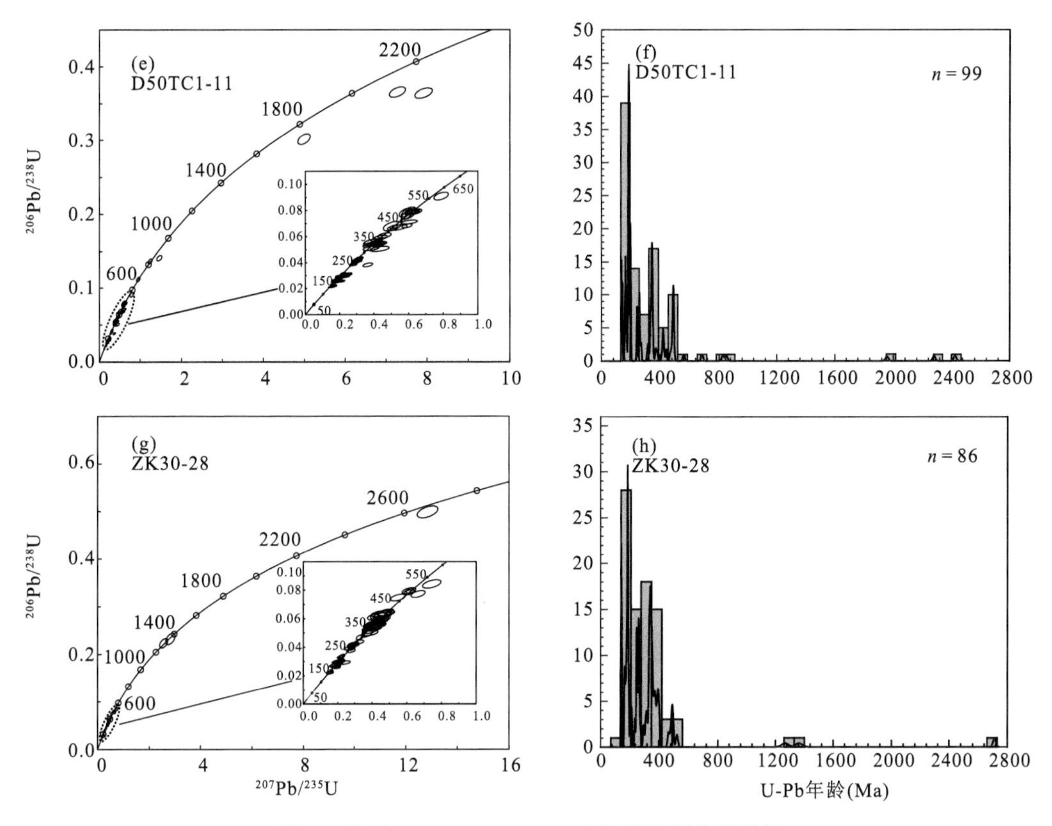

图 5 锆石 LA-ICP-MS U-Pb 定年谐和图和频谱图

Fig. 5 Zircon LA-ICP-MS U-Pb concordia diagrams and histograms.

表 1 二十二站组长石砂岩主量（%）和微量元素（$\times10^{-6}$）含量

Table 1 Major (%), trace and rare earth element concentrations ($\times10^{-6}$) of Ershi'erzhan Formation sandstone

样品号	ZK30-1-29	ZK30-1-30	ZK30-1-32	SBS-N1-B1	SBS-N1-B2	SBS-N1-B3	SBS-N1-B4	SBS-N1-B5	SBS-N1-B6	SBS-N1-B7	SBS-N1-B8	SBS-N1-B9	SBS-N1-B10
SiO_2	68.73	71.23	68.80	73.94	72.08	65.85	70.19	71.39	72.07	71.05	67.26	70.98	68.28
TiO_2	0.39	0.41	0.41	0.19	0.27	0.40	0.33	0.18	0.23	0.25	0.43	0.24	0.40
Al_2O_3	14.71	14.60	14.58	14.22	13.96	13.96	13.96	12.73	13.55	14.20	15.11	13.88	13.50
Fe_2O_3	0.39	0.70	0.66	1.91	1.79	3.73	3.13	2.06	1.70	2.47	3.40	2.26	2.90
FeO	2.81	2.74	2.71	0.96	1.36	2.98	2.82	1.66	1.44	1.00	2.75	1.92	1.95
MnO	0.03	0.03	0.03	0.07	0.09	0.13	0.11	0.12	0.09	0.09	0.11	0.09	0.10
MgO	1.14	1.27	1.23	0.22	0.41	0.91	0.41	0.48	0.51	0.22	0.97	0.41	0.64
CaO	1.50	0.41	1.43	0.28	0.75	2.59	0.98	2.63	1.42	1.32	1.99	1.33	2.88
Na_2O	2.67	2.34	2.74	4.47	4.42	3.66	4.25	4.76	4.82	4.33	3.60	4.34	3.71
K_2O	3.31	2.75	3.07	3.43	4.39	3.61	3.54	2.54	2.78	3.02	3.66	3.23	2.89
P_2O_5	0.11	0.09	0.11	0.06	0.06	0.11	0.08	0.06	0.07	0.07	0.11	0.07	0.08
LOI	4.16	3.41	4.09	1.13	1.70	4.84	2.84	2.92	2.64	2.42	3.22	2.99	4.37
Total	95.80	96.57	95.77	100.88	101.28	102.80	102.70	101.50	101.30	100.90	102.60	101.70	101.70
Fe_2O_3T	3.20	3.44	3.37	2.98	3.30	7.04	6.26	3.90	3.30	3.58	6.46	4.39	5.07
SiO_2/Al_2O_3	4.67	4.88	4.72	5.20	5.16	4.72	5.03	5.61	5.32	5.00	4.45	5.11	5.06
K_2O/Na_2O	1.24	1.16	1.12	0.77	0.99	0.99	0.83	0.53	0.58	0.70	1.02	0.74	0.78
Fe_2O_3/K_2O	0.12	0.25	0.21	0.56	0.41	1.03	0.88	0.81	0.61	0.82	0.93	0.70	1.00
F1	−5.75	−4.25	−3.79	−1.12	−3.02	−0.82	−1.01	0.09	−0.82	−0.05	−0.97	−0.90	−0.24
F2	−0.76	−2.43	−1.29	0.46	1.77	−0.01	0.21	−0.08	−0.15	0.09	−0.15	0.22	−0.41

续表

样品号	ZK30-1-29	ZK30-1-30	ZK30-1-32	SBS-N1-B1	SBS-N1-B2	SBS-N1-B3	SBS-N1-B4	SBS-N1-B5	SBS-N1-B6	SBS-N1-B7	SBS-N1-B8	SBS-N1-B9	SBS-N1-B10
La	26.40	27.80	29.40	18.10	43.70	37.70	65.50	18.60	24.10	28.50	38.40	36.70	78.40
Ce	51.50	52.70	57.80	37.00	82.20	75.40	123.00	34.90	46.90	51.90	75.10	68.80	151.00
Pr	6.01	6.02	6.68	4.03	8.70	8.30	12.70	3.79	5.05	5.68	8.12	7.25	16.20
Nd	22.30	22.10	25.10	14.50	28.80	28.70	41.70	13.80	17.90	19.90	29.10	24.90	54.30
Sm	4.27	4.06	4.31	2.48	4.32	4.30	5.37	2.53	2.97	3.09	4.44	3.55	7.83
Eu	0.80	0.69	0.87	0.58	0.64	0.71	0.64	0.48	0.49	0.57	0.80	0.59	1.17
Gd	3.39	3.70	3.56	1.87	2.90	3.74	3.94	1.75	2.12	2.71	3.97	2.89	5.49
Tb	0.52	0.57	0.62	0.36	0.54	0.56	0.66	0.36	0.40	0.44	0.59	0.51	0.80
Dy	2.78	2.79	3.17	1.85	2.59	3.37	3.16	1.69	2.06	2.28	3.34	2.31	4.14
Ho	0.61	0.54	0.59	0.34	0.52	0.64	0.60	0.35	0.38	0.48	0.68	0.47	0.80
Er	1.70	1.81	1.65	0.81	1.27	1.68	1.61	0.94	1.00	1.19	1.74	1.26	1.98
Tm	0.23	0.26	0.25	0.16	0.25	0.29	0.28	0.18	0.19	0.23	0.34	0.23	0.38
Yb	1.50	1.58	1.57	1.00	1.47	1.90	1.73	1.06	1.22	1.35	2.06	1.46	2.42
Lu	0.27	0.28	0.27	0.14	0.21	0.28	0.28	0.15	0.17	0.22	0.29	0.21	0.36
Eu/Eu^*	0.62	0.54	0.66	0.83	0.55	0.54	0.42	0.69	0.59	0.60	0.58	0.56	0.55
Ce/Ce^*	0.96	0.95	0.97	1.06	1.03	1.05	1.05	1.02	1.04	1.00	1.04	1.03	1.04
ΣREE	122.27	124.91	135.80	83.20	178.00	168.00	261.00	80.60	105.00	119.00	169.00	151.00	325.00
LREE/HREE	10.12	9.83	10.63	11.70	17.30	12.50	20.30	11.50	12.90	12.30	12.00	15.20	18.90
$(La/Yb)_N$	12.62	12.62	13.43	13.00	21.30	14.20	27.20	12.60	14.20	15.10	13.40	18.00	23.20
$(Eu/Eu^*)_{UCC}$	0.99	0.84	1.04	1.27	0.84	0.83	0.65	1.06	0.91	0.92	0.90	0.86	0.83
$(La/Yb)_{UCC}$	1.29	1.29	1.37	1.33	2.18	1.46	2.78	1.29	1.45	1.55	1.37	1.84	2.38
Li	31.30	39.70	29.50	4.45	3.24	5.00	4.56	17.90	4.05	4.62	53.10	5.12	7.11
Be	1.63	1.45	2.24	2.02	1.83	2.46	1.98	1.32	1.37	1.81	1.92	1.80	1.80
Sc	7.14	7.57	7.30	2.96	3.71	6.35	5.87	3.05	3.21	5.13	6.27	3.76	6.12
V	41.90	51.00	49.50	15.50	17.70	37.70	29.40	17.80	16.80	22.70	37.70	20.60	34.70
Cr	212.00	250.00	241.00	15.20	19.00	27.50	19.50	23.90	7.50	8.70	16.50	15.80	12.40
Co	8.82	9.78	8.49	2.18	3.89	5.90	4.22	3.63	3.37	2.98	5.55	3.63	4.49
Ni	19.60	29.60	20.40	4.91	6.60	7.48	5.66	6.42	4.26	3.95	6.73	5.14	5.65
Ga	16.70	16.40	16.80	15.60	21.00	22.70	27.90	13.60	15.30	17.50	23.00	21.00	32.40
Rb	99.20	66.70	95.30	90.50	113.00	106.00	96.80	70.10	72.50	82.50	103.00	100.00	91.30
Sr	307.00	154.00	303.00	176.00	140.00	196.00	147.00	178.00	177.00	184.00	225.00	143.00	179.00
Y	15.20	14.80	16.30	9.06	14.10	17.70	17.10	10.40	11.10	13.20	19.20	13.10	22.00
Ba	1010.00	946.00	963.00	500.00	703.00	438.00	515.00	490.00	481.00	456.00	625.00	560.00	331.00
Pb	11.20	11.10	13.20	17.50	15.80	14.40	12.00	16.10	16.70	16.80	19.30	16.80	11.30
Th	7.01	7.41	8.52	4.75	9.89	10.80	15.20	4.68	5.47	7.39	11.10	9.04	18.40
U	1.65	1.81	1.73	1.07	2.31	1.98	1.94	1.18	1.03	1.56	2.15	1.31	2.62
Nb	6.44	6.68	7.28	5.22	7.62	9.55	9.60	5.04	6.03	6.18	10.50	6.70	12.70
Ta	0.48	0.51	0.55	0.36	0.58	0.61	0.77	0.37	0.40	0.48	0.71	0.48	1.30
Zr	122.00	167.00	165.00	46.40	54.90	87.80	69.70	39.00	43.20	54.60	92.30	56.70	121.00
Hf	3.16	4.22	3.83	1.45	1.87	2.74	2.34	1.33	1.37	1.83	2.86	1.79	3.81
Rb/Sr	0.32	0.43	0.31	0.51	0.81	0.54	0.66	0.39	0.41	0.45	0.46	0.70	0.51
Rb/Nb	15.40	9.99	13.09	17.30	14.80	11.10	10.10	13.90	12.00	13.30	9.81	14.90	7.19

注：主量元素单位为%，稀土、微量元素单位为 10^{-6}；$F1 = -1.773w(TiO_2) + 0.607w(Al_2O_3) + 0.76w(Fe_2O_3) - 1.5w(MgO) + 0.616w(CaO) + 0.509w(Na_2O) - 1.224w(K_2O) - 9.09$；$F2 = -0.445w(TiO_2) + 0.07w(Al_2O_3) - 0.25w(Fe_2O_3) - 1.142w(MgO) + 0.438w(CaO) + 0.475w(Na_2O) + 1.426w(K_2O) - 6.861$，据 Roser and Korsch（1988）；样品 ZK30-1-29、ZK30-1-30、ZK30-1-32 数据为本次测得，其余样品数据据李良等，2017。

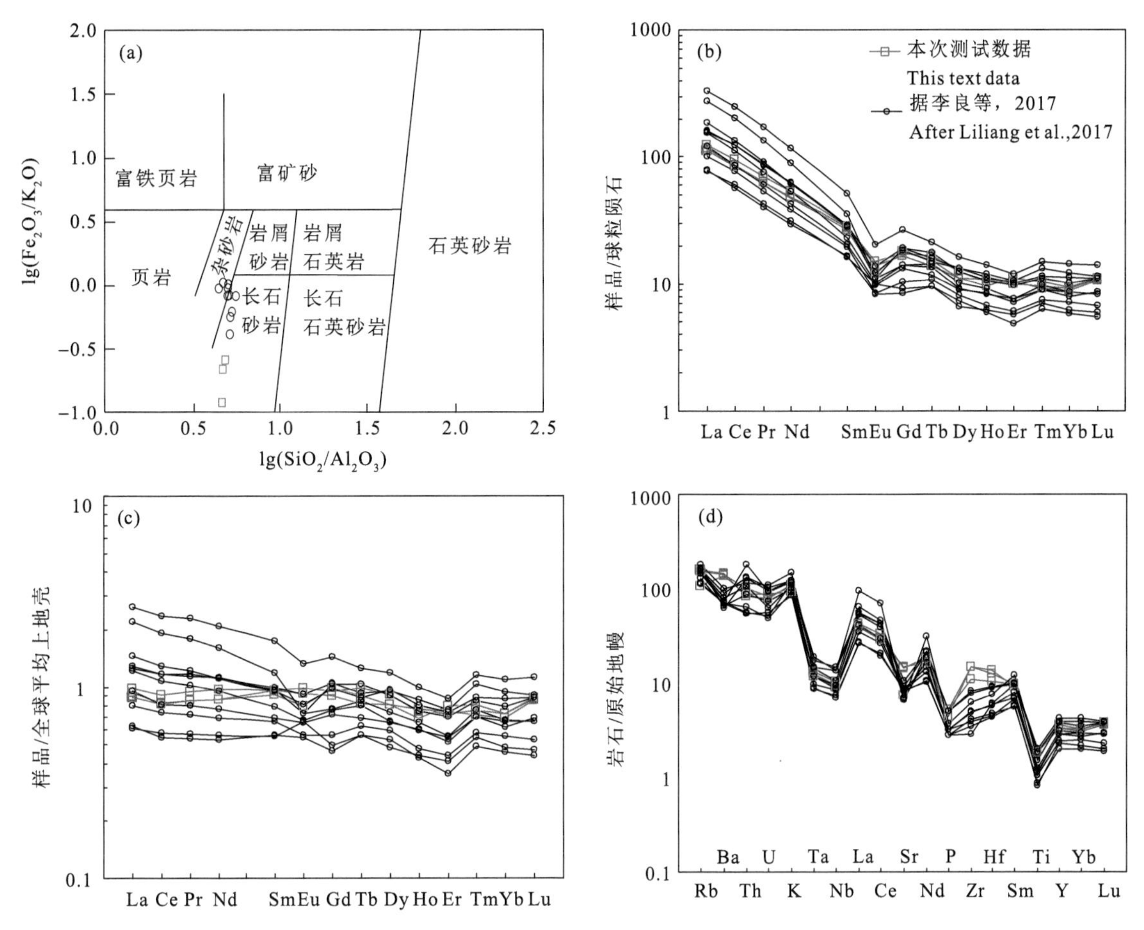

图6 （a）二十二站组砂岩岩石地球化学分类图解；（b）二十二站组砂岩球粒陨石；（c）全球大陆平均上地壳标准化稀土元素配分曲线；（d）二十二站组砂岩原始地幔标准化微量元素蛛蛛图（球粒陨石和原始地幔标准化值据文献 Sun et al.，1989；全球大陆平均上地壳标准化值据文献 Mclennan et al.，1993）

Fig. 6 (a) Geochemical classification diagrams of the Ershierzhan Formation sandstone; (b) Chondrite-normalized; (c) UCC-normalized rare earth element patterns of the Ershierzhan Formation sandstone; (d) Primitive mantle-normalized incompatible trace element variation diagrams of Ershierzhan Formation sandstone (normalized data of primitive mantle and chondrite from Sun et al., 1989; UCC data quote from Mclennan et al., 1995).

5 讨论

5.1 形成时代

长期以来，不同学者对二十二站组的形成时代存在争议。前人多根据二十二站组中被发现的生物化石推测其形成时代，Qiherige（1995）依据地层中发育的淡水底栖无脊椎动物（蠕虫）的生活遗迹化石将地层时代定为中侏罗世，Wu Heyong（2003b）依据二十二站组中的孢粉化石组合，认为其形成时代为晚侏罗世，Sun Guangrui（2002）依据地层中被发现的双壳类、介形动物群化石将其时代重新厘定为中-晚侏罗世 Bajocian-Tithoian 期。最新的碎屑锆石 LA-ICP-MS U-Pb 定年（Li Liang et al.，2017）将二十二站组沉积成岩时代定为晚侏罗世。本文主要通过对采自二十一站矿区二十二站组砂岩中碎屑锆石研究并结合搜集到的区域地质调查资料对其形成时代进行重新厘定。

本次研究从二十二站组 4 件砂岩样品中共获得 299 个可靠的年龄数据，结合它们的阴极发光图像特征以及 Th/U 比值，可以对二十二站组的形成年龄做出精确的限定。与最新锆石 U-Pb 测年结果（Li Liang et al.，2017；Zhen Miao，2017）不同的是，本次研究所测四个砂岩中锆石年龄分布范围均非常广，且表现出年龄越小锆石数量越多的特征，尤其是＞800Ma 的锆石仅零星分布，但是每件样品中均存在年龄老的锆石。样品 D50TC1-02 中最年轻锆石年龄为 144±1Ma（谐和度为 92%），样品 D50TC1-09 中最年轻锆石年龄为 139±2Ma 和 141±2Ma（谐和度均为 88%），样品 D50TC1-11 中最年轻锆石为 134±1Ma，136±1Ma 和

135±1Ma（谐和度分别为96%，91%，91%），样品Zk30-1-28中最年轻锆石为138±1Ma，谐和度为97%。这些锆石的阴极发光图像均具有震荡环带特征以及大于0.4的Th/U比值，因此可以判断这些锆石为岩浆成因的碎屑锆石，且未受到后期构造-热事件的改造作用而形成变质锆石（边）。

整体来看，此次测试发现最年轻锆石年龄为134±1Ma。此外，二十二站组泥岩中发现双壳类动物化石 *Sphaerium* ex gr. *anderssoni* (Grabau)，时代为J_3-K_1[①]，由此可将二十二站组形成时代的下限限定为早白垩世早期。

5.2 物源分析

关于二十二站组的物源也存在较大争议，许多学者认为二十二站组的物源来自北部的蒙古-鄂霍茨克造山带与南部的额尔古纳地块（He Zhengjun et al.，2003；Hou Wei，2006；Li Chunlei，2007；He Zhonghua et al.，2008a，2008b；Hou Wei et al.，2010a，2010b，2010c；Li Liang et al.，2017）；也有学者认为盆地沉积物主要来自南侧的陆块抬升基底，即古元古界兴华渡口群与寒武系兴隆群、古生代同碰撞和后碰撞花岗质岩石及早中生代中酸性火成岩（He Zhonghua et al.，2008a）；也有学者认为盆地的物源除来自南北两侧外，盆地内部也提供了部分物源（Sun Qiushi，2013）。

沉积物物源分析是盆地分析的重要内容，是再现沉积盆地演化、恢复古环境的重要依据（Xu Yajun，2007）。沉积物物源分析方法包括岩石学、地球化学、重矿物、同位素年代学等。稀土元素特征对于指示源岩具有重要意义，可作为鉴别砂岩物源的重要参数（Liu Junlai et al.，2003）。稀土元素配分模式可以用来指示物源岩性特征，因为源自基性岩石的稀土元素具有低的LREE/HREE比值，无Eu异常，而长英质岩石通常具有较高的LREE/HREE比值，且具Eu负异常。中酸性侵入岩（如花岗岩）和火山岩、长英质变质岩，以及来自大陆源区的沉积岩等Eu多显示负异常。二十二站组砂岩的稀土元素球粒陨石配分曲线[图6（b）]呈现轻稀土元素相对富集，重稀土元素相对亏损的特征，LREE/HREE平均比值为13.48，Eu/Eu^*比值范围为0.54～0.66，平均值为0.59，具明显的负Eu异常。稀土元素特征表明漠河盆地二十二站组沉积物物源主要来自上地壳。

在稀土元素配分模式相似的情况下，用稀土元素全球平均大陆上地壳成分(UCC)标准化图解可得到不同的分异形式，因此可以利用不同的分异形式对源岩特征做进一步区分（Zhang Pei et al.，2005）。二十二站组样品全球平均大陆上地壳成分（UCC）标准化稀土元素配分曲线[图6（c）]总体较为平缓，呈现出轻微的轻稀土元素富集，$(La/Yb)_{UCC}$平均值为1.66，且有些样品表现为Eu正异常，说明在不同时代的二十二站组沉积物发生了变化，可能有来自下地壳和幔源深部物质的加入。另外，在F1-F2物源区分析图上[图7（a）]，二十二站组砂岩样品大部分投在P3区域，有一个样品点落在P4区域，还有一个样品点落在P3、P4区域交界线上，说明二十二站沉积物源岩绝大部分来自大陆边缘弧长英质火成岩，少部分来自成熟大陆石英质物源区，可能为古老的地质体，克拉通或再旋回造山带（Roser and Korsch，1988）。砂岩宏观物质组成分析表明，二十二站组沉积物源岩性质主要代表了上地壳物质来源[图7（b）]，也可能有中基性物质或者相当于洋壳成分的碎屑。

为进一步揭示源岩属性，利用La-Th-Hf源岩属性判别图解（Gu X X et al.，2002）对二十二站组沉积物源岩属性进行分析。图8显示，大部分样品落入长英质物源区与安山质物源区之间的混合长英质/基性源区，个别样品落入安山质物源区。综合以上判别图解，二十二站组物源主要为上地壳大陆边缘弧长英质、安山质源区，并混有下地壳深部物质（基性岩），并且可能有成熟大陆石英质物源区的物质加入。漠河-上阿穆尔盆地周边均发育长英质活动大陆边缘物源区，因此仅仅依靠地球化学特征无法确定其物源究竟在哪里，即是盆地北部的西伯利亚板块物源区，还是盆地南部的额尔古纳地块北缘物源区，还是盆地基底，还是均兼而有之。同时也无法确定母岩组成，进而无法为讨论盆地性质和形成演化提供依据。利用沉积物中碎屑锆石年龄谱揭示源区并进行亲缘性对比是近年来流行的有效方法之一（Moecher and Samson，2006；Greentree and Li，2008；Wang et al.，2010；Yu et al.，2012；Dong et al.，2013；Kang Yu et al.，2018；Wang Yongsheng et al.，2019）。因此，本文通过对二十二站组沉积物中大量碎屑锆石的测试，在以上分析的基础上，进一步确定沉积物物源。

锆石属于超稳定重矿物，在沉积过程中，其成分和年龄不受影响（Morton et al.，1996），因此可以将其作为判断砂岩物源特征的有效证据（Sircombe，1999；Cawood and Nemchin，2000；Hallsworth et al.，2000；Li Yifei et al.，2018）。由于沉积岩的复杂性，本次研究四个样品中碎屑锆石 U-Pb 测年数据年龄分布宽泛，但是大多数年龄＜600Ma，总共只有 17 颗锆石年龄＞600Ma，显示出沉积物来源的复杂性和多源性。二十二站组沉积岩碎屑锆石 U-Pb 年龄数据为其物质源区提供了丰富的信息。本次研究共获得 299 组可靠的碎屑锆石 U-Pb 年龄且绝大多数碎屑锆石具岩浆锆石结构特征，主要来自于岩浆岩物源区或其再循环。本次测试在样品 D50TC1-09 中获得最年轻的锆石，其结构自形均一，具清晰的韵律环带，为岩浆锆石，年龄为 134±1Ma，属早白垩世早期，指示研究区存在早白垩世早期火成岩物源，同时反映漠河盆地在早白垩世之后的构造运动未在锆石上留下痕迹，而仅是盆内的填平补齐和对盆地内早白垩世之前沉积物的再分配。样品 ZK30-1-28 中具年龄最老的锆石，具微弱的环带结构，锆石颗粒半自形均一，其年龄为 2711±10Ma。该年龄反映了二十二站组沉积物具新太古代的物质来源。

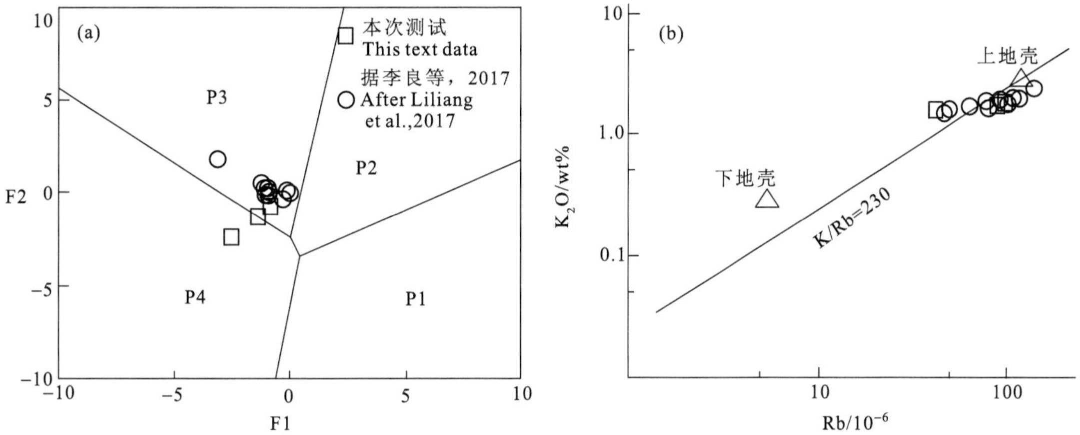

P1-镁铁质和少量中性火成岩源区；P2-主要中性火成岩源区（主要为安山岩）；P3-长英质火成岩源区（大陆边缘弧）；P4-古老的沉积体系或卡拉通/再旋回造山带

图 7　二十二站组砂岩源区母岩性质判别图

Fig. 7　Discrimination diagrams of mother rock for the Ershi'erzhan Formation sandstone.

（a）F1-F2 判别图解（底图据 Roser et al.，1988）；（b）K_2O-Rb 判别图解（底图据 Floyd et al.，1987）
(a) F1-F2 discrimination diagram (after Roser et al., 1988); (b) K_2O-Rb discrimination diagram (after Floyd et al.,1987)

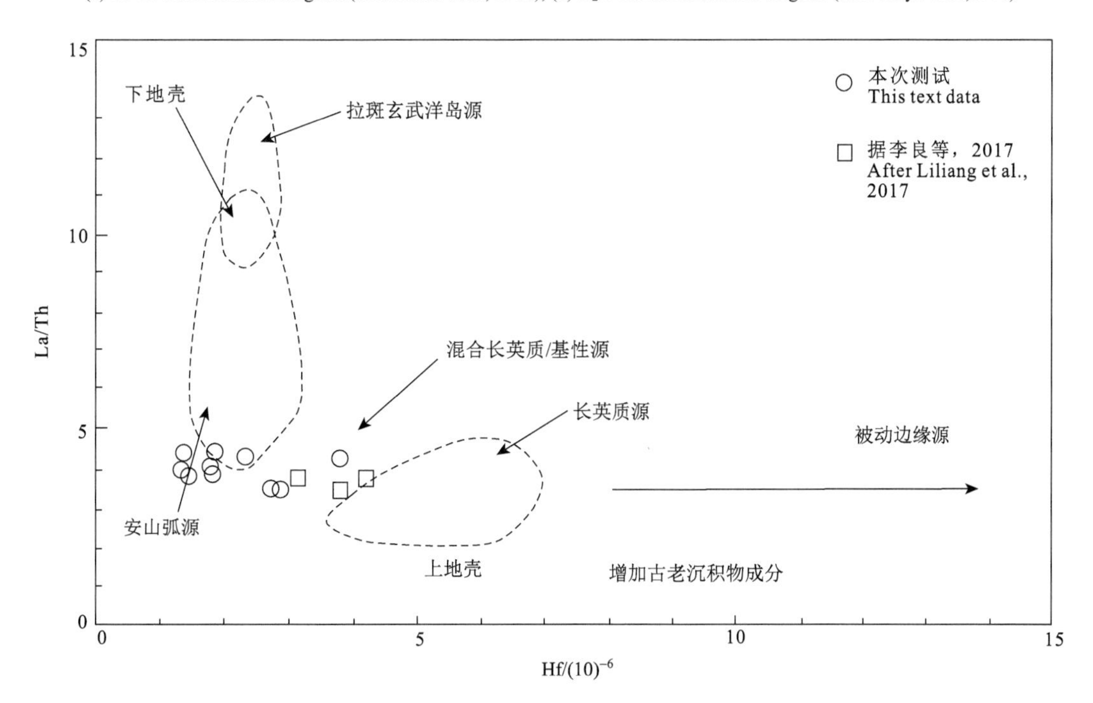

图 8　二十二站组砂岩 La/Th-Hf 判别图解（底图据 Gu X X et al.，2002）

Fig. 8　The diagrams of La/Th-Hf discrimination for the Ershi'erzhan Formation sandstone (after Gu X X et al., 2002).

从锆石的数量上分析（图 5），前寒武纪锆石数量最少（<10%），而侏罗纪碎屑锆石数量占大多数，有随着时代变新锆石保存数量增多的趋势。整体上，二十二站组碎屑锆石可被识别出较为明显的 4 期，表明其沉积物主要形成于四个时期（图 5）。

（1）2711±10Ma（新太古代）：俄罗斯研究者一般认为额尔古纳地块具有太古代的基底，俄罗斯的德茹洛利变质杂岩被认为晚太古代形成[③]。在我国，前人对额尔古纳地块中河流碎屑锆石的研究均未发现太古代时期形成的锆石。本次研究在 ZK30-1-28 样品中发现了一颗年龄为 2711±10Ma 的锆石，磨圆较好，Th/U 比值为 0.58，大小约 40μm×80μm，环带结构不清晰，为岩浆锆石。这颗锆石说明了额尔古纳地块存在晚太古代基底，其出露面积可能非常小，并且为二十二站组沉积物提供了极少的物源。

（2）1238～2428Ma（中元古代—古元古代）：本次研究各样品中获得的古元古代和中元古代年龄段的碎屑锆石（$N=11$）反映的时代和岩石类型与区域上的兴华渡口岩群相当，且在锆石形态上磨圆较好，表明经历了不止一次的搬运沉积过程，因此可以认为二十二站组沉积物中具有来自盆地基底的物质信息，即来自于盆地南缘的兴华渡口岩群。本次获得的古元古代锆石年龄（2428±9Ma）是兴华渡口岩群迄今为止最老的岩浆锆石年龄，为兴华渡口岩群的形成时代提供了新证据，同时也验证了地球化学数据分析认为二十二站组沉积物部分源区为古老的沉积岩系。此外，该年龄段的碎屑锆石数量少的原因可能是兴华渡口岩群分布面积小，因而被搬运到盆地中的物质较少。

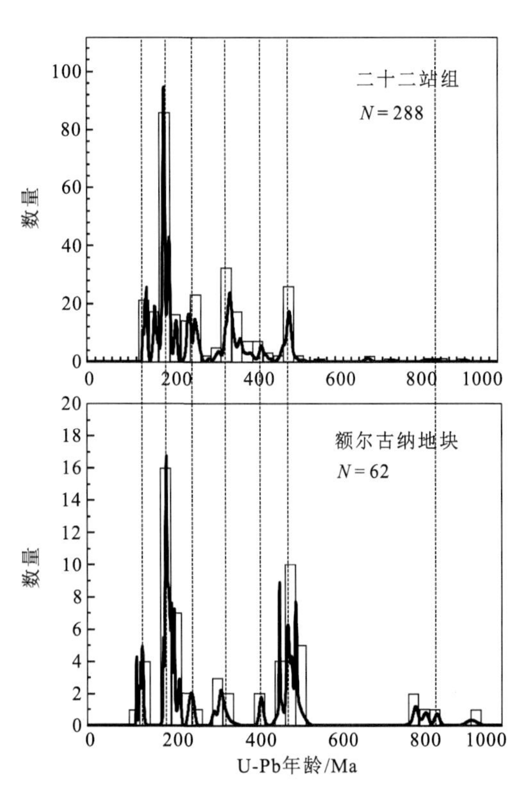

图 9　二十二站组碎屑锆石与额尔古纳地块岩浆锆石 U-Pb 定年结果对比

Fig. 9　Probability curves of ages for detrital zircons from the Ershi'erzhan Formation sandstone and magmatic zircons from the Erguna massif.

（$N=288$ 为碎屑锆石年龄数据个数；$N=62$ 为额尔古纳地块中侵入岩岩体年龄个数；额尔古纳地块侵入岩年龄统计数据据 Wu Fuyuan et al.2011）

（$N=288$ is the number of detrital zircon age data; $N=62$ is the number of the age of intrusive rock in Erguna block; the age statistics of intrusive rocks in Erguna block according to Wu Fuyuan et al., 2011）

（3）561～921Ma（新元古代）：除 ZK30-1-28 样品外，其他样品中均发现新元古代年龄段碎屑锆石，但数量很少（$N=7$）。这些新元古代的碎屑锆石年龄与额尔古纳地块新元古代侵入体年龄相吻合（Zhao Shuo et al.，2016）（图 9），是晋宁期古亚洲洋向额尔古纳-兴安地块俯冲形成大陆岩浆弧（活动大陆边缘）构造事件在研究区的记录（Wang Hongbo et al.，2013），Rodinia 超大陆演化的沉积响应（Zhao Shuo et al.，2016）。这些侵入体与兴华渡口岩群共同遭受了变质变形作用，且两者共同构成额尔古纳地块结晶基底（Wang Hongbo et al.，2013）。另外，有三个锆石年龄，分别为 561±7Ma，683±7Ma，684±7Ma，未能在额尔古纳新元古代侵入体中找到对应的年龄值，可能为前人未发现的侵入体年龄。

（4）134～540Ma（中生代—晚古生代）：将二十二站组中显生宙碎屑锆石年龄统计数据与额尔古纳地块中侵入体年龄统计数据（Wu Fuyuan et al.，2011）进行对比（图 9），漠河盆地二十二站组显生宙碎屑锆石年龄分布与额尔古纳地块花岗质侵入体年龄分布具有非常相似的峰值特征，说明二十二站组显生宙沉积物绝大部分来自于额尔古纳地块显生宙花岗质侵入体。这些花岗质侵入体的形成可能与蒙古-鄂霍茨克洋的俯冲、闭合有关（Wu Fuyuan et al.，2011）。此外，发现了 3 颗早白垩世早期的碎屑锆石，具有明显的环带特征，为岩浆锆石，其物源可能为早白垩世花岗岩，是加厚的岩石圈拆沉在地表形成火成岩的物质响应（Wu Fuyuan et al.，2011）。

5.3　构造背景

在(K_2O/Na_2O)-SiO_2判别图上［图 10（a）］，由于 K_2O/Na_2O 比值分布在 1 附近且 SiO_2 含量较高，漠河

盆地二十二站组砂岩大部分落入活动大陆边缘区域。在$(Al_2O_3/(CaO + Na_2O))$-$(Fe_2O_3 + MgO)$和(Al_2O_3/SiO_2)-$(Fe_2O_3 + MgO)$构造背景判别图上[图 10（b）、（c）]，绝大部分样品点均落在活动陆缘和大陆岛弧区域。二十二站组砂岩样品的主量元素构造图解显示出结果的一致性，表明其物源的大地构造背景属活动大陆边缘和大陆岛弧环境。

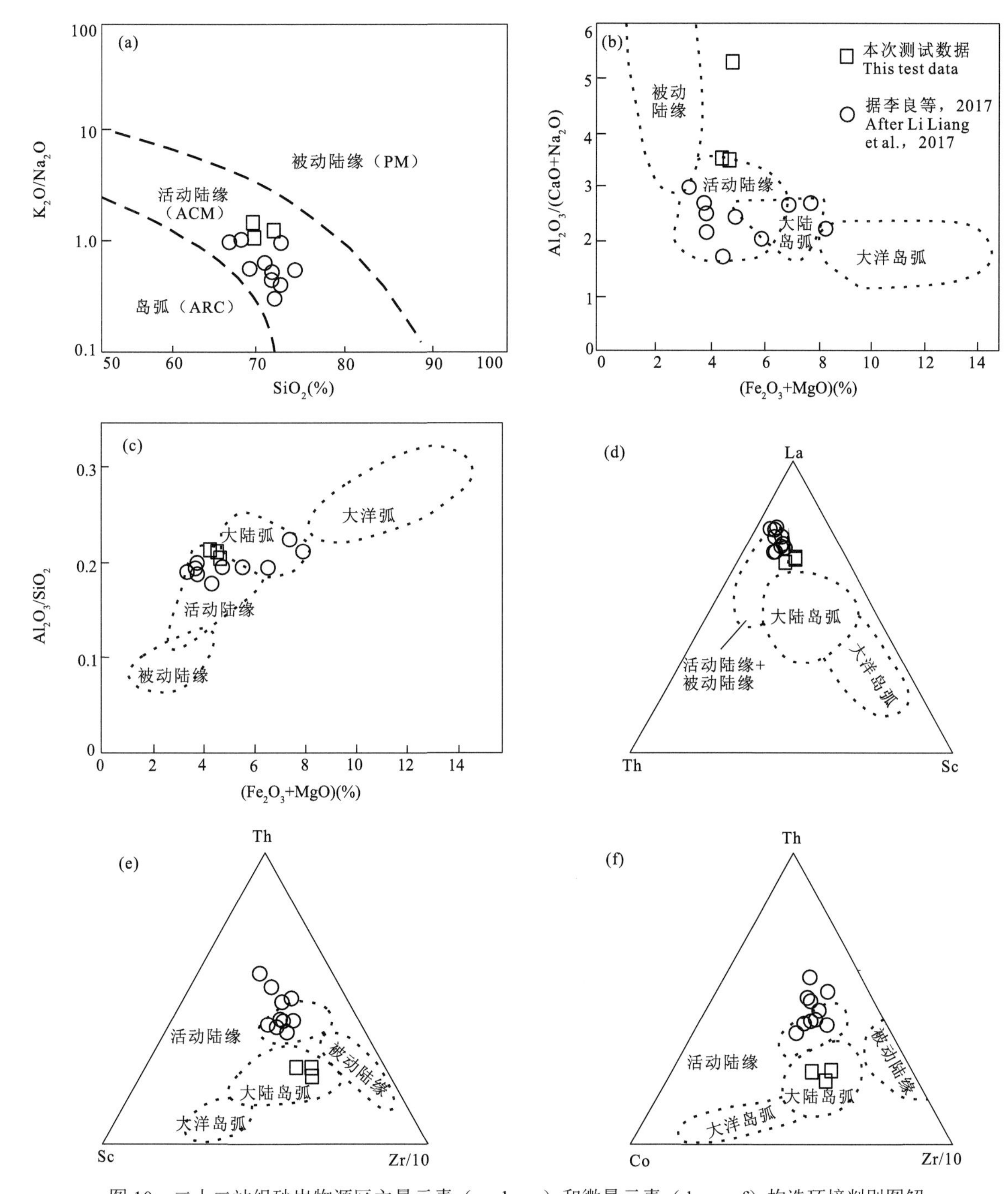

图 10　二十二站组砂岩物源区主量元素（a，b，c）和微量元素（d，e，f）构造环境判别图解

Fig. 10　Tectonic setting discrimination diagrams of the major elements（a，b，c）and trace elements（d，e，f）for the provenance from the Ershi’erzhan Formation sandstone.

（a）SiO_2-(K_2O/Na_2O)判别图解（底图据 Roser et al.，1986）；（b）(Fe_2O_3+MgO)-$(Al_2O_3/(CaO+Na_2O))$判别图解（底图据 Bhatia，1983）；（c）(Fe_2O_3+MgO)-(Al_2O_3/SiO_2)判别图解（底图据 Bhatia，1983）；（d）-（f）La-Th-Sc、Th-Sc-(Zr/10)、Th-Co-(Zr/10)判别图（底图据 Bhatia et al.，1986）

(a) SiO_2-(K_2O/Na_2O) discrimination diagram (after Roser et al., 1986); (b) (Fe_2O_3+MgO)-$(Al_2O_3/(CaO+Na_2O))$ discrimination diagram (after Bhatia, 1983); (c) (Fe_2O_3+MgO)-(Al_2O_3/SiO_2) discrimination diagram (after Bhatia, 1983); (d)-(f) La-Th-Sc、Th-Sc-(Zr/10)、Th-Co-(Zr/10) discrimination diagram (after Bhatia et al., 1986)

由于稀土元素和一些微量元素如 Th、Sc、Zr 和 Co 在一般水体中很难溶解，相对稳定，搬运、沉积过程中具非迁移性，能准确地反映物源区的地球化学性质和构造背景（Taylor and Mclennan，1985；Bhatia，1985；Girty et al.，1993）。因此，稀土和微量元素已被广泛的应用于沉积物物源和构造背景的研究（Bhatia and Crook，1986； McLennan et al.，1993）。将二十二站组砂岩样品微量元素测试数据投影到Bhatia和 Crook（1986）的La-Th-Sc，Th-Sc-Zr/10，Th-Co-Zr/10三角图上[图10（d）、（e）、（f）]，显示出样品点大部分均落在活动大陆边缘和大陆岛弧区域及其附近，与主量元素构造环境判别结果具有较高的一致性。另外，从次稳定的被动大陆边缘到非稳定的大陆岛弧区，∑REE，∑LREE/∑HREE 和 La/Yb 值明显降低（Bhatia，1985）。二十二站组砂岩稀土元素有关参数与Bhatia（1985）归纳总结的不同构造背景下沉积盆地杂砂岩REE特征值（表2）相比，与活动大陆边缘和大陆岛弧接近，与候伟等（2010c）的结论一致。

表2　各种构造背景下砂岩的REE参数表

Table 2　The REE geochemical parameter of sandstone in different tectonic setting

构造背景	物源区类型	La（$\times10^{-6}$）	Ce（$\times10^{-6}$）	∑REE（$\times10^{-6}$）	La/Yb	$(La/Yb)_N$	∑LREE/∑HREE	Eu/Eu^*	数据来源
海洋岛弧	未切割的岩浆弧	8±1.7	19±3.7	58±10	4.2±1.3	2.8±0.9	3.8±0.9	1.04±0.11	Bhatia,1985
大陆岛弧	切割的岩浆弧	27±4.5	59±8.2	146±20	11.0±3.6	7.5±2.5	7.7±1.7	0.79±0.13	
活动大陆边缘	上隆的基底	37	78	186	12.5	8.5	9.1	0.6	
被动大陆边缘	克拉通内部构造高低	39	85	210	15.9	10.8	8.5	0.56	
研究区	二十二站组	36	70	155	22.6	16	13	0.59	本文

上述不同的判别方法中，活动大陆边缘指安第斯型盆地，发育在或者靠近厚的大陆边缘；大陆岛弧指弧间前前弧或者后弧盆地，靠近火山弧，发育在厚的大陆地壳或者薄的大陆边缘上，二者均属于广义上的活动大陆边缘（Bhatia and Crook，1986）。结合区域地质资料，漠河-上阿穆尔盆地北部为西伯利亚板块南缘，具有活动大陆边缘性质，发育有岩浆弧（Bulukayefu and Natalin，1996）。盆地南部为大兴安岭北部，具有类似活动大陆边缘的性质（Bureau of Geology and Mineral Resources of Heilongjiang Province，1993；Sun Guangrui et al.，2002；Wu Guang et al.，2005）。因此，综合各种构造判别图解，二十二站组物源区的构造背景为活动大陆边缘的大陆岛弧环境。

6　结论

（1）二十二站组碎屑锆石大部分为岩浆结晶锆石，少部分锆石颗粒为增生-混合型锆石，显示出经历了后期构造-热事件改造的特征。此外，少部分锆石颗粒磨圆好，显示出其经历了多次搬运、沉积过程的特征，从而指示早先形成的古老沉积岩为二十二站组提供了物源。

（2）被发现的最古老锆石年龄为 2711±10Ma，说明额尔古纳地块存在晚太古宙的基底信息。被发现的最年轻锆石年龄为 134±1Ma，属早白垩世早期，指示研究区存在早白垩世早期火成岩物源，且前人发现了J_3-K_1时期的古生物化石，从而限定了二十二站组的沉积下限为早白垩世早期。

（3）二十二站组物源区的构造背景为活动大陆边缘的大陆岛弧环境。二十二站组物源主要为上地壳大陆边缘弧长英质、安山质源区，并混有下地壳深部物质（基性岩），并且可能有成熟大陆石英质物源区的物质加入。

（4）二十二站组碎屑锆石有随着时代变新锆石保存数量增多的趋势，其沉积物主要形成于四个时期：新太古代（2711±10Ma），说明额尔古纳地块存在晚太古宙的基底信息，并且为二十二站组沉积物提供了极少物源；中元古代—古元古代（1238～2428Ma），指示兴华渡口岩群为二十二站组沉积物提供了部分物源；新元古代（561～921Ma），是晋宁期古亚洲洋向额尔古纳-兴安地块俯冲形成大陆岩浆弧（活动大陆边缘）构造事件在研究区的记录，为二十二站组提供了部分物源；中生代—晚古生代（134～540Ma），是蒙古-鄂霍茨克洋俯冲、闭合过程中形成的花岗质岩浆在研究区的物质记录，且显生宙花岗岩质岩浆为二十二站组提供了最为丰富的物源。

致谢

野外工作得到马宝山工程师、郭宝军工程师、赫天枢工程师、何海平工程师的大力支持，锆石LA-ICP-MS U-Pb 定年得到自然资源部成矿作用与资源评价重点实验室侯可军研究员的帮助，样品岩石地球化学分析得到了核工业北京地质研究院测试研究中心刘牧工程师的帮助，在此一并表示衷心的感谢！

注释

①黑龙江省地质调查研究总院. 2014. 开库康幅、塔河县幅、新街基幅 1∶25 万区域地质调查报告。

②阎鸿铨，叶茂，孙维志. 2001. 大兴安岭满洲里和乌奴尔地区银、铅、锌和铜矿床预测研究. 沈阳：沈阳地质调查中心.

③洪骏男，马宝山. 2017. 黑龙江省塔河县二十一站河、主功河河谷地区岩金普查（北区）补充设计. 齐齐哈尔：齐齐哈尔地质勘查总院.

Reference

Anderson T. 2002. Correction of common lead in U-Pb analyses that do not report ^{204}Pb. Chemical Geology, 192 (1-2): 59-79.

Bhatia M R. 1985. Rare earth element geochemistry of Australian Paleozoic graywacke and mudrocks: Provenance and tectonic Control. Sedimentary Geology, 45 (1-2): 97-113.

Bhatia M R, Crook K AW. 1986. Trace element characteristics of graywacks and tectonic setting discrimination of sedimentary basins. Contributions to Mineralogy and Petrology, 92 (2): 181-193.

Bulukayefu, Natalin, Translated by Xu Guoguang. 1996. Adhibiting structures in southern of far eastern Russia. World Geology, 15(2): 35-38.

Bureau of Geology and Mineral Resources of Heilongjiang Province. 1993. Regional geology of Heilongjiang Province. Beijing: Geological Publishing House.

Cawood P A, Nemchin A A. 2000. Provenance record of a rift basin: U/Pb ages of detrital zircons from the Perth Basin, Western Australia. Sedimentary Geology, 134 (3-4):209-234.

Compston W, Williams I S, Kirschvink J L, Zhang Zichao, Ma Guogan. 1992. Zircon U-Pb ages for the early Cambrian time-scale. Journal of the geological society, 149 (2): 171-184.

Connelly J N. 2000. Degree of preservation of igneous zonation in zircon as a signpost for concordancy in U/Pb geochronology. Chemical Geology, 172: 25-39.

Corfu F, Hanchar J M, Hoskin P W O, Kinny P. 2003. Atlas of zircon textures. Reviews in Mineralogy & Geochemistry, 53 (1): 469-500.

Dong Yunpeng, Liu Xiaoming, Neubauer F, Zhang Guwei, Tao Ni, Zhang Yiguo, Zhang Xiaoning, Li Wei. 2013. Timing of Paleozoic amalgamation between the North China and South China Blocks: evidence from detrital zircon U-Pb ages. Tectonophysics, 586: 173-191.

Floyd P A, Leveridge B E. 1987. Tectonic environment of the Devonian Gramscatho basin, south Cornwall: framework mode and geochemical evidence from turbiditic sandstones. Journal of the Geological Society, 144: 531-542.

Girty G H, Hanson A D, Yoshinobu A S, Knaack C, Johnson D. 1993. Provenance of Paleozoic mudstones in a contact metamorphic aureole determined by rare earth element, Th, and Sc, analyses, Sierra Nevada California. Geology, 21(4): 363-366.

Greentree M R, Li Zhengxiang. 2008. The oldest known rocks in south-western China: SHRIMP U-Pb magmatic crystallisation age and detrital provenance analysis of the Paleoproterozoic Dahongshan group. Journal of Asian Earth Sciences, 33 (5-6): 289-302.

Guo Pei, Liu Chiyang, Wang Jianqiang, Li Changzhi. 2017. Considerations on the application of Detrital-Zircon geochronology to sedimentary provenance analysis. Acta Sedimentologica Sinica. 35(1):46-56(in Chinese with English abstract).

Gu X X, Liu J M, Zheng M H, Qi L. 2002. Provenance and tectonic setting of the Proterozoic turbidites in Hunan, South China: geochemical evidence. Journal of Sedimentary Research, 72(3): 393-407.

Hallsworth C R, Morton A C, Claoué-Long J, Fanning C M. 2000. Carboniferous sand provenance in the Pennine Basin, UK: Constraints from heavy mineral and detrital zircon age data. Sedimentary Geology, 137 (3-4): 147-185.

Hermann J, Rubatto D, Korsakov A, Shatsky V S. 2001. Multiple zircon growth during fast exhumation of diamondiferous, deeply subducted continental crust (Kokchetav Massif, Kazakhstan). Contributions to Mineralogy and Petrology, 2001, 141 (1): 66-82.

He Zhonghua, Liu Zhaojun, Guo Hongwei, Hou Wei, Dong Linsen. 2008a. Provenance analysis of middle Jurassic sediments and its geological significance in Mohe Basin. Journal of Jilin University (Earth Science Edition), 38 (3): 398-404(in Chinese with English abstract).

He Zhonghua, Wang Yufen, Hou Wei. 2008b. Geochemisry and provenance analysis of the middle Jurassic sandstones in the Mohe Basin, Heilongjiang. Sedimentary Geology and Tethyan Geology, 28 (4): 93-100 (in Chinese with English abstract).

He Zhengjun, Li Jinyi, Mo Shenguo, A. A. Sorokin. 2003. Geochemistry tectonic background and provenance analysis of the sandstones from the Mohe Foreland basin. Science in China (Series D), 33 (12): 1219-1226(in Chinese without English abstract).

Hou Wei. 2006. Evolution of sedimentary and provenance analysis in middle Jurassic of Mohe basin. Changchun: Jilin University (M. A thesis).

Hou Wei, Liu Zhaojun, He Yuping, He Zhonghua, Zhang Yueqiao, Zhang Lei. 2010a. Provenance analysis of upper Jurassic and its geological significances in Mohe basin. Geological Review, 56 (1): 71-81 (in Chinese with English abstract).

Hou Wei, Liu Zhaojun, He Yuping, He Zhonghua, Zhang Lei. 2010b. Sedimentary characteristics and tectonic setting of the upper Jurassic Mohe basin. Journal of Jilin University (Earth Science Edition), 40 (2): 286-297 (in Chinese with English abstract).

Hou Wei, Liu Zhaojun, He Yuping, He Zhonghua, Zhang Ziming, Bai Tong, Dong Desheng. 2010c. Application of REE geochemical characteristics of sandstone to study on provenance:A case from the middle Jurassic of Mohe basin in northeast China. Acta Sedimentologica Sinica, 28 (2): 285-293 (in Chinese with English abstract).

Kang Yu, Chen Gang, Xia Xiaoyu, Ren Shuaifeng, Zhang Weigang, Shi Pingpin. 2018. Detrial Zircon U-Pb geochronology and its geological implication of the Nancaode and Zhuanghegou formations in the southern margin Ordos basin. Acta Geologica Sinica, 91 (9): 1829-1842 (in Chinese with English abstract).

Li Chunlei. 2007. Structural characteristic, tectonic evolution and basin dynamics of Mohe basin. Beijing: China University of Geosciences (Beijing) (M. A thesis).

Liu Junhai, Yang Xianghua, Yu Shui, Wu Zhixuan, Jia Donghui. 2003. The REE geochemical characteristics of Paleocene-Eocene in the LiShui Sag of the DongHai basin. Geoscience, 17 (4): 421-427(in Chinese with English abstract).

Li Liang, Sun Fengyue, Li Bile, Xu Qinglin, Zhang Yajing, Lan Lishi. 2017. Geochronology of Ershi'erzhan formation sandstone in Mohe basin and tectonic environment of its provenance. Earth Science, 42 (1): 35-52 (in Chinese with English abstract).

Li Xiangwen, Bai Ling'an, Wang Keyong, Zhou Jinbo, Pang Long, Zhang Fei. 2017. Multi-source information metallogenic prognosis based on evidence weight method in upper Heilongjiang metallogenic belt. GOID, 38 (3): 19-24 (in Chinese with English abstract).

Liu Yongsheng, Hu Zhaochu, Zong Keqing, Gao Changgui, Gao Shan, Xu Juan, Chen Hailong. 2010. Reappraisement and refinement of zircon U-Pb isotope and trace element analyses by LA-ICP-MS. Chinese Science Bulletin, 55 (15): 1535-1546.

Li Yifei, Luo Jinhai, Xu Huan, You Jia, Chen Guanxu. 2018. Detrital Zircon U-Pb age, geochemical characteristics and geological significance of meta-sandstones form Boyang-Yuanlong area in the western Qinling orogenic belt. Geological Review: 64 (5): 1087-1102(in Chinese with English abstract).

Li Yun, Hou Xiaoyu, Yang Bo, Zhao Yuanyi, Chen Long. 2017. Indo-Chinese magmatic activity in the Duobaoshan district of Heilongjiang province and its geological significance. Acta Geologica Sinica, 91 (6): 2058-2077.

Ludwig K R. 2003. Users manual for Isoplot/Ex, version 3. 00: A Geochronological Toolkit for Microsoft Excel. Berkeley Geochronology Center Special Publication, No. 4: 1-70.

McLennan S M, Hemming S, McDaniel D K. 1993. Geochemical approaches to sedimentation, provenance, and tectonics. Special Paper of the Geological Society of America, 284: 21-40.

McLennan S M, Hemming S R, Taylor S R, Eriksson K A. 1995. Early Proterozoic crustal evolution: Geochemical and Nd-Pb isotopic evidence from metasedimentary rocks, southwestern North America. Geochimica et Coschimica Acta, 59 (6): 1153-1177.

Miao Zhongying, Zhao Xingmin, Deng Jian, Lu Cheng, Shi Shengbao, Wang Jing, Mao Liquan. 2014. Biomarker geochemistry of low grade metamorphic source rocks: A case from the Mohe formation of Mohe basin in northeast China. Acta Geologica Sinica, 88 (1): 134-143(in Chinese with English abstract).

Moecher D P, Samson S D. 2006. Differential zircon fertility of source terranes and natural bias in the detrital zircon record: Implications for sedimentary provenance analysis. Earth and Planetary Science Letters, 247 (3-4): 252-266.

Morton A C, Claoué-Long, J C, Berge C. 1996. SHRIMP constraints on sediment provenance and transport history in the Mesozoic Statfjord formation, North Sea. Journal of the Geological Society, 153 (6): 915-929.

Qiherige. 1995. Vestigiofossils of the middle Jurassic Ershierzhan Group in Mohe area, Heilongjiang province. Regional Geology of China, (3): 243-244(in Chinese without English abstract).

Roser B P, Korsch R J. 1986. Determination of tectonic setting of sandstone-mudstone suites using SiO_2 content and K_2O/Na_2O ratio. The Journal of Geology, 94 (5): 635-650.

Roser B P, Korsch R J. 1988. Provenance signatures of sandstone-mudstone suits determined using discriminant function analysis of major element data. Chemical Geology, 67 (1-2): 119-139.

Sircombe K N. 1999. Tracing provenance through the isotope ages of littoral and sedimentary detrital zircon, eastern Australia. Sedimentary Geology, 124 (1-4): 47-67.

Sun Guangrui, Li Yangchun, Zhang Yu. 2002. The basement tectonics of Ergun massif. Geology and Resources, 11 (3): 129-139(in Chinese with English abstract).

Sun Guangrui, Liu Xuguang, Han Zhenzhe, Liu Shiwei, Guo Kuicheng, Zhang Jinlian, Zhu Chunyan, Wang Deping. 2002. Stratigraphic division and age of the Mid-Upper Jurassic Ershierzhan group in the upper Heilongjiang River Basin. Geological Bulletin of China, 21 (3): 150-155 (in Chinese with English abstract).

Sun Qiushi. 2013. Study on the exhumation process from late Jurassic of Mohe basin. Changchun : Jilin University (M. A thesis).

Sun S S, McDonough W F. 1989. Chemical and isotopic systematics of oceanic Basalts: implications for mantle composition and process. In: Sauders A D and Norry M J (eds). Magmatism in the ocean basins. Geological Society. Special Publication, 42: 313-345.

Taylor S R, Mclennan S M. 1985. The continental crust:Its composition and evolution. Oxford: Blackwell Scientific Publication, 312.

Wang Hongbo, Yang Xiaoping. 2013. Main geological achievements and progress of the new round of national land and resources survey in north Daxinganling. Geological Bulletin of China, 32(2-3): 525-532(in Chinese with English abstract).

Wang Jian. 2007. The Geophysical research of structure characteristics in the western of the Mohe basin. Changchun : Jilin University (M. A thesis).

Wang Yuejun, Zhang Feifei, Fan Weiming, Zhang Guowei, Chen Shiyue, Cawood P A, Zhang Aimei. 2010. Tectonic setting of the south China block in the early Paleozoic: Resolving intracontinental and ocean closure models from detrital zircon U-Pb geochronology. Tectonics, 29: TC6020.

Wang Yongsheng, Tian Ziqiang, Hu Zhaoqi, Bai Qiao. 2019. Provenance analysis of the Yuantongshan formation in the Hefei basin and its geological signature: Evidence from detrital zircon dating. Earth Science Frontiers, 26 (2): 179-193(in Chinese with English abstract).

Wu Fuyuan, Sun Deyou, Ge Wenchun, Zhang Yanbin, Grant M L, Wilde S A, Jahn B. 2011. Geochronology of the Phanerozoic Granitoids in northeastern China. Journal of Asian Earth Sciences, 41 (1): 1-30.

Wu Genyao, Feng Zhiqiang, Yang Jianguo, Wang Zaijun, Zhang Liguo, Guo Qingxia. 2006. Tectonic setting and geological evolution of Mohe basin in Northeast China. Oil&Gas Geology, 27 (4): 528-535(in Chinese with English abstract).

Wu Guang, Sun Fengyue, Zhao Shengcai, Li Zhitong, Zhao Ailin, Pang Qingbang, Li Guangyuan. 2005. Discovery of early Paleozoic post-collision granites on the northern margin of the Erguna block and its geological significance. Science Bulletin, 50 (20): 2278-2288 (in Chinese without English abstract).

Wu Heyong, Wang Shihui, Yang Jianguo, Tang Zhenhai, Wang Zaijun, Zhang Qingshi. 2004. Analysis of exploration potential in surrounding basins of Daqing oilfield. Petroleum Geology, (4): 23-31(in Chinese with English abstract).

Wu Heyong, Xin Renchen, Yang Jianguo. 2003a. The middle Jurassic sedimentary evolution and Petroleum potential of the Mohe basin. Petroleum Geology & Experiment, 25 (2): 116-121(in Chinese with English abstract).

Wu Heyong, Yang Jianguo, Huang Qinghua, Liu Wenlong. 2003b. Sequence and Age of the Mesozoic strata in the Mohe basin. Journal of Stratigraphy, 27 (3): 193-198(in Chinese with English abstract).

Wu Yuanbao, Zheng Yongfei. 2004. Genesis of zircon and its constraints on interpretation of U-Pb age. Chinese Science Bulletin. 49 (15): 1554-1569.

Xin Renchen, Wu Heyong, Yang Jianguo. 2003. Upper Jurassic sequence-stratigraphic Framework of the Mohe basin. Journal of Stratigraphy, 27 (3): 199-204(in Chinese with English abstract).

Xu Yajun, Du Yuansheng, Yang Jianghai. 2007. Prospects of sediment provenance analysis. Geological Science and Technology Information, 26 (3): 26(in Chinese with English abstract).

Yang Jianguo, Wu Heyong, Liu Junlai. 2006. Stratigraphic correlation of the Mesozoic and Cenozoic in the outer basins of the Daqing exploration area, Heilongjiang, China. Geological Bulletin of China, 25 (9-10): 1088-1093(in Chinese with English abstract).

Yu Jinhai, O'Reilly S Y, Zhou M F, Griffin W L, Wang Lijuan. 2012. U-Pb geochronology and Hf-Nd isotopic geochemistry of the Badu Complex, southeastern China: Implications for the Precambrian crustal evolution and paleogeography of the Cathaysia Block. Precambrian Research, 222-223: 424-449.

Zhang Pei, Zheng Jianping, Zhang Ruisheng, Yu Chunmei. 2005. Rare earth elemental characteristics of Ordovician-Jurassic mudstone in Tabei uplift, Tarmi Basin. Acta Sedimentologica Sinica, 23 (4): 740-746(in Chinese with English abstract).

Zhang Shun, Lin Chunming, Wu ChaoDong, Yang Jianguo. 2003. Tectonic characteristics and basin evolution of the Mohe basin, Heilongjiang Province. Geological Journal of China Universities, 9 (3): 411-419(in Chinese with English abstract).

Zhao Liguo, Yang Xiaoping, Zhao Xingmin, Liu Yuan, Zhang Wenlong. 2014. LA-ICP-MS U-Pb geochronology of the sedimentary rock and volcano rock zircons from the Emoerhe group in the Mohe basin and its geological significance. Journal of Geomechanics, 20 (3): 285-291(in Chinese with English abstract).

Zhao Shuo, Xu Wenliang, Tang Jie, Li Yu, Guo Peng. 2016. Neoproterozoic magmatic events and tectonic attribution of the Erguna massif: Constraints from geochronological, geochemical and Hf isotopic data of intrusive rocks. Earth Science, 41 (11): 1803-1829(in Chinese with English abstract).

Zhen Miao. 2017. Materials evolution analysis in upper Jurassic of Mohe basin. Changchun: Jilin University (M. A thesis)

参考文献

布鲁卡耶夫, 纳塔林, 徐国光[译]. 1996. 俄罗斯远东南部的增生构造. 世界地质, 15 (2): 35-38.

郭佩, 刘池洋, 王建强, 李长志. 2017. 碎屑锆石年代学在沉积物源研究中的应用及存在问题, 沉积学报, 35 (1): 46-56.

和钟铧, 刘招君, 郭宏伟, 侯伟, 董林森. 2008a. 漠河盆地中侏罗世沉积源区分析及地质意义. 吉林大学学报 (地球科学版), 38 (3): 398-404.

和钟铧, 王玉芬, 侯伟. 2008b. 漠河盆地中侏罗统砂岩地球化学特征及物源属性分析. 沉积与特提斯地质, 28 (4): 93-100.

和政军, 李锦轶, 莫申国, A. A. Sorokin. 2003. 漠河前陆盆地砂岩岩石地球化学的构造背景和物源区分析. 中国科学 (D辑), 33 (12):1219-1226.

侯伟. 2006. 漠河盆地中侏罗世沉积演化与物源分析. 长春: 吉林大学 (硕士论文).

侯伟, 刘招君, 何玉平, 何钟铧, 张月巧, 张雷. 2010a. 漠河盆地上侏罗统物源分析及其地质意义. 地质论评, 56 (1): 71-81.

侯伟, 刘招君, 何玉平, 何钟铧, 张雷. 2010b. 漠河盆地上侏罗统沉积特征与构造背景. 吉林大学学报 (地球科学版), 40 (2): 286-297.

侯伟, 刘招君, 何玉平, 和钟铧, 张子明, 柏桐, 董德胜. 2010c. 砂岩稀土元素地球化学特征在沉积物源区分析中的应用: 以中国东北漠河盆地中侏罗统为例. 沉积学报, (2) 28: 285-293.

黑龙江省地质矿产局. 1993. 黑龙江省区域地质. 北京: 地质出版社.

康昱, 陈刚, 夏晓雨, 任帅锋, 张卫刚, 师平平. 2018. 鄂尔多斯盆地南缘南曹德组与庄河沟组碎屑锆石年代学及其地质意义. 地质学报, 91 (9): 1829-1842.

李春雷. 2007. 漠河盆地构造特征演化与成盆动力学研究. 北京: 中国地质大学 (北京) (硕士学位论文).

刘俊海, 杨香华, 于水, 吴志轩, 加东辉. 2003. 东海盆地丽水凹陷古新统沉积岩的稀土元素地球化学特征. 现代地质, 17 (4): 421-427.

李良, 孙丰月, 李碧乐, 许庆林, 张雅静, 兰理实. 2017. 漠河盆地二十二站组砂岩形成时代及物源区构造环境判别. 地球科学, 42 (1): 35-52.

李向文, 白令安, 王可勇, 周金波, 庞龙, 张飞. 2017. 上黑龙江成矿带基于证据权重法的金矿综合信息成矿预测. 黄金, 38 (3): 19-24.

李亦飞, 罗金海, 徐欢, 尤佳, 陈冠旭. 2018. 西秦岭伯阳-元龙地区中泥盆统变砂岩碎屑锆石 U-Pb 年龄和地球化学特征及地质意义. 地质论评, 64 (5): 1087-1102.

苗忠英, 赵省民, 邓坚, 陆程, 师生宝, 王晶, 毛立全. 2014. 浅变质烃源岩生物标志物地球化学——以漠河盆地漠河组为例. 地质学报, 88 (1): 134-143.

其和日格. 1995. 黑龙江省漠河地区中侏罗统二十二站组的遗迹化石. 中国区域地质, (3): 243-244.

孙广瑞, 李仰春, 张昱. 2002. 额尔古纳地块基底地质构造. 地质与资源, 11 (3): 129-139.

孙广瑞, 刘旭光, 韩振哲, 刘世伟, 郭奎城, 张金莲, 朱春燕, 王德平. 2002. 上黑龙江盆地中上侏罗统二十二站群的地层划分与时代. 地质通报, 21 (3): 150-155.

孙求实. 2013. 漠河盆地晚侏罗系以来剥露过程研究. 长春: 吉林大学 (硕士学位论文).

王洪波, 杨晓平. 2013. 大兴安岭北段新一轮国土资源大调查以来的主要基础地质成果与进展. 地质通报, 32 (2-3): :525-532.

王骞. 2007. 漠河盆地西部构造特征的地球物理研究. 长春: 吉林大学 (硕士论文).

王勇生, 田自强, 胡召齐, 白桥. 2019. 合肥盆地圆筒山组物源分析及其地质意义: 来自碎屑锆石的证据. 地学前缘, 26 (2): 179-193.

吴根耀, 冯志强, 杨建国, 汪在君, 张立国, 郭庆霞. 2006. 中国东北漠河盆地的构造背景和地质演化. 石油与天然气质, 27 (4): 528-535.

吴河勇, 王世辉, 杨建国, 唐振海, 汪在君, 张庆石. 2004. 大庆外围盆地勘探潜力. 石油地质, (4): 23-31.

吴河勇, 辛仁臣, 杨建国. 2003a. 漠河盆地中侏罗统沉积演化及含油气远景. 石油实验地质, 25 (2): 116-121.

吴河勇, 杨建国, 黄清华, 刘文龙. 2003b. 漠河盆地中生代地层层序及时代. 地层学杂志, 27 (3): 193-198.

武广, 孙丰月, 赵财胜, 李之彤, 赵爱琳, 庞庆帮, 李广远. 2005. 额尔古纳地块北缘早古生代后碰撞花岗岩的发现及其地质意义. 科学通报, 50 (20): 2278-2288.

辛仁臣, 吴河勇, 杨建国, 2003. 漠河盆地上侏罗统层序地层格架. 地层学杂志, 27 (3): 199-204.

徐亚军, 杜远生, 杨江海. 2007. 沉积物物源分析研究进展. 地质科技情报, 26 (3): 26.

杨建国, 吴河勇, 刘俊来. 2006. 大庆探区外围盆地中、新生代地层对比及四大勘探层系. 地质通报, 25 (9-10):1088-1093.

张沛, 郑建平, 张瑞生, 余淳梅. 2005. 塔里木盆地塔北隆起奥陶系—侏罗系泥岩稀土元素地球化学特征. 沉积学报, 23 (4): 740-746.

张顺, 林春明, 吴朝东, 杨建国. 2003. 黑龙江漠河盆地构造特征与成盆演化. 高校地质学报, 9 (3): 411-419.

赵立国, 杨晓平, 赵省民, 刘渊, 张文龙. 2014. 漠河盆地额木尔河群锆石 U-Pb 年龄及地质意义. 地质力学学报, 20 (3): 285-291.

赵硕, 许文良, 唐杰, 李宇, 郭鹏. 2016. 额尔古纳地块新元古代岩浆作用与微陆块构造属性: 来自侵入岩锆石 U-Pb 年代学、地球化学和Hf同位素的制约. 地球科学, 41 (11): 1803-1829.

甄淼. 2017. 漠河盆地上侏罗统物源演化分析. 长春: 吉林大学 (硕士论文).

Detrital zircon geochronology, geochemistry and geological significance of sandstone in the Ershierzhan Formation of the Ershiyizhan mining area, Mohe Basin

Yuanchao Wang[1, 2, 3], Yuanyi Zhao[1], Chunhua Liu[1], Xinfang Shui[1], Xianda Cheng[2, 4], Xin Gong[2, 5], Xuan Liu[2], Wei Tan[2] and Junnan Hong[6]

1. Insititute of Mineral Resources, Chinese Academy of Geological Sciences, Beijing 100037

2. School of Earth Sciences and Resources, China University of Geosciences, Beijing 100083
3. 12th Gold Team of Chinese People's Armed Police, Chengdu 610000
4. Xi'an Center of Mineral Resources Survey, CGS, Xi'an 710100
5. Geological and Mineral Exploration Institute, Guizhou Bureau of Geological exporation for Non-ferrous Metals and Nuclear Industry, Guiyang 550005
6. Heilongjiang Qiqihar General Institute of Geological Exploration, Qiqihar 161000

Abstract The Mohe Basin is located in the northern margin of the Erguna microplate in the Mongolia-Okhotsk fold belt. It is located in the collision zone between the Siberian plate and the North China plate. The Ershierzhan Formation is one of the Mesozoic sedimentary strata in the Mohe Basin. The predecessors discussed its formation age and source, but there is still much controversy. Based on the previous studies, the zircon chronology of detrital zircons and rock geochemistry again determines the age of its formation, and explores the tectonic setting of the source and source areas in this paper. The chronological study of detrital zircons shows that most of the detrital zircons in Ershierzhan Formation are magmatic zircons, and a small part of zircon particles are hyperplastic-mixed zircons, which shows that they have undergone post-structural-thermal event modification. In addition, a small portion of the zircon particles are rounded, indicating that they have undergone multiple handling and deposition processes, suggesting that the early formation of ancient sedimentary rocks provided a source for the Ershierzhan Formation. The youngest zircon age is 134±1Ma. In the previous regional geological survey report, the fossils of the paleontology in the J_3–K_1 period were found, and the sedimentary limit of Ershierzhan Formation was limited to the early Cretaceous, and it also explained the area of the early Cretaceous igneous rocks existed in the study area. The tectonic environment discrimination of the main and trace elements and provenance analysis reveal that the source of Ershierzhan Formation is mainly the felsic upper crust and the Andesitic source area of upper crust of the active continental margin and the continental island arc environment, and is mixed with the lower crust deep matter (Basite rock). The zircon LA-ICP-MS U-Pb dating results show that the detrital zircon in Ershierzhan Formation has a tendency to increase the amount of zircon preservation with the age, and its age distribution can be divided into four periods as a whole: Neoarchean (2711±10Ma), indicating the existence of the basement information of the Neoarchean in the Erguna massif;the Mesoproterozoic–Palaeozoic (1238～2428Ma), indicating that the Xinghuadukou rock group is the partial sediment source of Ershierzhan Formation; Neoproterozoic (561～921Ma) is a record of the continental magmatic arc (active continental margin) tectonic events in the study area during the subduction of the ancient Asian ocean to the Erguna-Xing'an block of the Jiningian period; the Mesozoic-Late Paleozoic (134～540Ma), a material record of the granitic magma formed during the subduction and closure of the Mongolian-Okhotsk in the study area, and the Phanerozoic granitic magma provided the most rich source for the Ershierzhan Formation.

Keywords Detrital zircon U-Pb chronology; Geochemistry; Sandstone; Provenance analysis; Ershierzhan Formation; Mohe Basin

附件 1 漠河盆地二十二站组碎屑锆石 LA-ICP-MS U-Pb 定年数据

Appendix 1 Detrital zircon LA-ICP-MS U-Pb data for metasedimentary rocks in Ershi'erzhan group, Mohe basin

测点点号	含量/10^{-6}			Th/U	同位素比值						年龄/Ma					
	Pb	Th	U		$^{206}Pb/^{207}Pb$	1σ	$^{207}Pb/^{235}U$	1σ	$^{206}Pb/^{238}U$	1σ	$^{207}Pb/^{206}Pb$	1σ	$^{207}Pb/^{235}U$	1σ	$^{206}Pb/^{238}U$	1σ
样品编号：D50TC1-02																
02-1	785	616	889	0.69	0.06203	0.00182	0.51024	0.01572	0.05941	0.00067	675	47	419	11	372	4
02-15	1665	1228	925	1.33	0.05972	0.00138	0.62761	0.01465	0.0764	0.00073	594	34	495	9	475	4
02-16	381	699	549	1.27	0.04949	0.00164	0.19988	0.00692	0.02935	0.00037	171	57	185	6	186	2
02-17	760	508	539	0.94	0.05854	0.00123	0.61495	0.01348	0.07639	0.00088	550	28	487	8	475	5
02-18	1335	942	1350	0.70	0.06432	0.00108	0.61661	0.01048	0.06956	0.00048	752	24	488	7	434	3

续表

测点点号	含量/10^{-6}			Th/U	同位素比值						年龄/Ma					
	Pb	Th	U		$^{206}Pb/^{207}Pb$	1σ	$^{207}Pb/^{235}U$	1σ	$^{206}Pb/^{238}U$	1σ	$^{207}Pb/^{206}Pb$	1σ	$^{207}Pb/^{235}U$	1σ	$^{206}Pb/^{238}U$	1σ
02-61	55	619	1338	0.46	0.05444	0.00145	0.25587	0.00695	0.03417	0.00037	389	41	231	6	217	2
02-62	26	381	533	0.71	0.05219	0.00202	0.27363	0.01039	0.03844	0.00054	294	61	246	8	243	3
02-64	52	751	1374	0.55	0.05313	0.00153	0.22747	0.00689	0.03111	0.0003	334	51	208	6	197	2
02-65	31	841	777	1.08	0.0484	0.0022	0.19379	0.00883	0.02927	0.00035	119	81	180	8	186	2
02-66	10	115	240	0.48	0.04922	0.00268	0.2262	0.01212	0.03409	0.00054	158	93	207	10	216	3
02-67	79	1185	2250	0.53	0.05076	0.00164	0.20435	0.00696	0.02932	0.0003	230	60	189	6	186	2
02-68	87	368	1077	0.34	0.0597	0.00212	0.55343	0.01927	0.06791	0.00075	593	56	447	13	424	5
02-69	75	2322	1145	2.03	0.05201	0.00196	0.27508	0.01097	0.03854	0.00041	286	72	247	9	244	3
02-70	119	1150	3511	0.33	0.04821	0.00176	0.19419	0.00773	0.02929	0.0003	109	72	180	7	186	2
02-72	88	1135	2522	0.45	0.05003	0.00178	0.20123	0.00757	0.02935	0.00029	196	69	186	6	186	2
02-74	66	1308	1617	0.81	0.04946	0.00163	0.21202	0.00731	0.03126	0.00032	170	61	195	6	198	2
02-76	60	881	1571	0.56	0.04966	0.00142	0.21296	0.00654	0.03124	0.00034	179	51	196	5	198	2
02-78	41	844	1282	0.66	0.05198	0.0019	0.18331	0.00728	0.02555	0.00029	284	70	171	6	163	2
02-79	187	2983	4792	0.62	0.05939	0.00156	0.25759	0.00822	0.03125	0.00029	581	53	233	7	198	2
02-80	23	587	586	1.00	0.05035	0.00221	0.20195	0.00915	0.02934	0.0004	211	80	187	8	186	3
02-81	110	766	1115	0.69	0.06077	0.00126	0.65871	0.01509	0.07876	0.0008	631	32	514	9	489	5
02-82	28	677	735	0.92	0.05186	0.002	0.20882	0.00823	0.0293	0.00037	279	67	193	7	186	2
02-83	32	707	842	0.84	0.0511	0.00173	0.20511	0.00698	0.02928	0.00032	245	58	189	6	186	2
02-84	236	3219	6347	0.51	0.05049	0.00068	0.21872	0.00355	0.03144	0.00031	217	20	201	3	200	2
02-85	19	372	492	0.76	0.0517	0.00236	0.22006	0.00966	0.03131	0.00045	272	74	202	8	199	3
02-86	62	298	686	0.43	0.05747	0.0015	0.6204	0.01634	0.07869	0.00084	510	39	490	10	488	5
02-87	33	647	817	0.79	0.05081	0.00153	0.22105	0.00732	0.03138	0.00034	232	56	203	6	199	2
02-88	112	1647	2786	0.59	0.05264	0.00114	0.24413	0.0055	0.03366	0.00031	313	34	222	4	213	2
02-90	98	1248	2591	0.48	0.05629	0.00126	0.24289	0.00545	0.03132	0.00028	464	34	221	4	199	2
02-91	100	1446	2428	0.60	0.05205	0.00103	0.24415	0.00492	0.03404	0.00032	287	29	222	4	216	2
02-92	24	577	577	1.00	0.05315	0.00204	0.22698	0.00849	0.03123	0.00044	335	59	208	7	198	3
02-94	94	234	750	0.31	0.06306	0.00123	0.97138	0.01947	0.11193	0.00117	710	25	689	10	684	7
02-95	54	508	1197	0.42	0.05419	0.00145	0.28738	0.008	0.03847	0.0004	379	44	256	6	243	2
02-96	127	1205	3888	0.31	0.05004	0.00078	0.20185	0.00348	0.02927	0.0003	197	22	187	3	186	2
02-97	53	612	1563	0.39	0.05318	0.00146	0.21393	0.00566	0.02931	0.00029	337	42	197	5	186	2
02-98	772	888	1703	0.52	0.12294	0.0011	6.17317	0.07798	0.36372	0.0036	1999	10	2001	11	2000	17
02-99	109	418	1231	0.34	0.06101	0.00114	0.64175	0.01226	0.07642	0.0008	640	24	503	8	475	5
02-100	60	1657	2004	0.83	0.05353	0.0015	0.16559	0.00439	0.02259	0.00022	351	42	156	4	144	1
02-103	51	660	1388	0.48	0.05323	0.0013	0.22395	0.00521	0.03066	0.00029	339	36	205	4	195	2
02-104	41	802	1096	0.73	0.05156	0.00135	0.20792	0.00575	0.02927	0.00032	266	43	192	5	186	2
02-105	54	522	1478	0.35	0.04872	0.00139	0.21046	0.00597	0.03141	0.00034	134	46	194	5	199	2
02-106	49	1124	1503	0.75	0.04924	0.0013	0.17509	0.00538	0.02566	0.00033	159	48	164	5	163	2
02-107	42	842	1112	0.76	0.04859	0.00156	0.19655	0.00674	0.02935	0.00034	128	59	182	6	186	2
02-108	224	536	2577	0.21	0.05645	0.00072	0.61511	0.01027	0.07898	0.00089	470	18	487	6	490	5
02-111	230	43	1791	0.02	0.06451	0.00079	1.09121	0.01648	0.12269	0.00131	758	16	749	8	746	8
02-113	47	802	913	0.88	0.06208	0.00237	0.33214	0.01476	0.03842	0.00044	677	76	291	11	243	3
样品编号：D50TC1-09																
09-3	492	591	1084	0.55	0.12512	0.00277	5.9531	0.14129	0.34454	0.00316	2031	29	1969	21	1908	15
09-6	169	1178	4192	0.28	0.06647	0.00158	0.31513	0.01021	0.03414	0.00042	821	47	278	8	216	3
09-7	79	476	1754	0.27	0.05427	0.001	0.29882	0.00587	0.03993	0.0004	382	27	265	5	252	2

续表

测点点号	含量/10^{-6}			Th/U	同位素比值						年龄/Ma					
	Pb	Th	U		$^{206}Pb/^{207}Pb$	1σ	$^{207}Pb/^{235}U$	1σ	$^{206}Pb/^{238}U$	1σ	$^{207}Pb/^{206}Pb$	1σ	$^{207}Pb/^{235}U$	1σ	$^{206}Pb/^{238}U$	1σ
09-8	29	640	756	0.85	0.04763	0.00177	0.192	0.00711	0.02929	0.00038	81	60	178	6	186	2
09-9	61	628	1252	0.50	0.05654	0.00201	0.31419	0.01432	0.03977	0.00064	474	72	277	11	251	4
09-13	11	211	417	0.51	0.05445	0.00289	0.16545	0.00975	0.02177	0.00034	390	105	155	8	139	2
09-14	15	115	221	0.52	0.05882	0.00263	0.44251	0.02095	0.05489	0.00086	560	76	372	15	344	5
09-16	27	514	966	0.53	0.04826	0.00176	0.15033	0.00533	0.02275	0.00025	112	62	142	5	145	2
09-17	29	502	805	0.62	0.05288	0.0017	0.21371	0.00712	0.02932	0.00033	323	55	197	6	186	2
09-18	152	500	1632	0.31	0.0682	0.00143	0.74554	0.01905	0.07886	0.00069	875	38	566	11	489	4
09-20	37	851	944	0.90	0.05199	0.0016	0.2098	0.00694	0.02926	0.00035	285	54	193	6	186	2
09-21	1022	395	2394	0.16	0.13858	0.00155	6.67616	0.10449	0.34847	0.0031	2210	15	2069	14	1927	15
09-22	18	274	587	0.47	0.0542	0.00273	0.18336	0.00851	0.02485	0.00029	379	84	171	7	158	2
09-23	42	912	1023	0.89	0.0664	0.0017	0.26787	0.00713	0.02934	0.00032	819	38	241	6	186	2
09-26	46	401	689	0.58	0.05681	0.00189	0.43813	0.01573	0.05594	0.00066	484	58	369	11	351	4
09-27	42	1118	1001	1.12	0.06477	0.00197	0.25582	0.00825	0.02871	0.00034	767	48	231	7	182	2
09-28	25	436	635	0.69	0.05292	0.00191	0.22616	0.00829	0.03119	0.0004	325	60	207	7	198	3
09-29	64	590	871	0.68	0.05465	0.00119	0.43902	0.01071	0.05823	0.00055	398	38	370	8	365	3
09-32	170	436	2473	0.18	0.06086	0.00114	0.57224	0.01601	0.06796	0.00126	634	31	459	10	424	8
09-33	29	853	1013	0.84	0.0554	0.00183	0.16847	0.00622	0.02205	0.00029	429	58	158	5	141	2
09-35	23	513	678	0.76	0.05491	0.00238	0.19638	0.00837	0.02609	0.00037	408	70	182	7	166	2
09-36	29	398	717	0.56	0.06787	0.00209	0.29674	0.00892	0.03189	0.00034	865	45	264	7	202	2
09-37	103	540	1098	0.49	0.05838	0.00098	0.63714	0.01312	0.07907	0.00094	544	25	501	8	491	6
09-40	42	213	442	0.48	0.06842	0.00204	0.72587	0.02332	0.07714	0.00116	881	42	554	14	479	7
09-43	677	119	1968	0.06	0.12317	0.00109	5.19099	0.05843	0.30515	0.00268	2003	9	1851	10	1717	13
09-46	65	231	1400	0.17	0.05253	0.00112	0.30727	0.00677	0.0424	0.00043	309	32	272	5	268	3
09-48	64	1466	1563	0.94	0.0551	0.00155	0.23199	0.00685	0.0305	0.00034	416	46	212	6	194	2
09-49	143	38	1641	0.02	0.06404	0.0012	0.72335	0.01393	0.08186	0.00071	743	26	553	8	507	4
09-50	35	366	684	0.53	0.0528	0.00151	0.3121	0.00944	0.0428	0.0005	320	48	276	7	270	3
09-52	44	264	517	0.51	0.04931	0.00173	0.21367	0.00774	0.03139	0.00035	162	64	197	6	199	2
09-53	73	286	394	0.73	0.05516	0.00262	0.4173	0.01868	0.05509	0.00059	419	81	354	13	346	4
09-54	106	876	794	1.10	0.05047	0.00168	0.21242	0.00869	0.03034	0.00044	217	68	196	7	193	3
09-57	140	417	997	0.42	0.05395	0.00103	0.42653	0.00991	0.05711	0.00067	369	31	361	7	358	4
09-59	135	811	1283	0.63	0.05494	0.00139	0.25202	0.00692	0.03321	0.00038	410	41	228	6	211	2
09-60	21	149	228	0.65	0.05468	0.00265	0.21892	0.01041	0.02934	0.00049	399	76	201	9	186	3
09-61	936	1080	1764	0.61	0.08167	0.00108	1.90625	0.04128	0.1681	0.00215	1238	23	1083	14	1002	12
09-62	202	931	1555	0.60	0.06714	0.0021	0.37071	0.01403	0.03943	0.00048	842	58	320	10	249	3
09-64	331	853	1313	0.65	0.06201	0.00109	0.67104	0.01069	0.07889	0.00084	675	17	521	6	489	5
09-66	84	705	629	1.12	0.04995	0.00185	0.2017	0.00776	0.02933	0.00034	193	68	187	7	186	2
09-67	85	571	357	1.60	0.05833	0.00213	0.31266	0.01064	0.03938	0.00051	542	52	276	8	249	3
09-68	130	971	1107	0.88	0.05202	0.00111	0.22307	0.00497	0.0312	0.00037	286	30	204	4	198	2
09-69	113	446	541	0.82	0.06013	0.00147	0.4573	0.01097	0.05549	0.00068	608	31	382	8	348	4
09-71	325	1559	1153	1.35	0.05569	0.00094	0.43614	0.00942	0.05675	0.0007	440	27	368	7	356	4
09-72	54	432	468	0.92	0.05001	0.00178	0.20136	0.00714	0.02935	0.00037	196	59	186	6	186	2
09-73	123	440	618	0.71	0.05656	0.00141	0.4635	0.01265	0.05929	0.00054	474	44	387	9	371	3
09-74	56	393	606	0.65	0.05417	0.00168	0.21786	0.00678	0.02936	0.00038	378	46	200	6	187	2
09-75	54	433	409	1.06	0.06184	0.00239	0.24569	0.00862	0.02928	0.00037	668	53	223	7	186	2
09-76	24	55	102	0.54	0.06421	0.00319	0.70273	0.03832	0.07841	0.00134	749	86	540	23	487	8

续表

测点点号	含量/10^{-6}			Th/U	同位素比值						年龄/Ma					
	Pb	Th	U		$^{206}Pb/^{207}Pb$	1σ	$^{207}Pb/^{235}U$	1σ	$^{206}Pb/^{238}U$	1σ	$^{207}Pb/^{206}Pb$	1σ	$^{207}Pb/^{235}U$	1σ	$^{206}Pb/^{238}U$	1σ
09-77	21	80	118	0.68	0.05762	0.00292	0.433	0.02067	0.0554	0.00094	515	75	365	15	348	6
09-78	210	909	1860	0.49	0.05452	0.00083	0.30952	0.00494	0.04127	0.00044	392	18	274	4	261	3
09-79	124	880	1164	0.76	0.04991	0.00113	0.21531	0.00549	0.03126	0.00039	191	37	198	5	198	2
09-81	394	1009	1711	0.59	0.0603	0.00099	0.66096	0.01487	0.07907	0.00097	614	28	515	9	491	6
09-83	679	381	656	0.58	0.12312	0.00123	5.88905	0.07505	0.34658	0.00315	2002	11	1960	11	1918	15
09-84	171	927	1548	0.60	0.06057	0.0015	0.28162	0.00873	0.03347	0.00038	624	47	252	7	212	2
09-85	412	103	1917	0.05	0.0703	0.00068	1.48998	0.01891	0.15365	0.00145	937	12	926	8	921	8
09-86	189	959	2177	0.44	0.05328	0.00075	0.25078	0.00388	0.03418	0.00034	341	18	227	3	217	2
09-87	101	817	794	1.03	0.05879	0.00172	0.234	0.00682	0.02897	0.00027	559	47	213	6	184	2
09-89	105	354	439	0.81	0.0539	0.00148	0.50276	0.01431	0.06777	0.00071	367	45	414	10	423	4
09-90	144	991	1490	0.67	0.05141	0.00103	0.20739	0.00438	0.02929	0.0003	259	30	191	4	186	2
09-93	82	488	604	0.81	0.05103	0.00155	0.26946	0.00802	0.03848	0.00042	242	48	242	6	243	3
09-95	79	784	980	0.80	0.0509	0.00153	0.15748	0.0049	0.02259	0.00029	236	48	148	4	144	2
09-96	66	408	389	1.05	0.04974	0.00198	0.26934	0.01089	0.03956	0.00051	183	70	242	9	250	3
09-98	542	1611	1801	0.89	0.0579	0.0008	0.62832	0.00998	0.07883	0.00079	526	18	495	6	489	5
09-99	99	819	781	1.05	0.053	0.00165	0.21324	0.0067	0.02929	0.00031	329	52	196	6	186	2
09-100	212	1428	1864	0.77	0.05008	0.001	0.23401	0.00523	0.03393	0.00038	199	31	214	4	215	2
09-101	200	953	539	1.77	0.05711	0.00136	0.46486	0.01179	0.05924	0.00064	496	37	388	8	371	4
09-102	87	547	443	1.24	0.04982	0.00175	0.28137	0.01034	0.04126	0.00059	187	59	252	8	261	4
09-103	97	465	636	0.73	0.05669	0.00174	0.33866	0.01344	0.0428	0.00057	479	64	296	10	270	4
样品编号：D50TC1-11																
11-2	81	322	456	0.71	0.05735	0.00177	0.42764	0.01483	0.05401	0.00079	505	51	361	11	339	5
11-3	172	257	986	0.26	0.06263	0.00099	0.78833	0.01705	0.09086	0.00122	696	24	590	10	561	7
11-4	59	548	826	0.66	0.05134	0.00169	0.15762	0.00531	0.02228	0.00026	256	56	149	5	142	2
11-5	137	347	625	0.56	0.05787	0.0014	0.63377	0.01775	0.07897	0.00087	525	42	498	11	490	5
11-6	52	420	566	0.74	0.05249	0.00211	0.21199	0.00908	0.02914	0.00039	307	73	195	8	185	2
11-7	178	1992	1493	1.33	0.05098	0.00121	0.17985	0.00463	0.02559	0.00039	240	32	168	4	163	2
11-9	61	316	510	0.62	0.05364	0.00144	0.30612	0.00883	0.04118	0.00055	356	41	271	7	260	3
11-15	272	1677	2318	0.72	0.07331	0.0048	0.36491	0.01159	0.03837	0.00061	1023	39	316	9	243	4
11-22	249	630	1639	0.38	0.05535	0.0009	0.51402	0.00898	0.06737	0.00053	426	25	421	6	420	3
11-24	197	445	1174	0.38	0.05818	0.00255	0.60135	0.01772	0.07801	0.00187	537	29	478	11	484	11
11-25	189	466	1055	0.44	0.06303	0.00129	0.5906	0.01213	0.06816	0.00062	709	28	471	8	425	4
11-26	192	2143	1128	1.90	0.04904	0.00158	0.19674	0.00634	0.02927	0.0003	150	56	182	5	186	2
11-27	294	670	1764	0.38	0.0561	0.00093	0.58578	0.01175	0.07592	0.00112	456	21	468	8	472	7
11-28	21	83	145	0.57	0.06247	0.00318	0.4327	0.02273	0.05058	0.0008	690	85	365	16	318	5
11-30	201	344	1374	0.25	0.06188	0.00193	0.60469	0.01919	0.07113	0.00061	670	53	480	12	443	4
11-31	152	978	1793	0.55	0.05698	0.0017	0.23083	0.00743	0.02938	0.00026	491	55	211	6	187	2
11-32	52	476	534	0.89	0.05342	0.00215	0.19456	0.00817	0.02647	0.0003	347	74	181	7	168	2
11-33	84	889	1172	0.76	0.05058	0.0018	0.14658	0.00547	0.02105	0.00022	222	67	139	5	134	1
11-34	141	857	1744	0.49	0.05064	0.00109	0.22259	0.00509	0.03192	0.00028	225	37	204	4	203	2
11-35	375	543	1132	0.48	0.06578	0.00089	1.23838	0.02041	0.13658	0.00135	799	19	818	9	825	8
11-37	77	787	609	1.29	0.05995	0.00221	0.21111	0.00804	0.0256	0.00029	602	63	194	7	163	2
11-38	72	640	508	1.26	0.05683	0.00205	0.23155	0.00851	0.02964	0.00039	485	57	211	7	188	2
11-39	60	306	484	0.63	0.05277	0.00184	0.30604	0.01116	0.04221	0.00063	319	56	271	9	267	4
11-41	504	4782	3080	1.55	0.05169	0.00089	0.22246	0.00414	0.03121	0.00029	272	26	204	3	198	2

续表

测点点号	含量/10^{-6}			Th/U	同位素比值						年龄/Ma					
	Pb	Th	U		$^{206}Pb/^{207}Pb$	1σ	$^{207}Pb/^{235}U$	1σ	$^{206}Pb/^{238}U$	1σ	$^{207}Pb/^{206}Pb$	1σ	$^{207}Pb/^{235}U$	1σ	$^{206}Pb/^{238}U$	1σ
11-42	84	915	1075	0.85	0.05612	0.0022	0.16814	0.0063	0.02185	0.00026	457	62	158	5	139	2
11-43	52	229	244	0.94	0.05661	0.00261	0.42585	0.02008	0.05463	0.00067	477	83	360	14	343	4
11-44	66	328	573	0.57	0.05399	0.00188	0.30523	0.01067	0.04106	0.00039	371	61	270	8	259	2
11-45	156	1339	1611	0.83	0.048	0.00102	0.19179	0.00398	0.02914	0.00033	99	28	178	3	185	2
11-46	66	660	1036	0.64	0.05355	0.00155	0.15634	0.00463	0.02125	0.00023	352	47	147	4	136	1
11-47	29	123	196	0.63	0.055	0.00256	0.38422	0.01907	0.05102	0.00074	412	85	330	14	321	5
11-48	50	241	211	1.14	0.05573	0.00265	0.41086	0.01866	0.05425	0.00075	442	76	349	13	341	5
11-49	123	293	1079	0.27	0.0545	0.00108	0.41371	0.0089	0.05508	0.00055	392	30	352	6	346	3
11-50	135	893	1332	0.67	0.05793	0.00185	0.25135	0.00833	0.03148	0.00033	527	54	228	7	200	2
11-51	188	1467	1733	0.85	0.05727	0.00122	0.23925	0.00537	0.03035	0.0003	502	32	218	4	193	2
11-52	330	1558	1514	1.03	0.05627	0.00105	0.42552	0.00852	0.0549	0.00055	463	27	360	6	345	3
11-55	68	657	1024	0.64	0.05643	0.00178	0.17018	0.00541	0.02195	0.00023	469	52	160	5	140	1
11-56	214	1167	635	1.84	0.05415	0.00182	0.40693	0.01397	0.05472	0.00055	377	59	347	10	343	3
11-57	257	1061	1821	0.58	0.05655	0.00108	0.41604	0.00824	0.05334	0.00041	474	30	353	6	335	2
11-58	68	179	612	0.29	0.05324	0.00155	0.40298	0.01209	0.05505	0.00064	339	47	344	9	345	4
11-59	189	250	843	0.30	0.06182	0.00107	0.95161	0.01805	0.11169	0.00122	668	23	679	9	683	7
11-60	100	865	1344	0.64	0.05697	0.00152	0.20124	0.00521	0.02572	0.00025	490	40	186	4	164	2
11-61	210	553	1059	0.52	0.05602	0.00096	0.61016	0.01196	0.07889	0.00078	453	26	484	8	489	5
11-62	767	2761	2341	1.18	0.05883	0.00074	0.64085	0.00951	0.07892	0.00064	561	18	503	6	490	4
11-65	254	321	796	0.40	0.07491	0.0011	1.45785	0.02557	0.14097	0.0015	1066	19	913	11	850	9
11-66	214	2287	1497	1.53	0.05102	0.00119	0.20578	0.0049	0.02929	0.0003	242	36	190	4	186	2
11-68	24	107	133	0.80	0.05193	0.00285	0.39668	0.02132	0.05604	0.00089	282	94	339	16	351	5
11-69	51	186	299	0.62	0.05559	0.00201	0.46123	0.01582	0.06081	0.00084	436	52	385	11	381	5
11-71	87	770	940	0.82	0.05359	0.0016	0.21493	0.00663	0.02908	0.00029	354	52	198	6	185	2
11-72	1434	531	2011	0.26	0.15737	0.00139	7.91256	0.08714	0.3642	0.00286	2428	9	2221	10	2002	14
11-73	119	685	412	1.66	0.05415	0.00175	0.41178	0.01305	0.05534	0.0006	377	52	350	9	347	4
11-74	45	437	411	1.06	0.04952	0.002	0.19831	0.0076	0.02938	0.00044	172	61	184	6	187	3
11-75	70	731	989	0.74	0.05383	0.00139	0.1678	0.00452	0.02262	0.00022	364	43	158	4	144	1
11-76	90	642	914	0.70	0.05148	0.00149	0.21825	0.006	0.03093	0.0003	262	45	200	5	196	2
11-77	219	2331	1374	1.70	0.05179	0.00127	0.21005	0.00568	0.02934	0.00027	276	45	194	5	186	2
11-78	63	680	955	0.71	0.05347	0.00189	0.15584	0.00551	0.02118	0.00021	349	62	147	5	135	1
11-79	236	2314	2039	1.14	0.05103	0.00113	0.20255	0.00468	0.02879	0.00023	242	38	187	4	183	1
11-80	67	700	534	1.31	0.05066	0.00214	0.18817	0.00756	0.0272	0.0003	225	72	175	6	173	2
11-81	47	257	190	1.35	0.051	0.00236	0.38651	0.01915	0.05545	0.00101	241	80	332	14	348	6
11-82	159	324	1026	0.32	0.05406	0.00107	0.58222	0.015	0.07825	0.00134	373	29	466	10	486	8
11-83	143	752	1574	0.48	0.05414	0.00099	0.28952	0.00561	0.03883	0.0004	377	25	258	4	246	2
11-84	60	643	590	1.09	0.04889	0.00201	0.17254	0.00689	0.02582	0.00037	143	67	162	6	164	2
11-85	301	514	2076	0.25	0.05838	0.0008	0.63368	0.00898	0.07886	0.00071	544	16	498	6	489	4
11-86	189	586	710	0.83	0.0571	0.00121	0.621	0.01368	0.07898	0.00074	495	32	490	9	490	4
11-87	242	1995	2490	0.80	0.05117	0.00097	0.2178	0.0045	0.03087	0.00028	248	31	200	4	196	2
11-89	264	2554	1811	1.41	0.04974	0.00101	0.21393	0.00453	0.03124	0.00029	183	32	197	4	198	2
11-90	120	311	605	0.51	0.05793	0.00132	0.62839	0.01445	0.07892	0.00081	527	32	495	9	490	5
11-91	40	370	317	1.16	0.05746	0.0025	0.24433	0.01124	0.03075	0.00037	509	80	222	9	195	2
11-92	198	1388	2712	0.51	0.0493	0.0011	0.19441	0.00412	0.02874	0.00025	162	33	180	3	183	2
11-93	136	1188	1342	0.89	0.05093	0.00123	0.20621	0.00541	0.02933	0.00029	238	42	190	5	186	2

续表

测点点号	含量/10^{-6}			Th/U	同位素比值						年龄/Ma					
	Pb	Th	U		$^{206}Pb/^{207}Pb$	1σ	$^{207}Pb/^{235}U$	1σ	$^{206}Pb/^{238}U$	1σ	$^{207}Pb/^{206}Pb$	1σ	$^{207}Pb/^{235}U$	1σ	$^{206}Pb/^{238}U$	1σ
11-94	87	539	596	0.90	0.05265	0.00161	0.30101	0.00956	0.0416	0.00048	314	51	267	7	263	3
11-96	52	470	764	0.62	0.05318	0.00164	0.1648	0.00497	0.0226	0.00029	336	45	155	4	144	2
11-97	132	798	874	0.91	0.05279	0.00143	0.31554	0.00939	0.04337	0.00047	320	48	278	7	274	3
11-98	81	405	355	1.14	0.05398	0.0018	0.40684	0.01344	0.055	0.00065	370	53	347	10	345	4
11-99	66	642	594	1.08	0.05214	0.00175	0.20577	0.00693	0.02877	0.00037	291	53	190	6	183	2
11-100	64	562	657	0.86	0.05006	0.00177	0.20231	0.00736	0.0293	0.00032	198	64	187	6	186	2
11-101	44	167	220	0.76	0.05494	0.00221	0.50596	0.02113	0.06788	0.0015	410	55	416	14	423	9
11-102	534	26	1293	0.02	0.11995	0.00113	4.98949	0.0623	0.30182	0.00293	1955	10	1818	11	1700	14
11-105	263	2363	2088	1.13	0.05282	0.00096	0.22767	0.00435	0.03132	0.00032	321	26	208	4	199	2
11-106	152	1625	1170	1.39	0.05154	0.00167	0.20478	0.00698	0.02883	0.00031	265	58	189	6	183	2
11-107	27	256	340	0.75	0.05285	0.00268	0.1801	0.00892	0.02495	0.00032	322	89	168	8	159	2
11-108	43	172	196	0.88	0.0584	0.00254	0.53462	0.02327	0.06687	0.00088	545	72	435	15	417	5
11-109	21	116	354	0.33	0.04996	0.00249	0.18394	0.00978	0.0266	0.00037	193	97	171	8	169	2
11-111	23	179	292	0.61	0.05782	0.00272	0.23251	0.01111	0.02929	0.00042	523	80	212	9	186	3
11-112	89	636	1123	0.57	0.051	0.00127	0.21941	0.00562	0.03121	0.0003	241	41	201	5	198	2
11-113	102	818	1220	0.67	0.04872	0.00134	0.19481	0.00548	0.02899	0.00026	135	49	181	5	184	2
11-114	91	862	888	0.97	0.05371	0.00202	0.21285	0.00894	0.02852	0.00027	359	78	196	7	181	2
11-115	238	2224	1792	1.24	0.05853	0.00156	0.24013	0.00691	0.02965	0.00023	550	49	219	6	188	1
11-116	141	877	1153	0.76	0.05189	0.00122	0.27895	0.0067	0.03908	0.00042	281	35	250	5	247	3
11-117	45	236	252	0.94	0.05032	0.00184	0.37008	0.01385	0.05355	0.00069	210	63	320	10	336	4
11-118	107	607	1045	0.58	0.05151	0.00121	0.27232	0.00634	0.03843	0.00039	264	35	245	5	243	2
11-119	31	71	805	0.09	0.05135	0.00168	0.17991	0.006	0.02544	0.00027	257	57	168	5	162	2
11-120	84	359	472	0.76	0.05396	0.00163	0.4107	0.01175	0.05562	0.00064	369	44	349	8	349	4
11-121	179	602	2342	0.26	0.05083	0.00086	0.28887	0.00561	0.04117	0.00039	233	27	258	4	260	2
11-122	18	43	415	0.10	0.05031	0.00219	0.19472	0.0081	0.02849	0.00037	209	72	181	7	181	2
11-123	218	731	750	0.97	0.05632	0.00118	0.60846	0.0124	0.07885	0.001	465	24	483	8	489	6
11-124	38	308	439	0.70	0.04842	0.00195	0.19518	0.00821	0.02937	0.00037	120	73	181	7	187	2
11-126	73	238	449	0.53	0.05401	0.00193	0.43918	0.01586	0.05929	0.0007	371	60	370	11	371	4
11-127	1211	409	1620	0.25	0.14427	0.00123	7.27581	0.07995	0.36566	0.00287	2279	9	2146	10	2009	14
11-128	274	1564	934	1.67	0.05537	0.00123	0.42441	0.00997	0.05566	0.00055	427	35	359	7	349	3
样品编号：ZK30-1-28																
28-1	129	544	337	1.61	0.05455	0.00201	0.47237	0.0165	0.06347	0.0009	394	53	393	11	397	5
28-2	17	41	96	0.43	0.05356	0.00295	0.44848	0.02343	0.06243	0.00123	353	82	376	16	390	7
28-3	51	230	232	0.99	0.05481	0.00269	0.39914	0.0191	0.05297	0.00068	405	84	341	14	333	4
28-5	89	952	1128	0.84	0.05012	0.00165	0.14903	0.00492	0.02162	0.00021	201	58	141	4	138	1
28-6	106	1030	1178	0.87	0.05053	0.00152	0.15826	0.00464	0.02286	0.00029	220	44	149	4	146	2
28-8	122	589	382	1.54	0.05323	0.00175	0.39193	0.01291	0.0536	0.0007	339	51	336	9	337	4
28-9	63	612	821	0.75	0.04967	0.00188	0.15367	0.00605	0.02246	0.00027	180	69	145	5	143	2
28-10	74	454	485	0.93	0.04926	0.00195	0.26371	0.00996	0.03914	0.00047	160	66	238	8	248	3
28-12	44	111	374	0.30	0.05528	0.00148	0.42052	0.01156	0.0554	0.00069	424	39	356	8	348	4
28-13	59	559	415	1.35	0.04889	0.00268	0.18365	0.01013	0.02758	0.0004	143	99	171	9	175	3
28-14	48	145	279	0.52	0.05597	0.002	0.46201	0.01733	0.05994	0.00089	451	57	386	12	375	5
28-15	78	223	608	0.37	0.05703	0.00159	0.44424	0.01266	0.0568	0.00082	493	38	373	9	356	5
28-18	137	583	1403	0.42	0.05161	0.00128	0.28496	0.00769	0.04005	0.00043	268	42	255	6	253	3
28-19	235	590	1824	0.32	0.05722	0.00104	0.46421	0.01099	0.05856	0.00059	500	34	387	8	367	4

续表

测点点号	含量/10^{-6}			Th/U	同位素比值						年龄/Ma					
	Pb	Th	U		$^{206}Pb/^{207}Pb$	1σ	$^{207}Pb/^{235}U$	1σ	$^{206}Pb/^{238}U$	1σ	$^{207}Pb/^{206}Pb$	1σ	$^{207}Pb/^{235}U$	1σ	$^{206}Pb/^{238}U$	1σ
28-20	85	510	472	1.08	0.04986	0.00188	0.28407	0.01065	0.0418	0.00065	189	58	254	8	264	4
28-21	35	314	303	1.04	0.05331	0.00291	0.19164	0.01039	0.0262	0.00041	342	94	178	9	167	3
28-22	109	285	912	0.31	0.05336	0.00131	0.4175	0.01126	0.05693	0.00074	344	38	354	8	357	5
28-23	35	136	131	1.04	0.05341	0.00309	0.46228	0.02776	0.06358	0.00098	346	108	386	19	397	6
28-24	73	341	578	0.59	0.05106	0.00181	0.29375	0.01051	0.04193	0.00051	244	60	262	8	265	3
28-25	34	113	174	0.65	0.05483	0.00271	0.45897	0.02163	0.06213	0.00091	405	79	384	15	389	6
28-26	104	455	307	1.48	0.05383	0.0021	0.43954	0.01676	0.05953	0.00081	364	61	370	12	373	5
28-27	100	591	1161	0.51	0.04914	0.00135	0.21799	0.00609	0.03225	0.00031	155	48	200	5	205	2
28-29	59	257	285	0.90	0.05735	0.00205	0.418	0.0149	0.05321	0.00068	505	56	355	11	334	4
28-30	126	902	1306	0.69	0.04995	0.00145	0.19992	0.00627	0.02905	0.00033	193	52	185	5	185	2
28-31	65	260	334	0.78	0.0548	0.00189	0.42819	0.01532	0.05713	0.00094	404	51	362	11	358	6
28-32	109	737	558	1.32	0.05348	0.00177	0.27721	0.01032	0.03749	0.00052	349	59	248	8	237	3
28-33	44	178	161	1.10	0.0512	0.00267	0.42923	0.02283	0.06159	0.00103	250	92	363	16	385	6
28-34	76	664	1187	0.56	0.05367	0.00152	0.16375	0.00493	0.02219	0.00027	357	46	154	4	142	2
28-35	73	340	287	1.19	0.0527	0.00209	0.38877	0.01556	0.05415	0.00078	316	65	333	11	340	5
28-36	18	146	164	0.89	0.0604	0.00323	0.24183	0.01339	0.02922	0.0005	618	90	220	11	186	3
28-37	85	456	907	0.50	0.04857	0.00139	0.22711	0.0064	0.03413	0.0004	127	44	208	5	216	3
28-39	32	226	367	0.62	0.04779	0.00259	0.1934	0.01085	0.02965	0.00055	89	90	180	9	188	3
28-40	76	364	542	0.67	0.05386	0.00197	0.32312	0.01227	0.04356	0.00058	365	62	284	9	275	4
28-41	266	1848	1144	1.62	0.05156	0.00135	0.28209	0.00782	0.03976	0.00046	266	43	252	6	251	3
28-43	162	598	1268	0.47	0.05087	0.00115	0.32925	0.00881	0.04691	0.00069	235	35	289	7	296	4
28-44	64	474	607	0.78	0.05063	0.00199	0.20865	0.00856	0.03008	0.0004	224	70	192	7	191	3
28-46	584	2815	1383	2.04	0.05732	0.00117	0.46547	0.01162	0.0594	0.00116	504	25	388	8	372	7
28-47	134	1153	990	1.16	0.04971	0.0014	0.19869	0.00607	0.02906	0.00039	182	46	184	5	185	2
28-48	72	672	610	1.10	0.05582	0.00229	0.19679	0.00789	0.02575	0.00029	445	69	182	7	164	2
28-50	79	600	1539	0.39	0.04989	0.00144	0.1515	0.00458	0.02204	0.00022	190	52	143	4	141	1
28-52	56	318	330	0.96	0.0509	0.00245	0.28925	0.01444	0.04125	0.00048	236	94	258	11	261	3
28-53	65	480	532	0.90	0.05195	0.00229	0.21728	0.00925	0.03066	0.00045	283	70	200	8	195	3
28-54	133	1074	1095	0.98	0.04979	0.00134	0.19344	0.00518	0.02827	0.00029	185	43	180	4	180	2
28-55	70	288	346	0.83	0.05297	0.002	0.4029	0.01611	0.05513	0.00065	328	69	344	12	346	4
28-56	48	330	591	0.56	0.04914	0.00206	0.19527	0.00834	0.02888	0.00034	154	78	181	7	184	2
28-57	23	100	114	0.87	0.05427	0.0023	0.38186	0.01606	0.05108	0.00063	382	72	328	12	321	4
28-58	168	352	639	0.55	0.06354	0.0016	0.74378	0.02174	0.08436	0.0012	726	38	565	13	522	7
28-59	51	475	425	1.12	0.04762	0.00276	0.17749	0.01022	0.02722	0.00036	80	102	166	9	173	2
28-60	51	227	266	0.85	0.05595	0.00246	0.38829	0.01856	0.04993	0.00076	451	79	333	14	314	5
28-61	55	325	473	0.69	0.0483	0.00207	0.21842	0.00845	0.0332	0.00048	114	63	201	7	211	3
28-62	109	799	980	0.82	0.05419	0.00164	0.22462	0.00733	0.02999	0.00038	379	51	206	6	190	2
28-63	141	732	1076	0.68	0.05244	0.00137	0.28394	0.00747	0.03925	0.0004	304	41	254	6	248	2
28-65	80	613	817	0.75	0.05368	0.00191	0.2096	0.00772	0.02823	0.00032	358	63	193	6	179	2
28-66	158	647	760	0.85	0.05583	0.00136	0.42045	0.01086	0.05447	0.00055	446	39	356	8	342	3
28-69	120	1025	855	1.20	0.05146	0.00144	0.20904	0.00632	0.02941	0.00032	262	49	193	5	187	2
28-70	178	1072	827	1.30	0.05201	0.00147	0.28108	0.00818	0.03951	0.00069	286	36	252	6	250	4
28-71	502	1583	1397	1.13	0.05679	0.00099	0.62435	0.01135	0.07983	0.0008	483	23	493	7	495	5
28-72	59	531	821	0.65	0.04795	0.00174	0.15001	0.00569	0.02273	0.00029	97	63	142	5	145	2
28-73	72	326	1663	0.20	0.05121	0.00129	0.16289	0.00433	0.02315	0.00031	250	37	153	4	148	2

续表

测点点号	含量/10^{-6}			Th/U	同位素比值						年龄/Ma					
	Pb	Th	U		$^{206}Pb/^{207}Pb$	1σ	$^{207}Pb/^{235}U$	1σ	$^{206}Pb/^{238}U$	1σ	$^{207}Pb/^{206}Pb$	1σ	$^{207}Pb/^{235}U$	1σ	$^{206}Pb/^{238}U$	1σ
28-75	69	195	391	0.50	0.05403	0.00186	0.4803	0.01848	0.06439	0.00094	372	60	398	13	402	6
28-77	59	544	517	1.05	0.05088	0.00248	0.18279	0.0085	0.02635	0.00032	235	85	170	7	168	2
28-78	237	172	531	0.32	0.0829	0.0014	2.55885	0.06508	0.22343	0.00406	1267	24	1289	19	1300	21
28-79	149	838	817	1.03	0.05156	0.00156	0.29596	0.00893	0.04178	0.00048	266	48	263	7	264	3
28-80	29	85	117	0.72	0.05341	0.00161	0.54865	0.01733	0.07477	0.00097	346	48	444	11	465	6
28-81	84	62	688	0.09	0.06236	0.00136	0.66521	0.01668	0.07731	0.00099	687	32	518	10	480	6
28-82	58	279	174	1.61	0.05648	0.00262	0.42939	0.02062	0.05532	0.00086	471	79	363	15	347	5
28-83	89	481	584	0.82	0.05093	0.00201	0.29557	0.01197	0.0421	0.00051	238	71	263	9	266	3
28-84	148	607	593	1.02	0.05169	0.00161	0.39741	0.01332	0.05585	0.00076	272	52	340	10	350	5
28-85	152	288	897	0.32	0.05672	0.00119	0.6143	0.01398	0.07852	0.00074	481	34	486	9	487	4
28-86	102	473	376	1.26	0.05217	0.00165	0.38845	0.01285	0.05409	0.00068	293	53	333	9	340	4
28-87	71	342	239	1.43	0.0569	0.00226	0.43169	0.01797	0.05533	0.00081	488	66	364	13	347	5
28-88	103	636	556	1.14	0.05103	0.00196	0.28722	0.01104	0.04108	0.00051	242	66	256	9	260	3
28-89	167	764	682	1.12	0.05411	0.00171	0.3692	0.01289	0.04937	0.00092	376	46	319	10	311	6
28-90	41	47	1075	0.04	0.05232	0.00177	0.20608	0.00797	0.02887	0.00074	300	45	190	7	183	5
28-91	69	299	296	1.01	0.05257	0.00229	0.40919	0.01846	0.05655	0.00083	310	76	348	13	355	5
28-94	1022	409	708	0.58	0.18643	0.00183	12.84299	0.16443	0.49925	0.00507	2711	10	2668	12	2611	22
28-95	802	594	1433	0.41	0.0871	0.00098	2.81806	0.07488	0.23248	0.00486	1363	24	1360	20	1347	25
28-96	143	496	919	0.54	0.05447	0.00109	0.40665	0.00798	0.05438	0.00063	391	24	346	6	341	4
28-97	91	141	595	0.24	0.05578	0.0014	0.612	0.01624	0.07959	0.00089	444	39	485	10	494	5
28-98	69	554	534	1.04	0.04867	0.00206	0.19349	0.00772	0.02911	0.00038	132	69	180	7	185	2
28-99	29	150	471	0.32	0.04998	0.00232	0.2029	0.00926	0.02967	0.00038	194	82	188	8	188	2
28-100	136	703	383	1.83	0.05372	0.00176	0.39622	0.01358	0.05332	0.00058	359	58	339	10	335	4
28-101	83	621	784	0.79	0.0518	0.00184	0.21711	0.00771	0.03044	0.00041	277	56	199	6	193	3
28-102	60	268	295	0.91	0.05266	0.0023	0.38216	0.01616	0.05294	0.00072	314	71	329	12	333	4
28-103	72	351	242	1.45	0.05236	0.00259	0.39888	0.0197	0.05523	0.00078	301	87	341	14	347	5
28-104	25	248	333	0.74	0.04896	0.00231	0.15049	0.00689	0.02226	0.00027	146	83	142	6	142	2

柴达木西部南翼山构造地表混积岩岩石学特征及沉积环境讨论*

于冬冬[1]，张永生[1]，邢恩袁[1]，左智峰[2]，侯献华[1]，王琳霖[1]，赵为永[3]

1. 中国地质科学院矿产资源研究所，国土资源部盐湖资源与环境重点实验室，北京 100037
2. 中国石油长庆油田分公司勘探部，陕西西安 751500
3. 中国石油青海油田分公司勘探开发研究院，甘肃敦煌 736202

内容提要 南翼山构造地表狮子沟组发育混积岩，作为一种特殊类型的沉积岩，对其研究具有一定的实际意义和科学价值。本文在野外观察和镜下鉴定的基础上，对南翼山混积岩样品进行了 X 衍射、主微量元素及碳氧同位素测试，分析了混积岩的岩石学特征，考虑到其成分和成因的复杂性，通过数据分析排除了多种干扰因素，最后利用有效指标综合判断了其沉积环境。结果显示：南翼山构造地表混积岩中陆源碎屑平均含量为58.7%，碳酸盐矿物含量为 31.0%，属于碳酸盐质陆源碎屑岩，主要为钙质泥岩，其中碎屑矿物以细粉砂级石英为主，碳酸盐矿物主要为泥晶方解石，黏土矿物组合为伊利石和有序伊蒙混层，且三者呈均匀混合的特征。数据分析表明 Ca、Na、Mn、Sr、Ba 等元素主要来源于自生矿物，稀土元素及碳氧同位素组成未受成岩作用影响，可以作为判断沉积时水体盐度和氧化还原条件的指标。根据 Sr/Cu 比值和 K_2O/Al_2O_3 比值，结合黏土矿物组合及伊利石结晶度和化学指数，判断地表混积岩沉积时为寒冷干旱的气候条件；Sr/Ba 比值和 Z 值表明古水体介质为咸水环境；U/Th 比值、自生 U 含量和 Ce 轻微负异常等指标综合判定其形成于弱氧化环境。以上研究表明，南翼山构造地表混积岩主要为陆源碎屑和碳酸盐矿物均匀混合的钙质泥岩，形成于寒冷干旱气候条件下的弱氧化陆相咸水湖泊环境中。

关键词 混积岩；岩石学特征；地球化学；沉积环境；南翼山构造；柴达木西部

混积岩为混合沉积的产物，其概念于 20 世纪 80 年代提出，指同一岩层内陆源碎屑与碳酸盐组分相混合的一类沉积岩（Mount，1984）。混积岩的成分介于碎屑岩和碳酸盐岩之间，可形成于滨海、滨浅湖、浅海陆棚、斜坡等环境（Tirsgaard，1996；Feng Jinlai et al.，2011a）。之前混积岩的研究并未给予足够的重视，只有少数学者在分类命名、形成机制和沉积特征等方面对其开展过研究，而且多集中在滨海相、浅海陆棚相和淡水湖泊环境（Yang Chaoqing et al.，1990；Zhang Xionghua，2000；Dong Guiyu et al.，2007；Zhang Jinliang et al.，2007）。近几年，随着混积岩重要性的日益凸显，咸水湖泊成为关注的新领域，发现该环境下形成的混积岩可以充当油气的优质储层（Feng Jinlai et al.，2011b；Xu Wei et al.，2014）。混积岩的形成受多因素控制，导致了其研究难度相较于碎屑岩或碳酸盐岩更大，尤其是在恢复沉积环境方面，但目前并没有相当成熟的方法，只能从多方面进行考虑，排除非沉积期因素的干扰，利用多种指标综合分析才能比较准确地恢复沉积时期的古环境。

南翼山构造古近系和新近系含有丰富的油气资源，长期以来其油气成藏特征一直是研究的重点（Wei Chengzhang et al.，1999；Gan Guiyuan et al.，2002；Zhang Ningsheng et al.，2006；Feng Jinlai et al.，2011b），而地表沉积岩由于缺少研究价值，很少有人关注。但近年来，在南翼山构造发现了富含钾锂硼的卤水资源，下一步开展盐田建设需要了解地表沉积岩的岩石学特征，所以明确地表混积岩的矿物组成显得尤为重要。本文以野外观察和镜下鉴定为基础，结合地球化学测试，分析了南翼山地表上新统狮子沟组混积岩的岩石学特征和沉积环境，以期为盐田建设提供地质依据，同时为咸水湖盆混积岩的地球化学研究以及柴西北地区上新世末期的古环境恢复提供一定的参考。

---

* 本文发表在：地质学报，2018，92（10）：2068-2080

1　地质概况

南翼山构造位于柴达木盆地西部茫崖凹陷内，其西北为红沟子和小梁山构造，东北为尖顶山构造，东南为油泉子构造，西南为咸水泉构造，为 NW-SE 向的大型箱状背斜[图 1（a）]。背斜核部宽缓，翼部地层变陡，且地层厚度较稳定，其中核部出露上新统狮子沟组，翼部出露第四系，地表多见风蚀残丘（Wei Chengzhang et al.，1999；Gan Guiyuan et al.，2002；Zhang Ningsheng et al.，2006）。南翼山构造沉积了巨厚的新生代地层，厚度达 5000 余米，自下而上发育路乐河组（E_{1+2}）、下干柴沟组（E_3）、上干柴沟组（N_1）、下油砂山组（N_2^1）、上油砂山组（N_2^2）、狮子沟组（N_2^3）和第四系（Q）。

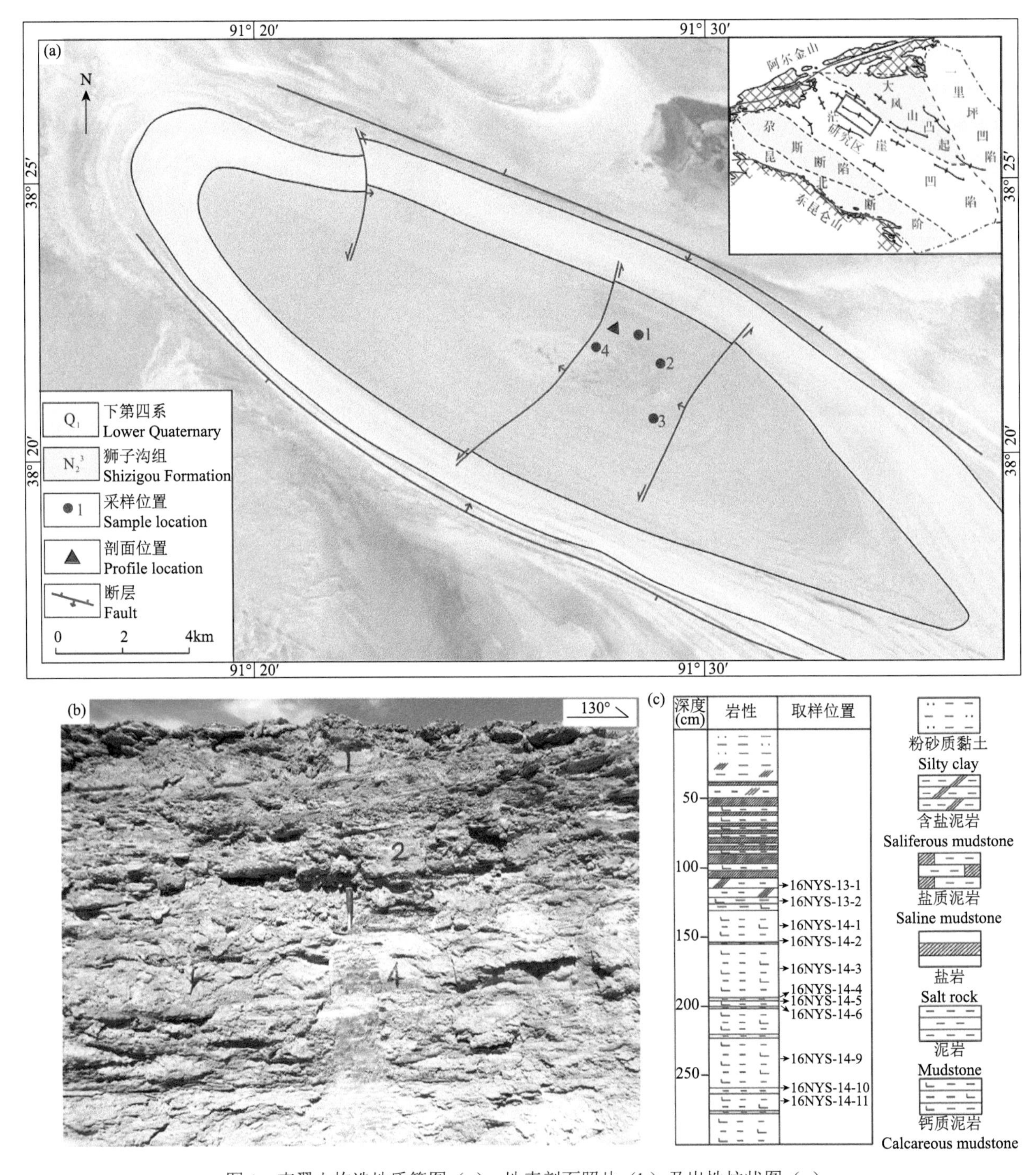

图 1　南翼山构造地质简图（a）、地表剖面照片（b）及岩性柱状图（c）

Fig. 1　Schematic geological map (a), photograph (b) and lithological column (c) of surface section of Nanyishan Structure.

上新世时期随着阿尔金山和东昆仑山的隆起，柴达木西部湖盆面积不断萎缩，沉积中心从一里坪逐渐迁至柴东三湖地区（Fu Suotang et al.，2013），此时南翼山构造开始隆起，发育背斜雏形。柴西地区上新世中晚期从山前向盆内依次发育泛滥平原、滨湖、浅湖和半深湖沉积，沉积物也有变细的趋势，山前地区主要为中粗砂岩、砾岩等粗碎屑沉积，盆内由粉砂岩、粉砂质泥岩和泥岩组成（Dang Yuqi et al.，2004）。钻井资料显示，南翼山地区上新统狮子沟组以灰色泥岩为主夹少量白色盐岩和石膏，第四系七个泉组只在翼部有出露，岩性与狮子沟组类似，粒度稍有变粗的趋势，以砂质泥岩为主，含有砂岩、盐岩和泥灰岩。

2 样品采集及测试方法

2.1 采样位置

样品采集自南翼山构造地表出露的狮子沟组[图 1(a)]，包括 1 个剖面及几个分散采样点，涉及 20 件样品，其中主微量元素和碳氧同位素测试样品取自地表剖面，X 衍射分析样品取自采样点 1～4。剖面位于南翼山构造北翼（38°22′14.62″N，91°28′4.12″E），是在地表露头上挖出的剖面，总厚度 280cm，产状 40°∠12°[图 1（b）]。按照沉积特征，将该剖面分为 4 大层 15 小层，其岩性特征自上而下具体描述如下：第一大层为土黄色粉砂质黏土层，向下逐渐含盐，硬度增大，厚度约 46cm；第二大层为浅灰色-灰白色盐岩夹浅黄色泥质条带，厚度约 65cm；第三大层为浅黄色-浅绿色钙质泥岩夹灰白色薄层岩盐，厚度 22cm；第四大层为厚层棕色钙质泥岩夹灰白色泥质条带，细分成 12 小层，其中 2、4、6、8、10、12 小层为灰白色泥质条带，厚度 1～2cm，其余均为棕色钙质泥岩，滴稀盐酸起泡强烈，本大层总厚度 147.5cm，未见底[图 1（c）]。第三和第四大层为研究目的层。

2.2 测试方法

本次研究对南翼山地表样品进行了全岩和黏土矿物 X 衍射分析、主微量元素和碳氧同位素分析，测试工作均在核工业北京地质研究所实验室完成。X 衍射分析使用 Panalytical X'Pert PRO X 射线衍射仪，执行标准为 SY/T 5163—2010；主量元素测试工作使用 AxiosmAX 型 X 射线荧光光谱仪，执行标准为 GB/T 14506.28—2010，分析误差 2%～3%；微量元素测试工作所用仪器为 NexION300D 等离子体质谱仪，执行标准为 GB/T 14506.30—2010，分析误差＜5%；碳氧同位素测试工作所用仪器为 MAT-253 气体同位素质谱仪，精度±0.2‰，执行标准为 DZ/T 0184.17—1997。

3 测试结果

研究区样品全岩和黏土矿物 X 衍射分析结果见表 1。全岩分析结果显示，样品中矿物主要有石英、斜长石、方解石、石盐和黏土矿物，部分样品含有文石、白云石和钾长石，经统计碎屑矿物成分占 15.3%～35.1%，碳酸盐矿物占 14.9%～42.9%，黏土矿物占 17.2%～43.8%。黏土矿物分析结果显示，伊利石和伊蒙混层是主要的黏土矿物类型，分别占 42%～63%和 22%～41%，绿泥石含量为 5%～16%，蒙脱石和高岭石极少。

样品的主量、微量及稀土元素测试数据见表 2。与上地壳（UCC）氧化物和微量元素含量相比（Rudnick et al.，2003），研究区样品中 Ca、Na、Sr 明显富集，其余元素相近或略微亏损[图 2（a）]。其中 Sr 元素含量为 632×10^{-6}～3372×10^{-6}，平均为 1266.9×10^{-6}，相较于上地壳平均含量（320×10^{-6}）明显富集。稀土元素用球粒陨石含量进行标准化处理（Taylor et al.，1985），绘制稀土元素的配分模式图[图 2（b）]。Ce 异常计算公式为 $\delta\mathrm{Ce} = \mathrm{Ce_N}/(0.5\mathrm{La_N} + 0.5\mathrm{Pr_N})$，Eu 异常计算公式为 $\delta\mathrm{Eu} = \mathrm{Eu_N}/(0.5\mathrm{Sm_N} + 0.5\mathrm{Gd_N})$。结果显示，稀土元素总量∑REE 为 87.51×10^{-6}～152.39×10^{-6}，平均为 128.28×10^{-6}，低于北美页岩（NASC）和澳大利亚后太古宙页岩（PAAS）的稀土总量；δCe 值为 0.91～0.93，平均值为 0.92，显示了轻微的负异常，δEu 介于 0.58～0.67 之间，平均值为 0.62，显示了明显的负异常。样品的稀土元素配分模式表现了稳定的一致性，表明了样品具有相同的物源。

表 1　南翼山构造地表样品全岩和黏土矿物 X 衍射（%）分析结果

Table 1　Whole rock and clay minerals X-ray diffraction of surface samples in Nanyishan Structure

序号	样品编号	矿物种类与含量							黏土矿物	黏土矿物相对含量				
		石英	钾长石	斜长石	方解石	白云石	文石	石盐		蒙脱石	伊蒙混层	伊利石	高岭石	绿泥石
1	16NYS-5	24. 1	—	9.1	20.7	—	—	5.9	40.2	2	35	51	4	8
2	16NYS-6	10.8	4.5	—	31.1	—	—	36.4	17.2	—	41	48	—	11
3	16NYS-16-1	18.4	—	7.8	22.4	2.3	17.5	1.2	30.4	—	34	52	—	14
4	16NYS-16-2	11.9	—	6.3	22.1	2.1	18.7	2.1	36.8	3	28	60	4	5
5	16NYS-17	15.4	—	6.3	12.9	8	14.6	3.5	39.3	—	27	60	4	9
6	16NYS-18	13.7	4.5	4.7	18.1	—	—	31.1	27.9	2	22	62	—	14
7	16NYS-31	16.1	—	5.3	20.3	18.6	—	4.8	34.9	—	—	—	—	—
8	16NYS-34	18.7	—	8.9	24.3	—	10	1.6	36.5	1	41	42	—	16
9	16NYS-35	25.3	—	9.8	14.9	—	—	6.2	43.8	—	24	63	6	7

表 2　南翼山构造地表样品主量、微量、稀土元素含量和碳氧同位素值分析结果

Table 2　Major, trace, rare earth element concentration and carbon, oxygen isotope values of surface samples in Nanyishan Structure

编号	16NYS-13-1	16NYS-13-2	16NYS-14-1	16NYS-14-2	16NYS-14-3	16NYS-14-4	16NYS-14-5	16NYS-14-6	16NYS-14-9	16NYS-14-10	16NYS-14-11
SiO_2	31.24	22.95	36.35	33.73	38.64	37.39	39.19	35.84	41.16	40.1	43.09
Al_2O_3	9.78	7.61	10.75	9.31	11.96	10.16	11.98	9.26	11.67	11.82	12.04
TFe_2O_3	4.36	3.28	4.41	3.54	4.52	3.64	5.01	3.34	4.67	4.69	4.77
MgO	3.25	2.4	3.1	2.75	3.07	2.81	3.13	2.83	2.93	2.91	2.85
CaO	16.3	20.98	15.87	14.55	13.8	14.63	13.3	13.57	12.14	11.67	11.7
Na_2O	6.15	5.04	4.34	6.03	3.8	4.43	3.48	6.66	4.48	4.47	3.81
K_2O	2.13	1.73	2.32	1.98	2.58	2.12	2.58	1.97	2.47	2.54	2.55
MnO	0.099	0.085	0.096	0.072	0.072	0.064	0.076	0.062	0.072	0.064	0.071
TiO_2	0.392	0.296	0.474	0.405	0.506	0.423	0.508	0.431	0.518	0.51	0.548
K_2O/Al_2O_3	0.22	0.23	0.22	0.21	0.22	0.21	0.22	0.21	0.21	0.21	0.21
Sc	10.6	8.94	12	11.3	13.6	11.7	13.9	10.6	13	12.9	13.2
V	72.8	60.3	82.9	69.4	90.2	79.2	93.3	68	89.5	92.5	87.3
Cr	57.7	47.9	66.3	54	71.4	62.4	73.3	55.7	68.1	70.8	68.3
Co	13.1	11.2	14.3	11.5	15.5	12.4	14.5	11.1	13.6	14.4	14.1
Ni	31.2	29.4	36.2	30.1	38.3	32.2	37.8	28.6	33.8	36.7	35.5
Cu	30.3	25.5	29.6	26	32.1	27.8	33.1	26.5	30.9	33.3	34
Rb	88.8	73.4	98.6	85.2	111	94.7	109	85.3	106	110	108
Sr	3372	1239	1043	632	788	678	726	568	868	1791	2231
Ba	968	433	453	374	495	426	491	391	468	516	508
Th	9.05	7.45	10.7	8.44	11.8	10.1	11.9	9.26	11.2	11.6	11.4
U	3.53	3.06	3.91	2.91	4.04	3.16	3.88	2.59	3.3	3.68	3.52
Sr/Cu	111.29	48.59	35.24	24.31	24.55	24.39	21.93	21.43	28.09	53.78	65.62
Sr/Ba	3.48	2.86	2.30	1.69	1.59	1.59	1.48	1.45	1.85	3.47	4.39
$Th_{自生}$	1.49	1.57	2.39	1.24	2.56	2.25	2.64	2.10	2.18	2.46	2.09
$U_{自生}$	1.93	1.81	2.15	1.38	2.08	1.49	1.91	1.07	1.38	1.74	1.54
U/Th	1.29	1.15	0.90	1.11	0.81	0.66	0.72	0.51	0.64	0.71	0.74
La	23.7	18.7	28.3	23.2	30.1	26.4	30.6	26.4	30.6	30.2	32.4
Ce	44.5	35.7	54.5	44.2	57.5	50.9	59.3	50	58.2	58.7	62.6
Pr	5.11	4.03	6.28	5.16	6.7	5.97	6.75	5.8	6.78	6.71	7.3
Nd	19.8	15.6	24.6	20	26.2	22.8	26	22.7	26.4	26.1	27.9
Sm	3.82	3.1	4.67	3.73	5.03	4.44	5.03	4.3	5.08	5.02	5.33
Eu	0.806	0.592	0.843	0.726	0.945	0.834	0.984	0.826	0.942	0.953	1.02
Gd	3.35	2.61	4.05	3.26	4.33	3.8	4.37	3.79	4.41	4.29	4.65
Tb	0.587	0.471	0.711	0.567	0.745	0.665	0.763	0.645	0.738	0.745	0.808

续表

编号	16NYS-13-1	16NYS-13-2	16NYS-14-1	16NYS-14-2	16NYS-14-3	16NYS-14-4	16NYS-14-5	16NYS-14-6	16NYS-14-9	16NYS-14-10	16NYS-14-11
Dy	3.13	2.64	3.75	3.05	3.98	3.53	3.98	3.4	3.95	3.9	4.06
Ho	0.608	0.506	0.712	0.575	0.759	0.674	0.775	0.662	0.777	0.761	0.787
Er	1.72	1.42	2.01	1.63	2.19	1.88	2.23	1.84	2.15	2.21	2.24
Tm	0.31	0.252	0.353	0.273	0.378	0.33	0.392	0.324	0.381	0.378	0.391
Yb	2.02	1.65	2.31	1.88	2.4	2.22	2.53	2.09	2.44	2.44	2.53
Lu	0.288	0.235	0.333	0.264	0.366	0.312	0.376	0.29	0.346	0.348	0.37
∑REE	109.75	87.51	133.42	108.52	141.62	124.76	144.08	123.07	143.19	142.76	152.39
LREE/HREE	8.14	7.94	8.38	8.44	8.35	8.30	8.35	8.44	8.43	8.47	8.62
La_N/Yb_N	7.93	7.66	8.28	8.34	8.48	8.04	8.17	8.54	8.47	8.36	8.65
δEu	0.67	0.62	0.58	0.62	0.60	0.61	0.63	0.61	0.59	0.61	0.61
δCe	0.91	0.93	0.93	0.92	0.92	0.92	0.93	0.91	0.92	0.93	0.92
δ^{13}C	−1.7	1.1	−0.9	−0.3	0.1	−0.2	−0.1	−0.4	−0.1	−0.1	−0.3
δ^{18}O	−0.4	0.1	−1	−1.6	−1.4	−1.7	−1.3	−2.4	−2	−2	−2.6
Z 值	123.62	129.60	124.96	125.89	126.81	126.04	126.45	125.29	126.10	126.10	125.39

注：主量元素氧化物含量单位为%，微量、稀土元素含量单位为$\times 10^{-6}$，碳氧同位素值单位为‰，下同。

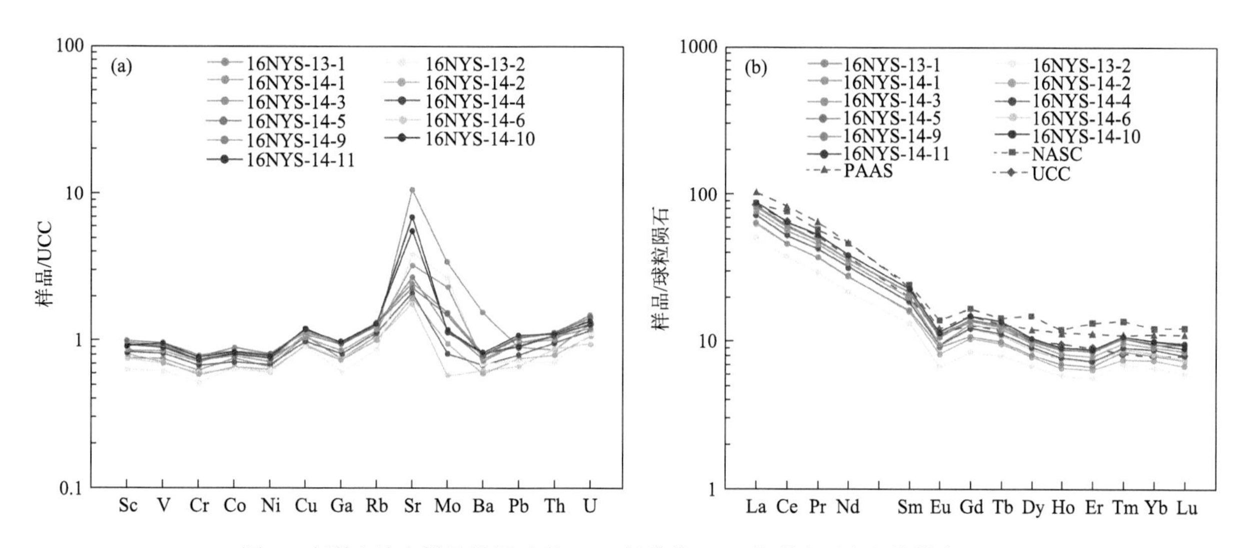

图 2 南翼山地表样品微量元素 UCC 标准化（a）和稀土元素配分模式（b）

Fig. 2 UCC-nomalized trace element and chondrite-nomalized REE pattern of surface samples in Nanyishan Structure.

UCC 数据来自 Rudnick 等（2003）；球粒陨石和 PAAS 数据分别来自 Taylor 等（1985）；NASC 数据来自 Haskin 等，（1968）

UCC values from Rudnick et al. (2003), chondrite and PAAS values from Taylor et al. (1985), NASC values from Haskin et al. (1968)

样品碳氧同位素值较稳定，δ^{13}C 值为−1.7‰～＋1.1‰，平均值为−0.26‰，δ^{18}O 值为−2.6‰～＋0.1‰，平均值为−1.48‰（均采用 PDB 标准），δ^{13}C 和 δ^{18}O 的变化幅度均较小（表 2）。对比典型的湖相碳酸盐岩碳氧同位素组成（Yuan Jianying et al.，2015），碳同位素未发生偏移，氧同位素呈正偏移。

4 岩石学特征

X 衍射结果统计显示（表 1），样品中碎屑矿物、碳酸盐矿物和黏土矿物三个端元的含量均小于 50%，为典型的混积岩。目前关于混积岩的分类命名尚未有统一的方案，大多借鉴碎屑岩和碳酸盐岩的分类方法，有学者将黏土、陆源碎屑和碳酸盐作为三端元进行划分（Yang Chaoqing et al.，1990；Zhang Xionghua，2000），也有陆源碎屑（包括黏土）和碳酸盐的两端元法（Dong Guiyu et al.，2007；Yuan Jianying et al.，2016）。本文采用两端元法，将碎屑矿物和黏土矿物归为一类。经统计陆源碎屑占 32.5%～78.9%，平均为 58.7%，碳酸盐矿物占 14.9%～42.9%，平均为 31.0%，从成分上看，南翼山地表混积岩属于碳酸盐质陆源碎屑岩。结合野外观察与薄片鉴定，将南翼山地表混积岩分为两类：钙质泥岩、含砂钙质泥岩。

（1）钙质泥岩：该类型在野外最为常见，整体呈棕色块状，中间夹有灰白色条带[图 3（a）、（b）]。

粒度细，手触摸有滑腻感，断口处切面光滑，滴稀盐酸起泡较剧烈，未见明显生物化石。镜下观察到碎屑矿物以细粉砂级石英为主，未见到自生现象，主要是母岩风化的产物，碳酸盐矿物中主要为泥晶方解石，未见明显的方解石颗粒及重结晶现象，可能形成于原地物理化学沉淀。碎屑矿物、碳酸盐矿物与黏土矿物均匀混合，棕色泥岩中黏土矿物含量较高[图 3（d）]，灰白色泥岩方解石含量相对升高[图 3（e）]，未见生物碎屑。

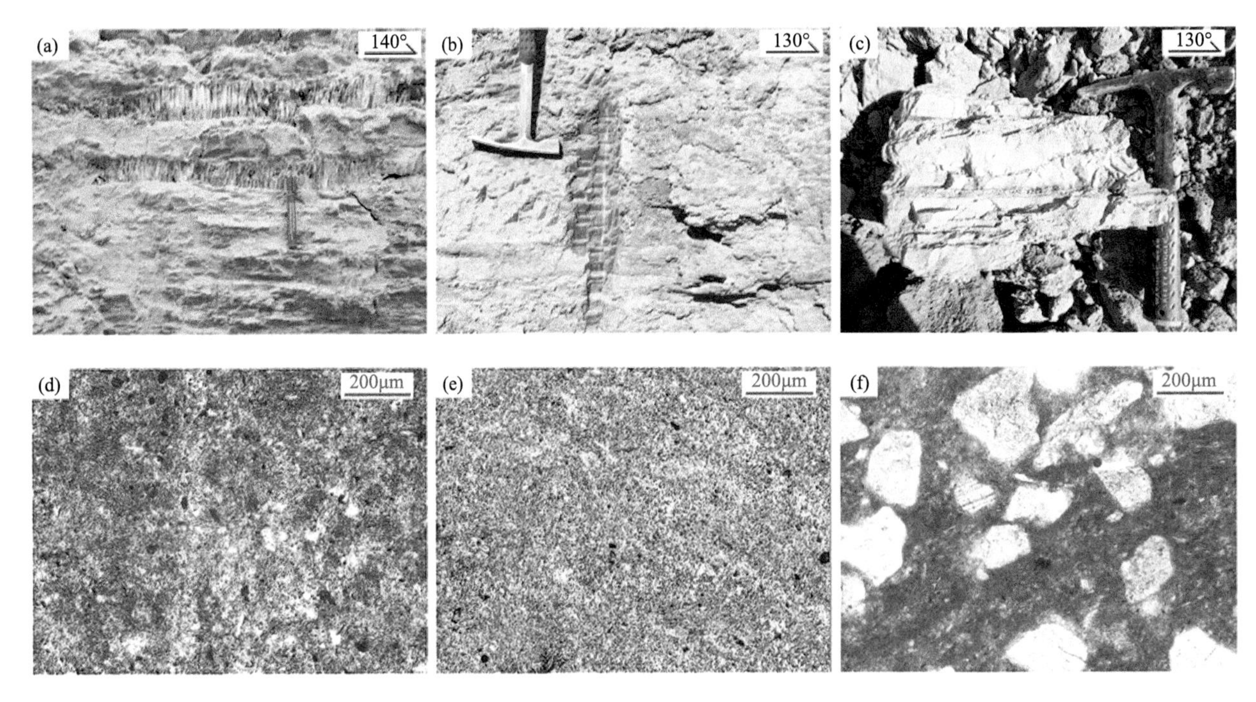

图 3　南翼山构造地表混积岩野外及显微镜下特征

Fig. 3　The characteristics of field and microscope of surface mixed rocks in Nanyishan Structure.

（a）棕色钙质泥岩，中间夹纤维状石盐；（b）棕色钙质泥岩夹灰白色条带；（c）灰色含砂钙质泥岩，夹有纹层状黄色条带，裂隙充填石盐；（d）钙质泥岩，碎屑矿物、方解石、黏土矿物均匀混合，黏土矿物含量高，单偏光；（e）钙质泥岩，碎屑矿物、方解石、黏土矿物均匀混合，方解石含量相对较高，单偏光；（f）含砂钙质泥岩，见明显石英、斜长石，单偏光

(a) brown calcareous mudstone with interbedded fibrous halite; (b) brown calcareous mudstone with interbedded of white bands; (c) grey sandy calcareous mudstone with lamellar yellow bands, fractures are filled with halite; (d) calcareous mudstone, clastic minerals, calcite and clay minerals are homogeneous mixed，the content of clay minerals is high, polarized light; (e) calcareous mudstone，clastic minerals, calcite and clay minerals are homogeneous mixed, the content of calcite is relatively high, polarized light; (f) sandy calcareous mudstone with obvious quartz and plagioclase，polarized light

（2）含砂钙质泥岩：仅在南翼山构造局部地区可见，颜色相对较浅，中间夹有薄层石盐[图 3（c）]。成分仍以陆源碎屑为主，含量在 60%以上，碎屑矿物明显增多，可见细砂级石英和长石颗粒，黏土矿物与泥晶方解石呈均匀混合的状态[图 3（f）]。

可以看出，南翼山地表混积岩以钙质泥岩为主，岩石中矿物呈粒度细、混合均匀的特点，反映了当时水动力较弱，以垂向沉降为主，是典型的安静低能湖泊环境下形成的混积岩（Xu Wei et al.，2014）。混积岩中黏土矿物主要为伊利石，其遇水不膨胀和不可塑的性质以及混积岩粒度细、成分均匀混合的特点都表明地表狮子沟组混积岩有利于储存卤水，开展盐田建设。

5　沉积环境讨论

5.1　数据有效性检验

相对于普通的碎屑岩和碳酸盐岩来说，混积岩的成分更加复杂，影响因素更加多样，许多指标不能直接用于判断沉积环境，必须加以分析，考虑各种因素对分析结果的影响。比如在讨论水体盐度及氧化还原条件时，就必须剔除陆源碎屑的影响（Chang Huajin et al.，2009），而且稀土元素、碳氧同位素、黏土矿物在成岩过程中都可能发生变化（Shields et al.，2001；Huang Sijing et al.，2003；Zhao Ming et al.，2006），所以在应用这些数据恢复沉积环境之前，对数据的有效性进行检验十分必要。

5.1.1 主微量元素的来源

从主微量元素的测试结果中可以看出，不同元素之间具有一定的相关性。Al_2O_3 与 TiO_2、SiO_2、K_2O、P_2O_5、Sc、Rb、Th、V、Co、Cr、∑REE 等具有良好的正相关关系，暗示它们主要来自陆源碎屑，而与 CaO、Na_2O、MnO 呈现负相关性，表明 CaO、MnO 主要与碳酸盐沉积有关，Na_2O 与石盐生成有关（图 4）。MgO 与 Al_2O_3 呈较强的正相关，与 CaO 呈弱负相关（图 4），表明 MgO 主要为陆源碎屑来源，部分来自白云石，这与全岩 X 衍射的结果一致。Sr 的异常富集是值得注意的地方，沉积物中的 Sr 主要有两个来源，一是陆源碎屑组分，

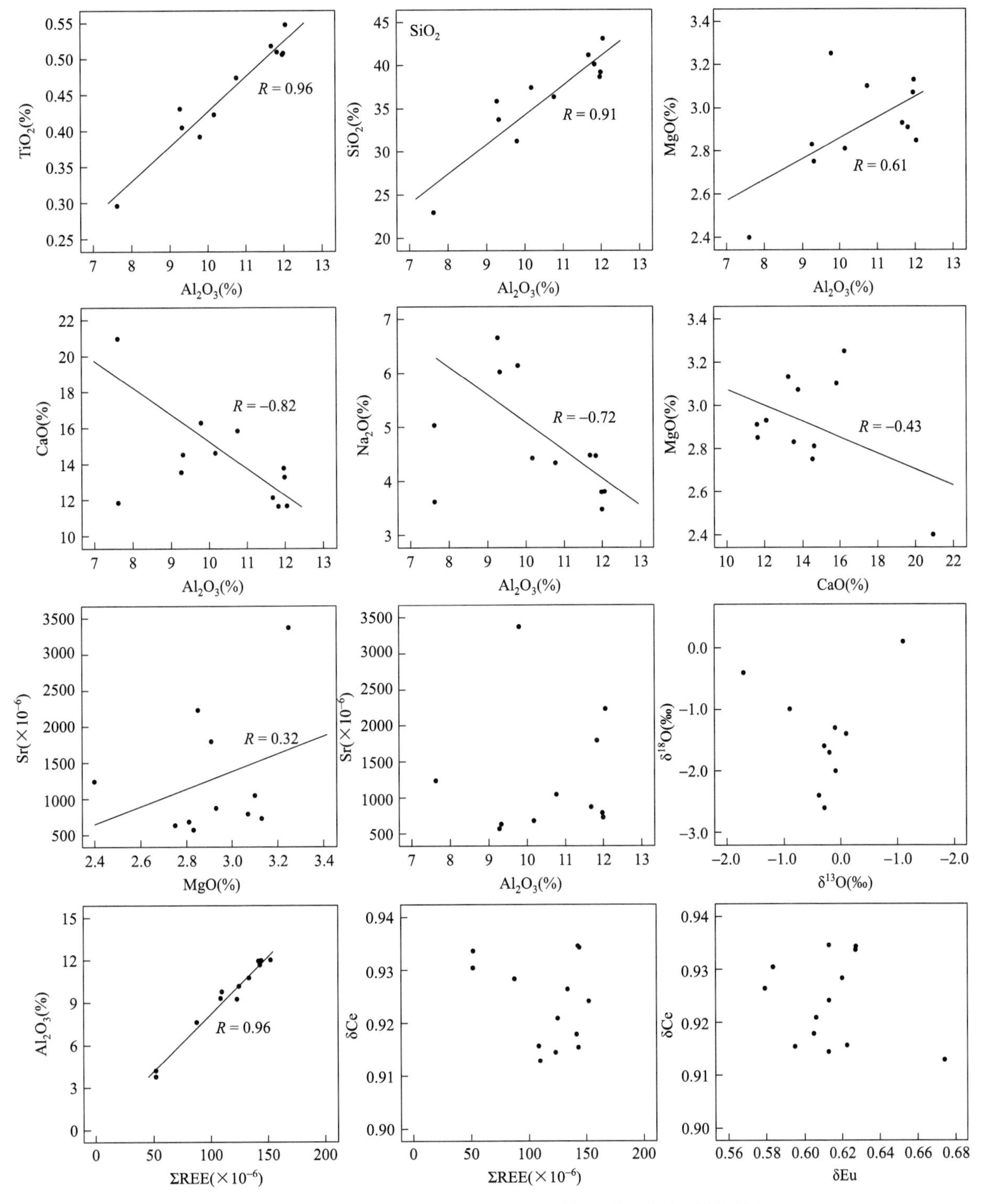

图 4 样品中主量、微量、稀土元素和碳氧同位素的相关性图解

Fig. 4 Correlation diagrams between major, trace, rare earth element and carbon, oxygen isotope in surface samples.

二是水体中通过物理化学作用沉积下来的，本次样品中 Sr 与 Al_2O_3、TiO_2、SiO_2 等不具相关性，与 MgO、MnO、CaO 呈中等相关，暗示 Sr 可能主要来源于碳酸盐矿物。通过以上相关性分析可以得出，来源于陆源碎屑的元素有 Al、Si、K、Ti、Sc、Rb、Th、V、Co 和 Cr 等，这些元素主要反映母岩的化学特征。自生矿物中的元素有 Ca、Na、Mn、Sr、Ba 等，它们对沉积时水体的盐度和氧化还原条件具有直接的指示意义。

5.1.2　成岩作用的影响

黏土矿物在沉积和成岩作用中经常发生转变，可以根据其结晶度和矿物组合判断成岩阶段。研究区伊利石的结晶度（$\Delta 2\theta$）0.40°～0.52°，平均值为（$\Delta 2\theta$）0.45°（表 3），根据 Kübler（1964）的成岩阶段划分标准，大于（$\Delta 2\theta$）0.42°处于成岩作用阶段，未发生变质。样品黏土矿物组合为伊利石 + 有序伊蒙混层（图 5），且伊蒙混层中伊利石含量 90%以上（表 3），可以判断其处于成岩晚期阶段。在埋藏和成岩过程中，蒙脱石会向伊蒙混层和伊利石转化，并且随着埋深的加大，转化程度越大，体现为伊利石含量和结晶程度增大（Zhao Ming et al.，2006）。样品中蒙脱石含量微乎其微，可能是在成岩过程中转化为伊蒙混层，并最终转化为伊利石，表明样品黏土矿物中存在成岩自生的伊利石。同时，伊利石含量和伊利石的结晶度并无相关性，所以黏土矿物中的伊利石可能是来源于成岩自生和沉积搬运两个方面。

表 3　样品中伊利石结晶度和化学指数

Table 3　Crystallinity and chemical index of illite in surface samples

样品编号	伊利石含量（%）	伊利石结晶度（°）	伊利石化学指数	伊蒙混层比（%I）
16NYS-5	51	0.52	0.23	92
16NYS-6	48	0.45	0.18	93
16NYS-16-1	52	0.46	0.14	91
16NYS-16-2	60	0.47	0.11	92
16NYS-17	60	0.47	0.18	94
16NYS-18	62	0.45	0.15	94
16NYS-34	42	0.42	0.12	92
16NYS-35	63	0.40	0.12	94

注：伊利石化学指数根据乙二醇曲线中 5Å/10Å 面积比计算，结晶度测量自 10Å 处的半峰宽。

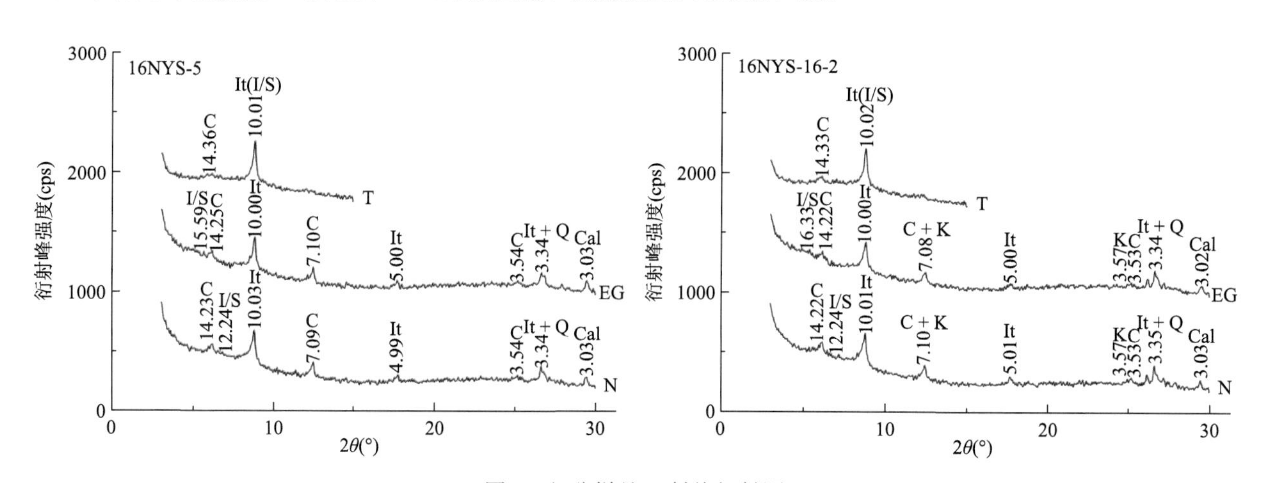

图 5　部分样品 X 射线衍射图

Fig. 5　X-ray diffraction diagrams of some samples.

N-自然风干样；EG-乙二醇处理样；T-加热样；C-绿泥石；It-伊利石；I/S-伊蒙混层；K-高岭石；Cal-方解石；Q-石英

N-air-dried samples; EG-glycolated samples; T-Heated samples; C-chlorite; It-illite; I/S-illite/smectite; K-kaolinite; Cal-calcite; Q-quartz

成岩过程中化学元素和同位素组成可能会发生变化，稀土元素 Eu 和 Ce 受到氧化电位的影响，在孔隙流体存在情况下会改变价而与其他稀土元素发生分异，样品中∑REE 与 Al_2O_3 显示强正相关性，δCe 和 δEu、

∑REE 无相关性（图 4），表明稀土元素也未受后期成岩作用的影响（Shields et al.，2001）。

同样，在应用碳氧同位素讨论地质意义时应先判断其是否能代表原始的碳氧同位素组成。Mn/Sr 比值常用来判断碳氧同位素是否受到成岩蚀变的影响，一般来说，碳酸盐岩成岩过程是丢失 Sr 获取 Mn 的过程，所以成岩蚀变较弱的岩石具有较高的锶含量和较低的锰含量（Huang Sijing et al.，2003），Mn/Sr<2～3 表示保留了原始碳氧同位素的组成。样品中的 Mn 平均含量 586.62×10^{-6}，Sr 平均含量 1266.91×10^{-6}，其比值仅为 0.46，表明受到成岩作用影响较小。另外，$\delta^{13}C$ 和 $\delta^{18}O$ 不具有明显的相关性，$\delta^{18}O>-10‰$也能够指示碳氧同位素未受成岩破坏。图 4 显示混积岩样品碳氧同位素不具有明显的相关性，且 $\delta^{18}O$ 介于 −2.6‰～＋0.1‰之间，以上指标都表明同位素组成在成岩过程中没有或者很少受到影响。

5.2 沉积环境

5.2.1 古气候

在不同的自然环境下，元素具有不同的分解、迁移和富集特征，其在沉积物中的含量变化可以反映当时的古气候条件，岩石中主量、微量元素含量及比值都可以作为判别古气候条件的有效手段，常用来反映古气候的指标主要有化学风化指数（CIA）、长石蚀变指数（PIA）、Sr/Cu、Mg/Ca、FeO/MnO、K_2O/Al_2O_3 等（Nesbitt et al.，1982；Zhang Tianfu et al.，2016）。由于样品中 CaO 主要是碳酸盐沉积形成的，Na_2O 与石盐形成有关，所以 CIA、PIA 等指标在这里并不适用。另外，像 Mg/Ca、FeO/MnO 等指标并没有定量化，只能反映相对的变化，所以也不能准确判断单一沉积时期的古气候。本文主要采用 Sr/Cu 比值、黏土矿物特征和 K_2O/Al_2O_3 来讨论古气候条件。微量元素中 Sr 和 Cu 的比值通常作为古气候变化研究的参数，Sr/Cu 比值介于 1.3～5 之间指示湿润气候，大于 5 指示干旱气候（Lermanm，1978）。黏土矿物的组合常作为气候的指示剂，温暖潮湿环境有利于高岭石形成和保存，而寒冷干旱气候对伊利石和绿泥石的形成和保存有利（Tang Yanjie et al.，2002）。此外，伊利石结晶度和化学指数是气候变化的灵敏反映指标，伊利石结晶度小（结晶程度大），代表寒冷干旱的气候条件（Ehrmann，1998），化学指数小于 0.5 代表富 Fe-Mg 伊利石，为物理风化产物（Esquevin，1969）。Al 通常形成于温暖湿润条件下的强烈化学风化作用，K 反映与冷干气候有关的弱化学风化作用，所以 K_2O/Al_2O_3 比值也可以反映气候，K_2O/Al_2O_3 高值（>0.2）反映冷干气候（Beckmann et al.，2005；Roy et al.，2013）。

研究区混积岩样品 Sr/Cu 比值介于 21.43～111.19 之间（图 6），明显大于 5，表明气候十分干旱。黏土矿物组合为伊利石、伊蒙混层和少量绿泥石，前文已经分析过伊利石有成岩自生和沉积搬运两方面的来源，且自生伊利石来自于蒙脱石的转化，推断沉积时黏土矿物组合以伊利石和蒙脱石为主，这恰恰是在寒冷干旱条件下易于形成的。样品中伊利石结晶度介于（$\Delta2\theta$）0.40°～0.52°，化学指数 0.11～0.23（表 3，图 5），且有着良好的相关性（$R=0.74$），表明结晶度与化学指数反映了相同的气候的环境，共同指示了冷干环境。样品中 Al 和 K 主要来源于陆源碎屑，其比值主要反映物源区的气候条件，样品的 K_2O/Al_2O_3 比值为 0.21～0.23（图 6），物源区主要为阿尔金山和东昆仑山，考虑到研究区的地理位置，也从侧面印证了沉积时为寒冷干旱的气候。

5.2.2 古盐度

Sr/Ba 比值可以作为古盐度判别的灵敏标志，一般来说淡水沉积物中 Sr/Ba<1，咸水沉积物中 Sr/Ba>1，且与古盐度呈正相关（Deng Hongwen et al.，1993）。样品分析结果显示，Sr/Ba 值在 1.45～4.39 之间，平均 2.38（图 6），反映了当时古水体介质为咸水环境。碳氧同位素也可以用来进行古水体介质盐度的定性判断，一般采用 Keith 等（1964）的经验公式：$Z=2.048\times(\delta^{13}C+50)+0.498\times(\delta^{18}O+50)$，当 Z 大于 120 时为海相灰岩，小于 120 时为淡水灰岩。近来研究表明，现代陆相咸水湖泊的 Z 值也大于 120，所以本文也采用这个指标来进一步判断古盐度。经计算，研究区样品 Z 值介于 123.62～129.60 之间，说明当时为陆相咸水湖泊环境，而且氧同位素的正偏移也在一定程度上反映了水平面下降，盐度升高。

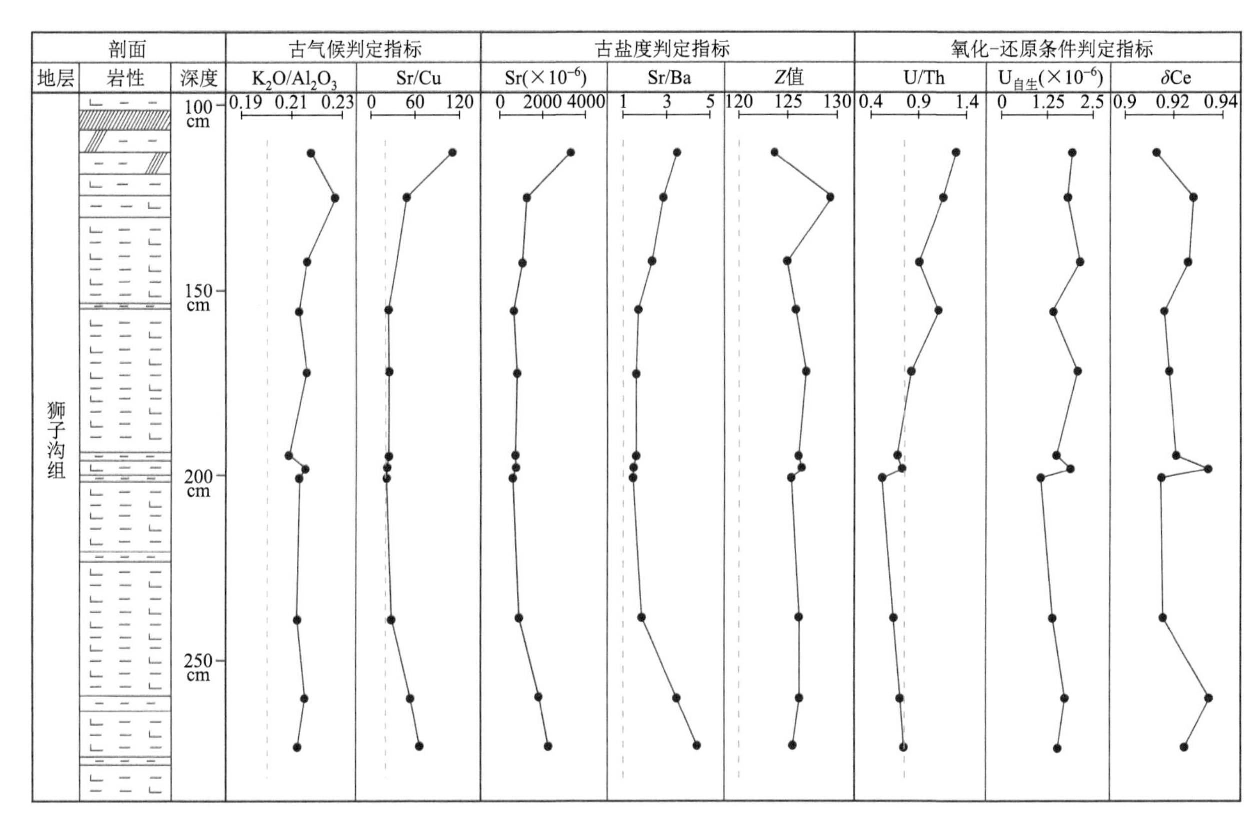

图 6　南翼山地表混积岩中地球化学指标曲线图

Fig. 6　Geochemical indicators curves of surface mixed rocks in Nanyishan Structure.

5.2.3　古氧化还原条件

水体的氧化还原条件影响多种元素的溶解度、赋存状态，所以可以借助这些微量元素在沉积物中的特征，恢复古水体的氧化还原环境。U、V、Co、Ni、Cr、Mo 等元素对氧化还原环境比较敏感，可以用来重建古环境（Tribovillard et al.，2006；Chang Huajin et al.，2009）。国内外许多学者先后提出了多种评价指标，如 DOP、V/（V + Ni）、U/Th、Ni/Co、V/Cr 等（Hatch et al.，1992；Rimmer，2004）（表 4），在不同地区不同时期的氧化还原环境判别中得到了良好的应用（Zhang Tianfu et al.，2016；Feng Yangwei et al.，2017）。Jones et al.（1994）认为 DOP、U/Th、自生 U、Ni/Co 和 V/Cr 是最可靠的指标，但是混积岩样品中 Th、Ni、Co、V 和 Cr 主要来源于陆源碎屑，不能直接用于判断水体氧化还原条件，必须剔除陆源碎屑的影响。目前常用的方法是根据 PAAS 的元素值进行校正，先计算样品中元素 X 受陆源碎屑影响的含量，经验公式为：$X_{碎屑} = (X/\mathrm{Al})_{平均页岩} \times \mathrm{Al}_{样品}$，那么自生的部分为 $X_{自生} = X_{总} - X_{碎屑}$（Reitz et al.，2004；Chang Huajin et al.，2009）。经计算，样品中自生 Th 含量为 1.24×10^{-6}～2.65×10^{-6}，自生 U 为 1.07×10^{-6}～2.15×10^{-6}，而 Ni、Co、V 和 Cr 基本全部来源于陆源碎屑，所以本文主要采用了 U/Th 和自生 U 来判断当时水体的氧化还原条件。混积岩样品的 U/Th 值为 0.5～1.29，自生 U 值均小于 5×10^{-6}（图 6，表 4），表明南翼山地区当时古水体为氧化-弱氧化环境。

表 4　氧化还原条件地球化学常用判定指标（据 Hatch et al.，1992；Jones et al. ，1994）

Table 4　Geochemical indicates of redox environment (after Hatch et al., 1992; Jones et al., 1994)

环境	DOP	U/Th	自生 U（$\times10^{-6}$）	Ni/Co	V/Cr	V/(V + Ni)
还原	＞0.75	＞1.25	＞12	＞7	＞4.25	＞0.77
贫氧	0.42～0.75	0.75～1.25	5～12	5～7	2～4.25	0.6～0.77
富氧	＜0.42	＜0.75	＜5	＜5	＜2	＜0. 6

另外，稀土元素记录了湖泊的环境变化信息，可以作为湖泊系统气候环境变化的代用指标，其中 Ce

异常可以用来判别氧化还原状态和水位变化情况，而且在泥岩中应用效果良好。在含氧的水体中，Ce^{3+}被氧化成易沉淀的Ce^{4+}，而造成 Ce 负异常（Guo Qingjun et al.，2007）。研究表明 δCe 值越高，湖泊水体越深（Zhang Hucai et al.，2009），本文利用 δCe 值进一步判断古水体氧化还原环境。样品的 δCe 均值为 0.92，显示了轻微负异常，反映当时的沉积环境为弱氧化性，所以综合判断沉积时为弱氧化环境。

6 结论

（1）X 衍射分析和薄片鉴定显示，南翼山构造地表狮子沟组混积岩中陆源碎屑占 58.7%，碳酸盐矿物占 31.0%，属于碳酸盐质陆源碎屑岩，主要为钙质泥岩。其中碎屑矿物以细粉砂级石英为主，碳酸盐矿物主要为泥晶方解石，黏土矿物组合为伊利石 + 有序伊蒙混层，并且三者呈均匀混合的状态，反映其形成于平静低能湖泊环境下。从混积岩矿物组成来看，南翼山地表有利于开展盐田建设。

（2）数据分析表明 Ca、Na、Mn、Sr、Ba 等元素主要来源于自生矿物，稀土元素及碳氧同位素组成未受成岩作用影响，它们可以用来判断沉积时古水体的盐度和氧化还原条件。黏土矿物中伊利石有沉积搬运和成岩自生两方面来源，在判断古气候时应结合结晶度和化学指数综合分析。

（3）地球化学指标 Sr/Cu＞20、K_2O/Al_2O_3＞0.2，结合黏土矿物组合及伊利石结晶度和化学指数，综合判定南翼山混积岩沉积时为寒冷干旱气候；Sr/Ba＞1.45 和 Z 值＞123 显示为咸水条件；U/Th 为 0.5～1.29、自生 U＜2.15×10^{-6} 和 Ce 轻微负异常等指标反映其形成于弱氧化沉积环境中。结合岩石学特征，推断南翼山地表混积岩形成于气候寒冷干旱、弱氧化的陆相咸水湖泊环境中。

References

Beckmann B, Flögel S, Hofmann P, Schulz M, Wagner T. 2005. Orbital forcing of Cretaceous river discharge in tropical Africa and ocean response. Nature, 437 (7056): 241-244.

Chang Huajin, Chu Xuelei, Feng Lianjun, Huang Jing, Zhang Qirui. 2009. Redox sensitive trace elements as paleoenvironments proxies. Geological Review, 55 (1): 91-99 (in Chinese with English abstract).

Dang Yuqi, Yin Chengming, Zhao Dongsheng. 2004. Sedimentary facies of the paleogene and neogene in western qaidam basin. Journal of Palaeogeography, 6 (3), 297-306 (in Chinese with English abstract).

Deng Hongwen, Qian Kai. 1993. Sedimentary Geochemistry and Environmental Analysis. Lanzhou: Gansu Science and Technology Press, 1-154 (in Chinese without English abstract).

Dong Guiyu, Chen Hongde, He Youbin, Qing Zhiyong, Luo Linxiong, Xin Changjing. 2007. Some Problems on the Study of the Mixed Siliciclastic-Carbonate Sediments. Advances in Earth Science, 22 (9): 931-939 (in Chinese with English abstract).

Ehrmann W. 1998. Implications of late Eocene to early Miocene clay mineral assemblages in McMurdo Sound (Ross Sea, Antarctica) on paleoclimate and ice dynamics. Palaeogeography Palaeoclimatology Palaeoecology, 139 (3-4): 213-231.

Esquevin J. 1969. Influence de la composition chimique des illites surcrystallite. Bull Centre Rech Pau-SNPA, 3 (1): 147-154.

Feng Jinlai, Hu Kai, Cao Jian, Chen Yan, Wang Longgang, Zhang Ying, Wang Mu, Zhao Jian. 2011a. A Review on Mixed Rocks of Terrigenous Clastics and Carbonates and Their Petroleum-Gas Geological Significance. Geological Journal of China Universities, 17 (2): 297-307 (in Chinese with English abstract).

Feng Jinlai, Cao Jian, Hu Kai, Chen Yan, Yang Shapyong, Liu Yuntian, Bian Lizeng, Zhang Guoqing. 2011b. Forming mechanism of middle-deep mixed rock reservoir in the Qaidam basin. Acta Petrologica Sinica, 27 (8): 2461-2472 (in Chinese with English abstract).

Feng Yangwei, Jiang Ting, Song Bo, Niu Yazhuo. 2017. Geochemical discrimination of Middle Permian Sedimentary Environment of the Yili Area, the Border between China and Kazakhstan. Acta Geologica Sinica, 91 (4): 942-953 (in Chinese with English abstract).

Fu Suotang, Zhang Daowei, Xue Jianqin, Zhang Xiaobao. 2013. Exploration potential and geological conditions of tight oil in the Qaidam Basin. Acta Sedimentologica Sinica, 31 (4): 672-682 (in Chinese with English abstract).

Gan Guiyuan, Wei Chengzhang, Chang Qingping, Yan Xiaolan, Cui Jun, Wang Anmin, Chen Dengqian. 2002. Characteristics and forming conditions of lake-facies carbonate-rock oil and gas pools in the nanyishan structure of the qaidam basin. Experimental Petroleum Geology, 24 (5), 413-417 (in Chinese with English abstract).

Guo Qingjun, Shields G A, Liu Congqing, Strauss H, Zhu maoyan, Pi Daohui, Goldber P, Yang Xinglian. 2007. Trace element chemostratigraphy of two Ediacaran-Cambrian successions in South China: Implications for organosedimentary metal enrichment and silicification in the Early Cambrian. Palaeogeography Palaeoclimatology Palaeoecology, 254 (1-2): 194-216.

Haskin L A, Haskin M A, Frey F A, Wildeman TR. 1968. Relative and absolute terrestrial abundances of the rare earths. Origin and Distribution of the Elements,

889-912.

Hatch J R, Leventhal J S. 1992. Relationship between inferred redox potential of the depositional environment and geochemistry of the Upper Pennsylvanian (Missourian) Stark Shale Member of the Dennis Limestone, Wabaunsee County, Kansas, U.S.A. Chemical Geology, 99 (1/3): 65-82.

Huang Sijing, Shi He. Mao Xiaodong, Zhang Meng, Shen Licheng, Wu Wenhui. 2003. Diagenetic alteration of earlier palaeozoic marine carbonate and preservation for the information of sea water. Journal of Chengdu University of Technology, 30 (1), 9-18 (in Chinese with English abstract).

Jones B, Manning D A C. 1994. Comparison of geochemical indices used for the interpretation of palaeoredox conditions in ancient mudstones. Chemical Geology, 111 (111): 111-129.

Keith M L, Weber J N. 1964. Carbon and oxygen isotopic composition of selected limestones and fossils. Geochimica Et Cosmochimica Acta, 28 (10-11): 1787-1816.

Kübler B. 1964. Les argiles, indicateurs de mè tamorphisme. Revue Institutė de la Francais de Pétrole, 19: 1093-1112.

Lermanm, A. 1978. Lakes: Chemistry, Geology, Physics. Berlin: Springer-Verlag: 79-83.

Mount J F. 1984. Mixing of siliciclastic and carbonate sediments in shallow shelf environments. Geology, 12 (7): 432-435.

Nesbitt H W, Young G M. 1982. Early proterozoic climates and plate motions inferred from major element chemistry of lutites. Nature, 299 (5885): 715-717.

Reitz A, Pfeifer K, Lange G J D, Klump J. 2004. Biogenic barium and the detrital ba/al ratio: a comparison of their direct and indirect determination. Marine Geology, 204 (3-4): 289-300.

Rimmer S M. 2004. Geochemical paleoredox indicators in Devonian-Mississippian black shales, Central Appalachian Basin (USA). Chemical Geology, 206 (3-4): 373-391.

Roy D K, Roser B P. 2013. Climatic control on the composition of Carboniferous-Permian Gondwana sediments, Khalaspir basin, Bangladesh. Gondwana Research, 23 (3): 1163-1171.

Rudnick R L, Gao S. 2003, Composition of the Continental Crust. Treatise on Geochemistry 3: 1-64.

Shields G, Stille P. 2001. Diagenetic constraints on the use of cerium anomalies as palaeoseawater redox proxies: an isotopic and REE study of Cambrian phosphorites. Chemical Geology, 175 (1-2): 29-48.

Tang Yanjie, Jia Jianye, Xie Xiande. 2002. Environment significance of clay minerals. Earth Science Frontiers, 9 (2): 337-344 (in Chinese with English abstract).

Taylor S R, Mclennan S M. 1985. The Continental Crust: Its Composition and Evolution. Blackwell, London, 57-72.

Tirsgaard H. 1996. Cyclic sedimentation of carbonate and siliciclastic deposits on a Late Precambrian ramp: The Elisabeth Bjerg Formation (Eleonore Bay supergroup), East Greenland. Journal of Sedimentary Research, 66 (4): 699-712.

Tribovillard N, Algeo T J, Lyons T, Riboulleau A. 2006. Trace metals as paleoredox and paleoproductivity proxies: An update. Chemical Geology, 232 (1-2): 12-32.

Wei Chengzhang, Li Zhongchun, Wu Changji. 1999. Characteristics of the fractured reservoir of condensate gas pool of nanyishan oil field in chaidamu basin. Natural Gas Industry, 19 (4): 5-7 (in Chinese with English abstract).

Xu Wei, Chen Kaiyuan, Cao Zhenglin, Xue Jianxun, Xiao Peng, Wang Wentao. 2014. Original mechanism of mixed sediments in the saline lacustrine basin. Acta Petrologica Sinica, 30 (6): 1804-1816 (in Chinese with English abstract).

Yang Chaoqing, Sha Qingan. 1990. Sedimentary environment of the middle Devonian Qujing Formation, Qujing, Yunnan province: a kind of mixing sedimentation of terrigenous clastics and carbonate. Acta Sedimentologica Sinica, 8 (2): 59-66 (in Chinese with English abstract).

Yuan Jianying, Huang Chenggang, Xia Qingsong, Cao Zhenglin, Zhao Fan, Wan Chuanzhi, Pan Xing. 2016. The characteristics of carbonate reservoir, and formation mechanism of pores in the saline lacustrine basin: a case study of the lower eocene ganchaigou formation in western qaidam basin. Geological Review, 62 (1): 111-126 (in Chinese with English abstract).

Yuan Jianying, Huang Chenggang, Cao Zhenglin, LI Zhiyong, WAN Chuangzhi, XU Li, PAN Xing, WU Lirong. 2015. Carbon and oxygen isotopic composition of saline lacustrine dolomite and its palaeoenvironmental significance: A case study of Lower Eocene Ganchaigou Formation in western Qaidam Basin. Geochimica, 44 (3): 254-266 (in Chinese with English abstract).

Zhao Ming, Chen Xiaoming, Ji Junfeng, ZHangyun, Zhangzhe. 2006. Diagenetic and paleogeothermal evolution of the clay minerals in the Paleogene Changwei prototype basin of Shandong province, China. Acta Petrologica Sinica, 22 (8): 2195-2204 (in Chinese with English abstract).

Zhang Hucai, Zhang Wenxiang, Chang Fengqin, Yang Lunqing, Lei Guoliang, Yang Mingsheng, Pu Yang, Lei Yanbin. 2009. Geochemical fractionation of rare earth elements in lacustrine deposits from Qaidam Basin. Sci Chin Ser D-Earth Sci, 39 (8): 1160-1168 (in Chinese without English abstract).

Zhang Jinliang, Si Xueqiang. 2007. Mixed siliciclastic—carbonate sediment in rift lacustrine basin-a case on the upper part of the fourth member of the eogene shahejie formation in jinjia area, dongying depression. Geological Review, 53 (4): 448-453 (in Chinese with English abstract).

Zhang Ningsheng, Ren Xiaojuan, Wei Jinxing, Kang Youxin, Zhang Cunhou. 2006. Rock types of mixosedimentite reservoirs and oil-gas distribution in Nanyishan of Qaidam basin. Acta Petrolei Sinica, 27 (1): 42-46 (in Chinese with English abstract).

Zhang Tianfu, Sun Lixin, Zhang Yun, Cheng Yinhang, Li Yanfeng. 2016. Geochemical characteristics of the Jurassic yan' man and Zhiluo formations in the northern margin of Ordos basin and their paleoenvironmental implications. Acta Geologica Sinica, 90 (12): 3454-3472 (in Chinese with English abstract).

Zhang Xionghua. 2000. Classification and origin of mixosedimentite. Geological Scienceand Technology Information, 19 (4): 31-34 (in Chinese with English abstract).

参考文献

常华进, 储雪蕾, 冯连君, 黄晶, 张启锐. 2009. 氧化还原敏感微量元素对古海洋沉积环境的指示意义. 地质论评, 55 (1): 91-99.

党玉琪, 尹成明, 赵东升. 2004. 柴达木盆地西部地区古近纪与新近纪沉积相. 古地理学报, 6 (3): 297-306.

邓宏文, 钱凯. 1993. 沉积地球化学与环境分析. 甘肃: 甘肃科学技术出版社, 1-54.

董桂玉, 陈洪德, 何幼斌, 秦志勇, 罗进雄, 辛长静. 2007. 陆源碎屑与碳酸盐混合沉积研究中的几点思考. 地球科学进展, 22 (9): 931-939.

冯进来, 胡凯, 曹剑, 陈琰, 王龙刚, 张英, 王牧, 赵健. 2011. 陆源碎屑与碳酸盐混积岩及其油气地质意义. 高校地质学报, 17 (2): 297-307.

冯进来, 曹剑, 胡凯, 陈琰, 杨少勇, 刘云田, 边立曾, 张国卿. 2011. 柴达木盆地中深层混积岩储层形成机制. 岩石学报, 27 (8): 2461-2472.

冯杨伟, 姜亭, 宋博, 牛亚卓. 2017. 中哈边境伊犁地区中二叠统沉积环境的地球化学判别. 地质学报, 91 (4): 942-953.

付锁堂, 张道伟, 薛建勤, 张晓宝. 2013. 柴达木盆地致密油形成的地质条件及勘探潜力分析. 沉积学报, 31 (4): 672-682.

甘贵元, 魏成章, 常青萍, 严晓兰, 崔俊, 王爱民, 陈登钱. 2002. 柴达木盆地南翼山湖相碳酸盐岩油气藏特征及形成条件. 石油实验地质, 24 (5): 413-417.

黄思静, 石和, 毛晓冬, 张萌, 沈立成, 武文慧. 2003. 早古生代海相碳酸盐的成岩蚀变性及其对海水信息的保存性. 成都理工大学学报 (自科版), 30 (1), 9-18.

汤艳杰, 贾建业, 谢先德. 2002. 黏土矿物的环境意义. 地学前缘, 9 (2): 337-344.

魏成章, 李忠春, 吴昌吉. 1999. 柴达木盆地南翼山凝析气藏裂缝性储层特征. 天然气工业, 19 (4): 5-7.

徐伟, 陈开远, 曹正林, 薛建勤, 肖鹏, 王文涛. 2014. 咸化湖盆混积岩成因机理研究. 岩石学报, 30 (6): 1804-1816.

杨朝青, 沙庆安. 1990. 云南曲靖中泥盆统曲靖组的沉积环境一种陆源碎屑. 沉积学报, 8 (2): 59-66.

袁剑英, 黄成刚, 夏青松, 曹正林, 赵凡, 万传治, 潘星. 2016. 咸化湖盆碳酸盐岩储层特征及孔隙形成机理——以柴西地区始新统下干柴沟组为例. 地质论评, 62 (1): 111-126.

袁剑英, 黄成刚, 曹正林, 李智勇, 万传治, 徐丽, 潘星, 吴丽荣. 2015. 咸化湖盆白云岩碳氧同位素特征及古环境意义: 以柴西地区始新统下干柴沟组为例. 地球化学, 44 (3): 254-266.

赵明, 陈小明, 季峻峰, 张耘, 张哲. 2006. 山东昌潍古近系原型盆地黏土矿物的成岩演化与古地温. 岩石学报, 22 (8): 2195-2204.

张虎才, 张文翔, 常凤琴, 杨伦庆, 雷国良, 杨明生, 蒲阳, 类延斌. 2009. 稀土元素在湖相沉积中的地球化学分异——以柴达木盆地贝壳堤剖面为例. 中国科学: 地球科学, 39 (8): 1160-1168.

张金亮, 司学强. 2007. 断陷湖盆碳酸盐与陆源碎屑混合沉积——以东营凹陷金家地区古近系沙河街组第四段上亚段为例. 地质论评, 53 (4): 448-453.

张宁生, 任晓娟, 魏金星, 康有新, 张存厚. 2006. 柴达木盆地南翼山混积岩储层岩石类型及其与油气分布的关系. 石油学报, 27 (1): 42-46.

张天福, 孙立新, 张云, 程银行, 李艳峰, 马海林, 鲁超, 杨才, 郭根万. 2016. 鄂尔多斯盆地北缘侏罗纪延安组、直罗组泥岩微量、稀土元素地球化学特征及其古沉积环境意义. 地质学报, 90 (12): 3454-3472.

张雄华. 2000. 混积岩的分类和成因. 地质科技情报, 19 (4): 31-34.

Petrological characteristics and sedimentary environment of the surface mixed rocks in Nanyishan structure, western Qaidam basin

Dongdong Yu[1], Yongsheng Zhang[1], Enyuan Xing[1], Zhifeng Zuo[2], Xianhua Hou[1], Linlin Wang[1] and Weiyong Zhao[3]

1. MLR Key Laboratory of Saline Lake Resources and Environments, Institute of Mineral Resources, Chinese Academy of Geological Sciences, Beijing 100037
2. Division of Exploration, Qinghai Oilfield Company, PetroChina, Xi'an, Shaanxi 751500
3. Research Institute of Exploration and Development, Qinghai Oilfield Company, PetroChina, Dunhuang, Gansu 736202

Abstract Mixed rocks were developed in the Shizigou Formation of Nanyishan Structure surface. As a special type of the sedimentary rocks, its research has some practical significance and scientific value. Based on field observation and microscopic identification, this paper analyzed petrological characteristics of the mixed rock in Nanyishan using X-ray diffraction, major and trace elements, carbon and oxygen isotope test of mixed rock. Considering the complexity of compositions and genesis, this paper employed the effective indicators to determine the sedimentary environment plus excluding a variety of interference factors through data analysis. The results show that the average contents of terrigenous clastics and carbonate minerals in the mixed rocks are 58.7% and 31.0%, respectively, implying that the mixed rocks belong to carbonatic terrigenous clastic rock. Among them, the terrigenous clastic rock is dominantly calcareous mudstones, with detrital minerals consisting mainly of fine silt grade quartz; the carbonate minerals are mainly micritic calcite and clay minerals assemblages are illite and ordered illite/smectite mixed layer; and above three rokcs are mixed homogeneously. Data analysis shows that elements (such as Ca, Na, Mn, Sr, Ba et al.) mainly derived from authigenic minerals. REE, carbon and oxygen isotopic compositions, which were not affected by diagenesis, can be used to determine the salinity and redox conditions of waters during deposition. Combined with assemblage of clay minerals, crystallinity of illite, and chemical index, Sr/Cu and K_2O/Al_2O_3 ratios suggest a cold and arid climate for the depositional period of the mixed rocks; Sr/Ba ratio and Z value indicate a salt water environment; and U/Th ratio, authigenic U and Ce negative anomaly synthetically show a weak oxidizing environment. The results above show that the mixed rocks in Nanyishan are dominantly calcareous mudstones consisting of terrigenous clastic and carbonate minerals, which formed in weak oxidizing saline lacustrine basin under the conditions of cold and dry climate.

Keywords Mixed rocks; Petrological characteristics; Geochemistry; Sedimentary environment; Nanyishan structure; Western Qaidam Basin

柴西南翼山构造上新统狮子沟组混积岩地球化学特征及物源指示意义*

于冬冬[1]，张永生[1]，邢恩袁[1]，王琳霖[1]，李凯[2]，黄围霖[3]

1. 中国地质科学院矿产资源研究所国土资源部盐湖资源与环境重点实验室，北京 100037
2. 中国地质大学（北京）地球科学与资源学院，北京 100083
3. 中国石油青海油田分公司勘探开发研究院，甘肃 敦煌 736202

摘要 柴西地区南翼山构造上新统混积岩中含有大量盐类矿物，其物源分析对盐矿物质来源研究有一定的参考价值。本文在野外观察和镜下鉴定的基础上，对南翼山构造上新统狮子沟组混积岩的地球化学特征进行了分析，并结合周缘造山带花岗岩的稀土元素数据，探讨了混积岩中陆源碎屑的物质来源。结果显示：南翼山构造狮子沟组混积岩中碎屑矿物、碳酸盐矿物和黏土矿物呈均匀混合的状态，元素含量与矿物组分密切相关，除 Ca、Na、Sr 元素外，其余元素含量普遍低于上地壳平均含量，其中 CaO 和 Na_2O 受碳酸盐和石盐形成影响，不能用于判别构造背景，而 Sc、Th、Zr、Hf 等微量元素和稀土元素较完整地保留了源岩的地球化学信息，可用于源区构造背景的判别及物源示踪，同时稀土元素配分模式一致性较好，表明混积岩中陆源碎屑具有相同的物源特征。稀土元素配分模式、La-Th-Sc、Th-Sc-Zr/10、La/Th-Hf 图解以及 Al_2O_3/TiO_2 比值共同指示南翼山混积岩的陆源碎屑组分来源于大陆岛弧背景下的长英质火成岩，通过与周缘造山带花岗岩稀土配分模式对比，推断晚志留世和晚二叠世花岗岩是混积岩中陆源碎屑和盐类矿物的主要源岩，南部的祁漫塔格是主要的物源区，且西部的阿尔金南段也有部分物源贡献。

关键词 混积岩；地球化学；构造背景；物源；南翼山构造；柴达木西部

混积岩是介于碎屑岩和碳酸盐岩之间的一种过渡类型沉积岩，在古代和现代沉积环境中都相当常见[1-3]。由于混积岩在油气运移和储集方面发挥着重要作用，其研究价值逐渐受到重视，近 20 年来的研究工作取得了令人瞩目的成果，主要集中在：（1）混积岩的分类命名[4-7]；（2）混积岩的沉积特征、沉积模式和成因机理[8-11]；（3）混积岩型储层特征与油气地质意义[12-16]。混积岩通常由同一沉积环境中外来陆源碎屑和盆内碳酸盐颗粒相互作用产生，沉积过程不同、海平面和气候变化都会导致不同比例和规模的混合，比纯碎屑岩或碳酸盐岩具有更加复杂的沉积模式[3]，所以混积岩的研究需要采取多学科结合的方法。前期研究主要从混积岩的岩石学特征出发，研究其沉积环境、形成原因和油气成藏作用[8-16]。地球化学方法近年来才开始应用到混积岩的研究中，由于混积岩成分的复杂性，常规地球化学指标并不能直接用于混积岩，在进行物源示踪或沉积环境分析之前必须进行化学元素有效性的检验，目前多采用相关性分析判断化学元素来源于陆源碎屑还是自生矿物[17, 18]。利用此方法，孙林华等[19]和伏美燕等[20]分别进行过混积岩的物源示踪和古环境分析，但总体上相关研究报道仍较少，因此本文有望对混积岩的地球化学研究提供一定的借鉴意义。

新生代时期南翼山构造处于湖相沉积环境，咸水湖泊中广泛发育混积岩，其中古近系和新近系混积岩作为油气良好的储层一直是前期研究的重点[14-16]，但近地表的上新统狮子沟组混积岩很少有人关注。近年来，柴达木西部新生代地层中发现了富钾锂硼卤水资源[21]，查明钾锂硼的物质来源对下一步找矿突破具有重要意义，研究工作亟待开展。南翼山构造混积岩中伴生有大量盐类矿物，作为柴达木西部卤水开发的重点区域，混积岩的物源分析有望成为摸清卤水物质来源的突破口。本文以野外观察和镜下鉴定为基础，结合地球化学测试，分析了南翼山狮子沟组混积岩的地球化学特征

* 本文发表在：地学前缘，2018，25（4）：65-75

和物源属性，以期为柴达木西部上新世以来盐矿的物质来源研究和咸水湖盆混积岩的地球化学研究提供一定的参考。

1　地质背景及样品采集

南翼山构造位于柴达木盆地西部，为一大型箱状背斜，整体呈 NW 向展布，背斜核部出露上新统狮子沟组，两翼被第四系覆盖，地表以风蚀残丘为主（图 1）[22]。南翼山沉积了约 6km 厚的新生代地层，其中狮子沟组为一套滨浅湖相沉积，混合沉积特征明显，地层中多见石盐和石膏夹层[23, 24]。南翼山构造形成演化受柴西地区周缘造山带活动控制，上新世时期随着阿尔金山和东昆仑山的隆升，柴西地区湖盆面积不断萎缩，沉积中心从一里坪地区逐渐迁至柴东三湖地区[25]，此时南翼山构造开始隆起，发育背斜雏形[26]。第四纪受阿尔金断裂强烈左行运动以及东昆仑和祁连山向盆内俯冲的影响，NE-SW 向挤压应力进一步加强[27]，南翼山形成“两断夹一隆”的背斜，同时不断遭受风化剥蚀，最终呈现了现今的构造形态[26]。

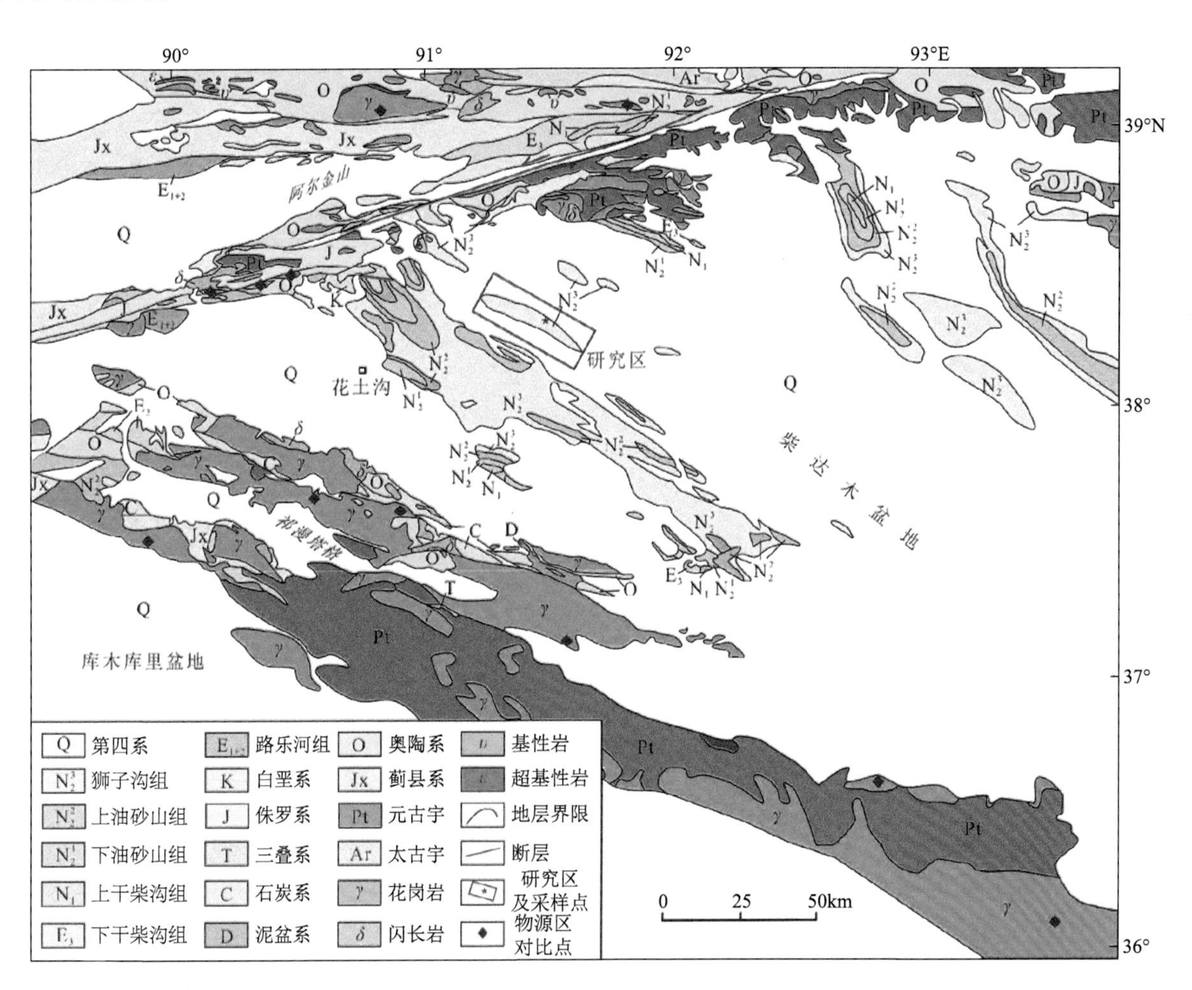

图 1　柴西地区地质简图（据文献[22]修改）

Fig. 1　Schematic geological map of the western Qaidam Basin.

样品采集自南翼山构造狮子沟组地层，距地表 2～4m，包括 3 个采样点，共涉及 13 件样品（图 1）。样品以棕色块状为主，夹有薄层石盐，粒度较细，切面光滑，遇酸起泡剧烈[图 2（a）]。在显微镜下可见到明显的混合沉积特征，其中碎屑矿物主要为细粉砂级石英颗粒，碳酸盐矿物以泥晶方解石为主，二者与黏土矿物均匀混合在一起[图 2（b）]。X 衍射分析结果表明，黏土矿物主要为伊利石和伊蒙混层，含有少量绿泥石，碎屑矿物、碳酸盐矿物和黏土矿物大致呈 1∶1.26∶1.38 的比例混合，为典型的混积岩（图 3）。按照前人的分类方案[7]，将碎屑矿物和黏土矿物统一归为陆源碎屑，其中陆源碎屑占 55.00%～78.90%，平均为 63.61%，碳酸盐矿物占 14.90%～42.90%，平均为 32.77%，南翼山狮子沟组混积岩属于碳酸盐质陆源碎屑岩，主要为钙质泥岩。

图 2　南翼山构造狮子沟组混积岩岩性特征

Fig. 2　Lithological characteristics of mixed rocks from the Shizigou Formation in the Nanyishan structure.

(a) 棕色混积岩夹薄层石盐，镜头方向 305°；(b) 混积岩，碎屑矿物、泥晶方解石和黏土矿物均匀混合，正交偏光×100

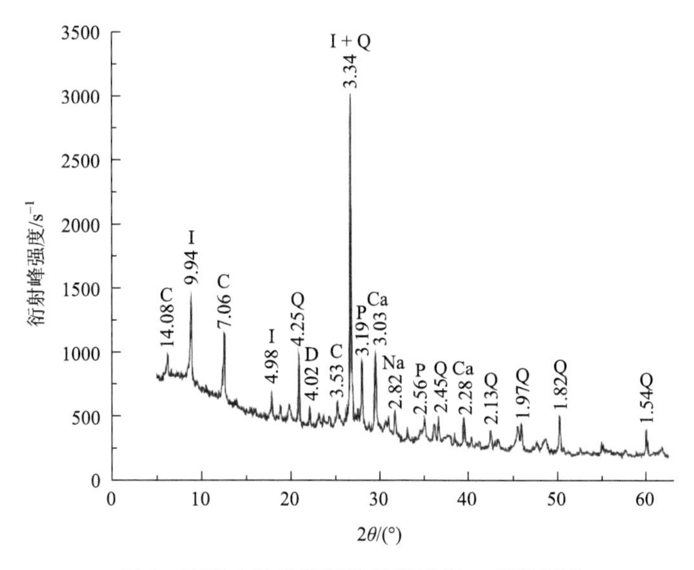

图 3　南翼山构造狮子沟组混积岩 X 衍射图谱

Fig. 3　X-ray diffraction spectrum of mixed rocks from the Shizigou Formation in the Nanyishan structure.

C-绿泥石；I-伊利石；Q-石英；Na-石盐；P-斜长石；Ca-方解石；D-白云石

2　测试方法及结果

样品测试工作委托核工业北京地质研究所分析测试中心完成，包括全岩主量、微量和稀土元素分析，其中主量元素在 Axios^{mAX} 型 X 射线荧光光谱仪上进行测试，微量和稀土元素在 NexION300D 等离子体质谱仪上进行测试，执行标准分别为 GB/T 14506.28—2010 和 GB/T 14506.30—2010，分析误差均＜5%。

2.1　主、微量元素

主量和微量元素测试数据见表 1。混积岩样品的主要化学成分为 SiO_2、CaO、Na_2O、Al_2O_3、MgO、K_2O、FeO 和 Fe_2O_3，平均含量分别为 28.19%、14.54%、13.91%、8.16%、2.82%、1.90%、1.77%和 1.51%，MnO、TiO_2 和 P_2O_5 含量低于 1%。烧失量（LOI）介于 18.90%～35.07%，表明样品中可能含有较多的有机质和碳酸盐。与上地壳（UCC）氧化物含量相比[28]，CaO 明显富集，表明样品中方解石含量较高，与前文分析结果一致。含盐混积岩中 Na_2O 含量很高，表明其与石盐的形成有关。其余元素呈略微亏损的状态[图 4（a）]。微量元素分析结果显示，大部分元素平均含量相较于上地壳平均含量轻微亏损，只有 Sr 元素富集现象十分明显，其含量为（279～3372）$\times10^{-6}$，平均为 1021×10^{-6}[图 4（b）]。

表 1　南翼山构造狮子沟组混积岩样品主量元素和微量元素测试分析结果

Table 1　Results of Major and trace elemental analyses of mixed rocks from the Shizigou Formation in the Nanyishan structure

样品号	岩性	ω_B/%											
		SiO_2	Al_2O_3	Fe_2O_3	MgO	CaO	Na_2O	K_2O	MnO	TiO_2	P_2O_5	FeO	LOI
N-11	含盐混积岩	16.25	3.78	0.89	3.43	9.36	36.4	1.34	0.024	0.214	0.053	0.76	
N-12	含盐混积岩	16.57	4.19	1.32	2.96	8.13	36.98	1.45	0.023	0.217	0.047	1.12	
N-13-1	混积岩	31.24	9.78	2.08	3.25	16.3	6.15	2.13	0.099	0.392	0.049	2.28	25.8
N-14-1	混积岩	36.35	10.75	2.06	3.1	15.87	4.34	2.32	0.096	0.474	0.04	2.35	22.16
N-14-2	混积岩	33.73	9.31	1.23	2.75	14.55	6.03	1.98	0.072	0.405	0.056	2.31	26.13
N-14-3	混积岩	38.64	11.96	1.97	3.07	13.8	3.8	2.58	0.072	0.506	0.08	2.55	20.87
N-20	含盐混积岩	14.28	3.44	0.17	1.28	6	34.56	0.62	0.024	0.133	0.12	0.94	
N-21-1	含盐混积岩	18.68	5.7	0.77	2.02	9.62	25.47	1.12	0.033	0.199	0.101	1.34	
N-21-2	混积岩	12.58	3.71	0.86	1.5	38.16	1.55	0.86	0.044	0.17	0.113	1.42	35.07
N-21-3	混积岩	27.38	7.06	1.5	4.09	16.52	15.68	2.48	0.057	0.297	0.116	1.34	
N-21-5	混积岩	38.35	11.32	2.41	2.82	14.66	3.9	2.44	0.078	0.478	0.116	1.95	21.42
N-22	混积岩	38.9	12.18	2.04	2.92	14.84	3.61	2.67	0.078	0.505	0.119	2.13	19.96
N-34	混积岩	43.47	12.88	2.28	3.52	11.26	2.37	2.75	0.075	0.561	0.119	2.57	18.90
样品号	岩性	$w_B/10^{-6}$											
		Sc	V	Cr	Co	Ni	Rb	Sr	Ba	Th	U	Zr	Hf
N-11	含盐混积岩	8.8	37.9	23.1	4.57	12.8	34.5	559	230	3.44	1.4	50.04	1.46
N-12	含盐混积岩	6.83	43.9	26.6	5.75	14.6	40.6	736	181	4.4	1.88	57.24	1.62
N-13-1	混积岩	10.6	72.8	57.7	13.1	31.2	88.8	3372	968	9.05	3.53	116.82	3.45
N-14-1	混积岩	12	82.9	66.3	14.3	36.2	98.6	1043	453	10.7	3.91	134.28	3.98
N-14-2	混积岩	11.3	69.4	54	11.5	30.1	85.2	632	374	8.44	2.91	103.86	3.3
N-14-3	混积岩	13.6	90.2	71.4	15.5	38.3	111	788	495	11.8	4.04	147.24	4.61
N-20	含盐混积岩	12.2	37.9	22.1	3.79	11	32.5	279	173	3.57	1.26	45.9	1.41
N-21-1	含盐混积岩	10.6	50.9	35.3	6.64	18.8	55.8	551	299	5.76	2.43	72	2.13
N-21-2	混积岩	5.28	29.4	22.2	7.43	25.7	34.6	2478	473	3.84	3.64	53.28	1.53
N-21-3	混积岩	10.1	53.8	42.7	9.39	25.6	64.1	734	367	6.9	3.22	94.68	2.81
N-21-5	混积岩	12.3	82.8	65.6	12.1	32.7	102	592	503	10.6	4.04	145.26	4.37
N-22	混积岩	13.2	92.4	72.1	14.6	37.6	112	640	553	11.5	4.63	156.96	4.54
N-34	混积岩	13.7	98.1	74.6	15.1	38.3	113	869	438	11.5	3.36	158.76	4.81

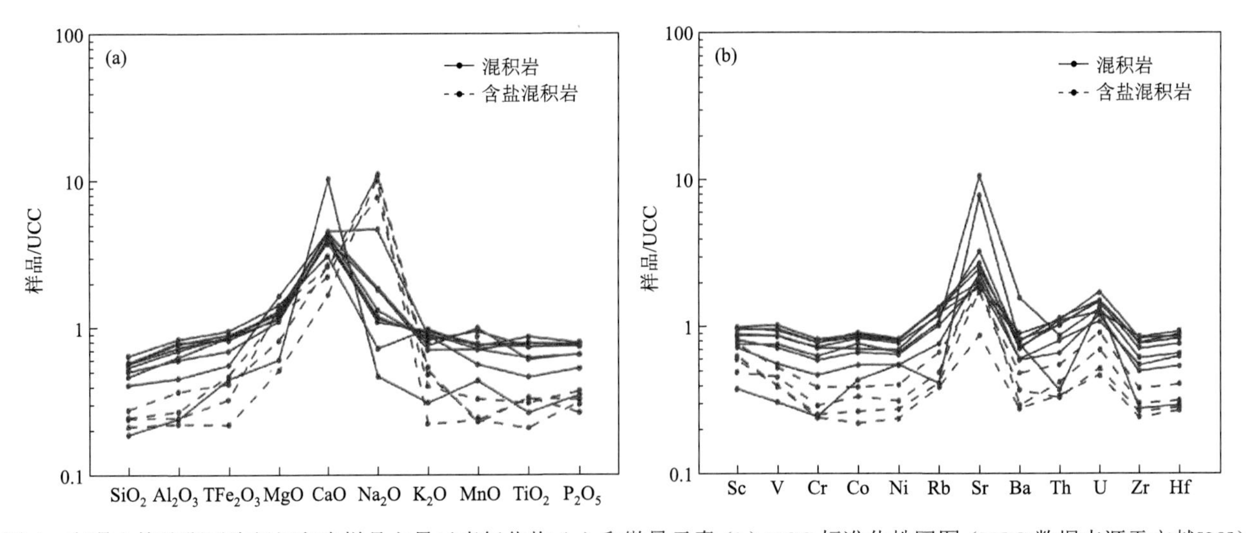

图 4　南翼山构造狮子沟组混积岩样品主量元素氧化物（a）和微量元素（b）UCC 标准化蛛网图（UCC 数据来源于文献[28]）

Fig. 4　Spider diagrams of UCC-nomalized major (a) and trace elements (b) of mixed rocks from the Shizigou Formation in the Nanyishan structure.

2.2 稀土元素

稀土元素测试分析结果见表 2。根据球粒陨石含量进行标准化处理[29]，得到稀土元素的配分模式，并与 UCC、北美页岩（NASC）和澳大利亚后太古宙页岩（PAAS）稀土元素的配分模式进行对比[28-30]（图 5）。Ce 和 Eu 异常计算公式分别为 $\delta Ce = Ce_N/(0.5La_N + 0.5Pr_N)$和 $\delta Eu = Eu_N/(0.5Sm_N + 0.5Gd_N)$。结果表明：混积岩稀土元素总量∑REE 为（53.09～141.75）$\times 10^{-6}$，平均为 115.48$\times 10^{-6}$（不包括含盐混积岩），低于 UCC 的稀土总量（148.14$\times 10^{-6}$）[28]、NASC 的稀土总量（173.21$\times 10^{-6}$）[30]和 PAAS 的稀土总量（184.76$\times 10^{-6}$）[29]；轻重稀土元素比值 LREE/HREE 平均为 8.31，$(La/Yb)_N$ 平均为 8.14，样品中轻重稀土分异明显。δCe 值介于 0.91～0.93，平均为 0.92，具有轻微的负异常。δEu 值介于 0.58～0.67，平均为 0.62，低于 NASC（0.68）、PAAS（0.63）和 UCC（0.69），具有较明显的负异常。整体上，混积岩稀土元素配分模式一致性较好，呈左高右低、整体平缓的趋势，略低于 NASC 和 PAAS，与 UCC 的配分模式较接近，表明所有样品的物源具有相似的特征。

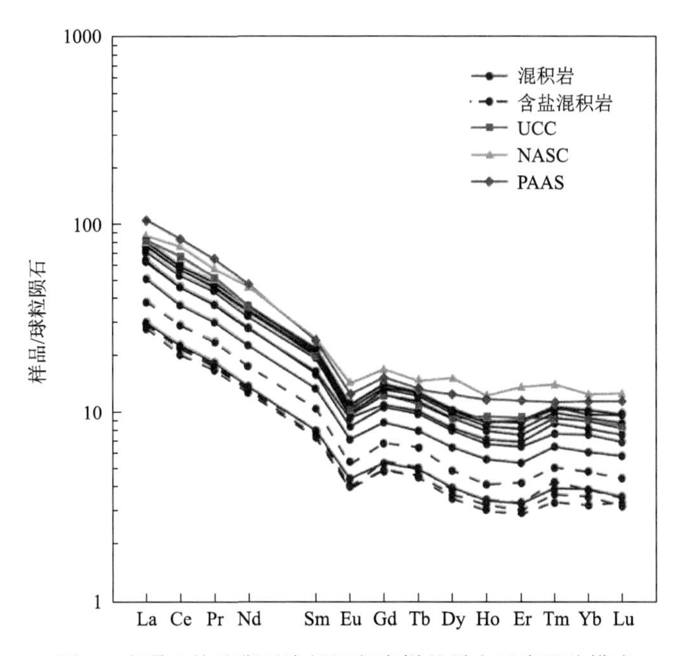

图 5 南翼山构造狮子沟组混积岩样品稀土元素配分模式

（UCC 数据来源于文献[28]；PAAS 数据来源于文献[29]；NASC 数据来源于文献[30]）

Fig. 5 Chondrite-nomalized REE patterns of mixed rocks from the Shizigou Formation in the Nanyishan structure.

表 2 南翼山构造狮子沟组混积岩样品稀土元素测试分析和计算结果

Table 2 Results of analysis and calculation of rare earth elements in mixed rocks from the Shizigou Formation in the Nanyishan structure

样品号	岩性	$w_B/10^{-6}$										
		La	Ce	Pr	Nd	Sm	Eu	Gd	Tb	Dy	Ho	Er
N-11	含盐混积岩	11.1	21.50	2.45	9.32	1.73	0.34	1.51	0.27	1.39	0.27	0.75
N-12	含盐混积岩	10.8	20.90	2.4	9.46	1.87	0.34	1.67	0.29	1.47	0.29	0.82
N-13-1	混积岩	23.7	44.50	5.11	19.8	3.82	0.81	3.35	0.59	3.13	0.61	1.72
N-14-1	混积岩	28.3	54.50	6.28	24.6	4.67	0.84	4.05	0.71	3.75	0.71	2.01
N-14-2	混积岩	23.2	44.20	5.16	20	3.73	0.73	3.26	0.57	3.05	0.58	1.63
N-14-3	混积岩	30.1	57.50	6.70	26.2	5.03	0.95	4.33	0.75	3.98	0.76	2.19
N-20	含盐混积岩	10.3	19.50	2.29	9.05	1.69	0.34	1.50	0.26	1.32	0.25	0.72

续表

样品号	岩性	$w_B/10^{-6}$										
		La	Ce	Pr	Nd	Sm	Eu	Gd	Tb	Dy	Ho	Er
N-21-1	含盐混积岩	14.3	27.70	3.24	12.5	2.42	0.46	2.09	0.38	1.86	0.35	1.05
N-21-2	混积岩	11.1	21.70	2.52	9.76	1.85	0.38	1.64	0.29	1.51	0.29	0.81
N-21-3	混积岩	18.8	35.60	4.13	16.1	3.11	0.62	2.69	0.46	2.48	0.48	1.34
N-21-5	混积岩	27.1	51.80	6.14	23.5	4.51	0.87	3.97	0.68	3.53	0.72	1.96
N-22	混积岩	28.7	55	6.41	24.9	4.84	0.91	4.14	0.74	3.86	0.75	2.12
N-34	混积岩	30	57.7	6.74	26.3	4.95	0.95	4.30	0.73	3.88	0.76	2.18
样品号	岩性	$\omega_B/10^{-6}$							LREE/HREE	$(La/Yb)_N$	δCe	δEu
		Tm	Yb	Lu	Y	∑REE	LREE	HREE				
N-11	含盐混积岩	0.13	0.88	0.12	7.6	51.76	46.44	5.32	8.73	8.52	0.93	0.63
N-12	含盐混积岩	0.15	0.96	0.13	8.2	51.57	45.77	5.79	7.90	7.59	0.93	0.58
N-13-1	混积岩	0.31	2.02	0.29	17.4	109.75	97.74	12.01	8.14	7.93	0.91	0.67
N-14-1	混积岩	0.35	2.31	0.33	20	133.42	119.19	14.23	8.38	8.28	0.93	0.58
N-14-2	混积岩	0.27	1.88	0.26	16.2	108.52	97.02	11.50	8.44	8.34	0.92	0.62
N-14-3	混积岩	0.38	2.40	0.37	21.9	141.62	126.48	15.15	8.35	8.48	0.92	0.60
N-20	含盐混积岩	0.12	0.79	0.13	7.25	48.26	43.17	5.08	8.49	8.78	0.91	0.65
N-21-1	含盐混积岩	0.18	1.19	0.17	10.1	67.89	60.62	7.27	8.34	8.12	0.92	0.61
N-21-2	混积岩	0.14	0.96	0.14	8.33	53.09	47.31	5.77	8.2	7.78	0.93	0.66
N-21-3	混积岩	0.23	1.52	0.22	13.7	87.78	78.36	9.42	8.32	8.36	0.91	0.64
N-21-5	混积岩	0.35	2.29	0.32	19.9	127.74	113.92	13.82	8.25	8.00	0.91	0.62
N-22	混积岩	0.38	2.52	0.37	21.3	135.63	120.76	14.87	8.12	7.70	0.92	0.61
N-34	混积岩	0.37	2.53	0.37	20.6	141.75	126.64	15.11	8.38	8.01	0.92	0.61

3　讨论

3.1　构造背景

构造背景限定了沉积盆地的物源属性，是进行物源分析的基础，目前地球化学元素分析被广泛应用于构造背景和物源的讨论[31, 32]。沉积物中 Al、Ti 含量常作为陆源碎屑贡献量的指标[33]，混积岩样品的 Al_2O_3 含量与 TiO_2、SiO_2、Sc、Th、Zr 等呈现了很强的正相关性，说明它们主要来自陆源碎屑；而与 CaO 和 Na_2O 呈现不相关或负相关性，说明 CaO 主要与碳酸盐沉积有关，Na_2O 主要来自石盐。另外，稀土元素总量∑REE 与 Al_2O_3 也表现出强正相关性，而 δCe 与∑REE、δEu 无相关性，说明稀土元素主要来源于陆源碎屑，且未受后期成岩作用的影响（图 6）。综上可以看出，混积岩中 Cao 和 Na_2O 并不适合判断构造背景，而 Sc、Th、Zr、Hf 等微量元素和稀土元素与 Al_2O_3 呈良好的正相关性，表明受成岩作用影响较弱，较完整地保留了源区岩石的地球化学信息，可以用来判断源区的构造背景以及进行物源示踪。

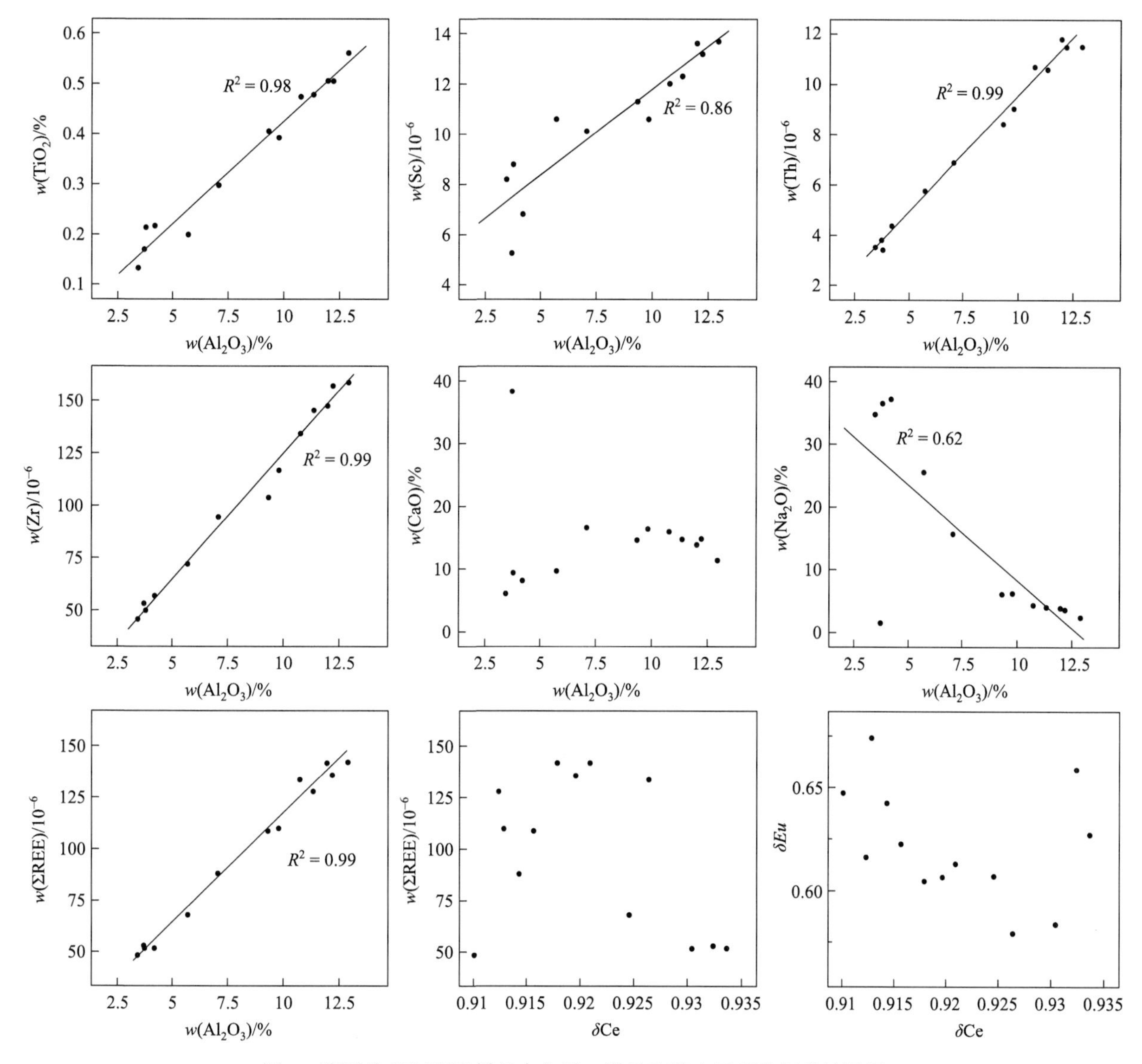

图 6 狮子沟组混积岩样品中主量、微量和稀土元素的相关性图解

Fig. 6 Correlations of major, trace elements and rare earth elements in mixed rocks from the Shizigou Formation.

Bhatia[31]归纳总结出大洋岛弧、大陆岛弧、活动大陆边缘和被动大陆边缘 4 类构造环境，利用 La-Th-Sc、Th-Co-Zr/10、Th-Sc-Zr/10 和 La/Sc-Ti/Zr 等图解可以进行判别[32]。南翼山狮子沟组混积岩样品在 La-Th-Sc 和 Th-Sc-Zr/10 图解中比较一致，几乎全部落在大陆岛弧区域中（图 7）。同时，混积岩样品的稀土元素特征值也更接近大陆岛弧（表 3），所以认为南翼山构造地表混积岩的陆源碎屑组分主要来自大陆岛弧环境中。

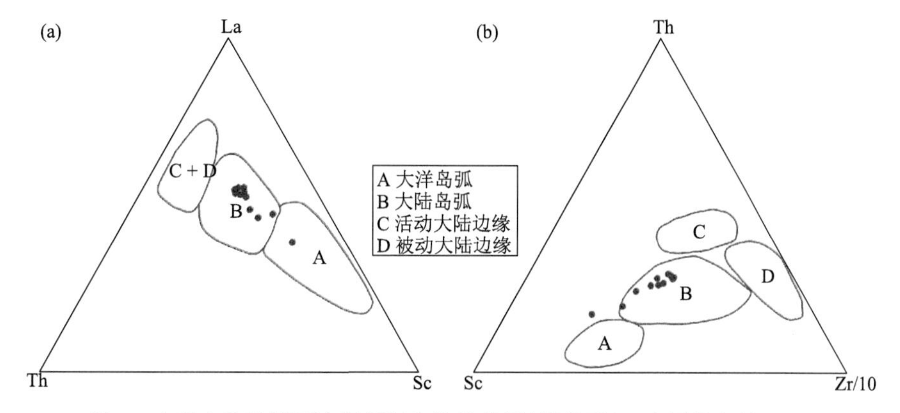

图 7 南翼山构造狮子沟组混积岩构造背景判别图解（底图据文献[32]）

Fig. 7 Discrimination diagrams for tectonic settings of mixed rocks from the Shizigou Formation in the Nanyishan structure.

表 3　南翼山构造狮子沟组混积岩样品与不同构造背景杂砂岩稀土元素对比表[31]

Table 3　Comparison of REE characteristics of mixed rocks from the Shizigou Formation in the Nanyishan structure and graywackes under different tectonic settings[31]

构造背景	ω(La)/10^{-6}	ω(Ce)/10^{-6}	δEu	$\omega(\sum REE)/10^{-6}$	LREE/HREE	$(La/Yb)_N$
大洋岛弧	8±1.7	19±3.7	1.04±0.11	58±10	3.8±0.9	2.8±0.9
大陆岛弧	27±4.5	59±8.2	0.79±0.13	146±20	7.7±1.7	7.5±2.5
主动大陆边缘	37	78	0.6	186	9.1	8.5
被动大陆边缘	39	85	0.56	210	8.5	10.8
南翼山混积岩	24.56	46.94	0.62	115.48	8.31	8.14

3.2　物源确定

稀土元素的分布模式可以反映盆地沉积物的物源性质，通常轻稀土富集、重稀土亏损和 Eu 负异常的特征暗示源岩主要为长英质岩，混积岩样品中的稀土元素配分模式正是呈现此特征。利用 La/Th-Hf 图解进一步判断源岩属性[34]，在 La/Th-Hf 图中，混积岩样品主要落在长英质岛弧源区，表明源岩以长英质火成岩为主（图 8）。另外，主量元素氧化物也能提供一定的物源信息，前人研究认为碎屑岩的 Al_2O_3/TiO_2 的比值可以用来判断源岩类型[35]。相关性分析表明，混积岩样品中 Al_2O_3 和 TiO_2 主要来源于陆源碎屑，可以作为物源判别指标，其 Al_2O_3/TiO_2 比值介于 17.67～28.64，平均为 23.24，指示研究区的源岩可能主要是长英质火成岩[36]。

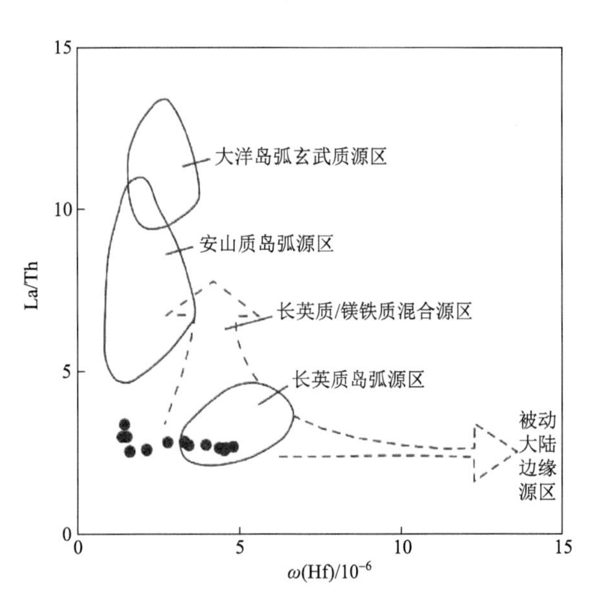

图 8　南翼山构造狮子沟组混积岩 La/Th-Hf 图解（底图据文献[34]）

Fig. 8　La/Th-Hf diagram of mixed rocks from the Shizigou Formation in the Nanyishan structure.

由此可见，南翼山构造狮子沟组混积岩的源区可能为大陆岛弧环境，陆源碎屑物质主要来自花岗岩。南翼山构造西部为阿尔金造山带，南部为东昆仑造山带，上新世末期受青藏高原隆升的影响，阿尔金和东昆仑造山带都比较活跃，这些造山带的隆升剥蚀为盆内沉积提供了物源[37]。前期地质调查发现，阿尔金山和东昆仑山遍布有大量花岗岩，从早古生代至早中生代均有发育[38]，显然想要准确地确定混积岩的物源区，必须对柴西及周缘地区古生代和中生代的构造背景加以分析。研究表明，震旦纪柴达木地块及邻区开始形成大陆裂谷盆地，至寒武纪已发育成熟的洋盆。奥陶纪开始发生俯冲消减作用，至晚志留世—早泥盆世，洋盆最终闭合[38, 39]，在此过程中形成了诸多火山岛弧，包括北祁连—北阿尔金火山岛弧、柴北缘—南阿尔金火山岛弧、祁漫塔格火山岛弧和中昆仑火山岛弧等，早古生代阿尔金—祁连—东昆仑地区处于多地体、多洋盆、多岛弧的古构造环境中[40]。早石炭世东昆仑地区开始发育新的洋盆，早二叠世松潘—甘孜洋向昆仑—柴达木地块下俯冲，在柴达木南缘形成陆缘弧和弧后盆地[41]，中三叠世东昆仑地区完成由洋壳俯冲到碰撞造山的转换，之后进入陆内演化阶段[40, 42]。因此，奥陶纪—早泥盆世和早二叠世—中三叠世形成的花岗岩有可能是研究区的源岩。

在上述基础上，收集了阿尔金和祁漫塔格早古生代至早中生代花岗岩的稀土元素数据，具体包括阿尔金北段喀腊大湾早奥陶世花岗岩和阿北中志留世花岗岩[43, 44]，阿尔金南段茫崖早奥陶世花岗岩、柴水沟早泥盆世花岗岩和阿克提山晚二叠世花岗岩[45, 46]，祁漫塔格西段十字沟晚志留世花岗岩、阿雅克库木湖北晚二叠世花岗岩和豹子沟早三叠世花岗岩[47, 48]，祁漫塔格东段那棱郭勒河晚志留世花岗岩、向阳沟晚二叠世花岗岩和虎头崖中三叠世花岗岩[42, 49, 50]（图 1）。将它们稀土元素平均值的配分模式与南翼山混积岩进行对比，发现混积岩样品的稀土配分模式与晚志留世和晚二叠世花岗岩相似性最好，而它们主要分布在研究区南部的祁漫塔格西段和东段，部分分布在西部的阿尔金南段（图 9）。岩石地球化学特征显示，晚志留世花岗岩形成于后碰撞阶段，但同时具有大陆岛弧性质[47, 49]，前人通过“滞后型弧火山岩”这一概念进行了合理解释，即后碰撞阶段岩浆原岩可能是由俯冲阶段产生而储存于下地壳或软流圈的原岩产生的[51]。晚二叠世花岗岩为典型的洋壳俯冲形成的大陆岛弧花岗岩[42, 46]，与构造背景分析结果相符。由此推断南翼山混积岩中陆源碎屑主要来自晚志留世和晚二叠世花岗岩，其中南部的祁漫塔格是主要的物源区，也有部分陆源碎屑来自于西部的阿尔金南段。

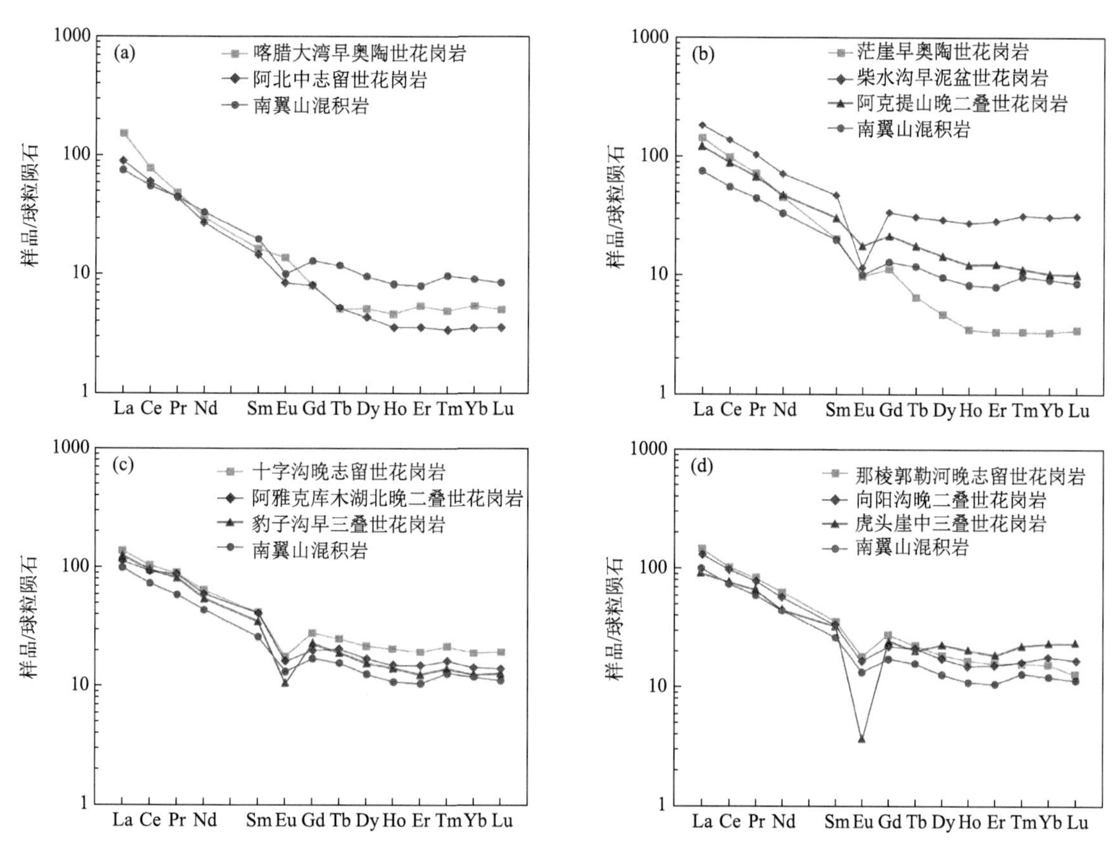

图 9　南翼山构造狮子沟组混积岩及物源区花岗岩稀土元素配分模式

Fig. 9　Chondrite-nomalized REE patterns in mixed rocks from the Shizigou Formation in the Nanyishan structure and in granites from source regions.

(a) 阿尔金北段；(b) 阿尔金南段；(c) 祁漫塔格西段；(d) 祁漫塔格东段

南翼山狮子沟组混积岩中盐类矿物主要出现在层间裂缝中，呈层状展布，其上下的混积岩隔水性较好，且断裂和裂缝发育较少[图 2（a）]，表明卤水并非来源于大气降水渗流和深部卤水的上涌，而是封存于层间的沉积卤水，其与混积岩的陆源碎屑具有相同的物源。

4　结论

（1）南翼山构造狮子沟组混积岩为碳酸盐质陆源碎屑岩，其中碎屑矿物、碳酸盐矿物和黏土矿物大致呈 1∶1.26∶1.38 的比例均匀混合。相较于上地壳平均含量，混积岩中 Ca、Na、Sr 元素明显富集，其余元素普遍相对亏损。稀土元素具有总量较低、轻稀土富集、重稀土亏损、Ce 和 Eu 负异常的特点，其配分模式左高右低、整体平缓，呈现了稳定的一致性，表明南翼山混积岩具有相同的物源特征。

（2）相关性分析表明，高含量氧化物 Cao 和 Na_2O 分别与碳酸盐和石盐的形成有关，不适合用于混积岩构造背景的判别，而 Sc、Th、Zr、Hf 等微量元素和稀土元素主要来源于陆源碎屑，且未受成岩作用影响，保留了源岩的地球化学信息，可以作为源区构造背景判断及物源示踪的指标。

（3）La-Th-Sc 和 Th-Sc-Zr/10 图解表明南翼山混积岩的陆源碎屑组分来源于大陆岛弧环境中，稀土元素配分模式、La/Th-Hf 图解以及 Al_2O_3/TiO_2 比值共同指示混积岩的源岩为长英质火成岩。对比周缘造山带酸性花岗岩的稀土元素配分模式，发现与晚志留世和晚二叠世花岗岩相似性最好，其主要分布在祁漫塔格，部分在阿尔金南段。混积岩中陆源碎屑和盐类矿物具有相同的物源，研究区南部的祁漫塔格是主要的物源区，西部的阿尔金南段是次要物源区。

致谢

感谢中国石油青海油田分公司梅建森、高雪峰、袁伟等工程师在野外工作中提供的指导与支持，感谢河北区域地质矿产调查研究所和核工业北京地质研究所工作人员在样品加工和测试工作中的帮助，感谢评审专家和编辑在审稿过程中提出的宝贵意见。

参考文献

MOUNT J F. Mixing of siliciclastic and carbonate sediments in shallow shelf environments[J]. Geology, 1984, 12 (12): 432-435.

TIRSGAARD H. Cyclic sedimentation of carbonate and siliciclastic deposits on a Late Precambrian ramp: the Elisabeth Bjerg Formation (Eleonore Bay supergroup), East Greenland[J]. Journal of Sedimentary Research, 1996, 66 (4): 699-712.

CHIARELLA D, LONGHITANO S G, TROPEANO M. Types of mixing and heterogeneities in siliciclastic-carbonate sediments[J]. Marine and Petroleum Geology, 2017, 88: 617-627.

张雄华. 混积岩的分类和成因[J]. 地质科技情报, 2000, 19 (4): 31-34.

杨朝青, 沙庆安. 云南曲靖中泥盆统曲靖组的沉积环境: 一种陆源碎屑与海相碳酸盐的混合沉积[J]. 沉积学报, 1990, 8 (2): 59-66.

郭福生, 严兆彬, 杜杨松. 混合沉积、混积岩和混积层系的讨论[J]. 地学前缘, 2003, 10 (3): 68-68.

董桂玉, 陈洪德, 何幼斌, 等. 陆源碎屑与碳酸盐混合沉积研究中的几点思考[J]. 地球科学进展, 2007, 22 (9): 931-939.

张金亮, 司学强. 断陷湖盆碳酸盐与陆源碎屑混合沉积: 以东营凹陷金家地区古近系沙河街组第四段上亚段为例[J]. 地质论评, 2007, 53 (4): 448-453.

陈世悦, 张顺, 刘惠民, 等. 湖相深水细粒物质的混合沉积作用探讨[J]. 古地理学报, 2017, 19 (2): 271-284.

赵会民. 辽河西部凹陷雷家地区古近系沙四段混合沉积特征研究[J]. 沉积学报, 2012, 30 (2): 283-290.

徐伟, 陈开远, 曹正林, 等. 咸化湖盆混积岩成因机理研究[J]. 岩石学报, 2014, 30 (6): 1804-1816.

MCNEILL D F, CUNNINGHAM K J, GUERTIN L A, et al. Depositional themes of mixed carbonate-siliciclastics in the South Florida Neogene: application to ancient deposits [J]. AAPG Memoir, 2004, 80: 23-43.

BARNABY R J, WARD W B. Outcrop analog for mixed siliciclastic-carbonate ramp reservoirs-stratigraphic hierarchy, facies architecture, and geologic heterogeneity: Grayburg Formation, Permian Basin, U.S.A.[J]. Journal of Sedimentary Research, 2007, 77 (1): 34-58.

张宁生, 任晓娟, 魏金星, 等. 柴达木盆地南翼山混积岩储层岩石类型及其与油气分布的关系[J]. 石油学报, 2006, 27 (1): 42-46.

冯进来, 曹剑, 胡凯, 等. 柴达木盆地中深层混积岩储层形成机制[J]. 岩石学报, 2011, 27 (8): 2461-2472.

FENG J L, CAO J, HU K, et al. Dissolution and its impacts on reservoir formation in moderately to deeply buried strata of mixed siliciclastic-carbonate sediments, Northwestern Qaidam Basin, Northwest China[J]. Marine and Petroleum Geology, 2013, 39 (1): 124-137.

邵磊, 刘志伟, 朱伟林. 陆源碎屑岩地球化学在盆地分析中的应用[J]. 地学前缘, 2000, 7 (3): 297-304.

常华进, 储雪蕾, 冯连君, 等. 氧化还原敏感微量元素对古海洋沉积环境的指示意义[J]. 地质论评, 2009, 55 (1): 91-99.

孙林华, 桂和荣, 陈松, 等. 皖北新元古代贾园组混积岩物源和构造背景的地球化学示踪[J]. 地球学报, 2010, 31 (6): 833-842.

伏美燕, 李娜, 黄茜, 等. 滨岸浅海混合沉积对海平面与气候变化的响应: 以塔里木盆地巴麦地区石炭系为例[J]. 沉积学报, 2017, 35 (6): 1110-1120.

郑绵平, 张雪飞, 侯献华, 等. 青藏高原晚新生代湖泊地质环境与成盐成藏作用[J]. 地球学报, 2013 (2): 129-138.

陈宣华, 党玉琪, 尹安, 等. 柴达木盆地及其周缘山系盆山耦合与构造演化[M]. 北京: 地质出版社, 2010: 1-437.

甘贵元, 魏成章, 常青萍, 等. 柴达木盆地南翼山湖相碳酸盐岩油气藏特征及形成条件[J]. 石油实验地质, 2002, 24 (5): 413-417.

党玉琪, 尹成明, 赵东升. 柴达木盆地西部地区古近纪与新近纪沉积相[J]. 古地理学报, 2004, 6 (3): 297-306.

付锁堂, 张道伟, 薛建勤, 等. 柴达木盆地致密油形成的地质条件及勘探潜力分析[J]. 沉积学报, 2013, 31 (4): 672-682.

于冬冬, 张永生, 侯献华, 等. 柴西南翼山构造形成演化及其对油气成藏的控制[J]. 断块油气田, 2017, 24 (6): 740-744.

张涛, 宋春晖, 王亚东, 等. 柴达木盆地西部地区晚新生代构造变形及其意义[J]. 地学前缘, 2012, 19 (5): 312-321.

RUDNIK R L, GAO S. Composition of the continental crust [M]//Treatise on geochemistry. Amsterdam: Elsevier, 2003, 3: 1-64.

TAYLOR S R, MCLENNAN S M. The continental crust: its composition and evolution[M]. Oxford: Blackwell, 1985: 57-72.

HASKIN L A, HASKIN M A, FREY F A, et al. Relative and absolute terrestrial abundances of the rare earths[M]//AHRENS L H. Origin and distribution of the elements. Oxford: Pergamon, 1968: 889-911.

BHATIA M R. Rare earth element geochemistry of Australian Paleozoic graywackes and mudrocks: provenance and tectonic control[J]. Sedimentary Geology,

1985, 45 (1): 97-113.
BHATIA M R, CROOK K A W. Trace element characteristics of greywackes and tectonic setting discrimination of sedimentary basins[J]. Contributions to Mineralogy and Petrology, 1986, 92 (2): 181-193.
贺子丁, 刘志飞, 李建如, 等. 南海西部 54 万年以来元素地球化学记录及其反映的古环境演变[J]. 地球科学进展, 2012, 27 (3): 327-336.
FLOYD P A, LEVERIDGE B E. Tectonic environment of the Devonian Gramscatho Basin, south Cornwall: framework mode and geochemical evidence from turbidite sandstones[J]. Journal of the Geological Society, 1987, 144 (4): 531-542.
HAYASHI K I, FUJISAWA H, HOLLAND H D, et al. Geochemistry of approximately 1.9 Ga sedimentary rocks from northeastern Labrador, Canada[J]. Geochimica et Cosmochimica Acta, 1997, 61 (19): 4115-4137.
KESKIN S. Geochemistry of Çamardı Formation sediments, central Anatolia (Turkey): implication of source area weathering, provenance, and tectonic setting[J]. Geosciences Journal, 2011, 15 (2): 185-195.
关平, 简星. 青藏高原北部新生代构造演化在柴达木盆地中的沉积记录[J]. 沉积学报. 2013, 31 (5): 824-833.
许志琴, 杨经绥, 李海兵, 等. 造山的高原: 青藏高原的地体拼合、碰撞造山及隆升机制[M]. 北京: 地质出版社, 2007: 37-202.
汤良杰, 金之钧, 张明利, 等. 柴达木震旦纪—三叠纪盆地演化研究[J]. 地质科学, 1999, 34 (3): 289-300.
许志琴, 杨经绥, 李海兵, 等. 青藏高原与大陆动力学: 地体拼合、碰撞造山及高原隆升的深部驱动力[J]. 中国地质, 2006, 33 (2): 221-238.
赵文津, 吴珍汉, 史大年, 等. 昆仑山深部结构与造山机制[J]. 中国地质, 2014, 41 (1): 1-18.
王秉璋, 陈静, 罗照华, 等. 东昆仑祁漫塔格东段晚二叠世—早侏罗世侵入岩岩石组合时空分布、构造环境的讨论[J]. 岩石学报, 2014, 30 (11): 3213-3228.
吴玉, 陈正乐, 陈柏林, 等. 北阿尔金喀腊大湾南段二长花岗岩地球化学、SHRIMP 锆石 U-Pb 年代学、Hf 同位素特征及其对壳-幔相互作用的指示[J]. 地质学报, 2017, 91 (6): 1227-1244.
孟令通, 陈柏林, 王永, 等. 北阿尔金早古生代构造体制转换的时限: 来自花岗岩的证据[J]. 大地构造与成矿学, 2016, 40 (2): 295-307.
康磊, 校培喜, 高晓峰, 等. 茫崖二长花岗岩、石英闪长岩的年代学、地球化学及岩石成因: 对阿尔金南缘早古生代构造-岩浆演化的启示[J]. 岩石学报, 2016, 32 (6): 1731-1748.
吴才来, 郜源红, 雷敏, 等. 南阿尔金茫崖地区花岗岩类锆石 SHRIMP U-Pb 定年、Lu-Hf 同位素特征及岩石成因[J]. 岩石学报, 2014, 30 (8): 2297-2323.
王秉璋. 祁漫塔格地质走廊域古生代—中生代火成岩岩石构造组合研究[D]. 北京: 中国地质大学 (北京), 2012: 71-201.
李光明, 沈远超, 刘铁兵. 东昆仑祁漫塔格地区华力西期花岗岩地质地球化学特征[J]. 地质与勘探, 2001, 37 (1): 73-78.
郝娜娜, 袁万明, 张爱奎, 等. 东昆仑祁漫塔格晚志留世—早泥盆世花岗岩: 年代学、地球化学及形成环境[J]. 地质论评, 2014, 60 (1): 201-215.
李侃, 高永宝, 钱兵, 等. 东昆仑祁漫塔格虎头崖铅锌多金属矿区花岗岩年代学、地球化学及 Hf 同位素特征[J]. 中国地质, 2015, 42 (3): 630-645.
李伍平, 路凤香. 钙碱性火山岩构造背景的研究进展[J]. 地质科技情报, 1999, 18 (2): 15-18.

Geochemical Characteristics and implication for provenance of mixed rocks from the Pliocene Shizi gou Formation in the Nanyishan structure of the western Qaidam Basin

Dongdong Yu[1], Yongsheng Zhang[1], Enyuan Xing[1], Linlin Wang[1], Kai Li[2] and Youlin Huang[3]

1. MLR Key Laboratory of Saline Lake Resources and Environments，Institute of Mineral Resources，Chinese Academy of Geological Sciences, Beijing 100037, China
2. School of Earth Sciences and Resources, China University of Geosciences (Beijing), Beijing 100083, China
3. Research Institute of Exploration and Development，Petro China Qinghai Oilfield Company, Dunhuang 736202, China

Abstract The Pliocene mixed rocks contain large amounts of salt minerals in the Nanyishan structure of the western Qaidam Basin, and their provenance analyses can be valuable in finding the origin of the salt minerals. Based on field observation and microscopic identification, we analyzed the geochemical characteristics of mixed rocks from the Pliocene Shizigou Formation in the Nanyishan structure. Combined with rare earth element (REE) characteristics of granites from the surrounding orogenic belts, we discussed the provenance of terrigenous clastics in mixed rocks. The results showed that the detrital, carbonate and clay minerals in mixed rocks were homogeneously mixed, and the elemental contents of the rocks were closely related to the mineral composition. Except for Ca, Na, and Sr, the mixed rocks had generally lower mineral contents than those of the upper continental crust. Considering that CaO and Na_2O are affected by the formation of carbonate and halite, Ca and Na are not suitable for identifying tectonic settings. In contrast, trace elements and REE such as Sc, Th, Co, Zr and Hf, retain mostly the geochemical information of source rocks and therefore can be used to evaluate the tectonic settings of the source area for provenance analysis. The REE patterns showed good consistency, indicating the terrigenous clastics of mixed rocks have the same provenance characteristics. Together with La-Th-Sc, Th-Sc-Zr/10, La/Th-Hf diagrams and Al_2O_3/TiO_2 ratio, our data suggest that the terrigenous clastics components of mixed rocks were derived from the felsic igneous rocks of continental island arc. In comparison with REE patterns in granites from the surrounding orogenic belts, we consider the Late Silurian and Late Permian granites to be the main source rocks for terrigenous clastics and salt minerals in the mixed rock. In addition, we hypothesize that the south part of the study area, Qimantag, is the main source area, and South Altyn Tagh in the west also takes part in the sedimentary source recharge.

Keywords Mixed rocks; Geochemistry; Tectonic setting; Provenance; Nanyishan structure; Western Qaidam Basin

青藏高原南部地热型锂资源*

王晨光[1]，郑绵平[1]，张雪飞[1]，叶传永[1]，伍倩[1]，陈双双[3]，黎明明[1,2]，丁涛[1,2]，杜少荣[1]

1. 中国地质科学院地质矿产资源研究所，自然资源部盐湖资源与环境重点实验室，北京 100037
2. 中国矿业大学（北京）地球科学与测绘工程学院，北京 100083
3. 中山大学地球科学与工程学院，广东省地球动力作用与地质灾害重点实验室，广州 510275

摘要 高温地热水中含有丰富的锂资源，世界各国对其中锂资源的开发利用的研究越来越多。对青藏高原丰富的地热资源中富锂地热资源进行了综述，得出其具有以下特点：（1）构造控制强烈，地热型锂资源主要分布在雅江缝合带两侧及其南部地区强烈活动的高温地热田中，受到沿近NS向正断层发育的裂谷或地堑盆地的强烈控制；（2）品质好，锂含量可高达239mg/L；mg/Li非常低，多数富锂地热系统Mg/Li介于0.03～1.48；Li/TDS相对较高且介于0.25%～1.14%（扎布耶富锂盐湖为0.19%；玻利维亚乌尤尼盐湖为0.08%～0.31%）；持续稳定排放数十年，部分达到工业品位（32.74mg/L）；伴生可以综合利用的B、Cs和Rb元素等；（3）规模大，据不完全统计，当前锂含量达到或超过19mg/L的富锂温泉至少有19处，年排出金属锂约4281t，折合碳酸锂 25686t，并且最新钻孔数据表明地热田深部潜力巨大；（4）属于非火山型，火山岩缺失；（5）深部来源成因，富锂地热系统的形成与印度和欧亚大陆碰撞引起的地壳深部部分熔融密切相关，深部熔融岩浆为富锂地热系统提供了稳定的热源，富锂的母地热流体沿着青藏高原南部广泛发育的断裂带上涌至地表形成高温富锂热泉。由此，认为青藏高原南部广泛发育的高温富锂地热资源是一种非常有价值、值得开发利用的地热型锂资源，随着地热水中锂提取技术的不断提升，青藏高原南部地热型锂资源有望成为一种可有效开发利用的锂矿床新类型——地热型锂矿床。

关键词 青藏高原南部；地热；锂资源

近几年，随着电动汽车的推广和普及，世界锂资源的需求不断增长，面对锂价格的不断上涨[1,2]，全球范围内锂矿的勘查保持持续的活跃态势[3]。富锂地热水中的锂资源越来越受到关注[4,5]，有学者甚至指出地热水中的锂资源量可达2Mt[6,7]。盐湖卤水锂为世界锂的需求贡献了3/4的资源量，多数学者都认为地热水是富锂盐湖卤水中锂的重要来源[4,8-16]。因此，许多学者开展了从富锂地热水中提取和利用锂原料的研究工作，认为富锂地热水将会成为满足锂资源需求不断上涨的有效途径之一[17-21]。中国西藏地热资源丰富，研究历史悠久，针对富锂地热资源的研究较早。1960年，郑绵平就在西藏的班戈湖南部发现了富锂热泉，并对其成因进行过初步研究，此外，还对羊八井地热田中的锂资源量进行了初步估算，获得锂资源量39万t[22]。近几十年，西藏地热资源中的热能得到了有效的开发和利用，但对其中的锂资源研究与利用没有给予足够的重视。随着中国经济产业转型，高新技术产业的迅猛发展，对锂资源的需求量正在逐年快速增加，加之国际锂需求市场的不断扩大和富锂的矿物原料的不足，从任何富锂水中提取锂都是有效的解决办法[23]。因此，在近年国内外锂资源的勘探、开发和市场动态变化的背景下，为了提高中国西藏高温富锂地热田中锂资源的研究、开发和利用程度，郑绵平等基于其几十年对青藏高原盐湖资源评价工作经验[14,16,24,25]及对世界范围内其他高温富锂地热田开发利用现状的了解，提出了地热水也是锂矿的构想，并针对这一构想开展了有针对性的野外和室内研究工作，在这些工作的基础上提出地热型锂矿资源的概念，认为随着提取技术的不断进步，地热型锂资源有希望成为一种新的矿床成因类型。

本文通过对前人相关研究资料的吸收和总结[13,22,26-41]，加上野外重点地区（古堆等地热田）的调查研究，对青藏高原南部富锂地热资源的分布、规模、成因类型和开发利用前景作一简要汇总。

* 本文发表在：科技导报，2020，38（15）：24-36

1　青藏高原南部高温富锂地热资源特点

青藏高原南部西藏高温富锂地热资源多分布在雅鲁藏布江缝合带两侧及南北向裂谷两侧主断裂与次级断裂的交叉部位（图 1），与世界其他高温富锂地热田相比，周围地层火山岩缺失，地表泉口温度多高于 70℃，接近当地水的沸腾温度，多属于高温非火山型地热田，并且受到小规模次级断裂的强烈控制，如羊八井地热田[35]、古堆地热田(图 2)[42]。天然水中锂含量非常低，世界主要河流中锂的含量不到 0.01×10^{-6}[42]，古堆地区普通冷泉水锂含量为 0.017×10^{-6}[43]。中国发布的盐湖卤水矿最低工业品位为 LiCl 200mg/L（折合

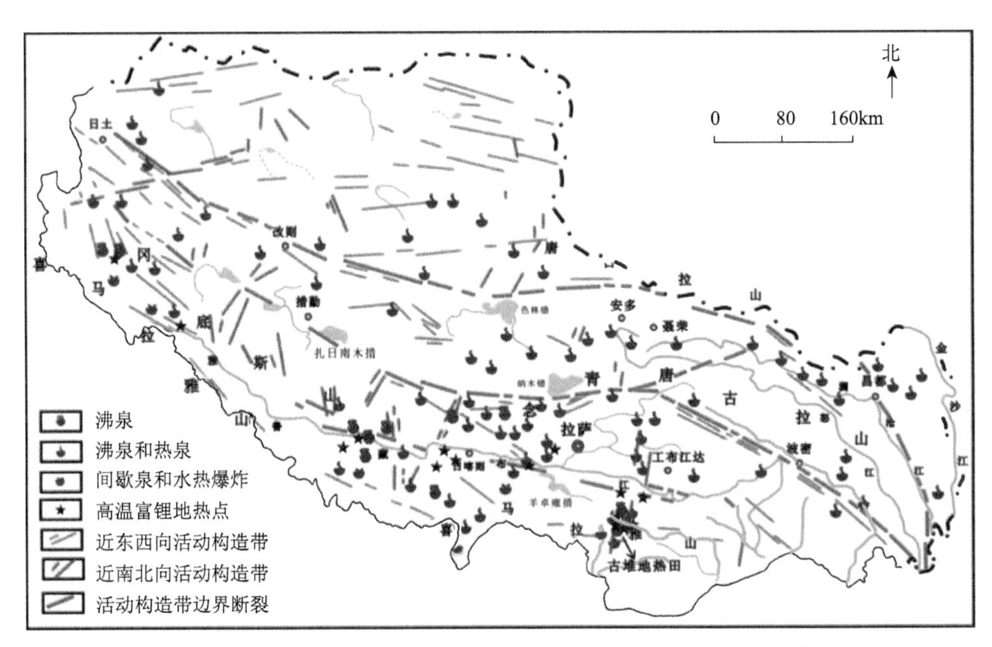

图 1　青藏高原南部西藏热泉点及富锂地热系统（点）分布[25]

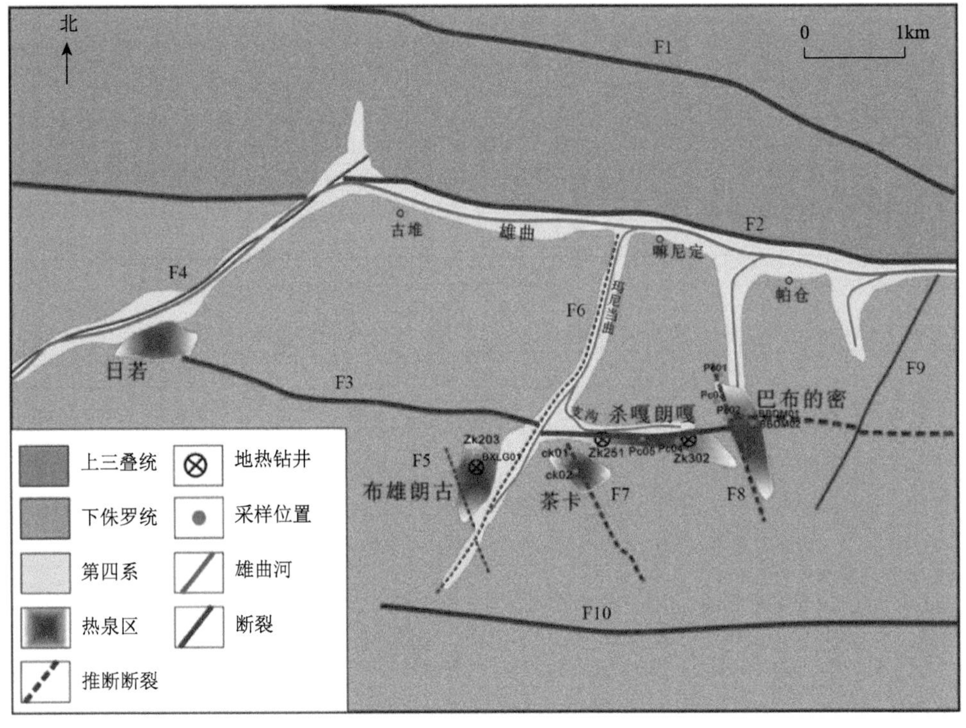

图 2　西藏古堆地热田地质简图及样品分布[40, 41, 43, 44]

Li 离子 32.74mg/L），边界品位为 150mg/L（折合 Li 离子 24.56mg/L）。相比之下，青藏高原南部部分地热水中锂含量非常之高，部分达到可利用的工业品位，如碱海子热泉锂离子含量高达 239mg/L（考察发现其有后期蒸发富集作用发生），竹墨沙热泉 Li 离子含量高达 65.40mg/L（野外考察发现其发育于雅鲁藏布江边，雨季江水多淹没泉区），莫落江沸泉 Li 离子含量高达 50mg/L 等（表 1）。此外，中国青藏高原南部高温富锂地热水多具有 Mg/Li 非常低，介于 0.03～1.48，平均值为 0.43 的特点。这种特征非常有利于工业化开采利用，虽然大部分高锂地热水的总矿化度（total dissolved solids，TDS）都不高，但其 Li/TDS 却并不低，有的高达 1.14%，而西藏最富锂的盐湖扎布耶富锂盐湖 Li/TDS 也不过为 0.19%，世界范围内最富锂的玻利维亚乌尤尼（Uyuni）盐湖的 Li/TDS 也不过为 0.08%～0.31%，因此，青藏高原南部西藏高温富锂地热水中锂含量所占比例非常高，并且据郑绵平等[14]、佟伟等[20]报道，青藏高原南部大部分的富锂地热水都数十年稳定的排出（图 3），具有稳定的热源和锂源，但至今，多数富锂地热水资源都汇入地表径流，造成很大的浪费。此外，近些年施工的羊八井钻孔深部地热水锂含量具有明显增高的趋势，表明在高温富锂地热田深部有希望找到更加富锂的地热资源。

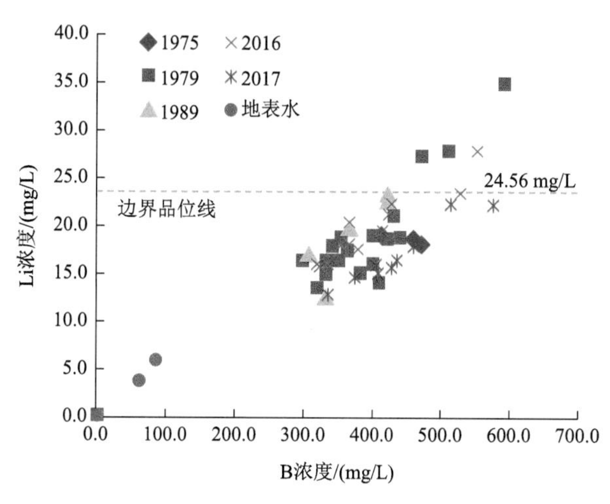

图 3　西藏古堆地热田 1975～2017 年硼锂变化趋势及关系图[37, 40, 43]

从 1956 年起，许多学者先后在西藏开展了大范围的地热地质调查，并发现雅鲁藏布地热带地热水富含 B、Li、Cs、Rb 等元素[25]。1995 年，郑绵平通过对西藏地热田泉华的研究中发现了泉华中铯的异常富集，并在随后系统的研究后提出了西藏水热成矿新类型——铯硅化矿床新类型[25]。通过总结前人的研究成果，发现富锂地热系统中也都同样发育有高浓度的 B、Cs、Rb 等可以综合利用的资源，无疑又增加了青藏高原南部高温富锂地热资源的开发利用价值。此外，青藏高原南部部分高温富锂地热水还大量排出 As（色米南岸 6.6mg/L[25, 38, 43]，最高纪录为青藏高原南部莫落江沸泉 126mg/L[37]）、Sb（色米 1.40mg/L[25, 38, 43]）、W（塔格架、色米和古堆的含量变化范围为 289.1～1103μg/L[31]）、Hg（0.006mg/L[25, 38, 43]）等有害元素，通过高温地热田这些有害元素地球化学特征和分布规律的研究，不仅可以有效控制有害元素对中国三江源头的污染，甚至可能将其综合利用起来。在不久的将来，高温富锂地热田中这些金属元素开发利用所产生的价值，有可能与地热田热能的开发利用产生的价值相媲美。

2　青藏高原南部西藏高温富锂地热资源的分布和规模

中国高温富锂地热资源主要分布在西藏地区，尤其以青藏高原南部地区最为丰富（图 1）。自 20 世纪 50 年代初期，西藏地矿局等先后在这些地区进行了大范围与地热地质有关或其他矿产地质调查研究之后，其大规模的地热活动越来越受到科学界的广泛重视，调查发现西藏的现代中、高温地热资源主要分布在青藏高原南部，泉华常见硅华，水化学类型以 Na-Cl 为主（表 1），西藏已发现各类地热显示区（点）近 700 处[37-41]，高温地热系统（温度≥150℃）共 57 处，中高温地热系统的地热水中普遍具有较高的锂含量，尤其是雅鲁藏布江两侧及青藏高原南部地热水富含 B、Cs、Li、Rb 等元素[25, 38, 43]，之后在对塔格架地热田深入研究中，郑绵平首次指出西藏多处地热水的 Cs、Li、Rb 和 B 的含量已达到单项综合利用指标（如昂仁县色米地热显示区温泉中 Li 含量高达 35mg/L[25, 38, 43]），并对其资源量做了初步的估算[16]。Grimaud 等[45]详细研究了西藏的 300 个热泉，并发现许多地热田产出富硼、锂和铯的地热水。尽管世界许多地热田的地热水中都富集有大量的锂元素，但在青藏高原，许多地热水显示出 B、Rb、Cs 等元素的异常富集达到了经济利用水平[45]。李振清[46]和雒洋冰等[13]认为西藏的热水活动与相应的活动构造基本一致，其中的青藏高原南部高温热水活动区从西到东又可以划分为当热雄错-古错水热带、申扎-定结水热带、那曲-亚东水热带和桑日-错那水热带（图 1），这与郑绵平等[14]关于西藏第四纪盆地及线性构造略图的表述基本一致。

表 1　青藏高原南部西藏富锂地热系统（点）特征

序号	温泉名称	样品编号	海拔/m	温度/℃	Mg/(mg/L)	Li/(mg/L)	Mg/Li	TDS/(mg/L)	Li/TDS/%	HBO_2/(mg/L)	Rb/(mg/L)	Cs/(mg/L)	主要水化学类型	参考文献
1	莫落江沸泉	—	4900	86	1.50	50.00	0.03	6240	0.80	1917.00	1.85	30.60	$Cl-CO_3-Na$	
2	多果曲热泉	—	4860	72	12.50	20.30	0.62	4330	0.47	1180.00	1.25	23.00	$HCO_3-Cl-Na$	
3	色米沸泉	—	4400	86	4.17	35.00	0.12	4340	0.81	1750.00	2.45	51.50	$Cl-CO_3-HCO_3-Na$	
4	拉布朗沸泉	—	4400	85	nd	23.75	—	4470	0.53	1965.00	2.10	58.00	$Cl-CO_3-HCO_3-Na$	
5	拉旺孜热泉	—	3910	63	7.28	21.30	0.34	2610	0.82	565.00	0.70	2.40	$Cl-HCO_3-Na-Ca$	
6	卡乌曲灿	—	4700	62	2.08	19.80	0.11	2496	0.79	500.00	3.62	27.20	$Cl-HCO_3-CO_3-Ca$	佟伟等[37]
7	查托岗温泉	—	4066	37	6.00	19.80	0.30	3280	0.60	550.00	2.90	23.80	—	
8	金噶曲灿	—	4462	57	14.40	23.80	0.61	2910	0.82	315.00	2.70	0.93	$Cl-HCO_3-Na-Ca$	
9	查巴曲珍	—	4000	41	10.40	27.80	0.37	4180	0.67	585.00	2.25	14.10	$Cl-CO_3-Na$	
11	竹墨沙热泉	—	3520	45	17.10	65.40	0.26	10611	0.62	2045.00	13.50	99.80	—	
12	日若沸泉	—	4440	10	11.20	35.00	0.32	3070	1.14	590.00	1.90	9.60	—	
16	碱海子	—	4100	na	34.38	239.00	0.14	95330	0.25	1478.00	6.95	0.53	$Cl-CO_3-Na$	
19	巴布的密沸泉	PC01	4449	74	3.8	16.1	0.24	2182.4	0.74	455.58368	0.7	3.4	Na-Cl	本文
20	巴布的密沸泉	PC02	4605	81	3.6	15.7	0.23	2206.4	0.71	481.88472	0.8	3.5	Na-Cl	本文
21	巴布的密沸泉	PC03	4605	70	3	16.5	0.18	2255.4	0.73	490.68936	0.8	3.6	Na-Cl	本文
27	巴布的密沸泉	BBDM01	4520	80	8.6	12.7	0.68	2316.6	0.55	376.79344	0.8	3.6	$Na-HCO_3-Cl$	本文
28	巴布的密沸泉	BBDM02	4520	20	9.2	14.6	0.63	2282.6	0.64	420.59088	0.7	2.6	Na-Cl	本文
22	杀嘎朗嘎沸泉	PC04	4645	85	3.4	18	0.19	2649.6	0.68	516.9904	0.9	3.4	Na-Cl	本文
23	杀嘎朗嘎沸泉	PC05	4630	84	3.6	22.2	0.16	2981.1	0.74	648.38272	1.2	5.2	Na-Cl	本文
16	杀嘎朗嘎地热钻井	ZK302	4630～4700	204	32.3	27.8	1.16	3324.9	0.84	621.06576	—	—	Na-Cl	王思琪[43]
17	杀嘎朗嘎地热钻井	ZK251	4630～4700	163	23.4	22.4	1.04	2798.3	0.80	480.75592	—	—	Na-Cl	王思琪[43]
24	布雄朗古沸泉	BXLG01	4488	83	4	22.2	0.18	2996.1	0.74	578.28424	0.9	3.9	Na-Cl	本文
18	布雄朗古地热钻井	ZK203	4630～4700	175.5	30.1	20.4	1.48	2902.5	0.70	412.12488	0	0	$Na-Cl-HCO_3$	王思琪[43]
25	茶卡热泉	CK01	4433	70	6.8	15	0.45	2472.8	0.61	459.986	0.8	3.6	$Na-HCO_3-Cl$	本文
26	茶卡热泉	CK02	4501	68	5.6	15.2	0.37	2303.3	0.66	459.986	0.8	3.3	Na-Cl	本文

而早在 1960 年，郑绵平就在对班戈湖南部含锂温泉的研究中，认为西藏各个地热带迄今仍有大量的 B、Li、K、F 携出，在此基础上，又对富锂温泉与火山、浅成岩浆的关系进行了探讨[47]。根据佟伟等[37]所编著的《西藏温泉志》对西藏高锂地热水进行了统计（表 1），根据其不完全统计，对有流量报道且锂含量达到或超过 19mg/L 的 19 处富锂地热系统 1 年的锂和硼排出量进行了计算，结果显示 1 年的锂排出量有 4281t 之巨，折合碳酸锂 25686t，硼排出量有 91882t 之多（表 2），这足以说明西藏高温富锂地热水的巨大开发利用潜力。国内外众多学者均认为富锂盐湖中的锂也是来自地热水[9, 17, 48]，这进一步说明了地热水中锂资源的重要性和推动地热锂资源研究、开发和利用的迫切性。此外，西藏多数高锂地热水常常伴生高含量的 Rb、Cs、B 等可以综合利用的资源（表 1），这将进一步提高地热型锂资源的开发利用价值。

表 2 青藏高原南部部分高温富锂地热系统 Li、B 资源量估算

序号	编号	温泉名称	海拔/m	涌水量/(L/s)	Li/(mg/L)	HBO_2/(mg/L)	锂排放/(t/a)	硼排放/(t/a)
1	ZGR10	莫落江沸泉	4900	1	50	1917	38	3023
2	ZBR8	多果曲热泉	4860	5	20.3	1180	77	3777
3	ZNR5	色米沸泉	4400	3	35	1750	79	5795
4	ZNR6	拉布朗沸泉	4400	—	23.75	1965	0	0
5	ZXTM11	拉旺孜热泉	3910	1.5	21.3	565	24	569
6	ZS'G1	卡乌曲灿	4700	20	19.8	500	300	6244
7	ZS'G2	查托岗温泉	4066	3	19.8	550	45	1030
8	ZGYZ1	金噶曲灿	4462	10	23.8	315	180	2364
9	ZRB2	查巴曲珍	4000	5	27.8	585	105	2564
10	ZNMA4	绒玛热泉	4900	—	20	—	0	0
11	ZNQ3	谷露间歇喷泉	4720	8	25.2	205	153	1303
12	ZDX14	羊八井地热田	4200	—	28.6	273	0	0
13	ZND1	卡布桑臭水	4280	0.01	19.4	385	0	2
14	ZQS1	竹墨沙热泉	3520	—	65.4	2045	0	0
15	ZCM4	日若沸泉	4440	68	35	590	1801	44283
16	ZCM5-1	布雄朗古沸泉	4400	10	27.9	510	211	4487
17	ZCM5-4	杀嘎朗嘎沸泉	4600	70	21.1	304	1118	14160
18	ZCM5-5	巴布的密沸泉	4630	10	19.8	365	150	2279
19	ZBX16-4	碱海子	4100	—	239	1478	0	0
			合计				4281	91882

通过对前人研究资料的综合分析发现，据不完全统计，西藏共发育达到边界品位的富锂地热点有 7 处，而锂含量达到或超过 19mg/L 的富锂温泉至少有 19 处（图 1），而锂离子含量达到 15mg/L 以上的热泉就有 26 处，分别是莫落江沸泉、多果曲热泉、色米沸泉、拉布朗沸泉、拉旺孜热泉等，这些高温富锂地热系统发育位置见图 1，其水化学特征如表 1 所示；通过对其中的布雄朗古沸泉、杀嘎朗嘎沸泉、巴布的密沸泉开展了野外调查与研究工作，发现古堆地热田除了发育有这几个热泉之外，还有茶卡热泉和日诺沸泉，并受到构造的强烈控制，热泉区发育在断裂交汇最近位置（图 2）。对古堆地热田 42 年以来 B、Li 数据进行投图发现（图 3），最近 42 年以来，其地热水 B、Li 含量基本稳定，地表水含量与地热水差别明显。此外，还发现其地热水硼、锂含量与地表水基本呈线性关系，表明了二元混合的特征，说明地表热泉是深部富含锂、硼的母地热流体与地表水混合的结果，进一步说明深部富锂母地热流体的存在。从总体上看，不仅青藏高原南部西藏高温富锂地热资源规模较大，且富含 B、Rb、Cs 等可以综合开发利用的元素，深部潜在的高温富锂地热资源进一步增加和提升了其开发利用价值。

3 青藏高原南部西藏高温富锂地热资源的成因

世界锂资源供给的 2/3 都是依靠富锂卤水中锂的提取，而多数的富锂地热系统都赋存在这些富锂盐湖卤水的周边，并对陆相富锂盐湖的形成起着非常重要的作用，如世界著名的智利阿塔卡马（Atacama）盐湖、玻利维亚乌尤尼盐湖、阿根廷翁布雷穆埃尔托（Hombre Muerto）盐湖、美国银峰（Silver Peak）盐湖、美国西尔斯（Searles）盐湖等。Bradley 等[49]在对构成陆相富锂卤水成矿模型的基本特征总结中，重点强调了地热活动的重要作用，Munk 等[9]进一步将地热活动对陆相富锂盐湖的控制作用总结为以下 5 点：（1）地热水提供了从富锂源岩中淋滤锂所需要的热水来源；（2）地热水可能直接提供了来自浅部岩浆热液或岩浆活动的富锂热液；（3）地热水可以通过浅部地下的蒸馏和沸腾作用来提高水中锂的浓度；（4）地热水引起的水对流循环可能是促使锂从源区到聚集区运移的有效方式；（5）地热水可以形成富锂的黏土矿物，若其后期在富黏土区发生了热液的淋滤和运移，便提供了形成富锂卤水的潜在来源。可见，地热活动对富锂盐湖卤水成矿的重要作用。因此，近年来，许多学者都对高温富锂地热流体的形成开展了广泛的研究。通过大量的文献调研发现，控制地热田形成的关键因素一是热源，二是构造形成的对流或地热水上升的通道，而形成地热资源的热源又有两种形式：首先是近代活动的活火山，例如南美安第斯山脉普遍发育的活火山，其周边往往发育有众多地热田；其次是深部熔融岩浆热源，这种热源在世界地热田中也广泛发育，例如美国的索尔顿海地热田[50-53]、中国的玛旁雍错地热田、羊八井地热田[26]、羊易地热田、古堆地热田等[31-32]，地质历史上强烈的构造活动造成的区域性深大断裂构成了质量热量传递通道，使深部热能迁移至较浅层位，形成地热田和地热显示区，尤其是不同方向的断裂相互切割的地区[43]，构造作用和火山作用提供了通道系统和驱动地下水对流循环的热量，是从大量岩石中浸出锂的必要条件[8]，大的地热田往往是沿着大的深断裂带分布的，尤其是多期次断裂的交叉点位置，如现今美国加利福尼亚州的索尔顿海地热田位于活跃的圣安德烈亚斯断层之上[50]，并叠加了之后的 SAFIF 走滑伸展[51]，其与玻利维亚和智利最具生产力的盐湖一样富含锂[4]，中国西藏地区的高温地热显示区多分布于深大断裂附近，尤其是近东西向与近南北向深大断裂相互切割交汇的地段（图 1、图 2），此外新西兰的 Taupo Geothermal Field 地热田、美国的黄石公园地热田（Yellowstone Geothermal Field）、冰岛的 Geysir geothermal field 地热田[54-56]等也都具有此类特征。在合适的热源和活动构造条件下，高温地热水的形成有两种路径，一种是大气降水或冰川融水沿着断裂带下渗，随着下渗深度的逐渐增加，越来越靠近深部高温岩体，并逐渐被加热，加热后的冷水浮力增大，沿着断裂上升地表，形成热泉，密集发育的热泉形成了地热田；另一种途径是大气降水或冰川融水沿着断裂带下渗到一定深度后，再与上升的岩浆热水混合，形成高温深部热储，混合的热储水再沿着断裂上升到地表形成热泉或热田，这两种不同方式形成的地热水在水化学特征上具有明显的区别，Chowdury 等[57]指出 B、F、As、Li 和 Cs 与岩浆结晶演化的最后一个阶段密切相关，基于实验研究[58, 59]和对花岗岩中微量元素的分析[45]，有学者指出富含 K、Rb、Cs、B、Li、F 和 Cl 的岩浆水可以在花岗岩演化的最后阶段释放出来，从青藏高原南部高温富锂地热水特征来看，其多为中性或碱性的 Na-Cl 型水，富含 B、Li、Rb、Cs、F 等特征性元素，同时热田区多有酸性地热水发育且富含 SO_4^{2-}，此外，Cl 离子含量与其他特征性离子多具有很好的线性关系，古堆地热田即有如此特征[31-33]。表明青藏高原南部高温富锂地热田多有深部岩浆流体的混入，众多地球物理的研究结果也证明了深部可能有岩浆岩体或岩浆熔融体的存在[60-66]。

世界范围内所有已经形成的地热田或热泉基本都受到以上因素所控制，但并不是所有的地热田都可以产生大量的富锂地热流体，富锂的地热田有其独有的特征，其唯一的区别就是锂的来源，全球各国不同的学者大都对这一问题进行过研究，虽仍有争议，但多数学者认为其来源基本有两个。

1）高温热水与富锂岩石（多数为凝灰岩等火山岩，如南美的 Vide 盐湖、美国的银峰湖）发生水岩反应，淋滤出其中的锂，如玻利维亚的乌尤尼盐沼，其周围广泛分布与第四纪流纹岩火山岩有关的热泉[8]，其热泉水含有异常高的锂含量，并被认为是卤水中锂的主要来源[4, 8, 10]；而早在 1983 年，郑绵平就在其研究“论西藏的盐湖”中提到了其 1960 年西藏野外工作中发现的班戈湖南部的含锂热泉的成因问题，其中就提到伦坡拉盆地火山岩沉积层的溶滤水中锂硼含量很高，并认为西藏中南部的新生代酸性火山岩是本区盐湖硼锂组分的重要来源[14]。2014 年，Araoka 等[67]通过对内华达州多个富锂盆地（Silver Peak playa，Clayton Valley brine field，Alkali Lake，Columbus Salt Marsh）的系统锂同位素研究，发现内华达州盆地锂的富集主要是通过当地地热活动相关的高温热水与富锂凝灰岩或火山玻璃的水岩反应淋滤其中的锂形成的，其周边的地热水中锂的含量高达 36×10^{-6}，这表明了高温地热水与富锂岩石的水岩反应淋滤作用对富锂地热水形成的重要性。

2）深部熔融岩浆分异结晶后期高温汽水溶液携出（由于锂在岩浆系统中是中度不相容的元素，因此在岩浆分异过程中通常聚集在残余熔体中[68]），如美国的索尔顿海地热田[50-53]、青藏高原南部措美县的古堆地热田[33, 69]、羊八井地热田[26]等。青藏高原南部地区广泛发育的高温高锂地热田，未见相关火山岩发育，即可能与 Francheteau 等[70]报道的引起青藏高原南部高热流的侵入岩有关[48]，Tan 等[71]利用氢氧同位素数据解释了西藏主要高温地热系统地下热水的循环过程，主要受到地下水的快速循环及残余岩浆水的升流作用，且在多数情况下，在距离热泉不远的地方都可以观察到花岗岩露头[48]，如古堆地热田东南部数十公里处发育有成岩时代为 12～17Ma 的富电气石花岗岩[72]。强烈活动的地热水流体通过在长英质富锂岩石地层中如花岗岩而发生水岩反应，淋滤出了其中的锂或直接与富锂的岩浆期后流体混合形成了青藏高原南部的高温富锂地热水，高温的地热水流体比传统的低温流体更能溶解岩石中的锂[73]，郑绵平通过对青藏高原全区的湖水成分趋势面分析，发现高原现代湖泊 B、Li、K 和 Cs 等元素丰度具有以冈底斯-雅鲁藏布为高值中心向外递减的特征，构成了青藏高原 B、Li、Cs 地球化学异常区，其成因与班怒构造带和雅鲁藏布西段聚敛带深部岩浆熔融体为中心通过火山和地热水向表层扩散有关[16]。

近年来，随着研究工作的不断深入和通过研究国内外近几十年的文献资料，并与南美“锂三角”周边正在形成的高温富锂地热水对比，从整体上看，西藏富锂盐湖的形成经过了多次的“预富集”过程，而高温富锂地热水的发育是其中重要一环，早期富锂洋壳在俯冲过程中熔融形成富锂岩浆完成了初始的富集，富锂岩浆沿着深大断裂上升过程中逐渐冷却结晶，在结晶分异晚期锂进一步在汽水热液中富集，随着印度与欧亚大陆的进一步碰撞抬升，在不同的地质条件下，部分富锂岩浆热液形成了富电气石-锂辉石花岗岩，部分富锂岩浆热液沿断裂带上升并与下渗地表水混合，沿着断裂带渗出地表，形成富锂地热水，部分地热水汇入盐湖，通过进一步地表的蒸发浓缩富集作用形成富锂盐湖，大部分富锂地热水汇入地表径流。

4 青藏高原南部西藏高温富锂地热资源开发利用前景

世界各国学者针对地热水中锂的提取和开发都做了许多研究，例如 1975 年，Hazen Research Incorporated 公司就对地热水中金属离子的提取和回收进行了初步的研究[34]；1978 年，Reno Metallurgy Research Center[74]，SRI International 公司[75]对地热水提取锂进行了中试研究；1984 年，Schultze 和 Bauer[76]报道了其可以将 Salton Sea 地热发电厂废水中 99%的锂提取出来；Salton Sea 地热发电厂甚至报道了其从地热水提取锂的中试工厂的建立及其商业开发计划[7, 47]；波兰的 RabkaZdroj 地热田其地热水中锂的含量并不是最高的，其锂含量范围为 10～16mg/L[77]，但是该国对从富锂地热水中提取锂的研究却较为领先，截至目前，已有报道的有两种方法，分别是混合电容去电离法（hybrid capacitive deionization）和应用有聚丙烯酸的天然或合成沸石法（natural and synthetic zeolites applying poly acrylic acid）[77]；Ziya 等[78]对土耳其不同类型地热水中锂含量及其提取方法进行了综合对比，并认为直接将富锂地热水中的锂沉淀为锂盐、通过膜滤器分离以及通过离子交换树脂捕获形式进行提取锂比较适合；中国学者也在不断推进地热水中锂的提取技术研究，并且已经取得了一定的研究成果，Sun 等[79]提出了一种基于新型磷酸铁锂电化学技术从地热水中绿色回收锂的方法，经过 8 个吸附-解吸循环，Li^+的回收率高达 90.65%，另外，在锂回收的整个过程中仅消耗了电能，没有使用或产生有机溶剂或其他有毒试剂，所有这些特性使上述方法成为从地热水中回收 Li^+ 的绿色且非常有前景的方法，Wang 等[80]报道了通过 EGDE 交联的球形 CTS/LMO（EGDE cross-linked spherical CTS/LMO）技术来对低锂含量的地热水中的锂进行选择性提取，并认为这种方法具有非常好的稳定性和选择性，并且在 5 次循环后吸附容量的衰减不超过 1.1%，以上从地热水中提取锂的方法，总结起来包括电化学、溶剂萃取、结晶沉淀[81]、离子交换[82]、吸附[2, 83]、含无机吸附剂的聚合物膜[84]或以上多种方法的组合[19, 85, 86]，在所有上述方法中，溶剂萃取和离子交换吸附得到了最广泛的深入研究[79]。但是截至目前，针对富锂地热流体中锂的提取技术的研究仍然有待加强，郑绵平基于在西藏铯硅化矿床中铯的提取经验，通过前期对富锂地热水蒸发浓缩实验过程中锂的浓度和其他化学成分的变化规律，并结合青藏高原南部高温富锂地热区地形地质等特征，认为单纯的蒸发浓缩效率太低，不能满足当前的开发利用需求。前期的探索研究，认为针对青藏高原南部的高温富锂地热水，膜法和吸附法仍然是非常有前景和值得考虑的，尤其是铝系物质或沸石类吸附剂等天然无污染的吸附剂，甚至还可以考虑天然物质和人工合成物质相结合的吸附剂，不仅可以绿色无污染，还可以增加人为的可控性，根据不同的地热水化学组分和类型，进行适当的调整，提高提取锂效率的同时降低能耗。此外，郑绵平[87]通过对索尔顿海地热田（Salton Sea geothermal）、克什米尔的 Puga 地热田、新西兰的 Taupo 地热田和中国的西藏现在正在发育的低盐度 Li、B、Cs、Rb 等稀碱金属地热流体的对比研究，认为低盐度的地热流体也是一种特殊的成矿流体，也是成矿作用一个新的领域。

5　结论

地热型锂资源是一种新型的锂资源，世界范围内广泛分布，潜在储量巨大，各国学者相继对地热型锂资源开展了研究，应引起中国学者的广泛重视。中国地热型锂资源主要分布在青藏高原南部西藏地区，受到南北向伸展构造和近东西向雅鲁藏布江缝合带的明显控制，中国高温富锂地热水具有 Mg/Li 低，Li/TDS 较高，数十年持续稳定的排出，部分达到工业品位，伴生高 B、Cs、Rb 等可综合利用元素的特点。此外，从总体上看，当前中国青藏高原南部高温富锂地热水每年排出金属锂规模较大，深部潜在资源丰富，具有较大的开发利用价值。雅鲁藏布江两侧及青藏高原南部地区火山活动发育较弱，深部上地壳部分熔融引起的岩浆作用对高温富锂地热水的发育至关重要。随着当代技术的不断进步，从富锂地热水中提取锂的技术逐渐多样，方法逐渐成熟，将来对高温富锂地热水中锂等资源的开发不仅会产生其应有的经济价值，而且会有利于降低其中有害元素对环境的污染。综上所述，中国青藏高原南部广泛发育的高温富锂地热资源是一种非常有价值的、值得开发利用的地热型锂资源，并且近地表富锂稀碱等矿质地热流体也是一种现代的低盐度热液成矿流体，通过对其成矿作用的研究，有助于加深对低温热液成矿作用的理解和区域深部找矿勘查。

参考文献

[1] Martin G, Rentsch L, Höck M, et al. Lithium market research—Global supply, future demand and price development[J]. Energy Storage Materials, 2016, 6: 171-179.

[2] Park J K. Principles and applications of lithium secondary batteries[M]. New York: Wiley, 2012.

[3] 刘丽君, 王登红, 高娟琴, 等. 国外锂矿找矿的新突破（2017～2018 年）及对我国关键矿产勘查的启示[J]. 地质学报, 2019, 93 (6): 1479-1488.

[4] Campbell M G. Battery lithium could come from geothermal waters[J]. New Scientist, 2009, 204 (2738): 23-23.

[5] Tomaszewska B, Szczepaå S A. Possibilities for the effi cient utilisation of spent geothermal waters[J]. Environmental Science & Pollution Research, 2014, 21 (19): 11409-11417.

[6] Gruber P W, Medina P A, Keoleian G A, et al. Global lithium availability—A constraint for electric vehicles？[J] Journal of Industrial Ecology, 2011, 15: 760-775.

[7] Kesler S E, Gruber P W, Medina P A, et al. Global lithium resources: Relative importance of pegmatite, brine and other deposits[J]. Ore Geology Reviews, 2012, 48 (5): 55-69.

[8] Ericksen G E, Vine J D, Ballón A R. Chemical composition and distribution of lithium-rich brines in salar de Uyuni and nearby salars in southwestern Bolivia[J]. Energy, 1978, 3 (3): 355-363.

[9] Munk L A, Bradley D C, Hynek S A, et al. Origin and evolution of Li-rich brines at Clayton Valley, Nevada, USA[C]//11th SGA Biennial Meeting. Antofagasta: SGA. 2011: 217-219.

[10] Shcherbakov A V, Dvorov V I. Thermal waters as a source for extraction of chemicals[J]. Geothermics, 1970, 2 (2): 1636-1639.

[11] Tan H, Chen J, Rao W, et al. Geothermal constraints on enrichment of boron and lithium in salt lakes: An example from a river-salt lake system on the northern slope of the eastern Kunlun Mountains, China[J]. Journal of Asian Earth Sciences, 2012, 51 (12): 21-29.

[12] Yu J Q, Gao C L, Cheng A Y, et al. Geomorphic, hydroclimatic and hydrothermal controls on the formation of lithium brine deposits in the Qaidam Basin, northern Tibetan Plateau, China[J]. Ore Geology Reviews, 2013, 50 (50): 171-183.

[13] 雒洋冰, 郑绵平, 任雅琼. 青藏高原特种盐湖与深部火山-地热水的相关性[J]. 科技导报, 2017, 35 (12): 44-48.

[14] 郑绵平, 刘文高, 向军, 等. 论西藏的盐湖[J]. 地质学报, 1983, 57 (2): 184-194.

[15] 郑绵平, 向军, 魏新俊, 等. 青藏高原盐湖[M]. 北京: 北京科学技术出版社, 1989.

[16] 郑绵平, 郑元, 刘杰. 青藏高原盐湖及地热矿床的新发现[J]. 中国地质科学院院报, 1990 (1): 151.

[17] Cetiner Z S, Özgür D, Özdilek G, et al. Toward utilising geothermal waters for cleaner and sustainable production: Potential of Li recovery from geothermal brines in Turkey[J]. International Journal of Global Warming, 2015, 7 (4): 439.

[18] Hano T, Matsumoto M, Ohtake T. Recovery of lithium from geothermal water by solvent extraction technique [J]. Solvent Extraction & Ion Exchange, 1992, 10: 195-206.

[19] Jeongeon P, Hideki S, Syouhei N, et al. Lithium recovery from geothermal water by combined adsorption methods[J]. Solvent Extraction & Ion Exchange, 2012, 30: 398-404.

[20] Krotscheck E, Smith R A. Separation and recovery of lithium from geothermal water by sequential adsorption process with l-$MnO_2$and TiO_2[J]. Ion Exchange Letters, 2012, 32: 2219-2233.

[21] Yanagase K, Yoshinaga T, Kawano K, et al. The recovery of lithium from geothermal water in the Hatchobaru area of Kyushu, Japan[J]. Bulletin of The Chemical Society of Japan, 1983, 56: 2490-2498.

[22] 郑绵平, 刘文高. 西藏发现富锂镁硼酸盐矿床[J]. 地质论评, 1982, 28 (3): 263-266.

[23] Guo Q, Wang Y, Liu W. Hydrogeochemistry and environmental impact of geothermal waters from Yangyi of Tibet, China[J]. Journal of Volcanology & Geothermal Research, 2009, 180 (1): 9-20.

[24] 郑绵平, 刘文高. 新的锂矿物——扎布耶石[J]. 矿物学报, 1987, 7 (3): 221-226.

[25] 郑绵平. 水热成矿新类型[M]. 北京: 地质出版社, 1995.

[26] Guo Q, Wang Y, Liu W. Major hydrogeochemical processes in the two reservoirs of the Yangbajing geothermal field, Tibet, China[J]. Journal of Volcanology & Geothermal Research, 2007, 166 (3): 255-268.

[27] Tan H, Su J, Xu P, et al. Enrichment mechanism of Li, B and K in the geothermal water and associated deposits from the Kawu area of the Tibetan plateau: Constraints from geochemical experimental data[J]. Applied Geochemistry, 2018, 93: 60-68.

[28] Guo Q H, Wang Y X, Liu W. O, H, and Sr isotope evidences of mixing processes in two geothermal fluid reservoirs at Yangbajing, Tibet, China[J]. Environmental Earth Sciences, 2010, 59: 1589-1597.

[29] Guo Q, Wang Y. Geochemistry of hot springs in the Tengchong hydrothermal areas, Southwestern China[J]. Journal of Volcanology & Geothermal Research, 2012, 215-216: 61-73.

[30] Guo Q, Liu M, Li J, et al. Fluid geochemical constraints on the heat source and reservoir temperature of the Banglazhang hydrothermal system, Yunnan-Tibet Geother mal Province, China[J]. Journal of Geochemical Exploration, 2017, 172: 109-119.

[31] Guo Q, Li Y, Luo L. Tungsten from typical magmatic hydrothermal systems in China and its environmental transport[J]. Science of The Total Environment, 2019, 657: 1523-1534.

[32] Guo Q, Planer-Friedrich B, Liu M, et al. Magmatic fluid input explaining the geochemical anomaly of very high arsenic in some southern Tibetan geothermal waters[J]. Chemical Geology, 2019, 513: 32-43.

[33] Wang C G, Zheng M P. Hydrochemical characteristics and evolution of hot fluids in the Gudui geothermal field in Comei County, Himalayas[J]. Geothermics, 2019, 81: 243-258.

[34] Zheng W, Tan H, Zhang Y, et al. Boron geochemistry from some typical Tibetan hydrothermal systems: Origin and isotopic fractionation[J]. Applied Geochemistry, 2015, 63: 436-445.

[35] 多吉. 典型高温地热系统——羊八井热田基本特征[J]. 中国工程科学, 2003, 5 (1): 42-47.

[36] 李建康, 刘喜方, 王登红. 中国锂矿成矿规律概要[J]. 地质学报, 2014, 88 (12): 2269-2283.

[37] 佟伟, 廖志杰. 西藏温泉志[M]. 北京: 科学出版社, 2000.

[38] 佟伟, 章铭陶, 张知非, 等. 西藏地热[M]. 北京: 科学出版社, 1981.

[39] 张知非, 沈敏子, 赵凤三. 西藏古堆高温水热系统的地下状况[M]//地热专辑 (第二辑). 北京: 地质出版社, 1989: 134-140.

[40] 郑绵平, 刘喜方. 青藏高原盐湖水化学及其矿物组合特征[J]. 地质学报, 2010, 84 (11): 1585-1600.

[41] 郑淑蕙, 张知非, 倪葆龄, 等. 西藏地热水的氢氧稳定同位素研究[J]. 北京大学学报 (自然科学版), 1982 (1): 99-106.

[42] Morozov N P. Geochemistry of the alkali metals in rivers [J]. Geokhimiya, 1969, 6 (3): 729-739.

[43] 王思琪. 西藏古堆高温地热系统水文地球化学过程与形成机理[D]. 北京: 中国地质大学 (北京), 2017.

[44] 刘昭, 陈康, 男达瓦. 西藏古堆地热田地下热水水化学特征[J]. 地质论评, 2017 (Suppl 1): 353-354.

[45] Grimaud D, Huang S, Michard G, et al. Chemical study of geothermal waters of Central Tibet (China) [J]. Geothermics, 1985, 14 (1): 35-48.

[46] 李振清. 青藏高原碰撞造山过程中的现代热水活动[D]. 北京: 中国地质科学院, 2002.

[47] Evans K R. Lithium—Chapter 10[M]//Gunn G. 2014-Critical metals handbook. New Jersey: Wiley-Blackwell, 2014.

[48] Munk L A, Hynek S A, Bradley D, et al. Lithium brines: A global perspective[J]. Review Economic Geology, 2016, 18: 339-365.

[49] Bradley D, Munk L, Jochens H, et al. A preliminary deposit model for lithium brines[R]. Reston, Virginia: U.S. Geological Survey, 2013.

[50] Brothers D S, Driscoll N W, Kent G M, et al. Tectonic evolution of the Salton Sea inferred from seismic reflection data[J]. Nature Geoscience, 2009, 2 (8): 581-584.

[51] Karakas O, Dufek J, Mangan M T, et al. Thermal and petrologic constraints on lower crustal melt accumulation under the Salton Sea Geothermal Field[J]. Earth and Planetary Science Letters, 2017, 467: 10-17.

[52] Lachenbruch A H, Sass J, Galanis S. Heat flow in southernmost California and the origin of the Salton Trough [J]. Journal of Geophysical Research-Solid Earth, 1985, 90: 6709-6736.

[53] Schmitt A K, Hulen J B. Buried rhyolites within the active, high-temperature Salton Sea geothermal system[J]. Journal of Volcanology and Geothermal Research, 2008, 178: 708-718.

[54] Elderfield H, Greaves M J. Strontium isotope geochemistry of icelandic geothermal systems and implications for sea water chemistry[J]. Geochimica et Cosmochimica Acta, 1981, 45: 2201-2212.

[55] Jones B, Renaut R W, Torfason H, et al. The geological history of Geysir, Iceland: A tephrochronological approach to the dating of sinter[J]. Journal of the Geological Society, 2007, 164 (6): 1241-1252.

[56] Geilert S, Vroon P Z, Keller N S, et al. Silicon isotope fractionation during silica precipitation from hot-spring waters: Evidence from the Geysir geothermal field, Iceland[J]. Geochimica et Cosmochimica Acta, 2015, 164: 403-427.

[57] Chowdhury A N, Handa B K, Das A K. High lithium, rubidium and cesium contents of thermal spring water, spring sediments and borax deposits in Puga Valley, Kashmir, India[J]. Geochemical Journal, 1974, 8: 61-65.

[58] Fuge R. On the behaviourof fluorine and chlorine during magmatic differentiation[J]. Contributions to Mineralogy & Petrology, 1977, 61 (3): 245-249.

[59] Webster E A, Holloway J R. The partitioning of REE's, Rb and Cp between silicic meh and a CI fluid[OL]. EOS, 1980, 61: 1152.

[60] Brown L D, Zhao W, Nelson K D, et al. Bright spots, structure, and magmatism in southern tibet from indepth seismic reflection profiling[J]. Science, 1996, 274: 1688-1690.

[61] Chen L, Booker J R, Jones A G, et al. Electrically conductive crust in Southern Tibet from INDEPTH magnetotelluric surveying[J]. Science, 1996, 274: 1694-1696.

[62] Kind R, Ni J, Zhao W, et al. Evidence from earthquake data for a partially molten crustal layer in Southern Tibet [J]. Science, 1996, 274: 1692-1694.

[63] Makovsky Y, Klemperer S L, Ratschbacher L, et al. INDEPTH wide-angle reflection observation of P-wave-to-S-wave conversion from crustal bright spots in Tibet[J]. Science, 1996, 274: 1690-1691.

[64] Nelson K D, Zhao W, Brown L D, et al. Partially molten middle crust beneath southern Tibet: Synthesis of project INDEPTH results[J]. Science, 1996, 274: 1684-1688.

[65] Wei W B, Jin S, Ye G F, et al. Conductivity structure and rheological property of lithosphere in Southern Tibet inferred from super-broadband magmetotulleric sounding [J]. Science in China (Earth Sciences), 2010, 53: 189-202.

[66] 谭捍东, 魏文博, Martyn U, 等. 西藏高原南部雅鲁藏布江缝合带地区地壳电性结构研究[J]. 地球物理学报, 2004, 47 (4): 685-690.

[67] Davis J R, Friedman I, Gleason J D. Origin of the lithium-rich brine, Clayton Valley, Nevada[J]. U.S. Geological Survey Bulletin, 1986, 1622: 131-138.

[68] Zhang L, Chan L H, Gieskes J M. Lithium isotope geochemistry of pore waters from Ocean Drilling Program Sites 918/919, Irminger Basin[J]. Geochimica et Cosmochimica Acta, 1998, 62 (14): 2437-2450.

[69] Wang C G, Zheng M P, Zhang X F, et al. O, H, and Sr isotope evidence for origin and mixing processes of the Gudui geothermal system, Himalayas, China[J]. Geoscience Frontiers, 2019, doi: 10.1016/j.gsf.2019.09.013.

[70] Francheteau J, Jaupart C, Shen X J, et al. High heat flow in southern Tibet[J]. Nature, 1984, 307 (5946): 32-36.

[71] Tan H, Zhang Y, Zhang W, Kong N, Zhang Q, Huang J. Understanding the circulation of geothermal waters in the Tibetan Plateau using oxygen and hydrogen stable isotopes[J]. Applied Geochemistry, 2014, 51: 23-32.

[72] Liu M L, Guo Q H, Wu G, et al. Boron geochemistry of the geothermal waters from two typical hydrothermal systems in Southern Tibet (China): Daggyai and Quzhuomu [J]. Geothermics, 2019, 82: 190-202.

[73] Chagnes A, Światowska J. Lithium Process Chemistry [M]. Amsterdam: Elsevier, 2015.

[74] Berthold C E. Magmamax No. 1 Geothermal minerals recovery pilot plant, engineering design[R]. Reho, Nevada: Hazen Research, Reno Metallurgy Research Center, 1978.

[75] Farley E P, Watson E L, Macdonald D D, et al. Recovery of heavy metals from high salinity geothermal brine [R]. Nevada: SRI International, 1980.

[76] Schultze L E, Bauer D J. Recovering lithium chloride from a geothermal brine[R]. Reston, Virginia: United States Bureau of Mines, Fort Meade in Maryland, 1984.

[77] Małgorzata W, Gracja F, Iwona O, et al. Investigations of the possibility of lithium acquisition from geothermal water using natural and synthetic zeolites applying poly (acrylic acid) [J]. Journal of Cleaner Production, 2018, 195: 821-830.

[78] Ziya S C, Özgür D, Göksel Ö, et al. Toward utilizing geothermal waters for cleaner and sustainable production: Potential of Li recovery from geothermal brines in Turkey[J]. International Journal of Global Warming, 2015, 7 (4): 439.

[79] Sun S, Yi X P, Li M L, et al. Green recovery of lithium from geothermal water based on a novel lithium iron phosphate electrochemical technique[J]. Journal of Cleaner Production, 2020, 247: 119178.

[80] Wang H S, Cui J, Li M L, et al. Selective recovery of lithium from geothermal water by EGDE cross-linked spherical CTS/LMO[J]. Chemical Engineering Journal, 2020, 389: 124410.

[81] Pauwels H, Brach M, Fouillac C. Lithium recovery from geothermal waters of Cesano (Italy) and Cronembourg (Alsace, France) [C]//12th New Zealand Geothermal Workshop. Orléans: Bureau de Recherches Géologiques et Minières, 1990: 117-123.

[82] Nishihama S, Onishi K, Yoshizuka K. Selective recovery process of lithium from seawater using integrated ion exchange methods[J]. Solvent Extraction and Ion Exchange, 2011, 29 (3): 421-431.

[83] Miyai Y, Ooi K, Katoh S. Recovery of lithium from seawater using a new type of ion-sieve adsorbent based on$MgMn_2O_4$[J]. Separation Science and Technology, 1998, 23 (1-3): 179-191.

[84] Chung K S, Lee J C, Kim W K, et al. Inorganic adsorbent containing polymeric membrane reservoir for the recovery of lithium from seawater[J]. Journal of Membrane Science, 2008, 325 (2): 503-508.

[85] Flexer V, Baspineiro C F, Galli C L. Lithium recovery from brines: a vital raw material for green energies with a potential environmental impact in its mining and processing[J]. Science Total Environment, 2018, 639: 1188-1204.

[86] Song J F, Nghiem L D, Li X M, et al. Lithium extraction from Chinese salt-lake brines: Opportunities, challenges, and future outlook[J]. Environmental Science-Water Research & Technology, 2017, 3: 593-597.

[87] Zheng M P. Preliminary discussion of low-salinity hydrothermal fluid mineralization[J]. Chinese Science Bulletin, 1999, 44 (Suppl 2): 141-143.

Geothermal-type lithium resources in southern Tibetan Plateau

Chenguang Wang[1], Mianping Zheng[1], Xuefei Zhang[1], Chuanyong Ye[1], Qian Wu[1], Shuangshuang Chen[3], Mingming Li[1, 2], Tao Ding[1, 2] and Shaorong Du[1]

1. MNR Key Laboratory of Saline Lake Resources and Environments, Institute of Mineral Resources, Chinese Academy of Geological Sciences, Beijing 100037, China
2. College of Geoscience and Surveying Engineering, China University of Mining and Technology (Beijing), Beijing 100083, China
3. Guangdong Provincial Key Laboratory of Geodynamics and Geohazards, School of Earth Sciences and Engineering, Sun Yat-sen University, Guangzhou 510275, China

Abstract The high-temperature geothermal water contains abundant lithium resources, and the development and the utilization of the geothermal-type lithium resources are increasingly paid attention around the world. This paper reviews the lithium-rich geothermal resources among the geothermal resources on the Qinghai-Tibet Plateau, and it is concluded that these resources have the following characteristics: (1) strong structural control: the lithium-rich geothermal spots in Southern Tibetan Plateau are often found in the intensely active high-temperature geothermal fields and are distributed on both sides of the Yarlung Zangbo suture zone and its southern part, strongly controlled by north-south trending rifts or grabens formed by E-W extension; (2) good quality: the lithium concentration is up to 239mg/L; the Mg/Li ratio is extremely low and ranges from 0.03 to 1.48 for most of the lithium-rich geothermal fluid; the Li/TDS value is relatively high and ranges from 0.25%–1.14% (Zhabuye lithium-rich salt lake: 0.19%; Salar de Uyuni (Bolivia): 0.08–0.31%); the continuous discharge is stable at least for several decades, in some parts reaches the industrial grades (32.74mg/L: according to the industrial grades of Salt lake brine), and in addition, the elements such as B, Cs, and Rb are rich and can be comprehensively utilized; (3) large scale: according to incomplete statistics, there are at least 16 lithium-rich hot springs with lithium concentration of 19mg/L or more, and the total discharge of lithium metal is about 4281 tons every year, equivalent to 25686 tons of lithium carbonate, moreover, drilling data show that the depth is still very promising; (4) lack of volcanism (non-volcanic geothermal system); (5) deep origin: the formation of lithium-rich geothermal resources are closely related to the deep crust partially melting caused by the collision of the Indo-Asia continent, the deep molten magma provides a stable heat source for the high-temperature lithium-rich geothermal field and the lithiumrich parent geothermal fluid rushes to the surface to form hot springs along the extensively developed tectonic fault zones in Southern Tibetan Plateau. Therefore, the widely developed high-temperature lithium-rich geothermal resources in the southern Qinghai-Tibet Plateau are valuable and worthy lithium resources. With the continuous advancement of the lithium extraction technologies on lithium-rich geothermal fluid, the lithium resource in Southern Tibetan Plateau is becoming a promising new type of mineral deposit—the geothermal-type lithium deposit and will be effectively exploited.

Keywords Southern Tibetan Plateau; Geothermal; Lithium resources

川宣地 1 井发现厚层海相可溶性“新型杂卤石钾盐”工业矿层*

张永生[1,2]，郑绵平[1,2]，邢恩袁[1,2]，左璠璠[1,2]，彭渊[1,2]，仲佳爱[1,2]，桂宝玲[1,2]，牛新生[1,2]，苏奎[1,2]，商雯君[1,2]，麻乾坤[1,2]，崔新宇[1,2]

1. 中国地质科学院矿产资源研究所自然资源部成矿作用与资源评价重点实验室，北京 100037

2. 中国地质科学院矿产资源研究所自然资源部盐湖资源与环境重点实验室，北京 100037

1 研究目的

近年来，中国地质科学院矿产资源研究所郑绵平院士团队在四川盆地东北部宣汉普光地区下三叠统嘉陵江组四—五段（简称“嘉四—五段”）发现了一种与石盐共生的碎屑颗粒杂卤石，此种分布于石盐基质中的碎屑颗粒杂卤石易溶于水，便于采用水溶法低成本、规模化开采，因而被命名为“新型杂卤石钾盐矿”，是一种全新的海相可溶性优质硫酸盐型钾盐矿床类型。

为进一步查明新型杂卤石钾盐矿的沉积分布特征与含钾性等基本地质参数，结合区域物探资料和野外地质调查结果，优选钻探靶区，以嘉四—五段为主要目的层系，设计部署锂钾综合地质调查井——川宣地 1 井，力争取得海相可溶性新型杂卤石钾盐找矿突破，并为该区钾盐资源评价提供可靠依据。

2 研究方法

川宣地 1 井于 2020 年 8 月完钻（井深 3797m），累计取心 837.25m，岩心采取率 98.08%，在井深 2900～3400m 的嘉四—五段发现多层海相富锂钾卤水和厚层新型杂卤石钾盐工业矿层。有关富锂钾卤水将另文论述，本文重点讨论新型杂卤石钾盐的岩矿特征、矿石品位、矿层厚度及其成果意义。

针对川宣地 1 井发现的厚层新型杂卤石钾盐，系统开展了岩心观察、镜下鉴定及电子探针等研究工作，并通过高密度采样（样品间隔 10cm 左右），采用等离子光谱仪和等离子质谱仪测试了该层段 475 个粉末样品的主微量元素。上述测试工作在国家地质实验测试中心和青海地质矿产测试应用中心完成，并对测试结果进行了对比校正。

3 结果

3.1 岩石矿物特征

宏观上，新型杂卤石钾盐主要分为两类：第一类以含硬石膏/硬石膏质杂卤石碎屑颗粒呈细粒-巨砾等不同粒级镶嵌在石盐基质中为主要特征[图 1（a）]，颗粒含量 20%～80%，岩心表面呈褐红色-肉红色，断面以浅灰色-灰白色为主，可见暗色条纹，石盐基质则表现为半透明烟灰色-灰黑色。第二类表现为块状含硬石膏/硬石膏质杂卤石层被石盐脉切割破碎[图 1（b）]，被切割边界形态仍保留完整，经拼接可复原其原始形态。

微观上，含硬石膏/硬石膏质杂卤石碎屑可见杂卤石与硬石膏晶体共生，杂卤石晶体在单偏光下无色透明，在正交偏光下具二级蓝绿干涉色，斜消光，可见粒状、柱状及放射状结构，常见聚片双晶[图 1（c）]；

* 本文发表在：中国地质，2021，48（1）：343-344

硬石膏在单偏光下突起较杂卤石高，在正交偏光下平行消光，干涉色可达三级绿，聚片双晶少见。胶结物石盐晶体在正交偏光下全消光，因此极易识别[图 1（c）]。此外，通过电子探针的识别，杂卤石晶体与石膏、石盐边界清晰，少见交代、穿插现象[图 1（d）、（e）]，应为同时期原生沉积产物。

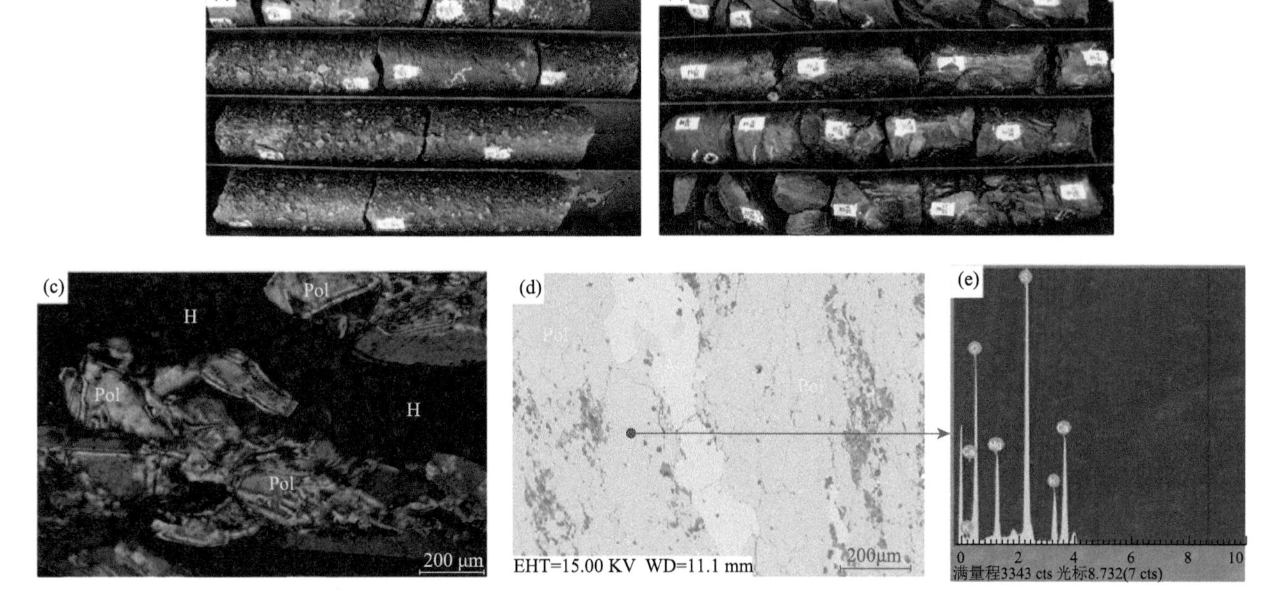

图 1 川宣地 1 井新型杂卤石钾盐矿岩心及显微特征

Fig. 1 Core and microscopic characteristics of the new type of polyhalite potassic salt ore in CXD1 well.

（a）、（b）新型杂卤石钾盐矿宏观特征；（c）正交偏光下，杂卤石被石盐晶体胶结；（d）电子探针下，杂卤石与硬石膏共生；（e）电子探针下，杂卤石能谱曲线；Pol-杂卤石，H-石盐，An-硬石膏

(a)、(b) Macroscopic characteristics of the new type of polyhalite potassic salt ore; (c) Polyhalite grains cemented by halite crystals (crossed nicols); (d) Polyhalite and anhydrite crystals under EPMA; (e) Polyhalite power spectral curve under EPMA; Pol-Polyhalite, H-Halite, An-Anhydrite

3.2 矿层特征

主微量元素测试分析表明，在川宣地 1 井嘉四—五段井深 3000.67～3387.65m 范围内，发现海相新型杂卤石钾盐含钾盐系共计 387m，其中赋存上、下两个主力钾盐组，井深分别为 3000.67～3062.67m、3376.1～3387.65m，钾盐含量（KCl 当量）在 3%～20.5%的新型杂卤石钾盐工业矿层厚度分别为 52.58m 和 10.23m，累计厚达 62.81m。在上钾盐组，高品位（KCl＞8%）的钾盐矿层厚达 32.86m，占该钾盐组矿层总厚度 62.5%。

关于新型杂卤石钾盐矿的矿床规模、分布规律和成矿模式的研究正在全面推进中。

4 结论

川宣地 1 井在下三叠统嘉四—五段发现累计厚达 62.81m 的新型杂卤石钾盐工业矿层，可划分为上、下两个钾盐组，明确了新型杂卤石钾盐矿的垂向分布特征，宣汉地区新型杂卤石钾盐矿预测资源量（KCl 当量）达数亿吨，取得了海相可溶性固体钾盐找矿的重大突破，有望形成连片突破的示范效应，并率先推动宣汉建设我国首个亿吨级大型海相可溶性固体钾盐资源基地。

5 致谢

川宣地 1 井是由中国地质科学院矿产资源研究所与四川巴人新能源有限公司共同出资，施工单位四川省地质矿产勘查开发局四〇五地质队保障钻探工程的顺利实施，达州市恒成能源（集团）有限责任公司为前期研究提供了基础资料，并得到宣汉县人民政府和中国石化勘探分公司的大力支持，在此一并感谢。

川东北普光地区发现新型杂卤石钾盐矿*

郑绵平　张永生　商雯君　邢恩袁　仲佳爱　桂宝玲　彭渊

中国地质科学院矿产资源研究所，北京 100037

1　研究目的

杂卤石（$K_2SO_4·MgSO_4·2CaSO_4·2H_2O$）是一种常见的钾镁硫酸盐矿物，因其较难溶于水，通常被视为重要的找钾标志。由于杂卤石是钾、镁、钙的硫酸盐复盐，其本身也是一种缓释性优质无氯复合钾镁肥。

全球杂卤石资源相对集中产于北纬25°～50°的纬度带（现今）内，产出的地质时代有二叠纪、三叠纪、古近纪、新近纪和第四纪，可概括为“一带五期”，主要产于前第四纪古盐盆中。中国杂卤石产出的地质时代集中在三叠纪、第四纪和古近纪，产出的盐盆地有四川盆地、柴达木盆地、江汉盆地等，其中以四川盆地三叠系杂卤石分布最广、潜在资源量最大。据原西南石油地质局第二地质大队钾盐勘查报告，四川盆地杂卤石远景资源量（折合 K_2O）约百亿吨以上。本次是在前人针对产于硬石膏岩中或与硬石膏岩呈薄互层杂卤石研究的基础上，对近年来在普光地区新发现的、分布于石盐基质中的盐晶胶结碎屑颗粒杂卤石（本文称之为“盐晶颗粒杂卤石岩”——一种新型杂卤石钾盐矿）的盐矿特征及其可利用性作初步探讨，以期引起业界的重视。

2　研究方法

本文以近年来川东北宣汉县普光地区恒成2井、恒成3井等钻探工作为基础，通过岩心观察、镜下鉴定、扫描电镜、能谱分析等手段，初步探讨这种新型固体杂卤石钾盐矿的特征及其可利用性。

3　研究结果

3.1　盐晶颗粒杂卤石岩的宏观特征

新型杂卤石钾盐矿赋存于埋深超过 3000m 的下、中三叠统嘉陵江组—雷口坡组海相蒸发岩层系中，内碎屑颗粒杂卤石呈星点状、不规则团块状或似条带状分布于石盐基质中，大小不一，细粒（<1mm）、粗粒（1～2mm）至砾屑（>2mm），局部见巨砾级颗粒（3～7cm），杂卤石团块有近似等轴的似圆状—似方状到长条状—椭球状—不规则状等不同形状[图 1（a）]，似条带状杂卤石具明显的揉皱和破碎现象[图 1（b）]。杂卤石呈灰白色或肉红色，发育暗色条纹，具贝壳状断口，粉晶—细晶结构，其中灰白色—黑色杂卤石不透明、呈土状光泽或光泽不明显，可能是杂卤石团块中含其他杂质较多造成；肉红色杂卤石半透明、蜡状光泽、结构细腻致密。鉴于内碎屑颗粒杂卤石含量一般为20%～30%，部分大于50%，胶结物多为石盐晶体，本文将之命名为“盐晶颗粒杂卤石岩”。另外，可见杂卤石呈薄层赋存于此类杂卤岩中。

* 本文发表在：中国地质，2018，45（5）：1074-1075

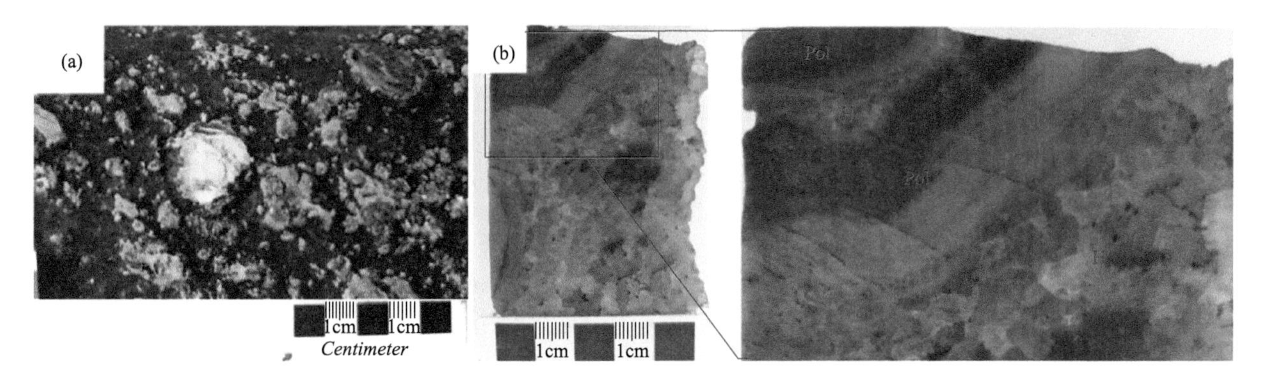

图1 恒成3井杂卤石手标本特征

Fig. 1 Surface features of polyhalite in Heng Cheng No.3 well.

3.2 盐晶颗粒杂卤石岩的显微特征

单偏光下，杂卤石晶体略显黄绿色，半自形—他形粒状、柱状结构，常见晶棱，突起较石盐高；正交光下，杂卤石晶体具二级蓝绿干涉色，斜消光，消光角为15°～20°，镜下杂卤石颗粒为二轴晶负光性，光轴角为60°～70°，胶结物石盐晶体呈全消光[图2（a）]。晶体集合具一定的方向性，中部晶体细小，一般为微晶-细晶，紧密、互相叠置生长，边部晶体颗粒较大，可能为次生晶体。在多处杂卤石中发现钾盐包体，或呈半自形-自形粒状均质体，负突起，交代杂卤石，含量1%～3%。

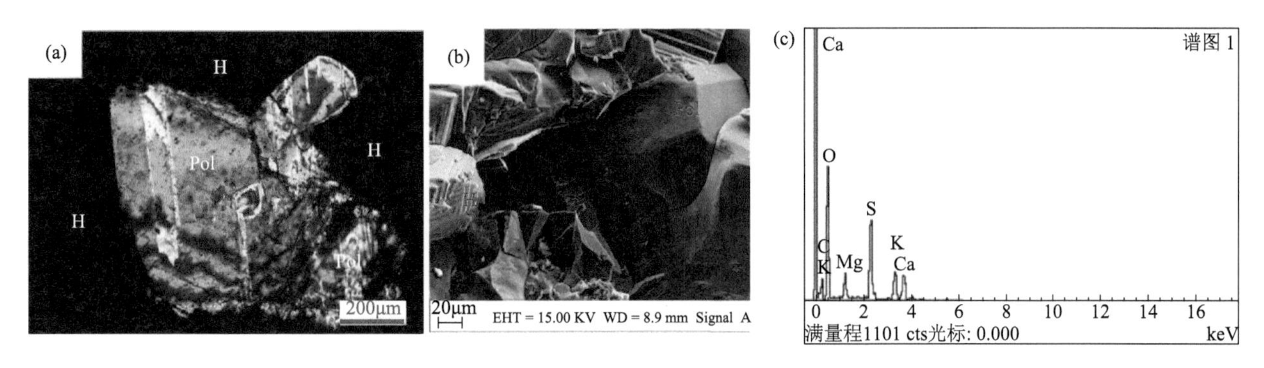

图2 盐晶颗粒杂卤石——新型杂卤石钾盐矿的显微特征

Fig. 2 Microscopic characteristics of polyhalite, a new type of polyhalite potash salt deposit.

（a）正交偏光下，被石盐晶体胶结的杂卤石颗粒：Pol-杂卤石，H-石盐；（b）扫描电镜下，较完整的杂卤石晶体，与石盐晶体边界清晰；（c）扫描电镜下，杂卤石能谱曲线；Pol-杂卤石，H-石盐

(a) Polyhalite grain cemented by halite crystal (crossed nicols); Pol-Polyhalit, H-Halite; (b) Fairly complete polyhalite crystal, showing clear boundary with halite; (c) Polyhalite power spectral curve under SEM. Pol-Polyhalite; H-Halite

扫描电镜下可见杂卤石自形晶体呈三斜晶系，贝壳状断口，解理不明显，晶体互相叠置紧密生长，杂卤石与石盐边界清晰，未见交代、穿插现象，结构简单，形态单一[图2（b）]，应为同期沉积的产物，为同生或原生杂卤石。

3.3 盐晶颗粒杂卤石岩的水溶特征

杂卤石团块、集合体等大小不等的内碎屑颗粒散布于石盐基质中[图2（b）]，在注入淡水后，作为胶结物的石盐基质迅速溶解，杂卤石颗粒失去支撑进入卤水溶液中，处于随机运动状态，并被进一步溶解于水中，成为可溶性内碎屑颗粒杂卤石。这些盐晶胶结的内碎屑颗粒杂卤石与钾石盐、光卤石等可溶性钾盐矿相当，便于水溶法开采，可通过对接井的方式进行注水溶采，生产成本大大降低，生产效率大为提高。溶解有内碎屑颗粒杂卤石的富钾卤水，可直接用于生产优质硫酸钾型钾肥或复合钾镁肥。因此，本文将这类“盐晶颗粒杂卤石岩”称为“一种新型杂卤石钾盐矿”。

4 结论

川东北普光地区发现的“盐晶颗粒杂卤石岩”——一种新型杂卤石钾盐矿，内碎屑颗粒中的杂卤石晶体结构简单，形态单一，与石盐、石膏、黏土矿物边界清晰，未见交代、穿插等关系，系同期沉积产物，为同生或原生杂卤石。大小不等的杂卤石内碎屑颗粒散布于石盐基质中，可直接采用水溶法溶采，成为能经济利用的大型整装优质硫酸钾（K_2SO_4）型钾盐矿，潜在经济价值巨大，有望成为中国新的大型海相固体钾盐战略基地。四川盆地三叠系赋存的这种“盐晶颗粒杂卤石岩”——一种新型杂卤石钾盐矿将成为中国海相钾盐勘查的主攻方向之一。

5 致谢

本文为国家重点研发计划课题“重点含钾盆地富钾规律、战略选区与深部探测技术示范”（2017YFC0602806）和中国地质调查项目资助的成果。感谢恒成公司王宁军副总、唐兵副总等在岩心编录过程中给予的支持和帮助！

第三部分

盐湖化工

Solid-liquid equilibria of KCl in polyethylene glycol 6000-H_2O mixed solvent at 288.2, 298.2, and 308.2 K: experiment and correlation*

Xudong Yu[1, 2], Qin Huang[2], Miangping Zheng[1], Lin Wang[2] and Maolan Li[2]

1. MNR Key Laboratory of Saline Lake Resources and Environments, Institute of Mineral Resources, CAGS, Beijing 100037, P. R. China
2. College of Materials and Chemistry & Chemical Engineering, Chengdu University of Technology, Chengdu 610059, P. R. China

Abstract The present study investigates the measurement of the phase equilibria of ternary systems KCl-PEG6000 (PEG with molecular weight 6000)-H_2O at 288.2, 298.2 and 308.2 K. All the experimental data concluding the solubility of KCl in PEG6000-H_2O mixed solvents, corresponding density, and refractive index were determined. According to the results, the phase diagrams at 288.2, 298.2 and 308.2 K are divided into the three parts of regions: unsaturated homogeneous liquid (L), one liquid and one solid KCl (L + S), one liquid and two solids KCl and PEG6000 (L + 2S); the area of (L + 2S) decreases with an increase of temperature, while the areas of (L) and (L + S) increase with an increase of temperature. With an increment of PEG6000 in mixed solvents, the solubilities of KCl and the density decreased; inversely, the refractive index increased at three temperatures. Comparing the diagrams of KCl-PEG1000/4000/6000-H_2O at 288.2, 298.2 and 308.2 K, the sizes of regions of (L) and (L + S) decreased and that of (L + 2S) increased with the increase of the molecular weight of PEG. A modified Pitzer model and Pitzer activity coefficient model were applied to simulate the equilibrium thermodynamics of the KCl-PEG6000-H_2O system at 288.2, 298.2, and 308.2 K. Comparing the experimental data and calculated dada, the thermodynamic models can give precise results with little error.

Keywords Polyethylene Glycol 6000; Solubility; KCl; Phase equilibria; Pitzer equation

Introduction

Saline lakes are valuable natural resources, and potassium-bearing saline lakes have become the most important source of potash production in Jordan, Israel, and China. There are 835 salt lakes with an area of $>1\ km^2$ on the Qinghai-Tibet Plateau. According to the method developed by Valyashk, the chemical types of saline lakes can be divided into three types: carbonate, sulfate, and chloride type (Zheng, 2014; Zheng et al., 2018). The salt lakes in Qaidam Basin mostly belong to the system of chloride type, and the coexistence of many ions in chloride type brine makes the relationship between salts more complicated; carnallite type double salt $MCl \cdot MgCl_2 \cdot 6H_2O$ ($M = K^+$, Rb^+, NH_4^+, $Li(H_2O)^+$), and solid solution [(K, Rb)Cl] and [(K, NH_4)Cl] were found in a previous work (Yu et al., 2012, 2015; Guo et al., 2017; Li et al., 2019a), which increases the difficulty of separation and extraction of potassium chloride form the brine. Commonly, the dissolution and crystallization behavior of salts in solution (pure liquids and mixed solvents) can be obtained through the phase equilibria study of the complex system at multiple temperatures, which are of significant importance for extracting KCl from liquid minerals. Therefore, a large number of studies of phase equilibria of potassium chloride in pure liquids have been done (Yu et al., 2012, 2014, 2015, 2017; Li et al., 2013, 2016, 2019a, 2019b; Guo et al., 2017; Gao et al., 2018).

It is worth mentioning that in the mixed solution of water-soluble organic solvents/ionic liquids + salt + water, the ionic liquids or water-soluble organic solvents and salt compete with each other for

* 本文发表在：Journal of Chemical Engineering of Japan, 2020, 53 (6): 229-236

the ability to dissolve in water. Therefore, an addition of new miscible component (organic solvents or ionic liquids) in the original aqueous solution may cause the salting-out effect and decrease the solubility of the salt, which can be applied to the crystallization for producing supersaturated solution. In the systems containing water-soluble organic solvents, salt, and water, organic solvents including methanol, ethanol, ethylene glycol, propanediol, and the soluble polymers such as polyethylene glycol (PEG) and dextran are often widely used in the mixed solution preparation. Meanwhile, the phase equilibria of salts in different mixed solvents have been carried out (Jimenez et al., 2009; Zafarani-Moattar et al., 2012; Zhao et al., 2014; Chen et al., 2015; Deyhimi et al., 2015; Zhong et al., 2016).

Among the above-mentioned organic solvents or soluble polymer, PEG is considered as one of the most applicable for the purpose of preparing the mixed solution, as it is low cost, nontoxic and non-pollution. Accordingly, phase equilibria of the system composed of salts-PEG-H_2O (salts: KCl, CsCl, Li_2SO_4, Na_2SO_4, $MgSO_4$) were determined (Taboada et al., 2005; Lovera et al., 2012; González-Amado et al., 2016; Sosa et al., 2017; Rodrigues Barreto et al., 2019; Yu et al., 2019a). The results show that the temperature, salt type, and the PEG molar mass can influence the phase equilibrium of the system. As a result, the systems KCl-PEG1000/4000-H_2O at three different temperatures were studied in our previous work (Huang et al., 2019; Yu et al., 2019b). The results show that the solubility of KCl in PEG-H_2O mixed solvent decrease with an increase in molecular weight of PEG due to the effect of steric hindrance getting stronger, which further leads to the change of phase region of KCl. Therefore, as a supplement of thermodynamic and a continuation of the previous work, the solid-liquid equilibrium for ternary system composed of PEG6000-potassium chloride-water at 288.2, 298.2 and 308.2 K was experimentally determined to study the influence of temperature and the PEG molar mass on the solid-liquid equilibrium (SLE). Meanwhile, the modified Pitzer equation and the original Pitzer activity coefficient model were applied to correlate the solid-liquid equilibrium data.

1 Experimental

1.1 Chemicals and instruments

Double-deionized water ($\kappa \leqslant 5.5 \times 10^{-6} S \cdot m^{-1}$) obtained from a water purification system (UPT-II-20T, Chengdu Ultrapure Technology Co., Ltd.) was used in all experiments. KCl and PEG6000 were obtained from Sinopharm Chemical Reagent Co., Ltd. (Shanghai) at 99.5%, and 99.0% purity, respectively. All materials were used without further purification.

The solubility measurement was carried out with a thermostatic evaporator (Inborn SHH250, China) having temperature stability of ± 0.1 K. An analytical balance (BSA124S, Sartorius) with an accuracy of ± 0.0001 g was employed to weigh the solutions. The refractive index (n_D) and density (ρ) were determined by the Abbe refractometer (WYA, China) with a precision of ± 0.0001 and the specific gravity method with a precision of ± 0.0002 $g \cdot cm^{-3}$.

1.2 Experimental methods

The phase equilibrium was achieved by mixing known masses of PEG and water with excess potassium chloride. A series of PEG6000-H_2O mixed solutions whose masses were known were prepared in clean 50 mL glass bottles with the PEG6000 mass fraction varying from 0 to saturation by an analytical balance. Then, excess salt KCl was dissolved in each PEG6000-H_2O mixed solvent to form the mixtures containing PEG6000, KCl, and H_2O. Placing magnets into glass bottles and stirring solutions on magnetic stirrers in the thermostatic evaporator at 288.2, 298.2 and 308.2 K for 72 h accelerated the equilibration. The samples were analyzed at regular intervals to determine whether the system achieved equilibrium. This was established as the point at which the composition of the sample remained constant. Once equilibria was reached, mixing was interrupted and samples were left to stand for at least 24 h. Subsequently, the clear liquid was collected by syringe and filtered for further determinations of concentration, density and refractive index. The contents of solid were measured by Schreinemakers' wet residue method (Fosbl et al., 2009).

1.3 Analytical methods

By means of measuring the concentration of the Cl^- ion to obtain the composition of KCl, this work adopts the $AgNO_3$ titration method (Institute of Qinghai Salt-Lake of Chinese Academy of Sciences, 1988).

The composition of PEG6000 was determined from the refractive index measurements with Eq. (1) below (Cheluget et al., 1994).

$$n_D = a_0 + a_1 \times w_{KCl} + a_2 \times w_{PEG} \tag{1}$$

Herein, a_0, a_1, a_2 denote the correlation coefficients. a_0 represents the refractive index of double-deionized water at 298.2 K; a_1, a_2 can be obtained by fitting the calibration curves where a_1 is 0.1335, a_2 is 0.1361. The refractive index was measured for a wide range of aqueous solutions [PEG 1000: 0–35% (w/w); KCl: 0–6% (w/w)].

Each sample was measured about three times until the values remained constant and the average of the three values was used to produce the final deterministic value.

2 Results and discussion

In our previous work (Huang et al., 2019; Yu et al., 2019b), the solubility of KCl in pure water at 288.2, 298.2 and 308.2 K was measured first to verify the accuracy and reproducibility of our experimental procedure, and the results showed that the experimental data are in good agreement with literature data.

The solubilities, densities (ρ), refractive indices (n_D), and composition of wet solids of the ternary system KCl-PEG6000-H_2O at 288.2, 298.2 and 308.2 K are given in Table 1 and Fig. 1 to 3. As depicted in Fig. 1 to 3, the following points, curves, and regions were obtained.

Table 1 Solubilities, densities (ρ), and refractive indices (n_D) for the KCl-PEG6000-H_2O ternary system at 288.2, 298.2, and 308.2 K[a]

No.	Density (ρ)	Refractive index	Composition of equilibrated solution		Composition of wet solid phase		Equilibrated solid phase
	[$g \cdot cm^{-3}$]	n_D	w_{KCl}	$w_{PEG6000}$	w_{KCl}	$w_{PEG6000}$	
				T = 288.2 K			
1A	1.1671	1.3680	0.2485	0.0000	—	—	KCl
2	1.1633	1.3716	0.2344	0.0383	0.6593	0.0170	KCl
3	1.1607	1.3758	0.2184	0.0782	0.6910	0.0309	KCl
4	1.1553	1.3806	0.1983	0.1203	0.7498	0.0375	KCl
5	1.1517	1.3852	0.1816	0.1637	0.6836	0.0633	KCl
6	1.1483	1.3888	0.1629	0.2093	0.8165	0.0459	KCl
7	1.1446	1.3937	0.1454	0.2564	0.6128	0.1162	KCl
8	1.1422	1.3993	0.1250	0.3063	0.7728	0.0795	KCl
9	1.1421	1.4048	0.1137	0.3545	0.6551	0.1380	KCl
10	1.1398	1.4099	0.0928	0.4082	0.5769	0.1904	KCl
11	1.1393	1.4159	0.0761	0.4619	0.2864	0.3568	KCl

Continued

No.	Density (ρ)	Refractive index	Composition of equilibrated solution		Composition of wet solid phase		Equilibrated solid phase
	[g·cm^{-3}]	n_D	w_{KCl}	$w_{PEG6000}$	w_{KCl}	$w_{PEG6000}$	
12	1.1348	1.4212	0.0547	0.5199	0.3592	0.3525	KCl
13B	1.1106	1.4244	0.0000	0.5961	—	—	—
				T = 298.2 K			
14C	1.1798	1.3704	0.2639	0.0000	—	—	KCl
15	1.1724	1.3726	0.2464	0.0377	0.6367	0.0182	KCl
16	1.1689	1.3765	0.2314	0.0769	0.6444	0.0356	KCl
17	1.1655	1.3812	0.2164	0.1175	0.8374	0.0244	KCl
18	1.1621	1.3851	0.2006	0.1599	0.6842	0.0632	KCl
19	1.1585	1.3899	0.1849	0.2038	0.8611	0.0347	KCl
20	1.1560	1.3946	0.1683	0.2495	0.6533	0.1040	KCl
21	1.1532	1.3997	0.1512	0.2971	0.6641	0.1176	KCl
22	1.1510	1.4051	0.1344	0.3462	0.6533	0.1387	KCl
23	1.1506	1.4113	0.1208	0.3956	0.5703	0.1934	KCl
24	1.1477	1.4164	0.1001	0.4500	0.5424	0.2288	KCl
25	1.1470	1.4228	0.0884	0.5014	0.4572	0.2985	KCl
26	1.1455	1.4289	0.0718	0.5569	0.3446	0.3933	KCl
27D	1.1077	1.4274	0.0000	0.6451	—	—	—
				T = 308.2 K			
28E	1.1847	1.3692	0.2851	0.0000	—	—	KCl
29	1.1785	1.3725	0.2646	0.0368	0.9611	0.0019	KCl
30	1.1735	1.3774	0.2512	0.0749	0.8173	0.0183	KCl
31	1.1694	1.3806	0.2350	0.1147	0.8649	0.0203	KCl
32	1.1634	1.3855	0.2162	0.1568	0.8404	0.0319	KCl
33	1.1607	1.3898	0.2027	0.1993	0.7037	0.0741	KCl
34	1.1572	1.3945	0.1841	0.2448	0.6623	0.1013	KCl
35	1.1543	1.3995	0.1678	0.2913	0.6667	0.1167	KCl
36	1.1513	1.4050	0.1510	0.3396	0.6199	0.1520	KCl
37	1.1494	1.4100	0.1309	0.3911	0.6031	0.1786	KCl
38	1.1484	1.4160	0.1187	0.4406	0.5380	0.2310	KCl
39	1.1466	1.4220	0.1051	0.4922	0.3406	0.3626	KCl
40	1.1454	1.4283	0.0884	0.5470	0.3608	0.3835	KCl
41	1.1418	1.4385	0.0709	0.6039	0.3929	0.3946	KCl
42F	1.1086	1.4403	0.0000	0.7304	—	—	—

a. Standard uncertainties u are $u(T) = 0.10$ K; $u_r(\rho) = 2.0 \cdot 10^{-4}$; $u_r(n_D) = 1.0 \cdot 10^{-4}$; u_r(KCl) = 0.0027; u_r(PEG6000) = 0.0005.

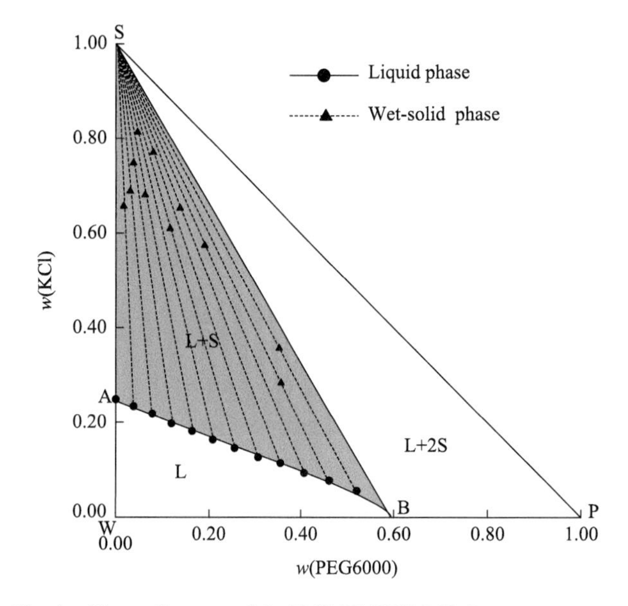

Fig. 1 Phase diagram of the KCl-PEG6000-H_2O ternary system at 288.2 K.

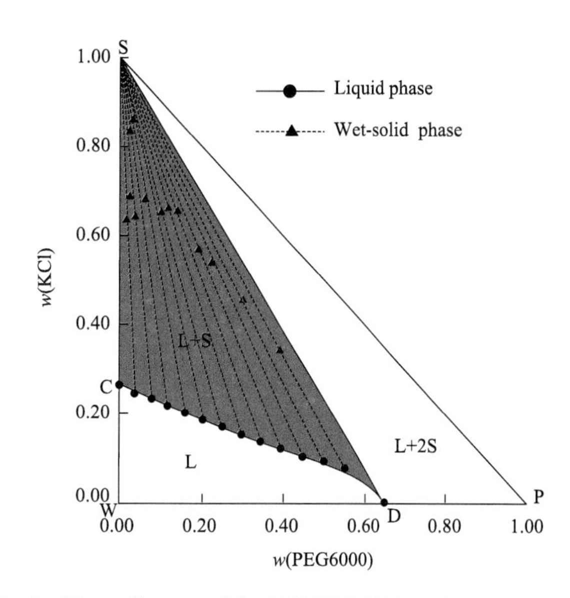

Fig. 2 Phase diagram of the KCl-PEG6000-H_2O ternary system at 298.2 K.

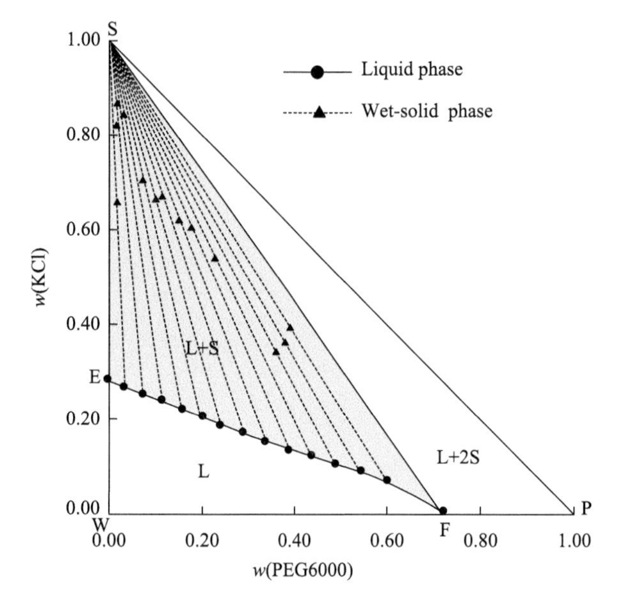

Fig. 3 Phase diagram of the KCl-PEG6000-H_2O ternary system at 308.2 K.

(1) Points A, C, and E represent the solubility of KCl in pure water at 288.2, 298.2 and 308.2 K; considering the crystallization behavior of PEG6000 has been studied previously (Zhang et al., 1997; Verheyen et al., 2001; Kidokoro et al., 2003; Hatakeyama et al., 2007; Cheng and Fan, 2009), and showed that PEG6000 crystallizes forming lamellae with chains either fully extended or folded once or twice, and chain-folding seems to be related to the crystallization procedure applied. Therefore, PEG6000 precipitated from mixed solution is assumed to be solid in this paper, and thus, points B, D, and F indicate the saturation points of PEG6000 in pure water at 288.2, 298.2, and 308.2 K;

(2) Curves AB, CD, and EF are saturation solubility curves of KCl in PEG6000-H_2O mixed solvents at 288.2, 298.2 and 308.2 K;

(3) The phase diagrams of the ternary system KCl-PEG6000-H_2O at 288.2, 298.2 and 308.2 K are split into three parts: areas WABW, WCDW, and WEFW are homogeneous areas of unsaturated liquid (L) containing KCl, PEG6000 and water; in areas ASBA, CSDC, and ESFE, one saturated liquid and one solid KCl (L + S) coexist; BSPB, DSPD, and FSPF regions are with the presence of one saturated liquid and two kinds of solids KCl and PEG6000 (L + 2S);

(4) The solubility of KCl in the PEG6000-H_2O mixed solvent decreased with the addition of PEG6000, indicating that PEG6000 has salting out effect on KCl.

The influence of the temperature on the phase relation ship of KCl-PEG6000-H_2O is expressed in Fig. 4. The results reveal that the area of (L + 2S) decreases with an increase of temperature, while the areas of (L) and (L + S) increase with an increase of temperature.

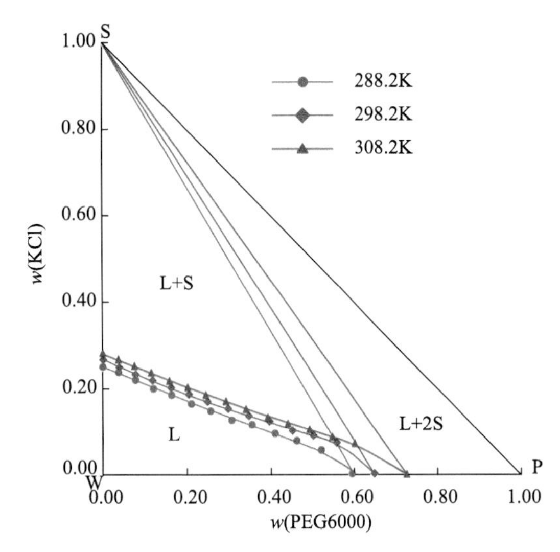

Fig. 4 Phase diagram of the KCl-PEG6000-H_2O ternary system at 288.2, 298.2 and 308.2 K.

The phase diagram of the KCl-PEG1000/4000 (PEG of molecular weight 1000 or 4000)-H_2O ternary system at 288.2, 298.2 and 308.2 K has been reported by our research group (Huang et al., 2019; Yu et al., 2019b). A comparison of the phase diagrams of the KCl-PEG1000-H_2O, KCl-PEG4000-H_2O, and KCl-PEG6000-H_2O systems at 288.2, 298.2 and 308.2 K (Fig. 5) indicates that: (1) in the system KCl-PEG1000-H_2O at 308.2 K, PEG1000 is in a liquid state and completely miscible with water, and thus the area of L + 2S disappears, while for other systems, the phase diagrams all comprise three areas: L, L + S, and L + 2S; (2) the sizes of regions of L and L + S decrease and that of L + 2S increases with the increase of the molecular weight of PEG when at the same temperature.

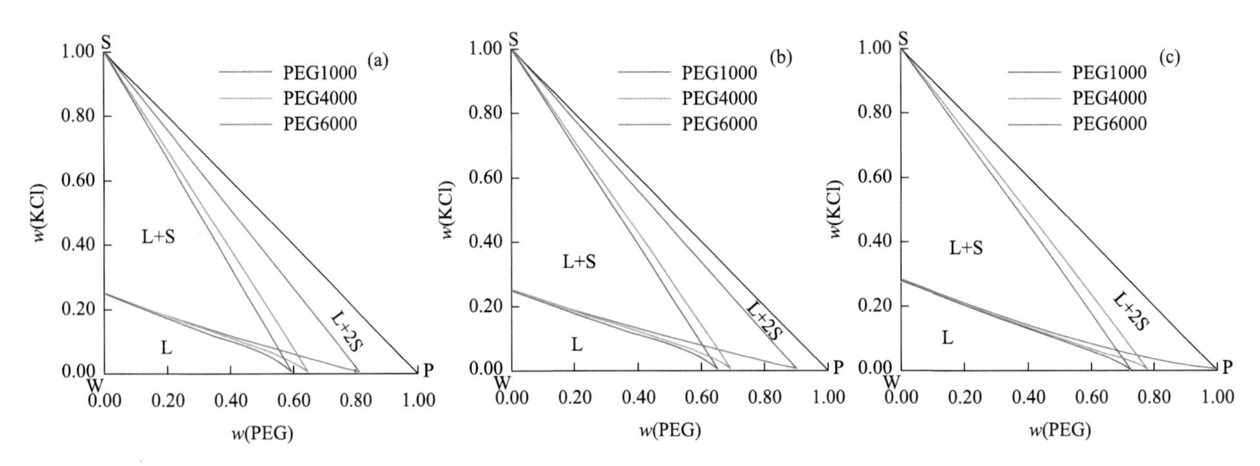

Fig. 5 Comparison of the phase diagrams of KCl-PEG1000/4000/6000-H_2O at 288.2, 298.2 and 308.2 K.

The diagrams of density vs composition (Fig. 6) and refractive index vs composition (Fig. 7) are plotted based on the experimental data in Table 1. From the figures above, with an increment of PEG6000 in mixed solvents, the solubility of KCl decreased, and the density of solution decreased; inversely, the refractive index increased at 288.2, 298.2 and 308.2 K.

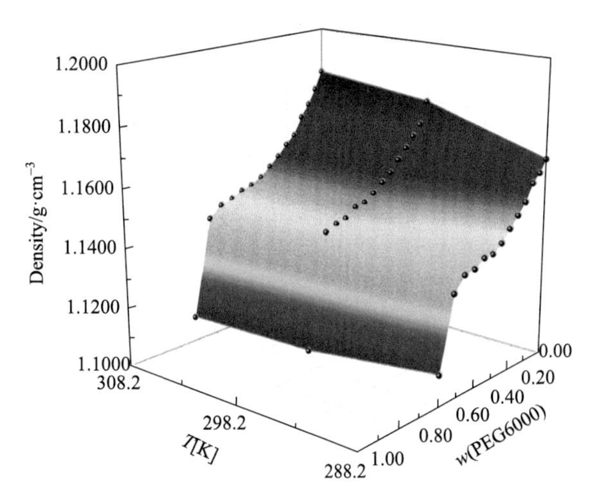

Fig. 6 Density vs composition diagram for the KCl-PEG6000-H_2O ternary system at 288.2, 298.2 and 308.2 K.

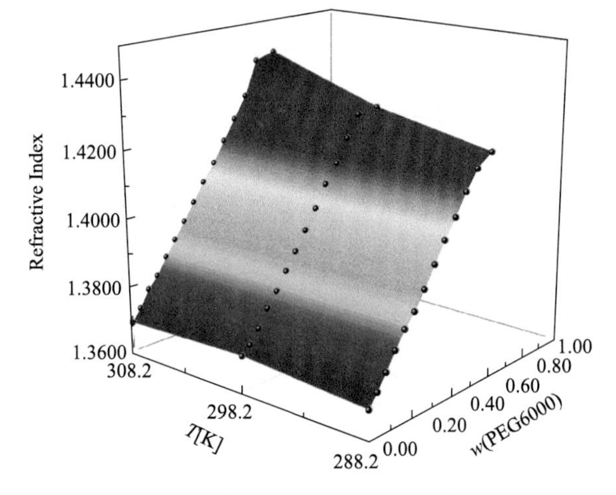

Fig. 7 Refractive index vs composition diagram for the KCl-PEG6000-H_2O ternary system at 288.2, 298.2 and 308.2 K.

3 Thermodynamic model

3.1 Original Pitzer activity coefficient model

An original Pitzer activity coefficient model (Pitzer and Mayorga, 1973) of electrolyte containing one component was considered to correlate thermodynamics data. Further, it had been used to predict the solubility of $NaCl/K_2SO_4$ in different mixed solvents (Kan et al., 2003; Jimenez et al., 2009). The main expressions of mean activity coefficient are as follows.

$$\ln\gamma_{\pm MX} = |Z_M Z_X| f^{\gamma} + \left(|\frac{2v_M \cdot v_X}{v}\right) m \cdot B_{MX}^{\gamma} + \left(\frac{2(v_M \cdot v_X)^{3/2}}{v}\right) m^2 \cdot C_{MX}^{\gamma} \tag{2}$$

Where

$$f^{\gamma} = -A^{\varphi}\left[I^{1/2}/(1+bI^{1/2}) + \frac{2}{b}\ln(1+bI^{1/2})\right] \tag{3}$$

$$I = \frac{1}{2}\sum m_i Z_i^2 \tag{4}$$

$$B_{MX}^{\gamma} = 2\beta_{MX}^{(0)} + \beta_{MX}^{(1)} \cdot g(\alpha_1 \cdot I^{1/2}) \tag{5}$$

$$g(x) = 2\left[1-\left(1+x-\frac{1}{2}x^2\right)\exp(-x)\right]/x^2 \tag{6}$$

$$C_{MX}^{\gamma} = 1.5C_{MX}^{\varphi} \tag{7}$$

Herein, subscripts M and X represent cation and anion, v_M, v_X, Z_X, Z_M are the number and charge number of cation and anion in one electrolyte, respectively; and m and I are molarity and ionic strength. In this study, α_1 and b are constants having values of 2.0, and 1.2, respectively. $\beta^{(0)}$, $\beta^{(1)}$, and C^{φ} are single salt parameters for the binary system KCl-H_2O, which can be calculated according to the relation with temperature from the literature (Guo et al., 2017) and are listed in Table 2. A^{φ} is the Debye-Hükel constant with the value of 0.3851, 0.3915, and 0.3985 at 288.2, 298.2, and 308.2 K, respectively, given by literature (Holmes et al., 1997).

Table 2 Pitzer binary parameters of the system KCl-H_2O at 288.2, 298.2 and 308.2 K

T [K]	$\beta^{(0)}$	$\beta^{(1)}$	C^{φ}	Ref.
288.2	0.0435	0.2043	−0.0003	Guo et al. (2017)
298.2	0.0480	0.2168	−0.0008	Guo et al. (2017)
308.2	0.0525	0.2293	−0.0013	Guo et al. (2017)

In this work, the effect of PEG6000 on salt in aqueous solution would be considered because it belonged to the type of non-electrolyte and existed in the ternary system KCl + PEG6000 + H_2O. A simple Born-type empirical expression had been given (Kan et al., 2003) for the effect of PEG6000 on salt as follows.

$$K_{sp} = (m_{KCl}\gamma_{\pm KCl}\gamma_N)^2 \tag{8}$$

Where

$$\gamma^N = 10^{\left(a+\frac{b}{T}+cI\right)x_N + d\cdot x_N^2} \tag{9}$$

$$OF = \sum(m_{exp} - m_{cal})^2 \tag{10}$$

Herein, K_{sp} represents solubility product constants of KCl which are available from literature (Huang et al., 2019; Yu et al., 2019b), and they are 5.5063, 7.9725 and 9.7737 at 288.2, 298.2 and 308.2 K, respectively. γ_N

represents the PEG6000 effect on salt, x_N is the mole fraction of PEG without salt, and $\gamma_N = 1$ when $x_{PEG} = 0$. Parameters a, b, c, and d can be acquired by minimizing the objective function OF. Parameters a, b, c, and d at different temperatures are presented in Table 3. According to the parameters in Tables 2 and 3, the solubility of KCl in mixed solvents PEG6000-H_2O at 288.2, 298.2 and 308.2 K can be calculated.

Table 3 Parameters a, b, c, and d of Born-type equation for the system KCl-PEG6000-H_2O at 288.2, 298.2 and 308.2 K

T [K]	a	b	c	d
288.2	131.0587	1.4487	−5.5742	2.1674
298.2	25.4914	1.0822	4.7457	1.2738
308.2	8.3243	1.0821	4.5793	1.2768

3.2 Modified Pitzer model

A modified Pitzer model which takes consideration of the long-range and short-range contribution was used to correlate the phase equilibrium data of ternary systems $(NH_4)_2SO_4$-PEG1000/4000-H_2O, Na_2CO_3-PEG1000/8000-H_2O and Na_2SO_4-PEG1000/8000-H_2O at 298.2 K (Wu et al., 1996). Further, the modified Pitzer model was also used to correlate the solubility of KCl in PEG1000/4000-H_2O mixed solvents at 288.2, 298.2, and 308.2 K in our previous work (Huang et al., 2019; Yu et al., 2019b). Accordingly, the modified Pitzer model was applied to correlate the phase equilibria data of the system KCl-PEG6000-H_2O at 288.2, 298.2 and 308.2 K.

The cross parameters $\beta_{12}^{(0)}$, $\beta_{12}^{(1)}$, and C_{112} and C_{122} of ternary systems KCl-PEG6000-H_2O were fitted by the solubility data in this study and listed in are Table 4. The physical properties of PEG6000 and H_2O at 288.2, 298.2 and 308.2 K are presented in Table 5. Based on the equations used in our previous work (Huang et al., 2019; Yu et al., 2019b), the calculation of the systems KCl-PEG6000-H_2O at 288.2, 298.2 and 308.2 K can be done.

Table 4 Binary and cross parameters for the KCl-PEG6000-H_2O ternary system at 288.2, 298.2 and 308.2 K[a]

Binary parameters					
T [K]	$B_{11}\times10^2$	$C_{111}\times10^5$	$\beta_{22}^{(0)}\times10^3$	$\beta_{22}^{(1)}\times10^2$	$C_{222}^{\varphi}\times10^3$
288.2	0.7088	1.296	−0.444	−1.0938	0.7356
298.2	0.7088	1.296	7.5231	−0.8338	0.9664
308.2	0.7088	1.296	9.9501	−1.0526	1.9808
Cross parameters					
T [K]	$\beta_{12}^{(0)}$	$\beta_{12}^{(1)}$	C_{112}	C_{122}	
288.2	0.0012031	0.0009665	0.0027536	0.0119737	
298.2	0.0272985	0.0103233	0.0003166	0.0097801	
308.2	0.117131	0.1468224	−0.000661	−0.002283	

a. The value of $B_{11}\times10^2$ and $C_{111}\times10^5$ were given by literature (Wu et al., 1996).

Table 5 Physical properties of the pure substances at the three studied temperatures

T [K]	Components	M [g·mol^{-1}]	V [m^3·mol^{-1}]	ρ [g·cm^{-3}]	r_i	D
288.2	H_2O	18.02	18.04×10^{-6}	0.9991	1	81.95
	PEG6000	6000	4.92×10^{-3}	1.1106	136	2.204
298.2	H_2O	18.02	18.07×10^{-6}	0.9970	1	78.34
	PEG6000	6000	4.93×10^{-3}	1.1077	136	2.204
308.2	H_2O	18.02	18.13×10^{-6}	0.9940	1	74.83
	PEG6000	6000	4.95×10^{-3}	1.1086	136	2.204

3.3 Solubility calculation

Based on Pitzer and Modified Pitzer models, the solubility of KCl in mixed solvents PEG6000-H_2O at three different temperatures (288.2, 298.2 and 308.2 K) can be calculated. The comparisons of experimental and calculated data are shown in Fig. 8–10. Meanwhile, the accuracy of the model was evaluated by the mean square deviation (σ) of experimental and calculated data, which can be calculated according to Eq. (11), and their results are presented in Table 6.

$$\sigma = \sqrt{\frac{\sum (w_{\text{exp}} - w_{\text{cal}})^2}{n}} \tag{11}$$

Herein, w_{exp} is the experimental value, w_{cal} represent the calculated value, n denote the number of samples. From Table 6 and Fig. 8–10, it can be found that the calculated solubility with the Pitzer model give better precision than the modified Pitzer model at 298.2 and 308.2 K, while the modified Pitzer model presented better accuracy than the Pitzer model at 288.2 K.

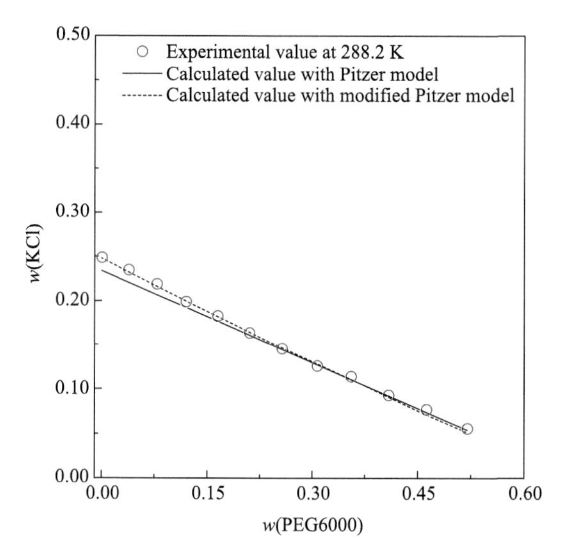

Fig. 8 The experimental and calculated solubility for ternary system KCl-PEG6000-H_2O at 288.2 K.

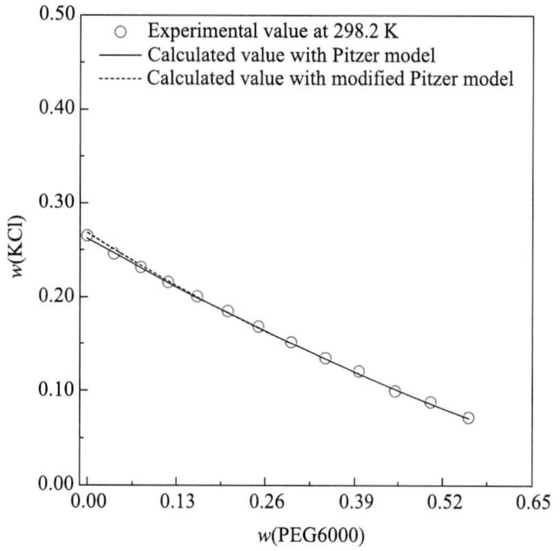

Fig. 9 The experimental and calculated solubility for ternary system KCl-PEG6000-H_2O at 298.2 K.

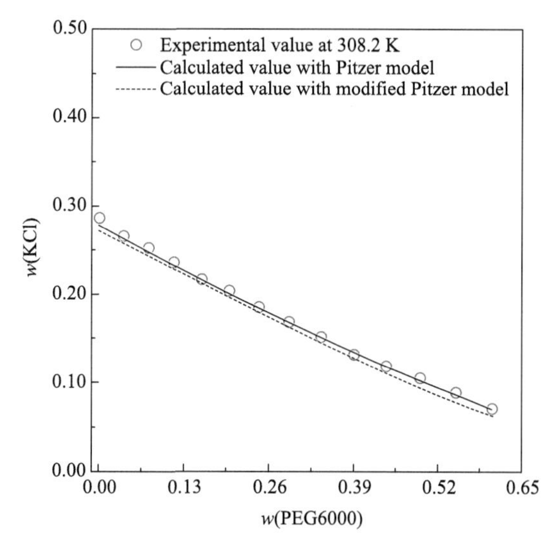

Fig. 10 The experimental and calculated solubility for ternary system KCl-PEG6000-H_2O at 308.2 K.

Table 6 The mean square deviation (σ) calculated with two model and experimental solubility

T [K]	288.2	298.2	308.2
Pitzer	0.0065	0.0013	0.0026
Modified Pitzer	0.0029	0.0026	0.0078

Conclusion

The solid-liquid phase equilibria of the system KCl-PEG6000-H_2O at 288.2, 298.2 and 308.2 K were carried out by using an isothermal dissolution method. The solubilities of KCl and the density decreased; inversely, the refractive index increased. These three phase diagrams all contain three areas: (L), (L + S), and (L + 2S); the area of (L + 2S) decreases with an increase of temperature, while the areas of (L) and (L + S) increase with an increase of temperature; the sizes of regions of (L) and (L + S) decrease and that of (L + 2S) increases with the increase of the molecular weight of PEG. Both the modified Pitzer model and Pitzer activity coefficient model are applied to simulate the solubility of KCl in PEG6000-H_2O mixed solvents. It shows that the calculated value agrees well with the experimental value.

Acknowledgement

This project was supported by the Research Fund from the NSFC (U1507111) and China Geological Survey (DD20201115).

Literature Cited

Cheluget, E. L., S. Gelinas, J. H. Vera and M. E. Weber; "Liquid-Liquid Equilibrium of Aqueous Mixtures of Poly (Propylene Glycol) with NaCl, " *J. Chem. Eng. Data*, **39,** 127-130 (1994)

Chen, J. X., Y. L. Zhong, J. Han, M. Su and X. N. Shi; "Liquid-Liquid Equilibria for Water + 1-Propanol (or 1-Butanol) + Potassium Chloride + Ammonium Chloride Quaternary Systems at 298.15 K, "*Fluid Phase Equilib.*, **397,** 50-57 (2015)

Cheng, G. W. and X. D. Fan; "Isothermal Crystallization Kinetics of Polyethylene Glycol, " *Polym. Mater. Sci. Eng. (Chengdu, China)*, **25,** 101-104 (2009)

Deyhimi, F., M. Abedi and Z. Karimzadeh; "Modeling of the Quaternary NaCl + KCl + CH_3OH + H_2O Mixed Electrolyte System: Binary and Ternary Mixing (Ion + Ion) and (Ion + Nonelectrolyte) Interaction Parameters, " *J. Chem. Thermodyn.*, **88,** 15-21 (2015)

Fosbl, P. L., K. Thomsen and E. H. Stenby; "Reverse Schreinemakers Method for Experimental Analysis of Mixed-Solvent Electrolyte Systems, " *J. Solution Chem.*, **38,** 1-14 (2009)

Gao, Y. Y., C. Ye, W.-Y. Zhang, L. Yang, S. H. Sang, and Y. Huang; "Phase Equilibria in the Ternary System $CaCl_2$-$SrCl_2$-H_2O and the Quaternary System KCl-$CaCl_2$-$SrCl_2$-H_2O at 373 K, " *J. Chem. Eng. Data*, **63,** 2738-2742 (2018)

González-Amado, M., E. Rodil, A. Arce, A. Soto and O. Rodríguez; "The Effect of Temperature on Polyethylene Glycol (4000 or 8000) - (Sodium or Ammonium) Sulfate Aqueous Two Phase Systems, " *Fluid Phase Equilib.*, **428,** 95-101 (2016)

Guo, L. J., H. J. Han, O. Y. Dong and Y. Yao; "Thermodynamics and Phase Equilibrium of the High Concentration Solid Solution-Aqueous Solution System KCl-RbCl-H_2O from T = 298.15 K to T = 323.15 K, " *J. Chem. Thermodyn.*, **106,** 285-294 (2017)

Hatakeyama, Y., H. Kasuga, M. Tanaka and H. Hatakeyama; "Cold Crystallization of Poly (Ethylene Glycol) -Water Systems, " *Thermochim. Acta*, **465,** 59-66 (2007)

Holmes, H. F., J. M. Simonson and R. E. Mesmer; "Additions and Corrections Aqueous Solutions of the Alkaline Earth Metal Chlorides. Corrected Constants for the Ion-Interaction Model, " *J. Chem. Thermodyn.*, **29,** 1363-1373 (1997)

Huang, Q., L. Wang, M. Li, P. Hu, X. Yu, H. Deng and Y. Zeng; "Measurements and Simulation of the Polyethylene Glycol 1000-Water-KCl Ternary System at 288.2, 298.2, and 308.2 K, " *J. Chem. Eng. Japan*, **52,** 325-332 (2019)

Institute of Qinghai Salt-Lake of Chinese Academy of Sciences; Analytical Methods of Brines and Salts, 2nd ed. (in Chinese), pp. 69-72, Chinese Science Press, Beijing, China (1988)

Jimenez, Y. P., M. E. Taboada and H. R. Galleguillos; "Solid-Liquid Equilibrium of K_2SO_4in Solvent Mixtures at Different Temperatures, " *Fluid Phase Equilib.*, **284,** 114-117 (2009)

Kan, A. T., G. Fu and M. B. Tomson; "Effect of Methanol and Ethylene Glycol on Sulfates and Halite Scale Formation, " *Ind. Eng. Chem. Res.*, **42,** 2399-2408 (2003)

Kidokoro, M., K. Sasaki, Y. Haramiishi and N. Matahira; "Effect of Crystallization Behavior of Polyethylene Glycol 6000 on the Properties of Granules Prepared by Fluidized Hot-Melt Granulation (FHMG), " *Chem. Pharm. Bull.*, **51,** 487-493 (2003)

Li, Z. Q., X. D. Yu, Q. H. Yin and Y. Zeng; "Thermodynamics Metastable Phase Equilibria of Aqueous Quaternary System LiCl + KCl + RbCl + H_2O at 323.15 K, " *Fluid Phase Equilib.*, **358,** 131-136 (2013)

Li, D. D., D. W. Zeng, X. Yin, H. J. Han, L. J. Guo and Y. Yao; "Phase Diagrams and Thermochemical Modeling of Salt Lake Brine Systems. II. NaCl + H_2O, KCl + H_2O, $MgCl_2$ + H_2O, and $CaCl_2$ + H_2O Systems, " *Calphad*, **53,** 78-89 (2016)

Li, C., B. Zhao, S. Wang, C. Y. Xue, H. F. Guo, D. Y. Wang and J. L. Cao; "Phase Diagrams of the Quinary System K^+, NH_4^+, Mg^{2+}// SO_4^{2-}, Cl^- -H_2O at 273.15 K and 298.15 K and Their Application, " *Fluid Phase Equilib.*, **499,** 112238 (2019a)

Li, D., Y. H. Liu, L. Z. Meng, Y. F. Guo, T. L. Deng and L. Yang; "Phase Diagrams and Thermodynamic Modeling of Solid-Liquid Equilibria in the System NaCl-KCl-$SrCl_2$-H_2O and Its Application in Industry, "*J. Chem. Thermodyn.*, **136,** 1-7 (2019b)

Lovera, J. A., A. P. Padilla and H. R. Galleguillos; "Correlation of the Solubilities of Alkali Chlorides in Mixed Solvents: Polyethylene Glycol + H_2O and Ethanol + H_2O, "*Calphad*, **38,** 35-42 (2012)

Pitzer, K. S. and G. Mayorga; "Thermodynamics of Electrolytes. II. Activity and Osmotic Coefficients for Strong Electrolytes with One or Both Ions Univalent, " *J. Phys. Chem.*, **77,** 2300-2308 (1973)

Rodrigues Barreto, C. L., S. de Sousa Castro, E. Cardozo de Souza Júnior, C. M. Veloso, L. A. Alcântara Veríssimo, V. S. Sampaio, O. R. Ramos Gandolfi, R. da Costa Ilhéu Fontan, I. C. Oliveira Neves and R. C. Ferreira Bonomo; "Liquid-Liquid Equilibrium Data and Thermodynamic Modeling for Aqueous Two-Phase System Peg1500 + Sodium Sulfate + Water at Different Temperatures, " *J. Chem. Eng. Data*, **64,** 810-816 (2019)

Sosa, F. H. B., D. de Araujo Sampaio, F. O. Farias, A. B. G. Bonassoli, L. Igarashi-Mafra and M. R. Mafra; "Measurement and Correlation of Phase Equilibria in Aqueous Two-Phase Systems Containing Polyethyleneglycol (2000, 4000, and 6000) and Sulfate Salts (Magnesium Sulfate and Copper Sulfate) at Different Temperatures (298.15, 318.15, and 338.15 K), "*Fluid Phase Equilib.*, **449,** 68-75 (2017)

Taboada, M. E., H. R. Galleguillos, T. A. Graber and S. Bolado; "Compositions, Densities, Conductivities, and Refractive Indices of Potassium Chloride or/and Sodium Chloride + PEG4000 + Water at 298.15 K and Liquid-Liquid Equilibrium of Potassium Chloride or Sodium Chloride + PEG4000 + Water at 333.15 K, "*J. Chem. Eng.* Data, **50,** 264-269 (2005)

Verheyen, S., P. Augustijns, R. Kinget and G. Van den Mooter; "Melting Behavior of Pure Polyethylene Glycol 6000 and Polyethylene Glycol 6000 in Solid Dispersions Containing Diazepam or Temazepam: A DSC Study, "*Thermochim. Acta*, **380,** 153-164 (2001)

Wu, Y. T., D. Q. Lin, Z. Q. Zhu and L. H. Mei; "Prediction of Liquid-Liquid Equilibria of Polymer-Salt Aqueous Two-Phase Systems by Modified Pitzer's Virial Equation, "*Fluid Phase Equilib.*, **124,** 67-79 (1996)

Yu, X. D., Y. Zeng and Z. X. Zhang; "Solid-Liquid Metastable Phase Equilibria in the Ternary Systems KCl + NH_4Cl + H_2O and NH_4Cl + $MgCl_2$ + H_2O at 298.15 K, "*J. Chem. Eng. Data*, **57,** 1759-1765 (2012)

Yu, X. D., Q. H. Yin, D. B. Jiang and Y. Zeng; "Metastable Equilibrium for the Quaternary System Containing with Lithium + Potassium + Magnesium + Chloride in Aqueous Solution at 323 K, "*Korean J. Chem. Eng.*, **31,** 1065-1069 (2014)

Yu, X. D., Y. Zeng, P. T. Mu, Q. Tan and D. B. Jiang; "Solid-Liquid Equilibria in the Quinary System LiCl-KCl-RbCl-$MgCl_2$-H_2O at T = 323 K, "*Fluid Phase Equilib.*, **387,** 88-94 (2015)

Yu, X. D., L. Wang, J. Chen and M. L. Li; "Salt-Water Phase Equilibria in Ternary Systems K + (Mg^{2+}), NH_4^+ //Cl^--H_2O at T = 273 K, "*J. Chem. Eng. Data*, **62,** 1427-1432 (2017)

Yu, X. D., L. Wang, M. L. Li, Q. Huang, Y. Zeng and Z. Lan; "Phase Equilibria of CsCl-Polyethylene Glycol (PEG) -H_2O at 298.15 K: Effect of Different Polymer Molar Masses ($PEG_{1000/4000/6000}$), "*J. Chem. Thermodyn.*, **135,** 45-54 (2019a)

Yu, X. D., Q. Huang, L. Wang, M. L. Li, H. Zheng and Y. Zeng; "Measurements and Simulation for Ternary System KCl-PEG4000-H_2O at T = (288, 298, and 308) K, "*J. Chem. Ind. Eng (China)*, **70,** 830-839 (2019b)

Zafarani-Moattar, M. T., E. Nemati-Kande and A. Soleimani; "Study of the Liquid-Liquid Equilibrium of 1-Propanol + Magnesium Sulphate and 2-Propanol + Lithium Sulphate Aqueous Two-Phase Systems at Different Temperatures: Experiment and Correlation, "*Fluid Phase Equilib.*, **313,** 107-113 (2012)

Zhang, X. X., H. Zhang, X. C. Wang, L. Hu and J. J. Niu; "Crystallization and Low Temperature Heat-Storage Behavior of PEG, "*J. Tianjin Inst. Text. Sci. Technol*, **16,** 11-14 (1997)

Zhao, D. D., S. N. Li, Q. G. Zhai, Y. C. Jiang and M. C. Hu; "Solid-Liquid Equilibrium (SLE) of the *N*, *N*-Dimethylacetamide (DMA) + MCl (M = Na, K, Rb, and Cs) + Water Ternary Systems at Multiple Temperatures, "*J. Chem. Eng. Data*, **59,** 1423-1434 (2014)

Zheng, M. P.; Saline Lakes and Salt Basin Deposits in China, pp. 41-52, Science Press, Beijing, China (2014)

Zheng, M. P., T. L. Deng and O. Aharon; Introduction to Salt Lake Sciences, pp. 24-27, Science Press, Beijing, China (2018)

Zhong, Y. L., J. X. Chen, J. Han, M. Su, Y. H. Li and A. D. Lu; "The Influence of Temperature on Liquid-Liquid-Solid Equilibria for (Water + 2-Propanol + KCl + NH_4Cl) Quaternary System, "*Fluid Phase Equilib.*, **425,** 158-168 (2016)

Experimental study of the Tibetan Dangxiong Co salt lake brine during isothermal evaporation at 25 ℃*

Zhen Nie[1, 2], Qian Wu[1, 2], Lingzhong Bu[1, 2], Yunsheng Wang[1, 2] and Mianping Zheng[1, 2]

1. MNR Key Laboratory of Saline Lake Resources and Environments, Institute of Mineral Resources, Chinese Academy of Geological Sciences, Beijing 100037, China
2. R&D Center for Saline Lakes and Epithermal Deposits, Chinese Academy of Geological Sciences, Beijing 100037, China

Abstract Production of lithium carbonate from brines has become the dominate trend in the world since the beginning of the century. Dangxiong Co, located in the interior of the Tibetan Plateau, China, is a carbonate-type lithium salt lake. The lake, rich in Li, B, K and other valuable elements, is of great economic value. The concentration rules of these elements and the salt crystallization paths in the brine were studied in an isothermal evaporation experiment at 25 ℃. The sequence in sedimentation of the primary salts, which crystallized from the brine during evaporation experiment at 25 ℃, was halite (NaCl)-trona ($Na_2CO_3 \cdot NaHCO_3 \cdot 2H_2O$)-zabuyelite ($Li_2CO_3$)-Glaserite ($3K_2SO_4 \cdot Na_2SO_4$)-sylvite (KCl)-borax ($Na_2B_4O_7 \cdot 10H_2O$). This is some different from what one may conclude with the metastable phase diagram of the quinary system Na-K-CO_3-SO_4-Cl-H_2O at 25 ℃. Lithium precipitation was a continuous process that occurred throughout the whole experiment. But, it was difficult to obtain high-grade lithium salt during the evaporation operation at 25 ℃. Potash was precipitated as Glaserite and sylvite in the experiment with high grade, which made the Dangxiong Co salt lake brine suitable to produce potash. Borax was precipitated in the late stage. High-grade borax could be obtained from Dangxiong Co salt lake brine. The experiment results indicate that the lithium carbonate exploiting technology that is being used on the Zabuye salt lake could be applicable to the Dangxiong Co salt lake.

Keywords Dangxiong Co salt lake; Isothermal evaporation; Lithium carbonate; Concentration characteristic; Crystallization path

Introduction

Exploitation of lithium resources has become more and more attractive recently because of its wide usage (Averill and Olson, 1978; Epstein et al., 1981; Hamzaoui et al., 2003; Hamzaoui et al., 2008; Hoshino, 2013), especially in portable devices and new energy production (Ebensperger et al., 2005; Chen et al., 2005; Wen et al., 2014). Production of lithium carbonate from brines is now rapidly increasing in China (Li et al., 2005; Zhu et al., 2008; Huang et al., 2008; Nie et al., 2011). There are more than 1500 salt lakes in China, which distribute in most provinces of China (Zheng et al., 2002); while, the lithium-rich salt lakes only distribute in Qinghai-Tibet Plateau (Zheng et al., 1988, 1989; Zheng and Liu, 2009). Table 1 shows the main lithium salt lakes in China. Among them, the Zabuye and Dangxiong Co are two typical carbonate-type lithium salt lakes in Tibet, and other salt lakes are located in Qinghai province.

* 本文发表在：Garbonates and Evaporites, 2019, 35: 5

Table 1 Chemical composition of main lithium salt lake brines in China (Song, 2000)

Salt lake	Chemical composition (%)								Mg/Li	ρ (g·cm^{-3})	Area (km^2)
	Na^+	K^+	Mg^{2+}	Ca^{2+}	Li^+	Cl^-	SO_4^{2-}	B_2O_3			
Qarham	2.37	1.25	4.890	0.051	0.031	18.80	0.44	0.009	1577	1.197	4224
Da Qadam	6.92	0.71	2.140	—	0.016	14.64	4.05	0.062	134	1.174	90
East Taij	5.13	1.47	2.990	0.020	0.085	14.95	4.78	0.110	35.2	1.266	201
West Taij	8.26	0.69	1.990	0.031	0.022	16.17	1.14	0.018	61.0	1.223	188
Yiliping	2.58	0.91	1.280	0.016	0.021	14.97	2.88	0.031	90.5	1.215	280
Zabuye	10.01	3.16	0.002	—	0.080	12.06	2.98	0.260	0.0025	1.295	247
Dangxiong Co	3.72	0.80	0.007	—	0.033	6.21	0.64	0.250	0.21	1.106	55

Dangxiong Co salt lake is located on the southwest side of the hinterland of the northern Tibetan Plateau, China, about 1100 km northwest of Lhasa, the capital of Tibet. The geographical coordinates of the lake are 86°38′00″–86°49′00″E, 31°30′00″–31°40′00″ N (Fig. 1). The elevation of the lake is currently at 4475 m above sea level. Its area is approximately 55 km^2. The brine of Dangxiong Co salt lake has a salinity of 130–180 g/L, which is unsaturated (Zheng et al., 1988; Wu et al., 2012). Dangxiong Co salt lake is a largescale deposit for lithium and boron as well as a middle-scale one for potassium. Moreover, the brine also contains rubidium, cesium and bromine. Therefore, the lake is very economically important.

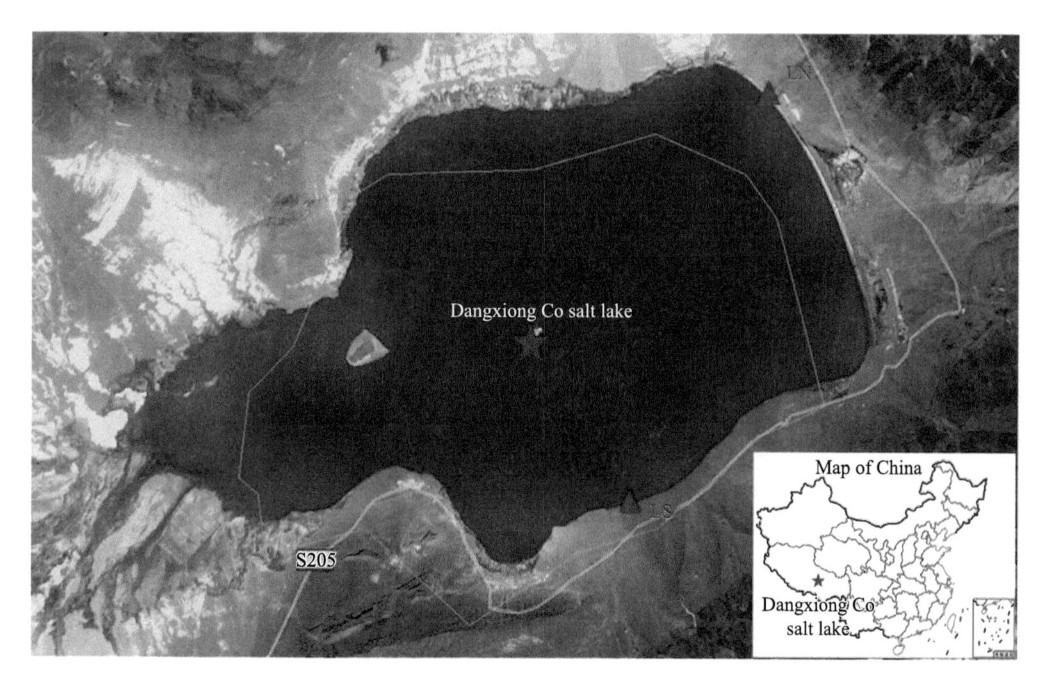

Fig. 1 Location of Tibetan Dangxiong Co salt lake, China and two sampling points on the north and south sides of the Dangxiong Co salt lake (LN—Sampling point on the north side of the Dangxiong Co salt lake, LS—Sampling point on the south side of the Dangxiong Co salt lake).

The solar pond technology takes the lake brine as the raw material and the solar energy as the heat energy. It uses the vast lakeside beach to build multi-level solar ponds, concentrates the brine and obtains the salts in stages, and has good economic and environmental effects on the utilization of salt lake resources. The local climatic condition with abundant solar energy, strong evaporation and low precipitation, is very suitable for the exploitation of salt lake brine by the solar pond technology in Qinghai and Tibet Plateau. To design a better solar pond technology, it is necessary to know the evaporative concentration characteristics and the mineral crystallization paths of the lake brines by doing the isothermal evaporation experiments (Menta and Dhar, 1979; Luis and Nancy, 1983).

To develop technology for exploitation of the mineral resources, isothermal evaporation experiments were carried with the brines of the Zabuye and Taijinaier salt lakes at 25 ℃ (Chen et al., 1980; Zhang et al., 1994; Yang et al., 1996; Zheng et al., 2007; Nie et al., 2010). Brine in Dangxiong Co, a carbonate-type salt lake, has some unique characteristics that make it different from the classical seawater system. Some field evaporation and cooling evaporation experiments had been done with the Dangxiong Co salt lake brine in Tibet and in lab, respectively (Wu et al., 2012; Wang et al., 2011). In spite of this, there is still a lack of basic information on the isothermal evaporation process of Dangxiong Co salt lake brine. To make full use of the local superior natural resources and the separation of useful mineral resources from salt lake brine, it is very necessary to summarize the evaporation and crystallization regularity of the salt lake brine under different temperatures. The results of this study lay the foundation of process route for the comprehensive development and utilization of Dangxiong Co salt lake brine.

In this paper, the results of the isothermal evaporation experiment at 25 ℃ with brine of the Dangxiong Co salt lake were described. The evaporative concentration characteristics and mineral crystallization paths were studied, and the difference of the brine with other carbonate-type lithium salt lakes was also discussed. The recommendations of the exploitation technology for the Dangxiong Co salt lake were provided.

Materials and methods

The brine samples for the experiment were collected from Dangxiong Co salt lake surface brine in October 2015 by 25-L bucket and transported to Beijing in a sealed container. The brine has a pH of 9.33, and a specific gravity of 1.106 at 25 ℃. The chemical composition of the brine sample labeled as Dxc-1 is shown in Table 3.

A field scientific observation station has been established in the Dangxiong Co salt lake area for more than 15 years, and the lake brine samples have been regularly observed and analyzed every month. According to the previous data analysis, the chemical composition of the salt lake brine is relatively stable and has not changed greatly except that a small range of desalination is caused by the global warming and glaciers melting. Therefore, the chemical components of the water sample (Dxc-1) in this study are representative, and the experimental results could be repeatable. The analysis results of chemical compositions of lake water before and after sampling ‘Dxc-1’ are shown in Table 2, and two sampling points on the north and south sides of the Dangxiong Co salt lake are shown in Fig. 1.

Table 2 Analysis of chemical compositions of lake water before and after sampling ‘Dxc-1’ (LN—Sampling point on the north side of the Dangxiong Co salt lake, LS—Sampling point on the south side of the Dangxiong Co salt lake)

Sample no.	Na^+ (mg/L)	K^+ (mg/L)	CO_3^{2-} (g/L)	Cl^- (g/L)	SO_4^{2-} (g/L)	Li^+ (mg/L)	B_2O_3 (mg/L)
Dxc-1	55,200	10,300	16.9	75.0	5.62	320	2580
LN2015-2-1	58,389	10,885	11.0	76.2	5.74	325	2616
LN2015-6-1	57,636	10,839	11.0	75.7	5.63	310	2326
LN2016-2-1	48,039	9787	10.5	65.9	5.75	301	2118
LN2016-6-1	47,880	8915	10.8	66.5	5.64	291	2069
LN2016-10-1	46,757	8816	11.4	65.0	5.56	310	2147
LS2015-2-1	57,552	10,986	11.4	76.4	5.59	315	2675
LS2015-6-1	56,117	10,237	10.5	72.1	5.59	291	2415
LS2016-2-1	47,403	8790	11.1	66.3	5.54	338	2389
LS2016-6-1	46,414	7886	10.1	58.7	4.96	312	2196
LS2016-10-1	45,396	8095	10.0	60.6	5.05	294	2077

The isothermal evaporation experiment was carried out in a constant temperature lab, which has an area of about 10 m^2. The temperature of the lab can be controlled at a constant value between −30 ℃ and 40 ℃ and has a precision of ± 3 ℃. The ambient temperatures at the salt lake generally range from −15 ℃ in January to 28 ℃ in late summer. The isothermal evaporation experiment was performed in the constant temperature lab at 25 ℃, and the temperature is within the range of the local temperature fluctuations.

A 10-kg brine sample was placed in an experimental plastic rectangle vessel of 15-cm height. An electric fan was used continuously at a constant wind speed (2.5–3 m/s) as an imitation of the natural wind to accelerate the speed of the brine evaporation. And a dehumidifier was used to control the humidity of the lab under 50%. Double deionized water with a conductivity of less than 1×10^{-4}s/m was used during the experiment.

During the experiment, the property of the brine and the precipitated salts, if there were any, was examined daily. When a new mineral was observed or after evaporating 3–4 days, liquid and solid phases in the brine were separated, and the physicochemical properties of the brine were measured. A 25-mL sample of the clarified solution was removed from the brine into a 250-mL volumetric flask with a filter pipette. The sample was then diluted to determine the composition of the liquid phase with the double deionized water. The salt sample was also analyzed and determined with an Olympus digital polarizing microscope (Olympus Company, Japan). The liquid samples were numbered after the sequence they sampled, while the solid samples were numbered responding to the liquid samples. The remainder of the brine was put in the tank to continue to evaporate. Repeat as it is mentioned, until the brine evaporates completely (Wang et al. 2011).

The content of K^+ and Li^+ ions in the liquid and solid phases was analyzed by atomic absorption spectrophotometry (WFX-120 atomic absorption spectrophotometer, Beijing Rayleigh Analytical Instrument Corp., Beijing, China). The concentration of Cl^- ion was determined by silver nitrate volumetric titration. The SO_4^{2-} ion was analyzed by barium sulfate gravimetric method. The $B_4O_7^{2-}$ and CO_3^{2-} ions were analyzed by acid-base titration in the presence or in the absence of mannitol. The Na^+ion concentration was determined by total ions balance calculation (Analysis Group, Qinghai Saline Lake Institute, 1988). The minerals in the solid phases were identified using the Olympus digital polarizing microscope.

Results

Chemical composition of brine and salts in experiment

After the experiment, we got the data of salt and corresponding equilibrium brine during evaporation. The chemical compositions of liquid and solid phases at 25 ℃ with the Jänecke index which is the mole percent of the various ions relative to 100 mol of total dry salts are listed in Tables 3 and 4. The percentages of salt minerals in solid phases precipitated are shown in Table 5.

Table 3 Chemical composition of Dangxiong Co salt lake brine during 25 ℃ isothermal evaporation

Sampling date	Sample no.	Specific gravity	Chemical composition (g/L)										Salinity	Jänecke index		
			Li^+	Na^+	K^+	Mg^{2+}	Ca^{2+}	Cl^-	HCO_3^-	CO_3^{2-}	SO_4^{2-}	$B_4O_7^{2-}$		2K	SO_4	CO_3
2016.6.25	Dxc-1	1.106	0.32	55.2	10.3	0.086	<0.002	75.0	4.22	16.9	5.62	2.87	170.52	27.92	12.40	59.69
2016.6.29	Dxc-2	1.131	0.40	70.7	13.6	0.110	<0.002	89.2	3.35	12.4	6.94	3.63	200.33	38.41	15.96	45.63
2016.7.5	Dxc-3	1.177	0.56	100.0	18.8	0.120	0.008	123.0	4.68	16.7	9.16	4.92	277.94	39.15	15.53	45.32
2016.7.8	Dxc-4	1.206	0.64	115.2	22.2	0.088	<0.002	145.2	3.80	20.6	11.60	2.91	322.32	37.96	16.15	45.90
2016.7.12	Dxc-5	1.227	0.72	125.8	24.8	0.096	<0.002	162.0	2.84	24.4	12.76	3.34	356.85	37.02	15.51	47.47
2016.7.19	Dxc-6	1.238	0.74	132.0	25.8	0.100	<0.002	172.8	1.26	26.2	13.26	3.45	375.71	36.47	15.26	48.27
2016.7.23	Dxc-7	1.241	0.78	134.8	27.0	0.100	<0.002	177.6	3.18	25.6	13.96	3.55	386.67	37.64	15.84	46.51
2016.7.27	Dxc-8	1.240	0.76	130.4	25.0	0.098	<0.002	179.6	1.60	44.8	13.90	3.52	399.77	26.40	11.95	61.65

Continued

Sampling date	Sample no.	Specific gravity	Chemical composition (g/L)										Salinity	Jänecke index		
			Li^+	Na^+	K^+	Mg^{2+}	Ca^{2+}	Cl^-	HCO_3^-	CO_3^{2-}	SO_4^{2-}	$B_4O_7^{2-}$		2K	SO_4	CO_3
2016.8.1	Dxc-9	1.229	0.74	126.6	24.8	0.092	<0.002	164.4	3.08	26.0	13.82	3.52	363.14	35.46	16.09	48.45
2016.8.6	Dxc-10	1.239	0.76	131.6	25.4	0.088	<0.002	170.6	3.22	25.4	13.58	3.48	374.22	36.52	15.89	47.59
2016.8.9	Dxc-11	1.244	0.80	137.8	26.4	0.068	<0.002	181.8	3.20	27.2	15.16	3.70	396.19	35.59	16.64	47.78
2016.8.13	Dxc-12	1.249	0.80	135.4	26.6	0.019	<0.002	181.0	1.20	27.6	15.94	3.73	392.31	35.21	17.18	47.61
2016.8.17	Dxc-13	1.251	0.82	136.2	27.0	0.010	<0.002	184.2	2.84	27.4	17.16	3.77	399.41	35.21	18.22	46.57
2016.8.25	Dxc-14	1.258	0.92	135.0	30.2	<0.002	<0.002	188.4	3.46	30.6	17.42	4.27	410.27	35.84	16.83	47.33
2016.9.3	Dxc-15	1.272	1.10	131.8	36.6	<0.002	<0.002	173.8	4.32	63.8	20.60	5.28	437.30	26.81	12.28	60.90
2016.9.9	Dxc-16	1.290	1.30	140.4	47.4	<0.002	<0.002	176.0	5.14	47.4	28.60	6.75	452.99	35.79	17.58	46.63
2016.9.15	Dxc-17	1.318	1.30	137.8	72.2	<0.002	<0.002	176.2	–	61.6	30.60	11.60	491.30	40.70	14.04	45.25
2016.9.21	Dxc-18	1.372	1.62	176.6	69.4	<0.002	<0.002	147.2	–	98.4	27.20	25.24	545.66	31.58	10.08	58.35
2016.9.30	Dxc-19	1.376	1.18	179.8	68.6	<0.002	<0.002	134.2	–	106.2	22.20	22.94	535.12	30.48	8.03	61.49
2016.10.14	Dxc-20	1.380	0.80	185.6	67.0	<0.002	<0.002	126.8	–	114.2	22.20	16.48	533.08	28.65	7.73	63.63
2016.10.22	Dxc-21	–	0.82	181.8	66.4	<0.002	<0.002	125.4	–	111.2	20.60	17.45	523.67	29.11	7.35	63.53
2016.11.1	Dxc-22	–	0.68	185.6	70.0	<0.002	<0.002	133.2	–	193.6	21.00	8.94	613.02	20.63	5.04	74.34

Table 4 Chemical compositions of salts precipitated with its corresponding brine concentration rate and salt deposition rate during 25 ℃ isothermal evaporation of Dangxiong Co salt lake brine

Sampling date	Sample no.	Mass percentage of each ion accounting to total mass of anion and cation (%)										Brine concentration rate (%)	Salt deposition rate (%)
		Li^+	Na^+	K^+	Mg^{2+}	Ca^{2+}	Cl^-	$B_4O_7^{2-}$	HCO_3^-	CO_3^{2-}	SO_4^{2-}		
2016.8.13	SDxc-12	0.016	40.73	0.43	0.180	0.019	57.10	0.108	0.00	0.00	0.11	42.36	0.16
2016.8.17	SDxc-13	0.027	36.65	0.47	0.630	0.012	54.44	0.233	6.15	1.08	0.30	40.94	0.56
2016.8.25	SDxc-14	0.017	38.52	0.59	0.140	0.016	56.14	0.154	2.60	0.06	0.19	36.52	1.74
2016.9.3	SDxc-15	0.023	40.22	0.68	0.037	0.013	56.89	0.154	1.71	0.01	0.23	26.11	5.13
2016.9.9	SDxc-16	0.120	38.67	0.87	0.120	0.018	55.53	0.233	1.54	1.39	0.34	21.93	7.64
2016.9.15	SDxc-17	0.075	34.45	5.22	<0.005	0.032	47.20	0.330	1.98	1.59	7.37	12.35	11.00
2016.9.21	SDxc-18	0.220	25.30	16.20	<0.005	0.012	41.11	0.718	2.72	6.23	5.24	5.29	13.86
2016.9.30	SDxc-19	0.340	28.19	10.77	<0.005	<0.005	23.59	8.113	0.00	14.82	4.53	3.05	15.04
2016.10.14	SDxc-20	0.260	27.23	11.37	<0.005	0.008	23.16	10.016	0.00	13.71	3.42	1.81	15.55
2016.10.22	SDxc-21	0.120	31.57	11.11	<0.005	0.008	20.25	4.200	1.75	26.30	3.86	0.96	15.88
2016.11.1	SDxc-22	0.140	30.31	10.95	<0.005	0.010	20.97	5.170	0.00	20.34	3.46	0.11	16.03

Table 5 Mass percentage of salt mineral in solid phase precipitated (%)

Sample no.	Li_2CO_3	$Na_2SO_4\cdot3K_2SO_4$	$Na_2B_4O_7\cdot10H_2O$	$Na_2CO_3\cdot NaHCO_3\cdot H_2O$	$Na_2CO_3\cdot H_2O$	KCl	NaCl
SDxc-12	0.08	0.18	0.25	0.00	0.00	0.66	98.82
SDxc-13	0.14	0.52	0.57	22.83	0.00	0.55	75.38
SDxc-14	0.09	0.32	0.37	9.48	0.00	0.89	88.84
SDxc-15	0.12	0.38	0.36	6.01	0.00	0.98	92.16
SDxc-16	0.62	0.57	0.56	5.57	0.00	1.23	91.43
SDxc-17	0.39	12.49	0.79	7.19	0.00	1.35	77.78
SDxc-18	1.22	9.43	1.83	10.48	0.01	25.78	51.26

Continued

Sample no.	Li_2CO_3	$Na_2SO_4·3K_2SO_4$	$Na_2B_4O_7·10H_2O$	$Na_2CO_3·NaHCO_3·H_2O$	$Na_2CO_3·H_2O$	KCl	NaCl
SDxc-19	2.18	9.44	24.01	0.00	0.01	18.38	45.98
SDxc-20	1.61	6.90	28.70	0.00	0.01	20.64	42.13
SDxc-21	0.99	10.30	15.91	10.00	0.01	25.74	37.05
SDxc-22	1.07	8.56	18.15	0.00	0.01	24.09	48.12

Minerals precipitated during 25 ℃ isothermal evaporation process

During 25 ℃ isothermal evaporation experiment of Dangxiong Co salt lake brine, the minerals precipitated successively as below:

Halite

Halite + Trona

Halite + Trona + Zabuyelite

Halite + Trona + Glaserite + Zabuyelite

Halite + Glaserite + Sylvite + Zabuyelite

Halite + Broax + Glaserite + Sylvite + Zabuyelite

Discussion

Crystallization path

Theoretical crystallization path

The ions of Na^+, Cl^-, K^+, CO_3^{2-} and SO_4^{2-} together make up 95 wt% of the total salt content (Table 4). To explain the concentration and crystallization behaviors of the Dangxiong Co salt lake brine, the metastable phase diagram of the quinary system Na-K-CO_3-SO_4-Cl-H_2O at 25 ℃ (Fang et al., 1991) was used in this paper. There are five solid-phase regions saturated with NaCl in the phase diagram, such as sylvite, glaserite, sodium carbonate heptahydrate, thenardite, and burkeite, respectively. The compositional point of the original brine for the experiment, marked as point L_0shown in Fig. 2, is located in the glaserite region of the phase diagram. When the brine is saturated, halite precipitates first. Then, glaserite crystallized from the brine next. The liquid-phase point of the brine moves to point F along with line GF, which is an extension of line GL_0, and crosses the common crystallization line of glaserite and sylvite at point F. When the liquid-phase point gets to point F, sylvite begins to crystallize with glaserite together. Then, the liquid-phase point moves to point E, which is an invariant point of the quinary system, and the glaserite, sylvite and sodium carbonate heptahydrate precipitate together here.

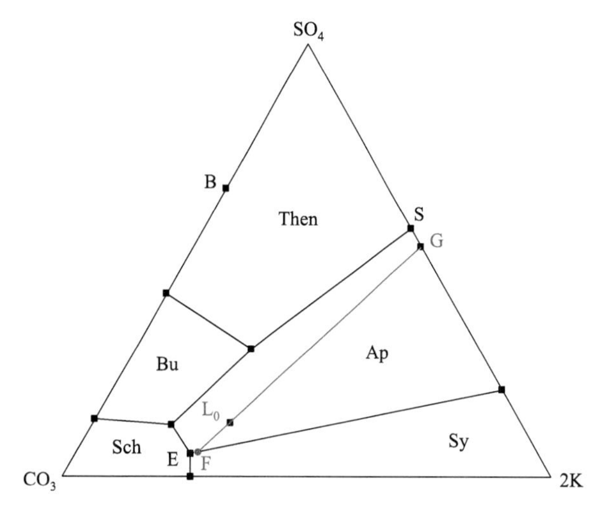

Fig. 2 Theoretical crystallization path of experimental brine in quinary water-salt system metastable phase diagram Na-K-CO_3- SO_4-Cl-H_2O (at 25 ℃) (Fang et al., 1991) (Sy-KCl, Ap-$3K_2SO_4·Na_2SO_4$, Then-Na_2SO_4, Bu-$2Na_2SO_4·Na_2CO_3$, Sch- $Na_2CO_3·7H_2O$).

The evaporation of the brine ends at point E. From the phase diagram, the conclusion can be reached that the sequence of mineral precipitation from the Dangxiong Co salt lake brine during the isothermal evaporation at 25 ℃ is NaCl, $3K_2SO_4·Na_2SO_4$, KCl and $Na_2CO_3·7H_2O$.

Real crystallization path

The original brine of the Dangxiong Co salt lake is unsaturated. Therefore, the quaternary water-salt system metastable phase diagram of Na-K-SO_4-Cl-H_2O at 25 ℃ can be used to explain the mineral crystallization path in the early stage of brine evaporation. The real chemical compositions of the brine during the evaporation process were all plotted in the phase diagram as shown in Fig. 3. The compositional point of the unsaturated original brine is located in the halite region. Therefore, the brine concentrates to saturation at the first stage of the evaporation. Then, the halite precipitates firstly. The second salt crystallized is the glaserite. Finally, the sylvite precipitates with the halite and glaserite together. It could be seen in Fig. 3 that the evaporation will end at the point H in the phase diagram.

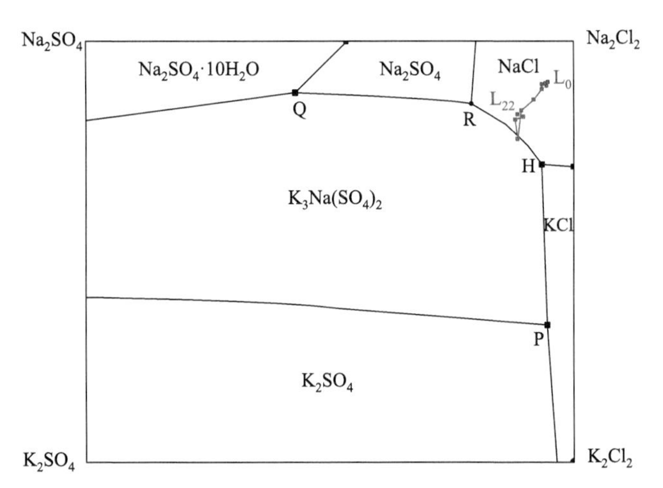

Fig. 3 Crystallization path of experimental brine in quaternary water-salt system metastable phase diagram Na-K-SO_4-Cl-H_2O (at 25 ℃).

But when the carbonate was saturated in the brine with evaporation, the quaternary water-salt system metastable phase diagram Na-K-SO_4-Cl-H_2O cannot respond the concentration behavior of the brine, and the metastable phase diagram of the quinary water-salt system Na-K-CO_3-SO_4-Cl-H_2O at 25 ℃ should be used. The real chemical compositions of the brine during the evaporation were plotted in the quinary water-salt system metastable phase diagram Na-K-CO_3-SO_4-Cl-H_2O at 25 ℃ as shown in Fig. 4. The compositional point of the original brine for the experiment is located in the glaserite region of the metastable phase diagram. It is halite but not glaserite that precipitated firstly from the brine during the isothermal evaporation process, because the brine is saturated with NaCl firstly when it is concentrated. Then, due to the high concentration of carbonate in the brine, trona becomes saturated as the evaporation proceeds. Along with the precipitation of trona from brine, the liquid-phase equilibrium points basically move far from the Na_2CO_3 solid-phase point of the phase diagram. Then, Li_2CO_3 becomes saturated and zabuyelite is precipitated together with halite and trona. After that, potassium becomes saturated in the brine, and it precipitates as glaserite from the brine. Later, KCl reached its saturation, and sylvite is precipitated. In the last stage of the evaporation process, borax precipitates in large amounts. The sequence of mineral precipitation from the brine is, thus, halite, trona, zabuyelite, glaserite, sylvite and borax.

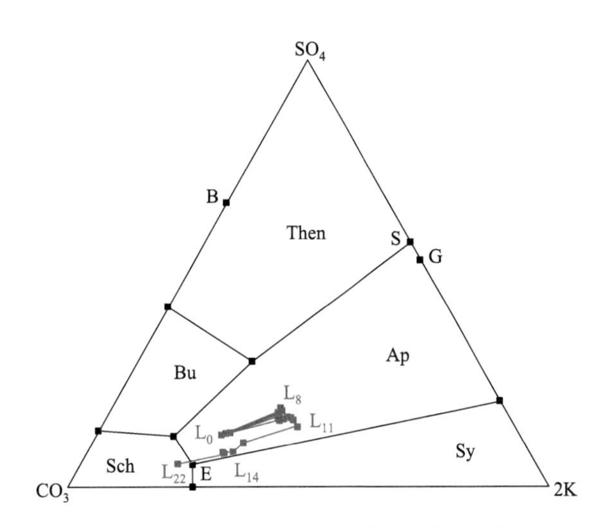

Fig. 4 Crystallization path of experimental brine in quinary water-salt system metastable phase diagram Na-K-CO_3-SO_4-Cl-H_2O (at 25 ℃).

It can be seen from Fig. 4 that the minerals precipitated from Dangxiong Co salt lake brine during isothermal evaporation at 25 ℃ are in accordance with those concluded from the real crystallization path drawn in the metastable phase diagram of the Na-K-CO_3-SO_4-Cl-H_2O quinary system at 25 ℃. Because of the effect of high concentration of carbonate in brine, the whole brine concentration rule and salt precipitation path of the brine during its 25 ℃ isothermal evaporation are essentially different from those deduced from the metastable phase diagram as shown in Fig. 2. The trona appeared very early in the evaporation; it is the key point which makes the difference. Other researchers also found this (Wang et al. 2011; Wu et al. 2012).

The concentration and precipitation rule of the main elements

The concentration and precipitation behavior of lithium in the Dangxiong Co salt lake brine at the temperature of 25 ℃ and its relation to the evaporation rate are shown in Fig. 5. As unsaturated in the brine, lithium is concentrated in the brine in the early and middle stages of the evaporation process. And the lithium concentration in the brine rises up with the specific gravity. When the brine concentration rate drops to 5.29%, lithium attains saturation in brine and begins to precipitate as mineral of lithium carbonate, which is also called zabuyelite. One peak of lithium precipitation was observed during the experiment when the brine concentration rate is 3.05%, and it responds to the decline of lithium concentration in the brine. It could be seen from Fig. 5 that the lithium precipitates only when the brine evaporates and condenses to a certain stage, but a small amount of lithium was entrained with the mother liquor in the early and middle stages of evaporation. It is difficult to obtain a high-grade lithium salt through evaporation and separation operation under 25 ℃, since the carbonate-type brine has the characteristics of high carbonate, lithium ions are not easily concentrated and the lithium carbonate could be easily dispersed and precipitated at various evaporation stages. Mixing salt with a lithium carbonate grade of only 2.18% could be obtained through the evaporation experiment.

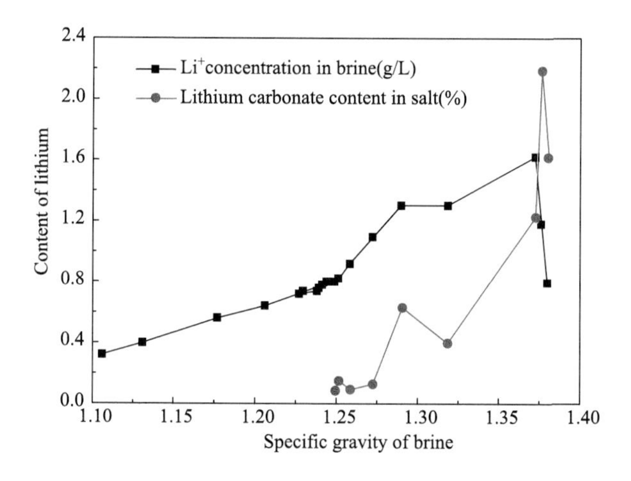

Fig. 5 Changes of lithium composition in brine and salts during 25 ℃ isothermal evaporation.

During the early and middle stages of evaporation, potassium is mainly precipitated as glaserite. In the late stage, it is precipitated as sylvite accompanied by minor amounts of glaserite. In the whole evaporation process of the experiment, potash started to precipitate as soon as potassium became saturated in the brine. High-grade potash can be obtained from Dangxiong Co salt lake brine through proper operations. The relation between the potassium mineral content in the solid phases and the evaporation rate is shown in Fig. 6. The potassium concentration in the liquid phases goes up progressively in the early stage, and then the growth rate has accelerated significantly after salt precipitation. The concentration change curve shows an inflection point at the specific gravity of 1.318, and the K^+ concentration in brine decreases after reaching a maximum

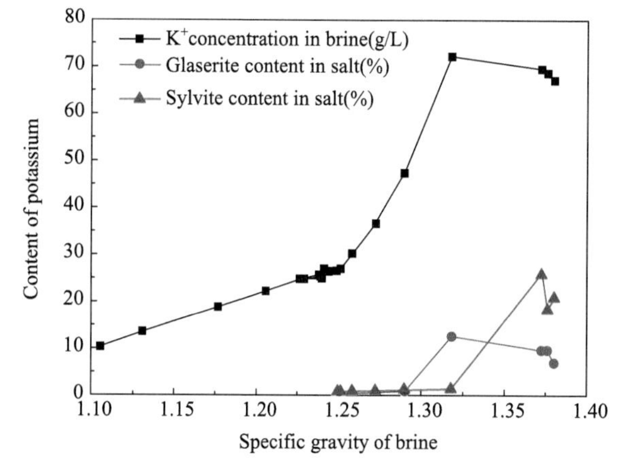

Fig. 6 Changes of potassium composition in brine and salts during 25 ℃ isothermal evaporation.

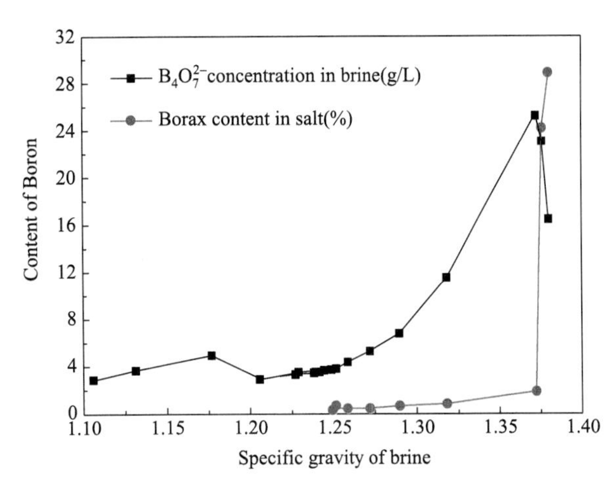

Fig. 7　Changes of boron composition in brine and salts during 25 ℃ isothermal evaporation.

value of 72.2 g/L, indicating that the potassium salts reach saturation and begin to crystallize. The glaserite precipitates slightly earlier, and the content of glaserite in salt is 12.49%. Then, the sylvite also begins to precipitate out, with a content of up to 25.78% at the specific gravity of 1.372. It can be seen that during the later period of brine evaporation, the sylvite and glaserite precipitate out and the change curve shows a fluctuation state, indicating that there is a mutual transformation between the sylvite and glaserite in the salt precipitation process. One of the salts dissolves, while the other precipitates.

Boron becomes highly concentrated during evaporation (Fig. 7). The concentration of boron in the liquid phases increases progressively with evaporation, especially in the late stage of the experiment. When the specific gravity of the brine gets to 1.372, boron becomes saturated in the brine and the borax begins to precipitate. In the evaporation experiment, almost no borax precipitates in the early stage of the experiment, and only when the specific gravity of the brine reached 1.38 is a large amount of boron precipitated as borax with a grade of almost 28.7%. Therefore, the bittern remaining in the intermediate and late stages of evaporation could be used to extract borax.

Comparison with other main lithium salt lakes

The Zabuye salt lake is the first one which had been reported that the lithium carbonate mineral could precipitate directly from the brine. The natural lithium carbonate crystal was found at this salt lake in 1987 (Zheng and Liu, 1987). The comparison of crystallization paths among some main lithium salt lake brines by isothermal evaporation experiment under 25 ℃ is shown in Table 6. It can be known that the concentration behavior of the ions and the crystallization paths of salts are similar between the Dangxiong Co and Zabuye salt lake brines. Halite, trona or thermonatrite, glaserite, sylvite, zabuyelite and borax precipitate from these two brines during the isothermal evaporation at 25 ℃. And the biggest difference between them is the precipitation time of trona, which precipitated just after halite at the early stage of salt precipitation with Dangxiong Co salt lake brine. But trona and thermonatrite will precipitate later during the evaporation of Zabuye salt lake brine (Zheng et al., 2007; Nie et al., 2010). It is mostly caused by the higher concentration of carbonate in Dangxiong Co salt lake brine, and this will be more conducive for the enrichment of Li^+ to obtain the better quality lithium-rich brine. In addition, the concentrations of Ca^{2+} and Mg^{2+} in Dangxiong Co salt lake brine are higher than that in Zabuye salt lake brine. During the evaporation process, Ca^{2+} and Mg^{2+} are precipitated in the form of double salt, and separated from Li_2CO_3, which facilitates the extraction and comprehensive utilization of lithium carbonate. Therefore, the technology of exploiting lithium carbonate from the Zabuye salt lake brine could be used in Dangxiong Co salt lake as an analogous, but the separation and utilization of sodium alkali salts should be considered specifically in the early stage of brine evaporation process.

Table 6　Crystallization paths of main lithium salt lake brines by isothermal evaporation experiment under 25 ℃

Salt lake	Crystallization path
Dangxiong Co	$NaCl$, $Na_2CO_3 \cdot NaHCO_3 \cdot 2H_2O$, Li_2CO_3, $Na_2SO_4 \cdot 10H_2O$, KCl
Zabuye	$NaCl$, $Na_2SO_4 \cdot K_2SO_4$, Li_2CO_3, $Na_2CO_3 \cdot NaHCO_3 \cdot 2H_2O$, $Na_2CO_3 \cdot H_2O$, KCl, $Na_2B_4O_7 \cdot 10H_2O$, $Na_2CO_3 \cdot 7H_2O$
East Taij	$NaCl$, $K_2SO_4 \cdot MgSO_4 \cdot 6H_2O$, $MgSO_4 \cdot 7H_2O$, KCl, $MgSO_4 \cdot 6H_2O$, $KCl \cdot MgSO_4 \cdot 2.75H_2O$, $KCl \cdot MgCl_2 \cdot 6H_2O$, $Li_2SO_4 \cdot H_2O$, $MgCl_2 \cdot 6H_2O$
West Taij	$NaCl$, $MgSO_4 \cdot 7H_2O$, KCl, $KCl \cdot MgCl_2 \cdot 6H_2O$

Conclusion

After performing the isothermal evaporation experiment at 25 ℃ with the brine of Dangxiong Co salt lake, the following conclusion can be drawn. (1) Trona precipitates earlier in the experiment compared to the metastable phase diagram. (2) Lithium precipitation is a continuous process that occurred throughout the experiment. During the evaporation experiment at 25 ℃, it is difficult to obtain high-grade lithium salt. (3) Potash is precipitated as glaserite and sylvite during the evaporation. Glaserite is precipitated first, and then the sylvite and glaserite precipitate together. (4) Boron is enriched in brine in the early and middle stages of the experiment and high-grade borax is precipitated in the late stage. Taking into account all the above, we may conclude that the lithium carbonate exploiting technology using on the Zabuye salt lake may be applied on the Dangxiong Co salt lake. Meantime, the high-grade potash and borax could be obtained from Dangxiong Co Salt Lake brine.

Acknowledgements

This paper was financially supported by the Subject of National Key R&D Program of China (2017YFC0602806) and the Institute of Mineral Resources, CAGS Research Fund (KK1915, KK1917). The authors also appreciate Prof. Nikolai Shadrin for his careful revision of the paper.

References

Analysis Group, Qinghai Saline Lake Institute, Chinese Academy of Sciences (1988) Analysis methods of brine and salt, 2nd edn. Science Press, Beijing, pp 30-70

Averill WA, Olson DL (1978) A review of extractive processes for lithium from ores and brines. Energy 3: 305-313

Chen JQ, Liu ZQ, Fu YJ et al (1980) 25 C isothermal evaporation and natural evaporation experiments of the Dongtaijinarer salt lake brine. Salt Lake Sci Technol Data 21: 71-82

Chen ZF, Jiang YX, Zhuang QC et al (2005) Preparation and characterization of a novel composite microporous polymer electrolyte for Li-ion batteries. Chin Sci Bull 50: 1435-1440

Ebensperger A, Maxwell P, Moscoso C (2005) The lithium industry: its recent evolution and future prospects. Resour Policy 30: 218-231

Epstein JA, Feist EM, Zmora J (1981) Extraction of lithium from the Dead Sea. Hydrometallurgy 6: 269-275

Fang CH, Niu ZD, Liu ZQ et al (1991) Studies on the metastable phase diagram in the quinary system Na-K-CO_3-SO_4-Cl-H_2O at 25 C. Acta Chim Sin 49: 1062-1070

Hamzaoui AH, M'nif A, Hammi H et al (2003) Contribution to the lithium recovery from brine. Desalination 158: 221-224

Hamzaoui AH, Jamoussi B, M'nif A et al (2008) Lithium recovery from highly solutions: response surface methodology (RSM) process parameters optimization. Hydrometallurgy 90: 1-7

Hoshino T (2013) Preliminary studies of lithium recovery technology from seawater by electrodialysis using ionic liquid membrane. Desalination 317: 11-16

Huang WN, Sun ZN, Wang XK et al (2008) Progress in industrialization for lithium extraction from salt lake. Modern Chem Ind 28: 14-19

Li YY, Di XL, Gao J (2005) The status of saline lake lithium resources and its present situation of exploitation. J Sea Lake Salt Chem Ind 34: 31-35

Luis VE, Nancy PF (1983) Study of the phase chemistry of the Salar de Atacama brine. In: Sixth international symposium on Salt, pp 345-366

Menta SK, Dhar JK (1979) Salt from Tsokor lake-1, A study of crystallization of salts by solar evaporation. J Indian Chem Soc 56: 809-812

Nie Z, Bu LZ, Zheng MP et al (2010) Phase chemistry study on brine from the Zabuye carbonate-type salt Lake in Tibet. Acta Geol Sin 84: 587-592

Nie Z, Bu LZ, Zheng MP et al (2011) Experimental study of natural brine solar ponds in Tibet. Sol Energy 85: 1537-1542

Song PS (2000) Comprehensive utilization of salt lake and related resources (continuation l). J Salt Lake Sci 8: 33-58

Wang SQ, Guo YF, Zhang N et al (2011) Caloric evaporation of the brine in Zangnan salt lake. Front Chem Sci Eng 5: 343-348

Wen R, Yue J, Ma ZF et al (2014) Synthesis of $Li_4Ti_5O_{12}$nanostructural anode materials with high charge-discharge capability. Chin Sci Bull 59: 2162-2174

Wu Q, Zheng MP, Nie Z et al (2012) Natural evaporation and crystallization regularity of Dangxiongcuo carbonate-type salt lake brine in Tibet. Chin J Inorg Chem 28: 1895-1903

Yang JY, Zhang Y, Cheng WY et al (1996) 25 C isothermal evaporating research in winter brine of Zabuye salt lake in Tibet. Journal of Sea Lake Salt Chem Ind 25: 21-24

Zhang BQ, Liu ZT, Fu YJ et al (1994) Study of the phase chemistry of Dongtaijinarer salt lake brine (I) -isothermal evaporation at 25 C. J Salt Lake Res 2: 57-60

Zheng MP, Liu WG (1987) A new Li mineral-Zabuyelite. Acta Mineral Sin 7: 221-227

Zheng MP, Liu XF (2009) Hydrochemistry of salt lakes of the Qinghai-Tibet plateau, China. Aquat Geochem 15: 293-320

Zheng XY, Tang Y, Xu C (1988) Saline lakes of Tibet. Science Press, Beijing

Zheng MP, Xiang J, Wei XJ et al (1989) Saline lakes of Tibet and Qinghai. Science and Technology Press, Beijing

Zheng XY, Zhang MG, Xu T et al (2002) Salt lakes of China. Science Press, Beijing

Zheng MP, Deng YJ, Nie Z et al (2007) 25 C isothermal evaporation of autumn brines from the Zabuye Salt Lake, Tibet, China. Acta Geol Sin 81: 1742-1749

Zhu ZH, Zhu CL, Wen XM et al (2008) Progress in production process of lithium carbonate. J Salt Lake Res 16: 64-72

Phase equilibria of the ternary systems potassium sulfate + polyethylene glycol (PEG6000/10,000) + water at 288.2, 298.2 and 308.2 K: experimental determination and correlation*

Hong Zheng[1], Shuai Chen[1], Lin Wang[1], Qian Wu[2] and Xudong Yu[1, 2]

1. College of Materials and Chemistry & Chemical Engineering, Chengdu University of Technology, Chengdu 610059, People's Republic of China
2. Institute of Mineral Resources, MNR Key Laboratory of Saline Lake Resources and Environments, CAGS, Beijing 100037, People's Republic of China

Abstract The phase equilibrium data were determined for the ternary systems potassium sulfate + polyethylene glycol (PEG6000/10,000) + water at 288.2, 298.2 and 308.2 K. Only solid-liquid equilibrium exists for the ternary system K_2SO_4 + PEG6000/10,000 + H_2O at 288.2 K and 298.2 K, the solubility, density, and refractive index were measured by using the isothermal dissolution equilibrium method. The solid-liquid equilibrium and liquid-liquid equilibrium exist simultaneously for the ternary system K_2SO_4 + PEG6000/10,000 + H_2O at 308.2 K, the corresponding compositions of binodal curve and tie-line were obtained experimentally by using the turbidimetric method. Results show that the solubility of K_2SO_4 decreases as the content of PEG increases in the PEG-H_2O mixed solvents. Meanwhile, as the molar mass of PEG increased, the liquid-liquid equilibrium formed more easily and the liquid-liquid equilibrium area was larger. In addition, the modified Pitzer model was used to calculate the solid-liquid equilibrium data for systems K_2SO_4 + PEG6000/10,000 + H_2O at 288.2, 298.2 and 308.2 K.

Keywords Potassium sulfate; Polyethylene glycol; Phase equilibrium; ATPS; Pitzer

1 Introduction

Sylvite is an important raw material for industry and agriculture, mainly used in the production of potash fertilizer, which is of great significance to ensure food security and increasing crop yield in China. The salt lake brines and underground brines in China are rich in liquid potassium mineral resources. As we all know, the phase diagram is the pattern of complex multi-ion interactions, which provides theoretical basis for the separation and purification of mineral resources in brine. In recent years, the aqueous two-phase system (ATPS) as a new separation/enrichment technology has attracted increasing attention. ATPS has been used in the extraction of metallic ions [1, 2], extractive crystallization of inorganic salts [3, 4]; ATPS is also used to extract biological materials such as cells and proteins [5-8], the recovery of nanoparticles[9] and other fields. The salt-polymer aqueous system is more popular because of its low cost, low viscosity, low interfacial tension, rapid phase separation and easy amplification compared with the polymer-polymer aqueous systems[10]. Therefore, it is more meaningful to develop salt-polymer aqueous systems for salt development and purification.

For salt-polymer-water systems, Wysoczanska et al. [11, 12] studied the effects of different kinds of salts (potassium citrate, potassium sodium tartrate) and molar mass polyethylene glycol (4000 $g \cdot mol^{-1}$, 6000 $g \cdot mol^{-1}$, 8000 $g \cdot mol^{-1}$) on the aqueous two-phase system, it was found that when the molar mass increased, the binodal curves move to the origin. Oliveira et al. [13] studied the liquid-liquid equilibria of the system formed by

* 本文发表在：Jouomal of Solution Chemistry, 2020, 49: 1154-1169

PEG1500/4000/6000 and $Na_2S_2O_3 \cdot 5H_2O$. Yu et al. [14-16] studied the effect of different molar mass of PEG and temperature on the ternary system CsCl + PEG (1000/4000/6000) + H_2O at 298.15 K and KCl + PEG1000/4000 + H_2O at 288, 298, and 308 K. Govindarajan [17] studied the effect of temperature on the phase equilibria of triammonium citrate + PEG2000 + H_2O at 298.15, 308.15 and 318.15 K, results show that the salting-out capacity of the system increases with an increase of temperature. For the system composed of K_2SO_4-mixed solvent, Urréjola et al. [18] studied the phase equilibria of the ternary system K_2SO_4 + C_2H_5OH + H_2O (278.15, 288.15, 298.15, 308.15, 318.15) K. Taboada et al. [19-21] studied the solid-liquid equilibrium of the ternary system K_2SO_4 + 1-propanol/methanol/ethanol/acetone + H_2O at 288.15 and 308.15 K.

Through the analysis of the above literature, the main factors that affect the phase forming ability of salt in a mixed solvent system are the type of organic solvent, the molar mass of polymer and the temperature of the system. Until now, the equilibrium data of potassium sulfate in different organic solvents (methanol, ethanol, acetone, and 1-propanol) at temperatures have been reported. However, few studies have been done on the equilibrium data of K_2SO_4 + PEG6000/10,000 + H_2O systems. Therefore, the experimental data of K_2SO_4 + PEG6000/10,000 + H_2O at the temperatures (288.2, 298.2, 308.2) K are reported in this work to study the influence of PEG molar mass and temperature on the phase diagram. Meanwhile, the modified Pitzer model was used to correlate the data of solid-liquid equilibrium.

2 Experimental section

2.1 Materials and apparatus

Double-deionized water ($\kappa \leqslant 5.5 \times 10^{-6} s \cdot m^{-1}$) obtained from a water purification system (UPT-II-20T, Chengdu Ultrapure Technology Co., Ltd.) was used in all the experiments. The chemicals potassium sulfate ($K_2SO_4 \geqslant 0.995$) and polyethylene glycol (PEG) with molar mass of 6000 $g \cdot mol^{-1}$ and 10,000 $g \cdot mol^{-1}$ (PEG $\geqslant$ 0.990) in this study were purchased from the Sinopharm Chemical Reagent Co. Ltd. Before use, K_2SO_4 needs to be dried for at least 2 h at the temperature of 378.2–383.2 K, PEG needs to be dried for 5 days at 323.2 K, and then cooled in the desiccator to before use.

An analytical balance (PRACTM224-1CN, Sartorius, Germany) with an accuracy of ± 0.0001 g and uncertainty of mass determination 0.0002 g was employed to weigh the solutions. The thermostat (Inborn SHH250, China) with an uncertainty of ± 0.1 K, was used for keeping a desired temperature in this study. The refractive index (n_D) and density (ρ) were determined by an Abbe refractometer (WYA, China) with a precision of ± 0.0001 and an uncertainty of 0.0002 and the density meter (DA-130N, Japan) with a resolution of 0.0001 $g \cdot cm^{-3}$ and an uncertainty of ± 0.001 $g \cdot cm^{-3}$.

2.2 Experimental method

2.2.1 Solid-liquid equilibria

The solid-liquid equilibria were studied by the isothermal dissolution equilibrium method. Excess potassium sulfate was added to known mixed solutions with different mass ratios (PEG-H_2O) in 50 mL glass bottle. After sealing the glass bottle, the sample was magnetically stirred for 5–6 days, and then set aside about 2 days until the solution reached equilibrium at the working temperature. All the experiment processes were done in the SHH-250 type thermostat at the corresponding temperature. Then, the clarified solution and solid phase of equilibrium state were taken out and analyzed for their composition; meanwhile, the density and refractive index of the liquid phase were measured at the working temperature.

2.2.2 Liquid-liquid equilibria

The binodal curves of liquid-liquid equilibrium were determined by the turbidimetric method [22]. Details of the experimental procedure were previously described by Yu et al. [14]. A small amount of PEG (usually 0.02 g) was added to 20 g K_2SO_4 saturated solution each time. The mixture was stirred under controlled temperature until the characteristic change of turbidity appeared, indicating the formation of the two-liquid phase. After forming turbidity, the next step was adding a few drops of water to the turbid solution, recording the mass of PEG and water to calculate the composition of the binodal curve. A complete binodal curve emerged by the repeated process of adding PEG and water. All the samples were put in a controlled bath at the working temperature.

According to the binodal curves data, the tie-line data of the known overall composition is prepared in the liquid-liquid phase area (usually 20 g mixed solution is prepared in a glass bottle). Similarly, the sample was stirred in a SHH-250 type thermostat for about 5 days, then the two-phase solution was separated by a pipette after 2 days of isothermal precipitation, and then its composition, density and refractive index were determined.

2.3 Analytical methods

The composition of solid phase was identified by Schreinemakers' wet residue method [23]. The composition of K_2SO_4 was acquired with the SO_4^{2-} concentration, and the concentration of SO_4^{2-} was determined by alizarin red volumetric titration with an uncertainty of ± 0.003 mass fraction[24]. The composition of PEG was acquired by measuring the refractive index in the sample. The relationship is shown in Eq. (1) [25]:

$$n_D = a_0 + a_1 \times w_{K_2SO_4} + a_2 \times w_{PEG} \tag{1}$$

where n_D is refractive index of solution, a_0 is the refractive index of double-deionized water at 298.2 K measured as 1.3325. a_1, a_2 are fitting parameters, $a_1 = 0.1137$, $a_2 = 0.1373$ and 0.1324 at PEG6000 and PEG10000, respectively.

3 Results and discussion

3.1 Phase equilibria of the K_2SO_4 + PEG6000/10,000 + H_2O system at 288.2 and 298.2 K

The phase equilibria of K_2SO_4 + PEG6000 + H_2O and K_2SO_4 + PEG10000 + H_2O ternary system were studied at 288.2 and 298.2 K. Results show that there is only solid-liquid phase equilibrium in this system. The solubilities, density, refractive index of the equilibrium liquid phase and the composition of the wet-solid phase are listed in Table 1.

Table 1 The solubilities, densities (ρ), refractive indices (n_D) for the ternary system K_2SO_4 (s) + PEG6000/10,000 (p) + H_2O at (288.2, and 298.2) K and 94.77 kPa

No.	Density	Refractive index	Composition of equilibrated solution		Composition of wet solid phase		Equilibrated solid phase
	ρ (g·cm^{-3})	n_D	w (s)	w (p)	w (s)	w (p)	
			K_2SO_4 + PEG6000 + H_2O at 288.2 K				
1A_1	1.0715	1.3446	0.0903	0.0000	—	—	K_2SO_4
2	1.0674	1.3484	0.0727	0.0463	0.6523	0.0174	K_2SO_4
3	1.0640	1.3539	0.0591	0.0942	0.6687	0.0332	K_2SO_4
4	1.0622	1.3593	0.0467	0.1429	0.6682	0.0497	K_2SO_4
5	1.0621	1.3655	0.0343	0.1933	0.6139	0.0773	K_2SO_4
6	1.0637	1.3718	0.0256	0.2437	0.6535	0.0866	K_2SO_4
7	1.0671	1.3787	0.0191	0.2943	0.6291	0.1113	K_2SO_4
8	1.0721	1.3858	0.0134	0.3447	0.6445	0.1242	K_2SO_4

Continued

No.	Density	Refractive index	Composition of equilibrated solution		Composition of wet solid phase		Equilibrated solid phase
	ρ (g·cm^{-3})	n_D	w (s)	w (p)	w (s)	w (p)	
9	1.0812	1.3937	0.0094	0.3960	0.6105	0.1557	K_2SO_4
10	1.0886	1.4010	0.0064	0.4466	0.5877	0.1853	K_2SO_4
11	1.0970	1.4086	0.0045	0.4974	0.6405	0.1796	K_2SO_4
12	1.1051	1.4161	0.0030	0.5485	0.6351	0.2008	K_2SO_4
$13B_1$	1.1106	1.4244	0.0000	0.5961	—	—	—
			K_2SO_4 + PEG6000 + H_2O at 298.2 K				
$1C_1$	1.0861	1.3452	0.1068	0.0000	—	—	K_2SO_4
2	1.0786	1.3495	0.0891	0.0455	0.7963	0.0102	K_2SO_4
3	1.0730	1.3544	0.0730	0.0927	0.7485	0.0251	K_2SO_4
4	1.0692	1.3595	0.0578	0.1413	0.7675	0.0349	K_2SO_4
5	1.0670	1.3651	0.0451	0.1910	0.7737	0.0453	K_2SO_4
6	1.0667	1.3713	0.0328	0.2418	0.7633	0.0592	K_2SO_4
7	1.0682	1.3772	0.0248	0.2926	0.7469	0.0759	K_2SO_4
8	1.0718	1.3852	0.0162	0.3443	0.6795	0.1122	K_2SO_4
9	1.0772	1.3931	0.0113	0.3954	0.5930	0.1628	K_2SO_4
10	1.0836	1.3994	0.0074	0.4466	0.5733	0.1920	K_2SO_4
11	1.0903	1.4081	0.0049	0.4975	0.5944	0.2028	K_2SO_4
12	1.0970	1.4141	0.0031	0.5482	0.5393	0.2534	K_2SO_4
13	1.1041	1.4226	0.0020	0.5987	0.3673	0.3796	K_2SO_4
$14D_1$	1.1077	1.4274	0.0000	0.6451	—	—	—
			K_2SO_4 + PEG10000 + H_2O at 288.2 K				
$1A_2$	1.0715	1.3446	0.0903	0.0000	—	—	K_2SO_4
2	1.0691	1.3502	0.0763	0.0462	0.7196	0.0140	K_2SO_4
3	1.0659	1.3555	0.0622	0.0937	0.7459	0.0254	K_2SO_4
4	1.0640	1.3606	0.0503	0.1425	0.6936	0.0460	K_2SO_4
5	1.0633	1.3667	0.0390	0.1920	0.6961	0.0607	K_2SO_4
6	1.0645	1.3729	0.0268	0.2432	0.6600	0.0850	K_2SO_4
7	1.0675	1.3799	0.0195	0.2941	0.6683	0.0995	K_2SO_4
8	1.0735	1.3877	0.0142	0.3449	0.6508	0.1222	K_2SO_4
9	1.0805	1.3946	0.0094	0.3959	0.5456	0.1816	K_2SO_4
10	1.0876	1.4021	0.0063	0.4471	0.6477	0.1585	K_2SO_4
11	1.0967	1.4100	0.0044	0.4976	0.6295	0.1852	K_2SO_4
12	1.1041	1.4177	0.0027	0.5485	0.5729	0.2349	K_2SO_4
$13B_2$	1.1074	1.4217	0.0000	0.5750	—	—	—
			K_2SO_4 + PEG10000 + H_2O at 298.2 K				
$1C_2$	1.0861	1.3452	0.1068	0.0000	—	—	K_2SO_4
2	1.078	1.3495	0.0872	0.0456	0.7731	0.0113	K_2SO_4
3	1.0731	1.3545	0.0727	0.0927	0.7225	0.0278	K_2SO_4
4	1.0692	1.3597	0.0581	0.1413	0.7402	0.0390	K_2SO_4
5	1.0670	1.3659	0.0450	0.1910	0.6601	0.0680	K_2SO_4
6	1.0665	1.3711	0.0336	0.2416	0.7211	0.0697	K_2SO_4
7	1.0680	1.3777	0.0235	0.2929	0.6408	0.1078	K_2SO_4
8	1.0716	1.3847	0.0166	0.3442	0.5349	0.1628	K_2SO_4
9	1.0774	1.3920	0.0115	0.3954	0.4854	0.2058	K_2SO_4
10	1.0839	1.4000	0.0076	0.4466	0.3838	0.2773	K_2SO_4
11	1.0912	1.4072	0.0050	0.4975	0.5152	0.2424	K_2SO_4
12	1.0972	1.4147	0.0029	0.5484	0.4431	0.3063	K_2SO_4
$13D^2$	1.1073	1.4270	0.0000	0.6336	—	—	—

Standard uncertainties u are $u(T) = 0.10$ K; $u(p) = 0.50$ kPa; $u(\rho) = 0.001$ $g \cdot cm^{-3}$; $u(n^D) = 0.0001$; $u(w(K_2SO_4)) = 0.003$; $u(w(PEG)) = 0.003$.

According to the data in Table 1, the phase diagram of systems K_2SO_4 + PEG6000 + H_2O and K_2SO_4 + PEG10000 + H_2O at 288.2 and 298.2 K are plotted in Fig. 1. In order to observe the effect of changes in molar mass of polymer and temperature on the phase equilibrium of the system K_2SO_4 + PEG6000/10,000 + H_2O, the comparison of the phase diagrams is shown in Fig. 2.

From Figs. 1 and 2, the following points, curves, and regions were obtained:

1. Points A_1, $A_2$and C_1, C_2 represent the saturation solubility of K_2SO_4 in pure water at 288.2 and 298.2 K, respectively; points B_1, B_2 and D_1, D_2 are the saturation points of PEG6000/10,000 in pure water at 288.2 and 298.2 K; points S, P, and W mean pure salt K_2SO_4, pure PEG, and pure water.

2. Curves A_1B_1, A_2B_2, C_1D_1, and C_2D_2 represent the solubility of K_2SO_4 in PEG-H_2O mixed solvent at 288.2 and 298.2 K, respectively.

3. All of the phase diagrams of the ternary system K_2SO_4 + PEG6000/10,000 + H_2O at 288.2 and 298.2 K have the same crystalline phase region and are divided into three areas: WA_1B_1W, WC_1D_1W, WA_2B_2W and WC_2D_2W areas represent the unsaturated liquid phase areas (L); $A_1SB_1A_1$, $C_1SD_1C_1$, $A_2SB_2A_2$ and $C_2SD_2C_2$ areas correspond to the areas where one saturated liquid phase and one solid phase of K_2SO_4 coexist (L + S); SPB_1S, SPD_1S, SPB_2S and SPD_2S areas have one saturated liquid phase and two solid phases of K_2SO_4 and PEG6000/10,000 (L + 2S).

4. It can be seen from the phase diagram in Fig. 1 that the solubility of K_2SO_4 in the mixed solvent PEG6000/10,000-H_2O decreases with the increase of PEG6000/10,000, indicating that PEG6000/10,000 has an anti-solvent effect on K_2SO_4.

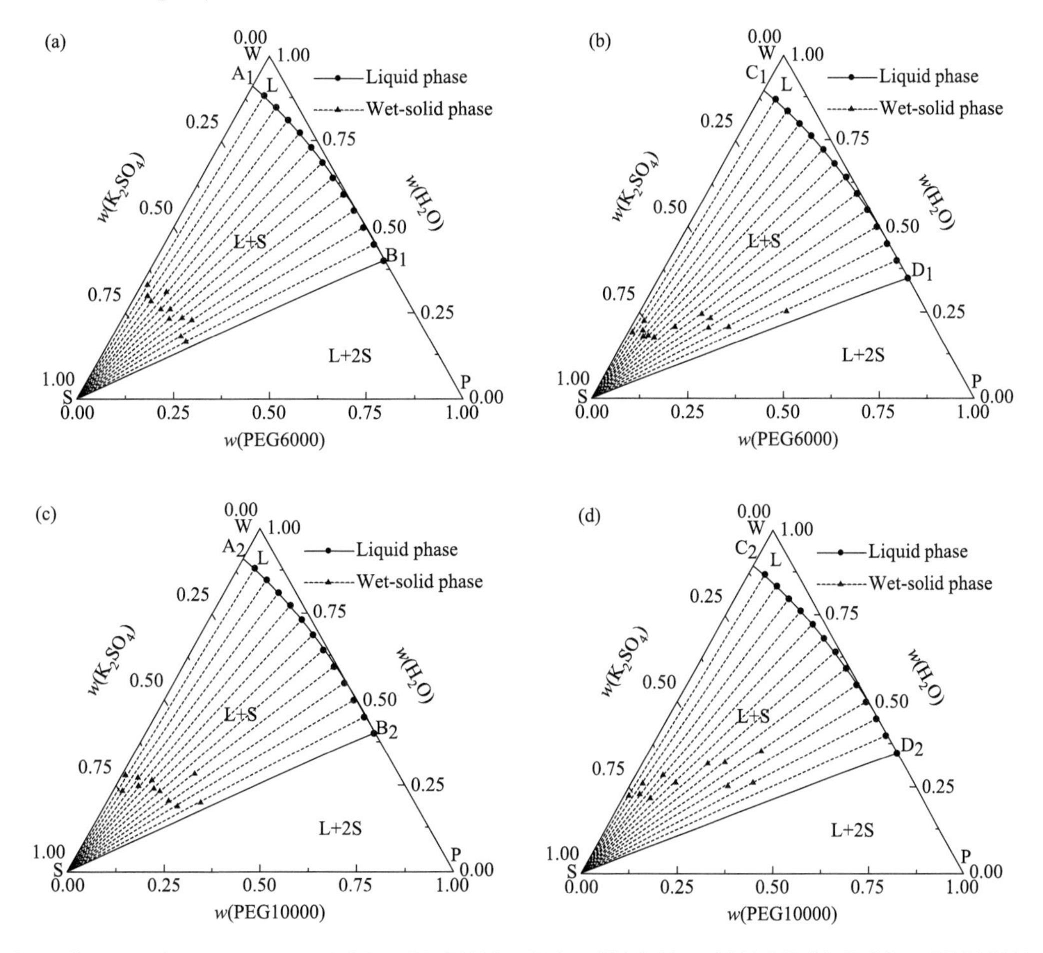

Fig. 1 Phase diagram of ternary systems K_2SO_4 + PEG6000 + H_2O at 288.2 (a) and 298.2 K (b), K_2SO_4 + PEG10000 + H_2O at 288.2 (c) and 298.2 K (d)

According to the comparison phase diagram of Fig. 2, when the research temperature is same, the areas of the phase region in the system K_2SO_4 + PEG + H_2O change slightly, this demonstrates that the molar mass of PEG has little effect on the phase equilibria of the system K_2SO_4 + PEG + H_2O at temperatures of 288.2 and 298.2 K; whereas the effect of temperature on the crystallization region is obvious, especially the region (L + S) and (L + 2S), the region (L + S) becomes larger, and the region (L + 2S) becomes smaller with the temperature increases.

In addition, it can be found from Table 1 that the density first decreases then increases, while the refractive index increases continuously with the increase of the PEG6000/10,000 content in the mixed solvent for the equilibrium liquid phase.

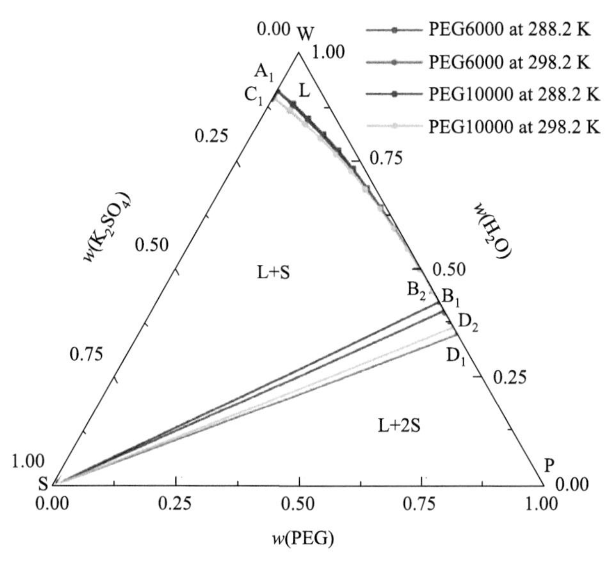

Fig. 2　Comparison of phase diagrams of ternary system K_2SO_4 + PEG6000/10,000 + H_2O at 288.2 K and 298.2 K.

3.2　Phase equilibria of the K_2SO_4 + PEG6000/10,000 + H_2O system at 308.2 K

For the ternary systems K_2SO_4 + PEG6000 + H_2O and K_2SO_4 + PEG10000 + H_2O at 308.2 K, both solid-liquid and liquid-liquid equilibria were included. The solid-liquid and liquid-liquid equilibria were investigated by using isothermal dissolution equilibrium and the turbidimetric method, respectively. The experimental data of solid-liquid phase equilibrium is shown in Table 2. The binodal curve data of the liquid-liquid phase equilibrium is given in Table 3. According to the data in Tables 2, and 3, the complete phase diagrams of K_2SO_4 + PEG6000 + H_2O and K_2SO_4 + PEG10000 + H_2O at 308.2 K are drawn as shown in Fig. 3.

Table 2　The solubilities, densities (ρ), refractive indices (n_D) for the ternary system K_2SO_4 (s) + PEG6000/10,000 (p) + H^2O at 308.2 K and 94.77 kPa

No.	Density	Refractive index	Composition of equilibrated solution		Composition of wet solid phase		Equilibrated solid phase
	ρ (g·cm^{-3})	n_D	w (s)	w (p)	w (s)	w (p)	
			K_2SO_4 + PEG6000 + H_2O at 308.2 K				
1F_1	1.0702	1.3657	0.0450	0.1851	0.7523	0.0825	K_2SO_4
2	1.0645	1.3794	0.0301	0.2829	0.7655	0.0684	K_2SO_4
3	1.0708	1.3919	0.0114	0.3882	0.7127	0.1128	K_2SO_4
4	1.0815	1.4049	0.0054	0.4681	0.7284	0.1279	K_2SO_4
5	1.0909	1.4154	0.0012	0.5706	0.7348	0.1515	K_2SO_4
6	1.1021	1.4289	0.0004	0.6657	0.7013	0.1990	K_2SO_4
7B_3	1.1086	1.4403	0.0000	0.7304	—	—	—
			K_2SO_4 + PEG10000 + H_2O at 308.2 K				
1F_2	1.0675	1.3371	0.0439	0.2389	0.7716	0.0571	K_2SO_4
2	1.0668	1.3380	0.0252	0.3247	0.6157	0.1280	K_2SO_4
3	1.0722	1.3387	0.0133	0.4062	0.6947	0.1257	K_2SO_4
4	1.0829	1.3398	0.0061	0.4962	0.7047	0.1474	K_2SO_4
5	1.0935	1.3413	0.0026	0.5830	0.5841	0.2431	K_2SO_4
6	1.1018	1.3422	0.0004	0.6641	0.6897	0.2061	K_2SO_4
7B_4	1.1060	1.4334	0.0000	0.7017	—	—	—

Standard uncertainties u are $u(T)$ = 0.10 K; $u(p)$ = 0.50 kPa; $u(\rho)$ = 0.001 g·cm^{-3}; $u(n_D)$ = 0.0001; $u(w(K_2SO_4))$ = 0.003; $u(w(PEG))$ = 0.003.

Table 3 Binodal curve data of the K_2SO_4 (s) + PEG6000/10,000 (p) + H_2O system at 308.2 K and 94.77 kPa

No.	*w* (s)	*w* (p)	No.	*w* (s)	*w* (p)	No.	*w* (s)	*w* (p)
				K_2SO_4 + PEG6000 + H_2O at 308.2 K				
1E_1	0.1134	0.0226	8	0.0872	0.0813	15	0.0698	0.1352
2	0.1094	0.0283	9	0.0841	0.0908	16	0.0698	0.1357
3	0.1052	0.0359	10	0.0813	0.0989	17	0.0667	0.1476
4	0.1011	0.0454	11	0.0785	0.1075	18	0.0636	0.1603
5	0.0974	0.0538	12	0.0767	0.1131	19	0.0602	0.1720
6	0.0938	0.0640	13	0.0743	0.1207	20F_1	0.0577	0.1851
7	0.0903	0.0742	14	0.0720	0.1281			
				K_2SO_4 + PEG10000 + H_2O at 308.2 K				
1E_2	0.1143	0.0090	9	0.0819	0.0727	17	0.0546	0.1731
2	0.1102	0.0122	10	0.0786	0.0853	18	0.0526	0.1869
3	0.1047	0.0197	11	0.0748	0.0968	19	0.0508	0.1936
4	0.0987	0.0298	12	0.0705	0.1114	20	0.0508	0.1934
5	0.0942	0.0398	13	0.0684	0.1178	21	0.0491	0.2089
6	0.0903	0.0497	14	0.0653	0.1300	22	0.0474	0.2174
7	0.0867	0.0584	15	0.0616	0.1441	23	0.0451	0.2299
8	0.0838	0.0668	16	0.0581	0.1568	24F_2	0.0439	0.2389

Standard uncertainties *u* are $u(T)$ = 0.10 K; $u(p)$ = 0.50 kPa; $u(w\ (K_2SO_4))$ = 0.001; $u(w$ (PEG)) = 0.001.

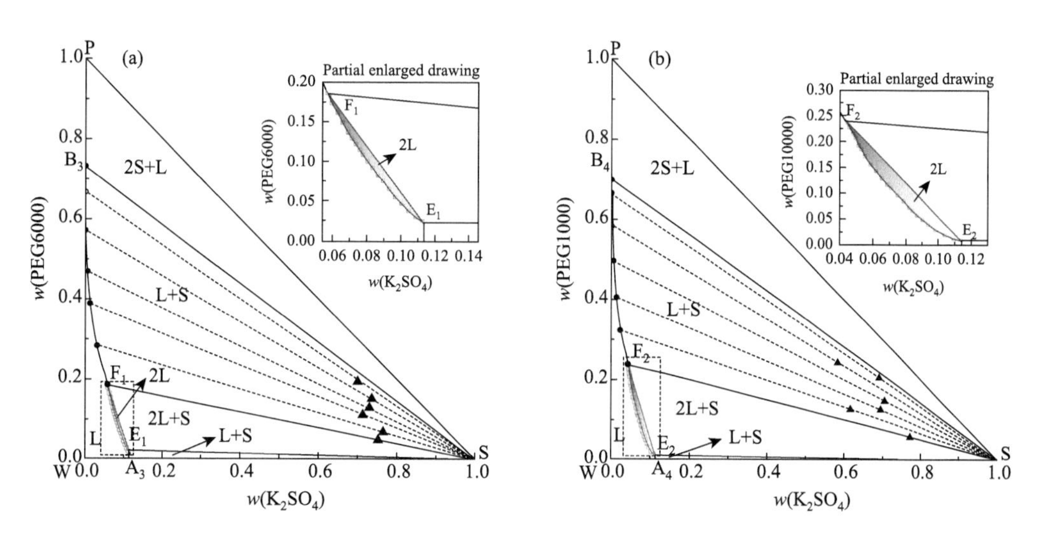

Fig. 3 Phase diagram of ternary system K_2SO_4 + PEG6000 + H_2O (a) and K_2SO_4 + PEG10000 + H_2O (b) at 308.2 K.

It can be seen from Tables 1, and 2 that PEG has an anti-solvent effect on the solubility of K_2SO_4 in solution. In this work, the yield (Y) which defined as Eq. (2) can be used to describe the ability of the anti-solvent effect of PEG.

$$Y = \frac{w_0 - w}{w_0} \tag{2}$$

where w_0 and w represent the solubility of K_2SO_4 in pure water and PEG-H_2O mixed solvent, respectively. The maximum yield of K_2SO_4 in PEG-H_2O mixed solvent at 288.2, 298.2 and 308.2 K are calculated as: PEG6000: 0.9668, 0.9813, 0.9968; PEG10000: 0.9701, 0.9728, 0.9968. This indicates that PEG has a good precipitation effect on K_2SO_4 and higher yields achieved with increasing temperature.

There is solid-liquid equilibrium at 288.2 and 298.2 K, both the solid-liquid and liquid-liquid equilibrium are included in the system K_2SO_4 + PEG6000/10,000 + H_2O at 308.2 K, which indicates that temperature increase is conducive to the formation of the liquid-liquid equilibrium. The reason for this phenomenon may be that: the mutual effect of hydrophobicity of PEG and the salting-out effect of salt is responsible for the two-phase formation [17]. As the temperature increases, the attraction between the PEG molecules increases, which makes it easier for the system to form a PEG-rich phase. In addition, the solubility of the salt increases as the temperature increases, and the salting out effect also increases, which makes it easier for the system to form a salt-rich phase.

As shown in Fig. 3, points S, P, and W mean pure salt, pure PEG, and coordinate origin; points A_3, A_4 and B_3, B_4 represent the solubility of saturated K_2SO_4 and PEG6000/10,000 in pure water at 308.2 K, respectively. The points E_1, E_2 and F_1, F_2 are the start and end points of the binodal curve line, and also the boundary points of the liquid-liquid equilibrium. Curves A_3E_1, A_4E_2 and F_1B_3, F_2B_4 represent the solubility of K_2SO_4 in PEG-H_2O mixed solvent at 308.2 K. From Fig. 3, there are 6 regions in the complete phase diagram: $A_3WB_3A_3$ and $A_4WB_4A_4$ are unsaturated liquid regions (L); regions SA_3E_1A, SA_4E_2S, SF_1B_3S and SF_2B_4S are composed of a saturated liquid phase and a solid phase of K_2SO_4 (L + S); $E_1F_1E_1$ and $E_2F_2E_2$ are the biphasic zone (2 L); SE_1F_1S and SE_2F_2S are regions composed of two saturated liquid phases and one solid phase of K_2SO_4 (2 L + S); SB_3PS and SB_4PS represent a region which composed of one liquid phase and two solid phases of K_2SO_4 and PEG6000/10,000 (L + 2S).

As shown in Figs. 1 and 3, the system only includes solid-liquid equilibrium at 288.2 and 298.2 K, and the area of (L + S) is the largest region of the phase diagram; the system included both solid-liquid and liquid-liquid equilibria at 308.2 K, the area of (L + S) is significantly larger than the area of (2 L). The larger phase area indicated that the pure salt can be prepared more easily in this region. Accordingly, it is beneficial to the precipitation of K_2SO_4 from the solution in the solid-liquid equilibrium phase region.

The binodal data were correlated using the following Eqs. (3) – (5) developed by Mistry et al. [26], Hu et al. [27], and Jayapal et al. [28]:

$$\ln(w_{\mathrm{PEG}}) = a + bw_{\mathrm{K_2SO_4}}^{0.5} + cw_{\mathrm{K_2SO_4}}^{3} \tag{3}$$

$$w_{\mathrm{PEG}} = a + bw_{\mathrm{K_2SO_4}}^{0.5} + cw_{\mathrm{K_2SO_4}} + dw_{\mathrm{K_2SO_4}}^{2} \tag{4}$$

$$w_{\mathrm{PEG}} = a + bw_{\mathrm{K_2SO_4}}^{0.5} + cw_{\mathrm{K_2SO_4}} \tag{5}$$

with $w_{\mathrm{K_2SO_4}}$, w_{PEG} being the mass fraction of K_2SO_4 and PEG6000/10,000, respectively. *a*, *b*, *c*, and *d* are the parameters of nonlinear equations. The parameter values and correlation coefficient R^2 are listed in Table 4, and R^2 is not less than 0.998. The results show that the binodal curve can be well correlated by the above three nonlinear equations.

Table 4 Parameters *a*, *b*, *c*, *d* of three nonlinear expressions for binodal curve in systems K_2SO_4 + PEG16000/10,000 + H_2O at 308.2 K

Equations	*a*	*b*	*c*	*d*	R^2
K_2SO_4 + PEG6000 + H_2O at 308.2 K					
$\ln(w_{\mathrm{PEG}}) = a + bw_{\mathrm{K_2SO_4}}^{0.5} + cw_{\mathrm{K_2SO_4}}^{3}$	−0.6407	−3.4236	−327.9818	—	0.9992
$w_{\mathrm{PEG}} = a + bw_{\mathrm{K_2SO_4}}^{0.5} + cw_{\mathrm{K_2SO_4}} + dw_{\mathrm{K_2SO_4}}^{2}$	0.4823	0.0057	−6.4119	20.5577	0.9997
$w_{\mathrm{PEG}} = a + bw_{\mathrm{K_2SO_4}}^{0.5} + cw_{\mathrm{K_2SO_4}}$	0.8962	−3.8866	3.8195	—	0.9996

Continued

Equations	a	b	c	d	R^2
	K_2SO_4 + PEG10000 + H_2O at 308.2 K				
$\ln(w_{PEG}) = a + bw_{K_2SO_4}^{0.5} + cw_{K_2SO_4}^3$	0.1428	−7.0026	−433.7408	—	0.9983
$w_{PEG} = a + bw_{K_2SO_4}^{0.5} + cw_{K_2SO_4} + dw_{K_2SO_4}^2$	1.1105	−5.6652	7.1817	−0.7409	0.9991
$w_{PEG} = a + bw_{K_2SO_4}^{0.5} + cw_{K_2SO_4}$	1.0984	−5.5444	6.8458	—	0.9991

The experimental tie-line data of the system K_2SO_4 (s) + PEG6000/10,000 (p) + H_2O at 308.2 K is given in Table 5. Furthermore, tie-line length (TLL) is a thermodynamic parameter, which represents the difference of strengthening thermodynamic function between top and bottom phases under constant pressure and temperature. It can be used to analyze the composition of top and bottom phases, and it is calculated with Eq. (6). The slope of tieline (STL) can be used to analyze the influence of temperature and molar mass of polymer on phase equilibrium, which is defined by Eq. (7). The data of TLL and STL are also listed in Table 5, and the corresponding experimental tie-line are also shown in Fig. 4

$$\mathrm{TLL} = \left[\left(w_{K_2SO_4}^{t} - w_{K_2SO_4}^{b}\right) + \left(w_{PEG}^{t} - w_{PEG}^{b}\right)\right]^{0.5} \tag{6}$$

$$\mathrm{STL} = \frac{(w_{PEG}^{t} - w_{PEG}^{b})}{\left(w_{K_2SO_4}^{t} - w_{K_2SO_4}^{b}\right)} \tag{7}$$

where, t and b represent the top phase and bottom phase.

Table 5 Experimental composition of tie-line for the K_2SO_4 (s) + PEG6000/10,000 (p) + H_2O system, densities (ρ) and refractive indices (n_D) at 308.2 K and 94.77 kPa

Top phase				Bottom phase				TLL	STL
w (s)	w (p)	ρ (g·cm^{-3})	n_D	w (s)	w (p)	ρ (g·cm^{-3})	n_D		
				K_2SO_4 + PEG6000 + H_2O at 308.2 K					
0.0699	0.1370	1.0736	1.3599	0.1032	0.0412	1.0841	1.3506	0.1014	−2.8731
0.0641	0.1578	1.0714	1.3626	0.1070	0.0341	1.0885	1.3492	0.1310	−2.8798
0.0612	0.1682	1.0717	1.3622	0.1091	0.0306	1.0876	1.3493	0.1457	−2.8738
				K_2SO_4 + PEG10000 + H_2O at 308.2 K					
0.0640	0.1347	1.0058	1.3365	0.0996	0.0306	1.0058	1.3349	0.1100	−2.9213
0.0561	0.1703	1.0056	1.3370	0.1079	0.0191	1.0063	1.3348	0.1599	−2.9212
0.0567	0.1676	1.0050	1.3366	0.1074	0.0197	1.0060	1.3347	0.1563	−2.9189
0.0510	0.1961	1.0055	1.3374	0.1135	0.0134	1.0069	1.3347	0.1931	−2.9244
0.0482	0.2112	1.0677	1.3684	0.1176	0.0102	1.0915	1.3466	0.2127	−2.8975

Standard uncertainties u are $u(T) = 0.10$ K; $u(p) = 0.50$ kPa; $u(\rho) = 0.001$ g·cm^{-3}; $u(n_D) = 0.0001$; u (w(K_2SO_4)) = 0.003; u(w(PEG)) = 0.003.

From Table 5; Fig. 4, the TLL value in system K_2SO_4 + PEG10000 + H_2O is greater than in system K_2SO_4 + PEG6000 + H_2O at 308.2 K. Results show that in the K_2SO_4 + PEG + H_2O system it is easier to separate phases when the molar mass of the polymer increases at the same temperature, the content of PEG in the top phase increases, and the content of K_2SO_4 in the bottom phase increases. However, the STL value in system K_2SO_4 + PEG10000 + H_2O is smaller than in system K_2SO_4 + PEG6000 + H_2O at 308.2 K. The results show that PEG moves into the top phase and K_2SO_4moves into the bottom phase, which makes the phase region of system K_2SO_4 + PEG10000 + H_2O at 308.2 K larger.

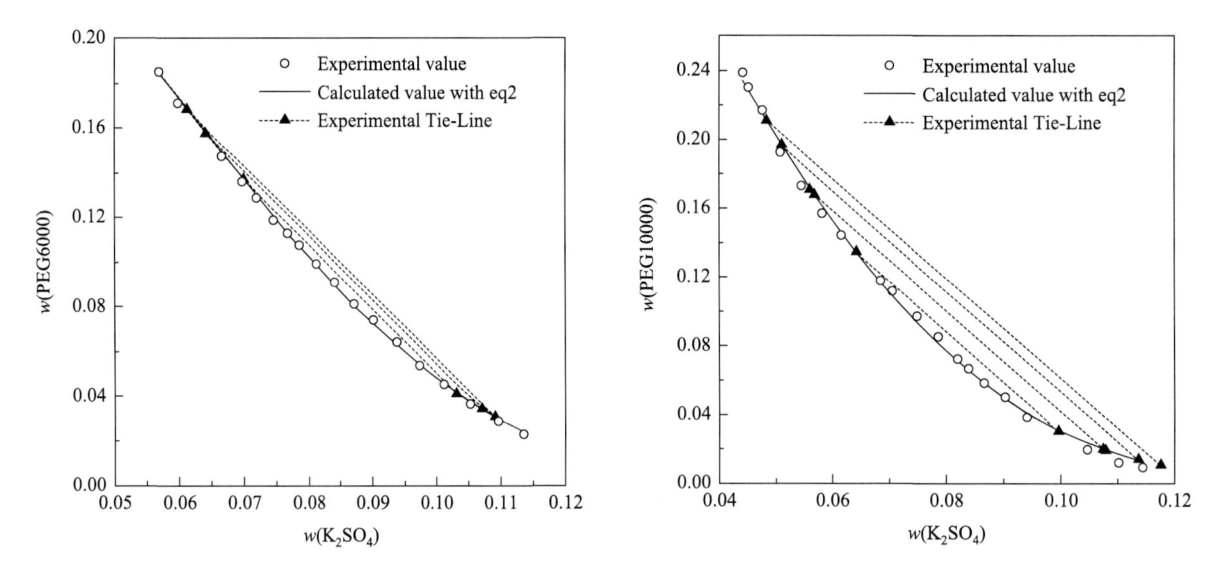

Fig. 4　The binodal curve and tie-line for ternary systems K_2SO_4 + PEG6000/10,000 + H_2O at 308.2 K.

3.3　Modified thermodynamic Pitzer model

For polymer-water mixed solvent systems, based on the Pitzer model, Wu et al. [29] proposed a modified Pitzer model, and successfully used it to correlate equilibrium data of mixed solvent systems PEG1000/4000-$(NH_4)_2SO_4$-H_2O, PEG1000/8000-Na_2CO_3-H_2O, PEG1000/8000-Na_2SO_4-H_2O. In our previous work [15, 16], the modified Pitzer model has been used to calculate the solubility of KCl in mixed solvents PEG1000/4000-H_2O at 288, 298 and 308 K. This model was also applied to correlate the solubility of K_2SO_4 in mixed solvents PEG6000/10,000-H_2O at 288.2, 298.2 and 308.2 K with the Eqs. (8) – (20).

$$\ln\gamma_{\pm}=-\frac{2A\sqrt{I}}{1+b\sqrt{I}}+\frac{1}{3}\left[B_{12}r_1m_1+8B_{22}m_2+I(B'_{12}r_1m_1+4B'_{22}m_2)+C_{112}r_1^2m_1^2+2C_{122}r_1m_1m_2+2\times2^{10.5}C_{222}^{\gamma}m_2^2\right]\tag{8}$$

$$\ln\gamma_3=-\frac{2AV_3\rho}{b^3}\left[1+b\sqrt{I}-\frac{1}{1+b\sqrt{I}}-2\ln(1+b\sqrt{I})\right]-\frac{M_3}{1000}\left[B_{11}r_1^2m_1^2+\left(B_{12}+\frac{n_w}{n_s}IB'_{12}\right)r_1m_1m_2\right.$$
$$\left.+4\left(B_{22}+\frac{n_w}{n_s}IB'_{12}\right)m_2^2+2C_{111}r_1^3m_1^3+2C_{112}r_1^2m_1^2m_2+2C_{122}r_1m_1m_2^2+2\times2^{10.5}C_{222}^{\phi}m_2^3\right]\tag{9}$$

$$B_{ij}=\beta_{ij}^{(0)}+\frac{2\beta_{ij}^{(1)}}{a_1^2I}\left[1-(1+a_1\sqrt{I})e^{-a_1\sqrt{I}}\right]\tag{10}$$

$$C_{222}^{\gamma}=\frac{3}{2}C_{222}^{\phi}=3C_{222}\tag{11}$$

$$I=\frac{1}{2}\sum_i m'_iz_i^2,m'_i=n_i/n_s\tag{12}$$

$$n_s=n_w+M_1n_1/1000\tag{13}$$

$$A=\frac{1.327757\times10^5}{DT}\sqrt{\frac{\rho}{DT}}\tag{14}$$

$$b=6.359696\sqrt{\frac{\rho}{DT}}\tag{15}$$

$$D=\sum_i\phi'_iD_i\tag{16}$$

$$\rho=\sum_i\phi'_i\rho_i\tag{17}$$

$$P_V = \sqrt{D_1 M_1} \tag{18}$$

$$\phi_i' = \frac{n_i V_i}{\sum_i n_i V_i} \tag{19}$$

$$V_1 = (2r_1 + 1) \cdot V_3 \tag{20}$$

where γ is the activity coefficients, n_w is the number of kilograms of water, and n_i, r_i m_i are the numbers of moles, segments and molality of components i, respectively. M is the molar mass. V_i denotes the molar volume of species i. I is ionic strength. A and b are Debye-Hückel constants, which can be estimated by the Eqs. (14) and (15). B_{ij} represents the second virial coefficients. B_{ij}' is the derivative of B_{ij} with respect to the ionic strength I. $\alpha_1 = 2$. D and ρ are the dielectric constant and the density of the mixed solvent. P_V is the polarity of PEG, and may be estimated from the summation of the group values given by Krevelen et al. [30]. ϕ is the salt-free volume fraction of non-ionic species i in the liquid phase.

The binary parameters of the model $\beta_{22}^{(0)}$, $\beta_{22}^{(1)}$, C_{222}^{ϕ} and B_{11}, C_{111} for the binary systems of K_2SO_4-H_2O and PEG6000/10,000-H_2O are listed in Table 6. The binary parameters $\beta_{22}^{(0)}$, $\beta_{22}^{(1)}$, C_{222}^{ϕ} of system K_2SO_4-H_2O can be calculated by modified Pitzer model according to the mean activity coefficient of K_2SO_4 in aqueous solution acquired by the Pitzer model [31]. The binary parameters B_{11}, C_{111} of PEG6000/10,000-H_2O can be obtained by literature[29]. Furthermore, the cross parameters $\beta_{12}^{(0)}$, $\beta_{12}^{(1)}$, and C_{112} and C_{122} of ternary systems K_2SO_4-PEG6000/10,000-H_2O can be fitted by the solid-liquid equilibrium solubility data in this study, and they are listed in Table 6, and the needed physical properties of PEG6000/10,000 and H_2O at 288.2, 298.2, 308.2 K are presented in Table 7.

Table 6 Binary and cross parameters of the ternary system K_2SO_4-PEG6000/10,000-H_2O at (288.2, 298.2, 308.2) K

T (K)	$B_{11}\times10^2$	$C_{111}\times10^5$	$\beta_{22}^{(0)}$	$\beta_{22}^{(1)}$	C_{222}^{ϕ}
		Binary parameters			
288.2	0.7088 [29]	1.2962 [29]	−0.7308	3.5620	0.6329
298.2	0.7088 [29]	1.2962 [29]	−0.5764	3.0898	0.4180
308.2	0.7088 [29]	1.2962 [29]	−0.5171	2.9602	0.3355
T (K)	$\beta_{12}^{(0)}$	$\beta_{12}^{(1)}$	$C_{112}\times10^3$		$C_{122}\times10^2$
		Cross parameters			
		K_2SO_4-PEG6000-H_2O			
288.2	0.3285	0.3149	−8.2388		7.6879
298.2	0.3474	0.2592	−7.0756		1.8945
308.2	0.3963	0.2038	−6.2245		−4.7421
		K_2SO_4-PEG10000-H_2O			
288.2	0.2435	0.1058	−2.4919		−2.3650
298.2	0.2301	0.1777	−2.9622		1.6402
308.2	0.1865	0.2830	−3.1124		7.5701

Table 7 Physical properties of pure chemicals at different temperatures

T (K)	Components	M (g·mol^{-1})	V (m^3·mol^{-1})	ρ (g·cm^{-3})	r_i	D
	PEG 6000	5706	4.6811×10^{-3}	1.2189	129.2723	2.204
288.2	PEG 10,000	10,251	8.4072×10^{-3}	1.2193	232.5677	2.204
	H_2O	18.02	1.8036×10^{-5}	0.9991	1	81.95

Continued

T (K)	Components	M (g·mol^{-1})	V (m^3·mol^{-1})	ρ (g·cm^{-3})	r_i	D
	PEG 6000	5706	4.6907×10^{-3}	1.2164	129.2723	2.204
298.2	PEG 10,000	10,251	8.4245×10^{-3}	1.2168	232.5677	2.204
	H_2O	18.02	1.8073×10^{-5}	0.9970	1	78.34
	PEG 6000	5706	4.7050×10^{-3}	1.2128	129.2723	2.204
308.2	PEG 10,000	10,251	8.4501×10^{-3}	1.2131	232.5677	2.204
	H_2O	18.02	1.8128×10^{-5}	0.9940	1	74.83

The unknown parameters are evaluated by the objective function Eq. (21)

$$\mathrm{OF} = (Q_{\exp} - Q_{\mathrm{cal}})^2 / Q_{\exp}^2 \tag{21}$$

where Q stands for any thermodynamic property.

K_2SO_4 is a 1–2 type electrolyte, so the expression of the solubility product constant K_{sp} of K_2SO_4 as Eq. (22).

$$K_{sp} = 4m_{K_2SO_4}^3\gamma_{\pm}^3 \tag{22}$$

The K_{sp} value can be calculated by Eq. (22) according to the solubility of K_2SO_4 in pure water at 288.2, 298.2 and 308.2 K measured by this work, the calculated $\ln K_{sp}$ values of K_2SO_4 are –4.3920, –4.0522, and –3.6811, respectively. In order to compare the degree of coincidence of experimental data and calculated data, we calculated the mean square deviation (σ) according to Eq. (23).

$$\sigma = \sqrt{\frac{\sum(w_{\exp} - w_{\mathrm{cal}})^2}{n}} \tag{23}$$

where $w_{\exp}$ is experimental value, w_{cal} represents calculated value, n denotes the number of samples, in this study, the calculated mean square deviation (σ) is 0.0005, 0. 0005, 0.0001 and 0.0006, 0.0009, 0.0018 for the K_2SO_4-PEG6000-H_2O and K_2SO_4-PEG10000-H_2O system at 288.2, 298.2, 308.2 K. Meanwhile, the comparison of experimental and calculated values is presented in Fig. 5; results show that calculated values agree well with experimental values.

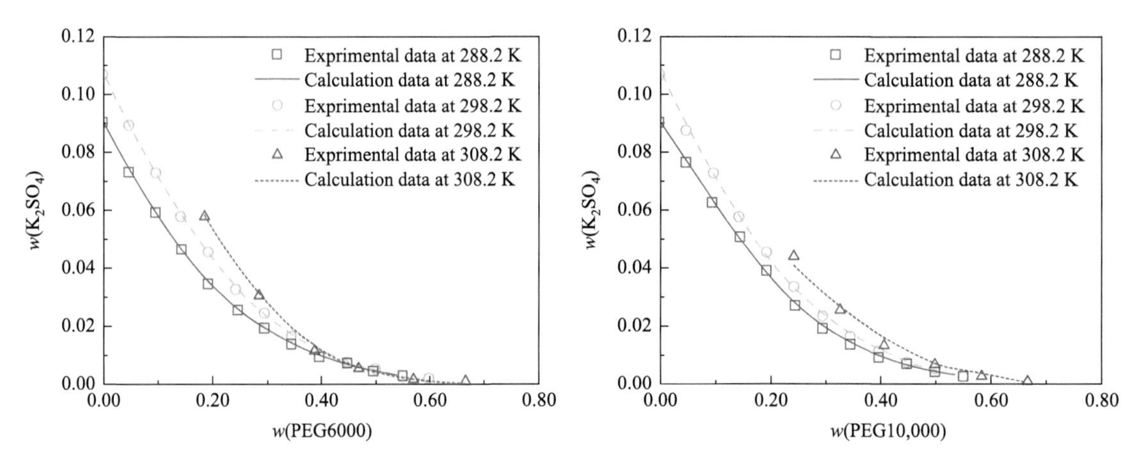

Fig. 5 The comparison of experimental and calculated values for ternary system K_2SO_4 + PEG6000/10,000 + H_2O at 288.2, 298.2, 308.2 K

4 Conclusions

1 The solubilities, densities, and refractive indices of the ternary system K_2SO_4 + PEG6000/10,000 + H_2O were measured at 288.2, 298.2, and 308.2 K. The corresponding complete phase diagram was drawn. The phase diagram consists of three phase regions: L, L + S, and L + 2S at 288.2 and 298.2 K, without the biphase region

formed. Both solid-liquid and liquid-liquid equilibrium exist in the system of K_2SO_4 + PEG6000/10,000 + H_2O at 308.2 K.

2 It can be seen from the phase diagram that all of the solubilities of K_2SO_4 at the three different temperatures decrease with an increase of the mass fraction of PEG6000/10,000, indicating that PEG6000/10,000 has anti-solvent effect on K_2SO_4.

3 The biphase region is formed at 308.2 K and the area of the biphase region increases with increasing molar mass, indicating that the ATPS prefer to be formed with higher temperature and lager molar mass of PEG.

4 The thermodynamic calculations of solid-liquid equilibrium data of system K_2SO_4 + PEG6000/10,000 + H_2O at 288.2, 298.2, and 308.2 K were done by using the modified Pitzer equation, and the calculated results agree well with experimental data.

Acknowledgements

This project was financially supported by the National Natural Science Foundation of China (U1507111).

References

[1] Silva, L. H. M., Silva, M. C. H., Júnior, A. J., Martins, J. P., Coimbra, J. S. R., Minim, L. A.: Hydrophobic effect on the partitioning of $[Fe(CN)_5(NO)]^{2-}$ and $[Fe(CN)_6]^{3-}$ anions in aqueous two-phase systems formed by triblock copolymers and phosphate salts. Sep. Purif. Technol. **60** (1), 103-112 (2008)

[2] Wang, P., Zhang, F., Li, P., Sun, T., Pan, Y. J., Zhang, Y. Q.: Partitioning performance of molybdenum in poly (ethyleneglycol) + sodium sulfate + water aqueous two-phase systems. J. Mol. Liq. **260**, 211-217 (2018)

[3] Taboada, M.E., Graber, T.A., Asenjo, J.A., Andrews, B.A.: Drowning-out crystallisation of sodium sulphate using aqueous two-phase systems. J. Chromatogr. B: Biomed. Sci. Appl. **743** (1), 101-105 (2000)

[4] Taboada, M.E., Palma, P.A., Graber, T.A.: Crystallization of potassium sulfate by cooling and saltingout using 1-propanol in a calorimetric reactor. Cryst. Res. Technol. **38** (1), 21-29 (2003)

[5] Jamehbozorg, B., Sadeghi, R.: Extractions of alkaloids codeine and caffeine with $[Bmim][BF_4]$/carbohydrate aqueous biphasic systems as a novel class of liquid-liquid extraction systems. J. Chem. Eng. Data **64** (3), 916-925 (2019)

[6] Clavijo, V., Torres-Acosta, M.A., Vives-Flórez, M.J., Rito-Palomares, M.: Aqueous two-phase systems for the recovery and purification of phage therapy products: Recovery of salmonella bacteriophageϕSan23 as a case study. Sep. Purif. Technol. **211**, 322-329 (2019)

[7] Silva, C.A.S., Coimbra, J.S.R., Rojas, E.E.G., Minim, L.A., Silva, L.H.M.: Partitioning of caseinomacropeptide in aqueous two-phase systems. J. Chromatogr. B **858** (1-2), 205-210 (2007)

[8] Rosa, P.A.J., Azevedo, A.M., Aires-Barros, M.R.: Application of central composite design to the optimisation of aqueous two-phase extraction of human antibodies. J. Chromatogr. A. **1141** (1), 50-60 (2007)

[9] Braas, G.M.F., Walker, S.G., Lyddiatt, A.: Recovery in aqueous two-phase systems of nanoparticulates applied as surrogate mimics for viral gene therapy vectors. J. Chromatogr. B: Biomed. Sci. Appl. **743** (1-2), 409-419 (2000)

[10] Graber, T.A., Gálvez, M.E., Galleguillos, H.R., Álvarez-Benedí, J.: Liquid-liquid equilibrium of the aqueous two-phase system water + PEG 4000 + lithium sulfate at different temperatures. experimental determination and correlation. J. Chem. Eng. Data **49** (6), 1661-1664 (2004)

[11] Wysoczanska, K., Do, H.T., Held, C., Sadowski, G., Macedo, E.A.: Effect of different organic salts on amino acids partition behaviour in PEG-salt ATPS. Fluid Phase Equilib. **456**, 84-91 (2018)

[12] Wysoczanska, K., Macedo, E.A.: Influence of the molecular weight of PEG on the polymer salt phase diagrams of aqueous two-phase systems. J. Chem. Eng. Data **61** (12), 4229-4235 (2016)

[13] Oliveira, A.C., Sosa, F.H.B., Costa, M.C., Filho, E.S.M., Ceriani, R.: Study of liquid-liquid equilibria in aqueous two-phase systems formed by poly (ethylene glycol) (PEG) and sodium thiosulfate pentahydrate ($Na_2S_2O_3 \cdot 5H_2O$) at different temperatures. Fluid Phase Equilib. **476**, 118-125 (2018)

[14] Yu, X.D., Wang, L., Li, M.L., Huang, Q., Zeng, Y., Lan, Z.: Phase equilibria of CsCl-polyethylene glycol (PEG) -H_2O at 298.15 K: effect of different polymer molar masses ($PEG_{1000/4000/6000}$). J. Chem. Thermodyn. **135**, 45-54 (2019)

[15] Huang, Q., Wang, L., Li, M.L., Hu, P.P., Yu, X.D., Deng, H., Zeng, Y.: Measurements and simulation of the polyethylene glycol 1000-water-KCl ternary system at 288.2, 298.2, and 308.2 K. J. Chem. Eng. Jpn. **52** (4), 325-332 (2019)

[16] Yu, X.D., Huang, Q., Wang, L., Li, M.L., Zheng, H., Zeng, Y.: Measurements and simulation for ternary system KCl-PEG4000-H_2O at T = 288, 298 and 308 K. CIESC J. **70** (3), 830-839 (2019)

[17] Govindarajan, R., Perumalsamy, M.: Phase equilibrium of PEG2000 + triammonium citrate + water system relating PEG molecular weight, cation, anion

with effective excluded volume, Gibbs free energy of hydration, size of cation, and type of anion at (298.15, 308.15, and 318.15) K. J. Chem. Eng. Data **58**, 2952-2958 (2013)

[18] Urréjola, S., Sanchez, A., Hervello, M.F.: Solubilities of sodium, potassium, and copper (II) sulfates in ethanol-water solutions. J. Chem. Eng. Data **56** (5), 2687-2691 (2011)

[19] Jimenez, Y.P., Taboada, M.E., Galleguillos, H.R.: Solid-liquid equilibrium of K_2SO_4in solvent mixtures at different temperatures. Fluid Phase Equilib. **284**, 114-117 (2009)

[20] Taboada, M.E., Véliz, D.M., Galleguillos, H.R., Graber, T.A.: Solubilities, densities, viscosities, electrical conductivities, and refractive indices of saturated solutions of potassium sulfate in water + 1-propanol at 298.15, 308.15, and 318.15 K. J. Chem. Eng. Data **47** (5), 1193-1196 (2002)

[21] Jimenez, Y.P., Taboada, M.E., Flores, E.K., Galleguillos, H.R.: Density, viscosity, and electrical conductivity in the potassium sulfate + water + 1-propanol system at different temperatures. J. Chem. Eng. Data **54** (6), 1932-1934 (2009)

[22] Jimenez, Y.P., Galleguillos, H.R.: (Liquid + liquid) equilibrium of ($NaNO_3$ + PEG4000 + H_2O) ternary system at different temperatures. J. Chem. Thermodyn. **43**, 1573-1578 (2011)

[23] Fosbl, P.L., Thomsen, K., Stenby, E.H.: Reverse Schreinemakers method for experimental analysis of mixed-solvent electrolyte systems. J. Solution Chem. **38** (1), 1-14 (2008)

[24] Lv, P., Zhong, Y., Meng, R.Y., Sun, B., Song, P.S.: Weighing titration analysis of the sulfate content. J. Salt Lake Res. **23** (3), 5-13 (2015)

[25] Cheluget, E.L., Gelinas, S., Vera, J.H., Weber, M.E.: Liquid-liquid equilibrium of aqueous mixtures of poly (propylene glycol) with NaCl. J. Chem. Eng. Data **39** (1), 127-130 (1994)

[26] Mistry, S.L., Kaul, A., Merchuk, J.C., Asenjo, J.A.: Mathematical modelling and computer simulation of aqueous two-phase continuous protein extraction. J. Chromatogr. A. **741** (2), 151-163 (1996)

[27] Hu, M.C., Zhai, Q.G., Jiang, Y.C., Jin, L.H., Liu, Z.H.: Liquid-liquid and liquid-liquid-solid equilibrium in PEG + Cs_2SO_4 + H_2O. J. Chem. Eng. Data **49** (5), 1440-1443 (2004)

[28] Jayapal, M., Regupathi, I., Murugesan, T.: Liquid-liquid equilibrium of poly (ethylene glycol) 2000 + potassium citrate + water at (25, 35, and 45) ℃. J. Chem. Eng. Data **52** (1), 56-59 (2007)

[29] Wu, Y.T., Lin, D.Q., Zhu, Z.Q., Mei, L.H.: Prediction of liquid-liquid equilibria of polymer-salt aqueous two-phase systems by a modified Pitzer's virial equation. Fluid Phase Equilib. **124**, 67-79 (1996)

[30] Krevelen, D.W.V., Nijehuis, K.T.: Properties of Polymers: Their Correlation with Chemical Structure; Their Numerical Estimation and Prediction from Additive Group Contributions, 4th edn. Elsevier, Amsterdam (2009)

[31] Greenberg, J.P., Møller, N.: The prediction of mineral solubilities in natural waters: a chemical equilibrium model for the Na-K-Ca-Cl-SO_4-H_2O system to high concentration from 0 to 250 ℃. Geochim. Cosmochim. Acta **53**, 2503-2518 (1989)

Solid-liquid equilibrium of quinary aqueous solution composed of lithium, potassium, rubidium, magnesium, and borate at 323.15 K*

Xudong Yu[1, 2], Mianping Zheng[1], Ying Zeng[1, 2] and Lin Wang[2]

1. MNR Key Laboratory of Saline Lake Resources and Environments, Institute of Mineral Resources, CAGS, Beijing 100037, China
2. College of Materials and Chemistry&Chemical Engineering, Chengdu University of Technology, Chengdu 610059, China

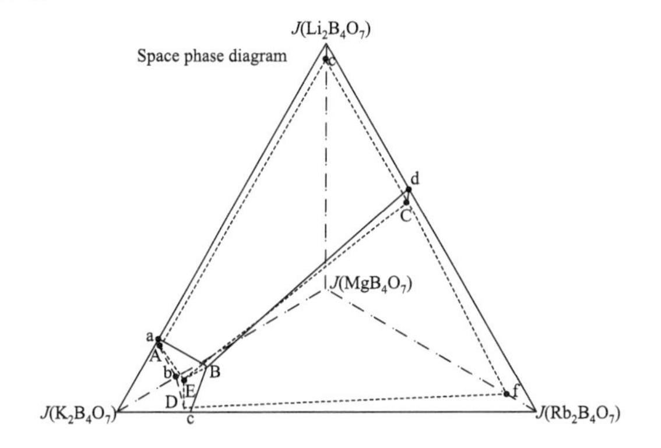

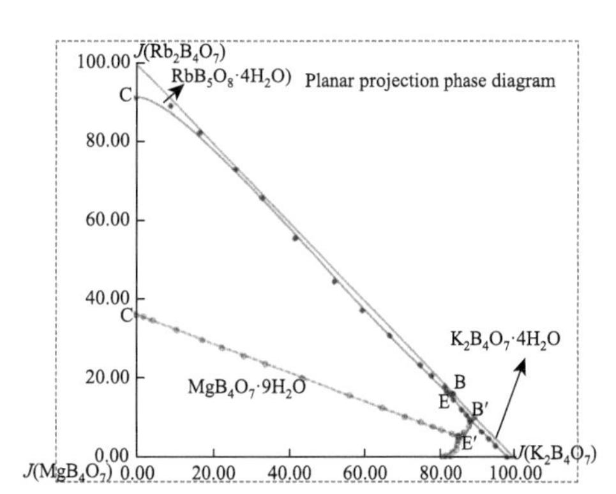

Abstract To understand the thermodynamic behaviors of the borate-containing system, the solid-liquid equilibrium data (solubilities, density, and refractive index) of quinary aqueous solution composed of lithium, potassium, rubidium, magnesium, and borate at 323.15 K were determined by using the isothermal dissolution method. The crystalloid forms of the solid phase were confirmed by the X-ray diffraction method. Results show that four single salts $Li_2B_4O_7·3H_2O$, $K_2B_4O_7·4H_2O$, $RbB_5O_8·4H_2O$, and $MgB_4O_7·9H_2O$ were formed at 323.15 K in the quinary invariant point, with no double salt or solid solution found. Under the condition of $Li_2B_4O_7$ saturation, the crystallization region of $MgB_4O_7·9H_2O$ occupied the largest crystallization region, and the salt $K_2B_4O_7·4H_2O$ had the smallest region. Among the four salts, the crystalline regions of the salts $MgB_4O_7·9H_2O$ and $RbB_5O_8·4H_2O$ changed obviously at 323.15 and 348.15 K, that of $MgB_4O_7·9H_2O$ enlarged at 323.15 K, and those of $RbB_5O_8·4H_2O$ and $K_2B_4O_7·4H_2O$ decreased.

1 Introduction

Saline lakes and underground brines are valuable natural resources. Large amounts of important raw materials for the chemical industry and metallurgical industry are now obtained from saline lakes and underground brines.[1] Saline lakes occur in XinJiang, Qinghai, Tibet, Inner Mongolia, and Shanxi. Underground brines are present at the Sichuan Basin, Jianghan Basin, and Qaidam Basin.[1] Of these, the underground brine in Sichuan Basin (Pingluo, Zigong, and Xuanda Salt Basin) is famous for its excellent quality with high concentrations of Na, K, B, Mg, and the rare elements Li and Rb. Among of them, the concentration of boron (4.99 $g·L^{-1}$) in Pingluo under-ground brine is higher than that in Searles Salt Lake (3.11 $g·L^{-1}$) and is 32.29 times the comprehensive utilization industrial grade.[2] The boron (by H_3BO_3) reserved in Pingluo is nearly about 2.987×10^7 t, which has great exploitation potential for meeting the huge demand for boron.[3-5] However, as mentioned above, the coexistence of many ions in underground brine makes the relationship between salts more complicated, which increases the difficulty of separation and extraction of boron form the brine.

* 本文发表在：Journal of Chemical & Engineering Data, 2019, 64: 5681-5687

Commonly, the dissolution and crystallization behavior of salts in solution can be obtained through the phase equilibria study of the complex salt−water system at multiple temperatures. Therefore, to understand the thermodynamic behaviors of the borates in aqueous solution, several phase diagrams of the system containing borates have been reported, for example, $Na^+//BO_2^-$, OH^-−H_2O at (263.15−323.15) K;[6-8] $Li^+//Cl^-$ (SO_4^{2-}), BO_2^--H_2O at 288.15 and 298.15 K;[9, 10] Na^+, $Ca^{2+}//Cl^-$, borate-H_2O and Li^+, Na^+, $Ca^{2+}//Cl^-$, borate-H_2O at 288.15 K;[11, 12] Li^+, $K^+//SO_4^{2-}$, $B_4O_7^{2-}$-H_2O at 288.15 K;[13] Li^+, Na^+, $Mg^{2+}//B_4O_7^{2-}$-H_2O at 273.15 K;[14] Li^+, Na^+, $K^+//CO_3^{2-}$, $B_4O_7^{2-}$-H_2O and Na^+, $K^+//Cl^-$, SO_4^{2-}, $B_4O_7^{2-}$-H_2O at 298.15 K;[15, 16] and the phase equilibria of quinary, quaternary, and ternary subsystems of system Li^+, K^+, Rb^+, $Mg^{2+}//Cl^-$, borate-H_2O.[17-24] After comprehension of the above research, three forms of boron ions such as BO_2^-, $B_4O_7^{2-}$, and $B_5O_8^-$ appeared in the solution, and the coexisting ions in the solution and temperature have a direct effect on the structure of boron ion.

Till now, the corresponding ternary and quaternary subsystems of quinary system lithium + potassium + rubidium + magnesium + borate at 323.15 K have been observed,[25-27] while there is no report about the phase equilibrium of the quinary system at 323.15 K. Consequently, the quinary solid-liquid equilibrium data of lithium + potassium + rubidium + magnesium + borate at 323.15 K is reported in detail.

2 Experimental section

2.1 Reagents

The ultrapure water was prepared by a UPT-II-20T Ulupure purification system (Chengdu Ultrapure Technology Co., Ltd.) with conductivity $\leqslant 5.5\times10^{-6}$ $S\cdot m^{-1}$, and the conductivity was measured by the Ulupure purification system with its own online conductometer. The chemicals hungchaoite ($MgB_4O_7\cdot9H_2O$) and rubidium pentaborate ($RbB_5O_8\cdot4H_2O$) were synthesized in our laboratory from an aqueous solution of MgO/Rb_2CO_3 and H_3BO_3.[28, 29] The information on $Li_2B_4O_7$ and $K_2B_4O_7\cdot5H_2O$ is presented in Table 1.

Table 1　Reagents used in this experiment

Regent	Initial purity (ω/ω%)	Purified method	Final purity (ω/ω%)	Analytical method[30]
$Li_2B_4O_7^a$	99.0	Recrystallization	99.5	Alkalimetry in the presence of mannitol for boron
$K_2B_4O_7\cdot5H_2O^b$	99.0	Recrystallization	99.0	
$MgB_4O_5(OH)_4\cdot7H_2O^c$	99.0		99.0	
$RbB_5O_6(OH)_4\cdot2H_2O^d$	99.0		99.0	

a. From the Chengdu Kelong Chemical Reagent Plant. b. From the Sinopharm Chemical Reagent Co., Ltd. c. Synthesized in our laboratory.[28] d. Synthesized in our laboratory.[29]

2.2 Apparatus and procedure

The quinary solid−liquid equilibrium was carried out via an isothermal dissolution method at 323.15 K under atmospheric pressure.[21] A series of mixtures were prepared by adding different ratios (from 0 to saturation) of a fourth new salt in the quaternary saturations corresponding to four quaternary subsystems. The sample was stirred in the thermostatic bath (THZ-82 type, China) with constant temperature at 323.15±0.2 K. Once the composition of the fourth salt in the system did not change, the equilibrium of the system was reached, and then stirrers were stopped. After the samples had remained static for 48−72 h, the density of the solution was measured by pycnometry method with the standard uncertainty of $\pm2.0\times10^{-4}g\cdot cm^{-3}$, the refractive index was determined by an Abbe refractometer (WYA type, China) with the standard uncertainty of $\pm1.0\times10^{-4}$. Meanwhile, the corresponding liquid phase were analyzed after the solid−liquid separation; the wet residue was obtained and dried at 323.15 K. The crystalloid forms of the solid phases were determined by the X-ray powder diffraction (DX-2700 type, China). To avoid the effect of the temperature change during sampling, the thermostat (Inborn SHH250, China) with an uncertainty of ±0.1 K was used for maintaining the desired temperature.

The total amount of boron was determined by using neutralization titration with standard uncertainty of ±0.3%.[30] The concentrations of $Li_2B_4O_7$, $K_2B_4O_7$, $Rb_2B_4O_7$, and MgB_4O_7 in this system were acquired with Li^+, K^+, Rb^+ (Inductively coupled plasma-optical emission spectrometry (ICP-OES), PerkinElmer, 5300 V-type), [31, 32] and Mg^{2+} (ethyl-enediaminetetraacetic acid (EDTA) complex metric titration method)[33] with standard uncertainty ±0.5%, respectively.

3 Results and discussion

There are different polymeric forms of boron B $(OH)_3$, $B(OH)_4^-$, BO_2^-, $B_4O_7^{2-}$, $B_5O_8^-$, etc. in the aqueous solution containing borate.[34, 35] Accordingly, to simplify the calculation, the traditional stoichiometric form $B_4O_7^{2-}$ was selected for the description of the concentration of boron in solution. The values of the invariant point of the corresponding subsystems of the quinary solid-liquid system lithium + potassium + rubidium + magnesium + borate at 323.15 K are presented in Table 2. The quinary solid-liquid equilibrium data (solubility, density, and refractive index) saturated with $Li_2B_4O_7$ are given in Table 3. The mass fraction w and the dry salt mole index J were used for the expression of the concentration of each solution component in Tables 2 and 3. Calculation formulas of the Janecke index in Table 2 (eqs 1−5) are as follows

$$\text{letting}[M]=\frac{w(Li_2B_4O_7)}{169.12}+\frac{w(K_2B_4O_7)}{233.44}+\frac{w(Rb_2B_4O_7)}{326.17}+\frac{w(MgB_4O_7)}{179.54}\text{mol} \tag{1}$$

$$J(Li_2B_4O_7)=\frac{w(Li_2B_4O_7)}{169.12\times[M]}\times 100 \tag{2}$$

$$J(K_2B_4O_7)=\frac{w(K_2B_4O_7)}{233.44\times[M]}\times 100 \tag{3}$$

$$J(Rb_2B_4O_7)=\frac{w(Rb_2B_4O_7)}{326.17\times[M]} \tag{4}$$

$$J(Mg\,B_4O_7)=\frac{w(MgB_4O_7)}{179.54\times[M]}\times 100 \tag{5}$$

Calculation formulas of the Janecke index in Table 3 (eqs 6−10) are as follows

$$\text{letting}[A]=\frac{w(K_2B_4O_7)}{233.44}+\frac{w(Rb_2B_4O_7)}{326.17}+\frac{w(MgB_4O_7)}{179.54}\text{mol} \tag{6}$$

$$J(K_2B_4O_7)=\frac{w(K_2B_4O_7)}{233.44\times[A]}\times 100 \tag{7}$$

$$J(Rb_2B_4O_7)=\frac{w(Rb_2B_4O_7)}{326.17\times[A]}\times 100 \tag{8}$$

$$J(MgB_4O_7)=\frac{w(MgB_4O_7)}{179.54\times[A]}\times 100 \tag{9}$$

$$J(H_2O)=\frac{w(H_2O)}{18.02\times[A]}\times 100 \tag{10}$$

Figure 1 presents the space phase diagram of the quinary system at 323.15 K; points a to f represent the saturation points of ternary subsystems; points A–D denote the saturation points of quaternary subsystems; and point E is the saturation point of the quinary system. Four univariant curves AE, BE, CE, and DE divided the space phase diagram into four parts, which correspond to four single salts. The results of space phase diagram (Fig. 1) and

X-ray diffraction analysis of point *E* (Fig. 2) show that salts $MgB_4O_7 \cdot 9H_2O$, $K_2B_4O_7 \cdot 4H_2O$, $RbB_5O_8 \cdot 4H_2O$, and $Li_2B_4O_7 \cdot 3H_2O$ coexist at point *E*. The X-ray diffraction pattern of points *A*, *D*, and 29 are given in Figures S1−S3 (Supporting Information).

Table 2 Composition of the subsystems' invariant points of the quinary system lithium + potassium + rubidium + magnesium + borate at 323.15 K and pressure p = 0.1 MPa[a]

No.	System	Composition of solution, w (B) $\times 10^2$				Jänecke index of dry salt $J(Li_2B_4O_7)+J(K_2B_4O_7)+J(Rb_2B_4O_7)+J(MgB_4O_7)=100$				Equilibrium solid phase
		$w(Li_2B_4O_7)$	$w(K_2B_4O_7)$	$w(Rb_2B_4O_7)$	$w(MgB_4O_7)$	$J(Li_2B_4O_7)$	$J(K_2B_4O_7)$	$J(Rb_2B_4O_7)$	$J(MgB_4O_7)$	
a	LiKB[25]	4.62	27.29	0.00	0.00	18.94	81.06	0.00	0.00	LiB + KB
b	KMgB	0.00	23.16	0.00	7.24	0.00	71.10	0.00	28.90	MB + KB
c	LiMgB	4.53	0.00	0.00	0.23	95.44	0.00	0.00	4.56	LiB + MB
d	LiRbB[25]	3.61	0.00	4.46	0.00	60.96	0.00	39.04	0.00	LiB + RB
e	KRbB[26]	0.00	25.14	7.73	0.00	0.00	81.96	18.04	0.00	RB + KB
f	RbMgB[26]	0.00	0.00	3.86	0.38	0.00	0.00	84.83	15.17	MB + RB
A	LiKMgB	4.63	25.11	0.00	0.37	19.98	78.51	0.00	1.51	LiB + MB + KB
B	LiKRbB[27]	3.23	25.28	6.83	0.00	12.88	73.00	14.12	0.00	LiB + KB + RB
C	LiRbMgB	3.63	0.00	4.87	0.25	56.80	0.00	39.51	3.69	LiB + MB + RB
D	KRbMgB	0.00	27.70	7.19	0.12	0.00	83.94	15.59	0.47	MB + KB + RB
E	LiKRbMgB	2.81	27.71	7.48	0.26	10.40	74.33	14.36	0.91	LiB + MB + KB + RB

a. Note: Standard uncertainties u are $u(T)$ = 0.20 K, $u_r(p)$ = 0.05; $u_r(w(Li_2B_4O_7))$ = 0.0030, $u_r(w(B))$ = 0.0050 (B = $K_2B_4O_7$, $Rb_2B_4O_7$, MgB_4O_7); LiB−$Li_2B_4O_7 \cdot 3H_2O$, KB−$K_2B_4O_7 \cdot 4H_2O$, RbB−$RbB_5O_8 \cdot 4H_2O$, MgB−$MgB_4O_7 \cdot 9H_2O$; Ternary subsystems: LiKB: Li^+, K^+//borate−H_2O, KMgB: K^+, Mg^{2+}//borate−H_2O, LiMgB: Li^+, Mg^{2+}//borate−H_2O, LiRbB: Li^+, Rb^+//borate−H_2O, KRbB: K^+, Rb^+//borate−H_2O, RbMgB: Rb^+, Mg^{2+}//borate−H_2O; Quaternary subsystems: LiKMgB: Li^+, K^+, Mg^{2+}//borate−H_2O, LiKRbB: Li^+, K^+, Rb^+//borate−H_2O, LiRbMgB: Li^+, Rb^+, Mg^{2+}//borate−H_2O, KRbMgB: K^+, Rb^+, Mg^{2+}//borate−H_2O; Quinary system: LiKRbMgB: Li^+, K^+, Rb^+, Mg^{2+}//borate−H_2O.

Table 3 Solubilities, densities, refractive indices of the quinary system lithium + potassium + rubidium + magnesium + borate at 323.15 K and pressure p = 0.1 MPa[a] (saturated with $Li_2B_4O_7$)

No.	Density (g·cm^{-3})	Refractiveindex (n_D)	Composition of solution, w(B) $\times 10^2$					Jänecke index of dry salt $J(K_2B_4O_7)+J(Rb_2B_4O_7)+J(MgB_4O_7)=100$				Equilibrium solid phase
			$w(Li_2B_4O_7)$	$w(K_2B_4O_7)$	$w(Rb_2B_4O_7)$	$w(MgB_4O_7)$	$w(H_2O)$	$J(K_2B_4O_7)$	$J(Rb_2B_4O_7)$	$J(MgB_4O_7)$	$J(H_2O)$	
1, A	1.3560	1.3856	4.63	25.11	0.00	0.37	69.89	98.12	0.00	1.88	3538	LiB + KB + MB
2	1.3544	1.3927	3.89	27.50	1.15	0.45	67.01	95.13	2.85	2.02	3003	LiB + KB + MB
3	1.3586	1.3935	4.01	27.70	1.97	0.40	65.92	93.48	4.76	1.76	2882	LiB + KB + MB
4	1.3608	1.3949	3.66	27.83	2.78	0.43	65.30	91.61	6.55	1.84	2785	LiB + KB + MB
5	1.3817	1.3965	3.83	26.47	3.93	0.44	65.33	88.66	9.42	1.92	2835	LiB + KB + MB
6	1.3896	1.3979	3.71	27.11	4.71	0.39	64.08	87.48	10.88	1.64	2679	LiB + KB + MB
7	1.3978	1.3986	3.57	26.85	5.31	0.41	63.86	86.10	12.19	1.71	2653	LiB + KB + MB
8	1.4006	1.3989	3.55	25.70	6.24	0.41	64.10	83.71	14.55	1.74	2705	LiB + KB + MB
9, E	1.5072	1.4046	2.81	27.71	7.48	0.26	61.74	82.96	16.03	1.01	2394	LiB + KB + RB + MB

Continued

No.	Density ($g \cdot cm^{-3}$)	Refractiveindex (n_D)	Composition of solution, $w(B) \times 10^2$					Jänecke index of dry salt $J(K_2B_4O_7)+J(Rb_2B_4O_7)+J(MgB_4O_7)=100$				Equilibrium solid phase
			$w(Li_2B_4O_7)$	$w(K_2B_4O_7)$	$w(Rb_2B_4O_7)$	$w(MgB_4O_7)$	$w(H_2O)$	$J(K_2B_4O_7)$	$J(Rb_2B_4O_7)$	$J(MgB_4O_7)$	$J(H_2O)$	
10, B	1.4471	1.3969	3.23	25.28	6.83	0.00	64.66	83.80	16.20	0.00	2776	LiB + KB + RB
11, E	1.5072	1.4046	2.81	27.71	7.48	0.26	61.74	82.96	16.03	1.01	2394	LiB + KB + RB + MB
12, C	1.1186	1.3535	3.63	0.00	4.87	0.25	91.25	0.00	91.47	8.53	31 022	LiB + RB + MB
13	1.1339	1.3519	2.97	0.40	5.51	0.06	91.06	9.05	89.19	1.76	26 680	LiB + RB + MB
14	1.1284	1.3528	3.16	0.76	5.24	0.03	90.81	16.71	82.43	0.86	25 859	LiB + RB + MB
15	1.1349	1.3550	2.94	1.45	5.64	0.03	89.94	26.24	73.05	0.71	21 086	LiB + RB + MB
16	1.1357	1.3559	2.78	2.11	5.87	0.06	89.18	33.03	65.75	1.22	18 082	LiB + RB + MB
17	1.1649	1.3601	2.78	3.57	6.64	0.16	86.85	41.85	55.71	2.44	13 189	LiB + RB + MB
18	1.1751	1.3620	2.76	5.17	6.16	0.23	85.68	52.34	44.63	3.03	11 237	LiB + RB + MB
19	1.1975	1.3655	2.78	7.09	6.19	0.27	83.67	59.72	37.32	2.96	9130	LiB + RB + MB
20	1.2194	1.3706	2.54	9.92	6.44	0.25	80.85	66.78	31.03	2.19	7051	LiB + RB + MB
21	1.2393	1.3745	2.58	13.36	5.84	0.24	77.98	74.84	23.41	1.75	5659	LiB + RB + MB
22	1.2739	1.3782	2.44	15.62	5.80	0.23	75.91	77.83	20.68	1.49	4900	LiB + RB + MB
23	1.3718	1.3858	2.57	20.33	6.30	0.10	70.70	81.42	18.06	0.52	3668	LiB + RB + MB
24	1.4530	1.3977	2.39	25.84	7.40	0.32	64.05	81.90	16.78	1.32	2630	LiB + RB + MB
25, E	1.5072	1.4046	2.81	27.71	7.48	0.26	61.74	82.96	16.03	1.01	2394	LiB + KB + RB + MB
26, D	1.4784	1.4006	0.00	27.70	7.19	0.12	64.99					KB + RB + MB
27	1.4792	1.3975	0.53	28.82	6.74	0.17	63.74					KB + RB + MB
28	1.4834	1.3975	0.75	28.66	7.19	0.14	63.26					KB + RB + MB
29	1.4907	1.4013	1.55	29.56	6.81	0.45	61.63					KB + RB + MB
30	1.4775	1.4019	1.18	30.23	6.51	0.50	61.58					KB + RB + MB
31	1.5062	1.4040	2.04	28.68	7.03	0.42	61.83					KB + RB + MB
32	1.5068	1.4045	2.08	29.43	7.15	0.35	60.99					KB + RB + MB
33, E	1.5072	1.4046	2.81	27.71	7.48	0.26	61.74					KB + RB + MB

a. Note: Standard uncertainties u are $u(T)=0.20$ K, $u_r(p)=0.05$; $u(\rho)=2.0\times10^{-4} g \cdot cm^{-3}$; $u(n_D)=1.0\times10^{-4}$; $u_r(w(Li_2B_4O_7))=0.0030$, $u_r(w(B))=0.0050$ (B = $K_2B_4O_7$, $Rb_2B_4O_7$, MgB_4O_7); LiB–$Li_2B_4O_7 \cdot 3H_2O$, KB–$K_2B_4O_7 \cdot 4H_2O$, RbB–$RbB_5O_8 \cdot 4H_2O$, MgB–$MgB_4O_7 \cdot 9H_2O$.

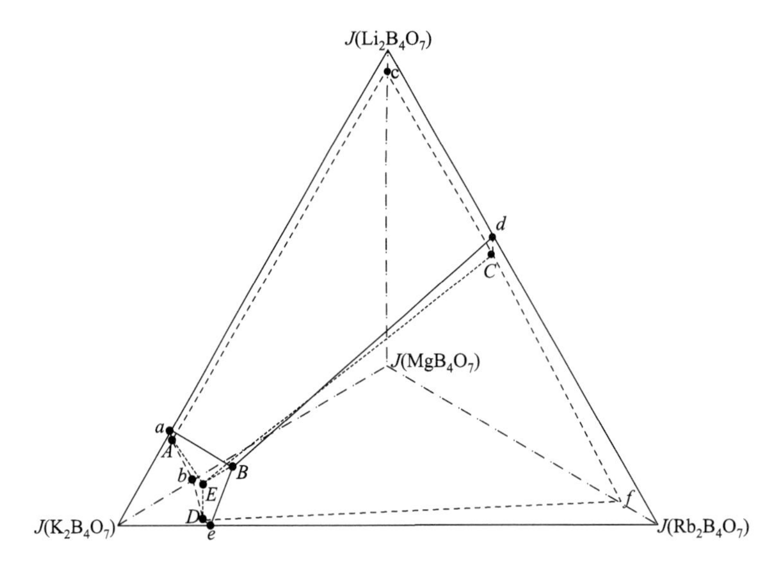

Fig. 1　Space diagram of the aqueous solution composed of lithium, potassium, rubidium, magnesium, and borate at 323.15 K.

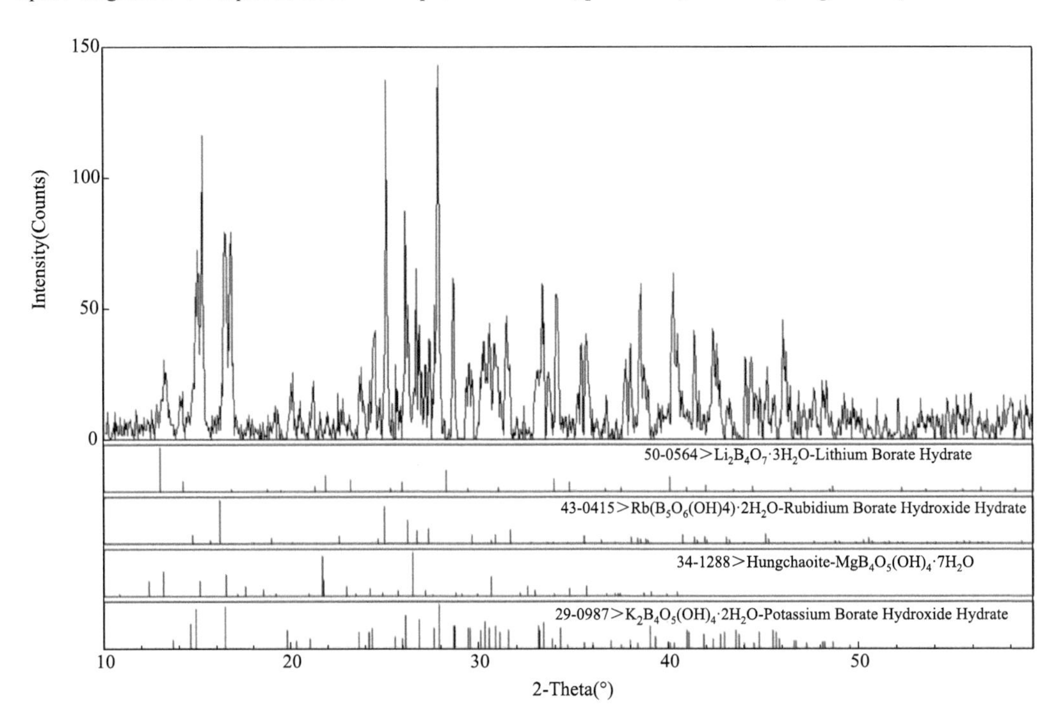

Fig. 2　X-ray diffraction pattern of the invariant point E in the aqueous solution composed of lithium, potassium, rubidium, magnesium, and borate at 323.15 K.

By analysis the result of the quinary solid−liquid system of lithium + potassium + rubidium + magnesium + chloride at 323.15 K,[36] there are various salt crystal forms, such as solid solution [(K, Rb)Cl], carnallite-type double salt $MCl·MgCl_2·6H_2O$ ($M = K^+$, Rb^+, Li $(H_2O)^+$), while when the anion changed from chloride to borate, coexist cation ions were still lithium, potassium, rubidium, magnesium, the crystals of solid solution and double salt disappeared, only four single salts formed. This may be due to the presence of two different borate forms layered structure tetrapolymer $B_4O_5(OH)_4^{2-}$ and double-ring structure pentapolymer $B_5O_6(OH)_4^-$ in the solution,[37] which makes the salt crystal strains different and does not have the condition for the formation of a solid solution or double salt.

Commonly, with an increasing number of components, the diagram becomes very complicated and is not very useful in practice. To make the phase diagram of a complex system easier to understand, the quinary system can be represented in two dimension when two projections (planar projection diagram and water content diagram) are combined. Fig. 3 presents the planar projection phase diagram (saturated with salt $Li_2B_4O_7$) of the quinary system at 323.15 K. The crystallization region of $K_2B_4O_7·4H_2O$ occupied the smallest crystallization regions, which means that it is difficult to separate from the mixture solution. The comparison of the phase diagrams at 323.15 and 348.15 K[21] of the quinary system (Fig. 4) shows that (1) both phase diagrams at 323.15 and 348.15 K are of simple type, no double salt or solid solution formed and (2) temperature has no effect on the crystalline form of salt, while it has obvious effect on the crystalline regions of salts, where the crystalline region of $MgB_4O_7·9H_2O$ enlarged at 323.15 K and those of $RbB_5O_8·4H_2O$ and $K_2B_4O_7·4H_2O$ decreased. Thus, at a relatively high temperature, pure salt $RbB_5O_8·4H_2O$ can be more easily obtained.

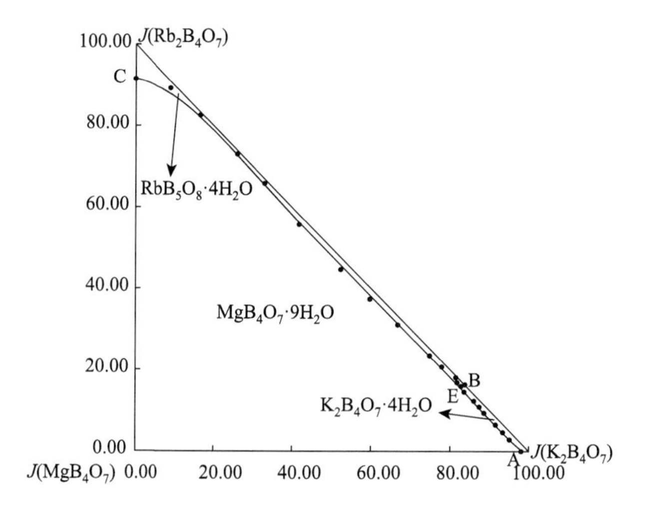

Fig. 3 Planar projection diagram of the aqueous solution composed of lithium, potassium, rubidium, magnesium, and borate at 323.15 K (saturated with $Li_2B_4O_7$).

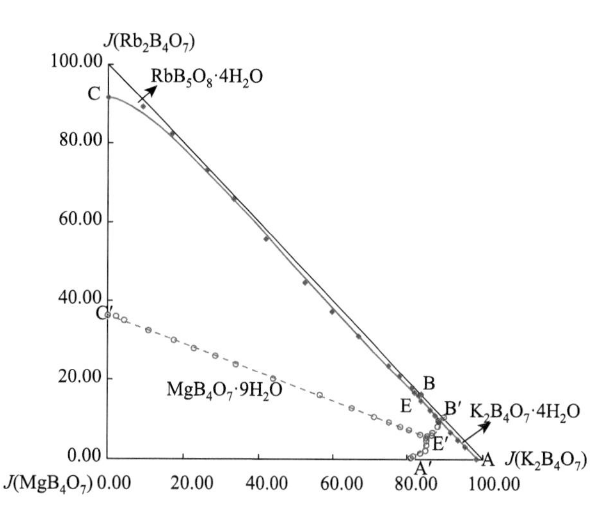

Fig. 4 Planar projection diagram of the aqueous solution composed of lithium, potassium, rubidium, magnesium, and borate at 323.15 K and 348.15 K[21] (saturated with $Li_2B_4O_7$) (red circle open, blue box solid, experimental value at 348.15 K[21] and 323.15 K; red line: solubility curve at 348.15 K; blue line: solubility curve at 323.15 K).

The water content diagram is plotted as shown in Fig. 5. The water content in curve CE decreases with the addition of $K_2B_4O_7$ in the solution until the value reaches the minimum at quinary invariant point *E*, while the water content changes in the opposite direction in curves AE and BE with the decrease of $K_2B_4O_7$ in the solution; the water content decreases until the value reaches the minimum at the quinary invariant point *E*.

The relationship between density or refractive index and composition at 323.15 K are presented in Figs. 6 and 7. The density and refractive index of the equilibrium solution are related to the content of salt in the solution. For the quinary system in this study, the solubility of salt $K_2B_4O_7$ is largest among the four salts, and it has a strong salting out effect on $Li_2B_4O_7$, RbB_5O_8, and MgB_4O_7. The values of density and refractive index of the solution at equilibrium have the same rule of change: the salt content in the solution increases with the addition of $K_2B_4O_7$ in the solution in curve CE (cosaturated with $Li_2B_4O_7·3H_2O$, $RbB_5O_8·4H_2O$, and $MgB_4O_7·9H_2O$); thus, the values of density and refractive index increase until the maximum values at

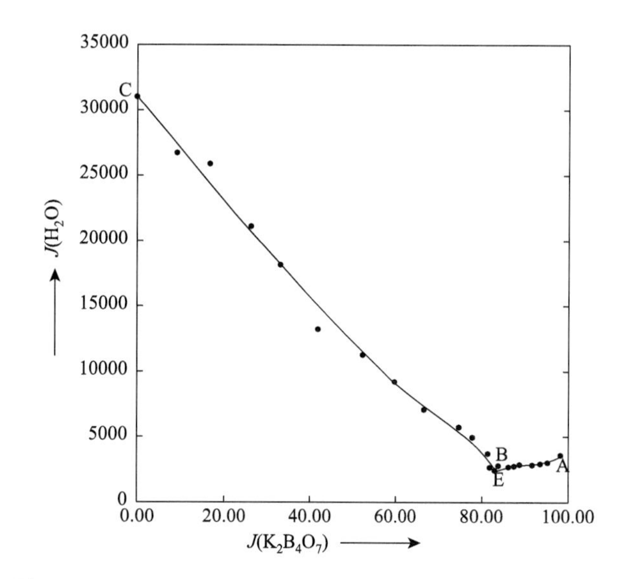

Fig. 5 Water content diagram of the aqueous solution composed of lithium, potassium, rubidium, magnesium, and borate at 323.15 K (saturated with $Li_2B_4O_7$).

point E is achieved; however, at curves AE (cosaturated with $Li_2B_4O_7·3H_2O$, $K_2B_4O_7·4H_2O$, and $MgB_4O_7·9H_2O$) and BE (cosaturated with $Li_2B_4O_7·3H_2O$, $K_2B_4O_7·4H_2O$, and $RbB_5O_8·4H_2O$), with increase in the Janecke index of $K_2B_4O_7$, the salting out effect on other salts is more obvious; thus, the content of the salts decrease and the values of density and refractive index decrease.

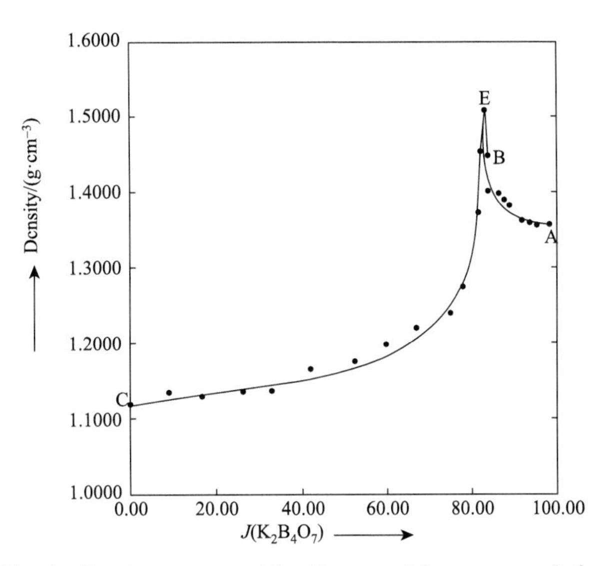

Fig. 6　Density vs composition diagram of the aqueous solution composed of lithium, potassium, rubidium, magnesium, and borate at 323.15 K (saturated with $Li_2B_4O_7$).

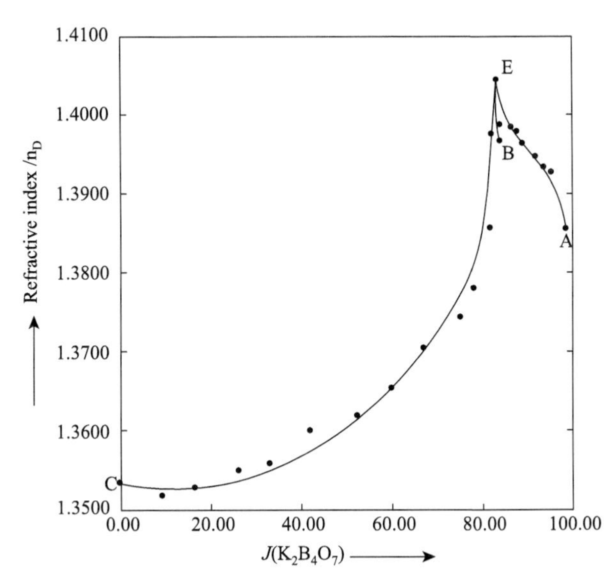

Fig. 7　Refractive index vs composition diagram of the aqueous solution composed of lithium, potassium, rubidium, magnesium, and borate at 323.15 K (saturated with $Li_2B_4O_7$).

4　Conclusions

(1) The solid-liquid equilibrium of quinary aqueous solution composed of lithium, potassium, rubidium, magnesium, and borate is a simple cosaturation type with three tetraborates $Li_2B_4O_7·3H_2O$, $K_2B_4O_7·4H_2O$, $MgB_4O_7·9H_2O$, and one pentaborate $RbB_5O_8·4H_2O$ formed at 323.15 K.

(2) At the quinary invariant point *E,* the water content reaches the minimum value, while the values of density and refractive index reach the maximum.

(3) The comparison of the phase diagrams at 323.15 K and 348.15 K of quinary system shows that temperature has no effect on the crystalline form of salt, while it has obvious effect on the crystalline regions of salts; the crystalline region of $MgB_4O_7·9H_2O$ enlarged at 323.15 K and that of $RbB_5O_8·4H_2O$ and $K_2B_4O_7·4H_2O$ decreased.

References

[1]　Zheng, M. P. *Saline Lakes and Salt Basin Deposits in China*; Sciences Press: Beijing, 2014.

[2]　Lin, Y. T.; Tang, Q. Distribution of Brine in Sichuan Basin and Its Prospects for Tapping. Geol. Chem. Miner. 1999, 21, 209-214.

[3]　Li, W.; Dong, Y. P.; Song, P. S. *The Development and Utilization of Salt Lake Brine Resource*; Chemical Industry Press: Beijing, 2012.

[4]　Teeple, J. E. *The Industrial Development of Searles Lake Brines with Equilibrium Data*; Chemical Catalog Company Inc: New York, 1929.

[5]　Yang, J. H., Xu, Y. X., Hou, X. D., Deng, X. L., Li, B. Y., Liu, Z. M. *Specifications for Salt-Lake, Salt Mineral Exploration* (*DZ/T 0212-2002*); Geological Publishing House: Beijing, China, 2003.

[6]　Vilarinho-Franco, T.; Teyssier, A.; Tenu, R.; Pecautc, J.; Delmas, J.; Heitzmann, M.; Capron, P.; Counioux, J. J.; Goutaudier, C. Solid-liquid equilibria in the ternary system $NaBO_2$-NaOH-H_2O thermal behavior of double salts. Fluid Phase Equilib. 2013, 360, 212-221.

[7]　Churikov, A. V.; Zapsis, K. V.; Khramkov, V. V.; Churikov, M. A.; Smotrov, M. P.; Kazarinov, I. A. Phase diagrams of the ternary systems $NaBH_4$ + NaOH + H_2O, KBH_4 + NaOH + H_2O, $NaBO_2$ + NaOH + H_2O, KBO_2 + NaOH + H_2O at-10 ℃. J. Chem. Eng. Data 2011, 56, 9-13.

[8]　Qin, S. Y.; Yin, B. W.; Zhang, Y. F.; Zhang, Y. Phase equilibria of the NaOH-$NaBO_2$-Na_2CO_3-H_2O system at 30 ℃, 60 ℃, and 100 ℃. J. Chem. Eng. Data 2015, 60, 3018-3023.

[9]　Gao, D. L.; Guo, Y. F.; Yu, X. P.; Wang, S. Q.; Deng, T. L. Solubilities, densities, and refractive indices in the salt-water ternary system

($LiCl + LiBO_2 + H_2O$) at T = 288.15 K and 298.15 K and p = 0.1 MPa. J. Chem. Eng. Data 2015, 60, 2594-2599.
[10] Gao, D. L.; Wang, Q.; Guo, Y. F.; Yu, X. P.; Wang, S. Q.; Deng, T. L. Solid-liquid phase equilibria in the aqueous ternary system $Li_2SO_4 + LiBO_2 + H_2O$ at T = 288.15 and 298.15 K. Fluid Phase Equilib.2014, 371, 121-124.
[11] Wang, M. X.; Lei, L. Y.; Guo, Y. F.; Meng, L. Z.; Wang, S. Q.; Deng, T. L. Phase equilibria of the reciprocal quaternary system (Na^+, Ca^{2+}//Cl^-, Borate-H_2O) at 288.15 K and 0.1 MPa. J. Chem. Eng. Data 2018, 63, 4005-4011.
[12] Chen, S. Q.; Guo, Y. F.; Li, L.; Zhang, S. S.; Lei, L. Y.; Li, M. L.; Duo, J.; Deng, T. L. Solid-liquid phase equilibria of the quinary system containing lithium, sodium, calcium, chloride, and borate ions at T = 288.15 K and p = 101.325 kPa. J. Chem. Eng. Data 2019, 64, 3050-3057.
[13] Sang, S. H.; Fu, C.; Zhang, T. T.; Zhang, X. P.; Xiao, L. J. Measurement of solid-liquid equilibria in quaternary system Li^+, K^+// SO_4^{2-} , $B_4O_7^{2-}$ -H_2O at 288 K. Chem. Res. Chin. Univ.2016, 32, 90-94.
[14] Yang, L.; He, X. F.; Gao, Y. Y.; Cui, R. Z.; Sang, S. H. Studies on phase equilibria in the quaternary systems $LiCl$-KCl-$MgCl_2$-H_2O and $Li_2B_4O_7$-$Na_2B_4O_7$-MgB_4O_7-H_2O at 273 K. J. Chem. Eng. Data 2018, 63, 1206-1211.
[15] Tursunbadalov, S.; Soliev, L. Investigation of phase equilibria in quinary water-salt systems. J. Chem. Eng. Data 2018, 63, 598-612.
[16] Tursunbadalov, S.; Soliev, L. Determination of phase equilibria and construction of comprehensive phase diagram for the quinary Na, K//Cl, SO_4, B_4O_7-H_2O system at 25 ℃. J. Chem. Eng. Data 2017, 62, 698-703.
[17] Yu, X. D.; Luo, Y. L.; Wu, L. T.; Cheng, X. L.; Zeng, Y. Solid-liquid equilibrium on the reciprocal aqueous quaternary system Li^+, Mg^{2+}//Cl^-, and Borate-H_2O at 323 K. J. Chem. Eng. Data 2016, 61, 3311-3316.
[18] Yin, Q. H.; Mu, P. T.; Tan, Q.; Yu, X. D.; Li, Z. Q.; Zeng, Y. Phase equilibria for the aqueous reciprocal quaternary system Rb^+, Mg^{2+}//Cl^-, borate-H_2O at 348 K. J. Chem. Eng. Data 2014, 59, 2235-2241.
[19] Guo, S. S.; Yu, X. D.; Zeng, Y. Phase equilibria for the aqueous reciprocal quaternary system K^+, Mg^{2+}//Cl^-, Borate-H_2O at 298 K. J. Chem. Eng. Data 2016, 61, 1566-1572.
[20] Yan, F. P.; Yu, X. D.; Yin, Q. H.; Zhang, Y. J.; Zeng, Y. The solubilities and physicochemical properties of the aqueous quaternary system Li^+, K^+, Rb^+//Borate-H_2O at 348 K. J. Chem. Eng. Data 2014, 59, 110-115.
[21] Yu, X. D.; Zeng, Y.; Guo, S. S.; Zhang, Y. J. Stable phase equilibrium and phase diagram of the quinary system Li^+, K^+, Rb^+, Mg^{2+}//Borate-H_2O at T = 348.15 K. J. Chem. Eng. Data 2016, 61, 1246-1253.
[22] Huang, Q.; Li, M. L.; Wang, L.; Yu, X. D.; Zeng, Y. The phase and physicochemical properties diagrams of systems Rb^+ (Mg^{2+}) //$Cl^-$and Borate-H_2O at 323 K. Russ. J. Phys. Chem. A 2019, 93, 211-217.
[23] Tan, Q.; Zeng, Y.; Mu, P. T.; Yu, X. D.; Zhang, Y. J. Stable phase equilibrium of aqueous quaternary system Li^+, K^+, Mg^{2+}//Borate-H_2O at 348 K. *J.* Chem. Eng. Data 2014, 59, 4173-4178.
[24] Zeng, Y.; Xie, G.; Wang, C.; Yu, X. D. Stable phase equilibrium in the aqueous quaternary system Rb^+, Mg^{2+}//Cl^-, borate-H_2O at 323 K. J. Chem. Eng. Data 2016, 61, 2419-2425.
[25] Feng, S.; Yu, X. D.; Cheng, X. L.; Zeng, Y. Phase diagrams and physicochemical properties of Li^+, K^+ (Rb^+) //Borate-H_2O systems at 323 K. Russ. J. Phys. Chem. A 2017, 91, 2149-2156.
[26] Yu, X. D.; Liu, M.; Wang, L.; Cheng, X. L.; Zeng, Y. Phase equilibria of potassium borate + rubidium borate + H_2O and rubidium borate + magnesium borate + H_2O aqueous ternary systems at 323 K. J. Chem. Eng. Chin. Univ.2017, 32, 514-521.
[27] Yu, X. D.; Zeng, Y.; Chen, P. J.; Li, L. G. Solid-liquid equilibrium of the quaternary system lithium, potassium, rubidium, and borate at T = 323 K. J. Chem. Eng. Data 2018, 63, 3125-3129.
[28] Jing, Y. A new synthesis method for magnesium borate. Sea-Lake Salt Chem. Ind. 2000, 29, 24-25.
[29] Zeng, Y.; Yu, X. D.; Liu, L. L.; Yin, Q. H. Method for Preparation Rubidium Pentaborate Tetrahydrate. China Patent CN103172078A2013.
[30] Institute of Qinghai Salt-lake of Chinese Academy of Sciences. *Analytical Methods of Brines and Salts,* 2nd ed.; Chines Science Press: Beijing, 1988.
[31] Shang, C. S., An, L. Y., Hu, Z. W. ICP-OES determination of rubidium in bitter. Chem. Res. Appl. 2012, 24, 642-645.
[32] Yuan, H. Z.; Zhu, Y. J.; Wu, L. P.; Zhang, X. Determination of high-content of lithium in natural saturated brines by inductively coupled plasma-atomic emission spectrometry. Rock Miner. Anal. 2011, 30, 87-89.
[33] Li, H. X.; Yao, Y.; Zeng, D. W. Measurement of concentrations of magnesium ion by masking method in presence of the interference lithium ion in salt water system. J. Salt Lake Res.2012, 20, 24-30.
[34] Ingri, N.; et al. Equilibrium studies of polyanions.10. On the first equilibrium steps in the acidification of $B(OH)_4^-$, an application of the self-medium method. Acta Chem. Scand. 1963, 17, 573-580.
[35] Gao, S. Y., Song, P. S., Xia, S. P., Zheng, M. P. *Salt Lake Chemistry, New Types of Borate & Lithium Salt Lake*; Science Press: Beijing, China, 2007.
[36] Yu, X. D.; Zeng, Y.; Mu, P. T.; Tan, Q.; Jiang, D. B. Solid-liquid equilibria in the quinary system $LiCl$-KCl-$RbCl$-$MgCl_2$-H_2O at T = 323 K. Fluid Phase Equilib.2015, 387, 88-94.
[37] Li, J.; Gao, S. Y. Chemistry of borates. J. Salt Lake Res. 1993, 1, 62-66.

Solid-liquid equilibrium of ternary system K_2SO_4-PEG4000-H_2O at T = (288, 298, and 308) K: measurements and correlation*

Xudong Yu[1, 2], Mianping Zheng[1], Hong Zheng[2], Shuai Chen[2] and Lin Wang[2]

1. MNR Key Laboratory of Saline Lake Resources and Environments, Institute of Mineral Resources, CAGS, Beijing 100037, China
2. College of Materials and Chemistry & Chemical Engineering, Chengdu University of Technology, Chengdu 610059, Sichuan, China

Abstract To obtain the solid-liquid equilibrium of ternary system K_2SO_4-Polyethylene glycol 4000 (PEG4000) -H_2O at (288, 298, and 308) K, the solubility of K_2SO_4 in PEG4000-H_2O mixed solvent, density, and refractive index for the ternary system were investigated by isothermal dissolution method. On the basis of the experimental data, the corresponding phase diagrams, and diagrams of density vs composition, refractive index vs composition at (288, 298, and 308) K were constructed. The phase diagrams of ternary system K_2SO_4-PEG4000-H_2O at 288, 298 and 308 K consist of one homogeneous area with unsaturated liquid phase (L), one saturated liquid phase and solid phase of K_2SO_4 (L + S), one area with one liquid phase and two solid phases (2S + L). The area of (2S + L) decreases with the increase of temperature, while the areas of (L) and (L + S) increase with the increase of temperature. Results show that with the addition of PEG4000 in the solution, the solubility of K_2SO_4 in the PEG4000-H_2O mixed solvent decreased, whereas the refractive index increased with the increase of PEG4000 in the mixed solvent. Moreover, the equilibrium thermodynamics of the K_2SO_4-PEG4000- H_2O system at 288, 298 and 308 K were calculated from the modified Pitzer equation. Finally, a comparison of the experimental and calculation diagrams revealed that the predictive solubilities agree well with the experimental values.

Keywords Phase equilibrium; K_2SO_4; Mixed solvent; Pitzer equation; PEG

1 Introduction

Salt lakes which contain a relatively high concentration of dissolved salts are widely distributed in the west of China. Large amounts of important raw materials for the chemical industry, agriculture, and metallurgical industry are now obtained from salt lakes. Since the 1970s, quaternary potassium-bearing salt lakes have become an important source of the world's potash resources[1, 2].

Salt lake is a solution system of multi-ion coexistence, and the interaction between salts is complex. The phase diagram can reflect the law of complex multi-ion interaction (dissolution and precipitation), and its research results can be used to explore the chemical production process and guide the formulation of the comprehensive utilization process of liquid mineral resources. This has led to numerous studies of the phase equilibria of ternary, quaternary, and quinary systems composed of salt and water/water-soluble organic solvents/ionic liquids[3-13]. It is worth mentioning that in the mixed solution of salt + water-soluble organic solvents + water, the salts and water-soluble organic solvents compete with each other for the ability to dissolve in water. Therefore, the addition of new miscible component in the original aqueous solution may cause the salting-out effect and decrease the solubility of the salts, which can be applied to the crystallization for producing supersaturated solution.

* 本文发表在：计算机与应用化学, 2020, 37 (1): 51-57

In the systems containing water-soluble organic solvents or soluble polymer, salt, and water, polyethylene glycol (PEG) is considered as one of the most applicable one for the purpose of preparing the mixed solution, as it is low cost, nontoxic and non-pollution. Till now, some papers focused on the system salt + PEG + water have been published [14-20]. However, to date, literature on the phase equilibrium of the K_2SO_4-PEG4000-H_2O ternary system is scarce. This study was done to investigate the equilibrium phase diagrams to evaluate the possible use of PEG4000 as a precipitant component of K_2SO_4 aqueous solutions. Accordingly, the completed phase diagrams of the K_2SO_4-PEG4000-H_2O system at 288, 298 and 308 K are presented herein. Moreover, the equilibrium thermodynamics data of the system K_2SO_4-PEG4000-H_2O at 288, 298 and 308 K were calculated by using the modified Pitzer equation.

2 Experimental section

2.1 Chemicals and instruments

Double-deionized water ($\kappa \leqslant 5.5 \times 10^{-6}$ S·m^{-1}) obtained from UPT water purification system was used in the experiments. Potassium sulfate (K_2SO_4, Molecular weight 174.26, CAS No. 7778-80-5), and polyethylene glycol 4000 (PEG4000, Average molecular weight 4000, CAS No. 25322-68-3) were obtained from Sinopharm Chmical Reagent Co., Ltd with analytical pure and used without further purification.

An analytical balance (Sartorius, Germany) with an accuracy of 0.0001 g was employed for the determination of the weight of solution. The thermostat (Inborn SHH250, China) with the temperature at 288/298/308±0.1 K was used for the phase equilibrium experiments. The refractive index (n_D) and density (ρ) were determined by the Abbe refractometer (WYA, China) with a precision of ±0.0001 and the pycnometry method with a precision of ±0.0002 g·cm^{-3}. Proper calibration requires a double-deionized water testing at 298 K.

2.2 Experimental methods

As reported in our previous work[15], the isothermal dissolution method was used for the solid-liquid equilibrium in this study. Firstly, the mixed solvent (PEG4000-H_2O) was placed in clean 50 mL with the mass fraction of PEG4000 varying from 0 to saturation. And then, the excess K_2SO_4 was dissolved in each PEG4000-H_2O mixed solvent. The sealed glass bottles with sample were put in a HY-5 Cyclotron Oscillator, which was taken into the SHH-250 type thermostat at T = 288, 298 and 308 K, with a constant oscillation frequency approximately 48 to 72 hours to accelerate equilibration. Once the equilibrium point of solution reached, the samples were kept on a static condition at working temperature for an additional 24 hours until the solution became clear, and then the clear liquid of each solution at equilibrium was taken out for analysis, meanwhile, the density (ρ) adopted the pycnometry method $\pm 2.0 \times 10^{-4}$ g·cm^{-3}, the refractive index (n_D) was measured by an WYA type Abbe refractometer with a precision of $\pm 1.0 \times 10^{-4}$. The composition of solid phase was identified by Schreinemakers' wet residue method [21].

The content of K_2SO_4 was acquired with the SO_4^{2-} concentration, and the concentration of SO_4^{2-} was determined by alizarin red volumetric titration with an uncertainty of ±0.0030 mass fraction[22].

The refractive index (n_D) of the mixed solution depends on the salt and PEG concentrations and is related to the mass fractions of K_2SO_4 and PEG4000 according to equation (1) [23]:

$$n_D = a_0 + a_1 \times w_{K_2SO_4} + a_2 \times w_{PEG4000} \tag{1}$$

where a_0 is the refractive index of double-deionized water at 298 K measured at 1.3325 and a_1 and a_2 are the fitting parameters estimated at 0.1125 and 0.1361, respectively. Accordingly, the concentration of

PEG4000 was determined by the refractive index measurements with an uncertainty of ± 0.0005 mass fraction. The refractive index was measured for a wide range of aqueous solutions [PEG 1000: 0–0.35 (*w/w*); K_2SO_4: 0–0.06 (*w/w*)].

3 Results and discussion

The mass fractions of the equilibrated solutions and wet solids, densities, and refractive indices of the K_2SO_4-PEG4000-H_2O ternary system at 288, 298 and 308 K are listed in Table 1, while Fig. 1 presents the respective solubility diagrams.

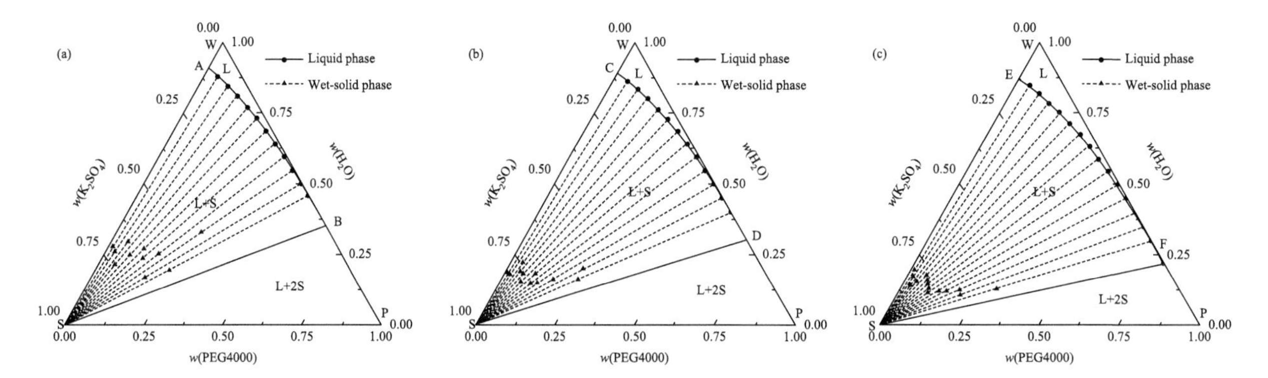

Fig. 1　Phase diagram of K_2SO_4-PEG4000-H_2O at 288, 298, and 308 K: (a) 288 K, (b) 298 K, (c) 308 K.

Table 1　The solubilities, densities (ρ), and refractive indices (n_D) for the ternary system K_2SO_4-PEG4000-H_2O at (288, 298, and 308) K

No.	Density (ρ) [$g \cdot cm^{-3}$]	Refractive index n_D	Composition of equilibrated solution		Composition of wet solid phase		Equilibrated solid phase
			w (K_2SO_4)	w (PEG4000)	w (K_2SO_4)	w (PEG4000)	
			$T = 288$ K				
1A	1.0715	1.3446	0.0903	0.0000	—	—	K_2SO_4
2	1.0671	1.3496	0.0750	0.0462	0.7066	0.0146	K_2SO_4
3	1.0635	1.3549	0.0604	0.0939	0.7104	0.0289	K_2SO_4
4	1.0610	1.3600	0.0471	0.1429	0.6516	0.0522	K_2SO_4
5	1.0621	1.3656	0.0368	0.1924	0.7328	0.0534	K_2SO_4
6	1.0638	1.3719	0.0272	0.2429	0.6670	0.0831	K_2SO_4
7	1.0685	1.3787	0.0193	0.2939	0.6137	0.1158	K_2SO_4
8	1.0730	1.3860	0.0142	0.3446	0.6350	0.1276	K_2SO_4
9	1.0800	1.3936	0.0097	0.3958	0.5759	0.1695	K_2SO_4
10	1.0870	1.4015	0.0064	0.4469	0.4043	0.2679	K_2SO_4
11	1.0948	1.4092	0.0045	0.4971	0.6645	0.1675	K_2SO_4
12	1.1031	1.4162	0.0032	0.5422	0.5718	0.2329	K_2SO_4
13B	1.1071	1.4281	0.0000	0.6485	—	—	—
			$T = 298$ K				
14C	1.0861	1.3452	0.1068	0.0000	—	—	K_2SO_4
15	1.0789	1.3494	0.0900	0.0455	0.8072	0.0096	K_2SO_4
16	1.0728	1.3539	0.0704	0.0930	0.8021	0.0198	K_2SO_4
17	1.0689	1.3594	0.0580	0.1413	0.7391	0.0391	K_2SO_4

Continued

No.	Density (ρ) [$g \cdot cm^{-3}$]	Refractive index n_D	Composition of equilibrated solution		Composition of wet solid phase		Equilibrated solid phase
			w (K_2SO_4)	w (PEG4000)	w (K_2SO_4)	w (PEG4000)	
18	1.0666	1.3649	0.0447	0.1910	0.7650	0.0470	K_2SO_4
19	1.0667	1.3716	0.0339	0.2416	0.7601	0.0600	K_2SO_4
20	1.0679	1.3772	0.0234	0.2929	0.7815	0.0655	K_2SO_4
21	1.0718	1.3846	0.0167	0.3441	0.7203	0.0979	K_2SO_4
22	1.0772	1.3914	0.0115	0.3954	0.7545	0.0982	K_2SO_4
23	1.0831	1.3991	0.0079	0.4465	0.7312	0.1210	K_2SO_4
24	1.0898	1.4074	0.0050	0.4975	0.6724	0.1638	K_2SO_4
25	1.0971	1.4153	0.0031	0.5483	0.5612	0.2414	K_2SO_4
26	1.1041	1.4219	0.0023	0.5986	0.5952	0.2429	K_2SO_4
27D	1.1118	1.4314	0.0000	0.6990	—	—	—
				T = 308 K			
28E	1.0938	1.3466	0.1257	0.0000	—	—	K_2SO_4
29	1.0859	1.3508	0.1038	0.0448	0.7942	0.0103	K_2SO_4
30	1.0797	1.3551	0.0874	0.0912	0.8054	0.0195	K_2SO_4
31	1.0741	1.3600	0.0714	0.1393	0.8310	0.0253	K_2SO_4
32	1.0702	1.3651	0.0569	0.1886	0.8047	0.0391	K_2SO_4
33	1.0677	1.3709	0.0434	0.2392	0.7624	0.0594	K_2SO_4
34	1.0673	1.3766	0.0307	0.2906	0.7706	0.0688	K_2SO_4
35	1.0688	1.3831	0.0217	0.3423	0.7728	0.0795	K_2SO_4
36	1.0723	1.3902	0.0152	0.3939	0.7783	0.0887	K_2SO_4
37	1.0799	1.3992	0.0089	0.4459	0.7867	0.0960	K_2SO_4
38	1.0834	1.4047	0.0064	0.4967	0.7509	0.1245	K_2SO_4
39	1.0895	1.4119	0.0043	0.5475	0.7269	0.1502	K_2SO_4
40	1.0957	1.4190	0.0027	0.5981	0.6857	0.1885	K_2SO_4
41	1.1012	1.4260	0.0015	0.6488	0.6924	0.1999	K_2SO_4
42	1.1062	1.4319	0.0008	0.6994	0.5668	0.3032	K_2SO_4
43F	1.1089	1.4445	0.0000	0.7789	—	—	—

From Fig. 1, the following points, curves, and regions were obtained:

(1) Points A, C, and E represent the solubility of K_2SO_4 in pure water at 288, 298, and 308 K, while points B, D, and F represent the saturation point of PEG4000 in water at 288, 298, and 308 K. The solubility of K_2SO_4 and PEG4000 in pure water both increase with temperature increasing.

(2) Curves AB, CD, and EF are the saturation curves of K_2SO_4 in a PEG4000-H_2O mixed solvents at 288, 298, and 308 K, respectively, the solubility of K_2SO_4 in the PEG4000-H_2O mixed solvent decreased with the addition of PEG4000 in the mixture.

(3) Areas WBAW, WDCW, and WFEW are the homogeneous areas of unsaturated liquid (L); areas BASB, DCSD, and FESF are the equilibrium areas containing solid phase K_2SO_4 and the saturated liquid phase (L + S); areas BSPB, DSPD, and FSPF are the areas with two solids (K_2SO_4 and PEG4000) and the saturated liquid (2S + L).

(4) The phase diagrams of ternary system K_2SO_4-PEG4000-H_2O at 288, 298 and 308 K consist of one homogeneous area with unsaturated liquid phase (L), one saturated liquid phase and solid phase of K_2SO_4 (L + S), one area with one liquid phase and two solid phase (2S + L), without the biphase region formed.

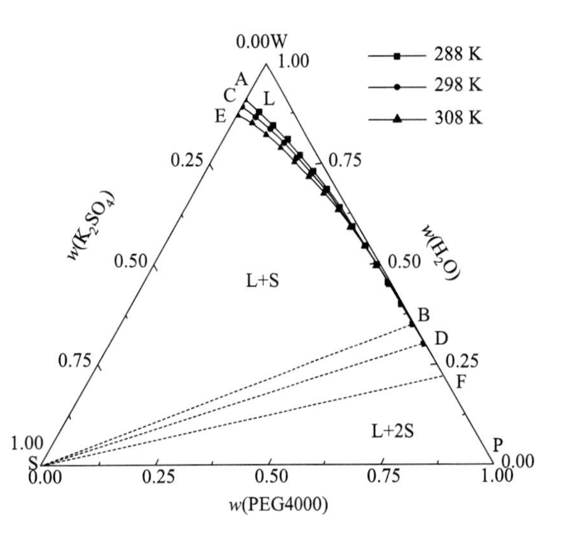

Fig. 2　Comparisons of the solubility of K_2SO_4 in PEG4000-H_2O mixed solvent at 288, 298, and 308 K.

The relationship of the temperature, density, and $w(K_2SO_4)$ is shown in Fig. 3. These three factors which influence the density are that: the mass fraction of K_2SO_4, the temperature, and the density of mixed solvent with PEG4000 saturated ($\rho_{288\ K} = 1.1071\ g \cdot cm^{-3}$, $\rho_{298\ K} = 1.1118\ g \cdot cm^{-3}$, $\rho_{308\ K} = 1.1089\ g \cdot cm^{-3}$). We can assume that the temperature has little effect on the density of mixed solvent with PEG4000 saturated, and the density of K_2SO_4 saturated solution changes from $1.0715\ g \cdot cm^{-3}$ to $1.0938\ g \cdot cm^{-3}$ with the rising of the temperature, thus the concentration of K_2SO_4 and PEG4000 in the solution both affect the density. Thus, the densities of the solution decrease first and then increase with the increase of the mass fraction of K_2SO_4 at 288, 298 and 308 K.

The relationship of the temperature, refractive index, and $w(K_2SO_4)$ is shown in Fig. 4. The refractive indices of the saturated solutions decrease with the mass fraction of K_2SO_4 increasing at 288, 298, and 308 K, the reason may be that the refractive index of saturated solution of PEG4000 is much larger than the refractive index of saturated solution of K_2SO_4 at three different temperatures, and with the addition of PEG4000 in the solution, the composition of K_2SO_4 decrease, thus the refractive index increase with the mass fraction of K_2SO_4 decreasing in the solution.

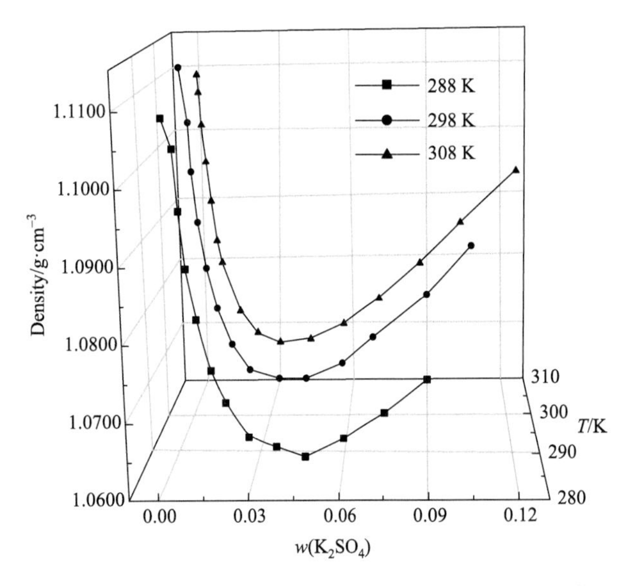

Fig. 3　Density vs composition of K_2SO_4-PEG4000-H_2O system at (288, 298, and 308) K.

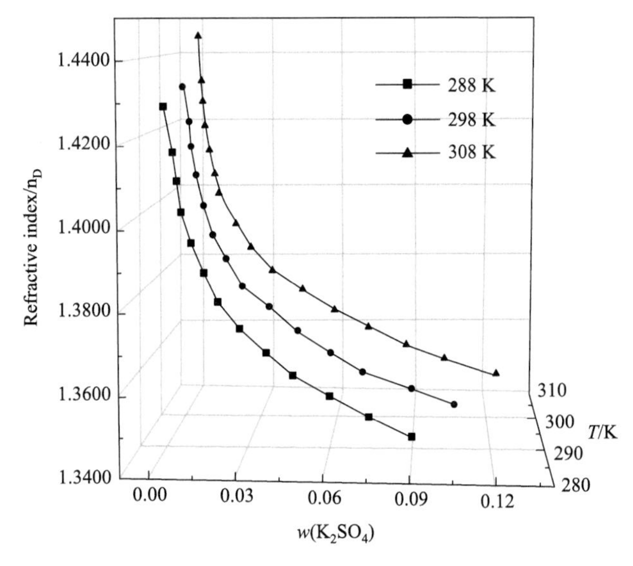

Fig. 4　Refractive index vs composition of K_2SO_4-PEG4000-H_2O system at (288, 298, and 308) K.

4　Solubility prediction

A modified Pitzer model considering long-range and short-range contribution given by Zhu et al.[24] had been successfully used to correlate equilibrium data of systems Na_2SO_4/Na_2CO_3-PEG1000/8000-H_2O and $(NH_4)_2SO_4$-PEG1000/4000-H_2O at 298 K. The calculations of system KCl-PEG1000/4000-H_2O at 288, 298, and 308 K by using the modified Pitzer model have been done by our previous work[15-16]. Therefore, this model was also applied to correlate the solubility of K_2SO_4 in mixed solvents PEG4000-H_2O at 288, 298, and 308 K with the following eq. (2) ~ eq. (14).

$$\begin{aligned}\ln\gamma_{\pm} = &-\frac{A\sqrt{I}}{1+b\sqrt{I}} + \frac{1}{2}[B_{12}r_1m_1 + 4B_{22}m_2 \\ &+ I(B'_{12}r_1m_1 + 2B'_{22}m_2) + C_{112}r_1^2m_1^2 \\ &+ 2C_{122}r_1m_1m_2 + 2C^{\gamma}_{222}m_2^2]\end{aligned} \tag{2}$$

$$\begin{aligned}ln\gamma_3 = &\frac{2AV_3d}{b^3}\left[1+b\sqrt{I} - \frac{1}{1+b\sqrt{I}} - 2ln(1+b\sqrt{I})\right] \\ &-\frac{M_3}{1000}\left[B_{11}r_1^2m_1^2 + \left(B_{12} + \frac{n_w}{n_s}IB'_{12}\right)r_1m_1m_2\right. \\ &+ 2\left(B_{22} + \frac{n_w}{n_s}IB'_{12}\right)m_2^2 + 2C_{111}r_1^3m_1^3 \\ &\left.+ 2C_{112}r_1^2m_1^2m_2 + 2C_{122}r_1m_1m_2^2 + 2C^{\phi}_{222}m_2^3\right]\end{aligned} \tag{3}$$

$$B_{ij} = \beta_{ij}^{(0)} + \frac{2\beta_{ij}^{(1)}}{\alpha_1^2 I}\left[1-(1+\alpha_1\sqrt{I})e^{-\alpha_1\sqrt{I}}\right] \tag{4}$$

$$C^{\gamma}_{222} = \frac{3}{2}C^{\phi}_{222} = 3C_{222} \tag{5}$$

$$I = \frac{1}{2}\sum_i m'_i z_i^2,\ m'_i = 1000n_i/(Mn) \tag{6}$$

$$M_n = M_3n_3 + M_1n_1 \tag{7}$$

$$A = \frac{1.327757\times10^5}{DT}\sqrt{\frac{\rho}{DT}} \tag{8}$$

$$b = 6.359696\sqrt{\frac{\rho}{DT}} \tag{9}$$

$$D = \sum_i \phi_i D_i \tag{10}$$

$$\rho = \sum_i \phi'_i \rho_i \tag{11}$$

$$P_V = \sqrt{D_1M_1} \quad \text{or} \quad P_{LL} = \frac{D_1-1}{D_1+2}V_1 \tag{12}$$

$$\phi_i = \frac{n_iV_i}{\sum_i n_iV_i} \tag{13}$$

$$V_1 = (2r_1+1)\cdot V_3 \tag{14}$$

Where n_i and r_i are the numbers of moles and segments of components i, respectively. M is the molar mass, B_{ij} represents the second virial coefficients. B'_{ij} is the derivative of B_{ij} to the ionic strength I. A and b are Debye-Hückel constants, which can be estimated by the equations (8) and (9). D and ρ are the dielectric constant and the density of the mixed solvent PEG4000-H_2O, which can be calculated approximately from the summation of the group values (P_V or P_{LL}) [25] as equations (10) to (14). ϕ_i is the salt-free volume, V_i denotes the molar volume of species i.

The binary parameters of the model $\beta_{22}^{(0)}$, $\beta_{22}^{(1)}$, C^{ϕ}_{222} and B_{11}, C_{111} for the binary systems of K_2SO_4-H_2O and PEG4000-H_2O were listed in Table 2, it should be mentioned that the binary parameters B_{11}, C_{111} of PEG4000-H_2O can be obtained by literature[24], the binary parameters $\beta_{22}^{(0)}$, $\beta_{22}^{(1)}$, C^{ϕ}_{222} of system K_2SO_4-H_2O can be calculated by modified Pitzer model according to the mean activity coefficient of K_2SO_4 in aqueous solution which can be acquired by Pitzer model[26], as shown in Fig. 5, it can be seen that the mean activity coefficient of K_2SO_4 can be accurately described by the modified Pitzer model and also indicated the binary parameters of this model is reliable. Furthermore, the cross parameters $\beta_{12}^{(0)}$, $\beta_{12}^{(1)}$, and C_{112} and C_{122} of ternary systems K_2SO_4-PEG4000-H_2O can be fitted by the solubility data in this study, and it was listed in Table 2, and the needed physical properties of PEG4000 and H_2O at 288, 298, 308 K are presented in Table 3.

Table 2 Binary parameters and cross parameters for ternary system K_2SO_4-PEG4000-H_2O at (288, 298, and 308) K

Binary parameters					
T/K	$B_{11}\times10^2$	$C_{111}\times10^5$	$\beta_{22}^{(0)}$	$\beta_{22}^{(1)}$	C_{222}^{ϕ}
288	0.7088[24]	1.296[24]	−0.7308	3.5620	0.6329
298	0.7088[24]	1.296[24]	−0.5764	3.0898	0.4180
308	0.7088[24]	1.296[24]	−0.5171	2.9602	0.3355

Cross parameters				
T/K	$\beta_{12}^{(0)}$	$\beta_{12}^{(1)}$	$C_{112}\times10^4$	$C_{122}\times10^4$
288	0.2866	0.0006	−0.8130	0.10005
298	0.3180	−0.0764	−0.4539	0.1003
308	0.3190	−0.0600	−0.4674	1.0011

Table 3 The physical properties of pure substances at different temperatures [24]

T/K	Components	M/g·mol^{-1}	V/m^3·mol^{-1}	ρ/g·cm^{-3}	r_i	D
288	H_2O	18.02	18.04×10^{-6}	0.9991	1	81.95
	PEG 4000	3750	3.1210×10^{-3}	1.2016	86	2.206
298	H_2O	18.02	18.07×10^{-6}	0.9970	1	78.34
	PEG 4000	3750	3.1261×10^{-3}	1.2010	86	2.206
308	H_2O	18.02	18.13×10^{-6}	0.9940	1	74.83
	PEG 4000	3750	3.1360×10^{-3}	1.1956	86	2.206

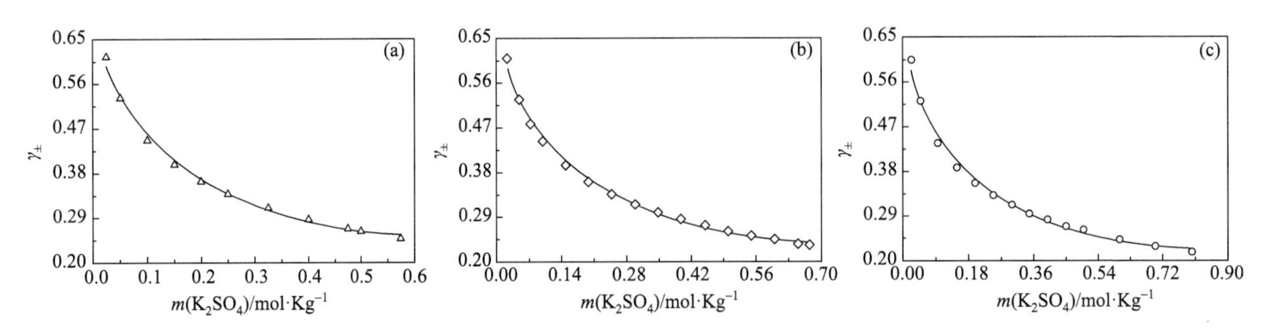

Fig. 5 The diagram of mean activity coefficient of K_2SO_4 for binary systems K_2SO_4-H_2O at 288 K (a), 298 K (b), 308 K (c). (△, ◇, ○: calculated value by Pitzer model; ——: calculated by modified Pitzer model in this study).

Based on above mentioned model parameters, the solubility calculation of K_2SO_4 in mixed solvent PEG4000-H_2O can be done by equations (2)~(15).

$$K_{sp} = 4m_{K_2SO_4}^3\gamma_{\pm}^3 \tag{15}$$

Where K_{sp} represents the solubility product of K_2SO_4, it can be calculated according to the solubility of K_2SO_4 in pure water at 288, 298, and 308 K, the corresponding lnK_{sp} values of K_2SO_4 are −4.3920, −4.0522, −3.6811, respectively; m is the molarity; $\gamma_{\pm}$ denote the mean activity coefficient of K_2SO_4. In order to compare the degree of coincidence of experimental data and calculated data, we calculated the mean square deviation (σ) according to equation (16).

$$\sigma = \sqrt{\frac{\sum(w_{exp} - w_{cal})^2}{n}} \tag{16}$$

Where w_{exp} is experimental value, w_{cal} represents calculated value, n denotes the number of samples, in this study, the calculated mean square deviation (σ) are 0.0036, 0.0027, and 0.0027 at 288, 298, 308 K. Meanwhile, the comparison of experimental and calculated values was presented in Fig. 6, results show that calculated value agree well with experimental value.

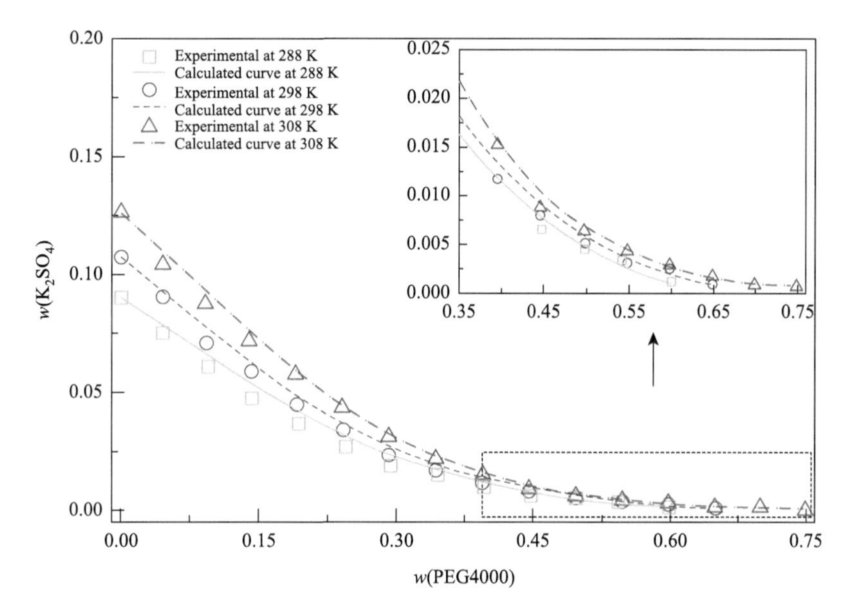

Fig. 6 The comparison of experimental and calculated value for ternary system K_2SO_4-PEG4000-H_2O at 288, 298, 308 K.

□, ○, △-experimental value at 288, 298, 308 K; ——, ------, -·-·-·-calculated curve by modified Pitzer model at 288, 298, 308 K

5 Conclusion

The solubilities, densities, and refractive indices of the ternary system K_2SO_4-PEG1000-H_2O were measured at 288, 298 and 308 K. The phase diagrams of K_2SO_4-PEG4000-H_2O at these three temperature consist of three areas: L, L + S, and 2S + L, without the biphase region formed. It can be observed that the solubility of K_2SO_4 at the three different temperatures all decreased with an increase in the mass fraction of PEG4000. The refractive indices increased with the addition of PEG4000 in the solution. The thermodynamics calculation of equilibrium data of system K_2SO_4-PEG4000-H_2O at 288, 298, and 308 K were carried out by using the modified Pitzer equation, and the calculation data agree well with the experimental data.

Acknowledgments

The studies were supported by the National Science Foundation of China (U1507111).

References

[1] Zheng Mianping. Saline Lakes and Salt Basin Deposits in China[M]. Beijing, Science Press, 2014.

[2] Zheng Mianping, Deng Tianlong, Aharon Oren. Introduction to Salt Lake Sciences[M]. Beijing, Science Press, 2018.

[3] Sohr Julia, Voigt Wolfgang, Zeng Dewen. IUPAC-NIST solubility data series. 104. lithium sulfate and its double salts in aqueous solutions[J]. J. Phys. Chem. Ref. Data., 2017, 46 (2): 023101.

[4] Zhao Xiaoling, Zhang Hongqi, Zhao Bin, et al. The phase diagram and the application for the quaternary system K^+, NH_4^+//Cl^-, SO_4^{2-}-H_2O at

80.0℃[J]. J. Solution. Chem., 2018, 47: 1419-1437.

[5] Hu Juanxin, Sang Shihua, Zhou Meifang, et al. Phase equilibria in the ternary systems KBr-$MgBr_2$-H_2O and NaBr-$MgBr_2$-H_2O at 348.15 K[J]. Fluid Phase Equilib., 2015, 392: 127-131.

[6] Meng Lingzong, Gruszkiewicz Miroslaw S, Deng Tianlong, et al. Isopiestic measurements of thermodynamic properties for the aqueous system LiBr-$CaBr_2$-H_2O at 373.15 K[J]. J. Chem. Thermodyn., 2019, 129: 83-91.

[7] Yu Xudong, Liu Min, Zheng Qiufeng, et al. Measurement and correlation of phase equilibria of ammonium, calcium, aluminum, and chloride in aqueous solution at 298.15 K[J]. J. Chem. Eng. Data., 2019, 64: 3514-3520.

[8] Wu Linxin, Zeng Ying, Chen Yu, et al. The stable phase equilibria of the ternary systems Na_2SO_4 + Rb_2SO_4 (Cs_2SO_4) + H_2O at 298.2 K[J]. J. Chem. Eng. Data., 2019, 64: 529-535.

[9] Jimenez Y P, Taboada M E, Galleguillos H R. Solid-liquid equilibrium of K_2SO_4 in solvent mixtures at different temperatures[J]. Fluid Phase Equilib., 2009, 284: 114-117.

[10] Urréjola S, Sánchez A, Hervello M F. Solubilities of sodium, potassium, and copper (II) sulfates in ethanol-water solutions[J]. J. Chem. Eng. Data., 2011, 56: 2687-2691.

[11] Taboada M E, Véliz D M, Galleguillos H R, et al. Solubilities, densities, viscosities, electrical conductivities, and refractive indices of saturated solutions of potassium sulfate in water + 1-propanol at 298.15, 308.15, and 318.15 K[J]. J. Chem. Eng. Data., 2002, 47: 1193-1196.

[12] Jimenez Y P, Taboada M E, Flores E K, et al. Densities, viscosity, and electrical conductivity in the potassium sulfate + water + 1-propanol system at different temperatures[J]. J. Chem. Eng. Data., 2009, 54: 1932-1934.

[13] Xie Juan, Liu Xiao, Pan Wen, et al. Phase equilibria in the system Na^+, K^+// SO_4^{2-} -$(CH_2OH)_2$-H_2O and Na^+, K^+//Cl^-, SO_4^{2-} -$(CH_2OH)_2$-H_2O at 328.15 K[J]. J. Chem. Thermodyn., 2017, 112: 155-165.

[14] Yu Xudong, Li Maolan, Wang Lin, et al. Solubilities, densitis, refractive indices of aqueous ternary systems PEG400-NaCl/KCl-H_2O at 298 K[J]. J. Salt. Lake. Res., 2019, 27 (2): 40-45.

[15] Huang Qin, Wang Lin, Li Maolan, et al. Measurements and simulation of the polyethylene glycol 1000-water-KCl ternary system at 288.2, 298.2, and 308.2 K[J]. J. Chem. Eng. Jpn., 2019, 52 (4): 325-332.

[16] Yu Xudong, Huang Qin, Wang Lin, et al. Measurements and simulation for ternary system KCl-PEG4000-H_2O at T = 288, 298 and 308 K[J]. CIESC J., 2019, 70 (3): 830-839.

[17] Yu Xudong, Wang Lin, Li Maolan, et al. Phase equilibria of CsCl-polyethylene glycol (PEG) -H_2O at 298.15 K: effect of different polymer molar masses (PEG1000/4000/6000) [J]. J. Chem. Thermodyn., 2019, 135: 45-54.

[18] Jahani F, Abdollahifar M, Haghnazari N. Thermodynamic equilibrium of the polyethylene glycol 2000 and sulphate salts solution[J]. J. Chem. Thermodyn., 2014, 69: 125-131.

[19] González-Amado M, Rodil E, Arce A, et al. The effect of temperature on polyethylene glycol (4000 or 8000) - (sodium or ammonium) sulfate aqueous two phase systems[J]. Fluid Phase Equilib., 2016, 428: 95-101.

[20] Hu Mancheng, Zhang Xiaolei, Li Shuni, et al. Phase equilibrium of ternary system PEG (Poly Ethylene Glycol) -Rb_2CO_3-H_2O[J]. Chin. J. Rare Met., 2006, 30 (6): 837-840.

[21] Fosbøl P L, Thomsen K, Stenby E H. Reverse schreinemakers method for experimental analysis of mixed-solvent electrolyte systems[J]. J. Solution Chem., 2009, 38 (1): 1-14.

[22] Lv Peng, Zhong Yuan, Meng Ruiying, et al. Weighing titration analysis of the sulfate content[J]. J. Salt. Lake. Res., 2015, 23 (3): 5-13.

[23] Cheluget E L, Gelinas S, Vera J H, et al. Liquid-liquid equilibrium of aqueous mixtures of poly (propylene glycol) with NaCl[J]. J. Chem. Eng. Data., 1994, 39: 127-130.

[24] Wu Youting, Lin Dongqiang, Zhu Ziqiang, et al. Prediction of liquid-liquid equilibria of polymer-salt aqueous two-phase systems by a modified Pitzer's virial equation[J]. Fluid Phase Equilib., 1996, 124: 67-79.

[25] Krevelen D W V, Nijehuis K T. Properties of Polymers: Their Correlation with Chemical Structure; Their Numerical Estimation and Prediction from Additive Group Contributions (4th ed) [M]. Amsterdam, Elsevier, 2009.

[26] Greenberg J P, Møller N. The prediction of mineral solubilities in natural waters: Achemical equilibrium model for the Na-K-Ca-Cl-SO_4-H_2O system to high concertration from 0 to 250℃ [J]. Geochim. Cosmochim. Acta., 1989, 53: 2503-2518.

盐湖卤水提锂技术及产业化发展*

丁涛[1]，郑绵平[2]，张雪飞[2]，伍倩[2]，张翔禹[1]

1. 中国矿业大学（北京）地球科学与测绘工程学院，北京 100083
2. 中国地质科学院地质矿产资源研究所，自然资源部盐湖资源与环境重点实验室，北京 100037

摘要 据美国地质调查局 2020 年最新统计，已查明的世界锂资源量为 8000 万 t，其中 59%锂资源分布在盐湖中，寻找不同类型盐湖的提锂技术亟待解决。根据盐湖卤水中锂和其他伴生离子赋存特征，综述了沉淀法、膜法、萃取法、盐梯度太阳池法、吸附法等盐湖提锂技术发展现状，总结了锂资源生产流程，探讨了各种方法对不同镁锂比盐湖的适应性及优势。研究发现，盐湖提锂不能按照单一方法进行，要根据不同盐湖赋存类型进行提锂工艺的选择。沉淀法已经在低镁锂比盐湖中经过工业化生产验证；吸附法是目前在高镁锂比盐湖中综合效果最为理想的提锂技术，其吸附容量高、一步直接提锂、循环性能高、稳定性强；天然矿物改性吸附法是未来盐湖提锂产业化应重点关注的方向。

关键词 盐湖锂资源；高镁锂比；提锂技术

新能源是能源革命的重要突破口，大力发展以锂资源为基础的锂电新能源，有利于突破资源瓶颈，转变发展方式，促进能源持续发展[1]。锂是稀有元素，它具有最轻的质量，因此具有独特的物理、化学性质，被广泛应用于军事、机械、化工、医药和新能源等行业。

随着全球锂消耗量的增加和锂产品价格的持续上涨促进了锂产业的迅猛发展。至今全球的锂生产主要集中在 SQM、FMC、Albemarle 和天齐锂业 4 家企业，其中除了天齐锂业以固态矿石提锂外，其他 3 家都是在盐湖卤水中进行提锂[2]。盐湖是一种综合性的宝贵的自然资源，人类已从盐湖中大量开采盐类资源，综合利用取得长足进步。表 1[3-12]是世界上主要富锂盐湖的化学组成。盐湖卤水中主要含有 Li^{+}、Na^{+}、K^{+}、Ca^{2+}、Mg^{2+}等阳离子和 SO_4^{2-}、Cl^{-}、CO_3^{2-} 等阴离子，按化学成分盐湖卤水分为碳酸盐型、硫酸盐型和氯化物型。盐湖提锂工艺和成本主要由盐湖类型控制，卤水中 Mg^{2+}、B^{3+}等伴生离子通常会给 Li^{+}分离带来麻烦。

目前，世界上已经被工业化开发的盐湖大多数都是低镁锂比盐湖（镁锂比低于 8），如智利 Atacama 湖、美国银峰（Silver Peak，USA）等。按照盐湖类型不同，目前比较成功的提锂方法有沉淀法、膜法、萃取法、吸附法和盐梯度太阳池法等[13-15]。针对盐湖卤水锂资源的赋存状态和特征，将盐湖提锂技术分为高镁锂比盐湖提锂技术和低镁锂比盐湖提锂技术，总结已经工业化生产提锂技术，在工业提锂的生产实践中，往往以单一方法为主，要根据自然资源的特点辅以其他方法综合应用。本文对吸附法中天然矿物改性作吸附剂进行机理预测和分析，提出各种提锂工艺在工业生产中面临的挑战，展望天然矿物改性作吸附剂的前景。

表 1 世界主要富锂盐湖卤水组成[3-12]

盐湖卤水来源	Mg^{2+}/Li^{+}	质量分数/%							
		Li^{+}	Na^{+}	K^{+}	B^{3+}	Mg^{2+}	Ca^{2+}	Cl^{-}	SO_4^{2-}
玻利维亚乌尤尼（Uyuni，Bolivia）	20.249	0.0321	7.06	1.17	0.071	0.65	0.0306	5	—
智利阿塔卡马（Atacama，Chile）	6.146	0.157	9.1	2.36	0.04	0.965	0.045	18.95	1.59
阿根廷翁布雷穆埃尔托（Hombre Muerto，Argentina）	1.371	0.062	9.789	0.617	0.035	0.085	0.053	15.80	0.853

* 本文发表在：科技导报，2020，38（15）：16-23

续表

盐湖卤水来源	Mg^{2+}/Li^{+}	质量分数/%							
		Li^{+}	Na^{+}	K^{+}	B^{3+}	Mg^{2+}	Ca^{2+}	Cl^{-}	SO_4^{2-}
美国瑟尔斯湖（Searles Lake，USA）	—	0.0054	11.08	2.53	—	—	0.0016	12.3	4.61
美国银峰（Silver Peak，USA）	6.667	0.006	6.2	0.8	—	0.04	0.05	10.06	0.71
以色列死海（Dead Sea，Israel）	2575.0	0.0012	3.01	0.56	0.03	3.09	1.29	16.1	0.061
中国扎布耶（Zabuye，China）	0.053	0.0489	7.29	1.66	—	0.0026	0.0106	9.53	—
中国西台吉乃尔（Taiji'naier，China）	65.161	0.031	5.63	0.44	—	2.02	0.02	13.42	3.41
中国一里坪（Yiliping，China）	60.95	0.021	2.58	0.91	0.031	1.28	0.016	14.97	2.88
中国察尔汗（Qarhan，China）	1577.4	0.0031	2.37	1.25	0.0087	4.89	0.051	18.8	0.44
中国大柴旦（Da Qaidam，China）	133.75	0.016	6.92	0.71	0.062	2.14	—	14.64	4.05

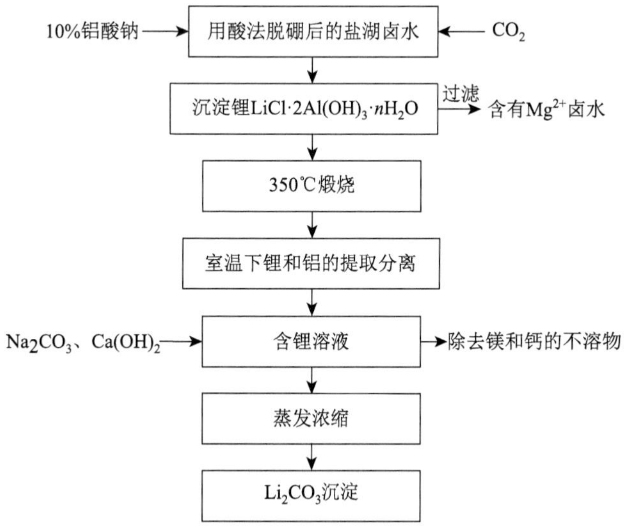

图 1　铝酸盐沉淀法提锂工艺流程图

1　低镁锂比盐湖提锂技术

1.1　沉淀法

沉淀法通过蒸发卤水将锂浓缩到一定浓度，然后利用化学沉淀反应，将 Li^{+}以沉淀形式从溶液中分离出来。因此，沉淀法主要包括 2 个方向，一是仅沉淀 Li + 的目标离子沉淀法，如铝酸盐沉淀法；二是 Li^{+}与伴生离子一起沉淀的共沉淀法，如碳酸盐沉淀法，硼理、硼镁共沉淀法。

1.1.1　铝酸盐沉淀法

铝酸盐沉淀法目前通用的两种盐类包括铝酸钠和铝酸钙，该方法先加入铝酸盐同时通入 CO_2，生成 $Al(OH)_3$，控制铝锂比在相应条件下得到铝锂沉淀物，经进一步处理得到 Li_2CO_3 产品，其工艺流程见图 1。

该方法的基本原理：

$$Al(OH)_3 + LiCl + nH_2O \longrightarrow LiCl\cdot 2Al(OH)_3\cdot nH_2O \text{（不溶物、沉淀锂）}$$

$$LiCl\cdot 2Al(OH)_3\cdot nH_2O + H_2O \longrightarrow xLiCl + (1-x)LiCl\cdot 2Al(OH)_3\cdot (n+1)H_2O \text{（洗脱锂）}$$

铝酸盐沉淀法将锂直接沉淀，其优点为锂和镁分离率高，Li_2CO_3 产品相对纯度高；其缺点是淡水消耗量大，蒸发能耗高，工序较多，生产成本高、周期长，并对 Li^{+} 浓度有一定要求。如何利用当地自然气候条件，并增加太阳能对工艺能量的供应解决高能耗问题，是铝酸盐沉淀法工业化生产应解决的问题。

1.1.2　碳酸盐沉淀法

碳酸盐沉淀法是目前工业应用最广泛的提锂技术，该方法流程是卤水先日晒蒸发析出 NaCl、KCl，再用酸或煤油将卤水脱硼，然后加入纯碱和石灰乳除去钙、镁，最后向 LiCl 溶液中加入纯碱得到 Li_2CO_3 产品，其工艺流程见图 2。Boryta 等[16]在专利中使用两阶段沉淀，第一阶段加入纯碱，第二阶段加入纯碱和 CaO，第一阶段滤出 $MgCO_3$，第二阶段 $Mg(OH)_2$、$CaCO_3$ 沉淀除去 Mg、Ca，固液分离后向溶液中加纯碱得到 Li_2CO_3 产品。Hamzaoui 等[17]分别用草酸盐和碳酸盐沉淀镁，并得出除镁效果与沉淀剂浓度、反应温度和持续时间 3 个条件有关。An 等[18]对玻利维亚乌尤尼盐湖卤水采用两步沉淀法提锂，加入石灰乳、草酸钠将 Ca、Mg 以 $CaSO_4·2H_2O$ 与 $Mg(OH)_2$ 的形式除去，最后加入纯碱，控制沉淀温度在 80～90℃时产生高纯度（99.55%）且结晶良好的 Li_2CO_3。因为碳酸盐沉淀法提锂过程中除镁需要大量碱，多用来处理低镁含量卤水，目前世界上低镁锂比硫酸盐型盐湖均采用此方法进行锂资源工业开发，包括智利的阿塔卡马盐湖、美国的西尔斯盐湖和银峰地下卤水等。

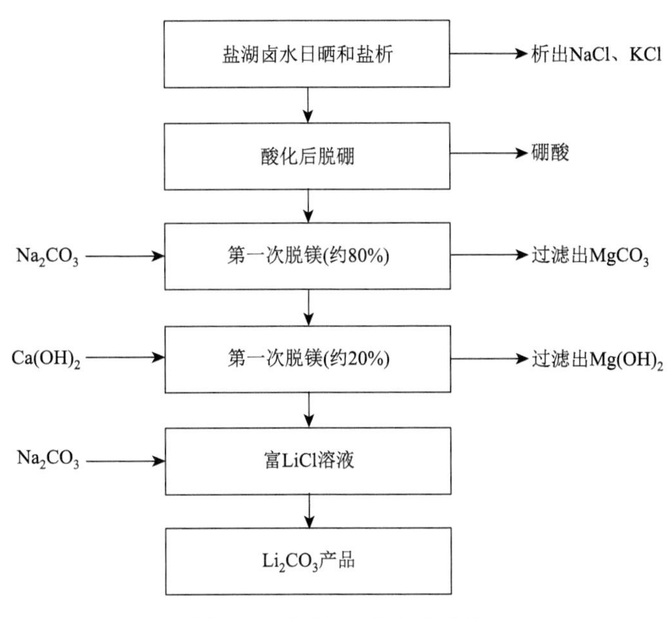

图 2 碳酸盐沉淀法提锂流程

近年来，已有一些学者将该方法改进，用于高镁锂比卤水提锂。陆增等[19]将盐湖晶间卤水进行三步蒸发，分别析出 NaCl、钾镁混盐和镁盐，得到的卤水加入适量纯碱并搅拌进一步除镁，控制纯碱溶液含量 14%～16%时，再加入 $Mg(OH)_2$ 溶液调整料液 pH 值为 9～10，固液分离后将 LiCl 溶液蒸发浓缩，富锂溶液中加入热纯碱并控制温度在 80～85℃得到 Li_2CO_3。

碳酸盐沉淀法利用太阳能日晒蒸发，整个提锂工艺可以生产多种盐类产品，但该方法生产周期长，锂回收率相对较低。目前主要在南美硫酸盐型低镁锂比盐湖中应用，未来在硫酸盐型低镁锂比盐湖应用中，应合理利用太阳能和其他能量，提高生产效率，缩短生产周期。

1.1.3 硼锂、硼镁共沉淀法

对于高镁锂比盐湖，魏新俊等[20]提出硼锂共沉淀的方法提取锂，工艺流程为一次冷冻、兑卤蒸发、一次蒸发、二次冷冻、二次蒸发、用过量硫酸沉淀硼锂；冷冻时温度控制在 0～25℃，最后用纯碱做沉淀剂制取 Li_2CO_3。钟辉等[21]提出另一种硼锂沉淀，先将卤水在盐田中析出钠钾混盐，再除 SO_4^{2-}，自然蒸发除镁，最后加纯碱制取 Li_2CO_3，该工艺流程见图 3，此方法锂的回收率可达 75%～85%，镁锂分离效果好，但工艺流程要控制酸性环境。

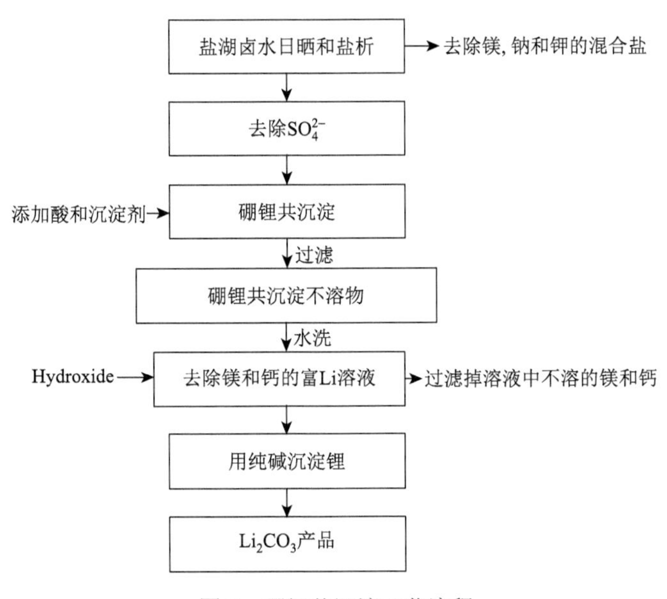

图 3 硼锂共沉淀工艺流程

钟辉等[22]针对硫酸镁亚型盐湖卤水开发了硼、镁共沉淀工艺，通过自然蒸发析出钾、镁混盐，再向卤水中加入碱性沉淀剂（控制 pH 值在 8～10），硼镁形成复盐或碳酸盐，实现镁锂初步分离，然后加入 NaOH 深度除镁，最后加 Na_2CO_3 得到 Li_2CO_3，该方法锂的回收率大于 80%，但沉淀过程中产生大量胶体，过滤过程锂损失率在 15%～20%。黄浩[23]用石灰乳做沉淀剂对硼镁共沉淀法进行改进，改进后锂损失率降到 7%，硼在此法中得到有效利用。未来共沉淀法应加以控制沉淀中锂的流失，或直接改变沉淀过程中沉淀剂形态，锂的流失问题减弱后工业共沉淀法应用有较好前景。

1.2　盐梯度太阳池法

扎布耶盐湖具有独特的水化学特性，除有在中度碳酸盐型的普通盐湖中常见的天然碱、石盐、芒硝、氯碳酸钠镁石和单斜钠钙石矿物外，还有大量硼砂、钾芒硝，其中锂镁比达 188.68，锂含量 1724mg/L，是中国少有的低镁锂比盐湖，表 2 为扎布耶盐湖与国内外产锂盐湖卤水特征对比。盐梯度太阳池法是针对低镁锂比盐湖，利用当地低温气候，先使 Li 接近饱和点并析出 $Na_2SO_4 \cdot 10H_2O$，得到的高锂混盐卤水（锂接近饱和点）利用太阳池技术不蒸发、升温条件得到富锂混盐和芒硝等沉淀，太阳池工作原理见图 4[24, 25]。目前该方法已被西藏扎布耶锂业高科技有限公司在扎布耶盐湖进行量产，年产量达 3000～5000t/a，是中国最大的锂工业生产基地[27]。虽然该方法已经工业化生产，但扎布耶盐湖的资源赋存情况世界少有，镁含量极低，且低温气候和当地自然条件都难以复制，盐梯度太阳池方法有其适用的局限性[26-28]，目前在中国藏北阿里地区和阿根廷部分地区已经采用此方法。

表 2　扎布耶盐湖与国内外产锂盐湖卤水特征对比

盐湖	$\frac{C_{Li^+}}{\Sigma 盐}/10^3$	Li^+/Mg^{2+}	$C_{Li^+}/(mg \cdot L^{-1})$	$C_{CO_3^{2-}}/(mg \cdot L^{-1})$	$C_{HCO_3^-}/(mg \cdot L^{-1})$
扎布耶盐湖南湖晶间卤水	3.93	188.68	1724	39829	399
大柴旦湖地表卤水	0.68	0.01	237	35.04	
美国瑟尔斯湖晶间卤水	0.24	—	81	27100	—
美国犹他州银峰地下卤水	2.23	1.0	400	—	—
美国大盐湖南湖地表卤水	0.25	0.01	58	—	473
智利阿塔卡马卤水	4.24	0.16	1570	—	230

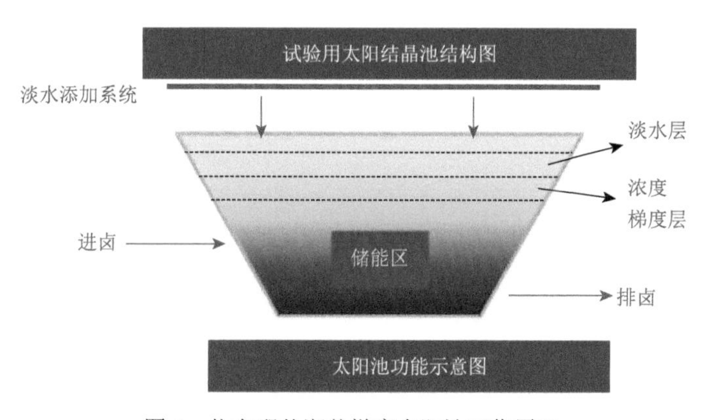

图 4　扎布耶盐湖盐梯度太阳池工作原理

2　高镁锂比盐湖提锂技术

2.1　纳滤与电渗析技术

膜分离技术主要包括膜吸附法、液膜萃取法、纳滤法、反渗透法和电渗析法。膜吸附法解决了离子筛吸附法的造粒问题，但吸附容量低；液膜萃取法和反渗透法需要压力驱动膜分离，所以主要介绍目前比较成熟的两种方法，纳滤法和电渗析法。

2.1.1　纳滤法

纳滤法提锂是根据纳滤膜的截留分子量和膜孔径对单价无机盐截留效果特异性对盐湖中镁、锂分离。纳滤法分离镁、锂效果可用表观截留率 R_{obs} 和分离因子 SF 表示

$$R_{obs} = (C_F - C_P)/C_F \times 100\%$$

$$SF = \frac{(C_{Mg^{2+}} / C_{Li^+})_P}{(C_{Mg^{2+}} / C_{Li^+})_F}$$

式中，C_P 和 C_F 分别为卤水原浓度和纳滤液浓度，mol/L；$C_{Mg^{2+}}$ 和 C_{Li^+} 分别为渗透液和原料液中 Mg^{2+} 和 Li^+ 的浓度，mol/L。SF 接近 1 时说明分离效果差，SF 越小说明分离效果越好[14]。

Shu 等[29]用 DL-2540 膜在东台吉乃尔和西台吉乃尔模拟卤水进行了镁锂的纳滤分离研究，研究表明 Mg、Li 分离与压力和温度有关，当温度升高 Mg、Li 分离率降低，当卤水流量增加，pH 值降低时 Mg^{2+}

的截留率提高，Li^+的截留率降低有利于镁锂分离。DL-2540 膜对 Mg^{2+}的截留率约为 60%，镁锂的分离因子 *SF* 在 0.35 左右。Wen 等[30]以东台吉乃尔盐湖卤水为原料，用 GE Osmonics 提供的 DL 2540C 膜进行纳滤分离研究，在 1.1MPa 的实验条件下，结果表明 DL 2540C 膜对硫酸盐的截留率高，SO_4^{2-} 分离效果好，当卤水中盐浓度增高，分离效果更差，DL 2540C 膜不适合此类型卤水提取锂。康为清等[31]采用 DK 纳滤膜验证纳滤法对盐湖提锂的可行性，用 3 种不同卤水为原料分别进行单级操作，实验结果原溶液中镁锂比分别从 48.5、42.31、28.3 降至 4.04、3.21、1.86，同时原卤水中全部 Ca^{2+}、SO_4^{2-} 被拦截，硼的截留率大于 73.81%，证明纳滤法可用于盐湖提锂。计超等[32]用 DK 纳滤膜分析其对镁锂分离效果，实验结果表明温度升高镁的截留率明显降低，而 pH 值降低增大 Mg^{2+}的截留率，当加入一价 Na^+、K^+会降低镁锂分离，并得出镁锂截留率关系

$$R_{\mathrm{Mg}}^{2+} = R_{\mathrm{Li}}^{+}\left(-1.06 + \frac{-7.26}{x - 0.77}\right)$$

成琪等[33]用 NT201（德国 Microdyn-Nadir 公司）纳滤膜，膜片有效直径为 0.07m，研究高镁锂比盐湖镁锂分离，实验结果显示低压，高镁锂比有利于镁锂分离，溶液中 Na^+、K^+会提高镁锂分离性能，而 SO_4^{2-} 明显影响镁锂分离。青海恒信融锂业使用纳滤法生产 Li_2CO_3，目前该公司中试已经能生产出电池级 Li_2CO_3，其生产流程见图 5。

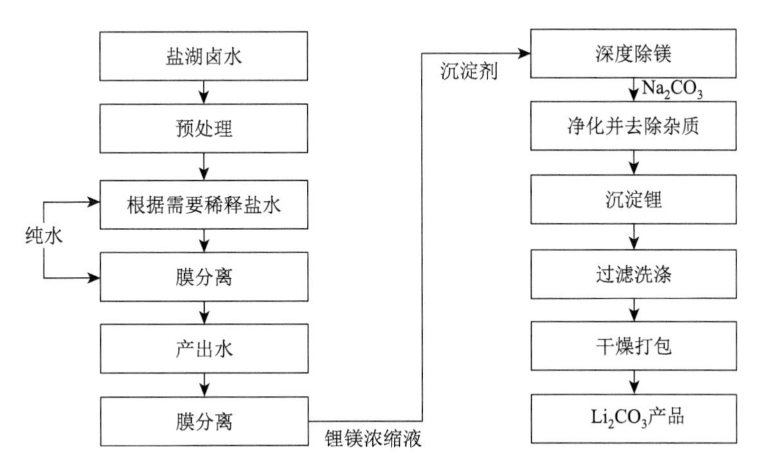

图 5 青海恒信融锂业使用纳滤法生产 Li_2CO_3 工艺流程[34]

据世界上已有的纳滤法对盐湖镁锂分离的研究，该方法对卤水总盐度要求高，卤水进行分离都要进行稀释，目前已有实验研究都是在低浓度的模拟卤水中进行，实验中镁铝比一般小于 30，因此纳滤技术目前难成为独立的提锂关键技术，但可以将纳滤技术与吸附法或电渗析技术结合，对镁锂分离技术进行优化。

2.1.2 电渗析法卤水提锂技术

电渗析法卤水提锂技术是在电场作用下溶液中带电粒子可以通过膜而迁移，如果用一价选择性离子交换膜，则可以将卤水中 Mg、Li 分离。该技术最早应用于海水淡化，21 世纪初开始应用在盐湖卤水 Mg、Li 分离。Bruggen 等[35]应用电渗析技术对 ED（AMV/CMV Selemion 膜和 ACS/CMS Tokuyama 膜）和 NF（NTR 7450 和 UTC-60）进行一价和二价阳离子分离评估，得出 UTC-60 膜最适合用于电渗析分离一价和二价阳离子。Jiang 等[36]用双极膜电渗析（EEDBM）技术尝试从盐湖卤水中提取 LiOH，用电流密度、进料浓度等参数评估提锂进程，评估 EEDBM 的过程成本约为 2.59 美元/kg，密度为 30mA/cm^2。马培华等[37]利用电渗析技术对镁锂化小于 300 的卤水进行分离，达到 Li^+回收率≥80%，Mg^{2+}脱除率≥95%，硼和 SO_4^{2-} 的脱除率≥99%的效果。Hoshino[38]将电渗析技术应用在海水提锂中，其离子液体膜可用于锂、钠分离。

青海锂业公司通过电渗析方法在东台吉乃尔盐湖用老卤生产 Li_2CO_3 产品，实际生产中发现电场作用会产生 H_2 和 OH^-，使用一段时间后产生的 $Mg(OH)_2$ 沉淀会覆盖离子交换膜影响膜的效率，因此用该方法生产需要经常拆洗膜，膜的维护成本高，生产维护工艺繁杂。

2.2 萃取法

萃取法提锂本质就是用某些对锂有特殊萃取性的有机溶剂萃取锂，在 1970 年前后该提锂方法就被提出[39]。美国锂公司 Nelli 等[40]最早提出用 20%磷酸三丁酯（TBP）、80%二异丁基酮（DIBK）和加入 $FeCl_3$ 进行高镁锂比盐湖提锂。Zhou 等[41-43]对此法进一步研究，并研发出成熟的盐湖卤水萃取锂工艺，在该萃取体系中 B 成为萃取 Li^+ 的有利条件。其萃取体系为 TBP/$FeCl_3$/MIBK，萃取机理为

$$FeCl_3(aq) + Cl^-(aq) == FeCl_4(aq)$$

$$Li^+(aq) + FeCl_4(aq) + n\mathrm{TBP}(org) == LiFeCl_4 \cdot n\mathrm{TBP}(org)$$

中性有机磷类萃取剂中，TBP/$FeCl_3$/煤油萃取体系是最有可能实现工业化的提锂技术，但实验 TBP 溶损问题严重，限制该方法进一步工业化，目前还未见工业化报道。除中性 TBP/$FeCl_3$/MIBK 萃取体系外，可用于 Li^+分离的萃取体系还有冠醚类化合物、β-双酮类、离子液体等。β-双酮类萃取体系萃取率高，萃取效果较好，但该体系中协萃剂价格昂贵，且碱性条件下溶损严重，目前工业化难度加大。冠醚类对锂的选择性较高，易萃取，但冠醚类化合物合成工艺繁杂，价格昂贵，未来还需进一步研究。离子液体萃取剂是一种新型绿色萃取剂，通过向离子液体结构中加入具有 Li^+亲和能力的官能团，可以使离子液体萃取锂，拥有一定应用前景，目前由于该方法需要加入大量官能团，溶液分子量大，造成黏度大对镁锂分离产生困难，未来设计合理功能化离子液体是主要研究问题。

2.3 吸附法

吸附法提锂是用天然或合成的化合物，制成可对卤水中锂离子进行选择性吸附，再用水或洗脱液将吸附剂中锂离子洗脱，使锂离子与杂质和伴生离子分离的方法。吸附法适用于高镁锂比盐湖，且提锂工艺简单，提锂过程中污染小，按照制备吸附剂材料不同分为有机吸附树脂吸附剂和无机吸附剂，无机吸附剂目前有层状吸附剂、无定型氢氧化物吸附剂、铝盐吸附剂、天然矿物吸附剂、离子筛吸附剂等。吸附剂提锂工艺在高镁锂比盐湖中应用目前存在如下问题：（1）吸附剂通用性差；（2）实际吸附容量远小于理论吸附容量；（3）吸附剂制备与成型工艺对吸附容量影响大；（4）吸附剂的复用频率低。

随着新能源技术快速发展，锂资源需求量将继续增加，而盐湖中锂资源的提取将成为锂资源发展的重心，高效地提取占比更大的高镁锂比盐湖中的锂更是一项重要的能源保障技术。在未来高镁锂比盐湖中应用锂的吸附剂进行工业化提取锂应重点关注以下研究方向。

1）铝盐吸附剂吸附速率快，且脱附过程不需要消耗酸，循环稳定性较好，已成功应用于工业化生产，在今后对铝盐吸附剂的研究要重点关注如何提高吸附容量，增强离子选择性，降低卤水总盐度对铝盐吸附剂吸附容量的影响。

2）锰系离子筛在吸附量和选择性方面均具有较大的优势，但目前脱附过程的锰溶损问题制约着其工业化应用，掺杂改性解决溶损问题、在制备过程中用镁代替锂制备离子筛前驱体是未来重点研究问题。

3 盐湖提锂产业化发展及存在问题

低镁锂比盐湖提锂技术相对成熟，沉淀法和盐梯度太阳池法等方法都已工业化应用多年，SQM 公司和 Albemarle 公司已经将沉淀法成熟的应用在低镁锂比盐湖提锂工业中。但沉淀法提锂工业生产成本高，提锂过程产生工业废料，环评压力大。今后在应用沉淀法进行工业化生产过程中，要将盐湖中的副产物变废为宝，如提锂同时将硼和钙合成为硼酸钙晶须材料。硼酸钙晶须材料为耐高温、高强度的军工材料，利用提锂的副产物既可以降低成本，又可以使提锂工艺更环保。

中国低镁锂比盐湖所占比例少，大部分盐湖为高镁锂比盐湖，所以高镁锂比盐湖提锂技术工业化应用面临问题更复杂。今后高镁锂比盐湖提锂技术工业化应用的实现要重点关注以下 3 方面研究方向。

1）纳滤法对卤水总盐度要求高，卤水进行分离要进行稀释，目前已有实验研究都是在低浓度的模拟

卤水中进行，实验中镁锂比一般小于30，因此纳滤技术目前难成为独立的提锂关键技术，但可以将纳滤技术与吸附法或电渗析技术结合，对镁锂分离技术进行优化。

2）吸附法中铝盐吸附剂吸附速率快，且脱附过程不需要消耗酸，循环稳定性较好，已成功应用于工业化生产，在今后研究中要提高吸附容量，增强离子选择性，降低卤水总盐度对铝盐吸附剂也有影响。锰系离子筛在吸附量和选择性方面均具有较大的优势，但目前脱附过程的锰溶损问题制约着其工业化应用，掺杂改性解决溶损问题，以及在制备过程中用镁代替锂制备离子筛前驱体是未来重点研究的问题。

3）天然矿物经过改性后，矿物孔径范围可控，对锂离子可以选择性吸附和洗脱，吸附过程中能源消耗更少，吸附锂工艺简单，产生废料少利于环保，因此针对盐湖气候条件和锂资源赋存的特点，研发出切实可行的利用天然矿物对液态锂提取新工艺，对加强盐湖、海水、地热水资源的综合利用和提高资源的利用效率有重大意义，但天然矿物选取要依据盐湖附近资源量进行成本计算，天然矿物的选取和机理的探索，矿物改性后成型造粒问题都是未来天然矿物改性作吸附剂的研究重点。

4 结论

目前工业化提锂技术中并没有一种技术能够适应所有的盐湖卤水类型，提锂技术适用性单一，所以工业化提锂发展可以将综合法作为日后关注的研究热点，将不同提锂技术进行综合应用，才能更高效地回收盐湖中锂资源。

参考文献（References）

[1] 王登红, 郑绵平, 王成辉, 等. 大宗急缺矿产和战略性新兴产业矿产调查工程进展与主要成果[J]. 中国地质调查, 2019, 6 (6): 1-11.

[2] Swain B. Recovery and recycling of lithium: A review[J]. Separation and Purification Technology, 2016, 172: 388-403.

[3] Zheng M, Deng T, Oren A. Introduction to salt lake sciences[M]. Beijing: Science Press, 2017.

[4] Zheng M, Liu X. Hydrochemistry of Salt Lakes of the Qinghai-Tibet Plateau, China[J]. Aquatic Geochemistry, 2009, 15 (1-2): 293-320.

[5] 郑绵平, 邓天龙, 阿哈龙·奥伦. 盐湖科学概论[M]. 北京: 科学出版社, 2017.

[6] 郑绵平, 刘喜方. 青藏高原盐湖水化学及其矿物组合特征[J]. 地质学报, 2010, 84 (11): 1585-1600.

[7] 郑绵平, 向军, 魏新俊, 等. 青藏高原盐湖[M]. 北京: 北京科学技术出版社, 1989.

[8] 郑绵平, 张永生, 刘喜方, 等. 中国盐湖科学技术研究的若干进展与展望[J]. 地质学报, 2016, 90 (9): 2123-2165.

[9] 郑绵平. 论中国盐湖[J]. 矿床地质, 2001, 20 (2): 181-189.

[10] 郑绵平. 青藏高原盐湖资源研究的新进展[J]. 地球学报, 2001, 22 (2): 97-102.

[11] 张雪飞, 郑绵平. 青藏高原盐类矿物研究进展[J]. 科技导报, 2017, 35 (12): 72-76.

[12] Jewell S, Kimball S M. Mineral commodity summaries 2019[R]. Reston: U.S. Geological Survey, 2019.

[13] 赵旭, 张琦, 武海虹, 等. 盐湖卤水提锂[J]. 化学进展, 2017, 29 (7): 796-808.

[14] 刘东帆, 孙淑英, 于建国. 盐湖卤水提锂技术研究与发展[J]. 化工学报, 2018, 69 (1): 141-155.

[15] 苏慧, 朱兆武, 王丽娜, 等. 从盐湖卤水中提取与回收锂的技术进展及展望[J]. 材料导报, 2019, 33 (13): 2119-2126.

[16] Boryta D A, Kullberg T F, Thurston A M. Production of lithium compounds directly from lithium containing brines: US7390466[P].2008-06-24.

[17] Hamzaoui A H, M'nif A, Hammi H, et al. Contribution to the lithium recovery from brine[J]. Desalination, 2003, 158: 221-224.

[18] An J W, Kang D J, Tran K T, et al. Recovery of lithium from Uyuni salar brine[J]. Hydrometallurgy, 2012, 117/118 (4): 64-70.

[19] 陆增, 胡适文, 袁建军. 从高镁锂比盐湖水中提取碳酸锂的方法: CN1398785[P].2003-02-26.

[20] 魏新俊, 王永浩, 保守君. 自卤水中同时沉淀硼锂的方法: CN1249272[P].2000-04-05.

[21] 钟辉, 杨建元, 张芃凡. 高镁锂比盐湖卤水中制取碳酸锂的方法: CN1335262[P].2002-02-13.

[22] 钟辉, 许惠. 一种硫酸镁亚型盐湖卤水镁锂分离方法: CN1454843[P].2003-11-12.

[23] 黄浩. 青海西台吉乃尔盐湖酸化老卤镁锂分离的技术研究[D]. 成都理工大学, 2009.

[24] 郑绵平, 郭珍旭, 张永生, 等. 从碳酸盐型卤水中提取锂盐方法: 中国 CN1270927[P].2000-10-25.

[25] 郑绵平. 利用太阳池从碳酸盐型卤水中结晶析出碳酸锂的方法: CN02129355[P].2003-02-26.

[26] 余疆江, 郑绵平, 唐力君, 等. 碳酸盐型卤水实验室模拟提锂和太阳池提锂的对比[J]. 化工进展, 2013, 32 (6): 1248-1252, 1260.

[27] 郑绵平, 刘喜方. 青藏高原盐湖水化学及其矿物组合特征[J]. 地质学报, 2010, 84 (11): 1585-1600.

[28] 乜贞, 卜令忠, 郑绵平. 中国盐湖锂资源的产业化现状——以西台吉乃尔盐湖和扎布耶盐湖为例[J]. 地球学报, 2010, 31 (1): 95-101.

[29] Shu Y S, Li J C, Xiao Y N. Separation of magnesium and lithium from brine using a Desal nanofiltration membrane. [J]. Journal of Water Process

Engineering, 2015, 7: 210-217.

[30] Wen X, Ma P, Zhu C, et al. Preliminary study on recovering lithium chloride from lithium-containing waters by nanofiltration[J]. Separation and Purification Technology, 2006, 49 (3): 230-236.

[31] 康为清，时历杰，赵有璟，等. 纳滤法用于盐湖卤水镁锂分离的初步实验[J]. 无机盐工业, 2014, 46 (12): 22-24.

[32] 计超，张杰，张志君，等. DK 纳滤膜对高镁锂比卤水的分离性能研究[J]. 膜科学与技术, 2014, 34 (3): 79-85.

[33] 成琪，关云山. 高镁锂比卤水的纳滤膜分离性能研究[J]. 无机盐工业, 2019, 51 (2): 35-39.

[34] 肖小玲，戴志锋，祝增虎，等. 吸附法盐湖卤水提锂的研究进展[J]. 盐湖研究, 2005, 13 (2): 66-69.

[35] Bruggen B V D, Koninckx A, Vandecasteele C. Separation of monovalent and divalent ions from aqueous solution by electrodialysis and nanofiltration[J]. Water Research, 2004, 38 (5): 1347-1353.

[36] Jiang C, Wang Y, Wang Q, et al. Production of lithium hydroxide from lake brines through electro-electrodialysis with bipolar membranes (EEDBM) [J]. Industrial &Engineering Chemistry Research, 2014, 53 (14): 6103-6112.

[37] 马培华，邓小川，温现民. 从盐湖卤水中分离镁和浓缩锂的方法: CN1626443A[P].2005-06-15.

[38] Hoshino T. Development of technology for recovering lithium from seawater by electrodialysis using ionic liquid membrane[J]. Fusion Engineering and Design, 2013, 88 (11): 2956-2959.

[39] Seeley F G, Baldwin W H. Extraction of lithium from neutral salt solutions with fluorinated β-diketones[J]. Journal of Inorganic & Nuclear Chemistry, 1976, 38 (5): 1049-1052.

[40] Nelli J R, Arthur T E. Recovery of lithium form bitterns: US 3537813[P].1970-06-26.

[41] Zhou Z, Qin W, Chu Y, et al. Elucidation of the structures of tributyl phosphate/Li complexes in the presence of $FeCl_3$via UV-visible, Raman and IR spectroscopy and the method of continuous variation[J]. Chemical Engineering Science, 2013, 101 (20): 577-585.

[42] Xiang W, Liang S, Zhou Z. Lithium recovery from salt lake brine by counter-current extraction using tributyl phosphate/$FeCl_3$in methyl isobutyl ketone[J]. Hydrometallurgy, 2017, 171: 27-32.

[43] Zhou Z, Liang S, Qin W, et al. Extraction equilibria of lithium with tributyl phosphate, diisobutyl ketone, acetophenone, methyl isobutyl ketone, and 2-heptanone in kerosene and $FeCl_3$[J]. Industrial & Engineering Chemistry Research, 2013, 52 (23): 7912-7917.

Development of lithium extraction technology and industrialization in brines of salt lake

Tao Ding[1], Mianping Zheng[2], Xuefei Zhang[2], Qian Wu[2] and Xiangyu Zhang[1]

1. College of Geoscience and Surveying Engineering, China University of Mining and Technology, Beijing 100083, China
2. MNR Key Laboratory of Saline Lake Resources and Environments, Institute of Mineral Resources, Chinese Academy of Geological Sciences, Beijing 100037, China

Abstract According to the latest statistics from the United States Geological Survey in 2020, the world has identified 80 million tons of lithium resources, of which 59% are distributed in salt lakes. This article reviews the development status of lithium extraction technology in salt lakes, and summarizes production processes of lithium resources. The adaptability and advantages of various extraction methods are explored with respect to different of magnesium to lithium ratios in salt lakes. It is found that lithium extraction in salt lakes cannot be carried out using a single method and selection of lithium extraction process should be based on different types of salt lakes. The precipitation method in low magnesium to lithium ratio salt lakes has been proven by industrial production while in high magnesium to lithium ratio salt lakes adsorption method is currently the most ideal extraction technology in terms of comprehensive effects, including high adsorption capacity, one-step direct extraction of lithium, high cycle performance and strong stability. Of all these methods, the natural mineral modified adsorption method should be the focus of future industrialization of lithium lake salt extraction.

Keywords Salt lake lithium resource; High magnesium-lithium ratio; Lithium extraction technology

黑北凹地富钾地下卤水自然蒸发实验研究*

彭玲玲[1,2]，魏学斌[3]，赵为永[3]，刘颖[1,2]，穆延宗[1]，乜贞[1]，王云生[1]

1. 中国地质科学院矿产资源研究所，自然资源部盐湖资源与环境重点实验室，北京 100037
2. 中国地质大学，北京 100083
3. 中国石油青海油田分公司勘探开发研究院，敦煌 817500

摘要 黑北凹地位于阿尔金山山前，其富钾地下卤水储量巨大，该卤水以钠、氯含量占绝对优势。富钾地下卤水化学组成简单，易于提取，具有很好的开发利用价值。以该富钾卤水为研究对象，在现场进行了自然蒸发实验。通过实验发现，卤水经过较长的石盐（NaCl）析出阶段后，分别达到光卤石（$KCl·MgCl_2·6H_2O$）和水氯镁石（$MgCl_2·6H_2O$）析出阶段，最后蒸干于南极石阶段，其析盐顺序为石盐—光卤石—水氯镁石—南极石；其中钾只以光卤石矿物析出，且析出阶段比较集中，在整个蒸发过程中氧化硼（B_2O_3）和锂则都在卤水中浓缩富集，没有析出。根据卤水蒸发过程中的析盐规律，其与 K^+，Na^+，Mg^{2+}∥Cl^--H_2O（25℃）介稳相图相符，蒸发过程中的体系变化趋势为液相体系点在石盐相区逐渐向远离石盐相点的方向移动，到达石盐与钾石盐共饱线后沿此线向石盐、钾石盐、光卤石共饱点移动，后沿石盐、光卤石共饱线移动，最终蒸干于石盐、光卤石、水氯镁石共饱点。经过对整个蒸发过程中卤水 pH 及密度的变化规律进行分析，指出可以通过 pH 及密度的变化来控制卤水的制卤过程。通过该蒸发实验研究，将为黑北凹地地下卤水的提钾工艺实验和后期开发提供科学依据。

关键词 黑北凹地；富钾地下卤水；自然蒸发；钾盐；相图；析盐规律

黑北凹地位于柴达木西北地区阿尔金山山前冲洪积扇群上，梁中凹地北东方向，察汗斯拉图以西。近年来，在梁中凹地-黑北凹地-察汗斯拉图一带发现大量砂砾石层型孔隙卤水，该卤水储量巨大，富水性较强，水位埋藏浅[1]，易抽取且抽取过程不易结盐。与南翼山等背斜构造第三系油田水富含 B、Li、Br、I 等不同[2-4]，黑北凹地砂砾石层型孔隙卤水化学组成简单，特征为高 Na，富含 K，低 B、Li 的氯化物型卤水。根据研究结果，此种类型卤水为高山融水溶滤 N_2 地层中的盐类而形成的溶滤型卤水[5]。卤水中 NaCl 含量达到单独开采工业品位，KCl 达到边界品位，由于地下卤水储量巨大，有望成为中国钾盐生产后备基地。现场蒸发实验对卤水中有用元素的可利用性、生产工艺及后期工业化生产有着重要的指导作用[6]。笔者以黑北凹地地下卤水为研究对象，借鉴前人对其他盐湖卤水蒸发实验研究的经验[7-11]，通过夏季小型现场蒸发实验，探究了卤水在蒸发浓缩过程中各元素的富集浓缩变化规律及析盐规律。

1 实验部分

1.1 实验原料

卤水取自黑 ZK02 井。初始卤水物化参数见表 1。

表 1 黑北凹地黑 ZK02 井卤水物化参数

项目	物理参数				各种离子质量浓度/($mg·L^{-1}$)		
	卤温/℃	密度/($g·cm^{-3}$)	pH	Br^-	Na^+	Ca^{2+}	Mg^{2+}
指标	22	1.2	6.454	40.3	106 660	6 312	6 075

* 本文发表在：无机盐工业，2019，51（6）：11-16

续表

项目	各种离子质量浓度/(mg·L^{-1})					
	K^+	Sr^{2+}	Li^+	Cl^-	SO_4^{2-}	B_2O_3
指标	2603	8.7	2.2	193 410	129.2	14.7

注：取样日期为2014年8月9日，气温为25℃，湿度为36.5%。

1.2　实验方法

蒸发实验在黑北凹地现场进行，蒸发初始所用容器为5个口径为50cm、深度为20cm的塑料盆。取卤水127.9kg分置于5个蒸发容器中，卤水深度为13cm。将装置放置在自然条件下蒸发，定期测定大气的温度、湿度，卤水的温度、密度、pH、蒸发量等物理参数；及时进行固液分离，并分别准确称量分离后的固体、液体质量，分别取固体、液体样品进行化学分析。为控制蒸发速度，蒸发过程中逐渐减少容器的个数，使卤水深度保持在10cm以上，当卤水深度较浅时进行合并继续蒸发。固、液分离频率根据卤水蒸发速率与析盐速率而定，直到最终无法进行固液分离为止。

1.3　分析方法

由中国科学院青海盐湖研究所盐湖化学分析测试部进行液、固相样品的分析测试[12]：Na^+、Ca^{2+}、Mg^{2+}、K^+、B_2O_3用等离子光谱仪分析；Sr^{2+}、Li^+用等离子质谱仪分析；Cl^-用以铬酸钾为指示剂的硝酸银容量法分析；SO_4^{2-}用盐酸联苯胺容量法分析；矿物种类鉴定用Rigaku MiniFlex 600型X射线衍射仪和JSM-5610LV型扫描电镜-能谱仪分析；每日常规测量中pH使用PHB-4型便携式pH计测量，湿度使用WS2080A型温湿度计测量，密度使用漂浮式比重计测量。

1.4　实验结果

蒸发实验从2014年8月9日开始，10月5日结束。期间进行了16次固、液分离，由于9月22日进行第13次固、液分离之后卤水浓度过大，固样难以晾干，并且会随着温度的变化在半固体与固体之间转化，故不再取固样进行溶解测试。蒸发实验观测数据见表2，固、液样品分析数据见表3、表4。根据配矿理论将分析数据进行配矿[13]，结果见表5。

表2　蒸发实验观测数据

取样日期	气温/℃	湿度/%	卤温/℃	密度/(g·cm^{-3})	pH	卤水质量/kg	成卤率/%	液样编号	固样编号
2014-08-09	25.0	36.5	22.0	1.200	6.45	127.90	100.00	HL0	—
2014-08-13	29.0	26.0	26.0	1.206	6.28	91.85	71.81	HL1	HS1
2014-08-17	34.5	22.0	30.1	1.207	6.08	56.90	44.49	HL2	HS2
2014-08-19	22.5	45.0	17.0	1.214	5.93	42.55	33.27	HL3	HS3
2014-08-23	27.0	28.0	25.0	1.220	5.67	28.80	22.52	HL4	HS4
2014-08-25	30.0	23.0	26.2	1.225	5.46	22.20	17.36	HL5	HS5
2014-08-28	23.0	45.0	19.0	1.239	5.14	15.65	12.24	HL6	HS6
2014-08-31	29.0	34.0	26.1	1.265	4.62	10.45	8.17	HL7	HS7
2014-09-03	28.0	39.0	30.0	1.307	4.04	7.15	5.59	HL8	HS8
2014-09-06	29.5	35.0	28.5	1.320	3.67	5.30	4.14	HL9	HS9
2014-09-09	25.5	51.0	34.0	1.350	3.08	4.10	3.21	HL10	HS10
2014-09-14	20.0	56.0	28.0	1.392	—	3.05	2.38	HL11	HS11
2014-09-20	29.1	41.0	32.5	1.409	2.26	2.25	1.76	HL12	HS12
2014-09-22	21.8	48.0	25.0	1.401	2.35	1.65	1.29	HL13	—
2014-09-26	18.0	61.0	16.5	1.404	2.16	1.10	0.86	HL14	—
2014-10-03	18.0	79.0	13.5	1.406	2.03	0.50	0.39	HL15	—
2014-10-05	25.0	42.0	21.0	1.400	—	0.40	0.31	HL16	—

注：1）取样时间均为16：00；2）成卤率计算方法为剩余卤水质量除以初始卤水质量[14]。

表 3 固样化学组成

固样编号	各种离子质量分数/%										相图指数/%		
	Na^+	Ca^{2+}	Mg^{2+}	K^+	Sr^{2+}	Li^+	Cl^-	SO_4^{2-}	B_2O_3	Br^-	KCl	NaCl	$MgCl_2$
HS1	34.790	0.378	0.178	0.153	0.003	0.003	54.586	0.483	0.006	0.002	0.326	98.894	0.780
HS2	35.833	0.501	0.257	0.193	0.004	0.003	56.621	0.563	0.005	0.003	0.398	98.513	1.089
HS3	34.018	0.489	0.403	0.234	0.005	0.003	54.509	0.231	0.008	0.004	0.504	97.712	1.783
HS4	35.874	0.485	0.360	0.222	0.005	0.003	57.179	0.298	0.007	0.004	0.455	98.029	1.516
HS5	33.105	0.722	0.711	0.452	0.011	0.003	54.701	0.105	0.016	0.006	0.982	95.847	3.172
HS6	30.155	1.135	1.138	0.441	0.018	0.003	52.031	0.218	0.023	0.008	1.026	93.535	5.439
HS7	27.806	1.565	1.715	0.669	0.024	0.003	51.090	0.180	0.031	0.015	1.621	89.841	8.537
HS8	21.805	1.745	3.076	3.219	0.025	0.003	48.478	0.130	0.032	0.020	8.338	75.297	16.366
HS9	2.990	1.685	7.276	8.638	0.027	0.002	36.601	0.035	0.035	0.027	31.331	14.459	54.210
HS10	3.258	2.438	7.068	6.539	0.037	0.002	35.833	0.028	0.050	0.027	25.743	17.100	57.157
HS11	1.953	3.144	7.944	3.710	0.065	0.002	35.076	0.027	0.055	0.040	16.393	11.505	72.102
HS12	0.228	4.876	8.339	0.154	0.096	0.003	33.324	0.007	0.069	0.060	0.876	1.728	97.396

注：相图指数指系统中各组分占干盐总量的质量分数，在本体系中 $w(KCl) + w(NaCl) + w(MgCl_2) = 100\%$。

表 4 液样化学组成

液样编号	各种离子质量浓度/ ($mg \cdot L^{-1}$)										相图指数/%		
	Na^+	Ca^{2+}	Mg^{2+}	K^+	Sr^{2+}	Li^+	Cl^-	SO_4^{2-}	B_2O_3	Br^-	KCl	NaCl	$MgCl_2$
HL0	106 660	6 312	6 075	260.3	8.7	2.2	193 410	129.2	14.7	40.3	0.168	91.778	8.054
HL1	104 180	8 152	7 749	3 060.0	137.4	9.5	199 370	1 242.0	192.3	63.5	1.938	87.979	10.082
HL2	93 260	11 350	12 080	4 691.0	259.6	15.0	202 440	1 006.0	323.6	102.7	3.049	80.821	16.130
HL3	84 970	13 920	14 920	5 673.0	306.9	16.1	203 460	847.1	383.2	125.8	3.792	75.722	20.486
HL4	71 920	19 250	20 550	7 910.0	443.6	20.8	211 270	652.6	583.1	179.8	5.418	65.671	28.911
HL5	60 140	23 660	25 440	9 797.0	601.6	27.4	216 840	709.7	725.8	227.3	6.888	56.372	36.740
HL6	42 490	30 730	33 450	12 870.0	759.9	31.6	228 370	480.4	942.2	260.5	9.311	40.982	49.708
HL7	22 440	42 190	45 610	17 260.0	1 002.0	39.1	257 180	307.2	1 255.0	385.0	12.253	21.238	66.509
HL8	10 490	54 400	58 030	21 060.0	1 331.0	49.3	299 910	217.2	1 706.0	458.3	13.653	9.067	77.280
HL9	8 828	67 500	60 550	9 863.0	1 670.0	60.9	315 270	163.9	2 097.0	592.5	6.755	8.061	85.184
HL10	3 843	81 290	68 870	4 502.0	1 296.0	73.8	350 660	142.5	2 549.0	800.8	2.980	3.391	93.629
HL11	2 479	100 590	72 820	1 835.0	2 325.0	89.1	392 280	95.0	3 180.0	916.3	1.186	2.136	96.678
HL12	2 459	108 390	72 450	1 993.0	2 327.0	102.2	405 340	127.3	3 719.0	926.3	1.293	2.128	96.579
HL13	2 329	110 860	69 320	1 616.0	1 470.0	124.0	397 970	116.9	4 650.0	1 008.0	1.098	2.111	96.791
HL14	2 276	108 190	70 890	1 627.0	976.0	145.1	398 730	124.0	5 529.0	1 118.0	1.083	2.019	96.898
HL15	1 424	111 060	69 370	942.3	398.5	161.2	397 350	91.4	5 204.0	1 503.0	0.648	1.306	98.045
HL16	1 559	102 950	73 090	974.0	226.3	180.0	392 740	98.9	5 790.0	1 468.0	0.636	1.357	98.007

注：相图指数指系统中各组分占干盐总量的质量分数，在本体系中 $w(KCl) + w(NaCl) + w(MgCl_2) = 100\%$。

表 5 固相中各盐类质量分数（%）

固样编号	w（$CaSO_4 \cdot 2H_2O$）	w（KCl）	w（NaCl）	w（$MgCl_2 \cdot 6H_2O$）	w（$CaCl_2 \cdot 6H_2O$）
HS1	0.33	0.26	98.68	0.48	0.26
HS2	0.37	0.31	98.26	0.67	0.39
HS3	0.16	0.40	97.73	1.10	0.62
HS4	0.19	0.36	97.98	0.93	0.54
HS5	0.07	0.77	96.10	1.95	1.10
HS6	0.16	0.81	93.85	3.35	1.83
HS7	0.14	1.28	90.55	5.28	2.75
HS8	0.11	6.86	79.01	10.54	3.48
HS9	0.05	31.92	18.79	43.25	5.99
HS10	0.04	25.33	21.46	44.04	9.12
HS11	0.05	16.22	14.52	55.87	13.34
HS12	0.02	0.83	2.08	71.91	25.17

2　数据处理与分析

2.1　相图分析

黑北凹地地下卤水中的阴离子主要为Cl^-，阳离子主要为Na^+、Ca^{2+}、Mg^{2+}、K^+等，其卤水类型按其化学性质分类为氯化物型，在蒸发实验中前期 Ca 基本不以矿物析出，因此可以采用 K^+，Na^+，Mg^{2+}//Cl^--H_2O（25℃）四元介稳相图；在实验后期，Ca 浓缩富集，达到饱和后以$CaCl_2 \cdot 6H_2O$矿物析出，因此可以选用 Na^+，Mg^{2+}，Ca^{2+}//Cl^--H_2O（25℃）四元介稳相图。根据整个蒸发过程中的固、液相中的离子组成及计算出的相图指数，绘得卤水蒸发结晶路线，见图 1。从图 1 可以看出，液相初始点位于 NaCl 相区，随着蒸发的进行液相点逐渐向 KCl 相区和远离 NaCl 相点的位置移动，直到 HL8 点到达 NaCl 与 KCl 共饱线，沿此共饱线向石盐、钾石盐、光卤石共饱点移动，因此共饱点为不稳定共饱点，之后随着蒸发液相点沿石盐与光卤石共饱线移动，最终蒸干于石盐、光卤石、水氯镁石共饱点。对应的固相点最初由 NaCl 相区缓慢朝 KCl 相区移动，这是由于卤水中的离子以Na^+、Cl^-占绝对优势，导致蒸发初始阶段析出物以 NaCl 为主，所以在相图中显示为长时间处于 NaCl 相区，HS8 点之后固相点迅速进入 KCl 相区，与液相点对比可知，液相点在 HL8 点到达 NaCl、KCl 共饱线，之后向石盐、钾石盐、光卤石共饱点移动，说明在 HL8 点之前液相中的 K 不断浓缩，而析出的固相以 NaCl 为主，到 HL8 点之后液相中的 KCl 开始析出。在石盐、钾石盐、光卤石共饱点钾石盐开始溶解，而石盐和光卤石共饱析出，至钾石盐回溶完成后，液相点随蒸发沿石盐、光卤石共饱线移动，在此相图中最终蒸干于石盐、光卤石、水氯镁石共饱点。结合相图及蒸发记录可知，在卤水密度达到 1.307g/cm^3，或者体积浓缩至原卤水体积的 5.13%时，KCl 便开始大量析出。

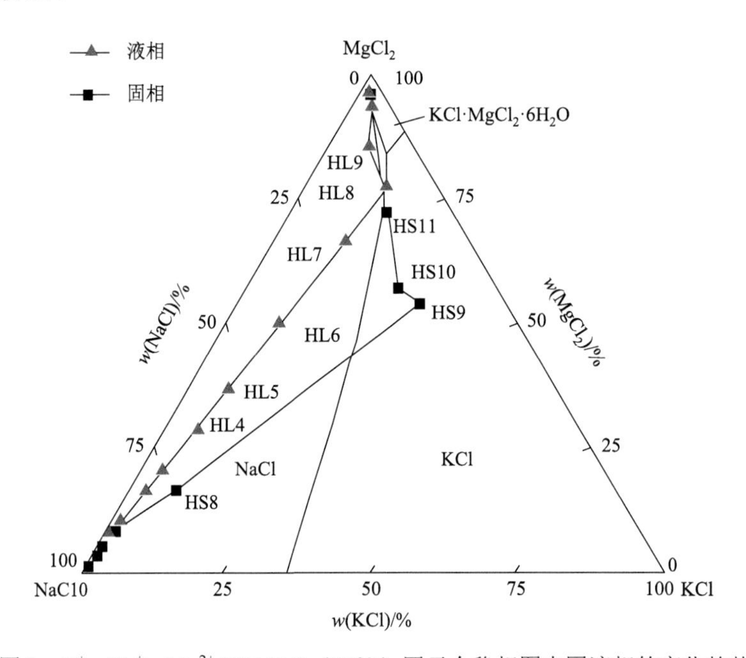

图 1　K^+，Na^+，Mg^{2+}//Cl^--H_2O（25℃）四元介稳相图中固液相的变化趋势

2.2　矿物析出次序

图 2 为固体样品 HS6、HS9、HS11、HS14 的扫描电镜-能谱仪分析结果；图 3 为固体样品 HS6 的 XRD 分析结果。从图 2、图 3 可以将实验过程中盐类的析出分为 4 个系列：1）石盐（HS1～HS6)；2）石盐 + 光卤石（HS7～HS10)；3）石盐 + 光卤石 + 水氯镁石（HS11～HS13)；4）石盐 + 光卤石 + 水氯镁石 + 南极石（HS14)。

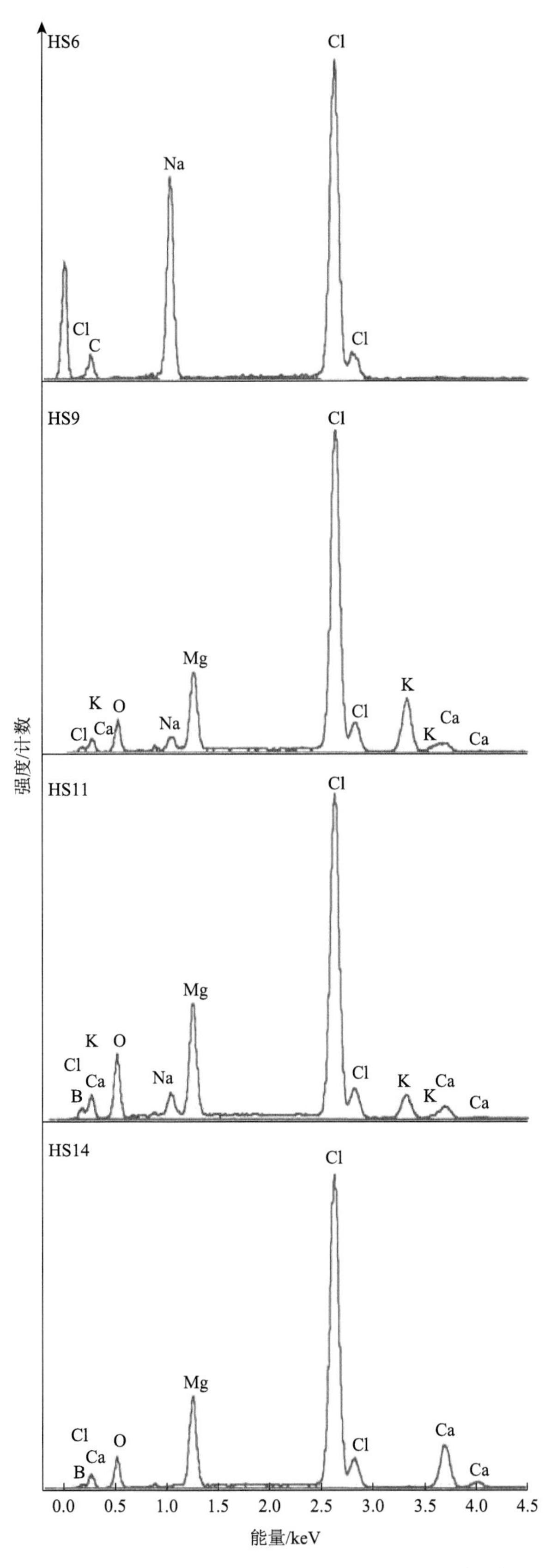

图 2 HS6、HS9、HS11、HS14 样品扫描电镜能谱分析结果

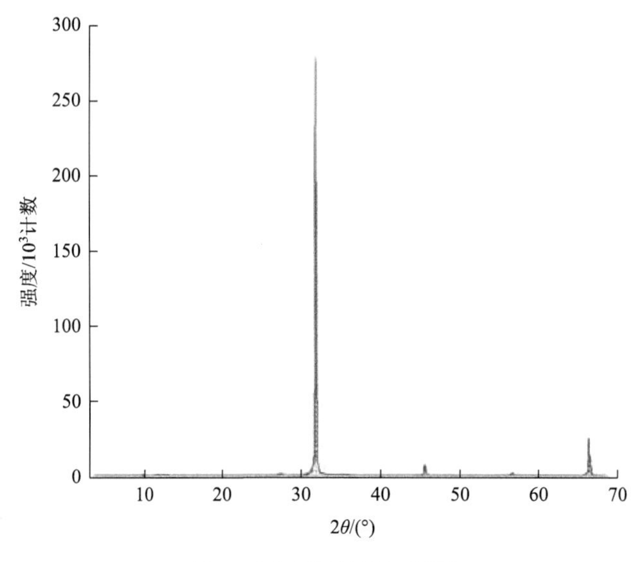

图 3　HS6 样品 XRD 谱图

2.3　离子富集规律分析

实验用卤水中 Na^+、Cl^-基本达到饱和，在卤水蒸发初始即有大量固相析出，而且析出的固相中石盐质量分数可达 98.68%。由蒸发记录（表 2）及固相配矿结果（表 5）可知，第七次固液分离后成卤率已达 8.17%，固相中石盐质量分数仍高达 90%以上，第八次固液分离后成卤率为 5.59%，固相中石盐质量分数有所下降（为 79.01%），其后的数次分离固相中石盐含量迅速下降。其他离子则经历了较长时间的富集过程，而后才开始成盐析出，这种结果印证了前述原卤离子含量单一的特征。

图 4 为固相中 KCl 质量分数与液相中 K^+ 质量浓度随着成卤率的变化趋势。从图 4 看出，蒸发初期液相中 K^+ 质量浓度缓慢增加，曲线较为平缓，从 HL0 到 HL3 成卤率从 100%降低到 33.27%，但是液相中 K^+ 质量浓度只从 2.603g/L 增加到 5.673g/L，仅增加至 2.2 倍，但是当到达 HL3 点之后，即成卤率达到 33.27%之后，K^+ 在液相中的富集速率逐渐加快，曲线斜率明显增大，直到 HL8 点即成卤率达到 5.59%时，液相中 K^+ 质量浓度达到极大值 21.060g/L，之后便迅速下降；与液相中 K^+ 质量浓度相对应，固相中 KCl 质量分数在 HS7 点之前近乎零，说明在 HS7 点之前，即成卤率到达 8.17%之前，液相中 K^+ 浓度一直未达到饱和，固相中微量的 KCl 含量来源于母液夹带，HS7 点之后，固相中的 KCl 含量迅速增加，直到 HS9 点，

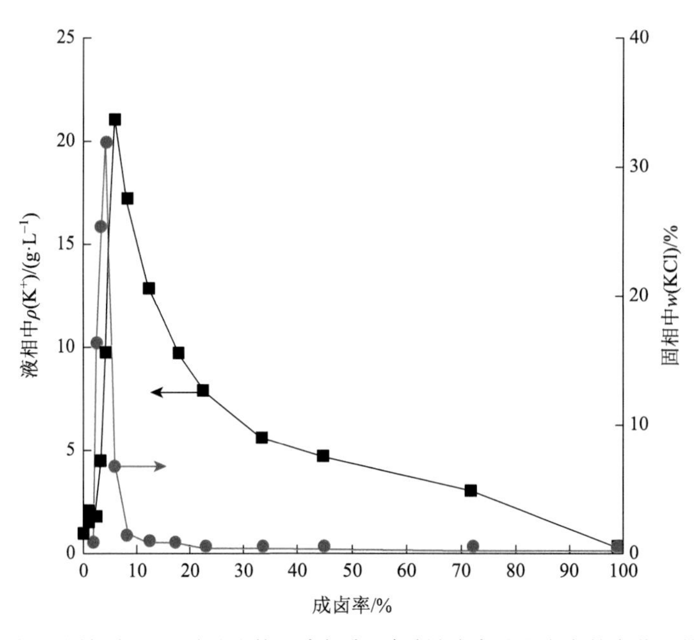

图 4　固相中 KCl 质量分数、液相中 K^+质量浓度随成卤率的变化规律

即成卤率为4.14%时，固相中KCl质量分数达到极大值31.92%，之后便迅速降低，说明该卤水中的钾盐在此阶段集中析出。固相中KCl质量分数从迅速增加到降低的过程，与液相中的K^+含量先短暂增加到极大值点后又迅速降低的变化规律相吻合。

图5为Mg在固、液相中的含量随成卤率的变化趋势。从图5看出，前期液相中Mg^{2+}质量浓度变化趋势与K^+类似，都是经历了较长时间的缓慢增加过程，在HL3点之前，即成卤率达到33.27%之前，曲线十分平缓，液相中Mg^{2+}质量浓度只从6.075g/L增加到14.920g/L，仅增加至2.5倍，HL3点之后液相中Mg^{2+}富集速率迅速增加，曲线斜率明显增大，直到HL16点，液相中Mg^{2+}质量浓度达到极大值73.090g/L，从HL11到HL16，液相中Mg^{2+}含量出现小幅的上下波动，说明其存在蒸发浓缩与析盐之间的一个平衡状态；固相中的Mg^{2+}质量分数，在HS4点之前，即成卤率达到22.52%之前，Mg^{2+}质量分数增加一直较为缓慢，曲线平缓并且略有波动，说明此时固相中的Mg^{2+}来源主要为母液夹带，从HS4点之后，固相中的Mg^{2+}富集速率开始逐渐加快，曲线的斜率逐渐增大，直到HS12，固相中的Mg^{2+}质量分数达到极大值8.339%。

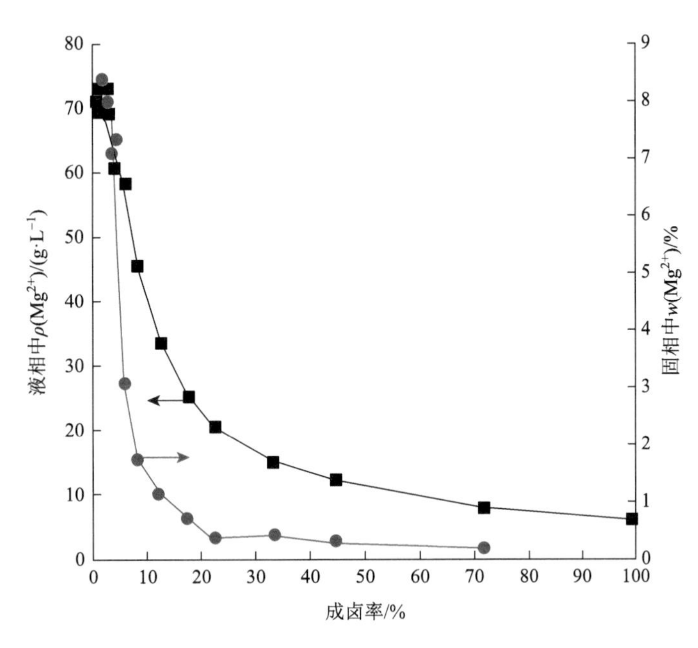

图5 Mg在固、液相中的含量随成卤率的变化规律

图6为液相中Li、B_2O_3随成卤率的变化趋势。从图6可以看出，二者在液相中随着成卤率的降低一直处于富集阶段，HL6点之前，卤量较大，Li、B_2O_3浓度低，蒸发浓缩速率较慢，HL6点之后，成卤率达到12.24%，二者开始迅速富集，但是最终并未出现下降趋势。从固样分析结果（表3）来看，随着成卤率的降低，Li、B_2O_3在固样中的质量分数并未出现大幅增长，说明其在整个蒸发过程中并未以盐的形式析出。根据浓缩倍数关系，在整个蒸发过程中，卤水浓缩了322.6倍，B_2O_3浓缩了393.9倍，而Li只浓缩了81.8倍，考虑到测试误差及实验过程中的母液夹带及损失，B_2O_3在整个蒸发过程中的富集倍数与卤水相符，而Li在蒸发过程中出现了损失，究竟以何种方式损失，还有待后续查证。

2.4 pH及密度变化规律分析

在蒸发实验过程中，连续测定了每次固液分离时卤水的pH及密度，根据测定结果绘制其随成卤率的变化规律见图7。从图7可以看出，随着成卤率的降低，pH及密度都呈现出规律性的变化：卤水pH不断降低，而密度不断增高，pH由初始的6.45降低到最终的2.03，密度由初始的1.200g/cm^{3}增加到最终的1.400g/cm^3。pH与密度的斜率（绝对值）均在HL8点开始迅速增大，其下降或上升速度明显加快，而此点正好对应KCl大量析出的点，因此此次实验pH及密度的变化规律可以作为卤水制卤过程的控制指标，可为后期卤水提钾实验和盐田实验提供相关数据支撑。

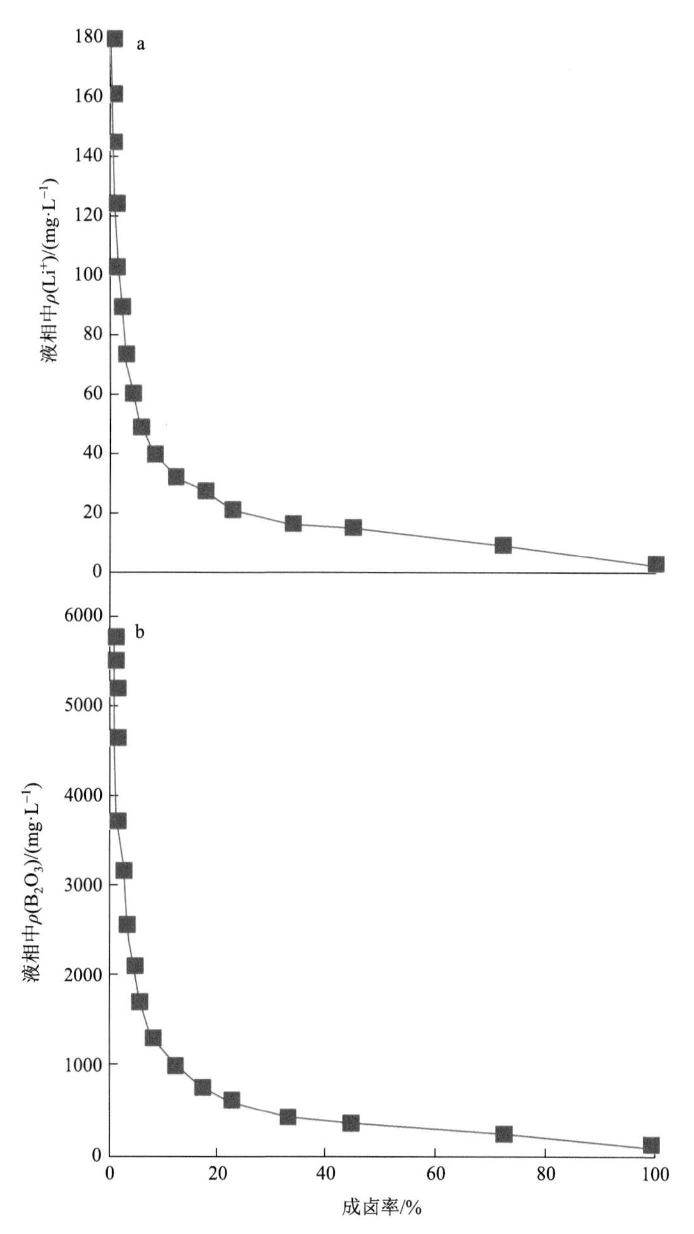

图 6　Li、B_2O_3 在液相中的质量浓度随成卤率的变化规律

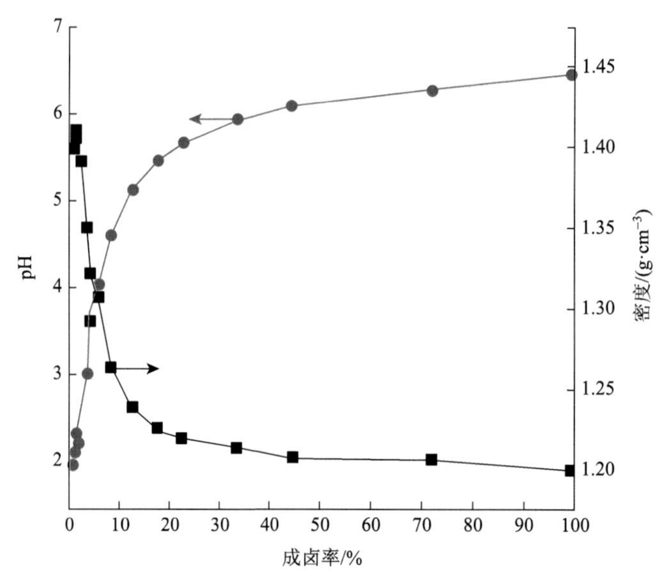

图 7　pH 及密度随成卤率的变化规律

3 结论

通过黑北凹地富钾卤水自然蒸发实验，可以得出以下结论：1）黑北凹地砂砾石层型孔隙卤水中 Na、Cl 离子含量远高于其他离子，在蒸发结晶过程中，经过了较长时间的石盐析出阶段，达到钾盐析出阶段，其析盐顺序为石盐—光卤石—水氯镁石—南极石；2）黑北凹地地下卤水在晒制过程中，其蒸发结晶路线与 K^+、Na^+、Mg^{2+}//Cl^--H_2O（25℃）介稳相图相符，原始卤水位于氯化钠相区，最终蒸干于石盐、光卤石、水氯镁石共饱点；3）黑北凹地地下卤水在蒸发过程中，钾以光卤石矿物析出，当成卤率达到 5.59%时，光卤石开始大量析出，且有一钾盐的集中析出阶段，便于钾盐的提取利用；4）B_2O_3、Li 在整个蒸发过程中一直在卤水中浓缩富集，没有结晶析出，其最终质量浓度分别为 5.79g/L 与 180mg/L，B_2O_3 浓缩了 393.9 倍，而 Li 只浓缩了 81.8 倍；5）蒸发过程中，卤水的 pH、密度数据直观反映了卤水的蒸发进度，可作为后续制卤阶段指示指标。

参考文献

[1] 李洪普，郑绵平，侯献华，等. 柴达木黑北凹地早更新世新型砂砾层卤水水化学特征与成因[J]. 地球科学——中国地质大学学报，2014，39 (10): 1333-1342.

[2] 樊启顺，马海州，谭红兵，等. 柴达木盆地西部典型地区油田卤水水化学异常及资源评价[J]. 盐湖研究，2007，15 (4): 6-12.

[3] 谭红兵，曹成东，李廷伟，等. 柴达木盆地西部古近系和新近系油田卤水资源水化学特征及化学演化[J]. 古地理学报，2007，9 (3): 313-320.

[4] 付建龙，于升松，李世金，等. 柴达木盆地西部第三系油田卤水资源可利用性分析[J]. 盐湖研究，2005，13 (3): 17-21.

[5] 郑绵平，侯献华，于长青，等. 成盐理论引领我国找钾取得重要进展[J]. 地球学报，2015，36 (2): 129-139.

[6] 姜旭，周保华，乜贞. 中国盐湖卤水蒸发实验研究进展[J]. 无机盐工业，2013，45 (6): 1-4.

[7] 乜贞，郑绵平. 西藏扎布耶盐湖夏季卤水盐田晒制研究[J]. 地球学报，2001，22 (3): 271-275.

[8] 乜贞，张永生，卜令忠，等. 西藏扎布耶盐湖卤水冬季制卤试验研究[J]. 地质通报，2005，24 (4): 386-390.

[9] 伍倩，郑绵平，乜贞，等. 西藏当雄错碳酸盐型盐湖卤水自然蒸发析盐规律研究[J]. 无机化学学报，2012，28 (9): 1895-1903.

[10] 伍倩，郑绵平，乜贞，等. 西藏当雄错盐湖卤水冬季日晒蒸发实验研究[J]. 地质学报，2013，87 (3): 433-440.

[11] 刘颖，王云生，乜贞，等. 西藏朋彦错盐湖碳酸盐型卤水 15℃等温蒸发实验研究[J]. 无机盐工业，2017，49 (8): 19-23.

[12] 中国科学院青海盐湖研究所分析室. 卤水和盐的分析方法[M]. 2 版. 北京：科学出版社，1988.

[13] 何法明，刘世昌，白崇庆，等. 盐类矿物鉴定工作方法手册[M]. 北京：化学工业出版社，1988.

[14] 高世扬，柳大纲. 大柴旦盐湖夏季组成卤水的天然蒸发（含硼海水型盐湖卤水的天然蒸发）[J]. 盐湖研究，1996 (Z1): 62-68.

Study on natural evaporation of underground potassium-rich brine in Heibei concave

Lingling Peng[1, 2], Xuebin Wei[3], Weiyong Zhao[3], Ying Liu[1, 2], Yanzong Mu[1], Zhen Nie[1] and Yunsheng Wang[1]

1. MNR Key Laboratory of Saline Lake Resources and Environments, Institute of Mineral Resources, Chinese Academy of Geological Sciences, Beijing 100037
2. China University of Geosciences, Beijing 100083
3. Exploration and Development Research Institute of PetroChina, Qinghai Oilfield Company, Dunhuang 817500

Abstract With huge reserves of underground potassium-rich brine, the Heibei concave is located in the piedmont of the Altun Mountains. The ion species in the brine are few except for Na or Cl whose content accounts for absolute advantage. Because of its simple chemical composition and easy extraction, the brine has a great value in industrial development and utilization. Taking the potassium-rich brine as the research object, the natural evaporation experiment was carried out on site. It was found through experiments that the brine reached the precipitation stage of carnallite ($KCl \cdot MgCl_2 \cdot 6H_2O$) and bischofite ($MgCl_2 \cdot 6H_2O$) after a long period of halite precipitation of (NaCl) , and finally dried out at the antarctic stage, and the sequence of the mineral crystallization during evaporation was halite, carnallite, bischofite and antarcticite. And potassium was only precipitated from carnallite minerals, and the precipitation stage was relatively concentrated. During the whole evaporation process, boron oxide (B_2O_3) and lithium were concentrated and enriched in the brine without precipitation. According to the crystallization law of brine during the evaporation process, it was consistent with the metastable phase diagram of K^+, Na^+, Mg^{2+} // Cl^--H_2O (25℃). The evolution trend of the brine during the evaporation process was that it moved from the NaCl point in the halite phase area to the sylvite and carnallite phase area and when reaching the univariant curve of halite and sylvite, it moved along the curve and arrives at the invariant point of halite, sylvite and carnallite. Then it moved along the univariant curve of halite and carnallite, and finally arrived at the invariant point of halite, carnallite and bischofite. After analyzing the variation of brine pH and density during the whole evaporation process, it was pointed out that the halogen production process of brine could be controlled by changes in pH and density. Through the evaporation experiment, it would provide a scientific basis for the potassium extraction process experiment and post- development of the underground brine in the Heibei concave.

Keywords Heibei concave; Underground brine; Natural evaporation; Potassium; Phase diagram; Crystallization regularity

第四部分

盐湖古气候与生态环境

Late Mio-Pliocene landscape evolution around the Chaka Basin in the NE Tibetan Plateau: insights from geochemical perspectives*

Yuanyuan Lü[1], Huiping Zhang[2], Binkai Li[3] and Qinghua Jin[4]

1. MLR Key Laboratory of Saline Lake Resources and Environments, Institute of Mineral Resources, CAGS, Beijing 100037, China
2. State Key Laboratory of Earthquake Dynamics, Institute of Geology, China Earthquake Administration, Beijing 100029, China
3. Qinghai Institute of Salt Lakes, Chinese Academy of Sciences, Xining 810008, China
4. Development and Research Center, China Geological Survey, Beijing 100037, China

Abstract The present NE Tibetan Plateau is characterized by a rugged mountain range and basin landforms. Knowledge of the landscape evolution, as recorded by basin sediments, will help to understand the growth history of the NE Tibetan Plateau. Our present contribution targeted the Neogene Chaka Basin south of the Qinghai Nan Shan and reports 40 geochemical Sr, Nd, trace element, and rare earth element (REE) results of the sediments. Values of ε_{Nd} vary from −15.6 to −11.6 for all samples, but ε_{Nd} is higher for samples younger than 5.5 Ma. $^{87}Sr/^{86}Sr$ ratios are all between 0.71441 and 0.71844, with the variation becoming less after 4.5 Ma. Little variation in the Sr-Nd isotopes and in the REE trend indicates that the source regions have very similar geochemical features. The contribution of aeolian loess into the basin sediments after ~3.5 Ma is also speculated from the trace element and REE results. These geochemical results reveal a change in the source region to the Qinghai Nan Shan around 4.5–5.5 Ma, which is consistent with previous litho face evidence. Since ~3.5 Ma, landscape and topographic relief were further developed; thus, aeolian loess was deposited onto the fluvial fans south of the Qinghai Nan Shan, while global cooling and aridification began to dominate. Collectively, our present geochemical study of the Chaka Basin sediments reveals stages of landscape evolution in NE Tibet since the late Miocene.

Keywords Neogene; Chaka Basin; Tibetan Plateau; Bulk geochemistry; Landscape evolution

1 Introduction

Since the Cenozoic Era, the Indian plate has been colliding with the Eurasian plate, and this continuous convergence has provided a dynamic environment for the outward and upward growth of the Tibetan Plateau (Molnar and Tapponnier, 1975a, 1975b). Northeastern Tibet is one of the youngest growing plateau margins and has been integrated into the plateau since the collision (Fig. 1; Tapponnier et al., 2001; Yuan et al., 2013). This plateau margin is also affected by the complex interaction between the Asian summer and winter monsoons and the Westerly jet. The development of the Loess Plateau in NE Tibet was often argued to be due to the uplift of the mountain ranges along the faults and to Cenozoic global cooling (An et al., 2001; Ding et al., 1995; Sun et al., 2020). Sediments eroded from the high-relief mountains mostly accumulated within adjacent basins (Dettman et al., 2003; Fan et al., 2007; Fang et al., 2005, 2007; Cheng et al., 2019). The study of these basin sediments helps us to gain a better understanding of the uplift and landscape evolution around the basin (Cheng et al., 2016; Wang et al., 2017). For example, the Guide Basin in the upper Yellow River region is confined by the Laji Shan (Shan means mountain in Chinese) to the north and by the West Qinling to the south (Fig. 1). Distribution of the zircon

* 本文发表在：Palaeogeography, Palaeoclimatology, Palaeoecology, 2020, 552: 109778-109786

U-Pb ages was reported by Lease et al. (2007) from the ~1200 m thick Neogene strata (~21–1.8 Ma; Fang et al., 2005). This showed a shift of the sediment source in the Guide Basin at approximately 8 Ma (Lease et al., 2007) and is supported by lithoface and provenance analysis.

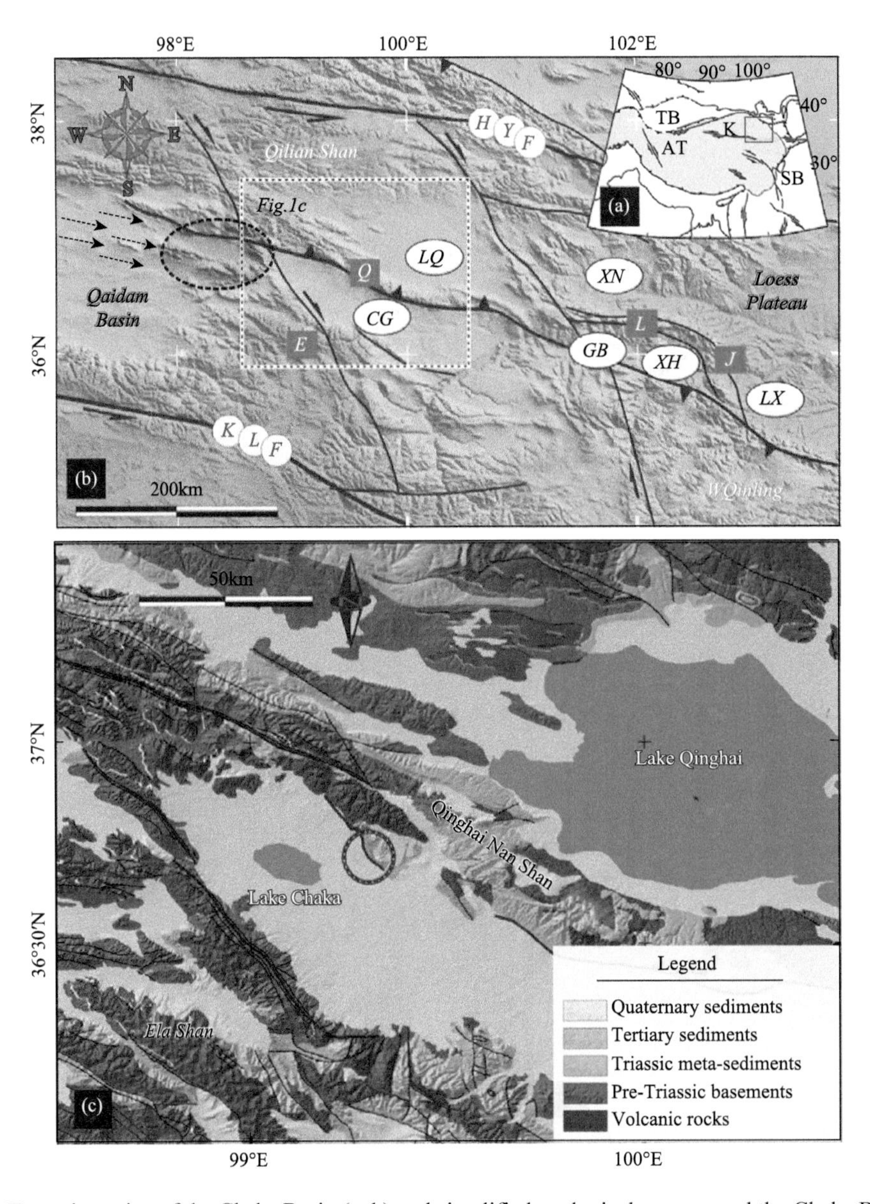

Fig. 1 Tectonic setting of the Chaka Basin (a, b) and simplified geological map around the Chaka Basin (c).

Inset map shows the location of the Chaka Basin relative to the plateau; TB, Tarim Basin; SB, Sichuan Basin; AT, Altyn Tagh Fault; KLF, Kunlun Fault; Abbreviations in b are as follows: CG, Chaka-Gonghe Basin; LQ, Lake Qinghai Basin; XN, Xining Basin; GB, Guide Basin; XH, Xunhua Basin; LX, Linxia Basin; E, Q, L, and J are Ela Shan, Qinghai Nan Shan, Laji Shan, and Jishi Shan, respectively; HYF, Haiyuan Fault; KLF, Kunlun Fault. The yellow ellipse marks the sample locations ($\varepsilon_{Nd}=-12$ to approximately -28) from Wan et al. (2006), and the red circle in c is the location of the Chaka section in this study. (For interpretation of the references to colour in this figure legend, the reader is referred to the web version of this article.)

With the exception of the means of sedimentology, isotopic signatures of the basin sediments can be used to identify mountain building and basin evolutionary processes. Because rare earth elements (REEs), for example, Nd, usually experience small mass-dependent fractionations during chemical weathering, sediment transport, and sedimentation, they are commonly used to trace the origin of detrital sediments and regional environmental changes (Goldstein and Jacobsen, 1988; Borg and Banner, 1996; Garzione et al., 2005; Tütken et al., 2002). Similarly, these geochemical tools have great potential to trace basin sediments and analyze the landscape evolution around the basin. Studies of the carbon and oxygen isotopic compositions of Neogene sediments in the Linxia and Xunhua basins in NE Tibet showed that the $\delta^{18}O$ values within the Xunhua Basin on the west side of

Jishi Shan are more positive than those of the Linxia Basin to the east after 11 Ma. This diversion in isotopic signature has been argued to be the consequence of the uplift and relief that increased the Jishi Shan from ~11 Ma onwards, blocking the transport of water vapor from east to west (Hough et al., 2011). Soon after, at ~280 km to the west of the Xunhua Basin, the Chaka Basin was separated from the Lake Qinghai Basin by the Qinghai Nan Shan [Fig. 1 (b)]. Sediment input into the basin is reported to have been deposited between ~11 Ma and ~3 Ma, or more recently (Zhang et al., 2012).

To decode the landscape evolution information recorded by the Neogene sediments within the Chaka Basin, our present study will analyze a series of geochemical compositions, such as Sr and Nd isotopes, trace elements, and REEs of basin sediments. We will reveal temporal changes of the sediment sources and thus the uplift of the Qinghai Nan Shan and the landscape evolution around the Chaka Basin during the past ~11 Myr.

2 Geological setting

The Chaka Basin is located in the southern Qilian orogenic belt, a transition zone between the Tibetan Plateau, Qaidam Basin, and the Loess Plateau (Fig. 1). Locally, the basin has been developed to the north of the Ela Shan adjacent to the Kunlun fault. As part of the Gonghe Basin, the Chaka Basin is separated from the Qaidam Basin to the west by the Ela Shan and from Lake Qinghai Basin by the Qinghai Nan Shan to the north. Bedrocks within the Ela Shan and the Qinghai Nan Shan are relatively similar; most of the lithology is Paleozoic and Triassic metamorphic rocks, such as gneiss, schist, and marble, and meta-sedimentary sandstone, siltstone, and slate rocks locally intruded by granite and granodiorite [Fig. 1 (c)].

On the southern side of the Qinghai Nan Shan, upper Tertiary strata were uplifted and exposed at the surface due to late Cenozoic tectonic activity (Zhang et al., 2012). Upper Miocene-Pliocene strata are continuous in the Chaka Basin and contain clear variations in sedimentary facies, similar to those in the Lake Qinghai Basin (Fu et al., 2013). Near Lake Qinghai, along the northern side of the Qinghai Nan Shan, the piedmont denudation surface is monotonically tilted towards Lake Qinghai (Su et al., 2017). Sedimentology, stratigraphy, and regional geology indicate that the Cenozoic strata in the Chaka Basin can be divided into four units (Fig. 2; Chang et al., 2008; Zhang et al., 2012). The lowest unit (N_1a; ~11.6–8.6 Ma; ~600–700 m thick) is characterized by gray, brownish- and greenish-gray, massive or laminated mudstone and siltstone interbedded with 50–100 cm thick, coarse sandstone, and pebble conglomerate lenses. The whole unit was braided river, floodplain, and shallow lake sediments (Zhang et al., 2012). The middle section ($N_{1\text{-}2}b$; 8.6–4.6 Ma; ~850–1000 m thick) is dominated by gray to gray-brown, medium-coarse grained sandstone interbedded with light yellow-green and gray siltstone. It has been interpreted as shallow lake, meandering river, and floodplain sedimentary environments (Zhang et al., 2012). The upper section (N_2c; 4.6–3.1 Ma; ~800 m thick) consists of massive, 2–10 m thick, yellow, brown, and light brown pebble/cobble, clast-supported conglomerates, with decimeter to meter thick, sandy, and silty lenses, indicating a fluvially dominated alluvial fan environment (Chang et al., 2008; Zhang et al., 2012). The top section, which is covered by Pleistocene strata (Qp_1), is ~4 m thick and mainly composed of gray gravel layers mixed with thin siltstones that represent fluvial/alluvial facies and a floodplain environment (Chang et al., 2008). The top of the eroded Cenozoic strata forms an angular unconformity which indicates the occurrence of significant uplift and erosion of the Qinghai Nan Shan since the late Miocene (Chang et al., 2008).

3 Sampling and analytical methods

Field samples were collected along the Chaka section following the locations in Zhang et al. (2012). A total of 39 sediment samples were obtained with 20–50 m intervals (CK01 to CK39, CK is the abbreviation of Chaka;

Fig. 2) and two loess samples (CK40 and CK41) were collected from the alluvial fan surface above the Neogene section (Table S1). Fifteen samples (CK01-15) were dated between ~9.2 and 6.0 Ma, and 24 samples (CK16-39) were dated between ~6.0 and ~2.5 Ma (Zhang et al., 2012). A 50 mg portion of each bulk sample was completely digested in a concentrated HF-$HClO_4$-HNO_3 solution in closed Teflon beakers on a hot plate set at 180–200 ℃. The samples were then dried, taken up in concentrated HNO_3, and dried once again. After this step, they were dissolved in 6 N HCl in closed Savillex Teflon beakers on a hotplate, and then dried. Then, they were taken up in 2.5 N HCl and loaded onto quartz columns to extract the Sr and Nd by conventional ion-exchange chromatographic techniques. The Sr and Nd isotopic analyses were conducted on a Neptune MC-ICP-MS at the Institute of Geology and Geophysics, Chinese Academy of Sciences (Yang et al., 2011, 2014). Reference materials SRM 987 and La Jolla Nd standard were used for quality control of Sr and Nd, respectively. The SRM 987 run along with samples yielded a mean $^{87}Sr/^{86}Sr$ ratio of 0.710217 ± 14 (2σ), while the La Jolla Nd standard run during the same period yielded a mean $^{143}Nd/^{144}Nd$ ratio of 0.510837 ± 6 (2σ). Total blanks (acid digestion plus column chemistry) for Nd and Sr were checked by ICP-MS and found to be negligible compared to the Nd and Sr amounts loaded onto the columns (less than 260 pg and 40 pg for Nd and Sr, respectively). The measured $^{143}Nd/^{144}Nd$ ratios are expressed as the fractional deviation in parts per 10^4 (units) from $^{143}Nd/^{144}Nd$ in a Chondritic Uniform Reservoir (CHUR), as measured at the present day:

$$\varepsilon_{Nd}(0) = [(^{143}Nd/^{144}Nd)_s / I_{CHUR}(0) - 1] \times 10^4 \tag{1}$$

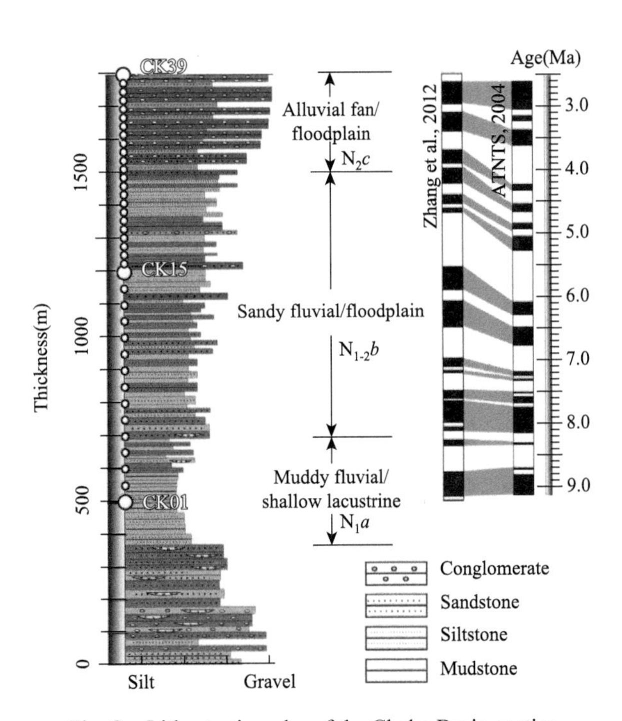

Fig. 2 Lithostratigraphy of the Chaka Basin section (modified from Zhang et al., 2012).
Samples CK01 to CK39 were collected along the section

where $(^{143}Nd/^{144}Nd)_s$ is the present-day ratio measured in the sample and I_{CHUR} (0) is the $^{143}Nd/^{144}Nd$ in the CHUR reference reservoir at present (I_{CHUR} (0) = 0.512638; Jacobsen and Wasserburg, 1980).

The trace element and REE concentrations were determined following Chinese Standard GB/T 14506.30—2010, using a Thermo Scientific Element High Resolution ICP-MS at the Beijing Research Institute of Uranium Geology, China.

4 Results

As shown in Table S1, the ε_{Nd} values of the bulk of samples from the Chaka Basin vary slightly from −15.6 to −11.6, with a mean value of −13.8. Interestingly, the samples above ~1300 m in this section [<5.5 Ma; Fig. 3 (a)] have relatively higher ε_{Nd} values than those of the older samples; the average values are −13.2 and −14.4, respectively [Fig. 3 (a)]. However, the $^{87}Sr/^{86}Sr$ ratios of the samples are distributed between 0.71441 and 0.71844 with a mean value of 0.71628. Unlike the ε_{Nd}, there is no obvious large difference across the whole section [Fig.3(b)]. Instead, the fluctuation of $^{87}Sr/^{86}Sr$ ratios shows an obviously more scattered pattern (0.71441–0.71844, <~1500 m) for the >4.5 Ma sample than those of <4.5 Ma (0.71548–0.71691). A similar pattern is also observed for the dispersion of Rb/Sr values, which are 0.43–0.79 with a mean of 0.58 [Fig. 3 (c); Table S1].

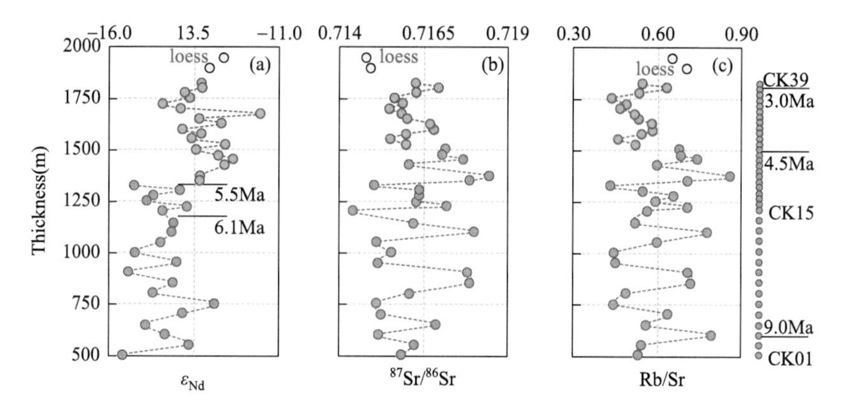

Fig. 3 Characteristics of Nd isotopes (a), Sr (b) isotopes, and Rb/Sr ratios (c) of the Neogene Chaka Basin sediments. Note change of geochemical signals ~5.4–4.5 Ma.

Trace elements in the samples are shown in Table S2. Considerable variations can be observed. Similarly, compared with the >4.5 Ma samples, the <4.5 Ma samples usually feature lower content of Li, Be, Co, Ni, and Sc, among others. REE concentrations are displayed in Fig. 4. One prominent feature can be distinguished: REE abundances are very constant from ~9.2 Ma to 4.5 Ma, they decrease from 4.5 Ma to 3.5 Ma, then increase from 3.5 Ma. However, REE contents at 3.5 Ma are still lower than those at >4.5 Ma.

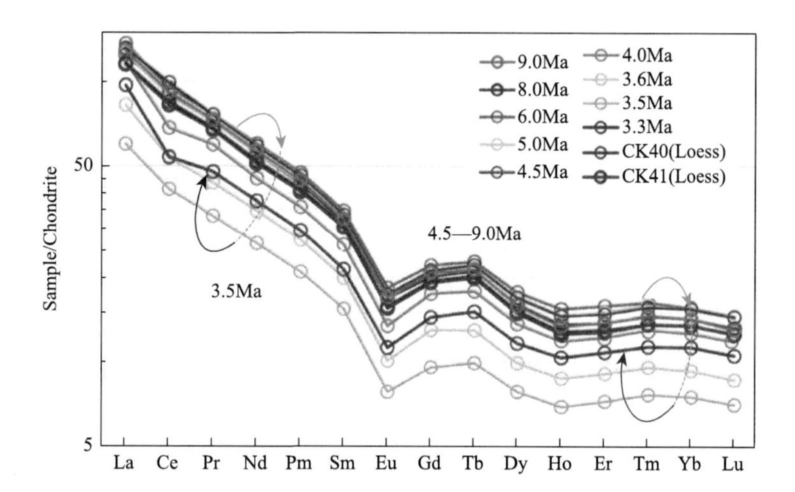

Fig. 4 Chondrite-normalized REE diagram for the Chaka sediments, showing very stable REE abundances from 9.0 Ma to 4.5 Ma, a decrease until 3.5 Ma, and then an increase at ~3.3 Ma. The two loess samples (CK40-41) show a similar pattern of abundance to the older Chaka section samples.

5 Discussion

5.1 Characteristics of ε_{Nd}, $^{87}Sr/^{86}Sr$ ratios and their implications

The averaged ε_{Nd} value increased from −14.4 to −13.2 after ~5.5 Ma [Fig. 3 (a)]. Because the ages of our studied section are young (<12 Ma) compared with the long half-life of ^{147}Sm decay ($T_{1/2} = 1.06\times10^{11}$; Sun, 2005), the variations in ε_{Nd} values in sedimentary rocks can be mainly attributed to a change in source materials. We infer that the variations of the ε_{Nd} values within the Chaka section are the result of a change in source material from the Ela Shan in the south to the Qinghai Nan Shan in the north. This finding is consistent with the available magnetostratigraphic results that indicate that the Qinghai Nan Shan began to be uplifted and provided sediments into the Chaka Basin around 6.1 Ma (Zhang et al., 2012). The Nd isotopic composition of the detrital sediments is commonly suggested to be the result of mechanical mixing of the older crust and younger detrital input; the latter usually has lower Nd isotopic composition (Kessarkar et al., 2003; Henry et al., 1997). Thus, we speculate that the higher ε_{Nd} values after 5.5 Ma are possibly due to the inputs of the relatively older crust materials in response to the Qinghai Nan Shan uplift, which may further imply that the crust materials in the Ela Shan are younger than those of the Qinghai Nan Shan.

Since Sr is more mobile than Nd, it can be released to solution, leading to the depletion of its abundance in the residual solids relative to the original rock (Singh et al., 2008). Additionally, Sr isotopic composition can be strongly affected by the processes of weathering, transportation (mineral sorting), and deposition (Faure, 1986; Sun, 2005; Tütken et al., 2002). Therefore, the variations in $^{87}Sr/^{86}Sr$ in sediments are often attributed to processes occurring during river transport (Singh et al., 2008). Within the Chaka section, we observed no evident difference in $^{87}Sr/^{86}Sr$ across the whole section, but larger fluctuations of $^{87}Sr/^{86}Sr$ ratios exist, for sample at >4.5 Ma [Fig. 3 (b)]. This large variation of $^{87}Sr/^{86}Sr$ ratios implies very complicated geochemical processes, most likely due to weathering and transportation under the depositional environments of shallow lacustrine and meandering river systems (Zhang et al., 2012). In other words, the complicated long-distance transportation and deposition processes from the Ela Shan to the south led to the large fluctuation of $^{87}Sr/^{86}Sr$ ratios before 4.5 Ma.

However, after 4.5 Ma, because the sediments in the section were mainly derived from the proximal Qinghai Nan Shan (Zhang et al., 2012), a simple sedimentary environment, the relatively shorter transport distance and residual time would all contribute to less variation in the $^{87}Sr/^{86}Sr$ ratio. However, the difference in Sr isotopes before and after 4.5 Ma is quite subtle, with average values of 0.71608 and 0.71638, respectively. This $^{87}Sr/^{86}Sr$ ratio feature thus indicates that the basement rocks in the Qinghai Nan Shan and the Ela Shan may have very similar $^{87}Sr/^{86}Sr$ ratios. Since the $^{87}Sr/^{86}Sr$ ratios were mainly determined from the bulk samples, Nd isotopic compositions can be related to the source rock, whereas Sr isotopic compositions are usually associated with many other complicated factors (such as the source rock, grain size, and chemical weathering; Rao et al., 2014).

It has been argued that the ratios of Rb/Sr in sediments reflect the degree of source-rock weathering (McLennan et al., 1993; Bhuiyan et al., 2011). Rb/Sr ratios of the Chaka samples are in the 0.43–0.79 range, with a mean of 0.58 [Figs. 3 (c) and 5; Table S1]. These values are higher than that of the average upper continental crust (0.32), but lower than the average post-Archean Australian shale (0.80; McLennan et al., 1993). Larger variations in Rb/Sr ratios, along with that of the $^{87}Sr/^{86}Sr$ ratio for the >4.5 Ma samples therefore indicate different degrees of weathering in the Ela Shan than in the Qinghai Nan Shan. The consequent transport, deposition, and weathering processes during landscape evolution led to the distinct $^{87}Sr/^{86}Sr$, Rb/Sr ratio fluctuation feature before and after ~4.5 Ma.

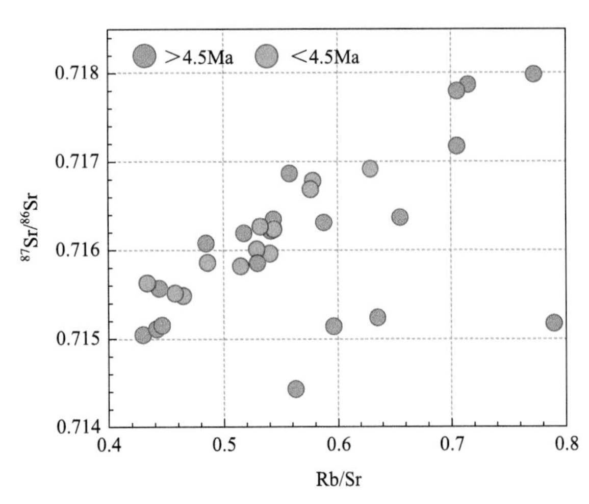

Fig. 5 Relationship between Rb/Sr and $^{87}Sr/^{86}Sr$ ratios in all samples.

Our present results of ε_{Nd} and the ratios of $^{87}Sr/^{86}Sr$ and Rb/Sr support the previous findings from lithoface, paleocurrent, and structural data (Zhang et al., 2012). However, there exists an ~0.6 Myr delay in the shift of ε_{Nd} relative to the uplift of the Qinghan Nan Shan, and a ~1.6 Myr delay in the shift of $^{87}Sr/^{86}Sr$ and Rb/Sr. These ~0.6–1.6 Myr delayed responses could be explained by a time lag between the initiation of the deformation at 6.1 Ma and the beginning of basement granite rock exposure at ~5.5 Ma and the completed basement exposure at ~4.5 Ma. It may have taken ~1.6 Myr to erode away the sediments on the top of the Qinghai Nan Shan [Fig. 3 (a)].

5.2 Insights from trace elements and REEs

Trace elements in the samples at <4.5 Ma are typically lower compared to those in the older samples. For example, the average Li content prior to 4.5 Ma (<1500 m) is 45.7 ppm, whereas that in the samples younger than 4.5 Ma (>1500 m) is 24.2 ppm. This feature implies that the trace elements in the rocks of the Qinghai Nan Shan are more depleted than those of the Ela Shan. The plot of La/Th against Hf is often used to discriminate between different source compositions (Floyd and Leveridge, 1987). In Fig. 6 (a), most of the Chaka section data are grouped around the felsic source; they tend to cluster around the average composition of the upper continental crust [Fig. 6 (a)]. Similarly, in the Co/Th vs. La/Sc diagram [Fig. 4 (b)], all data show low and constant Co/Th ratios with an average of 1.05, but relatively high and variable La/Sc ratios (3.2–6.0), suggesting that the Ela Shan and the Qinghai Nan Shan are dominated by felsic source rocks (McLennan and Taylor, 1984).

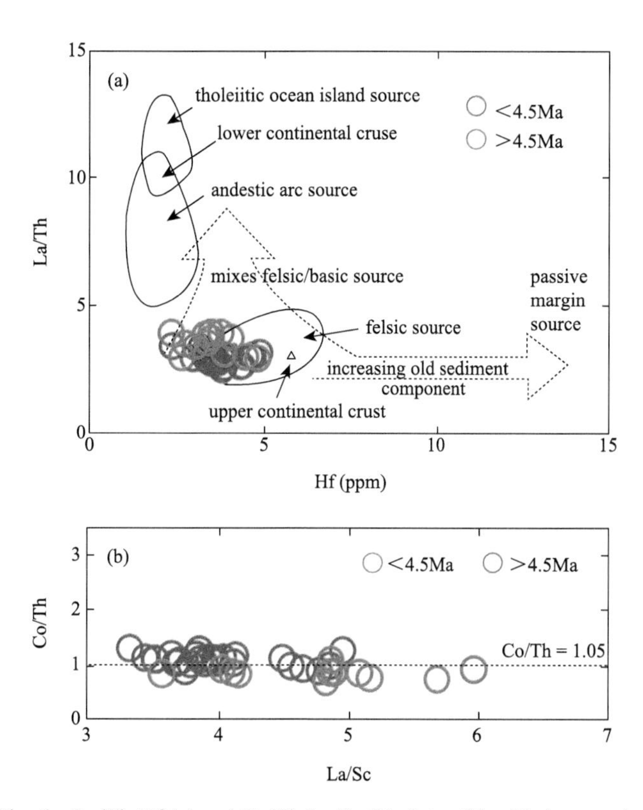

Fig. 6 La/Th-Hf (a) and Co/Th-La/Sc (b) plots of the Chaka samples.

All of the Chaka section samples (except for the CK40-41 loess samples) show higher Th/Sc ratios (0.8–1.3) and similar La/Sc ratios (2.0–5.0; Table S2) than the upper continental crust, suggesting a granite to granodiorite source. In the plot for La, Th, and Sc (Fig. 7), all sediments are plotted in silicic rocks between the granite and granodiorite compositions (as found in Bhatia and Crook, 1986; Girty et al., 1993; Singh, 2009). This may imply that the sediments were from granodiorite rocks in the Ela Shan before 4.5 Ma; however, increasing granite supply from the Qinghai Nan Shan since 4.5 Ma is consistent with the clast count analysis by Zhang et al. (2012).

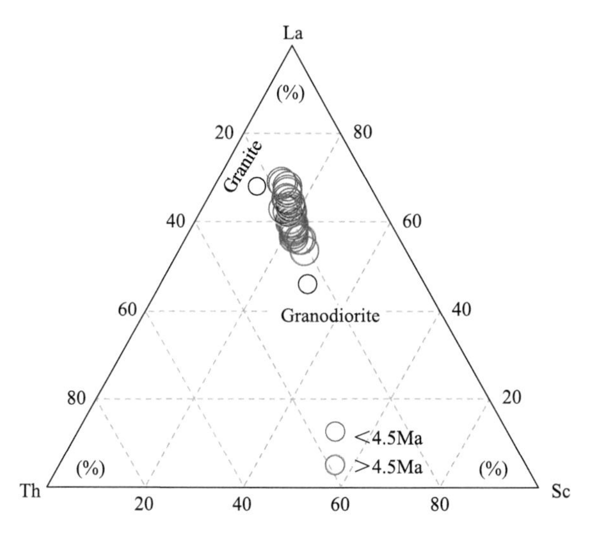

Fig. 7 Ternary discriminatory Sc-La-Th plot illustrating the position of the Chaka samples.

REE concentrations of the Chaka samples are listed in Table S2 and shown in Fig. 4. REE patterns are characterized by light REE enrichment and significant, negative Eu anomalies ($Eu/Eu^* = 0.61–0.68$), indicating that the main sedimentary source was continental crustal rock. All of the samples have relatively similar chondrite-normalized REE patterns, indicating that the REE geochemical characteristics of the Qinghai Nan Shan and the Ela Shan are similar. This again agrees with that indicated by the little variation in Nd and Sr isotopes and trace element geochemistry (Figs.3, 5 and 7). Similar to the features of Nd and Sr isotopes, REE abundances are relatively constant from ~9.2 Ma to 4.5 Ma, then decrease from 4.5 Ma to 3.5 Ma. The consistent

REE patterns from 9.2 Ma to 4.5 Ma reflects the REE characteristics of the source materials from the Ela Shan in the south. The decreasing trend of REE content between 4.5 Ma and 3.5 Ma is attributed to the input of the Qinghai Nan Shan materials.

Fractionation of Ce was previously known to occur during weathering and sedimentary processes (Borges et al., 2008). In the early stages, negative Ce anomalies are observed in weathering products (Braun et al., 1998). Positive Ce anomalies appear in intensely weathered lateritic profiles where soluble Ce^{3+} oxidizes to insoluble, thermodynamically stable Ce^{4+} and accumulates in secondary cerianite, Ce (IV) O_2 (Braun et al., 1998; Pan and Stauffer, 2000). Slightly negative Ce anomalies ($Ce/Ce^* = 0.74–0.99$; av. = 0.90; Fig. 8; Table S2) are observed in all Chaka samples. This negative Ce anomaly suggests that the chemical weathering around the Chaka Basin since the late Miocene through the Pliocene has not been very intense. The fluctuation in Ce anomalies reveals that there is a very stable trend from ~9 Ma to 4.5 Ma, but a more negative and remarkable scattered pattern after 4.5 Ma. Weathering from ~9 Ma to 4.5 Ma was possibly relatively more intense and stable than that of 4.5–3.0 Ma, similar to the observed worldwide increase in erosion rates and grain size due to climatic changes (Zhang et al., 2001). Thus, we speculate that Ce anomalies within the Chaka Basin were driven both by the change in sediment provenances and weathering intensity due to climate change from the late Miocene onwards. Further, a relatively more intensive weathering regime is usually favorable for the generation of much finer sediment grains and consequent deposition into the basin. During the formation of the Chaka Basin, finer sediments >4.5 Ma were derived from the distal Ela Shan. Thus, the siltstones and silty sandstones >4.5 Ma have less negative Ce anomalies than those in the younger coarse particles. Our Ce anomaly results thus agree well with the increase in grain sizes, indicating lithoface changes from shallow lake, meandering river, and floodplain sedimentary environments before 4.5 Ma into a dominated alluvial fan adjacent to the Qinghai Nan Shan (Fig. 2; Zhang et al., 2012).

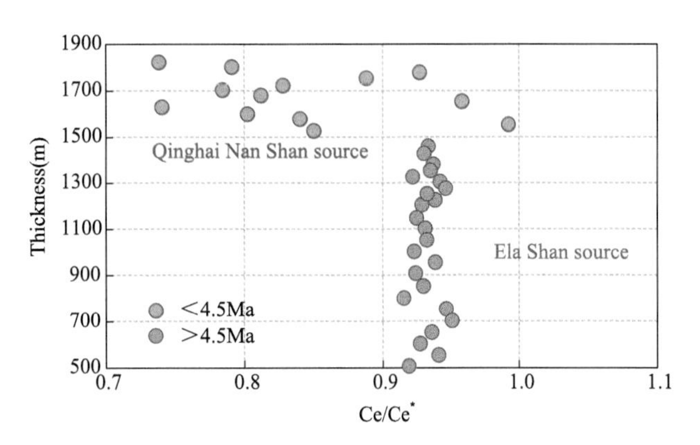

Fig. 8 Variation in Ce anomaly in the Chaka samples. Clearly different anomalies can be distinguished between the Ela Shan and Qinghai Nan Shan sources.

5.3 Possible loess input since the late Pliocene

Except for being relatively more negative, the fluctuation in Ce anomalies after 4.5 Ma is also much more scattered than what could be expected if all of the sediments were only from the Qinghai Nan Shan. As we observed, REE concentrations of the sediments from the Ela Shan >4.5 Ma are usually higher than those of <4.5 Ma sediments from the Qinghai Nan Shan (Table S2; Fig. 4). Interestingly, from 3.5 Ma onwards, REE contents increase to a level slightly lower than those >4.5 Ma. The REE concentrations of the two loess samples (CK40-41) on the alluvial fan are similar to those of the lacustrine sediments from 9.2 to 4.5 Ma. Therefore, we speculate that the increasing trend of REE concentrations since 3.5 Ma may be attributed to the input of the reworked fine grains with higher REE concentrations from the exposed underlying lacustrine sediments.

CK40 and CK41 are characterized by ΣREE enrichment (195 ppm and 191 ppm, respectively), which is in agreement with previous loess studies across central Asia (Taylor et al., 1983; Liu et al., 1993; Gallet et al., 1996; Rao et al., 2014; Sun et al., 2020). However, the ε_{Nd} values of these two loess samples (−12.6/−13.1, respectively) are much more negative than the results from the Chinese Loess Plateau, the Central Asian Orogen, the Northern Tibetan Plateau, and adjacent deserts (Sun et al., 2020), but they do overlap with the data from the Ela Shan and

Qinghai Nan Shan sediments (Fig. 9). Available ε_{Nd} values of the bedrocks west of the Chaka Basin mostly range from −12 to −18, with some negative values as low as approximately −28 (yellow ellipse in Fig. 1; Wan et al., 2006). We thus suggest that the loess within the Chaka Basin was mainly reworked from the nearby Ela Shan/Qinghai Nan Shan and the exposed Neogene Basin sediments.

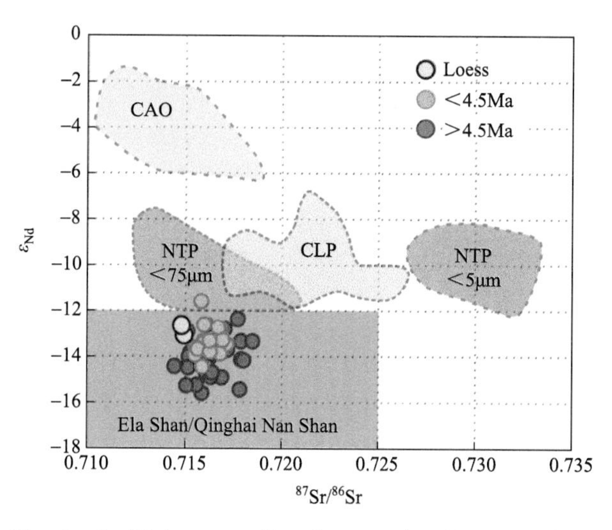

Fig. 9 Sr-Nd isotopes of aeolian samples from our Chaka section (this study), the northern Tibetan Plateau (NTP), Chinese Loess Plateau (CLP) and Central Asian Orogen (CAO) (modified after Sun et al., 2020).

The bottom left blue rectangle marks the ε_{Nd} concentrations of the potential source region in Fig. 1 (data are from Wan et al., 2006). (For interpretation of the references to colour in this figure legend, the reader is referred to the web version of this article.)

5.4 Landscape evolution around the Chaka Basin

Results of the Nd and Sr isotopes and REE data from the Neogene Chaka section enable us to reconstruct the landscape evolution around the Chaka Basin since ~11 Ma. The results can be summarized as follows, (1) The basin sediments were initially deposited at 11 Ma (Zhang et al., 2012) and continued until 5.5 Ma [Fig. 10 (a)]. During this time, the Chaka Basin was connected with the Lake Qinghai Basin to the north (Fu et al., 2013). The sediments before 5.5 Ma are mainly from granodiorite rock and are characterized by the relatively low ε_{Nd} values and scattered $^{87}Sr/^{86}Sr$ values [Fig. 3(a) and (b)]. Since the sediments were mainly from the Ela Shan to the south, the REE abundance remained stable during this period (Fig. 4) and the geochemical signatures are highly differentiated due to the complicated processes taking place during weathering, residence, and transport. (2) Basin sediments after 5.5 Ma had high ε_{Nd} values and very uniform $^{87}Sr/^{86}Sr$ values. This indicates that the sedimentary environments were relatively simple, since the sediments were transported within shorter distances from the Qinghai Nan Shan to the north. The Qinghai Nan Shan was initially uplifted and exhumed at ~6.1 Ma (Zhang et al., 2012), but began to provide granite sediments into the basin at 4.5 Ma. The growth and uplift of the Qinghai Nan Shan started to separate the Lake Qinghai and Chaka Basins [Fig. 10 (b)], and propagated along the strike of the Qinghai Nan Shan (Su et al., 2017). (3) The REE abundances began to decrease gradually until 3.5 Ma and then increase to a level lower than those of the >4.5 Ma samples. This represents the mixing and input of additional loess materials due to cooling during the Northern Hemisphere Glaciation. Uplift of the Qinghai Nan Shan raised the Neogene lacustrine sediments towards the surface and aided the reworking of windblown fine grains onto the alluvial fans in the pediment of the Qinghai Nan Shan [Fig. 10 (c)].

Late Miocene formation of the Chaka Basin, and uplift of the Qinghai Nan Shan mirror the overall feature of landscape evolution around the northeastern Tibetan Plateau. It has been suggested that south of the Chaka-Gonghe Basin, the Gonghe Nan Shan (Heka Shan) was uplifted at ~10–7 Ma (Craddock et al., 2011), separating the basin complex into the Gonghe and Guinan Basins. Approximately 180 km east of the Chaka section, the upward growth of Laji Shan at ~8 Ma began to confine the Guide Basin into an intermountain basin, with the exception of the Xining Basin to the north (Fang et al., 2005; Liu et al., 2013; Lease et al., 2007). Approximately 300 km further east, the uplift of the Jishi Shan started to break apart the Xunhua-Linxia Basin complex, and gradually trap the summer monsoon moisture into the Linxia Basin (Hough et al., 2011). The Miocene uplift of the range and formation of the basin was possibly due to the middle Miocene reorganization of deformation along the northeastern Tibetan Plateau (Lease et al., 2011), driven by the convection removal of mantle lithosphere from beneath central Tibet (Molnar and Stock, 2009).

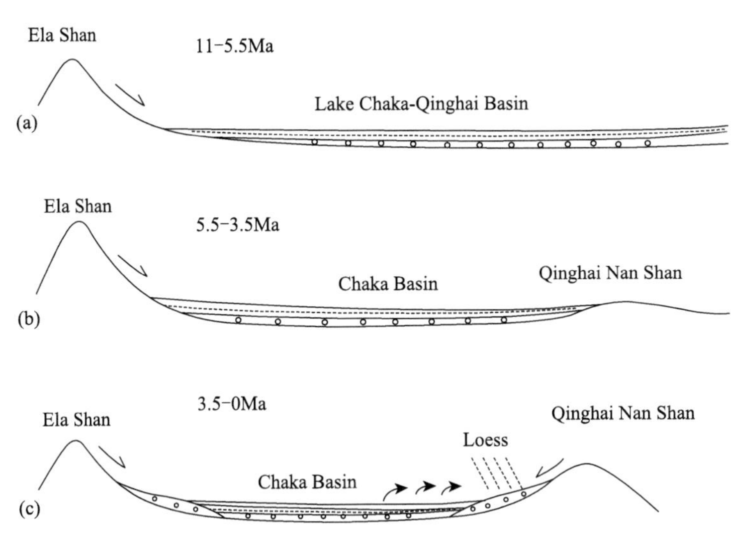

Fig. 10 Landscape evolution of the Chaka Basin since late Miocene.

(a) Sediments were eroded from the Ela Shan and deposited into the Chaka-Gonghe Basin; (b) from 5.5 to 3.5 Ma, growth of the Qinghai Nan Shan provided more and more sediments into the basin; (c) after 3.5 Ma, aeolian loess and reworked lacustrine fine grains were deposited onto the floodplains near the pediment of the Qinghai Nan Shan.

6 Conclusions

Temporal variations in Nd isotopic composition in the Chaka Basin indicate a change of provenance ~5.5 Myr, which is consistent with previous sedimentary evidence. The subtle Nd and Sr isotopic variations imply that the Qinghai Nan Shan and the Ela Shan have similar geochemical characteristics. The fluctuation of trace elements and REE concentrations further indicate that the Qinghai Nan Shan is more depleted in these elements than the Ela Shan is. We suggest a change in provenance from the Ela Shan to the Qinghai Nan Shan 5.5–4.5 Ma. Uplifted pediments in front of the Qinghai Nan Shan since 3.5 Ma began to accumulate aeolian loess and reworked wind-blown lacustrine silts. We infer that the multiple stages of landscape evolution of the Chaka Basin in NE Tibet was a result of late Cenozoic dynamic driving of the Tibetan Plateau.

Acknowledgments

We gratefully acknowledge Dr. Feng Cheng and two reviewers for the comments to improve our paper significantly. This work was supported by the NSF of China (No. 41673023; 41622204), the National Key Research and Development Program of China (No. 2017YFC0602704), Projects from the China Geological Survey (No. DD20190172; DD20201115), and the Second Tibetan Plateau Scientific Expedition and Research (STEP) (2019QZKK0704).

Appendix A. supplementary data

Supplementary data to this article can be found online at https: //doi.org/10.1016/j.palaeo.2020.109778.

References

An, Z., Kutzbach, J.E., Prell, W.L., Porter, S.C., 2001. Evolution of Asian monsoons and phased uplift of the Himalayan Tibetan plateau since late Miocene times. Nature 411, 62-66.

Bhatia, M.R., Crook, K.A.W., 1986. Trace element characteristics of graywackes and tectonic setting discrimination of sedimentary basins. Contrib. Mineral. Petr.92, 181-193.

Bhuiyan, M.A., Rahman, H., Dampare, M.J.J., Suzuki, S., 2011. Provenance, tectonics and source weathering of modern fluvial sediments of the Brahmaputra-Jamuna River, Bangladesh: Inference from geochemistry. J. Geochem. Explor.111 (3), 113-137.

Borg, L.E., Banner, J.L., 1996. Neodymium and strontium isotopic constraints on soil sources in Barbados, West Indies. Geochim. Cosmochim. Acta 60,

4193-4206.

Borges, J., Huh, B., Moon, Y., Noh, H., 2008. Provenance and weathering control on river bed sediments of the eastern Tibetan Plateau and the Russian Far East. Chem. Geol. 254, 52-72.

Braun, J.-J., Viers, J., Dupré, B., Polve, M., Ndam, J., Muller, J.-P., 1998. Solid/liquid REE fractionation in the lateritic system of Goyoum, East Cameroon: the implication for the present dynamics of the soil covers of the humid tropical regions. Geochim. Cosmochim. Acta 62 (2), 273-299.

Chang, Hong, Zhisheng, An, Wu, Feng, 2008. Major element records from the Qinghai Lake basin-Gonghe basin sediments indicating the Nanshan, Qinghai. Uplift Event 28 (5), 822-830.

Cheng, F., Fu, S., Jolivet, M., Zhang, C., Guo, Z., 2016. Source to sink relation between the Eastern Kunlun Range and the Qaidam Basin, northern Tibetan Plateau, during the Cenozoic. Geol. Soc. Am. Bull.128, 258-283.

Cheng, F., Garzione, C., Jolivet, M., Wang, W., Dong, J., Richter, F., Guo, Z., 2019. Provenance analysis of the Yumen Basin and northern Qilian Shan: Implications for the pre-collisional paleogeography in the NE Tibetan plateau and eastern termination of Altyn Tagh fault. Gondwana Res.65, 156-171.

Craddock, W., Kirby, E., Zhang, H., 2011. Late Miocene-Pliocene range growth in the interior of the northeastern Tibetan Plateau. Lithosphere 3, 420-438.

Dettman, D., Fang, L., Garzione, X., Li, J., 2003. Uplift-driven climate change at 12 Ma: a long $\delta^{18}O$ record from the NE margin of the Tibetan plateau. Earth Planet. Sci. Lett. 214 (1-2), 267-277.

Ding, Z., Liu, T., Rutter, N.W., Yu, Z., Guo, Z., Zhu, R., 1995. Ice-volume forcing of East Asian Winter Monsoon variations in the past 800,000 years. Quat. Res.44, 149-159.

Fan, M., Dettman, D., Song, L., Fang, C., Garzione, C.N., 2007. Climatic variation in the Linxia basin, NE Tibetan Plateau, from 13.1 to 4.3Ma: the stable isotope record. Palaeogeogr. Palaeocl. Palaeoecol.247 (3-4), 313-328.

Fang, X., Zhao, Z., Li, J., Yan, M., Pan, B., Song, C., Dai, S., 2005. Magnetostratigraphy of the late Cenozoic Laojunmiao anticline in the northern Qilian Mountains and its implications for the northern Tibetan Plateau uplift. Sci. China Ser. D 48 (7), 1040-1051.

Fang, X., Zhang, W., Meng, Q., Gao, J., Wang, X., King, J., Song, C., Dai, S., Miao, Y., 2007. High-resolution magnetostratigraphy of the Neogene Huaitoutala section in the eastern Qaidam Basin on the NE Tibetan Plateau, Qinghai Province, China and its implication on tectonic uplift of the NE Tibetan Plateau. Earth Planet. Sci. Lett.258 (1-2), 293-306.

Faure, G., 1986. Isotope Geology. John Wiley&Sons, New York (589 pp).

Floyd, P.A., Leveridge, B.E., 1987. Tectonic environment of the Devonian Gramscatho bssin, South Cornwall: framework mode and geochemical evidence from turbiditic sandstones. J. Geol. Soc. Lond.144, 531-542.

Fu, C., An, Z., Qiang, X., Bloemendal, J., Song, Y., Chang, H., 2013. Magnetostratigraphic determination of the age of ancient Lake Qinghai, and record of the East Asian monsoon since 4.63 Ma. Geology 41, 875-878.

Gallet, S., Jahn, B.-M., Torii, M., 1996. Geochemical characterization of the Luochuan loess-paleosol sequence, China, and paleoclimatic implications. Chem. Geol.133, 67-88.

Garzione, C.N., Ikari, M.J., Basu, A.R., 2005. Source of Oligocene to Pliocene sedimentary rocks in the Linxia basin in northeastern Tibet from Nd isotopes: Implications for tectonic forcing of climate. Geol. Soc. Am. Bull.117, 1156-1166.

Girty, G.H., Barber, R.W., Knaack, C., 1993. REE, Th, and SC evidence for the depositional setting and source rock characteristics of the Quanz Hill chert, Sierra Nevada, California. In: Johnsson, M.J., Basu, A. (Eds.), Processes Controlling the Composition of Clastic Sediments. Geol. Sot. Am. Spec. Pap 284. pp.109-l19.

Goldstein, S.J., Jacobsen, S.B., 1988. Nd and Sr isotopic systematics of river water suspended material: implications for crustal evolution. Earth Planet. Sci. Lett.87, 249-265.

Henry, P., Deloule, E., Michard, A., 1997. The erosion of the Alps: Nd isotopic and geochemical constraints on the sources of the periAlpine molasses sediments. Earth Planet. Sci. Lett.146, 627-644.

Hough, B.G., Garzione, C.N., Wang, Z., Lease, R.O., Burbank, D.W., Yuan, D., 2011. Stable isotope evidence for topographic growth and basin segmentation: Implications for the evolution of the NE Tibetan Plateau. Geol. Soc. Am. Bull.123, 168-185.

Jacobsen, S.B., Wasserburg, G.J., 1980. Sm-Nd isotopic evolution of chondrites. Earth Planet. Sci. Lett.50 (1), 139-155.

Kessarkar, P., Rao, M., Ahmad, V.P., Babu, G.A., 2003. Clay minerals and Sr-Nd isotopes of the sediments along the western margin of India and their implication for sediment provenanc. Mar. Geol.202, 55-69.

Lease, R.O., Burbank, D.W., Gehrels, G.E., Wang, Z., Yuan, D., 2007. Signatures of mountain building: Detrital zircon U/Pb ages from northeastern Tibet. Geology 35, 239-242.

Lease, R.O., Burbank, D.W., Clark, M.K., Farley, K.A., Zheng, D.W., Zhang, H.P., 2011. Middle Miocene reorganization of deformation along the northeastern Tibetan Plateau. Geology 39, 359-362.

Liu, C.-Q., Masuda, A., Okada, A., Yabuki, S., Zhang, J., Fan, Z.-L., 1993. A geochemical study of loess and desert sand in northern China: Implications for continental crust weathering and composition. Chem. Geol.106, 359-374.

Liu, S., Zhang, G., Pan, F., Zhang, H., Wang, P., Wang, P., Wang, Y., 2013. Timing of Xunhua and Guide basin development and growth of the northeastern

Tibetan Plateau, China. Basin Res.25 (1), 74-96.

McLennan, S.M., Taylor, S.R., 1984. Geochemistry of Archean metasedimentary rocks from West Greenland. Geochim. Cosmochim. Acta 48, 1-13.

McLennan, S., Hemming, M., McDennial, S., Hanson, G.N., 1993. Geochemical approaches to sedimentation, provenance and tectonics. Geol. Soc. Am. Spec. Pap.284, 21-40.

Molnar, P., Stock, J.M., 2009. Slowing of India's convergence with Eurasia since 20 Ma and its implications for Tibetan mantle dynamics. Tectonics 28, TC3001.

Molnar, P., Tapponnier, P., 1975a. Cenozoic tectonics of Asia: effects of a continental collision. Science 189 (4201), 419-426.

Molnar, P., Tapponnier, P., 1975b. Cenozoic tectonics of Asia: effects of a continental collision. Science 189, 419-426.

Pan, Y., Stauffer, M.R., 2000. Cerium anomaly and Th/U fractionation in the 1.85 Ga Flin Flon Paleosol: clues from REE-and U-rich accessory minerals and implications for paleoatmospheric rec onstruction. Am. Mineral.85, 898-911.

Rao, W., Chen, B., Tan, J., Weise, H.B., Wang, Y., 2014. Nd-Sr isotopic and REE geochemical compositions of late Quaternary deposits in the desert-loess transition, north-Central China: Implications for their provenance and past wind systems. Q. Int. 334-335, 197-212.

Singh, P., 2009. Major, trace and REE geochemistry of the Ganga River sediments: Influence of provenance and sedimentary processes. Chem. Geol.266, 242-255.

Singh, S.K., Rai, S.K., Krishnaswami, S., 2008. Sr and Nd isotopes in river sediments from the Ganga Basin: Sediment provenance and spatial variability in physical erosio. J. Geophys. Res.113, F03006.

Su, Q., Xie, H., Yuan, D.-Y., Zhang, H.-P., 2017. Along-strike topographic variation of Qinghai Nanshan and its significance for landscape evolution in the northeastern Tibetan Plateau. J. Asian Earth Sci.147, 226-239.

Sun, J.M., 2005. Nd and Sr isotopic variations in Chinese eolian deposits during the past 8 Ma: implications for provenance change. Earth Planet. Sci. Lett.240, 454-466.

Sun, Y., Yan, Y., Nie, J., Li, G., Shi, Z., Qiang, X., Chang, H., An, Z., 2020. Source-to-sink fluctuations of Asian aeolian deposits since the late Oligocene. Earth Sci. Rev.200, 102963.

Tapponnier, P., Xu, Z., Roger, Q., Meyer, F., Arnaud, B., Wittlinger, N., Yang, J., 2001. Oblique stepwise rise and growth of the Tibet plateau. Science 294 (5547), 1671-1677.

Taylor, S.R., McLennan, S.M., McCulloch, M.T., 1983. Geochemistry of loess, continental crustal composition and crustal model ages. Geochim. Cosmochim. Acta 47, 1897-1905.

Tütken, T., Eisenhauer, A., Wiegand, B., Hansen, B.T., 2002. Glacial-interglacial cycles in Sr and Nd isotopic composition of Arctic marine sediments triggered by the Svalbard/Barents Sea ice Sheet. Mar. Geol.182, 351-372.

Wan, Y., Zhang, J., Yang, J., Xu, Z., 2006. Geochemistry of high-grade metamorphic rocks of the North Qaidam mountains and their geological significance. J. Asian Earth Sci. 28, 174-184.

Wang, W., Zheng, W., Zhang, P., Li, Q., Kirby, E., Yuan, D., Zheng, D., Liu, C., Wang, Z., Zhang, H., 2017. Expansion of the Tibetan Plateau during the Neogene. Nat. Commun.8.

Yang, Y.H., Chu, Z.Y., Wu, F.Y., Xie, L.W., Yang, J.H., 2011. Precise and accurate determination of Sm, Nd concentrations and Nd isotopic compositions in geological samples by MC-ICP-MS. J. Anal. Atomic Spectrom.26, 1237-1244.

Yang, Y.H., Wu, F.Y., Xie, L.W., Chu, Z.Y., Yang, J.H., 2014. Reinvestigation of doubly charged ion of heavy rare earth elements interferences on Sr isotopic analysis using multi-collector inductively coupled plasma mass spectrometry. Spectrochim. Acta B At. Spectrosc.97, 118-123.

Yuan, D.-Y., Ge, W.-P., Chen, Z.-W., Li, C.-Y., Wang, Z.-C., Zhang, H.-P., Zhang, P.-Z., Zheng, D.-W., Zheng, W.-J., Craddock, W., Dayem, H., Duvall, K.E., Hough, A.R., Lease, B.G., Champagnac, R.O., Burbank, J.-D., Clark, D.W., Farley, M.K., Garzione, K.A., Kirby, C.N., Molnar, E., Roe, G.H., 2013. The growth of northeastern Tibet and its relevance to large-scale continental geodynamics: a review of recent studies. Tectonics 32 (5), 1358-1370.

Zhang, P., Molnar, P., Downs, W.R., 2001. Increased sedimentation rates and grain sizes 2-4 Myr ago due to the influence of climate change on erosion rates. Nature 410, 891-897.

Zhang, H., Craddock, P., Lease, W.H., Wang, R.O., Yuan, W.-T., Zhang, D.-Y., Molnar, P.-Z., Zheng, P., Zheng, W.-J., 2012. Magnetostratigraphy of the Neogene Chaka basin and its implications for mountain building processes in the north-eastern Tibetan Plateau. Basin Res.24, 31-50.

Grain size of surface sediments in Selin Co (central Tibet) linked to water depth and offshore distance*

Can Wang, Hailei Wang, Gao Song and Mianping Zheng

MNR Key Laboratory of Saline Lake Resources and Environment, Institute of Mineral Resources, CAGS, Beijing 100037, China

Abstract Grain size of lake sediments is often measured in paleolimnological studies, especially investigations of past lake-level changes. The paleo-hydrologic implications of such measures, however, remain unclear. We explored the relationship between grain-size characteristics of surface sediments in Selin Co, central Tibet (Median Diameter (Md), fine component percentage, and the grain-size frequency distribution curve), and both water depth and offshore distance. Under the same river/runoff transport and wind conditions, the Md value of grain size displays a significant negative correlation with water depth ($r = -0.767$, $N = 22$, $P < 0.001$) and offshore distance ($r = -0.633$, $N = 22$, $P = 0.002$), whereas the percentage of grains <10 μm has a significant positive correlation with water depth ($r = 0.689$, $N = 22$, $P <$ 0.001) and offshore distance ($r = 0.673$, $N = 22$, $P < 0.001$). The percentage of grains <4 μm was also positively correlated with water depth and offshore distance ($r = 0.549$, $N = 22$, $P < 0.01$ for both). We recommend that the grain sizes transported by river/runoff or wind be identified and eliminated from consideration before employing the Md value and <10 μm component of grain size for lake level reconstruction. The modal size of ~10 μm in the grain-size distribution curve is not affected by river/runoff or wind transportation, and is a reliable proxy for past lake level reconstruction, with smaller modal sizes associated with larger offshore distance [$r = -0.577$, $N = 22$, $P = 0.006$), larger lake area ($r = -0.786$, $N = 7$, $P = 0.036$) and higher annual precipitation ($r = -0.784$, $N = 8$, $P = 0.021$ for Bange (station 55279) and $r = -0.769$, $N = 8$, $P = 0.026$ for Shenzha (station 55472)].

Keywords Selin Co; Grain-size of lake sediments; Frequency distribution curve; Modal size at about 10 μm; Offshore distance; Water depth

Introduction

The grain size of clastic sediments is frequently used for paleoenvironmental reconstructions. The measure provides direct information on the source of material, transport-deposition processes and the sedimentary environment (Friedman and Sanders, 1978; McCave, 1978; Poizot et al., 2006; An et al., 2012; Wang et al., 2016). The frequency distribution curve of grain size has been employed to distinguish sediment types, which usually have different modal distributions and typical modal sizes.

For example, fluvial deposits usually consist of two grain size components, with modal sizes of 200-400 μm and 10–15 μm (Bennett and Best, 1995; Kranck et al., 1996; Påsse 1997; Xiao et al., 2013); loess deposits have a short-suspension component with dominant modal sizes of 16–32 μm, and a longsuspension component with dominant modal sizes of 2–6 μm (Tsoar and Pye, 1987; Sun et al., 2002). Grainsize distribution curve statistics on 60 samples from two extreme paleoenvironments (hyper-arid and humid), as well as comparison of records, indicated that the 40 μm mode represents the aeolian component, the 10–70 μm fraction of grain size is a valid proxy for the East Asian Winter Monsoon, and that the 70–650 μm fraction represents the intensity of dust storms (An et al., 2012).

---

* 本文发表在：Journal of Paleolimnology, 2019, 61: 217-229

Grain sizes of lacustrine sediments, however, can be difficult to interpret because they are affected by numerous factors and their distribution curve is strongly polymodal. Early studies suggested that deep-water deposits have a near-symmetrical unimodal distribution, whereas deposits controlled by two or more dynamic factors display slight negative unimodal or bimodal distributions (Sun et al., 2001). Surface sediments of Daihai Lake, China, displayed bimodal or trimodal grain-size distributions (Wang et al., 1990; Peng et al., 2005). Comparisons of lake sediment grain-size distribution curves in different areas indicated there are as many as six modes (median sizes <1, 2–10, 10–70, 70–150, 150–700, and >700 μm) (Yin et al., 2008). A grain-size study in Daihai Lake revealed that typical lacustrine sediments contain five distinct unimodal grain-size distributions, representing five grain-size components (Xiao et al., 2013).

Median Diameter (Md), the diameter at the 50th percentile of the distribution curve, is particularly sensitive to hydrodynamic conditions. Higher Md values represent coarser sediments in stronger hydrodynamic conditions, often interpreted to reflect lower lake level (Sun et al., 2001; Liu et al., 2003). Some research, however, suggested an alternate explanation, i.e. that greater rainfall enhances soil erosion and increases the transport capacity of streams and rivers, leading to more and coarser clastic materials transported by rivers and subsequently deposited in the central part of the lake (Kashiwaya et al., 1987; Peng et al., 2005; Jiang et al., 2010).

The fine component of the grain-size distribution is usually regarded as the main component of deposits in the deepest area of the lake. Both the <4 μm component (Graham and Rea, 1980; Shen et al., 2007; Jiang et al., 2010) and the <10 μm component (Peng et al., 2005; Yin et al., 2008) have been used in paleoenvironmental reconstructions, but it remains unclear which size class is more sensitive to lake level changes.

In this study we examined grain size of surface lake sediments in Selin Co, central Tibet, to explore relationships between grain-size characteristics and both water depth and offshore distance. Our goal was to explore how grain-size measurements in lake sediments could be used for paleoenvironmental reconstruction.

Study site

Selin Co is a brackish lake in the middle of Tibet (88°32′–89°22′E, 31°32′–32°07′N), about 90 km west of Bange County in Naqu District (Fig. 1). Total ion concentration in the lake is 11.07 g/L. The hydrochemical type is sodium sulfate. It is now the largest lake in Tibet, with an area of 2197 km^2 and an elevation of 4552 m a.s.l. (Lv et al., 2003; Laba et al., 2011). Selin Co and adjacent Lake Bange Co lie in a Cenozoic fault basin in the middle segment of the Bangonghu-Nujiang suture. Selin Co lies in the lowest area of the basin and has a catchment area of 42194 km^2. The lake is surrounded by highlands 200–500 m above the lake surface, which are covered with lush grass. The lake can be divided into east and west sectors. The east part is much larger, but shallower, and maximum water depth in the lake center is generally <35 m. The west part is smaller and deeper. A site 1.26 km from the shore has a water depth of 23.6 m, but this increases sharply to a depth of 35.4 m at a site 2.55 km offshore. The lake has a thermocline at 23–29 m depth, and the water temperature in the hypolimnion is 6–6.5 ℃.

Selin Co is a terminal lake that is drained by several rivers, including Zajia Zangbu, Zagen Zangbu, Ali Zangbu and Boqu Zangbu. The 409-km-long Zajia Zangbu, which is the longest inland river in Tibet, enters at the north shore of the eastern lake, and is the main perennial river that supplies Selin Co with water and sediment from glacier-covered Geladandong mountain. The water depth of the river in the wet season is ~2.5 m, whereas in spring it is only ~0.5–1 m deep. Annual average flow is 26.7 m^3/s (Wang et al., 1998).

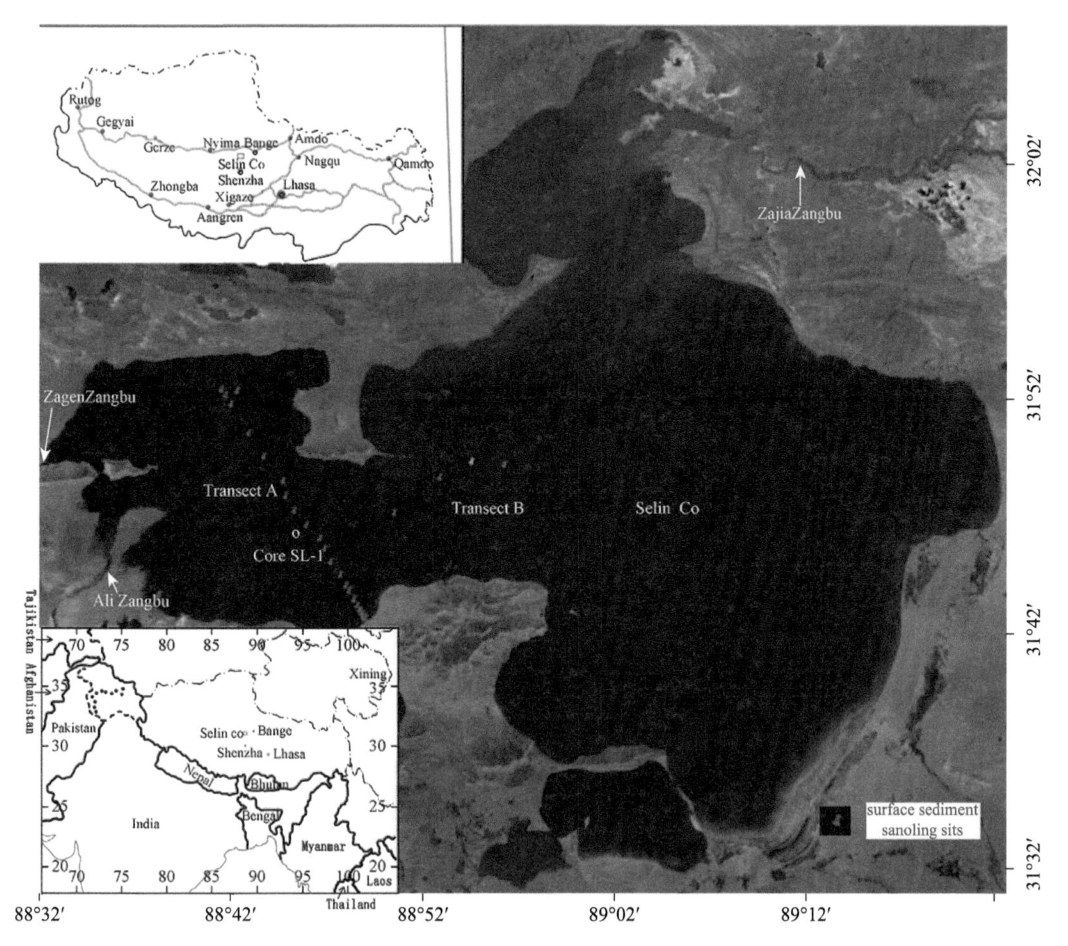

Fig. 1 Map showing the location of Selin Co and sediment sampling transects A and B in the western part of the lake.

Zagen Zangbu and Ali Zangbu enter at the west end of the lake, are intermittent rivers and generally supply Selin Co with water and sediments in the wet season, from June to October. Zagen Zangbu is shallow (~0.6 m) and the water is clear in the wet season. The annual average flow is small, and we estimated it as 5.6 m^3/s in August 2012. Ali Zangbu enters Selin Co in a small secondary lake of the western basin. It is shallower than Zagen Zangbu and the water is clear in the wet season. Boqu Zangbu was once an intermittent river that supplied Selin Co with water in the west, but it has dried completely.

The lake is in the semi-arid monsoon area of the Tibetan Plateau frigid zone, with very strong solar radiation. Annual precipitation is 290–321 mm and annual evaporation can be as high as 2176 mm. Annual mean air temperature is 0.8–1.0 ℃. The number of annual gale days (i.e. winds>62 km/hr) is 103–132 d (Dasang 2011). In winter, generally from December to March, precipitation falls mainly as snow rather than rain, given the very low air temperature. It does not, however, snow often. According to meteorological observations at the Selin-Bange Scientific Observation Station, between 2013 and 2015 it snowed <4 times a month, with a maximum snowfall of 5.7 mm.

Materials and methods

Surface sediments were collected with a dredge along two transects, A and B, in the western part of Selin Co (Fig. 1) in winter 2013, when the lake was covered with 30–40 cm of ice. Ice begins to form in late November and melts in early April. Transect A started at a site designated $A_0$1 (88°48′19.31″E, 31°42′14.45″N) and ended at site A18 (88°40′36.72″E, 31°52′5.93″N), with a total of 23 samples. Transect B started at site B1 (88°49′41.32″E, 31°46′42.85″N) and ended at B4 (88°55′31.21″E, 31°48′48.57″N), with a total of four samples. Water depth was measured from the ice surface using a metered rope with a weight. The offshore distance of

Fig. 2 The top 1 cm of sediment from each sample, (collected by a dredge) was carefully collected for grain-size analysis.

sampling sites was measured using maps in Google Earth and represents the shortest distance to the south or southwest shore of the peninsula. The top 1 cm of sediment from each grab sample was collected for grain-size analysis (Fig. 2). Snow samples on the ice were also collected along transect A to examine the aeolian component of the sediments.

Grain size analysis was conducted using a Mastersizer 2000 laser grain-size analyzer (Malvern Instruments, Malvern, UK) at the Key Laboratory of Saline Lake Resources and Environment, Ministry of Natural Resources, China. The detection range was 0.02–2000 μm. About 1 g of sample was boiled in 10 ml of 10% H_2O_2 for 10 min to remove organic matter. Then about 10 ml of 10% HCl was added for another 10 min to ensure that carbonate was completely removed. Next, the sample was rinsed in 1000 ml of distilled water for 24 h. A drop of 0.05 N $(NaPO_3)_6$ was then added after careful removal of the supernatant. The sample was dispersed for 10 min in an ultrasonic bath before measurement.

We compared the grain size of surface sediments in Selin Co with recent annual precipitation and lake area. Annual precipitation data are for the period 1973–2012 and come from Bange and Shenzha Counties, the two nearest counties to Selin Co. These two counties have a climate similar to the area around Selin Co, as they lie in the same Tibetan Plateau frigid zone. The lake area was calculated using Google Earth maps for the month of December, in years from 1988 to 2011. Historical grain-size data come from the top part of Core SL-1, taken in the center of the west lake basin of Selin Co, and span the period from 1973 to 2012, as determined by ^{137}Cs dating (Fig. 3). The largest ^{137}Cs activity peak in the core corresponds to the Chernobyl accident in 1986. In the Northern Hemisphere, the Chernobyl disaster was the largest nuclear leak of radionuclides from a plant, which were subsequently carried by the Westerlies to the Qinghai-Tibetan Plateau. The second largest ^{137}Cs activity

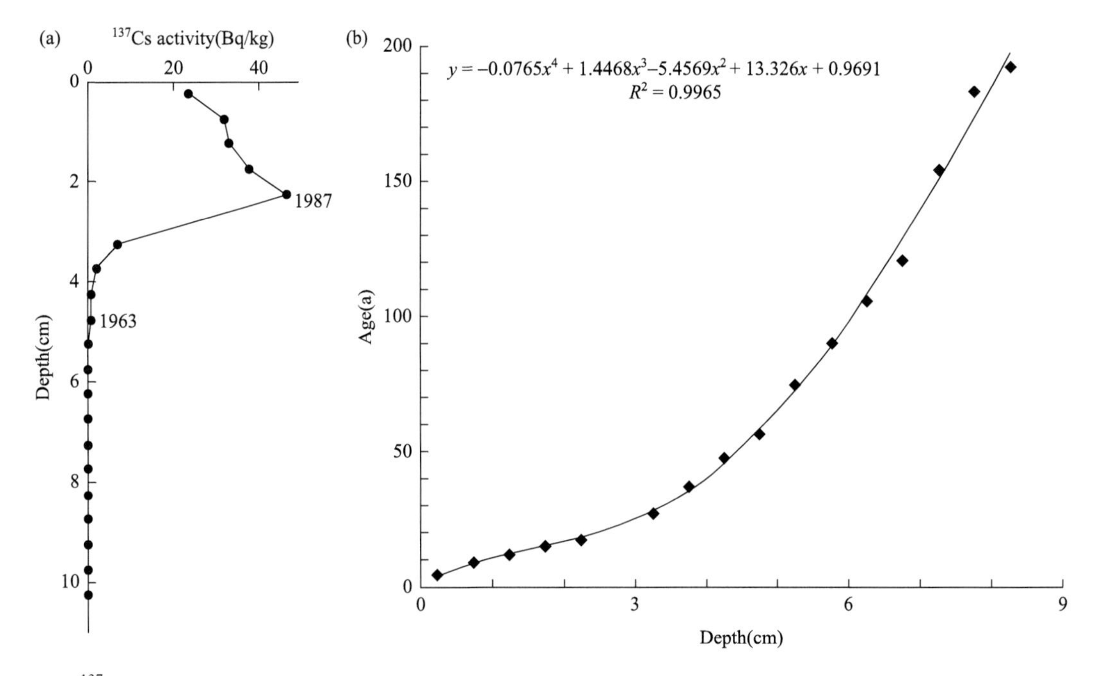

Fig. 3 ^{137}Cs activity and age-depth model for the top part of Core SL-1 from Selin Co (after Wang and Zheng 2014).

peak at a depth of 4.75 cm in the core corresponds to the time of greatest atmospheric nuclear weapons testing in 1963. See Wang and Zheng (2014) for further details on the core chronology. Relations between grain-size data, water depth, and offshore distance, were explored using Pearson correlation analysis in SPSS (IBM SPSS Statistics 19).

Results

Water depth and offshore distance at each sample site are shown in Table 1. Water depth at all sample sites ranged from 1.2 to 41.4 m, whereas offshore distance varied from 0.04 to 6.63 km. Fig. 4 shows a profile of the western lake floor, indicating the water depth and offshore distance of each sample site in the western lake. The lake deepens sharply within 2 km of the shore. Farther from shore, the lake bottom flattens, with a maximum water depth on transect A of 40.1 m.

Table 1 Water depth and offshore distance of each sample site

Sampling site	A_01	A_02	A_03	A_04	A_05	A1	A2	A3	A4	A5	A6	A7	A8	A9
Water depth (m)	1.2	2.6	3.4	7.5	14.9	23.6	25.5	35.4	38.2	40.1	37.4	39.6	28.9	36.4
Offshore distance (km)	0.04	0.13	0.29	0.52	0.85	1.26	1.78	2.55	3.09	4.05	5.37	6.52	6.63	5.87
Sampling site	A10	A11	A12	A13	A14	A15	A16	A17	A18	B1	B2	B3	B4	
Water depth (m)	35.2	30.6	30.3	37.2	35.3	31.4	30.4	35.4	32.6	41.4	38.5	35.9	36.4	
Offshore distance (km)	4.85	3.9	2.82	3.03	4.13	3.41	4.51	3.72	3.22	4.9	5.75	8.12	9.07	

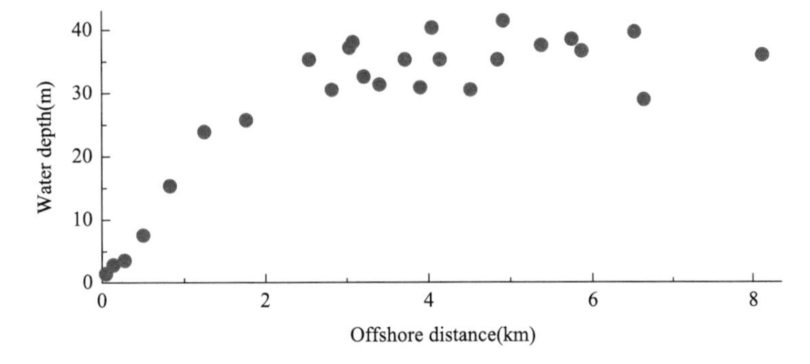

Fig. 4 Water depth and offshore distance of each site in Selin Co, showing the western lake floor.

Water depth at sites A_01–A_05 ranges from 1.2 to 14.9 m, and the offshore distance is only 0.04–0.85 km. Sediments at these sites are very dense and hard, or contain gravels with very thin, silty clay on the surface. The gravels are round and have high psephicity, with diameters >2 cm. They are clearly river sediments, indicating that these sites were at the lake shore in the past. The lake expanded rapidly in recent decades because of high rainfall and input of meltwater from glaciers and permafrost, as well as reduced evaporation (Zhuo et al., 2007). Site A1 is covered by much deeper water, 23.6 m, and sites A_01–A_05 must have been at the lake shore in recent decades, as there is sparse modern sediment deposited there. We eliminated these five sites in our analysis of lake sediments and water depth, and selected 22 samples, from sites A1–A18 and B1–B4, to investigate grain size. Sediments at these sites are mainly clay or silty clay, with the top 2–3 cm being tan and deeper deposits being dark gray or black. All the sediments are soft and hydrated. There are no diatoms in the sediments. The Md value of grain-size displays a significant negative correlation with water depth ($r = -0.767$, $N = 22$, $P < 0.0001$) and offshore distance ($r = -0.633$, $N = 22$, $P = 0.002$) (Fig. 5; Electronic Supplementary Material [ESM] Table 1).

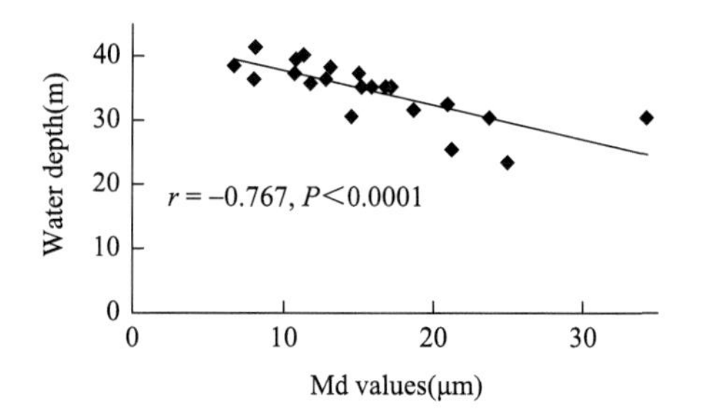

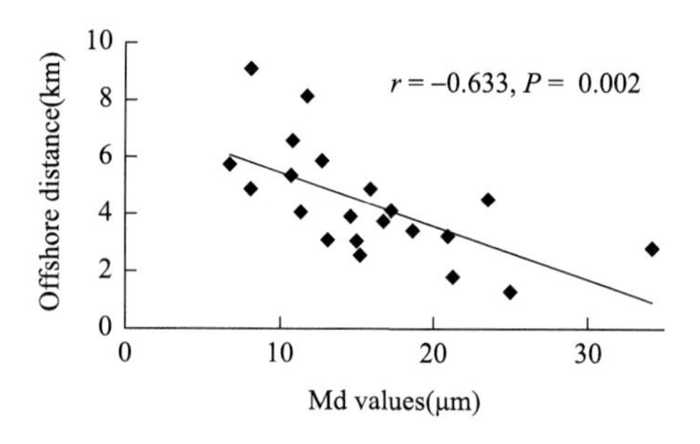

Fig. 5　Md value of grain size from surface sediment samples, plotted versus water depth (left) and offshore distance (right).

The percentage of <10 μm grain size has a more significant positive correlation with water depth ($r = 0.689$, $N = 22$, $P = 0.001$) and offshore distance ($r = 0.673$, $N = 22$, $P = 0.001$) than does the percentage of grains <4 μm ($r = 0.549$, $N = 22$, $P = 0.01$ for both water depth and offshore distance) (Fig. 6; ESM Table 1).

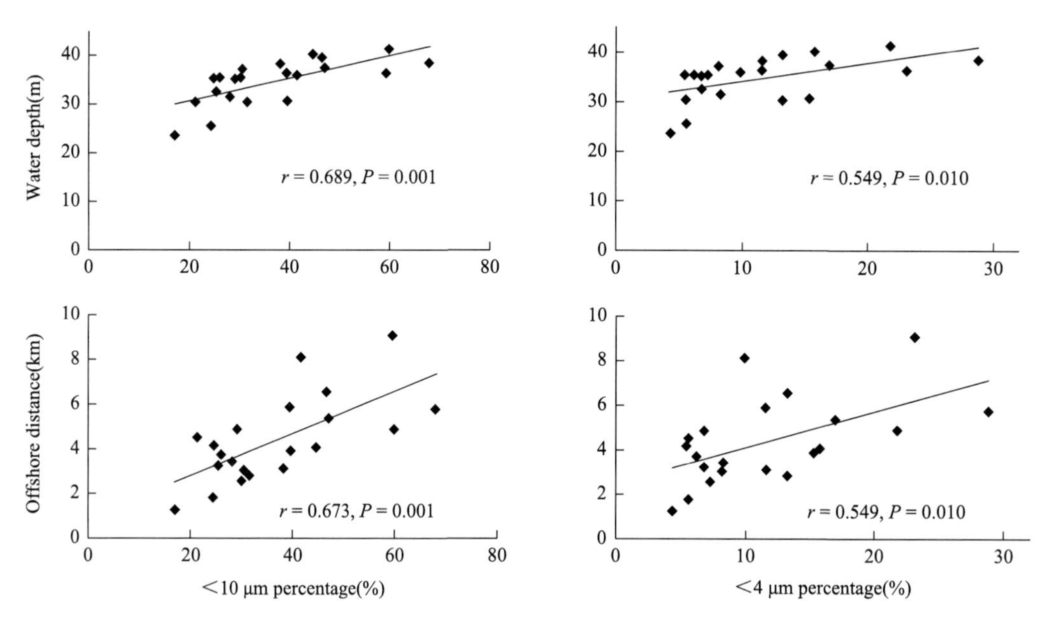

Fig. 6　Percentage of <10 μm (left) and percentage of <4 μm (right) grain sizes from surface sediment samples plotted versus water depth (upper) and offshore distance (lower).

Sample sites far from the shore show a typical bimodal distribution, with a primary modal size of about 10 μm and a secondary modal size of <1 μm [Fig. 7 (a)]. Sample sites with water depth <32.6 m or offshore distance <3.9 km (except site A16, with an offshore distance of 4.51 km) show a typical trimodal distribution, with a primary modal size of about 10 μm, a secondary coarse modal size of about 80 μm and a secondary finer modal size of <1 μm [Fig. 7 (b)].

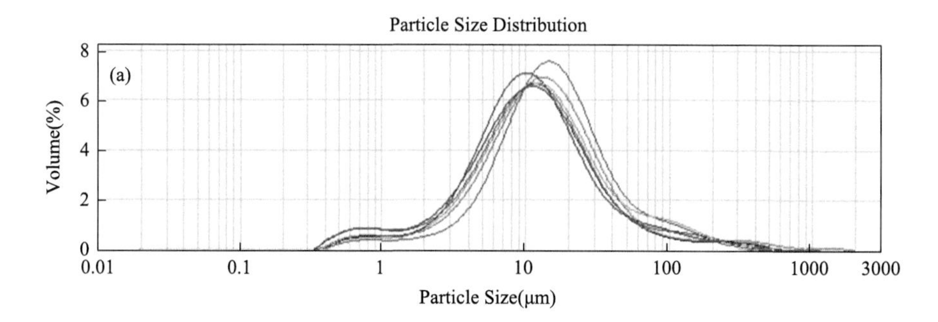

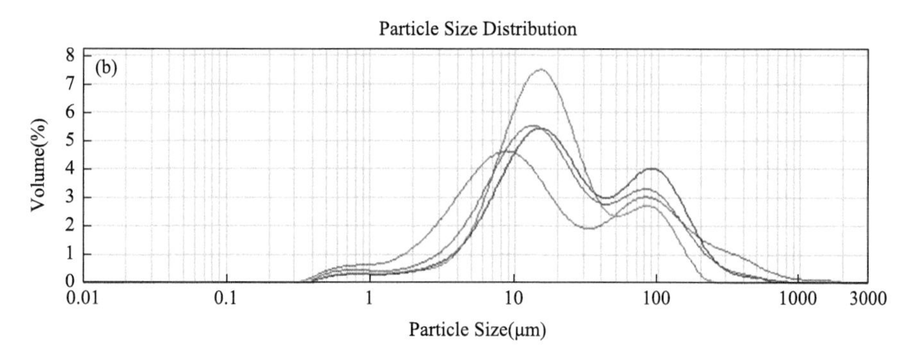

Fig. 7 Frequency distribution curve of grain size from sample sites that are far from the shore, showing a typical bimodal distribution (a) and from sample sites with water depth <32.6 m or offshore distance <3.9 km, showing a typical trimodal distribution (b).

The modal size at about 10 μm shows a weak negative correlation with water depth that is not significant at $P<0.05$ ($r=-0.412$, $N=22$, $P=0.064$) [Fig. 8 (a)], but a significant negative correlation with offshore distance ($r=-0.577$, $N=22$, $P=0.006$) [Fig. 8 (b); ESM Table 1].

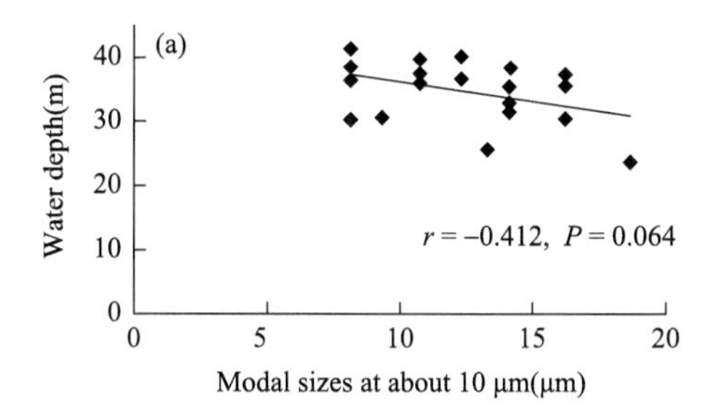

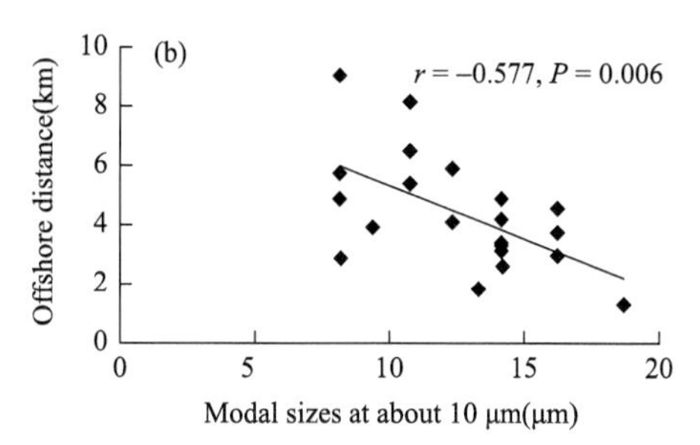

Fig. 8 The modal size at about 10 μm of surface sediment grain size in Selin Co plotted versus water depth (a) and offshore distance (b).

The lake area gradually enlarged from ~1787 km^2 in 1988 to ~2343 km^2 in 2011, indicating increasing lake level and greater offshore distances during this period. The modal size of ~10 μm shows a strong negative relationship with lake area ($r=-0.786$, $N=7$, $P=0.036$) [Fig. 9 (a), (b); ESM Table 1], but the Md correlation with lake area is not significant at $P<0.05$ ($r=-0.671$, $N=7$, $P=0.099$) [Fig. 9 (a), (d); ESM Table 1]. Likewise, correlation of the percentage of the <10 μm component with lake area was not significant at $P<0.05$ ($r=0.69$, $N=7$, $P=0.086$) [Fig. 9 (a), (c); ESM Table 1].

Ice-trapped aeolian dust of Selin Co has a coarser component induced by stronger wind conditions. The primary modal size of the material is ~70 μm (Fig. 10).

We explored the relationship between precipitation data and the modal grain sizes (Fig. 11 I; ESM Table 2) at about 10 μm (Fig. 11 Ia), the <10 μm component (Fig. 11 Ib) and the Md (Fig. 11 Ic) of grain-size. The modal size at about 10 μm is not significantly correlated with averaged 5-year precipitation for Bange ($r=-0.269$, $N=9$, $P=0.484$) or Shenzha ($r=-0.103$, $N=9$, $P=0.792$ for). Likewise, the correlation between the averaged 5-year precipitation and percentage of the <10 μm component was not correlated significantly for Bange ($r=-0.048$, $N=9$, $P=0.902$) or Shenza ($r=-0.227$, $N=9$, $P=0.556$). The same was true for correlations between averaged 5-year precipitation and the Md value ($r=0.022$, $N=9$, $P=0.955$ for Bange and $r=0.195$, $N=9$, $P=0.615$ for Shenzha). When a time lag of 4.2 years for precipitation data was applied (Fig. 11 II; ESM Table 2), however, the modal size at ~10 μm was negatively correlated with averaged 5-year precipitation for both locations (Fig. 11 IIa, $r=-0.784$, $N=8$, $P=0.021$ for Bange and $r=-0.769$, $N=8$, $P=0.026$ for Shenzha). The percentage of the <10-μm component positively correlates with averaged 5-year precipitation (Fig. 11 IIb, $r=0.734$, $N=8$, $P=0.038$ for Bange and $r=0.781$, $N=8$, $P=0.022$ for Shenzha), while the Md value is negatively correlated

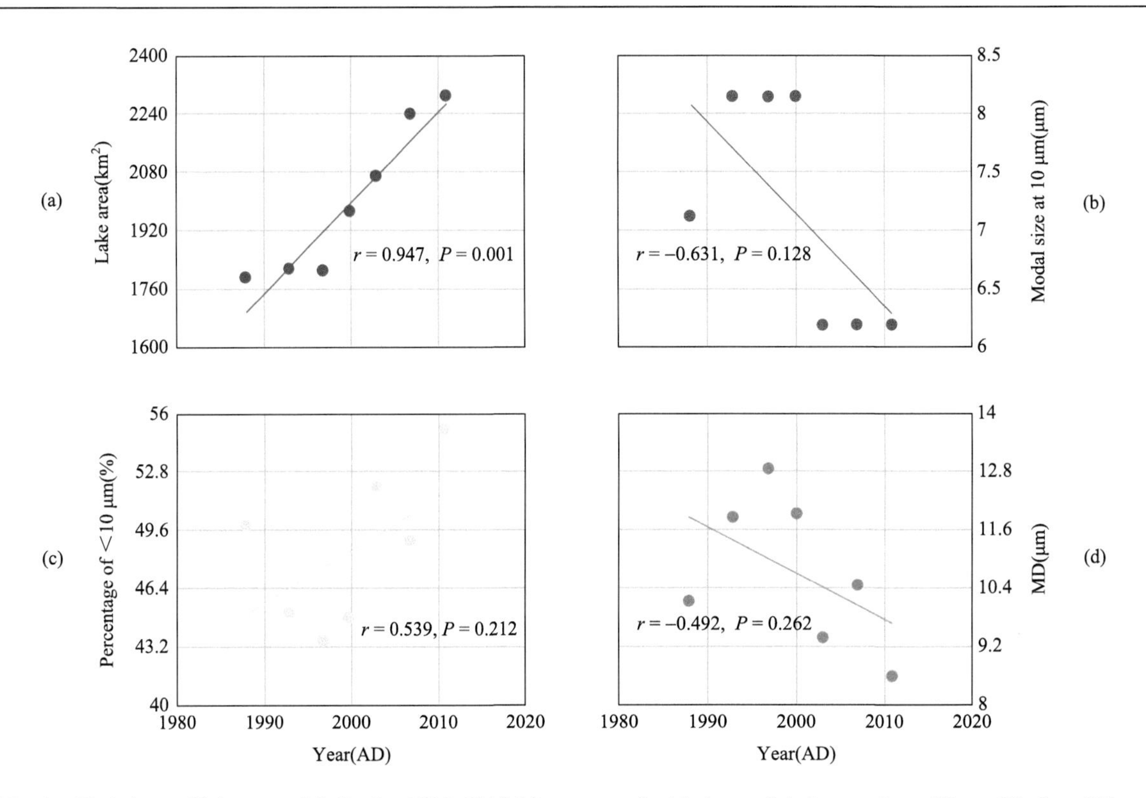

Fig. 9　Variations of lake area of Selin Co, 1988–2012 (a), compared with the modal sizes at about 10 μm (b), the <10 μm component (c) and the Md (d) of grain size in the Core SL-1 during the same period. Linear regression lines are shown.

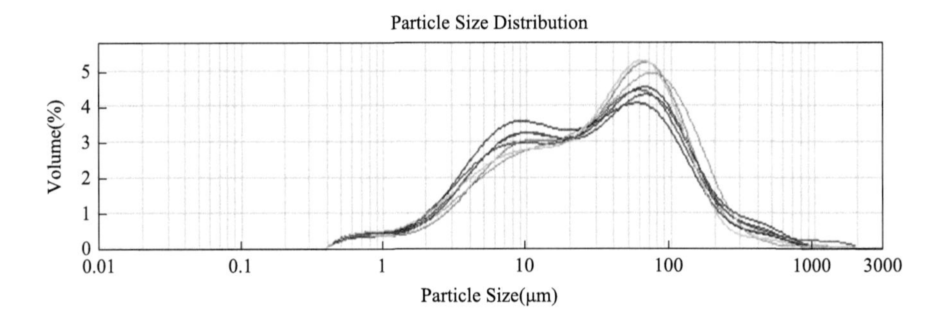

Fig. 10　Distribution curve of ice-trapped aeolian dust, showing a coarser aeolian component with a primary modal size of about 70 μm from snow samples on the ice along transect A in Selin Co.

with averaged 5-year precipitation (Fig. 11 IIc, $r = -0.746$, $N = 8$, $P = 0.034$ for Bange and $r = -0.786$, $N = 8$, $P = 0.021$ for Shenzha).

Discussion

The nature of clastic deposits in lake sediments is controlled by the hydraulic characteristics of the lake waters (Sly, 1978; Håkanson and Jansson, 1983). Median Diameter (Md) of grain size provides information on transport and deposition processes. For example, coarser particles are deposited under more dynamic conditions and lower lake level (Sun et al., 2001; Liu et al., 2003). Fine-grain sediments are usually the main component of deposits near the lake center. Fine components can be transported farther, and grain size decreases in the direction of transport (Self, 1977). Thus a larger fine-component percentage indicates higher lake level or greater offshore distance (Yin et al., 2008). Both Md and the fine component of grain size in loess and lacustrine deposits have applications as environmental indicators (Liu et al., 2003; Jiang et al., 2010; Xu and Wang, 2011).

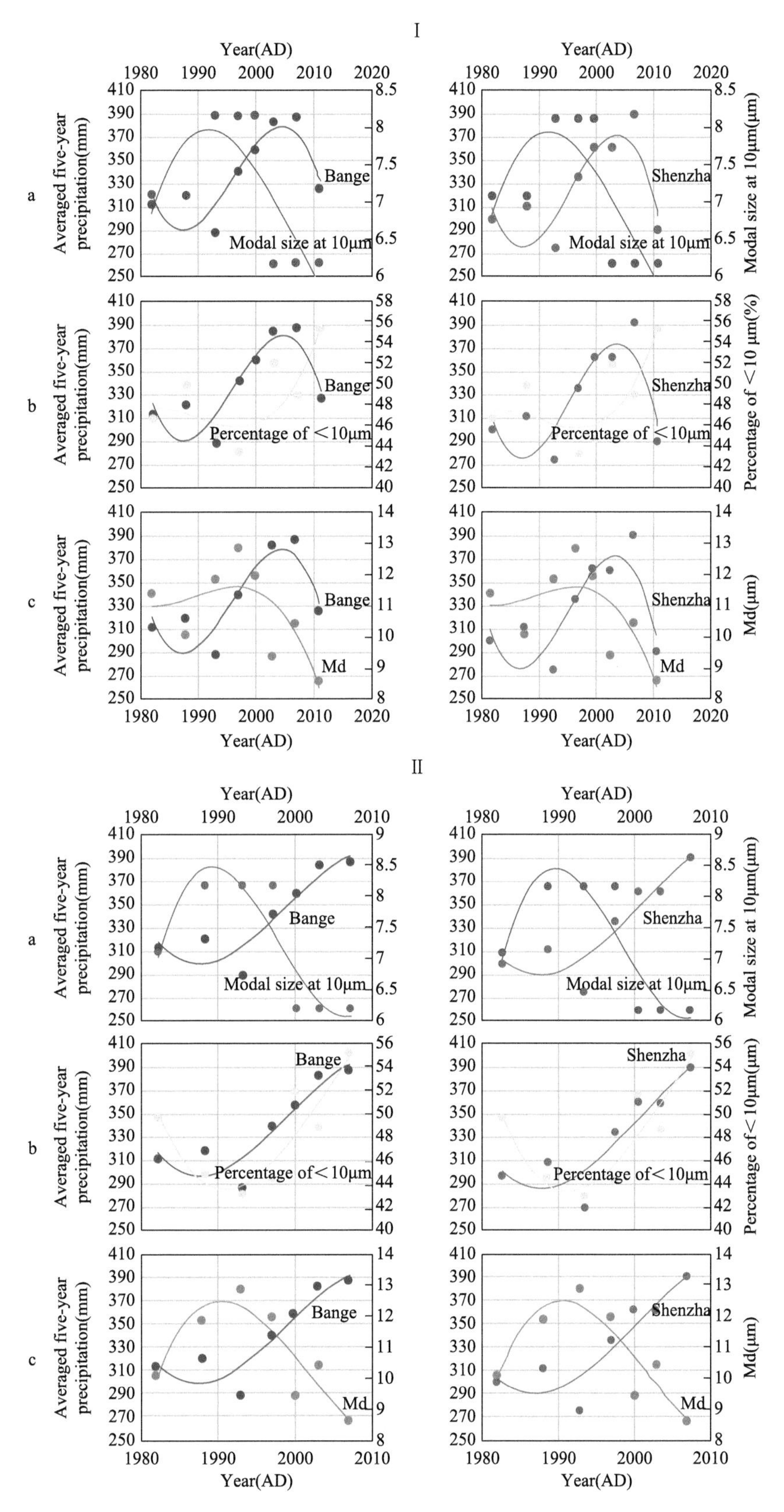

Fig. 11 I The modal sizes at about 10 μm (Ia), the <10 μm component (Ib) and the Md (Ic) of core sediment grain size plotted against averaged 5-year precipitation from Bangge (left) and Shenzha (right). Cubic regression lines are shown. There is no significant correlation between the three grain-size variables and averaged 5-year precipitation. II The modal sizes at about 10 μm (IIa), the <10 μm component (IIb) and the Md (IIc) of core sediment grain size plotted with averaged 5-year precipitation from Bangge (left) and Shenzha (right), with a 4.2-year time lag. Cubic regression lines are shown. The modal size at about 10 μm is negatively correlated with averaged 5-year precipitation, the percentage of <10 μm component positively correlates with averaged 5-year precipitation, whereas the Md value is negatively correlated with averaged 5-year precipitation.

Interpretation of grain size of lacustrine deposits, however, is more complicated when compared with interpretation of grain size in loess and deep-sea deposits. Clastic particles are deposited on lake bottoms mainly via three kinds of transport, i.e. river or runoff, wind (usually trapped on ice in winter and deposited in spring after ice melt), and long-term suspension maintained by wave action. This means that deposition of clastic particles is affected by multiple transport-deposition processes. Generally, river/runoff transport carries mainly coarse sediments with a modal size of 200–400 μm (Bennett and Best, 1995; Kranck et al., 1996; Påsse ,1997; Xiao et al., 2013). Surface sediments from sites A1–A18 and B1–B4 in this study show bimodal and trimodal frequency distribution curves, respectively, with no coarser modal size at 200–400 μm, suggesting that sediment particles at these sites are relatively unaffected by rivers. Indeed, these sites lie ＞15 km from the mouths of rivers that enter the lake. The Zajia Zangbu is the main river that supplies water and sediments to Selin Co, with a relatively large mean annual flow of 26.7 m^3/s, but it enters along the north shore of the eastern basin, ＞30 km from the study sites. Zagen Zangbu, an intermittent river, enters on the western side of the lake, some 15 km from the nearest sampling site, supplying Selin Co with water and sediments in the wet season from June to October. The river is shallow (～0.6 m) and the water is clear in the wet season. The annual average flow is small. Its influence on particle deposition in the studied sites is negligible. Ali Zangbu, also an intermittent river, enters a small secondary lake of the western basin, ＞15 km from the study sites. The river is shallower than Zagen Zangbu and the water is clear in the wet season. Its influence on particle deposition at the study sites is also negligible (Fig. 1).

The typical dominant component of ice-trapped aeolian dust displays a modal size of 50 μm in a lake under the influence of the East Asian Monsoon (Xiao et al., 2013). A modal size of ～40 μm in a lacustrine sediment core from the Qaidam Basin was attributed to aeolian origin (An et al., 2012). These aeolian components are coarser than the modal size of the offshore suspended component of lake deposits, mainly 2.1–10.7 μm (Xiao et al., 2013). In Selin Co, where winds are stronger, we discovered a coarser aeolian component, with a primary modal size of about 70 μm in ice-trapped aeolian dust (Fig. 10). In this study, sample sites far from the shore show a typical bimodal distribution with a primary modal size of about 10 μm and a secondary, finer modal size of＜1 μm [Fig. 7 (a)]. Sites near the shore show a trimodal distribution with a primary modal size of about 10 μm, a secondary coarser modal size of about 80 μm, and a secondary finer modal size of ＜1 μm [Fig. 7 (b)]. Selin Co is located in the middle of Tibet, where northwest airflow prevails in winter. Sediment transect A is closely aligned with this wind direction. The secondary coarser modal size of 80 μm at sample sites near the shore coincides well with the primary modal size of sediments from ice-trapped aeolian dust samples (～70 μm), suggesting that the component with a modal size of 80 μm in the distribution curve is an aeolian component. Consequently, the Md value and fine component of grain size at sites A1–A18 and B1–B4 in Selin Co may be affected by the coarser aeolian component and may not reflect lake level status. That is, a high Md value and low fine component may be attributed not only to low lake level, as indicated by some studies (Liu et al., 2003; Yin et al., 2008), but also by stronger wind conditions.

The relationship between Md value and the fine component percentage of grain size, and lake level status was, however, evident in this study. Md is negatively correlated with both water depth and offshore distance (Fig. 5), and the fine component percentage (both the ＜10 and ＜4 μm components) are positively correlated with both water depth and offshore distance (Fig. 6).

In this study, only the top 1 cm of surface sediment was sampled for grain-size analyses. The mean calculated sedimentation rate for 2.78 m Core SL-1 is 0.52 mm/a. Considering the top of the core has high water content and is not well compacted, the top 1 cm of sediment may represent ＜10 years, which is confirmed by a sediment collection experiment in the west part of Selin Co, carried out from January 2012 to June 2016. During that period, about 0.5 cm was collected in a sediment trap. Therefore, the top 1 cm of sediment was deposited within the last several years and can be regarded as having been deposited under relatively constant river/runoff transport and wind conditions. Under the same river/runoff transport and wind conditions, the Md value and fine component percentage of grain size should correspond with lake level, i.e. the Md value of grain size correlates negatively with both water depth and offshore distance, whereas the fine component percentages (both <10 and <4 μm) correlate positively with both water depth and offshore distance, confirming the consistency of particle deposition. We therefore recommend that the components of grain size transported by river/runoff or wind be distinguished and removed from consideration before employing the Md value and fine-component percentage for lake level reconstruction.

Additionally, among the fine components of lacustrine deposits, both the <4 μm and <10 μm components have been used widely in paleoenvironmental reconstruction (Graham and Rea, 1980; Shen et al., 2007; Jiang et al., 2010; Peng et al., 2005; Yin et al., 2008), but which variable is more sensitive to lake level remains unclear. In our study (Fig. 6), the <10 μm component shows a significant positive correlation with water depth ($r = 0.689$, $N = 22$, $P = 0.001$) and offshore distance ($r = 0.673$, $N = 22$, $P = 0.001$). These values are stronger than correlations between the <4 μm component ($r = 0.549$, $N = 22$, $P = 0.10$) for both for water depth and offshore distance.

The frequency distribution curve of grain sizes is especially useful for distinguishing different types of deposits or different components of a certain deposit (Visher, 1969; Sun et al., 2002; Yin et al., 2008; Xiao et al., 2013). Early studies suggested that grain size decreases in the direction of transport (Self 1977), which also applies to this investigation. Only sample sites with an offshore distance <3.9 km (except for the 4.51 km offshore distance at site A16) show a typical trimodal distribution, whereas sample sites farther from shore show a typical bimodal distribution (Fig. 7). This indicates that wind transport becomes less important toward the lake center. Besides, sample sites farther from the shore usually have greater water depth and particles there may be reworked by lake hydrodynamics.

The modal size at about 10 μm pertains to all samples [Fig. 7 (a), (b)], and is typical of the lacustrine component. This is supported by previous studies (Yin et al., 2008; An et al., 2012; Xiao et al., 2013). Thus, the modal size at about 10 μm can be distinguished easily from river/runoff and aeolian components and is not affected by river/runoff or wind transport. The modal size at about 10 μm shows a significant negative correlation with offshore distance ($r = -0.577$, $N = 22$, $P = 0.006$) [Fig. 8 (b)], indicating that the modal size of the grain-size distribution at about 10 μm is an effective proxy for offshore distance. The negative correlation with water depth, however, is not as strong ($r = -0.412$, $N = 22$, $P = 0.064$) [Fig. 8 (a)]. It is effective in Daihai Lake, central Inner Mongolia, where the offshore suspended component with modal size of 2.1–10.7 μm becomes finer towards the lake center (Xiao et al., 2013).

To verify the lake-level significance of the modal sizes at about 10 μm, the <10 μm component and the Md of grain size, we compared the lake area variations with grain size from Selin Co Core SL-1. The modal size at about 10 μm shows a negative relationship with lake area [Fig. 9 (a), (b)] though the Md value is not correlated with lake area at $P < 0.05$ [Fig. 9 (a), (d)]. The percentage of <10 μm component is not correlated significantly with lake area ($r = 0.69$, $N = 7$, $P = 0.086$) [Fig. 9 (a), (c)]. These relationships agree well with correlations between surface sediment grain size, and water depth and offshore distance. More importantly, the modal size at about 10 μm shows the strongest significant correlation with lake area among the three grain-size variables, further supporting the claim that the modal size at about 10 μm is an especially effective indicator for lake level change.

Moreover, using Core SL-1, we employed data on modern annual precipitation to verify the application of grain-size variables for lake level inference. The annual precipitation data were obtained from the Bange meteorological station (station 55279), 70 km east of Selin Co, and Shenzha (station 55472), 80 km south of Selin Co. Annual precipitation values from 1973 to 2011 were averaged over 5-year spans as the resolution of core samples for comparison is ~5 years. There is no significant correlation between the modal size at about 10 μm and averaged 5-year precipitation (Fig. 11 Ia). Likewise, the percentage of <10 μm component and Md values are not correlated significantly with averaged 5-year precipitation (Fig. 11 Ib, Ic).

Considering the large area of Selin Co and relatively small amount of precipitation and runoff into the lake, we assumed there is a time lag between precipitation and lake level change, i.e. high annual precipitation does not immediately cause the lake level to rise, and similarly, low annual rainfall does not cause an immediate drop in lake level. All three grainsize variables show significant relationships with precipitation data, as expressed by lake area, when a time lag of 4.2 years for precipitation data is applied (Fig. 11 II). The modal size at about 10 μm is negatively correlated with averaged 5-year precipitation (Fig. 11 IIa), the percentage of <10 μm component is positively correlated with averaged 5-year precipitation (Fig. 11 IIb), and the Md value is negatively correlated with averaged 5-year precipitation (Fig. 11 IIc), suggesting that precipitation influences grain size through the following mode: high precipitation leads to high lake level (with a time lag), which weakens the lake hydrodynamics, and consequently leads to finer particles being deposited on the lake floor. Therefore, the notion that coarser sediments occur under conditions of greater precipitation, higher lake levels and stronger river/stream transport, promoted by Peng et al. (2005) and Jiang et al. (2010), does not apply to large lakes like Selin Co.

Considering the resolution of $^{210}Pb/^{137}Cs$ dating, the 4.2-year time lag may result from the uncertainty of the core chronology. Furthermore, statistical error may also contribute, given the small number of samples.

The modal size at about 10 μm shows the strongest significant correlation with lake area (at the 0.05 level) among the three grain-size variables. It also shows significant correlations at $P<0.05$ with precipitation data from both Bangge and Shenzha, suggesting it is a reliable water depth and offshore distance indicator that can be used with confidence in paleo lake-level reconstructions and other paleolimnological research.

Conclusions

1. Under the same river/runoff transport and wind conditions, the Md value and fine-component percentage of the grain size of lacustrine deposits correspond well with lake level status, i.e. the Md value is negatively correlated with both water depth and offshore distance and the fine-component percentage is positively correlated with both water depth and offshore distance.

2. Among the fine components of lake deposits (<4 μm and <10 μm components), the <10 μm component is more sensitive to lake level than the<4 μm component.

3. We recommend that the components of grain-size be distinguished and that the components transported by river/runoff or wind be eliminated before employing the Md value and fine-component percentage of grain size in lake-level reconstruction.

4. The modal grain size at about 10 μm is easily distinguished from river/runoff and aeolian components and is not affected by river/runoff or wind transport. It shows a significant negative correlation with offshore distance ($P<0.01$), indicating that it is an effective proxy for past lake-level reconstruction.

5. Modern observation data confirm that the modal size of grain size at about 10 μm is an especially effective and reliable lake-level indicator, as it shows the strongest significant correlation with lake area among the three grain-size variables, as well as strong correlations with precipitation data from the Bangge and Shenzha stations ($P<0.05$).

Acknowledgements

This study was supported by the Fundamental Research Funds for Central Leveled Scientific Research Institutes (JYYWF20182701), the National Natural Science Foundation of China (41372179) and China Geological Survey (DD20160054). Professor Peter Hudson from South Australian Museum, Adelaide, kindly improved the English.

References

An FY, Ma HZ, Wei HC, Lai ZP (2012) Distinguishing aeolian signature from lacustrine sediments of the Qaidam Basin in northeastern Qinghai-Tibetan Plateau and its palaeoclimatic implications. Aeolian Res 4: 17-30

Bennett SJ, Best JL (1995) Mean flow and turbulence structure over fixed, two dimensional dunes: implications for sediment transport and bedform stability. Sedimentology 42: 491-513

Dasang (2011) The trends of air temperature and precipitation changes in the past 50 years in Selincuo Lake area, Tibet. Sci Technol Tibet 1: 42-45 (**in Chinese**)

Friedman GM, Sanders JE (1978) Principles of sedimentology. Wiley, New York, p 792

Graham EJ, Rea DK (1980) Grain size and mineralogy of sediment cores from western Lake Huron. J Great Lakes Res 6: 129-140

Håkanson L, Jansson M (1983) Principles of lake sedimentology. Springer, Berlin, p 316

Jiang QF, Shen J, Liu XQ, Ji JF (2010) Environmental changes recorded by lake sediments from Lake Jili, Xinjiang during the past 2500 years. J Lake Sci 22: 119-126 (**in Chinese, English abstract**)

Kashiwaya K, Yamamoto A, Fukuyama K (1987) Time variations of erosional force and grain size in Pleistocene lake sediments. Quat Res 28: 61-68

Kranck K, Smith PC, Milligan TG (1996) Grain-size characteristics of fine-grained unflocculated sediments II: 'multiround'distributions. Sedimentology 43: 597-606

Laba Chen T, Labazhuoma Cizhen (2011) Area change of Selincuo Lake and its forming reasons based on MODIS data. J Meteorol Environ 27: 69-72 (**in Chinese, English abstract**)

Liu XQ, Wang SM, Shen J (2003) The grainsize of the core QH-2000 in Qinghai Lake and its implication for paleoclimate and paleoenvironment. J Lake Sci 15: 112-117 (**in Chinese, English abstract**)

Lv P, Qu YG, Li WQ, Wang HS (2003) Shelincuo and Bangecuo extensional lake basins in the northern part of Tibet and present chasmic activities. Jilin Geol 22: 15-19 (**in Chinese, English abstract**)

McCave IN (1978) Grain size trends and transport along beaches: an example from eastern England. Mar Geol 28: 43-51

Påsse T (1997) Grain size distribution expressed as tanh-functions. Sedimentology 44: 1011-1014

Peng YJ, Xiao JL, Nakamura T, Liu BL, Inouchi Y (2005) Holocene East Asian monsoonal precipitation pattern revealed by grain-size distribution of core sediments of Daihai Lake in Inner Mongolia of north-central China. Earth Planet Sci Lett 233: 467-479

Poizot E, Mear Y, Thomas M, Garnaud S (2006) The application of geostatistics in defining the characteristic distance for grain size trend analysis. Comput Geosci 32: 360-370

Self RP (1977) Longshore variation in beach sands, Nautla area, Veracruz, Mexico. J Sediment Petrol 47: 1437-1443

Shen HY, Li SJ, Yu SB, Yao SC (2007) Grain-size characteristics of sediments from the Zigetang Co Lake, Tibetan Plateau and their environmental implication. Quat Sci 27: 613-619 (**in Chinese, English abstract**)

Sly PG (1978) Sedimentary processes in lakes. In: Lerman A (ed) Lakes: chemistry, geology, physics. Springer, Berlin, pp 65-89

Sun QL, Zhou J, Xiao JL (2001) Grain-size characteristics of Lake Daihai sediments and its paleaoenvironment significance. Mar Geol Quat Geol 21: 93-95 (**in Chinese, English abstract**)

Sun DH, Bloemendal J, Rea DK, Vandenberghe J, Jiang FC, An ZS, Su RX (2002) Grain-size distribution function of polymodal sediments in hydraulic and aeolian environments, and numerical partitioning of the sedimentary components. Sediment Geol 152: 263-277

Tsoar H, Pye K (1987) Dust transport and the question of desert loess formation. Sedimentology 34: 139-153

Visher GS (1969) Grain size distributions and depositional processes. J Sediment Petrol 39: 1074-1106

Wang HL, Zheng MP (2014) Lake level changes indicated by grain-size of Core SL-1 sediments since 5.33 ka BP in Selin Co, central Qinghai-Tibetan Plateau. Sci Technol Rev 32: 29-34

Wang SM, Yu YS, Wu RJ, Feng M (1990) The Daihai Lake: environment evolution and climate change. University of Science and Technology of China Press, Hefei, p 191 (**in Chinese**)

Wang SM, Dou HS, Chen KZ, Wang XC, Jiang JH (1998) China lakes record. Sci Press, Beijing, p 399 (**in Chinese**)

Wang HL, Zheng MP, Yuan HR (2016) Highstands in the interior of the Qinghai-Tibetan Plateau since the last interglacial: evidence from grain-size analysis of lacustrine core sediments in Zabuye Salt Lake. Geol J 51: 737-747

Xiao JL, Fan JW, Zhou L, Zhai DY, Wen RL, Qin XG (2013) A model for linking grain-size component to lake level status of a modern clastic lake. J Asian Earth Sci 69: 149-158

Xu SJ, Wang T (2011) Comparative study on the grain size characteristics of loess deposit both on Miaodao Islands and on the Laizhou Bay plain and its implications for provenance. Procedia Environ Sci 10: 1869-1875

Yin ZQ, Qin XG, Wu JS, Ning B (2008) Multimodal grain-size distribution characteristics and formation mechanism of lake sediments. Quat Sci 28: 345-353 (**in Chinese, English abstract**)

Zhuo G, Yang XH, Tang H (2007) Effects of climate change on lake acreage in Naqu region. Plateau Meteorol 26: 485-490 (**in Chinese, English abstract**)

High-resolution geochemical record for the last 1100 yr from Lake Toson, northeastern Tibetan Plateau, and its climatic implications*

Yuan Ling[1, 2], Xinqin Dai[3], Mianping Zheng[1], Qing Sun[4], Guoqiang Chu[5], Hailei Wang[1], Manman Xie[4] and Yabing Shan[6]

1. Institute of Mineral Resources, Chinese Academy of Geological Sciences, Beijing 100037, China
2. School of Ocean Sciences, China University of Geosciences, Wuhan 430074, China
3. College of Geoscience and Surveying Engineering, China University of Mining and Technology, Beijing 100083, China
4. National Research Center of Geoanalysis, Beijing 100037, China
5. Institute of Geology and Geophysics, Chinese Academy of Sciences, Beijing 100029, China
6. National Institute of Biological Sciences, Beijing 102206, China

Abstract Knowledge of the origin of past climatic variations on annual to decadal timescales is important for assessing the likelihood of the occurrence of similar variations in the future. Here we present a highresolution (82 μm) element record, obtained using Synchrotron Radiation X-ray Fluorescence, from the sediments of Toson Lake in the northeastern Tibetan Plateau. The record, which spans the last ~1100 yr, exhibits variability on an annual to decadal scale. The results of principal component analysis of the element dataset indicates that the first three principal components reflect variations in detrital inputs (via runoff and possibly eolian activity), drought conditions and temperature variations, respectively. Spectral analysis of samples scores on these components reveals strong El Niño-Southern Oscillation (ENSO) -like variability at periods of 3–8 years which indicates that the regional climatic variability was effected by ENSO events over the last~1100 yr. On the centennial time scale, the element data and TOC record suggest a warm and dry climate during the Medieval Warm Period (MWP) and a cold and wet Little Ice Age (LIA). The early 20th century was a warm interval, and drier than the LIA but wetter than the MWP.

Keywords High resolution; Element; Toson; Last millennia; Climate; Tibetan Plateau

1 Introduction

The Tibetan Plateau (TP), because of its elevation, size, geographical location and sensible heat flux, plays a key role in modulating regional atmospheric circulation patterns and climatic variability, on a regional as well as on a global scale (Liu et al., 2009; Grieβinger et al., 2011). Therefore, paleoclimatic reconstructions from the TP region are essential for improving our understanding of how complex forcing mechanisms can affect the regional climate (Liu et al., 2006). The climate of the TP during the last few millennia was characterized by significant variability, including the Medieval Warm Period (MWP) and the Little Ice Age (LIA), and knowledge of such intervals is important for evaluating the global warming trend of the past century (Xu, 2001; Holmes et al., 2009; Pu et al., 2013). During the late Holocene, the climate of the region is generally believed to have been characterized by warm-wet or cold-dry combinations, controlled by the Asian Summer Monsoon (An, 2000). However, in the Westerly-dominated regions of Asia, wet LIA conditions and dry MWP conditions have been identified (Qiang et al., 2005; Chen et al., 2010; Liu et al., 2011). Moreover, on short time scales, the relationship between air temperature and precipitation variations may be ambiguous (Pu et al., 2013). For example, Qiang et al. (2005) established the sequence of climatic changes at Lake Sugan, which consisted of warm-dry, cold-dry, and

* 本文发表在：Quaternary International, 2018, 487: 61-70

cold-wet combinations during the past two millennia. Toson Lake is in the transitional zone of the two regions and its climatic record is potentially important because of its marginal position in relation to the Asian Summer Monsoon. In addition, many studies suggest possible solar, ENSO and Pacific Decadal Oscillation (PDO) forcing of moisture oscillations in the TP during the last millennium (Zhao et al., 2009; Chu et al., 2011; Gou et al., 2014). Temperature variability over the TP is also affected by solar insolation, ENSO, North Atlantic Oscillation (NAO), Atlantic Multidecadal Oscillation (AMO), volcanism and greenhouse gas variations (Pu et al., 2013; Wang et al., 2014). However, the physical mechanisms responsible for these modes of variation are still not well understood, especially for the Asian continent, and further research needs to be undertaken (Fang et al., 2010).

Although previous investigations of paleoclimatic changes in the TP have been conducted, little is known about monsoon climate variability on annual to seasonal time scales (Chu et al., 2011), which impinges directly on people's lives (Yan et al., 2014). For example, ENSO, a coupled atmospheric-oceanic phenomenon, is the most significant factor causing global hydro-climatic variability (Allan, 2003; Fagel et al., 2008). It has been responsible for decreased rainfall and drought conditions in many regions of China, heavily impacting the livestock industry, water supplies and the overall economy. Therefore, it is crucial to understand the processes and mechanisms of high-frequency paleoclimatic variations in order to assess the probability of their occurrence in the future (Chu et al., 2013). However, one factor making it difficult to combine and compare paleoclimatic records with modern instrumental data is the limitations imposed on high-resolution paleoclimatic reconstruction imposed by traditional methods of geochemical analysis (Liu et al., 2014b). High-resolution proxy climate records from the TP are mainly derived from tree rings (Grieβinger et al., 2011; Shi et al., 2011; An et al., 2012; Xu et al., 2012; Sano et al., 2013; Liu et al., 2014a; Yang et al., 2014), ice cores (Thompson et al., 1989, 1997, 2000; Yao et al., 1997, 2002, 2006; Wang et al., 2003), and lacustrine sedimentary records (Zhou et al., 2007; Mischke et al., 2010; Chu et al., 2011; Wang et al., 2012). These archives provide various proxy records with an inherent annual-resolution chronology enabling high-resolution paleoclimatic reconstruction. However, few techniques can be used to obtain high-resolution data at annual to seasonal time scales. Recently, non-destructive analyses such as the micro X-ray Fluorescence (μ-XRF) core scanner (Rothwell et al., 2006; Das et al., 2008; Francus et al., 2009; Chawchai et al., 2015; Kelloway et al., 2015) and Synchrotron Radiation X-ray Fluorescence (SRXRF) have been developed. SRXRF is an effective method for analyzing element variations and the results can be correlated to climatic parameters such as seasonal or annual temperature and humidity (Kalugin et al., 2007, 2013; Chu et al., 2013; Ling et al., 2014; Xie et al., 2015; Sun et al., 2016).

In this study we present a high-resolution element dataset spanning the last ~1100 yr from the varved sediments of Lake Toson. The variations of K, Ca, Ti, Mn, Fe, Cu, Zn, Br, Rb, and Sr exhibit distinct inter-annual features, and we compare the results with multiple records derived from various proxies in order to characterize and interpret regional climatic variations over the last millennium.

2 Geological setting

Qaidam Basin, located in the northeastern TP, has an area of about 12×10^4 km^2 and an average altitude of about 2800 m. It is surrounded by the Kunlun Mountains to the south, the Altun Mountains to the west and the Qilian Mountains to the north and east (Zhao et al., 2007). The surrounding mountains rise to an elevation of >5000 m above sea level and the region has an extremely arid desert climate. Lake Toson (37°04′–37°13′N, 96°50′–97°03′E, altitude 2808 m, surface area of 165.9 km^2) is located on the northeastern edge of Qaidam Basin [Fig. 1 (a)]. Toson is a typical magnesium sulfate subtype lake with a salinity of 38.7 g/L. Western Hurleg Lake flows out through a small stream to feed Toson Lake, and snow melt water from the mountains to the north also enters the lake [Fig. 1(b) and (c)]. The maximum water depth is 25.7 m and the lake shore is surrounded by pluvial and alluvial sand-gravel beds, saline-alkali soil and aeolian sand dunes.

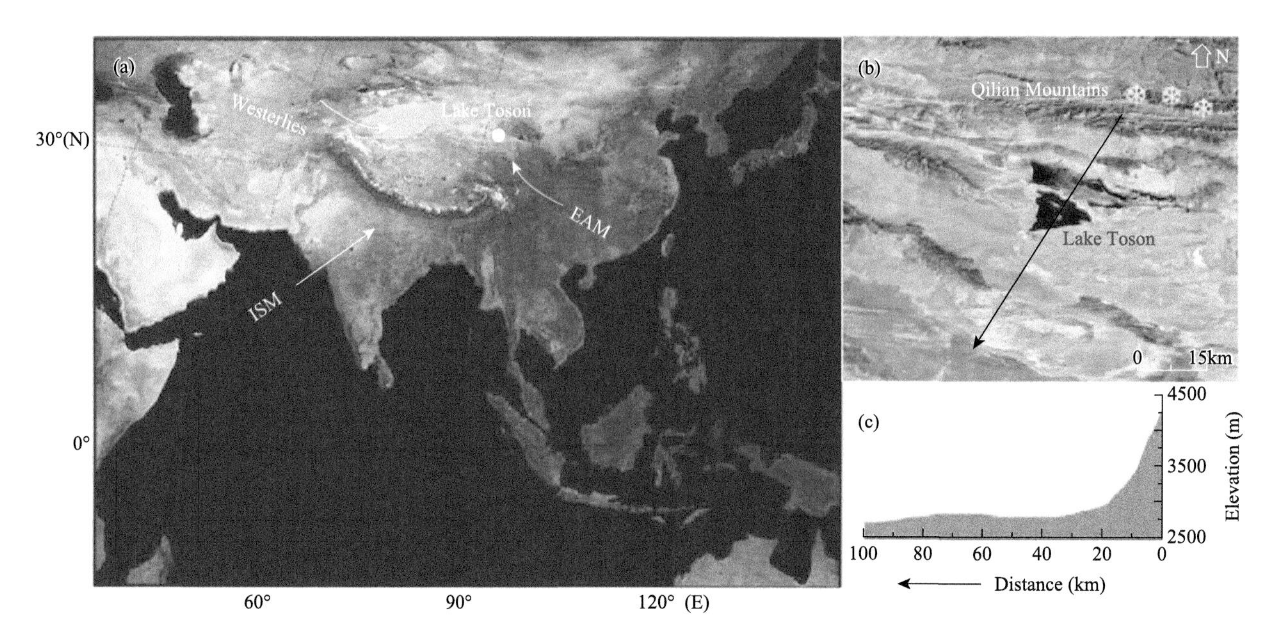

Fig. 1 (a) Location of Lake Toson within Asia; red arrows indicate the dominant directions of the East Asian monsoon (EAM), Indian Summer Monsoon (ISM) and the Westerlies. (b) Satellite photo of Toson Lake; the blue circle indicates the coring site, and the black line indicates the elevation profile shown in (c), from northeast to southwest. (For interpretation of the references to colour in this figure legend, the reader is referred to the web version of this article.)

Lake Toson is located on the margin of the region influenced by the East Asian summer monsoon (EAM) (Chen et al., 2008; Zhang et al., 2011), which results in a complex local climate system compared to other parts of the TP. Meteorological data from 1956 to 2004 from Delingha station indicate a mean annual temperature of ~4 ℃, a mean annual precipitation of ~160 mm, falling mainly during summer, and a mean annual potential evaporation of ~2000 mm (Zhao et al., 2010). The lake is ice-covered from the end of November until early April.

3 Methods

3.1 Sediment coring

In November 2015, a 60 cm sediment core was recovered from Toson Lake using a gravity corer developed at the Institute of Geology and Geophysics, Chinese Academy of Science. The coring site is shown in Fig. 1 (b). The core was gently lowered and slowly allowed to penetrate the sediments by gravity in order to recover an undisturbed record. The water in the upper part of the coring tube was carefully removed using a hose and syringe. The core tube was covered using columnar foam and kept vertical during transportation to the laboratory. In the laboratory, the core was dried further by inserting paper towels, photographed, and then split longitudinally and sampled (Chu et al., 2011).

3.2 Radiometric dating

A chronology was established by AMS ^{14}C dating of plant macrofossils from three depths in the core. We consider the plants to have been mainly aquatic, using both atmospheric carbon dioxide and bicarbonate dissolved in the lake water for photosynthesis. This combination of atmospheric carbon dioxide and dissolved carbon dioxide and bicarbonate in the lake water may result in heavier ^{14}C values consequently (Olsson and Kaup, 2001). Therefore, it is important to determine the reservoir ages for lakes of the TP in order to correct the dates for old carbon error. The carbon reservoir age (2010.3-year) of Toson Lake was determined by linear regression. All ^{14}C

dates were measured at Peking University and calibrated with Bacon software. The core spans the interval from 901–2015 AD. Fig. 2 shows the resulting age-depth model for the sediment core; the age-depth relationship is almost linear.

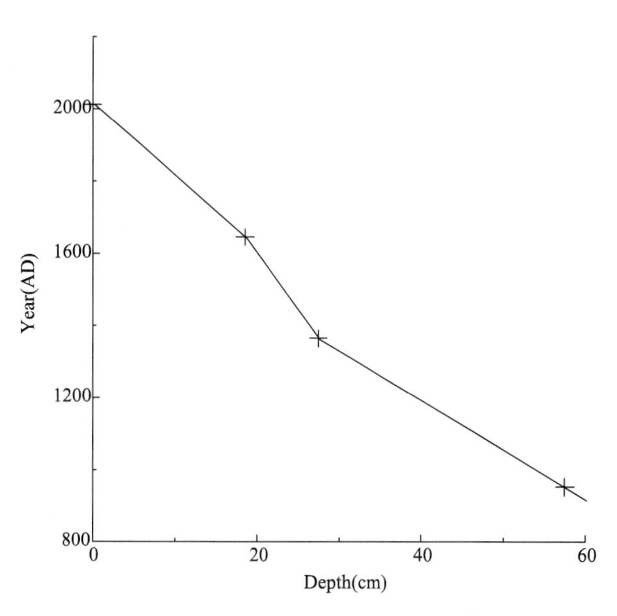

Fig. 2 Age-depth profile based on AMS ^{14}C dates from the sediment core from Lake Toson.

3.3 X-ray diffraction analyses

Eight subsamples were ground to pass a 200 mesh sieve. Subsequently, X-ray diffraction (XRD) analyses were performed using a Rigaku MiniFlex 600 Series X-ray powder diffractometer at the Key Laboratory of Saline Lake Resources and Environment, Ministry of Land and Resources, Beijing. A copper target was used and the operating conditions were 40 kV and 15 mA. The samples were scanned in the range of 3° to 70° (2θ) with a step of 0.02°. Identification and determination of minerals and their relative percentages were made using MDI Jade software. The calculation formula is:

$$X_i = \left(\frac{K_{i,n}}{I_{i,n}} \sum_{j-1}^{m} \frac{I_{j,n}}{K_{j,n}} \right)^{-1}$$

where $K_{i,n}$ and $K_{j,n}$ are the reference intensity ratios of mineral i and j in the diffraction ray n; $I_{i,n}$ and $I_{j,n}$ are the integrated intensities of mineral i and j in the diffraction ray n; m is the number of minerals in the sample. The determination error of the relative percentage composition of minerals is ±5%.

3.4 TOC analyses

Freeze-dried subsamples were ground to pass a 200 mesh sieve. The samples were then acidified with HCl and washed to neutral pH with distilled water. The concentrations of TOC were determined using a CHNSO elemental analyzer (Costech ECS 4010 Series) at the Institute of Geology, Chinese Academy of Geological Sciences. Blank samples and standards with known elemental composition were used for quality control. The precision of the method was 0.5%.

3.5 SRXRF analyses

SRXRF measurements were performed using the Hard X-ray beamline BL15U of the Shanghai Synchrotron Radiation Facility (SSRF). The sediment blocks (60×10×5 mm) were fixed on a sample stage, which was driven by a step motor. An incident beam energy of 20.05 keV was used for the determination. The beam size is 80 μm×300 μm. A Si (Li) detector was used to collect the XRF signal from samples with a scan time of 10 s. Chinese national standard material GBW07301 was used for quality control. Data were processed using program PyMca, and matrix effects were corrected using Compton scattered radiation. Normalized elemental intensities (K, Ca, Ti, Mn, Fe, Cu, Zn, Br, Rb and Sr) were thereby obtained. After standardization, the uncertainties of the intensities of elements are better than 5%, based on analysis of standard samples.

4 Results and discussions

4.1 Characteristics of Lake Toson sediments

Fig. 3　Microphotograph taken with a stereoscopic microscope under polarized and transmitted light.

The sediments in most of the core have a laminated structure (Fig. 4). In thin section, the laminae appear as a light-colored and dark-colored couplet (Fig. 3). The light-colored layer is composed mainly of detrital material, while the dark-colored layer mainly consists of organic matter. X-ray diffraction analysis shows that the minerals in the Lake Toson sediments are mainly detrital minerals (quartz, muscovite and chlorite) and carbonate minerals (calcite and aragonite). The detrital clay minerals are mainly muscovite and chlorite; muscovite, with an average content of 32%, is the best represented of the clay minerals. The average content of carbonate minerals is 38%. There are many large lithogenic (detrital) mineral grains in the laminated sediments that may have been of aeolian origin, as was suggested in the case of Lake Kusai in the TP (Liu et al., 2014b). The TOC content exhibits significant fluctuations during the past 1100 years, varying between 1.0 and 4.1%, and with high values during 937–1033 AD, 1283–1658 AD and 1870–2006 AD [Fig. 6 (h)].

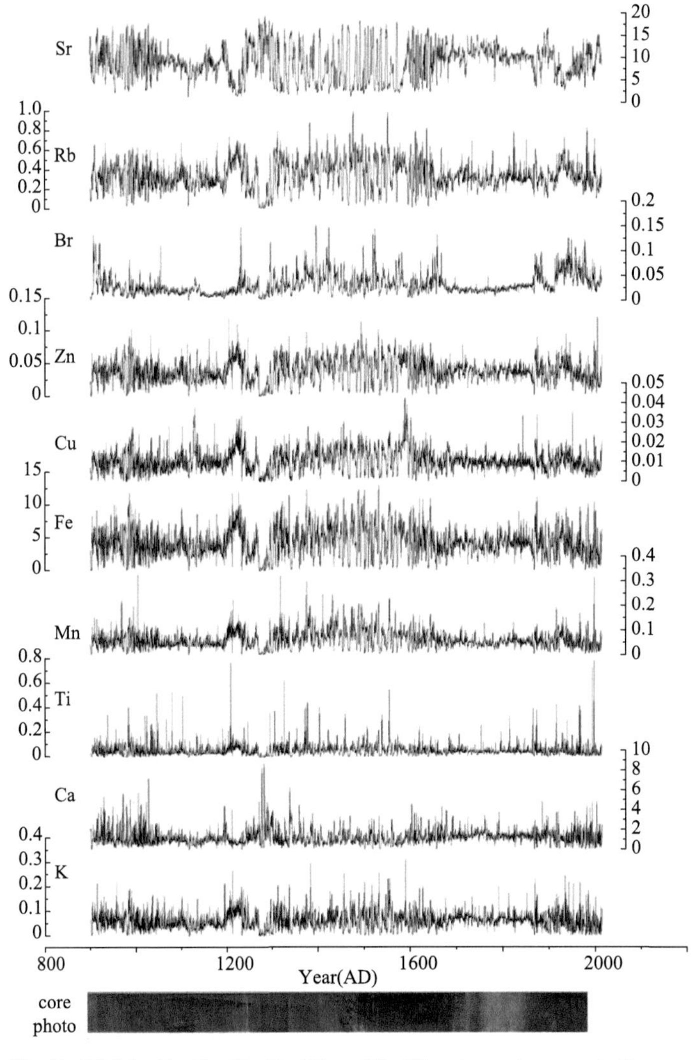

Fig. 4　Time series of K, Ca, Ti, Mn, Fe, Cu, Zn, Br, Rb and Sr. The element concentrations are normalized intensity (counts/ counts).

4.2 Principal components analysis

Fig. 4 shows time series of K, Ca, Ti, Mn, Fe, Cu, Zn, Br, Rb and Sr spanning the last ~1100 years. There is a broadly similar pattern of variation evident for K, Ti, Mn, Fe, Cu, Zn and Rb; and Ca and Sr also exhibit similar trends. There are systematic variations in the element assemblages with maxima in K, Ti, Mn, Fe, Cu, Zn and Rb coinciding with minima in Sr and Ca. Principal component analysis (PCA) was used to reduce the dimensionality of the large elemental dataset, in order to try to identify the main factors controlling the elemental distributions and to aid the interpretation of the geochemical behavior of specific elements. The technique has previously been applied successfully to geochemical datasets (Muller et al., 2008; Chu et al., 2013; Sun et al., 2016). The results are summarized in Table 1 and the variable loadings on the first three principal components are illustrated in Fig. 5. Three components explain 79.2% of the total variance. Component 1 (F1) accounts for 57.4%of the total and is characterized by high positive loadings (0.67–0.93) of K, Ti, Mn, Fe, Cu, Zn and Rb. F2 (12.7% of the total variance) is characterized by a high positive loading of Ca (0.93) and a negative loading of Sr (–0.53). F3 (9.0%) is characterized by a high positive loading of Br (0.96). Time series of sample scores on the first three principal components are illustrated in Fig. 6.

Table 1 Coefficients of determination and results of factor analysis for the Toson core

Element	K	Ca	Ti	Mn	Fe	Cu	Zn	Br	Rb	Sr	Zr	F1	F2	F3
K	1.00											0.87	0.13	0.02
Ca	0.06	1.00										–0.01	0.93	0.02
Ti	0.55	–0.10	1.00									0.73	–0.01	–0.04
Mn	0.71	–0.09	0.60	1.00								0.89	–0.08	0.17
Fe	0.77	–0.16	0.61	0.88	1.00							0.93	–0.18	0.15
Cu	0.54	–0.27	0.44	0.68	0.76	1.00						0.73	–0.36	0.21
Zn	0.66	–0.23	0.53	0.81	0.92	0.80	1.00					0.86	–0.29	0.24
Br	0.12	–0.09	0.14	0.25	0.24	0.27	0.29	1.00				0.10	–0.03	0.96
Rb	0.53	–0.31	0.40	0.63	0.71	0.64	0.80	0.33	1.00			0.67	–0.44	0.34
Sr	–0.44	0.38	–0.41	–0.60	–0.65	–0.63	–0.67	–0.33	–0.74	1.00		–0.59	–0.53	–0.33
Eigenvalues												5.8	1.4	0.9
PVE[a]												57.4	12.7	9.0
CPVE[b]												57.4	70.1	79.2

Notes: Extraction method: PCA; rotation method: varimax with Kaiser normalization.

a. PVE, percent of variance explained.

b. CPVE, cumulative percent of variance explained.

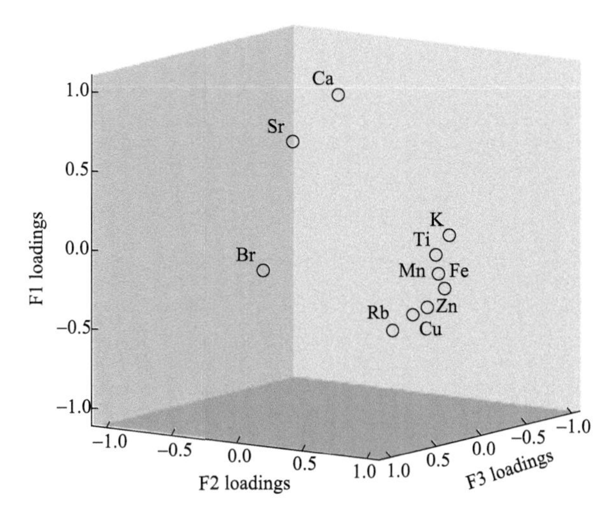

Fig. 5 Variable loadings on the first three components of a PCA of the analysed element concentrations.

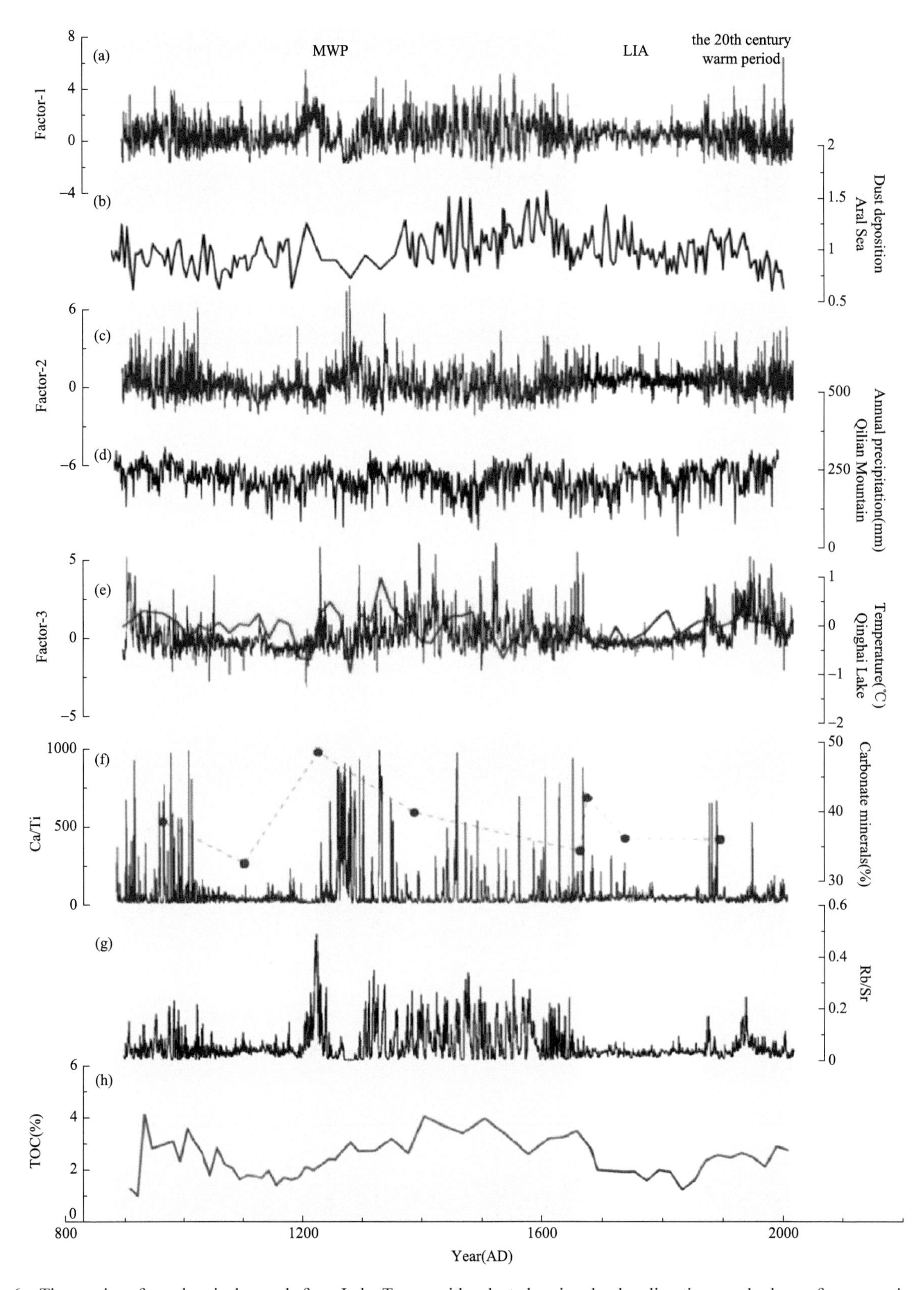

Fig. 6　Time series of geochemical records from Lake Toson, with selected regional paleoclimatic records shown for comparison.

(a) Sample scores on the 1st principal component of a PCA of the element record for Lake Toson. (b) Record of dust deposition from the Aral Sea (6–32μm and 2–6μm grain-size ratio). (c) Sample scores on the 2nd principal component of a PCA for Lake Toson. (d) Record of annual precipitation from the Qilian Mountains, China, based on the calibrated Qilian Mountain chronology (Yang et al., 2014). (e) Sample scores on the 3rd principal component of a PCA of the element record for Lake Toson; the red line is the alkenone-based temperature record from Lake Qinghai (Liu et al., 2006). (f) Ca/Ti ratio record for Lake Toson (very high values are removed); red curve is the % carbonate minerals in the sediments of Lake Toson. (g) Rb/Sr ratio record for Lake Toson. (h) TOC record for Lake Toson. (For interpretation of the references to colour in this figure legend, the reader is referred to the web version of this article.).

Lake sediments generally comprise two sources of material: allothogenic, derived from outside the lake; and authigenic which is produced within the lake water. Therefore, the proportions and flux of these components reflect changing catchment dynamics, including variations in run-off, soil erosion, weathering processes, and within-lake processes such as evaporative concentration or biological activity (Croudace, 2015). F1 represents more than 50% of the statistical variance and thus potentially contains significant information about these processes. The elements with high loadings on F1 (K, Ti, Mn, Fe, Cu, Zn and Rb) occur mainly in oxides and siliciclastic material which are typically land-derived through the erosion of lithogenic material and are indicative of surface runoff processes. The extent of surface runoff is dependent partly on climate and specifically the amount of precipitation. Thus higher F1 values may reflect increased catchment run-off, but it may also be influenced by aeolian deposition. Significantly, the variation of sample scores on F1 [Fig. 6 (a)] is generally consistent with the a dust deposition record from the Aral Sea inferred from grain-size ratios [Fig. 6 (b)]. Therefore, we conclude that F1 is broadly reflective of climatic variations, especially precipitation and aeolian processes. In addition, we consider the supply from surface runoff process is relatively limited, because the surface runoff often stop as the drought intensifying in the arid area. However, the sand storm run easily in this arid area on account of the regular strong winds, as well as the poor surface vegetation. So the supply of allothogenic elements from aeolian process might be more, which needs to be further investigated.

The second component is dominated by Sr and Ca. These elements have been shown to be sensitive paleoclimatic indicators and have been widely used to reconstruct various aspects of the lacustrine environment (Mischke et al., 2005; Kalugin et al., 2013; Zheng et al., 2016). Ca in the sediments of saline lakes is mainly controlled by lake water chemistry, temperature, evaporation and depositional processes. Generally, when the evaporative capacity is greater than precipitation, Ca^{2+} in the lake water becomes supersaturated, and carbonates start to crystallize and be deposited; therefore, evaporation is a major factor affecting carbonate deposition. As essential nutrients, Sr and Ca are subject to biogeochemical cycling (Margalef et al., 2014). Ca in lakes may have a biogenic origin which is often volumetrically substantial. Sr is an alkaline earth metal that is fixed by calcifying organisms at the same time as Ca, and is preferentially incorporated into aragonite. The positive correlation between Sr and Ca (Pearson correlation coefficient of 0.38; $p<0.001$; Table 1) suggests that Ca is also sourced from biogenic $CaCO_3$ with an additional small detrital source. In order to eliminate the detrital effect on the Ca concentration, we use the Ca/Ti ratio to identify Ca derived from within-lake processes such as evaporative concentration or biogenic production. Ti is useful for normalization since it is an unambiguous indicator of allochthonous inputs from the catchment (Croudace, 2015). The variation of the Ca/Ti ratio is also in accordance with the variation in carbonate minerals [Fig. 6 (f)]. High percentages of carbonate minerals correspond to a high Ca/Ti ratio, and vice versa. Therefore, we conclude that Ca mainly has an evaporative source, as well as being derived from biogenic $CaCO_3$. Clearly, Ca formed by biological action is also affected by precipitation- evaporation, because suitable salinity caused by precipitation-evaporation would favor the growth of specific organisms in the lake, thus the biogenic $CaCO_3$ might be increasing. Considering that F2 reflects the variation of chemically-mobile elements which are affected by precipitation-evaporation, it is possible the variations in F2 are more closely related to effective moisture, and that F2 is a drought-related factor since a high Ca input indicates a dry environment. Thus we interpret the variation of sample scores on F2 as a drought record, with the intervals of 901–1033 AD, 1283–1658 AD and 1870–2015 AD being relatively dry; and 1033–1283 AD and 1658–1870 AD being relatively wet.

F3 is dominated by Br. Since Br forms strong covalent bonds with organic molecules (Gilfedder et al., 2011), it has been proposed as a potential tool for identifying relative changes in organic bonds in lake sediments (Kalugin et al., 2007). Therefore, variations in Br may reflect biological activity and are an indirect source of environment information (Harvey, 1980; Leonova et al., 2011). In addition, in Lake Teletskoye, a positive correlation was observed between the Br concentration and annual temperature (Kalugin et al., 2007). Therefore, it is possible that F3 reflects temperature. If so, the Br records suggest that the intervals of 901–1033 AD, 1283–1658 AD and 1870–2015 AD were relatively warm; and those of 1033–1283 AD and 1658–1870 AD were relatively cold.

In summary, there are potentially two processes that influence the elemental distribution within the sediment core: (1) detrital influx from the catchment and atmospheric dust; (2) evaporative concentration and biogeochemical cycling of essential plant nutrient elements.

4.3 Element ratios and their paleoclimatic implications

Element ratios such as Ca/Ti and Rb/Sr can provide useful paleoclimatic information. The Ca content of lake sediments can be used as an index of changing of precipitation-evaporation, or effective moisture; the concentration is high when evaporation is stronger than precipitation. The variation of the Ca/Ti ratio (Fig. 6f) can be used to divide the past 1100 years into five distinct intervals: extremely high values from 901–1033 AD and 1283–1658 AD; relatively high values from 1870–2015 AD; and low values from 1033–1283 AD and 1658–1870 AD. Based on the Ca/Ti record, the driest interval was 901–1033 AD and 1283–1658 AD, followed by 1870–2015; and wet intervals occurred from 1033–1283 AD and 1658–1870 AD.

This interpretation is consistent with the variation of the Rb/Sr ratio. Rb and Sr exhibit a high negative correlation in the sediments of Lake Toson ($r = -0.74$, $p < 0.001$). They are related to their respective parent minerals: Rb mainly occurs in muscovite, chlorite and other minerals which are resistant to weathering; in contrast, Sr is chemically mobile and mainly occurs in carbonate minerals which are easily weathered, producing dissolved Sr^{2+} during the weathering process (Zeng et al., 2012). Therefore, the Rb/Sr ratio is an indicator of chemical weathering. Low Rb/Sr values in lake sediments indicate intense chemical weathering under a wet climate, and in contrast high Rb/Sr indicates a dry climate. The Rb/Sr ratio in the Lake Toson sediments is less than 0.2 from 901–1033 AD, decreases to less than 0.1 from 1033–1283 AD, increases to less than 0.5 from 1283–1658 AD, decreases to less than 0.1 from 1658–1870 AD, and increases to less than 0.2 from 1870–2015 AD. Therefore, its variation is highly coherent with the Ca/Ti ratio; both ratios indicate dry intervals from 901–1033 AD, 1283–1658 AD and 1870–2015 AD; and wet intervals from 1033–1283 AD and 1658–1870 AD (Fig. 6).

4.4 Paleoclimatic changes

Our results reveal distinct changes in element assemblages and selected element ratios in the sediments of Toson Lake, which we interpret as changes in the environment of the lake and its catchment which were ultimately driven by climate. The results of PCA of the element concentrations, together with TOC content and Ca/Ti and Rb/Sr ratios, clearly indicate distinct oscillations (Fig. 6). We infer that the intervals from 901–1658 AD and 1870–2015 AD were overall relatively warm and dry, which could reflect the MWP and 20th century warmth, respectively. However, we infer that the interval from 1033–1283 AD during the WMP was cold and wet. Compared with the MWP, the 20th century warm interval was wetter due to the significantly lower Ca/Ti and Rb/Sr ratios. The cold and wet interval from 1658–1870 AD corresponds to the LIA. However, TOC is high under the warm and dry conditions during 937–1033 AD, 1283–1658 AD and 1870–2006 AD, which indicates that the TOC may not be of terrestrial origin since the flourishing of terrestrial plants within the catchment would require warmer and wetter conditions. Instead, we speculate that high TOC values may reflect the contribution of organic matter which originated from within the lake.

During warm and dry conditions, high salinity would favor the growth within the lake of specific organisms such as *Artemia, Dunaliella salina* and halophilic bacteria. *Artemia* is a genus of anostracan crustaceans, which are favored by the salinity range of 2.6–340 g/L (Ma, 1993). In addition, a salinity range from 20 to 100 g/L constitutes optimum conditions for the growth of *Dunaliella salina*. The modern salinity of Toson lake is 38.7 g/L, and therefore we speculate that the flourishing of these halophilic organisms may have led to high sedimentary TOC values during intervals when the climate was warm and dry. However, further research is need to verify this possibility.

Fig. 6 (e) illustrates a comparison of the sample scores for PCA F3 from Lake Toson (black line) and the alkenone-based temperature record from Lake Qinghai (red line) (Liu et al., 2006), which is close to Lake Toson. There is a good correspondence between the two records, indicating that warm and cold intervals identified at Lake Toson are consistent with those previously identified in Lake Qinghai. Small discrepancies between the records may reflect uncertainties in sediment chronologies and local climatic differences.

In addition, a mean annual precipitation of ~160 mm in Lake Toson area is limited, and the contribution of snow melt water from the Qilian Mountains to the north of Lake Toson is much smaller. We estimate the real evaporation of this area to be ~200 mm using the equation of Yang et al. (2010). This relationship is given by: $E_s = K \times E_l$, with drier areas having lower K values. Since the K value of Badain Jaran Desert is ~0.1, and Lake Toson is wetter than the Badain Jaran Desert, we estimate the real evaporation using the K value to be 0.1. Therefore, we suggest that evaporation may have played a significant role in determining the humidity of this arid region. The MWP in the region would have been characterized by strong evaporation due to the occurrence of the warmest temperatures observed during the study interval; significant evaporation would also have occurred during the 20th century warm interval. Conversely, during the LIA, lower temperatures would have been accompanied by reduced evaporation, and thus the climate was humid. On the whole, the variations in temperature and effective moisture at Lake Toson during the last millennium indicate the occurrence of warm-dry or cold-wet associations, which is consistent with observations from mid-latitude Westerly-dominated regions. For example, a synthesized moisture curve based on both high-resolution and low-resolution records indicates that a wet climate prevailed during the typically cold LIA interval from ca 1500–1850 AD, while a dry climate occurred during the typically warm MWP (Chen et al., 2010; Liu et al., 2011).

4.5 Spectral analysis

Spectral analysis of PCA F1, which reflects variations in detrital inputs, reveals high-frequency quasi-periodicities (3, 4 and 7 years), and a low-frequency component with a periodicity of 56 years [Fig. 7 (a)]. These results are significant at the 95% confidence level. In the case of F2 (interpreted as a drought factor) a quasiperiodicity of 4 years is also detected at the 95% confidence interval [Fig. 7 (b)]. These periodicities are similar to those of ENSO. Previous studies have revealed classic ENSO periodicities of 3–8 years (Allan, 2003) and ENSO-related multi-decadal periodicities of 50–70 years (Macdonald and Case, 2005). El Niño events cause a warming (cooling) in the tropical Pacific and Indian Oceans that tends to suppress (enhance) the monsoon (Torrence and Webster, 1999; Krishnamurthy and Goswami, 2000). The contribution of ENSO to global air temperature variations in the past 150 years is 10–30% (Wang et al., 2003). F3 (interpreted as a temperature factor) exhibits high-frequency quasi-periodicities of 4 and 7 years, which also indicate an ENSO effect on temperature variations. For reference, instrumental data from 33 weather stations in Qinghai Province, including Delingha station, indicate 11 El Niño events and 8 La Nina events from 1958 to 2005 (Xu, 2013).

ENSO-like variability has been inferred from tree ring (Wang et al., 2008; Fang et al., 2010) and ice core records (Wang et al., 2003; Grigholm et al., 2009) from the TP region; and Wang et al. (2003) reported a strong ENSO influence on climate change on the Tibetan Plateau. Based on a chronology of tree ring indices in the

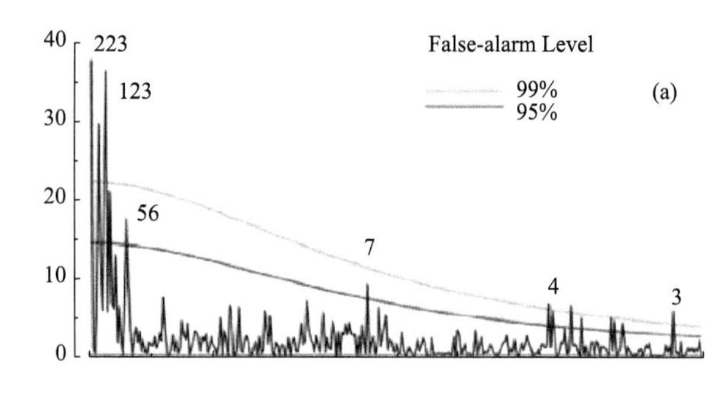

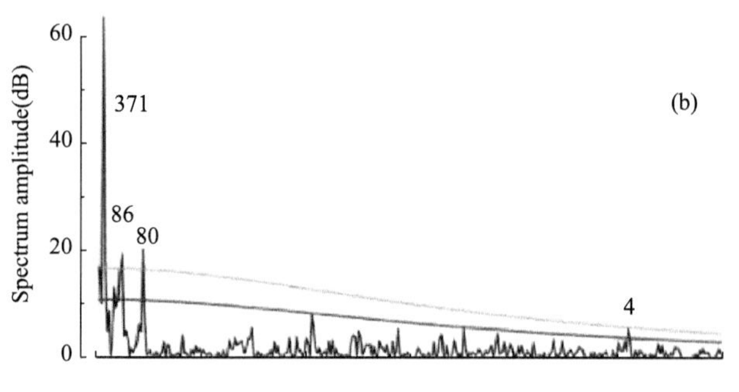

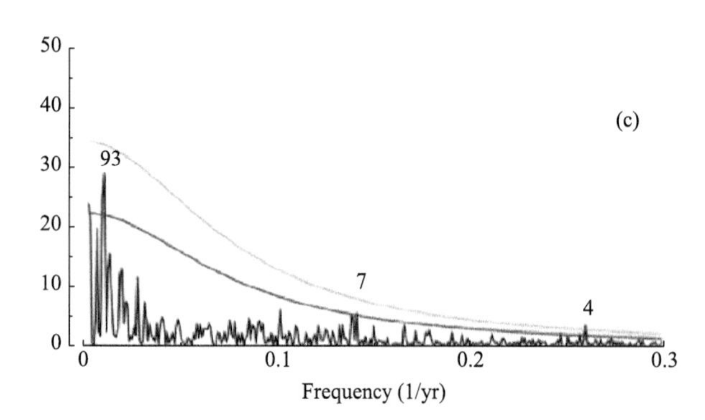

Fig. 7　Results of spectral analysis of the first three principal components (F1–F3, a–c) of a PCA of the element record for Lake Toson.

The spectra were estimated using the program REDFIT with OFAC = 4 and HIFAC = 1.1000 Monte Carlos simulations were used for the bias correction

southeastern TP, Fang et al. (2010) reconstructed local regional droughts and noted a correlation with ENSO during the last 568 years; and a tree ring index from the northern TP recorded five major drought events during the seventeenth century (1600–1700 AD) which are believed to have been a response to an interval of severe El Niño events (Wang et al., 2008). In addition, statistical analysis revealed a relationship between the Malan ice core $\delta^{18}O$ time series and the NAO (May-October) and SO (August-November) from 1887 to 1998. This indicates that variations in warm season air temperatures over the northern Tibetan Plateau were influenced by the strength of both the NAO and SO during the past century (Wang et al., 2003).

Spectral analysis also revealed periodicities of 223 and 123 years (99% confidence level) for PCA F1; periodicities of 37186 and 80 years (99% confidence level) for F2; and a periodicity of 93 years (95% confidence level) for F3. These periodicities could be linked with long-term changes in solar activity; previously reported periodicities of solar variations are 440, 360, 260, 230, 205, 200, 180, 168, 155, 147, 123, 106 and 88 years (Usoskin and Mursula, 2003; Ji et al., 2005; Zhao et al., 2009; Chu et al., 2013; Gou et al., 2014). Zhao et al. (2009) and Gou et al. (2014) reconstructed precipitation variations in the northeastern TP during the last millennium, and suggested a possible linkage with solar variations. Visual observation of a sediment redness record from Qinghai Lake also indicates cyclic variations in monsoonal moisture since the late glacial, and solar variations were interpreted to be an important cause of other cyclic environmental changes at Qinghai Lake (Ji et al., 2005).

5　Conclusions

SRXRF, which is non-destructive and yields an almost continuous record, has successfully been used to detect high-frequency paleoclimatic variations in the recent sediment record of Lake Toson. PCA was used to help interpret the large elemental dataset. PCA component 1 (F1) reflects detrital elements affected by precipitation and aeolian processes. F2 reflects mobile elements which are affected by evaporation-precipitation, and we suggest that its variations are related to drought conditions. F3 is driven by Br which may be a biologically-related element, and from its consistency with the alkenone-based temperature record from Lake Qinghai, we interpret it as reflecting temperature variations. Since there were no significant variations in precipitation during the study interval, evaporation is likely to be the main climatic variable controlled by temperature. A warm-dry interval correlative with the MWP and a cold-wet interval correlative with the LIA can be clearly identified. In addition, the 20th century warm period was drier than the LIA, but wetter the than the MWP. Finally, spectral analysis of the element record indicates that climate variability during the last millennium at Lake Toson was dominated by cyclical fluctuations with periodicities similar to those of ENSO and solar variability.

Acknowledgements

We thank the editors and two anonymous reviewers for their helpful comments. This project was supported by the Geological Survey Project of China (Grant No. 12120114048501). We are grateful to Aiguo Li, Chengwen Mao, Ke Yang and all members of the SSRF Synchrotron Radiation Facility.

Appendix A. Supplementary data

Supplementary data related to this article can be found at http: //dx.doi.org/10.1016/j.quaint.2017.03.067.

References

Allan, B.R.J., 2003. ENSO and Climatic Variability in the Past 150 Years. Cambridge University Press, Cambridge, U.K.

An, W.L., Liu, X.H., Leavitt, S.W., Ren, J.W., Sun, W.Z., Wang, W.Z., Wang, Y., Xu, G.B., Chen, T., Qin, D.H., 2012. Specific climatic signals recorded in earlywood and latewood $\delta^{18}O$ of tree rings in southwestern China. Tellus Ser. B-Chem. Phys. Meteorol. 64.

An, Z.S., 2000. The history and variability of the East Asian paleomonsoon climate. Quat. Sci. Rev.19 (1-5), 171-187.

Chawchai, S., Kylander, M.E., Chabangborn, A., Lö wemark, L., Wohlfarth, B., 2015. Testing commonly used X-ray fluorescence core scanning-based proxies for organic-rich lake sediments and peat. Boreas 45 (1), 180-189.

Chen, F.H., Chen, J.H., Holmes, J., Boomer, I., Austin, P., Gates, J.B., Wang, N.L., Brooks, S.J., Zhang, J.W., 2010. Moisture changes over the last millennium in arid central Asia: a review, synthesis and comparison with monsoon region. Quat. Sci. Rev.29, 1055-1068.

Chen, F.H., Yu, Z.C., Yang, M.L., Ito, E., Wang, S.M., Madsen, D.B., Huang, X.Z., Zhao, Y., Sato, T., John, B., Birks, H., Boomer, I., Chen, J.H., An, C.B., Wünnemann, B., 2008. Holocene moisture evolution in arid central Asia and its out-of-phase relationship with Asian monsoon history. Quat. Sci. Rev.27 (34), 351-364.

Chu, G.Q., Sun, Q., Li, S.Q., Ling, Y., Wang, X.H., Xie, M.M., Shang, W.Y., Li, A.G., Yang, K., 2013. Minor element variations during the past 1300 years in the varved sediments of Lake Xiaolongwan, north-eastern China. GFF 135 (3-4), 265-272.

Chu, G.Q., Sun, Q., Yang, K., Li, A.G., Yu, X.H., Xu, T., Yan, F., Wang, H., Liu, M.M., Wang, X.H., Xie, M.M., Ling, Y., 2011. Evidence for decreasing south Asian summer monsoon in the past 160 years from varved sediment in lake Xinluhai, Tibetan plateau. J. Geophys. Res. Atmos.116 (D2), 347-360.

Croudace, I.W., 2015. Micro-XRF studies of sediment cores. Dev. Paleoenviron. Res. 17.

Das, S.K., Routh, J., Roychoudhury, A.N., Klump, J.V., 2008. Major and trace element geochemistry in Zeekoevlei, South Africa: a lacustrine record of present and past processes. Appl. Geochem. 23 (8), 2496-2511.

Fagel, N., Boë s, X., Loutre, M.F., 2008. Climate oscillations evidenced by spectral analysis of Southern Chilean lacustrine sediments: the assessment of ENSO over the last 600 years. J. Paleolimnol.39 (2), 253-266.

Fang, K.Y., Gou, X.H., Chen, F.H., Li, J.B., D'Arrigo, R., Cook, E., Yang, T., Davi, N., 2010. Reconstructed droughts for the southeastern Tibetan Plateau over the past 568 years and its linkages to the Pacific and Atlantic Ocean climate variability. Clim. Dyn.35 (4), 577-585.

Francus, P., Lamb, H., Nakagawa, T., Marshall, M., Brown, E., Suigetsu 2006 project members, 2009. The potential of high-resolution X-ray fluorescence core scanning: applications in paleolimnology. Pages News 17 (3), 93-95.

Gilfedder, B.S., Petri, M., Wessels, M., Biester, H., 2011. Bromine species fluxes from Lake Constance's catchment, and a preliminary lake mass balance. Geochim. Cosmochim. Acta 75 (12), 3385-3401.

Gou, X.H., Yang, D., Chen, F.H., Yang, M.X., Gao, L.L., Nesje, A., Fang, K.Y., 2014. Precipitation variations and possible forcing factors on the Northeastern Tibetan Plateau during the last millennium. Quat. Res.81, 508-512.

Grießinger, J., Brä uning, A., Helle, G., Thomas, A., Schleser, G., 2011. Late Holocene Asian summer monsoon variability reflected by $\delta^{18}O$ in tree-rings from Tibetan junipers. Geophys. Res. Lett.38 (3), L03701-L03705.

Grigholm, B., Mayewski, P.A., Kang, S., Zhang, Y., Kaspari, S., Sneed, S.B., Zhang, Q., 2009. Atmospheric soluble dust records from a Tibetan ice core: possible climate proxies and teleconnection with the pacific decadal oscillation. J. Geophys. Res. Atmos.114 (D20), 311-311.

Harvey, G.R., 1980. A study of the chemistry of iodine and bromine in marine sediments. Mar. Chem.8 (4), 327-332.

Holmes, J.A., Cook, E.R., Yang, B., 2009. Climate change over the past 2000 years in Western China. Quat. Int.194 (1-2), 91-107.

Ji, J.F., Shen, J., Balsam, W., Chen, J., Liu, L.W., Liu, X.Q., 2005. Asian monsoon oscillations in the northeastern Qinghai-Tibet Plateau since the late glacial as interpreted from visible reflectance of Qinghai Lake sediments. Earth Planet. Sci. Lett.233 (1), 61-70.

Kalugin, I., Darin, A., Rogozin, D., Tretyakov, G., 2013. Seasonal and centennial cycles of carbonate mineralisation during the past 2500 years from varved sediment in Lake Shira, South Siberia. Quat. Int. 290-291 (2), 245-252.

Kalugin, I., Daryin, A., Smolyaninova, L., Andreev, A., Diekmann, B., Khlystov, O., 2007. 800-yr-long records of annual air temperature and precipitation over southern Siberia inferred from Teletskoye lake sediments. Quat. Res. 67 (3), 400-410.

Kelloway, S.J., Ward, C.R., Marjo, C.E., Wainwright, I.E., Cohen, D.R., 2015. Analysis of Coal Cores Using Micro-XRF Scanning Techniques, vol. 17 (6). Springer Netherlands, pp.721-731.

Krishnamurthy, V., Goswami, B.N., 2000. Indian monsoon-ENSO relationship on interdecadal timescale. J. Clim.13 (3), 579-595.

Leonova, G.A., Bobrov, V.A., Lazareva, E.V., Bogush, A.A., Krivonogov, S.K., 2011. Biogenic contribution of minor elements to organic matter of recent lacustrine sapropels (Lake Kirek as example). Lithol. Miner. Resour.46 (2), 99-114.

Ling, Y., Sun, Q., Zhu, Q.Z., Chen, D.L., Xu, W., Xie, M.M., Shan, Y.B., Wang, N., Chu, G.Q., 2014. Research on normalization method for element analysis of sediment with synchrotron radiation X-ray fluorescence (SRXRF) -An example of varved sediment in Lake Sihailongwan, Northeast China. Quat. Sci. 34 (6), 1327-1335 (In Chinese with English abstract).

Liu, W.G., Liu, Z.H., An, Z.S., Wang, X.L., Chang, H., 2011. Wet climate during the 'little ice age'in the arid Tarim basin, Northwestern China. Holocene 21 (3), 409-416.

Liu, X.H., Xu, G.B., Grießinger, J., An, W.L., Wang, W.Z., Zeng, X.M., Wu, G.J., Qin, D.H., 2014a. A shift in cloud cover over the southeastern Tibetan Plateau since 1600: evidence from regional tree-ring δ^{18}O and its linkages to tropical oceans. Quat. Sci. Rev.88, 55-68.

Liu, X.Q., Dong, H.L., Yang, X.D., Herzschuh, U., Zhang, E.L., Stuut, J.-B.W., Wang, Y.B., 2009. Late Holocene forcing of the Asian winter and summer monsoon as evidenced by proxy records from the northern Qinghai-Tibetan Plateau. Earth Planet. Sci. Lett.280 (1), 276-284.

Liu, X.Q., Yu, Z.T., Dong, H.L., Chen, H.-F., 2014b. A less or more dusty future in the Northern Qinghai-Tibetan Plateau? Sci. Rep.4, 6672-6672.

Liu, Z.H., Henderson, A.C.G., Huang, Y.S., 2006. Alkenone-based reconstruction of late-Holocene surface temperature and salinity changes in Lake Qinghai, China. Geophys. Res. Lett. 33 (9), 370-386.

Macdonald, G.M., Case, R.A., 2005. Variations in the Pacific decadal oscillation over the past millennium. Geophys. Res. Lett.32 (8), 93-114.

Margalef, O., Cortizas, A.M., Kylander, M., Pla-Rabes, S., Cañellas-Boltà, N., Pueyo, J.J., Séez, A., Valero-Garcés, B.L., Giralt, S., 2014. Environmental processes in Rano Aroi (Easter Island) peat geochemistry forced by climate variability during the last 70 kyr. Palaeogeogr. Palaeoclimatol. Palaeoecol.414, 438-450.

Ma, Z.Z., 1993. A study on the biogeography of brine shrimp. Mod. Fish. Inf. 8 (9), 20-26.

Mischke, S., Demske, D., Wünnemann, B., Schudack, M.E., 2005. Groundwater discharge to a Gobi desert lake during Mid and Late Holocene dry periods. Palaeogeogr. Palaeoclimatol. Palaeoecol. 225 (1), 157-172.

Mischke, S., Zhang, C., Börner, A., Herzschuh, U., 2010. Late glacial and Holocene variations of aeolian influx over the northeastern Tibetan Plateau recorded by laminated sediments of a saline meromictic lake. J. Quat. Sci. 25 (2), 162-177.

Muller, J., Kylander, M., Martinez-Cortizas, A., Wüst, R.A.J., Weiss, D., Blake, K., Coles, B., Garcia-Sanchez, R., 2008. The use of principle component analyses in characterising trace and major elemental distribution in a 55kyr peat deposit in tropical Australia: implications to paleoclimate. Geochim. Cosmochim. Acta 72 (2), 449-463.

Olsson, I.U., Kaup, E., 2001. The varying radiocarbon activity of some recent submerged estonian plants grown in the early 1990s. Radiocarbon 43 (2B), 809-820.

Pu, Y., Nace, T., Meyers, P.A., Zhang, H.C., Wang, Y.L., Zhang, C.L., Shao, X.H., 2013. Paleoclimate changes of the last 1000 yr on the eastern Qinghai-Tibetan Plateau recorded by elemental, isotopic, and molecular organic matter proxies in sediment from glacial Lake Ximencuo. Palaeogeogr. Palaeoclimatol. Palaeoecol. 379-380 (3), 39-53.

Qiang, M.R., Chen, F.H., Zhang, J.W., Gao, S.Y., Zhou, A.F., 2005. Climatic changes documented by stable isotopes of sedimentary carbonate in Lake Sugan, northeastern Tibetan Plateau of China, since 2 kaBP. Chin. Sci. Bull. 50 (17), 1930-1939.

Rothwell, R.G., Hoogakker, B., Thomson, J., Croudace, I.W., Frenz, M., 2006. Turbidite emplacement on the southern Balearic Abyssal Plain (western Mediterranean Sea) during marine isotope stages 1-3: an application of ITRAX XRF scanning of sediment cores to lithostratigraphic analysis. Geol. Soc. Lond. Spec. Publ. 267, 79-98.

Sano, M., Tshering, P., Komori, J., Fujita, K., Xu, C.X., Nakatsuka, T., 2013. May-September precipitation in the Bhutan Himalaya since 1743 as reconstructed from tree ring cellulose δ^{18}O. J. Geophys. Res. Atmos 118 (5), 8399-8410.

Shi, C., Daux, V., Risi, C., Hou, S.G., 2011. Reconstruction of southeast Tibetan Plateau summer cloud cover over the past two centuries using tree ring δ^{18}O. Clim. Past Discuss.7 (3), 1825-1844.

Sun, Q., Shan, Y., Sein, K., Su, Y.L., Zhu, Q.Z., Wang, L., Sun, J.M., Gu, Z.Y., Chu, G.Q., 2016. A 530-year-long record of the indian summer monsoon from carbonate varves in maar lake Twintaung, Myanmar. J. Geophys. Res. Atmos. 121 (10), 5620-5630.

Thompson, L.G., Mosley-Thompson, E., Davis, M.E., Bolzan, J.F., Dai, J., Yao, T.D., Gundestrup, N., Wu, X., Klein, L., Xie, Z., 1989. Holocene-late pleistocene climatic ice core records from Qinghai-tibetan plateau. Science 246, 474-477.

Thompson, L.G., Mosley-Thompson, E., Henderson, K.A., 2000. Ice-core palaeoclimate records in tropical south America since the last glacial maximum. J. Quat. Sci.15 (4), 377-394.

Thompson, L.G., Yao, T.D., Davis, M.E., Henderson, K.A., Mosley-Thompson, E., Lin, P.-N., Beer, J., Synal, H.-A., Cole-Dai, J., Bolzan, J.F., 1997. Tropical climate instability: the last glacial cycle from a Qinghai-tibetan ice core. Science 276 (5320), 1821-1825.

Torrence, C., Webster, P.J., 1999. Interdecadal changes in the ENSO-monsoon system. J. Clim.12 (8), 2679-2690.

Usoskin, I.G., Mursula, K., 2003. Long-term solar cycle evolution: review of recent developments. Sol. Phys.218 (218), 319-343.

Wang, J.L., Yang, B., Qin, C., Kang, S.Y., He, M.H., Wang, Z.Y., 2014. Tree-ring inferred annual mean temperature variations on the southeastern Tibetan Plateau during the last millennium and their relationships with the Atlantic multidecadal Oscillation. Clim. Dyn.43 (3-4), 627-640.

Wang, N.L., Thompson, L.G., Davis, M.E., Mosley-Thompson, E., Yao, T.D., Pu, J.C., 2003. Influence of variations in NAO and SO on air temperature over the northern Tibetan Plateau as recorded by $\delta^{18}O$ in the Malan ice core. Geophys Res. Lett.30 (22), 92-106.

Wang, X.C., Zhang, Q.B., Ma, K.P., Xiao, S.C., 2008. A tree-ring record of 500-year drywet changes in northern Tibet, China. Holocene 18 (4), 579-588.

Wang, Y.B., Liu, X.Q., Herzschuh, U., Yang, X.D., Zhang, E.L., 2012. Potential high resolution climate record on the northern Tibetan Plateau. In: PAGES 3rd Varve Working Group Workshop (21-24, April 2012), Manderscheid, Germany.

Xie, M.M., Sun, Q., Wang, N., Zhu, Q.Z., Shan, Y.B., Li, A.G., Yang, K., Mao, C.W., Wang, X.S., Chu, G.Q., Liu, J.Q., 2015. High resolution elements geochemical record during the past 1200 years in Huguangyan Maar Lake. Quat. Sci.35 (1), 152-159 (In Chinese with English abstract).

Xu, H., 2001. Advance in research on the Holocene climate fluctuations. Geol. Geochem. 29, 9-16 (In Chinese with English abstract).

Xu, H., Hong, Y.T., Hong, B., 2012. Decreasing Asian summer monsoon intensity after 1860 AD in the global warming epoch. Clim. Dyn. 39 (7-8), 2079-2088.

Xu, J., 2013. The Impact of ENSO Events to Droughts and Floods in Qinghai Province. Master thesis (In Chinese).

Yan, H., Shao, D., Wang, Y.H., Sun, L.G., 2014. Sr/Ca differences within and among three Tridacnidae species from the South China Sea: implication for paleoclimate reconstruction. Chem. Geol.390, 22-31.

Yang, B., Qin, C., Wang, J.L., He, M.H., Melvin, T.M., Osborn, T.J., Briffa, K.R., 2014. A 3,500-year tree-ring record of annual precipitation on the northeastern Tibetan Plateau. Proc. Natl. Acad. Sci. U. S. A.111 (8), 2903-2908.

Yang, X.P., Ma, N.N., Dong, J.F., Zhu, B.Q., Xu, B., Ma, Z.B., Liu, J.Q., 2010. Recharge to the inter-dune lakes and Holocene climatic changes in the Badain Jaran Desert, western China. Quat. Res.73 (1), 10-19.

Yao, T.D., Li, Z.X., Thompson, L.G., Mosley-Thompson, E., Wang, Y.Q., Tian, L.D., Wang, N.L., Duan, K.Q., 2006. $\delta^{18}O$ records from Tibetan ice cores reveal differences in climatic changes. Ann. Glaciol. 43 (1), 1-7.

Yao, T.D., Shi, Y.F., Thompson, L.G., 1997. High resolution record of paleoclimate since the little ice age from the Tibetan ice cores. Quat. Int. 37 (2), 19-23.

Yao, T.D., Thompson, L.G., Duan, K.Q., Xu, B.Q., Wang, N.L., Pu, J.C., Tian, L.D., Sun, W.Z., Kang, S.C., Xiang, Q., 2002. Temperature and methane records over the last 2 ka in Dasuopu ice core. Sci. China 45 (12), 1068-1074.

Zeng, Y., Chen, J., Zhu, Z., Li, J., Wang, J., Wan, G., 2012. The wet little ice age recorded by sediments in Huguangyan lake, tropical south China. Quat. Int. 263 (3), 55-62.

Zhang, J.W., Chen, F.H., Holmes, J.A., Li, H., Guo, X.Y., Wang, J.L., Li, S., Lü, Y.B., Zhao, Y., Qiang, M.R., 2011. Holocene monsoon climate documented by oxygen and carbon isotopes from lake sediments and peat bogs in China: a review and synthesis. Quat. Sci. Rev.30 (15-16), 1973-1987.

Zhao, C., Yu, Z.C., Zhao, Y., Emi, I., 2009. Possible orographic and solar controls of Late Holocene centennial-scale moisture oscillations in the northeastern Tibetan Plateau. Geophys. Res. Lett.36 (1), 272-277.

Zhao, C., Yu, Z.C., Zhao, Y., Ito, E., Kodama, K.P., Chen, F.H., 2010. Holocene millennialscale climate variations documented by multiple lake-level proxies in sediment cores from Hurleg Lake, Northwest China. J. Paleolimnol.44 (4), 995-1008.

Zhao, Y., Yu, Z., Chen, F., Ito, E., Zhao, C., 2007. Holocene vegetation and climate history at Hurleg Lake in the Qaidam Basin, northwest China. Rev. Palaeobot. Palynol.145 (3-4), 275-288.

Zheng, M.P., Zhang, Y.S., Liu, X.F., Nie, Z., Kong, F.J., Qi, W., Jia, Q.X., Pu, L.Z., Hou, X.H., Wang, H.L., Zhang, Z., Kong, W.G., Lin, Y.J., 2016. Acta Geol. Sin. 90 (4), 1195-1235.

Zhou, A., Chen, F., Qiang, M., Yang, M., Zhang, J., 2007. The discovery of annually laminated sediments (varves) from shallow Sugan Lake in inland arid China and their paleoclimatic significance. Sci. China Ser. D Earth Sci.50 (8), 1218-1224.

New understanding of lithium isotopic evolution from source to deposit—a case study of the Qaidam Basin*

Linlin Wang[1], Yongsheng Zhang[1], Enyuan Xing[1], Qingqiang Meng[2] and Dongdong Yu[1]

1. MLR Key Laboratory of Saline Lake Resources and Environments, Institute of Mineral Resources, Chinese Academy of Geological Sciences, Beijing 100037, China
2. Petroleum Exploration and Production Research Institute, Sinopec, Beijing 100083, China

Objective

The lithium-rich deep brine of the western Qaidam Basin is the successor of the lithium-rich salt lake on the surface. The identification of the lithium source has been a topic of intense interest among geologists. The dissolved lithium in lithium-rich protoliths transported to the basin through surface river water is considered as an important source of lithium in deep brine. The geochemical characteristics of lithium isotopes as they move from source to deposit are key to determining the recharge patterns of lithium in sedimentary basins in surface environment. However, until now, there have been few studies on the evolution of lithium isotopes in the process from source to deposit in the western Qaidam Basin. By studying the lithium traces of different types of sediments in source areas and the lithium isotope geochemistry of sediments and rivers in mountainous areas, alluvial plains, and salt lakes, we here summarize the laws of lithium isotope fractionation in surface processes of the western Qaidam Basin and determine the source of lithium in deep brines.

Methods

Samples were taken from the southern part of the Altun Mountains, the Qimantag Mountain in eastern Kunlun Mountains, and the western Qaidam Basin. The tests included lithium isotopes and lithium concentrations analysis, and completed by the ALS Laboratory.

Results

The lithium isotopic compositions of 12 protolith samples lie between 2.3‰ and 12.7‰ with Li concentrations ranging from 0.4 to 43.0 ppm (Appendix 1), which is located within the average Li isotopic composition of the upper mantle (δ^7Li = 4‰ ± 2‰) (Wan et al., 2017). The lithium isotopic compositions of the river waters lie between 12.5‰ and 14.9‰ with Li concentrations ranging from 0.022 to 0.130 ppm. The lithium isotopic composition of the brine in the Gasikule salt lake is 17.0‰ with Li concentration being 15.9 ppm. The lithium isotopic compositions of the channel and salt lake sediments lie between −5.5‰ and −4.0‰ with Li concentrations ranging from 34.7 ppm to 76.9 ppm. The Li concentrations of the channel and salt lake sediments were significantly higher than in the protolith, indicating the adsorption of Li^+ by native clay minerals and secondary clay minerals during weathering and erosion of the protolith. Against a background of low temperature, rapid and low weathering, the loss of Mg^{2+} during the water-rock reaction is small, and the isomorphism of Li^+ and Mg^{2+} increases the value of δ^7Li in the river water, thus forming a distinct lithium-isotope fractionation. In

* 本文发表在：Acta Geologica Sinica-English Edition, 2018, 92 (5): 2048-2049

the water-rock reaction process, ^{7}Li is preferentially enriched in the fluid, and ^{6}Li is preferentially enriched in the residual phase. Therefore, the value of δ^7Li decreased in the order of river, protolith and channel deposit (Fig. 1). Among them, the δ^7Li of the river water gradually increased from the uplifted area to the alluvial plain area, and the δ^7Li of the channel deposit gradually decreased from the headwaters to the downstream areas [Fig. 1 (a) and (b)]. In addition, the study found that the δ^7Li of the deep brine was 4.7‰, which was between the lithium isotopic compositions of the surface river water (with an average value of 13.8‰), Gasikule salt lake brine (17.0‰) and hot spring water (with an average value of −10.36‰), and cold spring water (with an average value of −5.81‰) (Xiao et al., 1994). This indicates that the lithium in the deep lithium-rich brine of the western Qaidam Basin originates from the dissolution of lithium in the lithium-rich protolith and deep fluid.

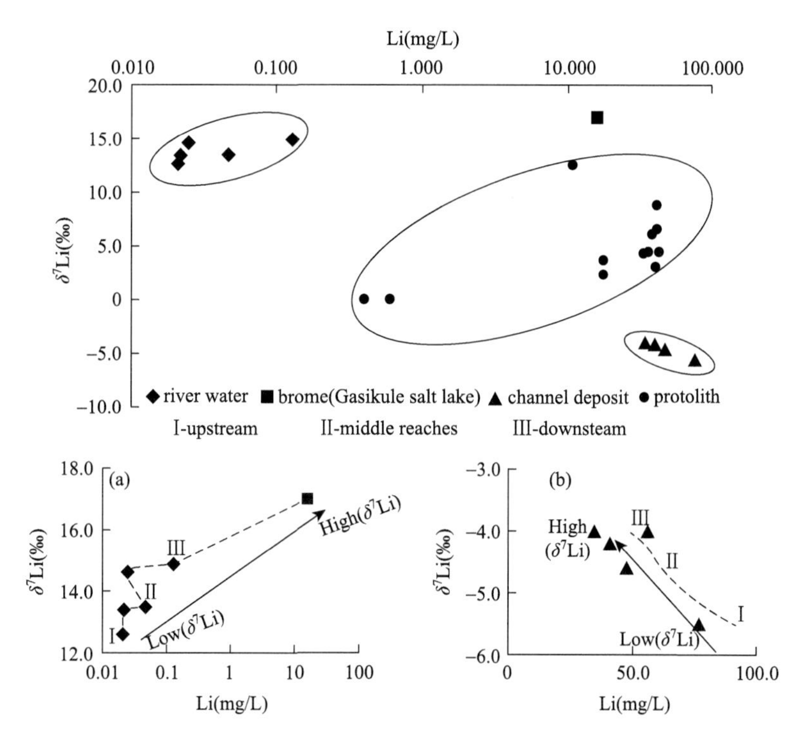

Fig. 1 Lithium isotopes and lithium concentrations in different media in western Qaidam Basin.

(a) Lithium isotopes and lithium concentrations in water at different locations between the source and the deposit; (b) Lithium isotopes and lithium concentrations in the channel sediments at different locations between the source and the deposit.

Conclusion

There is a distinct lithium-isotope fractionation from source to deposit in the brine of the western Qaidam Basin, and the value of δ^7Li decreased in the order of river, protolith and channel deposit. The lithium in the deep brine originates from the dissolution of lithium in the lithium-rich protolith and deep fluid.

Acknowledgments

This research was supported by the National Natural Science Foundation of China (grants No. 41572098 and 41541019).

References

Wan Hongqiong, Sun He, Liu Haiyang and Xiao Yilin, 2017. Lithium isotopic geochemistry in subduction zones: retrospects and prospects. *Acta Geologica Sinica* (English Edition), 91 (2): 688-710.

Xiao Yingkai, Qi Haiping, Wang Yunhui and Jin Lin, 1994. Lithium isotopic compositions of brine sediments and source water in Da Qaidam Lake, Qinghai, China. *Geochimica*, 23 (4): 329-338 (in Chinese with English abstract).

Appendix 1 Data of lithium isotope and Li concentrations in rivers, channel sediments, protoliths and Salt Lake in the western Qaidam Basin

Number	Sample	Li	δ^7Li (‰)	Number	Lithology of original rock	Li (ppm)	δ^7Li (‰)
R1	River water	0.021 (mg/L)	12.6	Q1	Granite	43.0	4.4
R2	River water	0.022 (mg/L)	13.4	Q3	Granite	41.0	3.0
R3	River water	0.047 (mg/L)	13.5	Q4	Granodiorite	41.6	6.6
R4	River water	0.025 (mg/L)	14.6	Q23	Diorite	17.7	2.3
R5	River water	0.130 (mg/L)	14.9	AL3	Andesite	33.8	4.3
L1	Brine of Gasikule Lake	15.90 (mg/L)	17.0	Q18	Hornfels	17.6	3.7
N16	Deep brine	241.59 (mg/L)	4.7	Q11	Marble	0.4	/
H2	Channel deposit	34.7 (ppm)	−4.0	AL5	Phyllite	41.2	8.8
H3	Channel deposit	41.0 (ppm)	−4.2	AL6	Mica schist	36.5	4.4
H4	Channel deposit	47.8 (ppm)	−4.6	Q14	Limestone	0.6	/
H5	Channel deposit	76.9 (ppm)	−5.5	Q15	Sandstone	38.1	6.1
HL1	Lakebed deposit	48.7 (ppm)	4.4	Q17	Sandstone	10.7	12.7

晚第四纪 MIS6 以来柴达木盆地成盐作用对冰期气候的响应*

陈安东[1]，郑绵平[1]，宋高[1]，王学锋[2]，李洪普[3]，韩光[3]，袁文虎[3]

1. 中国地质科学院矿产资源研究所，自然资源部盐湖资源与环境重点实验室，北京 100037
2. 中国科学院地质与地球物理研究所，新生代地质与环境重点实验室，北京 100029
3. 青海省柴达木综合地质矿产勘查院，青海格尔木 816099

内容提要 气候是控制柴达木盆地盐类沉积的主要因素之一，但是其作用机制尚待明确。作者以柴达木盆地察汗斯拉图盐湖的 3 个含盐剖面为研究对象，采用多接收电感耦合等离子质谱（MC-ICP-MS）铀系测年测定其沉积时代，并通过 X 射线粉晶衍射（XRD）分析测定其盐类矿物种类。铀系测年显示 D18 剖面石盐和芒硝层的沉积时代为 13.1±2.0ka BP～15.9±2.5ka BP，其中芒硝沉积年代属于末次冰期 MIS2 晚期；MXK2 剖面芒硝层的沉积时代分别为 131.7±39.5ka BP 和 158.3±10.8ka BP，D12 剖面芒硝层的沉积时代分别为 166.6±20.2ka BP 和 198.0±20.6ka BP，可以对应于倒数第二次冰期 MIS6。XRD 分析确定了 3 个剖面的盐类矿物主要为芒硝、石盐和石膏。综合多个盐湖晚第四纪成盐数据，本文认为倒数第二次冰期 MIS6 和末次冰期 MIS2 是柴达木盆地晚第四纪重要的成盐期，冰期的冷干气候有利于石盐和芒硝等盐类沉积。柴达木盆地"冰期成盐"的根本原因，是由于冰期环境下盆地周边山体冰川规模的扩张以及干冷的冰期气候，共同造成了盐湖补给水量的减少。此外，晚第四纪 MIS6 和 MIS2 的冰期降温也是盆地中冷相盐类沉积的直接原因。

关键词 柴达木盆地；盐湖；成盐期；倒数第二次冰期；末次冰期；铀系测年

世界上大型海相蒸发岩矿床通常是在干旱气候背景下沉积的，其中大多为"干热"气候（Herrero et al.，2015；Warren，2016）。因为盐类沉积通常是由于太阳能蒸发作用导致蒸发量超过补给水量，进而引起水体浓缩和盐类沉积，而"干热"的气候模式有利于这种成盐作用的进行（Herrero et al.，2015；Warren，2010，2016）。有研究表明加拿大麋鹿岬（Elk Point）盆地泥盆纪蒸发岩、美国新墨西哥盆地二叠纪蒸发岩、美国密执安盆地志留纪蒸发岩、欧洲蔡希斯坦盆地二叠纪蒸发岩矿床均形成于"干热"的气候条件下（钱自强等，1994；张永生等，2014）。此外，中国陕北奥陶纪、四川盆地三叠纪和新疆库车盆地古近纪等海相蒸发岩矿床均形成于高温气候背景下（张永生等，2014；赵艳军等，2015；刘成林等，2016）。

但是，在柴达木盆地和青藏高原等高海拔地区，却存在寒冷气候条件下的陆相盐类沉积。柴达木盆地各大盐湖区在第四纪期间均有盐类沉积，且大多具有冷相盐类组合特征（郑绵平等，2010）。而第四纪气候的主要特征是冰期与间冰期的交替发生（刘嘉麒等，2001），但对于冰期气候在柴达木盆地盐类沉积过程中所起的作用有待明确。沈振枢等（1993）提出柴达木盆地在冷期的盐层厚度明显大于暖期，黄麒等（2007）指出柴达木盆地各大盐湖区在末次冰期 MIS2 均有石盐沉积。陈安东等（2017）对中更新世以来柴达木盆地成盐期与青藏高原冰期序列进行了系统的对比，认为冰期环境有利于石盐和芒硝等盐类沉积。此外，对于气候因素在柴达木盆地盐类沉积过程中所起的作用，回答"高温成盐"还是"冰期成盐"这一科学问题，学术界尚存在不同的认识（郑绵平等，1989，1998；赵艳军等，2015；Chen Andong et al.，2017a；陈安东等，2017）。

2018 年作者在柴达木盆地西部察汗斯拉图盐湖开展矿产地质调查期间，采集了该盐湖的第四纪盐类矿物样品。本文以该盐湖近期开挖的卤水沟和露天矿坑的 3 个含芒硝剖面为研究对象，通过 XRD 分析和 MC-ICP-MS 铀系测年，测定其盐类矿物种类和沉积年代。通过本项研究及柴达木盆地多个盐湖晚第四纪

* 本文发表在：地质论评，2020，66（3）：611-624

成盐数据，本文提出 MIS6 以来成盐期与冰期和深海氧同位素阶段的对比方案，并对晚第四纪冰期气候在盐类沉积过程中所起的作用进行探讨，进一步阐明柴达木盆地“冰期成盐”机制。

1　区域概况

柴达木盆地位于青藏高原东北部 90°00′～98°20′E，35°55′～39°10′N 之间，面积 121000km^2，盆地内部海拔一般在 2675～3350m（图 1）。盆地周边被东昆仑山脉、祁连山脉和阿尔金山脉环绕。海拔 4800～5000m 以上的高山发育大陆型山岳冰川，加之昆仑山脉和祁连山脉腹地辽阔，因此构成了柴达木盆地主要的水源补给区（袁见齐等，1983；张彭熹，1987）。阿尔金山脉腹地相对较小，没有大型河流注入柴达木盆地。

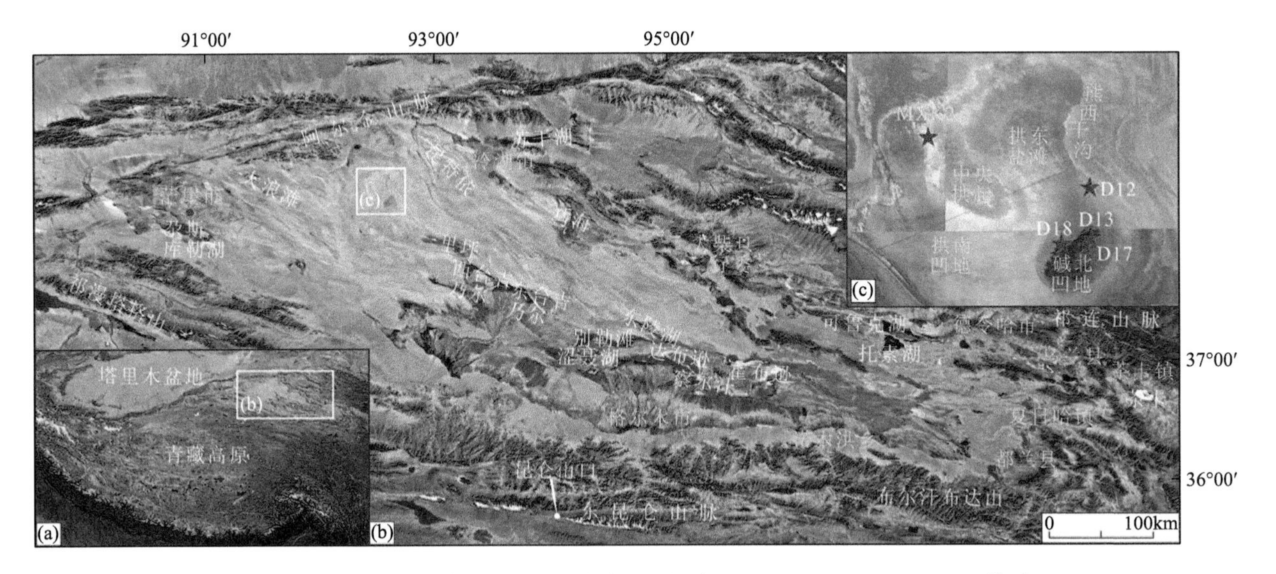

图 1　柴达木盆地西部察汗斯拉图盐湖位置图（据 Chen Andong et al.，2018 修改）

Fig. 1　Location of the study area. (modified after Chen Andong et al., 2018)

（a）青藏高原；（b）柴达木盆地；（c）察汗斯拉图盐湖；红色五角星示本文剖面位置

(a) Qinghai-Xizang (Tibetan) Plateau; (b) Qaidam Basin; (c) Chahansilatu Playa; Red star indicating the sampling sites

柴达木盆地干燥寒冷，少雨多风，具有高原大陆性气候特征。盆地内部月平均最低温和最高温分别为 −15～−10℃（1 月）和 15～17℃（7 月），盆地西部年均气温为 3℃（张彭熹，1987）。盆地西部年降水量低于 20mm，而蒸发量为 2000～3000mm（张彭熹，1987）。柴达木盆地极端干旱的气候有利于盐湖发育和盐类沉积。

本文研究区察汗斯拉图位于柴达木盆地的西北部，该盐湖西部为阿尔金山脉，东北方向为昆特依，西南方向为大浪滩，东南方向为一里坪（图 1）。该盐湖是一个干盐湖，地表主要出露下更新统—全新统的湖相和盐类化学沉积。该盐湖晶间卤水的水化学类型为硫酸镁亚型（袁见齐等，1995）。察汗斯拉图是柴达木盆地重要的芒硝矿区，芒硝矿层在该盐湖的中央拱起、拱南凹地和碱北凹地的中、上更新统地层中广泛分布。中央拱起是察汗斯拉图芒硝矿的主要采矿区，地表的芒硝层可厚达数米。碱北凹地的芒硝层中含有较多石盐，此外还是该盐湖钾镁盐沉积的核心部位。而拱南凹地的芒硝中含有一些石盐，但不含钾镁盐层。

本文的研究对象 D18 剖面位于碱北凹地卤水沟的最北端，采样位置均位于工业卤水的水位以上，避免了其他地层的卤水对样品造成污染；MXK2 剖面位于中央拱起西北部，是一个芒硝矿露天采矿坑；D12 剖面位于碱北凹地北部，是一个探矿坑。以上 3 个剖面均为含芒硝剖面，剖面照片及其中主要盐类矿物照片见图 2，剖面位置等信息见表 1。此外，作者还对碱北凹地中心 D13 和 D17 剖面开展了研究，但这两个剖面下部的盐层受到工业卤水污染，且剖面上部未找到原生石盐层，无法得出有效的铀系年代序列。

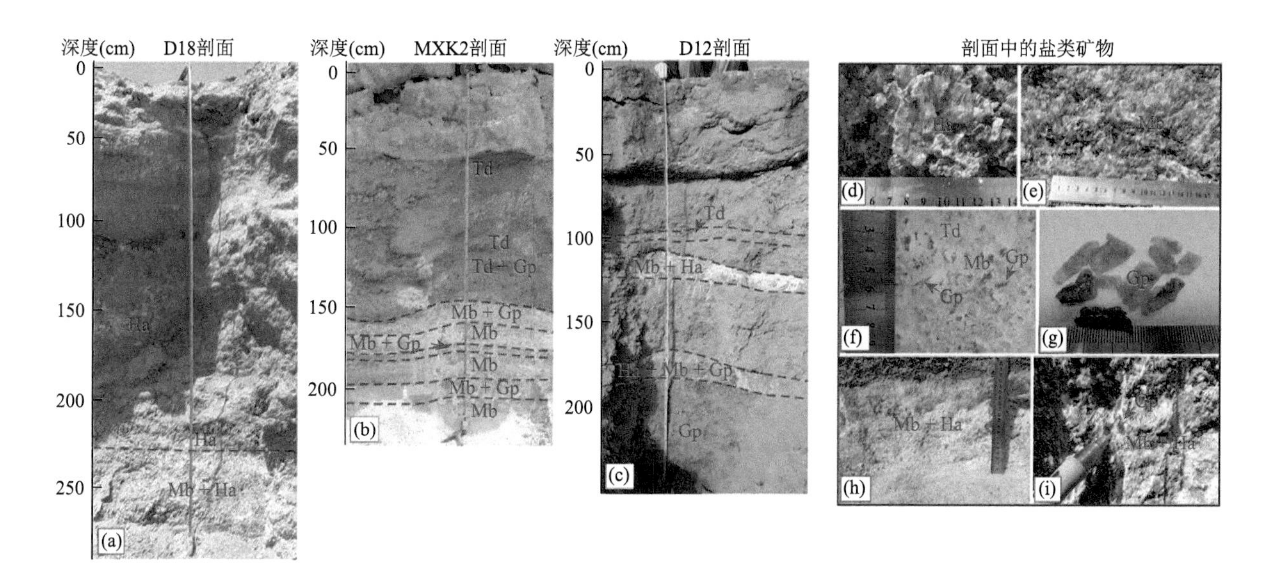

图2 柴达木盆地察汗斯拉图含盐地层剖面及其中主要盐类矿物

Fig. 2 Photos of salt-bearing profiles and main salt minerals in the Chahansilatu playa.

图中（a）、（b）和（c）分别为D18、MXK2和D12剖面；（d）为D18剖面石盐晶体；（e）为D18剖面底部的含石盐芒硝层；（f）为MXK2剖面的无水芒硝、芒硝和石膏；（g）为MXK2剖面石膏样品MXK2-3；（h）为D12剖面的含石盐芒硝层；（i）为D12剖面含石盐芒硝层中的石膏。Gp-石膏；Ha-石盐；Mb-芒硝；Td-无水芒硝

The profiles are D18 (a), MXK2 (b) and D12 (c); photos are showing halite crystals of D18 profile (d); mirabilite layer with halite of D18 profile (e); thenardite, mirabilite and gypsum of MXK2 profile (f); gypsum dating sample MXK2-3 of the MXK2 profile (g); mirabilite layer with halite of D12 profile (h); gypsum crystals in the mirabilite and halite layer of D12 (i). Gp-gypsum; Ha-halite; Mb-mirabilite; Td-thenardite

表1 柴达木盆地察汗斯拉图含盐剖面基本信息

Table 1 Information of salt-bearing profiles in the Chahansilatu playa

剖面名称	地理位置	坐标（E/N）	海拔高度（m）	主要盐类矿物
D18	碱北凹地西北部	92°35′2.9″/38°20′44.3″	2698	石盐、芒硝
MXK2	中央拱起西北部	92°24′43.3″/38°28′53.8″	2708	芒硝、石膏
D12	碱北凹地以北	92°37′15.3″/38°25′21.8″	2702	芒硝、石盐、石膏

2 样品处理和测试

2.1 XRD分析

为了准确地鉴定剖面中的盐类矿物种类，作者共选择了45个盐类矿物样品进行XRD分析。测试仪器为日本理学Rigaku MiniFlex 600型X射线粉晶衍射仪，测试条件：CuKa靶，电压/电流为40kV/15mA，扫描角度为3°～70°，步长为0.02°。盐类矿物样品均密封低温保存且未烘干，剔除晶体表面杂质后研磨至100目。XRD测试在自然资源部盐湖资源与环境重点实验室测试完成。

2.2 MC-ICP-MS铀系测年

在MXK2剖面中，测年样品选择地层中的原生石膏晶体，该类石膏晶体通常呈板柱状，无色透明或者含少量黑色杂质，长1～2cm，呈薄层状夹在芒硝层之间或芒硝层中[图2（f）和（g）]。依据石膏的形态及其在地层中的赋存状态，这种石膏通常被判定为原生石膏（Cody et al.，1988；Mees et al.，2012）。石膏样品共计3个，采样深度见表2。在样品前处理过程中，首先将石膏晶体在超声波清洗机中用酒精洗去杂质，碾碎后挑选其中无色透明的碎片研磨成200目以下的粉末。

表 2　察汗斯拉图剖面 ^{230}Th 测年结果（±2σ）

Table 2　^{230}Th dating results of the profiles（±2σ）

样品编号	测年矿物	采样深度（cm）	^{238}U（×10^{-9}）	^{232}Th（×10^{-12}）	$n(^{230}Th)/n(^{232}Th)$（×10^{-6}）	δ^{234}U（测量值）	$\frac{A(^{230}Th)}{A(^{238}U)}$	^{230}Th 年代（a）（未校正）	^{230}Th 年代（a）（校正）	$\delta^{234}U_{初始}$（校正）	^{230}Th 年代（ka BP）（校正）
D18-1	石盐	190～195	17.7±0.04	4178.2±84.0	11.9±0.2	253.8±2.5	0.1710±0.0009	15913.1±93.0	13157.5±1953.3	263.4±3.0	13.1±2.0
D18-2	石盐	215～220	15.7±0.03	4706.0±94.7	11.4±0.2	265.9±2.8	0.2070±0.0010	19352.7±112.4	15878.5±2463.4	278.1±3.5	15.9±2.5
D18-3	石盐	270	0.5±0.001	159.5±3.2	9.9±0.7	240.6±8.8	0.1898±0.0137	18007.4±1415.0	14265.6±2983.1	250.5±9.4	14.2±3.0
MXK2-3	石膏	155～160	69.9±0.24	116766.9±2368.0	9.4±0.2	139.2±3.2	0.9484±0.0061	181548.1±3223.4	131686.1±39462.5	201.9±22.8	131.7±39.5
MXK2-5	芒硝	198～211	5.49±0.01	3385.1±67.9	25.9±0.5	178.4±2.9	0.9677±0.0038	173257.5±1960.1	158304.7±10845.3	278.9±9.6	158.3±10.8
D12U2	石膏	90	28.4±0.05	31686.0±634.8	15.7±0.3	224.6±11.7	1.0594±0.0090	193255.8±6905.1	166644.9±20231.6	359.5±27.8	166.6±20.2
D12U3	石膏	110	44.0±0.07	51882.9±1039.7	15.7±0.3	226.4±5.5	1.1225±0.0056	225481.0±4995.0	197985.3±20609.9	395.9±25.0	198.0±20.6
GBW04412	-	-	10550.5±16.3	5569.4±112.3	33483.9±677.7	851.1±2.3	1.0720±0.0024	87068.1±327.2	87060.7±327.2	1088.1±3.1	87.0±0.3

注：标准物质 GBW04412 的年代为 85.00±4.00 ka（王立胜等，2016）；D18 剖面 ^{230}Th 年龄利用 $n(^{230}Th)/n(^{232}Th)$值为（2.2±1.1）×10^{-6}进行校正，参考柴达木盆地西部一里坪 15YZK01 深钻的碎屑物实测数据；MXK2 和 D12 剖面的 ^{230}Th 年龄利用 $n(^{230}Th)/n(^{232}Th)$值为（4.4±2.2）×10^{-6}，参考地壳平均值；BP（距今）中的“今”为公元 2000 年。$\frac{A(^{230}Th)}{A^{238}U}$ 为放射性活度比值。$\delta^{234}U=\left\{\left[\frac{A(^{234}U)}{A(^{238}U)}\right]-1\right\}\times1000$。

由于盐湖中沉积的芒硝易脱水且易溶化，系统封闭性较差，不宜作为铀系测年材料。芒硝晶体暴露在空气中容易脱去结晶水（$Na_2SO_4 \cdot 10H_2O \longrightarrow Na_2SO_4 + 10H_2O$），由无色透明晶体变成白色粉末[图 2（e），（f）和（h）]，在柴达木盆地的大风天气下，这一过程仅需要数十分钟就可以完成。而密封的、含石盐杂质的芒硝会随着温度升高而溶化，在夏季很容易完成这一过程。所以作者对采集的芒硝样品密封且低温（<10℃）保存，即便如此，在样品处理过程中也难免有部分芒硝脱水或者溶化。此外，石盐在地下卤水中可能会发生重结晶，系统封闭性也不如石膏。本文所采用的石盐测年材料，作者谨慎地选择 D18 剖面中薄层状的白色石盐，以及芒硝层中的石盐晶体。

作者并不试图强调石盐和芒硝也适合作为铀系测年材料，但是在有些盐湖剖面中没有其他更合适的测年材料可选的情况下，不妨采用石盐和芒硝作为尝试。而且，石盐和芒硝作为铀系测年材料，也已经取得了一系列成果（沈振枢等，1993；黄麒等，2007；侯献华等，2010；Han Wenxia et al.，2014）。本文用来测年的芒硝样品共计 1 个，石盐样品共计 3 个，采样深度见表 2。芒硝和石盐样品均在清理干净表面杂质后机械破碎，挑选无色透明的碎块研磨至 10 目以下。

本文用来测年的石膏、石盐和芒硝样品均利用 2mol/L 稀盐酸溶解，然后进行分离。采用多接收电感耦合等离子体质谱法（multiple collector inductively coupled plasma mass spectrometry，MC-ICP-MS）进行铀系定年，使用美国 Thermo-Fisher 公司的 Neptune Plus 型多接收电感耦合等离子体质谱仪测定样品中 U 和 Th 的同位素比值，最后计算出 ^{230}Th 年龄。在测试过程中，使用 GBW04412 标准物质控制测试误差。样品的分离纯化和质谱测量流程参照王立胜等（2016）的实验方法。测试实验在中国科学院地质与地球物理研究所铀系年代学实验室完成。

3 测试结果

3.1 主要盐类矿物

D18 剖面的盐类矿物以石盐（NaCl）和芒硝（$Na_2SO_4 \cdot 10H_2O$）为主，卤水沟底部有次生的石膏（$CaSO_4 \cdot 2H_2O$）晶体。作者在 D18 剖面未发现固体钾盐矿物，而对更靠近碱北凹地湖中心的 D13 剖面和 D17 剖面的细砂层中的盐类矿物进行 XRD 分析，发现其中的盐类矿物主要为钾石膏（$K_2Ca(SO_4)_2 \cdot H_2O$）、光卤石（$KMgCl_3 \cdot 6H_2O$）、钾石盐（KCl）和杂卤石（$K_2Ca_2Mg(SO_4)_4 \cdot 2H_2O$）。在 D13 和 D17 剖面中，钾石膏呈片状赋存在全新统细砂层中，钾石膏表面有白色粉末状的光卤石、杂卤石和钾石盐。在这两个剖面中，也有一些光卤石和杂卤石呈白色斑点状分布在细砂层中。在察汗斯拉图盐湖，地表的固体钾盐矿物大多分布在碱北凹地，而碱北凹地是该盐湖直到晚更新世末期—全新世才干涸的区域。

MXK2 剖面的芒硝相对较为纯净，芒硝层中除了石膏晶体，XRD 分析未检出其他盐类矿物。在 MXK2 剖面盐壳以下紧邻的粉砂层中，含有次生的石盐、石膏和芒硝（已脱水成无水芒硝）。D12 剖面的芒硝层中均含有较多的石盐，下部芒硝层中除了石盐还有石膏晶体。D12 剖面盐壳以下紧邻的粉砂层中，含有次生的石膏和芒硝（已脱水成无水芒硝）；底部的灰泥和粉砂层中含有次生的石膏晶体，灰泥中的碳酸盐矿物以文石和方解石为主，并含少量菱镁矿。

芒硝是典型的冷相盐类矿物，在低于 32℃的卤水中，Na_2SO_4 的溶解度随着温度降低而减小，因而芒硝更易在冰期的冷环境下析出，甚至在冰下析出（Last et al.，2005；Grasby et al.，2013；Herrero et al.，2015）。按照其成因，芒硝甚至不能被称为“真正的”蒸发岩矿物（Warren，2016）。芒硝作为冷相矿物，形成于年均温−7～−3℃以下且达到 7 个月以上的冷环境才能形成稳定的芒硝层（郑绵平等，1998，2016）。因此，芒硝对冰期冷环境带来的降温比较敏感。而石盐是广温相盐类矿物，石盐的沉积与卤水浓缩相关（郑绵平等，1998，2016）。

3.2 铀系测年结果

D18 剖面的石盐铀系年龄分别为 13.1±2.0ka BP（D18-1）、15.9±2.5ka BP（D18-2）和 14.2±3.0ka BP（D18-3）（表 2，图 3）。其中，D18-1 样品采自剖面 190～195cm 的石盐层；D18-2 采自 215～220cm 的石

盐层，该石盐层以深为含石盐的芒硝层；D18-3 采自剖面底部芒硝层中的石盐晶体。该剖面的芒硝沉积时代可以归属于末次冰期 MIS2。这 3 个样品的数据误差均超过 15%，实际上，对于柴达木盆地盐湖剖面顶部和钻孔浅部盐类矿物的铀系数据，以往在一里坪和东陵湖也出现过误差过大的问题，主要是由于样品中初始 Th 的含量较高导致（陈安东等，2017；Chen Andong et al.，2018）。D18 剖面 3 个样品的数据在误差范围内基本一致，均归属于晚更新世晚期，符合该区域的地质资料。前人对察汗斯拉图 ZK5025 钻孔近地表盐壳测定的 ^{14}C 数据也表明该盐湖在 MIS2 干涸（沈振枢等，1993；黄麒等，2007），与本文的铀系数据接近，表明我们的数据虽然误差比较大，但还是基本可靠的。

MXK2 剖面的石膏和芒硝铀系年代分别为 131.7±39.5ka BP（MXK2-3）和 158.3±10.8ka BP（MXK2-5）（图 3），其中 MXK2-3 出现较大的测年误差，如果不考虑测年误差的话，和 MXK2-5 均可归属于倒数第二次冰期 MIS6（130～191ka）。D12 剖面的石膏铀系年代分别为 166.6±20.2kaBP（D12U2）和 198.0±20.6kaBP（D12U3）（图 3），其中 D12U2 的年代属于 MIS6，但 D12U3 的年代略早于 MIS6，如果不考虑测年误差，甚至可以归属 MIS7。但是须知 MXK2 和 D12 剖面都是短剖面，而柴达木盆地盐湖的沉积速率往往较大，短剖面盐层的年代不可能相差太大。MXK2 剖面和 D12 剖面的铀系年代均归属于中更新世晚期，符合该区域的地质资料。

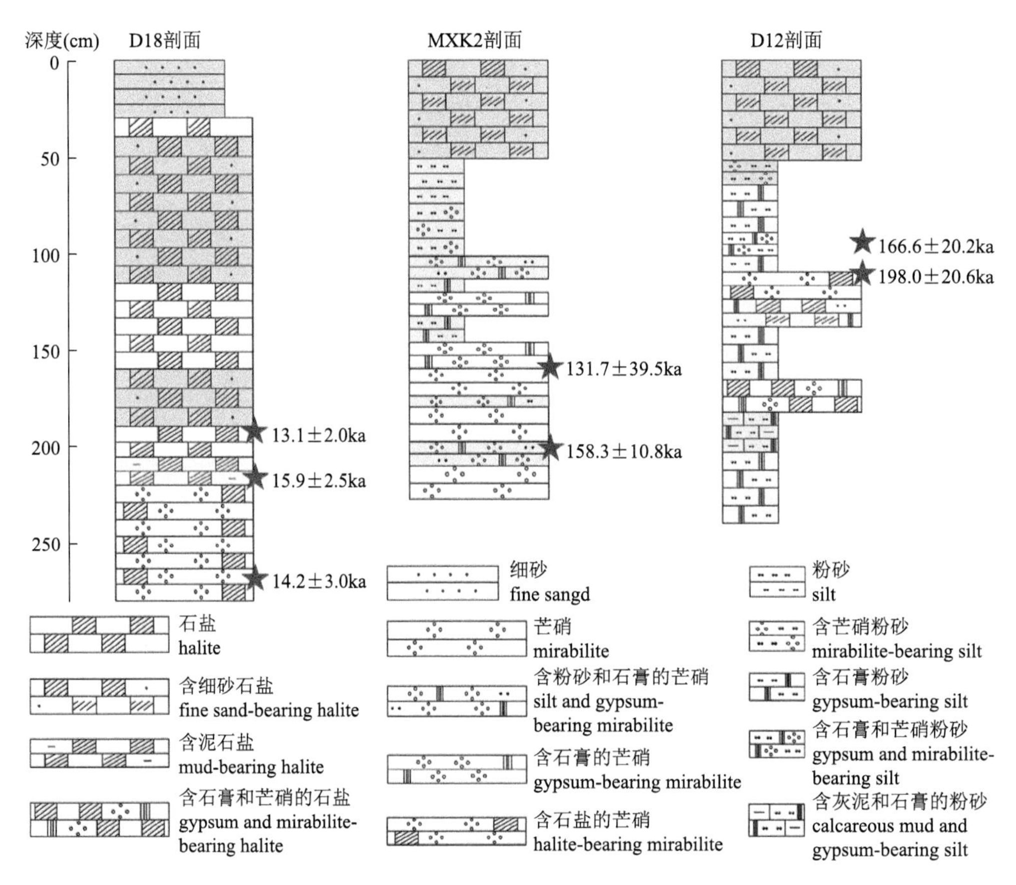

图 3　柴达木盆地察汗斯拉图剖面柱状图及其成盐时代

Fig. 3　Lithology and dating results of the profiles of the Chahansilatu playa.

4　分析与讨论

4.1　柴达木盆地盐湖 MIS6 以来成盐期与冰期的对比

中更新世以来的青藏高原冰期序列已经建立（施雅风，2002；易朝路等，2005；赵井东等，2011；崔之久等，2011），尤其是 MIS6 以来的冰期/间冰期温度和降水量的变化也已经取得了大量的研究成果（李吉均等，1991；潘保田等，1994，1997；施雅风，1998，2002；侯光良等，2019）。此外，近年来，柴达木盆地成盐期数据的不断积累（沈振枢等，1993；黄麒等，2007；Zhang Weilin et al.，2012；Wang Jiuyi et al.，2013；Chen Andong et al.，2017b；陈安东等，2017），使我们有条件将柴达木盆地成盐期与青藏高原第四纪冰期进行时间与空间上的系统对比。

为了将柴达木盆地成盐期与末次冰期 MIS2 进行对比，作者对多个盐湖在末次冰期的盐类沉积和测年数据进行了总结（表 3）。大量的同位素年代数据显示柴达木盆地西部盐湖在 MIS2 均有石盐沉积，个别盐湖还有芒硝和泻利盐沉积；在 MIS2 晚期盐湖面积萎缩，干盐滩面积扩张（表 3）。而柴达木盆地东部察尔汗盐湖成盐较晚，袁见齐等（1995）曾提出察尔汗盐湖的四次成盐期：31.0～25.8ka、24.7～20.0ka、19.0～16.5ka 和 15～8ka，其中大多数时间在 MIS2 范围内。在本文中，察汗斯拉图 D18 剖面的芒硝沉积年代对应于末次冰期 MIS2，与柴达木盆地东、西部盐湖在 MIS2 出现盐类沉积是可以对比的。所以，柴达木盆地东、西部盐湖在 MIS2 均有石盐沉积。

表 3　末次冰期早冰阶至新仙女木（YD）冰进期间柴达木盆地盐类沉积的测年数据

Table 3　Dating results of salt deposits in the Qaidam Basin since the Early Glacial Stage of the Last Glaciation to Younger Dryas event

盐湖	钻孔或采样点及位置（E/N）	盐类矿物及深度（m）	测年方法和数据（ka）	参考文献
大浪滩	ZK336 钻孔 91°25′15″/38°32′06″	石盐和芒硝（9.84）	^{14}C：31.26±1.90	沈振枢等，1993；黄麒等，2007
		石盐和芒硝（12.0）	^{230}Th：18.70±2.10	
	ZK312 钻孔 91°25′15″/38°32′06″	泻利盐（6.55）	^{230}Th：16.90±3.20	
		石盐（38.39）	^{230}Th：22.90±1.30	
		石盐和石膏（40.20）	^{230}Th：26.30±3.30	
		石盐和石膏（63.55）	^{14}C：26.77±0.60	
	ZK402 钻孔 91°21′25″/38°27′56″	石盐（1.35）	^{230}Th：27.90±9.10	
		石盐（3.35）	^{14}C：25.61±0.48	
		石盐（8.84）	^{14}C：31.30±3.1	
	梁 ZK02 钻孔 91°27′/38°33′	石盐（1.50）	^{230}Th：14.9±1.0	侯献华等，2010
		石盐（9.00）	^{230}Th：24.2±0.8	
		石盐（18.60）	^{230}Th：31.1±1.4	
察汗斯拉图	ZK5025 钻孔 92°36′35″/38°19′14″	石盐（1.70）	^{14}C：29.8±2.4	黄麒等，2007
		石盐（1.70）	^{14}C：29.79±2.42	沈振枢等，1993
	D18 剖面 92°35′2.9″/38°20′44.3″	石盐（1.90～1.95）	^{230}Th：13.1±2.0	本文
		石盐（2.15～2.20）	^{230}Th：15.9±2.5	
		石盐（2.70）	^{230}Th：14.2±3.0	
昆特依	地表盐壳	盐（0.10）	^{14}C：15.7±4.2	黄麒等，2007
		盐（0.10）	^{14}C：14.1±0.2	
	ZK3208 钻孔 93°02′03″/38°34′46″	盐（3.82）	^{230}Th：30.6±3.3	
		石膏（14.46）	^{230}Th：65.1±8.5	
一里坪	82CK1 钻孔	石盐（7.74）	^{14}C：12.1±0.65	黄麒等，2007
		石盐（15.20）	^{14}C：25.8±0.54	
		石盐（18.19）	^{14}C：32.5±0.65	
	15YZK01 钻孔 93°11′57″/37°58′03″	石盐（13.5～0.0）	32.5～2.0（据 ^{230}Th 年代计算）	陈安东等，2017
马海	地表盐壳	石盐	15（据 ^{230}Th 年代推断）	黄麒等，2007
尕斯库勒	地表盐壳	石盐	25 ka BP 以后（据 80CK1 钻孔和 ZK2605）	黄麒等，2007

续表

盐湖	钻孔或采样点及位置（E/N）	盐类矿物及深度（m）	测年方法和数据（ka）	参考文献
察尔汗盐湖区	CK6 钻孔 94°47′25″/37°06′31″	石盐（40.60）	^{14}C：24.8±0.94	沈振枢等，1993；黄麒等，2007
		石盐（52.22）	^{14}C：31.4±1.78	
	地表盐壳	石盐	^{230}Th：16.1±1.3	
		石盐	^{230}Th：12.2±1.1	
		石盐	^{230}Th：12±1.5	
	81CK1 钻孔达布逊东北湾	石盐（17.0）	^{230}Th：16.7±0.18	黄麒等，2007
		石盐（21.9）	^{230}Th：24.2±0.77	
		石盐（28.2）	^{230}Th：32.3±3.10	
	88CK1 钻孔别勒滩	石盐（14.21）	^{230}Th：16.0±0.80	
		石盐（24.33）	^{230}Th：32.5±1.40	
		石盐（17.2）	^{230}Th：22.4±1.7	
	89CK4 钻孔别勒滩	石盐（22.8）	^{230}Th：25.0±1.8	
		石盐（26.4）	^{230}Th：27.6±4.5	

倒数第二次冰期 MIS6 柴达木盆地西部盐湖均有石盐沉积，相关盐类沉积的测年数据已经有系统的总结（陈安东等，2017）。本文中，MXK2 剖面和 D12 剖面的芒硝沉积时代可以对应于倒数第二次冰期 MIS6，与前人在柴西多个盐湖取得的芒硝沉积年代是一致的。察汗斯拉图 SG-1 钻孔进尺 37.2～37.25m 和 37.60～37.65m 出现少量芒硝，依据古地磁获得的沉积速率计算其年代约为 185ka（Li Minghui et al.，2010；Zhang Weilin et al.，2012）。一里坪 15YZK01 钻孔进尺 72.0m 出现少量芒硝晶体，依据古地磁和 ^{230}Th 测年获得的沉积速率计算其年代为 173ka（Chen Andong et al.，2017b）。大浪滩的层状芒硝的 ^{230}Th 测年数据为 143～195ka（马妮娜等，2011）。以上数据表明柴西盐湖芒硝沉积时代在区域上可以相互佐证，柴西多个盐湖存在 MIS6 的芒硝沉积。

基于晚第四纪冰期与成盐期的对比关系，本文初步提出倒数第二次冰期 MIS6 以来的对比方案（表 4）。按照该对比方案，晚第四纪倒数第二次冰期 MIS6 和末次冰期 MIS2 是柴达木盆地两个重要的成盐期。末次冰期 MIS4 柴达木盆地西部盐湖也有盐沉积（黄麒等，2007），但不如 MIS2 普遍。末次间冰期 MIS5 柴达木盆地东部盐湖出现高湖面和泛湖期（樊启顺等，2010；Fan Qishun et al.，2010，2012），盐沉积较少。MIS3 为相对 MIS5 较弱的暖湿阶段，柴西部分盐湖出现盐沉积，而柴东湖泊出现高湖面记录（郑绵平等，2006）。MIS1 虽然总体上处于间冰期，但是仍然伴随着小冰期、新冰期和全新世早中期冰进等冷阶段，柴达木盆地出现干盐滩面积扩大的情况。而 MIS1 出现的全新世大暖期（8.5～3.0ka BP）中的稳定暖湿阶段（7.2～6.0ka BP）（施雅风等，1992），也与察尔汗盐湖洪泛期（8～5ka BP）基本重合（袁见齐等，1995），显示该暖湿阶段不利于盐沉积。

但是，这一对比方案还存在以下问题需要考虑：①蒸发岩沉积受到“气候-构造-物源”共同控制，而气候并不是控制成盐的唯一因素；②晚第四纪期间不受冰川融水补给的盐湖或者受冰川融水补给影响较小的盐湖可能存在不同的盐沉积特征；③同一个盐湖的河流入湖口、湖中心和湖边缘的盐沉积特征可能存在区别；④晚第四纪多个冰期的降温幅度不一致，而且间冰期升温过程中也伴随着冷期；⑤晚第四纪冰期/间冰期旋回还伴随着亚洲中心干旱化，柴达木盆地气候呈现出“振荡干化”且干旱程度加剧的特征。以上情况造成“冰期成盐”这一科学问题复杂化，还需要进一步完善。因此，我们的对比方案仅针对晚第四纪期间受冰川融水直接或者间接补给的大盐湖区，而单一的某个钻孔或剖面并不一定完全适用。

表 4 柴达木盆地 MIS6 以来成盐期与青藏高原冰期对比方案

Table 4 Comparison scheme between salt-forming periods in the Qaidam Basin and Quaternary glacial periods in the Qinghai-Xizang (Tibetan) Plateau since MIS6

年代（ka）	MIS	青藏高原冰期	柴达木盆地成盐期与盐类沉积
0～10	1	间冰期中包含小冰期，新冰期，全新世早中期冰进	干盐滩面积扩大，盐湖中沉积钾镁盐（黄麒等，2007）。察尔汗盐湖在全新世大暖期出现洪泛期（施雅风等，1992；袁见齐等，1995）
11～28	2	YD 冰进，近冰阶，末次冰盛期（LGM）	成盐期：盆地中几乎所有盐湖均有石盐沉积，部分盐湖有芒硝、泻利盐沉积。盐湖面积萎缩，在 15 ka BP 左右形成大面积干盐滩（黄麒等，2007）
32～58	3	3a 和 3c 暖期，3b 冰进	高湖面期：相对较弱，柴西部分盐湖有石盐沉积，柴东湖泊出现高湖面记录（郑绵平等，2006）
58～75	4	末次早冰阶	成盐期：相对较弱，柴西多个盐湖有石盐沉积
75～125	5	末次间冰期：5a、5c 和 5e 暖期，5b 和 5d 相对冷	高湖面和泛湖期：以碎屑岩和碳酸盐岩沉积为主，柴西盐湖有石膏沉积，石盐和芒硝沉积较少
130～191	6	倒数第二次冰期 MIS6	成盐期：柴西盐湖均有石盐沉积，多个盐湖出现芒硝沉积

4.2 晚第四纪气候对盐类沉积的控制作用

在通常情况下，高温可以加快卤水蒸发和浓缩，有利于盐类沉积（Warren，2010，2016）。赵艳军等（2015）计算获得柴达木第四纪盐湖石盐流体包裹体的均一温度为 17.9～38.2℃，平均值为 34.0℃。罗布泊上更新统石盐流体包裹体记录的温度最大值为 35.6～43℃，反映出罗布泊晚更新世末期处于干旱、炎热的环境（Sun Xiaohong et al.，2017）。此外，也有证据表明全新世以来，青海湖和茶卡盐湖随着温度的升高而导致古盐度升高，但在寒冷条件下古盐度降低（张彭熹等，1992；沈吉等，2001；刘兴起等，2007）。但是作者认为对于青藏高原和柴达木盆地受冰川融水补给的盐湖及其中的陆相盐沉积，在晚第四纪成盐过程中，冰期气候和冰川活动所起的作用是不容忽视的。

晚第四纪气候对柴达木盆地蒸发沉积矿床的控制作用，实质上是冰期/间冰期旋回对盐类沉积的控制作用。在柴达木盆地和青藏高原，凡是以冰川融水补给为主的盐湖，其盐类沉积必然受到冰川活动的影响，其成盐期必定受冰期控制。末次冰盛期（Last Glacial Maximum，LGM，24～18ka）青藏高原冰川面积是现代冰川的 7.5 倍，倒数第二次冰期 MIS6 冰川规模甚至比 LGM 还要大（李吉均等，1991；潘保田等，1994；施雅风等，1999）。鉴于柴达木盆地盐湖的水源补给主要依赖昆仑山脉、祁连山脉和阿尔金山脉的冰冻圈融水和大气降水，冰冻圈在冰期/间冰期旋回背景下的消长，将直接影响到盐湖的补给水量。在倒数第二次冰期 MIS6 和末次冰期 MIS2，作为柴达木盆地主要水源补给区的祁连山和昆仑山均有冰进的证据（赵井东等，2001；Zhou Shangzhe et al.，2002；Owen et al.，2006）。冰期冷环境下冰冻圈的规模扩张，汇入盆地的水量减少，最终引起盐湖水浓缩和盐度升高，也是盐类沉积的一个重要途径（Herrero et al.，2015；陈安东等，2017；Chen Andong et al.，2017b）。另外，在 MIS6 和 MIS2 期间的降温，也是盐湖中冷相矿物芒硝和泻利盐沉积的直接原因。按照芒硝形成和保存的条件推算，柴西地区在 MIS2 和 MIS6 降温至少达到 6～10℃。

此外，晚第四纪倒数第二次冰期 MIS6 和末次冰期 MIS2 气候均以“冷干”为特征（李吉均等，1991；潘保田等，1997；施雅风等，1999；侯光良等，2019），有利于盐类沉积。在倒数第二次冰期 MIS6，青藏高原东部气温相对比现在低 8～12℃，降水量减少（李吉均等，1991）。也有研究表明青藏高原东北部在 MIS6 降温 12℃以上（潘保田等，1994），气候相对末次冰期较湿（潘保田等，1994；赵井东等，2011）。在倒数第二次冰期 MIS6，柴达木盆地西部盐湖均有石盐沉积，一里坪、察汗斯拉图和大浪滩还有芒硝沉积（沈振枢等，1993；韩凤清等，1995；黄麒等，2007；马妮娜等，2011；Li Minghui et al.，2010；Zhang

Weilin et al.，2012；陈安东等，2017；Chen Andong et al.，2017b）。在末次冰盛期，青藏高原平均降温 7℃，降水量只有现代的 30%～70%（李吉均等，1991；施雅风等，1999）。在 MIS2 的冷干气候下，柴达木盆地几乎所有的盐湖均有石盐析出，并形成大面积的干盐滩（Chen and Bowler，1986；沈振枢等，1993；张保珍等，1995；黄麒等，2007；陈安东等，2017；Chen Andong et al.，2017b）。而古里雅冰芯记录显示末次冰期 MIS4 的降温不及 LGM，但是湿度稍大于 LGM（赵井东等，2011）。MIS4 的温度特征在深海氧同位素曲线上有明显的记录（Lisiecki et al.，2005），温度和湿度特征在青藏高原冰川规模上也有反映（赵井东等，2001，2002；Yang Jianqiang et al.，2006；王杰等，2012；Chevalier et al.，2019），所以在柴达木盆地 MIS4 的盐沉积不如 MIS2 普遍。

在此需要指出，即使在晚第四纪冰期降水量没有减少，仅依靠柴达木盆地内部的大气降水、深部水和油田水等也不足以维持盐湖的长期存在。柴西冷湖地区年均降水量低于 20 mm（张彭熹，1987），向东至都兰县夏日哈地区的降水量增加至 240 mm（Yu Lupeng et al.，2012）。盆地内部的深部水和油田水可以影响到盐湖的水化学组成（李润民，1983），但也很难维持一个盐湖区的长期存在。所以，柴达木盆地盐湖最重要的水源补给还是来自盆地周边山体（袁见齐等，1983；张彭熹，1987），通过地表和地下径流对盐湖进行补给。本文无法恢复晚第四纪冰川、河道和盐湖的演化过程，但是我们可以根据昆仑山现代冰川对三湖地区（台吉乃尔、涩聂和达布逊盐湖区）的补给情况将今论古（图 4，表 5）。

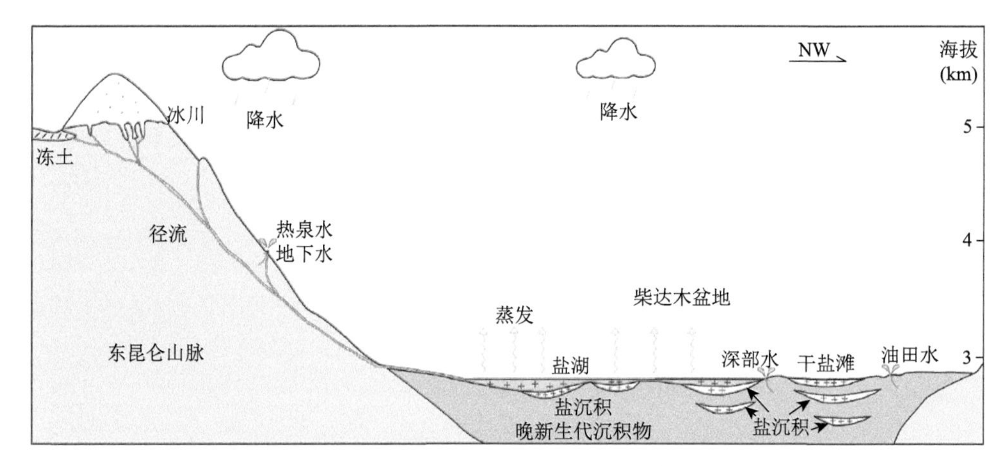

图 4 柴达木盆地盐湖与昆仑山冰川相互作用示意图

Fig. 4 Diagrammatic sketch of interactions between salt lakes in the Qaidam Basin and glaciers on the Kunlun Mountains.

表 5 柴达木盆地三湖地区盐湖及入湖河流

Table 5 Salt lakes in the Sanhu area and their inflow rivers

序号	盐湖区	主要固体盐类沉积	入湖河流
1	台吉乃尔（东台吉乃尔和西台吉乃尔湖）	石盐、芒硝和白钠镁矾	东台吉乃尔河和那陵格勒河
2	涩聂（涩聂湖和别勒滩）	石盐、光卤石和少量钾石盐	乌图美仁河、大灶火河、拖拉海河和清水河
3	达布逊（察尔汗和达布逊湖）	石盐、光卤石、水氯镁石和少量钾石盐	格尔木西河、格尔木东河和跃进河

三湖地区主要由发源于东昆仑山脉北坡的河流补给，河水主要来自山区的冰川融水和大气降水（袁见齐等，1983），还有少量的热泉水和地下水（周长进等，2002；高东林等，2006；余俊清等，2018）。在冰期随着冰冻圈规模扩张和雪线下降，这些水大部分以固体形式储存在山上。而更新世柴达木古湖由东南向西北补给，三湖地区是补给水流的主要起点之一（沈振枢等，1993；Chen Andong et al.，2017b）。按照这种补给方式，湖水可以由达布逊—涩聂—台吉乃尔——里坪—察汗斯拉图方向自东南向西北补给。因此，上述盐湖的盐沉积均与冰川活动有直接的联系。此外，大浪滩、马海、尕斯库勒和昆特依等盐湖在晚第四纪的成盐期也与冰期存在一定的对应关系（陈安东等，2017），但具体的补给过程有待进一步研究。

尽管有研究表明柴达木盆地在冰期出现“高湖面”（Phillips et al.，1993；张虎才等，2007；杨一博等，

2018)，但是本文认为至少 MIS6 和 MIS2 的冰期气候不利于“高湖面”的出现。虽然存在“温湿”气候模式的冰期，比如末次冰期 MIS3b 冰进，也有大量的数据表明在 MIS3，青藏高原色林错、扎布耶和纳木错以及柴达木盆地盐湖出现高湖面（郑绵平等，2006；刘俊英等，2007；Owen et al.，2006）。此外，末次间冰期 MIS5 柴达木盆地盐湖成盐数据也较少。目前已经有研究表明尕海、托素湖在末次间冰期 MIS5 出现高湖面，对应湖泊淡化期（樊启顺等，2010；Fan Qishun et al.，2010，2012）。而古里雅冰芯记录 MIS5e 青藏高原比现在升温 5℃以上（姚檀栋等，1997；施雅风等，1998），在 MIS3a 和 MIS3c 高出现在 3℃和 4℃（施雅风等，2002），均为暖期。以上研究表明 MIS5 和 MIS3 温度升高伴随着柴达木盆地湖泊水位升高，而“冷干”的 MIS6 和 MIS2 很难出现“高湖面”。

基于柴达木盆地盐类沉积与气候变化的关系，本文认为在短时间尺度上，全球变暖对于柴达木盆地盐类沉积是不利的。遥感数据显示近 40 年来青藏高原向暖湿气候转型，全球变暖导致高原冰川退化、冻土消融和降水量增加（闫丽娟等，2014；Yan Lijuan et al.，2015；Bibi et al.，2018；Li Huiying et al.，2019）。来自冰冻圈和大气圈的水体汇入盐湖水圈，造成了湖泊水位上涨、面积扩张和盐度下降。对于青藏高原和柴达木盆地受冰川融水补给的盐湖，全球变暖不利于其盐类沉积。更为直接的证据，来自 1989 年夏季气温升高引起高山冰川融化，导致柴达木盆地“特大洪水”事件，最终造成多个盐湖水位上涨和石盐溶解（袁见齐等，1995；黄麒等，2007）。不可否认随着冰川消融和降水量的增加，汇入盐湖的水量增加可以带来更多的盐类矿物，并促进分散的固体钾盐转为液相且向矿区低洼处汇集，但限于篇幅本文不拟赘述。

5 结论

（1）察汗斯拉图发育有倒数第二次冰期 MIS6 和末次冰期 MIS2 的芒硝和石盐层。中央拱起 MXK2 剖面和碱北凹地以北 D12 剖面芒硝的沉积年代可以对应于倒数第二次冰期 MIS6，碱北凹地 D18 剖面芒硝的沉积年代为末次冰期 MIS2 晚期。

（2）对于柴达木盆地受冰川融水补给的盐湖，其成盐期必定受到冰期的影响。倒数第二次冰期 MIS6 和末次冰期 MIS2 是柴达木盆地重要的成盐期，在盆地西部盐湖均有 MIS6 和 MIS2 的石盐沉积，个别盐湖还有芒硝和泻利盐沉积。末次间冰期 MIS5 温度升高不利于柴达木盆地盐类沉积。晚第四纪 MIS6 和 MIS2 柴达木盆地“冰期成盐”的根本原因，是由于冰期环境下冰川规模的扩张以及冷干的冰期气候，共同造成了盐湖补给水量的减少。此外，晚第四纪冰期 MIS6 和 MIS2 的降温也是导致冷相盐类（如芒硝和泻利盐）沉积的直接原因。

致谢

青海森盛矿业有限公司在野外矿产地质调查和采样工作中提供了支持，审稿专家和责任编辑为本文提供了宝贵的修改意见，在此一并表示感谢！

参考文献

(The literature whose publishing year followed by a “&” is in Chinese with English abstract; The literature whose publishing year followed by a“#” is in Chinese without English abstract)

陈安东, 郑绵平, 施林峰, 王海雷, 徐建明, 袁文虎. 2017. 柴达木盆地一里坪石膏 ^{230}Th 定年及成盐期与第四纪冰期和构造运动的关系. 地球学报, 38 (4): 494-504.

崔之久, 陈艺鑫, 张威, 周尚哲, 周力平, 张梅, 李川川. 2011. 中国第四纪冰期历史、特征及成因探讨. 第四纪研究, 31 (5): 749-764.

樊启顺, 赖忠平, 刘向军, 孙永娟, 隆浩. 2010. 晚第四纪柴达木盆地东部古湖泊高湖面光释光年代学. 地质学报, 84 (11): 1652-1660.

高东林, 马海洲, 张西营, 韩凤清, 周笃珺. 2006. 西台吉乃尔盐湖地下卤水的赋存特征. 盐湖研究, 14 (2): 1-6.

韩凤清, 黄麒, 王克俊, 王华安, 原力. 1995. 柴达木盆地昆特依盐湖的地球化学演化与古气候变化. 海洋与湖沼, 26 (5): 502-508.

侯献华, 郑绵平, 张成君, 施林峰, 王有德. 2010. 柴达木盆地西部大浪滩 140 ka 以来沉积特征与古环境. 地质学报, 84 (11): 1623-1630.

侯光良, 赖忠平, 刘向军, 鄂崇毅, 魏海成. 2019. 晚冰期以来青藏高原降水序列集成重建. 第四纪研究, 39 (3): 615-628.

黄麒, 韩凤清. 2007. 柴达木盆地盐湖演化与古气候波动. 北京: 科学出版社.

李吉均, 周尚哲, 潘保田. 1991. 青藏高原东部第四纪冰川问题. 第四纪研究, (3): 193-203.
李润民. 1983. 柴达木盆地察尔汗钾镁盐成矿地质条件. 地质论评, 29 (3): 262-268.
刘成林, 吴驰华, 王立成, 方小敏, 赵艳军, 颜茂都, 张永生, 曹养同, 张华, 吕凤琳. 2016. 中国陆块海相盆地成钾条件与预测研究进展综述. 地球学报, 37 (5): 581-606.
刘嘉麒, 倪云燕, 储国强. 2001. 第四纪的主要气候事件. 第四纪研究, 21 (3): 239-248.
刘俊英, 郑绵平, 袁鹤然, 刘喜方, 王海雷. 2007. 西藏扎布耶湖区 128～1.4 ka BP 的微体古生物与环境气候变化. 地质学报, 81 (12): 1618-1635.
刘兴起, 王永波, 沈吉, 王苏民, 杨波. 2007. 16000 a 以来青海茶卡盐湖的演化过程及其对气候的响应. 地质学报, 81 (6): 843-849.
马妮娜, 郑绵平, 马志邦, 陈文西, 孔凡晶, 施林峰. 2011. 柴达木盆地大浪滩地区表层芒硝的形成时代及环境意义. 地质学报, 85 (3): 433-444.
潘保田, 李吉均, 周尚哲. 1994. 青藏高原倒数第二次冰期冰楔的发现及其意义. 科学通报, 17: 1599-1602.
潘保田, 邬光剑. 1997. 青藏高原东北部最近两次冰期降温幅度的初步估算. 干旱区地理, 20 (2): 17-24.
钱自强, 曲一华, 刘群. 1994. 钾盐矿床. 北京: 地质出版社.
沈吉, 张恩楼, 夏威岚. 2001. 青海湖近千年来气候环境变化的湖泊沉积记录. 第四纪研究, 21 (6): 508-513.
沈振枢, 程果, 乐昌硕, 刘淑琴, 张发胜, 王强, 祁国柱, 张嘉尔, 葛同明, 雷世太. 1993. 柴达木盆地第四纪含盐地层划分及沉积环境. 北京: 地质出版社.
施雅风, 孔昭宸, 王苏民, 唐领余, 王富葆, 姚檀栋, 赵希涛, 张丕远, 施少华. 1992. 中国全新世大暖期的气候波动与重要事件. 中国科学 (B 辑), 22 (12): 1300-1308.
施雅风, 李吉均, 李炳元. 1998. 青藏高原晚新生代隆升与环境变化. 广东: 广东科技出版社.
施雅风, 李吉均, 李炳元, 姚檀栋, 王苏民, 李世杰, 崔之久, 王富保, 潘保田, 方小敏, 张青松. 1999. 晚新生代青藏高原的隆升与东亚环境变化. 地理学报, 54 (1): 10-20.
施雅风. 1998. 第四纪中期青藏高原冰冻圈的演化及其与全球变化的联系. 冰川冻土, 20 (3): 197-208.
施雅风. 2002. 中国第四纪冰期划分改进建议. 冰川冻土, 24 (6): 687-692.
施雅风, 姚檀栋. 2002. 中低纬度 MIS3b (54～44 ka BP) 冰期与冰川前进. 冰川冻土, 24 (1): 1-9.
王杰, 潘保田, 张国梁, 崔航, 曹泊, 耿豪鹏. 2012. 贡嘎山东坡中更新世晚期以来冰川作用年代学研究. 中国科学: 地球科学, 42 (12): 1889-1900.
王立胜, 马志邦, 程海, 段武辉, 肖举乐. 2016. MC-ICP-MS 测定铀系定年标样的 ^{230}Th 年龄. 质谱学报, 37 (3): 262-272.
闫丽娟, 郑绵平, 袁志洁. 2014. 近 40 年来气候变化对青海盐湖及其矿产资源开发的影响——以小柴旦湖为例. 矿床地质, 33 (5): 921-929.
姚檀栋, Thompson L G, 施雅风, 秦大河, 焦克勤, 杨志红, 田立德, Thompson E M. 1997. 古里雅冰芯中末次间冰期以来气候变化记录研究. 中国科学 (D 辑), 27 (5): 447-452.
杨一博, 方小敏, Albert Galy, 杨戎生. 2018. 柴达木盆地西部第四纪气候变化和流域风化. 第四纪研究, 38 (1): 76-85.
易朝路, 崔之久, 熊黑钢. 2005. 中国第四纪冰期数值年表初步划分. 第四纪研究, 25 (5): 609-619.
袁见齐, 霍承禹, 蔡克勤. 1983. 高山深盆的成盐环境——一种新的成盐模式的剖析. 地质论评, 29 (2): 159-165.
袁见齐, 杨谦, 孙大鹏, 霍承禹, 蔡克勤, 王文达, 刘训建. 1995. 察尔汗盐湖钾盐矿床的形成条件. 北京: 科学出版社.
余俊清, 洪荣昌, 高春亮, 成艾颖, 张丽莎. 2018. 柴达木盆地盐湖锂矿床成矿过程及分布规律. 盐湖研究, 26 (1): 7-14.
张保珍, 张彭熹, Lowenstein T K, Spencer R J. 1995. 青藏高原末次冰期盛冰阶的时限与干盐湖地质事件. 第四纪研究, (3): 193-201.
张虎才, 雷国良, 常凤琴, 樊红芳, 杨明生, 张文翔. 2007. 柴达木盆地察尔汗贝壳堤剖面年代学研究. 第四纪研究, 27 (4): 511-521.
张彭熹. 1987. 柴达木盆地盐湖. 北京: 科学出版社.
张彭熹, 张保珍, 钱桂敏, 李海军, 徐黎明. 1992. 青海湖全新世以来古环境参数的研究. 第四纪研究, (3): 225-238.
张永生, 邢恩袁, 陈文西. 2014. 国内外古陆表海盆成钾条件对比——兼论华北成钾的可能性. 矿床地质, 33 (5): 897-908.
赵井东, 周尚哲, 崔建新, 潘小多, 许刘兵, 张小伟. 2001. 摆浪河流域的 ESR 年代学与祁连山第四纪冰川新认识. 山地学报, 19 (6): 481-488.
赵井东, 周尚哲, 崔建新, 焦克勤, 业渝光, 许刘兵. 2002. 乌鲁木齐河源冰碛物的 ESR 测年研究. 冰川冻土, 24 (6): 737-743.
赵井东, 施雅风, 王杰. 2011. 中国第四纪冰川演化序列与 MIS 对比研究的新进展. 地理学报, 66 (7): 867-884.
赵艳军, 刘成林, 张华, Li Zhaoqi, 丁婷, 汪明泉. 2015. 古盐湖卤水温度对钾盐沉积的控制作用探讨. 岩石学报, 31 (9): 2751-2756.
郑绵平, 向军, 魏新俊, 郑元. 1989. 青藏高原盐湖. 北京: 北京科学技术出版社.
郑绵平, 袁鹤然, 赵希涛, 刘喜方. 2006. 青藏高原第四纪泛湖期与古气候. 地质学报, 80 (2): 169-180.
郑绵平, 刘喜方. 2010. 青藏高原盐湖水化学及其矿物组合特征. 地质学报, 84 (11): 1585-1600.
郑绵平, 张永生, 刘喜方, 齐文, 孔凡晶, 乜贞, 贾沁贤, 卜令忠, 侯献华, 王海雷, 张震, 孔维刚, 林勇杰. 2016. 中国盐湖科学技术研究的若干进展与展望. 地质学报, 90 (9): 2123-2165.
郑绵平, 赵元艺, 刘俊英. 1998. 第四纪盐湖沉积与古气候. 第四纪研究, (4): 297-307.
周长进, 董锁成. 2002. 柴达木盆地主要河流的水质研究及水环境保护. 资源科学, 24 (2): 37-41.
Bibi S, Wang Lei, Li Xiuping, Zhou Jing, Chen Deliang, Yao Tandong. 2018. Climatic and associated cryospheric, biospheric, and hydrological changes on the Tibetan Plateau: a review. International Journal of Climatology, DOI: 10.1002/joc.5411.
Chen Andong, Zheng Mianping, Shi Linfeng, Wang Hailei, Xu Jianming, Yuan Wenhu. 2017a. Gypsum ^{230}Th dating of the 15YZK01 drilling core in the Qaidam Basin: salt deposits and their link to Quaternary glaciation and tectonic movement. Acta Geoscientica Sinica, 38 (4): 494-504.

Chen Andong, Zheng Mianping, Shi Linfeng, Wang Hailei, Xu Jianming. 2017b. Magnetostratigraphy of deep drilling core 15YZK01 in the northwestern Qaidam Basin (NE Tibetan Plateau): tectonic movement, salt deposits and their link to Quaternary glaciation. Quaternary International, 436: 201-211.

Chen Andong, Zheng Mianping, Yao Haitao, Su Kui, Xu Jianming. 2018. Magnetostratigraphy and ^{230}Th dating of a drill core from the Southeastern Qaidam Basin: salt lake evolution and tectonic implications. Geosciences Frontiers, 9: 943-953.

Chen Kezao and Bowler J M. 1986. Late Pleistocene evolution of salt lakes in the Qaidam Basin, Qinghai Province, China. Palaeogeography, Palaeoclimatology, Palaeoecology, 54: 87-104.

Chevalier M L and Replumaz A. 2019. Deciphering old moraine age distributions in SE Tibet showing bimodal climatic signal for glaciations: Marine Isotope Stages 2 and 6. Earth and Planetary Science Letters, 507: 105-118.

Cody R D and Cody A M. 1988. Gypsum nucleation and crystal morphology in analog saline terrestrial environments. Journal of Sedimentary Petrology, 58: 247-255.

Cui Zhijiu, Chen Yinxin, Zheng Wei, Zhou Shangzhe, Zhou Liping, Zhang Mei, Li Chuanchuan. 2011&. Research history, glacial chronology and origins of Quaternary glaciations in China. Quaternary Sciences, 31 (5): 49-764.

Fan Qishun, Lai Zhongping, Liu Xiangjun, Sun Yongjuan, Long Hao. 2010&. Luminescence chronology of high lake levels of paleolakes in the late Quaternary eastern Qaidam Basin. Acta Geologica Sinica, 84 (11): 1652-1660.

Fan Qishun, Lai Zhongping, Long Hao, Sun Yongjuan, Liu Xiangjun.2010. OSL chronology for lacustrine sediments recording high stands of Gahai Lake in Qaidam Basin, northeastern Qinghai-Tibetan Plateau. Quaternary Geochronology, 5: 223-227.

Fan Qishun, Ma Haizhou, Cao Guangchao, Chen Zongyan, Cao Shengkui. 2012. Geomorphic and chronometric evidence for high lake level history in Gahai and Toson Lake of north-eastern Qaidam Basin, north-eastern Qinghai-Tibetan Plateau. Journal of Quaternary Science, 27 (8): 819-827.

Gao Donglin, Ma Haizhou, Zhang Xiying, Han Fengqing, Zhou Dujun. 2006&. Storage characteristics of ground brines of west Taijinar Salt Lake. Journal of Salt Lake Research. 14 (2): 1-6.

Grasby S E, Smith I R, Bell T, Forbes D L. 2013. Cryogenic formation of brine and sedimentary mirabilite in submergent coastal lake basins, Canadian Arctic. Geochimica et Cosmochimica Acta, 110: 13-28.

Han Fengqing, Huang Qi, Wang Kejun, Wang Huaan, Yuan Li. 1995&. Study of geochemical evolution and palaeoclimatic fluctuation of Kunteyi Salt Lake in the Qaidam Basin, Qinghai. Oceanologia et Limnologia Sinica, 26 (5): 502-508.

Han Wenxia, Ma Zhibang, Lai Zhongping, Appel Erwin, Fang Xiaomin, Yu Lupeng. 2014. Wind erosion on the north-eastern Tibetan Plateau: constraints from OSL and U-Th dating of playa salt crust in the Qaidam Basin. Earth Surface Process and Landforms, 39: 779-789.

Herrero M J, Escavy J I, Schreiber B C. 2015. Thenardite after mirabilite deposits as a cool climate indicator in the geological record: lower Miocene of central Spain. Climate of the Past, 11: 1-13.

Hou Guangliang, Lai Zhongping, Liu Xiangjun, E Chongyi, Wei Haicheng. 2019&. Synthetically reconstructed precipitation variability in the Qinghai-Tibet Plateau over the last 16 ka. Quaternary Sciences, 39 (3): 615-628.

Hou Xianhua, Zheng Mianping, Zhang Chengjun, Shi Linfeng, Wang Youde. 2010&. Sedimentary characteristics and paleoenvironmental of Dalangtan Salt Lake in western Qaidam Basin, since 140 ka BP. Acta Geologica Sinica, 84 (11): 1623-1630.

Huang Qi, Han Fengqing. 2007 #. Evolution of Salt Lakes and Palaeoclimate Fluctuation in Qaidam Basin. Beijing: Science Press.

Last W M and Ginn F M. 2005. Saline systems of the Great Plains of western Canada: an overview of the limnogeology and paleolimnology. Saline Systems, 10: 1-38.

Li Huiying, Mao Dehua, Li Xiaoyan, Wang Zongming, Wang Cuizhen. 2019. Monitoring 40-year lake area changes of the Qaidam Basin, Tibetan Plateau, using landsat time series. Remote Sensing, 11, 343: doi: 10.3390/rs11030343.

Li Jijun, Zhou Shangzhe, Pan Baotian. 1991&. The problems of Quaternary glaciations in the eastern part of Qinghai-Xizang Plateau. Quaternary Sciences, (3): 193-203.

Li Minghui, Fang Xiaomin, Yi Chaolu, Gao Shaopeng, Zhang Weilin, Galy A, 2010. Evaporite minerals and geochemistry of the upper 400 m sediments in a core from the Western Qaidam Basin, Tibet. Quaternary International, 218: 176-189.

Li Runming. 1983&. The geological conditions for the formation of the Qarhan K-Mg saline deposit in the Qaidam Basin. Geological Review, 29 (3): 262-268.

Lisiecki L E and Raymo M E. 2005. A Plio-Pleistocene stack of 57 globally distributed benthic δ^{18}O records. Paleoceanography, 20 (1): 1-16. DOI: 10.1029/2004PA001071.

Liu Chenglin, Wu Chihua, Wang Licheng, Fang Xiaomin, Zhao Yanjun, Yan Maodu, Zhang Yongsheng, Cao Yangtong, Zhang Hua, Lü Fenglin. 2016&. Advance in the study of forming condition and prediction of potash deposits of marine basins in China's small blocks: review. Acta Geoscientica Sinica, 37 (5): 581-606.

Liu Jiaqi, Ni Yunyan, Chu Guoqiang. 2001&. Main palaeoclimatic events in the Quaternary. Quaternary Sciences, 21 (3): 239-248.

Liu Junying, Zheng Mianping, Yuan Heran, Liu Xifang, Wang Hailei. 2007&. Microfossils and Climatic and Environmental Changes in the Zabuye Lake Area, Tibet, from 128 to 1.4 ka BP. Acta Geologica Sinica, 81 (12): 1618-1635.

Liu Xingqi, Wang Yongbo, Shen Ji, Wang Sumin, Yang Bo. 2007&. Evolution of Chaka Salt Lake during the Last 16000 years and its response to climatic Change. Acta Geologica Sinica, 81 (6): 843-849.

Ma Nina, Zheng Mianping, Ma Zhibang, Chen Wenxi, Kong Fanjing, Shi Linfeng. 2011. Forming age of surface mirabilite in Dalangtan, Qaidam Basin and its environmental significance. Acta Geographica Sinica, 85 (3): 433-444.

Mees F, Casteñeda C, Herrero J, Ranst E V. 2012. The nature and significance of variations in gypsum crystal morphology in dry lake basins. Journal of Sedimentary Research, 82: 41-56.

Owen L A, Robert F C, Ma Haizhou, Barnard P L. 2006. Late Quaternary landscape evolution in the Kunlun Mountains and Qaidam Basin, Northern Tibet: A framework for examining the links between glaciation, lake level changes and alluvial fan formation. Quaternary International, 154-155: 73-86.

Pan Baotian, Li Jijun, Zhou Shangzhe. 1994. Discovery and significance of ice wedges during Penultimate Glaciation in the Qinghai-Tibetan Plateau. Chinese Science Bulletin, 39: 578-582.

Pan Baotian, Wu Guangjian. 1997&. Preliminary estimation on the drop range in temperature during the last two glaciations in the northeastern Qinghai-Xizang Plateau. Arid Land Geography, 20 (2): 17-24.

Phillips F M, Zreda M G, Ku T L, Luo S D, Huang Qi, Elmore D, Kubik P W, Sharma P. 1993. $^{230}Th/^{234}U$ and ^{36}Cl dating of evaporite deposits from the western Qaidam Basin, China: implications for glacial-period dust export from Central Asia. Geological Society of America Bulletin, 105: 1606-1616.

Qian Ziqiang, Qu Yihua, Liu Qun. 1994#. Potash Deposits. Beijing: Geological Publishing House.

Shen Ji, Zhang Enlou, Xia Weilan. 2001&. Records from lake sediments of the Qinghai Lake to mirror climatic and environmental changes of the past about 1000 years. Quaternary sciences, 21 (6): 508-513.

Shen Zhenshu, Cheng Guo, Le Changshuo, Liu Shuqin, Zhang Fasheng, Wang Qiang, Qi Guozhu, Zhang Jiaer, Ge Tongming, Lei Shitai. 1993 #. The Division and Sedimentary Environment of Quaternary Salt-bearing Strata in Qaidam Basin. Beijing: Geological Publishing House.

Shi Yafeng, Kong Zhaozheng, Wang Sumin, Tang Lingyu, Wang Fubao, Yao Tandong, Zhao Xitao, Zhang Peiyuan, Shi Shaohua. 1994. The climatic fluctuation and important events of Holocene megathermal in China. Science in China (Series B), 37 (3): 353-365.

Shi Yafeng, Li Jijun, Li Bingyuan. 1998#. Uplift and Environmental Changes of Qinghai-Xizang (Tibetan) Plateau in the Late Cenozoic. Guangdong: Guangdong Science & Technology Press.

Shi Yafeng, Li Jijun, Li Bingyuan, Yao Tandong, Wang Sumin, Li Shijie, Cui Zhijiu, Wang Fubao, Pan Baotian, Fang Xiaomin, Zhang Qingsong. 1999&. Uplift of the Qinghai-Xizang (Tibetan) Plateau and East Asia environmental change during late Cenozoic. Acta Geographica Sinica, 54 (1): 10-20.

Shi Yafeng. 1998&. Evolution of the cryosphere in the Tibetan Plateau, China, and its Relationship with the global change in the Mid-Quaternary. Journal of Glaciology and Geocryology, 20 (3): 197-208.

Shi Yafeng. 2002&. A suggestion to improve the chronology of Quaternary glaciations in China. Journal of Glaciology and Geocryology, 24 (6): 687-692.

Shi Yafeng, Yao Tandong. 2002&. MIS3b (54～44 ka BP) cold period and glacial advance in middle and low latitudes. Journal of Glaciology and Geocryology, 24 (1): 1-9.

Sun Xiaohong, Zhao Yanjun, Liu Chenglin, Jiao Pengcheng, Zhang Hua, Wu Chihua. 2017. Paleoclimatic information recorded in fluid inclusions in halites from Lop Nur, Western China. Scientific Reports, 7: 16411. DOI: 10.1038/s41598-017-16619-4.

Wang Jie, Pan Baotian, Zhang Guoliang, Cui Hang, Cao Bo, Geng Haopeng. 2012. Late Quaternary glacial chronology on the eastern slope of Gongga Mountain, eastern Tibetan Plateau, China. Science China (Series D: Earth Sciences), 56 (3): 354-365.

Wang Jiuyi, Fang Xiaomin, Appel E, Zhang Weilin. 2013. Magnetostratigraphic and radiometric constraints on salt formation in the Qaidam Basin, NE Tibetan Plateau. Quaternary Science Reviews, 78: 53-64.

Wang Lisheng, Ma Zhibang, Cheng Hai, Duan Wuhui, Xiao Jule. 2016&. Determination of ^{230}Th dating age of Uranium-series standard samples by multiple collector inductively coupled plasma mass spectrometry. Journal of Chinese Mass Spectrometry Society, 37 (3): 262-272.

Warren J K. 2010. Evaporites through time: tectonic, climatic and eustatic controls in marine and nonmarine deposits. Earth-Science Reviews, 98: 217-268.

Warren J K. 2016. Evaporites: A Geological Compendium (Second Edition). Switzerland: Springer International Publishing Switzerland.

Yan Lijuan, Zheng Mianping, Yuan Zhijie. 2014&. Influence of climate change on salt lakes in Qinghai Province and their mineral resources exploitation in the past forty years: A case study of Xiao Qaidam Lake. Mineral Deposits, 33: 921-929.

Yan Lijuan, Zheng Mianping. 2015. Influence of climate change on saline lakes of the Tibet Plateau, 1973～2010. Geomorphology, 246: 68-78.

Yang Jianqiang, Zhang Wei, Cui Zhijiu, Yi Chaolu, Liu Kexin, Ju Yuanjiang, Zhang Xiaoyong. 2006. Late Pleistocene glaciations of the Diancang and Gongwang Mountains, southeast margin of the Tibetan plateau. Quaternary International, 154-155: 52-62.

Yang Yibo, Fang Xiaomin, Albert Galy, Yang Rongsheng. 2018&. Quaternary climate change and catchment weathering in the western Qaidam Basin. Quaternary Sciences, 38 (1): 76-85.

Yao Tandong, Thompson L G, Shi Yafeng, Qin Dahe, Jiao Keqin, Yang Zhihong, Tian Lide, Thompson E. M. 1997&. Climate variation since the last interglaciation recorded in the Guliya ice core. Science in China (Series D), 40 (6): 662-668.

Yi Chaolu, Cui Zhijiu, Xiong Heigang. 2005&. Numerical periods of Quaternary glaciations in China. Quaternary Sciences, 25 (5): 609-619.

Yuan Jianqi, Huo Chengyu, Cai Keqin. 1983&. The high mountain-deep basin saline environment—a new genetic model of salt deposits. Geological Review, 29 (2): 159-165.

Yuan Jianqi, Yang Qian, Sun Dapeng, Huo Chengyu, Cai Keqin, Wang Wenda, Liu Xunjian. 1995&. The Formation Conditions of the Potash Deposits in Charhan Saline Lake, Caidamu Basin, China. Beijing: Geological Publishing House.

Yu Lupeng, Lai Zhongping. 2012. OSL chronology and palaeoclimatic implications of aeolian sediments in the eastern Qaidam Basin of the northeastern Qinghai-Tibetan Plateau. Palaeogeography, Palaeoclimatology, Palaeoecology, 337-338: 120-129.

Yu Junqing, Hong Rongchang, Gao Chunliang, Cheng Aiying, Zhang Lisa. 2018&. Lithium brine deposits in Qaidam Basin: constraints on formation processes and distribution pattern. Journal of Salt Lake Research, 26 (1): 7-14.

Zhang Baozhen, Zhang Pengxi, Lowenstein T K, Spencer R J. 1995&. Time range of the Great Ice Age of the Last Glacial Stage and its related geological event of playa in the Qinghai-Xizang (Tibet) Plateau. Quaternary Sciences, (3): 193-201.

Zhang Hucai, Lei Guoliang, Chang Fengqin, Fan Hongfang, Yang Mingsheng, Zhang Wenxiang. 2007&. Age determination of shell bar section in salt lake Qarhan, Qaidam Basin. Quaternary Sciences, 27 (4): 511-521.

Zhang Pengxi. 1987#. Saline Lakes in the Qaidam Basin. Beijing: Science Press.

Zhang Pengxi, Zhang Baozhen, Qian Guimin, Li Haijun, Xu Liming. 1992&. The study of paleoclimatic parameter of Qinghai Lake since Holocene. Quaternary sciences, (3): 225-238.

Zhang Weilin, Appel E, Fang Xiaomin, Song Chunhui, Cirpka O. 2012. Magnetostratigraphy of deep drilling core SG-1 in the western Qaidam Basin (NE Tibetan Plateau) and its tectonic implications. Quaternary Research, 78: 139-148.

Zhang Yongsheng, Xing Enyuan, Chen Wenxi. 2014&. Comparative study of palaeo-epicontinental marine basin potash-forming conditions between China and foreign countries with special reference to potash-forming possibilities in North China. Mineral Deposits, 33 (5): 897-908.

Zhao Jingdong, Zhou Shangzhe, Cui Jianxin, Pan Xiaoduo, Xu Liubing, Zhang Xiaowei. 2001&. ESR chronology of Bailanghe valley and new understanding of Qilianshan Mountain's Quaternary glaciation. Journal of Mountain Science, 19 (6): 481-488.

Zhao Jingdong, Zhou Shangzhe, Cui Jianxin, Jiao Keqin, Ye Yuguang, Xu Liubing. 2002&. ESR dating of glacial tills at the headwaters of the Urumqi River in the Tianshan Mountains. Journal of Glaciology and Geocryology, 24 (6): 737-743.

Zhao Jingdong, Shi Yafeng, Wang Jie. 2011&. Comparison between Quaternary glaciations in China and the Marine Oxygen Isotope Stage (MIS): and improved schema. Acta Geographica Sinica, 66 (7): 867-884.

Zhao Yanjun, Liu Chenglin, Zhang Hua, Li Zhaoqi, Ding Ting, Wang Mingquan. 2015&. The controls of paleotemperature on potassium salt precipitation in ancient salt lakes. Acta Petrologica Sinica, 31 (9): 2751-2756.

Zheng Mianping, Xiang Jun, Wei Xinjun, Zheng Yuan. 1989#. Saline Lakes on the Qinghai-Xizang (Tibet) Plateau. Beijing: Beijing Science and Technology Press.

Zheng Mianping, Zhao Yuanyi, Liu Junying. 1998&. Quaternary saline lake deposition and paleoclimate. Quaternary Sciences, 4: 297-307.

Zheng Mianping, Liu Xifang. 2010&. Hydrochemistry and minerals assemblages of salt lakes in the Qinghai-Tibet Plateau, China. Acta Geologica Sinica, 84 (11): 1585-1600.

Zheng Mianping, Yuan Heran, Zhao Xitao, Liu Xiafang. 2006&. The Quaternary Pan-lake (overflow) period and paleoclimate on the Qinghai-Tibet Plateau. Acta Geologica Sinica, 80 (2): 169-180.

Zheng Mianping, Zhang Yongsheng, Liu Xifang, Nie Zhen, Kong Fanjing, Qi Wen, Jia Qinxian, Pu Lingzhong, Hou Xianhua, Wang Hailei, Zhang Zhen, Kong Weigang, Lin Yongjie. 2016. Progress and prospects of salt lake research in China. Acta Geologica Sinica (English Edition), 90: 1195-1235.

Zhou Changjin, Dong Suocheng. 2002&. Water quality of main rivers in the Qaidam Basin and water environmental protection. Resources Science, 24 (2): 37-41.

Zhou Shangzhe, Li Jijun, Zhang Shiqiang. 2002. Quaternary Glaciation of Bailang River Valley, Qilian Shan. Quaternary International, 97-98: 103-110.

Evaporite deposits in the Qaidam Basin and their response to Quaternary glacial climates since marine oxygen isotope stage 6 (MIS6)

Andong Chen[1], Mianping Zheng[1], Gao Song[1], Xuefeng Wang[2], Hongpu Li[3], Guang Han[3] and Wenhu Yuan[3]

1. MNR Key Laboratory of Saline Lake Resources and Environments, Institute of Mineral Resources, Chinese Academy of Geological Sciences, Beijing 100037
2. Key laboratory of Cenozoic Geology and Environment, Institute of Geology and Geophysics, Chinese Academy of Sciences, Beijing 100029
3. Qaidam Integrated Geological Exploration Institute of Qinghai Province, Golmud, Qinghai 816099

Objectives: Climate is one of the dominant factors which control evaporite deposits in Qaidam Basin, but its mechanism remains to be clarified. The aim of this paper is to make a comparison between the evaporite deposit period and glacial period since marine oxygen isotope stage 6 (MIS6), and clarify the role of late Quaternary glacial climate in evaporite deposit.

Methods: The authors took 3 salt-bearing profiles (D18, MXK2 and D12) from the western Qaidam Basin as the study object, applied multiple collector inductively coupled plasma mass spectrometry (MC-ICP-MS) to obtain the salt deposit age, and applied X-ray diffraction (XRD) to identify the salt minerals from the profiles.

Results: MC-ICP-MS U-series dating indicated the deposit ages of D18 profile are 13.1±2.0 ka BP～15.9± 2.5 ka BP, of which the mirabilite deposit belong to the late stage of last glacial MIS2; mirabilite deposit ages in MXK2 profile are 131.7±39.5 ka BP and 158.3±10.8 ka BP respectively, mirabilite deposit ages in D12 profile are 166.6±20.2 ka BP and 198.0 ± 20.6 ka BP, which can correspond to the penultimate glacial period MIS6. XRD analysis confirmed that salt minerals in the 3 profiles were mainly mirabilite, halite and gypsum.

Conclusions: Combining the study of this paper and evaporite deposits data of other salt lakes in the Qaidam Basin, this paper proposes that the penultimate glacial period MIS6 and the last glaciation MIS2 are two important salt-forming periods of the late Quaternary in the Qaidam Basin, and the cold and dry climate of the glacial period is favorable for salt deposits such as halite and mirabilite. The fundamental reason for salt deposit in the glacial environment in the Qaidam Basin is the expansion of glacier scale in the surrounding mountains, and the dry and cold glacial climate, resulting in the reduction of the recharge volume of the salt lakes in the Qaidam Basin. Besides, the temperature decrease during MIS6 and MIS2 is the direct genesis of the deposit of cold-phase salt minerals such as mirabilite and epsomite.

Keywords: Qaidam Basin; Salt lake; Evaporite deposit Period; Penultimate glacial period; Last glaciation; U-series dating

Dalangtan saline playa in a hyperarid region on Qinghai-Tibet Plateau-I: evolution and environments*

Fanjing Kong[1], Mianping Zheng[1], Bin Hu[1, 2], Alian Wang[3], Nina Ma[1] and Pablo Sobron[4, 5]

1. MLR Key Laboratory of Saline Lake Resources and Environments, Institute of Mineral Resources, Chinese Academy of Geological Sciences, Beijing, China
2. Institute of Geology and Geophysics, Chinese Academy of Sciences, Beijing, China
3. Department of Earth and Planetary Sciences, McDonnell Center for Space Sciences, Washington University in St. Louis, St. Louis, Missouri
4. SETI Institute, Mountain View, California
5. Impossible Sensing, St. Louis, Missouri

Abstract Since 2008, we have been studying a saline lake, Dalangtan (DLT) Playa, and its surroundings in a hyperarid region of the Qaidam Basin on the Tibetan Plateau as a potential Mars analog site. We describe the evolution of saline deposits in the Qaidam Basin (including DLT), based on investigative findings accumulated over the course of 60 years of geological surveys. In addition, we report regional meteorological patterns recorded for the past 32 years along with meteorological station recorded data at DLT since 2012. Overall, the DLT area on the Tibetan Plateau has low atmospheric pressure, high ultraviolet radiation, low annual mean temperatures (T) but large seasonal and diurnal T cycles, and extremely low relative humidity, all of which bear some similarities with the equatorial region on Mars. In addition, salt types similar to those found on Mars, such as magnesium sulfates, chlorides, and perchlorates, are found at the surface and subsurface in the DLT area (and the other two playas in the Qaidam Basin), thus supporting DLT as a Mars analogue in terms of mineralogy and geochemistry.

Keywords Mars analogue; Tibet Plateau; Saline playa; Evolution; Climatology

1 Introduction

Studies of terrestrial analog sites have played an essential role in understanding observations from past missions. Although no site on Earth is a perfect analogue to specific regions on Mars, individual analog sites on Earth do provide important scientific results that support fundamental science as applied to Mars and Mars missions.

Since 2008, our team has been studying the saline lake, Dalangtan (DLT) Playa, and its surroundings in the hyperarid region (Qaidam Basin) on the Tibetan Plateau in China as a potential Mars analog site (Kong et al., 2009, 2010; Mayer et al., 2009; Sobron et al., 2009; Wang and Zheng, 2009, 2011; Zheng et al., 2009; Wang et al., 2010, 2013; Anglés and Li, 2017a, 2017b; Cheng et al., 2017; Xiao et al., 2017). This article describes the evolution of saline deposits in the Qaidam Basin, based on the knowledge accumulated through 60 years of geological survey and detailed ground investigations (e.g., Zhang, 1987; Zhu et al., 1994; Zheng, 1997; Kong et al., 2014a, 2014b). In addition, we report on regional meteorological patterns averaged for the past 32 years. This is one of three articles based on our investigations of the DLT saline playa. In the second article, we report the analysis of the collected subsurface samples from two vertical stratigraphic cross sections at the DLT Playa, with an emphasis on mineralogy and geochemistry (Wang et al., 2018). In the third article of

* 本文发表在: Astrobiology, 2018, 18 (10): 1830

this series (Sobron et al., 2018), we report a multiscale mineralogy and geochemistry study of surface salts in an anticline in DLT (site 02).

This article discusses the salt precipitation history (through the Paleocene and to Pleistocene period) and distribution at different playas in Qaidam Basin along with the regional climate pattern. We note observed climatic, mineralogical, and geochemical similarities between the basin and some salt-rich areas on Mars.

2 Evolution of saline playas in Qaidam Basin

The Qaidam Basin is located in the northeast region of the Qinghai-Tibet (QT) Plateau. The QT Plateau has an average elevation of 4500 m and comprises five massive mountain chains that are east-west oriented and includes the Himalaya mountain chain (the highest among the five, average elevation of ~6100 m) to the south. This setting blocks the humid Indian monsoon to Qaidam Basin, which has the lowest average elevation (2676 m, the Senie Lake) on the QT Plateau. The Qaidam Basin is thus one of the driest places in nonpolar regions on Earth, with an aridity index (AI) of 0.008–0.04 recorded at specific locations (Zhu et al., 1994).

The Qaidam Basin is surrounded by the Qilian Mountains in the northeast, the Aljin Mountains in the northwest, and the Kunlun Mountains in the south (Fig. 1). This basin terrain began to subside, due to the opposite strike-slip of three deep faults that extend along each of the three mountain ranges, and then form the primary phase of the Qaidam sedimentary basin during the Mesozoic. The high elevation of the Qaidam Basin (2700–3300 m) results in low mean annual temperatures (T), large diurnal and seasonal T variations, and high ultraviolet radiation. These environmental conditions, in addition to hyperaridity, cause the evaporation of water bodies in the Qaidam Basin and the precipitation of various salts. The Qaidam Basin features four major saline playas that formed from brine evaporation in its internal depressions.

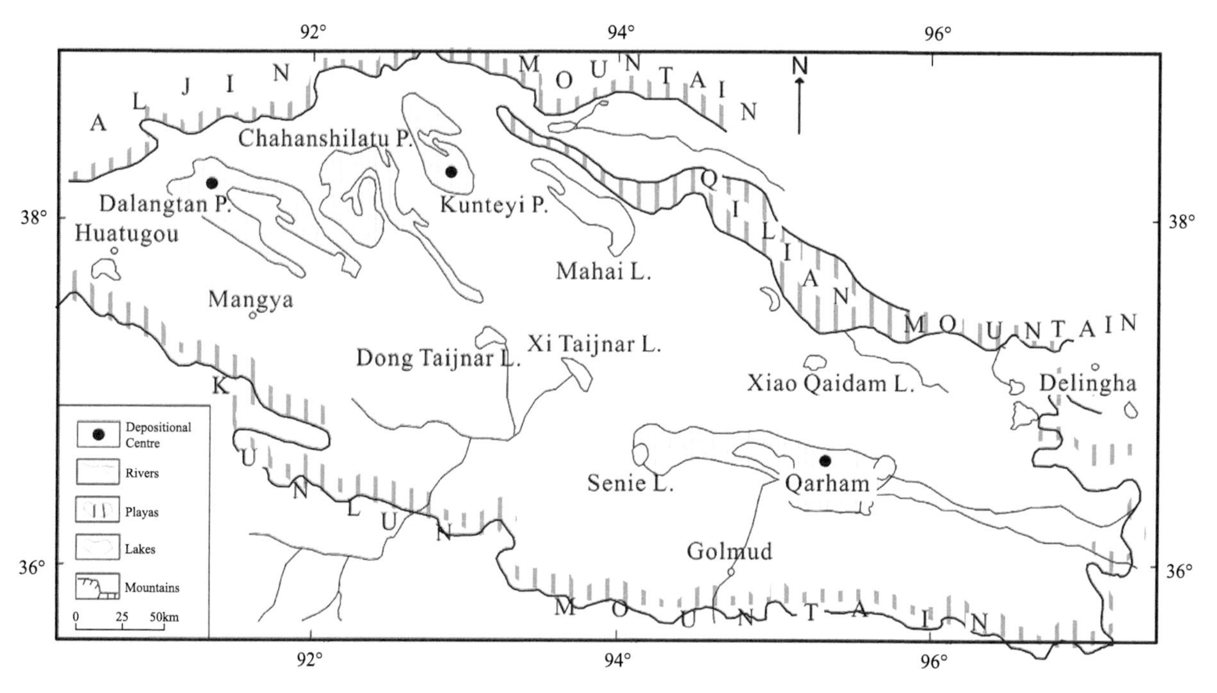

Fig. 1 Schematic map of Qaidam Basin.

The landforms of the Qaidam Basin have evolved since its formation. Since the Paleocene, starting with the collision between the Eurasia and India plates, the Qaidam Basin has experienced periods of subsidence, and a thick layer of sediments (up to 2600 m) has formed in the basin. In addition, regional tectonic activity, especially the uplift of the Tibetan Plateau, together with the deposition process itself, has altered the location of the depositional center several times within the basin.

The bases of the Qaidam Basin are thrust faults and strike slip faults, with main continental deposits in Mesozoic and Cenozoic. Triassic strata are absent over the basin, and deposits in Mesozoic, Jurassic, and Cretaceous are distributed in the northern margin of the basin. Jurassic strata is a set of river and lake facies, and coal-bearing swamp facies sediments with a thickness of 500 m, which indicates that the northern part of the Qaidam Basin was a freshwater lake in that time, with T consistent with a humid, subtropical climate. In the Cretaceous, a set of continental red clastic rock series developed that are mainly distributed in the northern margin of the basin. The sedimentary range was much more extensive than before. With the climate beginning to dry and water bodies changing from deep to shallow, the Qaidam Basin was essentially a collection of alluvial plains and shallow lakes.

The Cenozoic sedimentary strata developed most extensively in the Qaidam Basin and covered the whole basin. In addition, the main sedimentary formation from the Paleogene to the Neogene has been found in the western depression with a thickness of 1252–10,546 m. Sedimentary processes occurred from the upper Paleocene to the Pliocene, but the Quaternary sedimentary materials are located primarily in the eastern part of the basin. From the beginning of the early Pleistocene, a large number of salt deposits formed, most of them of 2000–3400 m thickness. Deposition occurred from the upper Paleocene to Pliocene, and gypsum salt sedimentation started to occur in the Oligocene; salt deposition was most prevalent during the Miocene.

In summary, the climate of the basin since the Cenozoic has become increasingly dry. Owing to the collision of the India plate and the Asian plate, the sedimentary center of the lake in the Qaidam Basin has migrated from the northwest to the east. By the late Pliocene, the center shifted from the northwest to the east of the basin, where the DLT Playa formed.

The Qaidam Basin is a closed drainage basin. River systems are mainly recharged by snowmelt from the surrounding mountains, whereas spring waters from deep faults and meteoritic water supply the modern lakes in the interior. Surface water bodies react with atmospheric water and with previously deposited salt-rich sediments, becoming salt ion rich. Therefore, the water chemistry of modern salt lakes in the Qaidam Basin is greatly controlled by the nature of local salt sediments. The long and complex history of salt deposits in the Qaidam Basin, especially the migration of the salt depositional center with time, has generated three major playas with different salt types: Mg-sulfate-rich DLT, chloride-rich Qarhan, and sulfate-chloride-rich Kunteyi.

2.1 Sulfate-rich sediments at the DLT Playa

As discussed previously, sulfate-rich sediments dominated during the first major salt forming period in the Pliocene and are distributed widely in the west and northwest parts of the Qaidam Basin. During this period, the sulfate-rich sediments first deposited at areas centered at the DLT secondary basin (aka DLT area, 38°0′–38°40′N, 91°10′–92°10′E), then extended to the northwest Qaidam Basin. In the late Pliocene, only a small amount of high-salinity sulfate brine was left in local depressions including the DLT depression, which eventually dried up during the Holocene to form the DLT Playa.

The DLT Playa is located in the center depression of the DLT area on the west margin of Qaidam Basin. The mineralogy of surface and shallow subsurface strata has been described in several studies (Zheng, 1997; Zheng et al., 2009; Wang et al., 2018, this volume). The predominant sulfate phase in the strata at the edge of the DLT area is mirabilite, a hydrated sodium sulfate, $Na_2SO_4 \cdot 10H_2O$. The strata at the center of the DLT Playa are composed mostly of hexahydrite, a hydrated magnesium sulfate, $MgSO_4 \cdot 6H_2O$. Bloedite, $Na_2Mg(SO_4)_2 \cdot 4H_2O$, was also identified in the DLT area (Table 1). This deposition trend from Na sulfate to Mg sulfate follows the normal geochemical evolution trend and the precipitation sequence of sulfate brines. The common occurrence of halite at the surface of the DLT and in samples collected from shallow subsurface strata at various locations of the playa supports the notion that the residual sulfate-rich brine from the first major salt forming period in the Pliocene reached the stage of halite saturation. In addition, ClO_4^- and ClO_3^- were detected in shallow subsurface regolith of the DLT Playa (Wang et al., 2018, this volume).

Table 1　Salt minerals found in three saline playas in qaidam basin and in mars

Minerals	Chemistry	DLT Playa	Kunteyi Playa	Qarhan Playa	Mars
Gypsum	$CaSO_4 \cdot 2H_2O$	Yes	Yes	Yes	Yes
Bassanite	$CaSO_4 \cdot 1/2H_2O$	Yes	Yes		
Anhydrite	$CaSO_4$	Yes		Yes	
Mirabilite	$Na_2SO_4 \cdot 10H_2O$	Yes	Yes		
Thenardite	Na_2SO_4	Yes	Yes		
Epsomite	$MgSO_4 \cdot 7H_2O$	Yes			
Hexahydrite	$MgSO_4 \cdot 6H_2O$	Yes	Yes		
Pentahydrite	$MgSO_4 \cdot 5H_2O$	Yes			
Starkeyite	$MgSO_4 \cdot 4H_2O$	Yes	Yes		Yes
Kieserite	$MgSO_4 \cdot H_2O$	Yes			Yes
Bloedite	$Na_2Mg(SO_4)_2 \cdot 4H_2O$	Yes	Yes		
Glauberite	$Na_2Ca(SO_4)_2$	Yes	Yes		
Glaserite	$K_3Na(SO_4)_2$	Yes			
Leonite	$K_2Mg(SO_4)_2 \cdot 4H_2O$	Yes			
Langbeinite	$K_2Mg_2(SO_4)_3$	Yes			
Loeweite	$Na_{12}Mg_7(SO_4)_{13} \cdot 15H_2O$	Yes	Yes		
Polyhalite	$K_2MgCa(SO_4)_3 \cdot 2H_2O$	Yes	Yes		
Picromerite	$K_2Mg(SO_4)_2 \cdot 6H_2O$	Yes			
Halite	$NaCl$	Yes	Yes	Yes	Yes
Sylvite	KCl	Yes	Yes	Yes	Yes
Carnallite	$KMgCl_3 \cdot 6H_2O$	Yes		Yes	
Bischofite	$MgCl_2 \cdot 6H_2O$		Yes	Yes	Yes
Antarcticite	$CaCl_2 \cdot 6H_2O$		Yes		
Dansite	$Na_{21}Mg(SO_4)_{10}Cl_3$	Yes			

Adapted from Zhu et al. (1994), Mohlmann et al. (2011), Kong et al. (2014b), and Wang et al. (2018). DLT, Dalangtan.

2.2 Chloride-rich sediments at Qarhan Playa

In the second major salt forming period of Qaidam Basin during the Pleistocene, chloride-rich sediments deposited at regions centered at the Qarhan lake area (Fig. 1). The chlorine-rich brine originated by way of migration from the west Qaidam Basin into the Qarhan depression, which eventually evaporated and formed the Qarhan Playa. Qarhan Playa is a large dry area in the Qarhan lake region (36°37′–37°12′N, 93°42′–96°14′E) at the center of the Qaidam Basin. Chloride-rich minerals, mainly halite NaCl, have been observed in the strata formed in the Pleistocene. Carnallite $KMgCl_3 \cdot 6H_2O$, sylvite KCl, and bischofite $MgCl_2 \cdot 6H_2O$ started to form in strata from the Holocene (Table 1), representing the very late evaporation stage of chlorine-rich brine.

2.3 Deposition transition from sulfates to chlorides at Kunteyi Playa

During the first major salt deposition period in the Pliocene, the deposition center of Qaidam Basin moved to the northwest, and the sulfate-chloride-bearing brine extended to the Kunteyi saline lake area (upper-middle in Fig. 1). As this migration went on, the brine chemistry evolved to an intermediate stage, transitioning from sulfate rich to chloride rich. Subsequently, these brines partially evaporated and formed the Kunteyi Playa and the Kunteyi Salt Lake.

The Kunteyi Playa has salt-bearing strata up to hundreds of meters thick. In this region, sulfates such as gypsum Ca-$SO_4 \cdot 2H_2O$, mirabilite, and glauberite $Na_2Ca(SO_4)_2$occur in the lower part of the strata, whereas hydrated chlorides, for example, bischofite $MgCl_2 \cdot 6H_2O$ and antarcticite $CaCl_2 \cdot 6H_2O$. Occur in the upper part of the strata or the surface of outcrops (Table 1).

3 Regional meteorological patterns

3.1 Three data sets

In this work, we use three data sets to describe the meteorological patterns of the DLT area. The first set is a 32-year record (1980–2011) of meteorology data collected at the Mangya meteorological station, located <50 km from the DLT Playa (Fig. 1). Meteorological data were provided by the Qinghai Meteorological Administration. This data set includes T, relative humidity (RH), atmospheric pressure at the surface, precipitation, evaporation, wind speed, and sunlight hours. These data were recorded each hour. The second set is a 2-year RH and T record collected by our team with a data logger RHT10 (Extech Co., Nashua, NH) at the surface of the Xiao Liang Shan anticline (6 km to the center of the DLT Playa) since 2010. These data were recorded each hour and were downloaded every 3-6 months. The third set is from a new meteorological station set up by us at the center of the DLT Playa in 2012, from which the data have been collected to the present and recorded each hour. All three playas in the Qaidam Basin, as discussed in Section 2, formed at local depressions and have quite similar elevations: 2690 m, 2670 m, and 2730 m for the DLT, Qarhan, and Kunteyi, respectively. Owing to their proximity and similar elevation, the three regions are subject to the same meteorological conditions. Although our data were only collected at, or near, the DLT Playa, the DLT measurements can be extrapolated to the other two playas.

3.2 Basic environmental characteristics

Figures 2, 4, 7–11 show the monthly average T (in ℃), RH (in %), lighting (in hours), wind speed (in m/s), precipitation (in mm), and evaporation (in mm) recorded at Mangya meteorological station (<50 km to DLT Playa). Fig. 3 and Fig. 5 show hourly variations of T and RH in two typical summer and winter days at logger RHT10.

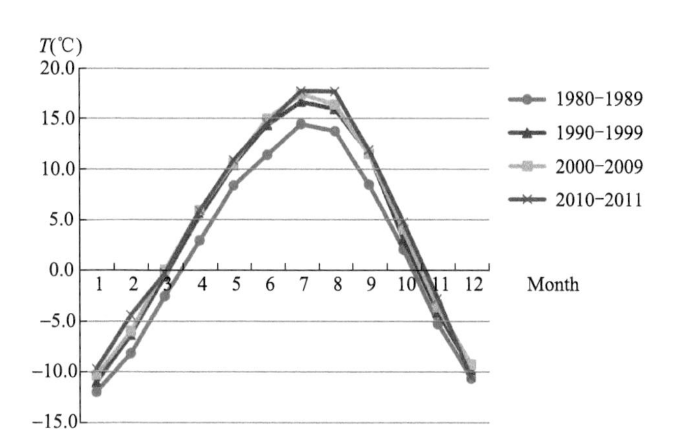

Fig. 2 Patterns of T (in ℃, monthly averaged) in 1980–2011, recorded at Mangya Meteorological Station (<50 km to DLT Playa). DLT, Dalangtan; T, temperature

Based on those data, we found that the annual average T in the DLT area is 3.5 ℃. The seasonal variation ΔT_{ave} is >30 ℃ (Fig. 2), but seasonal ΔT_{peak} is >55 ℃ (Fig. 3). The maximum monthly average air T was 16 ℃ in July, with T_{max}>30 ℃ occurring in the summer afternoon (Fig. 3). The minimum monthly average air T was −10.7 ℃ in January, with T_{min}<−25 ℃ occurring in the winter early morning (Fig. 3). The diurnal T variation in summer is about 20 ℃, and that value decreases in winter to about 15℃ (Fig. 3). The atmospheric pressure at the surface of the DLT area is about 70% of atmospheric pressure at sea level. The annual average surface pressure recorded is 709 mbar (Fig. 11). The surface pressure (717 mbar) in autumn is slightly higher than that in spring (706 mbar).

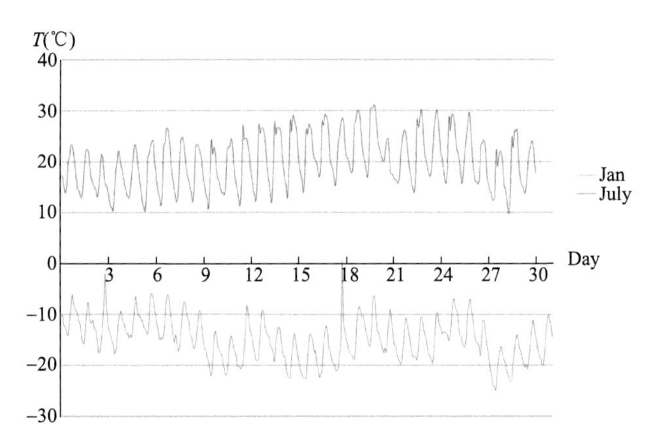

Fig. 3 Diurnal temperature variations during a typical summer month and a winter month recorded at the surface of XLS anticline from *T* to RH logger data (<6 km to the center of DLT Playa).
RH, relative humidity; XLS, Xiao Liang Shan.

RH is dependent on *T* and the partial pressure of water P_{H_2O}. The DLT area is extremely dry in spring and fall, but less dry in winter and summer when monthly averaged RH can rise to >35 %, although it never exceeds 40% (Fig. 4). The lowest RH values (<10%) were observed in early afternoon, with a variation of 15% between months (Fig. 5). The highest RH values (>60%) were observed in early morning, with a larger variation ~40% between months (Fig. 5). No RH value of 100% was found among our 2 years of recorded data (2010–2011) using an RHT10 logger, which is consistent with the lack of rain in our recording interval.

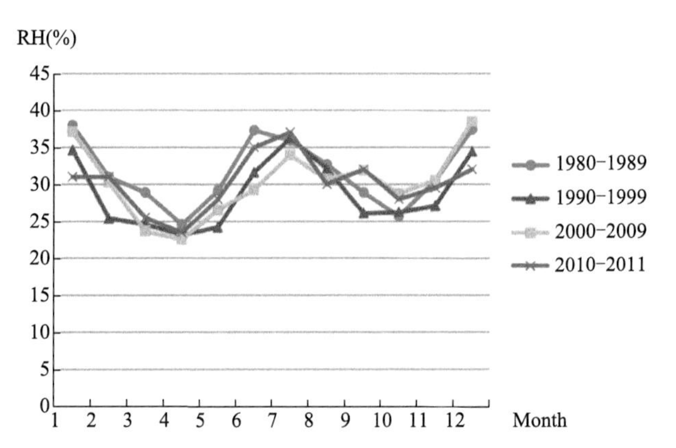

Fig. 4 Patterns of RH (%, monthly averaged) during 1980–2011, recorded at Mangya Meteorological Station (<50 km to DLT Playa).

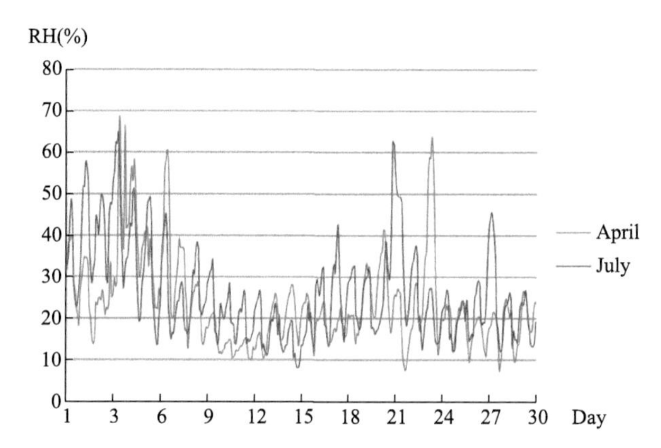

Fig. 5 Diurnal variations in RH during a typical spring month and a winter month recorded at the surface of XLS anticline from *T* to RH logger data (<6 km to the center of DLT Playa).

The changing pattern of water vapor pressures (P_{H_2O}) in 2011 is shown in Fig. 6. Data were recorded with the logger RHT10, and P_{H_2O} was calculated based on reference data of saturated water/ice vapor pressures (Haar et al., 1984; Wagner et al., 1994). The results indicate that the water vapor input is stronger from June to August.

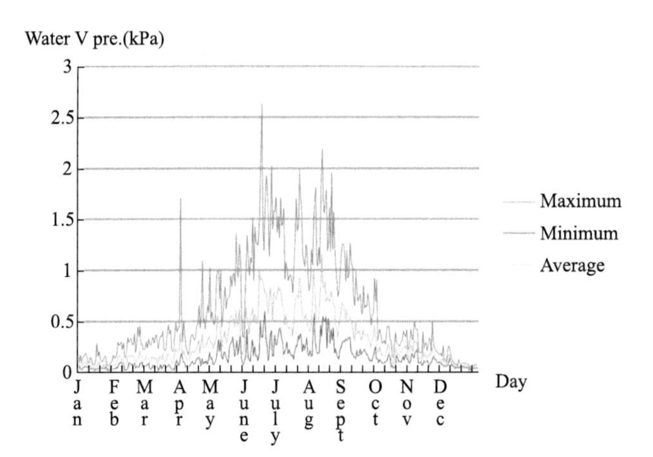

Fig. 6 Water vapor pressures variation at per hour density in 2011.

Water vapor transported from the west is the basic water vapor source of the DLT area. West wind fluctuations have a direct effect on the interannual change of the water vapor transport flux divergence over most of Northwest China (Wang et al., 2005). The Indian monsoon affects the southern and eastern parts of Northwest China by the southwest water vapor transport stream (Wang et al., 2005). For the Qaidam Basin, the vapor resources may have the same mechanism according to our observations.

The weather in the DLT Playa is sunny and windy. The monthly average hours of sunlight range from 220 during winter to 320 during summer (Fig. 7). The maximum monthly average wind speed occurs during spring (up to 3.5 m/s), and the minimum (about 1.5 m/s) occurs during winter (Fig. 8). The maximum wind speed can sometimes reach 20–22 m/s. A value of 20 m/s was recorded at the DLT Playa weather station.

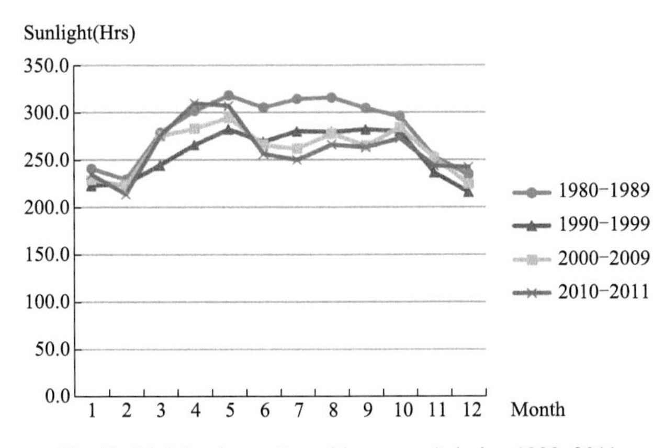

Fig. 7 Lighting hours (monthly average) during 1989–2011 recorded at Mangya Meteorological Station (<50 km to DLT Playa).

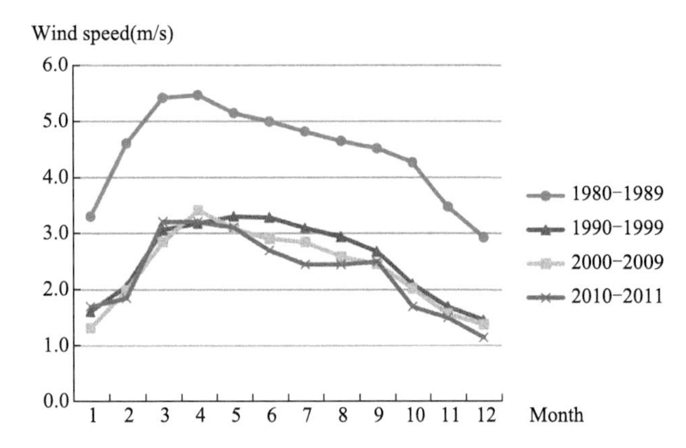

Fig. 8 Wind speed (monthly average) during 1989–2011 recorded at Mangya Meteorological Station (<50 km to DLT Playa).

Rainfall events in the DLT area are shown in Table 2. The annual total precipitation averaged for 32 years (1980–2011) is about 50 mm. Figure 9 shows that most precipitation (maximum <25 mm monthly average) happens in the summer months. In contrast, the evaporation rate of this region is extremely high. The strong wind, solar irradiation, and lack of vegetation all lead to an annual total evaporation of 2590 mm as averaged for 32 years. Fig. 10 shows that the strongest evaporation (maximum ~440 mm per month) occurs in the high T summer months. The AI of the DLT area, as defined by the annual total precipitation over annual total evaporation, is about 0.02 (variation over month is 0.01–0.05; Figs. 9 and 10), which classifies the DLT area as a hyperarid region on Earth.

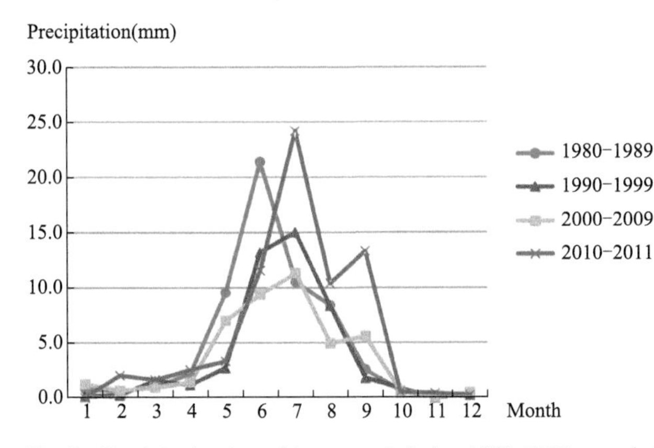

Fig. 9 Precipitation (monthly average) during 1989–2011 recorded at Mangya Meteorological Station (<50 km to DLT Playa).

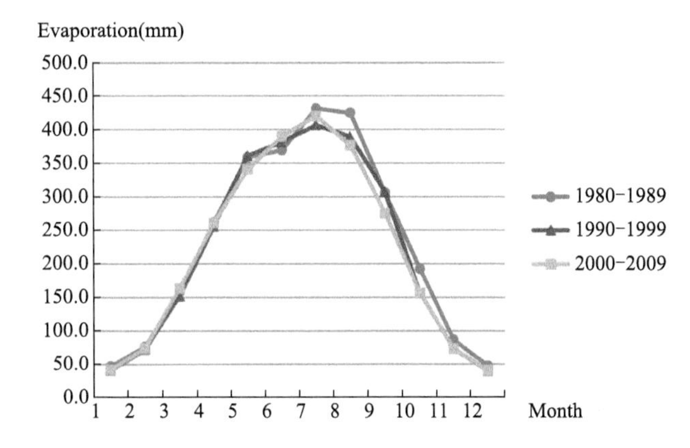

Fig. 10 Evaporation (monthly average) during 1989–2011 recorded at Mangya Meteorological Station (<50 km to DLT Playa).

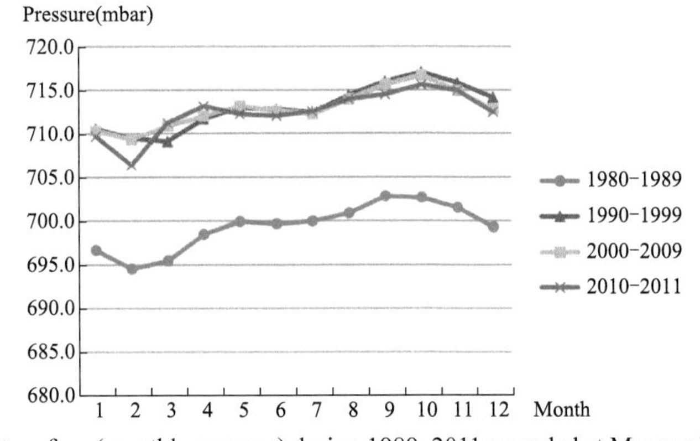

Fig. 11 Atmospheric pressure at surface (monthly average) during 1989–2011 recorded at Mangya Meteorological Station (<50 km to DLT Playa).

Table 2 Measured precipitation (in mm) from 1980 to 2011 at the mangya meteorological station, located <50 km from the dalangtan playa

Year	January	February	March	April	May	June	July	August	September	October	November	December	Total
1980	0.8	0.7	0.0	0.8	0.2	5.5	6.6	16.2	0.6	0.0	0.0	0.3	31.7
1981	0.0	1.1	0.0	2.9	1.0	8.8	17.8	26.2	0.9	0.5	0.0	0.1	59.3
1982	0.1	1.0	2.0	3.1	18.7	35.6	7.4	5.5	5.2	0.1	0.0	1.1	79.8
1983	0.5	0.4	1.1	0.0	2.2	16.7	28.7	6.3	2.2	2.1	0.0	0.0	60.2
1984	1.3	2.1	2.4	4.2	5.3	16.4	9.3	6.3	3.3	0.0	0.0	0.1	50.7
1985	0.2	0.0	0.3	0.0	7.8	8.5	6.8	0.7	2.4	0.0	0.1	1.9	28.7
1986	0.4	0.6	0.7	0.6	34.7	54.1	4.9	8.4	0.0	0.1	1.2	0.2	105.9
1987	0.0	0.2	1.4	5.7	5.3	24.0	9.9	0.7	1.7	2.9	0.3	0.3	52.4
1988	0.0	0.3	0.7	0.3	18.7	11.4	8.4	7.0	5.2	0.0	0.0	0.0	52.0
1989	4.6	0.0	0.8	4.8	1.5	32.9	5.3	6.8	4.0	0.5	0.0	0.5	61.7
1990	0.0	0.0	3.1	0.6	0.5	22.6	11.2	4.6	0.0	0.0	0.0	1.1	43.7
1991	0.0	0.0	0.9	4.9	0.8	3.8	7.1	14.7	0.0	0.0	0.0	0.0	32.2
1992	0.2	0.2	0.7	0.0	4.3	2.8	27.6	0.5	0.1	2.0	0.0	0.2	38.6
1993	0.6	0.9	1.6	0.0	1.8	45.0	27.1	8.9	0.0	0.0	0.0	1.1	87.0
1994	0.0	0.3	0.0	0.0	0.2	4.6	12.4	4.0	0.0	0.8	0.0	0.0	22.3
1995	0.1	1.0	0.0	5.2	0.0	4.1	2.1	3.6	12.0	4.7	0.0	0.0	32.8
1996	0.0	0.0	2.0	0.8	7.7	7.3	29.1	13.0	0.8	0.4	0.0	0.0	61.1
1997	0.0	0.0	0.0	0.0	11.0	8.8	8.3	18.2	0.0	0.0	0.0	0.0	46.3
1998	0.2	0.0	4.7	0.0	0.6	13.7	11.9	11.2	4.7	0.2	0.0	0.0	47.2
1999	0.0	0.0	2.4	0.0	0.3	19.0	13.2	4.9	0.0	0.2	0.0	0.0	40.0
2000	1.9	0.5	0.0	0.2	1.6	16.4	12.5	7.2	4.9	0.0	0.0	0.0	45.2
2001	0.0	0.4	0.0	0.0	0.8	3.9	1.1	5.8	0.1	0.0	0.0	0.0	12.1
2002	2.6	0.0	0.1	7.1	0.6	13.8	7.0	2.7	6.6	0.3	0.0	0.8	41.6
2003	0.0	1.4	3.1	0.9	4.1	21.0	14.7	0.0	2.6	0.3	0.0	0.2	48.3
2004	0.7	0.0	0.1	0.7	1.6	6.7	1.6	5.2	0.0	0.0	0.1	0.1	16.8
2005	0.0	2.2	5.4	0.0	32.4	2.6	22.5	21.3	0.6	1.3	0.0	0.0	88.3
2006	1.6	0.8	0.0	4.1	6.0	2.0	12.0	3.0	4.2	0.3	0.0	2.4	36.4
2007	0.0	0.0	0.1	0.9	2.7	17.8	14.6	0.3	11.9	1.8	0.0	0.0	50.1
2008	5.8	0.0	0.2	0.0	12.6	8.7	20.7	4.1	24.8	1.9	0.0	0.2	79.0
2009	0.0	1.2	0.3	0.0	7.7	1.0	6.3	0.0	0.2	0.0	0.0	1.1	17.8
2010	0	4.1	2.5	3.1	4.7	10.5	42.1	5.3	22.7	0.9	0	0.4	96.3
2011	0	0	0.8	2	2	12.6	6.3	15.5	3.9	0.1	0.8	0	44.0

4 DLT may provide some interesting case studies for Mars potential habitability

On the basis of the already reported occurrence of salt deposits in the three playas of Qaidam Basin (including the DLT Playa), and the regional meteorological patterns, we believe that the DLT area and the local region in general can serve as a Mars analog site that is consistent with desert-type terrain and applicable for investigation of the fundamental properties of hydrous salts in a natural setting under Mars-relevant environmental conditions, such as the precipitation of salts from a natural Na-K-Ca-Mg-Cl-SO_4-CO_3-H_2O system and their phase transformation or preservation under hyperarid conditions. Because similar types of hydrous salt deposits (magnesium-sulfates) at the surface of DLT were identified on Mars, the DLT area can also serve as an orbital remote sensing test region for the use of Vis-NIR spectroscopy, with known ground truth, and for the Operation Readiness Test of future landed missions to Mars (e.g., China's Mars 2020 mission).

The diverse salts with different chaotropic or kosmotropic properties in the Qaidam Basin (especially $MgCl_2 \cdot 6H_2O$ and $CaCl_i \cdot 6H_2O$ in the Kunteyi Playa) may provide some interesting case studies for Mars regarding potential habitability.

T and water activity are the most important physical parameters that influence microbial metabolism and growth (Beaty et al., 2006). A limit of $T \geqslant -25$ ℃ and water activity >0.5 has been recommended for considering Special Regions on Mars (Kminek et al., 2010). Future field exploration of the DLT region would enable detailed study of these parameters along with biomarker identification within salt deposits on-site and in laboratory study.

Acknowledgments

The authors thank Dr. Weigang Kong and Academician Wenjin Zhao for their help in preparing this article. This research was funded by the China Geological Survey program"Astrobiology on Mars analogous sites for saline environments" (Grant No.12120113019100) and "The comprehensive survey of lithium, and other resources in salt lakes on the Tibetan Plateau" (Grant Nos.121201103000150012 and DD20160025). A.W. and P.S. would like to express their great appreciation to the McDonnell Center for Space Science at Washington University in St. Louis for its generous financial support for the 2008 field expedition to Qaidam Basin, Tibetan Plateau, the laboratory sample analysis (2009–2012), and the article preparation (2014–2016). There are no competing financial interests existing for all authors of this article.

References

Anglés, A. and Li, Y. (2017a) The western Qaidam Basin as a potential Martian environmental analogue: an overview. *J Geophys Res Planets* 122: 856-888.

Anglés, A. and Li, Y.L. (2017b) Similar ring structures on Mars and Tibetan Plateau confirm recent tectonism on Martian northern polar region. *Int J Astrobiol* 16: 355-359.

Beaty, D., Buxbaum, K., Meyer, M., Barlow, N., Boynton, W., Clark, B., Deming, J., Doran, P.T., Edgett, K., Hancock, S., Head, J., Hecht, M., Hipkin, V., Kieft, T., Mancinelli, R., McDonald, E., McKay, C., Mellon, M., Newsom, H., Ori, G., Paige, D., Schuerger, A.C., Sogin, M., Spry, J.A., Steele, A., Tanaka, K., and Voytek, M. (2006) Findings of the Mars Special Regions Science Analysis Group. *Astrobiology* 6: 677-732.

Cheng, Z., Xiao, L., Wang, H., Yang, H., Li, J., Huang, T., Xu, Y., and Ma, N. (2017) Bacterial and archaeal lipids recovered from subsurface evaporites of Dalangtan Playa on the Tibetan Plateau and their astrobiological implications. *Astrobiology* 17: 1112-1122.

Haar, L., Gallagher, J.S., and Kell, G.S. (1984) *NBS/NRC Steam Tables*. Hemisphere Publishing Corp., New York, NY.

Kminek, G., Rummel, J.D., Cockell, C.S., Atlas, R., Barlow, N., Beaty, D., Boynton, W., Carr, M., Clifford, S., Conley, C.A., Davila, A.F., Debus, A., Doran, P., Hecht, M., Heldmann, J., Helbert, J., Hipkin, V., Horneck, G., Kieft, T.L., Klingelhoefer, G., Meyer, M., Newsom, H., Ori, G.G., Parnell, J., Prieur, D., Raulin, F., Schulze-Makuch, D., Spry, J.A., Stabekis, P.E., Stackebrand, E., Vago, J., Viso, M., Voytek, M., Wells, L., and Westall, F. (2010) Report of the COSPAR Mars special regions colloquium. *Adv Space Res* 46: 811-829.

Kong, F.J., Ma, N.N., Wang, A., and Amend, J. (2010) Isolation and identification of halophiles from evaporates in DaLangTan salt lake. *Acta Geologica Sinica (in Chinese)* 84: 1661-1667.

Kong, F.J., Zheng, M.P., Wang, A., and Ma, N.N. (2009) Endolithic halophiles found in evaporite salts on Tibet Plateau as a potential analog for Martian life in saline environment [abstract 1216]. In *40th Lunar and Planetary Science Conference Abstracts*, Lunar and Planetary Institute, Houston.

Kong, W.G., Zheng, M.P., and Kong, F.J. (2014a) Brine evolution in Qaidam Basin, northern Tibetan Plateau, and the formation of playas as Mars analogue site[abstract 1228]. In *45th Lunar and Planetary Science Conference Abstracts*, Lunar and Planetary Institute, Houston.

Kong, W.G., Zheng, M.P., Kong, F.J., and Chen, W.X. (2014b) Sulfate-bearing deposits at Dalangtan Playa and their implication for the formation and preservation of Martian salts. *Am Mineral* 99: 283-290.

Mayer, D.P., Arvidson, R.E., Wang, A., Sobron, P., and Zheng, M.P. (2009) Mapping minerals at a potential mars analog site on the Tibetan Plateau [abstract 1877]. In *40th Lunar and Planetary Science Conference Abstracts*, Lunar and Planetary Institute, Houston.

Mohlmann, D. and Thomsen, K. (2011) Properties of cryobrines on Mars. *Icarus* 212: 123-130.

Sobron, P., Freeman, J.J., and Wang, A. (2009) Field test of the water-wheel IR (WIR) spectrometer on evaporative salt deposits at Tibetan Plateau [abstract 2372]. In *40th Lunar and Planetary Science Conference Abstracts*, Lunar and Planetary Institute, Houston.

Sobron, P., Wang, A., Mayer, D., Bentz, J., Kong, F., and Zheng, M. (2018) DLT saline playa in a hyperarid region on Tibet Plateau-III: correlated multi-scale surface mineralogy and geochemistry survey. *Astrobiology* 18: 000-000.

Wagner, W., Saul, A., and Pruss, A. (1994) International equations for the pressure along the melting and along the sublimation curve of ordinary water substance. *J Phys Chem Ref Data* 23: 515-525.

Wang, A., Lu, Y.L., and Chou, I.M. (2013) Recurring slope lineae (RSL) and subsurface chloride hydrates on Mars [abstract 2606]. In *44th Lunar and Planetary Science Conference Abstracts*, Lunar and Planetary Institute, Houston.

Wang, A., Sobron, P., Kong, F., Zheng, M.P., and Zhao, Y. (2018) DLT saline playa in a hyperarid region on Tibet Plateau-II: preservation of salt with high hydration degrees in subsurface. *Astrobiology* 18: 000-000.

Wang, A. and Zheng, M.P. (2009) Evaporative salts from Saline Lakes on Tibetan Plateau: an analog for salts on Mars [abstract 1858]. In *40th Lunar and Planetary Science Conference Abstracts*, Lunar and Planetary Institute, Houston.

Wang, A. and Zheng, M.P. (2011) Saline Playa on Qinghai Tibet Plateau [abstract 6003] In *Analogue sites for the Mars Missions: MSL and Beyond*, Lunar and Planetary Institute, Houston.

Wang, A., Zheng, M.P., Kong, F.J., Sobron, P., and Mayer, D. (2010) Saline playas on Qinghai-Tibet Plateau as Mars analog for the formation-preservation of hydrous salts and biosignaturess[abstract for 2010]. *In Fall AGU*.

Wang, K.L., Jiang, H., and Zhao, H.Y. (2005) Atmospheric water vapor transport from westerly and monsoon over the Northwest China. *Adv Water Sci* 16: 432-438 (in Chinese).

Xiao, L., Wang, J., Dang, Y.N., Cheng, Z.Y., Huang, T., Zhao, J.N., Xu, Y., Huang, H., Xiao, Z.Y., and Komatsu, G. (2017) A new terrestrial analogue site for Mars research: the Qaidam Basin, Tibetan Plateau (NW China). *Earth Sci Rev* 164: 84-101.

Zhang, P.X. (1987) *Salt lakes on Qaidam Basin*. The Science Publishing House (in Chinese), Beijing.

Zheng, M. (1997) *An Introduction to Saline Lakes on the Qinghai-Tibet Plateau*. Springer, Netherlands.

Zheng, M.P., Wang, A., Kong, F.J., and Ma, N.N. (2009) Saline lakes on Qinghai-Tibet Plateau and salts on Mars [abstract 1454]. In *40th Lunar and Planetary Science Conference Abstracts*, Lunar and Planetary Institute, Houston.

Zhu, Y.T., Zhong, J.H., and Li, W.S. (1994) *The Neotectonic Movement and the Evolution of Saline Lakes of Qaidam Basin in Northwestern China*. Geological Publishing House (in Chinese), Beijing.

Edaphic characterization and plant zonation in the Qaidam Basin, Tibetan Plateau*

Xianjie Wang, Fanjing Kong, Weigang Kong and Wenning Xu

MLR Key Laboratory of Saline Lake Resources and Environments, Institute of Mineral Resources, Chinese Academy of Geological Sciences (CAGS), Beijing 100037, China.

This paper presents a study of edaphic characteristics and their relationship with plant distribution in the Qaidam Basin, Tibetan Plateau, and establishes a distribution model for plants in sandy gravel Gobi to dry salt lake areas. All of the communities in the study area were dominated by plants with strong saline-alkaline tolerance. In this area, salts appeared to migrate to the surface; the surface soil was striped, and the salt distribution varied from sandy gravel Gobi to dry salt lake areas. The salt composition mainly consisted of NaCl in the surface crust. In the subsurface layers, the salt composition was dominated by Ca^{2+}, Cl^- and SO_4^{2-}. The type of vegetation at the study site can be divided into two categories: salt-tolerant vegetation and weakly salt-tolerant vegetation. The salt-tolerant vegetation is influenced by Na^+, Cl^-, and the salinity. The soil of these vegetation communities had a higher salt and Na^+ concentration and a lower Ca^{2+} and K^+ concentration. The weakly salt-tolerant vegetation is mainly affected by the Ca^{2+}/Na^+ and K^+/Na^+ ratios. Based on the above results, a vegetation distribution model for saline lakes on the inland plateau was established.

Plant zonation is a common characteristic of salt marshes worldwide[1]. In low-stress environments, competition is an important factor for plant distribution, but in harsh physical conditions, the tolerance of species to an extreme soil environment determines the plant distribution[2]. Therefore, the soil type, especially the soil salinity, is a major factor in plant zonati on[3-6].

Solutes contributing to soil salinity include different ions, such as Na^+, K^+, Ca^{2+}, Mg^{2+}, Cl^-, SO_4^{2-}, CO_3^{2-}, and HCO_3^-. The relative proportions of Ca^{2+}, Na^+, Mg^{2+} and K^+ in the soil are considered to be critical factors for the development of plants in saline environments[7-10]. However, if the concentration of these ions is too high, the plant cells will suffer high osmotic pressure, which affects plant growth and development. Too much salt accumulation can also lead to a decrease in the osmotic pressure of plant roots, directly affecting the uptake of nutrients by roots, which can impair the growth of vegetation[11]. Furthermore, the concentrations of some ions do not necessarily increase when soil salinity increases. Therefore, determining the concentration of a specific ion or ion ratio is necessary when evaluating the relationship between soils and plants[1, 6]. Álvarez Rogel et al. reported that the ion ratios of K^+/Na^+, Ca^{2+}/Mg^{2+} and Ca^{2+}/Na^+ can best explain soil-vegetation relationships[1, 6, 12].

Many studies have determined the relationships between soil salinity, soil moisture and plant zonation on coast marshes[1, 6, 12-17]. For instance, the temporal gradients and topography of the salt marshes in Spain have been studied. A relationship model between soil and vegetation was established by analysing the correlation between the soil environment and specific vegetation types, allowing the vegetation to be used as an indicator of soil type[1, 6, 12, 14]. However, few studies on vegetation response to soils have been conducted in salt lakes. This paper focuses on the vegetation and soil characteristics on watershed of salt lake and a vegetation distribution model of saline lakes on an inland plateau is put forward.

There are many saline lakes in China; therefore, the total salt marsh area is large. The Qaidam Basin is located in

* 本文发表在：Scientific Reports, 2018, 8: 1822-1830

the northeast part of the Qinghai-Tibetan Plateau, which is the largest and highest plateau in the world. The basin is surrounded by the Kunlun Mountains, Altun Mountains and Qilian Mountains, forming a down-faulted closed inland basin. The elevations of the outer basin reach 4000–5000 m, and the lower parts of the basin are at 2700–3200 m. There are 51 lakes in the Qaidam Basin. Of these, Keluke Lake is a freshwater lake, and seven of the lakes have brackish water. The others are all saline lakes, and six of them are playas, which are rich in salt mineral resources[18].

The study site is located in the southern part of the Qaidam Basin, 22 km east of the city of Golmud. The site spans Tuanjie Lake to the diluvial plain on the north piedmont of the Kunlun Mountains. The plain (generally gravel Gobi or sand), which is mainly on the edge of alluvial plains, has some areas of fine soil and fluvial plain. Its dimensions range from approximately N36°37′–N36°25′, E95°11′10.5″–E95°11′13.6″, with an altitude of approximately 2750 m (Fig. 1).

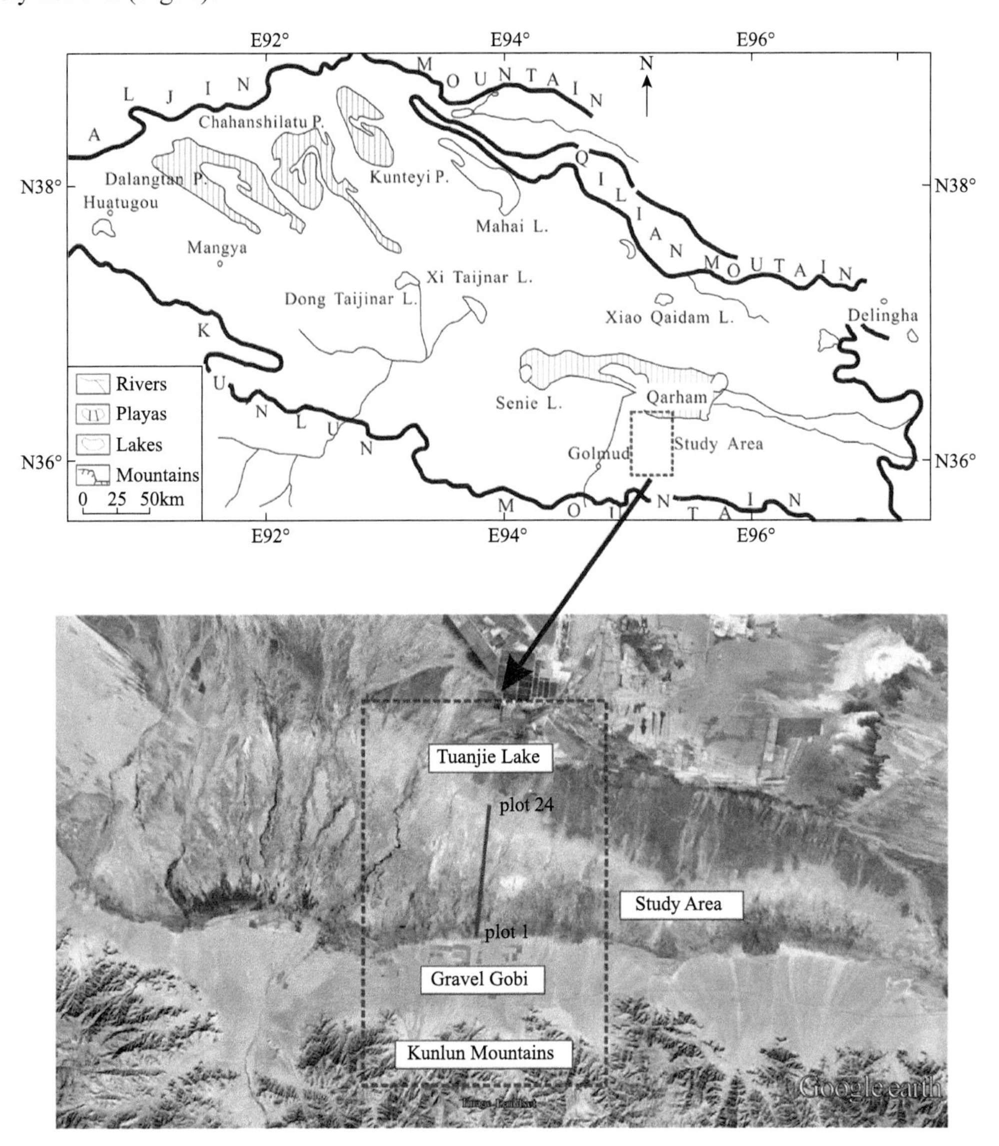

Fig. 1 The distribution of saline lakes in the Qaidam Basin and the location map of the study site.

Plots 1 to 24 are distributed from a gravel Gobi area to Tuanjie Lake. The upper part of the Fig. 1 was drawn by Xianjie Wang using Corel DRAW software (Corel DRAW X4 14.0.0.701), and the lower part was modified from map downloaded from Google Earth on May 2017 (Google Earth 7.1.5.1557)

Tuanjie Lake is formed atop the surrounding salt marshes of Qarhan Salt Lake and recharged by river water or groundwater. Phreatic water in the study site is close to the surface or overflowing, and the plant communities are dominated by *Tamarix* and salt-tolerant hygrophytes. The plant community structures are simple, having relatively few species. The predominant soil type is solonchak, according to the Food and Agriculture Organization classification[19].

The climate of the area is temperate and arid with a mean annual precipitation of 28.1 mm, but a mean annual evapotranspiration of 3456 mm[20]. The average temperature is 5.2 ℃. The total annual sunshine duration is long, exceeding 3100 h. The total solar radiation is 700 kJ/cm^2, and the average wind speed is 4.3 m/s[21]. In summary, the climate of the basin can be described as cold, dry and windy, with a large diurnal temperature difference, long sunshine duration and strong solar radiation.

Since the extreme harsh environment condition for the vegetation, the ecological environments here are fragile. The objective of this study was to investigate the relationship between soil and vegetation distribution, and establish a natural vegetation distribution model for saline lakes on an inland plateau. This model allows detection of environmental changes based on vegetation observations, which will help prevent environmental impacts and contribute to the management and conservation of this area[22]. Studies including both quantitative and qualitative environmental data from the area are needed to restore and preserve it. They can also provide basic data for the development of agriculture in the area, particularly for the saline lake agriculture.

Results

Vegetation characteristics. The plant types in the study site included shrubs and herbs. Shrub species included *Tamarix*, *Nitraria sibirica*, and *Lycium ruthenicum*. Herb species included *Phragmites australis*, *Leymus secalinus*, *Apocynum venetum*, *Salicornia europaea*, *Saussurea salsa*, *Kalidium gracile*, *Sphaerophysa salsula*, and *Glaux maritima*. The coverage of the vegetation is shown in Table 1. Spots 1 and 2 were located in the gravel Gobi vegetation on the diluvial clinoplain at the north piedmont of the Kunlun Mountains, and only a few reeds were distributed sparsely. With the lower water table, the surface soil transitioned to salt crust, where the dominant species were *Tamarix*, *Lycium ruthenicum*, *Apocynum venetum*, and *Phragmites australis* (spots 3 to 6). From spots 7 to 11, the soil moisture increased, and some salt-tolerance hygrophytes, such as *Leymus secalinus*, *Glaux maritima*, and *Triglochin maritimum* were present. Close to the playa centre, the vegetation was predominantly *Tamarix*, *Phragmites australis*, *Apocynum venetum*, *Saussurea salsa*, and *Nitraria sibirica*. In the lowest terrain of the playa, the only plants were *Phragmites australis*, and the coverage was very low (spots 22 and 23). The northern part next to spot 23 was a bare flat area covered with extensive salt crust (spot 24).

Table 1 Coverage of 10 species from 24 quadrats taken in the Tuanjie Lake salt marshes

	Phragmites australis	*Apocynum venetum*	*Lycium ruthenicum*	*Tamarix*	*Leymus secalinus*	*Saliconia europaea*	*Saussurea salsa*	*Kalidium gracile*	*Sphaerophysa salsula*	*Glaux maritime*
1	—	—	—	—	—	—	—	—	—	—
2	2	—	—	—	—	—	—	—	—	—
3	5	6	—	—	—	—	—	—	—	—
4	4	2	5	—	—	—	—	—	—	—
5	2	5	4	—	—	—	—	—	—	—
6	4	4	2	4	—	—	—	—	—	—
7	4	—	—	—	7	4	—	—	—	—
8	2	—	—	—	2	1	—	—	—	1
9	6	—	—	—	6	—	4	—	—	—
10	5	—	—	—	—	2	7	—	—	—
11	5	—	—	—	5	—	4	—	4	—
12	1	—	—	—	—	—	—	—	—	—

Continued

	Phragmites australis	*Apocynum venetum*	*Lycium ruthenicum*	*Tamarix*	*Leymus secalinus*	*Saliconia europaea*	*Saussurea salsa*	*Kalidium gracile*	*Sphaerophysa salsula*	*Glaux maritime*
13	4	—	—	—	—	—	—	—	—	—
14	2	—	—	2	—	—	—	—	—	—
15	2	5	—	—	—	—	—	—	—	—
16	2	—	—	—	—	—	1	—	—	—
17	4	5	—	—	—	—	—	—	—	—
18	4	1	—	5	—	—	—	—	—	—
19	1	1	—	5	—	—	—	—	—	—
20	1	4	—	—	—	—	—	—	—	—
21	1	4	—	2	—	—	—	4	—	—
22	1	—	—	—	—	—	—	—	—	—
23	1	—	—	—	—	—	—	—	—	—
24	—	—	—	—	—	—	—	—	—	—

Notes: Cover scale: -indicates 0% cover; 1 indicates <1% cover; 2 indicates 1–5% cover; 3 indicates 5% cover; 4 indicates 5–12.5% cover; 5 indicates 12.5–25% cover; 6 indicates 25–50% cover; 7 indicates 50–75% cover; 8 indicates >75% cover. Species coverage = π × (crown width/2)^2× quantity/25.

In the study site, *Phragmites australis* was distributed widely, with high coverage in almost every sampling spot, followed by *Apocynum venetum*, *Tamarix*, and *Lycium ruthenicum*. These plants mainly occurred in spots 3 to 6 and spots 14 to 21. *Saussurea salsa*, *Leymus secalinus*, *Salicornia europaea*, *Sphaerophysa salsula*, and *Glaux maritima* were concentrated in spots 7 to 11.

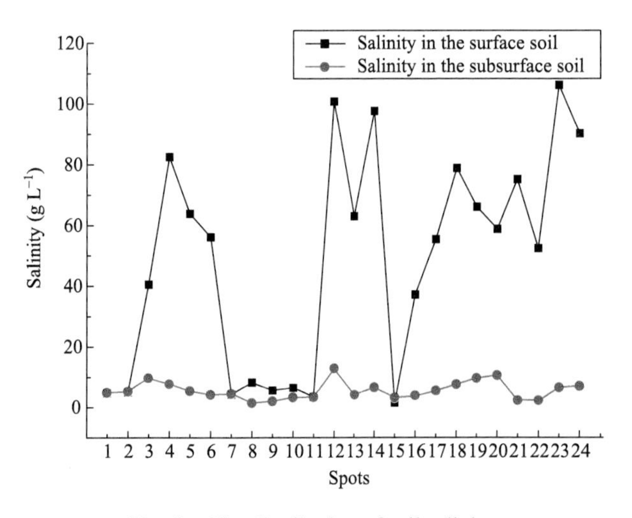

Fig. 2　The distribution of soil salinity.

Spatial variations in the physical and chemical parameters of the soil. Most soils in the study area were sandy and loamy. The salinity in the surface layer was greater than that in the deeper layers. The surface layer of most of the soils was a white salt crust, with a thickness of approximately 2–15 cm. However, the subsurface soil was mainly salty brown sand. The pH of the soils in the study site ranged from 8.0 to 9.0, indicating that the soils are alkaline. The physical and chemical characteristics were measured in the laboratory. The results showed that the salinity in most surface soils exceeded 92.875 g/kg, with a maximum value of 265.75 g/kg at spot 23. The salinity of all of the subsurface soil samples was below 32.2 g/kg; a minimum value of 3.608 g/kg occurred at spot10. These results suggest that there is a strong salt gradient in the vertical direction. Laterally, from the border of the north piedmont of the Kunlun Mountains to the Tuanjie Lake salt marshes, the surface soil had fluctuating salinity, but the salinity of the subsurface soil remained relatively constant (Fig. 2).

Except for spots 1, 2 and 7–11, the Na^+ concentration of the soil decreased greatly with depth. In contrast, the K^+, Ca^{2+}, and Mg^{2+}concentrations increased with depth in most of the sampled spots. The lateral distribution of the Na^+ concentration was similar to that of the salinity. The concentrations of Mg^{2+} and Ca^{2+} first increased and then decreased, and the K^+ concentration was relatively uniform (Fig. 3). The concentrations of Cl^- and

SO_4^{2-} were all higher in the surface than in the subsurface layers (Fig. 4). NaCl tended to be concentrated in the surface layer. In subsurface soil, the major soluble cation was Ca^{2+}, and the major anions were Cl^- and SO_4^{2-}.

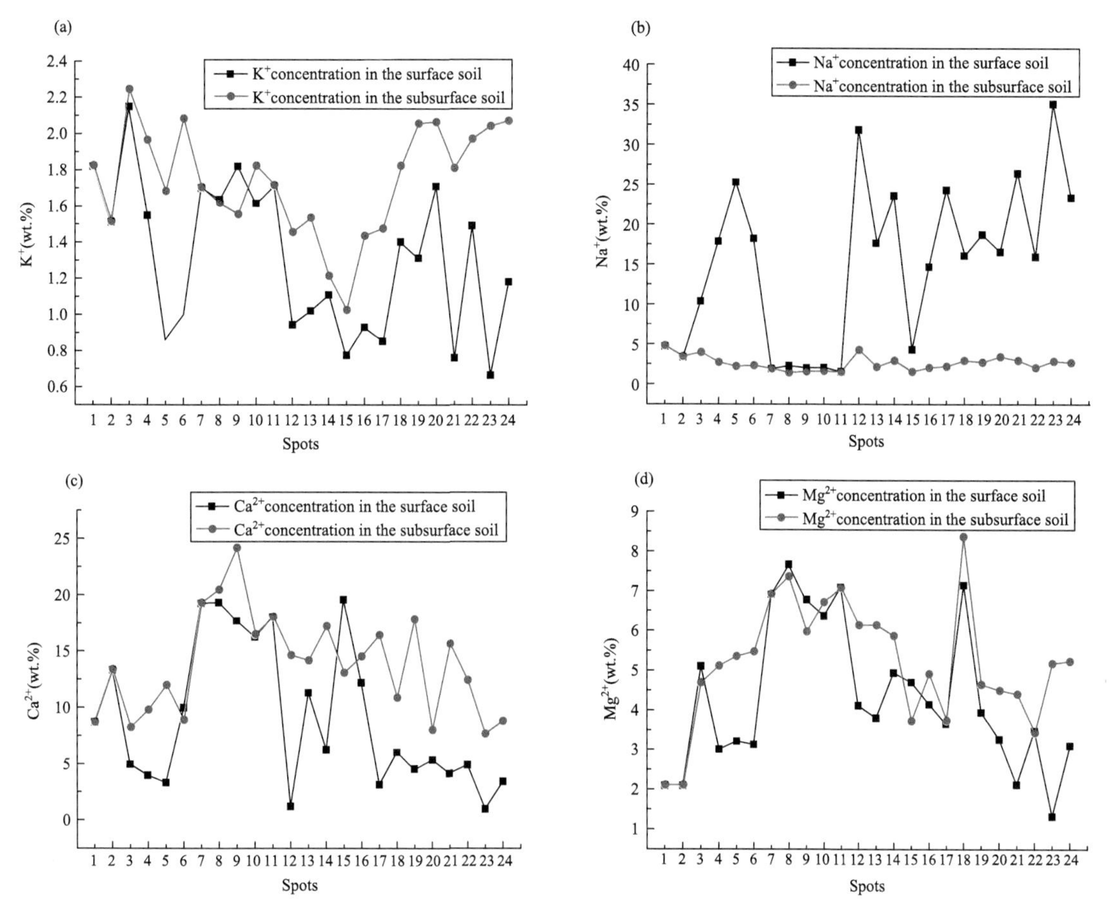

Fig. 3 The distributions of the main soil cations.

(a) K^+ concentration; (b) Na^+ concentration; (c) Ca^{2+} concentration; (d) Mg^{2+} concentration.

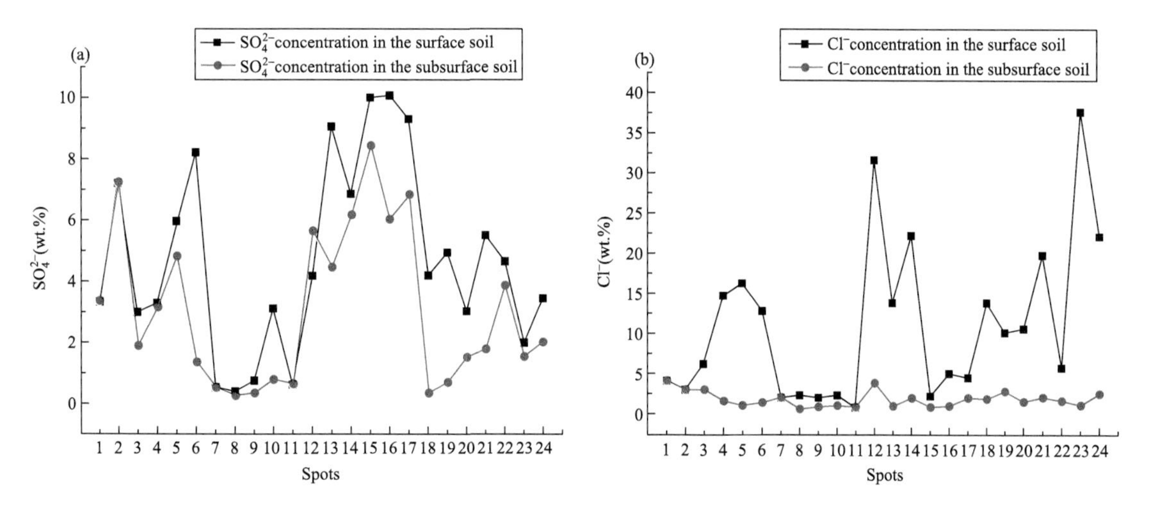

Fig. 4 The distributions of the main soil anions.

(a) SO_4^{2-} concentration; (b) Cl^- concentration.

Ion concentrations in the soils from spots 7 to 11 were relatively low and varied less than those from the other sampling spots; the ion concentrations did not vary much with depth. The low salinity could have been due to the dilution of salt in wet soils that were located in topographic lows with a shallow groundwater level. Furthermore, relatively weak surface accumulation of salt due to low capillary action could account for the lack of substantial vertical variation. Other key features of the soils from spots 7 to 11 include high concentrations of Ca^{2+} and Mg^{2+} and a Cl^- concentration that was higher than that of SO_4^{2-}.

The analysis of relationship between soil and vegetation by principal component analysis. To examine the plant-soil relationships, principal component analysis (PCA) was performed using the CANOCO 5 suite[23]. Dots represent the vegetation communities of the 24 sampling spots. The PCA results showed that the sum of the eigenvalues of the first and second axes was equal to 94% of the total sum of eigenvalues for all axes; as they account for most of the variance, we only considered the first and second axes in further analysis (Fig. 5).

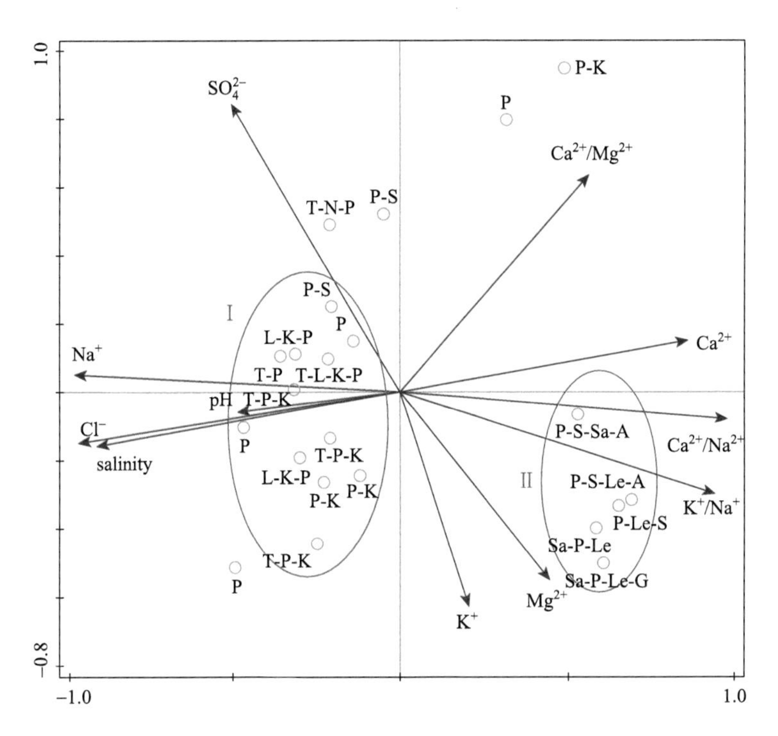

Fig. 5　Results of the PCA performed on the sample data.

P: *Phragmites australis* P-K: *Phragmites australis-Apocynum venetum* P-S: *Phragmites australis-Saussurea salsa* T-P: *Tamarix-Phragmites australis* L-K-P: *Lycium ruthenicum-Apocynum venetum-Phragmites australis* T-P-K: *Tamarix-Phragmites australis-Apocynum venetum* T-N-P: *Tamarix-Nitraria sibirica-Phragmites australis* Sa-P-Le: *Saliconia europaea-Phragmites australis-Leymus secalinus* P-Le-Sa: *Phragmites australis-Leymus secalinus-Saliconia europaea* T-L-K-P: *Tamarix-Lycium ruthenicum-Apocynum venetum-Phragmites australis* Sa-P-Le-G: *Saliconia europaea-Phragmites australis-Leymus secalinus-Glaux maritima* P-S-Sa-A: *Phragmites australis-Saussurea salsa-Saliconia europaea-Asparagus* L. P-S-Le-A: *Phragmites australis-Saussurea salsa-Leymus secalinus-Asparagus* L.

PCA axis 1, with an eigenvalue of 0.83, was best correlated with Na^+, Cl^-, salinity, Ca^{2+}, and the Ca^{2+}/Na^+ and K^+/Na^+ ratios. On the positive side of this axis are species that grow in soils with low concentrations of salt and Na^+ and high concentrations of Ca^{2+} and K^+. In contrast, the negative side includes species that grow best in soils with high salinity, high concentrations of Na^+ and Cl^-, and a low concentration of Ca^{2+}. The second axis, with an eigenvalue of 0.11, was primarily related to the SO_4^{2-} and K^+ concentrations. Since its eigenvalue was small, so it had small impact on the distribution of vegetation. As shown in Fig. 5, the vegetation at the study site can be divided into two categories: I) salt-tolerant vegetation communities including *Phragmites australis-Apocynum venetum*, *Tamarix-Phragmites australis-Apocynum venetum*, *Lycium ruthenicum-Phragmites australis-Apocynum venetum*, *Tamarix-Lycium ruthenicum-Phragmites australis-Apocynum venetum*, and II) weakly salt-tolerant hygrophytes, such as *Leymus secalinus*, *Glaux maritima*, and *Asparagus* L. The species of type I were influenced by Na^+, Cl^-, and the salinity. The soil corresponding to these vegetation communities had higher salinity and Na^+ concentrations and lower Ca^{2+} and K^+ concentrations. The species of type II were mainly controlled by the Ca^{2+}/Na^+ and K^+/Na^+ ratios. In these communities, the soil salinity and Na^+ concentrations were low, and the Ca^{2+} concentration was high.

Discussion

Spatiotemporal gradients of saline soil conditions in the Qaidam Basin. The Qaidam Basin is an inland arid plateau basin located in the north-eastern part of the Qinghai-Tibetan Plateau. The climate of coastal marshes alternates between periods of rainfall periods and drought. Salt is leached to the deepest soil horizons during the rainfall periods and rises to shallower horizons during the drought periods, resulting in the accumulation of salt near the surface[6, 12]. In contrast to the coastal marshes, the Qaidam Basin is surrounded by mountains that prevent warm, south-westerly air from entering the basin's interior, resulting in minimal precipitation[24]. In addition, the high elevation together with strong sunshine and solar radiation, result in the long-term accumulation of salt at the study site. The surface soil salinity is extremely high, and the seasonal variability of soil salinity is relatively low. Maritime salt marshes are periodically flooded by seawater[25]; thus, the contents of Na^+ and Cl^- are far higher than those of other ions. However, in inland areas, salt marshes are dominated by chlorides, sulphates, carbonates or bicarbonates. At the study site, the salt is mainly derived from leaching of rock salt. The salt marshes are dominated by chloride-type and sulphate-type soils.

The high gravel content in soil samples collected from the desert zone of the north piedmont of the Kunlun Mountains, together with the low groundwater level, would create a weak capillary effect, leading to the low soil salinity observed at spots 1 and 2. A higher groundwater level and soil sand content would enhance the capillary effect, resulting in increased soil salinity and surface salt accumulation (spots 3 to 6). At spots 7 to 11, the groundwater was near the surface in the lower terrain, and the soil salinity was very low. Close to the playa centre, the soil salinity was higher, reaching its maximum value near the centre of the playa lake (spots 22–24). Horizontally, the surface soil salinity was varied.

In the study area, the surface salt crust is very thick. The main factors leading to the formation of these crusts are aridity, high evaporation and salty groundwater[26]. When studying salt marshes in south-eastern Spain, Álvarez Rogel noted that with increased salinity, the relative percentages of Ca^{2+} and K^+ decreased, leading[26] to an imbalance in favour of the most toxic cations, such as Na^+ and Mg^{2+}. Similar results were also obtained in the present study. At the study site, moist soil had low salinity and a high concentration of Na^+ concentration and Ca^{2+}. Soil with less moisture had high salinity, with lower Ca^{2+} and higher Na^+ concentration, though the distributions of the K^+ and Mg^{2+} concentrations were not remarkable.

Many factors influenced the distribution of ions in the soil. For example, cations are absorbed by the soil's exchange complexes in the following sequence: Ca^{2+}, Mg^{2+}, K^+, and Na^+. Thus, during wet periods, divalent cations are more likely to be fixed in the surface horizons, increasing the Ca^{2+}/Na^+ and Ca^{2+}/Mg^{2+} ratios. However, salts such as NaCl and $MgCl_2$ are highly soluble and may be more mobile in the soil profile. In addition, changes in the concentration of NaCl affect the solubility of other salts[6].

Relationships between plant distributions and soils. In the study site, harsh climatic conditions and the unique geographical location cause considerable environmental stresses that affect vegetation growth; therefore, the soil salinity is an important factor affecting plant zonation. This result is consistent with previous studies that found that soil salinity was the decisive factor, especially in inland salt marshes[1, 6, 12-16]. However, not all ions increased with soil salinity. Soil salinity affects the overall vegetation distribution pattern, but specific ion concentrations or ion ratios can be further used to evaluate the relationship between soil and vegetation types[27].

Table 2 shows characteristics of the sampling spots dominated by *Apocynum venetum*, *Lycium ruthenicum*, *Saussurea salsa*, and *Tamarix*, including the soil salinity and ion ratio. *Saussurea salsa grew* mostly in soil with low salinity and high moisture, such as a low-humidity river bank, the lowland saline land at the edge of a saline lake, saline sand soil, or swamp meadow. The K^+/Na^+ ratio (0.96), Ca^{2+}/Na^+ ratio (9.13) and Ca^{2+}/Mg^{2+} ratio (2.51) in the soil dominated by *Saussurea salsa* were high. In other words, the K^+ and Ca^{2+} concentrations were high, while the Na^+ and Mg^{2+} contrations were low. *Saussurea salsa* can also grow in severely saline soil environments; however, the plants are short, and the coverage is low.

Table 2 The influence of the soil ion ratio on plant distribution

	Salinity	K^+/Na^+	Ca^{2+}/Na^+	Ca^{2+}/Mg^{2+}
Apocynum venetum	25.11	0.31	0.92	1.35
Lycium ruthenicum	45.06	0.17	0.67	1.69
Saussurea salsa	4.92	0.96	9.13	2.51
Tamarix	37.67	0.16	1.04	2.60

Apocynum venetum, *Tamarix* and *Lycium ruthenicum* grew in soil with high salinity. Compared to the other two salt-tolerant plants, *Lycium ruthenicum* favoured soils dominated by NaCl that had low K^+/Na^+ and Ca^{2+}/Na^+ ratios of 0.17 and 0.67, respectively. Thus, the Na^+ concentration was high, while the K^+ and Ca^{2+} concentrations were relatively low.

Apocynum venetum also grew in soil dominated by NaCl, as well as sulphate-type soil. The Ca^{2+}/Mg^{2+} ratio (1.35) was low, while the Mg^{2+} concentrations were high. The growth environment of *Lycium ruthenicum* was limited; it only appeared in spots 4 to 6. In contrast, *Apocynum venetum* was distributed in high salinity areas, showing its strong adaptability to hypersalinity.

Tamarix is salt-secreting plant that can grow in saline soil. *Tamarix* can excrete salt absorbed into its body through salt-secreting pores. The salt in the litter often adheres into blocks of sand to form a fixed sandbag[28, 29]. Therefore, *Tamarix* was the typical vegetation in soils with high surface soil salinity, low K^+/Na^+ (0.16) and high Ca^{2+}/Mg^{2+} (2.60). *Tamarix* preferred soils with lower magnesium concentrations compared to the other species.

Unlike soils in the coastal area of northern China, which have low salinity and high moisture content[30], soils in the study area are dry and have high salinity. *Phragmites australis* is distributed widely in the Tuanjie Lake playa, and it is present at almost every sampling point. *Phragmites australis* is resistant to extreme environmental factors, such as cold, drought, salt, wind, and preferred to high-humidity environment. Some scholars have noted that succulent halophytes seem to be the first colonizers of bare soil[1], but in this study site, the most salt-tolerant plant species was *Phragmites australis*.

Vegetation distribution model of saline lakes on an inland plateau. Based on the results of this study and the conclusions of previous literature, a vegetation distribution model of saline lakes on an inland plateau was established. The terrain from the surrounding mountains to the centre of the basin includes mountains, Gobi Desert areas, sand dunes, fine soil plains, swamps and saline lakes. The vegetation is distributed in five zones around the saline lakes or playas at the centre (Fig. 6).

The Qaidam Basin water resources mainly come from meltwater and precipitation in mountainous areas. The mountains around the basin are the sources of numerous rivers, but after those rivers flow through the edge of the alluvial fan, most runoff enters the ground. Groundwater is deep and difficult for vegetation to access, resulting in the formation of Gobi and other desert landscapes. While the groundwater level of the alluvial fan increases, the sparse vegetation begins to appear. The vegetation type is mainly psammophytes, the soil gravel content is high, and the organic matter and salinity are relatively low (Area A).

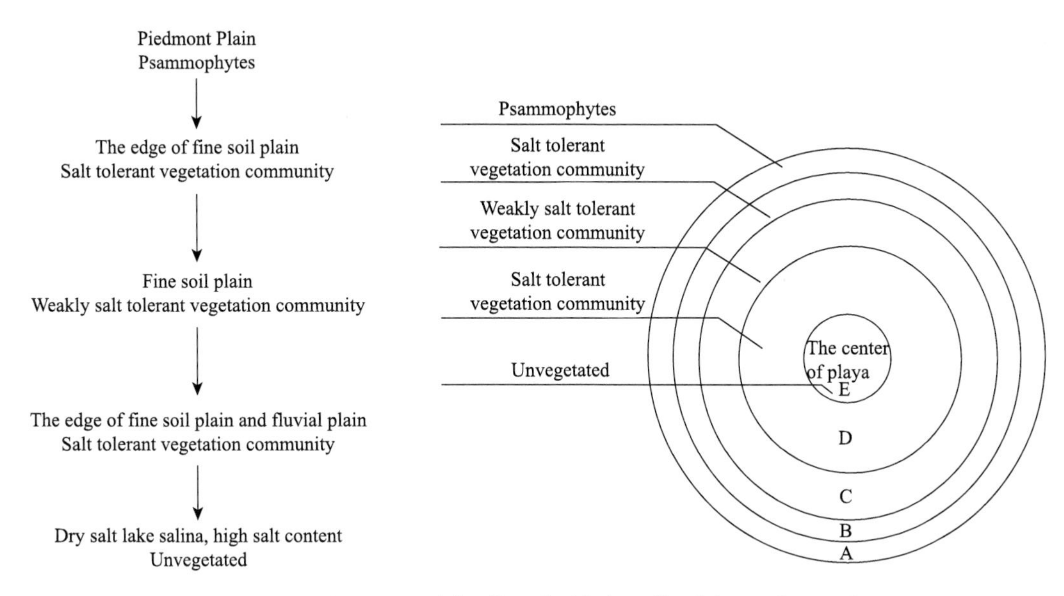

Fig. 6 The vegetation is zoned distributed with the saline lake or playa at the centre.

With the increase in the groundwater level, the soil sand content also increases; the salinity and Na^+ concentrations are high; and the Ca^{2+} concentration is relatively low. The surface soils have a salt crust, which may be very thick. This area is dominated by the salt-tolerant vegetation community, which includes *Tamarix*, *Nitraria sibirica*, *Phragmites australis*, *Apocynum venetum*, and *Lycium ruthenicum*. The roots of these salt-tolerant species can penetrate the surface of the heavy salt layer and absorb water and nutrients from the subsurface soil[31] (Area B).

In the fine soil plains, the soil texture is silt or clay and the groundwater level is close to the surface. The main vegetation type is weakly salt-tolerant hygrophytes, such as *Triglochin maritimum*, *Glaux maritima*, and *Leymus secalinus*. The soil has low salinity and Na^+ concentrations and a high Ca^{2+} concentration and water content. Plant growth pressure in this environment is low. The soil type is primarily loam and clay. The vegetation abundance and coverage are high, which creates a good pastoral area (Area C).

With increased proximity to the centre of a saline lake, the groundwater level decreases at the edge of the fine soil plain and lacustrine plain areas of dry saline lake; soil salinity and Na^+ concentration increase. The surface soil often forms a salt crust ranging from a few to over 10 cm thick. Throughout the zone, the vegetation associations are gradually dominated by halophytes, driven by salinity (Area D). The playa centre is bare saline land with no vegetation (Area E).

The model also reflects the relationship between the vegetation types and soil environment. For instance, increased coverage of *Triglochin maritimum*, *Glaux maritima*, *Leymus secalinus* and *Saussurea salsa* indicates a high water content, low salinity, high K^+ and Ca^{2+} concentrations and low Na^+ and Mg^{2+} concentrations. Increased coverage of *Tamarix*, *Lycium ruthenicum*, and *Apocynum venetum* indicates a low water content, high salinity, low K^+ and Ca^{2+} concentrations and high Na^+ and Mg^{2+} concentrations.

Based on the above description, the idealized distribution of vegetation in an inland plateau saline lake is zone. However, this zoned pattern is often disrupted by water systems and geological structures such as in the Tsagaan Us-Wutumeiren Basin and the Delingha Basin. Therefore, it is necessary to study the actual conditions of the vegetation and soil in each saline lake.

Materials and methods

Collection of soil samples and analysis. Field investigations were carried out from August to September 2015, and samples were collected from the gravel Gobi areas on the diluvial clinoplain in the north piedmont of the Kunlun Mountains to the Tuanjie Lake salt marshes (Fig. 1). There were 24 sampling sites within the 22 km transect. The vegetation data were recorded using 5 m × 5 m quadrats. The relative plant height, crown width and quantity of each species were measured, and the coverage of each species was calculated. Soil samples were taken layer by layer, and the sampling depth was 0–40 cm. Because sample plots 1, 2, 7 and 11 were not distinct, soil samples were collected from one layer. In the remaining sampling plots, samples were collected from two layers. A total of 44 soil samples were collected.

All of the soil samples collected in the field were analysed in the laboratory. The samples were dried at 65 ℃ and passed through a 2 mm sieve to remove coarse fragments before laboratory analysis. Salinity was measured in a 1 ∶ 2.5 soil water extraction using an electrical conductivity method, following an established procedure[32]. The concentrations of K^+, Na^+, Ca^{2+}, Mg^{2+}, Cl^-, and SO_4^{2-} were measured using a Perform X (Thermo Fisher Inc., U.S.A.) X-ray fluorescence spectrometer (XRF; effective diameter = 25 mm).

Statistical analysis. Principal component analysis (PCA) was performed using the programme CANOCO 5 to evaluate plant-soil relationships. When the cumulative variance of the analysis result was close to or greater than 80%, the PCA was considered useful. The results can be presented in ordination diagrams, which include arrows that represent environmental variables and points that represent species. A shorter vertical distance between a point and environmental variables indicates that the environmental variables have a greater effect on the distribution of vegetation[20].

References

[1] Álvarez-Rogel, J., José Martínez-Sánchez, J., Carrasco Blázquez, L. & Marín Semitiel, C. M. A conceptual model of salt marsh plant distribution in coastal dunes of southeastern Spain. Wetlands. 26, 703-717 (2006).

[2] Pennings, S. C. & Callaway, R. M. Salt marsh plant zonation: the relative importance of competition and physical factors. Ecology. 73, 681-690 (1992).

[3] García, L. V., Marañón, T., Moreno, A. & Clemente, L. Above-ground biomass and species richness in a Mediterranean salt marsh. J. Veg. Sci. 4, 417-424 (1993).

[4] Cantero, J. J., Cisneros, J. M. & Zobel, M. Environmental relationships of vegetation patterns in saltmarshes of central Argentina. Folia Geobot. 33, 133-145 (1998).

[5] Cantero, J. J., Leon, R. & Cisneros, J. Habitat structure and vegetation relationships in central Argentina salt marsh landscapes. Plant Ecol. 137, 79-100 (1998).

[6] Álvarez-Rogel, J., Ariza, F. A. & Silla, R. O. Soil salinity and moisture gradients and plant zonation in Mediterranean salt marshes of Southeast Spain. Wetlands. 20, 357-372 (2000).

[7] Grattan, S. R. & Grieve, C. M. Mineral element acquisition and growth response of plants grown in saline environments. Agr. Ecosyst. Environ. 38, 275-300 (1992).

[8] Reimann, C. & Brackle, S. W. Sodium relations in Chenopodiaceae: a comparative approach. Plant Cell Environ. 16, 323-328 (1993).

[9] Reimann, C. & Brackle, S. W. Salt tolerance and ion relations of Salsola kali L.: Differences between ssp. tragus (L.) Nyman and ssp. ruthenica (Iljin) Soó. New Phytol. 130, 37-45 (1995).

[10] Wang, X. Y. & Redmann, R. E. Adaptation to salinity in Hordeum jubatum L. populations studied using reciprocal transplants. Vegetatio. 123, 65-71 (1996).

[11] Li, L. Effects of soil-groundwater systems on natural vegetation growth: a case study at Dunhuang Basin. China University of Geosciences. (2013).

[12] Álvarez-Rogel, J., Silla, R. O. & Ariza, F. A. Edaphic characterization and soil ionic composition influencing plant zonation in a semiarid Mediterranean salt marsh. Geoderma. 99, 81-98 (2001).

[13] Piernik, A. Inland halophilous vegetation as indicator of soil salinity. Basic Appl. Ecol. 4, 525-536 (2003).

[14] Álvarez-Rogel, J., Jiménez-Cárceles, F. J., Roca, M. J. & Ortiz, R. Changes in soils and vegetation in a Mediterranean coastal salt marsh impacted by human activities. Estuar. Coast. Shelf S. 73, 510-526 (2007).

[15] Herrero, J. & Castañeda, C. Changes in soil salinity in the habitats of five halophytes after 20years. Catena. 109, 58-71 (2013).

[16] González-Alcaraz, M. N., Jiménez-Cárceles, F. J., Álvarez, Y. & Álvarez-Rogel, J. Gradients of soil salinity and moisture, and plant distribution, in a Mediterranean semiarid saline watershed: a model of soil-plant relationships for contributing to the management. Catena. 115, 150-158 (2014).

[17] Wang, J., Liu, M. S., Sheng, S., Xu, C., Liu, X. K. & Wang, H. J. Spatial distributions of soil water, salts and roots in an arid arbor-herb community. Acta Ecol. Sin. 28, 4120-4127 (2008).

[18] Zheng, X. Y., Zhang, M. G., Xu, Y. & Li, B. Salt Lakes of China (ed. Zheng) (Science Press, Beijing, 2002).

[19] FAO/UNESCO. Soil Map of the World. Revised legend. (1990).

[20] Zheng, M. P. *Saline Lakes on the Qinghai-Xizang (Tibet) Plateau* (ed. Zheng) (Beijing: Beijing Science & Technology Press, 1989).

[21] Yu, S. S., Liu, X. Q., Tan, H. B. & Cao, G. C. *Sustainable Utilization of Qarhan Salt Lake Resources* (ed. Yu) (Science Press, Beijing, 2009).

[22] Ababou, A., Chouieb, M., Khader, M., Mederbal, K. & Saidi, D. Using vegetation units as salinity predictors in the Lower Cheliff Algeria. Turk. J. Bot. 34, 73-82 (2010).

[23] Ter Braak, C. J. F. & Smilauer, P. *CANOCO Reference Manual and User's Guide: Software for Ordination* (ed. Ter Braak) (Microcomputer Power Ithaca, New York, 2012).

[24] Chen, L. Z. *China Flora and Vegetation Geography* (ed. Chen) (Science Press, Beijing, 2015).

[25] Chapman, V. J. *Salt Marshes and Salt Deserts of the World* 2nd edition (ed. Chapman) (Verlag Von J Cramer, Lehre, Germany, 1974).

[26] Koull, N. C. A. Soil-vegetation relationships of saline wetlands in north east of Algerian Sahara. Arid Land Res. Manag. 29, 72-84 (2015).

[27] Cebas-Csic, J. Á. R., Hernandez, J. & Silla, R. O. Patterns of spatial and temporal variations in soil salinity: example of a salt marsh in a semiarid climate. Arid Land Res. Manag. 11, 315-29 (1997).

[28] Yin, L. K. *Tamarix* spp. —the keyston species of desert ecosystem. Arid Zone Res. 12, 43-47 (1995).

[29] Yin, L. K. The ex-situ protection and the ecological adaptability of *Tamarix* L. Arid Zone Res. 19, 12-16 (2002).

[30] Li, W. Q., Liu, X. J., Khan, M. A. & Gul, B. Relationship between soil characteristics and halophytic vegetation in coastal region of North China. Pak. J. Bot. 40, 1081-1090 (2008).

[31] Zhou, D. J. & Chu, J. A. Development and use of bluish dogbane plant resources in Qaidam Basin. Qinghai Sci. Tech. 5, 47-48 (1998).

[32] Bao, S. D. The analysis of water soluble salt content of soil of referecing in Soil *Agro-chemistrical Analysis* 3rd edition (ed. Bao) Ch. 9, 183-187 (China Agriculture Press, Beijing, 2000) (In Chinese).

Influence of the regional climate variations on lake changes of Zabuye, Dangqiong Co and Bankog Co salt lakes in Tibet*

Yunsheng Wang[1,2], Mianping Zheng[1,2], Lijuan Yan[3], Lingzhong Bu[1,2] and Wen Qi[1,2]

1. MLR Key Laboratory of Saline Lake Resources and Environments, Institute of Mineral Resources, Chinese Academy of Geological Sciences, Beijing 100037, China
2. R&D Center for Saline Lakes and Epithermal Deposits, Chinese Academy of Geological Sciences, Beijing 100037, China
3. Chinese Academy of Geological Sciences, Beijing 100037, China

Abstract The lake hydrological and meteorological data of the Tibetan Plateau are not rich. This research reports the observed climatic data and measured water levels of saline lakes from the local meteorological stations in the Zabuye salt lake, the Dangqiong Co salt lake and the Bankog Co salt lake in recent two decades. Combining with satellite remote sensing maps, we have analyzed the changes of the water level of these three lakes in recent years and discussed the origins of the changes induced by the meteorological factors. The results show that the annual mean temperature and the water level reflect a general ascending trend in these three lakes during the observation period. The rising rates of the annual mean temperature were 0.08℃/a during 1991–2014 and 0.07℃/a during 2004–2014, and of the water level, were 0.032 m/a and 0.24 m/a, respectively. Analysis of changes of the meteorological factors shows the main cause for the increase of lake water quantity are the reduced lake evaporation and the increased precipitation in the lake basins by the rise of average temperature. Seasonal variation of lake water level is powered largely by the supply of lake water types and the seasonal change of regional climate.

Keywords Tibetan Plateau; Saline lake; Climate variation; Lake change

1 Introduction

With the development of the global warming, the research of environmental changes taking place in the Tibetan Plateau has become the focus of global concern. As a special body of water, the salt lake is closely related to the factors of atmosphere, soil and biology, and has obviously sensitive responses to the change of climate and environment system (Zheng et al., 2000, 2004; Liu et al., 2009; Kang et al., 2010; Song et al., 2014a). The alpine lakes which are in the natural state are less affected by human activities and can reflect the climate condition realistically. The Tibetan Plateau is a multi-lake distribution area in China. As the highest plateau and with the most abundant resources on earth, there are many saline lakes and salt water lake groups (Chang, 1987; Zheng, 2014). The research on plateau lakes is helpful to reveal the environmental evolution characteristics of the Tibetan Plateau and even the whole country.

Recent studies show that the water level of most of the lakes in Tibet region has risen (Qi et al., 2006a; Bianduo et al., 2009; Lei et al., 2013; Yan and Zheng, 2015a; Ma et al., 2016). The main reasons for this are the increase of annual precipitation, the amount of glacier melting and the melting of the frozen soil due to the increase of the annual mean temperature (Yao et al., 2007; Ye et al., 2008; Zhu et al., 2010; Wang et al., 2013; Ma et al., 2016). However, direct meteorological or hydrological observation data in the target saline lake are extremely scarce. The above research results are based on the interpretation of remote sensing images and the data

* 本文发表在：Journal of Geographical Sciences, 2019, 29 (11): 1895-1907

from the weather station of administrative region where saline lake is located (Jiang et al., 2017; Song et al., 2014b; Zhu et al., 2010). In fact, the northern Tibetan Plateau is vast in territory and complicated in topography. The impact of seasonal variation of climate in the saline lake area is also very significant. In order to get a thorough knowledge about the climate environment of the lake basin and the reasons for the change of the lake area, the climate change characteristics and the response of the water level and the lake area to meteorological factors in the central-south of the Qiangtang (Changtang) Plateau are analyzed in this paper. And datasets for the study on regional response of global change are provided.

2 Study area and data

2.1 Study area

The study area is located in central and western lake basin of northern Tibetan Plateau, including Zabuye, Dangqiong Co and Bankog Co (Fig. 1). Zabuye salt lake (31°14′–31°33′N, 83°52′–84°23′E) and Dangqiong Co (31°30′–31°40′N, 86°38′–86°49′E) lie on the western Tibetan Plateau, which belong to semi-arid climate zone of the Qiangtang Plateau. In the region, the climate shows strong radiation, low temperature, less precipitation and large evaporation. And the supply of the lake mainly depends on the precipitation and snow melt water. Zabuye Basin is a typical inland basin, and salt lake is the lowest point of confluence. It is divided into two lakes, north lake and south lake. The water area is 235 km^2. The Lunggar glacier sits close to the west of Zabuye. Water system in Zabuye area developed well, such as Luobujuqu, Jiaobuqu, Langmengaqu and Quanshui River and so on. The Luobujuqu is a perennial river, and the others are the stream rivers in rainy seasons and subsurface-flow in dry seasons. In addition, there is a lot of spring water around the lake. Zabuye Spring and Qiukuang Spring are two of them which have the maximum water inflow. The flow rate was not influenced by the season, having no direct relation with the rainy season precipitations. The area of Dangqiong Co is 56 km^2, its basin is high in southeast and low in northwest. The southern mountains are covered with snow all the year round. Geqianqiong, Qurebaima and Miangkangluoma are perennial rivers in Dangqiong Co basin. Bankog Co (31°42′–31°45′N, 89°29′–89°5′E) lies on the central Tibetan Plateau, and the supply mainly depends on surface rainfall and surface runoff. Bankog Co, Zabuye and Dangqiong Co are all carbonate type saline lakes. The latter two are rich in Li, K, B, Rb and Cs (Zheng et al., 2000, 2004; Zheng, 2014). They serve as important liquid mineral resources with great values of exploitation.

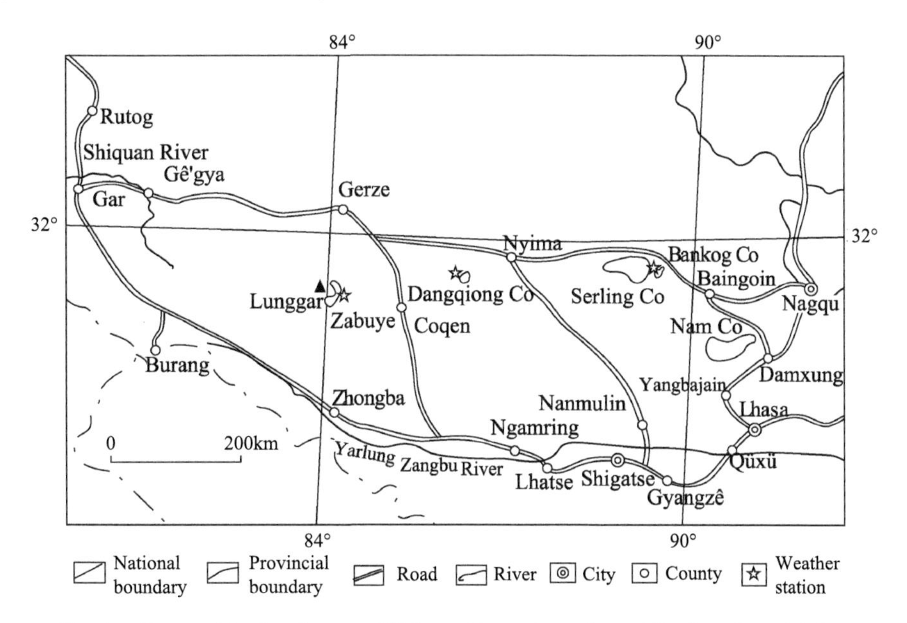

Fig. 1 Geographic sketch map of Zabuye, Dangqiong Co and Bankog Co salt lakes.

2.2 Data acquisition and preprocessing

Chinese Academy of Geological Sciences established three long-term saline lake meteorological observation stations in the above three saline lake districts in 1990, 2001 and 2004 respectively (Qi et al., 2006a). During the observation period, TRM-ZS2 high precision automatic meteorological station and LS206B propeller-type velocity meter were used. A large number of meteorological, hydrological and hydrochemical data were obtained for the first time in these districts. Although the running times of the three stations were different, the available measured data can reflect the change of lake area since the 1990s in general. It has scientific and practical significance for the study on dynamic change of lakes in the hinterland of the Tibetan Plateau and comprehensive utilization of lithium and potassium resources.

Water level observation points were located at the north and south sides of the three lakes. In general, water level was recorded once a week before 1998 and twice a month after 1998. In this paper we use the average monthly value to indicate water level. From January 1991 to December 2013, the number of missing data in the Zabuye salt lake was 33 and accounted for 14.1% of all data. From May 2004 to December 2013, the number for Dangqiong Co salt lake was 4 months and accounted for 5.6% of all data. And the data in Bankog Co salt lake was complete from May 2001 to October 2003. The velocities of rivers and springs around the salt lake were determined by velocity meter at fixed point. The flow rate was approxi mately equal to velocity multiplied by sectional area of riverbed. According to incomplete survey, the average annual runoff in Zabuye basin was 6.672×10^7 m^3/a during the period from 2001–2010. In Dangqiong Co basin, the average annual runoff was 3.6×10^6 m^3/a during the period 2004–2010. The main recharge source of Bankog Co is Kawazangbu River, and the supply failed to be counted because Bankog Co connects its adjacent mother lake.

In this paper, we used 120 Landsat MSS/TM/ETM images covering the study area to retrieve data on the saline lake surface extent for 1990–2013. The spatial resolution of images is 30 m. All of the images selected are cloud free or have only slight cloud cover (less than 5%).

3 Results

3.1 Annual mean temperature variation

The study shows that the change of annual mean air temperature is helpful to understand the characteristics of climate change and predict climate disasters, which has great significance to disaster prevention and reduction. The average value is used to analyze and discuss the meteorological elements in this section because it can better reflect the change of climate characteristics.

Several researchers argue that the general characteristics of climate change in the Tibetan Plateau are temperature rise, precipitation increase, potential evapotranspiration decrease and the trend from dry to humid status in most areas (Wu et al., 2005; Du, 2001; You et al., 2010). According to statistics, the annual mean temperature of the Zabuye salt lake began to rise gradually from 1991 and reached a high value in 1998, although there was a significant drop in 1997 (Fig. 2). The mean temperature change was smaller from 1998 to 2008, but it had a significant increase during the period 2009–2013. In the above three periods, the mean temperature of the Zabuye salt lake was 2.23, 2.71 and 3.67 ℃ respectively. The rise rate of annual mean temperature was 0.08 ℃/a from 1991 to 2013. Similar changes have taken place in the Dangqiong Co salt lake. The mean temperature raised from 3.2 ℃ in 2004 to 4.6 ℃ in 2014. The rise rate of annual mean temperature in the lake was 0.07 ℃/a. Based on the meteorological data of Bankog Co from June 2001 to

October 2002, the mean temperature was 0.1 ℃ and higher than the value provided by predecessors. The latter was −1 to −2 ℃. Meanwhile, the upward trend from 2000 to 2011 was more obvious (Zhao et al., 2004; Du et al., 2014). Generally, the highest temperature was usually observed in June and July, the lowest in December and January of the following year.

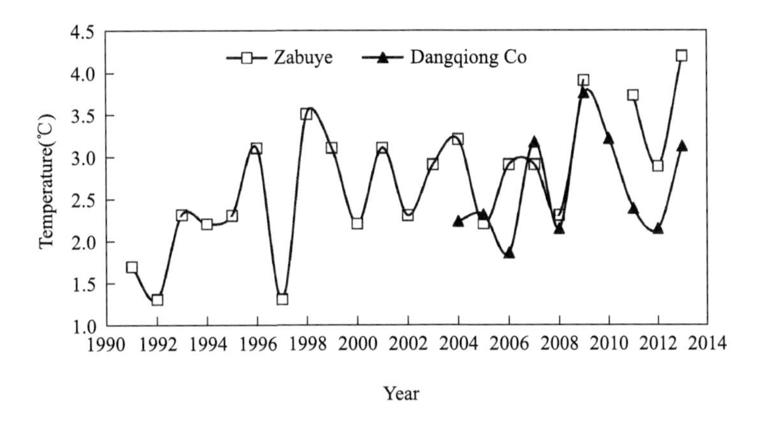

Fig. 2 The mean temperature of Zabuye and Dangqiong Co salt lakes.

Previous studies (Niu et al., 2004; Lu et al., 2005; Yao et al., 2006; Zheng et al., 2002) also showed that the 1960s was a relatively low temperature period in the Tibetan Plateau; the temperature began to rise in the 1970s, then came another relatively high temperature period in the mid-1980s, and the heating up was more obvious in the 1990s. In recent 50 years, the temperature has been increasing in fluctuation in the Tibetan Plateau, and the annual mean temperature in most areas rose and peaked in 1998 (Ma et al., 2003).

3.2 Annual evaporation, precipitation, and sunshine hours

3.2.1 Annual evaporation, precipitation

Lake evaporation is one of the main factors affecting the water level of saline lakes. According to statistics, the average annual evaporation of Zabuye salt lake was 2491.54 mm from 1991 to 2013. Dangqiong Co salt lake was 2361.19 mm from 2004 to 2013 (Fig. 3). And Bankog Co was 1978.1 mm from 2001 to 2002 (Zhao et al., 2004).

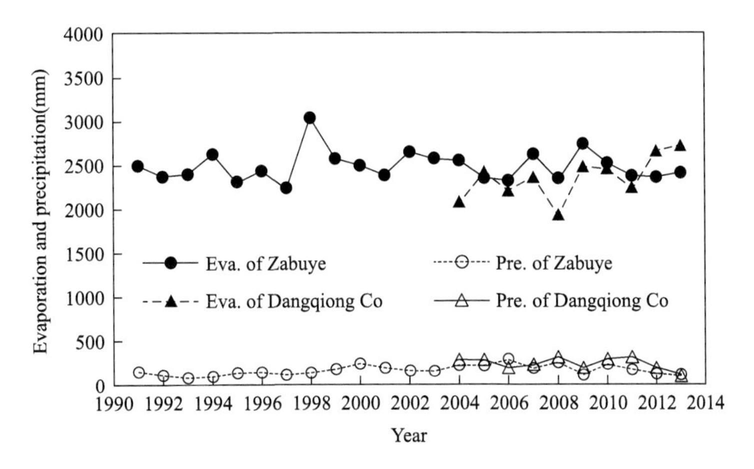

Fig. 3 Annual evaporation and precipitation of Zabuye and Dangqiong Co salt lakes.

From April to September, evaporation kept a high value and relatively low in the whole ice-free period.

Similar to temperature, the maximum evaporation emerged in June and July, the minimum emerged in December and January of the following year. The difference between the maximum and minimum evaporation of the two lakes varied from 300.6 to 420.7 mm and 187.6 to 486.4 mm respectively. The value of other months gradually changed from low to high or high to low. During the rainy season, evaporation was reduced because the air humidity was high. In addition, there was a certain relationship between evaporation and wind strength. In general, the evaporation can be increased with the wind getting strong.

According to the observation data in the last 19 years, the precipitation of Zabuye salt lake was more concentrated from the beginning of late July to the end of September. Rainfall predominated in the rainy season was more than 90% of the total annual rainfall. The annual mean precipitation was 170.2 mm, in which the rainfall was dominant and the amount of snow (hail) was relatively small. Similarly, the rainfall of Dangqiong Co salt lake was more concentrated in the third quarter, accounting for more than 80% of the total annual rainfall. The mean annual precipitation was 238.38 mm between 2004 and 2013 (Fig. 3). The rainfall of Bankog Co was 263.9 mm between 2002 and 2002. Overall, there was little precipitation in winter in the three lakes.

3.2.2 Annual sunshine hours

Those three salt lakes are very rich in solar energy, with the weather being mainly sunny or cloudy, rarely overcast and of thunderstorm. The results showed there was a significant decline before 2008 and then increased again in Zabuye salt lake area. The annual mean sunshine duration was 2883.0 hours (Fig. 4). There were 280 days of more than 7.2 hours sunshine duration, making up 76.7% of the annual sunshine amount. The mean annual sunshine hours of Bankog Co and Dangqiong Co were 2447.6 and 2158.2 hours (Fig. 4), and there were fewer days with more than 7.2 hours sunshine duration than in Zabuye salt lake. The distinction in sunshine duration between Zabuye and Dangqiong Co reflected in their salinity. Brine salinity of Zabuye salt lake was 30% and far more than Dangqiong Co. This illustrated that more sunshine was conducive to the enrichment of saline lake brine.

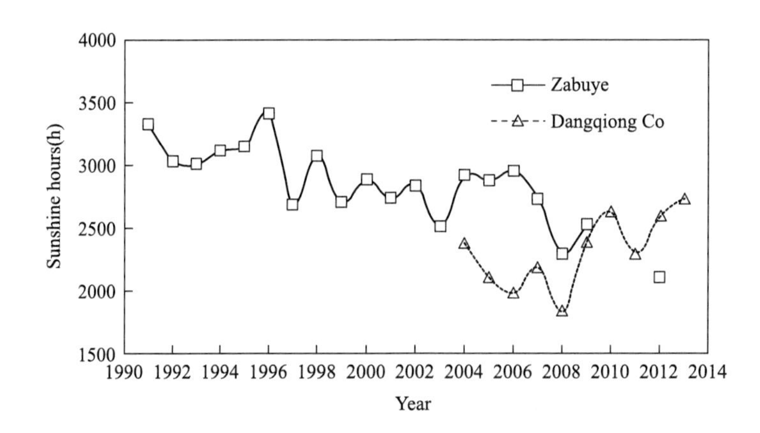

Fig. 4 The sunshine duration of Zabuye and Dangqiong Co salt lakes.

4 Discussion

4.1 Response of saline lake changes to climate fluctuation

As typical lake with glacial supplies in the central hinterland of the Tibetan Plateau, it is very important to study the law between the fluctuation of water level and climate change in Zabuye and Dangqiong Co.

The change of water level and its composition are highly sensitive and dependent on natural factors, and researchers can estimate the changes of climate according to the changes of the lake environment. For example, the water level in Qinghai Lake in recent years has dropped due to long-term climate changes and short-term fluctuations (Zhou et al., 1992; Shi et al., 2005), and the lake level of Ebinur (Aibi) and Aydingkol (Aiding) lakes in Xinjiang has been raised because of the increase of glacier melting and precipitation in the lake districts (Shi et al., 2007).

4.1.1 Dangqiong Co salt lake

Based on historical data of the Dangqiong Co over years, the change of water level was obviously influenced by temperature and precipitation. Because there was little precipitation in winter, water level dropped to the lowest point in January and February. As the day warmed up, the secondary peak was shown with the glaciers melted. After the maximum precipitation in June and July, the water level emerged the highest point in August and September. During the period from 2004 to 2014, the water level ascended to 2.2 m and the increase rate was 0.24 m/a (Fig. 5). The water level increased in two periods, 2004–2007 and 2010–2012. The increase rates were 0.144 m/a and 0.247 m/a respectively. It is worth noting that the mean temperature value of 2008 was the lowest during the period, and the water level reached the highest value in the rainy season with the highest-ever precipitation. Again the similar case occurred in 2010 and 2012.

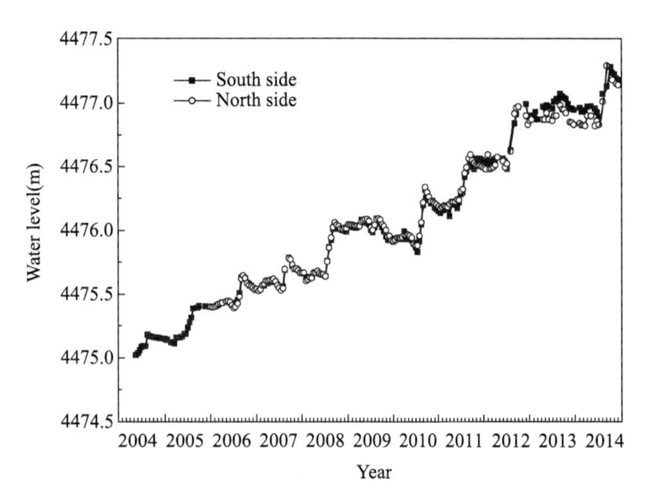

Fig. 5 Water level fluctuating recorded in Dangqiong Co salt lake during 2004–2014.

According to the result of remote sensing image interpretation, Dangqiong Co salt lake has a trend for expanding in both depth and extent from 2004 to 2014 (Fig. 6). Based on the meteorological data of Shenzha county, the nearest county to Dangqiong Co salt lake, the mean temperature in the period from 2001 to 2010 rose by 1.38 ℃ compared with the period from 1961 to 1970. The average growth rates of temperature and precipitation were 0.314 ℃/10a and 19.13 mm/10yr respectively, while the evaporation decreased based on the results of previous studies (Yan and Zheng, 2015a, 2015b; Ma et al., 2016). The temperature rise made the ice on the mountains of the south area melt, increasing the amount of water in rivers that drain into the lake. According to statistics, the area of glaciers has been reduced by about 24.35 km^2 from November 1990 to November 1999. The descending value was increased to 77.90 km^2 from November 1999 to November 2009 and the area of the lake increased by 12.55 km^2.

The enlargement in snow melt and rainfall has increased the supply of lake water. With the decrease of evaporation, lake area was larger. Dangqiong Co increased from 56.49 km^2 to 64.61 km^2 during the observation period from November 2001 to November 2013 (Fig. 6). It rose by 14.37% in the 12 years. This was consistent with the trend of global warming and gradual decrease of glaciers. Meanwhile, the evaporation on the surface of lakes was aggravated. The main reason for the large reduction in glacier area in 2009 was: the winter temperature has been second high in the Tibet region since 1971, leading to a large reduction in glaciers.

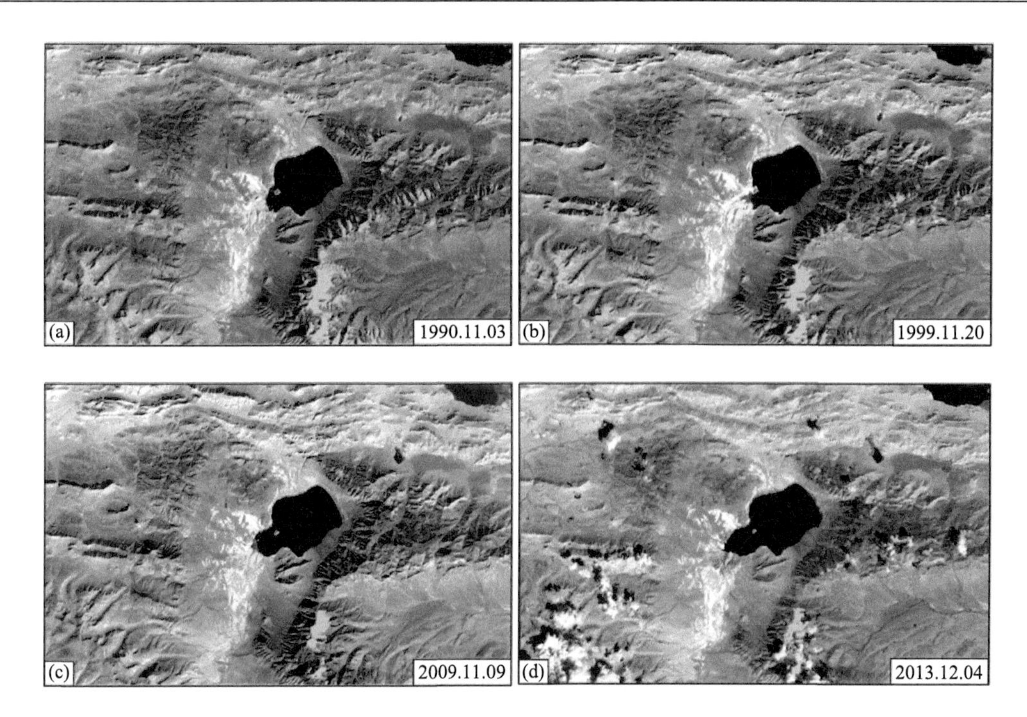

Fig. 6 Remote sensing images of Dangqiong Co salt lake.

4.1.2 Zabuye salt lake

Based on historical data of the Zabuye salt lake over years, the lowest point of water level in south lake appeared in July and August, and the highest point, in December or January of next year. The water level change of north lake is different from that of south lake. Before 1996, the lowest point of water level appeared in November and December, and the highest point in April. While after 1998, the lowest point of water level appeared in June and July, and the highest point in August and September (Qi et al., 2006a, 2006b), rising in a fluctuating way within this period. And the second peaks appeared in May and June.

The change characteristics in south and north lakes of Zabuye were different due to the supply types. The underground springs were the main supply of water in south lake and the precipitation in the catchment basin was the less prominent part. The main factors that determine the water level of saline lakes were air temperature and evaporation because the groundwater flow was relatively stable. It can be concluded from water level change curve that there was negative correlation between the water level of the south lake and the air temperature, and evaporation. Both the air temperature rising and the evaporation increasing caused a drop in the water level and vice versa.

Compared with south lake, there is positive correlation between water level of the north lake and the air temperature. Both the air temperature rising and the evaporation increasing caused a rise in the water level and vice versa. Geographically, the Lunggar glacier sits close to the west of Zabuye and its meltwater flows into the north lake. Based on the field observation results and previous studies (Yao et al., 2007; Ye et al., 2008), we believe glaciers meltwater was the main contribution factor besides atmospheric precipitation.

The average water level of the south lake dropped by 11 cm from 1991 to 2001, and for the north lake, by 9.7 cm from 1991 to 1996. The water level of the south and north lakes decreased because the mean temperature and evaporation of the saline lake basin increased continuously and rainfall decreased. It is a sign of regional arid climate in Zabuye saline lake basin. Before 1998, the water level of the south lake was higher than that of the north lake. The water level of the north lake jumped and exceeded that of the south lake in 1999. From 2002 to 2009, the water level of the south lake ascended by 48.5 cm, and for the north lake, by 69.8 cm from 1997 to 2009 (Fig. 7). Particularly, since 2009 the north lake has double peak which usually occurs in April and August. By comparison, the peak value of August is higher than that of April. The results also show that the precipitation in the rainy season has a faster and more obvious effect on supplying lake water, but only has a marginal effect on the general trend of Zabuye salt lake. After 2010, the water level of the north lake was higher than that of the south lake and both have a little change with seasons.

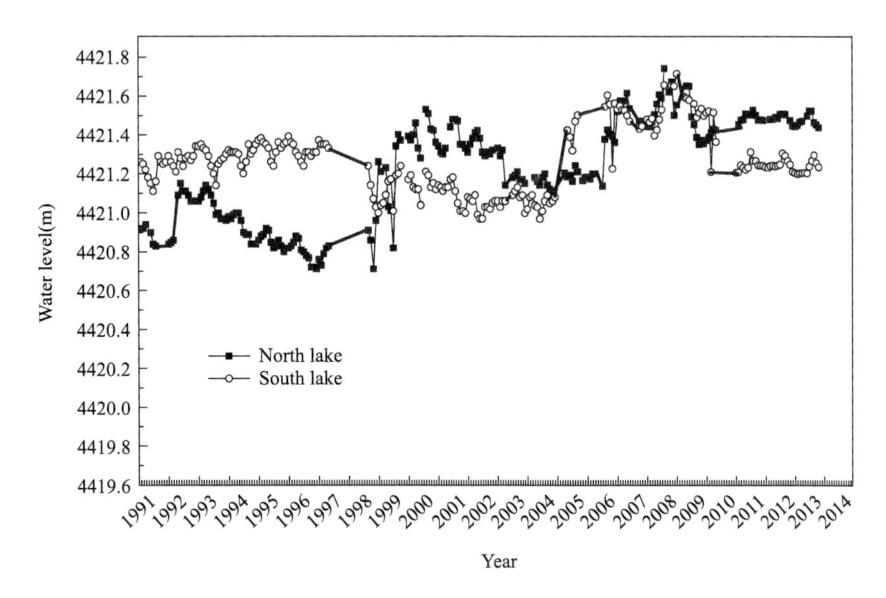

Fig. 7 Water level fluctuating recorded in Zabuye salt lake during 1991–2014.

Based on the result of remote sensing image interpretation, the area of Lunggar snow mountains which lie on the northwest side of Zabuye salt lake decreased by nearly 30 km^2 from 1976 to 2014, and the area of Zabuye salt lake in the same period increased from 137 km^2 to 235.97 km^2 (Fig. 8). The increasing extents of the south lake and the north lake were different because of the difference in the lake basin depth. The area of the north lake increased by 17.24 km^2, and for the south lake, by 81.74 km^2. The area of Zabuye salt lake increased by 72.25% over the past 23 years.

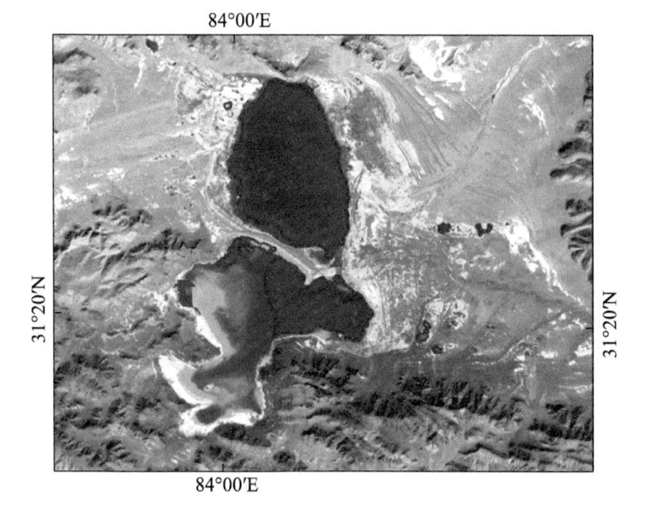

Fig. 8 Remote sensing images of Zabuye salt lake.

4.1.3 Bankog Co salt lake

Because of the periodic decline of the Siling Co surface, sand spits extended to the lake from both sides of the north and south connected at the final phase of the late Pleistocene. Bankog Co has been isolated and located in the east of the Siling Co. According to research, from 1959 to 2003, the lake surface of the Bankog Co was overall rising, reaching up to 1.75 m, although from 1959 to 1973 or a little later, there occurred a process of first falling, then rising and again slight falling, with a fall of 0.25 m, and an accompanied contraction of the lake area (Zhao et al., 2011). During the observation period from 2001 to 2003, annual variations of water level were 0.2 to 0.47 cm (Fig. 9). In August 2001, it reached the maximum value, 4522.11 m. During this period, the decline of water level in rainy season and dry season was 0.1 and 0.38 cm respectively (Zhao et al., 2006).

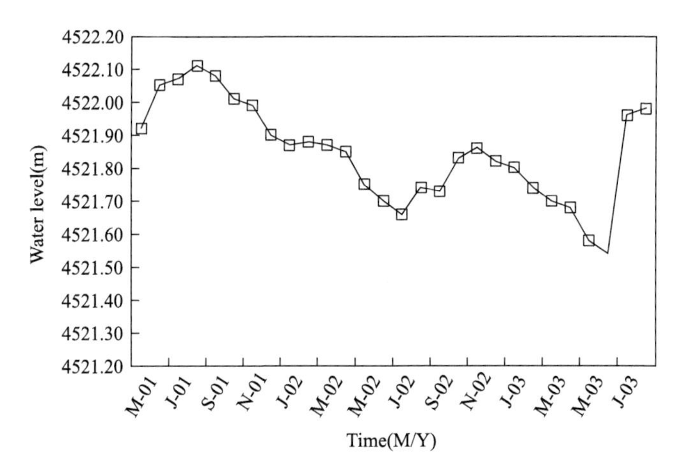

Fig. 9　Water level fluctuating recorded in Bankog Co salt lake during 2001–2003.

Based on previous studies and the result of remote sensing image interpretation, the area of Bankog Co has a trend of expansion from 1977 to 2010. The overall trend of mean temperature was increasing in Baingoin county after 1973. Mean temperature between 2001 and 2010 was 2.18 ℃ higher than the value between 1961 and 1970, and average speed of growth was 0.545 ℃/10a. Meanwhile, evaporation tended to decrease (Yan and Zheng, 2015a, 2015b; Ma et al., 2016). The enlargement in snow melt and rainfall has increased the supply of the basin. With the decrease of evaporation, lake area was bigger. The lake area increased from 41.4 km^2 in February 1977 to 131.82 km^2 in August 2010 (Fig. 10). The increased water volume in saline lake has led to lower salinity and ion concentration of lake water. The salinity of Bankog III dropped from 221.90 g/L in 1976 to 46.72 g/L in 2010 (Zheng et al., 2002; 2010). Siling Co located in the same basin became the largest lake in Tibet, the area has more than doubled and reached 2349.46 km^2. The area of each pull Dandong snow mountain dropped from 666.43 km^2 to 619.25 km^2, a reduction of 24.48 km^2.

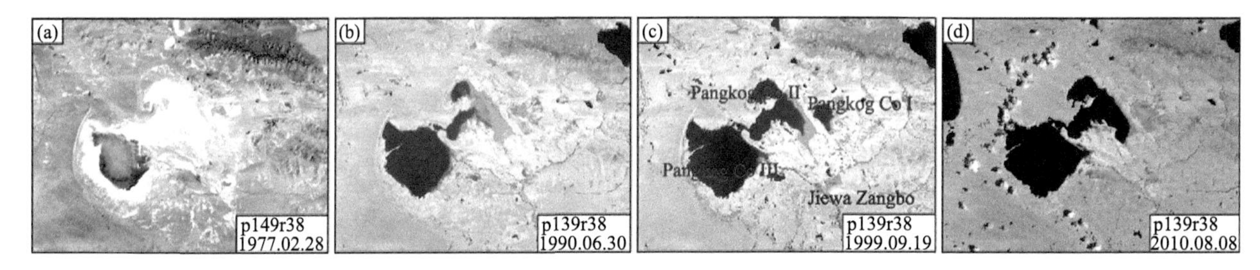

Fig. 10　Remote sensing images of Bankog Co salt lake.

As a whole the water level of the three lakes shows an upward trend. Due to the regional differences, the changing characteristics of the three lakes were different. Due to the large catchment area and shallow water, the influence of temperature on water level in Zabuye was stronger than that of precipitation. The peak caused by melt-water emerged in the May and June, and the secondary peak emerged with the rainfall in August and September. For Dangqiong Co, the influence of precipitation was more obvious, the peak emerged in rainy season and secondary peak emerged in June and July.

5　Conclusions

In this paper, climate change characteristics of Zabuye (1991–2014), Dangqiong Co (2004–2014) and Bankog Co (2001–2003) have been analyzed. The representativeness of weather parameters such as temperature, sunshine, evaporation and precipitation have been proved, considering the long-term meteorological data of field observational stations in Zabuye and Dangqiong Co salt lakes. Based on the analytical results, the climatic factors in the salt lake basin, especially the temperature, sunshine, evaporation, precipitation and wind determine the water level variation.

The following preliminary conclusions are obtained through analysis of the climate characteristics and dynamic changes of the water level in the salt lake basins.

(1) The annual mean temperature and precipitation in Zabuye, Dangqiong Co and Bankog Co basins increased gradually, but sunshine and evaporation decreased. Their water level ascended higher and higher, especially in the post-2000 period. Based on the relationship between the water changes and meteorological elements, analysis shows that water level had a positive correlation with air temperature and precipitation, and obviously a negative correlation with sunshine and evaporation.

(2) According to the analysis of climatic elements in salt lake regions and referring to the background of the climate change over the Tibetan Plateau, we can reasonably conclude that the climate in the observation area is changing to warming-wetting type and it is a partial consequence of water cycle increase by global warming. Water level rise has mainly been caused by increased precipitation, decreased evaporation and the increased meltwater of glacier retreating and permafrost degradation with the climate warming.

(3) Seasonal variation of water level is powered largely by the supply of lake water types and the seasonal change of regional climate. The key feature is the two-level surge in the water level within one year, particularly in shallow lakes.

(4) In the absence of continuous hydrological observation data, quantitative studies about glacial runoff and meltwater of frozen soil have not been carried on. The influence of the regional climate variations on the changes of salt lakes is shown in this paper. It is necessary to study on hydrological monitoring in the future, which will help to understand water cycle process in the high altitude inland river basins of the Tibetan Plateau.

Acknowledgments

We thank the constructive review from three anonymous reviewers. Since 1990, the following professional staff have attended the work of field observation. They are He Chaoxing, Zhang Fasheng, Gao Bingqi, Bianba, Huang Xinggen, Cheng Jiabai, Qing Guanghua, Baiduo, Gu fuqing, Wei Jianming, Fan Junfeng, Zhao Jianjun, Mima, Labaciren, Cidun, Queba, Danbazhaxi, Li Shenglu, Wang Guangcai, Ma Xiaoji, Ye Yugang, Ma Wanyun, Weigang Kong and so forth. Xiuyun Gu and the others provided logistics support. The meteorological and hydrological data of observation in this paper are the common fruit of everyone. Sincere thanks are given to all scholars who have devoted themselves to this work.

References

Bianduo, Bianbaciren, Li L *et al.*, 2009. The response of lake change to climate fluctuation in north Qinghai-Tibet Plateau in last 30 years. Journal of Geographical Sciences, 19 (2): 131-142.

Chang W Y B, 1987. Large lakes of China. Journal of Great Lakes Research, 13 (3): 235-249.

Du J, 2001. Change of temperature in Tibetan Plateau from 1961-2000. Acta Geographica Sinica, 56 (33): 232-239. (in Chinese)

Du J, Yang T B, He Y, 2014. Glaciers and lakes changes and climate response in the Selin Co Basin from 1990 to 2011. Journal of Arid Land Resources and Environment, 28 (12): 88-93. (in Chinese)

Jiang L G, Nielsen K, Andersen O B *et al.*, 2017. Monitoring recent lake level variations on the Tibetan Plateau using cryosat-2 sarin mode data. Journal of Hydrology, 544: 109-124.

Kang S C, Xu Y W, You Q L *et al.*, 2010. Review of climate and cryospheric change in the Tibetan Plateau. Environmental Research Letters, 5 (1): 1-8.

Lei Y B, Yao TD, Bird B W *et al.*, 2013. Coherent lake growth on the central Tibetan Plateau since the 1970s: Characterization and attribution. Journal of Hydrology, 483: 61-67.

Liu J S, Wang S Y, Yu S M *et al.*, 2009. Climate warming and growth of high-elevation inland lakes on the Tibetan Plateau. Global and Planetary Change, 67 (3/4): 209-217.

Lu A X, Yao T D, Wang L H, 2005. Study on the fluctuations of typical glaciers and lakes in the Tibetan Plateau using remote sensing. Journal of Glaciology and Geocryology, 27 (6): 783-792. (in Chinese)

Ma N, Szilagyi J, Niu G Y *et al*., 2016. Evaporation variability of Nam Co Lake in the Tibetan Plateau and its role in recent rapid lake expansion. Journal of Hydrology, 537: 27-35.

Ma X B, Li D L, 2003. Analyses on air temperature and its abrupt change over Qinghai-Xizang Plateau in modern age. Plateau Meteorology, 22 (5): 507-512. (in Chinese)

Niu T, Chen L X, Zhou Z J, 2004. The characteristics of climate change over the Tibetan Plateau in the last 40 years and the detection of climatic jumps. Advances in Atmospheric Sciences, 21 (2): 193-203.

Qi W, Zheng M P, 2006a. Initial research on water level fluctuation discipline of Zabuye Salt Lake in Tibet. Journal of Geographical Sciences, 22 (6): 693-699. (in Chinese)

Qi W, Zheng M P, 2006b. Winters and ARIMA model analyses of the lake level of salt Lake Zabuye, Tibetan Plateau. Journal of Lake Sciences, 18 (1): 21-28. (in Chinese)

Shi X H, Li L, Wang Q C *et al*., 2005. Climatic change and its influence on water level of Qinghai Lake. Meteorological Science and Technology, 33 (1): 58-62. (in Chinese)

Shi Y F, Shen Y P, Kang E *et al*., 2007. Recent and future climate change in northwest China. Climatic Change, 80 (3/4): 379-393.

Song C Q, Huang B, Ke L H *et al*., 2014a. Seasonal and abrupt changes in the water level of closed lakes on the Tibetan Plateau and implications for climate impacts. Journal of Hydrology, 514: 131-144.

Song C Q, Huang B, Ke L H *et al*., 2014b. Remote sensing of alpine lake water environment changes on the Tibetan Plateau and surroundings: A review. ISPRS Journal of Photogrammetry and Remote Sensing, 92: 26-37.

Wang X, Siegert F, Zhou A G *et al*., 2013. Glacier and glacial lake changes and their relationship in the context of climate change, central Tibetan Plateau 1972-2010. Global and Planetary Change, 111 (12): 246-257.

Wu S H, Yi Y H, Zheng D *et al*., 2005. Climate changes in the Tibetan Plateau during the last three decades. Acta Geologica Sinica, 60 (1): 3-l1. (in Chinese)

Yan L J, Zheng M P, 2015a. Influence of climate change on saline lakes of the Tibet Plateau, 1973-2010. Geomorphology, 246: 68-78.

Yan L J, Zheng M P, 2015b. The response of lake variations to climate change in the past forty years: A case study of the northeastern Tibetan Plateau and adjacent areas, China. Quaternary International, 371: 31-48.

Yao T D, Pu J C, Lu A X *et al*., 2007. Recent glacial retreat and its impact on hydrological processes on the Tibetan Plateau, China, and surrounding regions. Arctic Antarctic&Alpine Research, 39 (4): 642-650.

Yao T D, Zhu L P, 2006. The response of environmental changes on Tibetan Plateau to global changes and adaptation strategy. Advances in Earth Science, 21 (5): 459-464. (in Chinese)

Ye Q H, Yao T D, Chen F *et al*., 2008. Response of glacier and lake covariations to climate change in Mapam Yumco Basin on Tibetan Plateau during 1974-2003. Journal of China University of Geosciences, 19 (2): 135-145.

You Q L, Kang S C, Pepin N *et al*., 2010. Climate warming and associated changes in atmospheric circulation in the eastern and central Tibetan Plateau from a homogenized dataset. Global and Planetary Change, 72 (1/2): 11-24.

Zhao X T, Zhao Y Y, Zheng M P *et al*., 2011. Late Quaternary Lake Development and Denivellation of Bankog Co as well as lake evolution of southeastern North Tibetan Plateau during the Last Great Lake Period. Acta Geoscientica Sinica, 32 (1): 13-26. (in Chinese)

Zhao Y Y, Zhao X T, Zheng M P *et al*., 2006. The denivellation of Bankog Co in the past 50 years, Tibet. Acta Geologica Sinica, 80 (6): 876-884. (in Chinese)

Zhao Y Y, Zheng M P, Cai X G *et al*., 2004. Modern lake resources and environment on the western side of the Tibet section on the phase-II of the Qinghai-Tibet Railway. Geological Bulletin of China, 23 (7): 680-685. (in Chinese)

Zheng D, Lin Z Y, Zhang X Q, 2002. Progress in studies of Tibetan Plateau and global environmental change. Earth Sci. Front, 9 (1): 95-10. (in Chinese)

Zheng M P, 2014. Saline Lakes and Salt Basin Deposits in China. Beijing: Science Press.

Zheng M P, Liu X F, 2010. Hydrochemistry and minerals assemblages of salt lakes in the Qinghai-Tibet Plateau, China. Acta Geologica Sinica, 84 (11): 1585-1600. (in Chinese)

Zheng M P, Qi W, Jiang X F *et al*., 2004. Trend of salt lake changes in the background of global warming and tactics for adaptation to the changes. Acta Geologica Sinica, 78 (3): 795-807.

Zheng M P, Xiang J, Wei X J *et al*., 1989. Saline Lakes on the Qinghai-Xizang (Tibet) Plateau. Beijing: Science&Technology Press. (in Chinese)

Zheng M P, Zhao Y Y, Liu J Y, 2000. Palaeoclimatic indicators of China's Quaternary saline lake sediments and hydrochemistry. Acta Geologica Sinica, 74 (2): 259-265.

Zhou L S, Yang W D, 1992. Discussion of the climatic change in recent 600 years and dropping of Lake water-level, Qinghai Lake drainage basin. Journal of Lake Sciences, 4 (3): 25-31. (in Chinese)

Zhu L P, Xie M P, Wu Y H, 2010. Quantitative analysis of lake area variations and the influence factors from 1971 to 2004 in the Nam Co Basin of the Tibetan Plateau. Chinese Science Bulletin, 55 (13): 1294-1303.

生态区绿色勘查评价方法及评价标准探讨*

巩鑫[1,2]，赵元艺[3]，高知睿[4]，刘春花[3]

1. 中国地质大学（北京）地球科学与资源学院，北京 100083
2. 贵州有色金属和核工业地质勘查局地质矿产勘查院，贵州 贵阳 550005
3. 中国地质科学院矿产资源研究所，北京 100037
4. 中国地质调查局天津地质调查中心，天津 300170

摘要 绿色勘查为绿色发展在地质工作中的具体体现，我国在实施绿色勘查方面起步较晚，至今国家层面仍未出台统一的评价方法与评价标准，从而制约着对绿色勘查的认识及理解。通过查阅国家及地方关于地质勘查指导原则及规范标准，总结国内自实施绿色勘查以来取得的成就，分析国外矿业大国在不同生态区实行的地质勘查政策及评价标准。最终，结合我国地质勘查问题及现状、区域矿产资源分布与经济状况，提出不同生态区绿色勘查评价方法及准则。对地质勘查造成的影响进行量化，提出勘查工作在生态保护区、脆弱区、一般区分别达到 95 分、90 分、85 分以上者可称达到绿色勘查标准。并建议我国适当改变生态保护区、脆弱区地质工作政策，使环境保护与资源勘查、经济发展相协调。

关键词 绿色勘查；评价方法；评价标准；环境保护；生态区

0 引言

党的十九大报告提出将我国建设成为“富强、民主、文明、和谐、美丽”的现代化强国，如何处理好“金山银山”与“绿水青山”的关系，成为能否达到这一目标的关键因素。2017 年以来，甘肃、湖南、新疆等 15 个省（区）相继开展了生态保护区、脆弱区矿权清理工作。停止保护区、部分脆弱区内一切矿权勘查开发工作，一方面导致了我国地勘工作空间和范围严重压缩；另一方面也不利于资源丰富的贫困地区分享资源红利。

绿色勘查是一种先进的发展方式，正确运用绿色勘查理念指导找矿可以解决环境保护与地质勘查相冲突这一难题。绿色勘查是把保护生态环境放到地质勘查工作的首要位置，通过科学合理布局勘查工程，采用对环境影响小、可恢复的勘查技术手段，减少地质勘查对生态环境的影响，最终达到生态环境保护与矿山勘查开发、地方经济协同发展的目的[1, 2]。

1 绿色勘查的含义

绿色勘查指在地质勘查期间，以绿色发展理念为指导，以先进的、环保的技术工艺和勘查方法为基础，以及在勘查过程结束后，对生态环境造成的破坏进行恢复治理，最大限度地减少对生态环境造成的破坏或扰动，达到“绿色”与“勘查”协同发展的勘查技术手段[3-5]。绿色勘查是绿色发展在地质勘查领域的实践，是“创新、协调、绿色、开放、共享”五大发展理念的具体体现[6]。

2 绿色勘查的现状

2.1 国外绿色勘查的现状

国外主要矿业国（如澳大利亚、加拿大等）环保意识较强，实行绿色勘查起步较早。1999 年，澳大利

* 本文发表在：探矿工程（岩土钻掘工程），2019，46（3）：86-92

亚政府颁布了《环境和生物多样性保护法》(EPBC)，法规明确要求，对可能造成较大环境影响的投资项目必须在环境方面进行评估和审批。加拿大勘探开发者协会在2003年出台了《勘探工程卓越手册》，2004年编写成新版本的EES手册，EES手册为勘探者在地质勘查工程中提供了环境管理规范和要求。英国政府明确要求，在地质勘查完成后要对其进行标准化生态环境恢复[7-10]。

多数国外矿业国允许矿企、地勘单位在获得矿权开发许可证后进入生态区（生态保护区、生态脆弱区及生态一般区），利用绿色勘查理念对生态保护区内的矿产资源进行勘查开发，同时提高矿企、地勘单位进入门槛，加大政府、居民监督力度，最终实现生态保护与矿山勘查开发、地方经济协同发展[11-16]（表1）。

表1　国内外生态区矿山勘查要求对比

Table 1　Comparison of mine exploration requirements in ecological area at domestic and abroad

国家	生态保护区	生态脆弱区	生态一般区
中国	禁止一切矿企、地勘单位进行矿山勘查	允许适度矿山勘查，目前西部多数矿权被清退	允许矿山勘查
澳大利亚、加拿大、智利、南非、瑞典、冰岛、格陵兰、挪威、芬兰、俄罗斯	允许达到标准的矿企、地勘单位在获得矿业许可证后进入生态保护区、脆弱区、一般区进行矿山绿色勘查		

注：澳大利亚在保护区的核心区不允许进行矿山开发，但允许进行勘查工作。

2.2　国内绿色勘查的现状

我国《自然保护区条例》规定，生态环保区内禁止一切矿山勘查工作（表1），生态脆弱区允许适度进行勘查工作。据统计目前我国10%的重点成矿带位于已划定的生态保护区内，35%的重点成矿带位于生态脆弱区内。随着保护区、脆弱区数量、面积不断增多加大，地质行业的“倒逼”压力越来越重。旧式的地质勘查方式已不适合生态文明建设的要求，新式的地质勘查方式即绿色勘查应运而生[17, 18]。

我国绿色勘查尚处于起步阶段，2016年，青海省地勘局率先在青南多彩整装区实施绿色勘查，把地质勘查与生态保护、居民利益、社会和谐等因素有机结合，创造了“多彩模式”的勘查方法[19]；贵州西能集团启动了绿色勘查示范项目建设，组织了5家地质勘查单位7个地质勘查项目作为绿色勘查示范项目，取得了一定成效；甘肃省地矿局通过配备一批航空测量、物探及遥感高新设备，试图开展物探、化探无损方法研究[8]；中国地质调查局探矿工艺研究所加强绿色勘查新技术、新方法的研究，“一基多孔、一孔多支”新技术在若尔盖铀矿区成功利用，形成了一套生态脆弱区绿色钻探技术体系[20-22]。2016年5月10日我国发布了绿色勘查行动宣言，倡议地勘行业大力推进绿色勘查，为实现资源开发利用和生态环境保护协调发展做出更大的贡献。

绿色勘查的实施已初见成效，但绿色勘查理念未完全被地质工作者领悟，绿色勘查技术、方法尚未成熟，在判断现行的“绿色勘查”是否为真正的绿色勘查上缺少一套能够普遍适用并行之有效的评价方法及标准。

3　绿色勘查评价方法及影响因素

结合绿色勘查行为指南及我国不同生态区特征，提出绿色勘查评价指标，建立绿色勘查指标量化表，最终总结出不同生态区绿色勘查评价方法，并分析其影响因素。

3.1　我国生态区划分及特征

我国将生态区分为保护区、脆弱区及一般区。分析我国区域成矿带、区域经济发展状况具有以下特征[23-25]：保护区分布于我国大陆各个地区，分布零散，总面积较大；脆弱区主要分布于我国西北干旱—半干旱、青藏高原及西南地区，大部分位于胡焕庸线以西，总面积大，人口稀少，成矿带密集，潜在资源丰富，经济相对落后，多数成矿带中地质勘查工作即将面临被清退；一般区主要分布于我国东部湿润—半湿润地区，大部分位于胡焕庸线以东，人口密集，成矿带相对稀疏，人均资源相对匮乏，经济较发达。

3.2 绿色勘查评价方法

通过分析总结勘查工作对生态环境、社会的影响，提出绿色勘查评价指标，结合生态区的环境背景，提出生态保护区、脆弱区、一般区的绿色勘查评价方法[26-32]。将绿色勘查评价指标分自然生态环境、工程扰动、社会影响三大指标，16 项具体指标（表 2），并将每项指标予以具体分值来对绿色勘查的评定过程进行量化（若某地区未涉及某项具体指标，则按满分计算）。在此提出地质勘查工作在生态保护区、脆弱区、一般区分别达到95 分、90 分、85 分以上者可称为绿色勘查，认为达到了生态环境保护与资源勘查、经济发展相协调的目的。

3.3 评价指标影响因素

3.3.1 自然生态环境指标

（1）植被破坏率：指营地建设、工程施工、临时道路修建等开挖地表破坏的生态植被面积与矿区面积的比率，提出以下公式：

$$V = \frac{S_{营房} + S_{工程} + S_{道路}}{S_{矿区}} \tag{1}$$

其中：采用 $S_{矿区} = 2S_{勘查}$；在绿色勘查中 $S_{营房}$ 应采用“架空式”，对植被破坏可忽略不计；$S_{工程} = S_{钻探} + S_{槽深} + S_{硐探}$，$S_{硐探}$ 在勘查工作中采用较少且主要以尾矿对地表破坏，故在工程扰动指标中考虑。

按地质勘探阶段（对自然生态环境影响最大，下同）考虑，钻孔工程布局以网格状为例，钻孔平面间距以 50m×50m，钻孔平台面积 4m×4m（模块式钻机占地面积），槽探面积不大于钻孔平台面积即 4m×4m，道路修建面积以 50m×0.5m 为例，则代入公式（1）中得

$$V = \frac{0 + (4\times4 + 4\times4) + 50\times0.5}{2\times50\times50} \times 100\% = 2\%$$

故将植被破坏率以 2%、4%为界划分为 3 个等级，对其进行量化。

（2）土壤污染率：指勘查工作者生活、机台工作活动等产生的生活、生产垃圾所引起的土壤污染面积与矿区面积的比率。地质勘查中土壤污染多来自钻机机台对土壤的污染，提出土壤污染率不得大于钻机平台面积的 50%，即根据公式（1）得

$$V = \frac{4\times4}{2\times50\times50} \times 100\% \times 50\% \approx 0.2\%$$

故将土壤污染率以 0.2%、0.4%为界划分为 3 个等级，对其进行量化。

（3）土壤硬化率：指由营地建设及人为活动、机台施工、临时道路运输施工材料所引起的土质硬化面积与矿区面积的比率。地质勘查过程中引起的土壤硬化主要来源为机台硬化及临时道路硬化，提出土壤硬化率不得大于钻机平台及临时道路面积的 50%，即根据公式（1）得

$$V = \frac{4\times4 + 50\times0.5}{2\times50\times50} \times 100\% \times 50\% \approx 0.5\%$$

故将土壤硬化率以 0.5%、1%为界划分为 3 个等级，对其进行量化。

（4）水环境影响程度：指由工程施工、勘查工作者生活产生的垃圾引起勘查区内河流、湖泊的污染情况。对水体环境造成影响的因素主要由钻孔施工过程中泥浆的无规则排放，根据对水体的污染程度划分等级，进行量化。

表 2　绿色勘查评价指标量化表

Table 2　Quantitative index of green exploration evaluation

指标名称及分值		量化指标					
		级别及分值（分）			核心区	脆弱区	一般区
					≥95	≥90	≥85
自然生态环境指标（42）	植被破坏率（6）	≤2%（6）	（2～4）%（5～4）	≥4%（3～0）	≥40	≥38	≥36
	土壤污染率（6）	≤0.2%（6）	（0.2～0.4）%（5～4）	≥0.4%（3～0）			
	土壤硬化率（6）	≤0.5%（6）	（0.5～1）%（5～4）	≥1%（3～0）			
	水环境影响程度（6）	无（6）	轻微（5～4）	严重（3～0）			
	自然景观破坏（6）	无（6）	存在（0）				
	珍贵物种的破坏（6）	无（6）	存在（0）				
	垃圾处理率（6）	完全按要求处理（6）	存在少许未处理垃圾（5～4）	剩余垃圾对环境仍存在危害（3～0）			
工程扰动指标（32）	工程施工对生态破坏恢复率（8）	≥95%（8～7）	（95～85）%（6～4）	≤85%（3～0）	≥30	≥28	≥26
	地形地貌破坏恢复率（6）	≥90%（6）	（90～75）%（5～4）	≤75%（3～0）			
	临时道路恢复率（6）	≥90%（6）	（90～75）%（5～4）	≤75%（3～0）			
	尾矿及废石处理率（6）	≥95%（6）	（95～85）%（5～4）	≤85%（3～0）			
	地质次生灾害（6）	无（6）	微型且不存在发生灾害潜力（5～4）	存在发生灾害潜力（0）			
社会影响指标（26）	与居民和谐度：冲突事件发生次数（6）	0（6）	1（5）	≥2（0）	≥25	≥24	≥23
	与居民和谐度：对项目施工结果的满意度（6）	≥95%（6）	（95～80）%（5～4）	≤80%（3～0）			
	与居民和谐度：居民对施工噪声、废气、粉尘处理的满意度（6）	≥90%（6）	（90～75）%（5～4）	≤75%（3～0）			
	项目对当地经济、交通的影响（8）	利好作用（8～7）	较利好作用（6～4）	负面作用（0）			

（5）自然景观的破坏：指在地质勘查过程中，因工程施工、修建临时道路对勘查区已有的自然景观造成了破坏，使其失去了观赏价值。因自然景观破坏具难恢复性，故此指标实行一票否决制，存此种行为即未达到绿色勘查标准。

（6）珍贵物种的破坏：是指因勘查施工对勘查区内珍贵的植物、动物及动物栖息地造成破坏，甚者引起物种死亡及灭绝。珍贵物种的破坏具有不可逆性，故此指标实行一票否决制，存此种行为即未达到绿色勘查标准。

（7）垃圾处理率：指勘查工作结束后，对产生的生活、生产垃圾（废水、废泥浆材料及废油污）按标准处理的状况，根据处理情况划分等级，对其进行量化。

3.3.2　工程扰动指标

（1）工程施工对生态破坏恢复率：在勘查工程施工完成后，应及时对破坏的生态环境进行恢复，必要时进行人工增植、播散草籽等。项目验收时按生态环境恢复率进行划分等级，对其进行量化。

（2）地形地貌破坏恢复率：指机台平场、开挖槽探等勘查工程对勘查区地形地貌造成的破坏，在勘查工程结束后使其恢复的程度。该项指标中尤其槽探施工，对地形地貌破坏程度最大，故槽探施工完成后要及时将土回填。项目验收时根据原始地形地貌恢复率划分等级，对其进行量化。

（3）临时道路恢复率：在矿区因搬运施工机器、施工材料、生活用品而临时修建道路，对其生态植被、土壤硬度及地形地貌会造成一定的影响，项目验收时根据临时道路恢复率划分等级，对其进行量化。

（4）尾矿及废石处理率：硐探施工过程中所开采的尾矿及废石应尽量用于充填采空区，无法用于充填的尾矿及废石应进行合理处理，避免引发地质次生灾害。项目验收时根据尾矿及废石处理率划分等级，对其进行量化。

（5）地质次生灾害：机台及临时道路开挖、探槽施工和废石堆积等引起的滑坡、崩塌、泥石流、塌陷等地质次生灾害，项目验收时根据是否存在发生灾害的潜力及评估所带来的危害划分等级，对其进行量化。

3.3.3 社会影响指标

3.3.3.1 项目与居民的和谐度

（1）冲突事件发生次数：在勘查工作中因某种原因与当地居民发生冲突事件，根据事件次数划分等级，对其进行量化。一般认为冲突次数大于 2 次时，直接定级为未达到绿色勘查标准。

（2）对项目施工结果的满意度：通过调查统计当地政府、当地居民对项目施工结果的满意度，根据满意度划分等级，对其进行量化。

（3）居民对施工噪声、废气、粉尘处理的满意度：项目施工期间会对当地居民带来影响，通过调查在项目施工期间，居民对施工噪声、废气、粉尘处理的满意度，根据满意度划分等级，对其进行量化。

3.3.3.2 项目对当地经济、交通的影响

勘查项目结束后，评估此次勘查工作对当地经济、交通状况带来的作用，根据带来的作用划分等级。一般认为若勘查工作给当地经济、交通带来负面作用，直接定级为未达到绿色勘查标准。

4 建议

基于我国目前绿色勘查、经济现状及国家对保护区、脆弱区实行的政策，本文提出以下建议和策略。

（1）绿色勘查是当今形势下进行地质勘查的唯一路径，应将绿色勘查纳入地质勘查立项、设计、施工和验收中，主动调整工作布局、工作方法及工作顺序，把对生态环境的影响降至最低点。

（2）加快建立和完善国家层面的绿色勘查相关规范和评价标准，同时加强绿色勘查理念宣传，使绿色勘查在全国范围内得到顺利推广和应用。

（3）增加科技创新投入，加大绿色勘查新技术、新设备、新工艺的研究，使从技术、方法层面减少对环境的扰动。

（4）适当开放生态保护区（试验区、缓冲区）、生态脆弱区的地质勘查开发。提高矿企、地勘单位进入门槛，在获得环保资质、许可证条件，具有科学合理的生态保护、恢复治理方案及土地复垦方案的情况下，允许矿企、地勘单位进行矿山勘查开发活动。

（5）允许进入保护区、生态脆弱区的矿企、地勘单位必须缴纳环境保证金，占总项目经费一定比例，在项目结束后，按照绿色勘查评价指标进行评价、验收，验收达到绿色勘查标准即可返退环境保证金；对于达不到绿色勘查要求的，则动用环境保证金恢复生态。

（6）生态保护区、脆弱区施行“二不许”制度。即对区域地质环境影响严重且难以恢复的勘查开发不允许，绿色勘查不达标的勘查不允许。

5 结论

（1）实施绿色勘查是地质勘查领域贯彻落实中央生态文明战略的重要举措，是地质勘查行业持续健康发展的迫切要求，是实现保护生态环境和保障资源供给“双赢”的唯一途径。

（2）根据绿色勘查评价方法及量化指标，地质勘查工作影响指标在生态保护区、脆弱区、一般区赋分值分别达到95分、90分、85分以上者可称为绿色勘查，认为达到了生态环境保护与资源勘查、经济发展相协调的目的。

（3）适当改变生态保护区（试验区、缓冲区）、生态脆弱区的地质勘查政策，使部分达到标准和要求的矿企、地勘单位可进行矿权的勘查开发。运用绿色勘查方法，寻找、开发矿产资源，达到生态环境保护、资源勘查开发、地方经济协同发展，在保证生态环境不遭受破坏的前提下，又可使资源丰富的贫困地区享受到资源勘查开发带来的红利。

参考文献

[1] 付英, 黄贤营, 傅连珍, 等. 我国地质勘查行业发展现状与走向[J]. 中国国土资源经济, 2016, 29 (11): 11-14.
FU Ying, HUANG Xianying, FU Lianzhen, et al. Status and trend in development of geological exploration industry China [J]. Natural Resource Economics of China, 2016, 29 (11): 11-14.

[2] 赵元艺, 李小赛, 乔东海, 等. 西藏多龙矿集区绿色勘查与绿色矿山建议[J]. 地质论评, 2016, 62 (S1): 287-288.
ZHAO Yuanyi, LI Xiaosai, QIAO Donghai, et al. Suggestions of green exploration and green mining of Duolong Ore Concentration Are, Xizang (Tibet) [J]. Geological Review, 2016, 62 (S1): 287-288.

[3] 鞠建华, 强海洋. 中国矿业绿色发展的趋势和方向[J]. 中国矿业, 2017, 26 (2): 7-12.
JU Jianhua, QIANG Haiyang. The trend and direction of green development of the mining industry in China[J]. China Mining Magazine, 2017, 26 (2): 7-12.

[4] 杨俊鹏, 戴华阳, 张建伟. 新常态下我国绿色矿山建设面临问题与解决办法[J]. 中国矿业, 2017, 26 (1): 67-71.
YANG Junpeng, DAI Huayang, ZHANG Jianwei. The problems and solution of the construction of green mine in the new normal[J]. China Mining Magazine, 2017, 26 (1): 67-71.

[5] 王英超. 新常态下我国绿色勘查的发展探讨[J]. 地质论评, 2016, 62 (S2): 281-282.
WANG Yingchao. Discussion on the development of green exploration in our country under the new normal state[J]. Geological Review, 2016, 62 (S2): 281-282.

[6] 张新虎, 刘建宏, 黄万堂, 等. 绿色勘查理念: 认知、探索与实践[J]. 甘肃地质, 2017, 26 (1): 1-7.
ZHANG Xinhu, LIU Jianhong, HUANG Wantang, et al. Green exploration: cognition, explore and practice[J]. Gansu Geology, 2017, 26 (1): 1-7.

[7] 郑娟尔, 余振国, 冯春涛. 澳大利亚矿产资源开发的环境代价及矿山环境管理制度研究[J]. 中国矿业, 2010, 19 (11): 66-69, 84.
ZHENG Juaner, YU Zhenguo, FENG Chuntao. Environment cost caused by mineral resource exploitation and environment management system of Australia[J]. China Mining Magazine, 2010, 19 (11): 66-69, 84.

[8] Alexander W. Conservation and access to land for mining in protected areas: the conflict over mining in South Australia's Arkaroola Wilderness Sanctuary[J]. Journal of Environmental Law, 2014, 26 (2): 291-317.

[9] Monowar M, Nurlan O. Green governance and sustainability reporting in Kazak-hstan's oil, gas, and mining sector: evidence from a former USSR emerging economy[J]. Journal of Cleaner Production, 2017, 164 (10): 389-397.

[10] David R I, Pooe, Khomotso, et al. Exploring the challenges associated with the greening of supply chains in the South African manganese and phosphate mining industry: original research[J]. Journal of Transport and Supply Chain Management, 2014, 8 (1): 1-9.

[11] Samuel T K, Wilson, Hongtao, et al. The mining sector of Liberia: current practices and environmental challenges[J]. Environmental Science and Pollution Rese-arch, 2017, 24 (23): 18711-18720.

[12] Paul M, Kelvin M. Environmental injustice and post-colonial environmentalism: opencast coal mining, landscape and place[J]. Environment and Planning A, 2017, 49 (1): 29-46.

[13] Ellis J I, Clark M R, Rouse H L, et al. Environmental management frameworks for offshore mining: the New Zealand approach[J]. Marine Policy, 2017, 84 (17): 178-192.

[14] Juan O. Some criteria for evaluation of resources in a system of protected areas in Chile[J]. Environmental Conservation, 2015, 12 (2): 173-175.

[15] Anne M H, Frank V, Peter C, et al. Managing the social impacts of the rapidly-expanding extractive industries in greenland[J]. The Extractive Industries and Society, 2016, 3 (1): 25-33.

[16] Garcia, Leticia, Couto, et al. Brazil's worst mining disaster: corporations must be compelled to pay the actual environmental costs[J]. Ecological Applications, 2017, 27 (1): 5-9.

[17] 曹献珍. 国外绿色矿业建设对我国的借鉴意义[J]. 矿产环保与利用, 2011, 5 (6): 19-23.
CAO Xianzhen. Construction of green mining in foreign countries and reference mean to our country[J]. Conservation and Utilization of Mineral Resources, 2011, 5 (6): 19-23.

[18] 赵彦璞. 新常态下内蒙古矿产资源勘查开发思路[J]. 中国矿业, 2016, 25 (7): 28-31.

ZHAO Yanpu. The idea of mineral resources exploration and development in Inner Mongolia under the new normal condition[J]. China Mining Magazine, 2016, 25 (7): 28-31.

[19] 陈伯辉, 高元宏, 李玉胜, 等. 青海省绿色地勘技术及标准探讨[J]. 探矿工程 (岩土钻掘工程), 2016, 43 (10): 131-134.
CHEN Bohui, GAO Yuanhong, LI Yusheng, et al. Discussion on the green geological prospecting technique of Qinghai Province and the standard[J]. Exploration Engineering (Rock& Soil Drilling and Tunneling), 2016, 43 (10): 131-134.

[20] 吴金生, 李子章, 李政昭, 等. 绿色勘查中减少探矿工程对环境影响的技术方法[J]. 探矿工程 (岩土钻掘工程), 2016, 43 (10): 112-116.
WU Jinsheng, LI Zizhang, LI Zhengzhao, et al. Technological methods of reducing impact on environment by exploration in green exploration[J]. Exploration Engineering (Rock& Soil Drilling and Tunneling), 2016, 43 (10): 112-116.

[21] 刘海声, 穆元红, 刘鹏, 等. 绿色勘查技术在青海格尔木铜金山矿区钻探施工的应用分析[J]. 探矿工程 (岩土钻掘工程), 2017, 44 (3): 27-30.
LIU Haisheng, MU Yuanhong, LIU Peng, et al. Application analysis on green exploration technology in drilling construction in Tongjinshan Mining Area of Qinghai Province[J]. Exploration Engineering (Rock &Soil Drilling and Tunneling), 2017, 44 (3): 27-30.

[22] 贾占宏, 高元宏, 梁俭, 等. 绿色地质勘查综合技术应用分析[J]. 探矿工程 (岩土钻掘工程), 2017, 44 (4): 1-4.
JIA Zhanhong, GAO Yuanhong, LIANG Jian, et al. Application and analysis on comprehensive technology of green geological prospecting[J]. Exploration Engineering (Rock &Soil Drilling and Tunneling), 2017, 44 (4): 1-4.

[23] 刘军会, 邹长新, 高吉喜, 等. 中国生态环境脆弱区范围界定[J]. 生物多样性, 2015, 23 (6): 725-732.
LIU Junhui, ZOU Changxin, GAO Jixi, et al. Location determination of ecologically vulnerable regions in China[J]. Biodiversity Science, 2015, 23 (6): 725-732.

[24] 肖克炎, 邢树文, 丁建华, 等. 全国重要固体矿产重点成矿区带划分与资源潜力特征[J]. 地质学报, 2016, 90 (7): 1269-1280.
XIAO Keyan, XING Shuwen, DING Jianhua, et al. Division of major mineralization belts of China's key solid mineral resources and their mineral resource potential[J]. Acta Geologica Sinica, 2016, 90 (7): 1269-1280.

[25] 朱裕生, 王全明, 张晓华, 等. 中国成矿区带划分及有关问题[J], 地质与勘探, 1999, 35 (4): 1-4.
ZHU Yusheng, WANG Quanming, ZHANG Xiaohua, et al. Some problems on division of metallogenic belts in China[J]. Geology and Prospecting, 1999, 35 (4): 1-4.

[26] 王旭, 周爱国, 甘义群, 等. 青藏高原矿产资源开发与地质环境保护协调发展的对策讨论[J]. 干旱区资源环境, 2010, 24 (2): 69-73.
WANG Xu, ZHOU Aiguo, GAN Yiqun, et al. Study on the harmonize development of mineral resources exploration and the geo-environment protection countermeasures in Qinghai-Tibet[J]. Journal of Arid Land Resources and Environment, 2010, 24 (2): 69-73.

[27] 张文辉, 申文金. 关于绿色勘查标准化的思考[J]. 现代矿业, 2017, 9 (9): 8-11, 17.
ZHANG Wenhui, SHEN Wenjin. Considerations of standardization of green exploration[J]. Modern Mining, 2017, 9 (9): 8-11, 17.

[28] 黄敬军, 倪嘉曾, 赵永忠, 等. 绿色矿山创建标准及考评指标研究[J]. 中国矿业, 2008, 17 (7): 36-39.
HUANG Jingjun, NI Jiazeng, ZHAO Yongzhong, et al. Study on green mine construction standard and its check and evaluation index[J]. China Mining Magazine, 2008, 17 (7): 36-39.

[29] 闫志刚, 刘玉朋, 王雪丽. 绿色矿山建设评价指标与方法研究[J]. 中国煤炭, 2012, 38 (2): 116-120.
YAN Zhigang, LIU Yupeng, WANG Xueli. Evaluation criterion and method of green mine[J]. China Coal, 2012, 38 (2): 116-120.

[30] 尹伯悦, 赖明, 谢飞鸿, 等. 借鉴国外绿色建筑评估体系来研究我国绿色矿山建筑标准的建立和实施[J]. 中国矿业, 2006, 15 (6): 29-32.
YIN Boyue, LAI Ming, XIE Feihong, et al. The research assessment system for green building of the over main and our country mining green building standard model and implementation[J]. China Mining Magazine, 2006, 15 (6): 29-32.

[31] 张德明, 贾晓晴, 乔繁盛, 等. 绿色矿山评价指标体系的初步探讨[J]. 再生资源与循环经济, 2010, 3 (12): 11-13.
ZHANG Deming, JIA Xiaoqing, QIAO Fansheng, et al. Study on the index evaluation systems of green mines[J]. Recyclable Resources and Circular Economy, 2010, 3 (12): 11-13.

[32] 徐友宁, 袁汉春, 何芳, 等. 矿山环境地质问题综合评价指标体系[J]. 地质通报, 2003, 22 (10): 829-832.
XU Youning, YUAN Hanchun, HE Fang, et al. Comprehensive evaluation index system of the environmental geological problems of mines[J]. Geological Bulletin of China, 2003, 22 (10): 829-832.

Evaluation method and evaluation standard of green exploration in ecological area

Xin Gong[1, 2], Yuanyi Zhao[3], Zhirui Gao[4] and Chunhua Liu[3]

1. School of Earth Sciences and Resources，China University of Geosciences，Beijing 100083，China

2. Geological and Mineral Exploration Institute，Guizhou Bureau of Geological Exploration for Non-ferrous Metals and Nuclear Industry，Guiyang Guizhou 550005，China
3. Institute of Mineral Resources，Chinese Academy of Geological Sciences，Beijing 100037，China
4. Tianjin Center，China Geological Survey，Tianjin 300170，China

Abstract Green exploration is the concrete embodiment of green development in geological work. China began green exploration relatively late. Up to now, there has not been a unified evaluation method and standard yet at the national level, which restricts the understanding of green exploration. With review of the national and local guiding principles and standards for geological exploration, the advances are summarized since the implementation of green exploration in China, and analysis is made of the policies and evaluation standards for geological exploration implemented in different ecological zones by major mining countries abroad. Finally, in view of geological exploration problems and current situation, regional distribution of mineral resources and economic situation in China, green exploration evaluation methods and criteria are proposed for different ecological zones. The impact of geological exploration is quantified, and it is recommended that exploration work should reach 95 points, 90 points and 85 points respectively in ecological conservancy areas, fragile areas and general areas to meet the green exploration criteria. It is proposed that China should change the geological exploration policies for ecological conservancy areas and fragile areas, so as to keep ecological environmental protection in pace with resource exploration and economic development.
Keywords Green exploration; Evaluation methods; Evaluation standard; Environmental protection; Ecological area

西藏多不杂矿区土壤重金属特征及环境问题防治*

吴宇靓[1,2]，王松[1,2]，赵元艺[2]

1. 中国地质大学（北京），北京 100083
2. 中国地质科学院矿产资源研究所，自然资源部成矿作用与矿产资源评价重点实验室，北京 100037

内容提要 在矿产开发前，充分了解矿区的土壤重金属环境特征对矿产绿色开发和合理开采具有重要意义。位于西藏改则县的多不杂矿床是多龙矿集区内的重要矿床之一，目前已探明铜资源量达 234.44 万吨。本文以多不杂矿区为例，在地表选取 4 条剖面采集了 37 件土壤样品，开展了重金属元素（Cu、Pb、Zn、As、Cd、Cr、Hg）含量和 pH 值测试，对其中 10 件样品进行了重金属形态特征分析。采用单因子污染指数和内梅罗综合污染指数法分析研究了多不杂矿区内地表土壤重金属的含量、污染指数和形态特征，探讨了重金属元素来源及其对环境的影响程度。结果表明，多不杂矿区地表为碱性土壤，7 种重金属元素中，Cu 在所有样品中的含量均超过 GB 15618—1995Ⅲ级标准；Zn、Cd 仅有最大值超过Ⅲ级标准；Pb、As、Cr、Hg 含量均不超标。单因子污染指数呈现出 Cu＞Zn＞As＞Cd＞Cr＞Pb＞Hg 的分布特征，综合指数平均值为 11.60，表明重金属对环境的综合影响程度较高。Cu 主要以铁锰氧化态和残渣态存在，分别占 48.40%和 14.73%，Pb、Zn、As、Cr 主要以残渣态形式存在，分别占 90.27%、71.48%、79.59%、86.12%。Cd 主要以铁锰氧化态和碳酸盐态形式存在，分别占 40.05%和 22.42%，Hg 主要以残渣态和腐殖酸态存在，分别占 40.16%和 22.54%。7 种重金属元素来源主要有 3 类：Cu、Zn、Cd、Hg 源于黄铜矿的氧化淋滤；Pb、As 源于黄铁矿的氧化；Cr 源于岩石的自然风化。针对重金属相关的环境问题，提出了针对性的防治建议。

关键词 多不杂矿区；土壤重金属；环境特征；来源解析；生物有效性；防治意义

土壤质量直接影响食品安全，更与人体健康息息相关，近年来，土壤质量问题引起了各方面的广泛关注。2014 年全国土壤污染调查公报显示，我国有 16.1%的耕地土壤受到了污染，重金属是最主要的污染物，Cu、Pb、Zn、As、Cd、Cr 和 Hg 的含量分别超出标准值 2.1%、1.5%、0.9%、2.7%、7.0%、1.1%和 1.6%（Huang Ying et al.，2018）。矿体是有用元素相对富集的地段，矿区及其周边的人类活动更容易促使矿物暴露地表，导致土壤原有结构和状态发生重大变化，引起土壤重金属环境问题，因此，更应该重视在矿区周边开展土壤重金属研究工作。目前对于矿区的研究主要集中在矿区地质勘查、成矿机制、成矿预测等方面，对于矿山环境方面的研究也往往侧重于矿产开发后的生态恢复，开发前的环境属性研究较少。

多龙矿集区内目前发现了地堡那木岗、波龙、多不杂、荣那、拿若、尕尔勤等近十个铜金矿床（点），是我国最大的铜矿床，也是近年来我国发现的首个世界级超大型铜矿，被称为我国铜矿中的巨无霸，具有巨大的找矿潜力，被国家确定为矿产资源战略储备基地（Chen Hongqi et al.，2015）。多龙矿集区位于羌塘国家级自然保护区外围实验区的边缘，按照国家规定，实验区可以进入从事科学试验等活动，还包括一定范围的生产活动（Zhao Yuanyi et al.，2016）。多龙矿床目前尚未进行开发，本文以多不杂矿床为例，通过单因子污染指数、内梅罗综合污染指数法及重金属形态分析来研究矿区内地表土壤重金属的环境特征，并结合统计学方法进行重金属来源解析，提出环境问题防治建议，为将来矿床的绿色开发提供参考。

1 研究区概况

多龙矿集区位于西藏阿里地区改则县物玛乡境内，距县城北西方向约 100km，西藏自治区区道 301

* 本文发表在：地质学报，2019，93（11）：2985-2996

线，即安狮公路（安多至狮泉河）途径改则县。此外，从改则县城到新疆民丰正在修建柏油路面（216国道），正好穿过矿集区，交通较为便利，矿集区地理坐标范围为：83°23′00″E～83°27′00″E；32°47′00″N～32°50′00″N。多不杂矿床位于多龙矿集区的西部，海拔在 5km 左右，寒冷干旱，日照时间长，昼夜温差大，年降水量约 190mm，为高原亚寒带干旱季风气候。矿区内蒸发量大，沟谷常年无水或仅有季节性间歇溪流，生态环境十分脆弱[图 1（a）]，动植物生存条件差，环境恶劣，区内人口、植被稀少（Wang Song et al.，2016）[图 1（b）]，主要生长小薹草、小蒿草、白蒿、矮火绒草等矮小植物（Zhao Yuanyi et al.，2017）。

图 1　多不杂野外及镜下照片

Fig. 1　Field and mirror photos.

（a）多不杂地貌；（b）地表样品采集；（c）地表氧化带；（d）地表孔雀石化砂岩；（e）地表蓝铜矿化；（f）地表褐铁矿化砂岩；（g）黄铜矿（Ccp）与黄铁矿（Py）共生；（h）蓝辉铜矿（Dg）交代黄铜矿（Ccp）

(a) Duobuza's landform; (b) surface sample collection; (c) surface oxidation zone; (d) surface peacock petrified sandstone; (e) surface azurite; (f) surface limonite sandstone; (g) chalcopyrite (Ccp) symbiosis with pyrite (Py); (h) digenite (Dg) is metasomatic with chalcopyrite (Ccp)

多不杂矿床位于班公湖—怒江成矿带内，介于羌塘地块与冈底斯地块之间[图 2（a）]（Qiao Donghai et al.，2017）。矿区出露的地层较为简单（图 2），有下侏罗统曲色组一段（J_1q^1）、下侏罗统曲色组二段（J_1q^2）、下白垩统美日切组（K_1m）、渐新统康托组（E_3k）和第四系（Q）（Zhu Xiangping et al.，2012）。其中下侏罗统曲色组一段（J_1q^1）主要分布在矿区西部，约占矿区面积的 1/12，岩性为玄武岩；下侏罗统曲色组二段（J_1q^2），约占矿区面积的 1/3，岩性为浅绿灰色—浅黄褐色、薄—中厚层状变长石石英砂岩；下白垩统美日切组（K_1m），约占矿区面积的 1/12，呈不规则瘤状分布，岩性为紫红色安山质火山碎屑岩和安山玢岩；渐新统康托组（E_3k），约占矿区面积的 1/3，岩性为紫红色砂砾岩。第四系（Q）主要为坡积物及砂砾堆积层，主要分布在北部沟谷地区（Li Yubin et al.，2012；Ning Mohuan，2012；Zhu Xiangping et al.，2013）。多不杂矿区内岩浆活动强烈，侵入岩主要为花岗闪长斑岩、石英闪长玢岩和石英闪长岩等（Sun Jia et al.，

2017）。其中花岗闪长斑岩是主要的含矿岩石，主要金属矿物有黄铜矿、黄铁矿、褐铁矿、斑铜矿、磁铁矿、闪锌矿、蓝辉铜矿、孔雀石[图 1（c）、（d）、（e）、（f）、（g）、（h）]等；矿区内部分矿体已出露地表，地表孔雀石化较为严重[图 1（c）、（d）]，出现孔雀石化区域约长 300m，宽 50m，面积 15000m^2。矿区主要发育有近东西向和北东向两组断裂，东西向断裂早期为斑岩体的侵入提供通道，后期切割矿体。矿体整体呈厚板状，近北西向展布，向南倾斜，控制矿体长 1500m，延伸 200～600m，矿体厚度 200～500m。斑岩岩体中主要分布细脉状黄铜矿，矿体 Cu 品位为 0.27%～0.73%，平均品位为 0.52%（Chen Hongqi et al.，2015），铜金属资源总量 234.44 万吨（西藏自治区地质矿产勘查开发局第五地质大队，2011①）。矿体上部为细脉浸染状矿石，向深部逐渐过渡为稀疏浸染状矿石，铜含量相应降低（Li Jinxiang et al.，2008）。多不杂矿区矿体埋藏较浅，部分已出露地表，出露面积为 0.39km^2。斑岩体有黑云母化、钾长石化、硅化、绿泥石化、绢云母化等蚀变现象，且蚀变具有一定的分带性（Ning Mohuan，2012）。

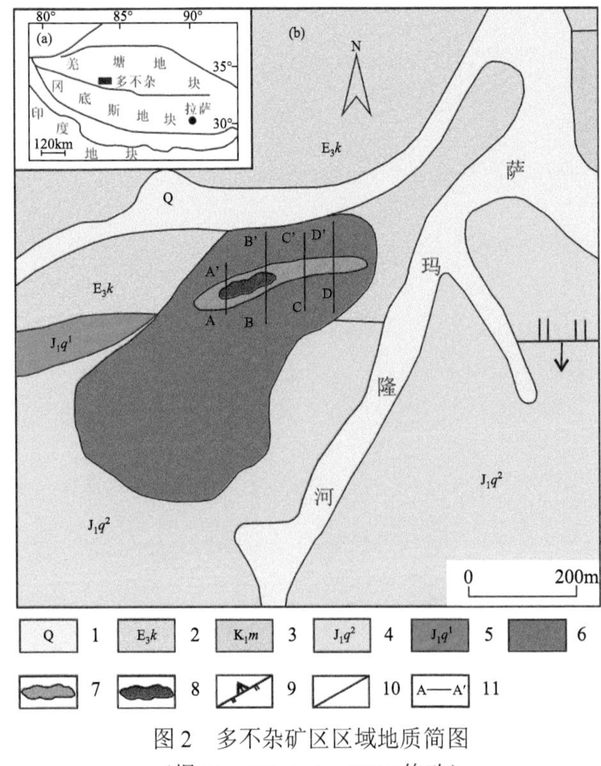

图 2　多不杂矿区区域地质简图

（据 Yang Yi et al.，2015 修改）

Fig. 2　Regional geological map of Duobuza mining area. (modified by Yang Yi et al., 2015)

（a）多不杂矿区大地构造位置图；（b）多不杂矿区地表土壤采样剖面位置；1-第四系（Q）；2-渐新统康托组（E_3k）；3-下白垩统美日切错组（K_1m）；4-下侏罗统曲色组二段（J_1q^2）；5-下侏罗统曲色组一段（J_1q^1）；6-蚀变和矿化区；7-孔雀石化；8-蓝铜矿化；9-逆断层；10-地层界限；11-采样剖面及编号

(a) Geotectonics map of in Duobuza mining area; (b) Location of surface soil sampling profile in Duobuza mining area; 1-Quaternary (Q); 2-Oligocene Kangtuo Formation (E_3k); 3-Lower Cretaceous Meiriqie Formation (K_1m); 4-Second Member of Lower Jurassic Quse Formation (J_1q^2); 5-First Member of Lower Jurassic Quse Formation (J_1q^1); 6-alteration and mineralization area; 7-peacock petrifaction; 8-azurite; 9-thrust fault; 10-stratigraphic boundary; 11-sampling profile and number

2　研究方法

2.1　样品的采集

本次工作在多不杂矿区地表孔雀石化分布区及邻区选取四条剖面（A-A′、B-B′、C-C′、D-D′），剖面间距 80～100m，这四条剖面既位于矿化区又穿过孔雀石化，用以控制矿区孔雀石化的范围。A-A′、B-B′剖面间可见少量蓝铜矿化[图 1（e）]，蓝铜矿化区域长约 110m，宽约 25m，面积约为 275m^2。在四条剖面上采集了土壤样品 37 件，其中 A-A′剖面长 100m，采集土壤样品 6 件；B-B′剖面长 200m，采集土壤样品 13 件；C-C′剖面长 150m，采集土壤样品 7 件；D-D′剖面长 200m，采集土壤样品 11 件。采样深度 0～20cm，采样位置见图 2（b），对其中 10 件样品进行重金属元素的形态测试。

2.2　样品测试方法

土壤样品测试由自然资源部合肥矿产资源监督检测中心完成，依据 DZ/T 0258—2014、DZ/T 0130.2—2006 测定土壤中重金属元素含量及 pH 值；测试仪器为原子荧光光谱仪、等离子体质谱仪、等离子体光谱仪等。依据 DD 2005-03、DZ/T 0130.2—2006 测定土壤中重金属元素形态；测试仪器为原子荧光光谱仪、等离子体质谱仪、等离子体光谱仪等。

① 西藏自治区地质矿产勘查开发局第五地质大队.西藏自治区改则县多不杂矿区铜矿普查报告.

2.3 研究方法

2.3.1 单因子污染指数和内梅罗综合污染指数评价

本文采用单因子污染指数和内梅罗综合污染指数法来研究多不杂矿区地表土壤重金属的含量特征。单因子污染指数法是将实测的土壤各参数值与国家标准值相比较，从而判定土壤污染级别；内梅罗综合污染指数法能够反映多种污染物对环境产生的综合影响程度，两种方法在评价土壤环境质量中都较为常用，评价标准采用 GB 15618—1995 三类土壤标准（表 1）。两种方法的评价标准见表 2、表 3，计算公式如下：

单污染指数评价法的公式为 $P_i = C_i/S_i$（State Environmental Protection Bureau，1995）。

表 1　土壤环境质量标准值（国家环境保护局，1995）（mg/kg）

Table 1　Standard values of soil environmental quality（State Environmental Protection Bureau, 1995）(mg/kg)

项目	土壤 pH 值				
	一级	二级			三级
	自然背景	<6.5	6.5～7.5	>7.5	>6.5
铜	≤35	50	100	100	400
砷	≤15	40	30	25	40
铬	≤90	150	200	250	300
铅	≤35	250	300	350	500
锌	≤100	200	250	300	500
镉	≤0.2	0.3	0.3	0.6	1
汞	≤0.15	0.3	0.5	1	1.5

表 2　土壤单污染指数评价划分标准（国家环境保护局，1995）（mg/kg）

Table 2　The criteria for evaluation and classification of soil single pollution index（State Environmental Protection Bureau, 1995）(mg/kg)

级别	无污染	轻度污染	中度污染	重度污染
P_i	P_i≤1	1<P_i≤5	5<P_i≤10	P_i>10

表 3　土壤质量级别划分标准（国家环境保护局，1995）（mg/kg）

Table 3　The criteria for classification of soil quality levels（State Environmental Protection Bureau, 1995）(mg/kg)

级别	优良	良好	较好	较差	极差
$P_{综}$	<0.7	0.7～（<1.0）	1.0～（<2.0）	2.0～（<3.0）	>3.0

P_i 为单污染评价指数，C_i 为第 i 种评价指数在环境中的观测值，S_i 为第 i 种评价指数的评价标准值，P_i 值越大就表示该参数在土壤环境中的污染程度越重（State Environmental Protection Bureau，1995）。

综合污染指数评价法公式为

$$P_{综}=\sqrt{\frac{\overline{P_i^2}+P_{i\max}^2}{2}} \tag{1}$$

式中：$\overline{P_i}$ 为所有单项污染指数的平均值，$P_{i\max}$ 为测得的各单项污染指数中的最大值（State Environmental Protection Bureau，1995）。

2.3.2 重金属元素形态分析

计算重金属元素各形态百分比，其中水溶态、离子交换态与碳酸盐态的重金属呈不稳定态，腐殖酸态、铁锰氧化态、强有机态为次稳定态，残渣态为稳定态（Gao Liancun et al.，1994）。在一定条件下呈次稳定态的重金属元素会向不稳定态转化（Gao Liancun et al.，1994；Zhang Jingru et al.，2017），将可能发生转化的这部分形态归入潜在可利用态。相关性强的形态间更容易发生转化，根据各元素形态间的相关性统计其潜在可利用态含量。累加各元素不稳定态（水溶态、离子交换态、碳酸盐态）与潜在可利用态的含量作为该元素的生物有效态比例，以此分析多不杂地表土壤重金属的生物有效性。

2.3.3 数据的统计与分析

本文采用 SPSS（22.0）统计软件对数据进行处理，对多不杂矿区地表土壤重金属的含量及 pH 值、重金属形态进行统计和聚类分析。

3 土壤重金属元素环境特征

3.1 重金属元素含量

通过对多不杂矿区地表 37 件土壤样品的重金属含量和 pH 值进行统计分析（表 4），可以得到：多不杂地表土壤的 pH 值在 8～9 之间，为碱性土壤，对矿区土壤重金属离子溶出的影响较小；从平均值来看，Cu、Pb、Zn、As、Cd、Cr、Hg 平均含量分别为 6475.69mg/kg、114.34mg/kg、313.01mg/kg、22.59mg/kg、0.54mg/kg、102.59mg/kg、0.04mg/kg，仅有 Cu 元素的平均含量超过国家Ⅲ级标准，其他元素的平均值均在正常范围内，Cu 元素平均含量为标准值的 16.19 倍，呈现出明显的异常；从最大值来看，仅有 Cu、Zn、Cd 三种元素的最大值超过了国家Ⅲ级标准，分别为标准值的 49.75 倍、1.54 倍、1.87 倍；从最小值来看，Cu 元素的最小值为标准值的 3.69 倍；从变异系数来看，7 种重金属元素变异系数由大到小的顺序为 Pb＞Cd＞Cu＞Zn＞Hg＞As＞Cr，即多不杂地表土壤中的 Pb 元素含量变化最不稳定，Cr 元素含量变化最稳定；从样品的超标情况来看，37 件样品中仅有 Cu、Zn、Cd 三种重金属被检测出超标，分别有 37 件、7 件、4 件样品超标，超标率分别为 100%、18.92%、10.81%。因此综合上述几项指标，地表土壤中含量异常的元素为 Cu、Zn、Cd 三种元素，其中 Cu 元素异常最为明显。

表 4 重金属含量统计表

Table 4 Heavy metal content statistics table

统计指标	Cu	Pb	Zn	As	Cd	Cr	Hg	pH
最大值（mg/kg）	19900.50	460.40	771.20	35.90	1.87	133.60	0.14	8.79
最小值（mg/kg）	1478.10	29.60	111.40	4.60	0.24	71.00	0.02	8.05
平均值（mg/kg）	6475.69	114.34	313.01	22.59	0.54	102.59	0.04	8.54
方差	16495681.97	8560.39	36384.76	59.75	0.14	200.46	0.00	0.03
标准差（mg/kg）	4061.49	92.52	190.75	7.73	0.38	14.16	0.02	0.18
变异系数（%）	62.72	80.92	60.94	34.22	70.48	13.80	58.83	2.08
Ⅲ级标准（mg/kg）	400	500	500	40	1	300	1.5	/
样品超标数（件）	37	0	7	0	4	0	0	/
样品超标率（%）	100	0	18.92	0	10.81	0	0	/

3.2 重金属元素污染指数

根据污染指数计算公式，计算出多不杂矿区地表 37 件土壤样品的单污染指数和综合污染指数并进行统计（表 5）。结果表明，Cu、Pb、Zn、As、Cd、Cr、Hg 的单污染指数分别在 3.695～49.751、0.059～0.921、0.223～1.542、0.115～0.898、0.235～1.867、0.237～0.445、0.013～0.091 之间。单污染指数平均值分别为 16.19、0.23、0.63、0.56、0.54、0.34、0.02，平均值由大到小的顺序为 Cu＞Zn＞As＞Cd＞Cr＞Pb＞Hg，

单污染指数代表重金属含量的超标倍数，单污染指数越大，即该重金属元素超标越严重。综合污染指数在2.668～35.576之间变化，平均值为11.60，甚至远远大于土壤质量为极差的标准值3，显示出研究区地表土壤受重金属影响程度很高。

表5 重金属污染指数统计表

Table 5 Heavy metal pollution index statistics table

参数	元素	最大值	最小值	平均值
P_i	Cu	49.751	3.695	16.19
	Pb	0.921	0.059	0.23
	Zn	1.542	0.223	0.63
	As	0.898	0.115	0.56
	Cd	1.867	0.235	0.54
	Cr	0.445	0.237	0.34
	Hg	0.091	0.013	0.02
$P_{综}$	/	35.576	2.668	11.60

3.3 重金属元素形态

多不杂矿区地表土壤中的Cu主要以铁锰氧化态和残渣态存在，分别占48.40%和14.73%，其他形态也占有一定比例；Pb、Zn、As、Cr主要以残渣态存在，所占比例分别为90.27%、71.48%、79.59%、86.12%；Cd主要以铁锰氧化态和碳酸盐态存在，两种形态分别占40.05%和22.42%；Hg主要以残渣态和腐殖酸态存在，分别占40.16%和22.54%（表6，表7，图3）。

表6 重金属元素形态统计结果（$\times10^{-6}$）

Table 6 Speciation statistic results of heavy metals（$\times10^{-6}$）

元素	水溶态	离子交换态	碳酸盐态	腐殖酸态	铁锰氧化态	强有机态	残渣态
Cu	34.1	1.7	839.7	828.4	3043.7	615.2	926.4
Pb	0.12	0.15	0.51	0.63	6.2	0.2	72.45
Zn	0.5	0.1	5.6	5.7	29.8	25.3	167.9
As	0.022	0.021	0.027	1.664	1.532	0.029	12.851
Cd	0.005	0.029	0.089	0.013	0.159	0.052	0.05
Cr	0.05	0.07	0.12	6.59	1.37	2.2	64.51
Hg	0.8	0.9	0.9	5.5	2.3	4.2	9.8

表7 重金属元素形态所占比例统计结果（%）

Table 7 The results of statistics on the proportion of heavy metal species（%）

元素	水溶态	离子交换态	碳酸盐态	腐殖酸态	铁锰氧化态	强有机态	残渣态
Cu	0.54	0.03	13.35	13.17	48.40	9.78	14.73
Pb	0.15	0.19	0.64	0.78	7.72	0.25	90.27
Zn	0.21	0.04	2.38	2.43	12.69	10.77	71.48
As	0.14	0.13	0.17	10.31	9.49	0.18	79.59
Cd	1.26	7.3	22.42	3.27	40.05	13.1	12.59
Cr	0.07	0.09	0.16	8.80	1.83	2.94	86.12
Hg	3.28	3.69	3.69	22.54	9.43	17.21	40.16

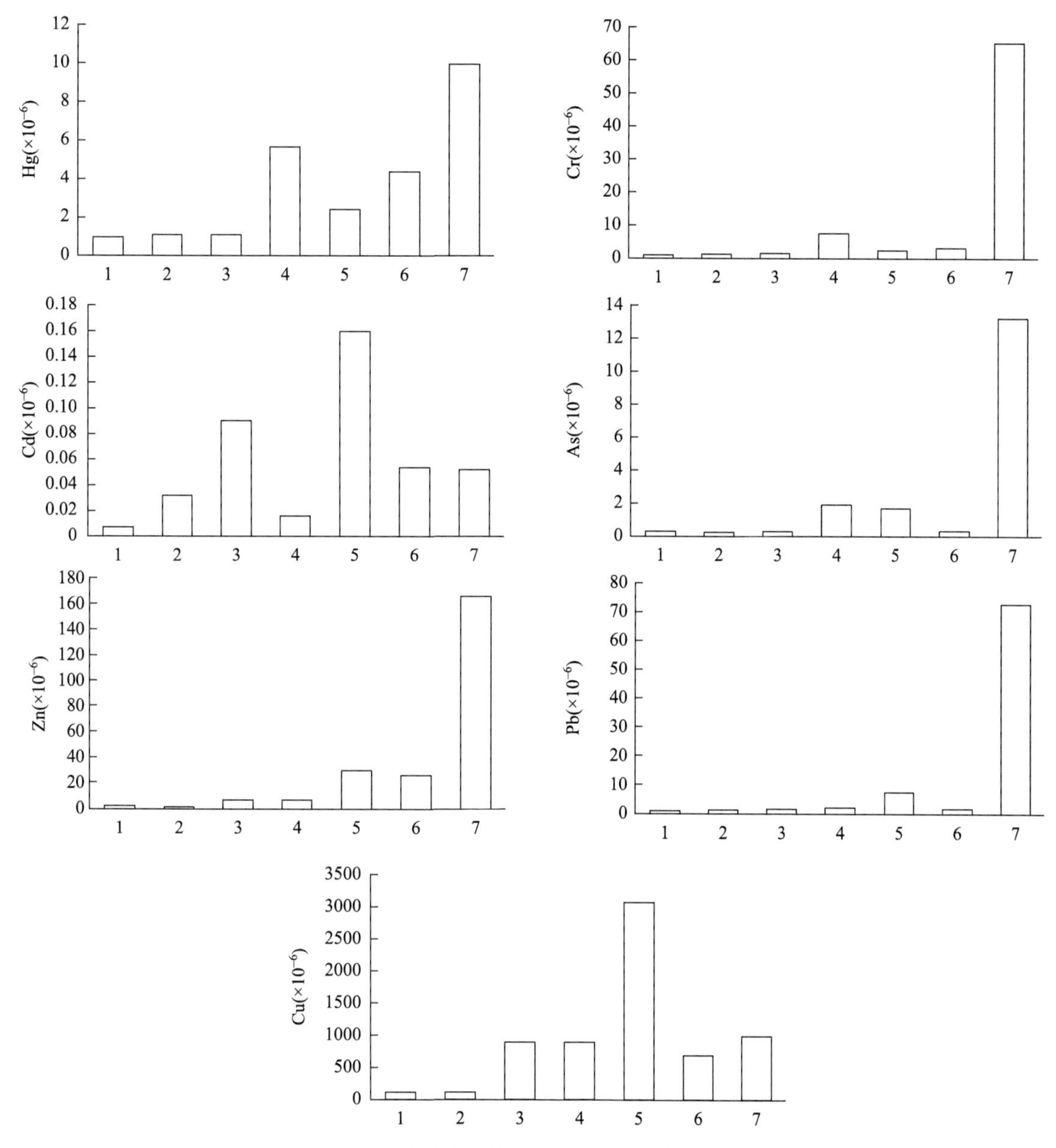

图 3 多不杂矿床地表土壤中 7 种重金属元素形态分布情况

Fig. 3 Speciation distribution of 7 heavy metal elements in surface soil of Duobuza deposit.

1-水溶态；2-离子交换态；3-碳酸盐态；4-腐殖酸态；5-铁锰氧化态；6-强有机态；7-残渣态

1-Water-dissolved state; 2-ion exchangeable state; 3-carbonate state; 4-humic acid state; 5-Fe-Mn oxidation state; 6-strong organic state; 7-residual state

4 讨论

4.1 重金属来源

源解析是预防和减少重金属污染的重要手段（Huang Ying et al.，2018）。目前所说的源解析包括两个层面，一是源识别，即定性地判断出污染源类型；二就是源解析，是在源识别的基础上定量地计算出各类污染源的贡献率大小（Li Jiao et al.，2018）。源解析的方法有很多，如利用地理信息系统（GIS）、主成分分析（PCA）、聚类分析（CA）和因子分析（FA）等来识别主要污染源（Francourí a et al.，2009；Li Xiaoping et al.，2010；Lin Yan et al.，2018）。聚类分析可以反映重金属元素之间的密切程度，聚类距离越小代表重金属元素之间的密切程度越高（Liu Qingping et al.，2018）。

根据重金属含量的聚类[图 4（a）]与统计结果（表 4），本文将八个变量因子分为两大类 4 组：①Cu、Zn、Cd、Hg；②Pb、As；③Cr；④pH。根据研究区的地质条件，结合上述统计数据，对多不杂矿区地表土壤重金属来源进行解析，具体分析如下。

4.1.1 Cu、Zn、Cd、Hg 的来源

聚类距离为 15 时，Cu、Zn、Cd、Cr、Hg 这五个元素聚在一组，由于 Cr 元素的变异系数与其他元素相差较大，以下单独列为一组。Cu、Zn、Cd、Hg 出现在一个分组中反映 Cu、Zn、Cd、Hg 元素之间具有相近的地球化学行为，容易一起形成共生矿物，Cd 元素具有很强的亲硫性，容易存在于硫化物矿物中，与 Cu、Zn、Hg 等元素伴生。单污染指数评价结果，地表剖面 37 件土壤样品中 Cu、Zn、Cd 元素被检测出超标（表 4），也佐证了这一观点，Cu、Zn、Cd 之间的富集具有很强的相关性。根据地质资料，多不杂矿床矿石中主要金属矿物为黄铜矿、黄铁矿[图 1（g）]、褐铁矿、斑铜矿、磁铁矿、闪锌矿、蓝辉铜矿[图 1（h）]、孔雀石等；赋存在深部的矿石矿物出露地表被氧化为孔雀石、蓝铜矿、褐铁矿等[图 1（c）、（d）、（e）、（f）]，地表孔雀石化较为严重[图 1（c）、（d）]。因此认为 Cu、Zn、Cd 和 Hg 一起赋存于硫化物矿物中，出露地表被氧化后经雨水冲刷，淋滤进地表土壤中。

4.1.2 Pb、As 的来源

Pb、As 出现在一个分组，表明 Pb、As 元素之间有较好的相关性，在矿物的形成过程中容易伴生在一起，都具有亲硫型，常出现在硫化物矿物中，同时这两个元素反映在一个分组中，说明多不杂矿床地表土壤中的 Pb、As 元素受到外界的干扰很小，保留着矿物中的存在形态。亲硫元素常常以类质同象的形式赋存在黄铁矿中（Qiao Donghai et al.，2017），黄铁矿出露地表被氧化过程中结构遭到破坏，Pb、As 元素被释放出来。由于这两种元素类质同象和迁移能力存在差异，因此它们在地表土壤中富集的程度也有一定差异，但这两种元素在研究区并不存在含量异常，Pb、As 的平均含量分别为 114.34mg/kg、22.59mg/kg 均不超过国家III级标准，且所有样品中最大值也低于标准值（表 4）。

4.1.3 Cr 的来源

Cr 元素的单污染指数介于 0.237～0.445 之间，平均值为 0.34（表 5），所有样品中 Cr 元素均不存在超标现象。Cr 的变异系数为 13.80%，变异系数远小于其他六种重金属元素（表 4），变化较小，说明并不存在使得 Cr 元素异常的特殊地质体。多不杂矿区尚未开发，生存条件差，无人居住，因此 Cr 主要是自然来源而非人为来源（Liu Qingping et al.，2018），故推测其来源是岩石的自然风化。

4.2 生物有效性

重金属的生物有效性是指生物对重金属吸收利用的程度。土壤中的重金属较难迁移，残留时间长、隐蔽性强、毒性较大，进入水、大气后会威胁人类的健康（周启星等，2001）。单从含量上来看，多不杂矿区地表土壤中的 Cu 元素平均含量为标准值的 16.19 倍，呈现出了明显的异常（表 4），对周边环境造成了严重的影响，地表明显可见大片孔雀石化和蓝铜矿化区域[图 1（c）、（d）、（e）、图 2（b）]。除此之外，重金属问题还有更深层次的潜在影响，直接影响人体健康。长期以来人们都认为土壤中的重金属含量越高，潜在的环境危害也就越大，但近年来的研究表明，重金属的毒性和它的含量之间没有很好的相关性，重金属在土壤中的存在形态和各种形态的比例才是决定它对环境及周围生态系统造成影响的关键（Davison et al.，1994；Quevauviller，1998；Peijnenburg et al.，2003；Li Zhibo et al.，2006；Liang Yanqiu et al.，2006；He Yuan et al.，2007；Cong Yuan et al.，2009；Violante et al.，2010；Zhang Chaoyang et al.，2012）。重金属元素的形态是指重金属元素在环境中的实际存在形式，不同的形态往往表现出不同的毒性和环境效应（Han Chunmei et al.，2005；He Zhongfa et al.，2009）。重金属的环境效应主要体现在重金属经水、土、植物迁移转化，最后对人体造成的影响（Chu Na et al.，2008）。水溶态、离子交换态与碳酸盐态的重金属在土壤中易于迁移转化，容易被植物吸收，对人类和环境的危害最大（Cong Yuan et al.，2009）。碳酸盐态在改进的 BCR 法中也被称为酸溶态重金属，对土壤的 pH 值反映比较敏感，在低 pH 值土壤条件下，形态容易发生

转化，释放出结合的重金属元素（Rauret et al.，1999）。强有机态很稳定，除非遭受微生物分解，才可释放出重金属元素（Cong Yuan et al.，2009）；残渣态又叫硅酸盐态，来源于土壤矿物，性质稳定，不易被植物吸收，在土壤中能够长时间保留（Chu Na et al.，2007；Wang Xiaoliang et al.，2013）；土壤中的铁锰氧化态、腐殖酸态也比较稳定，在氧化还原条件改变时才转变和释放（Cong Yuan et al.，2009）。

多不杂矿床目前尚未开发，矿区周边的经济方式以畜牧业为主。出露地表的矿石氧化淋滤出的重金属元素进入地表土壤中，由植物根系富集进入植物内，牛羊啃食后进一步富集进入牛羊体内，最终被人食用进入人体。其迁移路径为：矿石→水→土壤→植物→牛羊→人体。土壤中的重金属进入人体主要有三种途径：经口食入、皮肤接触和呼吸吸入，由农作物进入人体即为经口食入（Xu Youning et al.，2014），另外两种方式摄入的重金属数量很少，通过食物链进入人体是主要途径（Li Feng et al.，2008）。因此，重金属在迁移过程中被生物吸收利用的程度决定了人体重金属摄入量大小。土壤中重金属能够被吸收利用的形态所占的比例决定了它的生物有效性，重金属在自然界迁移时被生物吸收利用的部分即为生物有效态（Tao Wenjing et al.，2014）。水溶态、离子交换态、碳酸盐态的重金属元素极不稳定，容易迁移转化，被动植物吸收利用；残渣态极为稳定，可以不予考虑；腐殖酸态、铁锰氧化态、强有机态属于潜在可利用态，在一定条件下会转化为水溶态、离子交换态、碳酸盐态。

对土壤重金属形态进行聚类分析得到，Cu 元素碳酸盐态、腐殖酸态、水溶态、铁锰氧化态和强有机态之间具有较好的相关性[图 4（b）]，在一定条件下腐殖酸态、铁锰氧化态和强有机态能转化为碳酸盐态和水溶态；Pb 元素碳酸盐态和腐殖酸态之间，水溶态、离子交换态和强有机态之间相关性较好[图 4（c）]，在一定条件下腐殖酸态能转化为碳酸盐态，强有机态能转化为水溶态和离子交换态；Zn 元素碳酸盐态和铁锰氧化态之间，强有机态和水溶态之间相关性较好[图 4（d）]，在一定条件下铁锰氧化态能转化为碳酸盐态，强有机态能转化为水溶态；As 元素离子交换态、腐殖酸态和水溶态间相关性较好[图 4（e）]，在一定条件下腐殖酸态能转化为离子交换态和水溶态；Cd 元素碳酸盐态、铁锰氧化态和水溶态之间，离子交换态和强有机态之间相关性较好[图 4（f）]，在一定条件下铁锰氧化态能转化为碳酸盐态和水溶态，强有机态能转化为离子交换态；Cr 元素水溶态、离子交换态、铁锰氧化态和碳酸盐态间相关性较好[图 4（g）]，在一定条件下铁锰氧化态能转化为水溶态、离子交换态和碳酸盐态；Hg 元素水溶态、离子交换态、碳酸盐态、腐殖酸态、铁锰氧化态和强有机态间相关性较好[图 4（h）]，在一定条件下腐殖酸态、铁锰氧化态和强有机态能转化为水溶态、离子交换态和碳酸盐态。

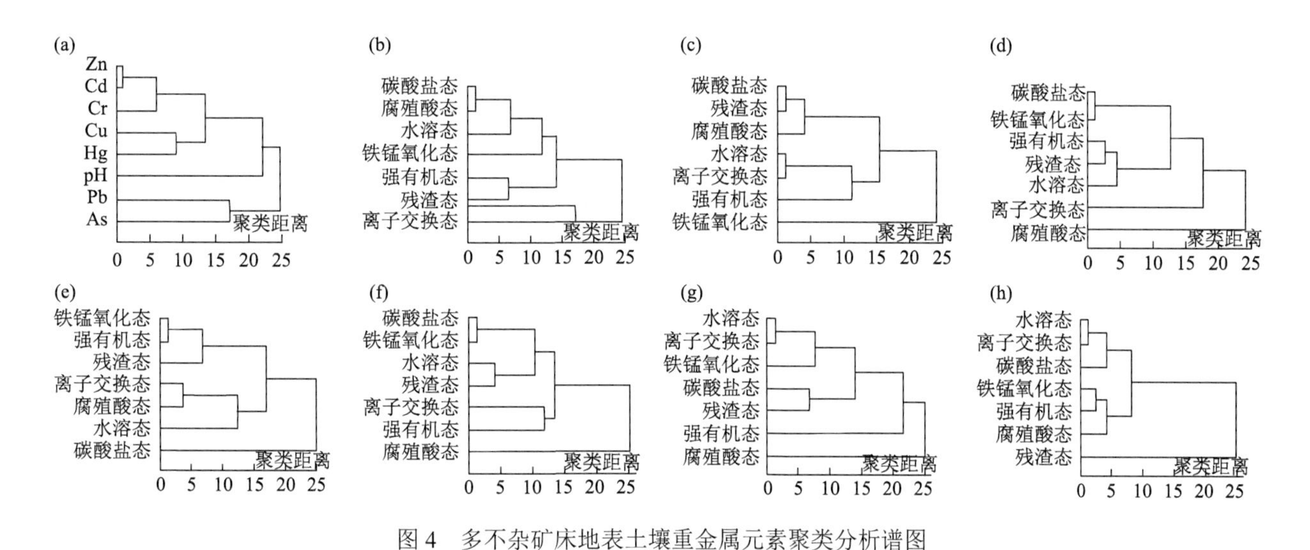

图 4 多不杂矿床地表土壤重金属元素聚类分析谱图

Fig. 4 Spectrogram of cluster analysis of heavy metal elements in surface soil of Duobuza deposit.

（a）地表剖面土壤重金属含量、pH 值聚类分析；（b）Cu 元素形态聚类分析谱图；（c）Pb 元素形态聚类分析谱图；（d）Zn 元素形态聚类分析谱图；（e）As 元素形态聚类分析谱图；（f）Cd 元素形态聚类分析谱图；（g）Cr 元素形态聚类分析谱图；（h）Hg 元素形态聚类分析谱图

(a) Soil heavy metal content, pH value cluster analysis; (b) Cu element morphological cluster analysis spectrum diagram; (c) Pb element morphological cluster analysis spectrum diagram; (d) Zn element morphological cluster analysis spectrum diagram; (e) As element morphological cluster analysis spectrum diagram; (f) Cd element morphological cluster analysis spectrum diagram; (g) Cr element morphological cluster analysis spectrum diagram; (h) Hg element morphology cluster analysis spectrum diagram

根据各元素形态间的相关性讨论各元素的潜在可利用态情况，由此计算出 7 种重金属的生物有效态比例（表 8），得出 Cu、Pb、Zn、As、Cd、Cr、Hg 生物有效态的比例分别为 85.27%、2.01%、26.09%、10.75%、84.13%、2.15%、59.84%。可知多不杂矿区地表土壤重金属元素最易迁移的顺序为 Cu＞Cd＞Hg＞Zn＞As＞Cr＞Pb，表明 Cu、Cd、Hg、Zn 的生物有效性较强，更容易被富集在植物体内，再通过食物链最终影响人体健康。重金属元素进入人体后难以排出，会在人体内积聚，沉积到器官或细胞内，轻者干扰器官及细胞的代谢，重则直接对机体产生毒性，导致死亡（Niu Lihua et al.，2009），另外，更有研究表明，重金属会积聚在人发中，靠近矿区居住的村民头发中的 Cu 含量明显高于远离矿区居住的村民（Ni Shanqin et al.，2012）。鉴于多不杂矿区地表土壤重金属异常对人类的生活可能产生一系列影响，有关部门必须给予足够的重视。

表 8　重金属元素生物有效态统计结果（%）

Table 8　The statistical results of bioavailability of heavy metal elements（%）

元素	水溶态	离子交换态	碳酸盐态	腐殖酸态	铁锰氧化态	强有机态	有效态
Cu	0.54	0.03	13.35	13.17	48.40	9.78	85.27
Pb	0.15	0.19	0.64	0.78	—	0.25	2.01
Zn	0.21	0.04	2.38	—	12.69	10.77	26.09
As	0.14	0.13	0.17	10.31	—	—	10.75
Cd	1.26	7.3	22.42	—	40.05	13.1	84.13
Cr	0.07	0.09	0.16	—	1.83	—	2.15
Hg	3.28	3.69	3.69	22.54	9.43	17.21	59.84

注：表格仅讨论了相关性强（聚类距离＜15）的次稳定态向不稳定态转化的情况，对相关性较弱（聚类距离＞15）的转化情况暂不予考虑，以“—”表示。

4.3　重金属问题防治建议

4.3.1　源头控制

重金属在任何环境下都不会降解，进入环境后，便成为污染物，在食物链的作用下进入人体，极少量就会威胁到人体健康（Luo Yuhu，2017）。重金属问题的防治包括源头预防和过程治理，污染预防远比污染治理重要得多，应以源头控制为主，杜绝重金属进入水体、土壤，有效降低重金属污染风险（Luo Yuhu，2017）。由本文重金属的来源可以看出，Cu、Zn、Cd、Hg 源于黄铜矿的氧化淋滤；Pb、As 源于黄铁矿的氧化；Cr 源于岩石的自然风化。因此，为了降低重金属对环境的影响，有效的方法是阻止黄铜矿与黄铁矿的氧化以及由于岩石的风化而释放重金属。要做到这一点，就需要加大草原保护力度，使草原尽量覆盖岩石和矿石，避免岩石和矿石裸露地表。这样，即可达到“源头控制”的目的。

4.3.2　严禁放牧

矿区地表可见大片孔雀石化和蓝铜矿化区域[图 1（c）、（d）、（e）、图 2]，说明此区域已受到重金属的影响，建议立即对这些地区修建围栏，严禁在此片区域放牧，避免牛羊因啃食了影响区的草料，使重金属通过食物链进入人体。尽可能地将矿区重金属的影响控制在对生态影响不到的区域。

4.3.3　改良土壤

对土壤重金属环境问题进行治理的途径主要有两种，一是直接从土壤中除去重金属；二是改变土壤的理化性质，抑制重金属向不稳定态转化，从而降低重金属的生物有效性（Zhong Zheng et al.，2015）。基于多不杂矿区土壤重金属的形态特征（表 7），矿区土壤重金属形态中 Cu 和 Cd 的碳酸盐态比例分别为 13.35% 和 22.42%，而碳酸盐态的重金属容易在酸性环境中释放至环境。因此建议在主要的牧草分布区土壤中，适当施加石灰，使重金属处于稳定状态（Shi Lei et al.，2018）。

5 结论

（1）多不杂矿区地表为碱性土壤，7 种重金属元素中，Cu 在所有样品中的含量均超过 GB 15618—1995III 级标准，最小值为标准值的 3.69 倍，平均含量为标准值的 16.19 倍，最大值为标准值的 49.75 倍；Zn、Cd 仅有最大值超过III级标准，分别是标准值的 1.54 倍、1.87 倍；Pb、As、Cr、Hg 含量均不超标。单因子污染指数平均值大小 Cu＞Zn＞As＞Cd＞Cr＞Pb＞Hg，土壤样品内梅罗综合指数平均值为 11.60，对环境的综合影响程度较高。地表土壤中的 Cu 主要以铁锰氧化态和残渣态存在，分别占 48.40%和 14.73%，Pb、Zn、As、Cr 主要以残渣态形式存在，分别占 90.27%、71.48%、79.59%、86.12%。Cd 主要以铁锰氧化态和碳酸盐态形式存在，分别占 40.05%和 22.42%，Hg 主要以残渣态和腐殖酸态存在，分别占 40.16%和 22.54%。

（2）土壤中的 7 种重金属元素来源主要有 3 类：①Cu、Zn、Cd、Hg 源于黄铜矿的氧化淋滤；②Pb、As 源于黄铁矿的氧化；③Cr 源于岩石的自然风化。矿区土壤重金属迁移顺序为 Cu＞Cd＞Hg＞Zn＞As＞Cr＞Pb，Cu、Cd、Hg、Zn 更容易被富集在植物体内，通过食物链进入人体。

（3）针对多不杂矿区土壤重金属相关的环境问题，提出了源头控制、严禁放牧、改良土壤共三方面的防治措施。

致谢

在野外工作中，西藏自治区自然资源厅、林业厅，阿里地区国土资源局、林业局、环保局，改则县国土资源局、林业局、环保局、生产安全局，改则县物玛乡人民政府，物玛乡萨玛隆村、本松村和扎多村委会，中铝矿产资源有限公司，西藏金龙矿业股份有限公司共 15 家单位提供了大力的支持；在样品的测试过程中合肥矿产资源监督检测中心也给予了极大帮助；论文写作过程中得到了福州大学的王力圆讲师、王少怀教授的指导，在此一并致以诚挚的谢意。

References

Chen Hongqi, Qu Xiaoming, Fan Shufang. 2015. Geological characteristics and metallogenic prospecting model of Duolong Porphyry copper gold ore concentration area in Gerze County, Tibet. Mineral Deposits, 34 (2): 321-332 (in Chinese with English abstract).

Chu Na, Zhao Yuanyi, Zhang Guangdi, Zhang Qin, Cai Jianhui, Xiong Qunyao, Li Dexian, Wang Jinsheng, Zhao Jinyan. 2007. Environmental trait of speciations of heavy metals in low-grade ore plot and soil of the Dawu river domain in the Dexing copper mine, Jiangxi province. Acta Geologica Sinica, 81 (5): 670-684 (in Chinese with English abstract).

Chu Na, Zhao Yuanyi, Zhang Guangdi, Yang Hui. 2008. Environmental effect of heavy metal elements in Dexing copper mine, Jiangxi province. Acta Geologica Sinica, 82 (4): 562 (in Chinese with English abstract).

Cong Yuan, Chen Yuelong, Yang Zhongfang, Hou Qingye, Hu Shengying, Guo Li. 2009. Chemical forms of heavy metals in soils and potential hazards to ecosystem in Beijing farmlands. Soils, 41 (1): 37-41 (in Chinese with English abstract).

Davison C M, Thomas R P, Mcvey S E, Perala R, Littlejohn D, Ure A M. 1994. Evaluation of a sequential extraction procedure for the speciation of heavy metals in sediments. Analytica Chimica Acta, 291 (3): 277-286.

Francouría A, Ló pezmateo C, Roca E, Ferná ndezmarcos M L. 2009. Source identification of heavy metals in pastureland by multivariate analysis in NW Spain. Journal of Hazardous Materials, 165 (1): 1008-1015.

Gao Liancun, He Guihua, Feng Suping, Wang Shuren, Cui Zhaojie. 1994. Solubilityand exchange of Cu, Pb, Zn, Cr species in simulant acid rain. Environmental Chemistry, 13 (5): 448-452 (in Chinese with English abstract).

Han Chunmei, Wang Linshan, Gong Zongqiang, Xu Huaxia. 2005. Chemical forms of soil heavy metals and their environmental significance. Chinese Journal of Ecology, 24 (12): 1499-1502 (in Chinese with English abstract).

He Yuan, Wang Xian, Chen Lidan, Zheng Shenghua, Cai Zhenzhen. 2007. Forms of heavy metals in the typical soils of Zoumadai of Quanzhou. Soils, 39 (2): 257-262 (in Chinese with English abstract).

He Zhongfa, Fang Zheng, Sun Yanwei, Li Jinzhu, Xia Chen, Wen Xiaohua, Zhang Zhuo, Liu Wenzhang, Jiang Simin. 2009. Preliminary study on chemical forms of Mercury in farmland soil. Shanghai Geology, 30 (1): 45-49 (in Chinese with English abstract).

Huang Ying, Deng Meihua, Wu Shaofu, Jan Japenga, Li Tingqiang, Yang Xiaoe, He Zhenli. 2018. A modified receptor model for source apportionment of heavy metal pollution in soil. Journal of Hazard Mater, 354: 161-169.

Li Feng, Zhang Xuexian, Dai Ruizhi. 2008. The bioavailability of heavy metal and environmental quality standard for soil. Guandong Weiliang Yuansu Kexue, 15 (1): 7-10 (in Chinese with English abstract).

Li Jiao, Wu Jin, Jiang Jinyuan, Teng Yanguo, He Lihuan, Song Liuting. 2018. Review on source apportionment of soil pollutants in recent ten years. Chinese Journal of Soil Science, 49 (1): 232-242 (in Chinese with English abstract).

Li Jinxiang, Li Guangming, Qin Kezhang, Xiao Bo. 2008. Geochemistry of porphyries and volcanic rocks and oreforming geochronology of Duobuza gold-rich porphyry copper deposit in Bangonghu belt, Tibet: Constraints on metallogenic tectonic settings. Acta Petrologica Sinica, 24 (3): 531-543 (in Chinese with English abstract).

Li Xiaoping, Feng Linna. 2010. Spatial distribution of hazardous elements in urban top soils surrounding Xi'an industrial areas, (NW, China): Controlling factors and contamination assessments. Journal of Hazardous Materials, 174 (1): 662-669.

Li Yubin, Duoji, Zhong Wanting, Li Yuchang, Qiangba Wangdui, Chen Hongqi, Liu Hongfei, Zhang Jinshu, Zhang Tianping, Xu Zhizhong, Fan Anhui, Suolang Wangqin. 2012. An exploration model of Dobuza porphyry copper-gold deposit in Gaize County, northern Tibet. Geology and Exploration, 48 (2): 274-287 (in Chinese with English abstract).

Li Zhibo, Luo Yongming, Song Jing. 2006. Study on soil environmental quality guidelines and standards II. health risk assessment of polluted soils. Acta Pedologica Sinica, 43 (1): 142-151 (in Chinese with English abstract).

Liang Yanqiu, Pan Wei, Liu Tingting, Xing Zhiqiang, Zang Shuliang. 2006. Speciation of heavy metals in soil from Zhangshi soil of Shenyang contaminated by industrial wastewater. Environmental Science and Management, 31 (2): 43-45 (in Chinese with English abstract).

Lin Yan, Ma Jin, Zhang Zhengdong, Zhu Yifang, Hou Hong, Zhao Long, Sun Zaijin, Xue Wenjuan, Shi Huading. 2018. Linkage between human population and trace elements in soils of the Pearl River Delta: Implications for source identification and riskassessment. Science of The Total Environment, 610-611: 944-950.

Liu Qingping, Deng Shiqiang, Zhao Yuanyi, Li Xiaosai. 2018. Sources analysis and element geochemical characteristics of regional soil heavy metals from the Ga'erqin ore deposit, Tibet, Acta Geoscientica Sinica, 39 (4): 481-490 (in Chinese with English abstract).

Luo Yuhu. 2017. The geochemical evolution and ecological response of natural sewage in Rongna River from the ore body, north-Tibet, China. Beijing: China University of Geosciences (Beijing) (in Chinese with English abstract).

Ni Shanqin, Li Ruiping, Wang Anjian. 2012. The distribution of heavy metal in the scalp hair of females near Dexing mine area, Jiangxi, China. Environmental Monitoring in China, 28 (2): 84-90 (in Chinese with Englishabstract).

Ning Mohuan. 2012. Geological characteristics and prospecting prediction of porphyry copper deposit in Dulong ore concentration area, Tibet. Chengdu: Chengdu University of Technology (in Chinese with English abstract).

Niu Lihua, Hu Qingdong, Li Guangwu, Wu Wenquan, Tian Yangchao. 2009. The relative property research about the neurotoxic effect of heavy metal ion copper, mercury and chrome to mice through ol-factory pathway and lavage. Jounal of Anhui Agriculture Science, 37 (36): 18384-18388 (in Chinese with English abstract).

Peijnenburg W J G M, Jager T. 2003. Monitoring approaches to assess bioaccessibility and bioavailability of metals: matrix issues. Ecotoxicology and Environment Safety, 56 (1): 63-77.

Qiao Donghai, Zhao Yuanyi, Wang Ao, LiYubin, Guo Shuo, Li Xiaosai, Wang Song. 2017. Geochronology, fluid inclusions, geochemical characteristics of Dibao Cu (Au) deposit, Duolong ore concentration area, Xizang (Tibet), and its genetic type. Acta Geologica Sinica, 91 (7): 1542-1564 (in Chinese with English abstract).

Quevauviller P. 1998. Operationally defined extraction procedures for soil and sediment analysis I. Standardization. Trends in Analytical Chemistry, 17 (5): 289-298.

Rauret G, Lopez-Sanchez J F, Sahuquillo A, Davidson C M, Ure A M, Quevauviller P H. 1999. Improvement of the BCR 3-step sequential extraction procedure prior to the certification of new sediment and soil reference materials. Journal of Environmental Monitoring, 1 (1): 57-61.

Shi Lei, Guo Zhaohui, Patch, Xiao Xiyuan, Xue Qinghua, Ran Hongzhen, Feng Wenli. 2018. Lime based amendments inhibiting uptake of cadmium in rice planted in contaminated soils. Transactions of the Chinese Society of Agricultural Engineering, 34 (11): 209-216 (in Chinese with English abstract).

State Environmental Protection Bureau. 1995. Environmental Quality Standard for Soil GB15618—1995. Beijing: China Standard Press (in Chinese).

Sun Jia, Mao Jingwen, Yao Fojun Duan Xianzhe. 2017. Relation between magmatic processes and porphyry copper-gold oreformation, the Duolong district, central Tibet. Acta Petrologica Sinica, 33 (10): 3217-3238 (in Chinese with English abstract).

Tao Wenjing, Chen Liya, Nie Quanxin, Huang Qin, Wang Jinyun. 2014. Study on analytic method of bioavailable form of heavy metalsin soil. Geology of Anhui, 24 (4): 300-303 (in Chinese with English abstract).

Violante A, Cozzolino V, Perelomov L, Caporale A G, Pigna M. 2010. Mobility and bioavailability of heavy metals and metalloids in soil environments. Journal of Soil Science and Plant Nutrition, 10 (3): 268-292.

Wang Song, Zhao Yuanyi, Li Xiaosai, Qiao Donghai. 2016. Characteristic of enverimental geology of Duobuza Cu (Au) deposit, Tibet. Geological Review, 62 (S1): 279-280 (in Chinese without English abstract).

Wang Xiaoliang, ZhaoYuanyi, Liu Jianping, Lu Lu, Yang Yongqiang, Chu Na. 2013. Characteristics and significance of Cadmium environmental geochemistry in soil of Dawu river basin in the Dexing copper orefeild. Geological Review, 59 (4): 781-788 (in Chinese with English abstract).

Xu Youning, Zhang Jianghua, Ke Hailing, Chen Huaqing, Qiao Gang, Liu Ruiping, Shi Yufei. 2014. Human health risk under the condition of farmland soil heavy metals pollution in a gold mining area. Geological Bulletin of China, 33 (8): 1239-1252 (in Chinese with English abstract).

Yang Yi, Zhang Zhi, Tang Juxing, Yu Chuan, Li Yubin, Wang Liqiang, Li Jianli, Gao Ke, Wang Qin, Yang Huanhuan. 2015. Mineralization, alteration and vein systems of the Bolong porphyry copper deposit in the Duolong ore concentration area, Tibet. Geology in China, 42 (3): 759-776 (in Chinese with English abstract).

Zhang Chaoyang, Peng Ping'an, Song Jianzhong, Liu Chengshuai, Peng Jue, Lu Puxiang. 2012. Utilization of modified BCR procedure for the chemical speciation of heavy metals in Chinese soil reference material. Ecology and Environmental Sciences, 21 (11): 1881-1884 (in Chinese with English abstract).

Zhang Jingru, Zhou Yongzhang, Ye Mai, Dou Lei, Li Xingyuan, Mo Liping. 2017. Bioavailability of heavy metal and transfer factors in a regional soil-to-crops system. Environmental Science and Technology, 40 (12): 256-266 (in Chinese with English abstract).

Zhao Yuanyi, Li Xiaosai, Qiao Donghai, Wang Song. 2016. Suggestions of green exploration and green mining of Duolong ore concentration area, Tibet. Geological Review, 62 (S1): 287-288 (in Chinese without English abstract).

Zhao Yuanyi, Wang Song, Qiao Donghai, Li Xiaosai, Li Ruiping, Wang Aiyun, Dai Jingjing, Jia Qinxian. 2017. The Giant copper deposit in China: the interpretation of theDuolong Copper Deposit. Scientific and Cultural Popularization of Land and Resources, (2): 14-17 (in Chinese without English abstract).

Zhong Zheng, He Guandi, He Tengbing. 2015. Advance in Prevention and Treatment of Heavy Metal Pollution in Soil. Guizhou Agricultural Sciences, 43 (6): 202-206 (in Chinese with English abstract).

Zhu Xiangping, Chen Huaan, Ma Dongfang, Huang Han xiao, Li Guangming, Wei Lujie, Liu Chaoqiang. 2012. Geology and alteration of the Duobuza porphyry copper-gold deposit in Tibet. Geology and Exploration, 48 (2): 199-206 (in Chinese with English abstract).

Zhu Xiangping, Chen Huaan, Ma Dongfang, Huang Hanxiao, Li Guangming, Li Yubin, Li Yuchang, Wei Lujie, Liu Chaoqiang. 2013. 40Ar/39Ar dating of hydrothermal K-feldspar and hydrothermal sericite from Bolong porphyry Cu-Au deposit in Tibet. Deposit Geology, 32 (5): 954-962 (in Chinese with English abstract).

参考文献

陈红旗, 曲晓明, 范淑芳. 2015. 西藏改则县多龙矿集区斑岩型铜金矿床的地质特征与成矿-找矿模型. 矿床地质, 34 (2): 321-332.

初娜, 赵元艺, 张光弟, 张勤, 蔡剑辉, 熊群尧, 李德先, 王金生, 赵金艳. 2007. 德兴铜矿低品位矿石堆浸场与大坞河流域土壤重金属元素形态的环境特征. 地质学报, 81 (5): 670-684.

初娜, 赵元艺, 张光弟, 杨慧. 2008. 江西省德兴铜矿矿区重金属元素的环境效应. 地质学报, 82 (4): 562-562.

丛源, 陈岳龙, 杨忠芳, 侯青叶, 胡省英, 郭莉. 2009. 北京市农田土壤重金属的化学形态及其对生态系统的潜在危害. 土壤, 41 (1): 37-41.

高连存, 何桂华, 冯素萍, 王淑仁, 崔兆杰. 1994. 模拟酸雨条件下降尘中 Cu, Pb, Zn, Cr 各形态的溶出和转化研究. 环境化学, 13 (5): 448-452.

国家环境保护局. 1995. 土壤环境质量标准 GB15618-1995. 北京: 标准出版社.

韩春梅, 王林山, 巩宗强, 许华夏. 2005. 土壤中重金属形态分析及其环境学意义. 生态学杂志, 24 (12): 1499-1502.

何园, 王宪, 陈丽丹, 郑盛华, 蔡真珍. 2007. 泉州走马埭典型土壤重金属的赋存形态分析. 土壤, 39 (2): 257-262.

何中发, 方正, 孙彦伟, 李金柱, 夏晨, 温晓华, 张琢, 刘文长, 江思珉. 2009. 农用地土壤中汞元素形态特征浅析. 上海地质, 30 (1): 45-49.

李娇, 吴劲, 蒋进元, 滕彦国, 何立环, 宋柳霆. 2018. 近十年土壤污染物源解析研究综述. 土壤通报, 49 (1): 232-242.

李金祥, 李光明, 秦克章, 肖波. 2008. 班公湖带多不杂富金斑岩铜矿床斑岩-火山岩的地球化学特征与时代: 对成矿构造背景的制约. 岩石学报, 24 (3): 531-543.

李玉彬, 多吉, 钟婉婷, 李玉昌, 强巴旺堆, 陈红旗, 刘鸿飞, 张金树, 张天平, 徐志忠, 范安辉, 索朗旺钦. 2012. 西藏改则县多不杂斑岩型铜金矿床勘查模型. 地质与勘探, 48 (2): 274-287.

李志博, 骆永明, 宋静. 2006. 土壤环境质量指导值与标准研究II-污染土壤的健康风险评估. 土壤学报, 43 (1): 142-151.

利锋, 张学先, 戴睿志. 2008. 重金属有效态与土壤环境质量标准制订. 广东微量元素科学杂志, 15 (1): 7-10.

梁彦秋, 潘伟, 刘婷婷, 邢志强, 臧树良. 2006. 沈阳张士污灌区土壤重金属元素形态分析. 环境科学与管理, 31 (2): 43-45.

刘青枰, 邓时强, 赵元艺, 李小赛. 2018. 西藏尕尔勤矿床区域性土壤重金属元素地球化学特征及来源解析. 地球学报, 39 (4): 481-490.

罗玉虎. 2017. 西藏荣那河中荣那矿体天然污水的地球化学演化与生态响应. 北京: 中国地质大学 (北京).

倪善芹, 李瑞萍, 王安建. 2012. 赣东北德兴矿区周边女性居民头发中重金属分布特征. 中国环境监测, 28 (2): 84-90.

宁墨奂. 2012. 西藏多龙矿集区斑岩铜矿床地质特征及找矿预测. 成都: 成都理工大学.

牛利华, 胡庆东, 李光武, 吴问全, 田杨超. 2009. 重金属铬、铜、汞经不同途径神经毒性的对比研究. 安徽农业科学杂志, 37 (36): 18384-18388.

乔东海, 赵元艺, 汪傲, 李玉彬, 郭硕, 李小赛, 王松. 2017. 西藏多龙矿集区地堡铜 (金) 矿床年代学、流体包裹体、地球化学特征及其成因类型研究. 地质学报, 91 (7): 1542-1564.

史磊, 郭朝晖, 彭驰, 肖细元, 薛清华, 冉洪珍, 封文利. 2018. 石灰组配土壤改良剂抑制污染农田水稻镉吸收. 农业工程学报, 34 (11): 209-216.

孙嘉, 毛景文, 姚佛军, 段先哲. 2017. 西藏多龙矿集区岩浆岩成因与成矿作用关系研究. 岩石学报, 33 (10): 3217-3238.

陶文靖, 程丽娅, 聂全新, 黄勤, 王金云. 2014. 土壤中重金属有效态分析方法研究. 安徽地质, 24 (4): 300-303.

王松, 赵元艺, 李小赛, 乔东海. 2016. 西藏多不杂铜 (金) 矿床环境地质特征. 地质论评, 62 (S1): 279-280.
王晓亮, 赵元艺, 柳建平, 路璐, 杨永强, 初娜. 2013. 德兴铜矿大坞河流域土壤中 Cd 的环境地球化学特征及意义. 地质论评, 59 (4): 781-788.
徐友宁, 张江华, 柯海玲, 陈华清, 乔冈, 刘瑞平, 史宇飞. 2014. 某金矿区农田土壤重金属污染的人体健康风险. 地质通报, 33 (8): 1239-1252.
杨毅, 张志, 唐菊兴, 毓川, 李玉彬, 王立强, 李建力, 高轲, 王勤, 杨欢欢. 2015. 西藏多龙矿集区波龙斑岩铜矿床蚀变与脉体系统. 中国地质, 42 (3): 759-776.
张朝阳, 彭平安, 宋建中, 刘承帅, 彭珏, 卢普相. 2012. 改进 BCR 法分析国家土壤标准物质中重金属化学形态. 生态环境学报, 21 (11): 1881-1884.
张景茹, 周永章, 叶脉, 窦磊, 李兴远, 莫莉萍. 2017. 土壤-蔬菜中重金属生物可利用性及迁移系数. 环境科学与技术, 40 (12): 256-266.
赵元艺, 李小赛, 乔东海, 王松. 2016. 西藏多龙矿集区绿色勘查与绿色矿山建议. 地质论评, 62 (S1): 287-288.
赵元艺, 王松, 乔东海, 李小赛, 李瑞萍, 王爱云, 代晶晶, 贾沁贤. 2017. 中国铜矿的巨无霸——多龙铜矿床解读. 国土资源科普与文化, (2): 14-17.
钟正, 何冠谛, 何腾兵. 2015. 土壤重金属污染防治研究进展. 贵州农业科学, 43 (6): 202-206.
周启星, 黄国宏. 2001. 环境生物地球化学及全球环境变化. 北京: 科学出版社, 1-256.
祝向平, 陈华安, 马东方, 黄瀚霄, 李光明, 卫鲁杰, 刘朝强. 2012. 西藏多不杂斑岩铜金矿床地质与蚀变. 地质与勘探, 48 (2): 199-206.
祝向平, 陈华安, 马东方, 黄瀚霄, 李光明, 李玉彬, 李玉昌, 卫鲁杰, 刘朝强. 2013. 西藏波龙斑岩铜金矿床钾长石和绢云母 40Ar/39Ar 年龄及其地质意义. 矿床地质, 32 (5): 954-962.

Characteristics of soil heavy metals and prevention of environmental problems in the Duobuza mining area, Tibet

Yujing Wu[1, 2], Song Wang[1, 2] and Yuanyi Zhao[2]

1. China University of Geosciences (Beijing), Beijing 100083
2. MRL Key Laboratory of Metallogeny and Mineral Assessment, Institute of Natural Resources, Chinese Academy of Geological Sciences, Beijing 100037

Abstract Before mineral exploitation, it is of great significance to fully understand the environmental characteristics of soil heavy metals in the mining area for the green exploitation and rational mining. The Duobuza deposit, located in Gaize County, Tibet, is one of the important deposits in the Dulong ore concentration area, and has been proved to contain 2.3444 million tons of copper resources at present. Taking the Duobuza mining area as an example, 37 soil samples were collected from four sections of the earth surface. The contents of heavy metal elements (Cu, Pb, Zn, As, Cd, Cr, Hg) and pH value were tested, and the speciation characteristics of heavy metals in 10 samples were analyzed. Single factor pollution index and Nemero comprehensive pollution index method were used to analyze the heavy metal contents, pollution index and speciation characteristics of surface soil in the Duobuza mining area. The sources of heavy metal elements and their extent of impact on the environment were also discussed. The results show that the surface soil is alkaline. Among the seven heavy metal elements, the contents of Cu in all samples are higher than the GB1 5618–1995 III standard, only the maximum value of the Zn, Cd exceed the grade III standard, the contents of Pb, As, Cr, Hg do not exceed the standard. The single pollution index shows Cu>Zn>As>Cd>Cr>Pb>Hg. The average value of the comprehensive pollution index is 11.60, which shows that heavy metals have a high degree of comprehensive impact on the environment. Cu mainly exists in the form of Fe-Mn oxidation and residual state, accounting for 48.40% and 14.73%, respectively. Pb, Zn, As, Cr exists mainly in the form of residuals, accounting for 90.27%, 71.48%, 79.59% and 86.12%, respectively. Cd in the forms of Fe-Mn oxidation and carbonate state, accounting for 40.05% and 22.42%, respectively. Hg in the forms of residual and humic acid state, accounting for 40.16% and 22.54%, respectively. There are three main sources of heavy metal elements: Cu, Zn, Cd, Hg derived from oxidation leaching of chalcopyrite; Pb and As derived from pyrite oxidation; Cr derived from natural weathering of rocks. Aiming at the environmental problems related to heavy metals, some suggestions were put forward for the prevention and control of heavy metals.

Keywords Duobuza Mining area; Soil heavy metals; Environmental characteristics; Source identification; Bioaccessibility; Prevention significance

Pollution, sources and environmental risk assessment of heavy metals in the surface AMD water, sediments and surface soils around unexploited Rona Cu deposit, Tibet, China*

Donghai Qiao[1, 2], Gaoshang Wang[2], Xiaosai Li[1], Song Wang[1] and Yuanyi Zhao[1]

1. MNR Key Laboratory of Metallogeny and Mineral Assessment, Institute of Mineral Resources, CAGS, Beijing 100037, China

2. Global Mineral Resources Strategic Research Center, Institute of Mineral Resources, CAGS, Beijing 100037, China

Highlights

- Unexploited Rona Cu deposit at Tibet has resulted in considerable environmental risk.
- pH and HMs in water, sediments and surface soils in the study area were investigated.
- The dominant pollution of HMs in sediments and surface soils came from Cu, Zn and As.
- The contents of Cu and Zn in acid water were 2114.0 and 1402.1 $\mu g \cdot L^{-1}$, respectively.
- Both total metal concentration and bioavailable metal fractions should be considered.

Graphical abstract

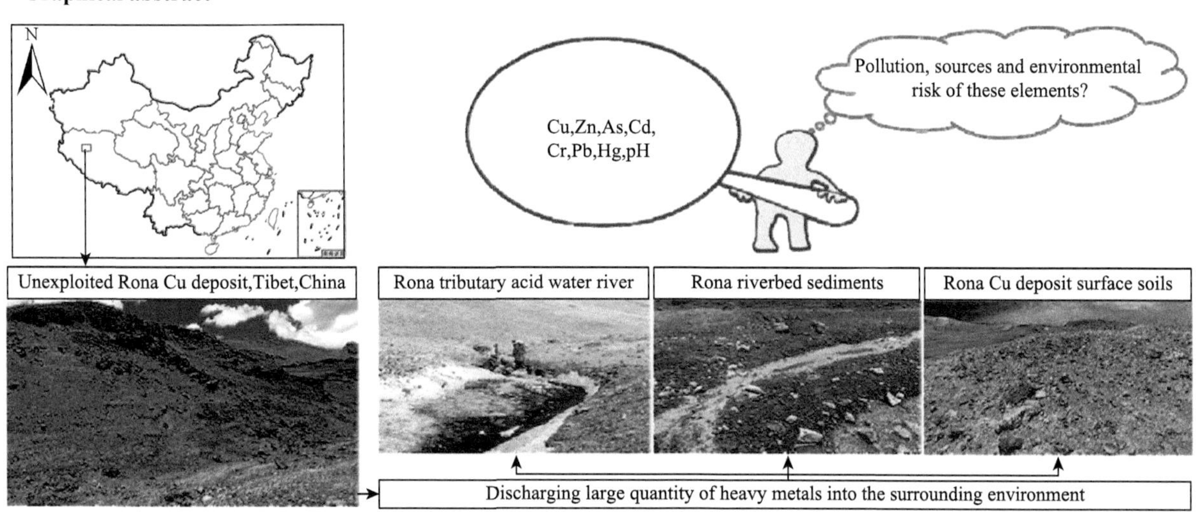

Abstract The pollution by heavy metals (HMs) of mining is a widespread problem in the world. However, the pollution by HMs around unexploited deposits (virgin fields) has been studied rarely, especially in Tibet, China. Water, sediments and surface soils were collected to investigate the concentrations of HMs around unexploited Rona Cu deposit in Tibet, China. Furthermore, geochemical fractions of these elements were also analyzed. Pollution and environmental risk introduced by HMs accumulation were assessed using pollution indices, geo-accumulation (I_{geo}), potential ecological risk index and risk assessment code (RAC). Results indicated that the pH values of Rona tributary river ranged from 2.70 to 3.08, and the average concentrations of Cu and Zn were 2114.00±65.89 and 1402.14±27.36 $\mu g \cdot L^{-1}$, respectively, exceeding their standard limits. The concentrations ($mg \cdot kg^{-1}$) of Cu, Zn and As ranged in 19.01–1763.10, 62.00–543.06 and 11.12–61.78 for sediments, respectively, and 154.60 –1489.35, 55.38–344.74 and 10.05– 404.03 for surface soils, respectively, exceeding their standard limits. According to RAC, almost all Cu, Zn and As near low risk status. However, Cd

* 本文发表在：Chemosphere, 2020, 248: 125988-126000

ranged from medium to very high risk in sediments, and low to high risk in surface soils. Statistical analysis suggested that Cu, Pb, Zn, As and Cd in sediments and surface soils may mainly derive from Rona deposit, whereas Cr and Hg may primarily originate from lithogenic sources. The results indicated that very high concentrations of HMs could be occurred in surface water, sediments and surface soils around unexploited deposits. Especially at high-altitude Tibet, the high environmental risk of HMs deserves more attention.

Keywords Heavy metals; Unexploited deposit; Environmental risk; Tibet; Geochemical fraction; Pollution assessment

1 Introduction

Human activities, such as mining, smelting, traffic and machinery manufacturing, have been recognized as major contributors to environmental pollution by HMs (Mostert et al., 2010). Mining activities, especially for non-ferrous metal mines, can generate and lead to the leaching of large quantities of HMs into environment inducing adverse influence on plants and human health (Ding et al., 2018). Previous researches have suggested that in most cases, water, sediments and soils in the vicinity of mines are polluted severely. For example, Local water supplies around Kilembe Cu mine in Western Uganda were contaminated with Co (Mwesigye et al., 2016). The concentrations of Pb in sediments around a Pb/Zn mine in NE Morocco, are expected to cause harmful effects on sediment-dwelling organisms (Azhari et al., 2017). Liu et al. (2013) reported that soils pollution by HMs was very high, and the concentrations of Cu and Cd were up to 426.15 and 2.55 $mg \cdot kg^{-1}$, respectively, around Dexing Cu mine in China. When HMs in sediments or soils have been transformed from stable fractions into susceptible, bioavailable and mobile existence, they could have caused threat to the health of animals and human beings (Madrid et al., 2002). For instance, Cd exposure can increase the chance of osteoporosis and lung cancer (Khan et al., 2015), while chronic exposure to As dust can lead to peripheral vascular disease (Rehman et al., 2019). The excessive intake of Pb adversely affect the central nervous system (Kaufmann et al., 2003). Zn may result infertility and kidney disease. Cu induces depression and Cr may cause tumor of respiratory organs (Sani et al., 2017). The toxicity of HMs is not obvious at the initial stage of accumulation, but it is difficult to eliminate when toxicity is shown (Lü et al., 2018). Due to their potential toxic, persistent and irreversible characteristic, the HMs such as Cu, Pb, Zn, As, Cd, Cr and Hg have been listed as priory control pollutants by the United States Environmental Protection Agency (USEPA) and caused more and more attention in many part of the world (Lei et al., 2008). China is one of the largest global producers and consumers of metals (Gunson and Jian, 2001). Especially, since the 21st century, continuously expanding industrialization has drastically increased extensive metal mining and smelting activities, causing the significant enrichment of HMs in soils and other environment medias (Kang et al., 2017). Therefore, assessment of HMs pollution is very essential to be carried out for chinese mines.

The HMs pollution around mines has been reported extensively, and most of the pollution sediments, soils or water column by HMs are attributed to the generation and release of acid mine drainage (AMD) from mines, containing high concentrations of HMs (Ramirez et al., 2005). Oxidation of sulphide ores, particularly those rich in pyrite, introduces high concentrations of HMs, sulphuric acid (hydrogen ions) and sulphate ions into waters and therefore generates AMD (Nordstrom et al., 2000). Anthropogenic activities such as mining and smelting of metal ores have greatly increased the generation of AMD (Mostert et al., 2010). During mining and smelting progresses, mine tailings, wastewaters and dusts could release a substantial amount of HMs to surrounding environment through surface runoff and aerial dust transport by wind (Rodríguez et al., 2009). Specifically, opencast mining and smelting activities have a relatively more serious environmental impact on soils and water streams, especially the pollution of environment by mine tailings (Yin et al., 2018), which was different from the causes of pollution in unexploited deposits (no tailings). For example, the concentrations of HMs in AMD can reach 41740 $\mu g \cdot L^{-1}$ for Cu, and 265 $\mu g \cdot L^{-1}$ for As in Jiama and Deerni Cu mines (being mined), respectively, in Tibetan plateau, China (Liu et al., 2018b), heavily polluted. Additionally, natural sources are also responsible for the production of HMs

(Shah et al., 2010). Previous research works have investigated the natural sources of HMs enrichment in ecosystems (Del R1'o et al., 2002). The results showed that the accumulation of HMs in sediments or soils depends on the type of weathered rocks and climatic environment in the area concerned (Kafayatullah et al., 2001; Khan et al., 2017). However, all of the previous researches are about the environmental pollution for non-mining areas or the deposits that have been mined or are being mined, and it is unclear for the generation of AMD for unexploited deposits, and the adverse influence on environment is rarely concerned.

According to the field geological work experience, ore body is the section where the available elements are abundant, and the primary halos of the surface are a good sign of ore prospecting, as well as an abnormal part of geochemistry. However, ore bodies with enriched useful elements and geochemical anomalies would produce pollution to the surrounding environment, which are the main sources of pollution. Therefore, even if these deposits are not mined, they also cause harm to the surrounding environment (Wang et al., 2006). Under favorable conditions, the sulphide ores in unexploited non-ferrous deposits, which are exposed, could be oxidized in natural condition as well to produce AMD (Nasrabadi et al., 2010). In consequence, it is significant to research the HMs pollution for unexploited deposits. Rona Cu deposit (unexploited) is polymetallic sulfide deposit, and the major ore minerals are pyrite (FeS_2), chalcopyrite ($CuFeS_2$), covellite (CuS), digenite ($4Cu_2S{\cdot}CuS$), bornite (Cu_5FeS_4), chalcocite (Cu_2S) and enargite (Cu_3AsS_4) (Tang et al., 2016). Rona river flows south of the Rona deposit and the flow is 39 L/s. A tributary river (natural spring water) in the middle of Rona deposit [Fig. 1 (a)] flows down slope and into Rona river [Fig. 1(b)] and the flow is 0.8 L/s, accompanied by a large amount of ore leaching materials, resulting in a noticeable impact on the surrounding environment. The seriousness of this environmental hazard can be seen from the obvious changes in riparian vegetation [Fig. 1 (c)]. Before tributary water inflows, the vegetation on both sides of Rona valley is lush, and after water inflows, there is almost no vegetation distribution on both sides of the valley. Field observation showed that the water in some sections of Rona river was yellowish brown [Fig. 1 (d)], with abundant yellow foam on ice sheet [Fig. 1 (e)], and the adjacent hillside were covered by large areas of oxidized ores [Fig. 1 (f)].

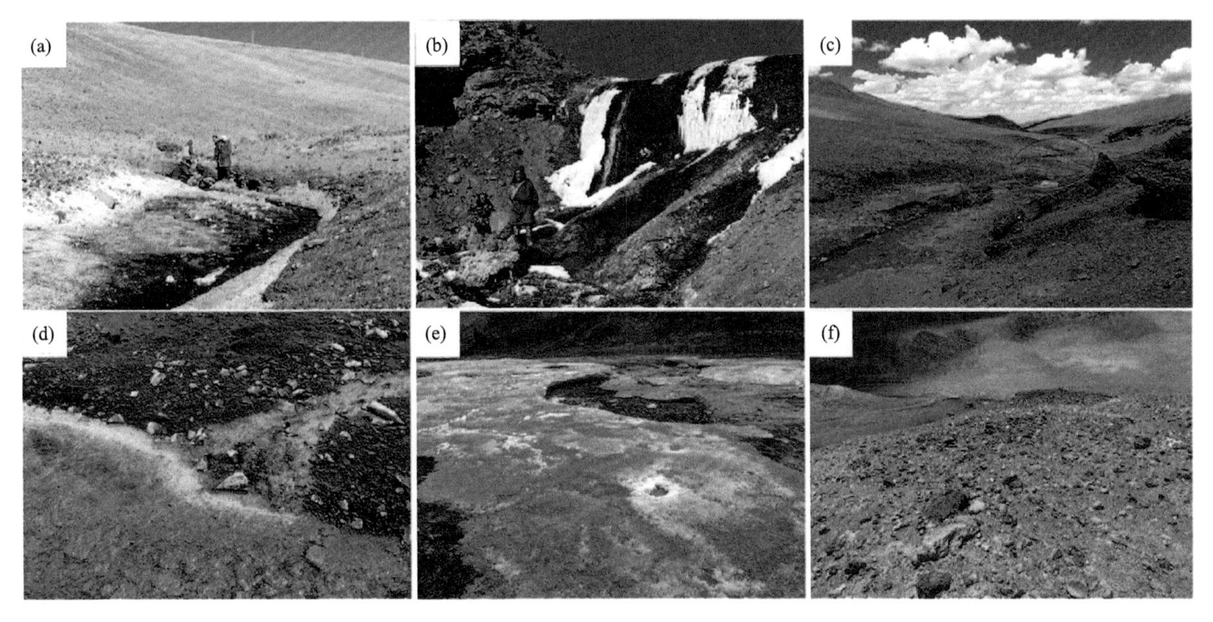

Fig. 1 Photographs of the unexploited Rona Cu deposit study area.

(a) Rona tributary river (natural spring water), (b) Tributary river water flows down slope and into Rona river, (c) Riparian vegetation changes before and after the inflow of tributary water (The red circle is the inlet position of tributary water), (d) Yellowish brown water of the Rona river, (e) Abundant yellow foam on the ice sheet, (f) Adjacent hillside covered with abundant oxidized ores. (For interpretation of the references to colour in this figure legend, the reader is referred to the Web version of this article.)

In this paper, the total concentrations of HMs and pH values of water, sediments and surface soils were determined to assess the pollution and environmental risk by pollution indices, I_{geo}, ecological risk indices and RAC. The objectives of this study were (i) to investigate the occurrence of Cu, Pb, Zn, As, Cd, Cr and Hg in water, sediments and surface soils and to determine the speciation of these elements using sequential extraction procedure; (ii) to assess the environmental risk of HMs pollution; and (iii) to define the sources of HMs preliminarily in the study area. The research results have important reference significance for the environmental protection and governance of Rona deposit. Given the fragile setting in the study area and its less external disturbance conditions, outcomes from this study can provide a unique perspective to understand the environmental pollution for unexploited deposits in other countries or regions with similar or the same geographical environment.

2 Materials and methods

2.1 Study area and sampling

Rona Cu deposit is located in the Tibet, China (N 32°47′00″–32°50′00″; E 83°23′00″-83°27′00″), about 100 km northwest of Gaize County, northeast of Samalong village (Fig. 2). At present, it is the first Porphyry-Epithermal deposit in Tibetan plateau and is currently the largest porphyry Cu deposit with single ore body in China. The Cu resources of the deposit are about 11 million tons, with an average Cu grade of 0.51% (Tang et al., 2016). In September 2012, Aluminum Corporation of China entered the study area to conduct ore prospecting, and discovered the Rona deposit in 2013. Rona deposit is currently in the scientific research and technical reserve stage, not mining. Compared with most parts of the China, the environmental characteristics on the study area is very special. The terrain of the study area is mainly mountain and valley, about 5000 m above sea level, and the average days with annual wind speed of 17 m/s are 200 d. The climate of the study area is Plateau subtropical arid monsoon climate and the annual mean temperature ranges from –0.1 ℃ to –2.5 ℃, with the highest and lowest temperature 26.0 ℃ and –44.6 ℃, respectively. Annual total precipitation in the study area is 308.3 mm, of which more than 70.0% concentrated in July and August (Qiao, 2018).

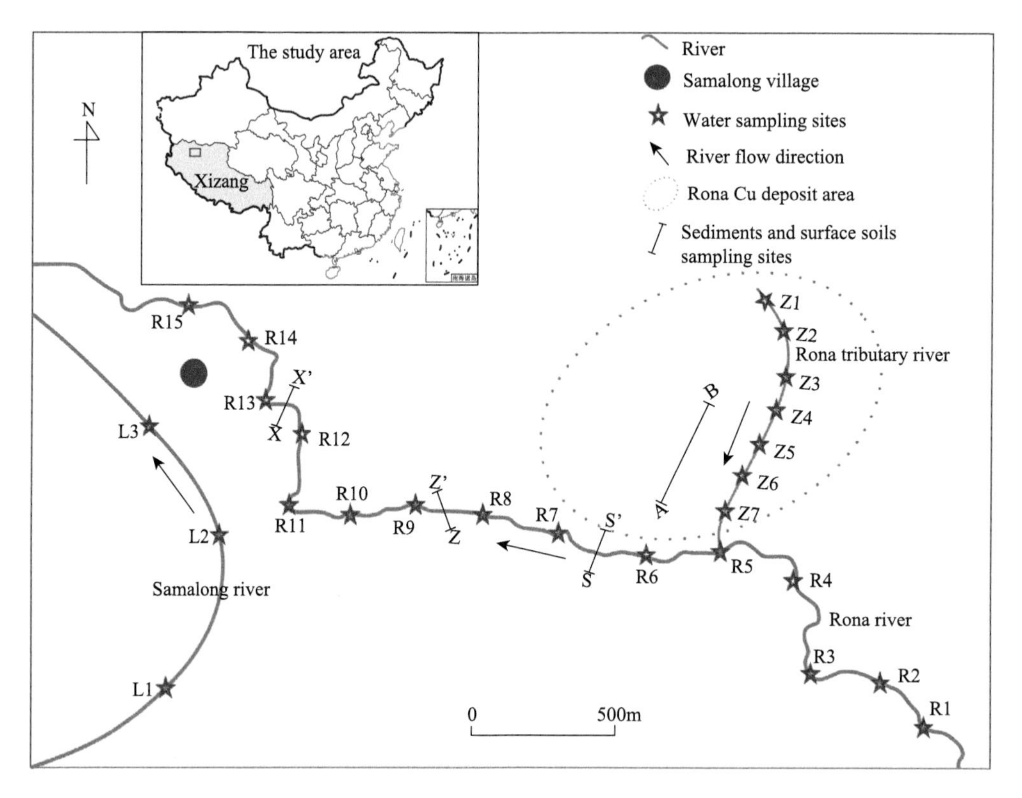

Fig. 2 Location of the study area and surface water, sediments and surface soils sampling sites.

A total of 22 surface water samples at depths of 0–5 cm were collected from Rona river (R1–R15) and its tributary (Z1-Z7) (Fig. 2). For locating the sample sites, a portable global positioning system (GPS) was used. All samples were filtered using pre-weighted glass fiber filters (0.45 μm) to remove suspended particles. Samples were acidified to a pH of <2 using nitric acid to analyze the concentration of elements. pH of surface water was measured using Hanna multiparameter instrument (Model HI9828, American) during sampling. All samples were kept at a temperature below 5 ℃ in sealed containers and transported back to laboratory. The Samalong river located in the adjacent valley of Rona river, is an ideal control river, and three surface water samples at depths of 0–5 cm were collected from the river (L1–L3) (Fig. 2). In addition, three well water samples (V1–V3) were collected from Samalong village.

A total of 58 sediment samples at depths of 30–50 cm were collected with the help of stainless steel hand spade (Fig. 2). From top to bottom, take a sediment sample every 10 cm deep, and 3–5 samples were collected at each point. For locating the sample sites, a portable global positioning system (GPS) was used. Sediment samples from Rona riverbed were taken from three sections of the riverbed: 20 samples from S-S′, 19 from Z-Z′ and 19 from X-X′, and each section was sampled from five points at a certain distance from each other (Fig. 2). The collected samples were air-dried, ground, sieved through 2 mm sized sieve to remove debris, stones and other coarse structures and carefully stored in clean polyethylene bags, labelled and then transported to the laboratory. The sampling locations of local background sediment samples from the upstream of Rona river are detailed in our previous study (Luo, 2017). Because of the hard bedrock beneath Rona tributary river, sediment samples cannot be collected.

A total of 22 surface soil samples at depths of 0–10 cm were collected from A-B hillside profile (Fig. 2). For locating the sample sites, a portable global positioning system (GPS) was used. Samples were collected from top of the mountain to a position near Rona river valley. The surface soil samples were air dried at room temperature, grounded and sieved through a 2 mm nylon mesh, and the coarse materials were removed. Then the samples were stored in zip-lock polyethylene bags for subsequent analysis. The sampling locations of local background surface soil samples are detailed in our previous study (Luo, 2017).

2.2 Chemical analysis

2.2.1 Analysis of surface and well water samples

In case of water samples, 100 mL from each water sample was taken in a beaker and 10 mL HNO_3 (65.0%) was added. The contents of the beakers were evaporated at 90 ℃ till the volume of the water to be digested remained 40 mL, filtered in 100 mL volumetric flask and volume was made up to the mark with deionized water. The concentrations of Cu, Pb, Zn, Cd and Cr were analyzed by inductively coupled plasma mass spectrometry system (ICP-MS) (Elan DCR-e, American), and As and Hg were analyzed by Atomic fluorescence spectrometry (AFS) (ASF 2202, China).

2.2.2 Analysis of sediment and surface soil samples

pH was measured (soil: water in the ratio 1 ∶ 2, agitated for about an hour) using a pH meter (Mettler Toledo FE20, Switzerland) (Saleem et al., 2018). The sieved sediment and surface soil samples was ground with mortar and pestle until fine particles (<200 μm). For metal concentration, the sediment and surface soil samples were weighed (0.25 g) into clean, dry, numbered and acid-washed poly tetrafluoroethylene (PTFE) beakers, respectively. HNO_3 (3.0 mL) followed by $HClO_4$ (3.0 mL) and then HF (10.0 mL) was added to each beaker and heated on hot plate at 180 ℃ for about 1–1.5 h (Cantle, 1986). HCl (10.0 mL) was added to each beaker and warmed gently. At last, the materials were transferred to 100 mL volumetric flask and made the volume up to the mark with double deionized water. After acid digestion, Graphite furnace atomic absorption spectroscopy (AA-1800E, China) was used to determine Cd concentration; flame atomic absorption spectrometry (AAS novAA 400, Germany) was used to determine the concentrations of Cr, Cu, Zn and Pb. Reduction gasification atomic fluorescence spectrometry

(PF6-3, China) was used to determine the concentrations of Hg and As.

A modified Tessier sequential extraction procedure (Tessier et al., 1979) was performed to separate HMs to water-soluble (F1), exchangeable (F2), carbonate-associated (F3), humic acid associated (F4), Fe-Mn oxide-associated (F5) and strong organic associated (F6), which were extracted by water, magnesium chloride solution (1.0 M), sodium acetate solution (1.0 M), sodium pyrophosphate solution (0.1 M), hydroxylamine hydrochloride solution (0.25 M) and hydrogen peroxide (30.0%), respectively. To give a check on the recovery of the whole procedure in comparison to the total concentration, the residual fraction (F7) was measured.

2.2.3 Quality control

In the experiment process, standard operating procedures, standard reference materials, reagent blanks, spiked samples recovery and analysis of replicates were implemented to check the consistency of the results. Precision and accuracy for HMs analysis are validated using standard reference materials from Center of National Standard Reference Material of China [water, GSB04-1767-2004; sediment, GBW07459 (GSS-22) and GBW07460 (GSS-25); soil, GBW07458 (GSS-17)]. Blanks and standard reference materials digestion and measurement was conducted as described for the above method. Accepted recovery rates for all HMs range from 93.0% to 109.0%. Differences in HMs concentrations between the determined and certified values are less than 10.0% and the deviation of duplicate samples was less than 5.0%, indicated that the results obtained were within the detectable range of the certified values. To ensure an adequate extraction efficiency, the standard reference material (GBW 07437) was carried out to test the accuracy of the sequential procedure. The recovery rates for all metals ranged from 80.4% to 104.9%, within the detectable range of the verification values.

2.3 Geochemical and environmental risk assessment

In recent years, different metal pollution indices have been applied for geochemical and environmental risk assessment (Praveena et al., 2007; Gong et al., 2008). In order to comprehensively evaluate pollution degree of water, sediments and surface soils, this paper employed different types of indices to determine the current pollution status of HMs in Rona Cu deposit area. In this study, five parameters including single pollution index (P_i), Nemrow integrated pollution index (P_n), I_{geo}, potential ecological risk of individual factor (E_{ir}) and Potential ecological risk index (RI) were calculated. These indices allow the assessment of metal pollution by individual elements (P_i and I_{geo}), holistic evaluation of the water, sediments and surface soils pollution (P_n) as well as the assessment of the ecological risk (E_{ir} and RI). Rodríguez et al. (2009) indicated that an adequate criterion for environmental risk assessment must include both total metal concentration and bioavailable metal fraction. Thus, RAC was used to assess the environmental risk of HMs. The selected indices are explained below.

2.3.1 Pollution indices

P_i (Hakanson, 1980) and P_n (Chen et al., 2015) were used to assess the pollution of HMs in water, sediments and surface soils. The P_i is defined as:

$$P_i = C_i / S_i \tag{1}$$

where C_i represents the concentration of metal i in samples, and S_i is its reference concentration. In this paper, the water, sediments, and surface soils reference concentrations are Chinese quality criterion of class Ⅲ for surface water (GB 3838–2002) and Soil quality Grade Ⅱ standards of China (CEPA, 1995), respectively. The P_n can be calculated using:

$$P_n = [0.5 \times (I_{\text{Avg}}^2 + I_{\text{Max}}^2)]^{1/2} \tag{2}$$

where I_{Avg} is the mean value of all pollution indices of the HMs considered, I_{Max} is the maximum value. The evaluated criteria of P_i and P_n are shown in Liu et al., 2018b and Chen et al. (2015).

2.3.2 Geo-accumulation index

I_{geo} is a geochemical criterion to evaluate pollution levels and has been used since the late 1960s (Muller, 1969). It can be calculated using:

$$I_{geo} = \log_2(C_n / 1.5B_n) \tag{3}$$

where C_n is the measured concentration of metal n in soils or sediments and B_n is the geochemical background value of the corresponding metal. In this study, the background values were calculated from the local average background values (ABVs). The factor 1.5 is introduced to minimize the possible variations in the background values due to lithogenic effects. The category of I_{geo} is shown in Chen et al. (2015).

2.3.3 Potential ecological risk index

To estimate the comprehensive risks from the HMs, the potential ecological risk index (RI) developed by Hakanson (1980) was applied in this study. RI reflects the collective effects of different pollutants in the sediments or soils and quantitatively measure the ecological risk caused by different pollutants. The RI is calculated by the following formulas:

$$C_f = C_n / C_b \tag{4}$$

$$E_{ir} = C_f \cdot T_r \tag{5}$$

$$RI = \sum E_{ir} \tag{6}$$

where C_f is the contamination factor of a HM, C_n is the concentration of metal n in the sediments or soils, and C_b is the baseline value of that metal in the sediments or soils. In this study, the local ABVs were used as the baseline values, since using the earth crust levels can be inappropriate due to geogenic differences (Matys Grygar and Popelka, 2016). E_{ir} is the potential risk of a HM, T_r is the toxic response factor for a HM. Finally, RI is the sum of the E_{ir} calculated for each HM. The T_r for Cu, Pb, Zn, As, Cd, Cr and Hg are 5, 5, 1, 10, 30, 2 and 40, respectively. The evaluated criteria of RI is shown in Hakanson (1980).

2.3.4 Risk assessment code

The RAC was useful to assess the environmental risk using sequential extraction as a characterization method (Liu et al., 2018a). Several authors have determined the speciation of sediments or soils HMs to better inform the environmental risk assessment (Perin et al., 1985; Jain, 2004). According to RAC, the metals in the sediments or soils are bound with different strengths to the different geochemical fractions, which leads to differences in the ability of metals to be released and enter into the environment. The RAC is assigned by taking into account the percentage of HMs associated with sediments or soils in the F2 + F3 fraction; thus, there is no risk (N) when the F2 + F3 is lower than 1.0%, there is low risk (L) for a range of 1.0%–10.0%, medium risk (M) for a range of 11.0%–30.0%, high risk (H) from 31.0% to 50.0% and very high risk (VH) for higher F2 + F3 percentages (Perin et al., 1985).

2.4 Statistical analysis

Statistical analyses were performed with the statistical software package SPSS version 22.0 software (IBM, Chicago, IL). Principal component analysis (PCA) and cluster analysis (CA) was employed to analyze the relationships and sources of HMs (Chabukdhara and Nema, 2012). The association coefficient between variables and factors reflect the degree of proximity between them. In the PCA, the principal components were calculated based on the correlation matrix, and VARIMAX normalized rotation was used.

3 Results and discussion

3.1 Descriptive statistical analysis of HM concentrations

3.1.1 Surface and well water

The concentrations of HMs and pH in surface and well water samples and permissible values of class III for surface water are shown in Table 1. The pH values of Rona tributary water ranged from 2.70 to 3.08, and fall outside the permissible values, acid water, whereas the values of all other samples ranged in 7.08–8.16. Surface water acidification may have various causes, but the main source so far is ore mining, which has been confirmed by a large number of literatures (Akabzaa et al., 2007). However, the Rona deposit has never been mined and the causes of surface water acidification would be discussed in the following section. Hammarstrom et al. (2003) showed that the diffusion degree of HMs pollution in mines was enhanced under acidic conditions, and acid water is an important carrier of HMs through diffusion. Therefore, the HMs pollution around Rona deposit may be affected seriously by acid water. The concentrations of Pb, Cd and Cr in Rona river and its tributary water all lower than their standard limits, and the concentrations of Hg and As all lower than their detection limits. Notably, the concentrations of Cu and Zn in tributary water ranged in 1977.00–2168.00 and 1361.00–1434.00 $\mu g \cdot L^{-1}$, respectively, exceeding their standard limits, indicating polluted by Cu and Zn. In addition, the maximum concentrations of Cu and Zn were 10.84 and 1.43 times than their standard limits set by Emission standard of pollutants for Cu, Ni and Co industry (GB 25467–2010), respectively.

In comparison with the ABVs, represented by Samalong river, the sites where Cu and Zn exceeding the maximum values of their corresponding ABVs are R5–R12 for Cu and R5–R10 for Zn, respectively, located downstream of the tributary acid water inflow (R5 is intersection) (Fig. 2), and maximum multiples are 8.35 and 15.78 for Cu and Zn, respectively, suggesting that Rona river was polluted by acid water. Noteworthily, the concentrations of Cd in Rona tributary water samples exceeded its ABV, and maximum multiple was 108 as many, indicating high environmental risk. For Samalong village well water, the concentrations of elements all lower than their standard limits set by Drinking Water Health Standards (GB 5749–2006), suggesting that the acid water had no effect on groundwater. Moreover, the mean concentrations of Cu, Pb and Zn in upper Rona river surface water (R1–R4) were 2.40, 0.18 and 2.78 $\mu g \cdot L^{-1}$, respectively, lower than their ABVs, indicating that the natural weathering of rocks could not pollute the surface water.

3.1.2 Sediments and surface soils

The descriptive statistics of HMs concentrations and pH of sediments and surface soils, ABVs (Luo, 2017) and the threshold values (Grade II) for protecting human health are presented in Table 1. The average concentrations of HMs in sediments and surface soils showed trends: Cu>Zn>Cr>Pb>As>Cd>Hg and Cu>Zn>Pb>As>Cr>Cd>Hg, respectively. The pH values of surface soil samples ranged from 5.00 to 8.65 with a mean value of 7.48±1.07. The soil samples in low pH may be impacted by hydrogen ions resulted from sulfide minerals oxidation. The solubility and migration of HMs are related to pH values and the HMs are generally more mobile at soil pH<7 as compared to high pH (Ma et al., 2016). Given that pH is crucial factor controlling bioavailability and mobility of HMs in sediments or soils, thus, the tested metal fractions in these samples would vary (Hu et al., 2008). The pH values in sediment samples ranged from 7.10 to 9.52 with a mean value of 8.50±0.42, less affected by acid water, which may be attributed to mineral characteristics (e.g. Clay minerals neutralize acids) in sediments (Rodríguez et al., 2009).

Table 1 Summary of the background values, guideline values and basic statistics of the HMs and pH in the water, sediment and surface soil samples in the study area

Sample	Location		HM concentrations (μg L^{-1} for water, mg kg^{-1} for sediments and surface soils, n = number)							pH
			Cr		Zn	Cd	Pb	Hg	As	
Water	Tributary (n = 7)	Range	1.49–4.54	1977.00 –2168.00	1361.00 –1434.00	0.98–1.08	0.32–6.39	–	–	2.70–3.08
		Mean	2.85±1.09	2114.00±65.89	1402.14±27.36	1.04±0.04	1.87±2.55	–	–	2.89±0.15
	Rona river (n = 15)	Range	0.38–12.50	1.87–55.60	2.24–110.00	0.002–0.12	0.04–0.23	–	–	7.08–7.92
		Mean	6.19±4.30	15.78±16.64	19.48±27.71	0.04±0.03	0.14±0.07	–	–	7.66±0.26
	Smalong river (n = 3)	Range	4.62–5.92	2.46–6.66	1.40–6.97	0.007–0.01	0.32–0.64	–	0.003–0.01	7.65–8.16
		Mean	5.29±0.65	4.08±2.26	4.46±2.82	0.009±0.002	0.50±0.16	–	0.005±0.004	7.93±0.26
	Samalong well water	Range	1.91–4.85	0.82–1.46	0.94–3.98	–	–	–	0.0027–0.003	7.98–8.01
	(n = 3)	Mean	3.76±1.61	1.11±0.32	2.70±1.58	–	–	–	0.003±0.0002	8.00±0.02
	standard limits of GB 25467–2010		–	200	1000	20	200	10	100	6.00–9.00
	standard limits of GB 5749–2006		50	1000	1000	5	10	1	10	6.50–8.50
	Class Ⅲ		50	1000	1000	5	50	0.1	50	6.00–9.00
Sediments	S-S′ (n = 20)	Range	19.62–65.54	110.58–1763.10	104.24–543.06	0.17–0.41	17.53–66.58	0.01–0.03	11.12–61.78	7.10–8.91
		Mean	37.28±14.23	656.95±525.35	229.75±133.01	0.23±0.07	29.85±14.18	0.01±0.01	28.02±14.01	8.27±0.55
	Z-Z′ (n = 19)	Range	37.55–97.49	19.01–1166.50	62.00–451.33	0.16–0.48	17.21–50.55	0.02–0.05	11.32–43.14	8.20–9.52
		Mean	64.29±18.82	307.86±340.64	178.65±112.00	0.27±0.10	32.22±11.37	0.02±0.01	23.75±10.09	8.65±0.35
	X-X′ (n = 19)	Range	65.29–145.68	51.78–1063.75	95.00–464.35	0.21–0.45	27.16–59.03	0.01–0.04	23.63–51.45	8.28–8.82
		Mean	99.00±22.06	445.24±397.26	205.37±123.21	0.31±0.08	42.77±9.10	0.03±0.01	34.54±7.97	8.59±0.14
	Mean		66.35±31.40	473.24±446.88	205.02±122.88	0.27±0.09	34.86±12.88	0.02±0.01	28.76±11.72	8.50±0.42
	ABVs		79.30	46.00	131.00	0.35	43.30	0.03	25.40	7.10
Surface soils	A-B (n = 22)	Range	17.57–65.90	154.60–1489.35	55.38–344.74	0.10–0.31	20.01–270.17	0.01–0.09	10.05–404.03	5.00–8.65
		Mean	37.77±11.61	785.38±343.79	123.99±76.41	0.17±0.06	86.09±69.45	0.03±0.02	62.29±84.59	7.48±1.07
	ABVs		52.73	14.85	49.27	0.13	13.88	0.05	12.97	-
	Grade Ⅱ		200	200	250	0.6	300	0.5	30	6.50–9.00

-Not detected or no data is available (detection limits: Cd<0.002, Pb<0.002, Hg<0.10 and As<0.60 μg L^{-1}).

The concentrations of Pb, Cd, Cr and Hg in sediments and surface soils all lower than their standard limits (Grade Ⅱ). However, the concentrations of Cu, Zn and As in sediments and surface soils all exceeded their corresponding standard limits (Table 1). The proportions of sediments and surface soils in which the concentrations exceeded the standard limits of Grade II were 56.9% and 95.5% for Cu, 34.5% and 9.1% for Zn, and 50.0% and 59.1% for As, respectively, indicating polluted by Cu, Zn and As. Compared with previous researches on the soils from Jiama and Deerni Cu deposits with similar geographic and geological characteristics in China, it is significantly apparent that the average concentrations of Cu, As and Zn in the soils of Rona deposit are higher than that of the Jiama (141.50, 30.40 and 89.80 $mg \cdot kg^{-1}$, respectively) and Deerni Cu deposits (47.30, 24.50 and 89.50 $mg \cdot kg^{-1}$, respectively) (Liu et al., 2018b), indicating polluted by Cu, Zn and As seriously in the study area. In comparison with the ABVs, the mean concentrations of Cu, Zn and As in sediments and surface soils all significant higher than their ABVs. Tremendous amounts of HMs have been added to the sediments and surface soils in the study area, suggesting that all these metals are heavily impacted by Rona Cu deposit. The concentrations of Cr and Hg in the sediments and surface soils are lower than their ABVs, suggested controlled by natural factors. It can be seen that the total concentrations of Cu, Zn and As were partly different between the sediments and surface soils. This is because they suffered from pollution of HMs to different extents. Sediments were polluted with acid water, whereas HMs in surface soils come from the oxidation of surface sulfide ores.

The highest concentration sites for Cu in sediments and surface soils are S11 and T7 (position not shown), 8.82 and 7.45 times than its standard limit, respectively; S11 and T7 for Zn, 2.17 and 1.38 times than its standard limit, respectively; and S11 and T21 for As (position not shown), 2.06 and 13.50 times than its standard limit, respectively. Accordingly, S11 is the most polluted site, located in the center of riverbed, suggesting that the sediments pollution by acid water is mainly concentrated near the riverbed. The orders of pollution for the highest concentration sites for Cu in sediments were S11>Z16>X9, likely because adsorption and deposition of Cu ions in water gradually fade away along the flow direction (Gaur et al., 2005). The highest concentration of As is 13.50 times higher than its standard limit, which may be related to the geological characteristics of Rona deposit. Rona deposit is rich in enargite (Cu_3AsS_4) (Li et al., 2015), exposure of the mineral to atmospheric oxygen and moisture resulting in release of high level of As.

Although total metal concentration is useful in identifying the pollution source and the potential for pollution, the mobility and bioavailability of metals not only depend on the total concentration of metals, but also on their specific chemical fraction (Tessier et al., 1979). For example, the fractions of F1, F2 and F3 are considered as the most bioavailable. HMs in F4, F5 and F6 fractions may be potentially bioavailable, while F7 fraction is not bioavailable (Tessier et al., 1979; Rauret et al., 1999). The percentages of HMs fractions in sediments and surface soils are shown in Fig. 3. As the figure shows, there were different patterns of fraction distributions of the HMs in sediments and surface soils. It was observed that in the sediments, the F1 + F2 + F3 proportions of Pb, As and Cr were small, with mean percentages of 2.07% for Pb, 0.38% for As and 0.63% for Cr. In contrast, large amounts of Pb, As and Cr were mainly associated with the F7 fraction, 80.7% for Pb, 86.7% for As and 86.3% for Cr of the total fractions, respectively. In the surface soils, the F1 + F2 + F3 proportions of Cu, Pb, Zn, As and Cr were small, with mean percentages of 1.13% for Cu, 1.07% for Pb, 1.04% for Zn, 0.25% for As and 0.41% for Cr. However, the F7 fraction was predominant for these HMs, 84.0% for Cu, 87.7% for Pb, 87.0% for Zn, 91.6% for As and 84.0% for Cr of the total fractions, respectively. Although the concentrations of Cu, Zn and As in surface soils were as high as 1489.35, 344.74 and 404.03 $mg \cdot kg^{-1}$, respectively, and As in sediments was as high as 61.78 $mg \cdot kg^{-1}$, these HMs would be unlikely to pose a direct and significant threat to the surrounding environment since HMs in F7 fraction are held within crystal structures of some primary and secondary minerals (Rodríguez et al., 2009), and elements in F7 fraction are difficult to leach out (Lei et al., 2008), suggesting that the environmental risk of Pb, As and Cr in sediments and Cu, Pb, Zn, As and Cr in surface soils may be low in the study area. However, as mentioned above, dominant pollution of HMs in sediments and surface soils came from Cu, Zn and As, therefore, the environmental risks of these elements still have to be highly focused.

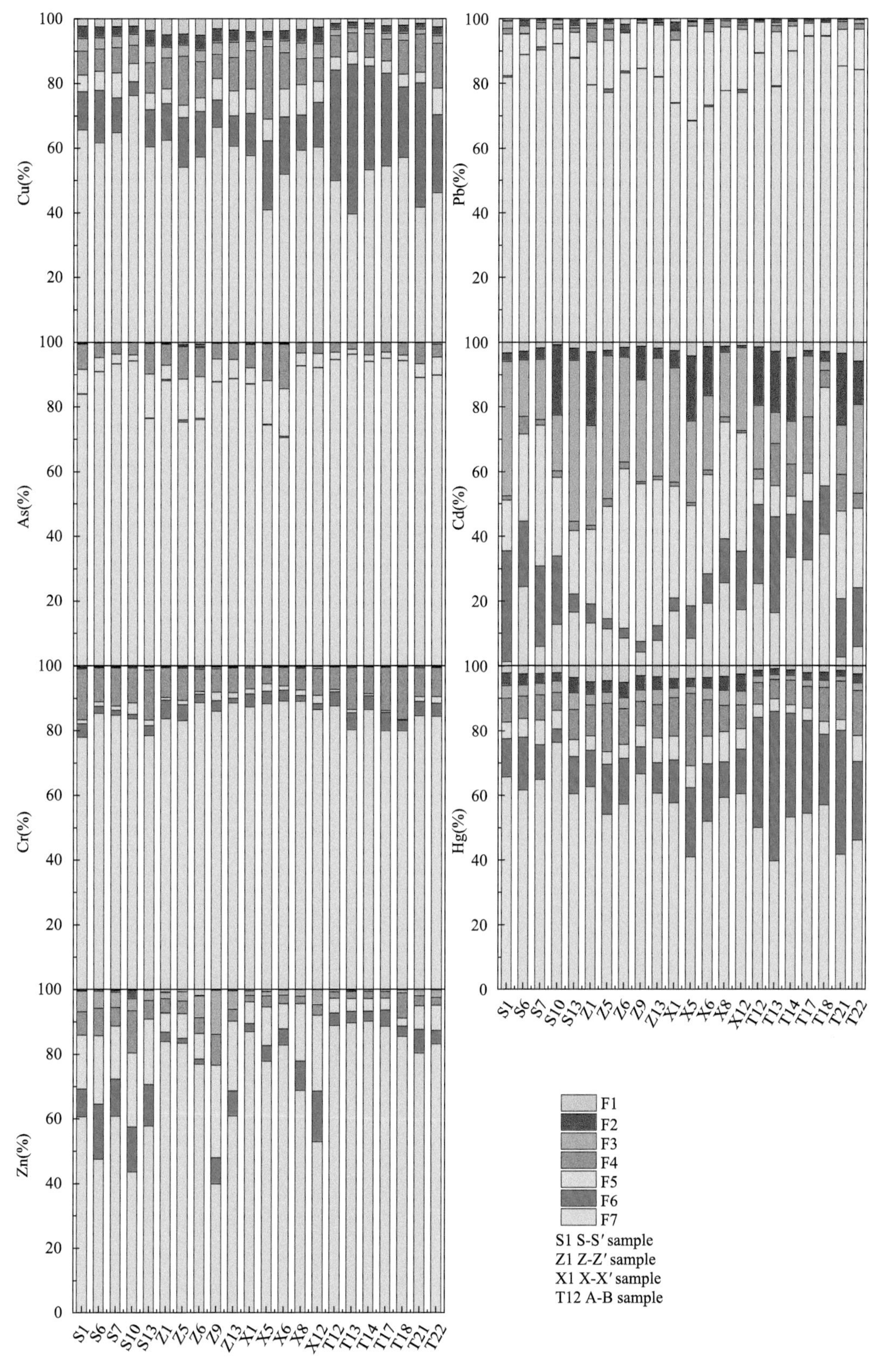

Fig. 3 The percentage contents of HM fractions in selected sediment and surface soil samples.

In sediments, the average Cu fractions followed the order of F7>F5>F4>F6>F3>F1>F2; the order of Zn was F7>F5>F6>F4>F3>F2>F1; and the order of Hg was F7>F6>F4>F5>F2>F3>F1. In surface soils, the order of Hg was F7>F6>F4>F5>F2>F3>F1. With respect to these HMs, although the F7 proportions of Cu, Zn and Hg in sediments and Hg in surface soils were high, the percentages of F4 + F5 + F6 were significant (40.8% for Cu, 35.9% for Zn and 30.4% for Hg in sediments, respectively; 46.2% for Hg in surface soils). Cu, Zn and Hg in F4 + F5 + F6 could be released when the redox potential was changed (Rinklebe and Shaheen, 2017), implying potentially environmental risk. Compared with surface soils (7.0%), Cu related to organic matter (F4 + F6) in sediments (16.6%) were significantly higher (Lei et al., 2008). In addition, compared with surface soils, the average percentage contents of Cu and Zn in sediments in F7 were significantly low, indicating relatively high mobility and bioavailability. As shown in Fig. 3, the total proportions of Cd in F1 + F2 + F3 were 38.4% and 33.7% for sediments and surface soils, respectively, and F1, F2 and F3 fractions of elements were of stronger mobility and bioavailability (Ma et al., 2016); additionally, for Cd mobility and bioavailability, Elliott et al. (1986) also observed that activity of Cd tends to bind to more mobile and bioavailable fractions, indicating high environmental risk. However, the concentrations of Cd in sediments and surface soils were slightly lower or close to their ABVs, environmental risk of Cd do not need to pay too much attention. Moreover, the F3 fraction of HMs can become exchangeable easily with conditions such as pH change (Gimeno-Garcia et al., 1995) and Kong and Bitton (2003) reported that HMs in F3 fraction are more toxic compared with other fractions. However, the F3 fraction in the study area was relatively low, probably because of hence absence of carbonates in the study area. Besides, surface soil sample T13 (pH = 5.40) has a lower percentage of F3 fraction than other surface soil samples (pH>7) due to lower pH. This results agreed with that of Wang and Wei (1995), who found that when acidic condition increased after acid water inflow, the HMs in F3 fraction would be released, causing greater harm to the environment.

3.2 Pollution characteristics

Fig. S1(a) indicated the numbers of different P_i levels of each element according to the category of the P_i values. The 100% of Pb, Cd and Cr in Rona river and its tributary water were unpolluted. The numbers of Cu reached the pollution levels of unpolluted, slightly polluted and moderately polluted were 15, 1 and 6, respectively, and Zn reached the levels of unpolluted and slightly polluted were 15 and 7, respectively. Fig. S1(b) indicated that the range of P_n was from 0.006 to 1.63. According to the category of P_n, 15 samples were unpolluted, and 7 samples were slightly polluted in tributary water.

Fig. S2 indicated the numbers of different P_i levels of each element in sediments and surface soils according to the category of the P_i values. The 100%samples of Pb, Cd, Cr and Hg in sediments and surface soils were unpolluted. The numbers of sediment and surface soil samples for Cu in heavily polluted were 22 and 15, respectively; for Zn in moderately polluted and slightly polluted were 2 and 2, respectively; and As in moderately polluted and heavily polluted were 1 and 4, respectively. Fig. 4 indicated that the range of P_n was 0.30–9.78. According to the category of P_n, 13 samples were safety; 18 were precaution level; 8 were slightly polluted; 18 were moderately polluted; and 23 samples were heavily polluted.

Fig. S3 indicated the proportions of different I_{geo} levels of each element in all sediments and surface soils according to the category of the I_{geo} values. More than half of the sediment and surface soil samples were moderately polluted (29.3% and 0%), moderately to heavily polluted (5.2% and 4.6%), heavily polluted (32.8% and 4.6%), heavily to extremely polluted (6.9% and 31.8%) or extremely polluted (0% and 59.1%) by Cu, respectively. The pollution level of Zn in sediment and surface soil samples with 8.6% and 18.2% of moderately polluted and 0% and 4.6% of moderately to heavily polluted, respectively. The As pollution in surface soil samples reached the levels of moderately polluted, moderately to heavily polluted, heavily to extremely polluted which took up 9.1%, 18.2% and 4.6% of all samples, respectively. The pollution of Cu, Zn, As, Cd, Cr and Hg in all other remaining sediment and surface soil samples remained unpolluted to moderately polluted or practically

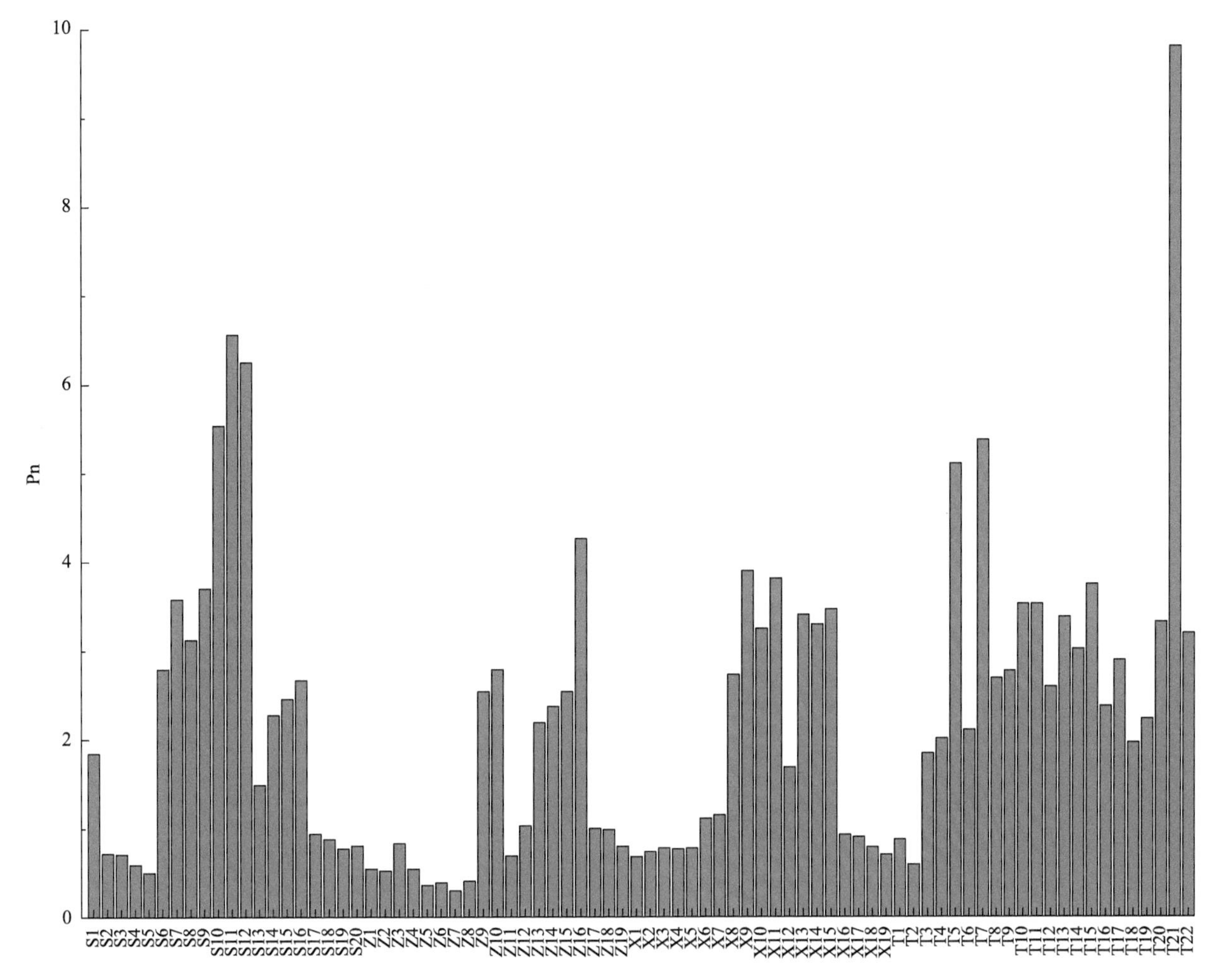

Fig. 4 The results of different P_n values for sediment and surface soil samples.

unpolluted. Accordingly, it could be concluded that the dominant pollution of HMs in sediments and surface soils came from Cu, Zn and As. Notably, Pb pollution in surface soil samples reached the levels of moderately polluted, moderately to heavily polluted and heavily polluted which took up 40.9%, 18.2% and 13.6% of all samples, respectively, different from the assessment results using P_i, and it may be the reason of that the I_{geo} considered all HMs as pollution contribution, overestimating the real pollution levels (Li and Yang, 2008).

3.3 Identification of pollution sources

3.3.1 Principal component analysis

The results of PCA for the HMs concentration in sediments are tabulated in Table S1. Two rotated principal components (PCs) were extracted with eigenvalues >1, accounting for 85.8% of the total variance. Indeed, the PCA (Fig. S4(a) and Table S1) showed that Cd, As, Pb, Zn and Cu were clustered to PC1, which explains 53.4% of the total variance, suggesting that Cd, As, Pb, Zn and Cu may have a common source or similar sources and moving together (Chabukdhara and Nema, 2012). As mentioned above, the dominant pollution of HMs in sediments came from Cu, Zn and As, and Cu, Zn and As are often associated with Pb and Cd (Alloway, 1995). Hence, these elements may derive from Rona tributary acid water. PC2 explained 32.4% of the total variance, and the loadings of Hg and Cr are 90.6% and 89.4%, of which concentrations were slightly lower or close to their ABVs. Moreover, Cr was lithogenic component (Hanesch et al., 2001). Therefore, Cr and Hg may primarily originate from lithogenic sources seemed to be controlled by parent rocks (Mico et al., 2006).

The results of PCA for the HMs concentration in surface soils are tabulated in Table S2. Three rotated principal components (PCs) were extracted with eigenvalues >1, accounting for 83.5% of the total variance. The PCA (Fig. S4b and Table S2) showed that As, Hg and Pb were clustered to PC1 accounting for 36.3% of the variance, and Cd, Zn and Cu were clustered to PC2 accounting for 30.2% of the variance. The mean concentrations of Cu, Pb, Zn, As and Cd were all higher than their ABVs, which may derive from the surface sulfide ores oxidation. Notably, the content of Hg was lower than its ABV, came from lithogenic sources. Cr was clustered to PC3 accounting for 17.1% of the variance, of which concentration was slightly lower than its ABV, which may also originate from lithogenic sources.

3.3.2 Cluster analysis

CA was used to group the HMs having homologous characteristics, which could give more details to further verify the results of above PCA analysis (Chabukdhara and Nema, 2012). CA examines distances between HMs and the most similar HMs are grouped forming one cluster and the process is repeated until all HMs belong to one cluster (Danielsson et al., 1999). The lower the value on the distance cluster, the more significant was the association. A criterion for the distance cluster of between 15 and 20 was used in this analysis.

In the sediments, two distinct clusters can be identified [Fig. S5(a)]. Cluster 1 contained Cu, Zn, Cd, Pb and As. Due to high content, these elements probably came from acid water. Cluster 2 contained Cr and Hg. The contents of these elements are below their ABVs and may originate from lithogenic sources. In the surface soils, three distinct clusters can be identified [Fig. S5(b)]. Cluster 1 contained Zn, Cd and Cu. The concentrations of these elements are higher than their ABVs and may come from surface sulfide ores oxidation. Cluster 2 contained Hg, As and Pb. The contents of As and Pb are higher than their ABVs and may derive from sulfide ores oxidation. Due to low content, Hg may originate from lithogenic sources. Cluster 3 contained only Cr. Due to low content, the Cr may originate from lithogenic sources. In general, the results of the CA agreed well with that of the PCA analysis. The Rona Cu deposit HMs inputs into the surrounding environment caused significant enrichments of HMs, such as Cu, Pb, Zn, As and Cd in the sediments and surface soils. Overall, the elements of Cu, Pb, Zn, As and Cd in sediments and surface soils enriched beyond their ABVs are seriously affected by Rona Cu deposit, whereas Cr and Hg may primarily originate from lithogenic sources. The causes of HMs formation are described as follows.

Rona tributary acid water formed in the middle of Rona Cu deposit may be caused by the North-South tectonic fault of the Tibetan plateau. The destruction of the original stable protection layer for metal sulfide ores and surrounding rocks caused the exposure of sulfur-bearing minerals in Rona deposit to atmospheric oxygen and moisture, under appropriate environmental conditions, sulfides are quickly oxidized and many associated elements, such as Cu, Pb, Zn, As and Cd are released and separated from sulfur, resulting in a large amount of AMD (Duruibe et al., 2007). Harrison et al. (1992) showed that the North-South tectonic fault in the study area formed in 7–9 Ma, later than the mineralization age of the deposit in 116–120 Ma (Tang et al., 2016). Furthermore, field observation showed that the tributary water distributed along the North-South tectonic fault [Fig. 1(a)]. Compared with other sulfurbearing minerals, pyrite is more likely to produce AMD (Obreque-Contreras et al., 2015) and the formation can be represented by the following equations:

$$2FeS_2(s)+2H_2O+7O_2 \longrightarrow 2Fe^{2+}+4SO_4^{2-}+4H^+ \tag{7}$$

$$2Fe^{2+}+2H^++1/2O_2 \longrightarrow 2Fe^{3+}+H_2O \tag{8}$$

$$2Fe^{3+}+6H_2O \longleftrightarrow 2Fe(OH)_3(s)+6H^+ \tag{9}$$

$$FeS_2(s)+14Fe^{3+}+8H_2O \longrightarrow 15Fe^{2+}+2SO_4^{2-}+16H^+ \tag{10}$$

The oxidation of chalcopyrite produced a large amount of soluble Cu^{2+}, which was the main source of Cu pollution in the study area, and the formation can be represented by the following equations:

$$2CuFeS_2+8O_2+H_2SO_4 \longrightarrow 2CuSO_4+Fe_2(SO_4)_3+H_2O \tag{11}$$

$$Fe_{1-x}+\left(2-\frac{x}{2}\right)O_2+H_2O \longrightarrow (1-x)Fe^{2+}+SO_4^{2-}+2H^+ \tag{12}$$

$$Fe^{2+}+1/4O_2+5/2H_2O \longleftrightarrow Fe(OH)_{3(s)}+2H^+ \tag{13}$$

3.4 Environmental risk assessment

Fig. S6 illustrated the potential ecological risks of the HMs in the sediments and surface soils. Cu in sediments posed a low, moderate, high and serious potential ecological risk in approximately 55.2%, 17.2%, 22.4% and 5.2% of the study area, respectively. In addition, Cd and Hg posed a moderate risk in approximately 1.7% and 22.4%, respectively, whereas all of the sediment samples were free from ecological risk posed by other elements. Cu in surface soils posed a moderate, serious and severe ecological risk in approximately 9.1%, 63.6% and 27.3% of the study area, respectively, and As posed a low, moderate, high and serious ecological risk in approximately 72.7%, 9.1%, 13.6% and 4.6%, respectively. Additionally, Hg posed a moderate risk in approximately 13.6%, and Pb posed a moderate and high risk in approximately 13.6% and 9.1%, respectively, whereas all of the surface soils were free from ecological risk posed by Zn and Cr. Due to the easy dissolution and transport of major chemical fractions of Cd in surface soils, 31.8% of moderate ecological risk of Cd were observed. In summary, the results indicated that pollution control should be carried out in some locations, especially for Cu and As.

As presented in Fig. 5, the RI values of the elements in the sediments ranged from 50.40 to 282.77. According to the category of RI, 17 samples were classified into moderate risk and 41 samples were classified into low risk. The RI values of the elements in the surface soils ranged from 139.07 to 982.53. According to the category of RI, 2 samples were classified into very high risk; 15 samples were classified into considerable risk; 4 samples were classified into moderate risk; and 1 sample was classified into low risk. The distribution of RI values in Fig. 5 showed an obviously corresponding relationship and similar changing trend with the distribution of P_n values in Fig. 4. Therefore, more reliable results can be obtained by using these two evaluation methods simultaneously.

The results of RAC assessment are tabulated in Table 2. Cu and Pb ranged from no risk to low risk in all sediments and surface soils; and Zn ranged from low to medium risk in sediment samples, from no risk to low risk in surface soil samples. Except sample Z6 (low risk), the element As was no risk in all sediments and surface soils. Except sample S13 (low risk), Cr was no risk in all samples. Moreover, the Hg was low risk in all sediments and surface soils. Notably, the Cd ranged from medium to very high risk in sediments, and low to high risk in surface soils. Obviously, there was disagreement between the RAC and the P_i analysis shown in Table 2. There are sediment and surface soil samples considered to be moderately or heavily polluted that ranged from no risk to low risk according to the RAC, e.g. S1, S6, S7, S10, S13, Z9, Z13, X8, X12, T12, T13, T14, T17, T18, T21 and T22 for Cu; and the samples considered to be slightly or heavily polluted are no risk according to the RAC, e.g. S10, Z13, X5, X6, T12, T13, T14, T17, T21 and T22 for As. On the contrary, samples considered to be unpolluted are low to very high risk according to the RAC for Cd. The all above analysis showed that there are obviously different environmental risk assessments among different evaluation methods, such as P_i, I_{geo}, E_{ir} and RAC, which may be caused by different emphasis of evaluation. Accordingly, when evaluating environmental risk of HMs, both total metal concentration and bioavailable metal fraction should be considered.

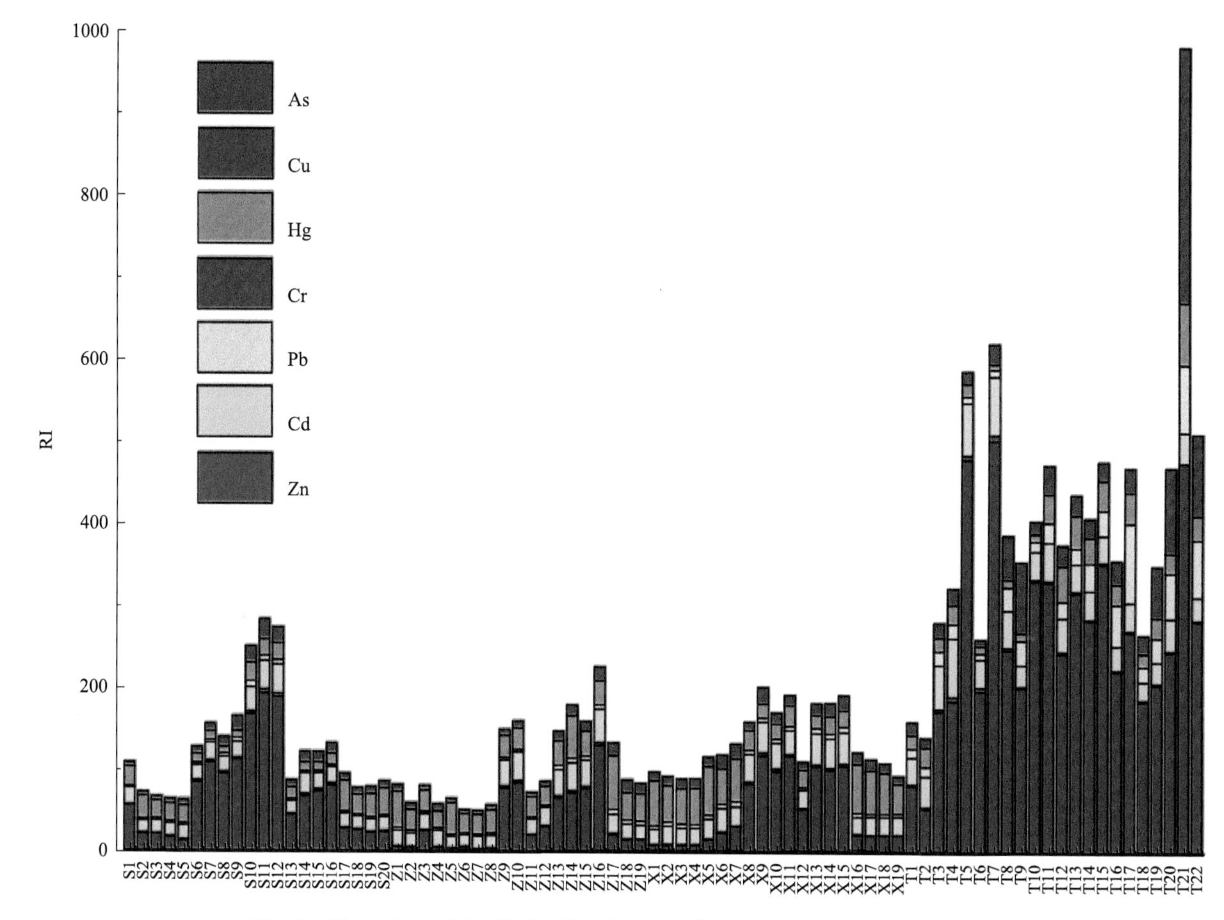

Fig. 5　The values of the RI for HMs in the sediment and surface soil samples.

Table 2　Comparison table of P_i and RAC assessment results for Cu, As and Cd

Sample name	Cu		As		Cd	
	P_i	RAC	P_i	RAC	P_i	RAC
S1	Moderately polluted	L	Safety	N	Safety	H
S6	Heavily polluted	L	Safety	N	Safety	M
S7	Heavily polluted	L	Safety	N	Safety	M
S10	Heavily polluted	L	Slightly polluted	N	Safety	H
S13	Moderately polluted	L	Safety	N	Safety	VH
Z1	Safety	N	Safety	N	Safety	VH
Z5	Safety	L	Safety	N	Safety	H
Z6	Safety	L	Safety	L	Safety	H
Z9	Heavily polluted	L	Safety	N	Safety	H
Z13	Moderately polluted	L	Slightly polluted	N	Safety	H
X1	Safety	L	Safety	N	Safety	H
X5	Safety	L	Slightly polluted	N	Safety	H
X6	Safety	L	Slightly polluted	N	Safety	H
X8	Heavily polluted	L	Safety	N	Safety	M
X12	Moderately polluted	L	Safety	N	Safety	M
T12	Heavily polluted	N	Slightly polluted	N	Safety	H
T13	Heavily polluted	N	Slightly polluted	N	Safety	M
T14	Heavily polluted	N	Slightly polluted	N	Safety	H
T17	Heavily polluted	N	Slightly polluted	N	Safety	M
T18	Moderately polluted	N	Safety	N	Safety	L
T21	Heavily polluted	L	Heavily polluted	N	Safety	H
T22	Heavily polluted	L	Heavily polluted	N	Safety	H

4 Conclusions and recommendations

The results of this paper showed that environmental pollution in mines was not entirely caused by anthropogenic activities. Unexploited deposits in natural background also had strong environmental risk. Natural metallic acid water discharge and the oxidation of surface sulfide ores in Rona Cu deposit have resulted in severe HMs pollution of the study area in Tibet, China. The elements Cu and Zn were two pollutants in acid water, and Rona river was obviously affected by acid water. The acid water had no effect on groundwater. The dominant pollution of HMs in sediments and surface soils came from Cu, Zn and As. Due to relatively high residual fraction, the environmental risk of Cu, Zn and As may be low. Cd in sediments and surface soils near high risk status. However, due to high content, the environmental risk of Cu, Zn and As still have to be highly focused. Due to low content, the environmental risk of Cd do not need to pay too much attention. The source of Cu, Pb, Zn, As and Cd in sediments and surface soils may mainly derive from Rona Cu deposit, whereas Cr and Hg may primarily originate from lithogenic sources. The treatment for Rona natural acid water pollution only needs to cut off the sewage source, thus, the mining of Rona Cu deposit may be the best way to solve the pollution. Unexploited Rona Cu deposit had produced considerable environmental risk for the surrounding environment, thus, it is necessary to pay more attention to the environmental risk of more unexploited deposits.

Declaration of interest statement

We declare that we have no financial and personal relationships with other people or organizations that can inappropriately influence our research work. There is no professional or other personal interest of any nature or kind in any product, service and/or company that could be construed as influencing the manuscript entitled"Pollution, sources and environmental risk assessment of heavy metals in the surface AMD water, sediments and surface soils around unexploited Rona Cu deposit, Tibet, China".

Acknowledgements

This work was financially supported by Investigation and Evaluation of the Whole Industrial Chain of Core Minerals in Emerging Industries (No. DD20190676) and Integrated Evaluation of Technology Economy and Environment of Duolong Deposit in Tibet (No. DD20160330).

Appendix A. Supplementary data

Supplementary data to this article can be found online at https://doi.org/10.1016/j.chemosphere.2020.125988.

References

Akabzaa, T.M., Armah, T.E.K., Baneong-Yakubo, B.K., 2007. Prediction of acid mine drainage generation potential in selected mines in the Ashaanti Metallogenic Belt using static geochemical methods. Environ. Geol.52, 957-964.

Alloway, B., 1995. Heavy Metals in Soils. Springer, London.

Azhari, A.E., Rhoujjati, A., Hachimi, L.M.E., Ambrosi, J.P., 2017. Pollution and ecological risk assessment of heavy metals in the soil-plant system and the sediment-water column around a former Pb/Zn mining area in NE Morocco. Ecotoxicol. Environ. Saf.144, 464-474.

Cantle, J.E., 1986. Atomic Absorption Spectrometry 5. Elsevier.

CEPA, 1995. Environmental Quality Standard for Soils (In Chinese) (GB 15618-1995).

Chabukdhara, M., Nema, A.K., 2012. Assessment of heavy metal contamination in Hindon River sediments: a chemometric and geochemical approach. Chemosphere 87, 945-953.

Chen, H.Y., Teng, Y.G., Lu, S.J., Wang, Y.C., Wang, J.S., 2015. Contamination features and health risk of soil heavy metals in China. Sci. Total Environ.

512-513, 143-153.

Danielsson, Å., Cato, I., Carman, R., Rahm, L., 1999. Spatial clustering of metals in the sediments of Skagerrak/Kattegat. Appl. Geochem.14, 689-706.

Del Rl'o, M., Font, R., Almela, C., Ve'lez, D., Montoro, R., Bailo'n, A.D.H., 2002. Heavy metals and arsenic uptake by wild vegetation in the Guadiamar river area after the toxic spill of the Aznalco'llar mine. J. Biotechnol.98, 125-137.

Ding, Z.W., Li, Y., Sun, Q.Y., Zhang, H.J., 2018. Trace elements in soils and selected agricultural plants in the Tongling mining Area of China. Int. J. Environ. Res. Publ. Health 15, 202.

Duruibe, J.O., Ogwuegbu, M.O.C., Egwurugwu, J.N., 2007. Heavy metal pollution and human biotoxic effects. Int. J. Phys. Sci.2 (5), 1942-1950.

Elliott, H.A., Liberati, M.R., Huang, C.P., 1986. Competitive adsorption of heavy metals by soil. J. Environ. Qual.15, 214-219.

Gaur, V.K., Gupta, S.K., Pandey, S.D., Gopal, K., Misra, V., 2005. Distribution of heavy metals in sediment and water of River Gomti. Environ. Monit. Assess. 102, 419-433.

Gimeno-Garcia, E., Andreu, V., Boluda, R., 1995. Distribution of heavy metals in rice farming soils. Arch. Environ. Contam. Toxicol. 29, 476-483.

Gong, Q.J., Deng, J., Xiang, Y.C., Wang, Q.F., Yang, L.Q., 2008. Calculating pollution indices by heavy metals in ecological geochemistry assessment and a case study in parks of Beijing. J. China Univ. Geosci.19 (3), 230-241.

Gunson, A.J., Jian, Y., 2001. Artisanal Mining in the People's Republic of China. International Institute of Environment and Development.

Hakanson, L., 1980. An ecological risk index for aquatic pollution control; A sedimentological approach. Water Res.14, 975-1001.

Hammarstrom, J.M., Seal, R.R., Jackson, J.C., 2003. Weathering of sulfidic shale and copper mine waste: secondary minerals and metal cycling in Great Smoky Mountains National Park, ennessee, and North Carolina, USA. Environ. Geol.45, 35-57.

Hanesch, M., Scholger, R., Dekkers, M.J., 2001. The application of fuzzy C-means cluster analysis and non-linear mapping to a soil data set for the detection of polluted sites. Earth and Geodesy 26, 885-891.

Harrison, T.M., Copeland, P., Kidd, W.S.F., Yin, A., 1992. Raising Tibet. Science 255, 1663-1670.

Hu, S., Chen, X., Shi, J., Chen, Y., Lin, Q., 2008. Particle-facilitated lead and arsenic transport in abandoned mine sites soil influenced by simulated acid rain. Chemosphere 71, 2091-2097.

Jain, C.K., 2004. Metal fractionation study on bed sediments of river Yamuna, India. Water Res.38, 569-578.

Kafayatullah, Q., Shah, M.T., Irfan, M., 2001. Biogeochemical and environmental study of the chromite-rich ultramafic terrain of Malakand area, Pakistan. Environ. Geol.40, 1482-1486.

Kang, X., Song, J., Yuan, H., Duan, L., Li, X.G., Li, N., Liang, X.M., Qu, B.X., 2017. Speciation of heavy metals in different grain sizes of Jiaozhou Bay sediments: bioavailability, ecological risk assessment and source analysis on a centennial timescale. Ecotoxicol. Environ. Saf.143, 296-306.

Kaufmann, R.B., Staes, C.J., Matte, T.D., 2003. Deaths related to lead poisoning in the United States, 1979-1998. Environ. Res.91, 78-84.

Khan, A., Khan, S., Khan, M.A., Qamar, Z., Waqas, M., 2015. The uptake and bioaccumulation of heavy metals by food plants, their effects on plants nutrients, and associated health risk: a review. Environ. Sci. Pollut. Res. 22 (18), 13772-13799.

Khan, M.A., Khan, S., Khan, A., Alam, M., 2017. Soil contamination with cadmium, consequences and remediation using organic amendments. Sci. Total Environ. 601-602, 1591-1605.

Kong, I.C., Bitton, G., 2003. Correlation between heavy metal toxicity and metal fractions of contaminated soils in Korea. Bull. Environ. Contam. Toxicol. 70, 557-565.

Lei, M., Liao, B.H., Zeng, Q.R., Qin, P.F., Khan, S., 2008. Fraction distributions of lead, cadmium, copper, and zinc in metal-contaminated soil before and after extraction with disodium ethylenediaminetetraacetic acid. Commun. Soil. Sci. Plan.39 (13-14), 1963-1978.

Li, M.S., Yang, S.X., 2008. Heavy metal contamination in soils and phytoaccumulation in a manganese mine wasteland, south China. Air Soil. Water Res.1, 31-41.

Li, G. M., Zhang, X.N., Qin, K.Z., Sun, X.G., Zhao, J.X., Yin, X.B., Li, J.X., Yuan, H.S., 2015. The telescoped porphyry-high sulfidation epithermal Cu (Au) mineralization of Rona deposit in Duolong ore cluster at the southern margin of Qiangtang Terrane, Central Tibet: integrated evidence from geology, hydrothermal alteration and sulfide assemblages. Acta Petrol. Sin.31, 2307-2324 (in Chinese).

Liu, G. N., Tao, L., Liu, X.H., Hou, J., Wang, A.J., Li, R.P., 2013. Heavy metal speciation and pollution of agricultural soils along Jishui River in non-ferrous metal mine area in Jiangxi Province, China. J. Geochem. Explor.132, 156-163.

Liu, G. N., Wang, J., Liu, X., Liu, X.H., Li, X.S., Ren, Y.Q., Wang, J., Dong, L.M., 2018a. Partitioning and geochemical fractions of heavy metals from geogenic and anthropogenic sources in various soil particle size fractions. Geoderma 312, 104-113.

Liu, R.P., Xu, Y.N., Zhang, J.H., Chen, H.Q., He, F., Qiao, G., Ke, H.L., Shi, Y.F., 2018b. A comparative study of the content of heavy metals in typical metallic mine rivers of the Tibetan Plateau. Geol. Bull. China 37 (12), 2154-2168 (in Chinese).

Lü, J., Jiao, W.B., Qiu, H.Y., Chen, B., Huang, X.X., Kang, B., 2018. Origin and spatial distribution of heavy metals and carcinogenic risk assessment in mining areas at You'xi County southeast China. Geoderma 310, 99-106.

Luo, Y.H., 2017. The geochemical evolution and ecological response of natural sewage in Rona river from the ore body, North-Tibet, China. Master dissertation of China University of Geosciences (Beijing) 1-73 (in Chinese).

Ma, X.L., Zuo, H., Tian, M.J., Zhang, L.Y., Meng, J., Zhou, X.N., Min, N., Chang, X.Y., Liu, Y., 2016. Assessment of heavy metals contamination in sediments from three adjacent regions of the Yellow River using metal chemical fractions and multivariate analysis techniques. Chemosphere 144, 264-272.

Madrid, L., Diaz-Barrientos, E., Madrid, F., 2002. Distribution of heavy metal contents of urban soils in parks of Seville. Chemosphere 49, 1301-1308.

Matys Grygar, T., Popelka, J., 2016. Revisiting geochemical methods of distinguishing natural concentrations and pollution by risk elements in fluvial sediments. J. Geochem. Explor.170, 39-57.

Mico, C., Recatala, L., Peris, M., Sanchez, J., 2006. Assessing heavy metal sources in agricultural soils of an European Mediterranean area by multivariate analysis. Chemosphere 65, 863-872.

MOHC, 2006. Standards for Drinking Water Quality (GB 5749-2006).

Mostert, M.M.R., Ayoko, G. A., Kokot, S., 2010. Application of chemometrics to analysis of soil pollutants. Trends Anal. Chem.29, 430-445.

Muller, G., 1969. Index of geo-accumulation in sediments of the rhine river. Geol. J. 2, 108-118.

Mwesigye, A.R., Young, S.D., Bailey, E.H., Tumwebaze, S.B., 2016. Population exposure to trace elements in the Kilembe copper mine area, Western Uganda: a pilot study. Sci. Total Environ.573, 366-375.

Nasrabadi, T., Bidhendi, G. N., Karbassi, A., Merdadi, N., 2010. Partitioning of metals in sediments of the haraz river (southern caspian sea basin). Environ. Earth. Sci. 59, 1111-1117.

NEPAC, 2002. Environmental Quality Standards for Surface Water (GB 3838-2002).

NEPAC, 2010. Emission Standard of Pollutants for Copper, Nickel, Cobalt Industry (GB 25467-2010).

Nordstrom, D.K., Alpers, C.N., Ptacek, C.J., Blowes, D.W., 2000. Negative pH and extremely acidic mine waters from Iron Mountain, California. Environ. Sci. Technol.34, 254-258.

Obreque-Contreras, J., Pérez-Flores, D., Gutiérrez, P., Chávez-Crooker, P., 2015. Acid mine drainage in Chile: an opportunity to apply bioremediation Technology. Hydrol. Curr. Res.6, 215.

Perin, G., Craboledda, L., Lucchese, M., Cirillo, R., Dotta, L., Zanette, M.L., Orio, A.A., 1985. Heavy metal speciation in the sediments of Northern Adriatic Sea-a new approach for environmental toxicity determination. In: Lekkas, T.D. (Ed.), Heavy Metal in the Environment, vol.2, pp.454-456.

Praveena, S.M., Radojevic, M., Abdullah, M.H., 2007. The assessment of mangrove sediment quality in mengkabong lagoon: an index analysis approach. Int. J. Environ. Sci. Educ.2 (3), 60-68.

Qiao, D.H., 2018. Environment Attribute Model of Rongna Deposit, Tibet. Master dissertation of China University of Geosciences (Beijing), pp.1-130 (in Chinese).

Ramirez, M., Massolo, S., Frache, R., Correa, J.A., 2005. Metal speciation and environmental impact on sandy beaches due to El Salvador copper mine, Chile. Mar. Pollut. Bull. 50, 62-72.

Rauret, G., Lopez-Sanchez, J.F., Sahuquillo, A., Davidson, C.M., Ure, A.M., Quevauviller, P.H., 1999. Improvement of the BCR 3-step sequential extraction procedure prior to the certification of new sediment and soil reference materials. J. Environ. Monit.1, 57-61.

Rehman, U., Khan, S., Muhammad, S., 2019. Ingestion of arsenic-contaminated drinking water leads to health risk and traces in human biomarkers (hair, nails, blood, and urine). Pakistan. Expo. Health. https: //doi.org/10.1007/s12403-019-00308-w.

Rinklebe, J., Shaheen, S.M., 2017. Geochemical distribution of Co, Cu, Ni, and Zn in soil profiles of fluvisols, luvisols, gleysols, and calcisols originating from Germany and Egypt. Geoderma 307, 122-138.

Rodríguez, L., Ruiz, E., Alonso-Azcárate, J., Rincón, J., 2009. Heavy metal distribution and chemical speciation in tailings and soils around a Pb-Zn mine in Spain. J. Environ. Manag. 90, 1106-1116.

Saleem, M., Iqbal, J., Akhter, G., Shah, M.H., 2018. Fractionation, bioavailability, contamination and environmental risk of heavy metals in the sediments from a freshwater reservoir, Pakistan. J. Geochem. Explor.184, 199-208.

Sani, H.A., Ahmad, M.B., Hussein, M.Z., Ibrahim, N.A., Musa, A., Saleh, T.A., 2017. Nanocomposite of ZnO with montmorillonite for removal of lead and copper ions from aqueous solutions. Process Saf. Environ. Protect.109, 97-105.

Shah, M.T., Begum, S., Khan, S., 2010. Pedo and biogeochemical studies of mafic and ultramafic rocks in the Mingora and Kabal areas, Swat, Pakistan. Environ. Earth Sci. 60, 1091-1102.

Tang, J.X., Song, Y., Wang, Q., Lin, B., Yang, C., Guo, N., Fang, X., Yang, H.H., Wang, Y.Y., Gao, K., Ding, S., Zhang, Z., Duan, J.L., Chen, H.Q., Su, D.K., Feng, J., Liu, Z.B., Wei, S.G., He, W., Song, J.L., Li, Y.B., Wei, L.J., 2016. Geological characteristics and exploration model of the tiegelongnan Cu (Au-Ag) deposit: the first ten million tons metal resources of a porphyry-epithermal deposit in Tibet. Acta Geosci. Sin.37, 663-690.

Tessier, A., Compbell, P. G. C., Bisson, M., 1979. Sequential extraction procedure for the speciation of particulate trace metals. Anal. Chem.51, 844-851.

Wang, Y., Wei, F.S., 1995. Environmental Chemistry of Soil Elements. China Environmental Science Press, pp. 58-150 (in Chinese).

Wang, M.Q., Wang, R.J., Chen, R.Y., Yan, G. S., 2006. A new suggestion to build a bridge between exploration and environmental assessment of mineral resources. Earth Environ.34 (2), 41-46 (in Chinese).

Yin, S.H., Wang, L.M., Kabwe, E., Chen, X., Yan, R.F., An, K., Zhang, L., Wu, A.X., 2018. Copper bioleaching in China: review and prospect. Minerals 8, 32.

第五部分

盐湖资源发展战略

创新开源，为全球锂电大产业发展提供资源保障*

郑绵平

中国地质科学院矿产资源研究所，自然资源部盐湖资源与环境重点实验室，北京 100037

随着世界范围引发的能源危机和环境污染加剧，新能源电动车乃至电动航空和移动储能设备等的迅速发展，锂在电池行业消费量从21世纪开始快速上升，尤其2015年以来，电动车产业迅猛发展，大力促进了锂的消费，至2018年，全球电池行业锂消费占比已增长至65%，达到了32万t碳酸锂当量（LCE），其余行业锂消费量占35%，仅有11.9万t LCE。预计全球锂资源需求量将保持15%～20%快速增长，锂资源的供给将是锂电产业持续发展的关键之一。

相关研究团队依据采用物质分析方法，预测了未来全球总的锂资源（自然锂矿产资源和城市锂矿产资源）的增长态势。若按照目前已知锂自然矿产储量1700万t（以下锂储量、资源量均以金属锂计算），假定无新增且锂的回收利用率可达100%，则锂自然矿产储量至少需要提高1倍，即增长至3300万t才能满足全球2080年以来对锂资源的需求。从中长期来看，二次资源（锂城市矿产利用）将发挥越来越重要作用。

全球锂自然资源，除已知锂储量3000万t外，资源量可达8000万t，后者为前者2.7倍。此外，全球还有待查明的巨量锂潜在资源，包括含锂盐湖、伟晶岩和含锂深部卤水、油田水、地热水和黏土等，以及待开拓的其他类型锂矿产。上述全球锂资源量和潜在资源，随着市场需求和科技发展，只要投入调查和工程量，将可能成为可开发的锂储量。如阿根廷广布的盐湖等，由于当地政府采取对外开放的探矿政策，2006年锂储量仅有20万t，至2018年猛增到640万t，增长了32倍。又如美国，曾经是伟晶岩和盐湖锂资源储量较少的国家，经多年调研，相继开发出锂蒙脱石（黏土型）、地热水、油田水等锂资源，而使美国在2005年锂储量3.8万t，资源量76万t，至2017年锂资源量达690万t，而2020年公布的锂储量达63万t。

经多年地质调查，中国已在青藏高原盐湖和新疆、四川等地伟晶岩型发现和评价一批锂资源和储量，截至2015年，已获锂储量约350万t，资源量约600万t，近年来由于中国地矿科技的进展，特别是近十余年来，相继在柴达木油田水、川东深部卤水、青藏高原-川东伟晶岩、西藏地热水及黔滇黏土等有大批锂资源的发现，经过多年科技攻关，破解了柴达木盐湖卤水Mg/Li特高的世界性难题，在2017年达到了年产3.5万t规模LCE。

以上说明，全球自然锂资源具有巨大远景，加之氢电池也是未来新能源汽车和储能另一潜力巨大资源，从而构成“三类资源”（全球自然锂资源、城市锂资源和氢能资源），支持未来全球新能源汽车和储能等发展所需资源是可以期待的。为了实现经济开发利用，无论是自然锂资源、城市锂资源，还是氢能资源，都存在其各自需要解决的关键科学技术难题和某些管理体制的制约。

锂资源发展的每一次进步既同相关国家管理与政策有关，更同科技进步密不可分。为了攻克上述盐类新能资源关键科技瓶颈，使中国在新能源锂电资源领域进入世界前列，亟须进一步完善相关管理机制政策，实行产学研政有效联合，推进跨部门多学科结合，协同创新，开源节流，以期为中国和全球锂电产业大发展提供资源保障。

* 本文发表在：科技导报，2020，38（15）：1

中国钾盐消费规律与需求预测*

牛新生[1,2]，王安建[1,3]，郑绵平[1,2]

1. 中国地质科学院矿产资源研究所，北京 100037
2. 自然资源部盐湖资源与环境重点实验室，北京 100037
3. 中国地质科学院全球矿产资源战略研究中心，北京 100037

摘要 钾盐是我国的紧缺矿种之一，从保障农业生产安全的战略角度出发，有必要对我国钾盐的中长期消费做出预测。本文对全球主要发达国家钾肥消费规律的研究表明，发达国家的人均钾肥消费量、单位耕地钾肥消费量和钾肥施用总量分别与人均 GDP 呈倒“S”形，具有明显的峰值特征。发达国家的钾肥施用总量轨迹可分为密集型、稀疏型和中间型三种。密集型国家国土面积中等，单位面积耕地钾肥施用量较大，人均 GDP 相对较低的 11000～16000 美元就达到钾肥消费的峰值；稀疏型国家国土面积较大，单位耕地面积钾肥施用量较低，单位耕地的粮食产量较低，人均 GDP 较高，19000～20000 美元才达到钾肥消费的峰值；中间型国家国土面积较大，人口众多，为农业大国、强国，钾肥消费总量较高，人均 GDP 18000 美元左右时达到钾肥消费的峰值。我国钾肥的消费特点属于中间型，在人均 GDP 达到 18000 美元时，将迎来钾盐消费的高峰，峰值约为 1400 万 t K_2O，时间在 2025～2028 年。

关键词 钾盐；钾肥；消费规律；需求预测

钾盐是可溶性的含钾盐类矿物，是钾肥和含钾肥料的最主要矿物原料，85%～95%的钾盐用于制作钾肥。截至 2018 年底，我国已成为全球第一大钾盐、钾肥消费国，因此，探讨钾盐、钾肥的消费规律并对国家的中期、长期钾盐需求做出预测，是指导钾盐未来产业发展的重要依据。

以往钾盐需求预测的主要方法有：①根据钾盐消费的年均增长率外推未来钾盐消费，如趋势预测法[1]；②根据人口粮食增长率预测钾肥需求量[2]；③根据氮、磷、钾肥的消费比例关系判断钾肥消费情况[3]；④数理统计法[4]；⑤谷物系数法[5]；⑥综合多种因素的作物体系-专家模型法[6]。以上方法的预测结果多局限于 2020 年之前，对中期、长期的预测结果相对较少；部分预测方法（如趋势预测法）的预测结果出现单向增长，长期来看，这并不符合钾盐的实际消费情况；此外，部分预测方法仅停留在理论探讨阶段，开展钾盐消费的中长期预测十分必要。

先期工业化国家的实际经验表明，人均能源和重要矿产资源的消费具有“S”形规律[7-13]，“S”形规律揭示了人类经济社会发展过程中资源消费的极限与周期问题[14]，已被广泛用于全球及中国一次能源、粗钢、铜、铝等矿产资源需求预测[9, 15-18]。与其他传统预测方法相比，这一方法基于对先期工业化国家资源消费规律的高度概括和总结，注重长周期的宏观消费规律分析，更加适用于资源需求的中期、长期预测。张艳等[19]初步论证了单位耕地面积化肥施用量随着经济的发展呈“S”形规律；唐尧[20]基于“S”形规律尝试了对钾盐的需求预测，认为中国与法国的单位耕地钾肥量消费量相近。在此基础上，本文拟运用资源消费“S”形规律的思路，对先期工业化国家的钾盐、钾肥消费规律做出多层次的总结，建立合适的预测模型，进而预测我国 2030 年之前的钾盐消费需求。

1 需求预测方法与模型

由于大部分的钾盐用作钾肥（本文取 90%），因此，钾盐的消费实际是钾肥的消费，对钾盐消费规律的分析实际是对钾肥消费规律的分析。下文中，所有涉及钾盐、钾肥的量纲均以折纯的 K_2O 计算，人均 GDP 均

* 本文发表在：中国矿业，2019，28（10）：6-12。

以1990年盖凯美元为计量单位。盖凯美元是可用于不同国家系统对比的一种通用国际元，是一个假定的货币单位，通常把1990年作为对比的基准年，下文中简称美元。我们分别从钾肥消费时间曲线、人均GDP-人均钾肥消费量、单位耕地粮食产量-单位耕地钾肥施用量、人均GDP-单位耕地钾肥施用量、人均GDP-钾肥施用总量五种关系对发达国家的钾肥消费规律进行探讨，同时，将中国、印度、巴西作为发展中国家的代表与主要发达国家进行对比分析。

1）钾肥消费时间曲线。发达国家的钾肥消费时间曲线表明，无论是人均钾肥消费的消费轨迹，还是单位耕地面积的钾肥消费轨迹，均具有明显的顶点现象（图1和图2）。西方主要发达国家均已度过峰值时刻并开始下降，但在具体时间节点上有所不同。

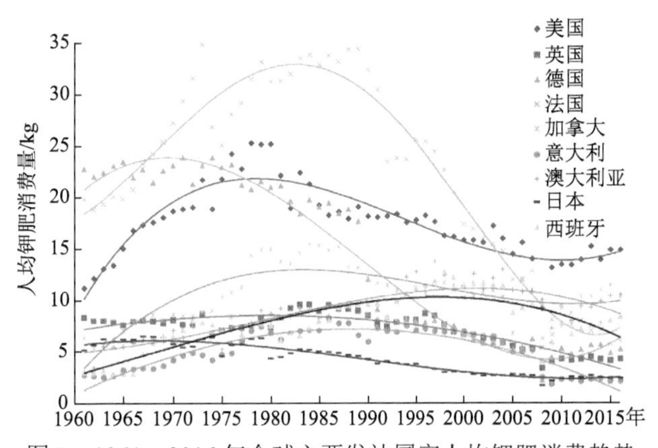

图1　1961～2016年全球主要发达国家人均钾肥消费趋势

Fig. 1　Per capita consumption trend of potassic fertilizer of developed countries from 1961 to 2016.

资料来源：文献[21]和文献[22]

人均钾肥消费方面，日本最早在20世纪60年代中期就已经达到了人均钾肥消费的顶点，而德国一直是钾肥生产大国，矿物钾肥的使用历史较长，钾盐资源丰富，其人均钾肥消费量在20世纪70年代初期达到了顶峰，多数发达国家如美国、法国、英国、加拿大、意大利等在20世纪70年代中期至80年代中期达到消费顶点，澳大利亚和西班牙直到2000年左右才迎来钾肥消费的峰值时刻（图1）。2000年以来，西方主要发达国家的钾肥消费趋势逐渐归于平缓，达到相对稳定的状态。

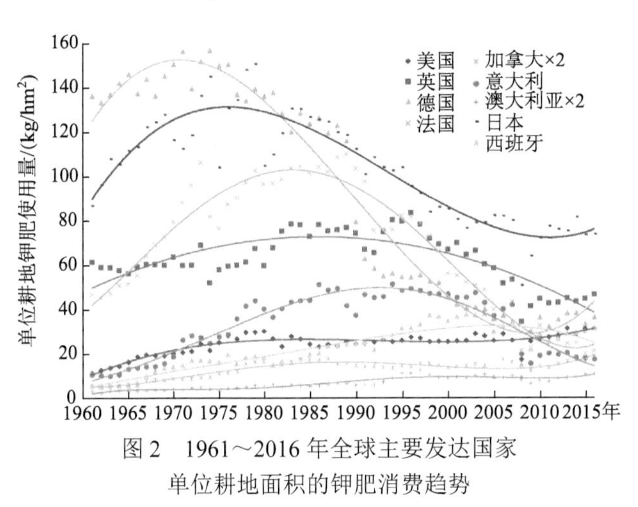

图2　1961～2016年全球主要发达国家单位耕地面积的钾肥消费趋势

Fig. 2　Per arable land potassic fertilizer consumption of developed countries from 1961 to 2016.

注：加拿大和澳大利亚的钾肥消费值进行了放大2倍处理；资料来源：文献[21]～[23]

单位耕地钾肥消费方面，主要发达国家也已先后达到钾肥消费的峰值（图2），峰值时刻呈波次递进的特点，并在峰值时刻之后逐步下降至相对稳定的水平上，其峰值区间介于1970～2010年。

与发达国家不同，主要发展中国家如中国、印度和巴西等国的钾肥消费均未达到钾肥消费的峰值阶段，人均钾肥消费量和单位耕地面积的钾肥消费量仍处于持续增长的阶段（图3和图4）。

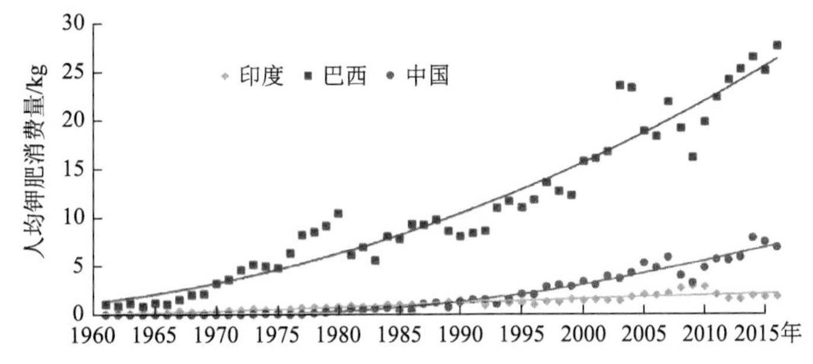

图3　1961～2016年主要发展中国家人均钾肥消费趋势

Fig. 3　Per capita consumption trends of potassic fertilizer of developing countries from 1961 to 2016.

资料来源：文献[21]～[23]

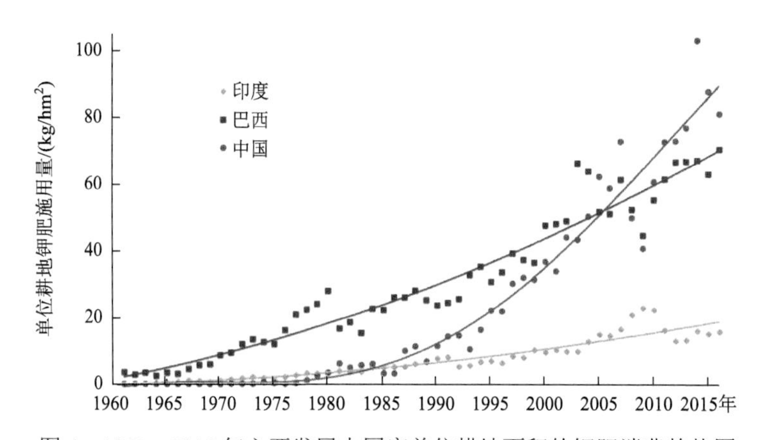

图 4　1961～2016 年主要发展中国家单位耕地面积的钾肥消费趋势图

Fig. 4　Per arable land potassic fertilizer consumption of developing countries from 1961 to 2016.

资料来源：文献[21]和文献[22]

2）人均 GDP-人均钾肥消费量。从人均 GDP 和人均钾肥消费量的关系来看，大部分发达国家的曲线大致呈“S”形（图 5）。各发达国家人均钾肥消费量达到顶点时的人均 GDP 并不一致，峰值时刻分布相对分散。其中，德国由于自己是主要钾盐生产国，因此其在人均 GDP 约 11000 美元的时候就达到了消费顶点；日本由于耕地奇缺，较为依赖化肥对耕地肥力的提升，也在人均 GDP 约 11000 美元的时候就达到了消费顶点。其他发达国家如美国、英国、法国、意大利、加拿大、澳大利亚、西班牙等国在人均 GDP 14000～20000 美元时达到消费顶点，之后呈持续下降趋势，在一定阶段时趋势变缓并逐渐稳定，“S”形轨迹非常明显。需要注意的是，图 5 中法国和德国曲线并不像美国、英国等呈现典型“S”形，而是略呈倒“U”形[19]，之所以呈现这种形态，一是 10000 美元以下时刻两国的人均钾肥消费量数据缺乏；二是两国的人均钾肥消费量峰值相对较高。造成这一现象的原因可能与两国人均钾肥量消费的意外降低有关，如德国所经历的东西德合并事件等因素，事实上，两国近十年的消费数据已经基本处于相对平稳的状态，因此，这种倒“U”形实际上是倒“S”形峰值拉高变形后的结果。

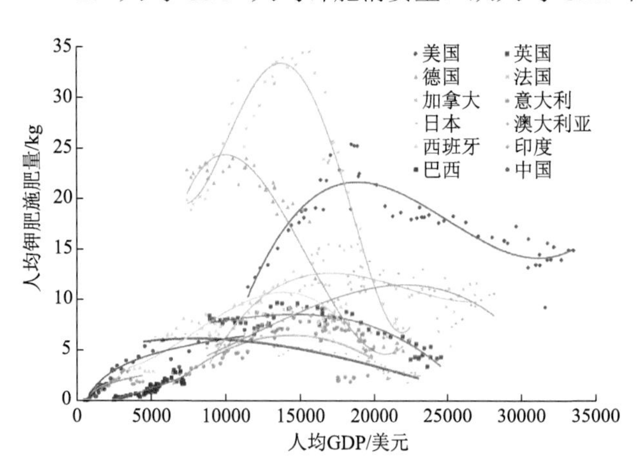

图 5　全球代表性国家人均 GDP 与人均钾肥消费量关系图

Fig. 5　Per capita GDP and per capita consumption trends of potassic fertilizer of typical countries.

资料来源：文献[21]和文献[22]

与发达国家不同，发展中国家中国、印度、巴西的钾肥消费仍处于增长过程中，中国已经接近钾肥消费的零增长点，巴西则处于钾肥消费的快速增长阶段（图 5）。

人均 GDP 和人均钾肥消费量的关系表明，当经济起飞时，人均钾肥消费量随人均 GDP 快速增长而同步增长，随着人均收入的增加，人均钾肥消费量增速趋缓，并在某一 GDP 值时增速为零，此时人均钾肥消费量达到顶点，其后开始逐渐降低。由于不同国家和地区经济发展水平、所处经济发展阶段和土壤肥力、种植结构等影响因素的不同，钾肥消费增长的起点、增长速率、峰值到来的时间点存在差异。

3）单位耕地粮食产量-单位耕地钾肥施用量。从单位耕地钾肥施用量与单位耕地粮食产量关系来看，主要发达国家美国、英国、法国、澳大利亚、德国、日本、意大利、加拿大、西班牙等国钾肥消费轨迹存在明显的顶点，整体轨迹呈倒“U”形（图 6），这个特点说明农作物对钾肥的需求不是传统认识上的“多多益善”，证明了钾肥应用的一个基本规律，即钾肥的使用不是粮食增产丰收的唯一影响因素。在早期的农业生产活动中，钾肥的投入对粮食作物的增产、增收至关重要，但随着科学技术的进步和种植条件的改善，粮食作物的丰收对钾肥投入的依赖性逐渐降低。发达国家已经走到了倒“U”形的右边，而发展中国家如中国、印度、巴西等仍处于“爬升”阶段，这也就解释了各发达国家钾肥消费的峰值现象。

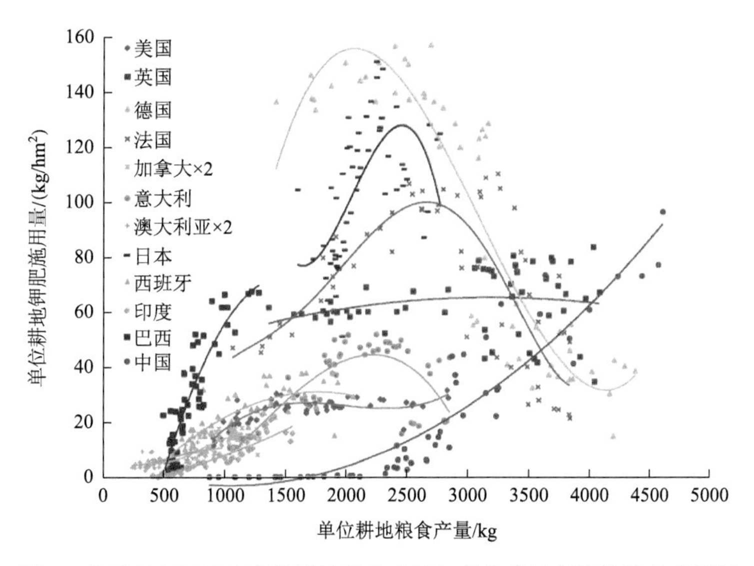

图 6 全球代表性国家单位耕地粮食产量与单位耕地钾肥施用量关系图

Fig. 6 Grain yield per arable land and per arable land potassic fertilizer consumption of typical counties.

注：加拿大和澳大利亚的钾肥消费值进行了放大 2 倍处理
资料来源：文献[21]和文献[22]

4）人均 GDP-单位耕地钾肥施用量。从人均 GDP 与单位耕地钾肥施用量关系图来看，主要发达国家单位耕地钾肥消费量的轨迹也是呈倒“S”形轨迹（图 7）。德国是工业化时期主要的钾肥供应国，钾肥在德国农业的发展过程中得到了充分的应用。因此钾肥消费的峰值时刻较早，在人均 GDP 略高于 11000 美元时，其单位耕地的钾肥施用量达到了顶点，日本、英国、意大利、法国、美国、加拿大和西班牙等大多数发达国家均在人均 GDP 12000～18000 美元时，单位耕地的钾肥施用量达到了最大值，之后开始持续下降，澳大利亚在人均 GDP 约 20000 美元时，单位耕地的钾肥施用量达到顶点。与之形成对比的是，由于中国、印度、巴西等国的经济发展程度仍远未达到发达国家水平，人均 GDP 相对较低，仍处于钾肥消费增长的快速阶段。

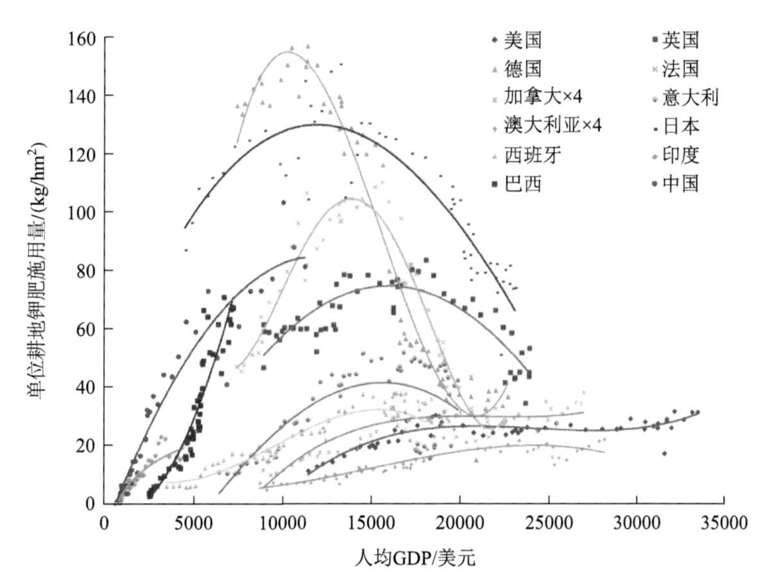

图 7 全球代表性国家单位耕地钾肥消费量与人均 GDP 关系轨迹图

Fig. 7 Per arable land potassic fertilizer consumption and per capita GDP of typical counties.

注：加拿大和澳大利亚的钾肥消费值进行了放大 4 倍处理
资料来源：文献[21]～[23]

5）人均 GDP-钾肥施用总量。与单位耕地钾肥施用量轨迹图相比，人均 GDP-钾肥施用总量关系图具有与之类似的特点，各国的钾肥消费总量轨迹均具有明显的顶点，大致呈倒“S”形。具体分析各发达国家钾肥消费的轨迹，其消费特点各有不同，根据单位耕地的钾肥施用量、单位耕地的粮食产量、钾肥消费峰值的达到时刻和国土面积等参数，可以进一步将发达国家分为三类消费类型。

第一类国家是钾肥消费的“密集型”国家。此类国家国土面积中等，单位面积耕地钾肥施用量较大，且单位耕地的粮食产量较高，人均 GDP 相对较低的 11000～16000 美元就达到了钾肥消费的峰值，代表性国家有英国、德国、法国、意大利、日本和西班牙（图 8），由于德国和法国钾肥消费量的“异常高”现象，并不具有普遍性，因此在预测模型中予以舍弃。

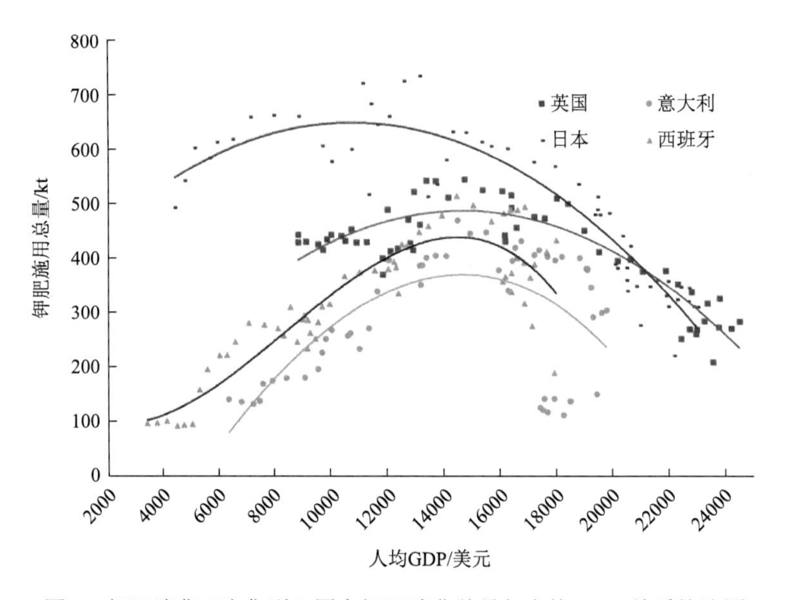

图 8　钾肥消费“密集型”国家钾肥消费总量与人均 GDP 关系轨迹图

Fig. 8　Total potassic consumption and per capita GDP of “intensive” countries.

资料来源：文献[21]和文献[23]

第二类国家是钾肥消费的“稀疏型”国家。此类国家国土面积较大，单位耕地面积钾肥施用量较小，单位耕地的粮食产量较小，人均 GDP 19000～20000 美元时才达到钾肥消费的峰值（图 9），以澳大利亚和加拿大两国为主要代表。

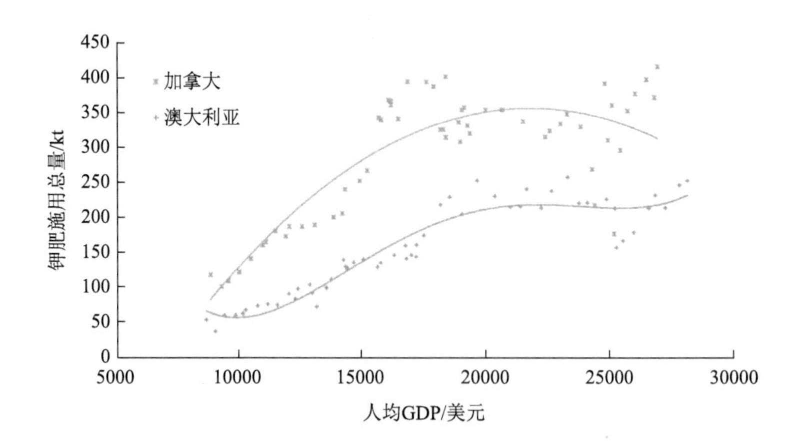

图 9　钾肥消费“稀疏型”国家钾肥消费总量与人均 GDP 关系轨迹图

Fig. 9　Total potassic consumption and per capita GDP of “sparse” countries.

资料来源：文献[21]和文献[23]

第三类国家是钾肥消费的“中间型”国家。以美国为主要代表，其特点是国土面积较大，耕地较多，且人口众多，为农业大国、强国，单位耕地钾肥消费量较大，钾肥消费的峰值时刻位于前两类国家之间，美国的峰值时刻人均 GDP 为 18000 美元。我们认为，中国、印度、巴西也属于此类国家，根据图 10 可知，中国、印度、巴西的钾肥消费总量仍远未到达顶点，中国相对巴西、印度更加接近峰值时刻。

需要指出的是，钾肥施用总量、单位耕地钾肥施用量和人均钾肥消费量三者的人均GDP分布图均呈倒“S”形轨迹，甚至峰值时刻也基本相同，但三者仍略有不同。相比人均钾肥消费量，单位耕地钾肥施用量轨迹分布趋同性更强，“S”形变化幅度更明显，原因在于钾肥直接作用于耕地，单位耕地钾肥施用量的表达更加客观，而钾肥施用总量轨迹的本质与单位耕地钾肥施用量相同，是钾肥消费总量曲线变化的直接表达，根据趋势变化，便于直接用来预测未来的钾肥需求情况。

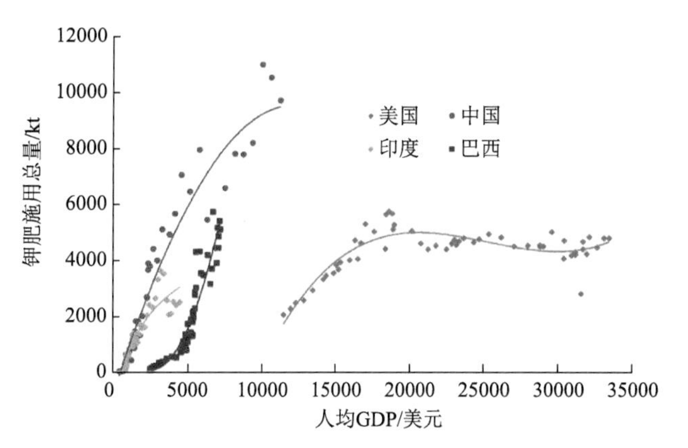

图 10 “中间型”国家钾肥消费总量与人均 GDP 关系轨迹图

Fig. 10 Total potassic consumption and per capita GDP of “medial” countries.

资料来源：文献[21]~[23]

2 需求预测方法、模型及指标体系

根据前述的钾盐、钾肥消费分析，我们认为资源消费预测的“S”形模型同样适用于钾盐、钾肥的消费预测，其规律适用于广大的发展中国家的需求预测，在总结钾肥消费规律的基础之上，建立了钾盐、钾肥需求预测的指标体系和方法。

1）需求预测的指标体系。①理论基础：前述发达国家的实例表明，人均钾肥消费量、单位耕地钾肥消费量和钾肥施用总量分别与人均 GDP 呈倒“S”形轨迹，依据国家消费类型的不同有三类不同情况，分别为密集型（以中等国土面积国家如法国、英国等国为代表）、中间型（以美国为代表）和稀疏型（以澳大利亚为代表），其峰值时刻分别对应人均 GDP 15000 美元、18000 美元和 22000 美元；发达国家单位耕地粮食产量与单位耕地钾肥消费量呈倒“U”形轨迹，即耕地的钾肥消费存在顶点，从理论上解释了钾肥消费峰值时刻形成的原因。②主要指标：人均 GDP、人均钾肥消费量、单位耕地钾肥施用量，预测体系（图 11）。③人均 GDP：包括国内生产总值和人口总数两项指标，这是决定钾肥需求的根本因素，人均 GDP 是衡量经济发展水平的重要指数，在定量的研究体系中，将人均 GDP 作为预测模型的自变量用于钾盐、钾肥需求预测中。④人均钾肥消费量：衡量不同国家和地区钾肥消费水平的重要指标，其峰值相对分散。⑤钾肥施用量：包括单位耕地钾肥施用量和钾肥施用总量，是衡量不同国家和地区单位耕地钾肥使用的重要指标，不仅可以体现不同农业发展阶段的钾肥消费规律，更能准确反映钾肥消费速度的变化轨迹。

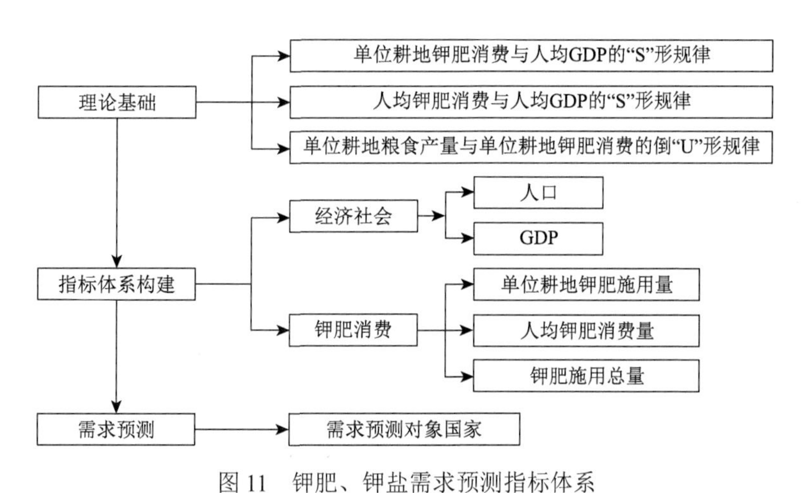

图 11 钾肥、钾盐需求预测指标体系

Fig. 11 Forecast index system of potash and potassic fertilizer demand.

2）预测过程。首先，建立预测国耕地钾肥或钾盐施用量与人均 GDP 的历史轨迹；其次，选取参照国类型，并选取参照国和预测国的转变点参数；再次，根据预测国经济发展规划或 GDP 增长趋势，确定预测区间人均 GDP 的增长方案；最后，按照 GDP 增长趋势，在预测国所属增长区内，按照其增长模式进行分段预测。

3　钾盐消费需求预测

综合考察发达国家的钾肥消费规律，我们认为中国的钾肥消费属于“中间型”，可与美国类比，理由有以下几方面：①我国国土面积广大，特别是与美国相比，地理纬度相当且同为农业大国，对钾肥的长远需求情况类似；②不同于法国、英国、日本等国，这些国家在达到钾肥的消费顶点后，需求量锐减，而我国是拥有 13 亿人口的消费大国，农业生产的安全事关 13 亿人口的饭碗，因此，即使将来我国的钾肥消费到达顶点，也不会出现钾肥需求量锐减的情况，而将逐步递减，并最终稳定在一定的需求水平上；③我国耕地虽然总体数量较大，但是人均耕地面积较少，这就需要单位耕地面积拥有较高的粮食产量，也决定了钾肥需求的必然性、高效性和长期性，因此钾肥消费的特点也不同于“稀疏型”国家。

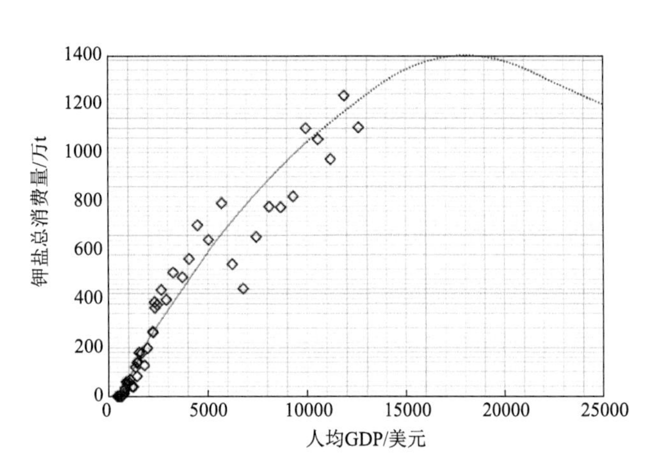

图 12　中国人均 GDP 与钾盐消费预测关系图

Fig. 12　Per capita GDP and potash consumption prediction in China.

由于钾肥几乎占据了钾盐产品的全部，钾盐与钾肥的消费轨迹十分相近，根据以上分析，得出中国钾盐的消费曲线，并对其未来的发展轨迹做出了趋势线推测（图 12）。由图 12 可知，在人均 GDP 为 5000 美元之前，我国钾盐消费处于快速上升区间，人均 GDP 达到 5000～18000 美元时钾盐消费处于减速度下降区间，人均 GDP 约 18000 美元时，我国钾盐消费将达到顶点，根据趋势线推测峰值为 1400 万 t K_2O，此后，我国的钾盐消费将呈逐渐下降的态势，推测至人均 GDP 约 25000 美元时，我国钾盐的年消费量约为 1200 万 t K_2O。

对于未来的人均 GDP 数值，按照中国经济发展增速分高、中、低三种情景预测，中国 2019～2030 年 GDP 年均增长率设定见表 1。

表 1　2019～2030 年人均 GDP 增速设定

Table 1　Per capita GDP set from 2019 to 2030

增速组别	2019 年	2020～2024 年	2026～2030 年
高速度/%	7.0	6.0	5.0
中速度/%	6.5	5.0	4.0
低速度/%	6.0	4.0	3.0

以 2015 年 GDP（采用美国咨商局的 GK 数据标准）和最新获得的 2016～2018 年数据为基准，再根据 2019～2030 年的人口预测，则可以计算得到三种不同增速情况下相应的人均 GDP 数据，然后根据此数据在图 12 读取相应的钾盐消费数据。据此预测，在高速度增长的情况下，我国最快在 2025 年迎来钾盐消费的峰值，中速度增长情况下，这一时间是 2026 年，低速度增长情况下，最迟将于 2028 年迎来钾盐消费的峰值。

4　结论和建议

1）钾肥消费的“S”形规律是客观存在的。钾盐的消费规律研究表明，钾盐、钾肥的消费具有与大宗矿产资源相类似的消费规律。在经济的快速发展阶段，与人类对其他矿产资源的需求规律类似，人类社会对营养物质的消费量也是逐步提升的，而这些营养物质主要是通过农业生产获得的，当经济发展到一定水平以后，营养物质的来源更加多样化，生产效率进一步提高，人类社会对钾盐、钾肥的消费将出现下降趋势，这种规律表现出来的形态就是明显的“S”形。

2）我国的钾盐消费也将遵循“S”形规律，即随着国民经济发展和人均 GDP 的不断提高，钾盐消费量将呈持续增长态势，目前仍处于增长阶段，在人均 GDP 达到 18000 美元时，我国的钾盐消费将达到峰值阶段，这一时间大约在 2025～2028 年间，峰值约为 1400 万 t K_2O，峰值年份取决于人均 GDP 的变化情况，度过峰值阶段之后，钾盐消费量将逐步下降，并最终稳定在一定的合理水平，形成“S”形的消费轨迹。

3）与美国类似，未来相当长的一段时间内，我国都将是全球最主要的钾盐消费国，即使度过消费峰值阶段，也将长期保持相对较高的消费量，对钾盐资源的供应形成较大压力。

4）面对钾盐资源匮乏的严峻现实，实现钾盐供应多元化是保障我国农业生产安全的必然趋势。首先，应立足国内，努力在境内实现钾盐找矿的较大突破，特别是海相钾盐找矿的突破；其次，我国在境外已经有若干企业在从事钾盐开采，应继续鼓励中资企业在国外开拓钾盐市场，不断提升我国在国际钾盐、钾肥市场的话语权；最后，实现钾盐进口国的多元化，避免受制于人。

参考文献

[1] 中国化学矿业协会. 钾盐供需形势分析及对策建议[J]. 化工矿物与加工, 2004, 33 (7): 1-2, 8.
China Chemical Mining Association. Analysis of the supply and demand of potash and its countermeasures[J]. Industrial Minerals &Processing, 2004, 33 (7): 1-2, 8.

[2] 鲍荣华. 钾盐资源可持续发展战略研究[R]. 2013.

[3] 林仟同. 钾盐供需态势与对策探讨[J]. 中国非金属矿工业导刊, 2005 (1): 45-47.
LIN Qiantong. The situation of supply and demand of potash salt and countermeasures[J]. China Non-Metallic Mining Industry Herald, 2005 (1): 45-47.

[4] 吴永娇, 马海州, 董锁成, 等. 化学需求模拟研究[J]. 盐湖研究, 2008, 16 (4): 42-47.
WU Yongjiao, MA Haizhou, DONG Suocheng, et al. Modelling the demand of potassium fertilizer in China[J]. Journal of Salt Lake Research, 2008, 16 (4): 42-47.

[5] 张卫峰, 王雁峰, 马文奇, 等. 中国化肥需求预测方法浅析[J]. 中国土壤与肥料, 2008 (2): 1-5.
ZHANG Weifeng, WANG Yanfeng, MA Wenqi, et al. A discussion on the method for fertilizer demand forecast of China [J]. Soil and Fertilizer Sciences in China, 2008 (2): 1-5.

[6] 张卫峰, 马文奇, 王雁峰, 等. 基于 CBEM 模型的 2010 年农田化肥需求预测[J]. 植物营养与肥料学报, 2008, 14 (3): 407-416.
ZHANG Weifeng, MA Wenqi, WANG Yanfeng, et al. Forecasting fertilizer demand of China in 2010using CBEM model [J]. Plant Nutrition and Fertilizer Science, 2008, 14 (3): 407-416.

[7] 王安建, 王高尚. 矿产资源与国家经济发展[M]. 北京: 地震出版社, 2002.

[8] 王安建, 王高尚, 陈其慎, 等. 能源与国家经济发展[M]. 北京: 地质出版社, 2008.

[9] 王安建, 王高尚, 陈其慎, 等. 矿产资源需求理论与模型预测[J]. 地球学报, 2010, 31 (2): 137-147.
WANG Anjian, WANG Gaoshang, CHEN Qishen, et al. The mineral resources demand theory and the prediction model [J]. Acta Geoscientia Sinica, 2010, 31 (2): 137-147.

[10] 王安建, 代涛, 刘固望. GDP 增速的“S”形演变轨迹——增速放缓背景下的中国矿产资源需求趋势[J]. 地球学报, 2016, 37 (5): 563-568.
WANG Anjian, DAI Tao, LIU Guwang. “S”-curve model of GDP growth rates: China’s demand trend for mineral resources in the background of slowdown GDP growth rates [J]. Acta Geoscientica Sinica, 2016, 37 (5): 563-568.

[11] 王高尚, 韩梅. 中国重要矿产资源的需求预测[J]. 地球学报, 2002, 23 (6): 483-490.
WANG Gaoshang, HAN Mei. The prediction of the demand on important mineral resources in China[J]. Acta Geoscientica Sinica, 2002, 23 (6): 483-490

[12] GHOSH S. Steel consumption and economic growth: evidence fromIndia[J]. Resources Policy, 2006, 31 (1): 7-11.

[13] DÖHRN R, KRÄTSCHELL K. Long-term trends in steelconsumption[J]. SSRN Electronic Journal, 2013, 27 (1): 43-49.

[14] 王安建, 王高尚, 周凤英. 能源和矿产资源消费增长的极限与周期[J]. 地球学报, 2017, 38 (1): 3-10.
WANG Anjian, WANG Gaoshang, ZHOU Fengying. The limits and cycles of the growth of energy and mineral resources consumption[J]. Acta Geoscientica Sinica, 2017, 38 (1): 3-10.

[15] 于汶加, 王安建, 王高尚. 中国能源消费“零增长”何时到来[J]. 地球学报, 2010, 31 (5): 635-644.
YU Wenjia, WANG Anjian, WANG Gaoshang. A prediction on the time of realizing zero growth of energy consumption in China[J]. Acta Geoscientica Sinica, 2010, 31 (5): 635-644.

[16] 徐铭辰, 王安建, 陈其慎, 等. 中国能源消费强度趋势分析[J]. 地球学报, 2010, 31 (5): 720-726.

XU Mingchen, WANG Anjian, CHEN Qishen, et al. Trend analysis of China's energy consumption intensity[J]. Acta Geoscientica Sinica, 2010, 31 (5): 720-726.

[17] WANG A J, WANG G S, CHEN Q S, et al. S-curve model of relationship between energy consumption and economic development[J]. Natural Resources Research, 2014, 24 (1): 53-64.

[18] 王高尚, 代涛, 柳群义. 全球矿产资源需求周期与趋势[J]. 地球学报, 2017, 38 (1): 11-16.
WANG Gaoshang, DAI Tao, LIU Qunyi. Cycles and trends of global mineral resources demand[J]. Acta Geoscientica Sinica, 2017, 38 (1): 11-16.

[19] 张艳, 于汶加, 陈其慎, 等. 化肥消费规律及中国化肥矿产需求趋势预测[J]. 资源科学, 2015, 37 (5): 977-987.
ZHANG Yan, YU Wenjia, CHEN Qishen, et al. Fertilizer consumption rule and prediction of China's fertilizer-related resource minerals demand[J]. Resources Science, 2015, 37 (5): 977-987.

[20] 唐尧. 我国钾盐资源概况及需求预测分析[J]. 化肥工业, 2015, 42 (4): 75-78.
TANG Yao. Overview of potash resource in China and demand forecasting andanalysis[J]. Chemical Fertilizer Industry, 2015, 42 (4): 75-78.

[21] International Fertilizer Association. Statistics database[EB/OL]. [2017-07-19]. http://ifadata.fertilizer.org/ucSearch.aspx.

[22] The Conference Board. The conference board total economy database[EB/OL]. [2015-06-08]. http://www.conferenceboard.org/data/economydatabase/.

[23] Food and Agriculture Organization of the United Nations. Databases[EB/OL]. [2017-11-22]. http://www.fao.org/statistics/databases/en/.

The consumption rules and demand prediction of potash in China

Xinsheng Niu[1, 2], Anjian Wang[1, 3] and Mianping Zheng[1, 2]

1. Institute of Mineral Resources, Chinese Academy of Geological Sciences, Beijing 100037, China
2. Key Laboratory of Saline Lake Resources and Environment, Ministry of Natural Resources, Beijing 100037, China
3. Research Center for Strategy of Global Mineral Resources, Chinese Academy of Geological Sciences, Beijing 100037, China

Abstract Potash is one of the scarce strategic mineral resources in China, it is necessary to make a prediction of the medium and long-term consumption of potash in China from a strategic point of view of maintaining the safety of agricultural production. The research on the potassic fertilizer consumption rules of major developed countries reveals that capita consumption of potassic fertilizer, consumption of potassic fertilizer per unit of arable land and total amount of potassic fertilizer consumption show inverted "S" shape with capita GDP, which have obvious peak characteristics. The trajectories of total amount of potassic fertilizer consumption of developed countries can be divided into three types: intensive type, sparse type and intermediate type. The intensive type countries have medium land area and higher consumption of potassic fertilizer per unit of arable land and reach consumption peak when capita GDP reaches \$11000–16000. The sparse type countries have larger land area and lower consumption of potassic fertilizer per unit of arable land and reach consumption peak when capita GDP reaches \$19000–20000. The intermediate countries have larger area and a large population. They are the largest agriculture countries and reach consumption peak when capita GDP reaches \$18000. China's consumption of potassic fertilizer is intermediate type and will reach consumption peak of potash when the capita GDP reaches \$18000. The peak value is about 14 million tons of K_2O per year, which will be from 2025 to 2028.

Keywords Potash; Potassic fertilizer; Consumption rule; Demand prediction

盐湖农业及其发展战略研究*

孔凡晶[1]，郑绵平[1]，张洪霞[2]，李真[3]，王利伟[1]

1. 自然资源部盐湖资源与环境重点实验室，中国地质科学院矿产资源研究所，北京 100037
2. 鲁东大学农学院，山东烟台 264025
3. 南京农业大学资源与环境学院，南京 210095

摘要 我国是多盐湖国家，拥有丰富的盐沼带及咸水资源，发展盐湖农业对于荒漠化治理、生态环境保护，对发展西部落后地区经济，开拓具有干旱、半干旱地区特色农业，具有现实和长远战略意义。随着全球人口增长，粮食短缺，淡水资源缺乏，充分利用咸水资源及盐碱地资源发展盐湖农业对保障人类食品安全十分必要。本文梳理了近年来在盐湖农业方面的认识及发展成果，指出发展盐湖农业存在的问题，并提出把盐湖农业列入国家科技规划、对盐湖流域盐碱地进行功能区划等创新驱动盐湖农业发展建议。

关键词 盐湖农业；盐湖生态系统；盐湖生物

一、前言

“盐湖农业”是从事盐湖流域盐水域与盐碱地系统生产食物和多种材料的科技与生产活动。盐湖及其生态环境可发展成为一种新型的农业，它既为一种盐水域水产养殖业，又与盐水域周缘盐沼泽耐盐生物群落相关联，从而构成水产-农牧业研究发展的新领域[1]。我国是多盐湖国家，拥有丰富的盐沼带及咸水资源，发展盐湖农业对于荒漠化治理、生态环境保护，对发展西部落后地区经济，开拓具有干旱、半干旱地区特色农业，具有现实和长远的战略意义。近年来，全球气候变化引起湖泊干涸，盐湖流域荒漠化趋势严重加剧，盐湖周边频繁发生的沙尘暴，严重影响着人们的健康和环境。随着全球人口增长，粮食短缺，淡水资源缺乏，充分利用咸水资源及盐碱地资源发展盐湖农业对保障人类食品安全十分必要。人民科学家钱学森先生在给郑绵平的信中指出，盐湖农业是 21 世纪的产业。本研究依托中国工程院与国家自然科学基金联合资助的咨询项目“盐湖流域盐碱地利用和盐湖农业发展战略研究”，对我国盐湖资源及技术现状进行了调研分析，对取得的认识和成果进行了梳理，提出“创新驱动盐湖农业发展”的建议，在盐湖水域和盐碱地以科技利用盐环境中的盐碱地、盐水域以及盐生物资源，发展种植业、养殖业以及相关多种工业、科学材料高值化产业，优化盐境生态环境。利用现代科学技术发展环境友好型、资源节约型、多领域多学科综合型盐境生物产业新领域，具有重要的战略意义。

二、盐湖农业研究的战略地位与应用价值

（一）发展盐湖农业具有重要的生态学意义

受近期气候干暖、湖面下降、湖泊盐化、湖区牧场退化等不利因素的影响，盐湖流域荒漠化趋势严重加剧。例如，北京周边的安固里淖、查干淖尔、乌拉盖高毕等湖泊正在急速干涸，湖盆流域荒漠化严重；内蒙古自治区的湖泊总面积从 2001 年至 2010 年前后，出现了大面积萎缩现象；京津冀风沙源区形成超过 2000km^2 的盐碱湖盆。裸露的湖盆逐渐盐漠化、沙漠化，成为新的化学物尘源或沙尘源。研究表明，干涸盐湖粉尘贡献占总量的 27%，却提供了 96.1%的盐尘。盐尘在污染、腐蚀、毒害等方面危害极大，盐尘暴中含有密度很高粒径很细的硫酸盐、氯化物及 Mn、As、Rb、Pb、Sr、Cr 等有害重金属元素，极大地污

* 本文发表在：中国工程科学，2019，21（1）：148-152

染空气、土壤、水质、食物，并腐蚀设备，引发疾病，导致受害区生态与自然环境的恶化，成为治理的难点[2]。因地制宜地发展盐湖农业，可使荒漠区恢复生机、绿化环境，从而降低粉尘污染外围城市环境，对于荒漠化治理、生态环境保护，具有重要的现实意义和长远的战略意义。

（二）发展盐湖农业对保障我国粮食安全具有重要的意义

当今世界人口迅速增长，食品短缺，特别是廉价的高质量蛋白质不足，农业生产已不能满足一些国家，特别是第三世界国家人口迅速增长的需要。我国也面临着耕地减少，粮食自给率降低的严峻形势。对此，习近平总书记要求“我们自己的饭碗主要要装自己生产的粮食”。积极研究开发盐湖农业技术，大力发展盐湖农业，利用盐湖水域发展水产养殖业；将广袤的盐湖、盐滩、盐田变为“良田”，扩大粮食等种植，能够增加粮食等食物供给，对我国粮食安全，保障 1.8×10^9 亩（1 亩≈$666.667m^2$）耕地红线具有重要的作用。

（三）发展盐湖农业具有重要的经济价值

盐湖系统中蕴含着丰富的盐生生物资源，通过发展盐湖农业可以实现其经济价值。盐水域已知最具有实用意义的盐生生物有盐藻、卤虫、螺旋藻、轮虫、嗜盐菌和嗜碱菌等。嗜盐的杜氏藻内胡萝卜素可高达 8%～10%（干重），并富含 30% 左右甘油和 30%～40% 蛋白质以及脂肪酸、叶绿素和四烯油等（干重），具有重要的营养和保健价值，盐藻在市场上十分畅销[3, 4]。盐卤虫干重蛋白质含量达 57%～60%、脂肪约为 18%，并含有多种氨基酸、不饱和脂肪酸、维生素等，特别是含鱼虾蟹生长所需要的 EPA 和矿物质，可作为性腺发育激素和抗病害载体，能使鱼虾早成熟，因而成为对虾、蟹幼体和高档鱼的优质饲料之一[5]。对这些生物资源的开发利用具有重要的医药和营养价值，经济前景广阔。

盐沼带有多种盐生植物分布，多为牛羊喜吃的饲料，也可作为能源植物。盐湖盐生植物有很大的经济价值，富含营养和多种元素。仅禾本科高品质优等牧草就有 11 种，如星星草、短芒大麦草、传统药材肉苁蓉、锁阳。有的还可作为中药材，具有重要的经济价值。当地草原用药值得开发，碱蓬含共轭亚油酸，可以食用，有医疗价值，也可以作为旅游景观植物，每到秋季，盐湖滩地呈现大片红色景观。盐碱地植被的梯度比变化，耐盐植物的独特习性等，都是进行科研和生态旅游的最佳选择[6]。

（四）盐湖生物具有重要的科学价值

盐湖、盐碱湖是一类重要的但非常脆弱的生态系统，还赋存有大量的各种具有重要科学和实用意义的盐湖生物，如目前已开展大量盐生经济植物种植，某些嗜碱细菌已开始用于工业生产，嗜盐菌（视）紫膜是理想的光电转化材料，具有广泛的应用前景；一些极端嗜盐微生物能代谢产生可以降解的聚羟基脂肪酸酯新型塑料；盐生植物和嗜盐菌及嗜碱菌中的抗盐基因可以作为特殊基因库，可以培育抗盐或抗盐碱的新的生物品种等[7]。

总之，发展盐湖农业，可以保护盐湖生物资源，改善和恢复恶化的盐湖生态环境；盐湖流域盐碱地还可以作为粮食后备种植基地，保障粮食安全，增加当地农牧民收入；发展盐湖农业具有良好的社会效益、生态效益和经济效益。

三、盐湖农业已取得的认识和成果

（一）发展盐湖农业的资源基础

1. 众多的盐湖资源

中国为多盐湖国家，据考察统计，面积大于 $1km^2$ 的现代内陆盐湖有 813 个，占全国 $1km^2$ 以上天然湖泊总数量的 29.04%；盐湖面积为 $4\times10^4km^2$，约占全国湖泊总面积的一半以上。我国盐湖具有水化学类型多样等特点[8]。特别是在我国西部内陆盐湖和东部滨海、与其相邻的广袤盐碱土和盐咸水域，总量分别为 1.5×10^9 亩和 $4.926\times10^{10}m^3$，这是值得引起重视的潜在土地资源和盐类矿产。地球上的第四纪盐湖主要分布于近代第四纪早期干旱和半干旱气候带，大致可将全球盐湖分布区划分为二带和二区；将我国盐湖带划分为四个盐湖区：青藏高原盐湖区（Ⅰ）、西北部盐湖区（Ⅱ）、东北部盐湖区（或称中北部盐湖区）（Ⅲ）和东部分散盐湖区（Ⅳ）。在盐湖区内再分亚区和小区[9]。

2. 盐湖农业资源：盐生生物

盐湖是湖泊中一种极端的类型，是一种特殊的水生和陆生生物生态环境。超过一般生物种群的耐盐度，在此极端的高盐环境，仍有个别生物种与其种群的多数成员不同，能够适应而繁衍，这种生物称为盐生生物。

随着盐湖湖水含盐度的增高，盐生生物的属种越来越稀少，那些能适应高（超）盐度的盐生生物，由于寡有天敌，反而更加孳生、繁盛，在条件适宜时其繁殖范围可扩展至整个盐湖（盐田）。盐湖生态系统包括盐沼带（Ⅰ）和盐水域（Ⅱ）两个亚系统。盐沼带（Ⅰ）以多种盐生物为主，如盐蒿、盐生藜科、田菁、红柳、紫穗槐等，往往构成重要的牧场。

盐生生物不仅有大量微生物、藻类，还有一些较高等的动植物，如端目纲、甲壳类、单子叶植物、昆虫以及脊椎动物（火烈鸟、罗非鱼等）[1, 7]。

3. 盐湖农业资源：盐碱地资源

盐湖流域盐碱地资源及人口规模：我国是多盐湖的国家，盐湖主要分布在西部干旱或半干旱地区，经济欠发达。盐湖流域主要涉及青海、西藏、新疆、内蒙古，以及黑龙江、吉林、甘肃、宁夏八个省（区）；盐湖流域盐碱地面积为 $5.236\times10^6km^2$，占全国盐碱地的55.9%[10]。

以柴达木盆地为例，柴达木盆地共有湖泊51个，除可鲁克湖为淡水湖和7个半咸水—咸水湖外，其他皆为盐湖，湖表卤水及晶间卤水矿化度都超过50g/L。其中有6个面积较大的干盐湖，盐类沉积储量居我国之冠。盆地中6大干盐湖各自的面积，名列世界现代盐湖面积前列，再加上柴达木盆地干旱少雨的气候条件使得柴达木盆地具有大面积的盐碱地。有资料显示含盐量较高、未开发利用的盐碱地约占总土地面积的11.7%，但是柴达木地区农业用地、草地、林地中盐碱土面积占比也很大，因此该区盐碱地资源非常丰富，具体数据目前尚未统计。柴达木盆地盐碱土类型以荒漠盐土、草甸盐土、沼泽盐土以及湖滨盐滩为主。盐碱土壤由于其不适于传统作物的生长，鉴于以往国内外盐碱地利用的研究，多注重采取人工改良土壤工程措施来适应“淡水”作物生长，但是对土壤进行改良需要大量的淡水进行洗盐，并且这些改良的土壤在耕作过程中很容易发生次生盐渍化，因此柴达木地区大量的盐碱地资源几乎未被利用。大面积的盐碱地资源为发展盐湖农业提供了大量的土地资源[11, 12]。

（二）盐湖农业进展成果

1. 利用盐生生物进行内蒙古干涸盐湖区生态治理

近年来，随着全球气候变化，湖泊干涸，盐湖流域荒漠化趋势严重加剧。比如在北京周边的六大湖泊安固里淖、查干淖尔、乌拉盖高毕等急速干涸，张北县的安固里诺尔，距北京200km，在2004年干涸，湖盆流域盐碱地荒漠化。北京每年春季频繁发生的沙尘暴，严重影响着人们的健康和环境。2008年以来在干涸盐湖先锋植物碱蓬治理技术上取得重要突破，目前在查干淖尔、安固里淖种植面积达到 $10000hm^2$，受到国内外相关领域的重视。取得了良好的生态效益、社会效益和经济效益。

2. 利用种植盐生经济植物进行青海经济转型

研究表明，种植盐生经济植物是进行青海经济转型的重要的经济形式：①尽管格尔木察尔汗盐湖区盐生植物类型比较单一，但其经济价值和生态价值相当可观；②气候整体变湿，盆地地下水位上升，为发展生态盐湖农业提供了最重要的水资源保障；③目前黑、红枸杞种植已在盆地形成产业，河东农场、河西农场、大格勒乡、诺木洪等地已大面积种植，枸杞产值已非常可观，合理的生态盐湖农业完全可以使当地致富。除枸杞外，白刺、罗布麻、芦苇等当地植物的高值化皆有潜力，此外还可引入与当地环境适应的物种，提升生态盐湖农业产值。

四、发展盐湖农业存在的问题

（1）盐湖农业是现代农业发展的新业态，具有高科技、高投入、高效益、高风险的特性，必须正视和解决目前存在的认识不足、措施不力等诸多问题，加强具有独具特色的盐湖农业发展之路，显得十分迫切和必要。

（2）盐湖农业是一个涉及多领域、多部门、多学科的新兴研究领域，在理论体系和实践上都需要进行完善。由于基础相对薄弱，土地、政策、税收、金融、人才、投入等配套政策不健全，发展合力尚未形成，加之商业化开发利用不够，也制约了盐湖农业的发展。

（3）目前盐湖开发主要生产无机盐，有的地区盐湖生态环境和生物资源受到破坏，如柴达木盆地，仅有个别的盐湖还保留着原始的生态和微观层次的微微生物资源，亟待进行保护和查明“达尔文之树黑洞”，抢救濒危的生物基因资源[13]。

五、盐湖农业创新驱动发展的建议

（1）加强盐湖农业科技创新，把盐湖农业列入国家科技规划。集中多学科人才，逐步创建与发展我国盐湖全国性和地区性以至企业性专门研究开发机构，逐步建立具有我国特色的盐湖农业创新体系。

（2）对盐湖流域气候、水资源、土壤和植被条件进行了调查分析，把盐湖流域盐碱地进行功能区划，盐碱地划分为：①可改良盐碱地；②自然发展的盐碱地；③混合发展的盐碱地。盐水域可划分为：①已有可利用盐生物资源（盐藻、螺旋藻、卤虫等）的盐水域；②尚未发现盐生物资源的盐水域。干盐滩依据盐类的类型划分微生物群落类别。通过现代生物学与技术等开展多学科研究，实现学科发展和资源的科学利用，提升农业潜力。

（3）建立盐湖区域土壤、水文、气象、生物种群和种质优良与品系五个基础数据库；基于这五个数据库，结合不同地域的生物、水文水利和土壤技术进行农业改造。需要重点支撑的重大工程和需要重点研发的项目有：①基于水文技术的节水土壤去盐化处理方案；②西部盐湖地区周边土壤和水文数据库建设；③盐湖与周边地下水的交互影响；④盐湖周边地区农业土壤改造中土壤肥力的提升；⑤基于矿物的盐湖淡水化处理方案；⑥盐湖以及周边水资源的合理调配以及系统水利工程建设；⑦深入开展耐盐/嗜盐生物种质的生理、生态、行为、病理实验研究。

（4）在青海、新疆、内蒙古、沿海等不同的生态区设置不同的盐湖农业示范基地，建立起国家级一二三产业融合的盐湖农业基地。

（5）政府应对盐湖农业取得的成效给予奖励和税收优惠，推进盐湖农业市场化和产业化机制的模式创新，调动企业及广大农牧民的积极性。

参考文献

[1] 郑绵平. 论“盐湖农业”[J]. 地球学报, 1995, 16 (4): 404-418.
Zheng M P. Study on “salt lake agriculture”[J]. Acta Geoscientica Sinica, 1995, 16 (4): 404-418.

[2] 韩同林. 京津地区沙尘暴与盐碱尘暴浅析[J]. 科学（上海), 2008, 60 (1): 46-49.
Han T L. Initial analysis on dust storm and salt dust storm in Beijing-Tianjin region [J]. Science (Shanghai), 2008, 60 (1): 46-49.

[3] 孔凡晶, 郑绵平. 盐湖杜氏藻研究的回顾与展望 [J]. 盐业与化工, 2007, 36 (5): 27-33.
Kong F J, Zheng M P. Retrospect and prospect of study on dunaliellain salt lake [J]. Journal of Salt Science and Chemical Industry, 2007, 36 (5): 27-33.

[4] 陈峰. 微藻生物技术 [M]. 北京: 中国轻工业出版社, 1999.
Chen F. Microalgae biotechnology [M]. Beijing: China Light Industry Press, 1999.

[5] 马志珍, 陈汇远. 中国盐湖卤虫的生物学特性及其在对虾育苗中的应用 [J]. 渔业信息与战略, 1994 (11): 14-19.

Ma Z Z, Chen H Y. Studies on Biological characteristics of brine shrimps (ar-lemia spp.) from saline lakes of China and the nauplii as a feed for penaeid shrimp, penaeus chinensis osbeck larvae [J]. Fishery Information and Strategy, 1994 (11): 14-19.
[6] Glenn E P, Brown J J, Blumwald E. Salt tolerance and crop potential of halophytes [J]. Critical Reviews in Plant Sciences, 1999, 18 (2): 227-255.
[7] 孔凡晶, 郑绵平. 盐湖生物学研究进展——第二届"盐湖生物学及嗜盐生物与油气生成学术研讨会"综述 [J]. 地球学报, 2007, 28 (6): 603-608.
Kong F J, Zheng M P. Research progress in saline lake biology: A review of the 2nd conference of saline lake biology and its relationship with petroleum generation [J]. Acta Geoscientica Sinica, 2007, 28 (6): 603-608.
[8] Zheng M P, Tang J, Lin J, et al. China salt lakes [J]. Hydrobiologia, 1993, 267: 23-26.
[9] 郑绵平. 论中国盐湖 [J]. 矿床地质, 2001, 20 (2): 181-189.
Zheng M P. On saline lakes of China [J]. Mineral Deposits, 2001, 20 (2): 181-189.
[10] 王遵亲, 祝寿泉, 俞仁培, 等. 中国盐渍土 [M]. 北京: 科学出版社, 1993.
Wang Z Q, Zhu S Q, Yu R P, et al. The saline soil in China [M]. Beijing: China Science Publishing &Media Ltd (CSPM), 1993.
[11] 王现洁, 孔凡晶, 孔维刚, 等. 发展柴达木盆地盐湖农业的资源基础 [J]. 科技导报, 2017, 35 (10): 93-98.
Wang X J, Kong F J, Kong W G, et al. Resource base of developing saline lake agriculture in Qaidam Basin [J]. Science &Technology Review, 2017, 35 (10): 93-98.
[12] Wang X J, Kong F J, Kong W G, et al. Edaphic characterization and plant zonation in the Qaidam Basin, Tibetan Plateau [J]. Scientific Reports, 2018, 8: 1822.
[13] 王家利, 储立民. 青藏高原盐湖微型和微微型真核浮游生物 [J]. 科技导报, 2017, 35 (12): 32-38.
Wang J L, Chu L M. Eukaryotic pico-and nano-plankton comm-unity in Qinghai-Tibetan Plateau saline lakes [J]. Science & Technology Review, 2017, 35 (12): 32-38.

Salt Lake Agriculture and Its Development Strategy

Fanjing Kong[1], Mianping Zheng[1], Hongxia Zhang[2], Zhen Li[3] and Liwei Wang[1]

1. Key Lab of Salt Lake Resources and Environments, Ministry of Natural Resources of the PRC, Institute of Mineral Resources, Chinese Academy of Geological Sciences, Beijing 100037, China
2. College of Agronomy, Ludong University, Yantai 264025, Shandong, China
3. College of Resources and Environmental Sciences, Nanjing Agricultural University, Nanjing 210095, China

Abstract China has numerous salt lakes, and the development of salt lake agriculture is of realistic and strategic significance in desert control, ecological environment protection, economic growth in the west, and development of featured agricultures in semi-arid and arid regions. As world population grows, food and fresh water shortage is aggravated, and thus developing the salt lake agriculture becomes essential in guaranteeing human food security. Strategic studies on the salt lake agriculture becomes especially urgent. This paper depicts recent achievements of the salt lake agriculture and identifies the problems in salt lake agriculture development. Finally, suggestions are proposed on innovation-driven development of the salt lake agriculture, including enrolling it into national science and technology planning, and zoning saline-alkali land in salt lake basins by functions.

Keywords Salt lake agriculture; Salt lake ecosystem; Salt lake organisms

青海盐湖资源综合开发利用及可持续发展战略

郑绵平　侯献华　樊　馥

中国地质科学院矿产资源研究所 自然资源部盐湖资源与环境重点实验室，北京 100037

一、青海盐湖资源综合开发及可持续发展的战略意义

青海是我国西部边疆的大后方，柴达木盆地是名副其实的聚宝盆。柴达木盆地盐湖资源综合开发利用，对于促进青海经济发展、社会进步、民生改善、支藏维疆，具有极其重要的战略地位。

（1）为保障国家粮食安全发挥支撑作用。青海盐湖资源极其丰富，尤其是柴达木盆地钾肥生产和销售在全国占有较大比重。以 2012 年为例，青海柴达木盆地钾肥生产（以氯化钾计）约 570 万 t，约占全国钾肥产量的 87%。目前，我国每年表观消费约 1900 万 t 氯化钾。加强青海盐湖资源的综合开发利用，对于保障我国粮食安全具有重要意义。

（2）为保障边疆地区稳定发挥基础作用。柴达木盆地盐湖资源开发利用是青海经济发展的重要支撑，以综合开发利用为主导，发展青海特色盐类化工产业，将进一步带动地区基础设施建设和工业基础条件的提升，除钾盐产业开发外，还有今后以锂电动车为主线的盐湖锂产业链以及硼、铷、铯、锶、溴、碘的盐湖特色产业链，对于改善人民生活、促进社会进步、确保边疆稳定、建设世界盐湖产业基地意义重大。

（3）为维护我国生态环境安全提供可靠屏障。柴达木盆地毗邻祁连山冰川与水源涵养生态功能区、三江源草原草甸湿地生态功能区和阿尔金草原荒漠化防治生态功能区。加快青海柴达木盐湖资源综合开发利用和以盐湖化工为核心的循环经济产业体系建设，是协调经济社会发展与生态环境矛盾的现实选择，是维护区域生态环境稳定和资源持续开发利用的必由之路。

二、青海盐湖资源综合开发利用面临的机遇和挑战

柴达木盐湖资源的综合开发利用，不仅可持续保障我国大农业的稳定发展，也必将成为我国经济发展新的增长点，在“一带一路”的布局中具有重大意义。当前，柴达木盆地以盐湖资源为切入点，依托多种资源优势，建设国家级循环经济先行区，面临诸多问题，必须要着力解决好六个突出问题。

（1）盐湖资源综合开发利用管理有待改善。目前，柴达木盆地循环经济体系尚未有效建立，盐湖资源综合开发利用缺乏科学有序的相关规章制度，致使盐湖资源开发各自为政。有关规划与实际情况存在较大差距。以盐湖镁为原料生产金属镁为例，其与内地白云石皮江法炼镁相比，缺乏市场竞争优势，规划金属镁产能，与国内市场需求脱节。盐湖生产的氯化镁、氯化钙及氯气等“副产物”缺乏统一有效监测，以致造成盆地出现“镁害”。柴达木盆地生态脆弱，大量利用淡水溶采钾盐导致部分河湖湿地萎缩与土地荒漠化、沙漠化扩张风险加剧，对湖区生态和城区人居环境影响较大。

（2）盐湖资源科技创新支撑严重不足。目前，青海盐湖资源开发缺乏有力的科技支撑载体，企业技术创新力量普遍薄弱，盐湖综合研发利用产业化若干关键技术瓶颈长期未能得到根本解决，以至不能较快产出推动盐湖资源产业发展的科研创新成果。

（3）盐湖资源综合开发利用程度低，整体效益尚未形成。柴达木盐湖有用组分多，且多为稀缺资源，综合开发价值巨大。但现状是，盆地各盐湖以开发钾盐为主，锂、硼战略新能源虽具有相当储量，但由于组分 Mg/Li 高、卤水提硼工艺尚不成熟，仍未能进行批量生产。

（4）钾盐后备资源储备不足。柴达木盐类资源总量巨大，但是钾资源储备面临严峻挑战。尤其是近十多年来，钾盐处于高强度开发，若继续保持目前开发强度，其储量将难以为继，并影响全国钾肥可持续供给。

（5）盐湖资源高强度、“单一”开发，将使盐类资源利用和绿色发展难以持续。盐湖资源开发的重点区域也是生态服务功能重要的尾闾湖泊湿地和防风固沙关键区域，盐湖资源开发与生态安全关系密切。我国钾肥每年的 KCl 需求量在 1500 万 t 以上，而生产 1t KCl 约产出 8～10t $MgCl_2$、15t NaCl。开发生产出一种产品时，如何将富含其他成分的中间产物、老卤、尾矿等储存或保管好，是一个很难解决的问题，这成为巨大的挑战。

（6）高水平、本地化人才资源缺乏。青海是西部发展、援藏稳疆的重要战略支撑点，但是中高级人员匮乏，亟须引起高度重视建立留住人才和多形式引智的长效机制，这是青海盐湖资源综合利用及可持续发展的根本保障。

三、走青海特色的盐湖综合开发利用及可持续发展道路

根据柴达木资源禀赋、经济技术水平和环境承载条件，研究认为，青海盐湖资源综合开发利用的总原则是：实施资源节约和生态保护优先战略，以环境承载力优化盐湖资源开发利用布局。控制钾肥生产规模，重点开发高价值的锂资源，加快硼、溴、碘资源开发，积极推进盐湖镁钠氯资源综合利用，加快中高端盐湖资源产品研发和生产，提高盐湖资源综合开发利用的质量和效益，推动盐湖产业转型升级。

到 2025 年，形成与盐湖资源开发利用相关的东部盐化工产业群和西部深层卤水钾锂硼锶溴碘铯联合产业集群，预计年产钾肥 700 万～900 万 t（KCl）、碳酸锂 8 万～10 万 t、硼酸 30 万 t、金属镁 5 万～10 万 t、纯碱 340 万 t、工业盐 250 万 t 以及 PVC 等产品，产值在 600 亿元以上。

到 2030 年，柴达木东西部均形成较完善的盐化工产业体系，重点发展价值较高的中高端原料和产品，预计年产碳酸锂 10 万～12 万 t、硼酸 30 万～35 万 t、溴素 5000t、碘 50t 以及 PVC 等产品，据柴达木生态保护形势，钾肥年产建议控制在 800 万～1000 万 t（高质量钾肥）以内，而金属镁亟须根据市场发展来确定其生产规模，以上产品总产值在 700 亿元以上。

同时，加强湿地、绿洲保护，在盐湖周边建造候鸟湿地生态示范区和优化咸水域盐湖生态环境，建设格尔木盐湖农业试验区、德令哈和大柴旦湿地生态示范区和绿色城镇群，使柴达木盆地成为全国盐湖盐沼利用绿色示范区，成为世界一流的综合高效和环保优化的盐湖明珠（图 1）。

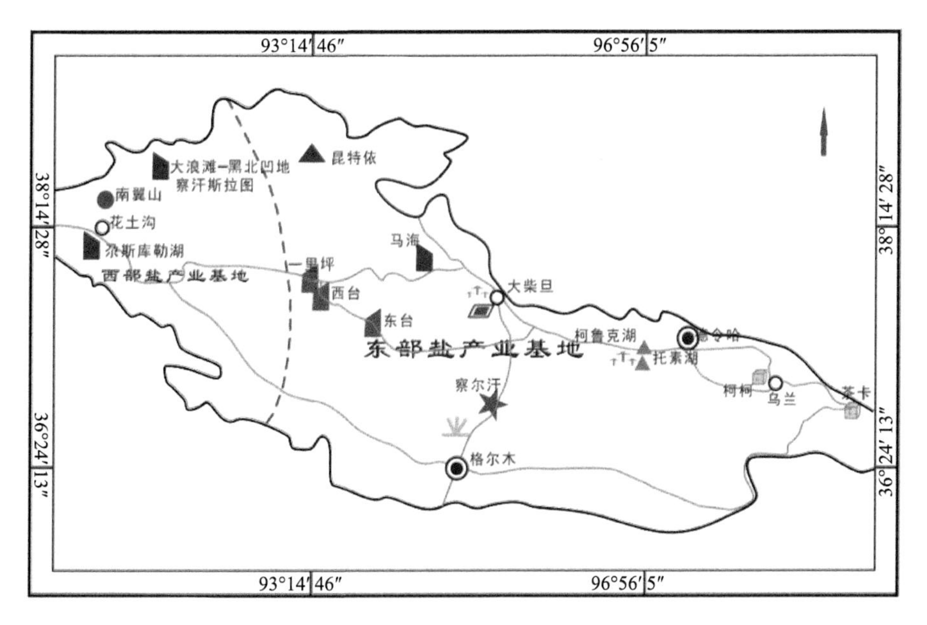

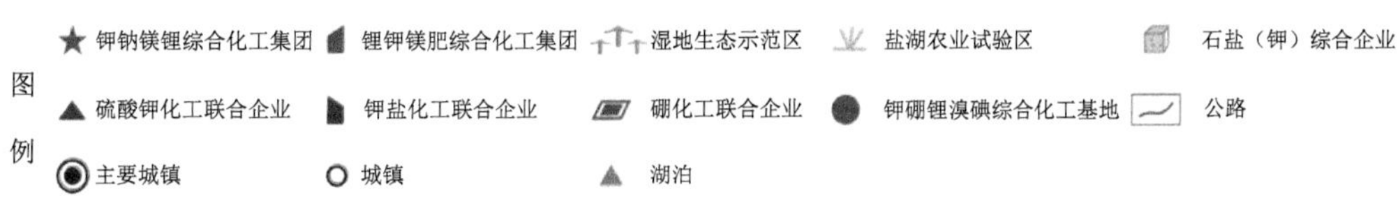

图 1　柴达木盐湖资源综合开发布局简图

四、青海盐湖资源综合开发利用及产业发展的技术路线

总的路线是：从市场需求出发，以创新驱动发展，先易后难。以钾为龙头，重点发展高价值的锂资源，加快开发硼、溴资源，积极推进大宗镁钠利用，加快中高端盐湖资源产品研发和生产。

（一）钾资源开发利用产业技术路线

①柴达木盆地东部，受淡水资源和氯化钾基础储量约束，到2025年察尔汗地区维持在550万～700万t高品质氯化钾/年产量较为合适，不再扩大柴达木东部钾肥的产量和产能。如果未来市场扩大和国际氯化钾供需形势变化，可以柴达木西部深层富钾卤水和其他盐湖为主，酌情逐步扩大钾盐生产规模（在此基础上增加高品位氯化钾100万～300万t）。②适度开采盐湖卤水资源，保持采补平衡，保证钾肥生产的长期稳定。③大力研发高品质的钾肥和中高端钾盐类产品（如硝酸钾等），提高资源产品附加值。

（二）锂、硼资源开发利用技术路线

①继续加强盐湖（深层卤水）锂、硼液体矿和固体硼矿的综合勘查和评价。②开展高镁锂比盐湖卤水提锂工艺的完善和创新。③发展锂硼精细化工产品，推动锂硼中高值产品开发，以大幅度提高经济效益。④继续完善贫锂、贫硼矿和含硼卤水提取、加工工艺和设备的研究。

（三）镁资源开发利用产业技术路线

①盐湖副产氯化镁科学排放储存，如按察尔汗已采氯化钾量计，已产出氯化镁2.6亿多吨，在未来11年，预计还将产出氯化镁约8亿t以上，如不采取有效措施，将造成生态环境的破坏、 氯化镁资源的损失和导致钾盐矿床的变质。②大力推动和支持盐湖氯化镁在镁基建材等行业的大批量产品的直接应用。③根据我国镁资源和镁产品的实际，循序渐进，积极推进盐湖镁氯资源的开发利用和深入研究镁合金、镁盐精细化工等高效利用。

五、保障措施与建议

第一，加强盐湖资源开发利用的领导和协调，推动盐湖资源管理创新。

20世纪60年代初，曾由国家科委、地质部、中科院、化工部联合成立国家青藏高原盐湖勘察与综合利用组，挂靠地质部。上述部委各派 1 人，成立“盐湖综合办公室”，曾对推动我国盐湖发展起到重大推动作用，特别是推动了察尔汗钾盐产业技术进步。新时期，建议由国家发改委或工信部成立“国家盐湖资源综合利用办公室”（简称“国家盐湖办”）。科技部、工信部、财政部以及青海、西藏、新疆等有关方面为成员单位，专职负责青海以及全国盐湖（盐类）资源开发及综合利用的统筹规划和组织制定相关规章制度，构建国家层面上的高效管理服务组织，推动盐湖管理服务创新。

与此相应，建议青海省加强盐湖资源管理局职能，赋予其综合管理和执法权，负责组织青海盐湖资源综合开发利用战略规划和统筹管理，全面监督盐湖资源开发与综合利用规划落实。

第二，建立国家盐湖综合研发创新中心和青海分中心，落实创新驱动发展战略。

我国盐类资源丰富，但缺乏盐湖资源开发利用的综合性创新、研发平台，制约了盐湖资源综合利用发展。为推进企业成为创新主体，加快产学研深度融合，建议以中国地质科学院盐湖资源环境研究中心为基础，联合青海盐湖集团、青海柴达木综合地质勘查院、中科院盐湖所、中国环境科学研究院、青海油田、天津科大等青海省内外相关单位，建立产学研结合的国家盐湖综合评价与利用创新中心，总部设立在北京，由“国家盐湖办”主管。同时，以青海盐湖集团等企业为牵头单位、中科院青海盐湖所等为科技支撑，设立若干盐湖产业研发青海创新分中心，由青海省盐湖管理局主管。

创新中心属于产业共性科技创新机构，通过协同创新，整合各方优势资源，着力开展：①柴达木盐湖资源应用基础研究；②深层含钾卤水资源综合评价与利用规范研究；③中高端产品研发以及科技成果推广和转化；④盐湖地质矿产、气象、水文和监测系统建设；⑤盐湖及周边地区的自然生态保护和盐湖资源综合开发利用的污染防控等研究。

第三，设立国家盐湖生态资源科技重大专项，集中开展盐湖资源开发利用的关键共性技术创新。

针对制约我国盐湖资源综合开发利用的重大关键性科技瓶颈，尤其是开发全国性高镁锂比盐湖提锂工艺、有效利用大宗盐湖镁以及建立柴达木水资源长期监测合理利用体系等，建议国家发改委、科技部、财政部设立“国家盐湖生态资源科技重大专项”，纳入国家“环境综合治理重大工程”之中，以应用示范和产业化为重点，凝聚国内相关领域的优势资源，形成支持盐湖资源综合开发利用共性技术研究、重大产品开发、应用示范、湖区生态保护与产业污染防控体系、产业化创新平台建设、人才培训为一体的专项计划。

第四，构建柴达木盆地水、盐资源调查综合监测网络，加强盐湖资源勘查和评价。

建议根据柴达木盆地淡水资源和盐业开发特点，以重点矿区为单元，建立柴达木盐湖资源与水资源的长期监测和评价体系；建立盐湖资源可持续利用和循环经济发展的监测、评价和绩效考核体系；除国家财政加大投入外，应鼓励企业积极参与盐湖资源（包括柴达木西部深层含钾、锂、硼等卤水）风险勘查和评价。

第五，建设盐湖绿色化工产业园区，加快构筑柴达木循环经济体系。

建设以盐湖资源综合开发利用为核心的循环经济示范园区，合理利用钾等盐类和淡水资源，推动盐湖资源利用的产业链延伸与耦合；优化盐湖资源综合开发利用基地的差异化布局，提高盐类资源的综合利用率，推动盐湖资源综合利用；研究建立以盐类产品质量和盐湖产业清洁生产水平以及盐湖资源综合开发利用水平为核心的行业准入机制、退出机制和企业考核机制；推广盐湖资源综合利用的示范机制，并逐步建立行业规范，推动盐湖化工产业基地的循环化改造和绿色化转型；积极探索金属镁一体化的低成本、低污染产业发展途径。

第六，创新盐湖资源开发体制机制，强化柴达木盆地生态环境保护。

制定并实施资源利用与生态环境保护红线管理制度，探索编制生态资产负债表，建立资源利用、环境绩效、生态资产考核管理与退出机制；实施差别化的资源环境管理机制，推动以盐湖资源开发为核心的循环化工产业差异化布局和示范园区创建；研究盐湖资源综合开发利用的生态补偿机制和环境税制，研究建立保障主要河-湖生态流量过程和生态需水的新机制；着力建设可鲁克湖-托素湖湿地、大柴旦湖湿地和鸭湖湿地生态示范区；在察尔汗设立“盐湖农业”试点；采取贴息贷款、财政补贴、税赋优惠等政策，支持节水技改和废弃物综合利用与污染物治理。

第七，大力加强青海盐湖资源综合开发利用高端人才保障工程建设。

从维护国家长治久安、巩固西部战略高地出发，从国家层面积极推进“智力援青工程”（如国内名校在青海设立分校等）。大力弘扬爱国、创业的献身精神，在全国范围营造到西部建功立业光荣的氛围；建立人才保障工程专项基金，完善人才激励机制，大幅提高高原工资待遇以及改善人才子女上学、就业等，为当地和援青人才发展创造宽松环境和安心创业条件。